ALGEBRA

Exponents and Radicals

$$x^a x^b = x^{a+b} \qquad \frac{x^a}{x^b} = x^{a-b} \qquad x^{-a} = \frac{1}{x^a} \qquad (x^a)^b = x^{ab} \qquad \left(\frac{x}{y}\right)^a = \frac{x^a}{y^a}$$

$$x^{1/n} = \sqrt[n]{x} \qquad x^{m/n} = \sqrt[n]{x^m} = \left(\sqrt[n]{x}\right)^m \qquad \sqrt[n]{xy} = \sqrt[n]{x}\sqrt[n]{y} \qquad \sqrt[n]{x/y} = \sqrt[n]{x}/\sqrt[n]{y}$$

Factoring Formulas

$$a^2 - b^2 = (a - b)(a + b) \qquad\qquad a^2 + b^2 \text{ does not factor over real numbers}$$
$$a^3 - b^3 = (a - b)(a^2 + ab + b^2) \qquad a^3 + b^3 = (a + b)(a^2 - ab + b^2)$$
$$a^n - b^n = (a - b)(a^{n-1} + a^{n-2}b + a^{n-3}b^2 + \cdots + ab^{n-2} + b^{n-1})$$

Binomials

$$(a \pm b)^2 = a^2 \pm 2ab + b^2 \qquad (a \pm b)^3 = a^3 \pm 3a^2b + 3ab^2 \pm b^3$$

Binomial Theorem

$$(a + b)^n = a^n + \binom{n}{1}a^{n-1}b + \binom{n}{2}a^{n-2}b^2 + \cdots + \binom{n}{n-1}ab^{n-1} + b^n,$$

$$\text{where } \binom{n}{k} = \frac{n(n-1)(n-2)\cdots(n-k+1)}{k(k-1)(k-2)\cdots 3\cdot 2\cdot 1} = \frac{n!}{k!(n-k)!}$$

Quadratic Formula

The solutions of $ax^2 + bx + c = 0$ are $x = \dfrac{-b \pm \sqrt{b^2 - 4ac}}{2a}$

GEOMETRY

Parallelogram	Triangle	Trapezoid	Circle	Sector

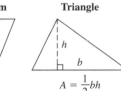
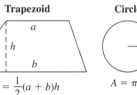
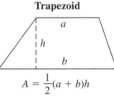
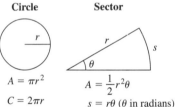

$A = bh$	$A = \frac{1}{2}bh$	$A = \frac{1}{2}(a + b)h$	$A = \pi r^2$	$A = \frac{1}{2}r^2\theta$
			$C = 2\pi r$	$s = r\theta$ (θ in radians)

Cylinder	Cone	Sphere

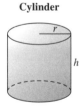

Equations of Lines and Circles

$$m = \frac{y_2 - y_1}{x_2 - x_1} \qquad \text{slope of line through } (x_1, y_1) \text{ and } (x_2, y_2)$$

$$y - y_1 = m(x - x_1) \qquad \text{point-slope form of line through } (x_1, y_1) \text{ with slope } m$$

$$y = mx + b \qquad \text{slope-intercept form of line with slope } m \text{ and } y\text{-intercept } (0, b)$$

$$(x - h)^2 + (y - k)^2 = r^2 \qquad \text{circle of radius } r \text{ with center } (h, k)$$

$V = \pi r^2 h$	$V = \frac{1}{3}\pi r^2 h$	$V = \frac{4}{3}\pi r^3$
$S = 2\pi rh$	$S = \pi rl$	$S = 4\pi r^2$
(lateral surface area)	(lateral surface area)	

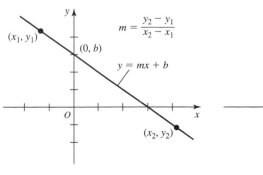

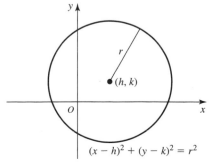

TRIGONOMETRY

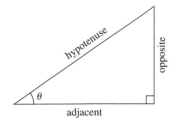

$$\cos\theta = \frac{\text{adj}}{\text{hyp}} \quad \sin\theta = \frac{\text{opp}}{\text{hyp}} \quad \tan\theta = \frac{\text{opp}}{\text{adj}}$$

$$\sec\theta = \frac{\text{hyp}}{\text{adj}} \quad \csc\theta = \frac{\text{hyp}}{\text{opp}} \quad \cot\theta = \frac{\text{adj}}{\text{opp}}$$

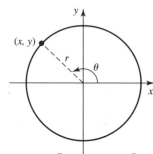

$$\cos\theta = \frac{x}{r} \qquad \sec\theta = \frac{r}{x}$$

$$\sin\theta = \frac{y}{r} \qquad \csc\theta = \frac{r}{y}$$

$$\tan\theta = \frac{y}{x} \qquad \cot\theta = \frac{x}{y}$$

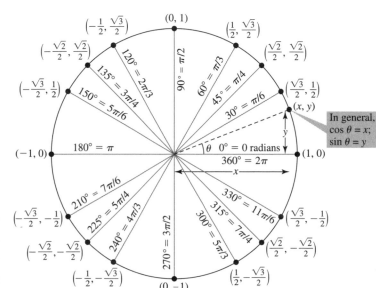

In general, $\cos\theta = x$; $\sin\theta = y$

Reciprocal Identities

$$\tan\theta = \frac{\sin\theta}{\cos\theta} \quad \cot\theta = \frac{\cos\theta}{\sin\theta} \quad \sec\theta = \frac{1}{\cos\theta} \quad \csc\theta = \frac{1}{\sin\theta}$$

Pythagorean Identities

$$\sin^2\theta + \cos^2\theta = 1 \quad \tan^2\theta + 1 = \sec^2\theta \quad 1 + \cot^2\theta = \csc^2\theta$$

Sign Identities

$$\sin(-\theta) = -\sin\theta \quad \cos(-\theta) = \cos\theta \quad \tan(-\theta) = -\tan\theta$$
$$\csc(-\theta) = -\csc\theta \quad \sec(-\theta) = \sec\theta \quad \cot(-\theta) = -\cot\theta$$

Addition Formulas

$$\sin(\alpha + \beta) = \sin\alpha\cos\beta + \cos\alpha\sin\beta \qquad \sin(\alpha - \beta) = \sin\alpha\cos\beta - \cos\alpha\sin\beta$$
$$\cos(\alpha + \beta) = \cos\alpha\cos\beta - \sin\alpha\sin\beta \qquad \cos(\alpha - \beta) = \cos\alpha\cos\beta + \sin\alpha\sin\beta$$
$$\tan(\alpha + \beta) = \frac{\tan\alpha + \tan\beta}{1 - \tan\alpha\tan\beta} \qquad\qquad \tan(\alpha - \beta) = \frac{\tan\alpha - \tan\beta}{1 + \tan\alpha\tan\beta}$$

Double-Angle Identities

$$\sin 2\theta = 2\sin\theta\cos\theta \qquad \cos 2\theta = \cos^2\theta - \sin^2\theta$$
$$= 2\cos^2\theta - 1$$
$$\tan 2\theta = \frac{2\tan\theta}{1 - \tan^2\theta} \qquad\qquad = 1 - 2\sin^2\theta$$

Half-Angle Formulas

$$\cos^2\theta = \frac{1 + \cos 2\theta}{2} \qquad \sin^2\theta = \frac{1 - \cos 2\theta}{2}$$

Graphs of Trigonometric Functions and Their Inverses

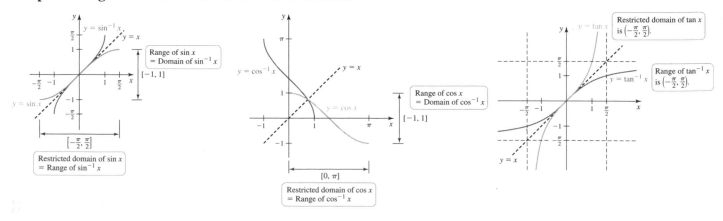

Calculus

EARLY TRANSCENDENTALS

SINGLE VARIABLE

WILLIAM BRIGGS
University of Colorado, Denver

LYLE COCHRAN
Whitworth University

with the assistance of
BERNARD GILLETT
University of Colorado, Boulder

Addison-Wesley

Boston Columbus Indianapolis New York San Francisco Upper Saddle River
Amsterdam Cape Town Dubai London Madrid Milan Munich Paris Montréal Toronto
Delhi Mexico City São Paulo Sydney Hong Kong Seoul Singapore Taipei Tokyo

Editor in Chief:	Deirdre Lynch
Senior Acquisitions Editor:	William Hoffman
Associate Editor:	Caroline Celano
Senior Project Editor:	Rachel Reeve
Associate Project Editor:	Leah Goldberg
Executive Director of Development:	Carol Trueheart
Development Editors:	Elka Block, David Chelton, Roberta Lewis, and Frank Purcell
Senior Managing Editor:	Karen Wernholm
Senior Production Project Manager:	Kathleen A. Manley
Digital Assets Manager:	Marianne Groth
Media Producer:	Lin Mahoney
Software Development:	Eileen Moore (Math XL and MyMathLab) and Mary Durnwald (TestGen)
Market Development Manager:	Dona Kenly
Executive Marketing Manager:	Jeff Weidenaar
Marketing Assistant:	Kendra Bassi
Senior Author Support/Technology Specialist:	Joe Vetere
Senior Prepress Supervisor:	Caroline Fell
Senior Manufacturing Manager:	Evelyn Beaton
Senior Media Buyer:	Ginny Michaud
Production Coordination and Composition:	Pre-Press PMG
Illustrations:	Scientific Illustrators
Cover and Interior Design:	Barbara T. Atkinson
Cover photo:	Road Song Copyright ©Pete Turner, Inc.

For Julie, Susan, Sally, Katie, Jeremy, Elise, Mary, Claire, and Katie, whose support, patience, and encouragement made this book possible.

For permission to use copyrighted material, grateful acknowledgment has been made to the copyright holders listed on p. xvii, which is hereby made part of the copyright page.

Many of the designations used by manufacturers and sellers to distinguish their products are claimed as trademarks. Where those designations appear in this book, and Addison-Wesley was aware of a trademark claim, the designations have been printed in initial caps or all caps.

Library of Congress Cataloging-in-Publication Data
Briggs, William L.
  Calculus: early transcendentals / William L. Briggs, Lyle Cochran;
with contributions by Bernard Gillett.
    p. cm.
  ISBN 978-0-321-66414-3
 1. Calculus—Textbooks.   2.   Transcendental functions—Textbooks.
 I. Cochran, Lyle.   II. Gillett, Bernard.   III. Title.
  QA303.2.B75 2010
  515—dc22   2009041075

    3 4 5 6 7 8 9 10—WC—13 12 11 10

Addison-Wesley
is an imprint of

www.pearsonhighered.com

ISBN-13 978-0-321-66414-3
ISBN-10   0-321-66414-0

Contents

Preface

This textbook supports a three-semester or four-quarter calculus sequence typically taken by students in mathematics, engineering, and the natural sciences. Our approach is based on many years of teaching calculus at diverse institutions using the best teaching practices we know.

Throughout the book, a concise and lively narrative motivates the ideas of calculus. Reviewers and class testers have consistently told us that the book mirrors the course they teach. Equally important, we believe that students will actually read the book. Topics are introduced through concrete examples, applications, and analogies rather than through abstract arguments. We appeal to students' intuition and geometric instincts to make calculus natural and believable. Once this intuitive foundation is established, generalizations and abstractions follow. We include informative proofs in the text, but less transparent proofs appear at the end of the sections or in Appendix B.

Pedagogical Features

Exercises

The exercises at the end of each section are one of the strongest features of the text. They are graded, varied, and original. In addition, they are labeled and carefully organized into groups.

- Each exercise set begins with *Review Questions* that check students' conceptual understanding of the essential ideas from the section.

- *Basic Skills* exercises are confidence-building problems that provide a solid foundation for the more challenging exercises to follow. Each example in the narrative is linked directly to a block of *Basic Skills* exercises via *Related Exercises* references at the end of the example solution.

- *Further Explorations* exercises expand on the *Basic Skills* exercises by challenging students to think creatively and to generalize newly acquired skills.

- *Applications* exercises connect skills developed in previous exercises to applications and modeling problems that demonstrate the power and utility of calculus.

- *Additional Exercises* are generally the most difficult and challenging problems; they include proofs of results cited in the narrative.

Each chapter concludes with a comprehensive set of *Review Exercises*.

Figures

Given the power of graphics software and the ease with which many students assimilate visual images, we devoted considerable time and deliberation to the figures in this book. Whenever possible, we let the figures communicate essential ideas using annotations

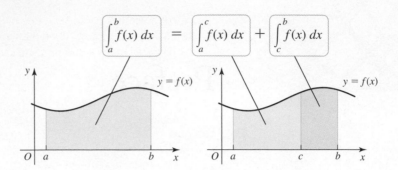

FIGURE 5.29

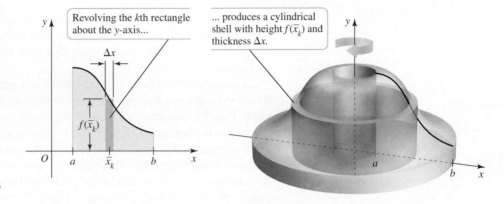

FIGURE 6.37

reminiscent of an instructor's voice at the board. Readers will quickly find that the figures facilitate learning in new ways.

Quick Check and *Margin Notes*

The narrative is interspersed with *Quick Check* questions that encourage students to read with pencil in hand and that resemble the kinds of questions instructors pose in class. Answers to the *Quick Check* questions are found at the end of the section in which they occur. *Margin Notes* offer reminders, provide insight, and clarify technical points.

Guided Projects

The *Instructor's Resource Guide and Test Bank* contains 78 *Guided Projects*. These projects allow students to work in a directed, step-by-step fashion, with various objectives: to carry out extended calculations, to derive physical models, to explore related theoretical topics, or to investigate new applications of calculus. The *Guided Projects* vividly demonstrate the breadth of calculus and provide a wealth of mathematical excursions that go beyond the typical classroom experience. A list of suggested *Guided Projects* is included at the end of each chapter.

Technology

We believe that a calculus text should help students strengthen their analytical skills *and* demonstrate how technology can extend (not replace) those skills. The exercises and examples in this text emphasize this balance. Calculators and graphing utilities are additional tools in the kit, and students must learn when and when not to use them. Our goal is to accommodate the different policies about technology that various instructors may use.

Throughout the book, exercises marked with ▦ indicate that the use of technology—ranging from plotting a function with a graphing utility to carrying out a calculation using a computer algebra system—may be needed.

Interactive Figures

The textbook is supported by a groundbreaking electronic book, created by Eric Schulz of Walla Walla Community College. This "live book" contains the complete text of the print book plus interactive animated versions of approximately 700 figures. Instructors can use these animations in the classroom to illustrate the important ideas of calculus, and students can explore the interactive animations while they are reading the textbook. In each case, these animations will help build students' geometric intuition of calculus. Available only within MyMathLab, the eBook provides instructors with powerful new teaching tools that expand and enrich the learning experience for teachers.

Content Highlights

In writing this text, we identified content in the calculus curriculum that consistently presents challenges to our students. We made organizational changes to the standard presentation of these topics or slowed the pace of the narrative to facilitate students' comprehension of material that is traditionally difficult. Two noteworthy modifications appear in the material for Calculus II and Calculus III, as outlined below.

Often appearing near the end of the term, the topics of sequences and series are the most challenging in Calculus II. By splitting this material into two chapters, we have given these topics a more deliberate pace and made them more accessible without adding significantly to the length of the narrative.

There *is* a clear and logical path through multivariate calculus, which is not apparent in many textbooks. We have carefully separated functions of several variables from vector-valued functions, so that these ideas are distinct in the minds of students. The book culminates when these two threads are joined in the last chapter, which is devoted to vector calculus.

Accuracy Assurance

One of the challenges we face with a first edition is ensuring the book meets the high standards of accuracy that instructors expect. More than 200 mathematicians reviewed the manuscript for accuracy, level of difficulty, and effective pedagogy. Additionally, nearly 1000 students participated in class-testing this book before publication. A team of mathematicians carefully examined each example, exercise, and figure in multiple rounds of editing, proofreading, and accuracy checking. From the beginning and throughout development, our goal has been to craft a textbook that is mathematically precise and pedagogically sound.

Text Versions

Calculus: Early Transcendentals

Complete (Chapters 1–14) ISBN 0-321-57056-1 | 978-0-321-57056-7
Single Variable Calculus (Chapters 1–10) ISBN 0-321-66414-0 | 978-0-321-66414-3
Multivariable Calculus (Chapters 8–14) ISBN 0-321-66415-9 | 978-0-321-66415-0

Calculus

Complete (Chapters 1–15) ISBN 0-321-33611-9 | 978-0-321-33611-8
Single Variable Calculus (Chapters 1–11) ISBN 0-321-66407-8 | 978-0-321-66407-5
Multivariable Calculus (Chapters 9–15) ISBN 0-321-66415-9 | 978-0-321-66415-0

Print Supplements

Instructor's Resource Guide and Test Bank

0-321-69173-3 | 978-0-321-69173-6
Bernard Gillett, University of Colorado at Boulder
Anthony Tongen, James Madison University

This guide represents significant contributions by the textbook authors and contains a variety of classroom support materials for instructors.

• Seventy-eight *Guided Projects*, correlated to specific chapters of the text, can be assigned to students for individual or group work. The *Guided Projects* vividly demonstrate the breadth of calculus and provide a wealth of mathematical excursions that go beyond the typical classroom experience.

• *Lecture Support Notes* give an *Overview* of the material to be taught in each section of the text, helpful classroom *Teaching Tips*, and a list of the *Interactive Figures* from the eBook. *Connections* among various sections of the text are also pointed out, and *Additional Activities* are provided.

• *Quick Quizzes* for each section in the text consist of multiple-choice questions that can be used as in-class quiz material or as Active Learning Questions.

• *Chapter Reviews* provide a list of key concepts from each chapter, followed by a set of chapter review questions.

• *Chapter Test Banks* consist of between 25 and 30 questions that can be used for in-class exams, take-home exams, or additional review material.

• *Student Study Cards*, consisting of key concepts for both single-variable and multivariable Calculus, are included for instructors to photocopy and distribute to their students as convenient study tools.

• *Answers* are provided at the back of the manual for all exercises in the manual, including the *Guided Projects*.

Instructor's Solutions Manuals

Mark Woodard, Furman University
Single Variable Calculus (Chapters 1–10) ISBN 0-321-66408-6 | 978-0-321-66408-2
Multivariable Calculus (Chapters 8–14) ISBN 0-321-66405-1 | 978-0-321-66405-1

The *Instructor's Solutions Manual* contains complete solutions to all the exercises in the text.

Student's Solutions Manuals

Mark Woodard, Furman University
Single Variable Calculus (Chapters 1–10) ISBN 0-321-66410-8 | 978-0-321-66410-5
Multivariable Calculus (Chapters 8–14) ISBN 0-321-66411-6 | 978-0-321-66411-2

The *Student's Solutions Manual* is designed for the student and contains complete solutions to all the odd-numbered exercises in the text.

Just-in-Time Algebra and Trigonometry for Early Transcendentals Calculus, Third Edition

ISBN 0-321-32050-6 | 978-0-321-32050-6
Guntram Mueller and Ronald I. Brent, University of Massachusetts—Lowell

Sharp algebra and trigonometry skills are critical to mastering calculus, and *Just-in-Time Algebra and Trigonometry for Early Transcendentals Calculus* is designed to bolster these skills while students study calculus. As students make their way through calculus, this text is with them every step of the way, showing them the necessary algebra or trigonometry topics and pointing out potential problem spots. The easy-to-use table of contents has algebra and trigonometry topics arranged in the order in which students will need them as they study calculus.

Media and Online Supplements

Technology Resource Manuals

Maple Manual by James Stapleton, North Carolina State University
Mathematica Manual by Marie Vanisko, Carroll College
TI-Graphing Calculator Manual by Elaine McDonald-Newman, Sonoma State University

These manuals cover Maple™ 13, Mathematica® 7, and the TI-83 Plus/TI-84 Plus and TI-89, respectively. Each manual provides detailed guidance for integrating a specific software package or graphing calculator throughout the course, including syntax and commands. These manuals are available to instructors and students through the Pearson Math and Stats Resources page, **www.pearsonhighered.com/mathstatsresources**, and MyMathLab®.

MyMathLab® Online Course (access code required)

MyMathLab is a text-specific, easily customizable online course that integrates interactive multimedia instruction with textbook content. MyMathLab gives you the tools you need to deliver all or a portion of your course online, whether your students are in a lab setting or working from home.

- **eBook featuring Interactive Figures** that can be manipulated to shed light on difficult-to-convey concepts. Instructors can use these animations in the classroom to illustrate the important ideas of calculus, and students can try out the interactive animations while they are using MyMathLab. In each case, these animations help build geometric intuition of calculus.

- **Interactive homework exercises,** correlated to your textbook at the objective level, are algorithmically generated for unlimited practice and mastery. Most exercises are free response and provide guided solutions, sample problems, and tutorial learning aids for extra help.

- **"Getting Ready" chapter** includes hundreds of exercises that address prerequisite skills in algebra and trigonometry. Each student can receive remediation for those skills with which he or she needs help.

- **Personalized Study Plan,** generated when students complete a test or quiz, indicates which topics have been mastered and links to tutorial exercises for topics students have not mastered. You can customize the Study Plan so that the topics available match your course content, or so that students' homework results also determine mastery.

- **Multimedia learning aids,** such as video lectures, Java applets, animations, and a complete interactive eBook, help students independently improve their understanding and performance. You can assign these multimedia learning aids as homework to help your students grasp the concepts.

- **Homework and Test Manager** lets you assign homework, quizzes, and tests that are automatically graded. Select just the right mix of questions from the MyMathLab exercise bank, instructor-created custom exercises, and/or TestGen® test items.

- **Gradebook,** designed specifically for mathematics and statistics, automatically tracks students' results, lets instructors stay on top of student performance, and gives them control over how to calculate final grades. You can also add offline (paper-and-pencil) grades to the gradebook.

- **MathXL® Exercise Builder** allows you to create static and algorithmic exercises for your online assignments. You can use the library of sample exercises as an easy starting point, or you can edit any course-related exercise.

- **Pearson Tutor Center (www.pearsontutorservices.com)** access is automatically included with MyMathLab. The Tutor Center is staffed by qualified math instructors who provide textbook-specific tutoring for students via toll-free phone, fax, email, and interactive Web sessions.

Students do their assignments in the Flash®-based MathXL Player, which is compatible with almost any browser (Firefox®, Safari™, or Internet Explorer®) on almost any platform (Macintosh® or Windows®). MyMathLab is powered by CourseCompass™, Pearson Education's online teaching and learning environment, and by MathXL, our online homework, tutorial, and assessment system. MyMathLab is available to qualified adopters. For more information, visit **www.mymathlab.com** or contact your Pearson representative.

MathXL® Online Course (access code required)

MathXL is an online homework, tutorial, and assessment system that accompanies Pearson's textbooks in mathematics or statistics.

- **Interactive homework exercises,** correlated to your textbook at the objective level, are algorithmically generated for unlimited practice and mastery. Most exercises are free response and provide guided solutions, sample problems, and learning aids for extra help.

- **"Getting Ready" chapter** includes hundreds of exercises that address prerequisite skills in algebra and trigonometry. Each student can receive remediation for those skills with which he or she needs help.

- **Personalized Study Plan,** generated when students complete a test, quiz, or homework, indicates which topics have been mastered and links to tutorial exercises for topics students have not mastered. Instructors can customize the available topics in the study plan to match their course concepts.

- **Multimedia learning aids,** such as video lectures, Java applets, and animations, help students independently improve their understanding and performance. These are assignable as homework, to further encourage their use.

- **Gradebook,** designed specifically for mathematics and statistics, automatically tracks students' results, lets you stay on top of student performance, and gives you control over how to calculate final grades.

- **MathXL Exercise Builder** allows you to create static and algorithmic exercises for your online assignments. You can use the library of sample exercises as an easy starting point or use the Exercise Builder to edit any of the course-related exercises.

- **Homework and Test Manager** lets you create online homework, quizzes, and tests that are automatically graded. Select just the right mix of questions from the MathXL exercise bank, instructor-created custom exercises, and/or TestGen test items.

The new, Flash®-based MathXL Player is compatible with almost any browser (Firefox®, Safari™, or Internet Explorer®) on almost any platform (Macintosh® or Windows®). MathXL is available to qualified adopters. For more information, visit our website at **www.mathxl.com**, or contact your Pearson representative.

TestGen®

TestGen (**www.pearsoned.com/testgen**) enables instructors to build, edit, print, and administer tests using a computerized bank of questions developed to cover all the objectives of the text. TestGen is algorithmically based, allowing instructors to create multiple but equivalent versions of the same question or test with the click of a button. Instructors can also modify test bank questions or add new questions. The software and test bank are available for download from Pearson Education's online catalog.

Video Lectures With Optional Captioning

The Video Lectures With Optional Captioning feature an engaging team of mathematics instructors who present comprehensive coverage of topics in the text. The lecturers' presentations include illustrative examples and exercises and support an approach that emphasizes visualization and problem solving. Available only through MyMathLab and MathXL.

PowerPoint® Lecture Slides

These PowerPoint slides contain key concepts, definitions, figures, and tables from the textbook. These files are available to qualified instructors through the Pearson Instructor Resource Center, **www.pearsonhighered/irc**, and MyMathLab.

Acknowledgments

We would like to express our thanks to the people who made many valuable contributions to this edition as it evolved through its many stages:

Development Editors

Elka Block

David Chelton

Roberta Lewis

Frank Purcell

Accuracy Checkers

Greg Friedman

Robert Pierce

Thomas Polaski

John Sammons

Joan Saniuk

Marie Vanisko

Diana Watson

Thomas Wegleitner

Roman Zadov

Reviewers, Class Testers, Focus Group Participants

Mary Kay Abbey, *Montgomery College*

J. Michael Albanese, *Central Piedmont Community College*

Michael R. Allen, *Tennessee Technological University*

Dale Alspach, *Oklahoma State University*

Alvina J. Atkinson, *Georgia Gwinnett College*

Richard Avery, *Dakota State University*

Rebecca L. Baranowski, *Estrella Mountain Community College*

Cathy Bonan-Hamada, *Mesa State College*

Michael J. Bonanno, *Suffolk County Community College*

Lynette Boos, *Trinity College*

Nathan A. Borchelt, *Clayton State University*

Mario B. Borha, *Moraine Valley Community College*

Michael R. Brewer, *California University of Pennsylvania*

Paul W. Britt, *Louisiana State University*

Tim Britt, *Jackson State Community College*

David E. Brown, *Utah State University*

Kirby Bunas, *Santa Rosa Junior College*

Chris K. Caldwell, *University of Tennessee at Martin*

Elizabeth Carrico, *Illinois Central College*

Tim Chappell, *Penn Valley Community College*

Karin Chess, *Owensboro Community & Technical College*

Ray E. Collings, *Georgia Perimeter College*

Carlos C. Corona, *San Antonio College*

Kyle Costello, *Salt Lake Community College*

Robert D. Crise, Jr., *Crafton Hills College*

Randall Crist, *Creighton University*

Joseph W. Crowley, *Community College of Rhode Island*

Patrick Cureton, *Hillsborough Community College*

Alberto L. Delgado, *Bradley University*

Amy Del Medico, *Waubonsee Community College*

Alicia Serfaty deMarkus, *Miami Dade College*

Joseph Dennin, *Fairfield University*

Emmett C. Dennis, *Southern Connecticut State University*

Andrzej Derdzinski, *Ohio State University*

Nirmal Devi, *Embry Riddle Aeronautical University*

Gary DeYoung, *Dordt College*

David E. Dobbs, *University of Tennessee*

Dr. Alvio Dominguez, *Miami-Dade College (Wolfson Campus)*

Christopher Donnelly, *Macomb Community College*

Anne M. Dougherty, *University of Colorado, Boulder*

Paul Drelles, *West Shore Community College*

Jerrett Dumouchel, *Florida State College at Jacksonville*

Sean Ellermeyer, *Kennesaw State University*

Dr. Amy H. Erickson, *Georgia Gwinnett College*

Robert Farinelli, *College of Southern Maryland*

Judith H. Fethe, *Pellissippi State Technical Community College*

Elaine B. Fitt, *Bucks County Community College*

Walden Freedman, *Humboldt State University*

Greg Friedman, *Texas Christian University*

Randy Gallaher, *Lewis & Clark Community College*

Javier Garza, *Tarleton State University*

Jürgen Gerlach, *Radford University*

Homa Ghaussi-Mujtaba, *Lansing Community College*

Tilmann Glimm, *Western Washington University*

Marvin Glover, *Milligan College*

Belarmino Gonzalez, *Miami-Dade College (Wolfson Campus)*

David Gove, *California State University, Bakersfield*

Phil Gustafson, *Mesa State College*

Aliakbar Montazer Haghighi, *Prairie View A&M University*

Mike Hall, *Arkansas State University*

Donnie Hallstone, *Green River Community College*

Sami Hamid, *University of North Florida*

Don L. Hancock, *Pepperdine University*

Keven Hansen, *Southwestern Illinois College*

David Hartenstine, *Western Washington University*

Kevin Hartshorn, *Moravian College*

Robert H. Hoar, *University of Wisconsin—LaCrosse*

Richard Hobbs, *Mission College*

Leslie Bolinger Horton, *Quinsigamond Community College*

Costel Ionita, *Dixie State College*

Stanislav Jabuka, *University of Nevada, Reno*

Mic Jackson, *Earlham College*

Tony Jenkins, *Northwestern Michigan College*

Jennifer Johnson-Leung, *University of Idaho*

Jack Keating, *Massasoit Community College*

Robert Keller, *Loras College*

Dan Kemp, *South Dakota State University*

Leonid Khazanov, *Borough of Manhattan Community College*

Gretchen Koch, *Goucher College*

Nicole Lang, *North Hennepin Community College*

Mary Margarita Legner, *Riverside City College*

Aihua Li, *Montclair State University*

John B. Little, *College of the Holy Cross*

Jean-Marie Magnier, *Springfield Technical Community College*

Shawna L. Mahan, *Pikes Peak Community College*

Tsun Zee Mai, *University of Alabama*

Nachimuthu Manickam, *DePauw University*

Tammi Marshall, *Cuyamaca College*

Lois Martin, *Massasoit Community College*

Chris Masters, *Doane College*

April Allen Materowski, *Baruch College*

Daniel Maxin, *Valparaiso University*

Mike McAsey, *Bradley University*

Stephen McDowall, *Western Washington University*

Mike McGrath, *Louisiana School for Math, Science, and the Arts*

Ken Mead, *Genesee Community College*

Jack Mealy, *Austin College*

Richard Mercer, *Wright State University*

Elaine Merrill, *Brigham Young University—Hawaii*

Juan Molina, *Austin Community College*

Kathleen Morris, *University of Arkansas*

Carrie Muir, *University of Colorado, Boulder*

Keith A. Nabb, *Moraine Valley Community College*

Paul O'Heron, *Broome Community College*

Michael Oppedisano, *Onondaga Community College*

Leticia M. Oropesa, *University of Miami*

Altay Ozgener, *Manatee Community College*

Shahrokh Parvini, *San Diego Mesa College*

Fred Peskoff, *Borough of Manhattan Community College*

Debra Pharo, *Northwestern Michigan College*

Philip Pickering, *Genesee Community College*

Jeffrey L. Poet, *Missiouri Western State University*

Tammy Potter, *Gadsden State Community College*

Jason Pozen, *Moraine Valley Community College*

Elaine A. Previte, *Bristol Community College*

Stephen Proietti, *Northern Essex Community College*

Suman Sanyal, *Clarkson University*

Brooke P. Quinlan, *Hillsborough Community College*

Douglas Quinney, *Keele University*

Traci M. Reed, *St. Johns River Community College*

Libbie H. Reeves, *Mitchell Community College*

Linda Reist, *Macomb Community College*

Harriette Markos Roadman, *New River Community College*

Kenneth Roblee, *Troy University*

Andrea Ronaldi, *College of Southern Maryland*

William T. Ross, *University of Richmond*

Behnaz Rouhani, *Georgia Perimeter College*

Eric Rowley, *Utah State University*

Ned W. Schillow, *Lehigh Carbon Community College*

Friedhelm Schwarz, *University of Toledo*

Randy Scott, *Santiago Canyon College*

Carl R. Seaquist, *Texas Tech University*

Deepthika Senaratne, *Fayetteville State University*

Dan Shagena, *New Hampshire Technical Institute*

Luz V. Shin, *Los Angeles Valley College*

Nándor Sieben, *Northern Arizona University*

Mark A. Smith, *Miami University*

Shing So, *University of Central Missouri*

Cindy Soderstrom, *Salt Lake Community College*

David St. John, *Malcolm X College*

Zina Stilman, *Community College of Denver*

Eleanor Storey, *Front Range Community College*

Jennifer Strehler, *Oakton Community College*

Linda Sturges, *SUNY—Maritime College*

Richard Sullivan, *Georgetown Unversity*

Donna M. Szott, *Community College of Allegheny County—South Campus*

Elena Toneva, *Eastern Washington University*

Anthony Tongen, *James Madison University*

Michael Tran, *Antelope Valley College*

John Travis, *Mississippi College*

Amitabha Tripathi, *SUNY at Oswego*

Preety N. Tripathi, *SUNY at Oswego*

Ruth Trygstad, *Salt Lake Community College*

David Tseng, *Miami Dade College—Kendall Campus*

Enefiok Umana, *Georgia Perimeter College*

Alexander Vaninsky, *Hostos Community College*

Linda D. VanNiewaal, *Coe College*

Anthony J. Vavra, *West Virginia Northern Community College*

Somasundaram Velummylum, *Claflin University*

Jim Voss, *Front Range Community College*

Yajni Warnapala-Yehiya, *Roger Williams University*

Leben Wee, *Montgomery College*

William Wells, *University of Nevada at Las Vegas*

Darren White, *Kennedy King College*

Bruno Wichnoski, *University of North Carolina—Charlotte*

Dana P. Williams, *Dartmouth College*

David B. Williams, *Clayton State University*

G. Brock Williams, *Texas Tech University*

Nicholas J. Willis, *George Fox University*

Mark R. Woodard, *Furman University*

Kenneth Word, *Central Texas College*

Zhanbo Yang, *University of the Incarnate Word*

Taeil Yi, *University of Texas at Brownsville*

David Zeigler, *California State University, Sacramento*

Hong Zhang, *University of Wisconsin, Oshkosh*

Credits

Note to Students

We offer several practical suggestions about how to gain the most from this book and from your calculus course.

1. Our experience in teaching calculus over many years tells us that the greatest obstacle to learning calculus is not the new ideas of calculus, which are often easily understood. Rather, students find a greater struggle with prerequisite skills—most notably algebra and trigonometry. Your progress with calculus will be far less difficult if you have a solid understanding of algebra and trigonometry before you begin Chapter 2. Take advantage of the material in Chapter 1 and Appendix A, as well as the review that your instructor may provide, so that your prerequisite skills are strong *before* you embark on the study of calculus.

2. An old saying is worth repeating: *Mathematics is not a spectator sport.* No one can expect to learn calculus merely by reading the book and listening to lectures. Your participation and engagement are essential. Read the book actively with a pencil and paper nearby. Use the margins for your notes, answer the Quick Check questions as you go, and work as many exercises as possible. Working exercises will accelerate your learning of calculus more than anything else you do.

3. The use of graphing calculators and computer software is a major issue in teaching and learning calculus. Instructors differ in their emphasis on technology, so it is important to understand your instructor's approach to the use of technology and to become proficient with the required technology as quickly as possible. You should strive for a balance between the use of technology and the use of what are called *analytical*, or pencil-and-paper, methods. Technology should be used to extend and check your analytical skills but never to replace them.

With these thoughts in mind, it is time to begin the calculus journey. We hope it is as exciting for you as it is for us every time we teach calculus.

William Briggs

Lyle Cochran

Bernard Gillett

1

Functions

Chapter Preview The goal of this chapter is to ensure that you begin your calculus journey fully equipped with the tools you will need. Here, you will see the entire cast of functions needed for calculus, which includes polynomials, rational functions, algebraic functions, exponential and logarithmic functions, and the trigonometric functions, along with their inverses. It is imperative that you work hard to master the ideas in this chapter and refer to it when questions arise.

1.1 Review of Functions

Mathematics is a language with an alphabet, a vocabulary, and many rules. If you are unfamiliar with set notation, intervals on the real number line, absolute value, the Cartesian coordinate system, or equations of lines and circles, please refer to Appendix A. Our starting point in this book is the fundamental concept of a function.

Everywhere around us we see relationships among quantities, or **variables**. For example, the consumer price index changes in time and the temperature of the ocean varies with latitude. These relationships can often be expressed by mathematical objects called **functions**. Calculus is the study of functions, and because we use functions to describe the world around us, calculus is a universal language for human inquiry.

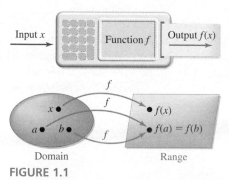

Domain   Range

FIGURE 1.1

> If the domain is not specified, we take it to be the set of all values of x for which f is defined. We will see shortly that the domain and range of a function may be restricted by the context of the problem.

DEFINITION Function

A **function** f is a rule that assigns to each value x in a set D a *unique* value denoted $f(x)$. The set D is the **domain** of the function. The **range** is the set of all values of $f(x)$ produced as x varies over the domain (Figure 1.1).

The **independent variable** is the variable associated with the domain; the **dependent variable** belongs to the range. The **graph** of a function f is the set of all points (x, y) in the xy-plane that satisfy the equation $y = f(x)$. The **argument** of a function is the expression on which the function works. For example, x is the argument when we write $f(x)$. Similarly, 2 is the argument in $f(2)$ and $x^2 + 4$ is the argument in $f(x^2 + 4)$.

QUICK CHECK 1 If $f(x) = x^2 - 2x$, find $f(-1)$, $f(x^2)$, $f(t)$, and $f(p - 1)$. ◄

The requirement that a function must assign a *unique* value of the dependent variable to each value in the domain is expressed in the vertical line test (Figure 1.2).

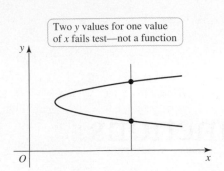

Two y values for one value of x fails test—not a function

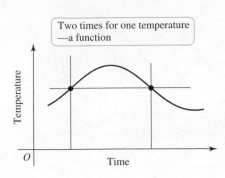

Two times for one temperature —a function

FIGURE 1.2

Vertical Line Test

A graph represents a function if and only if it passes the **vertical line test**: Every vertical line intersects the graph at most once. A graph that fails this test does not represent a function.

> ▶ A set of points or a graph that does *not* correspond to a function represents a **relation** between the variables. All functions are relations, but not all relations are functions.

EXAMPLE 1   Identifying functions State whether each graph in Figure 1.3 corresponds to a function.

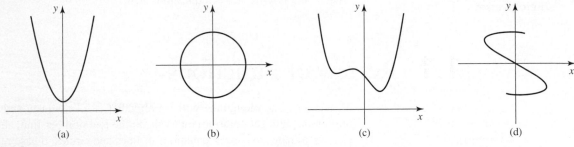

(a)   (b)   (c)   (d)

FIGURE 1.3

SOLUTION The vertical line test indicates that only graphs (a) and (c) represent functions. In graphs (b) and (d), it is possible to draw vertical lines that intersect the graph more than once. Equivalently, it is possible to find values of x that correspond to more than one value of y. Therefore, graphs (b) and (d) do not pass the vertical line test and do not represent functions.

Related Exercises 11–12 ◀

> ▶ A graphing window of $[a, b] \times [c, d]$ means $a \leq x \leq b$ and $c \leq y \leq d$.

EXAMPLE 2   Domain and range Graph each function with a graphing utility using the given window. Then state the domain and range of the function.

a. $y = f(x) = x^2 + 1;\quad [-3, 3] \times [-1, 5]$

b. $z = g(t) = \sqrt{4 - t^2};\quad [-3, 3] \times [-1, 3]$

c. $w = h(u) = \dfrac{1}{u - 1};\quad [-3, 5] \times [-4, 4]$

SOLUTION

a. Figure 1.4 shows the graph of $f(x) = x^2 + 1$. Because f is defined for all values of x, its domain is the set of all real numbers, or $(-\infty, \infty)$, or **R**. Because $x^2 \geq 0$ for all x, it follows that $x^2 + 1 \geq 1$ and the range of f is $[1, \infty)$.

b. When n is even, functions involving nth roots are defined provided the quantity under the root is nonnegative. In this case, the function g is defined provided $4 - t^2 \geq 0$, which means $t^2 \leq 4$, or $-2 \leq t \leq 2$. Therefore, the domain of g is $[-2, 2]$. By the

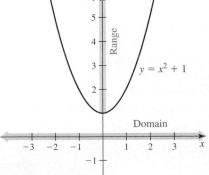

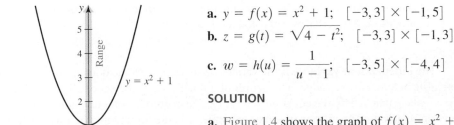

$y = x^2 + 1$

FIGURE 1.4

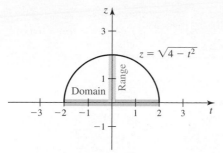

FIGURE 1.5

> The dashed vertical line $u = 1$ in Figure 1.6 indicates that the graph of $w = h(u)$ approaches a *vertical asymptote* as u approaches 1 and that w becomes large in magnitude for u near 1. Vertical and horizontal asymptotes are discussed in detail in Chapter 2.

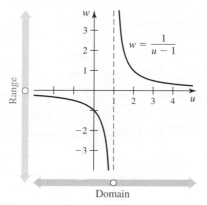

FIGURE 1.6

definition of the square root, the range consists only of nonnegative numbers. When $t = 0$, z reaches its maximum value of $g(0) = \sqrt{4} = 2$, and when $t = \pm 2$, z attains its minimum value of $g(\pm 2) = 0$. Therefore, the range of g is $[0, 2]$ (Figure 1.5).

c. The function h is undefined at $u = 1$, so its domain is $\{u : u \neq 1\}$ and the graph does not have a point corresponding to $u = 1$. We see that w takes on all values except 0; therefore, the range is $\{w : w \neq 0\}$. A graphing utility does *not* represent this function accurately if it shows the vertical line $u = 1$ as part of the graph (Figure 1.6).

Related Exercises 13–18 ◄

EXAMPLE 3 Domain and range in context At time $t = 0$ a stone is thrown vertically upward from the ground at a speed of 30 m/s. Its height above the ground in meters (neglecting air resistance) is approximated by the function $h = f(t) = 30t - 5t^2$. Find the domain and range of this function as they apply to this particular problem.

SOLUTION Although f is defined for all values of t, the only relevant times are between the time the stone is thrown ($t = 0$) and the time it strikes the ground. Solving the equation $h = 30t - 5t^2 = 0$, we find that

$$30t - 5t^2 = 0$$
$$5t(6 - t) = 0 \qquad \text{Factor.}$$
$$5t = 0 \quad \text{or} \quad 6 - t = 0 \qquad \text{Set each factor equal to 0.}$$
$$t = 0 \quad \text{or} \quad t = 6 \qquad \text{Solve.}$$

Therefore, the stone leaves the ground at $t = 0$ and returns to the ground at $t = 6$ s. An appropriate domain that fits the context of this problem is $\{t : 0 \leq t \leq 6\}$. The range consists of all values of $h = 30t - 5t^2$ as t varies over $[0, 6]$. The largest value of h occurs when the stone reaches its highest point at $t = 3$ s, which is $h = f(3) = 45$ m. Therefore, the range is $[0, 45]$. These observations are confirmed by the graph of the height function (Figure 1.7). Note that this graph is *not* the trajectory of the stone; the stone moves vertically.

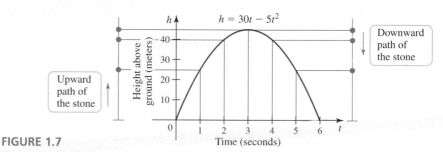

FIGURE 1.7

Related Exercises 19–20 ◄

QUICK CHECK 2 What are the domain and range of $f(x) = (x^2 + 1)^{-1}$? ◄

Composite Functions

> In the composition $y = f(g(x))$, f is called the *outer function* and g is the *inner function*.

Functions may be combined using sums $(f + g)$, differences $(f - g)$, products (fg), or quotients (f/g). The process called *composition* also produces new functions.

DEFINITION Composite Functions

Given two functions f and g, the composite function $f \circ g$ is defined by $(f \circ g)(x) = f(g(x))$. It is evaluated in two steps: $y = f(u)$, where $u = g(x)$. The domain of $f \circ g$ consists of all x in the domain of g such that $u = g(x)$ is in the domain of f (Figure 1.8).

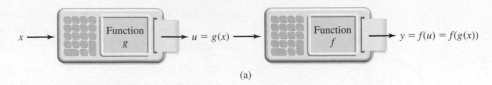

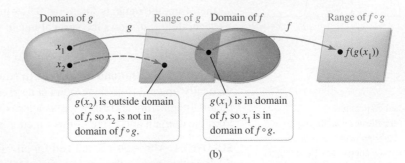

Domain of g Range of g Domain of f Range of $f \circ g$

$g(x_2)$ is outside domain of f, so x_2 is not in domain of $f \circ g$.

$g(x_1)$ is in domain of f, so x_1 is in domain of $f \circ g$.

FIGURE 1.8

(b)

EXAMPLE 4   Composite functions and notation Let $f(x) = 3x^2 - x$ and $g(x) = 1/x$. Simplify the following expressions.

a. $f(5p + 1)$ **b.** $g(1/x)$ **c.** $f(g(x))$ **d.** $g(f(x))$

SOLUTION In each case, the functions work on their arguments.

a. The argument of f is $5p + 1$, so

$$f(5p + 1) = 3(5p + 1)^2 - (5p + 1) = 75p^2 + 25p + 2.$$

b. Because g requires taking the reciprocal of the argument, we take the reciprocal of $1/x$ and find that $g(1/x) = 1/(1/x) = x$.

c. The argument of f is $g(x)$, so

$$f(g(x)) = f\left(\frac{1}{x}\right) = 3\left(\frac{1}{x}\right)^2 - \left(\frac{1}{x}\right) = \frac{3 - x}{x^2}.$$

d. The argument of g is $f(x)$, so

$$g(f(x)) = g(3x^2 - x) = \frac{1}{3x^2 - x}.$$

Related Exercises 21–30 ◀

EXAMPLE 5   Working with composite functions Identify possible choices for the inner and outer functions in the following composite functions. Give the domain of the composite function.

a. $h(x) = \sqrt{9x - x^2}$ **b.** $h(x) = \dfrac{2}{(x^2 - 1)^3}$

SOLUTION

a. An obvious outer function is $f(x) = \sqrt{x}$, which works on the inner function $g(x) = 9x - x^2$. Therefore, h can be expressed as $h = f \circ g$ or $h(x) = f(g(x))$. The domain of $f \circ g$ consists of all values of x such that $9x - x^2 \ge 0$. Solving this inequality gives $\{x : 0 \le x \le 9\}$ as the domain of $f \circ g$.

b. A good choice for an outer function is $f(x) = 2/x^3 = 2x^{-3}$, which works on the inner function $g(x) = x^2 - 1$. Therefore, h can be expressed as $h = f \circ g$ or $h(x) = f(g(x))$. The domain of $f \circ g$ consists of all values of $g(x)$ such that $g(x) \ne 0$, which is $\{x : x \ne \pm 1\}$.   *Related Exercises 31–34* ◀

EXAMPLE 6 More composite functions Given $f(x) = \sqrt[3]{x}$ and $g(x) = x^2 - x - 6$, find **(a)** $g \circ f$ and **(b)** $g \circ g$, and their domains.

SOLUTION

a. We have

$$(g \circ f)(x) = g(f(x)) = (\underbrace{\sqrt[3]{x}}_{f(x)})^2 - \underbrace{\sqrt[3]{x}}_{f(x)} - 6 = x^{2/3} - x^{1/3} - 6.$$

Because the domains of f and g are $(-\infty, \infty)$, the domain of $f \circ g$ is also $(-\infty, \infty)$.

b. In this case, we have the composition of two polynomials:

$$(g \circ g)(x) = g(g(x))$$
$$= g(x^2 - x - 6)$$
$$= \underbrace{(x^2 - x - 6)}_{g(x)}^2 - \underbrace{(x^2 - x - 6)}_{g(x)} - 6$$
$$= x^4 - 2x^3 - 12x^2 + 13x + 36$$

The domain of the composition of two polynomials is $(-\infty, \infty)$.

Related Exercises 35–44 ◄

QUICK CHECK 3 If $f(x) = x^2 + 1$ and $g(x) = x^2$, find $f \circ g$ and $g \circ f$. ◄

EXAMPLE 7 Using graphs to evaluate composite functions Use the graphs of f and g in Figure 1.9 to find the following values.

a. $f(g(5))$ **b.** $f(g(3))$ **c.** $g(f(3))$ **d.** $f(f(4))$

SOLUTION

a. According to the graphs, $g(5) = 1$ and $f(1) = 6$; it follows that $f(g(5)) = f(1) = 6$.

b. The graphs indicate that $g(3) = 4$ and $f(4) = 8$, so $f(g(3)) = f(4) = 8$.

c. We see that $g(f(3)) = g(5) = 1$. Observe that $f(g(3)) \neq g(f(3))$.

d. In this case, $f(f(4)) = f(\underbrace{8}_{8}) = 6$.

Related Exercises 45–46 ◄

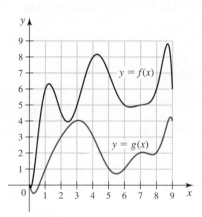

FIGURE 1.9

Symmetry

The word *symmetry* has many meanings in mathematics. Here we consider symmetries of graphs and the relations they represent. Taking advantage of symmetry often saves time and leads to insights.

DEFINITION Symmetry in Graphs

A graph is **symmetric with respect to the y-axis** if whenever the point (x, y) is on the graph, the point $(-x, y)$ is also on the graph. This property means that the graph is unchanged when reflected about the y-axis (Figure 1.10a).

A graph is **symmetric with respect to the x-axis** if whenever the point (x, y) is on the graph, the point $(x, -y)$ is also on the graph. This property means that the graph is unchanged when reflected about the x-axis (Figure 1.10b).

A graph is **symmetric with respect to the origin** if whenever the point (x, y) is on the graph, the point $(-x, -y)$ is also on the graph (Figure 1.10c). Symmetry about both the x- and y-axes implies symmetry about the origin, but not vice versa.

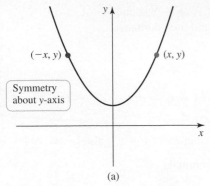

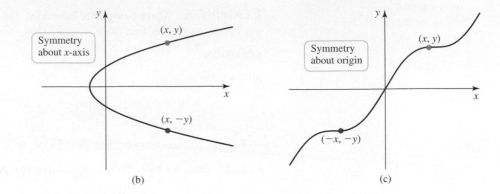

(a)

(b)

(c)

FIGURE 1.10

> **DEFINITION   Symmetry in Functions**
>
> An **even function** f has the property that $f(-x) = f(x)$ for all x in the domain. The graph of an even function is symmetric about the y-axis. Polynomials consisting of only even powers of the variable (of the form x^{2n}, where n is a nonnegative integer) are even functions.
>
> An **odd function** f has the property that $f(-x) = -f(x)$ for all x in the domain. The graph of an odd function is symmetric about the origin. Polynomials consisting of only odd powers of the variable (of the form x^{2n+1}, where n is a nonnegative integer) are odd functions.

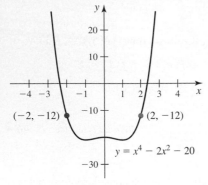

Even function—if (x, y) is on the graph, then $(-x, y)$ is on the graph.

FIGURE 1.11

QUICK CHECK 4 Explain why the graph of a nonzero function cannot be symmetric with respect to the x-axis. ◄

EXAMPLE 8   Identifying symmetry in functions Identify the symmetry, if any, in the following functions.

a. $f(x) = x^4 - 2x^2 - 20$ **b.** $g(x) = x^3 - 3x + 1$ **c.** $h(x) = \dfrac{1}{x^3 - x}$

SOLUTION

a. The function f consists of only even powers of x (where $20 = 20 \cdot 1 = 20x^0$ and x^0 is considered an even power). Therefore, f is an even function (Figure 1.11). This fact is verified by showing that $f(-x) = f(x)$:

$$f(-x) = (-x)^4 - 2(-x)^2 - 20 = x^4 - 2x^2 - 20 = f(x)$$

b. The function g consists of two odd powers and one even power (again, $1 = x^0$ is considered an even power). Therefore, we expect that the function has no symmetry about the y-axis or the origin (Figure 1.12). Note that

$$g(-x) = (-x)^3 - 3(-x) + 1 = -x^3 + 3x + 1,$$

so $g(-x)$ equals neither $g(x)$ nor $-g(x)$, and the function has no symmetry.

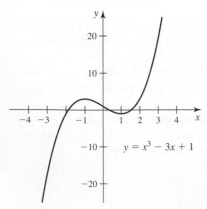

No symmetry—neither an even nor odd function.

FIGURE 1.12

▶ The symmetry of compositions of even and odd functions is considered in Exercises 65–71.

c. In this case, h is a composition of an odd function $f(x) = 1/x$ with an odd function $g(x) = x^3 - x$. Note that

$$h(-x) = \frac{1}{(-x)^3 - (-x)} = -\frac{1}{x^3 - x} = -h(x).$$

Because $h(-x) = -h(x)$, h is an odd function (Figure 1.13).

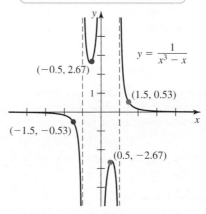

Odd function—if (x, y) is on the graph, then $(-x, -y)$ is on the graph.

$y = \dfrac{1}{x^3 - x}$

$(-0.5, 2.67)$

$(1.5, 0.53)$

$(-1.5, -0.53)$

$(0.5, -2.67)$

FIGURE 1.13

Related Exercises 47–54 ◀

SECTION 1.1 EXERCISES

Review Questions

1. Use the terms *domain, range, independent variable*, and *dependent variable* to explain how a function relates one variable to another variable.

2. Does the independent variable of a function belong to the domain or range? Does the dependent variable belong to the domain or range?

3. Explain how the vertical line test is used to detect functions.

4. If $f(x) = 1/(x^3 + 1)$, what is $f(2)$? What is $f(y^2)$?

5. Which statement about a function is true? (i) For each value of x in the domain, there corresponds one value of y; (ii) for each value of y in the range, there corresponds one value of x. Explain.

6. If $f(x) = \sqrt{x}$ and $g(x) = x^3 - 2$, find the compositions $f \circ g, g \circ f, f \circ f$, and $g \circ g$.

7. If $f(\pm 2) = 2$ and $g(\pm 2) = -2$, evaluate $f(g(2))$ and $g(f(-2))$.

8. Explain how to find the domain of $f \circ g$ if you know the domain and range of f and g.

9. Sketch a graph of an even function and give the function's defining property.

10. Sketch a graph of an odd function and give the function's defining property.

Basic Skills

11–12. Vertical line test *Decide whether graph A, graph B, or both graphs represent functions.*

11.

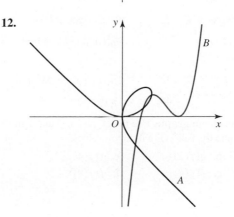

12.

13–18. Domain and range *Graph each function with a graphing utility using the given window. Then state the domain and range of the function.*

13. $f(x) = 3x^4 - 10;$ $[-2, 2] \times [-10, 15]$

14. $g(y) = \dfrac{y + 1}{y^2 - y - 6};$ $[-4, 6] \times [-3, 3]$

15. $f(x) = \sqrt{4 - x^2};$ $[-4, 4] \times [-4, 4]$

16. $F(w) = \sqrt[4]{2 - w};$ $[-3, 2] \times [0, 2]$

17. $h(u) = \sqrt[3]{u - 1};$ $[-7, 9] \times [-2, 2]$

18. $g(x) = (x^2 - 4)\sqrt{x + 5};$ $[-5, 5] \times [-10, 50]$

19–20. Domain in context *Determine an appropriate domain of each function. Identify the independent and dependent variables.*

19. A stone is thrown vertically upward from the ground at a speed of 40 m/s at time $t = 0$. Its distance d (in m) above the ground (neglecting air resistance) is approximated by the function $f(t) = 40t - 5t^2$.

20. The average production cost for a company to make n bicycles is given by the function $c(n) = 120 - 0.25n$.

21–30. Composite functions and notation *Let $f(x) = x^2 - 4$, $g(x) = x^3$, and $F(x) = 1/(x - 3)$. Simplify or evaluate the following expressions.*

21. $f(10)$ **22.** $f(p^2)$ **23.** $g(1/z)$ **24.** $F(y^4)$

25. $F(g(y))$ **26.** $f(g(w))$ **27.** $g(f(u))$ **28.** $\dfrac{f(2 + h) - f(2)}{h}$

29. $F(F(x))$ **30.** $g(F(f(x)))$

31–34. Working with composite functions *Find possible choices for outer and inner functions f and g such that the given function h equals $f \circ g$. Give the domain of h.*

31. $h(x) = (x^3 - 5)^{10}$ **32.** $h(x) = 2/(x^6 + x^2 + 1)^2$

33. $h(x) = \sqrt{x^4 + 2}$ **34.** $h(x) = \dfrac{1}{\sqrt{x^3 - 1}}$

35–40. More composite functions *Let $f(x) = |x|$, $g(x) = x^2 - 4$, $F(x) = \sqrt{x}$, and $G(x) = 1/(x - 2)$. Determine the following composite functions and give their domains.*

35. $f \circ g$ **36.** $g \circ f$ **37.** $f \circ G$

38. $f \circ g \circ G$ **39.** $G \circ g \circ f$ **40.** $F \circ g \circ g$

41–44. Missing piece *Let $g(x) = x^2 + 3$ and find a function f that produces the given composition.*

41. $(f \circ g)(x) = x^4 + 6x^2 + 9$ **42.** $(f \circ g)(x) = x^4 + 6x^2 + 20$

43. $(g \circ f)(x) = x^4 + 3$ **44.** $(g \circ f)(x) = x^{2/3} + 3$

45. **Composite functions from graphs** Use the graphs of f and g in the figure to determine the following function values.

 a. $f(g(2))$ **b.** $g(f(2))$ **c.** $f(g(4))$
 d. $g(f(5))$ **e.** $f(g(7))$ **f.** $f(f(8))$

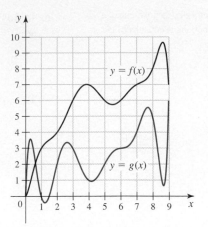

46. **Composite functions from tables** Use the table to evaluate the given compositions.

x	-1	0	1	2	3	4
$f(x)$	3	1	0	-1	-3	-1
$g(x)$	-1	0	2	3	4	5
$h(x)$	0	-1	0	3	0	4

 a. $h(g(0))$ **b.** $g(f(4))$ **c.** $h(h(0))$
 d. $g(h(f(4)))$ **e.** $f(f(f(1)))$ **f.** $h(h(h(0)))$
 g. $f(h(g(2)))$ **h.** $g(f(h(4)))$ **i.** $g(g(g(1)))$
 j. $f(f(h(3)))$

47–52. Symmetry *Determine whether the graphs of the following equations and functions have symmetry about the x-axis, the y-axis, or the origin. Check your work by graphing.*

47. $f(x) = x^4 + 5x^2 - 12$ **48.** $f(x) = 3x^5 + 2x^3 - x$

49. $f(x) = x^5 - x^3 - 2$ **50.** $f(x) = 2|x|$

51. $x^{2/3} + y^{2/3} = 1$ **52.** $x^3 - y^5 = 0$

53. **Symmetry in graphs** State whether the functions represented by graphs A, B, and C in the figure are even, odd, or neither.

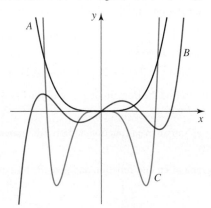

54. Symmetry in graphs State whether the functions represented by graphs A, B, and C in the figure are even, odd, or neither.

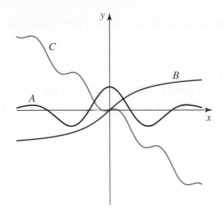

Further Explorations

55. Explain why or why not Determine whether the following statements are true and give an explanation or counterexample.

   a. The range of $f(x) = 2x - 38$ is all real numbers.
   b. The relation $f(x) = x^6 + 1$ is *not* a function because $f(1) = f(-1) = 2$.
   c. If $f(x) = x^{-1}$, then $f(1/x) = 1/f(x)$.
   d. In general, $f(f(x)) = (f(x))^2$.
   e. In general, $f(g(x)) = g(f(x))$.
   f. In general, $f(g(x)) = (f \circ g)(x)$.
   g. If $f(x)$ is an even function, then $cf(ax)$ is an even function, where a and c are real numbers.
   h. If $f(x)$ is an odd function, then $f(x) + d$ is an odd function, where d is a real number.
   i. If f is both even *and* odd, then $f(x) = 0$ for all x.

56. Range of power functions Using words and figures, explain why the range of $f(x) = x^n$, where n is a positive odd integer, is all real numbers. Explain why the range of $g(x) = x^n$, where n is a positive even integer, is all nonnegative real numbers.

57. Absolute value graphs Use the definition of absolute value to graph the equation $|x| - |y| = 1$. Use a graphing utility only to check your work.

58. Even and odd at the origin

   a. If $f(0)$ is defined and f is an even function, is it necessarily true that $f(0) = 0$? Explain.
   b. If $f(0)$ is defined and f is an odd function, is it necessarily true that $f(0) = 0$? Explain.

59–62. Polynomial composition *Determine a polynomial f that satisfies the following properties. (Hint: Determine the degree of f; then substitute a polynomial of that degree and solve for its coefficients.)*

59. $f(f(x)) = 9x - 8$ **60.** $(f(x))^2 = 9x^2 - 12x + 4$

61. $f(f(x)) = x^4 - 12x^2 + 30$ **62.** $(f(x))^2 = x^4 - 12x^2 + 36$

Applications

63. Launching a rocket A small rocket is launched vertically upward from the edge of a cliff 80 ft off the ground at a speed of 96 ft/s. Its height above the ground is given by the function $h(t) = -16t^2 + 96t + 80$, where t represents time measured in seconds.

   a. Assuming the rocket is launched at $t = 0$, what is an appropriate domain for h?
   b. Graph h and determine the time at which the rocket reaches its highest point. What is the height at that time?

64. Draining a tank (Torricelli's law) A cylindrical tank with a cross-sectional area of 100 cm^2 is filled to a depth of 100 cm with water. At $t = 0$, a drain in the bottom of the tank with an area of 10 cm^2 is opened allowing water to flow out of the tank. The depth of water in the tank at time $t \geq 0$ is $d(t) = (10 - 2.2t)^2$.

   a. Check that $d(0) = 100$, as specified.
   b. What is an appropriate domain for d?
   c. At what time is the tank first empty?

Additional Exercises

65–71. Combining even and odd functions *Let E be an even function and O be an odd function. Determine the symmetry, if any, of the following functions.*

65. $E + O$ **66.** $E \cdot O$ **67.** E/O **68.** $E \circ O$

69. $E \circ E$ **70.** $O \circ O$ **71.** $O \circ E$

72–75. Working with function notation *Consider the following functions and simplify the expressions* $\dfrac{f(x) - f(a)}{x - a}$ *and* $\dfrac{f(x + h) - f(x)}{h}$.

72. $f(x) = 3 - 2x$ **73.** $f(x) = 4x - 3$

74. $f(x) = 4x^2 - 1$ **75.** $f(x) = 1/(2x)$

QUICK CHECK ANSWERS

1. $3, x^4 - 2x^2, t^2 - 2t, p^2 - 4p + 3$
2. Domain is all real numbers; range is $\{y : 0 < y \leq 1\}$
3. $(f \circ g)(x) = x^4 + 1$ and $(g \circ f)(x) = (x^2 + 1)^2$.
4. If the graph were symmetric with respect to the x-axis, it would not pass the vertical line test. ◄

1.2 Representing Functions

We consider four different approaches to defining and representing functions: formulas, graphs, tables, and words.

Using Formulas

The following list is a brief catalog of the families of functions that are introduced in this chapter and studied systematically throughout this book; they are all defined by *formulas*.

> ➤ One version of the Fundamental Theorem of Algebra states that a nonconstant polynomial of degree n has exactly n roots, counting each root up to its multiplicity.

1. **Polynomials** are functions of the form

$$f(x) = a_n x^n + a_{n-1} x^{n-1} + \cdots + a_1 x + a_0,$$

   where the **coefficients** $a_0, a_1, \ldots, a_n$ are real numbers with $a_n \neq 0$ and the nonnegative integer n is the **degree** of the polynomial. The domain of any polynomial is the set of all real numbers. An nth-degree polynomial can have as many as n real **zeros** or **roots**—values of x at which $f(x) = 0$, which correspond to points at which the graph of f intersects the x-axis.

2. **Rational functions** are ratios of the form $f(x) = p(x)/q(x)$, where p and q are polynomials. Because division by zero is prohibited, the domain of a rational function is the set of all real numbers except those for which the denominator is zero.

3. **Algebraic functions** are constructed using the operations of algebra: addition, subtraction, multiplication, division, and roots. Examples of algebraic functions are $f(x) = \sqrt{2x^3 + 4}$ and $f(x) = x^{1/4}(x^3 + 2)$. In general, if an even root (square root, fourth root, and so forth) appears, then the domain does not contain points at which the quantity under the root is negative (and perhaps other points).

4. **Exponential functions** have the form $f(x) = b^x$, where the base $b \neq 1$ is a positive real number. Closely associated with exponential functions are **logarithmic functions** of the form $f(x) = \log_b x$, where $b > 0$ and $b \neq 1$. An exponential function has a domain consisting of all real numbers. Logarithmic functions are defined for positive real numbers.

   The most important exponential function is the **natural exponential function** $f(x) = e^x$, with base $b = e$, where $e \approx 2.71828\ldots$ is one of the fundamental constants of mathematics. Associated with the natural exponential function is the **natural logarithm function** $f(x) = \ln x$, which also has the base $b = e$.

5. The **trigonometric functions** are $\sin x$, $\cos x$, $\tan x$, $\cot x$, $\sec x$, and $\csc x$; they are fundamental to mathematics and many areas of application. Also important are their relatives, the **inverse trigonometric functions**.

6. Trigonometric, exponential, and logarithmic functions are a few examples of a large family called **transcendental functions**. Figure 1.14 shows the organization of these functions, all of which are explored in detail in upcoming chapters.

QUICK CHECK 1 Are all polynomials rational functions? Are all algebraic functions polynomials? ◄

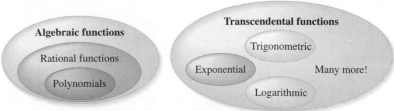

FIGURE 1.14

Using Graphs

Although formulas are the most compact way to represent many functions, graphs often provide the most illuminating representations. Two of countless examples of functions and their graphs are shown in Figure 1.15. Much of this book is devoted to creating and analyzing graphs of functions.

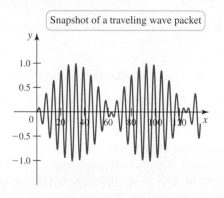

Snapshot of a traveling wave packet

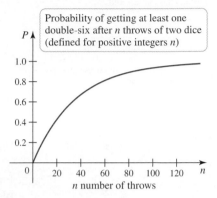

Probability of getting at least one double-six after n throws of two dice (defined for positive integers n)

n number of throws

FIGURE 1.15

There are two approaches to graphing functions.

- Graphing calculators and software are easy to use and powerful. Such **technology** easily produces graphs of most functions encountered in this book. We assume you know how to use a graphing utility.

- Graphing calculators, however, are not infallible. Therefore, you should also strive to master **analytical methods** (pencil-and-paper methods) in order to analyze functions and make accurate graphs by hand. Analytical methods rely heavily on calculus and are presented throughout this book.

The important message is this: Both technology and analytical methods are essential and must be used together in an integrated way to produce accurate graphs.

Linear Functions   One form of the equation of a line (see Appendix A) is $y = mx + b$, where m and b are constants. Therefore, the function $f(x) = mx + b$ has a straight-line graph and is called a **linear function**.

EXAMPLE 1   Linear functions and their graphs   Determine the function represented by the line in Figure 1.16.

SOLUTION   From the graph, we see that the y-intercept is $(0, 6)$. Using the points $(0, 6)$ and $(7, 3)$, the slope of the line is

$$m = \frac{3 - 6}{7 - 0} = -\frac{3}{7}.$$

Therefore, the line is described by the function $f(x) = -3x/7 + 6$.

Related Exercises 11–12 ◄

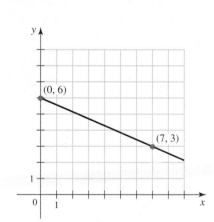

FIGURE 1.16

EXAMPLE 2   Demand function for CDs   After studying sales for several months, the owner of a large CD retail outlet knows that the number of new CDs sold in a day (called the *demand*) decreases as the retail price increases. Specifically, her data indicate that at a price of $14 per CD an average of 400 CDs are sold per day, while at a price of $17 per CD an average of 250 CDs are sold per day. Assume that the demand d is a *linear* function of the price p.

a. Find and graph the demand function $d = f(p) = mp + b$.

b. According to this model, how many CDs (on average) are sold at a price of $20?

SOLUTION

a. Two points on the graph of the demand function are given: $(p, d) = (14, 400)$ and $(17, 250)$. Therefore, the slope of the demand line is

$$m = \frac{400 - 250}{14 - 17} = -50 \text{ CDs per dollar.}$$

It follows that the equation of the linear demand function is

$$d - 250 = -50(p - 17).$$

Expressing d as a function of p, we have $d = f(p) = -50p + 1100$ (Figure 1.17).

b. Using the demand function with a price of \$20, the average number of CDs that could be sold per day is $f(20) = 100$.   *Related Exercises 13–14* ◀

> The units of the slope have meaning: For every dollar that the price is reduced, 50 more CDs can be sold.

Piecewise Functions   A function may have different definitions on different parts of its domain. For example, income tax is levied in tax brackets that have different tax rates. Functions that have different definitions on different parts of the domain are called **piecewise functions**. If all of the pieces are linear, the function is **piecewise linear**. Here are some examples.

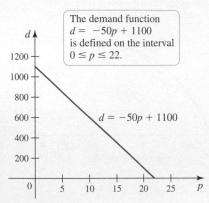

FIGURE 1.17

EXAMPLE 3   **Defining a piecewise function**   The graph of a piecewise linear function g is shown in Figure 1.18. Find a formula for the function.

SOLUTION   For $x < 2$, the graph is linear with a slope of 1 and a y-intercept of $(0, 0)$; its equation is $y = x$. For $x > 2$, the slope of the line is $-\frac{1}{2}$ and it passes through $(4, 3)$, so an equation of this piece of the function is

$$y - 3 = -\frac{1}{2}(x - 4) \quad \text{or} \quad y = -\frac{1}{2}x + 5.$$

For $x = 2$, we have $f(2) = 3$. Therefore,

$$g(x) = \begin{cases} x & \text{if } x < 2 \\ 3 & \text{if } x = 2 \\ -\dfrac{1}{2}x + 5 & \text{if } x > 2 \end{cases}$$

Related Exercises 15–16 ◀

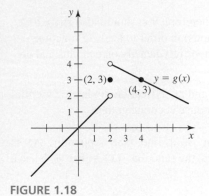

FIGURE 1.18

EXAMPLE 4   **Graphing piecewise functions**   Graph the following functions.

a. $f(x) = \begin{cases} \dfrac{x^2 - 5x + 6}{x - 2} & \text{if } x \neq 2 \\ 1 & \text{if } x = 2 \end{cases}$

b. $f(x) = |x|$, the absolute value function

SOLUTION

a. The function f is simplified by factoring and then canceling $x - 2$, assuming $x \neq 2$:

$$\frac{x^2 - 5x + 6}{x - 2} = \frac{(x - 2)(x - 3)}{x - 2} = x - 3.$$

Therefore, the graph of f is identical to the graph of the line $y = x - 3$ when $x \neq 2$. We are given that $f(2) = 1$ (Figure 1.19).

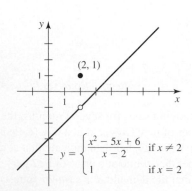

FIGURE 1.19

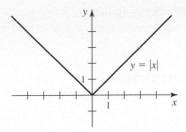

FIGURE 1.20

b. The absolute value of a real number is defined as

$$f(x) = |x| = \begin{cases} x & \text{if } x \geq 0 \\ -x & \text{if } x < 0 \end{cases}$$

Graphing $y = -x$ for $x < 0$ and $y = x$ for $x \geq 0$ produces the graph in Figure 1.20.

Related Exercises 17–20 ◄

Power and Root Functions

1. Power functions are a special case of polynomials; they have the form $f(x) = x^n$, where n is a positive integer. When n is an even integer, the function values are nonnegative and the graph passes through the origin, opening upward (Figure 1.21). For odd integers, the power function $f(x) = x^n$ has values that are positive when x is positive and negative when x is negative (Figure 1.22).

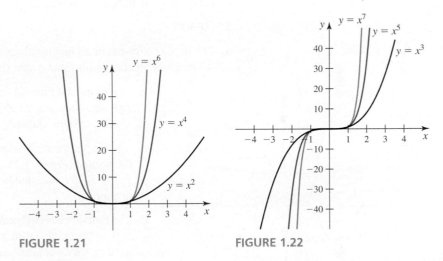

QUICK CHECK 2 What is the range of $f(x) = x^7$? What is the range of $f(x) = x^8$? ◄

FIGURE 1.21

FIGURE 1.22

▶ Recall that if n is a positive integer, then $x^{1/n}$ is the nth root of x; that is, $f(x) = x^{1/n} = \sqrt[n]{x}$.

2. Root functions are a special case of algebraic functions; they have the form $f(x) = x^{1/n}$, where $n > 1$ is a positive integer. Notice that when n is even (square roots, fourth roots, and so forth), the domain and range consist of nonnegative numbers. Their graphs begin steeply at the origin and then flatten out as x increases (Figure 1.23).

By contrast, the odd root functions (cube roots, fifth roots, and so forth) are defined for all real values of x; their range also consists of all real numbers. Their graphs pass through the origin, open upward for $x < 0$ and downward for $x > 0$, and flatten out as x increases in magnitude (Figure 1.24).

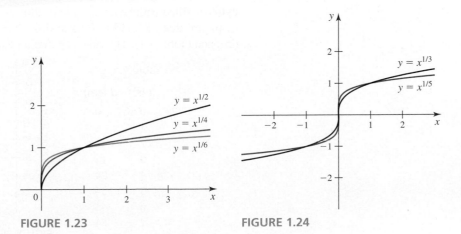

QUICK CHECK 3 What are the domain and range of $f(x) = x^{1/7}$? What are the domain and range of $f(x) = x^{1/10}$? ◄

FIGURE 1.23

FIGURE 1.24

Rational Functions   Rational functions figure prominently in this book and much is said later about graphing rational functions. The following example illustrates how analysis and technology work together.

EXAMPLE 5   Technology and analysis Consider the rational function

$$f(x) = \frac{3x^3 - x - 1}{x^3 + 2x^2 - 6}.$$

a. What is the domain of f?

b. Find the roots (zeros) of f.

c. Graph the function using a graphing utility.

d. At what points does the function have peaks and valleys?

e. How does f behave as x grows large in magnitude?

SOLUTION

a. The domain consists of all real numbers except those at which the denominator is zero. A computer algebra system shows that the denominator has one real zero at $x \approx 1.34$.

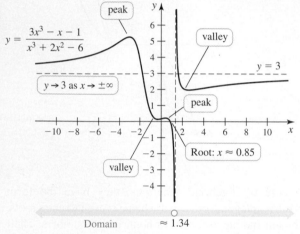

$$y = \frac{3x^3 - x - 1}{x^3 + 2x^2 - 6}$$

peak

valley

$y = 3$

$y \to 3$ as $x \to \pm\infty$

peak

Root: $x \approx 0.85$

valley

Domain ≈ 1.34

FIGURE 1.25

b. The roots of a rational function are the roots of the numerator, provided they are not also roots of the denominator. Using a computer algebra system, the only real root of the numerator is $x \approx 0.85$.

c. After experimenting with the graphing window, a reasonable graph of f is obtained (Figure 1.25). At the point where the denominator is zero, $x \approx 1.34$, the function becomes large in magnitude and f has a *vertical asymptote*.

d. The function has two peaks (soon to be called *local maxima*), one near $x = -3.0$ and one near $x = 0.4$. The function also has two valleys (soon to be called *local minima*), one near $x = -0.3$ and one near $x = 2.6$.

e. By zooming out, it appears that as x increases in the positive direction, the graph approaches the *horizontal asymptote* $y = 3$ from below, and as x becomes large and negative, the graph approaches $y = 3$ from above.

Related Exercises 21–24 ◄

Using Tables

Sometimes functions do not originate as formulas or graphs; they may start as numbers or data. For example, suppose you do an experiment in which a marble is dropped into a cylinder filled with heavy oil and is allowed to fall freely. You measure the total distance d, in cm, that the marble falls at times $t = 0, 1, 2, 3, 4, 5, 6,$ and 7 seconds after it is dropped (Table 1.1). The first step might be to plot the data points (Figure 1.26).

Table 1.1

t (s)	d (cm)
0	0
1	2
2	6
3	14
4	24
5	34
6	44
7	54

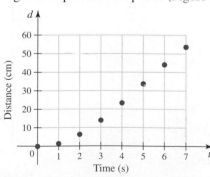

FIGURE 1.26

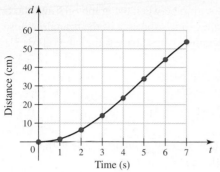

FIGURE 1.27

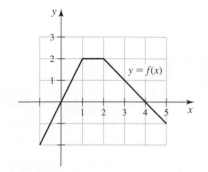

FIGURE 1.28

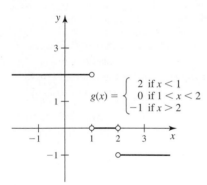

FIGURE 1.29

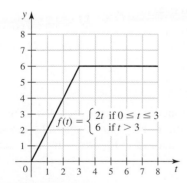

FIGURE 1.30

The data points suggest that there is a function $d = f(t)$ that gives the distance that the marble falls at *all* times of interest. Because the marble falls through the oil without abrupt changes, a smooth graph passing near the data points (Figure 1.27) is reasonable. Finding the best function that fits the data is a more difficult problem, which we discuss later in the text.

Using Words

Using words may be the least mathematical way to define functions, but it is often the way in which functions originate. Once a function is defined in words, it can often be tabulated, graphed, or expressed as a formula.

EXAMPLE 6   A slope function Let g be the **slope function** for a given function f. In words, this means that $g(x)$ is the slope of the curve $y = f(x)$ at the point $(x, f(x))$. Find and graph the slope function for the function f in Figure 1.28.

SOLUTION For $x < 1$, the slope of $y = f(x)$ is 2. The slope is 0 for $1 < x < 2$, and the slope is -1 for $x > 2$. At $x = 1$ and $x = 2$ the graph of f has a corner, so the slope is undefined at these points. Therefore, the domain of g is the set of all real numbers except $x = 1$ and $x = 2$, and the slope function is defined by the piecewise function (Figure 1.29)

$$g(x) = \begin{cases} 2 & \text{if } x < 1 \\ 0 & \text{if } 1 < x < 2 \\ -1 & \text{if } x > 2 \end{cases}$$

Related Exercises 25–26 ◄

EXAMPLE 7   An area function Let A be an **area function** for a positive function f. In words, this means that $A(x)$ is the area of the region between the graph of f and the t-axis from $t = 0$ to $t = x$. Consider the function (Figure 1.30)

$$f(t) = \begin{cases} 2t & \text{if } 0 \le t \le 3 \\ 6 & \text{if } t > 3 \end{cases}$$

a. Find $A(2)$ and $A(5)$.

b. Find a piecewise formula for the area function for f.

SOLUTION

a. The value of $A(2)$ is the area of the shaded region between the graph of f and the t-axis from $t = 0$ to $t = 2$ (Figure 1.31a). Using the formula for the area of a triangle,

$$A(2) = \frac{1}{2}(2)(4) = 4.$$

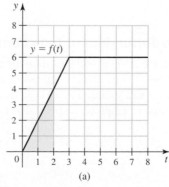

(a)

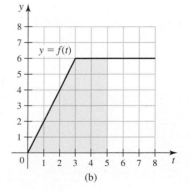

(b)

FIGURE 1.31

The value of $A(5)$ is the area of the shaded region between the graph of f and the t-axis on the interval $[0, 5]$ (Figure 1.31b). This area equals the area of the triangle whose base is the interval $[0, 3]$ plus the area of the rectangle whose base is the interval $[3, 5]$:

$$A(5) = \overbrace{\frac{1}{2}(3)(6)}^{\substack{\text{area of the}\\\text{triangle}}} + \overbrace{(2)(6)}^{\substack{\text{area of the}\\\text{rectangle}}} = 21$$

b. For $0 \le x \le 3$ (Figure 1.32a), $A(x)$ is the area of the triangle whose base is the interval $[0, x]$. Because the height of the triangle at $t = x$ is $f(x)$,

$$A(x) = \frac{1}{2}xf(x) = \frac{1}{2}x\underbrace{(2x)}_{f(x)} = x^2.$$

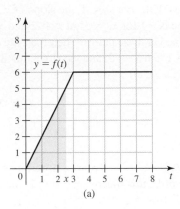

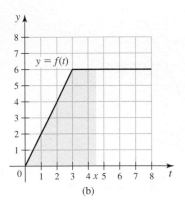

FIGURE 1.32

(a)   (b)

For $x > 3$ (Figure 1.32b), $A(x)$ is the area of the triangle on the interval $[0, 3]$ plus the area of the rectangle on the interval $[3, x]$:

$$A(x) = \overbrace{\frac{1}{2}(3)(6)}^{\substack{\text{area of}\\\text{the triangle}}} + \overbrace{(x - 3)(6)}^{\substack{\text{area of}\\\text{the rectangle}}} = 6x - 9$$

Therefore, the area function A (Figure 1.33) has the piecewise definition

$$y = A(x) = \begin{cases} x^2 & \text{if } 0 \le x \le 3 \\ 6x - 9 & \text{if } x > 3 \end{cases}$$

Related Exercises 27–28 ◄

$$A(x) = \begin{cases} x^2 & \text{if } 0 \le x \le 3 \\ 6x - 9 & \text{if } x > 3 \end{cases}$$

FIGURE 1.33

Transformations of Functions and Graphs

There are several ways to transform the graph of a function to produce graphs of new functions. Four transformations are common: *shifts* in the x- and y-directions and *scalings* in the x- and y-directions. These transformations, summarized in Figures 1.34–1.39, can save time in graphing and visualizing functions.

The graph of $y = f(x) + d$ is the graph of $y = f(x)$ shifted vertically by d units (up if $d > 0$ and down if $d < 0$).

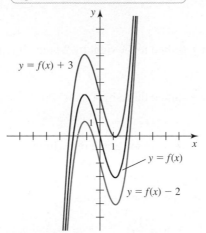

$y = f(x) + 3$

$y = f(x)$

$y = f(x) - 2$

FIGURE 1.34

The graph of $y = f(x - b)$ is the graph of $y - f(x)$ shifted horizontally by b units (right if $b > 0$ and left if $b < 0$).

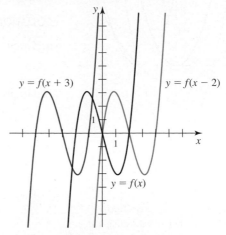

$y = f(x + 3)$

$y = f(x - 2)$

$y = f(x)$

FIGURE 1.35

For $c > 0$, the graph of $y = cf(x)$ is the graph of $y = f(x)$ scaled vertically by a factor of c (broadened if $0 < c < 1$ and steepened if $c > 1$).

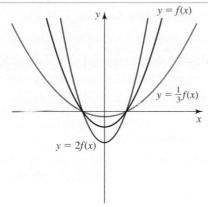

$y = f(x)$

$y = \frac{1}{3}f(x)$

$y = 2f(x)$

FIGURE 1.36

For $c < 0$, the graph of $y = cf(x)$ is the graph of $y = f(x)$ scaled vertically by a factor of $|c|$ and reflected across the x-axis (broadened if $-1 < c < 0$ and steepened if $c < -1$).

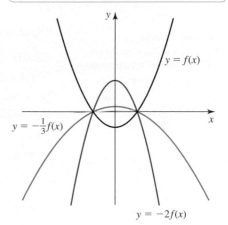

$y = f(x)$

$y = -\frac{1}{3}f(x)$

$y = -2f(x)$

FIGURE 1.37

For $a > 0$, the graph of $y = f(ax)$ is the graph of $y = f(x)$ scaled horizontally by a factor of a (broadened if $0 < a < 1$ and steepened if $a > 1$).

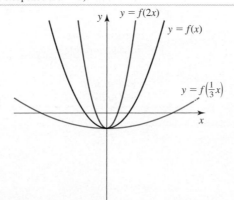

$y = f(2x)$

$y = f(x)$

$y = f\left(\frac{1}{3}x\right)$

FIGURE 1.38

For $a < 0$, the graph of $y = f(ax)$ is the graph of $y = f(x)$ scaled horizontally by a factor of $|a|$ and reflected about the y-axis (broadened if $-1 < a < 0$ and steepened if $a < -1$).

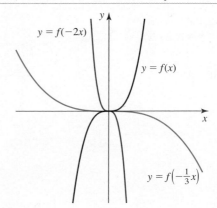

$y = f(-2x)$

$y = f(x)$

$y = f\left(-\frac{1}{3}x\right)$

FIGURE 1.39

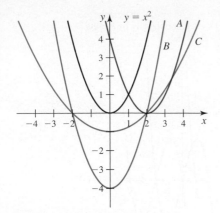

FIGURE 1.40

➤ You should verify that curve C also corresponds to a horizontal scaling and a vertical shift. It has the equation $y = f(ax) - 1$, where $a = \frac{1}{2}$.

EXAMPLE 8   Shifting parabolas The graphs A, B, and C in Figure 1.40 are obtained from the graph of $f(x) = x^2$ using shifts and scalings. Find the function that describes each graph.

SOLUTION

a. Graph A is the graph of f shifted to the right by 2 units. It represents the function

$$f(x - 2) = (x - 2)^2 = x^2 - 4x + 4.$$

b. Graph B is the graph of f shifted down by 4 units. It represents the function

$$f(x) - 4 = x^2 - 4.$$

c. Graph C is a broadened version of the graph of f shifted down by 1 unit. Therefore, it represents $cf(x) - 1 = cx^2 - 1$, for some value of c, with $0 < c < 1$ (because the graph is broadened). Using the fact that graph C passes through the points $(\pm 2, 0)$, we find that $c = \frac{1}{4}$. Therefore, the graph represents

$$y = \frac{1}{4}f(x) - 1 = \frac{1}{4}x^2 - 1.$$

Related Exercises 29–38 ◄

QUICK CHECK 4 How do you modify the graph of $f(x) = 1/x$ to produce the graph of $g(x) = 1/(x + 4)$? ◄

➤ Note that we can also write $g(x) = 2\left|x + \frac{1}{2}\right|$, which means the graph of g may also be obtained by a vertical scaling and a horizontal shift.

EXAMPLE 9   Scaling and shifting Graph $g(x) = |2x + 1|$.

SOLUTION We write the function as $g(x) = \left|2\left(x + \frac{1}{2}\right)\right|$ so it can be interpreted as a horizontal scaling and a horizontal shift. Letting $f(x) = |x|$, we have $g(x) = f\left(2\left(x + \frac{1}{2}\right)\right)$. Thus, the graph of g is obtained by scaling (steepening) the graph of f horizontally and shifting it $\frac{1}{2}$-unit to the left (Figure 1.41).

Related Exercises 29–38 ◄

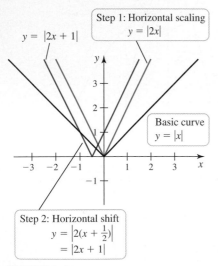

FIGURE 1.41

SUMMARY   Transformations

Given the real numbers a, b, c, and d and the function f, the graph of $y = cf(a(x - b)) + d$ is obtained from the graph of $y = f(x)$ in the following steps.

$$y = f(x) \xrightarrow[\text{by a factor of } |a|]{\text{horizontal scaling}} y = f(ax)$$

$$\xrightarrow[\text{by } b \text{ units}]{\text{horizontal shift}} y = f(a(x - b))$$

$$\xrightarrow[\text{by a factor of } |c|]{\text{vertical scaling}} y = cf(a(x - b))$$

$$\xrightarrow[\text{by } d \text{ units}]{\text{vertical shift}} y = cf(a(x - b)) + d$$

SECTION 1.2 EXERCISES

Review Questions

1. Give four ways that functions may be defined and represented.

2. What is the domain of a polynomial?

3. What is the domain of a rational function?

4. Describe what is meant by a piecewise linear function.

5. Sketch a graph of $y = x^5$.

6. Sketch a graph of $y = x^{1/5}$.

7. If you have the graph of $y = f(x)$, how do you obtain the graph of $y = f(x + 2)$?

8. If you have the graph of $y = f(x)$, how do you obtain the graph of $y = -3f(x)$?

9. If you have the graph of $y = f(x)$, how do you obtain the graph of $y = f(3x)$?

10. Given the graph of $y = x^2$, how do you obtain the graph of $y = 4(x + 3)^2 + 6$?

Basic Skills

11–12. Graphs of functions *Find the linear functions that correspond to the following graphs.*

11.

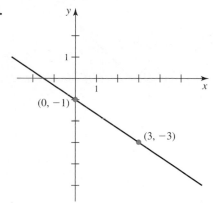

12.

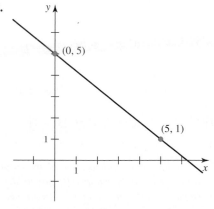

13. **Demand function** Sales records indicate that if DVD players are priced at \$250, then a large store sells an average of 12 units per day. If they are priced at \$200, then the store sells an average of

15 units pcr day. Find and graph the linear demand function for DVD sales. For what prices is the demand function defined?

14. **Fundraiser** The Biology Club plans to have a fundraiser for which \$8 tickets will be sold. The cost of room rental and refreshments is \$175. Find and graph the function $p = f(n)$ that gives the profit from the fundraiser when n tickets are sold. Notice that $f(0) = -\$175$; that is, the cost of room rental and refreshments must be paid regardless of how many tickets are sold. How many tickets must be sold to break even (zero profit)?

15–16. Graphs of piecewise functions *Write a definition of the functions whose graphs are given.*

15.

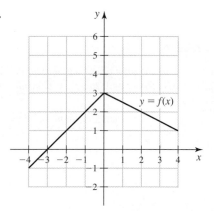

16.

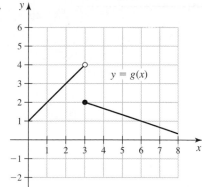

17–20. Piecewise linear functions *Graph the following functions.*

17. $f(x) = \begin{cases} 3x - 1 & \text{if } x \le 0 \\ -2x + 1 & \text{if } x > 0 \end{cases}$

18. $f(x) = \begin{cases} 3x - 1 & \text{if } x < 1 \\ x + 1 & \text{if } x \ge 1 \end{cases}$

19. $f(x) = \begin{cases} -2x - 1 & \text{if } x < -1 \\ 1 & \text{if } -1 \le x \le 1 \\ 2x - 1 & \text{if } x > 1 \end{cases}$

20. $f(x) = \begin{cases} 2x + 2 & \text{if } x < 0 \\ x + 2 & \text{if } 0 \le x \le 2 \\ 3 - x/2 & \text{if } x > 2 \end{cases}$

▣ 21–24. Graphs of functions

    a. *Use a graphing utility to produce a graph of the given function. Experiment with your choice of plot windows to see how the graph changes on different scales.*

    b. *Classify the function and give its domain.*

    c. *Discuss the interesting features of the function such as peaks, valleys, and intercepts (as in Example 5).*

21. $f(x) = x^3 - 2x^2 + 6$ **22.** $f(x) = \sqrt[3]{2x^2 - 8}$

23. $g(x) = \left| \dfrac{x^2 - 4}{x + 3} \right|$ **24.** $f(x) = \dfrac{\sqrt{3x^2 - 12}}{x + 1}$

25–26. Slope functions *Determine the slope function for the following functions.*

25. Use the figure for Exercise 15

26. Use the figure for Exercise 16

27–28. Area functions *Let $A(x)$ be the area of the region bounded by the t-axis and the graph of $y = f(t)$ from $t = 0$ to $t = x$. Consider the following functions and graphs.*

a. Find $A(2)$ **b.** Find $A(6)$ **c.** Find a formula for $A(x)$

27. $f(t) = 6$ (see figure)

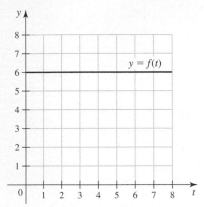

28. $y = f(t) = \begin{cases} -t + 2 & \text{if } t \leq 2 \\ 2t - 4 & \text{if } 2 < t < 4 \quad \text{(see figure)} \\ -\dfrac{1}{2}t + 6 & \text{if } t \geq 4 \end{cases}$

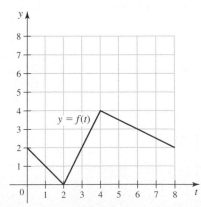

29. Transformations of $y = |x|$ The functions f and g in the figure were obtained by vertical and horizontal shifts and scalings

of $y = |x|$. Find formulas for f and g. Verify your answers with a graphing utility.

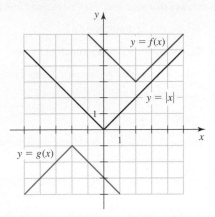

30. Transformations Use the graph of f in the figure to plot the following functions.

    a. $y = -f(x)$ **b.** $y = f(x + 2)$ **c.** $y = f(x - 2)$

    d. $y = f(2x)$ **e.** $y = f(x - 1) + 2$ **f.** $y = 2f(x)$

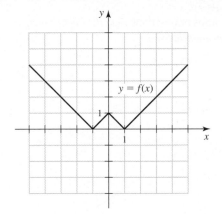

31. Transformations of $f(x) = x^2$ Use shifts and scalings to transform the graph of $f(x) = x^2$ into the graph of g. Use a graphing utility only to check your work.

    a. $g(x) = f(x - 3)$ **b.** $g(x) = f(2x - 4)$

    c. $g(x) = -3f(x - 2) + 4$ **d.** $g(x) = 6f\left(\dfrac{x - 2}{3}\right) + 1$

32. Transformations of $f(x) = \sqrt{x}$ Use shifts and scalings to transform the graph of $f(x) = \sqrt{x}$ into the graph of g. Use a graphing utility only to check your work.

    a. $g(x) = f(x + 4)$ **b.** $g(x) = 2f(2x - 1)$

    c. $g(x) = \sqrt{x - 1}$ **d.** $g(x) = 3\sqrt{x - 1} - 5$

▣ 33–38. Shifting and scaling *Use shifts and scalings to graph the given functions. Then check your work with a graphing utility. Be sure to identify an original function on which the shifts and scalings are performed.*

33. $g(x) = -3x^2$ **34.** $g(x) = 2x^3 - 1$

35. $g(x) = 2(x + 3)^2$ **36.** $p(x) = x^2 + 3x - 5$

37. $h(x) = -4x^2 - 4x + 12$ **38.** $h(x) = |3x - 6| + 1$

Further Explorations

39. Explain why or why not Determine whether the following statements are true and give an explanation or a counterexample.

    a. All polynomials are rational functions, but not all rational functions are polynomials.

    b. If f is a linear polynomial, then $f \circ f$ is a quadratic polynomial.

    c. If f and g are polynomials, then the degrees of $f \circ g$ and $g \circ f$ are equal.

    d. To graph $g(x) = f(x + 2)$, shift the graph of f two units to the right.

40–41. Intersection problems *Use analytical methods to find the following points of intersection. Use a graphing utility only to check your work.*

40. Find the point(s) of intersection between the parabola $y = x^2 + 2$ and the line $y = x + 4$.

41. Find the point(s) of intersection between the parabolas $y = x^2$ and $y = -x^2 + 8x$.

42–43. Functions from tables *Find a simple function that fits the data in the tables.*

42.

x	y
−1	0
0	1
1	2
2	3
3	4

43.

x	y
0	−1
1	0
4	1
9	2
16	3

44–47. Functions from words *Find a formula for a function describing the given situation. Graph the function and give a domain that makes sense for the problem. Recall that with constant speed, distance = speed · time elapsed or d = vt.*

44. A function $y = f(x)$ such that y is 1 less than the cube of x.

45. A function $y = f(x)$ such that if you run at a constant rate of 5 mi/hr for x hours, then you run y miles.

46. A function $y = f(x)$ such that if you ride a bike for 50 mi at x miles per hour, you arrive at your destination in y hours.

47. A function $y = f(x)$ such that if your car gets 32 mi/gal and gasoline costs $\$x$/gallon, then $100 is the cost of taking a y-mile trip.

48. Floor function The floor function, or greatest integer function, $f(x) = \lfloor x \rfloor$, gives the greatest integer less than or equal to x. Graph the floor function for $-3 \le x \le 3$.

49. Ceiling function The ceiling function, or smallest integer function, $f(x) = \lceil x \rceil$, gives the smallest integer greater than or equal to x. Graph the ceiling function for $-3 \le x \le 3$.

50. Sawtooth wave Graph the sawtooth wave defined by

$$f(x) = \begin{cases} \ \vdots \\ x + 1 & \text{for } -1 \le x < 0 \\ x & \text{for } 0 \le x < 1 \\ x - 1 & \text{for } 1 \le x < 2 \\ x - 2 & \text{for } 2 \le x < 3 \\ \ \vdots \end{cases}$$

51. Square wave Graph the square wave defined by

$$f(x) = \begin{cases} 0 & \text{for } x < 0 \\ 1 & \text{for } 0 \le x < 1 \\ 0 & \text{for } 1 \le x < 2 \\ 1 & \text{for } 2 \le x < 3 \\ \ \vdots \end{cases}$$

52–54. Roots and powers *Make a rough sketch of the given pairs of functions. Be sure to draw the graphs accurately relative to each other.*

52. $y = x^4$ and $y = x^6$ **53.** $y = x^3$ and $y = x^7$

54. $y = x^{1/3}$ and $y = x^{1/5}$

Applications

55. Bald eagle population Since DDT was banned and the Endangered Species Act was passed in 1973, the number of bald eagles in the United States has increased dramatically (see the figure). In the lower 48 states, the number of breeding pairs of bald eagles increased at a nearly linear rate from 1875 pairs in 1986 to 6471 pairs in the year 2000.

    a. Find a linear function $p(t)$ that models the number of breeding pairs from 1986 to 2000 $(0 \le t \le 14)$.

    b. Using the function in part (a), approximately how many breeding pairs were in the lower 48 states in 1995?

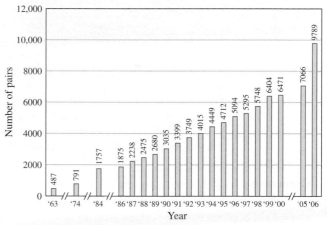

Source: U.S. Fish and Wildlife Service

56. Temperature scales

    a. Find the linear function $C = f(F)$ that gives the reading on the Celsius temperature scale corresponding to a reading on the Fahrenheit scale. Use the facts that $C = 0$ when $F = 32$ (freezing point) and $C = 100$ when $F = 212$ (boiling point).

    b. At what temperature are the Celsius and Fahrenheit readings equal?

57. Automobile lease vs. buy A car dealer offers a purchase option and a lease option on all new cars. Suppose you are interested in a car that can be bought outright for $25,000 or leased for a start-up fee of $1200 plus monthly payments of $350.

    a. Find the linear function $y = f(m)$ that gives the total amount you have paid on the lease option after m months.

b. With the lease option, after a 48-month (4-year) term, the car has a residual value of $10,000, which is the amount that you could pay to purchase the car. Assuming no other costs, should you lease or buy?

58–59. Functions from geometry

58. The surface area of a sphere of radius r is $S = 4\pi r^2$. Solve for r in terms of S and graph the radius function for $S \geq 0$.

T 59. A single slice through a sphere of radius r produces a *cap* of the sphere. If the thickness of the cap is h, then its volume is $V = \frac{1}{3}\pi h^2(3r - h)$. Graph the volume as a function of h for a sphere of radius 1. For what values of h does this function make sense?

T 60. Walking and rowing Kelly has finished a picnic on an island that is 200 m off shore, as shown in the figure. She wants to return to a beach house that is 600 m from the point P on the shore closest to the island. She plans to row a boat to a point on shore x meters from P and then jog along the (straight) shore to the house.

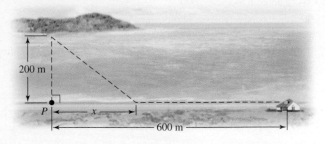

a. Let $d(x)$ be the total length of her trip as a function of x. Graph this function.

b. Suppose that Kelly can row at 2 m/s and jog at 4 m/s. Let $T(x)$ be the total time for her trip as a function of x. Graph $y = T(x)$.

c. Based on your graph in part (b), estimate the point on the shore at which Kelly should land in order to minimize the total time of her trip. What is that minimum time?

T 61. Optimal boxes Imagine a lidless box with height h and a square base whose sides have length x. The box must have a volume of 125 ft³.

a. Find and graph the function $S(x)$ that gives the surface area of the box for all values of $x > 0$.

b. Based on your graph in part (a), estimate the value of x that produces the box with a minimum surface area.

Additional Exercises

62. Composition of polynomials Let f be an nth-degree polynomial and let g be an mth-degree polynomial. What is the degree of the following polynomials?

a. $f \cdot f$ **b.** $f \circ f$ **c.** $f \cdot g$ **d.** $f \circ g$

63. Parabola vertex property Prove that if a parabola crosses the x-axis twice, the x-coordinate of the vertex of the parabola is halfway between the x-intercepts.

64. Parabola properties Consider the general quadratic function $f(x) = ax^2 + bx + c$, with $a \neq 0$.

a. Find the coordinates of the vertex in terms of a, b, and c.

b. Find the conditions on a, b, and c that guarantee that the graph of f crosses the x-axis twice.

65. Factorial function The factorial function is defined for positive integers as $n! = n(n-1)(n-2)\cdots3\cdot2\cdot1$.

a. Make a table of the factorial function for $n = 1, 2, 3, 4, 5$.

b. Graph these data points and then connect them with a smooth curve.

c. What is the least value of n for which $n! > 10^6$?

66. Sum of integers Let $S(n) = 1 + 2 + \cdots + n$, where n is a positive integer. It can be shown that $S(n) = n(n + 1)/2$.

a. Make a table of $S(n)$ for $n = 1, 2, \ldots, 10$.

b. How would you describe the domain of this function?

c. What is the least value of n for which $S(n) > 1000$?

67. Sum of squared integers Let $T(n) = 1^2 + 2^2 + \cdots + n^2$, where n is a positive integer. It can be shown that $T(n) = n(n + 1)(2n + 1)/6$.

a. Make a table of $T(n)$ for $n = 1, 2, \ldots, 10$.

b. How would you describe the domain of this function?

c. What is the least value of n for which $T(n) > 1000$?

QUICK CHECK ANSWERS

1. Yes; no **2.** $(-\infty, \infty), [0, \infty)$ **3.** Domain and range are $(-\infty, \infty)$. Domain and range are $[0, \infty)$. **4.** Shift the graph of f horizontally 4 units to the left. ◄

1.3 Inverse, Exponential, and Logarithmic Functions

Exponential functions are fundamental to all of mathematics. Many processes in the world around us are modeled by *exponential functions*, and they appear in finance, medicine, ecology, biology, economics, anthropology, and physics (among other disciplines). Every exponential function has an inverse function, which is a member of the family of *logarithmic functions*, also discussed in this section.

Exponential Functions

Exponential functions have the form $f(x) = b^x$, where the base $b \neq 1$ is a positive real number. An important question arises immediately: For what values of x can b^x be evaluated? We certainly know how to compute b^x when x is an integer. For example, $2^3 = 8$ and $2^{-4} = 1/2^4 = 1/16$. When x is rational, the numerator and denominator are interpreted as a power and root, respectively:

$$\overset{\text{power}}{16^{3/4}} = 16^{3/4} = \left(\sqrt[4]{16}\right)^3 = 8$$
$$\underset{\text{root}}{}$$

> $16^{3/4}$ can also be computed as $\sqrt[4]{16^3} = \sqrt[4]{4096} = 8$.

But what happens when x is irrational? How should 2^π be understood? Your calculator provides an approximation to 2^π, but where does the approximation come from? These questions will be answered eventually. For now we assume that b^x can be defined for all real numbers x and it can be approximated as closely as desired by using rational numbers as close to x as needed.

> **Exponent Rules**
>
> For any base $b > 0$ and real numbers x and y, we have
>
> **E1.** $b^x b^y = b^{x+y}$
>
> **E2.** $\dfrac{b^x}{b^y} = b^{x-y}$
>
> $\left(\text{which includes } \dfrac{1}{b^x} = b^{-x}\right)$
>
> **E3.** $(b^x)^y = b^{xy}$
>
> **E4.** $b^x > 0$ for all x

QUICK CHECK 1 Is it possible to raise a positive number b to a power and obtain a negative number? Is it possible to obtain zero? ◄

Properties of Exponential Functions $f(x) = b^x$

1. Because b^x is defined for all real numbers, the domain of f is $\{x: -\infty < x < \infty\}$. Because $b^x > 0$ for all values of x, the range of f is $\{y: 0 < y < \infty\}$.

2. For all $b > 0$, $b^0 = 1$, and thus $f(0) = 1$.

3. If $b > 1$, then f is an increasing function of x (Figure 1.42). For example, if $b = 2$, then $2^x > 2^y$ whenever $x > y$.

4. If $0 < b < 1$, then f is a decreasing function of x. For example, if $b = \frac{1}{2}$,

$$f(x) = \left(\frac{1}{2}\right)^x = \frac{1}{2^x} = 2^{-x}$$

and because 2^x increases with x, 2^{-x} decreases with x (Figure 1.43).

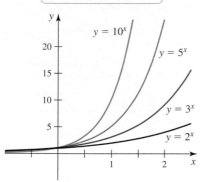

Larger values of b produce greater rates of increase in b^x if $b > 1$.

$y = 10^x$
$y = 5^x$
$y = 3^x$
$y = 2^x$

FIGURE 1.42

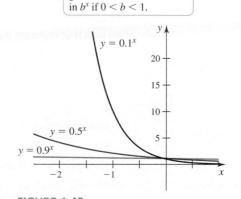

Smaller values of b produce greater rates of decrease in b^x if $0 < b < 1$.

$y = 0.1^x$
$y = 0.5^x$
$y = 0.9^x$

FIGURE 1.43

QUICK CHECK 2 Explain why $f(x) = (1/3)^x$ is a decreasing function. ◄

▷ The notation e was proposed by the Swiss mathematician Leonhard Euler (pronounced *oiler*) (1707–1783).

The Natural Exponential Function One of the bases used for exponential functions is special. For reasons that will become evident in upcoming chapters, the special base is e, one of the fundamental constants of mathematics. It is an irrational number with a value of $e = 2.718281828459\ldots$.

DEFINITION The Natural Exponential Function

The **natural exponential function** is $f(x) = e^x$, which has the base $e = 2.718281828459\ldots$.

The base e gives an exponential function that has the following valuable property. As shown in Figure 1.44, the graph of $y = e^x$ lies between the graphs of $y = 2^x$ and $y = 3^x$ (because $2 < e < 3$). At every point on the graph of $y = e^x$, it is possible to draw a *tangent line* (discussed in Chapter 2) that touches the graph only at that point. The natural exponential function is the only exponential function with the property that the slope of the tangent line at $x = 0$ is 1 (Figure 1.44); thus, e^x has both value and slope equal to 1 at $x = 0$. This property—minor as it may seem—leads to many simplifications when we do calculus with exponential functions.

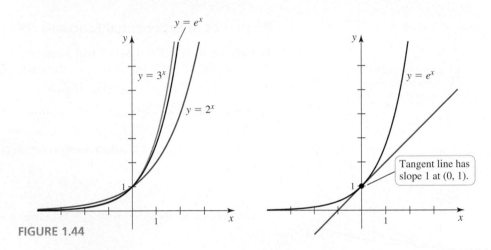

FIGURE 1.44

Inverse Functions

Consider the linear function $f(x) = 2x$, which takes any value of x and doubles it. The function that reverses this process by taking any value of $f(x) = 2x$ and mapping it back to x is called the *inverse function* of f, denoted f^{-1}. In this case, the inverse function is $f^{-1}(x) = x/2$. The effect of applying these two functions in succession looks like this:

$$x \xrightarrow{\ f\ } 2x \xrightarrow{\ f^{-1}\ } x$$

We now generalize this idea.

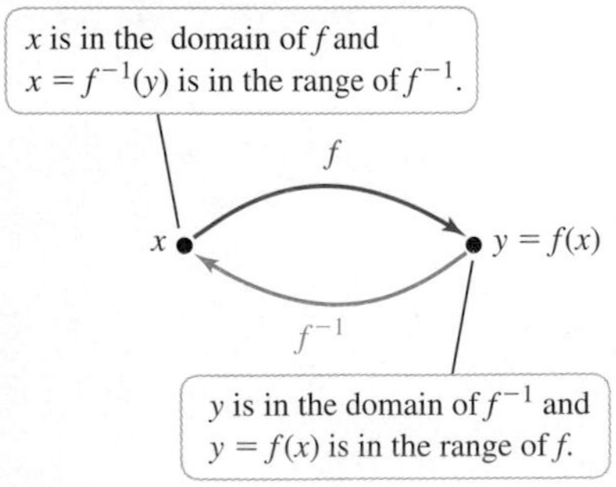

x is in the domain of f and $x = f^{-1}(y)$ is in the range of f^{-1}.

f

$x \bullet \qquad\qquad \bullet\, y = f(x)$

f^{-1}

y is in the domain of f^{-1} and $y = f(x)$ is in the range of f.

FIGURE 1.45

▷ The notation f^{-1} for the inverse can be confusing. The inverse is not the reciprocal; that is, $f^{-1}(x)$ is not $1/f(x) = (f(x))^{-1}$. We adopt the common convention of using simply *inverse* to mean *inverse function*.

DEFINITION Inverse Function

Given a function f, its inverse (if it exists) is a function f^{-1} such that whenever $y = f(x)$, then $f^{-1}(y) = x$ (Figure 1.45).

QUICK CHECK 3 What is the inverse of $f(x) = \frac{1}{3}x$? What is the inverse of $f(x) = x - 7$? ◁

Because the inverse "undoes" the original function, if we start with a value of x, apply f to it, and then apply f^{-1} to the result, we recover the original value of x; that is,

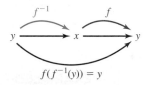

$$f^{-1}(f(x)) = x$$

Similarly, if we apply f^{-1} to a value of y and then apply f to the result, we recover the original value of y; that is,

$$f(f^{-1}(y)) = y$$

One-to-One Functions   We have defined the inverse of a function, but said nothing about when it exists. To ensure that f has an inverse on a domain, f must be *one-to-one* on that domain. This property means that every output of the function f must correspond to exactly one input. The one-to-one property is checked graphically by using the **horizontal line test**.

DEFINITION   One-to-One Functions and the Horizontal Line Test

A function f is **one-to-one** on a domain D if each value of $f(x)$ corresponds to exactly one value of x in D. More precisely, f is one-to-one on D if $f(x_1) \neq f(x_2)$ whenever $x_1 \neq x_2$, for x_1 and x_2 in D. The **horizontal line test** says that every horizontal line intersects the graph of a one-to-one function at most once (Figure 1.46).

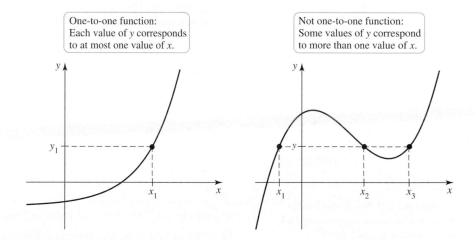

FIGURE 1.46

For example, in Figure 1.47, some horizontal lines intersect the graph of $f(x) = x^2$ twice. Therefore, f does not have an inverse function on the interval $(-\infty, \infty)$. However, if f is restricted to the interval $(-\infty, 0]$ or $[0, \infty)$, then it does pass the horizontal line test and it is one-to-one on these intervals.

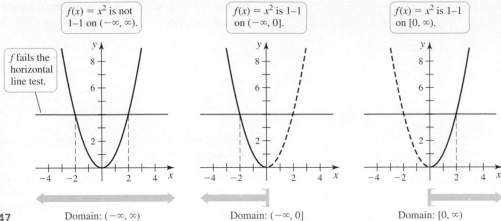

FIGURE 1.47

Domain: $(-\infty, \infty)$ Domain: $(-\infty, 0]$ Domain: $[0, \infty)$

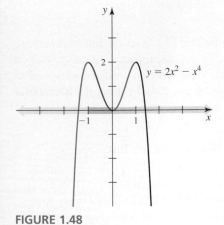

FIGURE 1.48

EXAMPLE 1   One-to-one functions Determine the (largest possible) intervals on which the function $f(x) = 2x^2 - x^4$ (Figure 1.48) is one-to-one.

SOLUTION The function is not one-to-one on the entire real line because it fails the horizontal line test. However, on the intervals $(-\infty, -1], [-1, 0], [0, 1]$, and $[1, \infty)$, f is one-to-one. The function is also one-to-one on any subinterval of these four intervals.   *Related Exercises 11–12* ◄

Existence of Inverse Functions Figure 1.49a illustrates the actions of a function f and its inverse f^{-1}. We see that f maps a value of x to a unique value of y. In turn, f^{-1} maps that value of y back to the original value of x. When f is *not* one-to-one, this procedure cannot be carried out (Figure 1.49b).

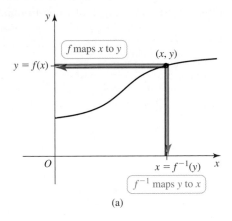

FIGURE 1.49

(a)   (b)

▷ The statement that a one-to-one function has an inverse may be plausible based on the graph of one-to-one functions. However, the proof of this theorem is fairly technical and is omitted.

THEOREM 1.1   Existence of Inverse Functions

Let f be a one-to-one function on a domain D with a range R. Then f has a unique inverse f^{-1} with domain R and range D such that

$$f^{-1}(f(x)) = x \quad \text{and} \quad f(f^{-1}(y)) = y$$

where x is in D and y is in R.

QUICK CHECK 4 The function that gives degrees Fahrenheit in terms of degrees Celsius is $F = 9C/5 + 32$. Explain why this function has an inverse. ◄

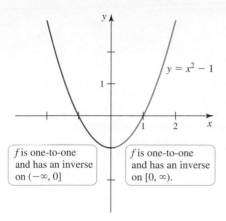

$y = x^2 - 1$

f is one-to-one and has an inverse on $(-\infty, 0]$

f is one-to-one and has an inverse on $[0, \infty)$.

FIGURE 1.50

> Once you find a formula for f^{-1} you can check your work by verifying that $f^{-1}(f(x)) = x$ and $f(f^{-1}(x)) = x$.

EXAMPLE 2   Does an inverse exist? Determine intervals on which $f(x) = x^2 - 1$ has an inverse function.

SOLUTION On the interval $(-\infty, \infty)$ the function does not pass the horizontal line test and is not one-to-one (Figure 1.50). However, if f is restricted to the intervals $(-\infty, 0]$ or $[0, \infty)$, it is one-to-one and an inverse exists.

Related Exercises 13–16 ◄

Finding Inverse Functions   The crux of finding an inverse for a function f is solving the equation $y = f(x)$ for x in terms of y. If it is possible to do so, then we have found a relationship of the form $x = f^{-1}(y)$. Interchanging x and y in $x = f^{-1}(y)$ so that x is the independent variable (which is the customary role for x), the inverse has the form $y = f^{-1}(x)$. Notice that if f is not one-to-one, this process leads to more than one inverse function.

PROCEDURE   Finding an Inverse Function

Suppose f is one-to-one on an interval I. To find f^{-1}:

1. Solve $y = f(x)$ for x. If necessary, choose the function that corresponds to I.

2. Interchange x and y and write $y = f^{-1}(x)$.

EXAMPLE 3   Finding inverse functions Find the inverse(s) of the following functions. Restrict the domain of f if necessary.

a. $f(x) = 2x + 6$ **b.** $f(x) = x^2 - 1$

SOLUTION

> A constant function (whose graph is a horizontal line) fails the horizontal line test and does not have an inverse.

a. Linear functions (except for constant linear functions) are one-to-one on the entire real line. Therefore, an inverse function for f exists for all values of x.

Step 1: Solve $y = f(x)$ for x: We see that $y = 2x + 6$ implies that $2x = y - 6$, or $x = (y - 6)/2$.

Step 2: Interchange x and y and write $y = f^{-1}(x)$:

$$y = f^{-1}(x) = \frac{x - 6}{2}$$

It is instructive to verify that the inverse relations $f(f^{-1}(x)) = x$ and $f^{-1}(f(x)) = x$ are satisfied:

$$f(f^{-1}(x)) = f\left(\frac{x - 6}{2}\right) = \underbrace{2\left(\frac{x - 6}{2}\right) + 6}_{f(x) = 2x + 6} = x - 6 + 6 = x.$$

$$f^{-1}(f(x)) = f^{-1}(2x + 6) = \underbrace{\frac{(2x + 6) - 6}{2}}_{f^{-1}(x) = (x - 6)/2} = x.$$

b. As shown in Example 2, the function $f(x) = x^2 - 1$ is not one-to-one on the entire real line; however, it is one-to-one on $(-\infty, 0]$ and on $[0, \infty)$. If we restrict our attention to either of these intervals, then an inverse function can be found.

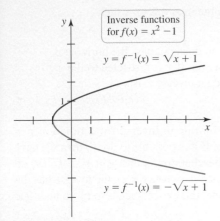

FIGURE 1.51

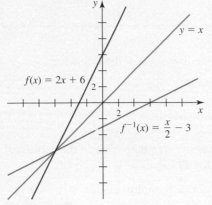

FIGURE 1.52

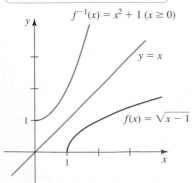

FIGURE 1.53

Step 1:   Solve $y = f(x)$ for x:

$$y = x^2 - 1$$
$$x^2 = y + 1$$
$$x = \begin{cases} +\sqrt{y+1} \\ -\sqrt{y+1} \end{cases}$$

Each branch of the square root corresponds to an inverse function.

Step 2:   Interchange x and y and write $y = f^{-1}(x)$:

$$y = f^{-1}(x) = \sqrt{x+1} \quad \text{or} \quad y = f^{-1}(x) = -\sqrt{x+1}$$

The interpretation of this result is important. Taking the positive branch of the square root, the inverse function $y = f^{-1}(x) = \sqrt{x+1}$ gives positive values of y; it corresponds to the branch of $f(x) = x^2 - 1$ on the interval $[0, \infty)$ (Figure 1.51). The negative branch of the square root, $y = f^{-1}(x) = -\sqrt{x+1}$, is another inverse function that gives negative values of y; it corresponds to the branch of $f(x) = x^2 - 1$ on the interval $(-\infty, 0]$.   *Related Exercises 17–24* ◄

QUICK CHECK 5 On what interval(s) does the function $f(x) = x^3$ have an inverse? ◄

Graphing Inverse Functions

The graphs of a function and its inverse have a special relationship, which is illustrated in the following example.

EXAMPLE 4   Graphing inverse functions Plot f and f^{-1} on the same coordinate axes.

a. $f(x) = 2x + 6$ **b.** $f(x) = \sqrt{x-1}$

SOLUTION

a. The inverse of $f(x) = 2x + 6$, found in Example 3, is

$$y = f^{-1}(x) = \frac{x-6}{2} = \frac{x}{2} - 3.$$

The graphs of f and f^{-1} are shown in Figure 1.52. Notice that both f and f^{-1} are increasing linear functions and they intersect at $(-6, -6)$.

b. The domain of $f(x) = \sqrt{x-1}$ is the set $\{x : x \geq 1\}$. On this domain f is one-to-one and has an inverse. It can be found in two steps:

Step 1:   Solve $y = \sqrt{x-1}$ for x:

$$y^2 = x - 1 \quad \text{or} \quad x = y^2 + 1$$

Step 2:   Interchange x and y and write $y = f^{-1}(x)$:

$$y = f^{-1}(x) = x^2 + 1$$

The graphs of f and f^{-1} are shown in Figure 1.53.   *Related Exercises 25–32* ◄

Looking closely at the graphs in Figure 1.52 and Figure 1.53, you see a symmetry that always occurs when a function and its inverse are plotted on the same set of axes. In each figure, one curve is the reflection of the other curve across the line $y = x$. These curves have *symmetry about the line $y = x$*, which means that the point (a, b) is on one curve whenever the point (b, a) is on the other curve (Figure 1.54).

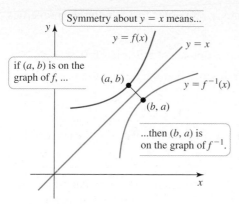

FIGURE 1.54

The explanation for the symmetry comes directly from the definition of the inverse. Suppose that the point (a, b) is on the graph of $y = f(x)$, which means that $b = f(a)$. By the definition of the inverse function, we know that $a = f^{-1}(b)$, which means that the point (b, a) is on the graph of $y = f^{-1}(x)$. This argument applies to all relevant points (a, b), so whenever (a, b) is on the graph of f, (b, a) is on the graph of f^{-1}. As a consequence, the graphs are symmetric about the line $y = x$.

Logarithmic Functions

Everything we learned about inverse functions is now applied to the exponential function $f(x) = b^x$. For any $b > 0$, with $b \neq 1$, this function is one-to-one on the interval $(-\infty, \infty)$. Therefore, it has an inverse.

> ▶ Logarithms were invented around 1600 for calculating purposes by the Scotsman John Napier and the Englishman Henry Briggs. Unfortunately, the word *logarithm*, derived from the Greek for reasoning (*logos*) with numbers (*arithmos*), doesn't help with the meaning of the word. **When you see** *logarithm*, **you should think** *exponent*.

DEFINITION   Logarithmic Function Base b

For any base $b > 0$, with $b \neq 1$, the **logarithmic function base b**, denoted $y = \log_b x$, is the inverse of the exponential function $y = b^x$. The inverse of the natural exponential function with base $b = e$ is the **natural logarithm function**, denoted $y = \ln x$.

The inverse relationship between logarithmic and exponential functions may be stated concisely in several ways. First, we have

$$y = \log_b x \quad \text{provided} \quad b^y = x.$$

Combining these two conditions results in two important relations.

> ▶ **Logarithm Rules**
>
> For any base $b > 0$ ($b \neq 1$) and positive real numbers x and y, the following relations hold:
>
> **L1.** $\log_b(xy) = \log_b x + \log_b y$
>
> **L2.** $\log_b\left(\dfrac{x}{y}\right) = \log_b x - \log_b y$
>
> $\left(\text{includes } \log_b \dfrac{1}{x} = -\log_b x\right)$
>
> **L3.** $\log_b(x^y) = y \log_b x$
>
> **L4.** $\log_b b = 1$

Inverse Relations for Exponential and Logarithmic Functions

For any base $b > 0$, with $b \neq 1$, the following inverse relations hold:

I1. $b^{\log_b x} = x$ (for all $x > 0$).

I2. $\log_b b^x = x$ (for all real values of x).

Properties of Logarithmic Functions   The graph of the logarithmic function is generated using the symmetry of the graphs of a function and its inverse. Figure 1.55 shows how the graph of $y = b^x$, for $b > 1$, is reflected across the line $y = x$ to obtain the graph of $y = \log_b x$.

The graphs of $y = \log_b x$ are shown (Figure 1.56) for several bases $b > 1$. Logarithms with fractional bases ($0 < b < 1$), although well defined, are generally not used. In fact, fractional bases can always be converted to bases with $b > 1$.

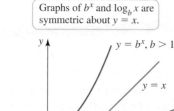

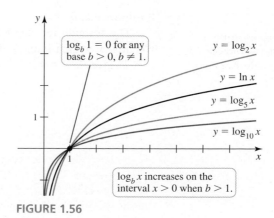

FIGURE 1.55

FIGURE 1.56

Logarithmic functions with base $b > 0$ satisfy properties that parallel the properties of the exponential functions given earlier.

1. Because the range of b^x is $\{y: 0 < y < \infty\}$, the domain of $\log_b x$ is $\{x: 0 < x < \infty\}$.

2. The domain of b^x is all real numbers, which implies that the range of $\log_b x$ is all real numbers.

3. Because $b^0 = 1$, it follows that $\log_b 1 = 0$.

4. If $b > 1$, then $\log_b x$ is an increasing function of x. For example, if $b = e$, then $\ln x > \ln y$ whenever $x > y$ (Figure 1.56).

> **QUICK CHECK 6** What is the domain of $f(x) = \log_b(x^2)$? What is the range of $f(x) = \log_b(x^2)$? ◄

EXAMPLE 5 Using inverse relations One thousand grams of a particular radioactive substance decays according to the function $m(t) = 1000e^{-t/850}$, where $t \geq 0$ measures time in years. When does the mass of the substance reach the safe level deemed to be 1 g?

SOLUTION Setting $m(t) = 1$, we solve $1000e^{-t/850} = 1$ by dividing both sides by 1000 and taking the natural logarithm of both sides:

$$\ln\left(e^{-t/850}\right) = \ln\left(\frac{1}{1000}\right)$$

> Because $\log_b x$ is a one-to-one function for $x > 0$, if $x = y$, then $\log_b x = \log_b y$. Therefore, we can take the $\log_b$ of both sides of an equation and produce a valid equation.

This equation is simplified by calculating $\ln(1/1000) \approx -6.908$ and observing that $\ln\left(e^{-t/850}\right) = -\dfrac{t}{850}$ (inverse property I2). Therefore,

$$-\frac{t}{850} \approx -6.908.$$

Solving for t, we find that $t \approx (-850)(-6.908) \approx 5872$ years.

Related Exercises 33–48 ◄

Change of Base

When working with logarithms and exponentials, it doesn't matter *in principle* which base is used. However, there are practical reasons for switching between bases. For example, most calculators have built-in logarithmic functions in just one or two bases. If you need to use a different base, then the change-of-base rules are essential.

Consider changing bases with exponential functions. Specifically, suppose you wish to express b^x (base b) in the form e^y (base e), where y must be determined. Taking the natural logarithm of both sides of $e^y = b^x$, we have

$$\underbrace{\ln e^y}_{y} = \underbrace{\ln b^x}_{x \ln b} \quad \text{which implies that } y = x \ln b.$$

It follows that $b^x = e^y = e^{x \ln b}$. For example, $4^x = e^{x \ln 4}$.

> A similar argument is used to derive more general formulas for changing from base b to any other positive base c.

The formula for changing from $\log_b x$ to $\ln x$ is derived in a similar way. We let $y = \log_b x$, which implies that $x = b^y$. Taking the natural logarithm of both sides of $x = b^y$ gives $\ln x = \ln b^y = y \ln b$. Solving for $y = \log_b x$ gives us the required formula:

$$y = \log_b x = \frac{\ln x}{\ln b}.$$

Change-of-Base Rules

Let b be a positive real number with $b \neq 1$. Then

$$b^x = e^{x \ln b}, \text{ for all } x \quad \text{and} \quad \log_b x = \frac{\ln x}{\ln b}, \text{ for } x > 0.$$

More generally, if c is a positive real number with $c \neq 1$, then

$$b^x = c^{x \log_c b}, \text{ for all } x \quad \text{and} \quad \log_b x = \frac{\log_c x}{\log_c b}, \text{ for } x > 0.$$

EXAMPLE 6   Changing bases

a. Express 2^{x+4} as an exponential function with base e.

b. Express $\log_2 x$ using base e and base 32.

SOLUTION

a. Using the change-of-base rule for exponential functions, we have

$$2^{x+4} = e^{(x+4)\ln 2}.$$

b. Using the change-of-base rule for logarithmic functions, we have

$$\log_2 x = \frac{\ln x}{\ln 2} \approx 1.44 \ln x.$$

To change from base 2 to base 32, we use the general change-of-base formula:

$$\log_2 x = \frac{\log_{32} x}{\log_{32} 2} = \frac{\log_{32} x}{1/5} = 5 \log_{32} x$$

The middle step follows from the fact that $2 = 32^{1/5}$, so $\log_{32} 2 = \frac{1}{5}$.

Related Exercises 49–56 ◄

SECTION 1.3 EXERCISES

Review Questions

1. For $b > 0$, what are the domain and range of $f(x) = b^x$?

2. Give an example of a function that is one-to-one on the entire real number line.

3. Explain why a function that is not one-to-one on an interval I cannot have an inverse function on I.

4. Explain with pictures why (a, b) is on the graph of f whenever (b, a) is on the graph of f^{-1}.

5. Sketch a function that is one-to-one and positive for $x \geq 0$. Make a rough sketch of its inverse.

6. Express the inverse of $f(x) = 3x - 4$ in the form $y = f^{-1}(x)$.

7. Explain the meaning of $\log_b x$.

8. Explain how the property $b^{x+y} = b^x b^y$ is related to the property $\log_b (xy) = \log_b x + \log_b y$.

9. For $b > 0$ with $b \neq 1$, what are the domain and range of $f(x) = \log_b x$ and why?

10. Express 2^5 using base e.

Basic Skills

11–12. One-to-one functions *Answer the questions in the following exercises using the graph of f.*

11. Find three intervals on which f is one-to-one, making each interval as large as possible.

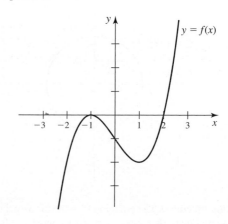

12. Find four intervals on which f is one-to-one, making each interval as large as possible.

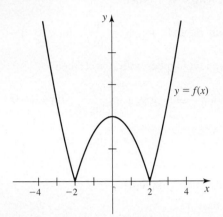

13–16. Where do inverses exist? *Use analytical and/or graphical methods to determine the intervals on which the following functions have an inverse (make each interval as large as possible).*

13. $f(x) = 3x + 4$

14. $f(x) = |2x + 1|$

15. $f(x) = 1/(x - 5)$

16. $f(x) = -(6 - x)^2$

17–22. Finding inverse functions

a. *Find the inverse of each function (on the given interval, if specified) and write it in the form $y = f^{-1}(x)$.*

b. *Verify the relationships $f(f^{-1}(x)) = x$ and $f^{-1}(f(x)) = x$.*

17. $f(x) = 6 - 4x$

18. $f(x) = 3x^3$

19. $f(x) = 3x + 5$

20. $f(x) = x^2 + 4$ for $x \geq 0$

21. $f(x) = \sqrt{x + 2}$

22. $f(x) = 2/(x^2 + 1)$ for $x \geq 0$

23. Splitting up curves The unit circle $x^2 + y^2 = 1$ consists of four one-to-one functions $f_1(x)$, $f_2(x)$, $f_3(x)$, and $f_4(x)$ (see figure).

a. Find a formula for each function and the domain of each function.

b. Find the inverse of each function and write it as $y = f^{-1}(x)$.

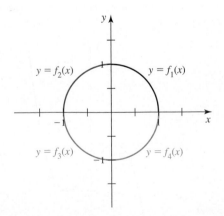

24. Splitting up curves The equation $y^4 = 4x^2$ is associated with four one-to-one functions $f_1(x)$, $f_2(x)$, $f_3(x)$, and $f_4(x)$ (see figure).

a. Find a formula for each function and the domain of each function.

b. Find the inverse of each function and write it as $y = f^{-1}(x)$.

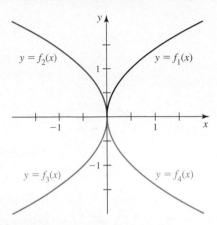

25–30. Graphing inverse functions *Find the inverse function (on the given interval, if specified) and graph both f and f^{-1} on the same set of axes. Check your work by looking for the required symmetry in the graphs.*

25. $f(x) = 8 - 4x$

26. $f(x) = 4x - 12$

27. $f(x) = \sqrt{x}$

28. $f(x) = \sqrt{3 - x}$

29. $f(x) = x^4 + 4$ for $x > 0$

30. $f(x) = 6/(x^2 - 9)$ for $x > 3$

31–32. Graphs of inverses *Sketch the graph of the inverse function.*

31.

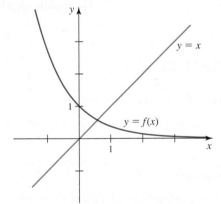

32.

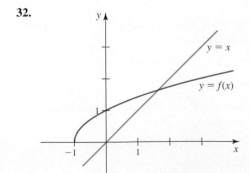

33–36. Solving logarithmic equations *Solve the following equations.*

33. $\log_{10} x = 3$

34. $\log_5 x = -1$

35. $\log_8 x = \frac{1}{3}$

36. $\log_b 125 = 3$

37–40. Properties of logarithms *Assume* $\log_b x = 0.36$, $\log_b y = 0.56$, *and* $\log_b z = 0.83$. *Evaluate the following expressions.*

37. $\log_b xz$

38. $\log_b \dfrac{\sqrt{xy}}{z}$

39. $\log_b \dfrac{\sqrt{x}}{\sqrt[3]{z}}$

40. $\log_b \dfrac{b^2 x^{5/2}}{\sqrt{y}}$

41–48. Solving equations *Without using a calculator, solve the following equations.*

41. $10^{x^2-4} = 1$

42. $3^{x^2-5x-5} = \frac{1}{3}$

43. $2^{|x|} = 16$

44. $9^x + 3^{x+1} - 18 = 0$

45. $7^x = 21$ 46. $2^x = 55$ 47. $3^{3x-4} = 15$ 48. $5^{3x} = 29$

49–52. Calculator base change *Write the following logarithms in terms of the natural logarithm. Then use a calculator to find the value of the logarithm, rounding your result to four decimal places.*

49. $\log_2 15$ 50. $\log_3 30$ 51. $\log_4 40$ 52. $\log_6 60$

53–56. Changing bases *Convert the following expressions to the indicated base.*

53. 2^x using base e

54. $3^{\sin x}$ using base e

55. $\ln |x|$ using $\log_5$

56. $\log_2 (x^2 + 1)$ using $\ln$

Further Explorations

57. **Explain why or why not** Determine whether the following statements are true and give an explanation or counterexample.

   a. If $y = 3^x$, then $x = \sqrt[3]{y}$.   b. $\dfrac{\log_b x}{\log_b y} = \log_b x - \log_b y$

   c. $\log_5 4^6 = 4 \log_5 6$ d. $2 = 10^{\log_{10} 2}$

   e. $2 = \ln 2^e$

   f. If $f(x) = x^2 + 1$, then $f^{-1}(x) = 1/(x^2 + 1)$.

   g. If $f(x) = 1/x$, then $f^{-1}(x) = 1/x$.

58. **Graphs of exponential functions** The following figure shows the graphs of $y = 2^x$, $y = 3^x$, $y = 2^{-x}$, and $y = 3^{-x}$. Match each curve with the correct function.

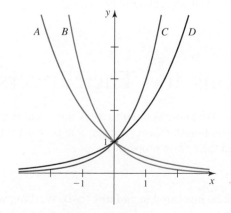

59. **Graphs of logarithmic functions** The following figure shows the graphs of $y = \log_2 x$, $y = \log_4 x$, and $y = \log_{10} x$. Match each curve with the correct function.

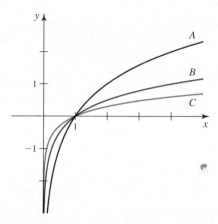

60. **Graphs of modified exponential functions** Without using a graphing utility, sketch the graph of $y = 2^x$. Then, on the same set of axes, sketch the graphs of $y = 2^{-x}$, $y = 2^{x-1}$, $y = 2^x + 1$, and $y = 2^{2x}$.

61. **Graphs of modified logarithmic functions** Without using a graphing utility, sketch the graph of $y = \log_2 x$. Then, on the same set of axes, sketch the graphs of $y = \log_2 (x - 1)$, $y = \log_2 x^2$, $y = (\log_2 x)^2$, and $y = \log_2 x + 1$.

62. **Large intersection point** Use any means to approximate the point(s) of intersection of the graphs of $f(x) = e^x$ and $g(x) = x^{123}$. Consider using logarithms.

63–66. Finding all inverses *Find all the inverses associated with the following functions and state their domains.*

63. $f(x) = (x + 1)^3$ 64. $f(x) = (x - 4)^2$

65. $f(x) = 2/(x^2 + 2)$ 66. $f(x) = 2x/(x + 2)$

Applications

67. **Population model** A culture of bacteria has a population of 150 cells when it is first observed. The population doubles every 12 hr, which means its population is governed by the function $p(t) = 150 \times 2^{t/12}$, where t is the number of hours after the first observation.

   a. Verify that $p(0) = 150$, as claimed.
   b. Show that the population doubles every 12 hr, as claimed.
   c. What is the population 4 days after the first observation?
   d. How long does it take the population to triple in size?
   e. How long does it take the population to reach 10,000?

68. **Charging a capacitor** A capacitor is a device that stores electrical charge. The charge on a capacitor accumulates according to the function $Q(t) = a(1 - e^{-t/c})$, where t is measured in seconds, and a and c are physical constants. The *steady-state charge* is the value that $Q(t)$ approaches as t becomes large.

   a. Graph the charge function for $t \geq 0$ using $a = 1$ and $c = 10$. Find a graphing window that shows the full range of the function.
   b. Vary the value of a holding c fixed. Describe the effect on the curve. How does the steady-state charge vary with a?

c. Vary the value of c holding a fixed. Describe the effect on the curve. How does the steady-state charge vary with c?

d. Find a formula that gives the steady-state charge in terms of a and c.

T 69. Height and time The height of a baseball hit straight up with an initial velocity of 64 ft/s is given by $h = f(t) = 64t - 16t^2$, where t is measured in seconds after the hit.

a. Is this function one-to-one on the interval $0 \le t \le 4$?

b. Find the inverse function that gives the time t at which the ball is at height h as the ball travels *upward*. Express your answer in the form $t = f^{-1}(h)$.

c. Find the inverse function that gives the time t at which the ball is at height h as the ball travels *downward*. Express your answer in the form $t = f^{-1}(h)$.

d. At what time is the ball at a height of 30 ft on the way up?

e. At what time is the ball at a height of 10 ft on the way down?

T 70. Velocity of a skydiver The velocity of a skydiver (in m/s) t seconds after jumping from the plane is $v(t) = 600(1 - e^{-kt/60})/k$, where k is a constant. The *terminal velocity* of the skydiver is the value that $v(t)$ approaches as t becomes large. Graph v with $k = 11$ and estimate the terminal velocity.

Additional Exercises

71. Reciprocal bases Assume that $b > 0$ and show that $\log_{1/b} x = -\log_b x$.

72. Proof of rule L1 Use the following steps to prove that $\log_b(xy) = \log_b x + \log_b y$.

a. Let $x = b^p$ and $y = b^q$. Solve these expressions for p and q, respectively.

b. Use property E1 for exponents to express xy in terms of b, p, and q.

c. Compute $\log_b(xy)$ and simplify.

73. Proof of rule L2 Modify Exercise 72 and use property E2 for exponents to prove that $\log_b(x/y) = \log_b x - \log_b y$.

74. Proof of rule L3 Use the following steps to prove that $\log_b(x^y) = y \log_b x$.

a. Let $x = b^p$. Solve this expression for p.

b. Use property E3 for exponents to express x^y in terms of b and p.

c. Compute $\log_b(x^y)$ and simplify.

T 75. Inverses of a quartic Consider the quartic polynomial $y = f(x) = x^4 - x^2$.

a. Graph f and find the intervals on which it is one-to-one. The goal is to find the inverse function on each of these intervals.

b. Make the substitution $u = x^2$ to solve the equation $y = f(x)$ for x in terms of y. Be sure you have included all possible solutions.

c. Write the inverse function in the form $y = f^{-1}(x)$ for each of the intervals found in part (a).

76. Inverse of composite functions

a. Let $g(x) = 2x + 3$ and $h(x) = x^3$. Consider the composite function $f(x) = g(h(x))$. Find f^{-1} directly and then express the inverse of f in terms of g^{-1} and h^{-1}.

b. Let $g(x) = x^2 + 1$ and $h(x) = \sqrt{x}$ for $x \ge 0$. Consider the composite function $f(x) = g(h(x))$. Find f^{-1} directly and then express the inverse of f in terms of g^{-1} and h^{-1}.

c. Explain why if h and g are one-to-one, the inverse of $f(x) = g(h(x))$ exists.

T 77–78. Inverses of (some) cubics *Finding the inverse of a cubic polynomial is equivalent to solving a cubic equation. A special case that is simpler than the general case is the cubic $f(x) = x^3 + ax$. Find the inverse of the following cubics using the substitution (known as Vieta's substitution) $x = z - a/(3z)$. Be sure to determine where the function is one-to-one.*

77. $f(x) = x^3 + 2x$ **78.** $f(x) = x^3 - 2x$

79. Nice property Prove that $(\log_b c)(\log_c b) = 1$, for $b > 0$ and $c > 0$.

QUICK CHECK ANSWERS

1. b^x is always positive (and never zero) for all x and for positive bases b.   **2.** Because $(1/3)^x = 1/3^x$ and 3^x increases as x increases, it follows that $(1/3)^x$ decreases as x increases.   **3.** $f^{-1}(x) = 3x$; $f^{-1}(x) = x + 7$.   **4.** For every Fahrenheit temperature, there is exactly one Celsius temperature, and vice versa. The given relation is also a linear function. It is one-to-one, so it has an inverse function.   **5.** The function $f(x) = x^3$ is one-to-one on $(-\infty, \infty)$, so it has an inverse for all values of x.   **6.** The domain of $\log_b(x^2)$ is all real numbers except zero (because x^2 is positive for $x \ne 0$). The range of $\log_b(x^2)$ is all real numbers. ◄

1.4 Trigonometric Functions and Their Inverses

This section is a review of what you need to know in order to study the calculus of trigonometric functions. Once the trigonometric functions are on stage, it makes sense to present the inverse trigonometric functions and their basic properties.

Radian Measure

Calculus typically requires that angles be measured in **radians** (rad). Working with a circle of radius r, the radian measure of an angle θ is the length of the arc associated with θ,

Degrees	Radians
0	0
30	$\pi/6$
45	$\pi/4$
60	$\pi/3$
90	$\pi/2$
120	$2\pi/3$
135	$3\pi/4$
150	$5\pi/6$
180	π

denoted s, divided by the radius of the circle r (Figure 1.57a). Working on a unit circle ($r = 1$), the radian measure of an angle is simply the length of the arc s associated with θ (Figure 1.57b). For example, the length of a full unit circle is 2π; therefore, an angle with a radian measure of π corresponds to a half circle ($\theta = 180°$) and an angle with a radian measure of $\pi/2$ corresponds to a quarter circle ($\theta = 90°$).

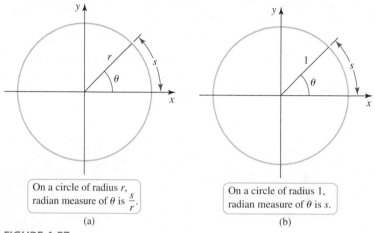

On a circle of radius r, radian measure of θ is $\dfrac{s}{r}$.

(a)

On a circle of radius 1, radian measure of θ is s.

(b)

FIGURE 1.57

QUICK CHECK 1 What is the radian measure of a 270° angle? What is the degree measure of a $5\pi/4$-rad angle? ◄

Trigonometric Functions

For acute angles, the trigonometric functions are defined as ratios of the sides of a right triangle (Figure 1.58). To extend these definitions to include all angles, we work in an xy-coordinate system with a circle of radius r centered at the origin. Suppose that $P(x, y)$ is a point on the circle. An angle θ is in **standard position** if its initial side is on the positive x-axis and its terminal side is the line segment OP between the origin and P. An angle is positive if it is obtained by a counterclockwise rotation from the positive x-axis (Figure 1.59). When the right-triangle definitions of Figure 1.58 are used with the right triangle in Figure 1.59, the trigonometric functions may be expressed in terms of x, y, and the radius of the circle, $r = \sqrt{x^2 + y^2}$.

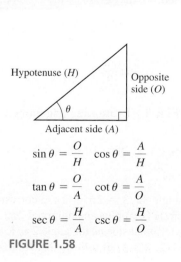

$$\sin\theta = \frac{O}{H} \quad \cos\theta = \frac{A}{H}$$

$$\tan\theta = \frac{O}{A} \quad \cot\theta = \frac{A}{O}$$

$$\sec\theta = \frac{H}{A} \quad \csc\theta = \frac{H}{O}$$

FIGURE 1.58

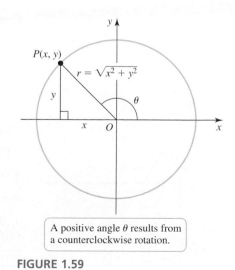

A positive angle θ results from a counterclockwise rotation.

FIGURE 1.59

▶ When working on a unit circle ($r = 1$), these definitions become

$$\sin\theta = y \qquad \cos\theta = x$$

$$\tan\theta = \frac{y}{x} \qquad \cot\theta = \frac{x}{y}$$

$$\sec\theta = \frac{1}{x} \qquad \csc\theta = \frac{1}{y}$$

> **DEFINITION   Trigonometric Functions**
>
> Let $P(x, y)$ be a point on a circle of radius r associated with the angle θ. Then
>
> $$\sin\theta = \frac{y}{r} \qquad \cos\theta = \frac{x}{r} \qquad \tan\theta = \frac{y}{x}$$
>
> $$\cot\theta = \frac{x}{y} \qquad \sec\theta = \frac{r}{x} \qquad \csc\theta = \frac{r}{y}$$

To find the trigonometric functions of the standard angles (multiples of 30° and 45°), it is helpful to know the radian measure of those angles and the coordinates of the associated points on the unit circle (Figure 1.60).

▶ Standard Triangles

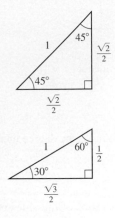

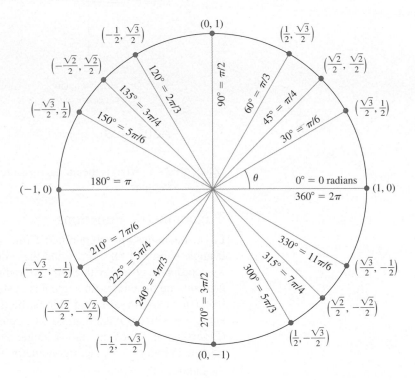

FIGURE 1.60

Combining the definitions of the trigonometric functions with the coordinates shown in Figure 1.60, we may evaluate these functions at any standard angle. For example,

$$\sin\frac{2\pi}{3} = \frac{\sqrt{3}}{2} \qquad \cos\frac{5\pi}{6} = -\frac{\sqrt{3}}{2} \qquad \tan\frac{7\pi}{6} = \frac{1}{\sqrt{3}}$$

$$\cot\frac{5\pi}{3} = -\frac{1}{\sqrt{3}} \qquad \sec\frac{7\pi}{4} = \sqrt{2} \qquad \csc\frac{3\pi}{2} = -1$$

EXAMPLE 1   Evaluating trigonometric functions Evaluate the following expressions.

a. $\sin(8\pi/3)$ **b.** $\csc(-11\pi/3)$

SOLUTION

a. The angle $8\pi/3 = 2\pi + 2\pi/3$ corresponds to a *counterclockwise* revolution of one full circle (2π) plus an additional $2\pi/3$ rad (Figure 1.61). Therefore, this angle has the same terminal side as the angle $2\pi/3$, and the corresponding point on the unit circle is $(-1/2, \sqrt{3}/2)$. It follows that $\sin(8\pi/3) = y = \sqrt{3}/2$.

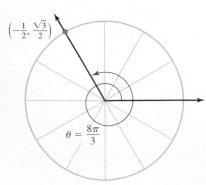

FIGURE 1.61

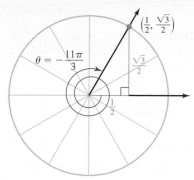

FIGURE 1.62

b. The angle $\theta = -11\pi/3 = -2\pi - 5\pi/3$ corresponds to a *clockwise* revolution of one full circle (2π) plus an additional $5\pi/3$ rad (Figure 1.62). Therefore, this angle has the same terminal side as the angle $\pi/3$. The coordinates of the corresponding point on the unit circle are $(1/2, \sqrt{3}/2)$, so $\csc(-11\pi/3) = 1/y = 2/\sqrt{3}$.

Related Exercises 15–22 ◄

QUICK CHECK 2 Evaluate $\cos(11\pi/6)$ and $\sin(5\pi/4)$. ◄

Trigonometric Identities

Trigonometric functions have a variety of properties, called **identities**, that are true for all angles in the domain. Here is a list of some commonly used identities.

Trigonometric Identities

Reciprocal Identities

$$\tan\theta = \frac{\sin\theta}{\cos\theta} \qquad \cot\theta = \frac{1}{\tan\theta} = \frac{\cos\theta}{\sin\theta}$$

$$\csc\theta = \frac{1}{\sin\theta} \qquad \sec\theta = \frac{1}{\cos\theta}$$

Pythagorean Identities

$$\sin^2\theta + \cos^2\theta = 1 \qquad 1 + \cot^2\theta = \csc^2\theta \qquad \tan^2\theta + 1 = \sec^2\theta$$

Double- and Half-Angle Formulas

$$\sin 2\theta = 2\sin\theta\cos\theta \qquad \cos 2\theta = \cos^2\theta - \sin^2\theta$$

$$\cos^2\theta = \frac{1 + \cos 2\theta}{2} \qquad \sin^2\theta = \frac{1 - \cos 2\theta}{2}$$

QUICK CHECK 3 Prove that $1 + \cot^2\theta = \csc^2\theta$. ◄

EXAMPLE 2 **Solving trigonometric equations** Solve the following equations.

a. $\sqrt{2}\sin x + 1 = 0$ **b.** $\cos 2x = \sin 2x$ where $0 \le x < 2\pi$.

SOLUTION

➤ By rationalizing the denominator, observe that $\dfrac{1}{\sqrt{2}} = \dfrac{1}{\sqrt{2}} \cdot \dfrac{\sqrt{2}}{\sqrt{2}} = \dfrac{\sqrt{2}}{2}$.

a. First, we solve for $\sin x$ to obtain $\sin x = -1/\sqrt{2} = -\sqrt{2}/2$. From the unit circle (Figure 1.60), we find that $\sin x = -\sqrt{2}/2$ if $x = 5\pi/4$ or $x = 7\pi/4$. Adding integer multiples of 2π produces additional solutions. Therefore, the set of all solutions is

$$x = \frac{5\pi}{4} + 2n\pi \quad \text{and} \quad x = \frac{7\pi}{4} + 2n\pi, \qquad n = 0, \pm 1, \pm 2, \pm 3, \ldots$$

b. Dividing both sides of the equation by $\cos 2x$ (assuming $\cos 2x \ne 0$), we obtain $\tan 2x = 1$. Letting $\theta = 2x$ gives us the equivalent equation $\tan\theta = 1$. This equation is satisfied by

$$\theta = \frac{\pi}{4}, \frac{5\pi}{4}, \frac{9\pi}{4}, \frac{13\pi}{4}, \frac{17\pi}{4}, \ldots$$

➤ Notice that the assumption $\cos 2x \ne 0$ is valid for these values of x.

Dividing by two and using the restriction $0 \le x < 2\pi$ gives the solutions

$$x = \frac{\theta}{2} = \frac{\pi}{8}, \frac{5\pi}{8}, \frac{9\pi}{8}, \text{ and } \frac{13\pi}{8}.$$

Related Exercises 23–34 ◄

Graphs of the Trigonometric Functions

Trigonometric functions are examples of **periodic functions**: Their values repeat over every interval of some fixed length. A function f is said to be periodic if $f(x + P) = f(x)$ for all x in the domain, where the **period** P is the smallest positive real number that has this property.

Period of Trigonometric Functions

The functions $\sin \theta$, $\cos \theta$, $\sec \theta$, and $\csc \theta$ have a period of 2π:

$$\sin (\theta + 2\pi) = \sin \theta \qquad \cos (\theta + 2\pi) = \cos \theta$$
$$\sec (\theta + 2\pi) = \sec \theta \qquad \csc (\theta + 2\pi) = \csc \theta$$

for all θ in the domain.

The functions $\tan \theta$ and $\cot \theta$ have a period of π:

$$\tan (\theta + \pi) = \tan \theta \qquad \cot (\theta + \pi) = \cot \theta$$

for all θ in the domain.

The graph of $y = \sin \theta$ is shown in Figure 1.63a. Because $\csc \theta = 1/\sin \theta$, these two functions have the same sign, but $y = \csc \theta$ is undefined with vertical asymptotes at $\theta = 0, \pm\pi, \pm 2\pi, \dots$. The functions $\cos \theta$ and $\sec \theta$ have a similar relationship (Figure 1.63b).

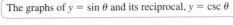

The graphs of $y = \sin \theta$ and its reciprocal, $y = \csc \theta$

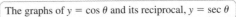

The graphs of $y = \cos \theta$ and its reciprocal, $y = \sec \theta$

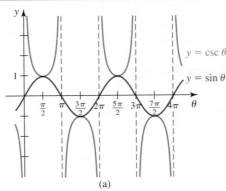

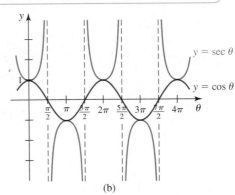

FIGURE 1.63   (a)   (b)

The graphs of $\tan \theta$ and $\cot \theta$ are shown in Figure 1.64. Each function has points, separated by π units, at which it is undefined.

The graph of $y = \tan \theta$ has period π.

The graph of $y = \cot \theta$ has period π.

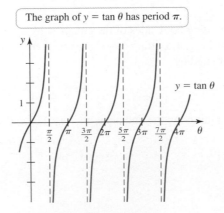

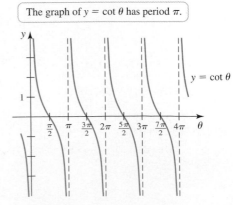

FIGURE 1.64

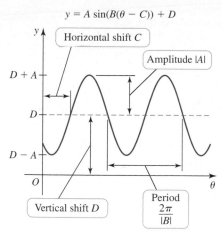

$$y = A \sin(B(\theta - C)) + D$$

Horizontal shift C

Amplitude |A|

Vertical shift D

Period $\dfrac{2\pi}{|B|}$

FIGURE 1.65

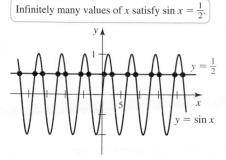

Daylight function gives length of day throughout the year.

$y = 12$

$D(t) = 2.8 \sin\!\left(\dfrac{2\pi}{365}(t - 81)\right) + 12$

| Jan 1 | Mar 21 | June 21 | Sep 21 | Dec 21 |

FIGURE 1.66

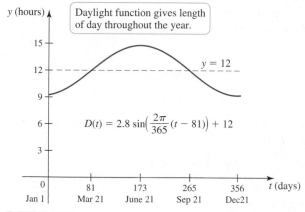

Infinitely many values of x satisfy $\sin x = \frac{1}{2}$.

$y = \frac{1}{2}$

$y = \sin x$

FIGURE 1.67

▷ The notation for the inverse trigonometric functions invites confusion: $\sin^{-1} x$ and $\cos^{-1} x$ do not mean the reciprocals of $\sin x$ and $\cos x$. The expression $\sin^{-1} x$ should be read *"angle whose sine is x,"* and $\cos^{-1} x$ should be read *"angle whose cosine is x."* The values of $\sin^{-1}$ and $\cos^{-1}$ are angles.

Transforming Graphs

Many physical phenomena, such as the motion of waves or the rising and setting of the sun, can be modeled using trigonometric functions; the sine and cosine functions are especially useful. With the transformation methods introduced in Section 1.3, we can show that the functions

$$y = A \sin\left(B(\theta - C)\right) + D \quad \text{and} \quad y = A \cos\left(B(\theta - C)\right) + D,$$

when compared to the graphs of $y = \sin\theta$ and $y = \cos\theta$, have a vertical stretch (or **amplitude**) of $|A|$, a **period** of $2\pi/|B|$, a horizontal shift (or **phase shift**) of C, and a **vertical shift** of D (Figure 1.65).

For example, at latitude 40° north (Beijing, Madrid, Philadelphia) there are 12 hr of daylight on the equinoxes (approximately March 21 and September 21), with a maximum of 14.8 hr of daylight on the summer solstice (approximately June 21) and a minimum of 9.2 hr of daylight on the winter solstice (approximately December 21). Using this information, it can be shown that the function

$$D(t) = 2.8 \sin\left(\frac{2\pi}{365}(t - 81)\right) + 12 \quad \text{(Figure 1.66)}$$

models the number of daylight hours t days after January 1 (Exercise 80). Notice that the graph of this function is obtained from the graph of $y = \sin t$ by (1) a horizontal scaling by a factor of $2\pi/365$, (2) a horizontal shift of 81, (3) a vertical scaling by a factor of 2.8, and (4) a vertical shift of 12.

Inverse Trigonometric Functions

The notion of inverse functions led from exponential functions to logarithmic functions (Section 1.3). We now carry out a similar procedure—this time with trigonometric functions.

Inverse Sine and Cosine   Our goal is to develop the inverses of the sine and cosine in detail. The inverses of the other four trigonometric functions then follow in an analogous way. So far, we have asked this question: Given an angle x, what is $\sin x$ or $\cos x$? Now we ask the opposite question: Given a number y, what is the angle x such that $\sin x = y$? Or, what is the angle x such that $\cos x = y$? These are inverse questions.

There are a few things to notice right away. First, these questions don't make sense if $|y| > 1$, because $-1 \le \sin x \le 1$ and $-1 \le \cos x \le 1$. Next, let's select an acceptable value of y, say $y = \frac{1}{2}$, and find the angle x that satisfies $\sin x = y = \frac{1}{2}$. It is apparent that infinitely many angles satisfy $\sin x = \frac{1}{2}$; all angles of the form $\pi/6 \pm 2n\pi$ and $5\pi/6 \pm 2n\pi$, where n is an integer, answer the inverse question (Figure 1.67). A similar situation occurs with the cosine function.

These inverse questions do not have unique answers because $\sin x$ and $\cos x$ are not one-to-one on their domains. To define their inverses, these functions must be restricted to intervals on which they are one-to-one. For the sine function, the standard choice is $[-\pi/2, \pi/2]$; for cosine, it is $[0, \pi]$ (Figure 1.68). Now when we ask for the angle x on the interval $[-\pi/2, \pi/2]$ such that $\sin x = \frac{1}{2}$, there is one answer: $x = \pi/6$. When we ask for the angle x on the interval $[0, \pi]$ such that $\cos x = -\frac{1}{2}$, there is one answer: $x = 2\pi/3$.

We define the **inverse sine**, or **arcsine**, denoted $y = \sin^{-1} x$ or $y = \arcsin x$, such that y is the angle whose sine is x, with the provision that y lies in the interval $[-\pi/2, \pi/2]$. Similarly, we define the **inverse cosine**, or **arccosine**, denoted $y = \cos^{-1} x$ or $y = \arccos x$, such that y is the angle whose cosine is x, with the provision that y lies in the interval $[0, \pi]$.

Restrict the domain of $y = \sin x$ to $\left[-\frac{\pi}{2}, \frac{\pi}{2}\right]$.

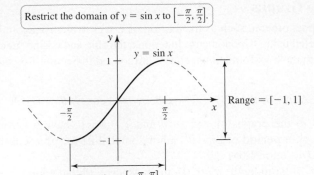

Restrict the domain of $y = \cos x$ to $[0, \pi]$.

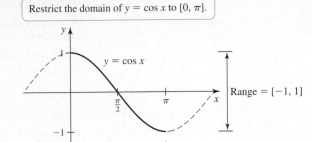

FIGURE 1.68

> **DEFINITION Inverse Sine and Cosine**
>
> $y = \sin^{-1} x$ is the value of y such that $x = \sin y$, where $-\pi/2 \le y \le \pi/2$.
> $y = \cos^{-1} x$ is the value of y such that $x = \cos y$, where $0 \le y \le \pi$.
> The domain of both $\sin^{-1} x$ and $\cos^{-1} x$ is $\{x: -1 \le x \le 1\}$.

Any invertible function and its inverse satisfy the properties

$$f(f^{-1}(y)) = y \quad \text{and} \quad f^{-1}(f(x)) = x.$$

These properties apply to the inverse sine and cosine, as long as we observe the restrictions on the domains. Here is what we can say:

- $\sin(\sin^{-1} x) = x$ and $\cos(\cos^{-1} x) = x$ for $-1 \le x \le 1$.
- $\sin^{-1}(\sin y) = y$ for $-\pi/2 \le y \le \pi/2$.
- $\cos^{-1}(\cos y) = y$ for $0 \le y \le \pi$.

QUICK CHECK 4 Explain why $\sin^{-1}(\sin 0) = 0$, but $\sin^{-1}(\sin 2\pi) \ne 2\pi$. ◄

EXAMPLE 3 Working with inverse sine and cosine Evaluate the following expressions.

a. $\sin^{-1}(\sqrt{3}/2)$ **b.** $\cos^{-1}(-\sqrt{3}/2)$ **c.** $\cos^{-1}(\cos 3\pi)$ **d.** $\sin\left(\sin^{-1}\left(\frac{1}{2}\right)\right)$

SOLUTION

a. $\sin^{-1}(\sqrt{3}/2) = \pi/3$ because $\sin(\pi/3) = \sqrt{3}/2$ and $\pi/3$ is in the interval $[-\pi/2, \pi/2]$.

b. $\cos^{-1}(-\sqrt{3}/2) = 5\pi/6$ because $\cos(5\pi/6) = -\sqrt{3}/2$ and $5\pi/6$ is in the interval $[0, \pi]$.

c. It's tempting to conclude that $\cos^{-1}(\cos 3\pi) = 3\pi$, but the result of an inverse cosine operation must lie in the interval $[0, \pi]$. Because $\cos(3\pi) = -1$ and $\cos^{-1}(-1) = \pi$, we have

$$\cos^{-1}\underbrace{(\cos 3\pi)}_{-1} = \cos^{-1}(-1) = \pi.$$

d. $\sin\left(\underbrace{\sin^{-1}\frac{1}{2}}_{\pi/6}\right) = \sin\frac{\pi}{6} = \frac{1}{2}.$

Related Exercises 35–40 ◄

Graphs and Properties Recall from Section 1.3 that the graph of the inverse f^{-1} is obtained by reflecting the graph of f about the identity line $y = x$. This operation produces the graphs of the inverse sine (Figure 1.69) and inverse cosine (Figure 1.70). The graphs make it easy to compare the domain and range of each function and its inverse.

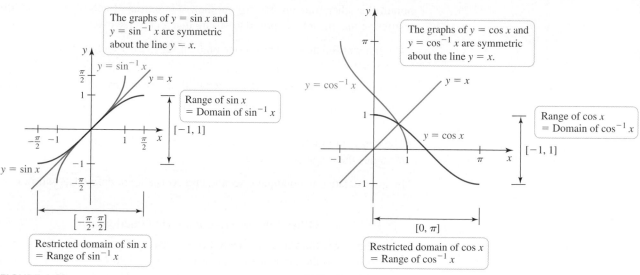

FIGURE 1.69

FIGURE 1.70

EXAMPLE 4   Right-triangle relationships

a. Suppose $\theta = \sin^{-1}(2/5)$. Find $\cos \theta$ and $\tan \theta$.

b. Find an alternative form for $\cot\left(\cos^{-1}(x/4)\right)$ in terms of x.

SOLUTION

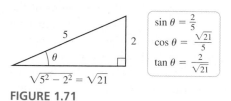

FIGURE 1.71

a. Relationships between the trigonometric functions and their inverses can often be simplified using a right-triangle sketch. The right triangle in Figure 1.71 satisfies the relationship $\sin \theta = \frac{2}{5}$, or, equivalently, $\theta = \sin^{-1}\left(\frac{2}{5}\right)$. We label the angle θ and the lengths of two sides; we then see the length of the third side is $\sqrt{21}$ (by the Pythagorean theorem). Now it is easy to read directly from the triangle:

$$\cos \theta = \frac{\sqrt{21}}{5} \quad \text{and} \quad \tan \theta = \frac{2}{\sqrt{21}}$$

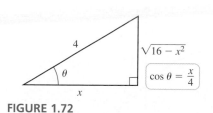

FIGURE 1.72

b. We draw a right triangle with an angle θ satisfying $\cos \theta = x/4$, or, equivalently, $\theta = \cos^{-1}(x/4)$ (Figure 1.72). The length of the third side of the triangle is $\sqrt{16 - x^2}$. It now follows that

$$\cot\left(\underbrace{\cos^{-1}\frac{x}{4}}_{\theta}\right) = \frac{x}{\sqrt{16 - x^2}}.$$

Related Exercises 41–46 ◄

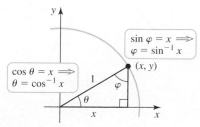

FIGURE 1.73

EXAMPLE 5   A useful identity Use right triangles to explain why $\cos^{-1} x + \sin^{-1} x = \pi/2$.

SOLUTION We draw a right triangle in a unit circle and label the acute angles θ and φ (Figure 1.73). These angles satisfy $\cos \theta = x$, or $\theta = \cos^{-1} x$, and $\sin \varphi = x$, or $\varphi = \sin^{-1} x$. Because θ and φ are complementary angles, we have

$$\frac{\pi}{2} = \theta + \varphi = \cos^{-1} x + \sin^{-1} x.$$

This result holds for $0 \leq x \leq 1$. An analogous argument extends the property to $-1 \leq x \leq 1$.

Related Exercises 47–48 ◄

Other Inverse Trigonometric Functions

The procedures that led to the inverse sine and inverse cosine functions can be used to obtain the other four inverse trigonometric functions. Each of these functions carries a restriction that must be imposed to ensure that an inverse exists:

- The tangent function is one-to-one on $(-\pi/2, \pi/2)$, which becomes the range of $y = \tan^{-1} x$.
- The cotangent function is one-to-one on $(0, \pi)$, which becomes the range of $y = \cot^{-1} x$.
- The secant function is one-to-one on $[0, \pi]$, excluding $x = \pi/2$; this set becomes the range of $y = \sec^{-1} x$.
- The cosecant function is one-to-one on $[-\pi/2, \pi/2]$, excluding $x = 0$; this set becomes the range of $y = \csc^{-1} x$.

The inverse tangent, cotangent, secant, and cosecant are defined as follows.

> ► Tables and books differ on the definition of the inverse secant and cosecant. In some books, $\sec^{-1} x$ is defined to lie in the interval $[-\pi, -\pi/2)$ when $x < 0$. Regardless of how it is defined, $\sec^{-1} x$ is not a continuous function.

DEFINITION   Other Inverse Trigonometric Functions

$y = \tan^{-1} x$ is the value of y such that $x = \tan y$, where $-\pi/2 < y < \pi/2$.
$y = \cot^{-1} x$ is the value of y such that $x = \cot y$, where $0 < y < \pi$.

The domain of both $\tan^{-1} x$ and $\cot^{-1} x$ is $\{x : -\infty < x < \infty\}$.

$y = \sec^{-1} x$ is the value of y such that $x = \sec y$, where $0 \leq y \leq \pi$, with $y \neq \pi/2$.
$y = \csc^{-1} x$ is the value of y such that $x = \csc y$, where $-\pi/2 \leq y \leq \pi/2$, with $y \neq 0$.

The domain of both $\sec^{-1} x$ and $\csc^{-1} x$ is $\{x : |x| \geq 1\}$.

The graphs of these inverse functions are obtained by reflecting the graphs of the original trigonometric functions about the line $y = x$ (Figures 1.74–1.77). The inverse secant and cosecant are somewhat irregular. The domain of the secant function (Figure 1.76) is restricted to the set $[0, \pi]$, excluding $x = \pi/2$, where the secant has a vertical asymptote. This asymptote splits the range of the secant into two disjoint intervals $(-\infty, -1]$ and $[1, \infty)$, which, in turn, splits the domain of the inverse secant into the same two intervals. A similar situation occurs with the cosecant.

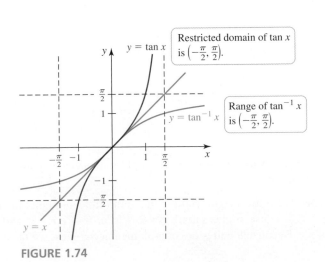

Restricted domain of $\tan x$ is $\left(-\frac{\pi}{2}, \frac{\pi}{2}\right)$.

Range of $\tan^{-1} x$ is $\left(-\frac{\pi}{2}, \frac{\pi}{2}\right)$.

FIGURE 1.74

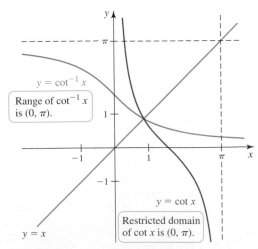

Range of $\cot^{-1} x$ is $(0, \pi)$.

Restricted domain of $\cot x$ is $(0, \pi)$.

FIGURE 1.75

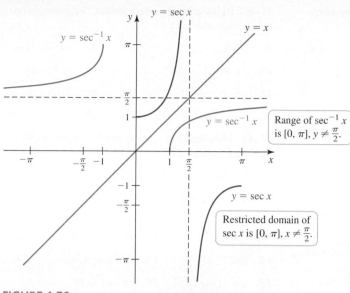

FIGURE 1.76

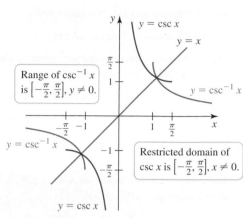

FIGURE 1.77

EXAMPLE 6   Working with inverse trigonometric functions Evaluate or simplify the following expressions.

a. $\tan^{-1}(-1/\sqrt{3})$ **b.** $\sec^{-1}(-2)$ **c.** $\sin(\tan^{-1}x)$

SOLUTION

a. The result of an inverse tangent operation must lie in the interval $(-\pi/2, \pi/2)$. Therefore,

$$\tan^{-1}\left(-\frac{1}{\sqrt{3}}\right) = -\frac{\pi}{6} \quad \text{because} \quad \tan\left(-\frac{\pi}{6}\right) = -\frac{1}{\sqrt{3}}.$$

b. The result of an inverse secant operation when $x \le -1$ must lie in the interval $(\pi/2, \pi]$. Therefore,

$$\sec^{-1}(-2) = \frac{2\pi}{3} \quad \text{because} \quad \sec\left(\frac{2\pi}{3}\right) = -2.$$

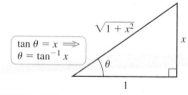

FIGURE 1.78

c. Figure 1.78 shows a right triangle with the relationship $x = \tan\theta$ or $\theta = \tan^{-1}x$, in the case that $0 \le \theta < \pi/2$. We see that

$$\sin\underbrace{(\tan^{-1}x)}_{\theta} = \frac{x}{\sqrt{1+x^2}}.$$

The same result follows if $-\pi/2 < \theta < 0$, in which case $x < 0$ and $\sin\theta < 0$.

Related Exercises 49–64 ◀

QUICK CHECK 5 Evaluate $\sec^{-1}1$ and $\tan^{-1}1$. ◀

SECTION 1.4 EXERCISES

Review Questions

1. Define the six trigonometric functions in terms of the sides of a right triangle.

2. Explain how a point $P(x, y)$ on a circle of radius r determines an angle θ and the values of the six trigonometric functions at θ.

3. Explain how the radian measure of an angle is determined.

4. Explain what is meant by the period of a trigonometric function. What are the periods of the six trigonometric functions?

5. What are the three Pythagorean identities for the trigonometric functions?

6. How are the sine and cosine functions used to define the other four trigonometric functions?

7. Where is the tangent function undefined?

8. What is the domain of the secant function?

9. Explain why the domain of the sine function must be restricted in order to define its inverse function.

10. Why do values of $\cos^{-1} x$ lie in the interval $[0, \pi]$?

11. Is it true that $\tan(\tan^{-1} x) = x$? Is it true that $\tan^{-1}(\tan x) = x$?

12. Sketch the graphs of $y = \cos x$ and $y = \cos^{-1} x$ on the same set of axes.

13. The function $\tan x$ is undefined at $x = \pm \pi/2$. How does this fact appear in the graph of $y = \tan^{-1} x$?

14. What are the domain and range of $\sec^{-1} x$?

Basic Skills

15–22. Evaluating trigonometric functions *Evaluate the following expressions by drawing the unit circle and the appropriate right triangle. Use a calculator only to check your work. All angles are in radians.*

15. $\cos(2\pi/3)$ **16.** $\sin(2\pi/3)$ **17.** $\tan(-3\pi/4)$

18. $\tan(15\pi/4)$ **19.** $\cot(-13\pi/3)$ **20.** $\sec(7\pi/6)$

21. $\cot(-17\pi/3)$ **22.** $\sin(16\pi/3)$

23–28. Trigonometric identities

23. Prove that $\tan^2 \theta + 1 = \sec^2 \theta$.

24. Prove that $\dfrac{\sin \theta}{\csc \theta} + \dfrac{\cos \theta}{\sec \theta} = 1$.

25. Prove that $\sec(\pi/2 - \theta) = \csc \theta$.

26. Prove that $\sec(x + \pi) = -\sec x$.

27. Find the exact value of $\cos(\pi/12)$.

28. Find the exact value of $\tan(3\pi/8)$.

29–34. Solving trigonometric equations *Solve the following equations.*

29. $\tan x = 1$ **30.** $2\theta \cos \theta + \theta = 0$

31. $\sqrt{2} \sin x - 1 = 0$

32. $\sin 3x = \sqrt{2}/2$, $0 \le x < 2\pi$

33. $\cos 3x = \sin 3x$, $0 \le x < 2\pi$

34. $\sin^2 \theta - 1 = 0$

35–40. Inverse sines and cosines *Without using a calculator, evaluate, if possible, the following expressions.*

35. $\sin^{-1}(\sqrt{3}/2)$ **36.** $\cos^{-1} 2$ **37.** $\cos^{-1}(-1/2)$

38. $\sin^{-1}(-1)$ **39.** $\cos(\cos^{-1}(-1))$ **40.** $\cos^{-1}(\cos 7\pi/6)$

41–46. Right-triangle relationships *Draw a right triangle to simplify the given expressions.*

41. $\cos(\sin^{-1} x)$ **42.** $\cos(\sin^{-1}(x/3))$

43. $\sin(\cos^{-1}(x/2))$ **44.** $\sin^{-1}(\cos \theta)$

45. $\sin(2 \cos^{-1} x)$ (*Hint:* Use $\sin 2\theta = 2 \sin \theta \cos \theta$.)

46. $\cos(2 \sin^{-1} x)$ (*Hint:* Use $\cos 2\theta = \cos^2 \theta - \sin^2 \theta$.)

47–48. Identities *Use right triangles to explain why the following identities are true.*

47. $\cos^{-1} x + \cos^{-1}(-x) = \pi$ **48.** $\sin^{-1} y + \sin^{-1}(-y) = 0$

49–56. Evaluating inverse trigonometric functions *Without using a calculator, evaluate or simplify the following expressions.*

49. $\tan^{-1} \sqrt{3}$ **50.** $\cot^{-1}(-1/\sqrt{3})$ **51.** $\sec^{-1} 2$

52. $\csc^{-1}(-1)$ **53.** $\tan^{-1}(\tan \pi/4)$ **54.** $\tan^{-1}(\tan 3\pi/4)$

55. $\csc^{-1}(\sec 2)$ **56.** $\tan(\tan^{-1} 1)$

57–62. Right-triangle relationships *Draw a right triangle to simplify the given expressions.*

57. $\cos(\tan^{-1} x)$ **58.** $\tan(\cos^{-1} x)$

59. $\cos(\sec^{-1} x)$ **60.** $\cot(\tan^{-1} 2x)$

61. $\sin\left(\sec^{-1}\left(\dfrac{\sqrt{x^2 + 16}}{4}\right)\right)$ **62.** $\cos\left(\tan^{-1}\left(\dfrac{x}{\sqrt{9 - x^2}}\right)\right)$

63–64. Right-triangle pictures *Express θ in terms of x using the inverse sine, inverse tangent, and inverse secant functions.*

63.   **64.**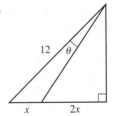

Further Explorations

65. Explain why or why not Determine whether the following statements are true and give an explanation or counterexample.

   a. $\sin(a + b) = \sin a + \sin b$

   b. The equation $\cos \theta = 2$ has multiple solutions.

   c. The equation $\sin \theta = \frac{1}{2}$ has exactly one solution.

   d. The function $\sin(\pi x/12)$ has a period of 12.

   e. Of the six basic trigonometric functions, only tangent and cotangent have a range of $(-\infty, \infty)$.

   f. $\dfrac{\sin^{-1} x}{\cos^{-1} x} = \tan^{-1} x$ **g.** $\cos^{-1}(\cos(15\pi/16)) = 15\pi/16$

   h. $\sin^{-1} x = 1/\sin x$

66–69. One function gives all six *Given the following information about one trigonometric function, evaluate the other five functions.*

66. $\sin \theta = -\frac{4}{5}$ and $\pi < \theta < 3\pi/2$ (Find $\cos \theta$, $\tan \theta$, $\cot \theta$, $\sec \theta$, and $\csc \theta$.)

67. $\cos \theta = \frac{5}{13}$ and $0 < \theta < \pi/2$

68. $\sec \theta = \frac{5}{3}$ and $3\pi/2 < \theta < 2\pi$

69. $\csc \theta = \frac{13}{12}$ and $0 < \theta < \pi/2$

70–73. Amplitude and period *Identify the amplitude and period of the following functions.*

70. $f(\theta) = 2 \sin 2\theta$

71. $g(\theta) = 3 \cos (\theta/3)$

72. $p(t) = 2.5 \sin \left(\frac{1}{2}(t - 3)\right)$

73. $q(x) = 3.6 \cos (\pi x/24)$

74–77. Graphing sine and cosine functions *Beginning with the graphs of $y = \sin x$ or $y = \cos x$, use shifting and scaling transformations to sketch the graph of the following functions. Use a graphing utility only to check your work.*

74. $f(x) = 3 \sin 2x$

75. $g(x) = -2 \cos (x/3)$

76. $p(x) = 3 \sin (2x - \pi/3) + 1$

77. $q(x) = 3.6 \cos (\pi x/24) + 2$

78–79. Designer functions *Design a sine function with the given properties.*

78. It has a period of 12 hr with a minimum value of -4 at $t = 0$ hr and a maximum value of 4 at $t = 6$ hr.

79. It has a period of 24 hr with a minimum value of 10 at $t = 3$ hr and a maximum value of 16 at $t = 15$ hr.

Applications

80. Daylight function for 40° N Verify that the function

$$D(t) = 2.8 \sin \left(\frac{2\pi}{365}(t - 81)\right) + 12$$

has the following properties, where t is measured in days and D is measured in hours.

a. It has a period of 365 days.

b. Its maximum and minimum values are 14.8 hr and 9.2 hr, respectively, which occur approximately at $t = 172$ and $t = 355$, respectively (corresponding to the solstices).

c. $D(81) = D(264) = 12$ (corresponding to the equinoxes).

81. Block on a spring A light block hangs at rest from the end of a spring when it is pulled down 10 cm and released. Assume the block oscillates with an amplitude of 10 cm on either side of its initial position, and with a period of 1.5 s. Find a function $d(t)$ that gives the displacement of the block t seconds after it is released, where $d(t) > 0$ represents downward displacement.

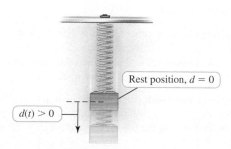

82. Approaching a lighthouse A boat approaches a 50-ft-high lighthouse whose base is at sea level. Let d be the distance between the boat and the base of the lighthouse. Let L be the distance between the boat and the top of the lighthouse. Let θ be the angle of elevation between the boat and the top of the lighthouse.

a. Express d as a function of θ.

b. Express L as a function of θ.

83. Ladders Two ladders of length a lean against opposite walls of an alley with their feet touching (see figure). One ladder extends h feet up the wall and makes a 75° angle with the ground. The other ladder extends k feet up the opposite wall and makes a 45° angle with the ground. Find the width of the alley in terms of a, h, and/or k. Assume the ground is horizontal and perpendicular to both walls.

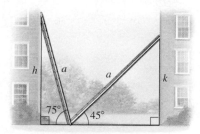

84. Pole in a corner A pole of length L is carried horizontally around a corner where a 3-ft-wide hallway meets a 4-ft-wide hallway. For $0 < \theta < \pi/2$, find the relationship between L and θ at the moment when the pole simultaneously touches both walls and the corner P. Estimate θ when $L = 10$ ft.

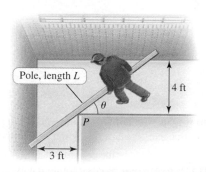

85. Little-known fact The shortest day of the year occurs on the winter solstice (near December 21) and the longest day of the year occurs on the summer solstice (near June 21). However, the latest sunrise and the earliest sunset do not occur on the winter solstice, and the earliest sunrise and the latest sunset do not occur on the summer solstice. At latitude 40° north, the latest sunrise occurs on January 4 at 7:25 A.M. (14 days after the solstice), and the earliest sunset occurs on December 7 at 4:37 P.M. (14 days before the solstice). Similarly, the earliest sunrise occurs on July 2 at 4:30 A.M. (14 days after the solstice) and the latest sunset occurs on June 7 at 7:32 P.M. (14 days before the solstice). Using sine functions, devise a function $s(t)$ that gives the time of sunrise t days after January 1 and a function $S(t)$ that gives the time of sunset t days after January 1. Assume that s and S are measured in minutes and $s = 0$ and $S = 0$ correspond to 4:00 A.M. Graph the functions. Then graph the length of the day function $D(t) = S(t) - s(t)$ and show that the longest and shortest days occur on the solstices.

86. Viewing angles An auditorium with a flat floor has a large flat-panel television on one wall. The lower edge of the television is 3 ft above the floor and the upper edge is 10 ft above the floor (see figure). Express θ in terms of x.

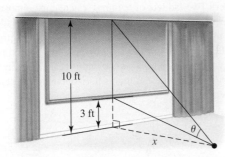

Additional Exercises

87. Area of a circular sector Prove that the area of a sector of a circle of radius r associated with a central angle θ (measured in rad) is $A = \frac{1}{2}r^2\theta$.

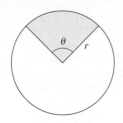

88. Law of cosines Use the figure to prove the law of cosines (which is a generalization of the Pythagorean theorem):
$$c^2 = a^2 + b^2 - 2ab\cos\theta.$$

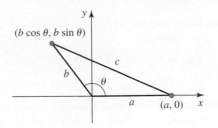

QUICK CHECK ANSWERS

1. $3\pi/2, 225°$ **2.** $\sqrt{3}/2; -\sqrt{2}/2$ **3.** Divide both sides of $\sin^2\theta + \cos^2\theta = 1$ by $\sin^2\theta$.   **4.** $\sin^{-1}(\sin 0) = \sin^{-1} 0 = 0$ and $\sin^{-1}(\sin(2\pi)) = \sin^{-1} 0 = 0$ **5.** $0, \pi/4$ ◄

CHAPTER 1   REVIEW EXERCISES

1. Explain why or why not Determine whether the following statements are true and give an explanation or counterexample.

a. A function could have the property that $f(x) = f(-x)$ for all x.
b. $\cos(a + b) = \cos a + \cos b$ for all a and b in $[0, 2\pi]$.
c. If f is a linear function of the form $f(x) = mx + b$, then $f(u + v) = f(u) + f(v)$ for all u and v.
d. The function $f(x) = 1 - x$ has the property that $f(f(x)) = x$.
e. The set $\{x: |x + 3| > 4\}$ can be drawn on the number line without lifting your pencil.
f. $\log_{10}(xy) = (\log_{10} x)(\log_{10} y)$.
g. $\sin^{-1}(\sin(2\pi)) = 0$.

2. Domain and range Find the domain and range of the following functions.

a. $f(x) = x^5 + \sqrt{x}$ **b.** $g(y) = \dfrac{1}{y - 2}$
c. $h(z) = \sqrt{z^2 - 2z - 3}$

3. Equations of lines Find an equation of the lines with the following properties. Graph the lines.

a. The line passing through the points $(2, -3)$ and $(4, 2)$
b. The line with slope $\frac{3}{4}$ and x-intercept $(-4, 0)$
c. The line with intercepts $(4, 0)$ and $(0, -2)$

4. Piecewise linear functions The parking costs in a city garage are $2.00 for the first half hour and $1.00 for each additional half hour. Graph the function $C = f(t)$ that gives the cost of parking for t hours, where $0 \le t \le 3$.

5. Graphing absolute value Consider the function $f(x) = 2(x - |x|)$. Express the function in two pieces without using the absolute value. Then graph the function by hand. Use a graphing utility only to check your work.

6. Function from words Suppose you plan to take a 500-mi trip in a car that gets 35 mi/gal. Find the function $C = f(p)$ that gives the cost of gasoline for the trip when gasoline costs $\$p$ per gallon.

7. **Graphing equations** Graph the following equations. Use a graphing utility only to check your work.

   a. $2x - 3y + 10 = 0$
   b. $y = x^2 + 2x - 3$
   c. $x^2 + 2x + y^2 + 4y + 1 = 0$
   d. $x^2 - 2x + y^2 - 8y + 5 = 0$

8. **Root functions** Graph the functions $f(x) = x^{1/3}$ and $g(x) = x^{1/4}$. Find all points where the two graphs intersect. For $x > 1$, is $f(x) > g(x)$ or is $g(x) > f(x)$?

9. **Root functions** Find the domain and range of the functions $f(x) = x^{1/7}$ and $g(x) = x^{1/4}$.

10. **Intersection points** Graph the equations $y = x^2$ and $x^2 + y^2 - 7y + 8 = 0$. At what point(s) do the curves intersect?

11. **Boiling-point function** Water boils at $212°$ F at sea level and at $180°$ F at an elevation of 6000 ft. Assume that the boiling point B varies linearly with altitude a. Find the function $B = f(a)$ that describes the dependence. Comment on whether a linear function gives a realistic model.

12. **Publishing costs** A small publisher plans to spend $1000 for advertising a paperback book and estimates the printing cost is $2.50 per book. The publisher will receive $7 for each book sold.

   a. Find the function $C = f(x)$ that gives the cost of producing x books.
   b. Find the function $R = g(x)$ that gives the revenue from selling x books.
   c. Graph the cost and revenue functions and find the number of books that must be sold for the publisher to break even.

13. **Shifting and scaling** Starting with the graph of $f(x) = x^2$, sketch the graph of the following functions. Use a graphing calculator only to check your work.

   a. $f(x + 3)$ b. $2f(x - 4)$ c. $-f(3x)$ d. $f(2(x - 3))$

14. **Shifting and scaling** The graph of $y = f(x)$ is shown in the figure. Sketch the graph of the following functions.

   a. $f(x + 1)$ b. $2f(x - 1)$ c. $-f(x/2)$ d. $f(2(x - 1))$

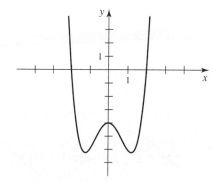

15. **Composite functions** Let $f(x) = x^3$, $g(x) = \sin x$, and $h(x) = \sqrt{x}$.

   a. Evaluate $h(g(\pi/2))$.
   b. Find $h(f(x))$.
   c. Find $f(g(h(x)))$.

   d. Find the domain of $g \circ f$.
   e. Find the range of $f \circ g$.

16. **Composite functions** Find functions f and g such that $h = f \circ g$.

   a. $h(x) = \sin(x^2 + 1)$
   b. $h(x) = (x^2 - 4)^{-3}$
   c. $h(x) = e^{\cos 2x}$

17. **Symmetry** Identify the symmetry in the graphs of the following equations.

   a. $y = \cos 3x$ b. $y = 3x^4 - 3x^2 + 1$ c. $y^2 - 4x^2 = 4$

18–19. Properties of logarithms and exponentials *Use properties of logarithms and exponentials, not a calculator, for the following exercises.*

18. Solve the equation $48 = 6e^{4k}$ for k.

19. Solve the equation $\log x^2 + 3 \log x = \log 32$ for x. Does the answer depend on the base of the log?

20. **Graphs of logarithmic and exponential functions** The figure shows the graphs of $y = 2^x$, $y = 3^{-x}$, and $y = -\ln x$. Match each curve with the correct function.

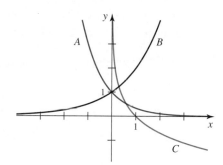

⚏ 21–22. Existence of inverses *Use analytical methods and/or graphing to determine the intervals on which the following functions have an inverse.*

21. $f(x) = x^3 - 3x^2$ 22. $g(t) = 2 \sin(t/3)$

⚏ 23–24. Finding inverses *Find the inverse on the specified interval and express it in the form $y = f^{-1}(x)$. Then graph f and f^{-1}.*

23. $f(x) = x^2 - 4x + 5$ for $x > 2$

24. $f(x) = 1/x^2$ for $x > 0$

25. **Degrees and radians**

   a. What is the radian measure of a $135°$ angle?
   b. What is the degree measure of a $4\pi/5$-rad angle?
   c. What is the length of the arc of a circle of radius 10 associated with an angle of $4\pi/3$ rad?

26. **Graphing sine and cosine functions** Use shifts and scalings to graph the following functions, and identify the amplitude and period.

   a. $f(x) = 4 \cos(x/2)$
   b. $g(\theta) = 2 \sin(2\pi\theta/3)$
   c. $h(\theta) = -\cos(2(\theta - \pi/4))$

27. Designing functions Find a trigonometric function f that satisfies each set of properties. Answers are not unique.

    a. It has a period of 6 hr with a minimum value of -2 at $t = 0$ hr and a maximum value of 2 at $t = 3$ hr.

    b. It has a period of 24 hr with a maximum value of 20 at $t = 6$ hr and a minimum value of 10 at $t = 18$ hr.

28. Graph to function Find a trigonometric function f represented by the graph in the figure.

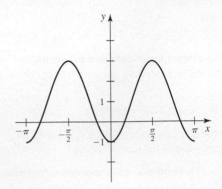

29. Matching Match each function with the corresponding graphs A–F.

    a. $f(x) = -\sin x$ **b.** $f(x) = \cos 2x$ **c.** $f(x) = \tan(x/2)$

    d. $f(x) = -\sec x$ **e.** $f(x) = \cot 2x$ **f.** $f(x) = \sin^2 x$

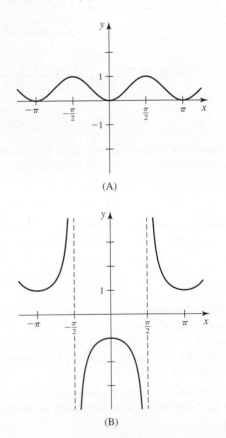

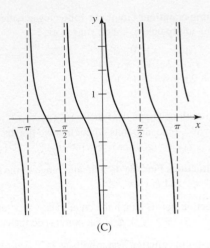

(C)

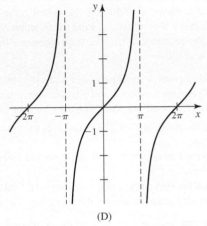

(D)

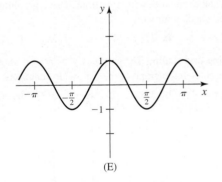

(E)

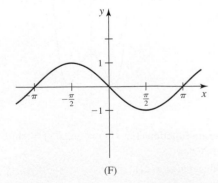

(F)

30–36. Inverse sines and cosines *Without using a calculator, evaluate or simplify the following expressions.*

30. $\sin^{-1}(\sqrt{3}/2)$ **31.** $\cos^{-1}(\sqrt{3}/2)$ **32.** $\cos^{-1}\left(-\frac{1}{2}\right)$

33. $\sin^{-1}(-1)$ **34.** $\cos(\cos^{-1}(-1))$ **35.** $\sin(\sin^{-1}x)$

36. $\cos^{-1}(\sin 3\pi)$

37. Right triangles Given that $\theta = \sin^{-1}\left(\frac{12}{13}\right)$, evaluate $\cos\theta$, $\tan\theta$, $\cot\theta$, $\sec\theta$, and $\csc\theta$.

38–45. Right-triangle relationships *Draw a right triangle to simplify the given expression. Assume $x > 0$ and $0 \leq \theta \leq \pi/2$.*

38. $\cos(\tan^{-1}x)$ **39.** $\sin(\cos^{-1}(x/2))$

40. $\tan(\sec^{-1}(x/2))$ **41.** $\cot^{-1}(\tan\theta)$

42. $\csc^{-1}(\sec\theta)$ **43.** $\sin^{-1}x + \sin^{-1}(-x)$

44. $\sin(2\cos^{-1}x)$ (*Hint:* Use $\sin 2\theta = 2\sin\theta\cos\theta$.)

45. $\cos(2\sin^{-1}x)$ (*Hint:* Use $\cos 2\theta = \cos^2\theta - \sin^2\theta$.)

46. Stereographic projections A common way of displaying a sphere (such as Earth) on a plane (such as a map) is to use a *stereographic projection*. Here is the two-dimensional version of the method, which maps a circle to a line. Let P be a point on the right half of a circle of radius R identified by the angle φ. Find the function $x = F(\varphi)$ that gives the x-coordinate ($x \geq 0$) corresponding to φ for $0 < \varphi \leq \pi$.

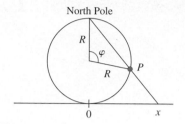

Chapter 1 Guided Projects

Applications of the material in this chapter and related topics can be found in the following Guided Projects. For additional information, see the Preface.

- Problem-solving skills
- Functions in action I
- Functions in action II

- Atmospheric CO_2
- Constant-rate problems
- Supply and demand

- Phase and amplitude
- Acid, noise, and earthquakes

2

Limits

Chapter Preview All of calculus is based on the idea of a *limit*. Not only are limits important in their own right, but they underlie the two fundamental operations of calculus: differentiation (calculating derivatives) and integration (evaluating integrals). Derivatives enable us to talk about the instantaneous rate of change of a function, which, in turn, leads to concepts such as velocity and acceleration, population growth rates, marginal cost, and flow rates. Integrals enable us to compute areas under curves, surface areas, and volumes. Because of the incredible reach of this single idea, it is essential to develop a solid understanding of limits. We first present limits intuitively by showing how they arise in computing instantaneous velocities and finding slopes of tangent lines. As the chapter progresses, we build more rigor into the definition of the limit, and we examine the different ways in which limits exist or fail to exist. The chapter concludes by introducing the important property called *continuity* and by giving the formal definition of a limit. By the end of the chapter, you will be ready to use limits when needed throughout the remainder of the book.

2.1 The Idea of Limits

This brief opening section illustrates how limits arise in two seemingly unrelated problems: finding the instantaneous velocity of a moving object and finding the slope of a line tangent to a curve. These two problems provide important insights into limits, and they reappear in various forms throughout the book.

Average Velocity

Suppose you want to calculate your average velocity as you travel along a straight highway. If you pass milepost 100 at noon and milepost 130 at 12:30 P.M., you travel 30 mi in a half-hour, so your **average velocity** over this time interval is $(30\,\text{mi})/(0.5\,\text{hr}) = 60\,\text{mi/hr}$. By contrast, even though your average velocity may be 60 mi/hr, it's almost certain that your **instantaneous velocity**, the speed indicated by the speedometer, varies from one moment to the next.

EXAMPLE 1 **Average velocity** A rock is launched vertically upward from the ground with a speed of 96 ft/s. Neglecting air resistance, a well-known formula from physics states that the position of the rock after t seconds is given by the function

$$s(t) = -16t^2 + 96t.$$

The position s is measured in feet with $s = 0$ corresponding to the ground. Find the average velocity of the rock between each pair of times.

a. $t = 1$ s and $t = 3$ s   **b.** $t = 1$ s and $t = 2$ s

SOLUTION Figure 2.1 shows the position of the rock on the time interval $0 \leq t \leq 3$.

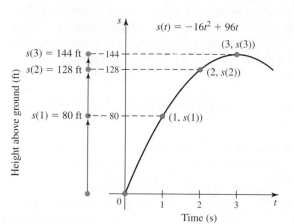

FIGURE 2.1

a. The average velocity of the rock over any time interval $[t_0, t_1]$ is the change in position divided by the elapsed time:

$$v_{av} = \frac{s(t_1) - s(t_0)}{t_1 - t_0}.$$

Therefore, the average velocity over the interval $[1, 3]$ is

$$v_{av} = \frac{s(3) - s(1)}{3 - 1} = \frac{144 \text{ ft} - 80 \text{ ft}}{3 \text{ s} - 1 \text{ s}} = \frac{64 \text{ ft}}{2 \text{ s}} = 32 \text{ ft/s}.$$

Here is an important observation: As shown in Figure 2.2a, the average velocity is simply the slope of the line joining the points $(1, s(1))$ and $(3, s(3))$ on the graph of the position function.

b. The average velocity of the rock over the interval $[1, 2]$ is

$$v_{av} = \frac{s(2) - s(1)}{2 - 1} = \frac{128 \text{ ft} - 80 \text{ ft}}{2 \text{ s} - 1 \text{ s}} = \frac{48 \text{ ft}}{1 \text{ s}} = 48 \text{ ft/s}.$$

Again, the average velocity is the slope of the line joining the points $(1, s(1))$ and $(2, s(2))$ on the graph of the position function (Figure 2.2b).

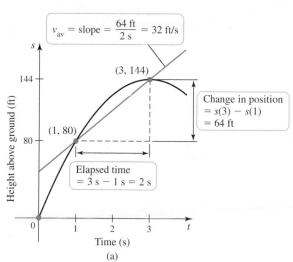

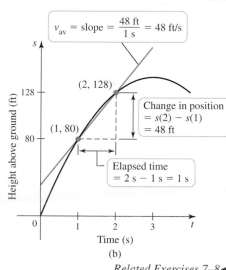

FIGURE 2.2   (a)   (b)

Related Exercises 7–8 ◀

A line joining two points on a curve is called a **secant line**. The slope of the secant line, denoted m_{sec}, for the position function in Example 1 on the interval $[t_0, t_1]$ is

$$m_{sec} = \frac{s(t_1) - s(t_0)}{t_1 - t_0}.$$

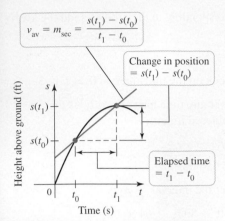

$$v_{av} = m_{sec} = \frac{s(t_1) - s(t_0)}{t_1 - t_0}$$

Change in position $= s(t_1) - s(t_0)$

Elapsed time $= t_1 - t_0$

Height above ground (ft)

Time (s)

FIGURE 2.3

Table 2.1

Time interval	Average velocity
$[1, 2]$	48 ft/s
$[1, 1.5]$	56 ft/s
$[1, 1.1]$	62.4 ft/s
$[1, 1.01]$	63.84 ft/s
$[1, 1.001]$	63.984 ft/s
$[1, 1.0001]$	63.9984 ft/s

➤ The same instantaneous velocity is obtained as t approaches 1 from the left (with $t < 1$) and as t approaches 1 from the right (with $t > 1$).

Example 1 demonstrates that the average velocity is the slope of a secant line on the graph of the position function; that is, $v_{av} = m_{sec}$ (Figure 2.3).

Instantaneous Velocity

To compute the average velocity, we use the position of the object at *two* distinct points in time. How do we compute the instantaneous velocity at a *single* point in time? As illustrated in Example 2, the instantaneous velocity at a point $t = t_0$ is determined by computing average velocities over intervals $[t_0, t_1]$ that decrease in length. As t_1 approaches t_0, the average velocities typically approach a unique number, which is the instantaneous velocity. This single number is called a **limit**.

EXAMPLE 2 Instantaneous velocity Estimate the *instantaneous velocity* of the rock in Example 1 at the *single* point $t = 1$.

SOLUTION We are interested in the instantaneous velocity at $t = 1$, so we compute the average velocity over smaller and smaller time intervals $[1, t]$ using the formula

$$v_{av} = \frac{s(t) - s(1)}{t - 1}.$$

Notice that these average velocities are also slopes of secant lines, several of which are shown in Table 2.1. We see that as t approaches 1, the average velocities appear to approach 64 ft/s. In fact, we could make the average velocity as close to 64 ft/s as we like by taking t sufficiently close to 1. Therefore, 64 ft/s is a reasonable estimate of the instantaneous velocity at $t = 1$.   *Related Exercises 9–12* ◄

In language to be introduced in Section 2.2, we say that the limit of v_{av} as t approaches 1 equals the instantaneous velocity v_{inst}, which is 64 ft/s. This statement is written compactly as

$$v_{inst} = \lim_{t \to 1} v_{av} = \lim_{t \to 1} \frac{s(t) - s(1)}{t - 1} = 64 \text{ ft/s}.$$

Figure 2.4 gives a graphical illustration of this limit.

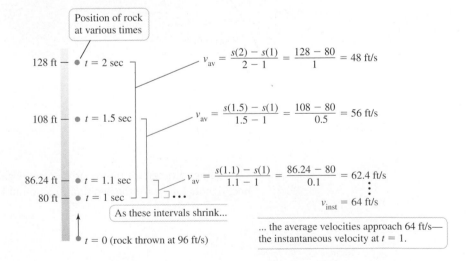

FIGURE 2.4

Position of rock at various times

128 ft — ● $t = 2$ sec $v_{av} = \dfrac{s(2) - s(1)}{2 - 1} = \dfrac{128 - 80}{1} = 48$ ft/s

108 ft — ● $t = 1.5$ sec $v_{av} = \dfrac{s(1.5) - s(1)}{1.5 - 1} = \dfrac{108 - 80}{0.5} = 56$ ft/s

86.24 ft — ● $t = 1.1$ sec $v_{av} = \dfrac{s(1.1) - s(1)}{1.1 - 1} = \dfrac{86.24 - 80}{0.1} = 62.4$ ft/s

80 ft — ● $t = 1$ sec

$v_{inst} = 64$ ft/s

As these intervals shrink...

... the average velocities approach 64 ft/s— the instantaneous velocity at $t = 1$.

● $t = 0$ (rock thrown at 96 ft/s)

Slope of the Tangent Line

> We define tangent lines carefully in Section 3.1. For the moment, imagine zooming in on a point P on a smooth curve. As you zoom in, the curve appears more and more like a line passing through P. This line is the *tangent line* at P.

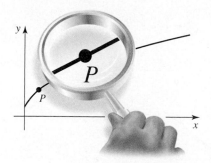

Several important conclusions follow from Examples 1 and 2. Each average velocity in Table 2.1 corresponds to the slope of a secant line on the graph of the position function (Figure 2.5). Just as the average velocities approach a limit as t approaches 1, the slopes of the secant lines approach the same limit as t approaches 1. Specifically, as t approaches 1, two things happen:

1. The secant lines approach a unique line called the **tangent line**.

2. The slopes of the secant lines m_{sec} approach the slope of the tangent line m_{tan} at the point $(1, s(1))$. Thus, the slope of the tangent line is also expressed as a limit:

$$m_{tan} = \lim_{t \to 1} m_{sec} = \lim_{t \to 1} \frac{s(t) - s(1)}{t - 1} = 64.$$

This limit is the same limit that defines the instantaneous velocity. Therefore, the instantaneous velocity at $t = 1$ is the slope of the line tangent to the position curve at $t = 1$.

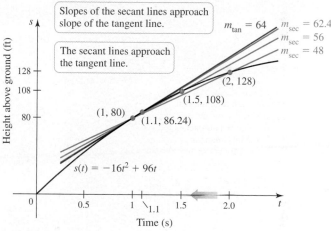

FIGURE 2.5

The parallels between average and instantaneous velocities, on one hand, and between slopes of secant lines and tangent lines, on the other, illuminate the power behind the idea of a limit. As $t \to 1$, slopes of secant lines approach the slope of a tangent line. And as $t \to 1$, average velocities approach an instantaneous velocity. Figure 2.6 summarizes these two parallel limit processes. These ideas lie at the foundation of what follows in the coming chapters.

AVERAGE VELOCITY ⟷ SECANT LINE

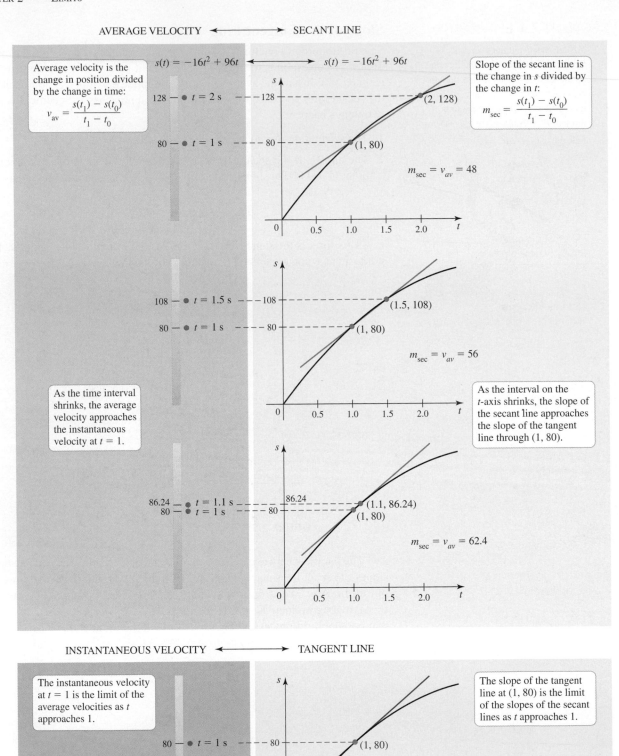

Average velocity is the change in position divided by the change in time:
$$v_{av} = \frac{s(t_1) - s(t_0)}{t_1 - t_0}$$

$s(t) = -16t^2 + 96t$ ⟷ $s(t) = -16t^2 + 96t$

Slope of the secant line is the change in s divided by the change in t:
$$m_{sec} = \frac{s(t_1) - s(t_0)}{t_1 - t_0}$$

128 — ● $t = 2$ s

80 — ● $t = 1$ s

(2, 128)

(1, 80)

$m_{sec} = v_{av} = 48$

As the time interval shrinks, the average velocity approaches the instantaneous velocity at $t = 1$.

108 — ● $t = 1.5$ s

80 — ● $t = 1$ s

(1.5, 108)

(1, 80)

$m_{sec} = v_{av} = 56$

As the interval on the t-axis shrinks, the slope of the secant line approaches the slope of the tangent line through (1, 80).

86.24 — ● $t = 1.1$ s
80 — ● $t = 1$ s

(1.1, 86.24)

(1, 80)

$m_{sec} = v_{av} = 62.4$

INSTANTANEOUS VELOCITY ⟷ TANGENT LINE

The instantaneous velocity at $t = 1$ is the limit of the average velocities as t approaches 1.

80 — ● $t = 1$ s

(1, 80)

$$v_{inst} = \lim_{t \to 1} \frac{s(t) - s(1)}{t - 1} = 64 \text{ ft/s}$$

The slope of the tangent line at (1, 80) is the limit of the slopes of the secant lines as t approaches 1.

$$m_{tan} = \lim_{t \to 1} \frac{s(t) - s(1)}{t - 1} = 64$$

Instantaneous velocity = 64 ft/s ⟷ Slope of the tangent line = 64

FIGURE 2.6

SECTION 2.1 EXERCISES

Review Questions

1. Suppose $s(t)$ is the position of an object moving along a line at time $t \geq 0$. What is the average velocity between the times $t = a$ and $t = b$?

2. Suppose $s(t)$ is the position of an object moving along a line at time $t \geq 0$. Describe a process for finding the instantaneous velocity at $t = a$.

3. What is the slope of the secant line between the points $(a, f(a))$ and $(b, f(b))$ on the graph of f?

4. Describe a process for finding the slope of the line tangent to the graph of f at $(a, f(a))$.

5. Describe the parallels between finding the instantaneous velocity of an object at a point in time and finding the slope of the line tangent to the graph of a function at a point.

6. Graph the parabola $f(x) = x^2$. Explain why the secant lines between the points $(-a, f(-a))$ and $(a, f(a))$ have slope zero. What is the slope of the tangent line at $x = 0$?

Basic Skills

7. **Average velocity** The position of an object moving along a line is given by the function $s(t) = -16t^2 + 128t$. Find the average velocity of the object over the following intervals.

    a. $[1, 4]$ b. $[1, 3]$
    c. $[1, 2]$ d. $[1, h]$, where $h > 0$ is a real number

8. **Average velocity** The position of an object moving along a line is given by the function $s(t) = -4.9t^2 + 30t + 20$. Find the average velocity of the object over the following intervals.

    a. $[0, 3]$ b. $[0, 2]$
    c. $[0, 1]$ d. $[0, h]$, where $h > 0$ is a real number

9. **Instantaneous velocity** Consider the position function $s(t) = -16t^2 + 128t$ (Exercise 7). Complete the following table with the appropriate average velocities. Then make a conjecture about the value of the instantaneous velocity at $t = 1$.

Time interval	$[1, 2]$	$[1, 1.5]$	$[1, 1.1]$	$[1, 1.01]$	$[1, 1.001]$
Average velocity					

10. **Instantaneous velocity** Consider the position function $s(t) = -4.9t^2 + 30t + 20$ (Exercise 8). Complete the following table with the appropriate average velocities. Then make a conjecture about the value of the instantaneous velocity at $t = 2$.

Time interval	$[2, 3]$	$[2, 2.5]$	$[2, 2.1]$	$[2, 2.01]$	$[2, 2.001]$
Average velocity					

11. **Instantaneous velocity** Consider the position function $s(t) = -16t^2 + 100t$. Complete the following table with the appropriate average velocities. Then make a conjecture about the value of the instantaneous velocity at $t = 3$.

Time interval	Average velocity
$[2, 3]$	
$[2.9, 3]$	
$[2.99, 3]$	
$[2.999, 3]$	
$[2.9999, 3]$	

12. **Instantaneous velocity** Consider the position function $s(t) = 3 \sin t$ that describes a block bouncing vertically on a spring. Complete the following table with the appropriate average velocities. Then make a conjecture about the value of the instantaneous velocity at $t = \pi/2$.

Time interval	Average velocity
$[\pi/2, \pi]$	
$[\pi/2, \pi/2 + 0.1]$	
$[\pi/2, \pi/2 + 0.01]$	
$[\pi/2, \pi/2 + 0.001]$	
$[\pi/2, \pi/2 + 0.0001]$	

Further Explorations

13–16. **Instantaneous velocity** *For the following position functions, make a table of average velocities similar to those in Exercises 9–12 and make a conjecture about the instantaneous velocity at the indicated time.*

13. $s(t) = -16t^2 + 80t + 60$ at $t = 3$

14. $s(t) = 20 \cos t$ at $t = \pi/2$

15. $s(t) = 40 \sin 2t$ at $t = 0$

16. $s(t) = 20/(t + 1)$ at $t = 0$

17–20. **Slopes of tangent lines** *For the following functions, make a table of slopes of secant lines and make a conjecture about the slope of the tangent line at the indicated point.*

17. $f(x) = 2x^2$ at $x = 2$ 18. $f(x) = 3 \cos x$ at $x = \pi/2$

19. $f(x) = e^x$ at $x = 0$ 20. $f(x) = x^3 - x$ at $x = 1$

21. **Tangent lines with zero slope**

    a. Graph the function $f(x) = x^2 - 4x + 3$.
    b. Identify the point $(a, f(a))$ at which the function has a tangent line with zero slope.
    c. Confirm your answer to part (b) by making a table of slopes of secant lines to approximate the slope of the tangent line at this point.

22. Tangent lines with zero slope

a. Graph the function $f(x) = 4 - x^2$.

b. Identify the point $(a, f(a))$ at which the function has a tangent line with zero slope.

c. Consider the point $(a, f(a))$ found in part (b). Is it true that the secant line between $(a - h, f(a - h))$ and $(a + h, f(a + h))$ has slope zero for any value of $h \neq 0$?

23. Zero velocity A projectile is fired vertically upward and has a position given by $s(t) = -16t^2 + 128t + 192$ for $0 \leq t \leq 9$.

a. Graph the position function for $0 \leq t \leq 9$.

b. From the graph of the position function, identify the time at which the projectile has an instantaneous velocity of zero; call this time $t = a$.

c. Confirm your answer to part (b) by making a table of average velocities to approximate the instantaneous velocity at $t = a$.

d. For what values of t on the interval $[0, 9]$ is the instantaneous velocity positive (the projectile moves upward)?

e. For what values of t on the interval $[0, 9]$ is the instantaneous velocity negative (the projectile moves downward)?

24. Impact speed A rock is dropped off the edge of a cliff and its distance s (in feet) from the top of the cliff after t seconds is $s(t) = 16t^2$. Assume the distance from the top of the cliff to the water below is 96 ft.

a. When will the rock strike the water?

b. Make a table of average velocities and approximate the velocity at which the rock strikes the water.

25. Slope of tangent line Given the function $f(x) = 1 - \cos x$ and the points $A(\pi/2, f(\pi/2))$, $B(\pi/2 + 0.05, f(\pi/2 + 0.05))$, $C(\pi/2 + 0.5, f(\pi/2 + 0.5))$, and $D(\pi, f(\pi))$ (see figure), find the slopes of the secant lines through A and D, A and C, and A and B. Use your calculations to make a conjecture about the slope of the line tangent to the graph of f at $x = \pi/2$.

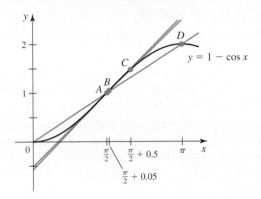

2.2 Definitions of Limits

Computing tangent lines and instantaneous velocities are just two of many important calculus problems that rely on limits. We now put these two problems aside until Chapter 3 and begin with a preliminary definition of the limit of a function.

> **DEFINITION   Limit of a Function (Preliminary)**
>
> Suppose the function f is defined for all x near a except possibly at a. If $f(x)$ is arbitrarily close to L (as close to L as we like) for all x sufficiently close (but not equal) to a, we write
>
> $$\lim_{x \to a} f(x) = L$$
>
> and say the limit of $f(x)$ as x approaches a equals L.

> ➤ The terms *arbitrarily close* and *sufficiently close* will be made precise when rigorous definitions of limits are given in Section 2.7.

Informally, we say that $\lim_{x \to a} f(x) = L$ if $f(x)$ gets closer and closer to L as x gets closer and closer to a from both sides of a. The value of $\lim_{x \to a} f(x)$ (if it exists) depends upon the values of f near a, but it does not depend on the value of $f(a)$. In some cases, the limit $\lim_{x \to a} f(x)$ equals $f(a)$. In other instances, $\lim_{x \to a} f(x)$ and $f(a)$ differ, or $f(a)$ may not even be defined.

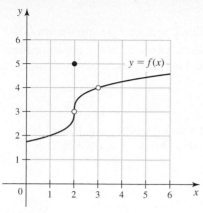

FIGURE 2.7

EXAMPLE 1 **Finding limits from a graph** Use the graph of f (Figure 2.7) to determine the following values, if possible.

a. $f(1)$ and $\lim_{x \to 1} f(x)$ **b.** $f(2)$ and $\lim_{x \to 2} f(x)$ **c.** $f(3)$ and $\lim_{x \to 3} f(x)$

SOLUTION

a. We see that $f(1) = 2$. As x approaches 1 from either side, the values of $f(x)$ approach 2 (Figure 2.8). Therefore, $\lim_{x \to 1} f(x) = 2$.

b. We see that $f(2) = 5$. However, as x approaches 2 from either side, $f(x)$ approaches 3 because the points on the graph of f approach the open circle at $(2, 3)$ (Figure 2.9). Therefore, $\lim_{x \to 2} f(x) = 3$ even though $f(2) = 5$.

c. In this case, $f(3)$ is undefined. We see that $f(x)$ approaches 4 as x approaches 3 from either side (Figure 2.10). Therefore, $\lim_{x \to 3} f(x) = 4$ even though $f(3)$ does not exist.

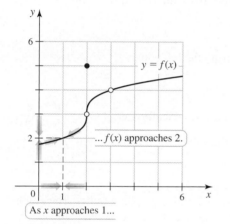

FIGURE 2.8

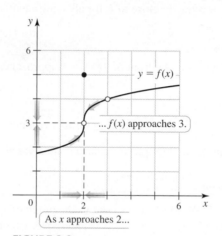

FIGURE 2.9

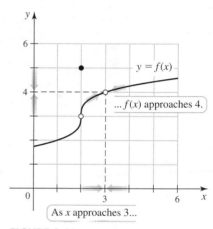

FIGURE 2.10

Related Exercises 7–10 ◄

QUICK CHECK 1 In Example 1, suppose we redefine the function at one point so that $f(1) = 1$. Does this change the value of $\lim_{x \to 1} f(x)$? ◄

EXAMPLE 2 **Finding limits from a table** Create a table of values of $f(x) = \dfrac{\sqrt{x} - 1}{x - 1}$ corresponding to values of x near 1. Then make a conjecture about the value of $\lim_{x \to 1} f(x)$.

> In Example 2, we have not stated with certainty that $\lim_{x \to 1} f(x) = 0.5$. But this is our best guess based upon the numerical evidence. Methods for calculating limits precisely are introduced in Section 2.3.

SOLUTION Table 2.2 lists values of f corresponding to values of x approaching 1 from both sides. The numerical evidence suggests that $f(x)$ approaches 0.5 as x approaches 1. Therefore, we make the conjecture that $\lim_{x \to 1} f(x) = 0.5$.

Table 2.2   ⟶ 1 ⟵

x	0.9	0.99	0.999	0.9999	1.0001	1.001	1.01	1.1
$f(x) = \dfrac{\sqrt{x} - 1}{x - 1}$	0.5131670	0.5012563	0.5001251	0.5000125	0.4999875	0.4998750	0.4987562	0.4880885

Related Exercises 11–14 ◄

One-Sided Limits

The limit $\lim_{x \to a} f(x) = L$ is referred to as a *two-sided* limit because $f(x)$ approaches L as x approaches a for values of x less than a *and* for values of x greater than a. For some functions, it makes sense to examine *one-sided* limits called left-hand and right-hand limits.

> As with two-sided limits, the value of a one-sided limit (if it exists) depends on the values of $f(x)$ near $x = a$ but not on the value of $f(a)$.

DEFINITION One-Sided Limits

1. **Right-hand limit** Suppose f is defined for all x near a with $x > a$. If $f(x)$ is arbitrarily close to L for all x sufficiently close to a with $x > a$, we write

$$\lim_{x \to a^+} f(x) = L$$

and say the limit of $f(x)$ as x approaches a from the right equals L.

2. **Left-hand limit** Suppose f is defined for all x near a with $x < a$. If $f(x)$ is arbitrarily close to L for all x sufficiently close to a with $x < a$, we write

$$\lim_{x \to a^-} f(x) = L$$

and say the limit of $f(x)$ as x approaches a from the left equals L.

> Computer-generated graphs and tables help us understand the idea of a limit. Keep in mind, however, that computers are not infallible and they may produce incorrect results, even for simple functions (see Example 5 and Exercises 37–38).

EXAMPLE 3 Examining limits graphically and numerically Let $f(x) = \dfrac{x^3 - 8}{4(x - 2)}$. Use tables and graphs to make a conjecture about the values of $\lim_{x \to 2^+} f(x)$, $\lim_{x \to 2^-} f(x)$, and $\lim_{x \to 2} f(x)$, if they exist.

SOLUTION Figure 2.11a shows the graph of f obtained with a graphing utility. The graph is misleading because $f(2)$ is undefined, which means there should be a hole in the graph at $(2, 3)$ (Figure 2.11b).

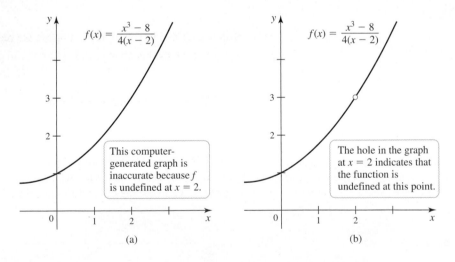

FIGURE 2.11   (a)   (b)

The graph in Figure 2.12a and the function values in Table 2.3 suggest that $f(x)$ approaches 3 as x approaches 2 from the right. Therefore, we write

$$\lim_{x \to 2^+} f(x) = 3,$$

which says the limit of $f(x)$ as x approaches 2 from the right equals 3.

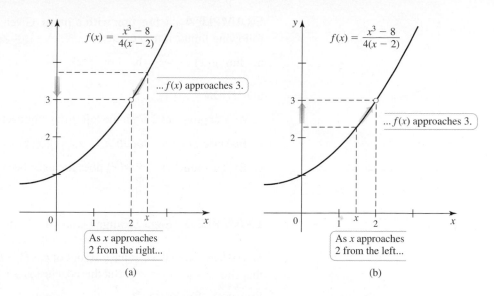

FIGURE 2.12

(a)   (b)

▷ Remember that the value of the limit does not depend upon the value of $f(2)$. In this case, $\lim_{x\to 2} f(x) = 3$ despite the fact that $f(2)$ is undefined.

Similarly, Figure 2.12b and Table 2.3 suggest that as x approaches 2 from the left, $f(x)$ approaches 3. So, we write

$$\lim_{x\to 2^-} f(x) = 3,$$

which says the limit of $f(x)$ as x approaches 2 from the left equals 3. Because $f(x)$ approaches 3 as x approaches 2 from either side, we write $\lim_{x\to 2} f(x) = 3$.

Table 2.3

x	1.9	1.99	1.999	1.9999	2.0001	2.001	2.01	2.1
$f(x) = \dfrac{x^3 - 8}{4(x - 2)}$	2.8525	2.985025	2.99850025	2.99985000	3.00015000	3.00150025	3.015025	3.1525

Related Exercises 15–16 ◀

Based upon the previous example, you might wonder whether the limits $\lim_{x\to a^-} f(x)$, $\lim_{x\to a^+} f(x)$, and $\lim_{x\to a} f(x)$ always exist and are equal. The remaining examples demonstrate that these limits sometimes have different values and in other cases, some or all of these limits do not exist. The following result is useful when comparing one-sided and two-sided limits.

▷ Recall that we write P *if and only if* Q when P implies Q and Q implies P.

THEOREM 2.1   Relationship Between One-Sided and Two-Sided Limits
Assume f is defined for all x near a except possibly at a. Then $\lim_{x\to a} f(x) = L$ if and only if $\lim_{x\to a^+} f(x) = L$ and $\lim_{x\to a^-} f(x) = L$.

A proof of Theorem 2.1 is outlined in Exercise 44 of Section 2.7. Using this theorem, it follows that $\lim_{x\to a} f(x) \neq L$ if either $\lim_{x\to a^+} f(x) \neq L$ or $\lim_{x\to a^-} f(x) \neq L$ (or both). Furthermore, if either $\lim_{x\to a^+} f(x)$ or $\lim_{x\to a^-} f(x)$ does not exist, then $\lim_{x\to a} f(x)$ does not exist. We put these ideas to work in the next two examples.

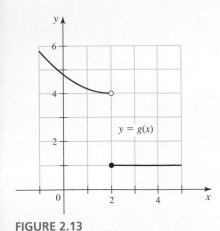

FIGURE 2.13

EXAMPLE 4   A function with a jump Given the graph of g in Figure 2.13, find the following limits, if they exist.

a. $\lim\limits_{x \to 2^-} g(x)$ **b.** $\lim\limits_{x \to 2^+} g(x)$ **c.** $\lim\limits_{x \to 2} g(x)$

SOLUTION

a. As x approaches 2 from the left, $g(x)$ approaches 4. Therefore, $\lim\limits_{x \to 2^-} g(x) = 4$.

b. Because $g(x) = 1$ for all $x \geq 2$, $\lim\limits_{x \to 2^+} g(x) = 1$.

c. By Theorem 2.1, $\lim\limits_{x \to 2} g(x)$ does not exist because $\lim\limits_{x \to 2^-} g(x) \neq \lim\limits_{x \to 2^+} g(x)$.

Related Exercises 17–20 ◄

EXAMPLE 5   Some strange behavior Examine $\lim\limits_{x \to 0} \cos(1/x)$.

SOLUTION From the first three values of $\cos(1/x)$ in Table 2.4, it is tempting to conclude that $\lim\limits_{x \to 0^+} \cos(1/x) = -1$. But this conclusion is not confirmed when we evaluate $\cos(1/x)$ for values of x closer to 0.

Table 2.4

x	$\cos(1/x)$
0.001	0.56238
0.0001	−0.95216
0.00001	−0.99936
0.000001	0.93675
0.0000001	−0.90727
0.00000001	−0.36338

> We might *incorrectly* conclude that $\cos(1/x)$ approaches −1 as x approaches 0 from the right.

The behavior of $\cos(1/x)$ near $x = 0$ is better understood by letting $x = 1/(n\pi)$, where n is a positive integer. In this case

$$\cos\left(\frac{1}{x}\right) = \cos n\pi = \begin{cases} 1 & \text{if } n \text{ is even} \\ -1 & \text{if } n \text{ is odd} \end{cases}$$

As n increases, the values of $x = 1/(n\pi)$ approach zero, while the values of $\cos(1/x)$ oscillate between −1 and 1 (Figure 2.14). Therefore, $\cos(1/x)$ does not approach one single number as x approaches 0 from the right. We conclude that $\lim\limits_{x \to 0^+} \cos(1/x)$ does *not* exist, which implies that $\lim\limits_{x \to 0} \cos(1/x)$ does not exist.

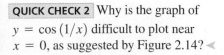

QUICK CHECK 2 Why is the graph of $y = \cos(1/x)$ difficult to plot near $x = 0$, as suggested by Figure 2.14? ◄

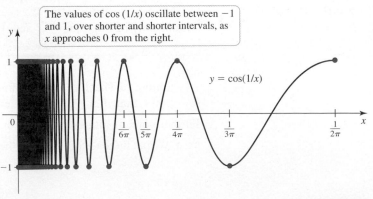

FIGURE 2.14

Related Exercises 21–22 ◄

Using tables and graphs to make conjectures for the values of limits worked well until Example 5. The limitation of technology in this example is not an isolated incident. For this reason, analytical techniques (paper-and-pencil methods) for finding limits are developed in the next section.

SECTION 2.2 EXERCISES

Review Questions

1. Explain in words the meaning of $\lim_{x \to a} f(x) = L$.

2. True or false: When $\lim_{x \to a} f(x)$ exists, it always equals $f(a)$. Explain.

3. Explain the meaning of $\lim_{x \to a^+} f(x) = L$.

4. Explain the meaning of $\lim_{x \to a^-} f(x) = L$.

5. If $\lim_{x \to a^-} f(x) = L$ and $\lim_{x \to a^+} f(x) = M$, where L and M are finite real numbers, then what must be true about L and M in order for $\lim_{x \to a} f(x)$ to exist?

6. What are the potential problems of using a graphing utility to determine $\lim_{x \to a} f(x)$?

Basic Skills

7. **Finding limits from a graph** Use the graph of h in the figure to find the following values, if they exist.

   a. $h(2)$ b. $\lim_{x \to 2} h(x)$ c. $h(4)$ d. $\lim_{x \to 4} h(x)$ e. $\lim_{x \to 5} h(x)$

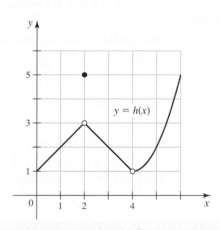

8. **Finding limits from a graph** Use the graph of g in the figure to find the following values, if they exist.

   a. $g(0)$ b. $\lim_{x \to 0} g(x)$ c. $g(1)$ d. $\lim_{x \to 1} g(x)$

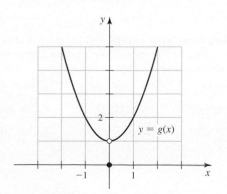

9. **Finding limits from a graph** Use the graph of f in the figure to find the following values, if they exist.

   a. $f(1)$ b. $\lim_{x \to 1} f(x)$ c. $f(0)$ d. $\lim_{x \to 0} f(x)$

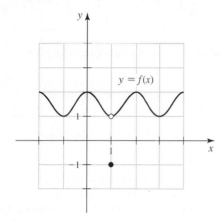

10. **Finding limits from a graph** Use the graph of f in the figure to find the following values, if they exist.

    a. $f(2)$ b. $\lim_{x \to 2} f(x)$ c. $\lim_{x \to 4} f(x)$ d. $\lim_{x \to 5} f(x)$

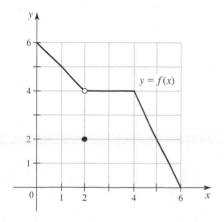

11. **Estimating a limit from tables** Let $f(x) = \dfrac{x^2 - 4}{x - 2}$.

    a. Calculate $f(x)$ for each value of x in the following table.

    b. Make a conjecture about the value of $\lim_{x \to 2} \dfrac{x^2 - 4}{x - 2}$.

x	1.9	1.99	1.999	1.9999
$f(x) = \dfrac{x^2 - 4}{x - 2}$				
x	2.1	2.01	2.001	2.0001
$f(x) = \dfrac{x^2 - 4}{x - 2}$				

12. Estimating a limit from tables Let $f(x) = \dfrac{x^3 - 1}{x - 1}$.

a. Calculate $f(x)$ for each value of x in the following table.

b. Make a conjecture about the value of $\displaystyle\lim_{x \to 1} \dfrac{x^3 - 1}{x - 1}$.

x	0.9	0.99	0.999	0.9999
$f(x) = \dfrac{x^3 - 1}{x - 1}$				
x	1.1	1.01	1.001	1.0001
$f(x) = \dfrac{x^3 - 1}{x - 1}$				

13. Estimating the limit of a function Let $g(t) = \dfrac{t - 9}{\sqrt{t} - 3}$.

a. Make two tables, one showing the values of g for $t = 8.9, 8.99,$ and 8.999 and one showing values of g for $t = 9.1, 9.01,$ and 9.001.

b. Make a conjecture about the value of $\displaystyle\lim_{t \to 9} \dfrac{t - 9}{\sqrt{t} - 3}$.

14. Estimating the limit of a function Let $f(x) = (1 + x)^{1/x}$.

a. Make two tables, one showing the values of f for $x = 0.01,$ $0.001, 0.0001,$ and 0.00001 and one showing values of f for $x = -0.01, -0.001, -0.0001,$ and -0.00001. Round your answers to five digits.

b. Estimate the value of $\displaystyle\lim_{x \to 0} (1 + x)^{1/x}$.

c. What mathematical constant does $\displaystyle\lim_{x \to 0} (1 + x)^{1/x}$ appear to equal?

15. One-sided and two-sided limits Let $f(x) = \dfrac{x^2 - 25}{x - 5}$. Use tables and graphs to make a conjecture about the values of $\displaystyle\lim_{x \to 5^+} f(x)$, $\displaystyle\lim_{x \to 5^-} f(x)$, and $\displaystyle\lim_{x \to 5} f(x)$, if they exist.

16. One-sided and two-sided limits Let $g(x) = \dfrac{x - 100}{\sqrt{x} - 10}$. Use tables and graphs to make a conjecture about the values of $\displaystyle\lim_{x \to 100^+} g(x)$, $\displaystyle\lim_{x \to 100^-} g(x)$, and $\displaystyle\lim_{x \to 100} g(x)$, if they exist.

17. One-sided and two-sided limits Use the graph of f in the figure to find the following values, if they exist. If a limit does not exist, explain why.

a. $f(1)$ b. $\displaystyle\lim_{x \to 1^-} f(x)$ c. $\displaystyle\lim_{x \to 1^+} f(x)$ d. $\displaystyle\lim_{x \to 1} f(x)$

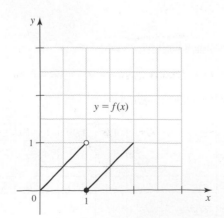

18. One-sided and two-sided limits Use the graph of g in the figure to find the following values, if they exist. If a limit does not exist, explain why.

a. $g(2)$ b. $\displaystyle\lim_{x \to 2^-} g(x)$ c. $\displaystyle\lim_{x \to 2^+} g(x)$

d. $\displaystyle\lim_{x \to 2} g(x)$ e. $g(3)$ f. $\displaystyle\lim_{x \to 3^-} g(x)$

g. $\displaystyle\lim_{x \to 3^+} g(x)$ h. $g(4)$ i. $\displaystyle\lim_{x \to 4} g(x)$

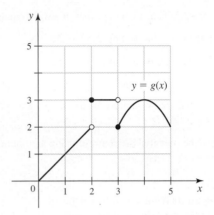

19. Finding limits from a graph Use the graph of f in the figure to find the following values, if they exist. If a limit does not exist, explain why.

a. $f(1)$ b. $\displaystyle\lim_{x \to 1^-} f(x)$ c. $\displaystyle\lim_{x \to 1^+} f(x)$

d. $\displaystyle\lim_{x \to 1} f(x)$ e. $f(3)$ f. $\displaystyle\lim_{x \to 3^-} f(x)$

g. $\displaystyle\lim_{x \to 3^+} f(x)$ h. $\displaystyle\lim_{x \to 3} f(x)$ i. $f(2)$

j. $\displaystyle\lim_{x \to 2^-} f(x)$ k. $\displaystyle\lim_{x \to 2^+} f(x)$ l. $\displaystyle\lim_{x \to 2} f(x)$

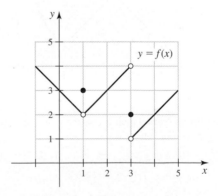

20. Finding limits from a graph Use the graph of g in the figure on the next page to find the following values, if they exist. If a limit does not exist, explain why.

a. $g(-1)$ b. $\displaystyle\lim_{x \to -1^-} g(x)$ c. $\displaystyle\lim_{x \to -1^+} g(x)$

d. $\displaystyle\lim_{x \to -1} g(x)$ e. $g(1)$ f. $\displaystyle\lim_{x \to 1} g(x)$

g. $\displaystyle\lim_{x \to 3} g(x)$ h. $g(5)$ i. $\displaystyle\lim_{x \to 5^-} g(x)$

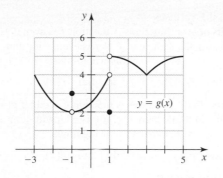

21. Strange behavior near $x = 0$

a. Create a table of values of $\sin\left(\dfrac{1}{x}\right)$ for $x = \dfrac{2}{\pi}, \dfrac{2}{3\pi}, \dfrac{2}{5\pi}, \dfrac{2}{7\pi}, \dfrac{2}{9\pi}$, and $\dfrac{2}{11\pi}$. Describe the pattern of values you observe.

b. Why does a graphing utility have difficulty plotting the graph of $y = \sin(1/x)$ near $x = 0$ (see figure)?

c. What do you conclude about $\lim\limits_{x \to 0} \sin(1/x)$?

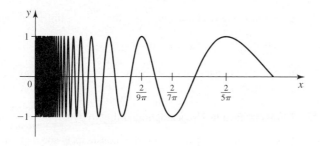

22. Strange behavior near $x = 0$

a. Create a table of values of $\tan(3/x)$ for $x = 1, 0.1, 0.01, 0.001, 0.0001$, and 0.00001. Describe the general pattern in the values you observe.

b. Use a graphing utility to graph $y = \tan(3/x)$. Why does a graphing utility have difficulty plotting the graph near $x = 0$?

c. What do you conclude about $\lim\limits_{x \to 0} \tan(3/x)$?

Further Explorations

23. Explain why or why not Determine whether the following statements are true and give an explanation or counterexample.

a. The value of $\lim\limits_{x \to 3} \dfrac{x^2 - 9}{x - 3}$ does not exist.

b. The value of $\lim\limits_{x \to a} f(x)$ can always be found by computing $f(a)$.

c. The value of $\lim\limits_{x \to a} f(x)$ does not exist if $f(a)$ is undefined.

24–25. Sketching graphs of functions *Sketch the graph of a function with the given properties. You do not need to find a formula for the function.*

24. $f(1) = 0, f(2) = 4, f(3) = 6,\ \lim\limits_{x \to 2^-} f(x) = -3,$
$\lim\limits_{x \to 2^+} f(x) = 5$

25. $g(1) = 0, g(2) = 1, g(3) = -2,\ \lim\limits_{x \to 2} g(x) = 0,$
$\lim\limits_{x \to 3^-} g(x) = -1, \lim\limits_{x \to 3^+} g(x) = -2$

26–29. Calculator limits *Estimate the value of the following limits by creating a table of function values for $h = 0.01, 0.001$, and 0.0001, and $h = -0.01, -0.001$, and -0.0001.*

26. $\lim\limits_{h \to 0} (1 + 2h)^{1/h}$

27. $\lim\limits_{h \to 0} (1 + 3h)^{2/h}$

28. $\lim\limits_{h \to 0} \dfrac{2^h - 1}{h}$

29. $\lim\limits_{h \to 0} \dfrac{\ln(1 + h)}{h}$

30. A step function Let $f(x) = \dfrac{|x|}{x}$ for $x \neq 0$.

a. Sketch a graph of f on the interval $[-2, 2]$.

b. Does $\lim\limits_{x \to 0} f(x)$ exist? Explain your reasoning after first examining $\lim\limits_{x \to 0^-} f(x)$ and $\lim\limits_{x \to 0^+} f(x)$.

31. The floor function For any real number x, the *floor function* (or *greatest integer function*), $\lfloor x \rfloor$, is defined to be the greatest integer less than or equal to x (see figure).

a. Compute $\lim\limits_{x \to -1^-} \lfloor x \rfloor$, $\lim\limits_{x \to -1^+} \lfloor x \rfloor$, $\lim\limits_{x \to 2^-} \lfloor x \rfloor$, and $\lim\limits_{x \to 2^+} \lfloor x \rfloor$.

b. Compute $\lim\limits_{x \to 2.3^-} \lfloor x \rfloor$, $\lim\limits_{x \to 2.3^+} \lfloor x \rfloor$, and $\lim\limits_{x \to 2.3} \lfloor x \rfloor$.

c. In general, for an integer a, state the values of $\lim\limits_{x \to a^-} \lfloor x \rfloor$ and $\lim\limits_{x \to a^+} \lfloor x \rfloor$.

d. In general, if a is not an integer, state the values of $\lim\limits_{x \to a^-} \lfloor x \rfloor$ and $\lim\limits_{x \to a^+} \lfloor x \rfloor$.

e. For what values of a does $\lim\limits_{x \to a} \lfloor x \rfloor$ exist? Explain.

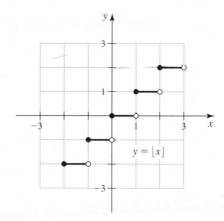

32. The ceiling function For any real number x, the *ceiling function*, $\lceil x \rceil$, is defined to be the least integer greater than or equal to x.

a. Graph the ceiling function $y = \lceil x \rceil$ for $-2 \le x \le 3$.

b. Evaluate $\lim\limits_{x \to 2^-} \lceil x \rceil$, $\lim\limits_{x \to 1^+} \lceil x \rceil$, and $\lim\limits_{x \to 1.5} \lceil x \rceil$.

c. For what values of a does $\lim\limits_{x \to a} \lceil x \rceil$ exist? Explain.

33. Limit by graphing Use the zoom and trace features of a graphing utility to approximate $\lim\limits_{x \to 1} \dfrac{\sqrt{2x - x^4} - \sqrt[3]{x}}{1 - x^{3/4}}$.

34. Limit by graphing Use the zoom and trace features of a graphing utility to approximate $\lim\limits_{x \to 0} \dfrac{6^x - 3^x}{2x}$.

Applications

35. Postage rates Assume that postage for sending a first-class letter in the United States is $0.42 for the first ounce (up to and including 1 oz) plus $0.17 for each additional ounce (up to and including each additional ounce).

a. Graph the function $p = f(w)$ that gives the postage p for sending a letter that weighs w ounces if $0 < w \le 5$.

b. Evaluate $\lim\limits_{w \to 3.3} f(w)$.

c. Interpret the limits $\lim\limits_{w \to 1^+} f(w)$ and $\lim\limits_{w \to 1^-} f(w)$.

d. Does $\lim\limits_{w \to 4} f(w)$ exist? Explain.

36. The Heaviside function The Heaviside function is used in engineering applications to simulate flipping a switch. It is defined as

$$H(x) = \begin{cases} 0 & \text{if } x < 0 \\ 1 & \text{if } x \ge 0 \end{cases}$$

a. Sketch a graph of H on the interval $[-1, 2]$.

b. Does $\lim\limits_{x \to 0} H(x)$ exist? Explain your reasoning after first examining $\lim\limits_{x \to 0^-} H(x)$ and $\lim\limits_{x \to 0^+} H(x)$.

37–38. Pitfalls of using a calculator *Suppose you want to estimate* $\lim\limits_{x \to a} \dfrac{f(x)}{g(x)}$. *If $g(x)$ is nearly equal to 0 when x is close to a, then a calculator utility may round the value of $g(x)$ to 0.*

37. Calculator limit Evaluate $\dfrac{\sin x^{20}}{x^{20}}$ for $x = 0.1, 0.01, \ldots, 0.00001$.

Based on the value given by your calculator, propose a value for $\lim\limits_{x \to 0^+} \dfrac{\sin x^{20}}{x^{20}}$. Using limit techniques introduced later in the text, it can be shown that $\lim\limits_{x \to 0^+} \dfrac{\sin x^{20}}{x^{20}} = 1$. Does your proposed value for the limit agree? Explain.

38. Calculator limit

a. Graph the function $y = \dfrac{x \sin x}{1 - \cos x}$ with the window $[-1, 1] \times [0, 3]$. Based upon this graph, estimate the value of $\lim\limits_{x \to 0} \dfrac{x \sin x}{1 - \cos x}$. What is the value of $\dfrac{x \sin x}{1 - \cos x}$ at $x = 0$?

b. Values of $\dfrac{x \sin x}{1 - \cos x}$ near 0, computed with a computer algebra system, are shown in the accompanying table. It appears that $\lim\limits_{x \to 0} \dfrac{x \sin x}{1 - \cos x}$ is undefined (does not exist). Using limit techniques introduced later in the text, it can be shown that $\lim\limits_{x \to 0} \dfrac{x \sin x}{1 - \cos x} = 2$. What do you think happened, and why?

x	$\pm 10^{-6}$	$\pm 10^{-7}$	$\pm 10^{-8}$	$\pm 10^{-9}$	$\pm 10^{-10}$
$\dfrac{x \sin x}{1 - \cos x}$	2	1	undefined	undefined	undefined

Additional Exercises

39. Limits of even functions A function f is even if $f(-x) = f(x)$ for all x in the domain of f. If f is even, with $\lim\limits_{x \to 2^+} f(x) = 5$ and $\lim\limits_{x \to 2^-} f(x) = 8$, find the following limits.

a. $\lim\limits_{x \to -2^+} f(x)$ b. $\lim\limits_{x \to -2^-} f(x)$

40. Limits of odd functions A function g is odd if $g(-x) = -g(x)$ for all x in the domain of g. If g is odd, with $\lim\limits_{x \to 2^+} g(x) = 5$ and $\lim\limits_{x \to 2^-} g(x) = 8$, find the following limits.

a. $\lim\limits_{x \to -2^+} g(x)$ b. $\lim\limits_{x \to -2^-} g(x)$

41. Limits by graphs

a. Use a graphing utility to estimate the values of $\lim\limits_{x \to 0} \dfrac{\tan 2x}{\sin x}$, $\lim\limits_{x \to 0} \dfrac{\tan 3x}{\sin x}$ and $\lim\limits_{x \to 0} \dfrac{\tan 4x}{\sin x}$.

b. Make a conjecture about the value of $\lim\limits_{x \to 0} \dfrac{\tan nx}{\sin x}$ for any real constant n.

42. Limits by graphs Graph $f(x) = \dfrac{\sin nx}{x}$ for $n = 1, 2, 3,$ and 4 (four graphs). Use the window $[-1, 1] \times [0, 5]$.

a. Estimate the values of $\lim\limits_{x \to 0} \dfrac{\sin x}{x}$, $\lim\limits_{x \to 0} \dfrac{\sin 2x}{x}$, $\lim\limits_{x \to 0} \dfrac{\sin 3x}{x}$, and $\lim\limits_{x \to 0} \dfrac{\sin 4x}{x}$.

b. Make a conjecture about the value of $\lim\limits_{x \to 0} \dfrac{\sin nx}{x}$ for any real constant n.

43. Limits by graphs Use a graphing utility to plot $y = \dfrac{\sin nx}{\sin mx}$ for at least three different pairs of nonzero constants m and n of your choice. Estimate $\lim\limits_{x \to 0} \dfrac{\sin nx}{\sin mx}$ in each case. Then use your work to make a conjecture about the value of $\lim\limits_{x \to 0} \dfrac{\sin nx}{\sin mx}$ for any nonzero values of m and n.

QUICK CHECK ANSWERS

1. The value of $\lim\limits_{x \to 1} f(x)$ depends only on the value of f *near* 1, not at 1. Therefore, changing the value of $f(1)$ will not change the value of $\lim\limits_{x \to 1} f(x)$.   **2.** A graphing device has difficulty plotting $y = \cos(1/x)$ near 0 because values of the function vary between -1 and 1 over shorter and shorter intervals as x approaches 0. ◄

2.3 Techniques for Computing Limits

Graphical and numerical techniques for estimating limits, like those presented in the previous section, provide intuition about limits. These techniques, however, occasionally lead to incorrect results. Therefore, we turn our attention to analytical methods for evaluating limits precisely.

Limits of Linear Functions

The graph of $f(x) = mx + b$ is a line with slope m and y-intercept b. From Figure 2.15, we see that $f(x)$ approaches $f(a)$ as x approaches a. Therefore, if f is a linear function we have $\lim_{x \to a} f(x) = f(a)$. It follows that for linear functions, $\lim_{x \to a} f(x)$ is found by direct substitution of $x = a$ into $f(x)$. This observation leads to the following theorem, which is proved in Exercise 28 of Section 2.7.

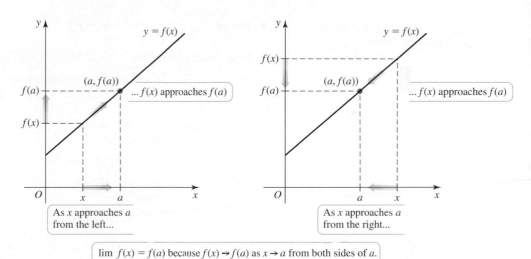

FIGURE 2.15

$$\lim_{x \to a} f(x) = f(a) \text{ because } f(x) \to f(a) \text{ as } x \to a \text{ from both sides of } a.$$

THEOREM 2.2   Limits of Linear Functions

Let a, b, and m be real numbers. For linear functions $f(x) = mx + b$,

$$\lim_{x \to a} f(x) = f(a) = ma + b.$$

EXAMPLE 1   Limits of linear functions Evaluate the following limits.

a. $\lim_{x \to 3} f(x)$, where $f(x) = \left(\frac{1}{2}x - 7\right)$

b. $\lim_{x \to 2} g(x)$, where $g(x) = 6$

SOLUTION

a. $\lim_{x \to 3} f(x) = \lim_{x \to 3} \left(\frac{1}{2}x - 7\right) = f(3) = -\frac{11}{2}.$

b. $\lim_{x \to 2} g(x) = \lim_{x \to 2} 6 = g(2) = 6.$

Related Exercises 11–16 ◀

Limit Laws

The following limit laws greatly simplify the evaluation of many limits.

> **THEOREM 2.3   Limit Laws**
> Assume $\lim\limits_{x \to a} f(x)$ and $\lim\limits_{x \to a} g(x)$ exist. The following properties hold, where c is a real number and $m > 0$ and $n > 0$ are integers.
>
> **1. Sum** $\lim\limits_{x \to a} [f(x) + g(x)] = \lim\limits_{x \to a} f(x) + \lim\limits_{x \to a} g(x)$
>
> **2. Difference** $\lim\limits_{x \to a} [f(x) - g(x)] = \lim\limits_{x \to a} f(x) - \lim\limits_{x \to a} g(x)$
>
> **3. Constant multiple** $\lim\limits_{x \to a} [cf(x)] = c \lim\limits_{x \to a} f(x)$
>
> **4. Product** $\lim\limits_{x \to a} [f(x)g(x)] = \left[\lim\limits_{x \to a} f(x)\right]\left[\lim\limits_{x \to a} g(x)\right]$
>
> **5. Quotient** $\lim\limits_{x \to a} \left[\dfrac{f(x)}{g(x)}\right] = \dfrac{\lim\limits_{x \to a} f(x)}{\lim\limits_{x \to a} g(x)}$, provided $\lim\limits_{x \to a} g(x) \ne 0$
>
> **6. Power** $\lim\limits_{x \to a} [f(x)]^n = \left[\lim\limits_{x \to a} f(x)\right]^n$
>
> **7. Fractional power** $\lim\limits_{x \to a} [f(x)]^{n/m} = \left[\lim\limits_{x \to a} f(x)\right]^{n/m}$, provided $f(x) \ge 0$ for x near a if m is even and n/m is reduced to lowest terms

> ▶ Law 6 is a special case of Law 7. Letting $m = 1$ in Law 7 gives Law 6.

A proof of Law 1 is outlined in Section 2.7. Laws 2–5 are proved in the Appendix B. Law 6 is proved from Law 4 as follows.

For a positive integer n, if $\lim\limits_{x \to a} f(x)$ exists, we have

$$\lim_{x \to a} [f(x)]^n = \lim_{x \to a} \underbrace{[f(x)\, f(x) \cdots f(x)]}_{n \text{ factors of } f(x)}$$

$$= \underbrace{\left[\lim_{x \to a} f(x)\right]\left[\lim_{x \to a} f(x)\right] \cdots \left[\lim_{x \to a} f(x)\right]}_{n \text{ factors of } \lim\limits_{x \to a} f(x)} \quad \text{Repeated use of Law 4}$$

$$= \left[\lim_{x \to a} f(x)\right]^n$$

> ▶ Recall that to take even roots of a number (for example, square roots or fourth roots), the number must be nonnegative if the result is to be real.

In Law 7, the limit of $[f(x)]^{n/m}$ involves the mth root of $f(x)$ when x is near a. If the fraction n/m is in lowest terms and m is even, this root is undefined unless $f(x)$ is nonnegative for all x near $x = a$, which explains the restrictions shown.

EXAMPLE 2   Evaluating limits Suppose $\lim\limits_{x \to 2} f(x) = 4$, $\lim\limits_{x \to 2} g(x) = 5$, and $\lim\limits_{x \to 2} h(x) = 8$. Use the limit laws in Theorem 2.3 to compute each limit.

a. $\lim\limits_{x \to 2} \dfrac{f(x) - g(x)}{h(x)}$ **b.** $\lim\limits_{x \to 2} [6f(x)g(x) + h(x)]$ **c.** $\lim\limits_{x \to 2} [g(x)]^3$

SOLUTION

a. $\lim\limits_{x \to 2} \dfrac{f(x) - g(x)}{h(x)} = \dfrac{\lim\limits_{x \to 2} [f(x) - g(x)]}{\lim\limits_{x \to 2} h(x)}$ Law 5

$$= \dfrac{\lim\limits_{x \to 2} f(x) - \lim\limits_{x \to 2} g(x)}{\lim\limits_{x \to 2} h(x)} \quad \text{Law 2}$$

$$= \dfrac{4 - 5}{8} = -\dfrac{1}{8}$$

b. $\lim_{x\to 2} [6f(x)g(x) + h(x)] = \lim_{x\to 2} [6f(x)g(x)] + \lim_{x\to 2} h(x)$ Law 1

$= 6 \cdot \lim_{x\to 2} [f(x)g(x)] + \lim_{x\to 2} h(x)$ Law 3

$= 6 \cdot \left[\lim_{x\to 2} f(x)\right] \cdot \left[\lim_{x\to 2} g(x)\right] + \lim_{x\to 2} h(x)$ Law 4

$= 6 \cdot 4 \cdot 5 + 8 = 128$

c. $\lim_{x\to 2} [g(x)]^3 = \left[\lim_{x\to 2} g(x)\right]^3 = 5^3 = 125$ Law 6   *Related Exercises 17–22* ◄

Limits of Polynomial and Rational Functions

The limit laws are now used to find the limits of polynomial and rational functions. For example, to evaluate the limit of the polynomial $p(x) = 7x^3 + 3x^2 + 4x + 2$ at an arbitrary point a, we proceed as follows:

$$\lim_{x\to a} p(x) = \lim_{x\to a} (7x^3 + 3x^2 + 4x + 2)$$

$$= \lim_{x\to a} (7x^3) + \lim_{x\to a} (3x^2) + \lim_{x\to a} (4x + 2) \quad \text{Law 1}$$

$$= 7 \lim_{x\to a} (x^3) + 3 \lim_{x\to a} (x^2) + \lim_{x\to a} (4x + 2) \quad \text{Law 3}$$

$$= 7\left(\underbrace{\lim_{x\to a} x}_{a}\right)^3 + 3\left(\underbrace{\lim_{x\to a} x}_{a}\right)^2 + \underbrace{\lim_{x\to a} (4x + 2)}_{4a + 2} \quad \text{Law 6}$$

$$= 7a^3 + 3a^2 + 4a + 2 = p(a) \quad \text{Theorem 2.2}$$

As in the case of linear functions, the limit of a polynomial is found by direct substitution; that is, $\lim_{x\to a} p(x) = p(a)$ (Exercise 87).

It is now a short step to evaluating limits of rational functions of the form $f(x) = p(x)/q(x)$, where p and q are polynomials. Applying Law 5, we have

$$\lim_{x\to a} \frac{p(x)}{q(x)} = \frac{\lim_{x\to a} p(x)}{\lim_{x\to a} q(x)} = \frac{p(a)}{q(a)} \quad \text{provided } q(a) \neq 0,$$

which shows that limits of rational functions are also evaluated by direct substitution.

> The conditions under which direct substitution $\left(\lim_{x\to a} f(x) = f(a)\right)$ can be used to evaluate a limit become clear in Section 2.6, when the important property of *continuity* is discussed.

THEOREM 2.4 Limits of Polynomial and Rational Functions
Assume p and q are polynomials and a is a constant.

a. Polynomial functions: $\lim_{x\to a} p(x) = p(a)$

b. Rational functions: $\lim_{x\to a} \frac{p(x)}{q(x)} = \frac{p(a)}{q(a)}$ provided $q(a) \neq 0$

QUICK CHECK 1 Evaluate $\lim_{x\to 2} (2x^4 - 8x - 16)$ and $\lim_{x\to -1} \frac{x-1}{x}$. ◄

EXAMPLE 3 Limit of a rational function Evaluate $\lim_{x\to 2} \frac{3x^2 - 4x}{5x^3 - 36}$.

SOLUTION Notice that the denominator of this function is nonzero at $x = 2$. Using Theorem 2.4b,

$$\lim_{x\to 2} \frac{3x^2 - 4x}{5x^3 - 36} = \frac{3(2^2) - 4(2)}{5(2^3) - 36} = 1.$$

Related Exercises 23–25 ◄

QUICK CHECK 2 Use Theorem 2.4b to compute $\lim_{x\to 1} \frac{5x^4 - 3x^2 + 8x - 6}{x + 1}$. ◄

EXAMPLE 4   **An algebraic function**   Evaluate $\lim\limits_{x \to 2} \dfrac{\sqrt{2x^3 + 9} + 3x - 1}{4x + 1}$.

SOLUTION Using Theorems 2.3 and 2.4, we have

$$\lim_{x \to 2} \frac{\sqrt{2x^3 + 9} + 3x - 1}{4x + 1} = \frac{\lim\limits_{x \to 2}\left(\sqrt{2x^3 + 9} + 3x - 1\right)}{\lim\limits_{x \to 2}(4x + 1)} \qquad \text{Law 5}$$

$$= \frac{\sqrt{\lim\limits_{x \to 2}(2x^3 + 9)} + \lim\limits_{x \to 2}(3x - 1)}{\lim\limits_{x \to 2}(4x + 1)} \qquad \text{Laws 1 and 7}$$

$$= \frac{\sqrt{(2(2)^3 + 9)} + (3(2) - 1)}{(4(2) + 1)} \qquad \text{Theorem 2.4}$$

$$= \frac{\sqrt{25} + 5}{9} = \frac{10}{9}$$

Notice that the limit at $x = 2$ equals the value of the function at $x = 2$.

Related Exercises 26–30 ◄

One-Sided Limits

Theorem 2.2, limit laws 1–6, and Theorem 2.4 also hold for left-sided and right-sided limits. In other words, these laws remain valid if we replace $\lim\limits_{x \to a}$ with $\lim\limits_{x \to a^+}$ or $\lim\limits_{x \to a^-}$. Law 7 must be modified slightly for one-sided limits, as shown below.

THEOREM 2.3 (CONTINUED)   **Limit Laws for One-Sided Limits**

Laws 1–6 hold with $\lim\limits_{x \to a}$ replaced by $\lim\limits_{x \to a^+}$ or $\lim\limits_{x \to a^-}$. Law 7 is modified as follows. Assume $m > 0$ and $n > 0$ are integers.

7. Fractional Power

a. $\lim\limits_{x \to a^+} [f(x)]^{n/m} = \left[\lim\limits_{x \to a^+} f(x)\right]^{n/m}$ provided $f(x) \geq 0$ for x near a with $x > a$, if m is even and n/m is reduced to lowest terms

b. $\lim\limits_{x \to a^-} [f(x)]^{n/m} = \left[\lim\limits_{x \to a^-} f(x)\right]^{n/m}$ provided $f(x) \geq 0$ for x near a with $x < a$, if m is even and n/m is reduced to lowest terms

---

EXAMPLE 5   **Calculating left- and right-sided limits**   Let

$$f(x) = \begin{cases} -2x + 4 & \text{if } x \leq 1 \\ \sqrt{x - 1} & \text{if } x > 1 \end{cases}$$

Find the values of $\lim\limits_{x \to 1^-} f(x)$, $\lim\limits_{x \to 1^+} f(x)$, and $\lim\limits_{x \to 1} f(x)$, or state that they do not exist.

SOLUTION The graph of f (Figure 2.16) suggests that $\lim\limits_{x \to 1^-} f(x) = 2$ and $\lim\limits_{x \to 1^+} f(x) = 0$. We verify this observation analytically by applying the limit laws. For $x \leq 1$, $f(x) = -2x + 4$; therefore,

$$\lim_{x \to 1^-} f(x) = \lim_{x \to 1^-}(-2x + 4) = 2. \quad \text{Theorem 2.2}$$

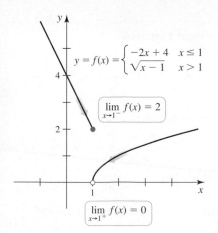

$$y = f(x) = \begin{cases} -2x + 4 & x \leq 1 \\ \sqrt{x - 1} & x > 1 \end{cases}$$

$\lim\limits_{x \to 1^-} f(x) = 2$

$\lim\limits_{x \to 1^+} f(x) = 0$

FIGURE 2.16

For $x > 1$, note that $x - 1 > 0$; it follows that

$$\lim_{x \to 1^+} f(x) = \lim_{x \to 1^+} (\sqrt{x - 1}) = 0. \quad \text{Law 7}$$

Because $\lim_{x \to 1^-} f(x) = 2$ and $\lim_{x \to 1^+} f(x) = 0$, $\lim_{x \to 1} f(x)$ does not exist by Theorem 2.1.

Related Exercises 31–36 ◄

Other Techniques

So far, we have evaluated limits by direct substitution. A more challenging problem is finding $\lim_{x \to a} f(x)$ when the limit exists, but $\lim_{x \to a} f(x) \neq f(a)$. Two typical cases are shown in Figure 2.17. In the first case, $f(a)$ is defined, but it is not equal to $\lim_{x \to a} f(x)$; in the second case, $f(a)$ is not defined at all.

FIGURE 2.17
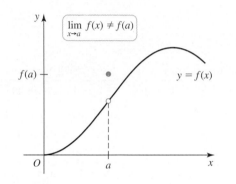
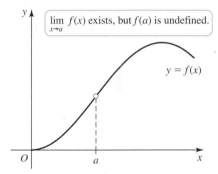

EXAMPLE 6 **Other techniques** Evaluate the following limits.

a. $\displaystyle\lim_{x \to 2} \frac{x^2 - 6x + 8}{x^2 - 4}$

b. $\displaystyle\lim_{x \to 1} \frac{\sqrt{x} - 1}{x - 1}$

SOLUTION

a. This limit cannot be found by direct substitution because the denominator is zero when $x = 2$. Instead, the numerator and denominator are factored; then, assuming $x \neq 2$, we cancel like factors:

> The argument used in this example is common. In the limit process, x approaches 2, but $x \neq 2$. Therefore, we may cancel like factors.

$$\frac{x^2 - 6x + 8}{x^2 - 4} = \frac{(x - 2)(x - 4)}{(x - 2)(x + 2)} = \frac{x - 4}{x + 2}$$

Because $\dfrac{x^2 - 6x + 8}{x^2 - 4} = \dfrac{x - 4}{x + 2}$ whenever $x \neq 2$, the two functions have the same limit as x approaches 2 (Figure 2.18). Therefore,

$$\lim_{x \to 2} \frac{x^2 - 6x + 8}{x^2 - 4} = \lim_{x \to 2} \frac{x - 4}{x + 2} = \frac{2 - 4}{2 + 2} = -\frac{1}{2}.$$

b. This limit was approximated numerically in Example 2 of Section 2.2; we conjectured that the value of the limit is $\frac{1}{2}$. Direct substitution fails in this case because the denominator is zero at $x = 1$. Instead, we first simplify the function by multiplying the numerator and denominator by the *algebraic conjugate* of the numerator. The conjugate of $\sqrt{x} - 1$ is $\sqrt{x} + 1$; therefore,

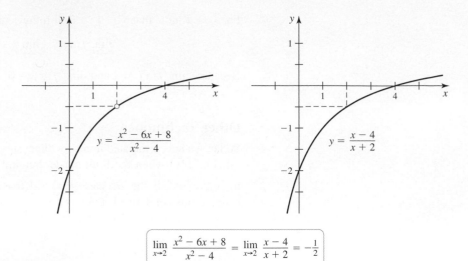

FIGURE 2.18

$$\lim_{x \to 2} \frac{x^2 - 6x + 8}{x^2 - 4} = \lim_{x \to 2} \frac{x - 4}{x + 2} = -\frac{1}{2}$$

$$\frac{\sqrt{x} - 1}{x - 1} = \frac{(\sqrt{x} - 1)(\sqrt{x} + 1)}{(x - 1)(\sqrt{x} + 1)} \qquad \text{Rationalize the numerator.}$$

$$= \frac{x + \sqrt{x} - \sqrt{x} - 1}{(x - 1)(\sqrt{x} + 1)} \qquad \text{Expand the numerator.}$$

$$= \frac{x - 1}{(x - 1)(\sqrt{x} + 1)} \qquad \text{Simplify.}$$

$$= \frac{1}{\sqrt{x} + 1} \qquad \text{Cancel like factors when } x \neq 1.$$

The limit can now be evaluated:

$$\lim_{x \to 1} \frac{\sqrt{x} - 1}{x - 1} = \lim_{x \to 1} \frac{1}{\sqrt{x} + 1} = \frac{1}{1 + 1} = \frac{1}{2}$$

Related Exercises 37–48 ◄

QUICK CHECK 3 Evaluate $\displaystyle\lim_{x \to 5} \frac{x^2 - 7x + 10}{x - 5}$. ◄

An Important Limit

Despite our success in evaluating limits using direct substitution, algebraic manipulation, and the limit laws, there are important limits for which these techniques do not work. One such limit arises when investigating the slope of a line tangent to the graph of an exponential function.

EXAMPLE 7   Slope of the line tangent to $f(x) = 2^x$ Estimate the slope of the line tangent to the graph of $f(x) = 2^x$ at the point $P(0, 1)$.

SOLUTION In Section 2.1, the slope of a tangent line was obtained by finding the limit of slopes of secant lines; the same strategy is employed here. We begin by selecting a point Q near P on the graph of f with coordinates $(x, 2^x)$. The secant line joining the points $P(0, 1)$ and $Q(x, 2^x)$ is an approximation to the tangent line. To compute the slope of the tangent line (denoted by m_{tan}) at $x = 0$, we look at the slope of the secant line $m_{\text{sec}} = (2^x - 1)/x$ and take the limit as x approaches 0.

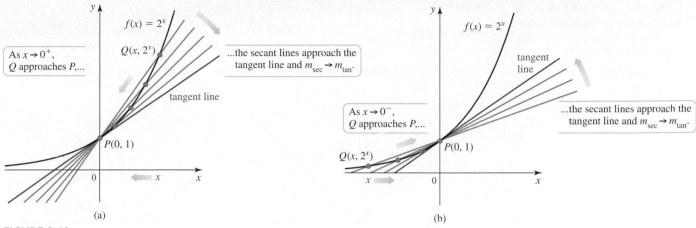

FIGURE 2.19

The limit $\lim\limits_{x \to 0} \dfrac{2^x - 1}{x}$ exists only if it has the same value as $x \to 0^+$ (Figure 2.19a) and as $x \to 0^-$ (Figure 2.19b). Because it is not an elementary limit, it cannot be evaluated using the limit laws of this section. Instead, we investigate the limit using numerical evidence. Choosing positive values of x near 0 results in Table 2.5.

Table 2.5

x	1.0	0.1	0.01	0.001	0.0001	0.00001
$m_{\mathrm{sec}} = \dfrac{2^x - 1}{x}$	1.000000	0.7177	0.6956	0.6934	0.6932	0.6931

> Example 7 shows that
> $$\lim_{x \to 0} \frac{2^x - 1}{x} \approx 0.693,\ \text{which is}$$
> approximately ln 2. The connection between the natural logarithm and slopes of lines tangent to exponential curves is made clear in Chapters 3 and 6.

We see that as x approaches 0 from the right, the slopes of the secant lines approach the slope of the tangent line, which is approximately 0.693. A similar calculation (Exercise 49) gives the same approximation for the limit as x approaches 0 from the left.

Because the left-hand and right-hand limits are the same, we conclude that $\lim\limits_{x \to 0} (2^x - 1)/x \approx 0.693$ (Theorem 2.1). Therefore, the slope of the line tangent to $f(x) = 2^x$ at $x = 0$ is approximately 0.693.   *Related Exercises 49–50* ◄

The Squeeze Theorem

> The Squeeze Theorem is also called the Pinching Theorem or the Sandwich Theorem.

The *Squeeze Theorem* provides another useful method for calculating limits. Suppose the functions f and h have the same limit L at a and assume the function g is trapped between f and h (Figure 2.20). The Squeeze Theorem says that g must also have the limit L at a. A proof of this theorem is outlined in Exercise 54 of Section 2.7.

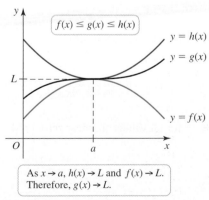

As $x \to a$, $h(x) \to L$ and $f(x) \to L$.
Therefore, $g(x) \to L$.

FIGURE 2.20

THEOREM 2.5   The Squeeze Theorem
Assume the functions f, g, and h satisfy $f(x) \le g(x) \le h(x)$ for all values of x near a, except possibly at a. If $\lim\limits_{x \to a} f(x) = \lim\limits_{x \to a} h(x) = L$, then $\lim\limits_{x \to a} g(x) = L$.

EXAMPLE 8   Sine and cosine limits   A geometric argument (Exercise 86) may be used to show that for $-\pi/2 < x < \pi/2$,

$$-|x| \le \sin x \le |x| \quad \text{and} \quad 0 \le 1 - \cos x \le |x|.$$

Use the Squeeze Theorem to confirm the following limits:

a. $\lim\limits_{x \to 0} \sin x = 0$ **b.** $\lim\limits_{x \to 0} \cos x = 1$

> The two limits in Example 8 play a crucial role in establishing fundamental properties of the trigonometric functions. They reappear in Section 2.6.

SOLUTION

a. Letting $f(x) = -|x|$, $g(x) = \sin x$, and $h(x) = |x|$, we see that g is trapped between f and h on $-\pi/2 < x < \pi/2$ (Figure 2.21a). Because $\lim_{x \to 0} f(x) = \lim_{x \to 0} h(x) = 0$ (Exercise 35), the Squeeze Theorem implies that $\lim_{x \to 0} g(x) = \lim_{x \to 0} \sin x = 0$.

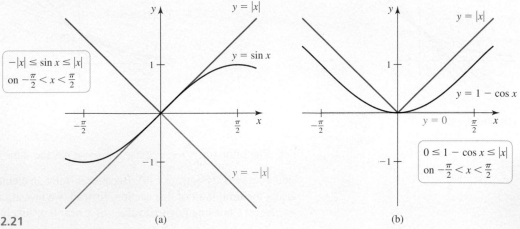

FIGURE 2.21

(a)   (b)

b. In this case, we let $f(x) = 0$, $g(x) = 1 - \cos x$, and $h(x) = |x|$ (Figure 2.21b). Because $\lim_{x \to 0} f(x) = \lim_{x \to 0} h(x) = 0$, the Squeeze Theorem implies that $\lim_{x \to 0} g(x) = \lim_{x \to 0} (1 - \cos x) = 0$. By the limit laws, it follows that $\lim_{x \to 0} 1 - \lim_{x \to 0} \cos x = 0$, or $\lim_{x \to 0} \cos x = 1$.   *Related Exercises 51–54* ◄

EXAMPLE 9 **Applying the Squeeze Theorem** Use the Squeeze Theorem to verify that $\lim_{x \to 0} x^2 \sin(1/x) = 0$.

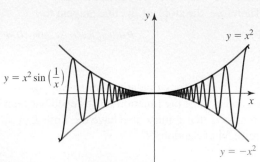

FIGURE 2.22

SOLUTION For any real number θ, $-1 \le \sin \theta \le 1$. Letting $\theta = 1/x$ for all $x \ne 0$, it follows that

$$-1 \le \sin\left(\frac{1}{x}\right) \le 1.$$

Noting that $x^2 > 0$ for $x \ne 0$, each term in this inequality is multiplied by x^2:

$$-x^2 \le x^2 \sin\left(\frac{1}{x}\right) \le x^2$$

These inequalities are illustrated in Figure 2.22. Because $\lim_{x \to 0} x^2 = \lim_{x \to 0} (-x^2) = 0$, the Squeeze Theorem implies that $\lim_{x \to 0} x^2 \sin(1/x) = 0$.

Related Exercises 51–54 ◄

QUICK CHECK 4 Suppose f satisfies $1 \le f(x) \le 1 + \dfrac{x^2}{6}$ for all values of x near zero. Find $\lim_{x \to 0} f(x)$, if possible. ◄

SECTION 2.3 EXERCISES

Review Questions

1. How is $\lim_{x \to a} f(x)$ calculated if f is a polynomial function?

2. How are $\lim_{x \to a^-} f(x)$ and $\lim_{x \to a^+} f(x)$ calculated if f is a polynomial function?

3. For what values of a does $\lim_{x \to a} r(x) = r(a)$ if r is a rational function?

4. Assume $\lim_{x \to 3} g(x) = 4$ and $f(x) = g(x)$ whenever $x \neq 3$.

   Evaluate $\lim_{x \to 3} f(x)$, if possible.

5. Explain why $\lim_{x \to 3} \dfrac{x^2 - 7x + 12}{x - 3} = \lim_{x \to 3} (x - 4)$.

6. If $\lim_{x \to 2} f(x) = -8$, find $\lim_{x \to 2} [f(x)]^{2/3}$.

7. Suppose p and q are polynomials. If $\lim_{x \to 0} \dfrac{p(x)}{q(x)} = 10$ and $q(0) = 2$, find $p(0)$.

8. If $\lim_{x \to 2} f(x) = \lim_{x \to 2} h(x) = 5$, find $\lim_{x \to 2} g(x)$ when $f(x) \leq g(x) \leq h(x)$ for all values of x.

9. Evaluate $\lim_{x \to 5} \sqrt{x^2 - 9}$.

10. Suppose
$$f(x) = \begin{cases} 4 & \text{if } x \leq 3 \\ x + 2 & \text{if } x > 3 \end{cases}$$

    Compute $\lim_{x \to 3^-} f(x)$ and $\lim_{x \to 3^+} f(x)$.

Basic Skills

11–16. Limits of linear functions *Evaluate the following limits.*

11. $\lim_{x \to 4} (3x - 7)$

12. $\lim_{x \to 1} (-2x + 5)$

13. $\lim_{x \to -9} (5x)$

14. $\lim_{x \to 2} (-3x)$

15. $\lim_{x \to 6} (4)$

16. $\lim_{x \to -5} (\pi)$

17–22. Applying limit laws *Assume $\lim_{x \to 1} f(x) = 8$, $\lim_{x \to 1} g(x) = 3$, and $\lim_{x \to 1} h(x) = 2$. Compute the following limits and state the limit laws used to justify your computations.*

17. $\lim_{x \to 1} [4f(x)]$

18. $\lim_{x \to 1} \left[\dfrac{f(x)}{h(x)} \right]$

19. $\lim_{x \to 1} \left[\dfrac{f(x)g(x)}{h(x)} \right]$

20. $\lim_{x \to 1} \left[\dfrac{f(x)}{g(x) - h(x)} \right]$

21. $\lim_{x \to 1} [h(x)]^5$

22. $\lim_{x \to 1} \sqrt[3]{f(x)g(x) + 3}$

23–30. Evaluating limits *Evaluate the following limits.*

23. $\lim_{x \to 1} (2x^3 - 3x^2 + 4x + 5)$

24. $\lim_{t \to -2} (t^2 + 5t + 7)$

25. $\lim_{x \to 1} \dfrac{5x^2 + 6x + 1}{8x - 4}$

26. $\lim_{t \to 3} \sqrt[3]{t^2 - 10}$

27. $\lim_{b \to 2} \dfrac{3b}{\sqrt{4b + 1} - 1}$

28. $\lim_{x \to 2} (x^2 - x)^5$

29. $\lim_{x \to 3} \dfrac{-5x}{\sqrt{4x - 3}}$

30. $\lim_{h \to 0} \dfrac{3}{\sqrt{16 + 3h} + 4}$

31. **One-sided limits** Let
$$f(x) = \begin{cases} x^2 + 1 & \text{if } x < -1 \\ \sqrt{x + 1} & \text{if } x \geq -1 \end{cases}$$

    Compute the following limits or state that they do not exist.

    a. $\lim_{x \to -1^-} f(x)$
    b. $\lim_{x \to -1^+} f(x)$
    c. $\lim_{x \to -1} f(x)$

32. **One-sided limits** Let
$$f(x) = \begin{cases} 0 & \text{if } x \leq -5 \\ \sqrt{25 - x^2} & \text{if } -5 < x < 5 \\ 3x & \text{if } x \geq 5 \end{cases}$$

    Compute the following limits or state that they do not exist.

    a. $\lim_{x \to -5^-} f(x)$
    b. $\lim_{x \to -5^+} f(x)$
    c. $\lim_{x \to -5} f(x)$
    d. $\lim_{x \to 5^-} f(x)$
    e. $\lim_{x \to 5^+} f(x)$
    f. $\lim_{x \to 5} f(x)$

33. **One-sided limits**

    a. Evaluate $\lim_{x \to 2^+} \sqrt{x - 2}$.

    b. Why don't we consider evaluating $\lim_{x \to 2^-} \sqrt{x - 2}$?

34. **One-sided limits**

    a. Evaluate $\lim_{x \to 3^-} \sqrt{\dfrac{x - 3}{2 - x}}$.

    b. Why don't we consider evaluating $\lim_{x \to 3^+} \sqrt{\dfrac{x - 3}{2 - x}}$?

35. **Absolute value limit** Show that $\lim_{x \to 0} |x| = 0$ by first evaluating $\lim_{x \to 0^-} |x|$ and $\lim_{x \to 0^+} |x|$. Recall that
$$|x| = \begin{cases} -x & \text{if } x < 0 \\ x & \text{if } x \geq 0 \end{cases}$$

36. **Absolute value limit** Show that $\lim_{x \to a} |x| = |a|$ for any real number. (*Hint:* Consider three cases, $a < 0$, $a = 0$, and $a > 0$.)

37–48. Other techniques *Evaluate the following limits, where a and b are fixed real numbers.*

37. $\lim_{x \to 1} \dfrac{x^2 - 1}{x - 1}$

38. $\lim_{x \to 3} \dfrac{x^2 - 2x - 3}{x - 3}$

39. $\lim_{x \to 4} \dfrac{x^2 - 16}{4 - x}$

40. $\lim_{t \to 2} \dfrac{3t^2 - 7t + 2}{2 - t}$

41. $\lim_{x \to b} \dfrac{(x - b)^{50} - x + b}{x - b}$

42. $\lim_{x \to -b} \dfrac{(x + b)^7 + (x + b)^{10}}{4(x + b)}$

43. $\lim_{x \to -1} \dfrac{(2x - 1)^2 - 9}{x + 1}$

44. $\lim_{h \to 0} \dfrac{\dfrac{1}{5 + h} - \dfrac{1}{5}}{h}$

45. $\lim_{x \to 9} \dfrac{\sqrt{x} - 3}{x - 9}$

46. $\lim_{t \to a} \dfrac{\sqrt{3t + 1} - \sqrt{3a + 1}}{t - a}$

47. $\lim_{h \to 0} \dfrac{\sqrt{16 + h} - 4}{h}$

48. $\lim_{x \to 0} \dfrac{a - \sqrt{a^2 - x^2}}{x^2}$

49. Slope of a tangent line

a. Sketch a graph of $y = 2^x$ and carefully draw three secant lines connecting the points $P(0, 1)$ and $Q(x, 2^x)$, for $x = -3, -2,$ and -1.

b. Find the slope of the line that joins $P(0, 1)$ and $Q(x, 2^x)$ for $x \neq 0$.

c. Complete the table and make a conjecture about the value of $\lim\limits_{x \to 0^-} \dfrac{2^x - 1}{x}$.

x	-1.0	-0.1	-0.01	-0.001	-0.0001	-0.00001
$\dfrac{2^x - 1}{x}$						

50. Slope of a tangent line

a. Sketch a graph of $y = 3^x$ and carefully draw four secant lines connecting the points $P(0, 1)$ and $Q(x, 3^x)$ for $x = -2, -1, 1,$ and 2.

b. Find the slope of the line that joins $P(0, 1)$ and $Q(x, 3^x)$ for $x \neq 0$.

c. Complete the table and make a conjecture about the value of $\lim\limits_{x \to 0} \dfrac{3^x - 1}{x}$.

x	-0.1	-0.01	-0.001	-0.0001	0.1	0.01	0.001	0.0001
$\dfrac{3^x - 1}{x}$								

51. Applying the Squeeze Theorem

a. Show that $-|x| \leq x \sin\left(\dfrac{1}{x}\right) \leq |x|$ for all $x \neq 0$.

b. Illustrate the inequality $-|x| \leq x \sin\left(\dfrac{1}{x}\right) \leq |x|$ with a graph.

c. Use the Squeeze Theorem to show that $\lim\limits_{x \to 0} x \sin\left(\dfrac{1}{x}\right) = 0$.

52. A cosine limit by the Squeeze Theorem It can be shown that $1 - x^2/2 \leq \cos x \leq 1$ for x near 0.

a. Illustrate the inequality $1 - x^2/2 \leq \cos x \leq 1$ with a graph.
b. Use the inequality in part (a) to find $\lim\limits_{x \to 0} \cos x$.

53. A sine limit by the Squeeze Theorem It can be shown that $1 - \dfrac{x^2}{6} \leq \dfrac{\sin x}{x} \leq 1$ for x near 0.

a. Illustrate the inequality $1 - \dfrac{x^2}{6} \leq \dfrac{\sin x}{x} \leq 1$ with a graph.

b. Use the inequality in part (a) to find $\lim\limits_{x \to 0} \dfrac{\sin x}{x}$.

54. A logarithm limit by the Squeeze Theorem

a. Draw a graph to verify that $-|x| \leq x^2 \ln x^2 \leq |x|$ for $-1 \leq x \leq 1$.

b. Use the Squeeze Theorem to determine $\lim\limits_{x \to 0} x^2 \ln x^2$.

Further Explorations

55. Explain why or why not Determine whether the following statements are true and give an explanation or counterexample. Assume a and L are finite numbers.

a. If $\lim\limits_{x \to a} f(x) = L$, then $f(a) = L$.

b. If $\lim\limits_{x \to a^-} f(x) = L$, then $\lim\limits_{x \to a^+} f(x) = L$.

c. If $\lim\limits_{x \to a} f(x) = L$ and $\lim\limits_{x \to a} g(x) = L$, then $f(a) = g(a)$.

d. The limit $\lim\limits_{x \to a} \dfrac{f(x)}{g(x)}$ does not exist if $g(a) = 0$.

e. If $\lim\limits_{x \to 1^+} \sqrt{f(x)} = \sqrt{\lim\limits_{x \to 1^+} f(x)}$, it follows that $\lim\limits_{x \to 1} \sqrt{f(x)} = \sqrt{\lim\limits_{x \to 1} f(x)}$.

56–63. Evaluating limits *Evaluate the following limits, where c and k are constants.*

56. $\lim\limits_{h \to 0} \dfrac{100}{(10h - 1)^{11} + 2}$

57. $\lim\limits_{x \to 2} (5x - 6)^{3/2}$

58. $\lim\limits_{x \to 5} (3x - 16)^{3/7}$

59. $\lim\limits_{x \to 1} \dfrac{\sqrt{10x - 9} - 1}{x - 1}$

60. $\lim\limits_{x \to 2} \left(\dfrac{1}{x - 2} - \dfrac{2}{x^2 - 2x} \right)$

61. $\lim\limits_{h \to 0} \dfrac{(5 + h)^2 - 25}{h}$

62. $\lim\limits_{x \to c} \dfrac{x^2 - 2cx + c^2}{x - c}$

63. $\lim\limits_{w \to -k} \dfrac{w^2 + 5kw + 4k^2}{w^2 + kw}$

64. Finding a constant Suppose

$$f(x) = \begin{cases} 3x + b & \text{if } x \leq 2 \\ x - 2 & \text{if } x > 2 \end{cases}$$

Determine a value of the constant b for which $\lim\limits_{x \to 2} f(x)$ exists and state the value of the limit, if possible.

65. Finding a constant Suppose

$$g(x) = \begin{cases} x^2 - 5x & \text{if } x \leq -1 \\ ax^3 - 7 & \text{if } x > -1 \end{cases}$$

Determine a value of the constant a for which $\lim\limits_{x \to -1} g(x)$ exists and state the value of the limit, if possible.

66–72. Useful factorization formula *Calculate the following limits using the factorization formula*
$$x^n - a^n = (x - a)(x^{n-1} + x^{n-2}a + x^{n-3}a^2 + \cdots + xa^{n-2} + a^{n-1}),$$
where n is a positive integer and a is a real number.

66. $\lim\limits_{x \to 2} \dfrac{x^5 - 32}{x - 2}$

67. $\lim\limits_{x \to 1} \dfrac{x^6 - 1}{x - 1}$

68. $\lim\limits_{x \to -1} \dfrac{x^7 + 1}{x + 1}$ (*Hint:* Use the formula for $x^7 - a^7$ with $a = -1$.)

69. $\lim\limits_{x \to a} \dfrac{x^5 - a^5}{x - a}$

70. $\lim\limits_{x \to a} \dfrac{x^n - a^n}{x - a}$, for any positive integer n

71. $\lim\limits_{x \to 1} \dfrac{\sqrt[3]{x} - 1}{x - 1}$
$\left(\textit{Hint: } x - 1 = \left(\sqrt[3]{x}\right)^3 - (1)^3 \right)$

72. $\lim\limits_{x \to 16} \dfrac{\sqrt[4]{x} - 2}{x - 16}$

73–76. Limits involving conjugates *Evaluate the following limits.*

73. $\lim\limits_{x\to 1} \dfrac{x-1}{\sqrt{x}-1}$

74. $\lim\limits_{x\to 1} \dfrac{x-1}{\sqrt{4x+5}-3}$

75. $\lim\limits_{x\to 4} \dfrac{3(x-4)\sqrt{x+5}}{3-\sqrt{x+5}}$

76. $\lim\limits_{x\to 0} \dfrac{x}{\sqrt{cx+1}-1}$, where c is a constant

77. Creating functions satisfying given limit conditions Give examples of functions f and g such that $\lim\limits_{x\to 1} f(x) = 0$ and $\lim\limits_{x\to 1} (f(x)\cdot g(x)) = 5$.

78. Creating functions satisfying given limit conditions Give an example of a function f satisfying $\lim\limits_{x\to 1}\left(\dfrac{f(x)}{x-1}\right) = 2$.

79. Finding constants Find constants b and c in the polynomial $p(x) = x^2 + bx + c$ such that $\lim\limits_{x\to 2} \dfrac{p(x)}{x-2} = 6$. Are the constants unique?

Applications

80. A problem from relativity theory Suppose a spaceship of length L_0 is traveling at a high rate of speed v relative to an observer. To the observer, the ship appears to have a smaller length given by the *Lorentz contraction formula*

$$L = L_0\sqrt{1-\dfrac{v^2}{c^2}}$$

where c is the speed of light.

a. What is the observed length L of the ship if it is traveling at 50% of the speed of light?

b. What is the observed length L of the ship if it is traveling at 75% of the speed of light?

c. In parts (a) and (b), what happens to L as the speed of the ship increases?

d. Find $\lim\limits_{v\to c^-} L_0\sqrt{1-\dfrac{v^2}{c^2}}$ and explain the significance of this limit.

81. Limit of the radius of a cylinder A right circular cylinder with a height of 10 cm and a surface area of S cm^2 has a radius given by

$$r(S) = \dfrac{1}{2}\left(\sqrt{100+\dfrac{2S}{\pi}} - 10\right).$$

Find $\lim\limits_{S\to 0^+} r(S)$ and interpret your result.

82. Torricelli's Law A cylindrical tank is filled with water to a depth of 9 m. At $t = 0$, a drain in the bottom of the tank is opened and water flows out of the tank. The depth of water in the tank (measured from the bottom of the tank) t seconds after the drain is opened is approximated by

$$d(t) = (3 - 0.015t)^2 \quad \text{for } 0 \le t \le 200.$$

Evaluate and interpret $\lim\limits_{t\to 200^-} d(t)$.

83. Electric field The magnitude of the electric field at a point x meters from the midpoint of a 0.1-m line of charge is given by $E(x) = \dfrac{4.35}{x\sqrt{x^2+0.01}}$ (in units of newtons per coulomb, N/C). Evaluate $\lim\limits_{x\to 10} E(x)$.

Additional Exercises
84–85. Limits of composite functions

84. If $\lim\limits_{x\to 1} f(x) = 4$, find $\lim\limits_{x\to -1} f(x^2)$.

85. Suppose $g(x) = f(1-x)$ for all x, $\lim\limits_{x\to 1^+} f(x) = 4$, and $\lim\limits_{x\to 1^-} f(x) = 6$. Find $\lim\limits_{x\to 0^+} g(x)$ and $\lim\limits_{x\to 0^-} g(x)$.

86. Two trigonometric inequalities Consider the angle θ in standard position in a unit circle where $0 \le \theta < \pi/2$ or $-\pi/2 \le \theta < 0$ (use both figures).

a. Show that $|AC| = |\sin\theta|$ for $-\pi/2 < \theta < \pi/2$. (*Hint:* Consider the cases $0 < \theta < \pi/2$ and $-\pi/2 < \theta < 0$ separately.)

b. Show that $|\sin\theta| < |\theta|$ for $-\pi/2 < \theta < \pi/2$. (*Hint:* The length of arc AB is θ if $0 < \theta < \pi/2$ and $-\theta$ if $-\pi/2 < \theta < 0$.)

c. Conclude that $-|\theta| \le \sin\theta \le |\theta|$ for $-\pi/2 < \theta < \pi/2$.

d. Show that $0 \le 1 - \cos\theta \le |\theta|$ for $-\pi/2 < \theta < \pi/2$.

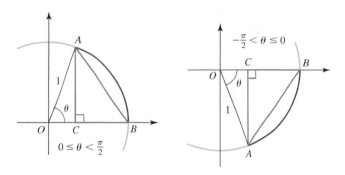

87. Theorem 2.4a Given the polynomial

$$p(x) = b_n x^n + b_{n-1}x^{n-1} + \cdots + b_1 x + b_0$$

prove that $\lim\limits_{x\to a} p(x) = p(a)$ for any value of a.

QUICK CHECK ANSWERS

1. 0, 2   **2.** 2   **3.** 3   **4.** 1

2.4 Infinite Limits

Two more limit scenarios are frequently encountered in calculus and are discussed in this and the following section. An *infinite limit* occurs when function values increase or decrease without bound near a point. The other type of limit, known as a *limit at infinity,* occurs when the independent variable x increases or decreases without bound. The ideas behind infinite limits and limits at infinity are quite different. Therefore, it is important to distinguish these limits and the methods used to calculate them.

An Overview

To illustrate the differences between limits at infinity and infinite limits, consider the values of $f(x) = 1/x^2$ in Table 2.6. As x approaches 0 from either side, $f(x)$ grows larger and larger. Because $f(x)$ does not approach a finite number as x approaches 0, $\lim_{x \to 0} f(x)$ does not exist. Nevertheless, we use limit notation and write $\lim_{x \to 0} f(x) = \infty$. The infinity symbol indicates that $f(x)$ grows arbitrarily large as x approaches 0. This is an example of an **infinite limit**; in general, the *dependent variable* becomes arbitrarily large in magnitude as the *independent variable* approaches a finite number.

With **limits at infinity**, the opposite occurs: The *dependent variable* approaches a finite number as the *independent variable* becomes arbitrarily large. In Table 2.7 we see that $f(x) = 1/x^2$ approaches 0 as x becomes arbitrarily large. In this case we write $\lim_{x \to \infty} f(x) = 0$.

Table 2.6

x	$f(x) = 1/x^2$
±0.1	100
±0.01	10,000
±0.001	1,000,000
↓	↓
0	∞

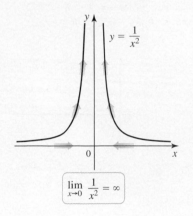

$$\lim_{x \to 0} \frac{1}{x^2} = \infty$$

Table 2.7

x	$f(x) = 1/x^2$
10	0.01
100	0.0001
1000	0.000001
↓	↓
∞	0

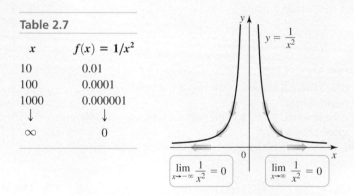

$$\lim_{x \to -\infty} \frac{1}{x^2} = 0 \qquad \lim_{x \to \infty} \frac{1}{x^2} = 0$$

A general picture of these two limit scenarios is shown in Figure 2.23.

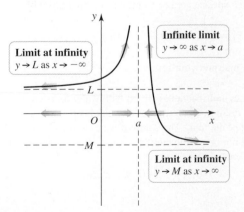

Limit at infinity
$y \to L$ as $x \to -\infty$

Infinite limit
$y \to \infty$ as $x \to a$

Limit at infinity
$y \to M$ as $x \to \infty$

FIGURE 2.23

Infinite Limits

The following definition of infinite limits is informal, but it is adequate for most functions encountered in this book. A precise definition is given in Section 2.7.

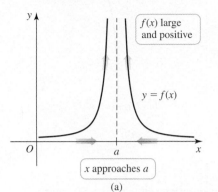

f(*x*) large and positive

y = *f*(*x*)

x approaches *a*

(a)

x approaches *a*

y = *f*(*x*)

f(*x*) large and negative

(b)

FIGURE 2.24

> **DEFINITION   Infinite Limits**
>
> Suppose f is defined for all x near a. If $f(x)$ grows arbitrarily large for all x sufficiently close (but not equal) to a (Figure 2.24a), we write
>
> $$\lim_{x \to a} f(x) = \infty.$$
>
> We say the limit of $f(x)$ as x approaches a is infinity.
>
> If $f(x)$ is negative and grows arbitrarily large in magnitude for all x sufficiently close (but not equal) to a (Figure 2.24b), we write
>
> $$\lim_{x \to a} f(x) = -\infty.$$
>
> In this case, we say the limit of $f(x)$ as x approaches a is negative infinity. In both cases, the limit does not exist.

EXAMPLE 1   Infinite limits Evaluate $\displaystyle\lim_{x \to 1} \frac{x}{(x^2 - 1)^2}$ and $\displaystyle\lim_{x \to -1} \frac{x}{(x^2 - 1)^2}$ using the graph of the function.

SOLUTION The graph of $f(x) = \dfrac{x}{(x^2 - 1)^2}$ (Figure 2.25) shows that as x approaches 1 (from either side), the values of f grow arbitrarily large. Therefore, the limit does not exist and we write

$$\lim_{x \to 1} \frac{x}{(x^2 - 1)^2} = \infty.$$

As x approaches -1, the values of f are negative and grow arbitrarily large in magnitude; therefore,

$$\lim_{x \to -1} \frac{x}{(x^2 - 1)^2} = -\infty.$$

Related Exercises 7–8 ◄

Example 1 illustrates *two-sided* infinite limits. As with finite limits, we also need to work with right-hand and left-hand infinite limits, which are both *one-sided*.

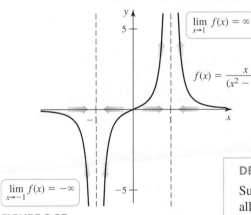

$\lim_{x \to 1} f(x) = \infty$

$f(x) = \dfrac{x}{(x^2 - 1)^2}$

$\lim_{x \to -1} f(x) = -\infty$

FIGURE 2.25

> **DEFINITION   One-sided Infinite Limits**
>
> Suppose f is defined for all x near a with $x > a$. If $f(x)$ becomes arbitrarily large for all x sufficiently close to a with $x > a$, we write $\displaystyle\lim_{x \to a^+} f(x) = \infty$ (Figure 2.26a).
> The one-sided infinite limits $\displaystyle\lim_{x \to a^+} f(x) = -\infty$ (Figure 2.26b), $\displaystyle\lim_{x \to a^-} f(x) = \infty$ (Figure 2.26c), and $\displaystyle\lim_{x \to a^-} f(x) = -\infty$ (Figure 2.26d) are defined analogously.

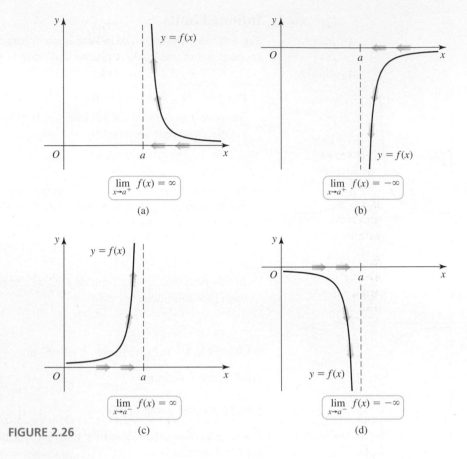

FIGURE 2.26

(a) $\lim\limits_{x\to a^+} f(x) = \infty$

(b) $\lim\limits_{x\to a^+} f(x) = -\infty$

(c) $\lim\limits_{x\to a^-} f(x) = \infty$

(d) $\lim\limits_{x\to a^-} f(x) = -\infty$

QUICK CHECK 1 Sketch the graph of a function and its vertical asymptote that satisfies the conditions $\lim\limits_{x\to 2^+} f(x) = -\infty$ and $\lim\limits_{x\to 2^-} f(x) = \infty$. ◄

In all the infinite limits illustrated in Figure 2.26, the line $x = a$ is called a *vertical asymptote*, a vertical line that is approached by the graph of f as x approaches a.

➤ The shorthand notation $\pm\infty$ means either $+\infty$ or $-\infty$.

> **DEFINITION   Vertical Asymptote**
>
> If $\lim\limits_{x\to a} f(x) = \pm\infty$, $\lim\limits_{x\to a^+} f(x) = \pm\infty$, or $\lim\limits_{x\to a^-} f(x) = \pm\infty$, the line $x = a$ is called a **vertical asymptote** of f.

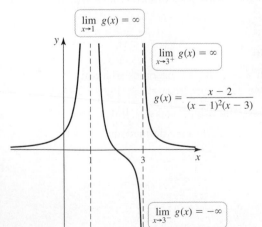

FIGURE 2.27

EXAMPLE 2   Determining limits graphically The vertical lines $x = 1$ and $x = 3$ are vertical asymptotes of the function $g(x) = \dfrac{x - 2}{(x - 1)^2 (x - 3)}$.

Use Figure 2.27 to determine the following limits, if possible.

a. $\lim\limits_{x\to 1} g(x)$ **b.** $\lim\limits_{x\to 3^-} g(x)$ **c.** $\lim\limits_{x\to 3} g(x)$

SOLUTION

a. The values of g grow arbitrarily large as x approaches 1 from either side. Therefore, $\lim\limits_{x\to 1} g(x) = \infty$.

b. The values of g are negative and grow arbitrarily large in magnitude as x approaches 3 from the left, so $\lim\limits_{x\to 3^-} g(x) = -\infty$.

c. Note that $\lim\limits_{x\to 3^+} g(x) = \infty$. Because the one-sided limits at $x = 3$ are not equal, $\lim\limits_{x\to 3} g(x)$ does not exist.

Related Exercises 9–16 ◄

Finding Infinite Limits Analytically

Many infinite limits are analyzed using a simple arithmetic property: The fraction a/b grows arbitrarily large in magnitude if b approaches 0 while a remains nonzero and relatively constant. For example, consider the fraction $(5 + x)/x$ for values of x approaching 0 from the right (Table 2.8).

We see that $\dfrac{5 + x}{x} \to \infty$ as $x \to 0^+$ because the numerator $5 + x$ approaches 5 while the denominator is positive and approaches 0. Therefore, we write $\lim\limits_{x \to 0^+} \dfrac{5 + x}{x} = \infty$.

Similarly, $\lim\limits_{x \to 0^-} \dfrac{5 + x}{x} = -\infty$ because the numerator approaches 5 while the denominator approaches 0 through negative values.

Table 2.8

x	$\dfrac{5 + x}{x}$
0.01	$\dfrac{5.01}{0.01} = 501$
0.001	$\dfrac{5.001}{0.001} = 5001$
0.0001	$\dfrac{5.0001}{0.0001} = 50{,}001$
$\downarrow$	$\downarrow$
0^+	∞

QUICK CHECK 2 Evaluate $\lim\limits_{x \to 0^+} \dfrac{x - 5}{x}$ and $\lim\limits_{x \to 0^-} \dfrac{x - 5}{x}$ by determining the sign of the numerator and denominator. ◄

EXAMPLE 3 **Evaluating limits analytically** Evaluate the following limits.

a. $\lim\limits_{x \to 3^+} \dfrac{2 - 5x}{x - 3}$

b. $\lim\limits_{x \to 3^-} \dfrac{2 - 5x}{x - 3}$

SOLUTION

a. As $x \to 3^+$, the numerator $2 - 5x$ approaches $2 - 5(3) = -13$ while the denominator $x - 3$ is positive and approaches 0. Therefore,

$$\lim\limits_{x \to 3^+} \dfrac{\overbrace{2 - 5x}^{\text{approaches } -13}}{\underbrace{x - 3}_{\substack{\text{positive and}\\\text{approaches } 0}}} = -\infty.$$

b. As $x \to 3^-$, $2 - 5x$ approaches $2 - 5(3) = -13$ while $x - 3$ is negative and approaches 0. Therefore,

$$\lim\limits_{x \to 3^-} \dfrac{\overbrace{2 - 5x}^{\text{approaches } -13}}{\underbrace{x - 3}_{\substack{\text{negative and}\\\text{approaches } 0}}} = \infty.$$

These limits imply that the given function has a vertical asymptote at $x = 3$.

Related Exercises 17–22 ◄

EXAMPLE 4 **Evaluating limits analytically** Evaluate $\lim\limits_{x \to -4^+} \dfrac{-x^3 + 5x^2 - 6x}{-x^3 - 4x^2}$.

SOLUTION First we factor and simplify, assuming $x \neq 0$:

$$\dfrac{-x^3 + 5x^2 - 6x}{-x^3 - 4x^2} = \dfrac{-x(x - 2)(x - 3)}{-x^2(x + 4)} = \dfrac{(x - 2)(x - 3)}{x(x + 4)}.$$

> We can assume that $x \neq 0$ because we are considering function values near $x = -4$.

As $x \to -4^+$, we find that

$$\lim\limits_{x \to -4^+} \dfrac{-x^3 + 5x^2 - 6x}{-x^3 - 4x^2} = \lim\limits_{x \to -4^+} \dfrac{\overbrace{(x - 2)(x - 3)}^{\text{approaches } 42}}{\underbrace{x(x + 4)}_{\substack{\text{negative and}\\\text{approaches } 0}}} = -\infty.$$

This limit implies that the given function has a vertical asymptote at $x = -4$.

Related Exercises 17–22 ◄

QUICK CHECK 3 Verify that $x(x + 4) \to 0$ through negative values as $x \to -4^+$. ◄

> Example 5 illustrates that $f(x)/g(x)$ might not grow arbitrarily large in magnitude if *both* $f(x)$ and $g(x)$ approach 0. Such limits are called *indeterminate forms* and are examined in detail in Section 4.7.

EXAMPLE 5   Location of vertical asymptotes Let $f(x) = \dfrac{x^2 - 4x + 3}{x^2 - 1}$. Evaluate the following limits and find the vertical asymptotes of f. Verify your work with a graphing utility.

a. $\lim\limits_{x \to 1} f(x)$ **b.** $\lim\limits_{x \to -1^-} f(x)$ **c.** $\lim\limits_{x \to -1^+} f(x)$

SOLUTION

a. Notice that as $x \to 1$, both the numerator and denominator of f approach 0, and the function is undefined at $x = 1$. To compute $\lim\limits_{x \to 1} f(x)$, we first factor:

> It is permissible to cancel the $x - 1$ factors in $\lim\limits_{x \to 1} \dfrac{(x - 1)(x - 3)}{(x - 1)(x + 1)}$ because x approaches 1, but is not equal to 1. Therefore, $x - 1 \neq 0$.

$$\lim_{x \to 1} f(x) = \lim_{x \to 1} \frac{x^2 - 4x + 3}{x^2 - 1}$$

$$= \lim_{x \to 1} \frac{(x - 1)(x - 3)}{(x - 1)(x + 1)} \qquad \text{Factor.}$$

$$= \lim_{x \to 1} \frac{(x - 3)}{(x + 1)} \qquad \text{Cancel like factors, } x \neq 1.$$

$$= \frac{1 - 3}{1 + 1} = -1 \qquad \text{Substitute } x = 1.$$

Therefore, $\lim\limits_{x \to 1} f(x) = -1$ (even though $f(1)$ is undefined). The line $x = 1$ is *not* a vertical asymptote of f.

b. In part (a) we showed that

$$f(x) = \frac{x^2 - 4x + 3}{x^2 - 1} = \frac{x - 3}{x + 1} \qquad \text{provided } x \neq 1.$$

We use this fact again. As x approaches -1 from the left, the one-sided limit is

$$\lim_{x \to -1^-} f(x) = \lim_{x \to -1^-} \frac{\overbrace{x - 3}^{\text{approaches } -4}}{\underbrace{x + 1}_{\substack{\text{negative and} \\ \text{approaches } 0}}} = \infty.$$

c. As x approaches -1 from the right, the one-sided limit is

$$\lim_{x \to -1^+} f(x) = \lim_{x \to -1^+} \frac{\overbrace{x - 3}^{\text{approaches } -4}}{\underbrace{x + 1}_{\substack{\text{positive and} \\ \text{approaches } 0}}} = -\infty.$$

The infinite limits $\lim\limits_{x \to -1^+} f(x) = -\infty$ and $\lim\limits_{x \to -1^-} f(x) = \infty$ each imply that the line $x = -1$ is a vertical asymptote of f. The graph of f generated by a graphing utility *may* appear as shown in Figure 2.28a. If so, two corrections must be made. A hole should appear in the graph at $(1, -1)$ because $\lim\limits_{x \to 1} f(x) = -1$, but $f(1)$ is undefined. It is also a good idea to replace the solid vertical line with a dashed line to emphasize that the vertical asymptote is not a part of the graph of f (Figure 2.28b).

➤ Graphing utilities vary in how they display vertical asymptotes. The errors shown in Figure 2.28 do *not* occur on all graphing utilities.

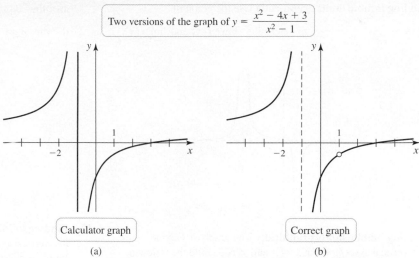

Two versions of the graph of $y = \dfrac{x^2 - 4x + 3}{x^2 - 1}$

| Calculator graph | Correct graph |

FIGURE 2.28

(a)   (b)

Related Exercises 23–26 ◄

QUICK CHECK 4 The line $x = 2$ is not a vertical asymptote of $y = \dfrac{(x - 1)(x - 2)}{x - 2}$. Why not? ◄

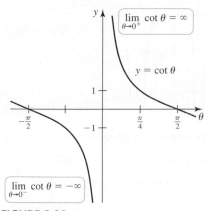

$\lim\limits_{\theta \to 0^+} \cot \theta = \infty$

$y = \cot \theta$

$\lim\limits_{\theta \to 0^-} \cot \theta = -\infty$

FIGURE 2.29

EXAMPLE 6 **Limits of trigonometric functions** Evaluate the following limits.

a. $\lim\limits_{\theta \to 0^+} \cot \theta$ **b.** $\lim\limits_{\theta \to 0^-} \cot \theta$

SOLUTION

a. Recall that $\cot \theta = \cos \theta / \sin \theta$. Furthermore (Example 8, Section 2.3), $\lim\limits_{\theta \to 0^+} \cos \theta = 1$ and $\sin \theta$ is positive and approaches 0 as $\theta \to 0^+$. Therefore, as $\theta \to 0^+$, $\cot \theta$ becomes arbitrarily large and positive, which means $\lim\limits_{\theta \to 0^+} \cot \theta = \infty$. This limit is confirmed by the graph of $\cot \theta$ (Figure 2.29), which has a vertical asymptote at $\theta = 0$.

b. In this case, $\lim\limits_{\theta \to 0^-} \cos \theta = 1$ and as $\theta \to 0^-$, $\sin \theta \to 0$ with $\sin \theta < 0$. Therefore, as $\theta \to 0^-$, $\cot \theta$ is negative and becomes arbitrarily large in magnitude. It follows that $\lim\limits_{\theta \to 0^-} \cot \theta = -\infty$, as confirmed by the graph of $\cot \theta$.   *Related Exercises 27–32* ◄

SECTION 2.4 EXERCISES

Review Questions

1. Use a graph to explain the meaning of $\lim\limits_{x \to a^+} f(x) = -\infty$.

2. Use a graph to explain the meaning of $\lim\limits_{x \to a} f(x) = \infty$.

3. What is a vertical asymptote?

4. Consider the function $F(x) = f(x)/g(x)$ with $g(a) = 0$. Does F necessarily have a vertical asymptote at $x = a$? Explain your reasoning.

5. Suppose $f(x) \to 100$ and $g(x) \to 0$ with $g(x) < 0$ as $x \to 2$. Determine $\lim\limits_{x \to 2} \dfrac{f(x)}{g(x)}$.

6. Evaluate $\lim\limits_{x \to 3^-} \dfrac{1}{x - 3}$ and $\lim\limits_{x \to 3^+} \dfrac{1}{x - 3}$.

Basic Skills

▣ 7. **Finding infinite limits numerically** Compute the values of $f(x) = \dfrac{x + 1}{(x - 1)^2}$ in the following table and use them to determine $\lim\limits_{x \to 1} f(x)$.

x	$\dfrac{x + 1}{(x - 1)^2}$	x	$\dfrac{x + 1}{(x - 1)^2}$
1.1		0.9	
1.01		0.99	
1.001		0.999	
1.0001		0.9999	

8. Finding infinite limits graphically Use the graph of

$$f(x) = \frac{x}{(x^2 - 2x - 3)^2}$$ to determine $\lim\limits_{x \to -1} f(x)$ and $\lim\limits_{x \to 3} f(x)$.

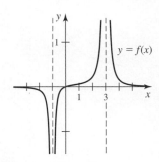

$y = f(x)$

9. Finding infinite limits graphically The graph of f in the figure has vertical asymptotes at $x = 1$ and $x = 2$. Find the following limits, if possible.

a. $\lim\limits_{x \to 1^-} f(x)$ b. $\lim\limits_{x \to 1^+} f(x)$ c. $\lim\limits_{x \to 1} f(x)$

d. $\lim\limits_{x \to 2^-} f(x)$ e. $\lim\limits_{x \to 2^+} f(x)$ f. $\lim\limits_{x \to 2} f(x)$

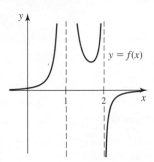

$y = f(x)$

10. Finding infinite limits graphically The graph of g in the figure has vertical asymptotes at $x = 2$ and $x = 4$. Find the following limits, if possible.

a. $\lim\limits_{x \to 2^-} g(x)$ b. $\lim\limits_{x \to 2^+} g(x)$ c. $\lim\limits_{x \to 2} g(x)$

d. $\lim\limits_{x \to 4^-} g(x)$ e. $\lim\limits_{x \to 4^+} g(x)$ f. $\lim\limits_{x \to 4} g(x)$

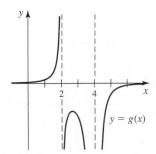

$y = g(x)$

11. Finding infinite limits graphically The graph of h in the figure has vertical asymptotes at $x = -2$ and $x = 3$. Find the following limits, if possible.

a. $\lim\limits_{x \to -2^-} h(x)$ b. $\lim\limits_{x \to -2^+} h(x)$ c. $\lim\limits_{x \to -2} h(x)$

d. $\lim\limits_{x \to 3^-} h(x)$ e. $\lim\limits_{x \to 3^+} h(x)$ f. $\lim\limits_{x \to 3} h(x)$

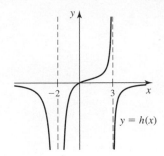

$y = h(x)$

12. Finding infinite limits graphically The graph of p in the figure has vertical asymptotes at $x = -2$ and $x = 3$. Find the following limits, if possible.

a. $\lim\limits_{x \to -2^-} p(x)$ b. $\lim\limits_{x \to -2^+} p(x)$ c. $\lim\limits_{x \to -2} p(x)$

d. $\lim\limits_{x \to 3^-} p(x)$ e. $\lim\limits_{x \to 3^+} p(x)$ f. $\lim\limits_{x \to 3} p(x)$

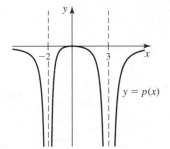

$y = p(x)$

13. Finding infinite limits graphically Graph the function

$$f(x) = \frac{1}{x^2 - x}$$ using a graphing utility with the window

$[-1, 2] \times [-10, 10]$. Use the graph to determine the following limits.

a. $\lim\limits_{x \to 0^-} f(x)$ b. $\lim\limits_{x \to 0^+} f(x)$ c. $\lim\limits_{x \to 1^-} f(x)$ d. $\lim\limits_{x \to 1^+} f(x)$

14. Finding infinite limits graphically Graph the function

$$f(x) = \frac{e^{-x}}{x(x + 2)^2}$$ using a graphing utility. (Experiment with your choice of a graphing window.) Use the graph to determine the following limits.

a. $\lim\limits_{x \to -2^+} f(x)$ b. $\lim\limits_{x \to -2} f(x)$ c. $\lim\limits_{x \to 0^-} f(x)$ d. $\lim\limits_{x \to 0^+} f(x)$

15. Sketching graphs Sketch a possible graph of a function f, together with vertical asymptotes, satisfying all of the following conditions.

$$f(1) = 0 \qquad f(3) \text{ is undefined} \qquad \lim\limits_{x \to 3} f(x) = 1$$

$$\lim\limits_{x \to 0^+} f(x) = -\infty \qquad \lim\limits_{x \to 2} f(x) = \infty \qquad \lim\limits_{x \to 4^-} f(x) = \infty$$

16. Sketching graphs Sketch a possible graph of a function g, together with vertical asymptotes, satisfying all of the following conditions.

$$g(2) = 1 \qquad g(5) = -1 \qquad \lim\limits_{x \to 4} g(x) = -\infty$$

$$\lim\limits_{x \to 7^-} g(x) = \infty \qquad \lim\limits_{x \to 7^+} g(x) = -\infty$$

17–22. Evaluating limits analytically *Evaluate the following limits, or state that they do not exist.*

17. a. $\lim\limits_{x\to 2^+} \dfrac{1}{x-2}$ **b.** $\lim\limits_{x\to 2^-} \dfrac{1}{x-2}$ **c.** $\lim\limits_{x\to 2} \dfrac{1}{x-2}$

18. a. $\lim\limits_{x\to 3^+} \dfrac{2}{(x-3)^3}$ **b.** $\lim\limits_{x\to 3^-} \dfrac{2}{(x-3)^3}$ **c.** $\lim\limits_{x\to 3} \dfrac{2}{(x-3)^3}$

19. $\lim\limits_{x\to 0} \dfrac{x^3-5x^2}{x^2}$ **20.** $\lim\limits_{t\to 5} \dfrac{4t^2-100}{t-5}$

21. $\lim\limits_{x\to 1^+} \dfrac{x^2-5x+6}{x-1}$ **22.** $\lim\limits_{z\to 4} \dfrac{z-5}{(z^2-10z+24)^2}$

23–26. Finding vertical asymptotes *Find all vertical asymptotes, $x=a$, of the following functions. For each value of a, evaluate* $\lim\limits_{x\to a^+} f(x)$, $\lim\limits_{x\to a^-} f(x)$, *and* $\lim\limits_{x\to a} f(x)$.

23. $f(x) = \dfrac{x^2-9x+14}{x^2-5x+6}$ **24.** $f(x) = \dfrac{\cos x}{x^2+2x}$

25. $f(x) = \dfrac{x+1}{x^3-4x^2+4x}$ **26.** $f(x) = \dfrac{x^3-10x^2+16x}{x^2-8x}$

27–30. Trigonometric limits *Evaluate the following limits.*

27. $\lim\limits_{\theta\to 0^+} \csc \theta$ **28.** $\lim\limits_{x\to 0^-} \csc x$

29. $\lim\limits_{x\to 0^+} (-10 \cot x)$ **30.** $\lim\limits_{\theta\to \pi/2^+} \dfrac{1}{3} \tan \theta$

T 31. Finding infinite limits graphically Graph the function $y = \tan x$ with the window $[-\pi, \pi] \times [-10, 10]$. Use the graph to determine the following limits.

a. $\lim\limits_{x\to \pi/2^+} \tan x$ **b.** $\lim\limits_{x\to \pi/2^-} \tan x$

c. $\lim\limits_{x\to -\pi/2^+} \tan x$ **d.** $\lim\limits_{x\to -\pi/2^-} \tan x$

T 32. Finding infinite limits graphically Graph the function $y = \sec x \tan x$ with the window $[-\pi, \pi] \times [-10, 10]$. Use the graph to determine the following limits.

a. $\lim\limits_{x\to \pi/2^+} \sec x \tan x$ **b.** $\lim\limits_{x\to \pi/2^-} \sec x \tan x$

c. $\lim\limits_{x\to -\pi/2^+} \sec x \tan x$ **d.** $\lim\limits_{x\to -\pi/2^-} \sec x \tan x$

Further Explorations

33. Explain why or why not Determine whether the following statements are true and give an explanation or counterexample.

a. The line $x = 1$ is a vertical asymptote of the function $f(x) = \dfrac{x^2-7x+6}{x^2-1}$.

b. The line $x = -1$ is a vertical asymptote of the function $f(x) = \dfrac{x^2-7x+6}{x^2-1}$.

c. If g has a vertical asymptote at $x = 1$ and $\lim\limits_{x\to 1^+} g(x) = \infty$, then $\lim\limits_{x\to 1^-} g(x) = \infty$.

34. Finding a function with vertical asymptotes Find polynomials p and q such that p/q is undefined at $x = 1$ and $x = 2$, but p/q has a vertical asymptote only at $x = 2$. Sketch a graph of your function.

35. Finding a function with infinite limits Give a formula for a function f that satisfies $\lim\limits_{x\to 6^+} f(x) = \infty$ and $\lim\limits_{x\to 6^-} f(x) = -\infty$.

36. Matching Match functions a–f with graphs A–F in the figure without using a graphing utility.

a. $f(x) = \dfrac{x}{x^2+1}$ **b.** $f(x) = \dfrac{x}{x^2-1}$

c. $f(x) = \dfrac{1}{x^2-1}$ **d.** $f(x) = \dfrac{x}{(x-1)^2}$

e. $f(x) = \dfrac{1}{(x-1)^2}$ **f.** $f(x) = \dfrac{x}{x+1}$

A.
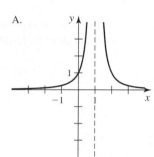

B.
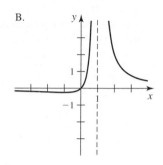

C.
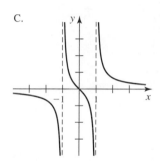

D.
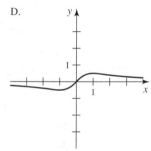

E.
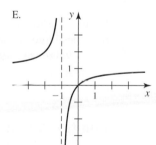

F.
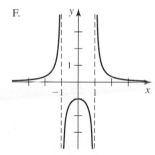

T 37–44. Asymptotes *Use analytical methods and/or a graphing utility to identify the vertical asymptotes (if any) of the following functions.*

37. $f(x) = \dfrac{x^2-3x+2}{x^{10}-x^9}$ **38.** $g(x) = 2 - \ln x^2$

39. $h(x) = \dfrac{e^x}{(x+1)^3}$ **40.** $p(x) = \sec\left(\dfrac{\pi x}{2}\right), |x| < 2$

41. $g(\theta) = \tan\left(\dfrac{\pi\theta}{10}\right)$ **42.** $q(s) = \dfrac{\pi}{s - \sin s}$

43. $f(x) = \dfrac{1}{\sqrt{x}\sec x}$ **44.** $g(x) = e^{1/x}$

Additional Exercises

45. Limits with a parameter Let $f(x) = \dfrac{x^2 - 7x + 12}{x - a}$.

  a. For what values of a, if any, does $\lim\limits_{x \to a^+} f(x)$ equal a finite number?

  b. For what values of a, if any, does $\lim\limits_{x \to a^+} f(x) = \infty$?

  c. For what values of a, if any, does $\lim\limits_{x \to a^+} f(x) = -\infty$?

46–47. Steep secant lines

  a. Given the graph of f in the following figures, find the slope of the secant line that passes through $(0, 0)$ and $(h, f(h))$ in terms of h for $h > 0$ and $h < 0$.

  b. Calculate the limit of the slope of the secant line found in part (a) as $h \to 0^+$ and $h \to 0^-$. What does this tell you about the tangent line to the curve at $(0, 0)$?

46. $f(x) = x^{1/3}$

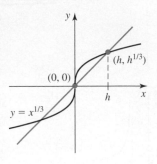

47. $f(x) = x^{2/3}$

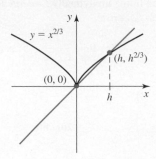

QUICK CHECK ANSWERS

1. Answers will vary, but all graphs should have a vertical asymptote at $x = 2$. **2.** $-\infty$; ∞ **3.** As $x \to -4^+$, $x < 0$ and $(x + 4) > 0$, so $x(x + 4) \to 0$ through negative values.

4. $\lim\limits_{x \to 2} \dfrac{(x - 1)(x - 2)}{x - 2} = \lim\limits_{x \to 2} (x - 1) = 1$, which is not an infinite limit, so $x = 2$ is not a vertical asymptote. ◄

2.5 Limits at Infinity

Limits at infinity—as opposed to infinite limits—occur when the independent variable becomes large in magnitude. For this reason, limits at infinity determine what is called the *end behavior* of a function. An application of these limits is to determine whether a system that evolves in time (such as an ecosystem or a large oscillating structure) reaches a steady state.

Limits at Infinity and Horizontal Asymptotes

Consider the function $f(x) = \tan^{-1} x$, whose domain is $(-\infty, \infty)$ (Figure 2.30). As x becomes arbitrarily large (denoted $x \to \infty$), $f(x)$ approaches $\pi/2$, and as x becomes arbitrarily large in magnitude and negative (denoted $x \to -\infty$), $f(x)$ approaches $-\pi/2$. These limits are expressed as

$$\lim_{x \to \infty} \tan^{-1} x = \frac{\pi}{2} \quad \text{and} \quad \lim_{x \to -\infty} \tan^{-1} x = -\frac{\pi}{2}.$$

The graph of f approaches the horizontal line $y = \pi/2$ as $x \to \infty$ and it approaches the horizontal line $y = -\pi/2$ as $x \to -\infty$. These lines are called *horizontal asymptotes*.

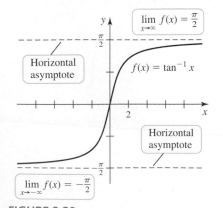

FIGURE 2.30

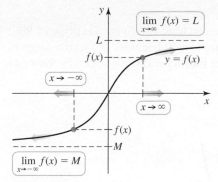

FIGURE 2.31

DEFINITION Limits at Infinity and Horizontal Asymptotes

If $f(x)$ becomes arbitrarily close to a finite number L for all sufficiently large and positive x, then we write

$$\lim_{x \to \infty} f(x) = L.$$

We say the limit of $f(x)$ as x approaches infinity is L. In this case the line $y = L$ is a **horizontal asymptote** of f (Figure 2.31). The limit at negative infinity, $\lim_{x \to -\infty} f(x) = M$, is defined analogously and in this case the horizontal asymptote is $y = M$.

QUICK CHECK 1 Evaluate $x/(x + 1)$ for $x = 10, 100$, and 1000. What is $\lim_{x \to \infty} \dfrac{x}{x + 1}$? ◄

EXAMPLE 1 Limits at infinity Evaluate the following limits.

a. $\lim_{x \to -\infty} \left(2 + \dfrac{10}{x^2} \right)$ **b.** $\lim_{x \to \infty} \left(5 + \dfrac{\sin x}{\sqrt{x}} \right)$

SOLUTION

> The limit laws of Theorem 2.3 and the Squeeze Theorem apply if $x \to a$ is replaced with $x \to \infty$ or $x \to -\infty$.

a. As x becomes large and negative, x^2 becomes large and positive; in turn, $10/x^2$ approaches 0. By the limit laws of Theorem 2.3,

$$\lim_{x \to -\infty} \left(2 + \dfrac{10}{x^2} \right) = \underbrace{\lim_{x \to -\infty} 2}_{\text{equals } 2} + \underbrace{\lim_{x \to -\infty} \left(\dfrac{10}{x^2} \right)}_{\text{equals } 0} = 2 + 0 = 2.$$

Notice that $\lim_{x \to \infty} \left(2 + \dfrac{10}{x^2} \right)$ is also equal to 2. Therefore, the graph of $y = 2 + 10/x^2$ approaches the horizontal asymptote $y = 2$ as $x \to \infty$ and as $x \to -\infty$ (Figure 2.32).

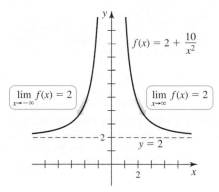

FIGURE 2.32

b. The numerator of $\sin x / \sqrt{x}$ is bounded between -1 and 1; therefore, for $x > 0$

$$-\dfrac{1}{\sqrt{x}} \le \dfrac{\sin x}{\sqrt{x}} \le \dfrac{1}{\sqrt{x}}.$$

As $x \to \infty$, $\sqrt{x}$ becomes arbitrarily large, which means that

$$\lim_{x \to \infty} \dfrac{-1}{\sqrt{x}} = \lim_{x \to \infty} \dfrac{1}{\sqrt{x}} = 0.$$

It follows by the Squeeze Theorem (Theorem 2.5) that $\lim_{x \to \infty} \dfrac{\sin x}{\sqrt{x}} = 0$.

Using the limit laws of Theorem 2.3,

$$\lim_{x \to \infty} \left(5 + \dfrac{\sin x}{\sqrt{x}} \right) = \underbrace{\lim_{x \to \infty} 5}_{\text{equals } 5} + \underbrace{\lim_{x \to \infty} \left(\dfrac{\sin x}{\sqrt{x}} \right)}_{\text{equals } 0} = 5.$$

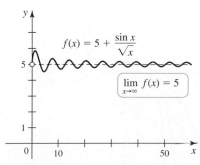

FIGURE 2.33

The graph of $y = 5 + \dfrac{\sin x}{\sqrt{x}}$ approaches the horizontal asymptote $y = 5$ as x becomes large (Figure 2.33). Note that the curve intersects its asymptote infinitely many times. *Related Exercises 9–14* ◄

Infinite Limits at Infinity

It is possible for a limit to be *both* an infinite limit and a limit at infinity. This type of limit occurs if $f(x)$ becomes arbitrarily large in magnitude as x becomes arbitrarily large in magnitude. Such a limit is called an **infinite limit at infinity** and is illustrated by the function $f(x) = x^3$ (Figure 2.34).

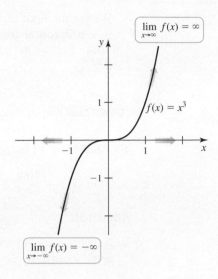

$$\lim_{x \to \infty} f(x) = \infty$$

$$f(x) = x^3$$

$$\lim_{x \to -\infty} f(x) = -\infty$$

FIGURE 2.34

DEFINITION   Infinite Limits at Infinity

If $f(x)$ becomes arbitrarily large as x becomes arbitrarily large, then we write

$$\lim_{x \to \infty} f(x) = \infty.$$

The limits $\lim_{x \to \infty} f(x) = -\infty$, $\lim_{x \to -\infty} f(x) = \infty$, and $\lim_{x \to -\infty} f(x) = -\infty$ are defined similarly.

Infinite limits at infinity tell us about the behavior of polynomials for large-magnitude values of x. First, consider power functions $f(x) = x^n$, where n is a positive integer. Figure 2.35 shows that when n is even, $\lim_{x \to \pm\infty} x^n = \infty$, and when n is odd, $\lim_{x \to \infty} x^n = \infty$ and $\lim_{x \to -\infty} x^n = -\infty$.

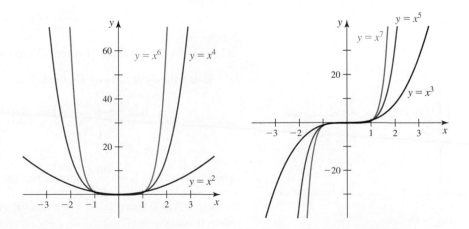

FIGURE 2.35

It follows that reciprocals of power functions $f(x) = 1/x^n = x^{-n}$, where n is a positive integer, behave as follows:

$$\lim_{x \to \infty} \frac{1}{x^n} = \lim_{x \to \infty} x^{-n} = 0 \quad \text{and} \quad \lim_{x \to -\infty} \frac{1}{x^n} = \lim_{x \to -\infty} x^{-n} = 0$$

QUICK CHECK 2 Describe the behavior of $p(x) = -3x^3$ as $x \to \infty$ and as $x \to -\infty$. ◄

From here, it is a short step to finding the behavior of any polynomial as $x \to \pm\infty$. Let $p(x) = a_n x^n + a_{n-1} x^{n-1} + \cdots + a_2 x^2 + a_1 x + a_0$. Notice that as $x \to \pm\infty$, the behavior of p is determined by the term $a_n x^n$ with the highest power of x.

THEOREM 2.6   Limits at Infinity of Powers and Polynomials

Let n be a positive integer and let p be the polynomial, where $a_n \neq 0$.
$p(x) = a_n x^n + a_{n-1} x^{n-1} + \cdots + a_2 x^2 + a_1 x + a_0.$

1. $\displaystyle \lim_{x \to \pm\infty} x^n = \infty$ when n is even

2. $\displaystyle \lim_{x \to \infty} x^n = \infty$ and $\displaystyle \lim_{x \to -\infty} x^n = -\infty$ when n is odd

3. $\displaystyle \lim_{x \to \pm\infty} \frac{1}{x^n} = \lim_{x \to \pm\infty} x^{-n} = 0$

4. $\displaystyle \lim_{x \to \pm\infty} p(x) = \infty$ or $-\infty$, depending on the degree of the polynomial and the sign of the leading coefficient a_n.

EXAMPLE 2   Limits at infinity Evaluate the limits as $x \to \pm\infty$ of the following functions.

a. $p(x) = 3x^4 - 6x^2 + x - 10$ **b.** $q(x) = -2x^3 + 3x^2 - 12$

SOLUTION

a. We use the fact that the limit is determined by the behavior of the leading term:

$$\lim_{x \to \infty} (3x^4 - 6x^2 + x - 10) = \lim_{x \to \infty} 3 \underbrace{x^4}_{\to \infty} = \infty$$

Similarly,

$$\lim_{x \to -\infty} (3x^4 - 6x^2 + x - 10) = \lim_{x \to -\infty} 3 \underbrace{x^4}_{\to \infty} = \infty.$$

b. Noting that the leading coefficient is negative, we have

$$\lim_{x \to \infty} (-2x^3 + 3x^2 - 12) = \lim_{x \to \infty} (-2 \underbrace{x^3}_{\to \infty}) = -\infty$$

$$\lim_{x \to -\infty} (-2x^3 + 3x^2 - 12) = \lim_{x \to -\infty} (-2 \underbrace{x^3}_{\to -\infty}) = \infty$$

Related Exercises 15–20 ◄

End Behavior

The behavior of polynomials as $x \to \pm\infty$ is an example of what is often called *end behavior*. Having treated polynomials, we now turn to the end behavior of rational, algebraic, and transcendental functions.

EXAMPLE 3   End behavior of rational functions   Determine the end behavior for the following rational functions and use a graph to confirm the results.

a. $f(x) = \dfrac{3x + 2}{x^2 - 1}$ **b.** $g(x) = \dfrac{40x^4 + 4x^2 - 1}{10x^4 + 8x^2 + 1}$ **c.** $h(x) = \dfrac{2x^2 + 6x - 2}{x + 1}$

SOLUTION

a. An effective approach for evaluating limits of rational functions at infinity is to divide both the numerator and denominator by x^n, where n is the largest power appearing in the denominator. This strategy forces the terms corresponding to lower powers of x to approach 0 in the limit. In this case, we divide by x^2:

$$\lim_{x \to \infty} \frac{3x + 2}{x^2 - 1} = \lim_{x \to \infty} \frac{\dfrac{3x + 2}{x^2}}{\dfrac{x^2 - 1}{x^2}} = \lim_{x \to \infty} \frac{\overbrace{\dfrac{3}{x} + \dfrac{2}{x^2}}^{\text{approaches 0}}}{\underbrace{1 - \dfrac{1}{x^2}}_{\text{approaches 0}}} = \frac{0}{1} = 0$$

> Recall that the *degree* of a polynomial is the highest power of x that appears.

A similar calculation gives $\displaystyle\lim_{x \to -\infty} \frac{3x + 2}{x^2 - 1} = 0$, and thus the graph of f has the horizontal asymptote $y = 0$. You should confirm that the zeros of the denominator are $x = -1$ and $x = 1$, which correspond to vertical asymptotes (Figure 2.36). In this example, the degree of the polynomial in the numerator is *less than* the degree of the polynomial in the denominator.

b. Again we divide both the numerator and denominator by the largest power appearing in the denominator, which is x^4:

$$\lim_{x \to \infty} \frac{40x^4 + 4x^2 - 1}{10x^4 + 8x^2 + 1} = \lim_{x \to \infty} \frac{\dfrac{40x^4}{x^4} + \dfrac{4x^2}{x^4} - \dfrac{1}{x^4}}{\dfrac{10x^4}{x^4} + \dfrac{8x^2}{x^4} + \dfrac{1}{x^4}}$$ Divide the numerator and denominator by x^4.

$$= \lim_{x \to \infty} \frac{40 + \overbrace{\dfrac{4}{x^2}}^{\text{approaches 0}} - \overbrace{\dfrac{1}{x^4}}^{\text{approaches 0}}}{10 + \underbrace{\dfrac{8}{x^2}}_{\text{approaches 0}} + \underbrace{\dfrac{1}{x^4}}_{\text{approaches 0}}}$$ Simplify.

$$= \frac{40 + 0 + 0}{10 + 0 + 0} = 4$$ Evaluate limits.

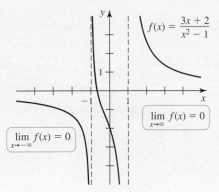

FIGURE 2.36

Using the same steps (dividing each term by x^4), it can be shown that $\displaystyle\lim_{x \to -\infty} \frac{40x^4 + 4x^2 - 1}{10x^4 + 8x^2 + 1} = 4$. This function has the horizontal asymptote $y = 4$ (Figure 2.37). Notice that the degree of the polynomial in the numerator *equals* the degree of the polynomial in the denominator.

c. We first divide the numerator and denominator by the largest power of x appearing in the denominator, which is x:

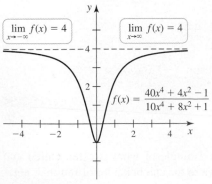

FIGURE 2.37

$$\lim_{x\to\infty}\frac{2x^2+6x-2}{x+1}=\lim_{x\to\infty}\frac{\dfrac{2x^2}{x}+\dfrac{6x}{x}-\dfrac{2}{x}}{\dfrac{x}{x}+\dfrac{1}{x}}$$

Divide the numerator and denominator by x.

$$=\lim_{x\to\infty}\frac{\overbrace{2x}^{\text{arbitrarily large}}+\overbrace{6}^{\text{constant}}-\overbrace{\dfrac{2}{x}}^{\text{approaches 0}}}{\underbrace{1}_{\text{constant}}+\underbrace{\dfrac{1}{x}}_{\text{approaches 0}}}$$

Simplify.

$$=\infty$$

Take limits.

A similar analysis shows that $\displaystyle\lim_{x\to-\infty}\frac{2x^2+6x-2}{x+1}=-\infty$. Because these limits are not finite, there are no horizontal asymptotes. In this case, the degree of the polynomial in the numerator is *greater than* the degree of the polynomial in the denominator.

There is more to be learned about the end behavior of this function. Using long division, the function h is written

$$h(x)=\frac{2x^2+6x-2}{x+1}=2x+4-\underbrace{\frac{6}{x+1}}_{\substack{\text{approaches 0}\\ \text{as }x\to\infty}}$$

As $x\to\infty$, the term $6/(x+1)$ approaches 0, and we see that the function h behaves like the linear function $\ell(x)=2x+4$. For this reason, the graphs of h and ℓ approach each other as $x\to\infty$ (Figure 2.38). A similar argument shows that the graphs of h and ℓ approach each other as $x\to-\infty$. The line described by ℓ is called a **slant**, or **oblique**, **asymptote** (Exercises 66–71).   *Related Exercises 21–26* ◄

The conclusions reached in Example 3 can be generalized for all rational functions. These results are summarized in Theorem 2.7 (see Exercise 72).

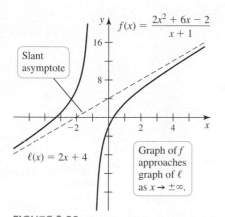

Slant asymptote

$f(x)=\dfrac{2x^2+6x-2}{x+1}$

$\ell(x)=2x+4$

Graph of f approaches graph of ℓ as $x\to\pm\infty$.

FIGURE 2.38

THEOREM 2.7   End Behavior and Asymptotes of Rational Functions

Suppose $f(x)=\dfrac{p(x)}{q(x)}$ is a rational function, where

$$p(x)=a_mx^m+a_{m-1}x^{m-1}+\cdots+a_2x^2+a_1x+a_0\quad\text{and}$$

$$q(x)=b_nx^n+b_{n-1}x^{n-1}+\cdots+b_2x^2+b_1x+b_0$$

with $a_m\neq0$ and $b_n\neq0$.

a. If $m<n$, then $\displaystyle\lim_{x\to\pm\infty}f(x)=0$, and $y=0$ is a horizontal asymptote of f.

b. If $m=n$, then $\displaystyle\lim_{x\to\pm\infty}f(x)=a_m/b_n$, and $y=a_m/b_n$ is a horizontal asymptote of f.

c. If $m>n$, then $\displaystyle\lim_{x\to\pm\infty}f(x)=\infty$ or $-\infty$, and f has no horizontal asymptote.

d. Assuming that $f(x)$ is in reduced form (p and q share no common factors), vertical asymptotes occur at the zeros of q.

QUICK CHECK 3 Use Theorem 2.7 to find the vertical and horizontal asymptotes of $y=\dfrac{10x}{3x-1}$. ◄

Although it isn't stated explicitly, Theorem 2.7 implies that a rational function can have at most one horizontal asymptote, and whenever there is a horizontal asymptote, $\lim\limits_{x\to\infty} \dfrac{p(x)}{q(x)} = \lim\limits_{x\to-\infty} \dfrac{p(x)}{q(x)}$. The same cannot be said of other functions, as the next examples show.

EXAMPLE 4   End behavior of an algebraic function   Examine the end behavior of
$$f(x) = \frac{10x^3 - 3x^2 + 8}{\sqrt{25x^6 + x^4 + 2}}.$$

SOLUTION   The square root in the denominator forces us to revise the strategy used with rational functions. First, consider the limit as $x \to \infty$. The highest power of the polynomial in the denominator is 6. The polynomial is under a square root, so we divide the numerator and denominator by $\sqrt{x^6} = x^3$ for $x \geq 0$. The limit is evaluated as follows:

$$\lim_{x\to\infty} \frac{10x^3 - 3x^2 + 8}{\sqrt{25x^6 + x^4 + 2}} = \lim_{x\to\infty} \frac{\dfrac{10x^3}{x^3} - \dfrac{3x^2}{x^3} + \dfrac{8}{x^3}}{\sqrt{\dfrac{25x^6}{x^6} + \dfrac{x^4}{x^6} + \dfrac{2}{x^6}}} \qquad \text{Divide by } \sqrt{x^6} = x^3.$$

$$= \lim_{x\to\infty} \frac{10 - \overbrace{\dfrac{3}{x}}^{\text{approaches }0} + \overbrace{\dfrac{8}{x^3}}^{\text{approaches }0}}{\sqrt{25 + \underbrace{\dfrac{1}{x^2}}_{\text{approaches }0} + \underbrace{\dfrac{2}{x^6}}_{\text{approaches }0}}} \qquad \text{Simplify.}$$

$$= \frac{10}{\sqrt{25}} = 2 \qquad \text{Evaluate limits.}$$

> **Recall that**
> $$\sqrt{x^2} = |x| = \begin{cases} x & \text{if } x \geq 0 \\ -x & \text{if } x < 0 \end{cases}$$
> Therefore,
> $$\sqrt{x^6} = |x^3| = \begin{cases} x^3 & \text{if } x \geq 0 \\ -x^3 & \text{if } x < 0 \end{cases}$$
> Because x is negative as $x \to -\infty$, we have $\sqrt{x^6} = -x^3$.

As $x \to -\infty$, x^3 is negative, so we divide numerator and denominator by $\sqrt{x^6} = -x^3$ (which is positive):

$$\lim_{x\to-\infty} \frac{10x^3 - 3x^2 + 8}{\sqrt{25x^6 + x^4 + 2}} = \lim_{x\to-\infty} \frac{\dfrac{10x^3}{-x^3} - \dfrac{3x^2}{-x^3} + \dfrac{8}{-x^3}}{\sqrt{\dfrac{25x^6}{x^6} + \dfrac{x^4}{x^6} + \dfrac{2}{x^6}}} \qquad \begin{array}{l}\text{Divide by}\\ \sqrt{x^6} = -x^3 > 0.\end{array}$$

$$= \lim_{x\to-\infty} \frac{-10 + \overbrace{\dfrac{3}{x}}^{\text{approaches }0} - \overbrace{\dfrac{8}{x^3}}^{\text{approaches }0}}{\sqrt{25 + \underbrace{\dfrac{1}{x^2}}_{\text{approaches }0} + \underbrace{\dfrac{2}{x^6}}_{\text{approaches }0}}} \qquad \text{Simplify.}$$

$$= \frac{-10}{\sqrt{25}} = -2 \qquad \text{Evaluate limits.}$$

The limits reveal two asymptotes, $y = 2$ and $y = -2$. Observe that the graph crosses both horizontal asymptotes (Figure 2.39). *Related Exercises 27–30* ◄

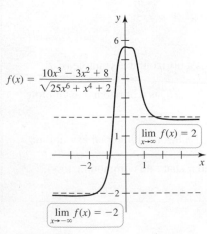

$$f(x) = \frac{10x^3 - 3x^2 + 8}{\sqrt{25x^6 + x^4 + 2}}$$

$$\lim_{x\to\infty} f(x) = 2$$

$$\lim_{x\to-\infty} f(x) = -2$$

FIGURE 2.39

EXAMPLE 5   End behavior of transcendental functions   Determine the end behavior of the following transcendental functions.

a. $f(x) = e^x$ and $f(x) = e^{-x}$ **b.** $g(x) = \ln x$ **c.** $f(x) = \cos x$

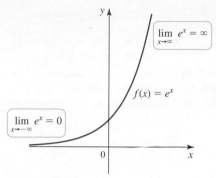

FIGURE 2.40

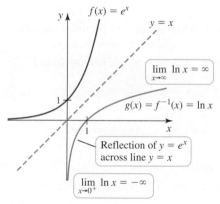

FIGURE 2.41

Table 2.9

x	$\ln x$
10	2.302
10^5	11.513
10^{10}	23.026
10^{50}	115.129
10^{99}	227.956
$\downarrow$	$\downarrow$
∞	???

QUICK CHECK 4 How do the functions e^{10x} and e^{-10x} behave as $x \to \infty$ and as $x \to -\infty$? ◄

SOLUTION

a. The graph of $f(x) = e^x$ (Figure 2.40) makes it clear that as $x \to \infty$, e^x increases without bound. All exponential functions b^x with $b > 1$ behave this way, because raising a number greater than 1 to ever-larger powers produces numbers that increase without bound. The figure also suggests that as $x \to -\infty$, the graph of e^x approaches the horizontal asymptote $y = 0$. This claim is confirmed analytically by recognizing that

$$\lim_{x \to -\infty} e^x = \lim_{x \to \infty} e^{-x} = \lim_{x \to \infty} \frac{1}{e^x} = 0.$$

Therefore, $\lim_{x \to \infty} e^x = \infty$ and $\lim_{x \to -\infty} e^x = 0$. Because $e^{-x} = 1/e^x$, it follows that $\lim_{x \to \infty} e^{-x} = 0$ and $\lim_{x \to -\infty} e^{-x} = \infty$.

b. The domain of $\ln x$ is $\{x : x > 0\}$, so we evaluate $\lim_{x \to 0^+} \ln x$ and $\lim_{x \to \infty} \ln x$ to determine end behavior. For the first limit, recall that $\ln x$ is the inverse of e^x (Figure 2.41), and the graph of $\ln x$ is a reflection across the line $y = x$ of the graph of e^x. The horizontal asymptote $(y = 0)$ of e^x is also reflected across $y = x$, becoming a vertical asymptote $(x = 0)$ for $\ln x$. These observations imply that $\lim_{x \to 0^+} \ln x = -\infty$.

It is not obvious whether the graph of $\ln x$ approaches a horizontal asymptote or whether the function grows without bound as $x \to \infty$. Furthermore, the numerical evidence (Table 2.9) is inconclusive because $\ln x$ increases very slowly. The inverse relation between e^x and $\ln x$ is again useful. The fact that the *domain* of e^x is $(-\infty, \infty)$ implies that the *range* of $\ln x$ is also $(-\infty, \infty)$. Therefore, the values of $\ln x$ lie in the interval $(-\infty, \infty)$, and it follows that $\lim_{x \to \infty} \ln x = \infty$.

c. The cosine function oscillates between -1 and 1 as x approaches infinity (Figure 2.42). Therefore, $\lim_{x \to \infty} \cos x$ does not exist. For the same reason, $\lim_{x \to -\infty} \cos x$ does not exist.

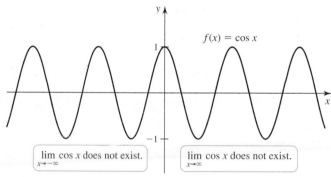

FIGURE 2.42

Related Exercises 31–36 ◄

THEOREM 2.8   End Behavior of e^x, e^{-x}, and $\ln x$
The end behavior for e^x and e^{-x} on $(-\infty, \infty)$ and $\ln x$ on $(0, \infty)$ is given by the following limits:

$$\lim_{x \to \infty} e^x = \infty \qquad \text{and} \qquad \lim_{x \to -\infty} e^x = 0$$

$$\lim_{x \to \infty} e^{-x} = 0 \qquad \text{and} \qquad \lim_{x \to -\infty} e^{-x} = \infty$$

$$\lim_{x \to 0^+} \ln x = -\infty \qquad \text{and} \qquad \lim_{x \to \infty} \ln x = \infty$$

SECTION 2.5 EXERCISES

Review Questions

1. Explain the meaning of $\lim\limits_{x \to -\infty} f(x) = 10$.

2. What is a horizontal asymptote?

3. Determine $\lim\limits_{x \to \infty} \dfrac{f(x)}{g(x)}$ if $f(x) \to 100{,}000$ and $g(x) \to \infty$ as $x \to \infty$.

4. Describe the end behavior of $g(x) = e^{-2x}$.

5. Describe the end behavior of $f(x) = -2x^3$.

6. Three cases arise when examining the end behavior of a rational function $f(x) = p(x)/q(x)$. State each case, and describe the associated end behavior.

7. Evaluate $\lim\limits_{x \to \infty} e^x$, $\lim\limits_{x \to -\infty} e^x$, and $\lim\limits_{x \to \infty} e^{-x}$.

8. Describe with a sketch the end behavior of $f(x) = \ln x$.

Basic Skills

9–14. Limits at infinity *Evaluate the following limits.*

9. $\lim\limits_{x \to \infty} \left(3 + \dfrac{10}{x^2} \right)$

10. $\lim\limits_{x \to \infty} \left(5 + \dfrac{1}{x} + \dfrac{10}{x^2} \right)$

11. $\lim\limits_{\theta \to \infty} \dfrac{\cos \theta}{\theta^2}$

12. $\lim\limits_{x \to \infty} \dfrac{3 + 2x + 4x^2}{x^2}$

13. $\lim\limits_{x \to -\infty} \dfrac{\cos x^5}{x}$

14. $\lim\limits_{x \to -\infty} \left(5 + \dfrac{100}{x} + \dfrac{\sin^4 x^3}{x^2} \right)$

15–20. Infinite limits at infinity *Determine the following limits.*

15. $\lim\limits_{x \to \infty} (3x^{12} - 9x^7)$

16. $\lim\limits_{x \to -\infty} (3x^7 + x^2)$

17. $\lim\limits_{x \to -\infty} (-3x^{16} + 2)$

18. $\lim\limits_{x \to -\infty} 2x^{-8}$

19. $\lim\limits_{x \to \infty} (-12x^{-5})$

20. $\lim\limits_{x \to -\infty} (2x^{-8} + 4x^3)$

21–26. Rational functions *Evaluate* $\lim\limits_{x \to \infty} f(x)$ *and* $\lim\limits_{x \to -\infty} f(x)$ *for the following rational functions. Then give the horizontal asymptote of f (if any).*

21. $f(x) = \dfrac{6x^2 - 9x + 8}{3x^2 + 2}$

22. $f(x) = \dfrac{4x^2 - 7}{8x^2 + 5x + 2}$

23. $f(x) = \dfrac{2x + 1}{3x^4 - 2}$

24. $f(x) = \dfrac{12x^8 - 3}{3x^8 - 2x^7}$

25. $f(x) = \dfrac{40x^5 + x^2}{16x^4 - 2x}$

26. $f(x) = \dfrac{-x^3 + 1}{2x + 8}$

27–30. Algebraic functions *Evaluate* $\lim\limits_{x \to \infty} f(x)$ *and* $\lim\limits_{x \to -\infty} f(x)$ *for the following functions. Then give the horizontal asymptote(s) of f (if any).*

27. $f(x) = \dfrac{4x^3 + 1}{2x^3 + \sqrt{16x^6 + 1}}$

28. $f(x) = \dfrac{4x^3}{2x^3 + \sqrt{9x^6 + 15x^4}}$

29. $f(x) = \dfrac{\sqrt[3]{x^6 + 8}}{4x^2 + \sqrt{3x^4 + 1}}$

30. $f(x) = 4x\left(3x - \sqrt{9x^2 + 1}\right)$

31–36. Transcendental functions *Determine the end behavior of the following transcendental functions by evaluating appropriate limits. Then provide a simple sketch of the associated graph, showing asymptotes if they exist.*

31. $f(x) = -3e^{-x}$

32. $f(x) = 2^x$

33. $f(x) = 1 - \ln x$

34. $f(x) = |\ln x|$

35. $f(x) = \sin x$

36. $f(x) = \dfrac{50}{e^{2x}}$

Further Explorations

37. **Explain why or why not** Determine whether the following statements are true and give an explanation or counterexample.

   a. The graph of a function can never cross one of its horizontal asymptotes.

   b. A rational function f can have both $\lim\limits_{x \to \infty} f(x) = L$ and $\lim\limits_{x \to -\infty} f(x) = \infty$.

   c. The graph of any function can have at most two horizontal asymptotes.

38–47. Horizontal and vertical asymptotes

   a. Evaluate $\lim\limits_{x \to \infty} f(x)$ and $\lim\limits_{x \to -\infty} f(x)$, and then identify the horizontal asymptotes.

   b. Find the vertical asymptotes. For each vertical asymptote $x = a$, evaluate $\lim\limits_{x \to a^-} f(x)$ and $\lim\limits_{x \to a^+} f(x)$.

38. $f(x) = \dfrac{x^2 - 4x + 3}{x - 1}$

39. $f(x) = \dfrac{2x^3 + 10x^2 + 12x}{x^3 + 2x^2}$

40. $f(x) = \dfrac{\sqrt{16x^4 + 64x^2} + x^2}{2x^2 - 4}$

41. $f(x) = \dfrac{3x^4 + 3x^3 - 36x^2}{x^4 - 25x^2 + 144}$

42. $f(x) = 16x^2\left(4x^2 - \sqrt{16x^4 + 1}\right)$

43. $f(x) = \dfrac{x^2 - 9}{x(x - 3)}$

44. $f(x) = \dfrac{x - 1}{x^{2/3} - 1}$

45. $f(x) = \dfrac{\sqrt{x^2 + 2x + 6} - 3}{x - 1}$

46. $f(x) = \dfrac{|1 - x^2|}{x(x + 1)}$

47. $f(x) = \sqrt{|x|} - \sqrt{|x - 1|}$

48–51. End behavior for transcendental functions

48. The central branch of $f(x) = \tan x$ is shown in the figure.

    a. Evaluate $\lim\limits_{x \to \pi/2^-} \tan x$ and $\lim\limits_{x \to -\pi/2^+} \tan x$. Are these limits infinite limits or limits at infinity?

    b. Sketch a graph of $g(x) = \tan^{-1} x$ by reflecting the graph of f over the line $y = x$, and use it to evaluate $\lim\limits_{x \to \infty} \tan^{-1} x$ and $\lim\limits_{x \to -\infty} \tan^{-1} x$.

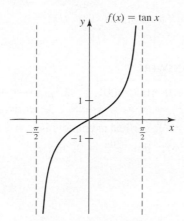

49. Graph $y = \sec^{-1} x$ and evaluate the following limits using the graph. Use the domain given in the text.

    a. $\lim\limits_{x \to \infty} \sec^{-1} x$ **b.** $\lim\limits_{x \to -\infty} \sec^{-1} x$

50. The **hyperbolic cosine function**, denoted $\cosh x$, is used to model the shape of a hanging cable (a telephone wire, for example). It is defined as $\cosh x = \dfrac{e^x + e^{-x}}{2}$.

    a. Determine its end behavior by evaluating $\lim\limits_{x \to \infty} \cosh x$ and $\lim\limits_{x \to -\infty} \cosh x$.

    b. Evaluate $\cosh 0$. Use symmetry and part (a) to sketch a plausible graph for $y = \cosh x$.

51. The **hyperbolic sine function** is defined as $\sinh x = \dfrac{e^x - e^{-x}}{2}$.

    a. Determine its end behavior by evaluating $\lim\limits_{x \to \infty} \sinh x$ and $\lim\limits_{x \to -\infty} \sinh x$.

    b. Evaluate $\sinh 0$. Use symmetry and part (a) to sketch a plausible graph for $y = \sinh x$.

52–53. Sketching graphs *Sketch a possible graph of a function f that satisfies all of the given conditions. Be sure to sketch all vertical and horizontal asymptotes.*

52. $f(-1) = -2, f(1) = 2, f(0) = 0, \lim\limits_{x \to \infty} f(x) = 1,$
$\lim\limits_{x \to -\infty} f(x) = -1$

53. $\lim\limits_{x \to 0^+} f(x) = \infty, \lim\limits_{x \to 0^-} f(x) = -\infty, \lim\limits_{x \to \infty} f(x) = 1,$
$\lim\limits_{x \to -\infty} f(x) = -2$

54. Asymptotes Find the vertical and horizontal asymptotes of $f(x) = e^{1/x}$.

55. Asymptotes Find the vertical and horizontal asymptotes of $f(x) = \dfrac{\cos x + 2\sqrt{x}}{\sqrt{x}}$.

Applications

56–61. Steady states *If a function f represents a system that varies in time, the existence of $\lim\limits_{t \to \infty} f(t)$ means that the system reaches a steady state (or equilibrium). For the following systems, determine if a steady state exists and give the steady-state value.*

56. The population of a bacteria culture is given by $p(t) = \dfrac{2500}{t + 1}$.

57. The population of a culture of tumor cells is given by
$$p(t) = \frac{3500t}{t + 1}.$$

58. The amount of drug (in mg) in the blood after an IV tube is inserted is $m(t) = 200(1 - 2^{-t})$.

59. The value of an investment is given by $v(t) = \$1000e^{0.065t}$.

60. The population of a colony of squirrels is given by
$$p(t) = \frac{1500}{3 + 2e^{-0.1t}}.$$

61. The amplitude of an oscillator is given by $a(t) = 2\left(\dfrac{t + \sin t}{t}\right)$.

62–65. Looking ahead to sequences *A sequence is an infinite ordered list of numbers that is often defined by a function. For example, the sequence $\{2, 4, 6, 8, \dots\}$ is specified by the function $f(n) = 2n$, where $n = 1, 2, 3, \dots$. The limit of such a sequence is $\lim\limits_{n \to \infty} f(n)$, provided the limit exists. All the limit laws for limits at infinity may be applied to limits of sequences. Find the limit of the following sequences, or state that the limit does not exist.*

62. $\left\{4, 2, \dfrac{4}{3}, 1, \dfrac{4}{5}, \dfrac{2}{3}, \dots\right\}$, which is defined by $f(n) = \dfrac{4}{n}$ for $n = 1, 2, 3 \dots$

63. $\left\{0, \dfrac{1}{2}, \dfrac{2}{3}, \dfrac{3}{4}, \dots\right\}$, which is defined by $f(n) = \dfrac{n - 1}{n}$ for $n = 1, 2, 3 \dots$

64. $\left\{\dfrac{1}{2}, \dfrac{4}{3}, \dfrac{9}{4}, \dfrac{16}{5}, \dots\right\}$, which is defined by $f(n) = \dfrac{n^2}{n + 1}$ for $n = 1, 2, 3 \dots$

65. $\left\{2, \dfrac{3}{4}, \dfrac{4}{9}, \dfrac{5}{16}, \dots\right\}$, which is defined by $f(n) = \dfrac{n + 1}{n^2}$ for $n = 1, 2, 3 \dots$

Additional Exercises

66–71. Oblique (slant) asymptotes *Suppose p/q is a rational function where the degree of p is 1 greater than the degree of q. Using polynomial long division, p/q can be written as*

$$\frac{p(x)}{q(x)} = mx + b + \frac{r(x)}{s(x)}$$

where r/s is a rational function with the property $r(x)/s(x) \to 0$ as $x \to \pm\infty$. This fact implies that $\dfrac{p(x)}{q(x)} \approx mx + b$ when x is large. The line $y = mx + b$ is an oblique (or slant) asymptote of p/q. Complete the following steps for the given functions.

    a. *Use polynomial long division to find the oblique asymptote of f.*
    b. *Find the vertical asymptotes of f.*
    c. *Graph f and all of its asymptotes with a graphing utility. Then sketch a graph of the function by hand, correcting any errors appearing in the computer-generated graph.*

66. $f(x) = \dfrac{x^2 - 1}{x + 2}$ **67.** $f(x) = \dfrac{x^2 - 3}{x + 6}$

68. $f(x) = \dfrac{3x^2 - 2x + 7}{2x - 5}$ **69.** $f(x) = \dfrac{x^2 - 2x + 5}{3x - 2}$

70. $f(x) = \dfrac{3x^2 - 2x + 5}{3x + 4}$ **71.** $f(x) = \dfrac{4x^3 + 4x^2 + 7x + 4}{1 + x^2}$

72. **End behavior of a rational function** Suppose $f(x) = \dfrac{p(x)}{q(x)}$ is a rational function, where

$$p(x) = a_m x^m + a_{m-1} x^{m-1} + \cdots + a_2 x^2 + a_1 x + a_0,$$
$$q(x) = b_n x^n + b_{n-1} x^{n-1} + \cdots + b_2 x^2 + b_1 x + b_0, a_m \neq 0,$$

and $b_n \neq 0$.

    a. Prove that if $m = n$, then $\displaystyle\lim_{x \to \pm\infty} f(x) = \dfrac{a_m}{b_n}$.

    b. Prove that if $m < n$, then $\displaystyle\lim_{x \to \pm\infty} f(x) = 0$.

QUICK CHECK **ANSWERS**

1. $10/11, 100/101, 1000/1001, 1$ **2.** $p(x) \to -\infty$ as $x \to \infty$ and $p(x) \to \infty$ as $x \to -\infty$ **3.** Horizontal asymptote is $y = \frac{10}{3}$; vertical asymptote is $x = \frac{1}{3}$

4. $\displaystyle\lim_{x \to \infty} e^{10x} = \infty$, $\displaystyle\lim_{x \to -\infty} e^{10x} = 0$, $\displaystyle\lim_{x \to \infty} e^{-10x} = 0$, $\displaystyle\lim_{x \to -\infty} e^{-10x} = \infty$ ◄

2.6 Continuity

The graphs of many functions encountered in this text contain no holes, jumps, or breaks. For example, if $L = f(t)$ represents the length of a fish t years after it is hatched, then the length of the fish changes gradually as t increases. Consequently, the graph of $L = f(t)$ contains no breaks (Figure 2.43a). Some functions, however, do contain abrupt changes in their values. Consider a parking meter that accepts only quarters and each quarter buys 15 min of parking. Letting $c(t)$ be the cost (in dollars) of parking for t min, the graph of c has breaks at integer multiples of 15 min (Figure 2.43b).

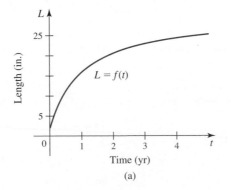

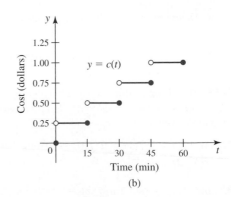

FIGURE 2.43   (a)   (b)

QUICK CHECK 1 For what values of t in $(0, 60)$ does the graph of $y = c(t)$ in Figure 2.43b have discontinuities? ◄

Informally, we say that a function f is *continuous* at $x = a$ if the graph of f contains no holes or breaks at $x = a$ (if the graph near $x = a$ can be drawn without lifting the pencil). If a function is not continuous at $x = a$, then a is a point of discontinuity.

Continuity at a Point

This informal description of continuity is sufficient for determining the continuity of simple functions, but it is not precise enough to deal with more complicated functions such as

$$h(x) = \begin{cases} x \sin \dfrac{1}{x} & \text{if } x \neq 0 \\ 0 & \text{if } x = 0 \end{cases}$$

It is difficult to determine whether the graph of h has a break at $x = 0$ because it oscillates rapidly as x approaches 0 (Figure 2.44). We need a better definition.

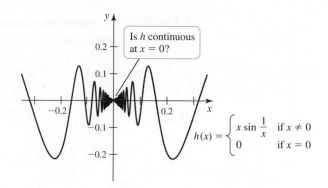

FIGURE 2.44

DEFINITION   Continuity at a Point

A function f is **continuous** at a if $\lim\limits_{x \to a} f(x) = f(a)$. If f is not continuous at a, then a is a point of discontinuity.

There is more to this definition than first appears. If $\lim\limits_{x \to a} f(x) = f(a)$, then $f(a)$ and $\lim\limits_{x \to a} f(x)$ must both exist, and they must be equal. The following checklist is helpful in determining whether a function is continuous at a.

Continuity Checklist

In order for f to be continuous at a, the following three conditions must hold:

1. $f(a)$ is defined (a is in the domain of f).

2. $\lim\limits_{x \to a} f(x)$ exists.

3. $\lim\limits_{x \to a} f(x) = f(a)$ (the value of f equals the limit of f at a).

If *any* item in the continuity checklist fails to hold, the function fails to be continuous at a. From this definition, we see that continuity has an important practical consequence:

> *If f is continuous at a, then $\lim\limits_{x \to a} f(x) = f(a)$, and direct substitution may be used to evaluate $\lim\limits_{x \to a} f(x)$.*

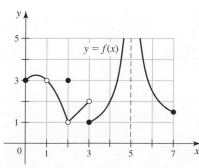

FIGURE 2.45

EXAMPLE 1   Points of discontinuity Use the graph of f in Figure 2.45 to identify values of x on the interval $(0, 7)$ at which f is not continuous.

> In Example 1, the discontinuities at $x = 1$ and $x = 2$ are called **removable discontinuities** because they can be removed by redefining the function at these points (in this case $f(1) = 3$ and $f(2) = 1$). The discontinuity at $x = 3$ is called a **jump discontinuity**. The discontinuity at $x = 5$ is called an **infinite discontinuity**. These terms are discussed in Exercises 83–89.

SOLUTION The function f has discontinuities at $x = 1, 2, 3,$ and 5 because the graph contains holes or breaks at each of these locations. These claims are verified using the continuity checklist.

- $f(1)$ is not defined.
- $f(2) = 3$ and $\lim_{x \to 2} f(x) = 1$. Therefore $f(2)$ and $\lim_{x \to 2} f(x)$ exist but are not equal.
- $\lim_{x \to 3} f(x)$ does not exist because the left-hand limit $\lim_{x \to 3^-} f(x) = 2$ differs from the right-hand limit $\lim_{x \to 3^+} f(x) = 1$.
- Neither $\lim_{x \to 5} f(x)$ nor $f(5)$ exists.   *Related Exercises 9–12* ◄

EXAMPLE 2 **Identifying discontinuities** Determine whether the following functions are continuous at a. Justify each answer using the continuity checklist.

a. $f(x) = \dfrac{3x^2 + 2x + 1}{x - 1}; \quad a = 1$

b. $g(x) = \dfrac{3x^2 + 2x + 1}{x - 1}; \quad a = 2$

c. $h(x) = \begin{cases} x \sin\left(\dfrac{1}{x}\right) & \text{if } x \neq 0 \\ 0 & \text{if } x = 0 \end{cases}; \, a = 0$

SOLUTION

a. The function f is not continuous at $x = 1$ because $f(1)$ is undefined.

b. Because g is a rational function and the denominator is nonzero at $x = 2$, it follows by Theorem 2.3 that $\lim_{x \to 2} g(x) = g(2) = 17$. Therefore, g is continuous at 2.

c. By definition, $h(0) = 0$. In Exercise 51 of Section 2.3, we used the Squeeze Theorem to show that $\lim_{x \to 0} x \sin\left(\dfrac{1}{x}\right) = 0$. Therefore, $\lim_{x \to 0} h(x) = h(0)$, which implies that h is continuous at 0.   *Related Exercises 13–18* ◄

The following theorems make it easier to test various combinations of functions for continuity at a point.

THEOREM 2.9   Continuity Rules
If f and g are continuous at a, then the following functions are also continuous at a. Assume c is a constant and $n > 0$ is an integer.

a. $f + g$	**b.** $f - g$
c. cf	**d.** fg
e. f/g, provided $g(a) \neq 0$	**f.** $[f(x)]^n$

To prove the first result, note that if f and g are continuous at a, then $\lim_{x \to a} f(x) = f(a)$ and $\lim_{x \to a} g(x) = g(a)$. From the limit laws of Theorem 2.3, it follows that

$$\lim_{x \to a} [f(x) + g(x)] = f(a) + g(a).$$

Therefore, $f + g$ is continuous at a. Similar arguments lead to the continuity of differences, products, quotients, and powers of continuous functions. The next theorem is a direct consequence of Theorem 2.9.

> **THEOREM 2.10   Polynomials and Rational Functions**
>
> **a.** A polynomial function is continuous for all x.
>
> **b.** A rational function (a function of the form $\dfrac{p}{q}$, where p and q are polynomials) is continuous for all x for which $q(x) \neq 0$.

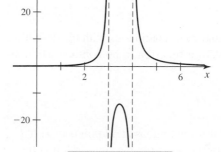

$f(x) = \dfrac{x}{x^2 - 7x + 12}$

Continuous everywhere except $x = 3$ and $x = 4$

FIGURE 2.46

EXAMPLE 3   Applying the continuity theorems For what values of x is the function
$$f(x) = \frac{x}{x^2 - 7x + 12} \text{ continuous?}$$

SOLUTION

a. Because f is rational, Theorem 2.10b implies it is continuous for all x at which the denominator is nonzero. The denominator factors as $(x - 3)(x - 4)$, so it is zero at $x = 3$ and $x = 4$. Therefore, f is continuous for all x except $x = 3$ and $x = 4$ (Figure 2.46). *Related Exercises 19–24* ◄

The following theorem allows us to determine when a composition of two functions is continuous at a point. Its proof is informative and is outlined in Exercise 90.

> **THEOREM 2.11   Continuity of Composite Functions at a Point**
> If g is continuous at a and f is continuous at $g(a)$, then the composite function $f \circ g$ is continuous at a.

The theorem says that under the stated conditions on f and g, the limit of their composition is evaluated by direct substitution; that is,
$$\lim_{x \to a} f(g(x)) = f(g(a)).$$

This result can be stated in another instructive way. Because g is continuous at a, we have $\lim_{x \to a} g(x) = g(a)$. Therefore
$$\lim_{x \to a} f(g(x)) = f(\underbrace{g(a)}_{\lim_{x \to a} g(x)}) = f\left(\lim_{x \to a} g(x)\right).$$

In other words, the order of a function evaluation and a limit may be switched for continuous functions.

QUICK CHECK 2 Evaluate $\lim_{x \to 4} \sqrt{x^2 + 9}$ and $\sqrt{\lim_{x \to 4} (x^2 + 9)}$. How do these results illustrate that the order of a function evaluation and a limit may be switched for continuous functions? ◄

EXAMPLE 4   Limit of a composition Evaluate $\displaystyle\lim_{x \to 0} \left(\frac{x^4 - 2x + 2}{x^6 + 2x^4 + 1}\right)^{10}$.

SOLUTION The rational function $\dfrac{x^4 - 2x + 2}{x^6 + 2x^4 + 1}$ is continuous for all x because its denominator is always positive (Theorem 2.10b). Therefore $\left(\dfrac{x^4 - 2x + 2}{x^6 + 2x^4 + 1}\right)^{10}$, which is the composition of the continuous function $f(x) = x^{10}$ and a continuous rational function, is continuous for all x by Theorem 2.11. By direct substitution,
$$\lim_{x \to 0} \left(\frac{x^4 - 2x + 2}{x^6 + 2x^4 + 1}\right)^{10} = \left(\frac{0^4 - 2 \cdot 0 + 2}{0^6 + 2 \cdot 0^4 + 1}\right)^{10} = 2^{10} = 1024.$$

Related Exercises 25–28 ◄

Continuity on an Interval

A function is *continuous on an interval* if it is continuous at every point in that interval. Consider the functions f and g whose graphs are shown in Figure 2.47. Both these functions are continuous for all x in (a, b), but what about the endpoints? To answer this question, we introduce the ideas of *left-continuity* and *right-continuity*.

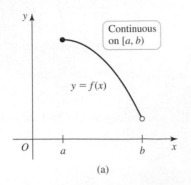

(a)

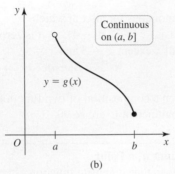

(b)

FIGURE 2.47

DEFINITION Continuity at Endpoints

A function f is **continuous from the left** (or **left-continuous**) at a if $\lim\limits_{x \to a^-} f(x) = f(a)$ and f is **continuous from the right** (or **right-continuous**) at a if $\lim\limits_{x \to a^+} f(x) = f(a)$.

Combining the definitions of left-continuous and right-continuous with the definition of continuity at a point, we define what it means for a function to be continuous on an interval.

DEFINITION Continuity on an Interval

A function f is **continuous on an interval I** if it is continuous at all points of I. If I contains its endpoints, continuity on I means continuous from the right or left at the endpoints.

To illustrate these definitions, consider again the functions in Figure 2.47. In Figure 2.47a, f is continuous from the right at a because $\lim\limits_{x \to a^+} f(x) = f(a)$; but it is not continuous from the left at b because $f(b)$ is not defined. Therefore, f is continuous on the interval $[a, b)$. The behavior of the function g in Figure 2.47b is the opposite: It is continuous from the left at b, but it is not continuous from the right at a. Therefore, g is continuous on $(a, b]$.

QUICK CHECK 3 Modify the graphs of the functions f and g in Figure 2.47 to obtain functions that are continuous on $[a, b]$. ◄

EXAMPLE 5 Intervals of continuity Determine the intervals of continuity for

$$f(x) = \begin{cases} x^2 + 1 & \text{if } x \le 0 \\ 3x + 5 & \text{if } x > 0 \end{cases}$$

SOLUTION This piecewise function consists of two polynomials that describe a parabola and a line (Figure 2.48). By Theorem 2.10, f is continuous for all $x \ne 0$. From its graph, it appears that f is left-continuous at $x = 0$. This observation is verified by noting that

$$\lim_{x \to 0^-} f(x) = \lim_{x \to 0^-} (x^2 + 1) = 1,$$

which means that $\lim\limits_{x \to 0^-} f(x) = f(0)$. However, because

$$\lim_{x \to 0^+} f(x) = \lim_{x \to 0^+} (3x + 5) = 5 \ne f(0),$$

we see that f is not right-continuous at $x = 0$. Therefore, f is continuous on $(-\infty, 0]$, and it is continuous on $(0, \infty)$. *Related Exercises 29–34* ◄

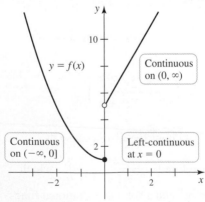

FIGURE 2.48

Functions Involving Roots

Recall that Limit Law 7 of Theorem 2.3 states

$$\lim_{x \to a} [f(x)]^{n/m} = \left[\lim_{x \to a} f(x) \right]^{n/m}$$

provided $f(x) \geq 0$ for x near a if m is even and n/m is reduced. Therefore, if m is odd and f is continuous at a, then $[f(x)]^{n/m}$ is continuous at a, because

$$\lim_{x \to a} [f(x)]^{n/m} = \left[\lim_{x \to a} f(x) \right]^{n/m} = [f(a)]^{n/m}.$$

When m is even, the continuity of $[f(x)]^{n/m}$ must be handled more carefully because this function is defined only when $f(x) \geq 0$. Exercise 59 of Section 2.7 establishes an important fact:

> If f is continuous at a and $f(a) > 0$, then f is positive for all values of x sufficiently close to a.

Combining this fact with Theorem 2.11 (the continuity of composite functions), it follows that $[f(x)]^{n/m}$ is continuous at a provided $f(a) > 0$. At points where $f(a) = 0$, the behavior of $[f(x)]^{n/m}$ varies. Often we find that $[f(x)]^{n/m}$ is left- or right-continuous at that point, or it may be continuous from both sides.

THEOREM 2.12   Continuity of Functions with Roots

Assume that m and n are positive integers with no common factors.
If m is an odd integer, then $[f(x)]^{n/m}$ is continuous at all points at which f is continuous.
If m is even, then $[f(x)]^{n/m}$ is continuous at all points a at which f is continuous and $f(a) > 0$.

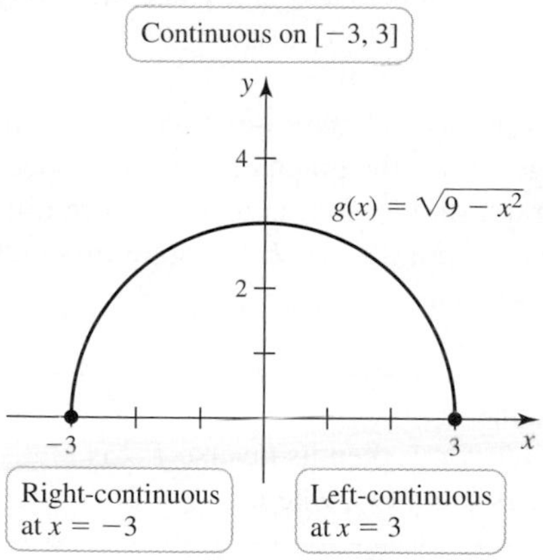

Continuous on $[-3, 3]$

Right-continuous at $x = -3$

Left-continuous at $x = 3$

FIGURE 2.49

EXAMPLE 6   Continuity with roots For what values of x are the following functions continuous?

a. $g(x) = \sqrt{9 - x^2}$ **b.** $f(x) = (x^2 - 2x + 4)^{2/3}$

SOLUTION

a. The graph of g is the upper half of the circle $x^2 + y^2 = 9$ (which can be verified by solving $x^2 + y^2 = 9$ for y). From Figure 2.49, it appears that g is continuous on $[-3, 3]$. To verify this fact, note that g involves an even root ($m = 2, n = 1$ in Theorem 2.12). If $-3 < x < 3$, then $9 - x^2 > 0$ and by Theorem 2.12, g is continuous for all x on $(-3, 3)$.

At the right endpoint, $\lim_{x \to 3^-} \sqrt{9 - x^2} = 0 = g(3)$ by Limit Law 7, which implies that g is left-continuous at 3. Similarly, g is right-continuous at -3 because $\lim_{x \to -3^+} \sqrt{9 - x^2} = 0 = g(-3)$. Therefore, g is continuous on $[-3, 3]$.

b. The polynomial $x^2 - 2x + 4$ is continuous for all x by Theorem 2.10a. Because f involves an odd root ($m = 3, n = 2$ in Theorem 2.12), f is continuous for all x.

Related Exercises 35–44 ◀

QUICK CHECK 4 On what interval is $f(x) = x^{1/4}$ continuous? On what interval is $f(x) = x^{2/5}$ continuous? ◀

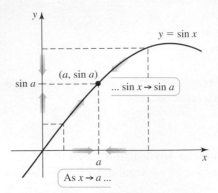

$(a, \sin a)$

$\sin a$

... $\sin x \to \sin a$

As $x \to a$ ...

FIGURE 2.50

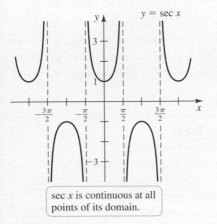

$y = \sec x$

sec x is continuous at all points of its domain.

FIGURE 2.51

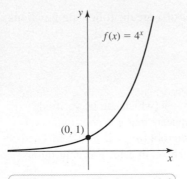

$f(x) = 4^x$

$(0, 1)$

f is continuous at all points provided 4^x is defined for both rational and irrational numbers.

FIGURE 2.52

Continuity of Transcendental Functions

The understanding of continuity that we have developed with algebraic functions may now be applied to transcendental functions.

Trigonometric Functions   In Example 8 of Section 2.3, we used the Squeeze Theorem to show that $\lim_{x \to 0} \sin x = 0$ and $\lim_{x \to 0} \cos x = 1$. Because $\sin 0 = 0$ and $\cos 0 = 1$, these limits imply that $\sin x$ and $\cos x$ are continuous at 0. The graph of $y = \sin x$ (Figure 2.50) suggests that $\lim_{x \to a} \sin x = \sin a$ for any value of a, which means that $\sin x$ is continuous everywhere. The graph of $y = \cos x$ also indicates that $\cos x$ is continuous for all x. Exercise 93 outlines a proof of these results.

With these facts in hand, we appeal to Theorem 2.9e to discover that the remaining trigonometric functions are continuous on their domains. For example, because $\sec x = 1/\cos x$, the secant function is continuous for all x for which $\cos x \neq 0$ (for all x except odd multiples of $\pi/2$) (Figure 2.51). Likewise, the tangent, cotangent, and cosecant functions are continuous at all points of their domains.

Exponential Functions   The continuity of exponential functions of the form $f(x) = b^x$, with $0 < b < 1$ or $b > 1$, raises an important question. Consider the function $f(x) = 4^x$ (Figure 2.52). Evaluating f is routine if x is rational:

$$4^3 = 4 \cdot 4 \cdot 4 = 64; \quad 4^{-2} = \frac{1}{4^2} = \frac{1}{16}; \quad 4^{3/2} = \sqrt{4^3} = 8; \quad \text{and} \quad 4^{-1/3} = \frac{1}{\sqrt[3]{4}}.$$

But what is meant by 4^x when x is an irrational number, such as $\sqrt{2}$? In order for $f(x) = 4^x$ to be continuous for all real numbers, it must also be defined when x is an irrational number. Providing a working definition for an expression such as $4^{\sqrt{2}}$ requires mathematical results that don't appear until Chapter 6. Until then, we assume without proof that the domain of $f(x) = b^x$ is the set of all real numbers and that f is continuous at all points of its domain.

Inverse Functions   Suppose a function f is continuous and one-to-one on an interval I. Reflecting the graph of f through the line $y = x$ generates the graph of f^{-1}. The reflection process introduces no discontinuities in the graph of f^{-1}, so it is plausible (and indeed, true) that f^{-1} is continuous on the interval corresponding to I. We state this fact without a formal proof.

> **THEOREM 2.13   Continuity of Inverse Functions**
> If a continuous function f has an inverse on an interval I, then its inverse f^{-1} is also continuous (on the interval consisting of the points $f(x)$, where x is in I).

Because all the trigonometric functions are continuous on their domains, they are also continuous when their domains are restricted for the purpose of defining inverse functions. Therefore, by Theorem 2.13, the inverse trigonometric functions are continuous at all points of their domains.

Logarithmic functions of the form $f(x) = \log_b x$ are continuous at all points of their domains for the same reason: They are inverses of exponential functions, which are one-to-one and continuous. Collecting all these facts together, we have the following theorem.

THEOREM 2.14 Continuity of Transcendental Functions
The following functions are continuous at all points of their domains.

Trigonometric		**Inverse Trigonometric**		**Exponential**	
$\sin x$	$\cos x$	$\sin^{-1} x$	$\cos^{-1} x$	b^x	e^x
$\tan x$	$\cot x$	$\tan^{-1} x$	$\cot^{-1} x$	**Logarithmic**	
$\sec x$	$\csc x$	$\sec^{-1} x$	$\csc^{-1} x$	$\log_b x$	$\ln x$

For each function listed in Theorem 2.14, we have $\lim\limits_{x \to a} f(x) = f(a)$, provided a is in the domain of the function. This means that limits involving these functions may be evaluated by direct substitution at points in the domain.

EXAMPLE 7 Limits involving transcendental functions Evaluate the following limits after determining the continuity of the functions involved.

a. $\lim\limits_{x \to 0} \dfrac{\cos^2 x - 1}{\cos x - 1}$ **b.** $\lim\limits_{x \to 1} \left(\sqrt[4]{\ln x} + \tan^{-1} x \right)$

SOLUTION

> Limits like the one in Example 7a are denoted 0/0 and are known as *indeterminate forms*, to be studied further in Section 4.7.

a. Both $\cos^2 x - 1$ and $\cos x - 1$ are continuous for all x by Theorems 2.9 and 2.14. However, the ratio of these functions is not continuous when $\cos x - 1 = 0$, which corresponds to all integer multiples of 2π. Note that both the numerator and denominator of $\dfrac{\cos^2 x - 1}{\cos x - 1}$ approach 0 as $x \to 0$. To evaluate the limit, we factor and simplify:

$$\lim_{x \to 0} \frac{\cos^2 x - 1}{\cos x - 1} = \lim_{x \to 0} \frac{(\cos x - 1)(\cos x + 1)}{\cos x - 1} = \lim_{x \to 0} (\cos x + 1)$$

(where $\cos x - 1$ may be canceled because it is nonzero as x approaches 0). The limit on the right is now evaluated using direct substitution:

$$\lim_{x \to 0} (\cos x + 1) = \cos 0 + 1 = 2$$

b. By Theorem 2.14, $\ln x$ is continuous on its domain, $(0, \infty)$. However, $\ln x > 0$ only when $x > 1$, so Theorem 2.12 implies $\sqrt[4]{\ln x}$ is continuous on $(1, \infty)$. At $x = 1$, $\sqrt[4]{\ln x}$ is right-continuous (Quick Check 5). The domain of $\tan^{-1} x$ is all real numbers, so it is continuous on $(-\infty, \infty)$. Therefore, $f(x) = \sqrt[4]{\ln x} + \tan^{-1} x$ is continuous on $[1, \infty)$. Because the domain of f does not include points with $x < 1$,

> **QUICK CHECK 5** Show that $f(x) = \sqrt[4]{\ln x}$ is right-continuous at $x = 1$. ◄

$\lim\limits_{x \to 1^-} \left(\sqrt[4]{\ln x} + \tan^{-1} x \right)$ does not exist, which implies that $\lim\limits_{x \to 1} \left(\sqrt[4]{\ln x} + \tan^{-1} x \right)$ does not exist.

Related Exercises 45–48 ◄

The Intermediate Value Theorem

A common problem in mathematics is finding solutions to equations of the form $f(x) = L$. Before attempting to find values of x satisfying this equation, it is worthwhile to determine whether a solution exists.

The existence of solutions is often established using a result known as the **Intermediate Value Theorem**. Given a function f and a constant L, we assume L lies between $f(a)$ and $f(b)$. The Intermediate Value Theorem says that if f is continuous on $[a, b]$, then the graph of $y = f(x)$ must cross the horizontal line $y = L$ at least once (Figure 2.53). Although this theorem is easily illustrated, its proof goes beyond the scope of this text.

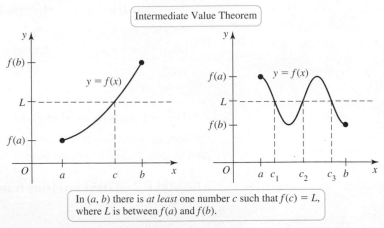

In (a, b) there is *at least* one number c such that $f(c) = L$, where L is between $f(a)$ and $f(b)$.

FIGURE 2.53

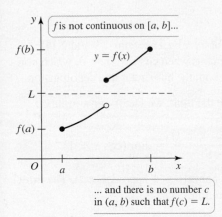

f is not continuous on $[a, b]$...

... and there is no number c in (a, b) such that $f(c) = L$.

FIGURE 2.54

> **THEOREM 2.15   The Intermediate Value Theorem**
> Suppose f is continuous on the interval $[a, b]$ and L is a number between $f(a)$ and $f(b)$. Then there is at least one number c in (a, b) satisfying $f(c) = L$.

The importance of continuity in Theorem 2.15 is illustrated in Figure 2.54, where we see a function f that is not continuous on $[a, b]$. For the value of L shown in the figure, there is no value of c in (a, b) satisfying $f(c) = L$.

EXAMPLE 8   Finding an interest rate Suppose you invest \$1000 in a special 5-year savings account with a fixed annual interest rate r, with monthly compounding. The amount of money A in the account after 5 years (60 months) is $A(r) = 1000\left(1 + \dfrac{r}{12}\right)^{60}$. Your goal is to have \$1400 in the account after 5 years.

a. Use the Intermediate Value Theorem to show there is a value of r in $(0, 0.08)$—that is, an interest rate between 0% and 8%—for which $A(r) = 1400$.

b. Use a graphing utility to illustrate your explanation in part (a), and then estimate the interest rate required to reach your goal.

SOLUTION

a. As a polynomial in r (of degree 60), $A(r) = 1000\left(1 + \dfrac{r}{12}\right)^{60}$ is continuous for all r.

Evaluating $A(r)$ at the endpoints of the interval $[0, 0.08]$, we have $A(0) = 1000$ and $A(0.08) = 1489.85$. Therefore

$$A(0) < 1400 < A(0.08)$$

and it follows, by the Intermediate Value Theorem, that there is a value of r in $(0, 0.08)$ for which $A(r) = 1400$.

b. The graphs of $y = A(r)$ and the horizontal line $y = 1400$ are shown in Figure 2.55; it is evident that they intersect between $r = 0$ and $r = 0.08$. Solving $A(r) = 1400$ algebraically or using a root finder reveals that the curve and line intersect at $r \approx 0.0675$. Therefore, an interest rate of approximately 6.75% is required for the investment to be worth \$1400 after 5 years.

Related Exercises 49–54 ◄

> **QUICK CHECK 6** Does the equation $f(x) = x^3 + x + 1 = 0$ have a solution on the interval $[-1, 1]$? Explain. ◄

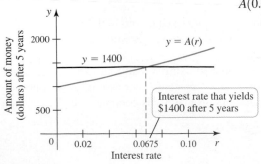

FIGURE 2.55

SECTION 2.6 EXERCISES

Review Questions

1. Which of the following functions are continuous for all values in their domain? Justify your answers.

   a. $a(t)$ = altitude of a skydiver t seconds after jumping from a plane

   b. $n(t)$ = number of quarters needed to park in a metered parking space for t minutes

   c. $T(t)$ = temperature t minutes after midnight in Chicago on January 1

   d. $p(t)$ = number of points scored by a basketball player after t minutes of a basketball game

2. Give the three conditions that must be satisfied by a function to be continuous at a point.

3. What does it mean for a function to be continuous on an interval?

4. We informally described a function f to be continuous at a if its graph contains no holes or breaks at a. Explain why this is not an adequate definition of continuity.

5. Complete the following sentences.

   a. A function is continuous from the left at a if _____.
   b. A function is continuous from the right at a if _____.

6. Describe the points (if any) at which a rational function fails to be continuous.

7. What is the domain of $f(x) = e^x/x$ and where is f continuous?

8. Explain in words and pictures what the Intermediate Value Theorem says.

Basic Skills

9–12. Discontinuities from a graph *Determine the points at which the following functions f have discontinuities. For each point state the conditions in the continuity checklist that are violated.*

9.
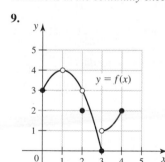

10.
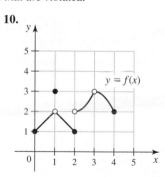

11.
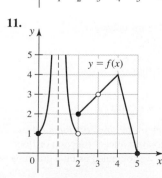

12.
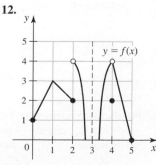

13–18. Continuity at a point *Determine whether the following functions are continuous at a. Use the continuity checklist to justify your answer.*

13. $f(x) = \sqrt{x - 2}$; $a = 1$

14. $g(x) = \dfrac{1}{x - 3}$; $a = 3$

15. $f(x) = \begin{cases} \dfrac{x^2 - 1}{x - 1} & \text{if } x \neq 1 \\ 3 & \text{if } x = 1 \end{cases}$; $a = 1$

16. $f(x) = \begin{cases} \dfrac{x^2 - 4x + 3}{x - 3} & \text{if } x \neq 3 \\ 2 & \text{if } x = 3 \end{cases}$; $a = 3$

17. $f(x) = \dfrac{5x - 2}{x^2 - 9x + 20}$; $a = 4$

18. $f(x) = \begin{cases} \dfrac{x^2 - x}{x + 1} & \text{if } x \neq -1 \\ 0 & \text{if } x = -1 \end{cases}$; $a = -1$

19–24. Continuity on intervals *Use Theorem 2.10 to determine the intervals on which the following functions are continuous.*

19. $p(x) = 4x^5 - 3x^2 + 1$

20. $g(x) = \dfrac{3x^2 - 6x + 7}{x^2 + x + 1}$

21. $f(x) = \dfrac{x^5 + 6x + 17}{x^2 - 9}$

22. $s(x) = \dfrac{x^2 - 4x + 3}{x^2 - 1}$

23. $f(x) = \dfrac{1}{x^2 - 4}$

24. $f(t) = \dfrac{t + 2}{t^2 - 4}$

25–28. Limits of compositions *Evaluate the following limits and justify your answers.*

25. $\displaystyle\lim_{x \to 0} (x^8 - 3x^6 - 1)^{40}$

26. $\displaystyle\lim_{x \to 2} \left(\dfrac{3}{2x^5 - 4x^2 - 50} \right)^4$

27. $\displaystyle\lim_{x \to 1} \left(\dfrac{x + 5}{x + 2} \right)^4$

28. $\displaystyle\lim_{x \to \infty} \left(\dfrac{2x + 1}{x} \right)^3$

29–32. Intervals of continuity *Determine the intervals of continuity for the following functions.*

29. The graph of Exercise 9

30. The graph of Exercise 10

31. The graph of Exercise 11

32. The graph of Exercise 12

33. **Intervals of continuity** Let

$$f(x) = \begin{cases} x^2 + 3x & \text{if } x \geq 1 \\ 2x & \text{if } x < 1 \end{cases}$$

   a. Use the continuity checklist to show that f is not continuous at 1.

   b. Is f continuous from the left or right at 1?

   c. State the interval(s) of continuity.

34. **Intervals of continuity** Let

$$f(x) = \begin{cases} x^3 + 4x + 1 & \text{if } x \leq 0 \\ 2x^3 & \text{if } x > 0 \end{cases}$$

a. Use the continuity checklist to show that f is not continuous at 0.

b. Is f continuous from the left or right at 0?

c. State the interval(s) of continuity.

35–40. Functions with roots *Determine the interval(s) on which the following functions are continuous. Be sure to consider right- and left-continuity at the endpoints.*

35. $f(x) = \sqrt{2x^2 - 16}$

36. $g(x) = \sqrt{x^4 - 1}$

37. $f(x) = \sqrt[3]{x^2 - 2x - 3}$

38. $f(t) = (t^2 - 1)^{3/2}$

39. $f(x) = (2x - 3)^{2/3}$

40. $f(z) = (z - 1)^{3/4}$

41–44. Limits with roots *Determine the following limits and justify your answers.*

41. $\displaystyle\lim_{x \to 2} \sqrt{\dfrac{4x + 10}{2x - 2}}$

42. $\displaystyle\lim_{x \to -1} \left(x^2 - 4 + \sqrt[3]{x^2 - 9} \right)$

43. $\displaystyle\lim_{x \to 3} \left(\sqrt{x^2 + 7} \right)$

44. $\displaystyle\lim_{t \to 2} \dfrac{t^2 + 5}{1 + \sqrt{t^2 + 5}}$

45–48. Continuity and limits with transcendental functions *Determine the interval(s) on which the following functions are continuous; then evaluate the given limits.*

45. $f(x) = \csc x;\quad \displaystyle\lim_{x \to \pi/4} f(x);\quad \lim_{x \to 2\pi^-} f(x)$

46. $f(x) = e^{\sqrt{x}};\quad \displaystyle\lim_{x \to 4} f(x);\quad \lim_{x \to 0^+} f(x)$

47. $f(x) = \dfrac{1 + \sin x}{\cos x};\quad \displaystyle\lim_{x \to \pi/2^-} f(x);\quad \lim_{x \to 4\pi/3} f(x)$

48. $f(x) = \dfrac{\ln x}{\sin^{-1} x};\quad \displaystyle\lim_{x \to 1^-} f(x)$

49. Intermediate Value Theorem and interest rates Suppose \$5000 is invested in a savings account for 10 years (120 months), with an annual interest rate of r, compounded monthly. The amount of money in the account after 10 years is $A(r) = 5000(1 + r/12)^{120}$.

a. Use the Intermediate Value Theorem to show there is a value of r in $(0, 0.08)$—an interest rate between 0% and 8%—that allows you to reach your savings goal of \$7000 in 10 years.

b. Use a graph to illustrate your explanation in part (a); then, approximate the interest rate required to reach your goal.

50. Intermediate Value Theorem and mortgage payments You are shopping for a \$150,000, 30-year (360-month) loan to buy a house. The monthly payment is

$$m(r) = \dfrac{150{,}000(r/12)}{1 - (1 + r/12)^{-360}}$$

where r is the annual interest rate. Suppose banks are currently offering interest rates between 6% and 8%.

a. Use the Intermediate Value Theorem to show there is a value of r in $(0.06, 0.08)$—an interest rate between 6% and 8%—that allows you to make monthly payments of \$1000 per month.

b. Use a graph to illustrate your explanation to part (a). Then determine the interest rate you need for monthly payments of \$1000.

51–54. Applying the Intermediate Value Theorem

a. *Use the Intermediate Value Theorem to show that the following equations have a solution on the given interval (a, b).*

b. *Use a graphing utility to find all the solutions to the equation on (a, b).*

c. *Illustrate your answers with an appropriate graph.*

51. $2x^3 + x - 2 = 0;\quad (-1, 1)$

52. $\sqrt{x^4 + 25x^3 + 10} = 5;\quad (0, 1)$

53. $x^3 - 5x^2 + 2x = -1;\quad (-1, 5)$

54. $-x^5 - 4x^2 + 2\sqrt{x} + 5 = 0;\quad (0, 3)$

Further Explorations

55. Explain why or why not Determine whether the following statements are true and give an explanation or counterexample.

a. If a function is left-continuous and right-continuous at a, then it is continuous at a.

b. If a function is continuous at a, then it is left-continuous and right-continuous at a.

c. If $a < b$ and $f(a) \le L \le f(b)$, then there is some value of c between a and b for which $f(c) = L$.

d. Suppose f is continuous on $[a, b]$. Then there is a point c in (a, b) such that $f(c) = [f(a) + f(b)]/2$.

56. Continuity of the absolute value function Prove that the absolute value function $|x|$ is continuous for all values of x. (*Hint:* Using the definition of the absolute value function, compute $\displaystyle\lim_{x \to 0^-} |x|$ and $\lim_{x \to 0^+} |x|$.)

57–60. Continuity of functions with absolute values *Use the continuity of the absolute value function (Exercise 56) to determine the interval(s) on which the following functions are continuous.*

57. $f(x) = |x^2 + 3x - 18|$

58. $g(x) = \left| \dfrac{x + 4}{x^2 - 4} \right|$

59. $h(x) = \left| \dfrac{1}{\sqrt{x} - 4} \right|$

60. $h(x) = |x^2 + 2x + 5| + \sqrt{x}$

61–70. Miscellaneous limits *Evaluate the following limits.*

61. $\displaystyle\lim_{x \to \pi} \dfrac{\cos^2 x + 3\cos x + 2}{\cos x + 1}$

62. $\displaystyle\lim_{x \to 5\pi/2} \dfrac{\sin^2 x + 6\sin x + 5}{\sin^2 x - 1}$

63. $\displaystyle\lim_{x \to \pi/2} \dfrac{\sin x - 1}{\sqrt{\sin x} - 1}$

64. $\displaystyle\lim_{\theta \to 0} \dfrac{\dfrac{1}{2 + \sin \theta} - \dfrac{1}{2}}{\sin \theta}$

65. $\displaystyle\lim_{x \to 0} \dfrac{\cos x - 1}{\sin^2 x}$

66. $\displaystyle\lim_{x \to 0^+} \cot x$

67. $\displaystyle\lim_{x \to \infty} \dfrac{\tan^{-1} x}{x}$

68. $\displaystyle\lim_{t \to \infty} \dfrac{\cos t}{e^{3t}}$

69. $\displaystyle\lim_{x \to 1^-} \dfrac{x}{\ln x}$

70. $\displaystyle\lim_{x \to 0^+} \dfrac{x}{\ln x}$

71. Pitfalls using technology The graph of the *sawtooth function* $y = x - \lfloor x \rfloor$, where $\lfloor x \rfloor$ is the greatest integer function or floor function (Exercise 31, Section 2.2), was obtained using a graphing utility (see figure). Identify any inaccuracies appearing in the graph and then plot an accurate graph by hand.

$$y = x - \lfloor x \rfloor$$

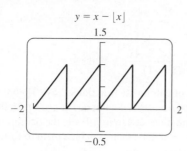

72. Pitfalls using technology Graph the function $f(x) = \dfrac{\sin x}{x}$ using a graphing window of $[-\pi, \pi] \times [0, 2]$.

a. Sketch a copy of the graph obtained with your graphing device and describe any inaccuracies appearing in the graph.

b. Sketch an accurate graph of the function. Is f continuous at 0?

73. Sketching functions

a. Sketch the graph of a function that is not continuous at 1, but is defined at 1.

b. Sketch the graph of a function that is not continuous at 1, but has a limit at 1.

74. An unknown constant Determine the value of the constant a for which the function

$$f(x) = \begin{cases} \dfrac{x^2 + 3x + 2}{x + 1} & \text{if } x \neq -1 \\ a & \text{if } x = -1 \end{cases}$$

is continuous at -1.

75. An unknown constant Let

$$g(x) = \begin{cases} x^2 + x & \text{if } x < 1 \\ a & \text{if } x = 1 \\ 3x + 5 & \text{if } x > 1 \end{cases}$$

a. Determine the value of a for which g is continuous from the left at 1.

b. Determine the value of a for which g is continuous from the right at 1.

c. Is there a value of a for which g is continuous at 1? Explain.

76–77. Applying the Intermediate Value Theorem *Use the Intermediate Value Theorem to verify that the following equations have three solutions on the given interval. Use a graphing utility to find the approximate roots.*

76. $x^3 + 10x^2 - 100x + 50 = 0$; $(-20, 10)$

77. $70x^3 - 87x^2 + 32x - 3 = 0$; $(0, 1)$

Applications

78. Parking costs Determine the intervals of continuity for the parking cost function c introduced at the outset of this section (see the figure). Consider $0 \leq t \leq 60$.

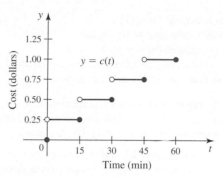

79. Investment problem Assume you invest \$250 at the end of each year for 10 years at an annual interest rate of r. The amount of money in your account after 10 years is $A = \dfrac{250[(1 + r)^{10} - 1]}{r}$. Assume your goal is to have \$3500 in your account after 10 years.

a. Use the Intermediate Value Theorem to show that there is an interest rate r in $(0.01, 0.10)$—between 1% and 10%—that allows you to reach your financial goal.

b. Use a calculator to estimate the interest rate required to reach your financial goal.

80. Applying the Intermediate Value Theorem Suppose you park your car at a trailhead in a national park and begin a 2-hr hike to a lake at 7 A.M. on a Friday morning. On Sunday morning, you leave the lake at 7 A.M. and start the 2-hr hike back to your car. Assume the lake is 3 mi from your car. Let $f(t)$ be your distance from the car t hours after 7 A.M. on Friday morning and let $g(t)$ be your distance from the car t hours after 7 A.M. on Sunday morning.

a. Evaluate $f(0)$, $f(2)$, $g(0)$, and $g(2)$.

b. Let $h(t) = f(t) - g(t)$. Find $h(0)$ and $h(2)$.

c. Use the Intermediate Value Theorem to show that there is some point along the trail that you will pass at exactly the same time of morning on both days.

81. The monk and the mountain A monk set out from a monastery in the valley at dawn. He walked all day up a winding path, stopping for lunch and taking a nap along the way. At dusk, he arrived at a temple on the mountaintop. The next day, the monk made the return walk to the valley, leaving the temple at dawn, walking the same path for the entire day, and arriving at the monastery in the evening. Must there be one point along the path that the monk occupied at the same time of day on both the ascent and descent? (*Hint:* The question can be answered without the Intermediate Value Theorem.) (*Source:* Arthur Koestler, *The Act of Creation.*)

Additional Exercises

82. Does continuity of $|f|$ imply continuity of f? Let

$$g(x) = \begin{cases} 1 & \text{if } x \geq 0 \\ -1 & \text{if } x < 0 \end{cases}$$

a. Write a formula for $|g(x)|$.
b. Is g continuous at $x = 0$? Explain.
c. Is $|g|$ continuous at $x = 0$? Explain.
d. For any function f, if $|f|$ is continuous at a, does it necessarily follow that f is continuous at a? Explain.

83–84. Classifying discontinuities *The discontinuities in graphs (a) and (b) are* removable discontinuities *because they disappear if we redefine f at a so that $f(a) = \lim\limits_{x \to a} f(x)$. The function in graph (c) has a* jump discontinuity *because left and right limits exist at a but are unequal. The discontinuity in graph (d) is an* infinite discontinuity *because the function has a vertical asymptote at a.*

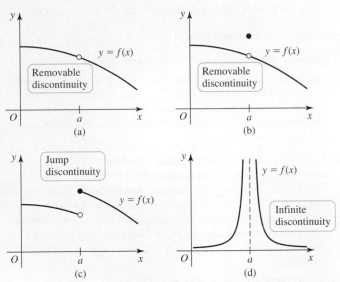

83. Is the discontinuity at a in graph (c) removable? Explain.

84. Is the discontinuity at a in graph (d) removable? Explain.

85–86. Removable discontinuities *Show that the following functions have a removable discontinuity at the given point.*

85. $f(x) = \dfrac{x^2 - 7x + 10}{x - 2}; \ x = 2$

86. $g(x) = \begin{cases} \dfrac{x^2 - 1}{1 - x} & \text{if } x \neq 1 \\ 3 & \text{if } x = 1 \end{cases}; \ x = 1$

87. Do removable discontinuities exist?

a. Does the function $f(x) = x \sin(1/x)$ have a removable discontinuity at $x = 0$?
b. Does the function $g(x) = \sin(1/x)$ have a removable discontinuity at $x = 0$?

88–89. Classifying discontinuities *Classify the discontinuities in the following functions at the given points.*

88. $f(x) = \dfrac{|x - 2|}{x - 2}; \ x = 2$

89. $h(x) = \dfrac{x^3 - 4x^2 + 4x}{x(x - 1)}; \ x = 0$ and $x = 1$

90. Continuity of composite functions Prove Theorem 2.11: If g is continuous at a and f is continuous at $g(a)$, then the composition $f \circ g$ is continuous at a. (*Hint:* Write the definition of continuity for f and g separately; then, combine them to form the definition of continuity for $f \circ g$.)

91. Continuity of compositions

a. Find functions f and g such that each function is continuous at 0, but $f \circ g$ is not continuous at 0.
b. Explain why examples satisfying part (a) do not contradict Theorem 2.11.

92. Violation of the Intermediate Value Theorem? Let $f(x) = \dfrac{|x|}{x}$. Then $f(-2) = -1$ and $f(2) = 1$. Therefore, $f(-2) < 0 < f(2)$, but there is no value of c between -2 and 2 for which $f(c) = 0$. Does this fact violate the Intermediate Value Theorem? Explain.

93. Continuity of $\sin x$ and $\cos x$

a. Use the identity $\sin(a + h) = \sin a \cos h + \cos a \sin h$ with the fact that $\lim\limits_{x \to 0} \sin x = 0$ to prove that $\lim\limits_{x \to a} \sin x = \sin a$, thereby establishing that $\sin x$ is continuous for all x. (*Hint:* Let $h = x - a$ so that $x = a + h$ and note that $h \to 0$ as $x \to a$.)
b. Use the identity $\cos(a + h) = \cos a \cos h - \sin a \sin h$ with the fact that $\lim\limits_{x \to 0} \cos x = 1$ to prove that $\lim\limits_{x \to a} \cos x = \cos a$.

QUICK CHECK ANSWERS

1. $t = 15, 30, 45$ **2.** Both expressions have a value of 5, showing that $\lim\limits_{x \to a} f(g(x)) = f\left(\lim\limits_{x \to a} g(x)\right)$.   **3.** Fill in the endpoints.   **4.** $[0, \infty); (-\infty, \infty)$ **5.** Note that

$$\lim_{x \to 1^+} \sqrt[4]{\ln x} = \sqrt[4]{\lim_{x \to 1^+} \ln x} = 0 \text{ and } f(1) = \sqrt[4]{\ln 1} = 0.$$

Because the limit from the right and the value of the function at $x = 1$ are equal, the function is right-continuous at $x = 1$.
6. The equation has a solution on the interval $[-1, 1]$ because f is continuous on $[-1, 1]$ and $f(-1) < 0 < f(1)$. ◄

2.7 Precise Definitions of Limits

The limit definitions already encountered in this chapter are adequate for most elementary limits. However some of the terminology used, such as *sufficiently close* and *arbitrarily large*, needs clarification. The goal of this section is to give limits a solid mathematical foundation by transforming the previous limit definitions into precise mathematical statements.

Moving Toward a Precise Definition

Assume the function f is defined for all x near a, except possibly at a. Recall that $\lim_{x \to a} f(x) = L$ means that $f(x)$ is arbitrarily close to L for all x sufficiently close (but not equal) to a. This limit definition is made precise by observing that the distance between $f(x)$ and L is $|f(x) - L|$ and that the distance between x and a is $|x - a|$. Therefore, we write $\lim_{x \to a} f(x) = L$ if we can make $|f(x) - L|$ arbitrarily small for any x, distinct from a, with $|x - a|$ sufficiently small. For instance, if we want $|f(x) - L|$ to be less than 0.1, then we must find a number $\delta > 0$ such that

$$|f(x) - L| < 0.1 \quad \text{whenever} \quad |x - a| < \delta \quad \text{and} \quad x \neq a.$$

> The phrase *for all x near a* means for all x in an open interval containing a.

> The Greek letters δ (delta) and ε (epsilon) represent small positive numbers when discussing limits.

> The two conditions $|x - a| < \delta$ and $x \neq a$ are written concisely as $0 < |x - a| < \delta$.

If, instead, we want $|f(x) - L|$ to be less than 0.001, then we must find *another* number $\delta > 0$ such that

$$|f(x) - L| < 0.001 \quad \text{whenever} \quad 0 < |x - a| < \delta.$$

For the limit to exist, it must be true that for *any* $\varepsilon > 0$, we can always find a $\delta > 0$ such that

$$|f(x) - L| < \varepsilon \quad \text{whenever} \quad 0 < |x - a| < \delta.$$

EXAMPLE 1 **Determining values of δ from a graph** Figure 2.56 shows the graph of a linear function f with $\lim_{x \to 3} f(x) = 5$. For each value of $\varepsilon > 0$, determine a value of $\delta > 0$ satisfying the statement

$$|f(x) - 5| < \varepsilon \quad \text{whenever} \quad 0 < |x - 3| < \delta.$$

a. $\varepsilon = 1$ **b.** $\varepsilon = \frac{1}{2}$

SOLUTION

a. With $\varepsilon = 1$, we want $f(x)$ to be less than 1 unit from 5, which means $f(x)$ is between 4 and 6. To determine a corresponding value of δ, draw the horizontal lines $y = 4$ and $y = 6$ (Figure 2.57a). Then sketch vertical lines passing through the points where the horizontal lines and the graph of f intersect (Figure 2.57b). We see that the vertical lines intersect the x-axis at $x = 1$ and $x = 5$. Note that $f(x)$ is less than 1 unit from 5 on the y-axis if x is within 2 units of 3 on the x-axis. So, for $\varepsilon = 1$, we let $\delta = 2$ or any smaller positive value.

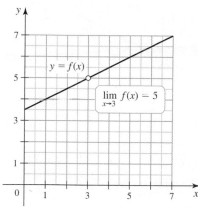

FIGURE 2.56

> The founders of calculus, Isaac Newton (1642–1727) and Gottfried Leibniz (1646–1716), developed the core ideas of calculus without using a precise definition of a limit. It was not until the 19th century that a rigorous definition was introduced by Louis Cauchy (1789–1857) and later refined by Karl Weierstrass (1815–1897).

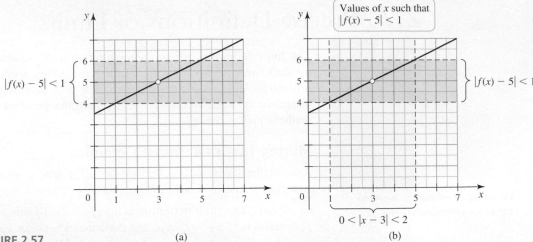

FIGURE 2.57   (a)   (b)

▶ Once an acceptable value of δ is found satisfying the statement

$$|f(x) - L| < \varepsilon \quad \text{whenever}$$

$$0 < |x - a| < \delta,$$

any smaller positive value of δ also works.

b. With $\varepsilon = \frac{1}{2}$, we want $f(x)$ to lie within a half-unit of 5 or, equivalently, $f(x)$ must lie between 4.5 and 5.5. Proceeding as in part (a), we see that $f(x)$ is within a half-unit of 5 on the y-axis if x is less than 1 unit from 3 (Figure 2.58). So for $\varepsilon = \frac{1}{2}$, we let $\delta = 1$ or any smaller positive number.

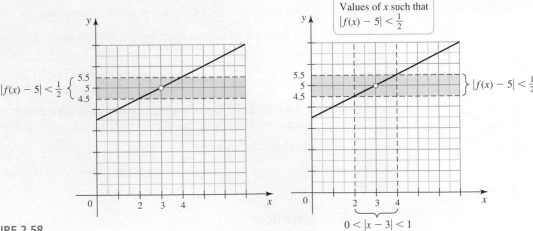

FIGURE 2.58

Related Exercises 9–12 ◀

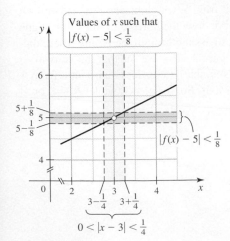

FIGURE 2.59

The idea of a limit, as illustrated in Example 1, may be described in terms of a contest between two people named Epp and Del. First, Epp picks a particular number $\varepsilon > 0$; then, he challenges Del to find a corresponding value of $\delta > 0$ such that

$$|f(x) - 5| < \varepsilon \quad \text{whenever} \quad 0 < |x - 3| < \delta. \tag{1}$$

To illustrate, suppose Epp chooses $\varepsilon = 1$. From Example 1, we know that Del will satisfy (1) by choosing $0 < \delta \le 2$. If Epp chooses $\varepsilon = \frac{1}{2}$, then (by Example 1) Del responds by letting $0 < \delta \le 1$. If Epp lets $\varepsilon = \frac{1}{8}$, then Del chooses $0 < \delta \le \frac{1}{4}$ (Figure 2.59). In fact, there is a pattern: For *any* $\varepsilon > 0$ that Epp chooses, no matter how small, Del will satisfy (1) by choosing a positive value of δ satisfying $0 < \delta \le 2\varepsilon$. Del has discovered a mathematical relationship: If $0 < \delta \le 2\varepsilon$ and $0 < |x - 3| < \delta$, then $|f(x) - 5| < \varepsilon$ for *any* $\varepsilon > 0$. This conversation illustrates the general procedure for proving that $\lim_{x \to a} f(x) = L$.

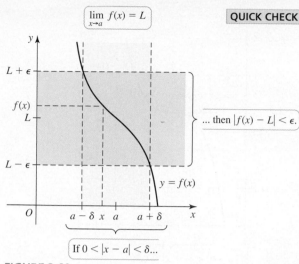

$$\lim_{x \to a} f(x) = L$$

... then $|f(x) - L| < \epsilon.$

If $0 < |x - a| < \delta$...

FIGURE 2.60

▶ The value of δ in the precise definition of a limit depends only on ε.

▶ Definitions of the one-sided limits $\lim_{x \to a^+} f(x) = L$ and $\lim_{x \to a^-} f(x) = L$ are discussed in Exercises 39–43.

QUICK CHECK 1 In Example 1, find a positive number δ satisfying the statement

$$|f(x) - 5| < \frac{1}{100} \quad \text{whenever} \quad 0 < |x - 3| < \delta. \blacktriangleleft$$

A Precise Definition

Example 1 dealt with a linear function, but it points the way to a precise definition of a limit for any function. As shown in Figure 2.60, $\lim_{x \to a} f(x) = L$ means that for *any* positive number ε, there is another positive number δ such that

$$|f(x) - L| < \varepsilon \quad \text{whenever} \quad 0 < |x - a| < \delta.$$

In all limit proofs, the goal is to find a relationship between ε and δ that gives an admissible value of δ, in terms of ε only. This relationship must work for any positive value of ε.

DEFINITION   Limit of a Function

Assume that $f(x)$ exists for all x in some open interval containing a, except possibly at a. We say that the **limit of $f(x)$ as x approaches a is L**, written

$$\lim_{x \to a} f(x) = L,$$

if for *any* number $\varepsilon > 0$ there is a corresponding number $\delta > 0$ such that

$$|f(x) - L| < \varepsilon \quad \text{whenever} \quad 0 < |x - a| < \delta.$$

EXAMPLE 2   Finding δ for a given ε using a graphing utility Let $f(x) = x^3 - 6x^2 + 12x - 5$ and demonstrate that $\lim_{x \to 2} f(x) = 3$ as follows. For the given values of ε, use a graphing utility to find a value of $\delta > 0$ such that

$$|f(x) - 3| < \varepsilon \quad \text{whenever} \quad 0 < |x - 2| < \delta.$$

a. $\varepsilon = 1$ **b.** $\varepsilon = \frac{1}{2}$

SOLUTION

a. The condition $|f(x) - 3| < \varepsilon = 1$ implies that $f(x)$ lies between 2 and 4. Using a graphing utility, we graph f and the lines $y = 2$ and $y = 4$ (Figure 2.61). These lines intersect the graph of f at $x = 1$ and at $x = 3$. We now sketch the vertical lines $x = 1$ and $x = 3$ and observe that $f(x)$ is within 1 unit of 3 whenever x is within 1 unit of 2 on the x-axis (Figure 2.61). Therefore, with $\varepsilon = 1$, we choose δ such that $0 < \delta \le 1$.

b. The condition $|f(x) - 3| < \varepsilon = \frac{1}{2}$ implies that $f(x)$ lies between 2.5 and 3.5 on the y-axis. We now find that the lines $y = 2.5$ and $y = 3.5$ intersect the graph of f at $x \approx 1.21$ and $x \approx 2.79$ (Figure 2.62). Observe that if x is less than 0.79 units from 2 on the x-axis, then $f(x)$ is less than a half-unit from 3 on the y-axis. Therefore with $\varepsilon = \frac{1}{2}$ we let $0 < \delta \le 0.79$.

    This procedure could be repeated for smaller and smaller values of $\varepsilon > 0$. For each value of ε, there exists a corresponding value of δ, proving that the limit exists.

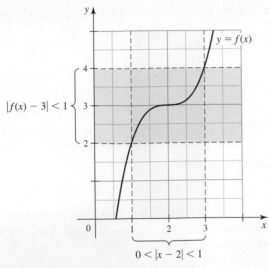

FIGURE 2.61

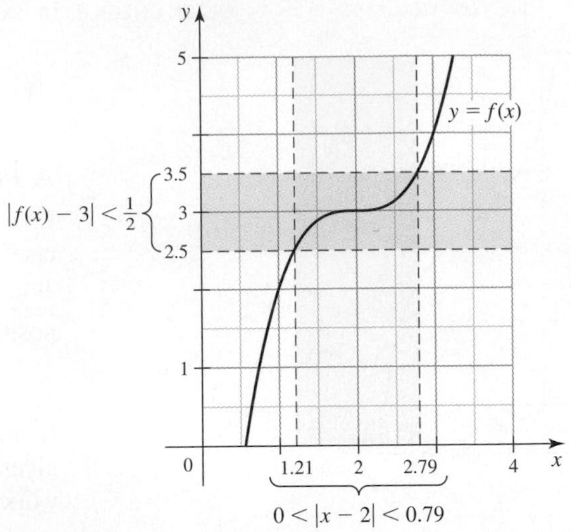

FIGURE 2.62

Related Exercises 13–14 ◄

QUICK CHECK 2 For the function f given in Example 2, estimate a value of $\delta > 0$ satisfying $|f(x) - 3| < 0.25$ whenever $0 < |x - 2| < \delta$. ◄

The inequality $0 < |x - a| < \delta$ means that x lies between $a - \delta$ and $a + \delta$ with $x \neq a$. We say that the interval $(a - \delta, a + \delta)$ is **symmetric about** a because a is the midpoint of the interval. Symmetric intervals are convenient, but Example 3 demonstrates that we don't always get symmetric intervals without a bit of extra work.

EXAMPLE 3   Finding a symmetric interval Figure 2.63 shows the graph of g with $\lim_{x \to 2} g(x) = 3$. For each value of ε, find the corresponding values of $\delta > 0$ that satisfy the condition

$$|g(x) - 3| < \varepsilon \quad \text{whenever} \quad 0 < |x - 2| < \delta.$$

a. $\varepsilon = 2$ **b.** $\varepsilon = 1$

c. For any given value of ε, make a conjecture about the corresponding values of δ that satisfy the limit condition.

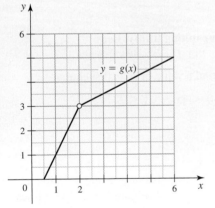

FIGURE 2.63

SOLUTION

a. With $\varepsilon = 2$, we need a value of $\delta > 0$ such that $g(x)$ is within 2 units of 3, which means between 1 and 5, whenever x is less than δ units from 2. The horizontal lines $y = 1$ and $y = 5$ intersect the graph of g at $x = 1$ and $x = 6$. Therefore $|g(x) - 3| < 2$ if x lies in the interval $(1, 6)$ with $x \neq 2$ (Figure 2.64a). However, we want x to lie in an interval that is *symmetric* about 2. We can guarantee that $|g(x) - 3| < 2$ only if x is less than 1 unit away from 2, on either side of 2 (Figure 2.64b). Therefore, with $\varepsilon = 2$ we take $\delta = 1$ or any smaller positive number.

b. With $\varepsilon = 1$, $g(x)$ must lie between 2 and 4 (Figure 2.65a). This implies that x must be within a half-unit to the left of 2 and within 2 units to the right of 2. Therefore, $|g(x) - 3| < 1$ provided x lies in the interval $(1.5, 4)$. To obtain a symmetric interval about 2, we take $\delta = \frac{1}{2}$ or any smaller positive number. Then we are guaranteed that $|g(x) - 3| < 1$ when $0 < |x - 2| < \frac{1}{2}$ (Figure 2.65b).

c. From parts (a) and (b), it appears that if we choose $\delta \leq \varepsilon/2$, the limit condition is satisfied for any $\varepsilon > 0$.

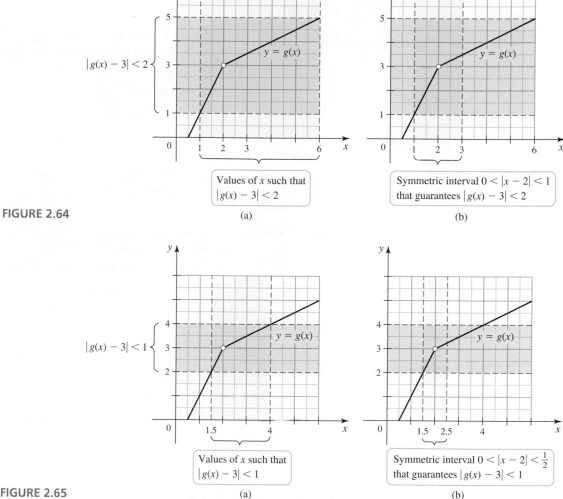

FIGURE 2.64

Values of x such that $|g(x) - 3| < 2$

(a)

Symmetric interval $0 < |x - 2| < 1$ that guarantees $|g(x) - 3| < 2$

(b)

FIGURE 2.65

Values of x such that $|g(x) - 3| < 1$

(a)

Symmetric interval $0 < |x - 2| < \frac{1}{2}$ that guarantees $|g(x) - 3| < 1$

(b)

Related Exercises 15–18 ◄

Limit Proofs

We use the following two-step process to prove that $\lim\limits_{x \to a} f(x) = L$.

> ➤ The first step of the limit-proving process is the preliminary work of finding a candidate for δ. The second step verifies that the δ found in the first step actually works.

Steps for proving that $\lim\limits_{x \to a} f(x) = L$

1. **Find δ.** Let ε be an arbitrary positive number. Use the inequality $|f(x) - L| < \varepsilon$ to find a condition of the form $|x - a| < \delta$, where δ depends only on the value of ε.

2. **Write a proof.** For any $\varepsilon > 0$, assume $0 < |x - a| < \delta$ and use the relationship between ε and δ found in Step 1 to prove that $|f(x) - L| < \varepsilon$.

EXAMPLE 4   **Limit of a linear function** Prove that $\lim\limits_{x \to 4} (4x - 15) = 1$ using the precise definition of a limit.

SOLUTION

Step 1:   *Find* δ. In this case, $a = 4$ and $L = 1$. Assuming $\varepsilon > 0$ is given, we use $|(4x - 15) - 1| < \varepsilon$ to find an inequality of the form $|x - 4| < \delta$. If $|(4x - 15) - 1| < \varepsilon$, then

$$|4x - 16| < \varepsilon$$
$$4|x - 4| < \varepsilon \quad \text{Factor } 4x - 16.$$
$$|x - 4| < \frac{\varepsilon}{4} \quad \text{Divide by 4 and identify } \delta = \varepsilon/4.$$

We have shown that $|(4x - 15) - 1| < \varepsilon$ implies $|x - 4| < \varepsilon/4$. Therefore, a plausible relationship between δ and ε is $\delta = \varepsilon/4$. We now write the actual proof.

Step 2:   *Write a proof.* Let $\varepsilon > 0$ be given and assume $0 < |x - 4| < \delta$ where $\delta = \varepsilon/4$. The aim is to show that $|(4x - 15) - 1| < \varepsilon$ for all x such that $0 < |x - 4| < \delta$. We simplify $|(4x - 15) - 1|$ and isolate the $|x - 4|$ term:

$$|(4x - 15) - 1| = |4x - 16|$$
$$= 4\underbrace{|x - 4|}_{\text{less than } \delta \, = \, \varepsilon/4}$$
$$< 4\left(\frac{\varepsilon}{4}\right) = \varepsilon$$

We have shown that for any $\varepsilon > 0$,

$$|f(x) - L| = |(4x - 15) - 1| < \varepsilon \quad \text{whenever} \quad 0 < |x - 4| < \delta$$

provided $0 < \delta \le \varepsilon/4$. Therefore, $\lim\limits_{x \to 4} (4x - 15) = 1$.

Related Exercises 19–24 ◄

Justifying Limit Laws

The precise definition of a limit is used to prove the limit laws in Theorem 2.3. Essential in several of these proofs is the triangle inequality, which states that

$$|x + y| \le |x| + |y| \quad \text{for all real numbers } x \text{ and } y.$$

EXAMPLE 5   **Proof of Limit Law 1** Prove that if $\lim\limits_{x \to a} f(x)$ and $\lim\limits_{x \to a} g(x)$ exist, then

$$\lim\limits_{x \to a} [f(x) + g(x)] = \lim\limits_{x \to a} f(x) + \lim\limits_{x \to a} g(x).$$

SOLUTION Assume that $\varepsilon > 0$ is given. Let $\lim\limits_{x \to a} f(x) = L$, which implies that there exists a $\delta_1 > 0$ such that

$$|f(x) - L| < \frac{\varepsilon}{2} \quad \text{whenever} \quad 0 < |x - a| < \delta_1.$$

Similarly, let $\lim\limits_{x \to a} g(x) = M$, which implies there exists a $\delta_2 > 0$ such that

$$|g(x) - M| < \frac{\varepsilon}{2} \quad \text{whenever} \quad 0 < |x - a| < \delta_2.$$

> The minimum value of a and b is denoted $\min\{a, b\}$. If $x = \min\{a, b\}$, then x is the smaller of a and b. If $a = b$, then x equals the common value of a and b. In either case, $x \leq a$ and $x \leq b$.

Let $\delta = \min\{\delta_1, \delta_2\}$ and suppose $0 < |x - a| < \delta$. Because $\delta \leq \delta_1$, it follows that $0 < |x - a| < \delta_1$ and $|f(x) - L| < \varepsilon/2$. Similarly, because $\delta \leq \delta_2$, it follows that $0 < |x - a| < \delta_2$ and $|g(x) - M| < \varepsilon/2$. Therefore,

$$|[f(x) + g(x)] - (L + M)| = |(f(x) - L) + (g(x) - M)| \quad \text{Rearrange terms.}$$
$$\leq |f(x) - L| + |g(x) - M| \quad \text{Triangle inequality}$$
$$< \frac{\varepsilon}{2} + \frac{\varepsilon}{2} = \varepsilon$$

> Proofs of other limit laws are outlined in Exercises 25–26.

We have shown that given any $\varepsilon > 0$, if $0 < |x - a| < \delta$ then $|[f(x) + g(x)] - (L + M)| < \varepsilon$, which implies that $\lim\limits_{x \to a} [f(x) + g(x)] = L + M = \lim\limits_{x \to a} f(x) + \lim\limits_{x \to a} g(x)$.

Related Exercises 25–28 ◀

Infinite Limits

> Notice that for infinite limits, N plays the role that ε plays for regular limits. It sets a tolerance or bound for the function values $f(x)$.

In Section 2.4, we stated that $\lim\limits_{x \to a} f(x) = \infty$ if $f(x)$ grows *arbitrarily large* as x approaches a. More precisely, this means that for any positive number N (no matter how large), $f(x)$ is larger than N if x is sufficiently close to a but not equal to a.

DEFINITION **Two-Sided Infinite Limit**

The **infinite limit** $\lim\limits_{x \to a} f(x) = \infty$ means that for any positive number N there exists a corresponding $\delta > 0$ such that

$$f(x) > N \quad \text{whenever} \quad 0 < |x - a| < \delta.$$

As shown in Figure 2.66, to prove that $\lim\limits_{x \to a} f(x) = \infty$, we let N represent *any* positive number. Then we find a value of $\delta > 0$, depending only on N, such that

$$f(x) > N \quad \text{whenever} \quad 0 < |x - a| < \delta.$$

This process is similar to the two-step process for finite limits.

> Precise definitions for $\lim\limits_{x \to a} f(x) = -\infty$, $\lim\limits_{x \to a^+} f(x) = -\infty$, $\lim\limits_{x \to a^+} f(x) = \infty$, $\lim\limits_{x \to a^-} f(x) = -\infty$, and $\lim\limits_{x \to a} f(x) = \infty$ are given in the exercises.

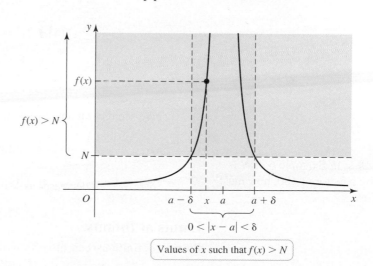

FIGURE 2.66

Values of x such that $f(x) > N$

Steps for proving that $\lim_{x \to a} f(x) = \infty$

1. **Find δ.** Let N be an arbitrary positive number. Use the statement $f(x) > N$ to find an inequality of the form $|x - a| < \delta$, where δ depends only on N.

2. **Write a proof.** For any $N > 0$, assume $0 < |x - a| < \delta$ and use the relationship between N and δ found in Step 1 to prove that $f(x) > N$.

EXAMPLE 6 **An Infinite Limit Proof** Let $f(x) = \dfrac{1}{(x - 2)^2}$. Prove that

$$\lim_{x \to 2} f(x) = \infty.$$

SOLUTION

Step 1: *Find $\delta > 0$.* Assuming $N > 0$, we use the inequality $\dfrac{1}{(x - 2)^2} > N$ to find δ, where δ depends only on N. Taking reciprocals of this inequality, it follows that

$$(x - 2)^2 < \frac{1}{N}$$

> Recall that $\sqrt{x^2} = |x|$.

$$|x - 2| < \frac{1}{\sqrt{N}} \quad \text{Take the square root of both sides.}$$

The inequality $|x - 2| < \dfrac{1}{\sqrt{N}}$ has the form $|x - 2| < \delta$ if we let $\delta = \dfrac{1}{\sqrt{N}}$. We now write a proof based on this relationship between δ and N.

Step 2: *Write a proof.* Suppose $N > 0$ is given. Let $\delta = \dfrac{1}{\sqrt{N}}$ and assume $0 < |x - 2| < \delta = \dfrac{1}{\sqrt{N}}$. Squaring both sides of the inequality $|x - 2| < \dfrac{1}{\sqrt{N}}$ and taking reciprocals, we have

$$(x - 2)^2 < \frac{1}{N} \quad \text{Square both sides.}$$

$$\frac{1}{(x - 2)^2} > N \quad \text{Take reciprocals of both sides.}$$

We see that for any positive N, if $0 < |x - 2| < \delta = \dfrac{1}{\sqrt{N}}$, then

$$f(x) = \frac{1}{(x - 2)^2} > N.$$ It follows that $\lim_{x \to 2} \dfrac{1}{(x - 2)^2} = \infty$. Note that because

QUICK CHECK 3 In Example 6, if N is increased by a factor of 100, how must δ change?

$\delta = \dfrac{1}{\sqrt{N}}$, δ decreases as N increases.   *Related Exercises 29–32* ◄

Limits at Infinity

Precise definitions can also be written for the limits at infinity $\lim_{x \to \infty} f(x) = L$ and $\lim_{x \to -\infty} f(x) = L$. For discussion and examples, see Exercises 52–53.

SECTION 2.7 EXERCISES

Review Questions

1. Suppose x lies in the interval $(1, 3)$ with $x \neq 2$. Find the smallest positive value of δ such that the inequality $0 < |x - 2| < \delta$ is true.

2. Suppose $f(x)$ lies in the interval $(2, 6)$. What is the smallest value of ε such that $|f(x) - 4| < \varepsilon$?

3. Which one of the following intervals is not symmetric about $x = 5$?

    a. $(1, 9)$ **b.** $(4, 6)$ **c.** $(3, 8)$ **d.** $(4.5, 5.5)$

4. Does the set $\{x: 0 < |x - a| < \delta\}$ include the point $x = a$? Explain.

5. State the precise definition of $\lim\limits_{x \to a} f(x) = L$.

6. Interpret $|f(x) - L| < \varepsilon$ in words.

7. Suppose $|f(x) - 5| < 0.1$ whenever $0 < x < 5$. Find all values of $\delta > 0$ such that $|f(x) - 5| < 0.1$ whenever $0 < |x - 2| < \delta$.

8. Give the definition of $\lim\limits_{x \to a} f(x) = \infty$ and interpret it using pictures.

Basic Skills

9. **Determining values of δ from a graph** The function f in the figure satisfies $\lim\limits_{x \to 2} f(x) = 5$. Determine a value of $\delta > 0$ satisfying each statement.

    a. If $0 < |x - 2| < \delta$, then $|f(x) - 5| < 2$.

    b. If $0 < |x - 2| < \delta$, then $|f(x) - 5| < 1$.

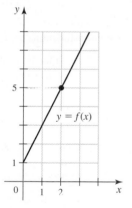

10. **Determining values of δ from a graph** The function f in the figure satisfies $\lim\limits_{x \to 2} f(x) = 4$. Determine a value of $\delta > 0$ satisfying each statement.

    a. If $0 < |x - 2| < \delta$, then $|f(x) - 4| < 1$.

    b. If $0 < |x - 2| < \delta$, then $|f(x) - 4| < 1/2$.

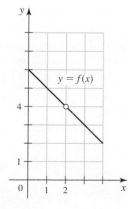

11. **Determining values of δ from a graph** The function f in the figure satisfies $\lim\limits_{x \to 3} f(x) = 6$. Determine a value of $\delta > 0$ satisfying each statement.

    a. If $0 < |x - 3| < \delta$, then $|f(x) - 6| < 3$.

    b. If $0 < |x - 3| < \delta$, then $|f(x) - 6| < 1$.

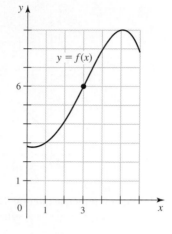

12. **Determining values of δ from a graph** The function f in the figure satisfies $\lim\limits_{x \to 4} f(x) = 5$. Determine a value of $\delta > 0$ satisfying each statement.

    a. If $0 < |x - 4| < \delta$, then $|f(x) - 5| < 1$.

    b. If $0 < |x - 4| < \delta$, then $|f(x) - 5| < 0.5$.

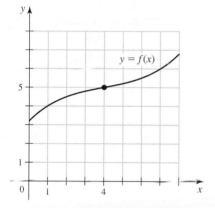

13. **Finding δ for a given ε using a graph** Let $f(x) = x^3 + 3$ and note that $\lim\limits_{x \to 0} f(x) = 3$. For each value of ε, use a graphing utility to find a value of $\delta > 0$ such that $|f(x) - 3| < \varepsilon$ whenever $0 < |x - 0| < \delta$. Sketch graphs illustrating your work.

    a. $\varepsilon = 1$ **b.** $\varepsilon = 0.5$

14. **Finding δ for a given ε using a graph** Let $g(x) = 2x^3 - 12x^2 + 26x + 4$ and note that $\lim\limits_{x \to 2} g(x) = 24$. For each value of ε, use a graphing utility to find a value of $\delta > 0$ such that $|g(x) - 24| < \varepsilon$ whenever $0 < |x - 2| < \delta$. Sketch graphs illustrating your work.

    a. $\varepsilon = 1$ **b.** $\varepsilon = 0.5$

15. **Finding a symmetric interval** The function f in the figure satisfies $\lim\limits_{x \to 2} f(x) = 3$. For each value of ε, find a value of $\delta > 0$ such that

$$|f(x) - 3| < \varepsilon \quad \text{whenever} \quad 0 < |x - 2| < \delta. \tag{2}$$

a. $\varepsilon = 1$ **b.** $\varepsilon = \frac{1}{2}$

c. For any given value of ε, make a conjecture about the corresponding value of δ satisfying (2).

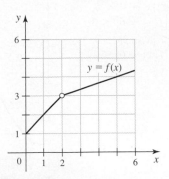

16. **Finding a symmetric interval** The function f in the figure satisfies $\lim_{x \to 3} f(x) = 4$. For each value of ε, find a value of $\delta > 0$ such that

$$|f(x) - 4| < \varepsilon \quad \text{whenever} \quad 0 < |x - 3| < \delta. \quad (3)$$

a. $\varepsilon = 2$ **b.** $\varepsilon = \frac{1}{2}$

c. For any given value of ε, make a conjecture about the corresponding value of δ satisfying (3).

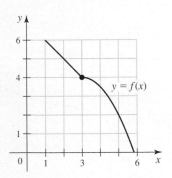

17. **Finding a symmetric interval** Let $f(x) = x^2$ and note that $\lim_{x \to 2} f(x) = 4$. For each value of ε, use a graphing utility to find a value of $\delta > 0$ such that $|f(x) - 4| < \varepsilon$ whenever $0 < |x - 2| < \delta$.

a. $\varepsilon = 1$ **b.** $\varepsilon = 0.5$

c. For any given value of ε, make a conjecture about the value of δ that satisfies the preceding inequality.

18. **Finding a symmetric interval** Let $f(x) = \dfrac{x^3 - 1}{x - 1}$ and note that $\lim_{x \to 1} f(x) = 3$. For each value of ε, use a graphing utility to find a value of $\delta > 0$ such that $|f(x) - 3| < \varepsilon$ whenever $0 < |x - 1| < \delta$.

a. $\varepsilon = 1.5$ **b.** $\varepsilon = 0.75$

c. For any given value of ε, make a conjecture about the value of δ that satisfies the preceding inequality.

19–24. **Limit proofs** *Use the precise definition of a limit to prove the following limits.*

19. $\lim_{x \to 1} (8x + 5) = 13$ **20.** $\lim_{x \to 3} (-2x + 8) = 2$

21. $\lim_{x \to 4} \dfrac{x^2 - 16}{x - 4} = 8$ (*Hint:* Factor and simplify.)

22. $\lim_{x \to 3} \dfrac{x^2 - 7x + 12}{x - 3} = -1$

23. $\lim_{x \to 0} x^2 = 0$ (*Hint:* Use the identity $\sqrt{x^2} = |x|$.)

24. $\lim_{x \to 3} (x - 3)^2 = 0$ (*Hint:* Use the identity $\sqrt{x^2} = |x|$.)

25. **Proof of Limit Law 2** Suppose $\lim_{x \to a} f(x) = L$ and $\lim_{x \to a} g(x) = M$. Prove that $\lim_{x \to a} [f(x) - g(x)] = L - M$.

26. **Proof of Limit Law 3** Suppose $\lim_{x \to a} f(x) = L$. Prove that $\lim_{x \to a} [cf(x)] = cL$, where c is a constant.

27. **Limit of a constant function and $f(x) = x$** Give proofs of the following theorems.

a. $\lim_{x \to a} c = c$ for any constant c

b. $\lim_{x \to a} x = a$ for any constant a

28. **Continuity of linear functions** Prove Theorem 2.2: If $f(x) = mx + b$, then $\lim_{x \to a} f(x) = ma + b$ for constants m and b. (*Hint:* For a given $\varepsilon > 0$, let $\delta = \varepsilon/|m|$.) Explain why this result implies that linear functions are continuous.

29–32. **Limit proofs for infinite limits** *Use the precise definition of infinite limits to prove the following limits.*

29. $\lim_{x \to 4} \dfrac{1}{(x - 4)^2} = \infty$ **30.** $\lim_{x \to -1} \dfrac{1}{(x + 1)^4} = \infty$

31. $\lim_{x \to 0} \left(\dfrac{1}{x^2} + 1 \right) = \infty$ **32.** $\lim_{x \to 0} \left(\dfrac{1}{x^4} - \sin x \right) = \infty$

Further Explorations

33. **Explain why or why not** Determine whether the following statements are true and give an explanation or counterexample. Assume a and L are finite numbers and assume $\lim_{x \to a} f(x) = L$.

a. For a given $\varepsilon > 0$, there is one value of $\delta > 0$ for which $|f(x) - L| < \varepsilon$ whenever $0 < |x - a| < \delta$.

b. The limit $\lim_{x \to a} f(x) = L$ means that given an arbitrary $\delta > 0$, we can always find an $\varepsilon > 0$ such that $|f(x) - L| < \varepsilon$ whenever $0 < |x - a| < \delta$.

c. The limit $\lim_{x \to a} f(x) = L$ means that for any arbitrary $\varepsilon > 0$, we can always find a $\delta > 0$ such that $|f(x) - L| < \varepsilon$ whenever $0 < |x - a| < \delta$.

d. If $|x - a| < \delta$, then $a - \delta < x < a + \delta$.

34. Finding δ algebraically Let $f(x) = x^2 - 2x + 3$.

    a. For $\varepsilon = 0.25$, find a corresponding value of $\delta > 0$ satisfying the statement

$$|f(x) - 2| < \varepsilon \quad \text{whenever} \quad 0 < |x - 1| < \delta.$$

    b. Verify that $\lim_{x \to 1} f(x) = 2$ by doing the following. For any $\varepsilon > 0$, find a corresponding value of $\delta > 0$ satisfying the statement

$$|f(x) - 2| < \varepsilon \quad \text{whenever} \quad 0 < |x - 1| < \delta.$$

35–38. Challenging limit proofs *Use the definition of a limit to prove the following results.*

35. $\lim\limits_{x \to 3} \dfrac{1}{x} = \dfrac{1}{3}$ (*Hint:* As $x \to 3$, eventually the distance between x and 3 will be less than 1. Start by assuming $|x - 3| < 1$ and show $\dfrac{1}{|x|} < \dfrac{1}{2}$.)

36. $\lim\limits_{x \to 4} \dfrac{x - 4}{\sqrt{x} - 2} = 4$ (*Hint:* Multiply the numerator and denominator by $\sqrt{x} + 2$.)

37. $\lim\limits_{x \to 1/10} \dfrac{1}{x} = 10$ (*Hint:* To find δ, you will need to bound x away from 0. So let $\left| x - \dfrac{1}{10} \right| < \dfrac{1}{20}$.)

38. $\lim\limits_{x \to 5} \dfrac{1}{x^2} = \dfrac{1}{25}$

39–43. Precise definitions for left- and right-hand limits *Use the following definitions.*

    Assume f exists for all x near a with x > a. We say that the **limit of $f(x)$ as x approaches a from the right of a is L** *and write* $\lim\limits_{x \to a^+} f(x) = L$, *if for any $\varepsilon > 0$ there exists $\delta > 0$ such that*

$$|f(x) - L| < \varepsilon \text{ whenever } 0 < x - a < \delta.$$

    Assume f exists for all values of x near a with x < a. We say that the **limit of $f(x)$ as x approaches a from the left of a is L** *and write* $\lim\limits_{x \to a^-} f(x) = L$, *if for any $\varepsilon > 0$ there exists $\delta > 0$ such that*

$$|f(x) - L| < \varepsilon \text{ whenever } 0 < a - x < \delta.$$

39. Comparing definitions Why is the last inequality in the definition of $\lim\limits_{x \to a} f(x) = L$, namely, $0 < |x - a| < \delta$, replaced with $0 < x - a < \delta$ in the definition of $\lim\limits_{x \to a^+} f(x) = L$?

40. Comparing definitions Why is the last inequality in the definition of $\lim\limits_{x \to a} f(x) = L$, namely, $0 < |x - a| < \delta$, replaced with $0 < a - x < \delta$ in the definition of $\lim\limits_{x \to a^-} f(x) = L$?

41. One-sided limit proofs Prove the following limits for

$$f(x) = \begin{cases} 3x - 4 & \text{if } x < 0 \\ 2x - 4 & \text{if } x \ge 0 \end{cases}$$

    a. $\lim\limits_{x \to 0^+} f(x) = -4$ **b.** $\lim\limits_{x \to 0^-} f(x) = -4$

    c. $\lim\limits_{x \to 0} f(x) = -4$

42. Determining values of δ from a graph The function f in the figure satisfies $\lim\limits_{x \to 2^+} f(x) = 0$ and $\lim\limits_{x \to 2^-} f(x) = 1$. Determine a value of $\delta > 0$ satisfying each statement.

    a. $|f(x) - 0| < 2$ whenever $0 < x - 2 < \delta$
    b. $|f(x) - 0| < 1$ whenever $0 < x - 2 < \delta$
    c. $|f(x) - 1| < 2$ whenever $0 < 2 - x < \delta$
    d. $|f(x) - 1| < 1$ whenever $0 < 2 - x < \delta$

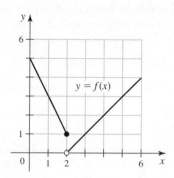

43. One-sided limit proof Prove that $\lim\limits_{x \to 0^+} \sqrt{x} = 0$.

Additional Exercises

44. The relationship between one-sided and two-sided infinite limits Prove the following statements to establish the fact that $\lim\limits_{x \to a} f(x) = L$ if and only if $\lim\limits_{x \to a^-} f(x) = L$ and $\lim\limits_{x \to a^+} f(x) = L$.

    a. If $\lim\limits_{x \to a^-} f(x) = L$ and $\lim\limits_{x \to a^+} f(x) = L$, then $\lim\limits_{x \to a} f(x) = L$.
    b. If $\lim\limits_{x \to a} f(x) = L$, then $\lim\limits_{x \to a^-} f(x) = L$ and $\lim\limits_{x \to a^+} f(x) = L$.

45. Definition of one-sided infinite limits We say that $\lim\limits_{x \to a^+} f(x) = -\infty$ if for each negative number N, there exists $\delta > 0$ such that

$$f(x) < N \quad \text{whenever} \quad a < x < a + \delta.$$

    a. Write an analogous formal definition for $\lim\limits_{x \to a^+} f(x) = \infty$.
    b. Write an analogous formal definition for $\lim\limits_{x \to a^-} f(x) = -\infty$.
    c. Write an analogous formal definition for $\lim\limits_{x \to a^-} f(x) = \infty$.

46–47. One-sided infinite limits *Use the definitions given in Exercise 45 to prove the following infinite limits.*

46. $\lim\limits_{x \to 1^+} \dfrac{1}{1 - x} = -\infty$ **47.** $\lim\limits_{x \to 1^-} \dfrac{1}{1 - x} = \infty$

48–49. Definition of an infinite limit *We write* $\lim\limits_{x \to a} f(x) = -\infty$ *if for any negative number M there exists a $\delta > 0$ such that*

$$f(x) < M \quad \text{whenever} \quad 0 < |x - a| < \delta.$$

Use this definition to prove the following statements.

48. $\lim\limits_{x \to 1} \dfrac{-2}{(x - 1)^2} = -\infty$ **49.** $\lim\limits_{x \to -2} \dfrac{-10}{(x + 2)^4} = -\infty$

50–51. Definition of a limit at infinity *The limit at infinity*
$\lim\limits_{x\to\infty} f(x) = L$ *means that for any $\varepsilon > 0$, there exists $N > 0$ such that*

$$|f(x) - L| < \varepsilon \quad \text{whenever} \quad x > N.$$

Use this definition to prove the following statements.

50. $\lim\limits_{x\to\infty} \dfrac{10}{x} = 0$

51. $\lim\limits_{x\to\infty} \dfrac{2x + 1}{x} = 2$

52–53. Definition of infinite limits at infinity *We say that*
$\lim\limits_{x\to\infty} f(x) = \infty$ *if for each positive number M, there is a corresponding $N > 0$ such that*

$$f(x) > M \quad \text{whenever} \quad x > N.$$

Use this definition to prove the following statements.

52. $\lim\limits_{x\to\infty} \dfrac{x}{100} = \infty$

53. $\lim\limits_{x\to\infty} \dfrac{x^2 + x}{x} = \infty$

54. Proof of the Squeeze Theorem Assume the functions f, g, and h satisfy the inequality $f(x) \le g(x) \le h(x)$ for all values of x near a, except possibly at a. Prove that if $\lim\limits_{x\to a} f(x) = \lim\limits_{x\to a} h(x) = L$, then $\lim\limits_{x\to a} g(x) = L$.

55. Limit proof Suppose f is defined for all values of x near a, except possibly at a. Assume for every integer $N > 0$ there is another integer $M > 0$ such that $|f(x) - L| < 1/N$ whenever $|x - a| < 1/M$. Prove that $\lim\limits_{x\to a} f(x) = L$ using the precise definition of a limit.

56–58. Proving that $\lim\limits_{x\to a} f(x) \neq L$ *Use the following definition for the nonexistence of a limit. Assume f is defined for all values of x near a, except possibly at a. We say that $\lim\limits_{x\to a} f(x) \neq L$ if for some $\varepsilon > 0$ there is no value of $\delta > 0$ satisfying the condition*

$$|f(x) - L| < \varepsilon \quad \text{whenever} \quad 0 < |x - a| < \delta.$$

56. For the following function, note that $\lim\limits_{x\to 2} f(x) \neq 3$. Find a value of $\varepsilon > 0$ for which the preceding condition for nonexistence is satisfied.

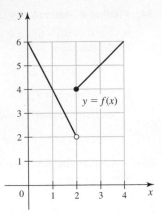

57. Prove that $\lim\limits_{x\to 0} \dfrac{|x|}{x}$ does not exist.

58. Let

$$f(x) = \begin{cases} 0 & \text{if } x \text{ is rational} \\ 1 & \text{if } x \text{ is irrational} \end{cases}$$

Prove that $\lim\limits_{x\to a} f(x)$ does not exist for any value of a. (*Hint:* Assume $\lim\limits_{x\to a} f(x) = L$ for some values of a and L and let $\varepsilon = \frac{1}{2}$.)

59. A continuity proof Suppose f is continuous at a and assume $f(a) > 0$. Show that there is a positive number $\delta > 0$ for which $f(x) > 0$ for all values of x in $(a - \delta, a + \delta)$. (In other words, f is positive for all values of x sufficiently close to a.)

QUICK CHECK ANSWERS

1. $\delta = \frac{1}{50}$ or smaller **2.** $\delta = 0.62$ or smaller **3.** δ must decrease by a factor of $\sqrt{100} = 10$ (at least).

CHAPTER 2 REVIEW EXERCISES

1. Explain why or why not Determine whether the following statements are true and give an explanation or counterexample.

  a. The rational function $\dfrac{x - 1}{x^2 - 1}$ has vertical asymptotes at $x = -1$ and $x = 1$.

  b. Numerical or graphical methods always produce good estimates of $\lim\limits_{x\to a} f(x)$.

  c. The value of $\lim\limits_{x\to a} f(x)$, if it exists, is found by calculating $f(a)$.

  d. If $\lim\limits_{x\to a} f(x) = \infty$ or $\lim\limits_{x\to a} f(x) = -\infty$, then $\lim\limits_{x\to a} f(x)$ does not exist.

  e. If $\lim\limits_{x\to a} f(x)$ does not exist, then either $\lim\limits_{x\to a} f(x) = \infty$ or $\lim\limits_{x\to a} f(x) = -\infty$.

  f. If a function is continuous on the intervals (a, b) and $[b, c)$, where $a < b < c$, then the function is also continuous on (a, c).

  g. If $\lim\limits_{x\to a} f(x)$ can be calculated by direct substitution, then f is continuous at $x = a$.

2.   Estimating limits graphically Use the graph of f in the figure to find the following values, if possible.

a. $f(-1)$ b. $\lim\limits_{x\to-1^-} f(x)$ c. $\lim\limits_{x\to-1^+} f(x)$ d. $\lim\limits_{x\to-1} f(x)$

e. $f(1)$ f. $\lim\limits_{x\to1} f(x)$ g. $\lim\limits_{x\to2} f(x)$ h. $\lim\limits_{x\to3^-} f(x)$

i. $\lim\limits_{x\to3^+} f(x)$ j. $\lim\limits_{x\to3} f(x)$

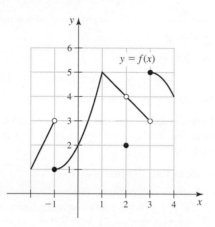

3.   Points of discontinuity Use the graph of f in the figure to determine the values of x in the interval $(-3, 5)$ at which f fails to be continuous. Justify your answers using the continuity checklist.

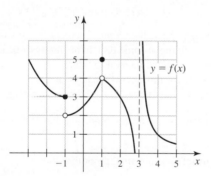

4.   Computing a limit graphically and analytically

a.   Graph $y = \dfrac{\sin 2\theta}{\sin \theta}$. Comment on any inaccuracies in the graph and then sketch an accurate graph of the function.

b.   Estimate $\lim\limits_{\theta\to0} \dfrac{\sin 2\theta}{\sin \theta}$ using the graph in part (a).

c.   Verify your answer to part (b) by finding the value of $\lim\limits_{\theta\to0} \dfrac{\sin 2\theta}{\sin \theta}$ analytically using the trigonometric identity $\sin 2\theta = 2 \sin \theta \cos \theta$.

5.   Computing a limit numerically and analytically

a.   Estimate $\lim\limits_{x\to\pi/4} \dfrac{\cos 2x}{\cos x - \sin x}$ by making a table of values of $\dfrac{\cos 2x}{\cos x - \sin x}$ for values of x approaching $\pi/4$. Round your estimate to four digits.

b.   Use analytic methods to find the value of $\lim\limits_{x\to\pi/4} \dfrac{\cos 2x}{\cos x - \sin x}$.

6.   Long-distance phone calls Suppose a long-distance phone call costs \$0.75 for the first min (or any part of the first min), plus \$0.10 for each additional min (or any part of a min).

a.   Graph the function $c = f(t)$ that gives the cost for talking on the phone for t minutes for $0 \le t \le 5$.

b.   Evaluate $\lim\limits_{t\to2.9} f(t)$.

c.   Evaluate $\lim\limits_{t\to3^-} f(t)$ and $\lim\limits_{t\to3^+} f(t)$.

d.   Interpret the meaning of the limits in part (c).

e.   For what values of t is f continuous? Explain.

7.   Sketching a graph Sketch the graph of a function f with all the following properties.

$$\lim_{x\to-2^-} f(x) = \infty \qquad \lim_{x\to-2^+} f(x) = -\infty \qquad \lim_{x\to0} f(x) = \infty$$

$$\lim_{x\to3^-} f(x) = 2 \qquad \lim_{x\to3^+} f(x) = 4 \qquad f(3) = 1$$

8–21.   Calculating limits *Calculate the following limits analytically.*

8. $\lim\limits_{x\to1000} 18\pi^2$

9. $\lim\limits_{x\to1} \sqrt{5x + 6}$

10. $\lim\limits_{h\to0} \dfrac{\sqrt{5x + 5h} - \sqrt{5x}}{h}$, where x is constant

11. $\lim\limits_{x\to1} \dfrac{x^3 - 7x^2 + 12x}{4 - x}$

12. $\lim\limits_{x\to4} \dfrac{x^3 - 7x^2 + 12x}{4 - x}$

13. $\lim\limits_{x\to1} \dfrac{1 - x^2}{x^2 - 8x + 7}$

14. $\lim\limits_{x\to3} \dfrac{\sqrt{3x + 16} - 5}{x - 3}$

15. $\lim\limits_{x\to3} \dfrac{1}{x - 3}\left(\dfrac{1}{\sqrt{x + 1}} - \dfrac{1}{2}\right)$

16. $\lim\limits_{t\to1/3} \dfrac{t - 1/3}{(3t - 1)^2}$

17. $\lim\limits_{x\to3} \dfrac{x^4 - 81}{x - 3}$

18. $\lim\limits_{p\to1} \dfrac{p^5 - 1}{p - 1}$

19. $\lim\limits_{x\to81} \dfrac{\sqrt[4]{x} - 3}{x - 81}$

20. $\lim\limits_{\theta\to\pi/4} \dfrac{\sin^2\theta - \cos^2\theta}{\sin \theta - \cos \theta}$

21. $\lim\limits_{x\to\pi/2} \dfrac{\dfrac{1}{\sqrt{\sin x}} - 1}{x + \pi/2}$

22.   **One-sided limits** Evaluate $\lim\limits_{x\to1^+} \sqrt{\dfrac{x - 1}{x - 3}}$ and $\lim\limits_{x\to1^-} \sqrt{\dfrac{x - 1}{x - 3}}$.

23.   **Applying the Squeeze Theorem**

a.   Use a graphing utility to illustrate the inequalities

$$\cos x \le \dfrac{\sin x}{x} \le \dfrac{1}{\cos x}$$

on $[-1, 1]$.

b.   Use part (a) and the Squeeze Theorem to explain why

$$\lim_{x\to0} \dfrac{\sin x}{x} = 1.$$

24.   **Applying the Squeeze Theorem** Assume the function g satisfies the inequality $1 \le g(x) \le \sin^2 x + 1$ for x near 0. Use the Squeeze Theorem to find $\lim\limits_{x\to0} g(x)$.

25–29.   Finding infinite limits *Determine the following infinite limits, if possible.*

25. $\lim\limits_{x\to5} \dfrac{x - 7}{x(x - 5)^2}$

26. $\lim\limits_{x\to-5^+} \dfrac{x - 5}{x + 5}$

27. $\lim\limits_{x\to3^-} \dfrac{x - 4}{x^2 - 3x}$

28. $\lim\limits_{u \to 0^+} \dfrac{u - 1}{\sin u}$ **29.** $\lim\limits_{x \to 0^-} \dfrac{2}{\tan x}$

T 30. Finding vertical asymptotes Let $f(x) = \dfrac{x^2 - 5x + 6}{x^2 - 2x}$.

a. Calculate $\lim\limits_{x \to 0^-} f(x)$, $\lim\limits_{x \to 0^+} f(x)$, $\lim\limits_{x \to 2^-} f(x)$, and $\lim\limits_{x \to 2^+} f(x)$.
b. Does the graph of f have any vertical asymptotes? Explain.
c. Graph f and then sketch the graph with paper and pencil, correcting any errors obtained with the graphing utility.

31–36. Limits at infinity *Evaluate the following limits.*

31. $\lim\limits_{x \to \infty} \dfrac{2x - 3}{4x + 10}$ **32.** $\lim\limits_{x \to \infty} \dfrac{x^4 - 1}{x^5 + 2}$

33. $\lim\limits_{x \to -\infty} (-3x^3 + 5)$ **34.** $\lim\limits_{z \to \infty} \left(e^{-2z} + \dfrac{2}{z} \right)$

35. $\lim\limits_{x \to \infty} (3 \tan^{-1} x + 2)$ **36.** $\lim\limits_{r \to \infty} \dfrac{1}{\ln r + 1}$

37–40. End behavior *Determine the end behavior of the following functions.*

37. $f(x) = \dfrac{4x^3 + 1}{1 - x^3}$ **38.** $f(x) = \dfrac{x + 1}{\sqrt{9x^2 + x}}$

39. $f(x) = 1 - e^{-2x}$ **40.** $f(x) = \dfrac{1}{\ln x^2}$

41–42. Vertical and horizontal asymptotes *Find all vertical and horizontal asymptotes of the following functions.*

41. $f(x) = \dfrac{1}{\tan^{-1} x}$ **42.** $f(x) = \dfrac{2x^2 + 6}{2x^2 + 3x - 2}$

43–46. Continuity at a point *Determine whether the following functions are continuous at $x = a$ using the continuity checklist to justify your answers.*

43. $f(x) = \dfrac{1}{x - 5};\ \ a = 5$

44. $g(x) = \begin{cases} \dfrac{x^2 - 16}{x - 4} & \text{if } x \neq 4 \\ 9 & \text{if } x = 4 \end{cases};\ a = 4$

45. $h(x) = \sqrt{x^2 - 9};\ \ a = 3$

46. $g(x) = \begin{cases} \dfrac{x^2 - 16}{x - 4} & \text{if } x \neq 4 \\ 8 & \text{if } x = 4 \end{cases};\ a = 4$

47–50. Continuity on intervals *Find the intervals on which the following functions are continuous. Specify right or left continuity at the endpoints.*

47. $f(x) = \sqrt{x^2 - 5}$ **48.** $g(x) = e^{\sqrt{x-2}}$

49. $h(x) = \dfrac{2x}{x^3 - 25x}$ **50.** $g(x) = \cos(e^x)$

51. Determining unknown constants Let

$$g(x) = \begin{cases} 5x - 2 & \text{if } x < 1 \\ a & \text{if } x = 1 \\ ax^2 + bx & \text{if } x \geq 1 \end{cases}$$

Determine values of the constants a and b for which g is continuous at $x = 1$.

52. Left and right continuity
a. Is $h(x) = \sqrt{x^2 - 9}$ left-continuous at $x = 3$? Explain.
b. Is $h(x) = \sqrt{x^2 - 9}$ right-continuous at $x = 3$? Explain.

53. Sketching a graph Sketch the graph of a function that is continuous on $(0, 1]$ and continuous on $(1, 2)$ but is not continuous on $(0, 2)$.

T 54. Intermediate Value Theorem
a. Use the Intermediate Value Theorem to show that the equation $x^5 + 7x + 5 = 0$ has a solution in the interval $(-1, 0)$.
b. Find a solution to $x^5 + 7x + 5 = 0$ in $(-1, 0)$ using a root finder.

T 55. Antibiotic dosing The amount of an antibiotic (in mg) in the blood t hours after an intravenous line is opened is given by

$$m(t) = 100(e^{-0.1t} - e^{-0.3t}).$$

a. Use the Intermediate Value Theorem to show the amount of drug is 30 mg at some time in the interval $[0, 5]$ and again at some time in the interval $[5, 15]$.
b. Estimate the times at which $m = 30$ mg.
c. Is the amount of drug in the blood ever 50 mg?

56. Limit proof Give a formal proof that $\lim\limits_{x \to 1} (5x - 2) = 3$.

57. Limit proof Give a formal proof that $\lim\limits_{x \to 5} \dfrac{x^2 - 25}{x - 5} = 10$.

58. Limit proofs
a. Assume $|f(x)| \leq L$ for all x near a and $\lim\limits_{x \to a} g(x) = 0$. Give a formal proof that $\lim\limits_{x \to a} [f(x)g(x)] = 0$.
b. Find a function f for which $\lim\limits_{x \to 2} [f(x)(x - 2)] \neq 0$. Why doesn't this violate the result stated in (a)?
c. The Heaviside function is defined as

$$H(x) = \begin{cases} 0 & \text{if } x < 0 \\ 1 & \text{if } x \geq 0 \end{cases}$$

Explain why $\lim\limits_{x \to 0} [xH(x)] = 0$.

59. Infinite limit proof Give a formal proof that $\lim\limits_{x \to 2} \dfrac{1}{(x - 2)^4} = \infty$.

Chapter 2 Guided Projects

Applications of the material in this chapter and related topics can be found in the following Guided Projects. For additional information, see the Preface.

• Fixed-point iteration

• Local linearity

3

Derivatives

Chapter Preview   Now that you are familiar with limits, the door to calculus stands open. The first task is to introduce the fundamental concept of the *derivative*. Suppose a function f represents a quantity of interest, say the variable cost of manufacturing an item, the population of a country, or the position of an orbiting satellite. The derivative of f is another function, denoted f', which gives the changing slope of the curve $y = f(x)$. Equivalently, the derivative of f gives the *instantaneous rate of change* of f at points in the domain. We use limits not only to define the derivative, but also to develop efficient rules for finding derivatives. The applications of the derivative—which we introduce along the way—are endless because almost everything around us is in a state of change, and derivatives describe change.

3.1 Introducing the Derivative

In this section we return to the problem of finding the slope of a line tangent to a curve, introduced at the beginning of Chapter 2. This concept is important for several reasons.

- We identify the slope of the tangent line with the *instantaneous rate of change* of a function (Figure 3.1).

- The slopes of the tangent lines as they change along a curve are the values of a new function called the *derivative*.

- If a curve represents the trajectory of a moving object, the line tangent to the curve at a point gives the direction of motion at that point (Figure 3.2).

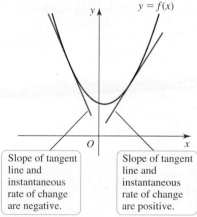

FIGURE 3.1

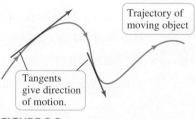

FIGURE 3.2

In Section 2.1 we gave an intuitive definition of a tangent line and used numerical evidence to estimate its slope. We now make these ideas precise.

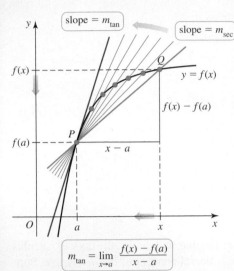

$$m_{tan} = \lim_{x \to a} \frac{f(x) - f(a)}{x - a}$$

FIGURE 3.3

➤ Figure 3.3 assumes $x > a$. Analogous pictures and arguments apply if $x < a$.

Tangent Lines and Rates of Change

Consider the curve $y = f(x)$ and a secant line intersecting the curve at the points $P(a, f(a))$ and $Q(x, f(x))$ (Figure 3.3). The difference $f(x) - f(a)$ is the change in the value of f on the interval $[a, x]$, while $x - a$ is the change in x. As discussed in Chapter 2, the slope of the secant line $\overleftrightarrow{PQ}$ is

$$m_{sec} = \frac{f(x) - f(a)}{x - a}$$

and it gives the **average rate of change** of f on the interval $[a, x]$.

Figure 3.3 also shows what happens as the variable point x approaches the fixed point a. Under suitable conditions, the slopes m_{sec} of the secant lines approach a unique number m_{tan} that we call the *slope of the tangent line*; that is,

$$m_{tan} = \lim_{x \to a} \frac{f(x) - f(a)}{x - a}.$$

The secant lines themselves approach a unique line that intersects the curve at P with slope m_{tan}; this line is the *tangent line at a*. The slope of the tangent line is also referred to as the **instantaneous rate of change** of f at a because it measures how quickly f changes at a. We summarize these observations as follows.

DEFINITION   Rates of Change and the Tangent Line

The **average rate of change** in f on the interval $[a, x]$ is the slope of the corresponding secant line:

$$m_{sec} = \frac{f(x) - f(a)}{x - a}$$

The **instantaneous rate of change** in f at $x = a$ is

$$m_{tan} = \lim_{x \to a} \frac{f(x) - f(a)}{x - a}, \tag{1}$$

which is also the **slope of the tangent line** at $x = a$, provided this limit exists. The **tangent line** at $x = a$ is the unique line through $(a, f(a))$ with slope m_{tan}. Its equation is

$$y - f(a) = m_{tan}(x - a).$$

QUICK CHECK 1 Sketch the graph of a function f near a point a. As in Figure 3.3, draw a secant line that passes through $(a, f(a))$ and a neighboring point $(x, f(x))$ with $x < a$. Use arrows to show how the secant lines approach the tangent line as x approaches a. ◄

➤ If x and y have physical units, then the average and instantaneous rates of change have units of (units of y)/ (units of x). For example, if y has units of m and x has units of s, the units of the rates of change are m/s.

EXAMPLE 1   Equation of a tangent line Let $f(x) = -16x^2 + 96x$ (the position function considered in Section 2.1) and consider the point $P(1, 80)$ on the curve.

a. Find the slope of the line tangent to the graph of f at P.

b. Find an equation of the tangent line in part (a).

SOLUTION

a. We use the definition of the slope of the tangent line with $a = 1$:

$$
\begin{aligned}
m_{tan} &= \lim_{x \to 1} \frac{f(x) - f(1)}{x - 1} && \text{Definition of slope of tangent line} \\[2mm]
&= \lim_{x \to 1} \frac{(-16x^2 + 96x) - 80}{x - 1} && f(x) = -16x^2 + 96x; f(1) = 80 \\[2mm]
&= \lim_{x \to 1} \frac{-16(x - 5)(x - 1)}{x - 1} && \text{Factor the numerator.} \\[2mm]
&= -16 \underbrace{\lim_{x \to 1} (x - 5)}_{-4} = 64 && \text{Cancel factors } (x \neq 1) \text{ and evaluate the limit.}
\end{aligned}
$$

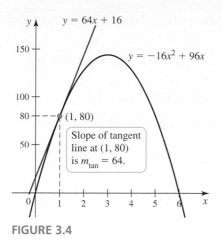

FIGURE 3.4

We have confirmed the conjecture made in Section 2.1 that the slope of the line tangent to the graph of $f(x) = -16x^2 + 96x$ at $(1, 80)$ is 64.

b. An equation of the line passing through $(1, 80)$ with slope $m_{tan} = 64$ is $y - 80 = 64(x - 1)$ or $y = 64x + 16$. The graph of f and the tangent line at $(1, 80)$ are shown in Figure 3.4. *Related Exercises 11–16* ◄

QUICK CHECK 2 In Example 1, is the slope of the tangent line at $x = 2$ greater than or less than the slope at $x = 1$? ◄

An alternative formula for the slope of the tangent line is helpful for future work. We now let $(a, f(a))$ and $(a + h, f(a + h))$ be the coordinates of P and Q, respectively (Figure 3.5). The difference in the x-coordinates of P and Q is $(a + h) - a = h$. Note that Q is located to the right of P if $h > 0$ and to the left of P if $h < 0$.

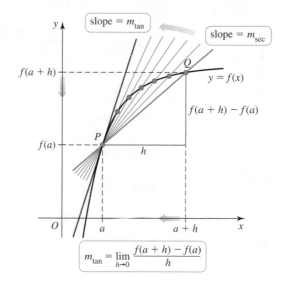

FIGURE 3.5

The slope of the secant line $\overleftrightarrow{PQ}$ using the new notation is $m_{sec} = \dfrac{f(a + h) - f(a)}{h}$.

As h approaches 0, the variable point Q approaches P and the slopes of the secant lines approach the slope of the tangent line. Therefore, the slope of the tangent line at $(a, f(a))$, which is also the instantaneous rate of change of f at a, is

$$m_{tan} = \lim_{h \to 0} \frac{f(a + h) - f(a)}{h}.$$

ALTERNATIVE DEFINITION   Rates of Change and the Tangent Line

The **average rate of change** in f on the interval $[a, a + h]$ is the slope of the corresponding secant line:

$$m_{sec} = \frac{f(a + h) - f(a)}{h}$$

The **instantaneous rate of change** in f at $x = a$ is

$$m_{tan} = \lim_{h \to 0} \frac{f(a + h) - f(a)}{h}, \qquad (2)$$

which is also the **slope of the tangent line** at $x = a$, provided this limit exists.

EXAMPLE 2 Equation of a tangent line Find an equation of the line tangent to the graph of $f(x) = x^3 + 4x$ at $x = 1$.

SOLUTION We let $a = 1$ in definition (2) and first find $f(1 + h)$. After expanding and collecting terms, we have

$$f(1 + h) = (1 + h)^3 + 4(1 + h) = h^3 + 3h^2 + 7h + 5.$$

Substituting $f(1 + h)$ and $f(1) = 5$, the slope of the tangent line is

$$m_{\tan} = \lim_{h \to 0} \frac{f(1 + h) - f(1)}{h} \qquad \text{Definition of } m_{\tan}$$

$$= \lim_{h \to 0} \frac{(h^3 + 3h^2 + 7h + 5) - 5}{h} \qquad \text{Substitute } f(1 + h) \text{ and } f(1) = 5.$$

$$= \lim_{h \to 0} \frac{h(h^2 + 3h + 7)}{h} \qquad \text{Simplify.}$$

$$= \lim_{h \to 0} (h^2 + 3h + 7) \qquad \text{Cancel } h \text{, noting } h \neq 0.$$

$$= 7 \qquad \text{Evaluate the limit.}$$

The tangent line has slope $m_{\tan} = 7$ and passes through the point $(1, 5)$ (Figure 3.6); its equation is $y - 5 = 7(x - 1)$ or $y = 7x - 2$. We could also say that the instantaneous rate of change of f at $x = 1$ is 7. *Related Exercises 17–22* ◄

> By the definition of the limit as $h \to 0$, notice that h approaches 0 but $h \neq 0$. Therefore, it is permissible to cancel h from the numerator and denominator of $\dfrac{h(h^2 + 3h + 7)}{h}$.

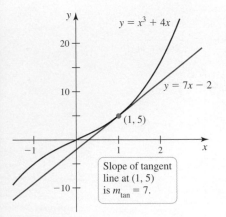

FIGURE 3.6

QUICK CHECK 3 Set up the calculation in Example 2 using definition (1) for the slope of the tangent line rather than definition (2). Does the calculation appear more difficult using definition (1)? ◄

The Derivative Function

So far we have computed the slope of the tangent line at one fixed point on the curve. If this point is moved along the curve, the tangent line also moves, and, in general, its slope changes (Figure 3.7). For this reason, the slope of the tangent line for the function f is itself a function of x, called the **derivative** of f.

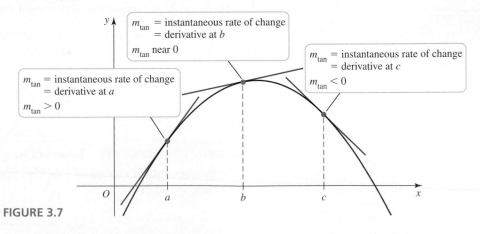

FIGURE 3.7

We let f' (read *f prime*) denote the derivative function for f, which means that $f'(a)$, when it exists, is the slope of the line tangent to the graph of f at $(a, f(a))$. Using definition (2) for the slope of the tangent line, we have

$$f'(a) = m_{\tan} = \lim_{h \to 0} \frac{f(a + h) - f(a)}{h}.$$

More generally, $f'(x)$, when it exists, is the slope of the tangent line (and the instantaneous rate of change) at the variable point $(x, f(x))$. Replacing a by the variable x in the expression for $f'(a)$ gives the definition of the **derivative function**.

> The process of finding f' is called *differentiation*, and to *differentiate* f means to find f'.

> Just as we have two definitions for the slope of the tangent line, we may also use the following definition for the derivative of f at a:
>
> $$f'(a) = \lim_{x \to a} \frac{f(x) - f(a)}{x - a}$$

DEFINITION   The Derivative

The **derivative** of f is the function

$$f'(x) = \lim_{h \to 0} \frac{f(x + h) - f(x)}{h}$$

provided the limit exists. If $f'(x)$ exists, we say f is **differentiable** at x. If f is differentiable at every point of an open interval I, we say that f is differentiable on I.

EXAMPLE 3   The slope of a curve Consider once again the function $f(x) = -16x^2 + 96x$ of Example 1 and find its derivative.

SOLUTION

> Notice that this argument applies for $h > 0$ and for $h < 0$; that is, the limit as $h \to 0^+$ and the limit as $h \to 0^-$ are equal.

$$f'(x) = \lim_{h \to 0} \frac{f(x + h) - f(x)}{h} \qquad \text{Definition of } f'(x)$$

$$= \lim_{h \to 0} \frac{\overbrace{-16(x + h)^2 + 96(x + h)}^{f(x+h)} - \overbrace{(-16x^2 + 96x)}^{f(x)}}{h} \qquad \text{Substitute.}$$

$$= \lim_{h \to 0} \frac{-16(x^2 + 2xh + h^2) + 96x + 96h + 16x^2 - 96x}{h} \qquad \begin{array}{l}\text{Expand the}\\ \text{numerator.}\end{array}$$

$$= \lim_{h \to 0} \frac{h(-32x + 96 - 16h)}{h} \qquad \begin{array}{l}\text{Simplify and}\\ \text{factor out } h.\end{array}$$

$$= \lim_{h \to 0} (-32x + 96 - 16h) = -32x + 96 \qquad \begin{array}{l}\text{Cancel } h \text{ and}\\ \text{evaluate the limit.}\end{array}$$

The derivative is $f'(x) = -32x + 96$, which gives the slope of the tangent line (equivalently, the instantaneous rate of change) at *any* point in the domain. For example, at the point $(1, 80)$, the slope of the tangent line is $f'(1) = -32(1) + 96 = 64$, confirming the calculation in Example 1. The slope of the tangent line at $(3, 144)$ is $f'(3) = -32(3) + 96 = 0$, which means the tangent line is horizontal at that point (Figure 3.8).

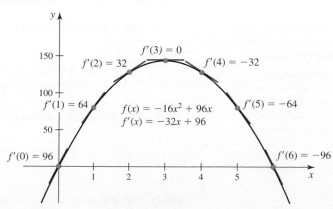

FIGURE 3.8

Related Exercises 23–32 ◄

QUICK CHECK 4 In Example 3, determine the slope of the tangent line at $x = 2$. ◄

Derivative Notation

For historical and practical reasons, several notations for the derivative are used. To see the origin of one notation, recall that the slope of the secant line $\overleftrightarrow{PQ}$ between two points $P(x, f(x))$ and $Q(x + h, f(x + h))$ on the curve $y = f(x)$ is $\dfrac{f(x + h) - f(x)}{h}$. The quantity h is the *change* in the x-coordinates in moving from P to Q. A standard notation for change is the symbol Δ (uppercase Greek letter delta). So, we replace h by Δx to represent the change in x. Similarly, $f(x + h) - f(x)$ is the change in y, denoted Δy (Figure 3.9). Therefore, the slope of $\overleftrightarrow{PQ}$ is

$$\frac{f(x + \Delta x) - f(x)}{\Delta x} = \frac{\Delta y}{\Delta x}.$$

> The notation $\dfrac{dy}{dx}$ is read *the derivative of y with respect to x* or *dy dx*. It does not mean dy divided by dx, but it is a reminder of the limit of $\Delta y / \Delta x$.

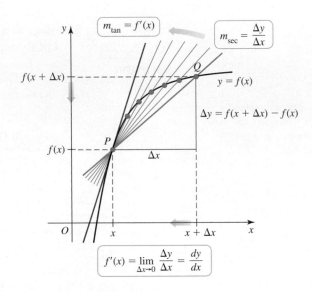

FIGURE 3.9

By letting $\Delta x \to 0$, the slope of the tangent line at $(x, f(x))$ is

$$f'(x) = \lim_{\Delta x \to 0} \frac{f(x + \Delta x) - f(x)}{\Delta x} = \lim_{\Delta x \to 0} \frac{\Delta y}{\Delta x} = \frac{dy}{dx}.$$

The new notation for the derivative is $\dfrac{dy}{dx}$; it reminds us that $f'(x)$ is the limit of $\dfrac{\Delta y}{\Delta x}$ as $\Delta x \to 0$.

> The derivative notation dy/dx was introduced by Gottfried Wilhelm von Leibniz (1646–1716), one of the coinventors of calculus. His notation is used today in its original form. The notation used by Sir Isaac Newton (1642–1727), the other coinventor of calculus, has fallen into disuse.

In addition to the notation $f'(x)$ and $\dfrac{dy}{dx}$, other common ways of writing the derivative include

$$\frac{df}{dx}, \qquad \frac{d}{dx}(f(x)), \qquad D_x(f(x)), \quad \text{and} \quad y'(x).$$

Each of the following notations represents the derivative of f evaluated at $x = a$.

$$f'(a), \qquad y'(a), \qquad \frac{df}{dx}\bigg|_{x=a}, \quad \text{and} \quad \frac{dy}{dx}\bigg|_{x=a}.$$

QUICK CHECK 5 What are some other ways to write $f'(3)$, where $y = f(x)$? ◄

EXAMPLE 4   A derivative calculation Let $y = f(x) = \sqrt{x}$.

> Example 4 gives the first of many derivative formulas to be presented in the text:
>
> $$\frac{d}{dx}(\sqrt{x}) = \frac{1}{2\sqrt{x}}.$$

a. Compute $\dfrac{dy}{dx}$.

b. Find an equation of the line tangent to the graph of f at $(4, 2)$.

SOLUTION

a.
$$\frac{dy}{dx} = \lim_{h \to 0} \frac{f(x + h) - f(x)}{h} \qquad \text{Definition of } \frac{dy}{dx} = f'(x)$$

$$= \lim_{h \to 0} \frac{\sqrt{x + h} - \sqrt{x}}{h} \qquad \text{Substitute } f(x) = \sqrt{x}.$$

$$= \lim_{h \to 0} \frac{\left(\sqrt{x + h} - \sqrt{x}\right)\left(\sqrt{x + h} + \sqrt{x}\right)}{h\left(\sqrt{x + h} + \sqrt{x}\right)} \qquad \begin{array}{l}\text{Multiply the numerator and} \\ \text{denominator by } \sqrt{x + h} + \sqrt{x}.\end{array}$$

$$= \lim_{h \to 0} \frac{1}{\sqrt{x + h} + \sqrt{x}} = \frac{1}{2\sqrt{x}} \qquad \text{Simplify and evaluate the limit.}$$

b. The slope of the tangent line at $x = 4$ is

$$\left.\frac{dy}{dx}\right|_{x=4} = \frac{1}{2\sqrt{4}} = \frac{1}{4}.$$

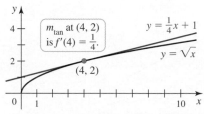

FIGURE 3.10

The tangent line at $(4, 2)$ has slope $m = \frac{1}{4}$ (Figure 3.10), so an equation is

$$y - 2 = \frac{1}{4}(x - 4) \text{ or } y = \frac{1}{4}x + 1. \qquad \textit{Related Exercises 33–34} \blacktriangleleft$$

QUICK CHECK 6 In Example 4, do the slopes of the tangent lines increase or decrease as x increases? Explain. ◄

If a function is given in terms of variables other than x and y, we make an adjustment to the derivative definition. For example, if $y = g(t)$, we replace f with g and x with t to obtain the *derivative of g with respect to t*:

$$g'(t) = \lim_{h \to 0} \frac{g(t + h) - g(t)}{h}.$$

QUICK CHECK 7 Express the derivative of $p = q(r)$ in three ways. ◄

Other notation for $g'(t)$ includes $\dfrac{dg}{dt}, \dfrac{d}{dt}(g(t)), D_t(g(t))$, and $y'(t)$.

EXAMPLE 5   Another derivative calculation Let $g(t) = 1/t^2$ and compute $g'(t)$.

SOLUTION

$$g'(t) = \lim_{h \to 0} \frac{g(t + h) - g(t)}{h} \qquad \text{Definition of } g'$$

$$= \lim_{h \to 0} \frac{1}{h}\left[\frac{1}{(t + h)^2} - \frac{1}{t^2}\right] \qquad \text{Substitute } g(t) = 1/t^2.$$

$$= \lim_{h \to 0} \frac{1}{h}\left[\frac{t^2 - (t + h)^2}{t^2(t + h)^2}\right] \qquad \text{Common denominator}$$

$$= \lim_{h \to 0} \frac{1}{h}\left[\frac{-2ht - h^2}{t^2(t + h)^2}\right] \qquad \text{Expand the numerator and simplify.}$$

$$= \lim_{h \to 0} \left[\frac{-2t - h}{t^2(t + h)^2}\right] \qquad h \neq 0; \text{ cancel } h.$$

$$= -\frac{2}{t^3} \qquad \text{Evaluate the limit.}$$

Related Exercises 35–38 ◄

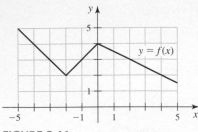

FIGURE 3.11

Graphs of Derivatives

The function f' is called the derivative of f because it is *derived* from f. The following examples illustrate how to *derive* the graph of f' from the graph of f.

EXAMPLE 6   **Graph of the derivative**   Sketch the graph of f' from the graph of f (Figure 3.11).

SOLUTION   The graph of f consists of line segments, which are their own tangent lines. Therefore, the slope of the curve $y = f(x)$ for $x < -2$ is -1; that is, $f'(x) = -1$ for $x < -2$. Similarly, $f'(x) = 1$ for $-2 < x < 0$ and $f'(x) = -\frac{1}{2}$ for $x > 0$ (Figure 3.12).

> In terms of limits at $x = -2$, we can write
>
> $$\lim_{h \to 0^-} \frac{f(-2 + h) - f(-2)}{h} = -1 \text{ and}$$
>
> $$\lim_{h \to 0^+} \frac{f(-2 + h) - f(-2)}{h} = 1. \text{ Because}$$
>
> the one-sided limits are not equal, $f'(-2)$ does not exist. The analogous one-sided limits at $x = 0$ are also unequal.

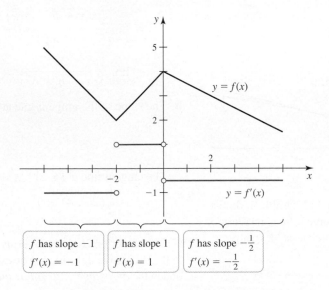

FIGURE 3.12

f has slope -1 $f'(x) = -1$

f has slope 1 $f'(x) = 1$

f has slope $-\frac{1}{2}$ $f'(x) = -\frac{1}{2}$

QUICK CHECK 8   In Example 6, why is the graph of f' not continuous at $x = -2$ and at $x = 0$? ◄

Notice that the slopes of the tangent lines change abruptly at $x = -2$ and $x = 0$. As a result, $f'(-2)$ and $f'(0)$ are undefined and the graph of the derivative is discontinuous at these points.   *Related Exercises 39–44* ◄

EXAMPLE 7   **Graph of the derivative**   Sketch the graph of g' using the graph of g (Figure 3.13).

SOLUTION   Without an equation for g, the best we can do is to find the general shape of the graph of g'. Here are the key observations.

1. First note that the lines tangent to the graph of g at $x = -3$, $x = -1$, and $x = 1$ have a slope of 0. Therefore,

$$g'(-3) = g'(-1) = g'(1) = 0,$$

which means the graph of g' has x-intercepts at these points (Figure 3.14).

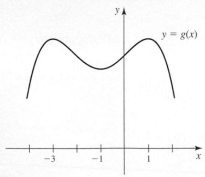

FIGURE 3.13

2. For $x < -3$, the slopes of the tangent lines are positive and decrease to 0 as x approaches -3 from the left. Therefore, $g'(x)$ is positive for $x < -3$ and decreases to 0 as x approaches -3.

3. For $-3 < x < -1$, $g'(x)$ is negative; it initially decreases as x increases and then increases to 0 at $x = -1$. For $-1 < x < 1$, $g'(x)$ is positive; it initially increases as x increases and then returns to 0 at $x = 1$.

4. Finally, $g'(x)$ is negative and decreasing for $x > 1$. Because the slope of g changes gradually, the graph of g' is continuous with no jumps or breaks.

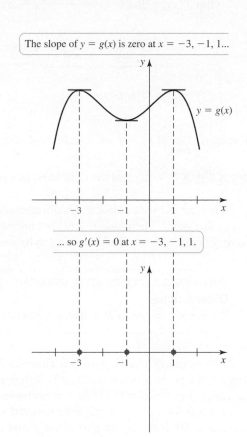

The slope of $y = g(x)$ is zero at $x = -3, -1, 1$...

... so $g'(x) = 0$ at $x = -3, -1, 1$.

FIGURE 3.14

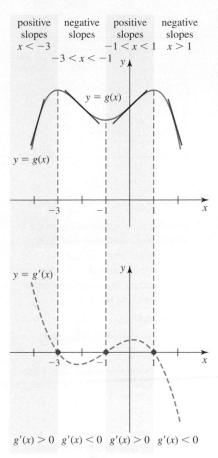

positive slopes	negative slopes	positive slopes	negative slopes
$x < -3$	$-3 < x < -1$	$-1 < x < 1$	$x > 1$

$g'(x) > 0 \quad g'(x) < 0 \quad g'(x) > 0 \quad g'(x) < 0$

Related Exercises 39–44 ◄

Continuity

We now return to the discussion of continuity (Section 2.6) and investigate the relationship between continuity and differentiability. Specifically, we show that if a function is differentiable at a point, then it is also continuous at that point.

THEOREM 3.1   Differentiable Implies Continuous
If f is differentiable at a, then f is continuous at a.

Proof Assume f is differentiable at a point a, which implies that

$$f'(a) = \lim_{x \to a} \frac{f(x) - f(a)}{x - a}$$

exists. To show that f is continuous at a, we must show that $\lim_{x \to a} f(x) = f(a)$. The key is the identity

$$f(x) = \frac{f(x) - f(a)}{x - a}(x - a) + f(a), \qquad x \neq a. \tag{3}$$

➤ Expression (3) is an identity because it holds for all values of $x \neq a$, which can be seen by canceling $x - a$ and simplifying.

Taking the limit as x approaches a on both sides of (3) and simplifying, we have

$$\lim_{x \to a} f(x) = \lim_{x \to a} \left[\frac{f(x) - f(a)}{x - a}(x - a) + f(a) \right] \quad \text{Use identity.}$$

$$= \lim_{x \to a} \underbrace{\left(\frac{f(x) - f(a)}{x - a} \right)}_{f'(a)} \underbrace{\lim_{x \to a} (x - a)}_{0} + \underbrace{\lim_{x \to a} f(x)}_{f(a)} \quad \text{Theorem 2.3}$$

$$= f'(a) \cdot 0 + f(a) \quad \text{Evaluate limits.}$$

$$= f(a) \quad \text{Simplify.}$$

Therefore, $\lim_{x \to a} f(x) = f(a)$, which means that f is continuous at a. ◄

QUICK CHECK 9 Verify that the right-hand side of (3) equals $f(x)$ if $x \neq a$. ◄

Theorem 3.1 tells us that if f is differentiable at a point, then it is necessarily continuous at that point. Therefore, if f is *not* continuous at a point, then f is *not* differentiable there (Figure 3.15). So, Theorem 3.1 can be stated in another way.

> **THEOREM 3.1 (ALTERNATIVE VERSION) Not Continuous Implies Not Differentiable**
> If f is not continuous at a, then f is not differentiable at a.

It is tempting to read more into Theorem 3.1 than what it actually states. If f is continuous at a point, f is *not* necessarily differentiable at that point. For example, consider the continuous function in Figure 3.16 and note the **corner point** at a. Ignoring the portion of the graph for $x > a$, we might be tempted to conclude that ℓ_1 is the line tangent to the curve at a. By ignoring the part of the graph for $x < a$, we might incorrectly conclude that ℓ_2 is the line tangent to the curve at a. The slopes of ℓ_1 and ℓ_2 are not equal. Because of the abrupt change in the slope of the curve at a, f is not differentiable at a: The limit that defines f' does not exist at a.

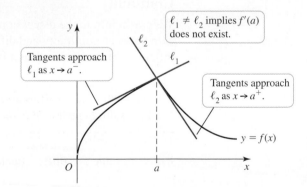

FIGURE 3.16

Another common situation occurs when the graph of a function f has a vertical tangent line at a. In this case, $f'(a)$ is undefined because the slope of a vertical line is undefined. A vertical tangent line may occur at a sharp point on the curve called a **cusp** (for example, the function $f(x) = \sqrt{|x|}$ in Figure 3.17a). In other cases, a vertical tangent line may occur without a cusp (for example, the function $f(x) = \sqrt[3]{x}$ in Figure 3.17b).

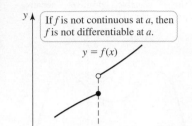

FIGURE 3.15

▶ The alternative version of Theorem 3.1 is called the *contrapositive* of the first statement of Theorem 3.1. A statement and its contrapositive are two equivalent ways of expressing the same statement. For example, the statement

If I live in Denver, then I live in Colorado

is logically equivalent to its contrapositive:

If I do not live in Colorado, then I do not live in Denver.

▶ To avoid confusion about continuity and differentiability, it helps to think about the function $f(x) = |x|$: It is continuous everywhere but not differentiable at 0.

▶ Continuity requires that $\lim_{x \to a} f(x) = f(a)$. Differentiability requires more:
$$\lim_{x \to a} \frac{f(x) - f(a)}{x - a} \text{ must exist.}$$

▶ See Exercises 61–64 for a formal definition of a vertical tangent line.

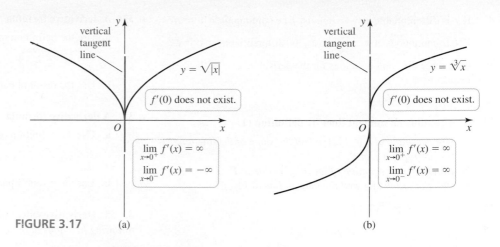

FIGURE 3.17   (a)   (b)

When Is a Function Not Differentiable at a Point?

A function f is *not* differentiable at a if at least one of the following conditions holds:

a. f is not continuous at a (Figure 3.15).

b. f has a corner at a (Figure 3.16).

c. f has a vertical tangent at a (Figure 3.17).

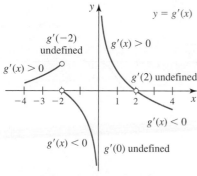

FIGURE 3.18

EXAMPLE 8   Continuous and differentiable Consider the graph of g in Figure 3.18.

a. Find the values of x in the interval $(-4, 4)$ at which g is not continuous.

b. Find the values of x in the interval $(-4, 4)$ at which g is not differentiable.

c. Sketch a graph of the derivative of g.

SOLUTION

a. The function g fails to be continuous at $x = -2$ (where the one-sided limits are not equal) and at $x = 2$ (where g is not defined).

b. Because it is not continuous at $x = \pm 2$, g is not differentiable at those points. Furthermore, g is not differentiable at $x = 0$, because the graph has a cusp at that point.

c. A rough sketch of the derivative (Figure 3.19) has the following features:

- $g'(x) > 0$ for $-4 < x < -2$ and $0 < x < 2$.
- $g'(x) < 0$ for $-2 < x < 0$ and $2 < x < 4$.
- $g'(x)$ approaches $-\infty$ as $x \to 0^-$ and $g'(x)$ approaches ∞ as $x \to 0^+$.
- $g'(x)$ approaches 0 as $x \to 2$ from either side, although $g'(2)$ does not exist.

Related Exercises 45–46 ◄

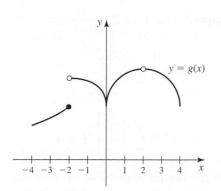

FIGURE 3.19

SECTION 3.1 EXERCISES

Review Questions

1. Use definition (1) for the slope of a tangent line to explain how slopes of secant lines approach the slope of the tangent line at a point.

2. Explain why the slope of a secant line can be interpreted as an average rate of change.

3. Explain why the slope of the tangent line can be interpreted as an instantaneous rate of change.

4. Given the function f, what does f' represent?

5. Given a function f and a point a in its domain, what does $f'(a)$ represent?

6. Explain the relationships among the slope of a tangent line, the instantaneous rate of change, and the value of the derivative at a point.

7. Why is the notation $\dfrac{dy}{dx}$ used to represent the derivative?

8. If f is differentiable at $x = a$, must f be continuous at $x = a$?

9. If f is continuous at $x = a$, must f be differentiable at $x = a$?

10. Give three different notations for the derivative of f with respect to x.

Basic Skills

11–16. Equations of tangent lines by definition (1)

  a. Use definition (1) (p. 122) to find the slope of the line tangent to the graph of f at P.
  b. Determine an equation of the tangent line at P.
  c. Plot the graph of f and the tangent line at P.

11. $f(x) = x^2 - 5$; $P(3, 4)$

12. $f(x) = -3x^2 - 5x + 1$; $P(1, -7)$

13. $f(x) = -5x + 1$; $P(1, -4)$ **14.** $f(x) = 5$; $P(1, 5)$

15. $f(x) = \dfrac{1}{x}$; $P(-1, -1)$ **16.** $f(x) = \dfrac{4}{x^2}$; $P(-1, 4)$

17–22. Equations of tangent lines by definition (2)

  a. Use definition (2) (p. 123) to find the slope of the line tangent to the graph of f at P.
  b. Determine an equation of the tangent line at P.

17. $f(x) = 2x + 1$; $P(0, 1)$ **18.** $f(x) = 3x^2 - 4x$; $P(1, -1)$

19. $f(x) = x^4$; $P(-1, 1)$ **20.** $f(x) = \dfrac{1}{2x + 1}$; $P(0, 1)$

21. $f(x) = \dfrac{1}{3 - 2x}$; $P\left(-1, \frac{1}{5}\right)$ **22.** $f(x) = \sqrt{x - 1}$; $P(2, 1)$

23–28. Derivatives and tangent lines

  a. For the following functions and points, find $f'(a)$.
  b. Determine an equation of the line tangent to the graph of f at $(a, f(a))$ for the given value of a.

23. $f(x) = 8x$; $a = -3$ **24.** $f(x) = x^2$; $a = 3$

25. $f(x) = 4x^2 + 2x$; $a = -2$ **26.** $f(x) = 2x^3$; $a = 10$

27. $f(x) = \dfrac{1}{\sqrt{x}}$; $a = 1/4$ **28.** $f(x) = \dfrac{1}{x^2}$; $a = 1$

29–32. Lines tangent to parabolas

  a. Find the derivative function f' for the following functions f.
  b. Find an equation of the line tangent to the graph of f at $(a, f(a))$ for the given value of a.
  c. Graph f and the tangent line.

29. $f(x) = 3x^2 + 2x - 10$; $a = 1$ **30.** $f(x) = 3x^2$; $a = 0$

31. $f(x) = 5x^2 - 6x + 1$; $a = 2$ **32.** $f(x) = 1 - x^2$; $a = -1$

33. A derivative formula

  a. Use the definition of the derivative to determine
  $\dfrac{d}{dx}(ax^2 + bx + c)$, where a, b, and c are constants.
  b. Use the result of part (a) to find $\dfrac{d}{dx}(4x^2 - 3x + 10)$.

34. A derivative formula

  a. Use the definition of the derivative to determine
  $\dfrac{d}{dx}(\sqrt{ax + b})$, where a and b are constants.
  b. Use the result of part (a) to find $\dfrac{d}{dx}(\sqrt{5x + 9})$.

35–38. Derivative calculations *Evaluate the derivative of the following functions at the given point.*

35. $y = 1/(t + 1)$; $t = 1$ **36.** $y = t - t^2$; $t = 2$

37. $c = 2\sqrt{s} - 1$; $s = 25$ **38.** $A = \pi r^2$; $r = 3$

39–40. Derivatives from graphs *Use the graph of f to sketch a graph of f'.*

39. 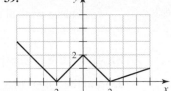   **40.**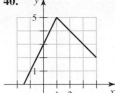

41. Matching functions with derivatives Match the functions (a)–(d) in the first set of figures with the derivative functions (A)–(D) in the next set of figures.

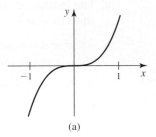
(a)

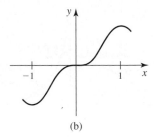
(b)

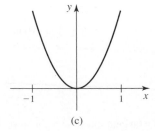
(c)

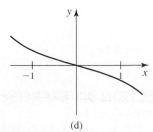
(d)

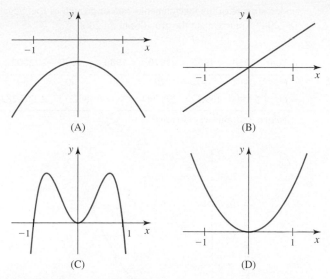

(A)   (B)

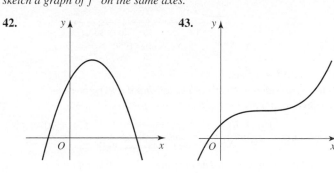

(C)   (D)

42–44. Sketching derivatives *Reproduce the graph of f and then sketch a graph of f' on the same axes.*

42.   **43.**

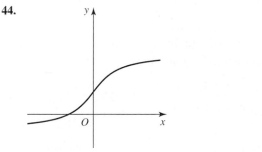

44.

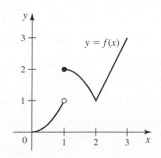

45. Where is the function continuous? Differentiable? Use the graph of f in the figure to do the following.

  a. Find the values of x in $(0, 3)$ at which f is not continuous.
  b. Find the values of x in $(0, 3)$ at which f is not differentiable.
  c. Sketch a graph of f'.

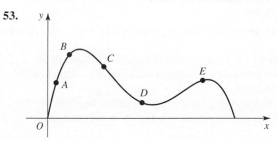

46. Where is the function continuous? Differentiable? Use the graph of g in the figure to do the following.

  a. Find the values of x in $(0, 4)$ at which g is not continuous.
  b. Find the values of x in $(0, 4)$ at which g is not differentiable.
  c. Sketch a graph of g'.

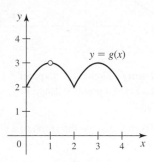

Further Explorations

47. Explain why or why not Determine whether the following statements are true and give an explanation or counterexample.

  a. For linear functions, the slope of any secant line always equals the slope of any tangent line.
  b. The slope of the secant line passing through the points P and Q is less than the slope of the tangent line at P.
  c. Consider the graph of the parabola $f(x) = x^2$. For $x > 0$ and $h > 0$, the secant line through $(x, f(x))$ and $(x + h, f(x + h))$ always has a greater slope than the tangent line at $(x, f(x))$.
  d. If the function f is differentiable for all values of x, then f is continuous for all values of x.

48. Slope of a line Consider the line $f(x) = mx + b$, where m and b are constants. Show that $f'(x) = m$ for all x. Interpret this result.

49–52. Calculating derivatives

  a. *For the following functions, find f' using the definition*
$$f'(x) = \lim_{h \to 0} \frac{f(x + h) - f(x)}{h}.$$
  b. *Determine an equation of the line tangent to the graph of f at $(a, f(a))$ for the given value of a.*

49. $f(x) = \sqrt{3x + 1}; \ a = 8$ **50.** $f(x) = \sqrt{x + 2}; \ a = 7$

51. $f(x) = \dfrac{2}{3x + 1}; \ a = -1$ **52.** $f(x) = \dfrac{1}{x}; \ a = -5$

53–54. Analyzing slopes *Use the points A, B, C, D, and E in the following graphs to answer these questions.*

  a. *At which point(s) is the slope of the curve negative?*
  b. *At which point(s) is the slope of the curve positive?*
  c. *Using A–E, list the slopes in decreasing order.*

53.

54.

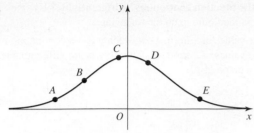

55. Finding f from f' Sketch the graph of $f'(x) = x$ (the derivative of f). Then, sketch a possible graph of f. Is there more than one possible graph?

56. Finding f from f' Create the graph of a continuous function $y = f(x)$, where

$$f'(x) = \begin{cases} 1 & \text{if } x < 0 \\ 0 & \text{if } 0 < x < 1 \\ -1 & \text{if } x > 1 \end{cases}$$

Is there more than one possible graph?

Applications

57. Power and energy Energy is the capacity to do work and power is the rate at which energy is used or consumed. Therefore, if $E(t)$ is the energy function for a system, then $P(t) = E'(t)$ is the power function. A unit of energy is the kilowatt-hour (1 kWh is the amount of energy needed to light ten 100-W light bulbs for an hour); the corresponding units for power are kW. The following figure shows the energy consumed by a small community over a 25-hr period.

  a. Estimate the power at $t = 10$ and $t = 20$ hr. Be sure to include units in your calculation.
  b. At what times on the interval $[0, 25]$ is the power zero?
  c. At what times on the interval $[0, 25]$ is the power a maximum?

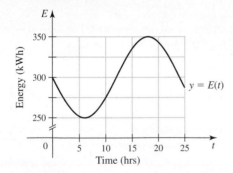

58. Population of Las Vegas Let $p(t)$ represent the population of the Las Vegas metropolitan area t years after 1950, as shown in the table and figure.

  a. Compute the average rate of growth of Las Vegas from 1970 to 1980.
  b. Explain why the average rate of growth calculated in part (a) is a good estimate of the instantaneous rate of growth of Las Vegas in 1975.
  c. Compute the average rate of growth of Las Vegas from 1990 to 2000. Is this average rate of growth an overestimate or

underestimate of the instantaneous rate of growth of Las Vegas in 2000? Approximate the growth rate in 2000.

Year	1950	1960	1970	1980	1990	2000
t	0	10	20	30	40	50
$p(t)$	59,900	139,126	304,744	528,000	852,737	1,563,282

Source: U.S. Bureau of Census

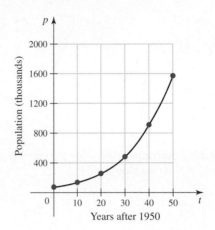

Additional Exercises

59–60. One-sided derivatives The *left-hand* and *right-hand* *derivatives* of a function at a point a are given by

$$f'_+(a) = \lim_{h \to 0^+} \frac{f(a + h) - f(a)}{h} \quad \text{and} \quad f'_-(a) = \lim_{h \to 0^-} \frac{f(a + h) - f(a)}{h}$$

provided these limits exist. The derivative $f'(a)$ exists if and only if $f'_+(a) = f'_-(a)$.

  a. *Sketch the following functions.*
  b. *Compute $f'_+(a)$ and $f'_-(a)$ at the given point a.*
  c. *Is f continuous at a? Is f differentiable at a?*

59. $f(x) = |x - 2|$; $a = 2$

60. $f(x) = \begin{cases} 4 - x^2 & \text{if } x \leq 1 \\ 2x + 1 & \text{if } x > 1 \end{cases}$; $a = 1$

61–64. Vertical tangent lines *If a function f is continuous at a and $\lim_{x \to a} |f'(x)| = \infty$, then the curve $y = f(x)$ has a vertical tangent line at a and the equation of the tangent line is $x = a$. If a is an endpoint of a domain, then the appropriate one-sided derivative (Exercises 59–60) is used. Use this definition to answer the following questions.*

61. Graph the following functions and determine the location of the vertical tangent lines.

  a. $f(x) = (x - 2)^{1/3}$ **b.** $f(x) = (x + 1)^{2/3}$
  c. $f(x) = \sqrt{|x - 4|}$ **d.** $f(x) = x^{5/3} - 2x^{1/3}$

62. The preceding definition of vertical tangent line includes four cases: $\lim_{x \to a^+} f'(x) = \pm\infty$ combined with $\lim_{x \to a^-} f'(x) = \pm\infty$ (for example, one case is $\lim_{x \to a^+} f'(x) = -\infty$ and $\lim_{x \to a^-} f'(x) = \infty$).

Make a rough sketch of a (continuous) function that has a vertical tangent line at $x = a$ in each of the four cases.

63. Verify that $f(x) = x^{1/3}$ has a vertical tangent line at $x = 0$.

64. Graph the following curves and determine the location of any vertical tangent lines.

    a. $x^2 + y^2 = 9$ **b.** $x^2 + y^2 + 2x = 0$

65–68. Find the function *The following limits represent the slope of a curve $y = f(x)$ at the point $(a, f(a))$. Determine a function f and a number a; then, calculate the limit.*

65. $\displaystyle\lim_{x \to 2} \dfrac{\dfrac{1}{x+1} - \dfrac{1}{3}}{x - 2}$ **66.** $\displaystyle\lim_{h \to 0} \dfrac{\sqrt{2+h} - \sqrt{2}}{h}$

67. $\displaystyle\lim_{h \to 0} \dfrac{(2+h)^4 - 16}{h}$ **68.** $\displaystyle\lim_{x \to 1} \dfrac{3x^2 + 4x - 7}{x - 1}$

69. Is it differentiable? Is $f(x) = \dfrac{x^2 - 5x + 6}{x - 2}$ differentiable at $x = 2$? Justify your answer.

70. Derivative of x^n Use the symbolic capabilities of a calculator to calculate $f'(x)$ using the definition $\displaystyle\lim_{h \to 0} \dfrac{f(x+h) - f(x)}{h}$ for the following functions.

    a. $f(x) = x^2$ **b.** $f(x) = x^3$
    c. $f(x) = x^4$ **d.** $f(x) = x^{100}$

    e. Based upon your answers to parts (a)–(d), propose a formula for $f'(x)$ if $f(x) = x^n$ where n is a positive integer.

71. Determining the unknown constant Let

$$f(x) = \begin{cases} 2x^2 & \text{if } x \le 1 \\ ax - 2 & \text{if } x > 1 \end{cases}$$

Determine a value of a (if possible) for which $f'(1)$ exists.

72. Graph of the derivative of the sine curve

    a. Use the graph of $y = \sin x$ (see figure) to sketch the graph of the derivative of the sine function.

    b. Based upon your graph in part (a), what function equals $\dfrac{d}{dx}(\sin x)$?

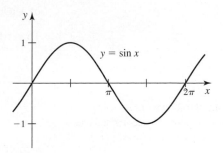

QUICK CHECK ANSWERS

2. The slope is less at $x = 2$.   **3.** Definition (1) requires factoring the numerator or long division in order to cancel $(x - 1)$.   **4.** 32   **5.** $\dfrac{df}{dx}\Big|_{x=3}, \dfrac{dy}{dx}\Big|_{x=3}, y'(3)$ **6.** The slopes of tangent lines decrease as x increases. The values of $f'(x) = \dfrac{1}{2\sqrt{x}}$ also decrease as x increases.

7. $\dfrac{dq}{dr}, \dfrac{dp}{dr}, D_r(q(r)), q'(r), p'(r)$ **8.** The slopes of the tangent lines change abruptly at $x = -2$ and $x = 0$. ◄

3.2 Rules of Differentiation

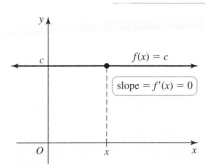

FIGURE 3.20

➤ We expect the derivative of a constant function to be 0 at every point because the values of a constant function do not change. This means the instantaneous rate of change is 0 at every point.

If you always had to use limits to evaluate derivatives, as we did in Section 3.1, calculus would be a tedious affair. The goal of this section is to establish rules and formulas for quickly evaluating derivatives—not just for individual functions but for entire families of functions.

The Constant and Power Rules for Derivatives

The graph of the **constant function** $f(x) = c$ is a horizontal line with a slope of 0 at every point (Figure 3.20). It follows that $f'(x) = 0$ or, equivalently, $\dfrac{d}{dx}(c) = 0$ (Exercise 70).

THEOREM 3.2   Constant Rule

If c is a real number, then $\dfrac{d}{dx}(c) = 0$.

QUICK CHECK 1 Find the values of $\frac{d}{dx}(5)$ and $\frac{d}{dx}(\pi)$. ◄

Next, consider power functions of the form $f(x) = x^n$, where n is a positive integer. If you completed Exercise 70 in Section 3.1, you found that

$$\frac{d}{dx}(x^2) = 2x, \qquad \frac{d}{dx}(x^3) = 3x^2, \quad \text{and} \quad \frac{d}{dx}(x^4) = 4x^3.$$

In each case, the derivative of x^n appears to be evaluated by placing the exponent n in front of x as a coefficient and decreasing the exponent by 1; in other words, $\frac{d}{dx}(x^n) = nx^{n-1}$. To verify this conjecture, we use the definition of the derivative in the form

$$f'(a) = \lim_{x \to a} \frac{f(x) - f(a)}{x - a}.$$

> Note that this formula agrees with familiar factoring formulas for differences of perfect squares and cubes:
>
> $x^2 - a^2 = (x - a)(x + a)$
>
> $x^3 - a^3 = (x - a)(x^2 + xa + a^2)$

If $f(x) = x^n$, then $f(x) - f(a) = x^n - a^n$. A factoring formula gives

$$x^n - a^n = (x - a)(x^{n-1} + x^{n-2}a + \cdots + xa^{n-2} + a^{n-1}).$$

Therefore,

$$
\begin{aligned}
f'(a) &= \lim_{x \to a} \frac{x^n - a^n}{x - a} && \text{Definition of } f'(a) \\[2mm]
&= \lim_{x \to a} \frac{(x - a)(x^{n-1} + x^{n-2}a + \cdots + xa^{n-2} + a^{n-1})}{x - a} && \text{Factor } x^n - a^n. \\[2mm]
&= \lim_{x \to a} (x^{n-1} + x^{n-2}a + \cdots + xa^{n-2} + a^{n-1}) && \text{Cancel common factors.} \\[2mm]
&= \underbrace{a^{n-1} + a^{n-2} \cdot a + \cdots + a \cdot a^{n-2} + a^{n-1}}_{n \text{ times } a^{n-1}} = na^{n-1} && \text{Evaluate the limit.}
\end{aligned}
$$

Replacing a by the variable x in $f'(a) = na^{n-1}$, we obtain the following result, known as the *Power Rule*.

> The $n = 0$ case of the Power Rule is the Constant Rule. You will see several versions of the Power Rule as we progress. It is extended first to integer powers, both positive and negative, then to rational powers, and, finally, to real powers.

THEOREM 3.3   Power Rule

If n is a positive integer, then $\frac{d}{dx}(x^n) = nx^{n-1}$.

EXAMPLE 1   Derivatives of power and constant functions Evaluate the following derivatives.

a. $\dfrac{d}{dx}(x^9)$ **b.** $\dfrac{d}{dx}(x)$ **c.** $\dfrac{d}{dx}(2^8)$

SOLUTION

a. $\dfrac{d}{dx}(x^9) = 9x^{9-1} = 9x^8$ Power Rule

b. $\dfrac{d}{dx}(x) = \dfrac{d}{dx}(x^1) = 1x^0 = 1$ Power Rule

QUICK CHECK 2 Use the graph of $y = x$ to give a geometric explanation of why $\frac{d}{dx}(x) = 1$. ◄

c. You might be tempted to use the Power Rule here, but $2^8 = 256$ is a constant. So, by the Constant Rule, $\dfrac{d}{dx}(2^8) = 0$.   *Related Exercises 7–12* ◄

Constant Multiple Rule

Consider the problem of finding the derivative of a constant c multiplied by a function f (assuming that f' exists). We apply the definition of the derivative in the form

$$f'(x) = \lim_{h \to 0} \frac{f(x+h) - f(x)}{h}$$

to the function cf:

$$\frac{d}{dx}[cf(x)] = \lim_{h \to 0} \frac{cf(x+h) - cf(x)}{h} \quad \text{Definition of the derivative of } cf$$

$$= \lim_{h \to 0} \frac{c[f(x+h) - f(x)]}{h} \quad \text{Factor out } c.$$

$$= c \lim_{h \to 0} \frac{f(x+h) - f(x)}{h} \quad \text{Theorem 2.3}$$

$$= cf'(x) \quad \text{Definition of } f'(x)$$

> ➤ Theorem 3.4 says that the derivative of a constant multiplied by a function is the constant multiplied by the derivative of the function.

THEOREM 3.4   Constant Multiple Rule

If f is differentiable at x and c is a constant, then

$$\frac{d}{dx}[cf(x)] = cf'(x).$$

EXAMPLE 2   Derivatives of constant multiples of functions Evaluate the following derivatives.

a. $\dfrac{d}{dx}\left(-\dfrac{7x^{11}}{8}\right)$ **b.** $\dfrac{d}{dt}\left(\dfrac{3}{8}\sqrt{t}\right)$

SOLUTION

a.
$$\frac{d}{dx}\left(-\frac{7x^{11}}{8}\right) = -\frac{7}{8} \cdot \frac{d}{dx}(x^{11}) \quad \text{Constant Multiple Rule}$$

$$= -\frac{7}{8} \cdot 11x^{10} \quad \text{Power Rule}$$

$$= -\frac{77}{8}x^{10} \quad \text{Simplify.}$$

> ➤ Recall from Example 4 of Section 3.1 that $\dfrac{d}{dt}(\sqrt{t}) = \dfrac{1}{2\sqrt{t}}$.

b.
$$\frac{d}{dt}\left(\frac{3}{8}\sqrt{t}\right) = \frac{3}{8} \cdot \frac{d}{dt}(\sqrt{t}) \quad \text{Constant Multiple Rule}$$

$$= \frac{3}{8} \cdot \frac{1}{2\sqrt{t}} \quad \text{Replace } \frac{d}{dt}(\sqrt{t}) \text{ by } \frac{1}{2\sqrt{t}}.$$

$$= \frac{3}{16\sqrt{t}} \qquad\qquad \textit{Related Exercises 13–18} ◄$$

Sum Rule

Many functions are sums of simpler functions. Therefore, it is useful to establish a rule for calculating the derivative of the sum of two or more functions.

▶ In words, Theorem 3.5 states that the derivative of a sum is the sum of the derivatives.

THEOREM 3.5   Sum Rule
If f and g are differentiable at x, then

$$\frac{d}{dx}[f(x) + g(x)] = f'(x) + g'(x).$$

Proof Let $F = f + g$, where f and g are differentiable at x and use the definition of the derivative:

$$\frac{d}{dx}[f(x) + g(x)] = F'(x)$$

$$= \lim_{h \to 0} \frac{F(x + h) - F(x)}{h} \qquad \text{Definition of derivative}$$

$$= \lim_{h \to 0} \frac{[f(x + h) + g(x + h)] - [f(x) + g(x)]}{h} \qquad \text{Replace } F \text{ with } f + g.$$

$$= \lim_{h \to 0} \left[\frac{f(x + h) - f(x)}{h} + \frac{g(x + h) - g(x)}{h} \right] \qquad \text{Regroup.}$$

$$= \lim_{h \to 0} \frac{f(x + h) - f(x)}{h} + \lim_{h \to 0} \frac{g(x + h) - g(x)}{h} \qquad \text{Theorem 2.3}$$

$$= f'(x) + g'(x) \qquad \text{Definition of } f' \text{ and } g' \qquad ◀$$

QUICK CHECK 3 If $f(x) = x^2$ and $g(x) = 2x$, what is the derivative of $f(x) + g(x)$? ◀

The Sum Rule can be extended to three or more differentiable functions, f_1, $f_2, \ldots, f_n$, to obtain the **Generalized Sum Rule**:

$$\frac{d}{dx}[f_1(x) + f_2(x) + \cdots + f_n(x)] = f_1'(x) + f_2'(x) + \cdots + f_n'(x)$$

The difference of two functions $f - g$ can be rewritten as the sum $f + (-g)$. By combining the Sum Rule with the Constant Multiple Rule, the **Difference Rule** is established:

$$\frac{d}{dx}[f(x) - g(x)] = f'(x) - g'(x)$$

EXAMPLE 3   Derivative of a polynomial Determine $\dfrac{d}{dw}(2w^3 + 9w^2 - 6w + 4)$.

SOLUTION

$$\frac{d}{dw}(2w^3 + 9w^2 - 6w + 4)$$

$$= \frac{d}{dw}(2w^3) + \frac{d}{dw}(9w^2) - \frac{d}{dw}(6w) + \frac{d}{dw}(4) \qquad \text{Generalized Sum Rule and Difference Rule}$$

$$= 2\frac{d}{dw}(w^3) + 9\frac{d}{dw}(w^2) - 6\frac{d}{dw}(w) + \frac{d}{dw}(4) \qquad \text{Constant Multiple Rule}$$

$$= 2 \cdot 3w^2 + 9 \cdot 2w - 6 \cdot 1 + 0 \qquad \text{Power Rule}$$

$$= 6w^2 + 18w - 6 \qquad \text{Simplify.}$$

Related Exercises 19–34 ◀

The technique used to differentiate the polynomial in Example 3 may be used for *any* polynomial. Much of the remainder of this chapter is devoted to discovering rules of differentiation for rational, exponential, logarithmic, algebraic, and trigonometric functions.

The Derivative of the Natural Exponential Function

The exponential function $f(x) = b^x$ was introduced in Chapter 1. Let's begin by looking at the graphs of two members of this family, $y = 2^x$ and $y = 3^x$ (Figure 3.21). The slope of the line tangent to the graph of $f(x) = b^x$ at $x = 0$ is given by

$$f'(0) = \lim_{h \to 0} \frac{f(0 + h) - f(0)}{h} = \lim_{h \to 0} \frac{b^h - b^0}{h} = \lim_{h \to 0} \frac{b^h - 1}{h}.$$

We investigate this limit numerically for $b = 2$ and $b = 3$. Table 3.1 shows values of $\dfrac{2^h - 1}{h}$ and $\dfrac{3^h - 1}{h}$ (which are slopes of secant lines) for values of h approaching 0 from the right.

Table 3.1

h	$\dfrac{2^h - 1}{h}$	$\dfrac{3^h - 1}{h}$
1.0	1.000000	2.000000
0.1	0.717735	1.161232
0.01	0.695555	1.104669
0.001	0.693387	1.099216
0.0001	0.693171	1.098673
0.00001	0.693150	1.098618

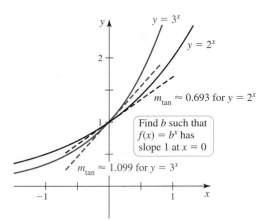

FIGURE 3.21

Exercise 60 gives similar approximations for the limit as h approaches 0 from the left. These numerical values suggest that

$$\lim_{h \to 0} \frac{2^h - 1}{h} \approx 0.693 \quad \text{Less than 1}$$

$$\lim_{h \to 0} \frac{3^h - 1}{h} \approx 1.099 \quad \text{Greater than 1}$$

These two facts, together with the graphs in Figure 3.21, suggest that there is a number b with $2 < b < 3$ such that the graph of $y = b^x$ has a tangent line with slope 1 at $x = 0$. This number b has the property that

$$\lim_{h \to 0} \frac{b^h - 1}{h} = 1.$$

We show in Chapter 4 that, indeed, such a number b exists. In fact, it is the number $e = 2.718281828459\ldots$ that was introduced in Chapter 1. Therefore, the exponential function whose tangent line has slope 1 at $x = 0$ is the *natural exponential function* $f(x) = e^x$ (Figure 3.22).

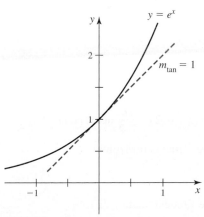

FIGURE 3.22

> The constant e was identified and named by the Swiss mathematician Leonhard Euler (1707–1783) (pronounced "oiler") and is also called Euler's constant.

DEFINITION   The Number e

The number $e = 2.718281828459\ldots$ satisfies

$$\lim_{h \to 0} \frac{e^h - 1}{h} = 1.$$

It is the base of the natural exponential function $f(x) = e^x$.

With the preceding facts in mind, the derivative of $f(x) = e^x$ is computed as follows:

$$\frac{d}{dx}(e^x) = \lim_{h \to 0} \frac{e^{x+h} - e^x}{h} \qquad \text{Definition of the derivative}$$

$$= \lim_{h \to 0} \frac{e^x \cdot e^h - e^x}{h} \qquad \text{Property of exponents}$$

$$= \lim_{h \to 0} \frac{e^x(e^h - 1)}{h} \qquad \text{Factor out } e^x.$$

$$= e^x \cdot \underbrace{\lim_{h \to 0} \frac{e^h - 1}{h}}_{1} \qquad e^x \text{ is constant as } h \to 0.$$

$$= e^x \cdot 1 = e^x$$

We have proved a remarkable fact: The derivative of the exponential function is itself; it is the only function (other than constant multiples of e^x and $f(x) = 0$) with this property.

> The Power Rule *cannot* be applied to exponential functions; that is,
> $$\frac{d}{dx}(e^x) \neq xe^{x-1}.$$

THEOREM 3.6   The Derivative of e^x

The function $f(x) = e^x$ is differentiable for all real numbers x and

$$\frac{d}{dx}(e^x) = e^x.$$

QUICK CHECK 4 Find the derivative of $f(x) = 4e^x - 3x^2$. ◄

Slopes of Tangent Lines

The derivative rules presented in this section allow us to determine slopes of tangent lines and rates of change for many functions.

EXAMPLE 4   Finding tangent lines

a. Write an equation of the line tangent to the graph of $f(x) = 2x - \dfrac{e^x}{2}$ at the point $\left(0, -\frac{1}{2}\right)$.

b. Find the point(s) on the graph of f where the tangent line is horizontal.

SOLUTION

a. To find the slope of the tangent line at $\left(0, -\frac{1}{2}\right)$, we first calculate $f'(x)$:

$$f'(x) = \frac{d}{dx}\left(2x - \frac{e^x}{2}\right)$$

$$= \frac{d}{dx}(2x) - \frac{d}{dx}\left(\frac{1}{2}e^x\right) \qquad \text{Difference Rule}$$

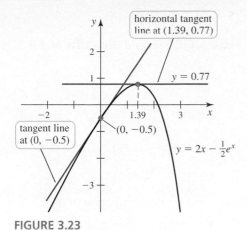

FIGURE 3.23

➤ Observe that the function has a maximum value of approximately 0.77 at the point where the tangent line has a slope of 0. We explore the importance of horizontal tangent lines in Chapter 4.

$$= 2\frac{d}{dx}(x) - \frac{1}{2}\cdot\frac{d}{dx}(e^x) \quad \text{Constant Multiple Rule}$$

$$\underbrace{\phantom{2\frac{d}{dx}(x)}}_{1} \quad \underbrace{\phantom{\frac{d}{dx}(e^x)}}_{e^x}$$

$$= 2 - \frac{1}{2}e^x \qquad \text{Evaluate derivatives.}$$

It follows that the slope of the tangent line at $x = 0$ is

$$f'(0) = 2 - \frac{1}{2}e^0 = \frac{3}{2}.$$

Figure 3.23 shows the tangent line passing through $\left(0, -\frac{1}{2}\right)$; it has the equation

$$y - \left(-\frac{1}{2}\right) = \frac{3}{2}(x - 0) \quad \text{or} \quad y = \frac{3}{2}x - \frac{1}{2}.$$

b. Because the slope of a horizontal tangent line is 0, our goal is to solve $f'(x) = 2 - \frac{1}{2}e^x = 0$. Multiplying both sides of this equation by 2 and rearranging gives the equation $e^x = 4$. Taking the natural logarithm of both sides, we find that $x = \ln 4$. Thus, $f'(x) = 0$ at $x = \ln 4 \approx 1.39$, and f has a horizontal tangent at $(\ln 4, f(\ln 4)) \approx (1.39, 0.77)$ (Figure 3.23).   *Related Exercises 35–41* ◄

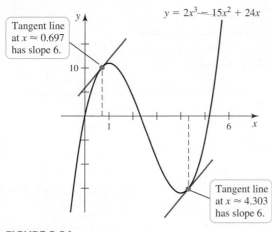

FIGURE 3.24

EXAMPLE 5   Slope of a tangent line Let $f(x) = 2x^3 - 15x^2 + 24x$. For what values of x does the line tangent to the graph of f have a slope of 6?

SOLUTION The tangent line has a slope of 6 when

$$f'(x) = 6x^2 - 30x + 24 = 6.$$

Subtracting 6 from both sides of the equation and factoring, we have

$$6(x^2 - 5x + 3) = 0.$$

Using the quadratic formula, the roots are

$$x = \frac{5 - \sqrt{13}}{2} \approx 0.697 \quad \text{and} \quad x = \frac{5 + \sqrt{13}}{2} \approx 4.303.$$

Therefore, the slope of the curve at these points is 6 (Figure 3.24).

Related Exercises 35–41 ◄

QUICK CHECK 5 Determine the point(s) at which $f(x) = x^3 - 12x$ has a horizontal tangent line. ◄

Higher-Order Derivatives

➤ Parentheses are placed around n to distinguish a derivative from a power. Therefore $f^{(n)}$ is the nth derivative of f and f^n is the function f raised to the nth power.

➤ The notation $\dfrac{d^2 f}{dx^2}$ comes from $\dfrac{d}{dx}\left(\dfrac{df}{dx}\right)$ and is read d 2 f dx squared.

Because the derivative of a function f is a function in its own right, we can take the derivative of f'. The result is the *second derivative of f*, denoted f'' (read f *double prime*). The derivative of the second derivative is the *third derivative of f*, denoted f''' or $f^{(3)}$ (f *triple prime*). For any positive integer n, $f^{(n)}$ represents the nth derivative of f. Other common notations for the nth derivative of $y = f(x)$ include $\dfrac{d^n f}{dx^n}$ and $y^{(n)}$. In general, derivatives of order $n \geq 2$ are called **higher-order derivatives**.

DEFINITION   Higher-Order Derivatives

Assuming f can be differentiated as often as necessary, the **second derivative** of f is

$$f''(x) = f^{(2)}(x) = \frac{d^2f}{dx^2} = \frac{d}{dx}[f'(x)].$$

For integers $n \geq 1$, the ***n*th derivative** is

$$f^{(n)}(x) = \frac{d^nf}{dx^n} = \frac{d}{dx}[f^{(n-1)}(x)].$$

EXAMPLE 6   Finding higher-order derivatives Find the third derivative of the following functions.

a. $f(x) = 3x^3 - 5x + 12$ **b.** $y = 3t + 2e^t$

> In Example 6, note that $f^{(4)}(x) = 0$, which means that all successive derivatives are also 0. In general, the nth derivative of an nth-degree polynomial is a constant, which implies that derivatives of order $k > n$ are 0.

SOLUTION

a.

$$f'(x) = 9x^2 - 5$$

$$f''(x) = \frac{d}{dx}(9x^2 - 5) = 18x$$

$$f'''(x) = 18$$

b. Here we use an alternative notation for higher-order derivatives:

$$\frac{dy}{dt} = \frac{d}{dt}(3t + 2e^t) = 3 + 2e^t$$

$$\frac{d^2y}{dt^2} = \frac{d}{dt}(3 + 2e^t) = 2e^t$$

$$\frac{d^3y}{dt^3} = \frac{d}{dt}(2e^t) = 2e^t$$

QUICK CHECK 6 With $f(x) = x^5$, find $f^{(5)}(x)$ and $f^{(6)}(x)$. With $g(x) = e^x$, find $g^{(100)}(x)$. ◄

In this case, $\dfrac{d^ny}{dt^n} = 2e^t$ for $n \geq 2$.   *Related Exercises 42–46* ◄

SECTION 3.2 EXERCISES

Review Questions

Assume the derivatives of f and g exist in Exercises 1–6.

1. If the limit definition of a derivative can be used to find f', then what is the purpose of using other rules to find f'?

2. In this section, it is shown that the rule $\dfrac{d}{dx}(x^n) = nx^{n-1}$ is valid for what values of n?

3. Give a nonzero function that is its own derivative.

4. How do you find the derivative of the sum of two functions, $f + g$?

5. How do you find the derivative of a constant multiplied by a function?

6. How do you find the fifth derivative of a function?

Basic Skills

7–12. Derivatives of power and constant functions *Find the derivative of the following functions.*

7. $y = x^5$ 8. $f(t) = t^{11}$ 9. $f(x) = 5$

10. $g(x) = e^3$ 11. $h(t) = t$ 12. $f(v) = v^{100}$

13–18. Derivatives of constant multiples of functions *Find the derivative of the following functions.*

13. $f(x) = 5x^3$ 14. $g(w) = \frac{5}{6}w^{12}$ 15. $p(x) = 8x$

16. $g(t) = 6\sqrt{t}$ 17. $g(t) = 100t^2$ 18. $f(s) = \dfrac{\sqrt{s}}{4}$

19–24. Derivatives of the sum of functions *Find the derivative of the following functions.*

19. $f(x) = 3x^4 + 7x$

20. $g(x) = 6x^5 - x$

21. $f(x) = 10x^4 - 32x + e^2$

22. $f(t) = 6\sqrt{t} - 4t^3 + 9$

23. $g(w) = 2w^3 + 3w + e^w$

24. $s(t) = 4\sqrt{t} - \frac{1}{4}t^4 + t + 1$

25–28. Derivatives of products *Find the derivative of the following functions by first expanding the expression. Simplify your answers.*

25. $f(x) = (2x + 1)(3x^2 + 2)$

26. $g(r) = (5r^3 + 3r + 1)(r^2 + 3)$

27. $h(x) = (x^2 + 1)^2$

28. $h(x) = \sqrt{x}(\sqrt{x} - 1)$

29–34. Derivatives of quotients *Find the derivative of the following functions by first simplifying the expression.*

29. $f(w) = \dfrac{w^3 - w}{w}$

30. $y = \dfrac{12s^3 - 8s^2 + 12s}{4s}$

31. $g(x) = \dfrac{x^2 - 1}{x - 1}$

32. $h(x) = \dfrac{x^3 - 6x^2 + 8x}{x^2 - 2x}$

33. $y = \dfrac{x - a}{\sqrt{x} - \sqrt{a}}$; a is a positive constant.

34. $y = \dfrac{x^2 - 2ax + a^2}{x - a}$; a is a constant.

35–38. Equations of tangent lines

   a. *Find an equation of the tangent line at $x = a$.*

   b. *Use a graphing utility to graph the curve and the tangent line on the same set of axes.*

35. $y = -3x^2 + 2$; $a = 1$

36. $y = x^3 - 4x^2 + 2x - 1$; $a = 2$

37. $y = e^x$; $a = \ln 3$

38. $y = \dfrac{e^x}{4} - x$; $a = 0$

39. Finding slope locations Let $f(x) = x^2 - 6x + 5$.

   a. Find the values of x for which the slope of the curve $y = f(x)$ is 0.

   b. Find the values of x for which the slope of the curve $y = f(x)$ is 2.

40. Finding slope locations Let $f(t) = t^3 - 27t + 5$.

   a. Find the values of t for which the slope of the curve $y = f(t)$ is 0.

   b. Find the values of t for which the slope of the curve $y = f(t)$ is 21.

41. Finding slope locations Let $f(x) = 2x^3 - 3x^2 - 12x + 4$.

   a. Find all points on the graph of f at which the tangent line is horizontal.

   b. Find all points on the graph of f at which the tangent line has slope 60.

42–46. Higher-order derivatives *Find $f'(x)$, $f''(x)$, and $f^{(3)}(x)$ for the following functions.*

42. $f(x) = 3x^3 + 5x^2 + 6x$

43. $f(x) = 5x^4 + 10x^3 + 3x + 6$

44. $f(x) = 3x^2 + 5e^x$

45. $f(x) = \dfrac{x^2 - 7x - 8}{x + 1}$

46. $f(x) = 10e^x$

Further Explorations

47. Explain why or why not Determine whether the following statements are true and give an explanation or counterexample.

   a. The derivative $\dfrac{d}{dx}(10^5)$ equals $5 \cdot 10^4$.

   b. The slope of a line tangent to $f(x) = e^x$ is never 0.

   c. $\dfrac{d}{dx}(4e^x) = 4xe^{x-1}$

   d. $\dfrac{d}{dx}(e^x) = xe^{x-1}$

   e. The nth derivative $\dfrac{d^n}{dx^n}(5x^3 + 2x + 5)$ equals 0 for any integer $n \geq 3$.

48. Tangent lines Suppose the derivative of the function f exists and assume $f(3) = 1$ and $f'(3) = 4$. Let $g(x) = x^2 + f(x)$ and $h(x) = 3f(x)$.

   a. Find an equation of the line tangent to $y = g(x)$ at $x = 3$.

   b. Find an equation of the line tangent to $y = h(x)$ at $x = 3$.

49. Derivatives from tangent lines Suppose the line tangent to the graph of f at $x = 2$ is $y = 4x + 1$ and suppose $y = 3x - 2$ is the line tangent to the graph of g at $x = 2$. Find an equation of the line tangent to the following curves at $x = 2$.

   a. $y = f(x) + g(x)$

   b. $y = f(x) - 2g(x)$

   c. $y = 4f(x)$

50–53. Derivatives from a graph *Let $F = f + g$ and $G = 3f - g$, where the graphs of f and g are shown in the figure. Find the following derivatives.*

50. $F'(1)$ **51.** $G'(1)$ **52.** $F'(5)$ **53.** $G'(5)$

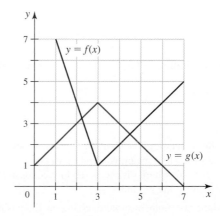

54–56. Derivatives from a table *Use the table to find the following derivatives.*

x	1	2	3	4	5
$f'(x)$	3	5	2	1	4
$g'(x)$	2	4	3	1	5

54. $\dfrac{d}{dx}[f(x) + g(x)]\Big|_{x=1}$

55. $\dfrac{d}{dx}[1.5f(x)]\Big|_{x=2}$

56. $\dfrac{d}{dx}[2x - 3g(x)]\Big|_{x=4}$

57–59. Derivatives from limits *The following limits represent $f'(a)$ for some function f and some value of a.*

a. *Find a function f and a number a.*
b. *Determine the value of the limit by finding $f'(a)$.*

57. $\displaystyle\lim_{h \to 0} \frac{\sqrt{9 + h} - \sqrt{9}}{h}$

58. $\displaystyle\lim_{h \to 0} \frac{(1 + h)^8 + (1 + h)^3 - 2}{h}$

59. $\displaystyle\lim_{x \to 1} \frac{x^{100} - 1}{x - 1}$

60. Important limits Complete the following table and give approximations for $\displaystyle\lim_{h \to 0^-} \frac{2^h - 1}{h}$ and $\displaystyle\lim_{h \to 0^-} \frac{3^h - 1}{h}$.

h	$\dfrac{2^h - 1}{h}$	$\dfrac{3^h - 1}{h}$
−1.0		
−0.1		
−0.01		
−0.001		
−0.0001		
−0.00001		

61–64. Calculator limits *Use a calculator to approximate the following limits.*

61. $\displaystyle\lim_{x \to 0} \frac{e^{3x} - 1}{x}$

62. $\displaystyle\lim_{n \to \infty} \left(1 + \frac{1}{n}\right)^n$

63. $\displaystyle\lim_{x \to 0^+} x^x$

64. $\displaystyle\lim_{x \to 0^+} \left(\frac{1}{x}\right)^x$

65. Calculating limits exactly Find a function f and a number a such that $f'(a) = \displaystyle\lim_{x \to 0} \frac{e^x - 1}{x}$.

Applications

66. Projectile trajectory The position of a small rocket that is launched vertically upward is given by $s(t) = -5t^2 + 40t + 100$ for $0 \le t \le 10$, where t is measured in seconds and s is measured in meters above the ground.

a. Find the instantaneous rate of change in the position (instantaneous velocity) of the rocket for $0 \le t \le 10$ s.

b. At what time is the instantaneous velocity zero?
c. At what time does the instantaneous velocity have the greatest magnitude for $0 \le t \le 10$.
d. Graph the position and instantaneous velocity for $0 \le t \le 10$.

67. Height estimate The distance an object falls (under the influence of Earth's gravity, neglecting air resistance) is given by $d(t) = 16t^2$, where d is measured in ft and t is measured in seconds. A rock climber sits on a ledge on a vertical wall and carefully observes the time it takes for a small stone to fall from the ledge to the ground.

a. Compute $d'(t)$. What units are associated with the derivative, and what does it measure?
b. If it takes 6 s for a stone to fall to the ground, how high is the ledge? How fast is the stone moving when it strikes the ground (in mi/hr)?

68. Cell growth When observations begin at $t = 0$, a cell culture has 1200 cells and continues to grow according to the function $p(t) = 1200\, e^t$, where p is the number of cells and t is measured in days.

a. Compute $p'(t)$. What units are associated with the derivative and what does it measure?
b. On the interval $[0, 4]$, when is the growth rate $p'(t)$ the least? When is it the greatest?

69. Gas mileage Starting with a full tank of gas, the distance traveled by a car is related to the amount of gas consumed by the function $D(g) = 0.05g^2 + 35g$, where D is measured in miles and g in gallons.

a. Compute dD/dg. What units are associated with the derivative and what does it measure?
b. Find dD/dg for $g = 0, 5$, and 10 gal (include units). What do your answers say about the gas mileage for this car?
c. What is the range of this car if it has a 12-gal tank?

Additional Exercises

70. Constant Rule Proof For the constant function $f(x) = c$, use the limit definition of a derivative to show that $f'(x) = 0$.

71. Alternative proof of the Power Rule The Binomial Theorem states that for any positive integer n,

$$(a + b)^n = a^n + na^{n-1}b + \frac{n(n - 1)}{2 \cdot 1} a^{n-2}b^2$$
$$+ \frac{n(n - 1)(n - 2)}{3 \cdot 2 \cdot 1} a^{n-3}b^3 + \cdots + nab^{n-1} + b^n.$$

Use this formula and the definition $f'(x) = \displaystyle\lim_{h \to 0} \frac{f(x + h) - f(x)}{h}$

to show that $\dfrac{d}{dx}(x^n) = nx^{n-1}$ for any positive integer n.

72. Looking ahead: Power Rule for negative integers Suppose n is a negative integer and $f(x) = x^n$. Use the following steps to prove that $f'(a) = na^{n-1}$, which means the Power Rule for positive integers extends to all integers. This result is proved in Section 3.3 by a different method.

a. Assume that $m = -n$, so that $m > 0$. Use the definition

$$f'(a) = \lim_{x \to a} \frac{x^n - a^n}{x - a} = \lim_{x \to a} \frac{x^{-m} - a^{-m}}{x - a}.$$

Simplify using the factoring rule

$$x^n - a^n = (x - a)(x^{n-1} + x^{n-2}a + \cdots + xa^{n-2} + a^{n-1})$$

until it is possible to take the limit.

b. Use this result to find $\dfrac{d}{dx}(x^{-7})$ and $\dfrac{d}{dx}\left(\dfrac{1}{x^{10}}\right)$.

73. Extending the Power Rule to $n = \frac{1}{2}, \frac{3}{2},$ and $\frac{5}{2}$ With Theorem 3.3 and Exercise 72, we have shown that the Power Rule,

$$\frac{d}{dx}(x^n) = nx^{n-1},$$ applies to any integer n. Later in the chapter, we extend this rule so that it applies to any rational number n.

a. Explain why the Power Rule is consistent with the formula

$$\frac{d}{dx}(\sqrt{x}) = \frac{1}{2\sqrt{x}}.$$

b. Prove that the Power Rule holds for $n = \frac{3}{2}$. (*Hint:* Use the definition of the derivative: $\dfrac{d}{dx}(x^{3/2}) = \lim\limits_{h \to 0} \dfrac{(x + h)^{3/2} - x^{3/2}}{h}$.)

c. Prove that the Power Rule holds for $n = \frac{5}{2}$.

d. Propose a formula for $\dfrac{d}{dx}(x^{n/2})$ for any positive integer n.

74. Computing the derivative of $f(x) = e^{-x}$

a. Use the definition of the derivative to show that

$$\frac{d}{dx}(e^{-x}) = e^{-x} \cdot \lim_{h \to 0} \frac{e^{-h} - 1}{h}.$$

b. Show that the limit in part (a) is equal to -1. Use the facts that $\lim\limits_{h \to 0} \dfrac{e^h - 1}{h} = 1$ and e^x is continuous for all x.

c. Use parts (a) and (b) to find the derivative of $f(x) = e^{-x}$.

75. Computing the derivative of $f(x) = e^{2x}$

a. Use the definition of the derivative to show that

$$\frac{d}{dx}(e^{2x}) = e^{2x} \cdot \lim_{h \to 0} \frac{e^{2h} - 1}{h}.$$

b. Show that the limit in part (a) is equal to 2. (*Hint:* Factor $e^{2h} - 1$.)

c. Use parts (a) and (b) to find the derivative of $f(x) = e^{2x}$.

76. Computing the derivative of $f(x) = x^2 e^x$

a. Use the definition of the derivative to show that

$$\frac{d}{dx}(x^2 e^x) = e^x \cdot \lim_{h \to 0} \frac{(x^2 + 2xh + h^2)e^h - x^2}{h}.$$

b. Manipulate the limit in part (a) to arrive at $f'(x) = e^x(x^2 + 2x)$. Use the fact that $\lim\limits_{h \to 0} \dfrac{e^h - 1}{h} = 1$.

QUICK CHECK ANSWERS

1. $\dfrac{d}{dx}(5) = 0$ and $\dfrac{d}{dx}(\pi) = 0$ because 5 and π are constants.

2. The slope of the curve $y = x$ is 1 at any point; therefore,

$$\frac{d}{dx}(x) = 1. \quad \textbf{3.} \ 2x + 2 \quad \textbf{4.} \ f'(x) = 4e^x - 6x$$

5. $x = 2$ and $x = -2$ **6.** $f^{(5)}(x) = 120, f^{(6)}(x) = 0,$ $g^{(100)}(x) = e^x$ ◄

3.3 The Product and Quotient Rules

The derivative of a sum of functions is the sum of the derivatives. So, you might be tempted to assume that the derivative of a product is the product of the derivatives. Consider, however, the functions $f(x) = x^3$ and $g(x) = x^4$. In this case, $\dfrac{d}{dx}[f(x)g(x)] = \dfrac{d}{dx}(x^7) = 7x^6$, but $f'(x)g'(x) = 3x^2 \cdot 4x^3 = 12x^5$. Therefore, $\dfrac{d}{dx}(f \cdot g) \neq f' \cdot g'$. Similarly, the derivative of a quotient is *not* the quotient of the derivatives. The purpose of this section is to develop rules for differentiating products and quotients of functions.

Product Rule

Here is an anecdote that suggests the formula for the Product Rule. Imagine running along a road at a constant speed. Your speed is determined by two factors: the length of your stride and the number of strides you take each second. Therefore,

$$\text{running speed} = \text{stride length} \cdot \text{stride rate}.$$

If your stride length is 3 ft and you take 2 strides/s, then your speed is 6 ft/s.

Now, suppose your stride length increases by 0.5 ft, from 3 to 3.5 ft. Then the change in speed is calculated as follows:

$$change \text{ in speed} = \text{change in stride length} \cdot \text{stride rate}$$

$$= 0.5 \cdot 2 = 1 \text{ ft/s}$$

Alternatively, suppose your stride length remains constant but your stride rate increases by 0.25 strides/s, from 2 to 2.25 strides/s. Then

$$change \text{ in speed} = \text{stride length} \cdot \text{change in stride rate}$$

$$= 3 \cdot 0.25 = 0.75 \text{ ft/s}$$

If both your stride rate and stride length change simultaneously, we expect two contributions to the change in your running speed:

$$change \text{ in speed} = (\text{change in stride length} \cdot \text{stride rate})$$

$$+ (\text{stride length} \cdot \text{change in stride rate})$$

$$= 1 \text{ ft/s} + 0.75 \text{ ft/s} = 1.75 \text{ ft/s}$$

This argument correctly suggests that the derivative (or rate of change) of a product of two functions has *two components*, as shown by the following rule.

> In words, Theorem 3.7 states that the derivative of the product of two functions equals the derivative of the first function multiplied by the second function plus the first function multiplied by the derivative of the second function.

THEOREM 3.7   Product Rule

If f and g are differentiable at x, then

$$\frac{d}{dx}[f(x)g(x)] = f'(x)g(x) + f(x)g'(x).$$

Proof We apply the definition of the derivative to the function fg:

$$\frac{d}{dx}[f(x)g(x)] = \lim_{h \to 0} \frac{f(x+h)g(x+h) - f(x)g(x)}{h}$$

A useful tactic is to add $-f(x)g(x+h) + f(x)g(x+h)$ (which equals 0) to the numerator, so that

$$\frac{d}{dx}[f(x)g(x)]$$

$$= \lim_{h \to 0} \frac{f(x+h)g(x+h) - f(x)g(x+h) + f(x)g(x+h) - f(x)g(x)}{h}$$

The fraction is now split and the numerators are factored:

$$\frac{d}{dx}[f(x)g(x)]$$

$$= \lim_{h \to 0} \frac{f(x+h)g(x+h) - f(x)g(x+h)}{h} + \lim_{h \to 0} \frac{f(x)g(x+h) - f(x)g(x)}{h}$$

$$= \lim_{h \to 0} \left[\overbrace{\frac{f(x+h) - f(x)}{h}}^{\text{approaches } f'(x) \text{ as } h \to 0} \cdot \underbrace{g(x+h)}_{\substack{\text{approaches} \\ g(x) \\ \text{as } h \to 0}} \right] + \lim_{h \to 0} \left[\underbrace{f(x)}_{\substack{\text{equals} \\ f(x) \text{ as} \\ h \to 0}} \cdot \overbrace{\frac{g(x+h) - g(x)}{h}}^{\text{approaches } g'(x) \text{ as } h \to 0} \right]$$

> As $h \to 0$, $f(x)$ does not change in value; it is independent of h.

$$= f'(x) \cdot g(x) + f(x) \cdot g'(x).$$

The continuity of g is used to conclude that $\lim_{h \to 0} g(x+h) = g(x)$. ◄

EXAMPLE 1 **Using the Product Rule** Find and simplify the following derivatives.

a. $\dfrac{d}{dv}[v^2(2\sqrt{v} + 1)]$ **b.** $\dfrac{d}{dx}(x^2 e^x)$

SOLUTION

> Recall from Example 4 of Section 3.1 that $\dfrac{d}{dv}(\sqrt{v}) = \dfrac{1}{2\sqrt{v}}$.

a. $\dfrac{d}{dv}[v^2(2\sqrt{v} + 1)] = \left[\dfrac{d}{dv}(v^2)\right](2\sqrt{v} + 1) + v^2\left[\dfrac{d}{dv}(2\sqrt{v} + 1)\right]$ Product Rule

$= 2v(2\sqrt{v} + 1) + v^2\left(2 \cdot \dfrac{1}{2\sqrt{v}}\right)$ Evaluate the derivatives.

$= (4v^{3/2} + 2v) + v^{3/2} = 5v^{3/2} + 2v$ Simplify.

QUICK CHECK 1 Find the derivative of $f(x) = x^5$. Then, find the same derivative using the Product Rule with $f(x) = x^2 x^3$. ◄

b. $\dfrac{d}{dx}(x^2 e^x) = \underbrace{2x}_{\frac{d}{dx}(x^2)} \cdot e^x + x^2 \cdot \underbrace{e^x}_{\frac{d}{dx}(e^x)} = (2x + x^2)e^x$

Related Exercises 7–16 ◄

Quotient Rule

Consider the quotient $q(x) = \dfrac{f(x)}{g(x)}$ and note that $f(x) = g(x)q(x)$. By the Product Rule, we have

$$f'(x) = g'(x)q(x) + g(x)q'(x).$$

Solving for $q'(x)$, we find that

$$q'(x) = \frac{f'(x) - g'(x)q(x)}{g(x)}.$$

Substituting $q(x) = \dfrac{f(x)}{g(x)}$ produces a rule for finding $q'(x)$:

$$q'(x) = \frac{f'(x) - g'(x)\dfrac{f(x)}{g(x)}}{g(x)}$$ Replace $q(x)$ with $\dfrac{f(x)}{g(x)}$.

$$= \frac{g(x)\left(f'(x) - g'(x)\dfrac{f(x)}{g(x)}\right)}{g(x) \cdot g(x)}$$ Multiply numerator and denominator by $g(x)$.

$$= \frac{g(x)f'(x) - f(x)g'(x)}{[g(x)]^2}$$ Simplify.

This calculation produces the correct result for the derivative of a quotient. However, there is one subtle point: How do we know that the derivative of f/g exists in the first place? A complete proof of the Quotient Rule is outlined in Exercise 76.

> In words, Theorem 3.8 states that the derivative of the quotient of two functions equals the denominator multiplied by the derivative of the numerator minus the numerator multiplied by the derivative of the denominator, all divided by the denominator squared.
>
> An easy way to remember the Quotient Rule is with
>
> $$\frac{LoD(Hi) - HiD(Lo)}{(Lo)^2}.$$

THEOREM 3.8 **The Quotient Rule**

If f and g are differentiable at x, then the derivative of f/g at x exists provided $g(x) \neq 0$ and

$$\frac{d}{dx}\left[\frac{f(x)}{g(x)}\right] = \frac{g(x)f'(x) - f(x)g'(x)}{[g(x)]^2}.$$

EXAMPLE 2   Using the Quotient Rule Find and simplify the following derivatives.

a. $\dfrac{d}{dx}\left[\dfrac{x^2 + 3x + 4}{x^2 - 1}\right]$ **b.** $\dfrac{d}{dx}(e^{-x})$

SOLUTION

> The Product and Quotient Rules are used on a regular basis throughout this text. Therefore, it is a good idea to memorize these rules (along with the other derivative rules and formulas presented in this chapter) so that you can evaluate derivatives quickly.

a.

$$\dfrac{d}{dx}\left[\dfrac{x^2 + 3x + 4}{x^2 - 1}\right] = \dfrac{\overbrace{(x^2 - 1)(2x + 3)}^{\substack{(x^2 - 1)\,\cdot\,\text{the derivative} \\ \text{of }(x^2 + 3x + 4)}} - \overbrace{(x^2 + 3x + 4)2x}^{\substack{(x^2 + 3x + 4)\,\cdot\,\text{the} \\ \text{derivative of }(x^2 - 1)}}}{\underbrace{(x^2 - 1)^2}_{\substack{\text{the denominator} \\ (x^2 - 1)\text{ squared}}}} \qquad \text{Quotient Rule}$$

$$= \dfrac{2x^3 - 2x + 3x^2 - 3 - 2x^3 - 6x^2 - 8x}{(x^2 - 1)^2} \qquad \text{Expand.}$$

$$= \dfrac{-3x^2 - 10x - 3}{(x^2 - 1)^2} \qquad \text{Simplify.}$$

b. We rewrite e^{-x} as $\dfrac{1}{e^x}$, and use the Quotient Rule:

$$\dfrac{d}{dx}\left(\dfrac{1}{e^x}\right) = \dfrac{e^x \cdot 0 - 1 \cdot e^x}{(e^x)^2} = \dfrac{-1}{e^x} = -e^{-x}.$$

Related Exercises 17–26 ◄

QUICK CHECK 2 Find the derivative of $f(x) = x^5$. Then find the same derivative using the Quotient Rule with $f(x) = x^8/x^3$. ◄

EXAMPLE 3   Finding tangent lines Find an equation of the line tangent to the graph of $f(x) = \dfrac{x^2 + 1}{x^2 - 4}$ at the point $(3, 2)$. Plot the curve and tangent line.

SOLUTION To find the slope of the tangent line, we compute f' using the Quotient Rule:

$$f'(x) = \dfrac{(x^2 - 4)\,2x - (x^2 + 1)\,2x}{(x^2 - 4)^2} \qquad \text{Quotient Rule}$$

$$= \dfrac{2x^3 - 8x - 2x^3 - 2x}{(x^2 - 4)^2} = \dfrac{-10x}{(x^2 - 4)^2} \qquad \text{Simplify.}$$

The slope of the tangent line at $(3, 2)$ is

$$m_{\text{tan}} = f'(3) = \dfrac{-10(3)}{(3^2 - 4)^2} = -\dfrac{6}{5}.$$

Therefore, an equation of the tangent line is

$$y - 2 = -\dfrac{6}{5}(x - 3), \quad \text{or} \quad y = -\dfrac{6}{5}x + \dfrac{28}{5}.$$

The graphs of f and the tangent line are shown in Figure 3.25.   *Related Exercises 27–30* ◄

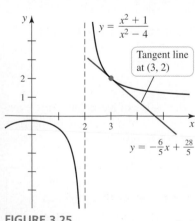

FIGURE 3.25

Extending the Power Rule to Negative Integers

The Power Rule in Section 3.2 says that $\dfrac{d}{dx}(x^n) = nx^{n-1}$ for nonnegative integers n. Using the Quotient Rule, we show that the Power Rule also holds if n is a negative integer. Assume n is a negative integer and let $m = -n$, so that $m > 0$. Then

$$\frac{d}{dx}(x^n) = \frac{d}{dx}\left(\frac{1}{x^m}\right) \qquad\qquad x^n = \frac{1}{x^{-n}} = \frac{1}{x^m}$$

$$= \frac{x^m\overbrace{\left[\dfrac{d}{dx}(1)\right]}^{\substack{\text{derivative of a}\\ \text{constant is 0}}} - 1\overbrace{\left(\dfrac{d}{dx}x^m\right)}^{\substack{\text{equals}\\ mx^{m-1}}}}{(x^m)^2} \qquad \text{Quotient Rule}$$

$$= \frac{-mx^{m-1}}{x^{2m}} \qquad\qquad \text{Simplify.}$$

$$= -mx^{-m-1} \qquad\qquad \frac{x^{m-1}}{x^{2m}} = x^{m-1-2m}$$

$$= nx^{n-1} \qquad\qquad \text{Replace } -m \text{ by } n.$$

THEOREM 3.9   Extended Power Rule

If n is any integer, then

$$\frac{d}{dx}(x^n) = nx^{n-1}.$$

QUICK CHECK 3 Find the derivative of $f(x) = 1/x^5$ in two different ways: using the Extended Power Rule and using the Quotient Rule. ◄

EXAMPLE 4   Using the Extended Power Rule Find the following derivatives.

a. $\dfrac{d}{dx}\left(\dfrac{9}{x^5}\right)$ **b.** $\dfrac{d}{dt}\left[\dfrac{3t^{16} - 4}{t^6}\right]$

SOLUTION

a. $\dfrac{d}{dx}\left(\dfrac{9}{x^5}\right) = \dfrac{d}{dx}(9x^{-5}) = 9(-5x^{-6}) = -45x^{-6} = -\dfrac{45}{x^6}$

b. The derivative of $\dfrac{3t^{16} - 4}{t^6}$ can be evaluated by the Quotient Rule, but an alternative method is to rewrite the expression using negative powers:

$$\frac{3t^{16} - 4}{t^6} = \frac{3t^{16}}{t^6} - \frac{4}{t^6} = 3t^{10} - 4t^{-6}$$

We now differentiate using the Extended Power Rule:

$$\frac{d}{dt}\left[\frac{3t^{16} - 4}{t^6}\right] = \frac{d}{dt}(3t^{10} - 4t^{-6}) = 30t^9 + 24t^{-7}$$

Related Exercises 31–36 ◄

The Derivative of e^{kx}

Consider the composite function $y = e^{2x}$, for which we presently have no differentiation rule. We rewrite the function and apply the Product Rule:

$$\frac{d}{dx}(e^{2x}) = \frac{d}{dx}(e^x \cdot e^x) \qquad\qquad e^{2x} = e^x \cdot e^x$$

$$= \frac{d}{dx}(e^x) \cdot e^x + e^x \cdot \frac{d}{dx}(e^x) \quad \text{Product Rule}$$

$$= e^x \cdot e^x + e^x \cdot e^x = 2e^{2x} \quad \text{Evaluate derivatives.}$$

In a similar fashion, $y = e^{3x}$ is differentiated by writing it as the product $y = e^x \cdot e^{2x}$. You should verify that $\frac{d}{dx}(e^{3x}) = 3e^{3x}$. Extending this strategy, it can be shown that $\frac{d}{dx}(e^{kx}) = ke^{kx}$ for positive integers k (Exercise 78 illustrates a proof by induction). The Quotient Rule is used to show that the rule holds for negative integers k (Exercise 79). Finally, we prove in Section 3.6 (Exercise 72) that the rule holds for all real numbers k.

THEOREM 3.10   The derivative of e^{kx}

For real numbers k,

$$\frac{d}{dx}(e^{kx}) = ke^{kx}.$$

EXAMPLE 5   Exponential derivatives Compute dy/dx for the following functions.

a. $y = xe^{5x}$ **b.** $y = 1000e^{0.07x}$

SOLUTION

a. We use the Product Rule and the fact that $\frac{d}{dx}(e^{kx}) = ke^{kx}$:

$$\frac{dy}{dx} = \underbrace{1}_{\frac{d}{dx}(x) = 1} \cdot e^{5x} + x \cdot \underbrace{5e^{5x}}_{\frac{d}{dx}(e^{5x}) = 5e^{5x}} = (1 + 5x)e^{5x}$$

b. Here we use the Constant Multiple Rule:

$$\frac{dy}{dx} = 1000 \cdot \frac{d}{dx}(e^{0.07x}) = 1000 \cdot 0.07e^{0.07x} = 70e^{0.07x}$$

QUICK CHECK 4 Find the derivative of $f(x) = 4e^{0.5x}$. ◄

Related Exercises 37–42 ◄

Rates of Change

The derivative provides information about the instantaneous rate of change of a function. The next example illustrates this concept.

EXAMPLE 6   Population growth rates The population of a culture of cells increases and approaches a constant level (often called the *steady state* or a *carrying capacity*) and is modeled by the function $p(t) = \dfrac{400}{1 + 3e^{-0.5t}}$, where $t \geq 0$ is measured in hr (Figure 3.26).

a. Compute and graph the instantaneous growth rate of the population for any $t \geq 0$.

b. At approximately what time is the instantaneous growth rate the greatest?

c. What is the steady-state population?

SOLUTION

a. The instantaneous growth rate is given by the derivative of the population function:

$$p'(t) = \frac{d}{dt}\left(\frac{400}{1 + 3e^{-0.5t}}\right)$$

$$= \frac{(1 + 3e^{-0.5t}) \cdot \overbrace{\frac{d}{dt}(400)}^{\text{equals } 0} - 400\frac{d}{dt}(1 + 3e^{-0.5t})}{(1 + 3e^{-0.5t})^2} \qquad \text{Quotient Rule}$$

$$= \frac{-400(-1.5e^{-0.5t})}{(1 + 3e^{-0.5t})^2} = \frac{600e^{-0.5t}}{(1 + 3e^{-0.5t})^2} \qquad \text{Simplify.}$$

> Methods for determining exactly when the growth rate is a maximum are discussed in Chapter 4.

The growth rate has units of cells per hour; its graph is shown in Figure 3.26.

b. The growth rate $p'(t)$ has a maximum at the point at which the population curve is steepest. Using a graphing utility, this point corresponds to $t \approx 2.20$ hr and the growth rate has a value of $p'(2.20) \approx 50$ cells/hr.

c. To determine whether the population approaches a fixed value after a long period of time (the steady-state population), we must investigate the limit of the population function as $t \to \infty$. In this case, the steady-state population exists and is

$$\lim_{t \to \infty} p(t) = \lim_{t \to \infty} \frac{400}{1 + \underbrace{3e^{-0.5t}}_{\text{approaches } 0}} = 400$$

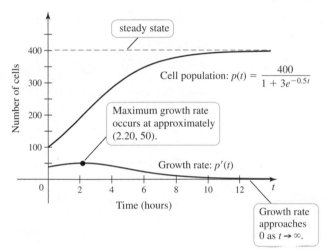

FIGURE 3.26

which is confirmed by the population curve in Figure 3.26. Notice that as the population approaches its steady state, the growth rate p' approaches zero.   *Related Exercises 43–48* ◄

Combining Derivative Rules

Some situations call for the use of multiple differentiation rules. This section concludes with one such example.

EXAMPLE 7 **Combining derivative rules** Find the derivative of

$$y = \frac{4xe^x}{x^2 + 1}.$$

SOLUTION In this case, we have the quotient of two functions, with a product $(4x \cdot e^x)$ in the numerator.

$$\frac{dy}{dx} = \frac{(x^2 + 1) \cdot \frac{d}{dx}(4xe^x) - (4xe^x) \cdot \frac{d}{dx}(x^2 + 1)}{(x^2 + 1)^2} \qquad \text{Quotient Rule}$$

$$= \frac{(x^2 + 1)(4e^x + 4xe^x) - (4xe^x)(2x)}{(x^2 + 1)^2} \qquad \begin{array}{l}\frac{d}{dx}(4xe^x) = 4e^x + 4xe^x \\ \text{by the Product Rule}\end{array}$$

$$= \frac{4e^x(x^3 - x^2 + x + 1)}{(x^2 + 1)^2}. \qquad \text{Simplify.}$$

Related Exercises 49–52 ◄

SECTION 3.3 EXERCISES

Review Questions

1. How do you find the derivative of the product of two functions that are differentiable at a point?

2. How do you find the derivative of the quotient of two functions that are differentiable at a point?

3. State the Extended Power Rule for differentiating x^n. For what values of n does the rule apply?

4. Give two ways to differentiate $f(x) = 1/x^{10}$.

5. What is the derivative of $y = e^{kx}$? For what values of k does this rule apply?

6. Give two ways to differentiate $f(x) = (x - 3)(x^2 + 4)$.

Basic Skills

7–12. Derivatives of products *Find the derivative of the following functions.*

7. $f(x) = 3x^4(2x^2 - 1)$

8. $g(x) = 6x - 2xe^x$

9. $h(x) = (5x^7 + 5x)(6x^3 + 3x^2 + 3)$

10. $f(x) = \left(1 + \dfrac{1}{x^2}\right)(x^2 + 1)$

11. $g(w) = e^w(w^3 - 1)$

12. $s(t) = 4e^t\sqrt{t}$

13–16. Derivatives by two different methods

a. Use the Product Rule to find the derivative of the given function. Simplify your result.
b. Find the derivative by expanding the product first. Verify that your answer agrees with part (a).

13. $f(x) = (x - 1)(3x + 4)$

14. $y = (t^2 + 7t)(3t - 4)$

15. $g(y) = (3y^4 - y^2)(y^2 - 4)$

16. $h(z) = (z^3 + 4z^2 + z)(z - 1)$

17–22. Derivatives of quotients *Find the derivative of the following functions.*

17. $f(x) = \dfrac{x}{x + 1}$

18. $f(x) = \dfrac{x^3 - 4x^2 + x}{x - 2}$

19. $y = (3t - 1)(2t - 2)^{-1}$

20. $h(w) = \dfrac{w^2 - 1}{w^2 + 1}$

21. $g(x) = \dfrac{e^x}{x^2 - 1}$

22. $y = (2\sqrt{x} - 1)(4x + 1)^{-1}$

23–26. Derivatives by two different methods

a. Use the Quotient Rule to find the derivative of the given function. Simplify your result.
b. Find the derivative by first simplifying the function. Verify that your answer agrees with part (a).

23. $f(w) = \dfrac{w^3 - w}{w}$

24. $y = \dfrac{12s^3 - 8s^2 + 12s}{4s}$

25. $y = \dfrac{x - a}{\sqrt{x} - \sqrt{a}}$; a is a positive constant.

26. $y = \dfrac{x^2 - 2ax + a^2}{x - a}$; a is a constant.

27–30. Equations of tangent lines

a. Find an equation of the line tangent to the given curve at $x = a$.
b. Use a graphing utility to graph the curve and the tangent line on the same set of axes.

27. $y = \dfrac{x + 5}{x - 1}$; $a = 3$

28. $y = \dfrac{2x^2}{3x - 1}$; $a = 1$

29. $y = xe^x$; $a = -1$

30. $y = \dfrac{e^x}{x}$; $a = 1$

31–36. Extended Power Rule *Find the derivative of the following functions.*

31. $f(x) = 3x^{-9}$

32. $y = \dfrac{4}{p^3}$

33. $g(t) = 3t^2 + \dfrac{6}{t^7}$

34. $y = \dfrac{w^4 + 5w^2 + w}{w^2}$

35. $g(t) = \dfrac{t^3 + 3t^2 + t}{t^3}$

36. $p(x) = \dfrac{4x^3 + 3x + 1}{2x^5}$

37–42. Derivatives with exponentials *Compute the derivative of the following functions.*

37. $f(x) = 15e^{3x}$

38. $y = 3x^2 - 2x + e^{-2x}$

39. $g(x) = \dfrac{x}{e^x}$

40. $f(x) = (1 - 2x)e^{-x}$

41. $y = \dfrac{2e^x + 3e^{-x}}{3}$

42. $A = 2500e^{0.075t}$

43–44. Population growth *Consider the following population functions.*

a. Find the instantaneous growth rate of the population for $t \geq 0$.
b. What is the instantaneous growth rate at $t = 5\,hr$?
c. At what time is the instantaneous growth rate the greatest?
d. Evaluate and interpret $\lim\limits_{t \to \infty} p'(t)$.
e. Use a graphing utility to graph the population and its growth rate for $0 \leq t \leq 200$.

43. $p(t) = \dfrac{200t}{t + 2}$

44. $p(t) = \dfrac{800}{1 + 7e^{-0.2t}}$

45. **Antibiotic decay** The half-life of an antibiotic in the bloodstream is 10 hr. If an initial dose of 20 mg is administered, the quantity left after t hours is modeled by $Q(t) = 20e^{-0.0693t}$ for $t \geq 0$.

a. Find the instantaneous rate of change of the amount of antibiotic in the bloodstream for $0 \leq t \leq 10$ hr.
b. How fast is the amount of antibiotic changing at $t = 0$ hr? At $t = 2$ hr?
c. Evaluate and interpret $\lim\limits_{t \to \infty} Q(t)$ and $\lim\limits_{t \to \infty} Q'(t)$.

46. **Bank account** A $200 investment in a savings account grows according to $A(t) = 200e^{0.0398t}$ for $t \geq 0$.

a. Find the balance of the account after 10 years.
b. How fast is the account growing (in dollars/year) at $t = 10$ years?
c. Use your answers to parts (a) and (b) to write the equation of the line tangent to the curve $A = 200e^{0.0398t}$ at the point $(10, A(10))$.

47. Finding slope locations Let $f(x) = xe^{2x}$.

    a. Find the values of x for which the slope of the curve $y = f(x)$ is 0.

    b. Explain the meaning of your answer to part (a) in terms of the graph of f.

48. Finding slope locations Let $f(t) = 100e^{-0.05t}$.

    a. Find the values of t for which the slope of the curve $y = f(t)$ is -5.

    b. Does the graph of f have a horizontal tangent line?

49–52. Combining rules *Compute the derivative of the following functions.*

49. $g(x) = \dfrac{x(3 - x)}{2x^2}$

50. $h(x) = \dfrac{(x - 1)(2x^2 - 1)}{(x^3 - 1)}$

51. $g(x) = \dfrac{4e^{x/4}}{(x^2 + x)(1 - x)}$

52. $h(x) = \dfrac{(x + 1)}{x^2 e^x}$

Further Explorations

53. Explain why or why not Determine whether the following statements are true and give an explanation or counterexample.

    a. The derivative $\dfrac{d}{dx}(e^5)$ equals $5 \cdot e^4$.

    b. The Quotient Rule must be used to evaluate $\dfrac{d}{dx}\left(\dfrac{x^2 + 3x + 2}{x}\right)$.

    c. $\dfrac{d}{dx}\left(\dfrac{1}{x^5}\right) = \dfrac{1}{5x^4}$

    d. $\dfrac{d^n}{dx^n}(e^{3x}) = 3^n \cdot e^{3x}$ for any integer $n \geq 1$.

54–57. Higher-order derivatives *Find $f'(x)$, $f''(x)$, and $f'''(x)$.*

54. $f(x) = \dfrac{1}{x}$

55. $f(x) = x^2 e^{3x}$

56. $f(x) = \dfrac{x}{x + 2}$

57. $f(x) = \dfrac{x^2 - 7x}{x + 1}$

58–61. Choose your method *Use any method to evaluate the derivative of the following functions.*

58. $f(x) = \dfrac{4 - x^2}{x - 2}$

59. $f(x) = 4x^2 - \dfrac{2x}{5x + 1}$

60. $f(z) = z^2(e^{3z} + 4) - \dfrac{2z}{z^2 + 1}$

61. $h(r) = \dfrac{2 - r - \sqrt{r}}{r + 1}$

62. Tangent lines Suppose the derivative of f exists, and assume that $f(2) = 2$ and $f'(2) = 3$. Let $g(x) = x^2 \cdot f(x)$ and $h(x) = \dfrac{f(x)}{x - 3}$.

    a. Find an equation of the line tangent to $y = g(x)$ at $x = 2$.

    b. Find an equation of the line tangent to $y = h(x)$ at $x = 2$.

63. The Witch of Agnesi The graph of $y = \dfrac{a^3}{x^2 + a^2}$, where a is a constant is called the *witch of Agnesi* (named after the 18th-century Italian mathematician Maria Agnesi).

    a. Let $a = 3$ and find the line tangent to $y = \dfrac{27}{x^2 + 9}$ at $x = 2$.

    b. Plot the function and the tangent line found in part (a).

64–69. Derivatives from a table *Use the following table to find the given derivatives.*

x	1	2	3	4	5
$f(x)$	5	4	3	2	1
$f'(x)$	3	5	2	1	4
$g(x)$	4	2	5	3	1
$g'(x)$	2	4	3	1	5

64. $\dfrac{d}{dx}[f(x)g(x)]\Big|_{x=1}$

65. $\dfrac{d}{dx}\left[\dfrac{f(x)}{g(x)}\right]\Big|_{x=2}$

66. $\dfrac{d}{dx}[xf(x)]\Big|_{x=3}$

67. $\dfrac{d}{dx}\left[\dfrac{f(x)}{(x + 2)}\right]\Big|_{x=4}$

68. $\dfrac{d}{dx}\left[\dfrac{xf(x)}{g(x)}\right]\Big|_{x=4}$

69. $\dfrac{d}{dx}\left[\dfrac{f(x)g(x)}{x}\right]\Big|_{x=4}$

70. Derivatives from tangent lines Suppose the line tangent to the graph of f at $x = 2$ is $y = 4x + 1$ and suppose $y = 3x - 2$ is the line tangent to the graph of g at $x = 2$. Find the line tangent to the following curves at $x = 2$.

    a. $y = f(x)g(x)$

    b. $y = \dfrac{f(x)}{g(x)}$

Applications

71. Electrostatic force The magnitude of the electrostatic force between two point charges Q and q of the same sign is given by $F(x) = \dfrac{kQq}{x^2}$, where x is the distance between the charges and $k = 9 \times 10^9\,\text{N} \cdot \text{m}^2/\text{C}^2$ is a physical constant (C stands for coulomb, the unit of charge; N stands for newton, the unit of force).

    a. Find the instantaneous rate of change of the force with respect to the distance between the charges.

    b. For two identical charges with $Q = q = 1\,\text{C}$, what is the instantaneous rate of change of the force at a separation of $x = 0.001$ m?

    c. Does the instantaneous rate of change of the force increase or decrease with the separation? Explain.

72. Gravitational force The magnitude of the gravitational force between two objects of mass M and m is given by $F(x) = -\dfrac{GMm}{x^2}$, where x is the distance between the centers of mass of the objects and $G = 6.7 \times 10^{-11}\,\text{N} \cdot \text{m}^2/\text{kg}^2$ is the gravitational constant (N stands for newton, the unit of force; the negative sign indicates an attractive force).

    a. Find the instantaneous rate of change of the force with respect to the distance between the objects.

    b. For two identical objects of mass $M = m = 0.1$ kg, what is the instantaneous rate of change of the force at a separation of $x = 0.01$ m?

    c. Does the instantaneous rate of change of the force increase or decrease with the separation? Explain.

Additional Exercises

73. Special Product Rule In general, the derivative of a product is not the product of the derivatives. Find nonconstant functions f and g such that the derivative of fg equals $f'g'$.

74. Special Quotient Rule In general, the derivative of a quotient is not the quotient of the derivatives. Find nonconstant functions f and g such that the derivative of f/g equals f'/g'.

75. Means and tangents Suppose f is differentiable on an interval containing a and b, and let $P(a, f(a))$ and $Q(b, f(b))$ be distinct points on the graph of f. Let c be the x-coordinate of the point at which the tangent lines to the curve at P and Q intersect, assuming that the tangent lines are not parallel (see figure).

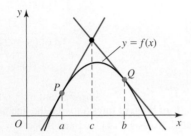

a. If $f(x) = x^2$, show that $c = (a + b)/2$, the arithmetic mean of a and b, for real numbers a and b.
b. If $f(x) = \sqrt{x}$, show that $c = \sqrt{ab}$, the geometric mean of a and b, for $a > 0$ and $b > 0$.
c. If $f(x) = 1/x$, show that $c = 2ab/(a + b)$, the harmonic mean of a and b, for $a > 0$ and $b > 0$.
d. Find an expression for c in terms of a and b for any (differentiable) function f whenever c exists.

76. Proof of the Quotient Rule Let $F = f/g$ be the quotient of two functions that are differentiable at x.

a. Use the definition of F' to show that
$$\frac{d}{dx}\left[\frac{f(x)}{g(x)}\right] = \lim_{h \to 0} \frac{f(x + h)g(x) - f(x)g(x + h)}{h \cdot g(x + h) \cdot g(x)}.$$

b. Now add $-f(x)g(x) + f(x)g(x)$ (which equals 0) to the numerator in the preceding limit to obtain
$$\lim_{h \to 0} \frac{f(x + h)g(x) - f(x)g(x) + f(x)g(x) - f(x)g(x + h)}{h \cdot g(x + h) \cdot g(x)}.$$

Use this limit to obtain the Quotient Rule.
c. Explain why $F' = (f/g)'$ exists, whenever $g(x) \neq 0$.

77. Product Rule for the second derivative Assuming the first and second derivatives of f and g exist, find a formula for
$$\frac{d^2}{dx^2}[f(x)g(x)].$$

78. Proof by induction: derivative of e^{kx} for positive integers k
Proof by induction is a method in which one begins by showing that a statement, which involves positive integers, is true for a particular value (usually $k = 1$). In the second step, the statement is assumed to be true for $k = n$, and the statement is proved for $k = n + 1$, which concludes the proof.

a. Show that $\dfrac{d}{dx}(e^{kx}) = ke^{kx}$ for $k = 1$.
b. Assume the rule is true for $k = n$ (that is, assume $\dfrac{d}{dx}(e^{nx}) = ne^{nx}$), and show this implies that the rule is true for $k = n + 1$. (*Hint:* write $e^{(n+1)x}$ as the product of two functions, and use the Product Rule.)

79. Derivative of e^{kx} for negative integers k Use the Quotient Rule and Exercise 78 to show that $\dfrac{d}{dx}(e^{kx}) = ke^{kx}$ for negative integers k.

80. Quotient Rule for the second derivative Assuming the first and second derivatives of f and g exist, find a formula for $\dfrac{d^2}{dx^2}\left[\dfrac{f(x)}{g(x)}\right]$.

81. Product Rule for three functions Assume that f, g, and h are differentiable at x.

a. Use the Product Rule (twice) to find a formula for
$$\frac{d}{dx}[f(x)g(x)h(x)].$$

b. Use the formula in (a) to find $\dfrac{d}{dx}[e^{2x}(x - 1)(x + 3)]$.

82. One of the Leibniz Rules One of several Leibniz Rules in calculus deals with higher-order derivatives of products. Let $(fg)^{(n)}$ denote the nth derivative of the product fg for $n \geq 1$.

a. Prove that $(fg)^{(2)} = gf'' + 2f'g' + fg''$.
b. Prove that, in general,
$$(fg)^{(n)} = \sum_{k=0}^{n}\binom{n}{k}f^{(k)}g^{(n-k)}$$

where $\dbinom{n}{k} = \dfrac{n!}{k!(n - k)!}$ are the binomial coefficients.
c. Compare the result of (b) to the expansion of $(a + b)^n$.

QUICK CHECK ANSWERS

1. $f'(x) = 5x^4$ by either method **2.** $f'(x) = 5x^4$ by either method **3.** $f'(x) = -5x^{-6}$ by either method
4. $f'(x) = 2e^{0.5x}$ ◄

3.4 Derivatives of Trigonometric Functions

From variations in market trends and ocean temperatures to daily fluctuations in tides and hormone levels, change is often cyclical or periodic. Trigonometric functions are well suited for describing such cyclical behavior. In this section, we investigate the derivatives of trigonometric functions and their many uses.

➤ Results stated in this section assume that angles are measured in *radians*.

Two Special Limits

Our principal goal is to determine derivative formulas for $\sin x$ and $\cos x$. In order to do this, we use two special limits.

THEOREM 3.11   Trigonometric Limits

$$\lim_{x \to 0} \frac{\sin x}{x} = 1 \qquad \lim_{x \to 0} \frac{\cos x - 1}{x} = 0$$

Table 3.2

x	$\dfrac{\sin x}{x}$
±0.1	0.9983341665
±0.01	0.9999833334
±0.001	0.9999998333

Note that these limits cannot be evaluated by direct substitution because in both cases, the numerator and denominator approach zero as $x \to 0$. We first examine numerical and graphical evidence supporting Theorem 3.11, and then we offer an analytic proof.

The values of $\dfrac{\sin x}{x}$, rounded to 10 digits, appear in Table 3.2. As x approaches zero from both sides, it appears that $\dfrac{\sin x}{x}$ approaches 1. Figure 3.27 shows a graph of $y = \dfrac{\sin x}{x}$, with a hole at $x = 0$, where the function is undefined. The graphical evidence also strongly suggests (but does not prove) that $\lim\limits_{x \to 0} \dfrac{\sin x}{x} = 1$. Similar evidence also indicates that $\dfrac{\cos x - 1}{x}$ approaches 0 as x approaches 0.

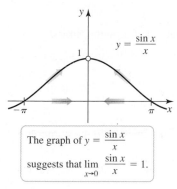

The graph of $y = \dfrac{\sin x}{x}$ suggests that $\lim\limits_{x \to 0} \dfrac{\sin x}{x} = 1$.

FIGURE 3.27

Using a geometric argument and the methods of Chapter 2, we now prove $\lim\limits_{x \to 0} \dfrac{\sin x}{x} = 1$. The proof that $\lim\limits_{x \to 0} \dfrac{\cos x - 1}{x} = 0$ is found in Exercise 61.

Proof Consider Figure 3.28, in which $\triangle OAD$, $\triangle OBC$, and the sector OAC of the unit circle (with central angle x) are shown. Observe that $0 < x < \pi/2$ and

$$\text{area of } \triangle OAD < \text{area of sector } OAC < \text{area of } \triangle OBC. \tag{1}$$

Because the circle in Figure 3.28 is a *unit* circle, $OA = OC = 1$. It follows that $\sin x = \dfrac{AD}{OA} = AD$, $\cos x = \dfrac{OD}{OA} = OD$, and $\tan x = \dfrac{BC}{OC} = BC$. From these observations, we conclude that

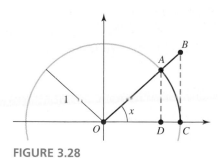

FIGURE 3.28

- area of $\triangle OAD = \dfrac{1}{2}(OD)(AD) = \dfrac{1}{2}\cos x \sin x$

- area of sector $OAC = \dfrac{1}{2} \cdot 1^2 \cdot x = \dfrac{x}{2}$

- area of $\triangle OBC = \dfrac{1}{2}(OC)(BC) = \dfrac{1}{2}\tan x$

➤ Area of the sector of a circle of radius r formed by a central angle θ:

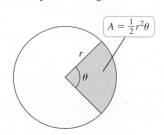

$$A = \tfrac{1}{2}r^2\theta$$

Substituting these results into (1), we have

$$\frac{1}{2}\cos x \sin x < \frac{x}{2} < \frac{1}{2}\tan x.$$

Replacing $\tan x$ with $\dfrac{\sin x}{\cos x}$ and multiplying the inequalities by $\dfrac{2}{\sin x}$ (which is positive) leads to the inequalities

$$\cos x < \frac{x}{\sin x} < \frac{1}{\cos x}.$$

When we take reciprocals and reverse the inequalities, we have

$$\cos x < \frac{\sin x}{x} < \frac{1}{\cos x} \tag{2}$$

for $0 < x < \pi/2$.

A similar argument may be used to show that the inequalities in (2) also hold for $-\pi/2 < x < 0$. Taking the limit as $x \to 0$ in (2), we find that

$$\underbrace{\lim_{x\to 0} \cos x}_{1} < \lim_{x\to 0} \frac{\sin x}{x} < \underbrace{\lim_{x\to 0} \frac{1}{\cos x}}_{1}.$$

The Squeeze Theorem (Theorem 2.5) now implies that $\displaystyle\lim_{x\to 0} \frac{\sin x}{x} = 1$.   ◄

EXAMPLE 1   Calculating trigonometric limits Evaluate the following limits.

a. $\displaystyle\lim_{x\to 0} \frac{\sin 4x}{x}$ **b.** $\displaystyle\lim_{x\to 0} \frac{\sin 3x}{\sin 5x}$

SOLUTION

a. To use the fact that $\displaystyle\lim_{x\to 0} \frac{\sin x}{x} = 1$, the argument of the sine function in the numerator must be the same as the denominator. Multiplying and dividing $\dfrac{\sin 4x}{x}$ by 4, we evaluate the limit as follows:

$$\lim_{x\to 0} \frac{\sin 4x}{x} = \lim_{x\to 0} \frac{4 \sin 4x}{4x} \qquad \text{Multiply and divide by 4.}$$

$$= 4 \lim_{t\to 0} \frac{\sin t}{t} \qquad \text{Factor out 4 and let } t = 4x; \ t \to 0 \text{ as } x \to 0.$$

$$= 4(1) = 4 \qquad \text{Theorem 3.11}$$

b. In order to obtain limits of the form $\displaystyle\lim_{x\to 0} \frac{\sin ax}{ax}$, the first step is to divide the numerator and denominator of $\dfrac{\sin 3x}{\sin 5x}$ by x:

$$\frac{\sin 3x}{\sin 5x} = \frac{(\sin 3x)/x}{(\sin 5x)/x}$$

As in part (a), we now divide and multiply $\dfrac{\sin 3x}{x}$ by 3 and divide and multiply $\dfrac{\sin 5x}{x}$ by 5. In the numerator, we let $t = 3x$, and in the denominator we let $u = 5x$. In each case, $t \to 0$ and $u \to 0$ as $x \to 0$. Therefore,

$$\lim_{x \to 0} \frac{\sin 3x}{\sin 5x} = \lim_{x \to 0} \frac{\dfrac{3 \sin 3x}{3x}}{\dfrac{5 \sin 5x}{5x}} \qquad \text{Multiply and divide by 3 and 5.}$$

$$= \frac{3}{5} \frac{\lim\limits_{t \to 0} (\sin t)/t}{\lim\limits_{u \to 0} (\sin u)/u} \qquad t = 3x \text{ in numerator and } u = 5x \text{ in denominator}$$

$$= \frac{3}{5} \cdot \frac{1}{1} = \frac{3}{5} \qquad \text{Both limits equal 1.} \qquad \textit{Related Exercises 7–14} \blacktriangleleft$$

QUICK CHECK 1 Evaluate $\lim\limits_{x \to 0} \dfrac{\tan 2x}{x}$. ◄

Derivatives of Sine and Cosine Functions

With the trigonometric limits of Theorem 3.11, the derivative of the sine function can be found. We start with the definition of the derivative

$$f'(x) = \lim_{h \to 0} \frac{f(x + h) - f(x)}{h}$$

with $f(x) = \sin x$ and then appeal to the sine addition identity:

$$\sin(x + h) = \sin x \cos h + \cos x \sin h$$

The derivative is

$$f'(x) = \lim_{h \to 0} \frac{\sin(x + h) - \sin x}{h} \qquad \text{Definition of derivative}$$

$$= \lim_{h \to 0} \frac{\sin x \cos h + \cos x \sin h - \sin x}{h} \qquad \text{Sine addition identity}$$

$$= \lim_{h \to 0} \frac{\sin x (\cos h - 1) + \cos x \sin h}{h} \qquad \text{Factor } \sin x.$$

$$= \lim_{h \to 0} \frac{\sin x (\cos h - 1)}{h} + \lim_{h \to 0} \frac{\cos x \sin h}{h} \qquad \text{Theorem 2.3}$$

$$= \sin x \left[\underbrace{\lim_{h \to 0} \frac{\cos h - 1}{h}}_{0} \right] + \cos x \left[\underbrace{\lim_{h \to 0} \frac{\sin h}{h}}_{1} \right] \qquad \begin{array}{l}\text{Both } \sin x \text{ and } \cos x \text{ are} \\ \text{independent of } h.\end{array}$$

$$= (\sin x)(0) + \cos x (1) \qquad \text{Theorem 3.11}$$

$$= \cos x \qquad \text{Simplify.}$$

We have proved the important result that $\dfrac{d}{dx}(\sin x) = \cos x$.

The fact that $\dfrac{d}{dx}(\cos x) = -\sin x$ is proved in a similar way using a cosine addition identity (Exercise 63).

THEOREM 3.12   Derivatives of Sine and Cosine

$$\frac{d}{dx}(\sin x) = \cos x \qquad \frac{d}{dx}(\cos x) = -\sin x$$

From a geometric point of view, these derivative formulas make sense. Because $f(x) = \sin x$ is a periodic function, we expect its derivative to be periodic. Observe that the horizontal tangent lines on the graph of $f(x) = \sin x$ (Figure 3.29a) occur at the zeros of $f'(x) = \cos x$. Similarly, the horizontal tangent lines on the graph of $f(x) = \cos x$ occur at the zeros of $f'(x) = -\sin x$ (Figure 3.29b).

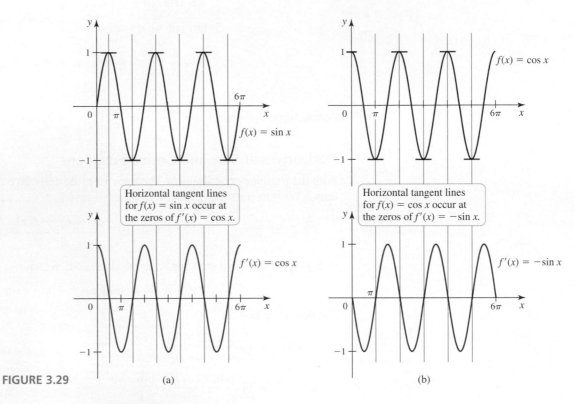

FIGURE 3.29   (a)   (b)

QUICK CHECK 2 At what points on the interval $[0, 2\pi]$ does the graph of $f(x) = \sin x$ have tangent lines with positive slopes? At what points on the interval $[0, 2\pi]$ is $\cos x > 0$? Explain the connection. ◄

EXAMPLE 2   Derivatives involving trigonometric functions Calculate dy/dx for the following functions.

a. $y = e^{2x}\cos x$ **b.** $y = \sin x - x\cos x$ **c.** $y = \dfrac{1 + \sin x}{1 - \sin x}$

SOLUTION

a.
$$\frac{dy}{dx} = \frac{d}{dx}(e^{2x}\cdot\cos x) = \overbrace{2e^{2x}\cos x}^{\substack{\text{derivative of } e^{2x} \\ \cdot \cos x}} + \overbrace{e^{2x}(-\sin x)}^{\substack{e^{2x}\cdot\text{ the} \\ \text{derivative of }\cos x}} \quad \text{Product Rule}$$

$$= e^{2x}(2\cos x - \sin x) \qquad\qquad \text{Simplify.}$$

b.
$$\frac{dy}{dx} = \frac{d}{dx}(\sin x) - \frac{d}{dx}(x\cos x) \qquad\qquad \text{Difference Rule}$$

$$= \cos x - [\underbrace{(1)\cos x}_{\substack{\text{derivative of} \\ x\,\cdot\,\cos x}} + \underbrace{x(-\sin x)}_{\substack{x\,\cdot\,\text{derivative of} \\ \cos x}}] \qquad \text{Product Rule}$$

$$= x\sin x \qquad\qquad\qquad\qquad \text{Simplify.}$$

$$\textbf{c.} \quad \frac{dy}{dx} = \frac{(1 - \sin x)\overbrace{(\cos x)}^{\substack{\text{derivative of} \\ 1 + \sin x}} - (1 + \sin x)\overbrace{(-\cos x)}^{\substack{\text{derivative of} \\ 1 - \sin x}}}{(1 - \sin x)^2} \qquad \text{Quotient Rule}$$

$$= \frac{\cos x - \cos x \sin x + \cos x + \sin x \cos x}{(1 - \sin x)^2} \qquad \text{Expand.}$$

$$= \frac{2 \cos x}{(1 - \sin x)^2} \qquad \text{Simplify.} \quad \textit{Related Exercises 15–22} \blacktriangleleft$$

Derivatives of Other Trigonometric Functions

The derivatives of $\tan x$, $\cot x$, $\sec x$, and $\csc x$ are obtained using the derivatives of $\sin x$ and $\cos x$ together with the Quotient Rule and trigonometric identities.

EXAMPLE 3 Derivative of the tangent function Calculate $\dfrac{d}{dx}(\tan x)$.

> Recall that $\tan x = \dfrac{\sin x}{\cos x}$, $\cot x = \dfrac{\cos x}{\sin x}$, $\sec x = \dfrac{1}{\cos x}$, and $\csc x = \dfrac{1}{\sin x}$.

SOLUTION Using the identity $\tan x = \dfrac{\sin x}{\cos x}$ and the Quotient Rule, we have

$$\frac{d}{dx}(\tan x) = \frac{d}{dx}\left(\frac{\sin x}{\cos x}\right)$$

$$= \frac{\cos x \overbrace{\cos x}^{\substack{\text{derivative of} \\ \sin x}} - \sin x \overbrace{(-\sin x)}^{\substack{\text{derivative of} \\ \cos x}}}{\cos^2 x} \qquad \text{Quotient Rule}$$

$$= \frac{\cos^2 x + \sin^2 x}{\cos^2 x} \qquad \text{Simplify numerator.}$$

$$= \frac{1}{\cos^2 x} = \sec^2 x \qquad \cos^2 x + \sin^2 x = 1$$

Therefore, $\dfrac{d}{dx}(\tan x) = \sec^2 x$.   *Related Exercises 23–25* ◄

The derivatives of $\cot x$, $\sec x$, and $\csc x$ are given in Theorem 3.13 (Exercises 23–25).

> One way to remember Theorem 3.13 is to learn the derivatives of the sine, tangent, and secant functions. Then, replace each function by its corresponding **cofunction** and put a negative sign on the right-hand side of the new derivative equation.
>
> $$\frac{d}{dx}(\sin x) = \cos x \quad \leftrightarrow$$
>
> $$\frac{d}{dx}(\cos x) = -\sin x$$
>
> $$\frac{d}{dx}(\tan x) = \sec^2 x \quad \leftrightarrow$$
>
> $$\frac{d}{dx}(\cot x) = -\csc^2 x$$
>
> $$\frac{d}{dx}(\sec x) = \sec x \tan x \quad \leftrightarrow$$
>
> $$\frac{d}{dx}(\csc x) = -\csc x \cot x$$

THEOREM 3.13 Derivatives of the Trigonometric Functions

$$\frac{d}{dx}(\sin x) = \cos x \qquad\qquad \frac{d}{dx}(\cos x) = -\sin x$$

$$\frac{d}{dx}(\tan x) = \sec^2 x \qquad\qquad \frac{d}{dx}(\cot x) = -\csc^2 x$$

$$\frac{d}{dx}(\sec x) = \sec x \tan x \qquad\qquad \frac{d}{dx}(\csc x) = -\csc x \cot x$$

QUICK CHECK 3 The formulas for $\dfrac{d}{dx}(\cot x)$, $\dfrac{d}{dx}(\sec x)$, and $\dfrac{d}{dx}(\csc x)$ can be determined using the Quotient Rule. Why? ◄

EXAMPLE 4   **Derivatives involving sec x and csc x**   Find the derivative of $y = \sec x \csc x$.

SOLUTION

$$\frac{dy}{dx} = \frac{d}{dx}(\sec x \cdot \csc x)$$

$$= \underbrace{\sec x \tan x \csc x}_{\text{derivative of }\sec x} + \sec x \underbrace{(-\csc x \cot x)}_{\text{derivative of }\csc x} \qquad \text{Product Rule}$$

$$= \underbrace{\frac{1}{\cos x}}_{\sec x} \cdot \underbrace{\frac{\sin x}{\cos x}}_{\tan x} \cdot \underbrace{\frac{1}{\sin x}}_{\csc x} - \underbrace{\frac{1}{\cos x}}_{\sec x} \cdot \underbrace{\frac{1}{\sin x}}_{\csc x} \cdot \underbrace{\frac{\cos x}{\sin x}}_{\cot x} \qquad \begin{array}{l}\text{Write functions in terms of}\\ \sin x \text{ and } \cos x.\end{array}$$

$$= \frac{1}{\cos^2 x} - \frac{1}{\sin^2 x} \qquad \text{Cancel and simplify.}$$

$$= \sec^2 x - \csc^2 x \qquad \begin{array}{l}\text{Definition of }\sec x \text{ and } \csc x\\ \textit{Related Exercises 26–32} \blacktriangleleft\end{array}$$

QUICK CHECK 4 Why is the derivative of $\sec x \csc x$ equal to the derivative of $\dfrac{1}{\cos x \sin x}$? ◄

Higher-Order Trigonometric Derivatives

Higher-order derivatives of the sine and cosine functions are important in many applications. A few higher-order derivatives of $y = \sin x$ reveal a pattern:

$$\frac{dy}{dx} = \cos x \qquad\qquad \frac{d^2 y}{dx^2} = \frac{d}{dx}\cos x = -\sin x$$

$$\frac{d^3 y}{dx^3} = \frac{d}{dx}(-\sin x) = -\cos x \qquad\qquad \frac{d^4 y}{dx^4} = \frac{d}{dx}(-\cos x) = \sin x$$

We see that the higher-order derivatives of $\sin x$ cycle back periodically to $\pm \sin x$. In general, it can be shown that $\dfrac{d^{(2n)} y}{dx^{(2n)}} = (-1)^n \sin x$, with a similar result for $\cos x$ (Exercise 68). This cyclic behavior in the derivatives of $\sin x$ and $\cos x$ does not occur with the other trigonometric functions.

QUICK CHECK 5 Find $\dfrac{d^2 y}{dx^2}$ and $\dfrac{d^4 y}{dx^4}$ when $y = \cos x$. Find $\dfrac{d^{40} y}{dx^{40}}$ and $\dfrac{d^{42} y}{dx^{42}}$ when $y = \sin x$. ◄

EXAMPLE 5   **Second-order derivatives**   Find the second derivative of $y = \csc x$.

SOLUTION By Theorem 3.13, $\dfrac{dy}{dx} = -\csc x \cot x$.

Applying the Product Rule gives the second derivative:

$$\frac{d^2 y}{dx^2} = \frac{d}{dx}(-\csc x \cot x)$$

$$= \left(\frac{d}{dx}(-\csc x)\right)\cot x - \csc x \frac{d}{dx}(\cot x) \qquad \text{Product Rule}$$

$$= (\csc x \cot x)\cot x - \csc x (-\csc^2 x) \qquad \text{Calculate derivatives.}$$

$$= \csc x (\cot^2 x + \csc^2 x) \qquad \text{Factor.}$$

$$\textit{Related Exercises 33–36} \blacktriangleleft$$

SECTION 3.4 EXERCISES

Review Questions

1. Why is it not possible to evaluate $\lim\limits_{x \to 0} \dfrac{\sin x}{x}$ by direct substitution?

2. How is $\lim\limits_{x \to 0} \dfrac{\sin x}{x}$ used in this section?

3. Explain why the Quotient Rule is used to determine the derivative of $\tan x$ and $\cot x$.

4. How can you use the derivatives $\dfrac{d}{dx}(\sin x) = \cos x$, $\dfrac{d}{dx}(\tan x) = \sec^2 x$, and $\dfrac{d}{dx}(\sec x) = \sec x \tan x$ to remember the derivatives of $\cos x$, $\cot x$ and $\csc x$?

5. If $f(x) = \sin x$, then what is the value of $f'(\pi)$?

6. Where does $\sin x$ have a horizontal tangent line? Where does $\cos x$ have a value of zero? Explain the connection between these two observations.

Basic Skills

7–14. Trigonometric limits *Use Theorem 3.11 to evaluate the following limits.*

7. $\lim\limits_{x \to 0} \dfrac{\sin 3x}{x}$

8. $\lim\limits_{x \to 0} \dfrac{\sin 5x}{3x}$

9. $\lim\limits_{x \to 0} \dfrac{\tan 5x}{x}$

10. $\lim\limits_{\theta \to 0} \dfrac{\cos^2 \theta - 1}{\theta}$

11. $\lim\limits_{x \to 0} \dfrac{\tan 7x}{\sin x}$

12. $\lim\limits_{\theta \to 0} \dfrac{\sec \theta - 1}{\theta}$

13. $\lim\limits_{x \to 2} \dfrac{\sin (x - 2)}{x^2 - 4}$

14. $\lim\limits_{x \to -3} \dfrac{\sin (x + 3)}{x^2 + 8x + 15}$

15–22. Calculating derivatives *Find dy/dx for the following functions.*

15. $y = \sin x + \cos x$

16. $y = 5x^2 + \cos x$

17. $y = e^{-x} \sin x$

18. $y = \sin x + 4e^{0.5x}$

19. $y = \sin x \cos x$

20. $y = \dfrac{(x^2 - 1)\sin x}{\sin x + 1}$

21. $y = \cos^2 x$

22. $y = \dfrac{x \sin x}{1 + \cos x}$

23–25. Derivatives of other trigonometric functions *Verify the following derivative formulas using the Quotient Rule.*

23. $\dfrac{d}{dx}(\cot x) = -\csc^2 x$

24. $\dfrac{d}{dx}(\sec x) = \sec x \tan x$

25. $\dfrac{d}{dx}(\csc x) = -\csc x \cot x$

26–32. Derivatives involving other trigonometric functions *Find the derivative of each of the following functions.*

26. $y = \tan x + \cot x$

27. $y = \sec x + \csc x$

28. $y = \dfrac{\tan w}{1 + \tan w}$

29. $y = \dfrac{\cot x}{1 + \csc x}$

30. $y = \dfrac{\tan t}{1 + \sec t}$

31. $y = \dfrac{1}{\sec z \csc z}$

32. $y = \csc^2 \theta - 1$

33–36. Second-order derivatives *Find y'' for each of the following functions.*

33. $y = \cot x$

34. $y = \tan x$

35. $y = \sec x \csc x$

36. $y = \cos \theta \sin \theta$

Further Explorations

37. **Explain why or why not** Determine whether the following statements are true and give an explanation or counterexample.

    a. $\dfrac{d}{dx}(\sin^2 x) = \cos^2 x$

    b. $\dfrac{d^2}{dx^2}(\sin x) = \sin x$

    c. $\dfrac{d^4}{dx^4}(\cos x) = \cos x$

    d. The function $\sec x$ is not differentiable at $x = \pi/2$.

38–43. Trigonometric limits *Evaluate the following limits or state that they do not exist.*

38. $\lim\limits_{x \to 0} \dfrac{\sin ax}{bx}$, where a and b are constants with $b \neq 0$

39. $\lim\limits_{x \to 0} \dfrac{\sin ax}{\sin bx}$, where a and b are constants with $b \neq 0$

40. $\lim\limits_{x \to \pi/2} \dfrac{\cos x}{x - (\pi/2)}$

41. $\lim\limits_{x \to 0} \dfrac{3 \sec^5 x}{x^2 + 4}$

42. $\lim\limits_{x \to \infty} \dfrac{\cos x}{x}$

43. $\lim\limits_{x \to \pi/4} 3 \csc 2x \cot 2x$

44–49. Calculating derivatives *Find dy/dx for the following functions.*

44. $y = \dfrac{\sin x}{1 + \cos x}$

45. $y = x \cos x \sin x$

46. $y = \dfrac{1}{2 + \sin x}$

47. $y = \dfrac{2 \cos x}{1 + \sin x}$

48. $y = \dfrac{x \cos x}{1 + x^3}$

49. $y = \dfrac{1 - \cos x}{1 + \cos x}$

T 50–53. Equations of tangent lines

   a. *Find the equation of the line tangent to the following curves at the given value of x.*

   b. *Use a graphing utility to plot the curve and the tangent line.*

50. $y = 4 \sin x \cos x$; $x = \pi/3$

51. $y = 1 + 2 \sin x$; $x = \pi/6$

52. $y = \csc x$; $x = \pi/4$

53. $y = \dfrac{\cos x}{1 - \cos x}$; $x = \pi/3$

54. Locations of tangent lines

    a. For what values of x does $g(x) = x - \sin x$ have a horizontal tangent line?

    b. For what values of x does $g(x) = x - \sin x$ have a slope of 1?

55. Locations of horizontal tangent lines For what values of x does $f(x) = x - 2\cos x$ have a horizontal tangent line?

56. Matching Match the graphs of the functions in (a)–(d) with the graphs of their derivatives in (A)–(D).

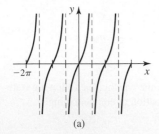

(a)

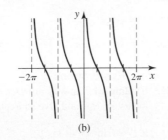

(b)

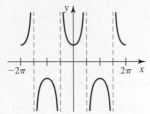

(c)

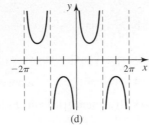

(d)

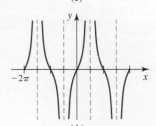

(A)

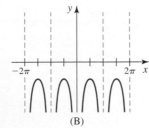

(B)

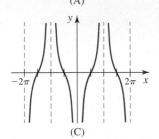

(C)

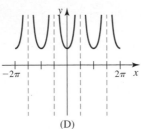

(D)

Applications

57. Velocity of an oscillator An object oscillates along a line, and its displacement in centimeters is given by $y(t) = 30(\sin t - 1)$, where $t \geq 0$ is measured in seconds and y is positive in the upward direction.

    a. Graph the position function for $0 \leq t \leq 10$.

    b. Find the velocity of the oscillator, $v(t) = y'(t)$.

    c. Graph the velocity function for $0 \leq t \leq 10$.

    d. At what times and positions is the velocity zero?

    e. At what times and positions is the velocity a maximum?

    f. The acceleration of the oscillator is $a(t) = v'(t)$. Find and graph the acceleration function.

58. Damped sine wave The graph of $f(t) = e^{-kt}\sin t$ with $k > 0$ is called a *damped* sine wave; it is used in a variety of applications, such as modeling the vibrations of a shock absorber.

    a. Use a graphing utility to graph f for $k = 1, \frac{1}{2}$, and $\frac{1}{10}$ to understand why these curves are called damped sine waves. What effect does k have on the behavior of the graph?

    b. Compute $f'(t)$ for $k = 1$, and use it to determine where the graph of f has a horizontal tangent.

    c. Evaluate $\lim_{t \to \infty} e^{-t}\sin t$ by using the Squeeze Theorem. What does the result say about the vibrations of a damped sine wave?

59. A differential equation A differential equation is an equation involving an unknown function and its derivatives. Consider the differential equation $y''(t) + y(t) = 0$.

    a. Show that $y = A\sin t$ satisfies the equation for any constant A.

    b. Show that $y = B\cos t$ satisfies the equation for any constant B.

    c. Show that $y = A\sin t + B\cos t$ satisfies the equation for any constants A and B.

Additional Exercises

60. Using identities Use the identity $\sin 2x = 2\sin x\cos x$ to find $\frac{d}{dx}(\sin 2x)$. Then use the identity $\cos 2x = \cos^2 x - \sin^2 x$ to express the derivative of $\sin 2x$ in terms of $\cos 2x$.

61. Proof of $\lim_{x \to 0} \dfrac{\cos x - 1}{x} = 0$ Use the trigonometric identity $\cos^2 x + \sin^2 x = 1$ to prove that $\lim_{x \to 0} \dfrac{\cos x - 1}{x} = 0$. (*Hint:* Begin by multiplying the numerator and denominator by $\cos x + 1$.)

62. Another method for proving $\lim_{x \to 0} \dfrac{\cos x - 1}{x} = 0$ Use the half-angle formula $\sin^2 x = \dfrac{1 - \cos(2x)}{2}$ to prove that $\lim_{x \to 0} \dfrac{\cos x - 1}{x} = 0$.

63. Proof of $\dfrac{d}{dx}(\cos x) = -\sin x$ Use the limit definition of the derivative and the trigonometric identity

$$\cos(x + h) = \cos x\cos h - \sin x\sin h$$

to prove that $\dfrac{d}{dx}(\cos x) = -\sin x$.

64. Continuity of a piecewise function Let

$$f(x) = \begin{cases} \dfrac{3\sin x}{x} & \text{if } x \neq 0 \\ a & \text{if } x = 0 \end{cases}$$

For what values of a is f continuous?

65. Continuity of a piecewise function Let

$$g(x) = \begin{cases} \dfrac{1 - \cos x}{2x} & \text{if } x \neq 0 \\ a & \text{if } x = 0 \end{cases}$$

For what values of a is g continuous?

66. Computing limits with angles in degrees Suppose your graphing calculator has two functions, one called $\sin x$, which calculates the sine of x when x is in radians and the other called $s(x)$, which calculates the sine of x when x is in degrees.

a. Explain why $s(x) = \sin\left(\dfrac{\pi}{180} x\right)$.

b. Evaluate $\lim\limits_{x \to 0} \dfrac{s(x)}{x}$. Verify your answer by estimating the limit on your calculator.

67. Derivatives of $\sin^n x$ Calculate the following derivatives using the Product Rule.

a. $\dfrac{d}{dx}(\sin^2 x)$ **b.** $\dfrac{d}{dx}(\sin^3 x)$ **c.** $\dfrac{d}{dx}(\sin^4 x)$

d. Based upon your answers to parts (a)–(c), make a conjecture about the value of $\dfrac{d}{dx}(\sin^n x)$, where n is a positive integer. Then, prove the result by induction.

68. Higher-order derivatives of $\sin x$ and $\cos x$ Prove that

$$\frac{d^{2n}}{dx^{2n}}(\sin x) = (-1)^n \sin x \quad \text{and} \quad \frac{d^{2n}}{dx^{2n}}(\cos x) = (-1)^n \cos x.$$

69–72. Identifying derivatives from limits *The following limits equal the derivative of a function f at a point a.*

a. *Find one possible f and a.*   **b.** *Evaluate the limit.*

69. $\lim\limits_{h \to 0} \dfrac{\sin\left(\dfrac{\pi}{6} + h\right) - \dfrac{1}{2}}{h}$

70. $\lim\limits_{h \to 0} \dfrac{\cos\left(\dfrac{\pi}{6} + h\right) - \dfrac{\sqrt{3}}{2}}{h}$

71. $\lim\limits_{x \to \pi/4} \dfrac{\cot x - 1}{x - \dfrac{\pi}{4}}$

72. $\lim\limits_{h \to 0} \dfrac{\tan\left(\dfrac{5\pi}{6} + h\right) + \dfrac{1}{\sqrt{3}}}{h}$

QUICK CHECK ANSWERS

1. 2   **2.** $0 < x < \frac{\pi}{2}$ and $\frac{3\pi}{2} < x < 2\pi$. The value of $\cos x$ is the slope of the line tangent to the curve $y = \sin x$.
3. The Quotient Rule is used because each function is a quotient when written in terms of the sine and cosine functions.

4. $\dfrac{1}{\cos x \sin x} = \dfrac{1}{\cos x} \cdot \dfrac{1}{\sin x} = \sec x \csc x$

5. $\dfrac{d^2 y}{dx^2} = -\cos x, \dfrac{d^4 y}{dx^4} = \cos x, \dfrac{d^{40}}{dx^{40}}(\sin x) = \sin x,$

$\dfrac{d^{42}}{dx^{42}}(\sin x) = -\sin x$ ◄

3.5 Derivatives as Rates of Change

The theme of this section is the *derivative as a rate of change*. Observing the world around us, we see that almost everything is in a state of change: The size of the Internet is increasing; your blood pressure fluctuates; as supply increases, prices decrease; and the universe is expanding. This section explores a few of the many applications of this idea and demonstrates why calculus is called the mathematics of change.

One-Dimensional Motion

Describing the motion of objects such as projectiles and planets was one of the challenges that led to the development of calculus in the 17th century. We begin by considering the motion of an object confined to one dimension; that is, the object moves along a line. This motion could be horizontal (for example, a car moving along a straight highway) or it could be vertical (such as a projectile launched vertically into the air).

> When describing the motion of objects, it is customary to use t as the independent variable to represent time. Generally, motion is assumed to begin at $t = 0$.

$$\text{Displacement } \Delta s = f(a + \Delta t) - f(a)$$

$s = 0 \quad s = f(a) \quad s = f(a + \Delta t) \quad s$

FIGURE 3.30

Position and Velocity   Suppose an object moves along a straight line and its location at time t is given by the **position function** $s = f(t)$. All positions are measured relative to a reference point, which is often the origin at $s = 0$. The **displacement** of the object between $t = a$ and $t = a + \Delta t$ is $\Delta s = f(a + \Delta t) - f(a)$, where the elapsed time is Δt units (Figure 3.30).

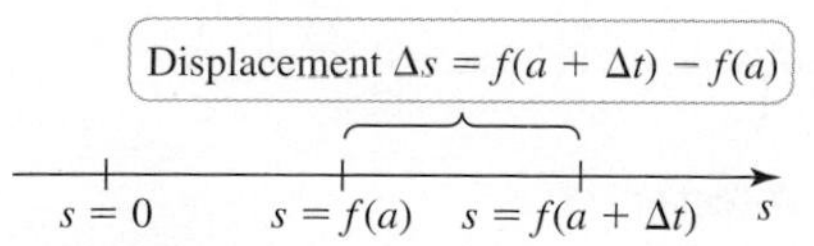

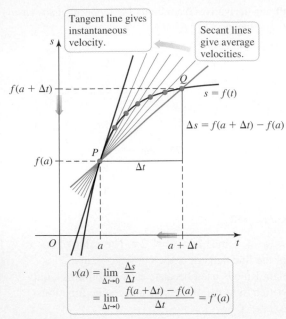

FIGURE 3.31

> Using the various derivative notations, the velocity is also written $v(t) = s'(t) = ds/dt$. If *average* or *instantaneous* is not specified, *velocity* is understood to mean instantaneous velocity.

QUICK CHECK 1 Does the speedometer in your car measure average or instantaneous velocity? ◄

Recall from Section 2.1 that the **average velocity** of the object over the interval $[a, a + \Delta t]$ is the displacement Δs of the object divided by the elapsed time Δt:

$$\frac{\Delta s}{\Delta t} = \frac{f(a + \Delta t) - f(a)}{\Delta t}$$

The average velocity is the slope of the secant line passing through the points $P(a, f(a))$ and $Q(a + \Delta t, f(a + \Delta t))$ (Figure 3.31).

As Δt approaches 0, the average velocity is calculated over smaller and smaller time intervals, and the limiting value of these average velocities is the **instantaneous velocity** at $t = a$. This is the same argument used to arrive at the derivative. The conclusion is that the instantaneous velocity at time a, denoted $v(a)$, is the derivative of the position function evaluated at a:

$$v(a) = \lim_{\Delta t \to 0} \frac{f(a + \Delta t) - f(a)}{\Delta t} = f'(a)$$

Equivalently, the instantaneous velocity at a is the instantaneous rate of change in the position function at a; it also equals the slope of the line tangent to the curve $s = f(t)$ at $P(a, f(a))$.

DEFINITION   Average and Instantaneous Velocity

Let $s = f(t)$ be the position function of an object moving along a line. The **average velocity** of the object over the time interval $[a, a + \Delta t]$ is the slope of the secant line between $(a, f(a))$ and $(a + \Delta t, f(a + \Delta t))$:

$$\frac{f(a + \Delta t) - f(a)}{\Delta t}$$

The **instantaneous velocity** at $t = a$ is the slope of the line tangent to the position curve, which is the derivative of the position function:

$$v(a) = \lim_{\Delta t \to 0} \frac{f(a + \Delta t) - f(a)}{\Delta t} = f'(a)$$

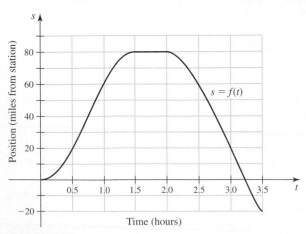

FIGURE 3.32

EXAMPLE 1   Position and velocity of a patrol car Assume a police station is located along a straight east-west freeway. At noon ($t = 0$), a patrol car leaves the station heading east. The position function of the car $s = f(t)$ gives the location of the car in miles east ($s > 0$) or west ($s < 0$) of the station t hours after noon (Figure 3.32).

a. Describe the location of the patrol car during the first 3.5 hr of the trip.

b. Calculate the average velocity of the car between noon and 2:00 P.M. ($0 \leq t \leq 2$).

c. Calculate the displacement and average velocity of the car between 2:00 P.M. and 3:30 P.M. ($2 \leq t \leq 3.5$).

d. At what time(s) is the instantaneous velocity greatest *as the car travels east*?

e. At what time(s) is the patrol car at rest?

SOLUTION

a. The graph of the position function indicates the car travels 80 mi east between $t = 0$ (noon) and $t = 1.5$ (1:30 P.M.). The position of the car does not change from $t = 1.5$ to $t = 2$, and therefore the car is at rest from 1:30 P.M. to 2:00 P.M. Starting at $t = 2$, the car's distance from the station decreases, which means the car travels west, eventually ending up 20 mi west of the station at $t = 3.5$ (3:30 P.M.) (Figure 3.33).

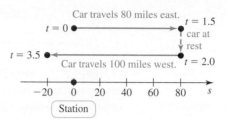

FIGURE 3.33

b. Using Figure 3.32, we find that $f(0) = 0$ mi and $f(2) = 80$ mi. Therefore, the average velocity during the first 2 hours is

$$\frac{\Delta s}{\Delta t} = \frac{f(2) - f(0)}{2 - 0} = \frac{80 \text{ mi}}{2 \text{ hr}} = 40 \text{ mi/hr}.$$

c. The position of the car at 3:30 P.M. is $f(3.5) = -20$ (the negative sign indicates the car is 20 miles *west* of the station), and the position of the car at 2:00 P.M. is $f(2) = 80$. Therefore, the displacement is

$$\Delta s = f(3.5) - f(2) = -20 \text{ mi} - 80 \text{ mi} = -100 \text{ mi}$$

during an elapsed time of $\Delta t = 3.5 - 2 = 1.5$ hr (the *negative* displacement indicates that the car moved 100 miles *west*). The average velocity is

$$\frac{\Delta s}{\Delta t} = \frac{-100 \text{ mi}}{1.5 \text{ hr}} \approx -66.7 \text{ mi/hr}.$$

d. The greatest eastward instantaneous velocity corresponds to points at which the graph has the greatest positive slope. The greatest slope appears to occur between $t = 0.5$ and $t = 1$. During this time interval, the car also has a nearly constant velocity because the curve is approximately linear. We conclude that the eastward velocity is largest from 12:30 to 1:00.

e. The car is at rest when the instantaneous velocity is zero. So, we look for points at which the slope of the curve is zero. These points occur at times between $t = 1.5$ and $t = 2$.   *Related Exercises 9–10* ◄

Speed and Acceleration   When only the magnitude of the velocity is of interest, we use **speed**, which is the absolute value of the velocity:

$$\text{speed} = |v|$$

For example, a car with an instantaneous velocity of -30 mi/hr travels with a speed of 30 mi/hr.

> Newton's first law of motion says that in the absence of external forces, a moving object has no acceleration, which means the magnitude and direction of the velocity are constant.

A more complete description of an object moving along a line includes its **acceleration**, which is the instantaneous rate of change of the velocity; that is, acceleration is the derivative of the velocity function with respect to time t. If the acceleration is positive, the object's velocity increases; if it is negative, the object's velocity decreases. Because velocity is the derivative of the position function, acceleration is the second derivative of the position. Therefore,

$$a = \frac{dv}{dt} = \frac{d^2 s}{dt^2}.$$

DEFINITION   Velocity, Speed, and Acceleration

Suppose an object moves along a line with position $s = f(t)$. Then,

velocity at time t: $v = \dfrac{ds}{dt} = f'(t)$

speed at time t: $|v| = |f'(t)|$

acceleration at time t: $a = \dfrac{dv}{dt} = \dfrac{d^2s}{dt^2} = f''(t)$

▸ The units of derivatives are consistent with the notation. If s is measured in meters and t is measured in seconds, the units of the velocity $\dfrac{ds}{dt}$ are m/s. The units of the acceleration $\dfrac{d^2s}{dt^2}$ are m/s².

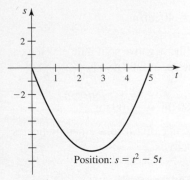

Position: $s = t^2 - 5t$

FIGURE 3.34

▸ Figure 3.34 gives the graph of the position function, not the path of the object. The motion is along a horizontal line.

EXAMPLE 2   Velocity and acceleration Suppose the position (in feet) of an object moving horizontally at time t (in seconds) is $s = t^2 - 5t$ for $0 \le t \le 5$ (Figure 3.34). Assume that positive values of s correspond to positions to the right of $s = 0$.

a. Graph the velocity function for $0 \le t \le 5$, and determine when the object is stationary, moving to the left, and moving to the right.

b. Graph the acceleration function for $0 \le t \le 5$, and determine the acceleration of the object when its velocity is zero.

c. Describe the motion of the object.

SOLUTION

a. The velocity is $v = s'(t) = 2t - 5$. The object is stationary when $v = 2t - 5 = 0$, or at $t = 2.5$ s. Solving $v = 2t - 5 > 0$, the velocity is positive (motion to the right) for $\frac{5}{2} < t < 5$. Similarly, the velocity is negative (motion to the left) for $0 \le t < \frac{5}{2}$. The graph of the velocity function (Figure 3.35) confirms these observations.

b. The acceleration is the derivative of the velocity or $a = v'(t) = s''(t) = 2$. This means that the acceleration is 2 ft/s² for $0 \le t \le 5$ (Figure 3.36).

c. Starting at an initial position of $s(0) = 0$, the object moves in the negative direction (to the left) with decreasing speed until it comes to rest momentarily at $s\left(\frac{5}{2}\right) = -\frac{25}{4}$. The object then moves in the positive direction (to the right) with increasing speed, reaching its initial position at $t = 5$. During this time interval, the acceleration is constant.

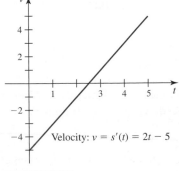

Velocity: $v = s'(t) = 2t - 5$

FIGURE 3.35

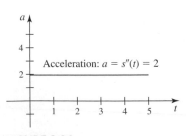

Acceleration: $a = s''(t) = 2$

FIGURE 3.36

Related Exercises 11–16 ◀

QUICK CHECK 2 Describe in words the velocity of an object that has a positive constant acceleration. Could an object have a positive acceleration and a decreasing speed? ◀

The acceleration due to Earth's gravitational field is denoted g. In metric units $g \approx 9.8 \text{ m/s}^2$ on the surface of Earth; in the U.S. Customary System (USCS), $g \approx 32 \text{ ft/s}^2$.

The derivation of the position function is given in Section 6.1. Once again we mention that the graph of the position function is not the path of the stone.

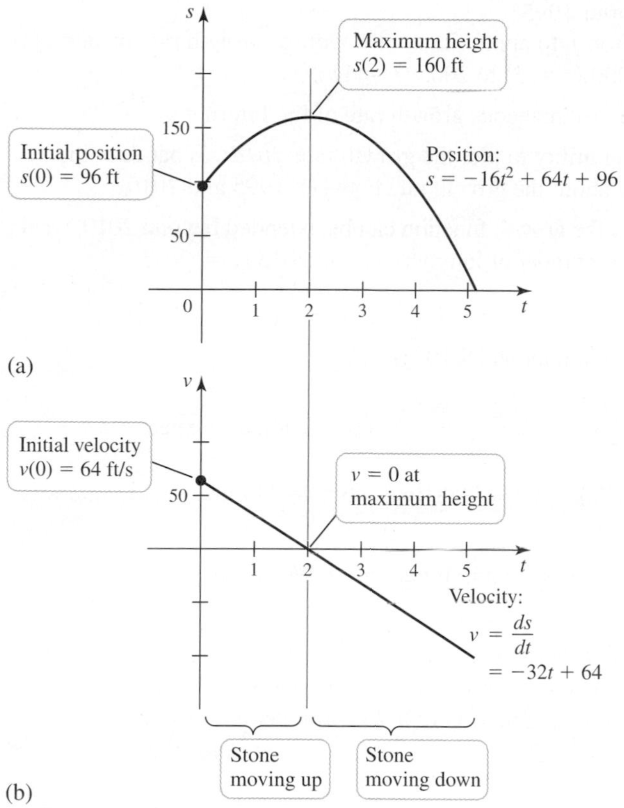

Free Fall We now consider problems in which an object moves vertically in Earth's gravitational field, assuming that no other forces (such as air resistance) are at work.

EXAMPLE 3 **Motion in a gravitational field** Suppose a stone is thrown vertically upward with an initial velocity of 64 ft/s from a bridge 96 ft above a river. By Newton's laws of motion, the position of the stone (measured as the height above the ground) after t seconds is

$$s(t) = -16t^2 + 64t + 96$$

where $s = 0$ is the level of the river (Figure 3.37a).

a. Find the velocity and acceleration functions.

b. What is the highest point above the river reached by the stone?

c. With what velocity will the stone strike the river?

SOLUTION

a. The velocity of the stone is the derivative of the position function, and its acceleration is the derivative of the velocity function. Therefore,

$$v = \frac{ds}{dt} = -32t + 64 \quad \text{and} \quad a = \frac{dv}{dt} = -32.$$

b. When the stone reaches its high point, its velocity is zero (Figure 3.37b). Solving $v(t) = -32t + 64 = 0$ yields $t = 2$ s, and thus the stone reaches its maximum height 2 s after it is thrown. Its height at that instant is

$$s(2) = -16(2)^2 + 64(2) + 96 = 160 \text{ ft}.$$

c. To determine the velocity at which the stone strikes the river, we first determine *when* it reaches the river. The stone strikes the river when $s(t) = -16t^2 + 64t + 96 = 0$. Dividing both sides of the equation by -16, we obtain $t^2 - 4t - 6 = 0$. Using the quadratic formula, the solutions are $t \approx 5.16$ or $t \approx -1.16$. Because the stone is thrown at $t = 0$, only positive values of t are of interest; therefore, the relevant root is $t \approx 5.16$ s. The velocity of the stone when it strikes the river is approximately

$$v(5.16) = -32(5.16) + 64 = -101.1 \text{ ft/s}.$$

Related Exercises 17–18 ◄

(a)

(b)

FIGURE 3.37

QUICK CHECK 3 In Example 3, does the rock have the greater speed at $t = 1$ or $t = 3$? ◄

Growth Models

Much of the change in the world around us can be classified as *growth*: Populations, prices, and computer networks all tend to increase in size. Modeling growth is important because it often leads to an understanding of underlying processes and allows for predictions.

We let $p = f(t)$ be the measure of a quantity of interest (for example, the population of a species or the consumer price index), where $t \geq 0$ represents time. The average growth rate of p between time $t = a$ and a later time $t = a + \Delta t$ is the change Δp divided by elapsed time Δt. Therefore, the **average growth rate** of p on the interval $[a, a + \Delta t]$ is

$$\frac{\Delta p}{\Delta t} = \frac{f(a + \Delta t) - f(a)}{\Delta t}.$$

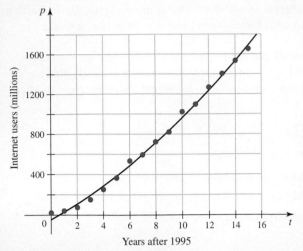

FIGURE 3.38

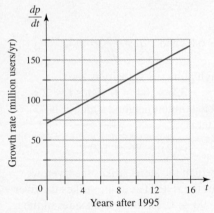

FIGURE 3.39

QUICK CHECK 4 Using the growth function in Example 4, compare the growth rates in 1996 and 2010. ◄

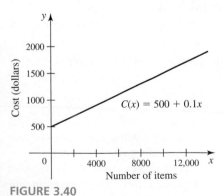

FIGURE 3.40

▶ Although x is a whole number of units, we treat it as a continuous variable, which is reasonable if x is large.

If we now let $\Delta t \to 0$, then $\dfrac{\Delta p}{\Delta t}$ approaches the derivative $\dfrac{dp}{dt}$, which is the **instantaneous growth rate** (or simply **growth rate**) of p with respect to time:

$$\frac{dp}{dt} = \lim_{\Delta t \to 0} \frac{\Delta p}{\Delta t}$$

EXAMPLE 4   Internet growth The number of worldwide Internet users between 1995 and 2010 is shown in Figure 3.38. A reasonable fit to the data is given by the function $p(t) = 3.0t^2 + 70.8t - 45.8$, where t measures years after 1995.

a. Use the function p to approximate the average growth rate of Internet users from 2000 ($t = 5$) to 2005 ($t = 10$).

b. What was the instantaneous growth rate of the Internet in 2006?

c. Use a graphing utility to plot the growth rate dp/dt. What does the graph tell you about the growth rate between 1995 and 2010?

d. Assuming that the growth function can be extended beyond 2010, what is the predicted number of Internet users in 2015 ($t = 20$)?

SOLUTION

a. The average growth rate over the interval $[5, 10]$ is

$$\frac{\Delta p}{\Delta t} = \frac{p(10) - p(5)}{10 - 5} \approx \frac{962 - 383}{5} \approx 116 \text{ million users/year.}$$

b. The growth rate at time t is $p'(t) = 6t + 70.8$. In 2006 ($t = 11$), the growth rate was $p'(11) \approx 137$ million users/year.

c. The graph of p' for $0 \le t \le 16$ is shown in Figure 3.39. We see that the growth rate is positive and increasing for $t \ge 0$.

d. A projection of the number of Internet users in 2015 is $p(20) \approx 2570$ million users, or about 2.6 billion users. This figure represents roughly one-third of the world's population, assuming a projected population of 7.2 billion people in 2015.

Related Exercises 19–20 ◄

Average and Marginal Cost

Our final example illustrates how derivatives arise in business and economics. As you will see, the mathematics of derivatives is the same in economics as it is in other applications. However, the vocabulary and interpretation used by economists are quite different.

Imagine a company that manufactures large quantities of a product such as mousetraps, DVD players, or snowboards. Associated with the manufacturing process is a **cost function** $C(x)$ that gives the cost of manufacturing x items of the product. A simple cost function might have the form $y = C(x) = 500 + 0.1x$, as shown in Figure 3.40. It includes a **fixed cost** of \$500 (setup costs and overhead) that is independent of the number of items produced. It also includes a **unit cost**, or **variable cost**, of \$0.10 per item produced. For example, the cost of producing 1000 items is $C(1000) = \$600$.

If the company produces x items at a cost of $C(x)$, then the **average cost** is $\dfrac{C(x)}{x}$ per item. For the cost function $C(x) = 500 + 0.1x$, the average cost is

$$\frac{C(x)}{x} = \frac{500 + 0.1x}{x} = \frac{500}{x} + 0.1.$$

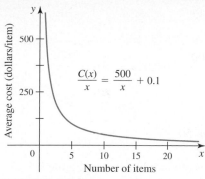

FIGURE 3.41

For example, the average cost of manufacturing 1000 items is

$$\frac{C(1000)}{1000} = \frac{\$600}{1000} = \$0.60/\text{unit.}$$

Plotting $C(x)/x$, we see that the average cost decreases as the number of items produced increases (Figure 3.41).

The average cost gives the cost of items already produced. But what about the cost of producing additional items? Having produced x items, the cost of producing another Δx items is $C(x + \Delta x) - C(x)$. Therefore, the average cost per item of producing those Δx additional items is

$$\frac{C(x + \Delta x) - C(x)}{\Delta x} = \frac{\Delta C}{\Delta x}.$$

> The average describes the past; the marginal describes the future.
> —Old saying

If we let $\Delta x \to 0$, we see that

$$\lim_{\Delta x \to 0} \frac{\Delta C}{\Delta x} = C'(x)$$

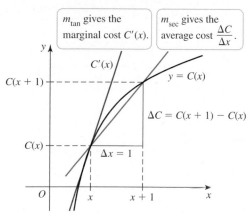

FIGURE 3.42

which is called the **marginal cost**. In reality, we cannot let $\Delta x \to 0$ because Δx represents whole numbers of items.

Here is a useful interpretation of the marginal cost. Suppose $\Delta x = 1$. Then, $\Delta C = C(x + 1) - C(x)$ is the cost to produce *one* additional item. In this case we write

$$\frac{\Delta C}{\Delta x} = \frac{C(x + 1) - C(x)}{1}.$$

If the *slope* of the cost curve does not vary significantly near the point x, then—as shown in Figure 3.42—we have

$$\frac{\Delta C}{\Delta x} \approx \lim_{\Delta x \to 0} \frac{\Delta C}{\Delta x} = C'(x).$$

> The approximation $\Delta C/\Delta x \approx C'(x)$ says that the slope of the secant line between $(x, C(x))$ and $(x + 1, C(x + 1))$ is approximately equal to the slope of the tangent line at $(x, C(x))$. This approximation is good if the cost curve is nearly linear over a one-unit interval.

Therefore, the cost of producing one additional item, having already produced x items, is approximated by the marginal cost $C'(x)$. In the preceding example, we have $C'(x) = 0.1$, so if $x = 1000$ items have been produced, then the cost of producing the 1001st item is approximately $C'(1000) = \$0.10$. With this simple linear cost function, the marginal cost tells us what we already know: The cost of producing one additional item is the variable cost of $\$0.10$. With more realistic cost functions, the marginal cost may be variable.

DEFINITION **Average and Marginal Cost**

The **cost function** $C(x)$ gives the cost to produce the first x items in a manufacturing process. The **average cost** to produce x items is $\overline{C}(x) = C(x)/x$. The **marginal cost** $C'(x)$ is the approximate cost to produce one additional item after producing x items.

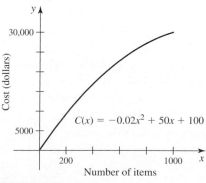

FIGURE 3.43

EXAMPLE 5 **Average and marginal costs** Suppose the cost of producing x items is given by the function (Figure 3.43)

$$C(x) = -0.02x^2 + 50x + 100 \quad \text{for} \quad 0 \le x \le 1000.$$

a. Determine the average and marginal cost functions.

b. Determine the average and marginal cost when $x = 100$ items and interpret these values.

c. Determine the average and marginal cost when $x = 900$ items and interpret these values.

SOLUTION

a. The average cost is

$$\overline{C}(x) = \frac{C(x)}{x} = \frac{-0.02x^2 + 50x + 100}{x} = -0.02x + 50 + \frac{100}{x}$$

and the marginal cost is

$$\frac{dC}{dx} = -0.04x + 50.$$

The average cost decreases as the number of items produced increases (Figure 3.44a). The marginal cost decreases linearly with a slope of -0.04 (Figure 3.44b).

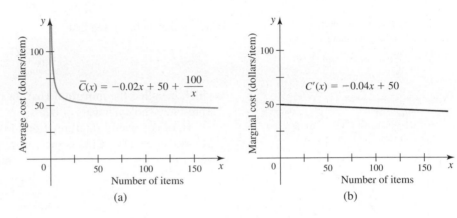

FIGURE 3.44   (a)   (b)

b. To produce $x = 100$ items, the average cost is

$$\overline{C}(100) = \frac{C(100)}{100} = \frac{-0.02(100)^2 + 50(100) + 100}{100} = \$49/\text{item}$$

and the marginal cost is

$$C'(100) = -0.04(100) + 50 = \$46/\text{item}.$$

These results mean that the average cost of producing 100 items is \$49 per item, but the cost of producing one additional item (the 101st item) is only \$46. Therefore, producing one more item is less expensive than the average cost of producing the first 100 items.

c. To produce $x = 900$ items, the average cost is

$$\overline{C}(900) = \frac{C(900)}{900} = \frac{-0.02(900)^2 + 50(900) + 100}{900} \approx \$32/\text{item}$$

and the marginal cost is

$$C'(900) = -0.04(900) + 50 = \$14/\text{item}.$$

The comparison with part (b) is revealing. The average cost of producing 900 items has dropped to \$32 per item. More striking is that the marginal cost (the cost of producing the 901st item) has dropped to \$14.   *Related Exercises 21–24* ◄

QUICK CHECK 5 In Example 5, what happens to the average cost as the number of items produced increases from $x = 1$ to $x = 100$? ◄

SECTION 3.5 EXERCISES

Review Questions

1. Use a graph to explain the difference between the average rate of change and the instantaneous rate of change of a function f.

2. Complete the following statement. If $\dfrac{dy}{dx}$ is large, then small changes in x will result in relatively _____ changes in the value of y.

3. Complete the following statement: If $\dfrac{dy}{dx}$ is small, then small changes in x will result in relatively _____ changes in the value of y.

4. What is the difference between the *velocity* and *speed* of an object moving in a straight line?

5. Define the acceleration of an object moving in a straight line.

6. An object moving along a line has a constant negative acceleration. Describe the velocity of the object.

7. Suppose the average cost of producing 200 gas stoves is $70 per stove and the marginal cost at $x = 200$ is $65 per stove. Interpret these costs.

8. Explain in your own words the adage: The average describes the past; the marginal describes the future.

Basic Skills

9. Highway travel A state patrol station is located on a straight north-south freeway. A patrol car leaves the station at 9:00 A.M. heading north with position function $s = f(t)$ that gives its location in miles t hours after 9:00 A.M. (see figure). Assume s is positive when the car is north of the patrol station.

a. Determine the average velocity of the car during the first 45 minutes of the trip.

b. Find the average velocity of the car over the interval $[0.25, 0.75]$. Is the average velocity a good estimate of the velocity at 9:30 A.M.?

c. Find the average velocity of the car over the interval $[1.75, 2.25]$. Estimate the velocity of the car at 11:00 A.M. and determine the direction in which the patrol car is moving.

d. Describe the motion of the patrol car relative to the patrol station between 9:00 A.M. and 12:00 P.M.

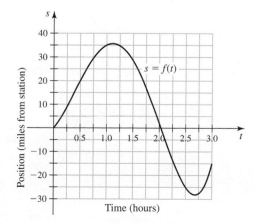

10. Airline travel The following figure shows the position function $s = f(t)$ of an airliner on an out-and-back trip from Seattle to Minneapolis, where $s = f(t)$ is the number of ground miles from Seattle t hours after take-off at 6:00 A.M. The plane returns to Seattle 8.5 hours later at 2:30 P.M.

a. Calculate the average velocity of the airliner during the first 1.5 hours of the trip ($0 \le t \le 1.5$).

b. Calculate the average velocity of the airliner between 1:30 P.M. and 2:30 P.M. ($7.5 \le t \le 8.5$).

c. At what time(s) is the velocity 0? Give a plausible explanation.

d. Determine the velocity of the airliner at noon ($t = 6$) and explain why the velocity is negative.

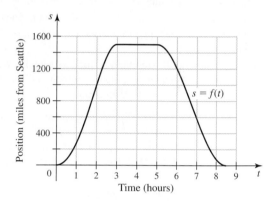

11–16. Position, velocity, and acceleration *Suppose the position of an object moving horizontally after t seconds is given by the following functions $s = f(t)$, where s is measured in feet, with $s > 0$ corresponding to positions right of the origin.*

a. *Graph the position function.*

b. *Find and graph the velocity function. When is the object stationary, moving to the right, and moving to the left?*

c. *Determine the velocity and acceleration of the object at $t = 1$.*

d. *Determine the acceleration of the object when its velocity is zero.*

11. $f(t) = t^2 - 4t; \ 0 \le t \le 5$

12. $f(t) = -t^2 + 4t - 3; \ 0 \le t \le 5$

13. $f(t) = 2t^2 - 9t + 12; \ 0 \le t \le 3$

14. $f(t) = 18t - 3t^2; \ 0 \le t \le 8$

15. $f(t) = 2t^3 - 21t^2 + 60t; \ 0 \le t \le 6$

16. $f(t) = -6t^3 + 36t^2 - 54t; \ 0 \le t \le 4$

17. A stone thrown vertically Suppose a stone is thrown vertically upward from the edge of a cliff with an initial velocity of 64 ft/s from a height of 32 ft above the ground. The height s (in feet) of the stone above the ground t seconds after it is thrown is $s = -16t^2 + 64t + 32$.

a. Determine the velocity v of the stone after t seconds.

b. When does the stone reach its highest point?

c. What is the height of the stone at the highest point?

d. When does the stone strike the ground?

e. With what velocity does the stone strike the ground?

18. A stone thrown vertically on Mars Suppose a stone is thrown vertically upward from the edge of a cliff on Mars (where the acceleration of gravity is only about 12 ft/s²) with an initial velocity of 64 ft/s from a height of 192 ft above the ground. The height s of the stone above the ground after t seconds is given by $s = -6t^2 + 64t + 192$.

   a. Determine the velocity v of the stone after t seconds.
   b. When does the stone reach its highest point?
   c. What is the height of the stone at the highest point?
   d. When does the stone strike the ground?
   e. With what velocity does the stone strike the ground?

⊤ 19. Population growth in Georgia The population of the state of Georgia (in thousands) from 1995 ($t = 0$) to 2005 ($t = 10$) is modeled by the polynomial $p(t) = -0.27t^2 + 101t + 7055$.

   a. Determine the average growth rate from 1995 to 2005.
   b. What was the growth rate for Georgia in 1997 ($t = 2$) and 2005 ($t = 10$)?
   c. Use a graphing utility to graph p' for $0 \le t \le 10$. What does this graph tell you about population growth in Georgia during the period of time from 1995 to 2005?

⊤ 20. Consumer price index The U.S. consumer price index (CPI) measures the cost of living based on a value of 100 in the years 1982–1984. The CPI for the years 1995–2010 (see figure) is modeled by the function $c(t) = 151e^{0.026t}$, where t represents years after 1995.

   a. Is the average growth rate greater between the years 1995–2000 or 2005–2010?
   b. Is the growth rate greater in 2000 ($t = 5$) or 2005 ($t = 10$)?
   c. Use a graphing utility to graph the growth rate for $0 \le t \le 15$. What does the graph tell you about growth in the cost of living during this time period?

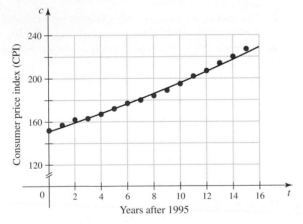

21–24. Average and marginal cost *Consider the following cost functions.*

   a. *Find the average cost and marginal cost functions.*
   b. *Determine the average and marginal cost when $x = a$.*
   c. *Interpret the values obtained in part (b).*

21. $C(x) = 1000 + 0.1x, \ 0 \le x \le 5000, \ a = 2000$

22. $C(x) = 500 + 0.02x, \ 0 \le x \le 2000, \ a = 1000$

23. $C(x) = -0.01x^2 + 40x + 100, \ 0 \le x \le 1500, \ a = 1000$

24. $C(x) = -0.04x^2 + 100x + 800, \ 0 \le x \le 1000, \ a = 500$

Further Explorations

25. Explain why or why not Determine whether the following statements are true and give an explanation or counterexample.

   a. If the acceleration of an object remains constant, then its velocity is constant.
   b. If the acceleration of an object moving along a line is always 0, then its velocity is constant.
   c. It is impossible for the instantaneous velocity at all times $a \le t \le b$ to equal the average velocity over the interval $a \le t \le b$.
   d. A moving object can have negative acceleration and increasing speed.

26. A feather dropped on the moon On the moon, a feather will fall to the ground at the same rate as a heavy stone. Suppose a feather is dropped from a height of 40 m above the surface of the moon. Then, its height s (in meters) above the ground after t seconds is $s = 40 - 0.8t^2$. Determine the velocity and acceleration of the feather the moment it strikes the surface of the moon.

27. Velocity of a bullet A bullet is fired vertically into the air at an initial velocity of 1200 ft/s. On Mars, the height s (in feet) of the bullet above the ground after t seconds is $s = 1200t - 6t^2$ and on Earth, $s = 1200t - 16t^2$. How much higher will the bullet travel on Mars than on Earth?

28. Velocity of a car The graph shows the position $s = f(t)$ of a car t hours after 5:00 P.M. relative to its starting point $s = 0$, where s is measured in miles.

   a. Describe the velocity of the car. Specifically, when is it speeding up and when is it slowing down?
   b. At approximately what time is the car traveling the fastest? The slowest?
   c. What is the approximate maximum velocity of the car? The approximate minimum velocity?

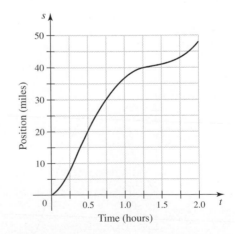

29. Velocity from position The graph of $s = f(t)$ represents the position of an object moving along a line at time $t \geq 0$.

  a. Assume the velocity of the object is 0 when $t = 0$. For what other values of t is the velocity of the object zero?

  b. When is the object moving in the positive direction and when is it moving in the negative direction?

  c. Sketch a graph of the velocity function.

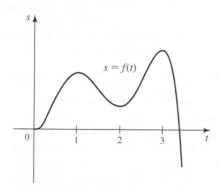

30. Fish length Assume the length L (in cm) of a particular species of fish after t years is modeled by the following graph.

  a. What does dL/dt represent and what happens to this derivative as t increases?

  b. What does the derivative tell you about how this species of fish grows?

  c. Sketch a rough graph of L' and L''.

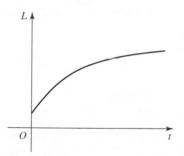

31–34. Average and marginal profit *Let $C(x)$ represent the cost of producing x items and $p(x)$ be the sale price per item if x items are sold. The profit $P(x)$ of selling x items is $P(x) = x\,p(x) - C(x)$ (revenue minus costs). The **average profit per item** when x items are sold is $P(x)/x$ and the **marginal profit** is dP/dx. The marginal profit approximates the profit obtained by selling one more item given that x items have already been sold. Consider the following cost functions C and price functions p.*

  a. *Find the profit function P.*

  b. *Find the average profit function and marginal profit function.*

  c. *Find the average profit and marginal profit if $x = a$ units have been sold.*

  d. *Interpret the meaning of the values obtained in part (c).*

31. $C(x) = -0.02x^2 + 50x + 100,\ p(x) = 100,\ a = 500$

32. $C(x) = -0.02x^2 + 50x + 100,\ p(x) = 100 - 0.1x,\ a = 500$

33. $C(x) = -0.04x^2 + 100x + 800,\ p(x) = 200,\ a = 1000$

34. $C(x) = -0.04x^2 + 100x + 800,\ p(x) = 200 - 0.1x,\ a = 1000$

Applications

35. Population growth of the United States Suppose $p(t)$ represents the population of the United States (in millions) t years after the year 1900. The graph of p' is shown in the figure.

  a. Approximately when (in what year) was the U.S. population growing most slowly between 1900 to 1990? Estimate the growth rate in that year.

  b. Approximately when (in what year) was the U.S. population growing most rapidly between 1900 and 1990? Estimate the growth rate in that year.

  c. In what years, if any, was p decreasing?

  d. In what years was the population growth rate increasing?

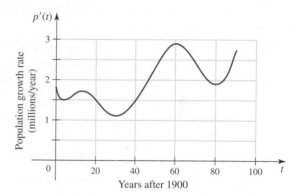

36. Average of marginal production Economists use *production functions* to describe how the output of a system varies with another variable such as labor or capital. For example, the production function $P(L) = 200L + 10L^2 - L^3$ gives the output of a system as a function of the number of laborers L. The *average product* $A(L)$ is the average output per laborer when L laborers are working; that is, $A(L) = P(L)/L$. The *marginal product* $M(L)$ is the approximate change in output when one additional laborer is added to L laborers; that is, $M(L) = \dfrac{dP}{dL}$.

  a. For the production function given here, compute and graph P, A, and L.

  b. Suppose the peak of the average product curve occurs at $L = L_0$, so that $A'(L_0) = 0$. Show that for a general production function, $M(L_0) = A(L_0)$.

37. Velocity of a marble The position (in meters) of a marble rolling up a long incline is given by $s = \dfrac{100t}{t + 1}$, where t is measured in seconds and $s = 0$ is the starting point.

  a. Graph the position function.

  b. Find the velocity function for the marble.

  c. Graph the velocity function and give a description of the motion of the marble.

  d. At what time is the marble 80 m from its starting point?

  e. At what time is the velocity 50 m/s?

38. Tree growth Let b represent the base diameter of a conifer tree and let h represent the height of the tree, where b is measured in centimeters and h is measured in meters. Assume the height is related to the base diameter by the function $h = 5.67 + 0.70b + 0.0067b^2$.

  a. Graph the height function.

  b. Plot and interpret the meaning of $\dfrac{dh}{db}$.

39. A different interpretation of marginal cost Suppose a large company makes 25,000 gadgets per year in batches of x items at a time. After analyzing setup costs to produce each batch and taking into account storage costs, it has been determined that the total cost $C(x)$ of producing 25,000 gadgets in batches of x items at a time is given by

$$C(x) = 1{,}250{,}000 + \frac{125{,}000{,}000}{x} + 1.5x.$$

a. Determine the marginal cost and average cost functions. Graph and interpret these functions.

b. Determine the average cost and marginal cost when $x = 5000$.

c. The meaning of average cost and marginal cost here is different than earlier examples and exercises. Interpret the meaning of your answer in part (b).

40. Diminishing returns A cost function of the form $C(x) = \frac{1}{2}x^2$ reflects *diminishing returns to scale*. Find and graph the cost, average cost and marginal cost functions. Interpret the graphs and explain the idea of diminishing returns.

41. Revenue function A store manager estimates that the demand for an energy drink decreases with increasing price according to the function $d(p) = \dfrac{100}{p^2 + 1}$, which means that at price p (in dollars), $d(p)$ units can be sold. The revenue generated at price p is $R(p) = p \cdot d(p)$.

a. Find and graph the revenue function.

b. Find and graph the marginal revenue $R'(p)$.

c. From the graph of the revenue function and its derivative, estimate the price that should be charged to maximize the revenue.

42. Fuel economy Suppose you own a fuel-efficient hybrid automobile containing a monitor on the dashboard that displays the mileage and gas consumption. The number of miles you can drive with g gallons of gas remaining in the tank on a particular stretch of highway is given by $m = 50g - 25.8g^2 + 12.5g^3 - 1.6g^4$, for $0 \le g \le 4$.

a. Graph and interpret the mileage function.

b. Graph and interpret the gas mileage m/g.

c. Graph and interpret dm/dg.

43. Spring oscillations A spring hangs from the ceiling at equilibrium with a mass attached to its end. Suppose you pull downward on the mass and release it 10 in below its equilibrium position. The distance x (in inches) of the mass from its equilibrium position after t seconds is given by the function $x(t) = 10 \sin t - 10 \cos t$, where x is positive when the mass is above the equilibrium position.

a. Graph and interpret this function.

b. Find $\dfrac{dx}{dt}$ and interpret the meaning of this derivative.

c. At what times is the velocity of the mass zero?

d. The function given here for x is a model for the motion of a spring. In what ways is this model unrealistic?

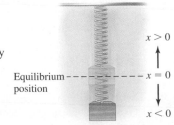

44. Pressure and altitude Earth's atmospheric pressure decreases with altitude from a sea level pressure of 1000 millibars (the unit of pressure used by meteorologists). Letting z be the height above Earth's surface (sea level) in km, the atmospheric pressure is modeled by $p(z) = 1000e^{-z/10}$.

a. Compute the pressure at the summit of Mt. Everest, which has an elevation of roughly 10 km. Compare the pressure on Mt. Everest to the pressure at sea level.

b. Compute the average change in pressure in the first 5 km above Earth's surface.

c. Compute the rate of change of the pressure at an elevation of 5 km.

d. Does $p'(z)$ increase or decrease with z? Explain.

e. What is the meaning of $\lim_{z \to \infty} p(z) = 0$?

45. A race Jean and Juan run a one-lap race on a circular track. Their angular positions on the track during the race are given by the functions $\theta(t)$ and $\phi(t)$, respectively, where $0 \le t \le 4$ min (see figure). These angles are measured in radians, where $\theta = \phi = 0$ represent the starting position and $\theta = \phi = 2\pi$ represent the finish position. The angular velocities of the runners are $\theta'(t)$ and $\phi'(t)$.

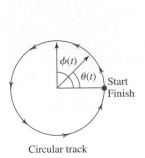

Circular track

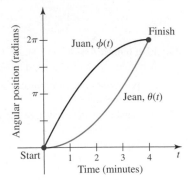

a. Compare in words the angular velocity of the two runners and the progress of the race.

b. Which runner has the greater average angular velocity?

c. Who wins the race?

d. Jean's position is given by $\theta(t) = \pi t^2/8$. What is her angular velocity at $t = 2$ and at what time is her angular velocity the greatest?

e. Juan's position is given by $\phi(t) = \pi t(8 - t)/8$. What is his angular velocity at $t = 2$ and at what time is his angular velocity the greatest?

46. Power and energy Power and energy are often used interchangeably, but they are quite different. **Energy** is what makes matter move or heat up. It is measured in units of **joules** or **Calories**, where 1 Cal = 4184 J. One hour of walking consumes roughly 10^6 J, or 240 Cal. On the other hand, **power** is the rate at which energy is used, which is measured in **watts**, where 1 W = 1 J/s. Other useful units of power are **kilowatts** (1 kW = 10^3 W) and **megawatts** (1 MW = 10^6 W). If energy is used at a rate of 1 kW for one hour, the total amount of energy used is 1 **kilowatt-hour** (1 kWh = 3.6×10^6 J). Suppose the cumulative energy used in a large building over a 24-hr period is given by $E(t) = 100t + 4t^2 - \dfrac{t^3}{9}$ kWh, where $t = 0$ corresponds to midnight.

a. Graph the energy function.

b. The power is the rate of energy consumption; that is, $P(t) = E'(t)$. Find the power over the interval $0 \le t \le 24$.

c. Graph the power function and interpret the graph. What are the units of power in this case?

47. Flow from a tank A cylindrical tank is full at time $t = 0$ when a valve in the bottom of the tank is opened. Torricelli's Law says that the volume of water in the tank after t hours is $V = 100(200 - t)^2$, measured in m³.

 a. Graph the volume function. What is the volume of water in the tank before the valve is opened?
 b. How long does it take for the tank to empty?
 c. Find the rate at which water flows from the tank and plot the flow rate function.
 d. At what time is the magnitude of the flow rate a minimum? A maximum?

48. Cell population The population of a culture of cells after t days is approximated by the function $P(t) = \dfrac{1600}{1 + 7e^{-0.02t}}$ for $t \geq 0$.

 a. Graph the population function.
 b. What is the average growth rate during the first 10 days?
 c. Looking at the graph, when does the growth rate appear to be a maximum?
 d. Differentiate the population function to determine the growth rate function $P'(t)$.
 e. Graph the growth rate. When is it a maximum and what is the population at the time that the growth rate is a maximum?

49. Bungee jumper A woman attached to a bungee cord jumps from a bridge that is 30 m above a river. Her height in meters above the river t seconds after the jump is $y(t) = 15(1 + e^{-t}\cos t)$ for $t \geq 0$.

 a. Determine her velocity at $t = 1$ and $t = 3$.
 b. Use a graphing utility to determine when she is moving downward and when she is moving upward during the first 10 s.
 c. Use a graphing utility to estimate the maximum upward velocity.

50. Spring runoff The flow of a small stream is monitored for 90 days between May 1 and August 1. The total water that flows past a gauging station is given by

$$V(t) = \begin{cases} \dfrac{4}{5}t^2 & \text{if } 0 \leq t < 45 \\[2mm] \dfrac{-4}{5}(t^2 - 180t + 4050) & \text{if } 45 \leq t < 90 \end{cases}$$

where V is measured in ft³ and t is measured in days, with $t = 0$ corresponding to May 1.

 a. Graph the volume function.
 b. Find the flow rate function $V'(t)$ and graph it. What are the units of the flow rate?
 c. Describe the flow of the stream over the 3-month period. Specifically, when is the flow rate a maximum?

51. Temperature distribution A thin copper rod, 4 m in length, is heated at its midpoint and the ends are held at a constant temperature of 0°. When the temperature reaches equilibrium, the temperature profile is given by $T(x) = 40x(4 - x)$, where $0 \leq x \leq 4$ m is the position along the rod. The **heat flux** at a point on the rod equals $-kT'(x)$, where k is a constant. If the heat flux is positive at a point, heat moves in the positive x-direction at that point, and if the heat flux is negative, heat moves in the negative x-direction.

 a. With $k = 1$, what is the heat flux at $x = 1$? At $x = 3$?
 b. For what values of x is the heat flux negative? Positive?
 c. Explain the statement that heat flows out of the rod at its ends.

QUICK CHECK ANSWERS

1. Instantaneous velocity **2.** An object has positive constant acceleration when its velocity is increasing. If the velocity is negative but increasing, then the acceleration is positive and the speed is decreasing. For example, the velocity may increase from -2 m/s to -1 m/s to 0 m/s. **3.** $v(1) = 32$ ft/s and $v(3) = -32$ ft/s, so the speed is 32 ft/s at both times. **4.** The growth rate in 1996 ($t = 1$) is approximately 77 million users/year. It is less than half of the growth rate in 2010 ($t = 15$), which is approximately 161 million users/year. **5.** As x increases from 1 to 100, the average cost decreases from \$150/item to \$49/item. ◄

3.6 The Chain Rule

The differentiation rules presented so far allow us to find derivatives of many functions. However, these rules are inadequate for finding the derivatives of most *composite functions*. Here is a typical situation. If $f(x) = x^3$ and $g(x) = 5x + 4$, then their composition is $f(g(x)) = (5x + 4)^3$. One way to find the derivative is by expanding $(5x + 4)^3$ and differentiating the resulting polynomial. Unfortunately, this strategy becomes prohibitive for functions such as $(5x + 4)^{100}$. We need a better approach.

QUICK CHECK 1 Explain why it is not practical to calculate $\dfrac{d}{dx}(5x + 4)^{100}$ by first expanding $(5x + 4)^{100}$. ◄

Chain Rule Formulas

An efficient method for differentiating composite functions, called the *Chain Rule*, is motivated by the following example. Suppose Yancey, Uri, and Xan pick apples. Let y, u, and x represent the number of apples picked in some period of time by Yancey, Uri, and Xan, respectively. Yancey picks apples three times faster than Uri, which means the rate at

> Expressions such as dy/dx should not be treated as fractions. Nevertheless, you can check symbolically that you have written the Chain Rule correctly by noting that du appears in the "numerator" and "denominator." If it were "canceled," the Chain Rule would have dy/dx on both sides.

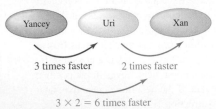

FIGURE 3.45

which Yancey picks apples with respect to Uri is $\dfrac{dy}{du} = 3$. Uri picks apples twice as fast as Xan, so $\dfrac{du}{dx} = 2$. Therefore, Yancey picks apples at a rate that is $3 \cdot 2 = 6$ times greater than Xan's rate, which means that $\dfrac{dy}{dx} = 6$ (Figure 3.45). Observe that

$$\frac{dy}{dx} = \frac{dy}{du} \cdot \frac{du}{dx} = 3 \cdot 2 = 6.$$

The equation $\dfrac{dy}{dx} = \dfrac{dy}{du} \cdot \dfrac{du}{dx}$ is one form of the Chain Rule. It is referred to as Version 1 of the Chain Rule in this text.

Alternatively, the Chain Rule may be expressed in terms of composite functions. Let $y = f(u)$ and $u = g(x)$, which means y is related to x through the composite function $y = f(u) = f(g(x))$. The derivative $\dfrac{dy}{dx}$ is now expressed as the product

$$\underbrace{\frac{d}{dx}[f(g(x))]}_{\frac{dy}{dx}} = \underbrace{f'(u)}_{\frac{dy}{du}} \cdot \underbrace{g'(x)}_{\frac{du}{dx}}.$$

Replacing u with $g(x)$ results in

$$\frac{d}{dx}[f(g(x))] = f'(g(x)) \cdot g'(x)$$

which we refer to as Version 2 of the Chain Rule.

> The two versions of the Chain Rule differ only in notation. Mathematically, they are identical. Version 2 of the Chain Rule states that the derivative of $y = f(g(x))$ is the derivative of f evaluated at $g(x)$ multiplied by the derivative of g evaluated at x.

THEOREM 3.14   The Chain Rule
Suppose g is differentiable at x and $y = f(u)$ is differentiable at $u = g(x)$. The composite function $y = f(g(x))$ is differentiable at x, and its derivative can be expressed in two equivalent ways:

$$\frac{dy}{dx} = \frac{dy}{du} \cdot \frac{du}{dx} \qquad \text{Version 1}$$

$$\frac{d}{dx}[f(g(x))] = f'(g(x)) \cdot g'(x) \quad \text{Version 2}$$

A proof of the Chain Rule is given at the end of this section. For now, it's important to learn how to use it. With the composite function $f(g(x))$, we refer to g as the *inner function* and f as the *outer function* of the composition. The key to using the Chain Rule is identifying the inner and outer functions. The following four steps simplify the differentiation process, although you will soon find that the procedure can be streamlined.

> There may be different ways to choose an inner function $u = g(x)$ and an outer function $y = f(u)$. Nevertheless, we refer to *the* inner and *the* outer function for the most obvious choices.

Guidelines for Using the Chain Rule
Assume the differentiable function $y = f(g(x))$ is given.

1. Identify the outer function f, the inner function g, and let $u = g(x)$.

2. Replace $g(x)$ by u to express y in terms of u:

$$y = f(\underbrace{g(x)}_{u}) \Rightarrow y = f(u)$$

QUICK CHECK 2 Identify an inner function (call it g) of $y = (5x + 4)^3$. Let $u = g(x)$ and express the outer function f in terms of u. ◄

3. Calculate the product $\dfrac{dy}{du} \cdot \dfrac{du}{dx}$.

4. Replace u by $g(x)$ in $\dfrac{dy}{du}$ to obtain $\dfrac{dy}{dx}$.

EXAMPLE 1 Version 1 of the Chain Rule For each of the following composite functions, find the inner function $u = g(x)$ and the outer function $y = f(u)$. Then, use Version 1 of the Chain Rule to find $\dfrac{dy}{dx}$.

a. $y = (5x + 4)^3$ **b.** $y = \sin^3 x$ **c.** $y = \sin x^3$

SOLUTION

a. The inner function of $y = (5x + 4)^3$ is $u = 5x + 4$, and the outer function is $y = u^3$. By Version 1 of the Chain Rule, we have

$$\frac{dy}{dx} = \frac{dy}{du} \cdot \frac{du}{dx} \qquad \text{Version 1}$$

$$= 3u^2 \cdot (5) \qquad y = u^3 \Rightarrow \frac{dy}{du} = 3u^2$$

$$\qquad\qquad\qquad\qquad u = 5x + 4 \Rightarrow \frac{du}{dx} = 5$$

$$= 3(5x + 4)^2 \cdot (5) \quad \text{Replace } u \text{ by } 5x + 4.$$

$$= 15(5x + 4)^2$$

> When using trigonometric functions, expressions such as $\sin^n (x)$ always mean $(\sin x)^n$, except when $n = -1$. In Example 1, $\sin^3 x = (\sin x)^3$.

b. Replacing the shorthand form $y = \sin^3 x$ with $y = (\sin x)^3$, we identify the inner function as $u = \sin x$. Letting $y = u^3$, we have

$$\frac{dy}{dx} = \frac{dy}{du} \cdot \frac{du}{dx} = 3u^2 \cdot \cos x = \underbrace{3 \sin^2 x}_{3u^2} \cos x.$$

QUICK CHECK 3 In Example 1a, we showed that
$$\frac{d}{dx}((5x + 4)^3) = 15(5x + 4)^2.$$
Verify this result by expanding $(5x + 4)^3$ and differentiating. ◄

c. Although $y = \sin x^3$ appears to be similar to the function $y = \sin^3 x$ in part (b), the inner function in this case is $u = x^3$ and the outer function is $y = \sin u$. Therefore,

$$\frac{dy}{dx} = \frac{dy}{du} \cdot \frac{du}{dx} = (\cos u) \cdot 3x^2 = 3x^2 \cos x^3.$$

Related Exercises 7–16 ◄

Version 2 of the Chain Rule, $\dfrac{d}{dx}[f(g(x))] = f'(g(x)) \cdot g'(x)$, is equivalent to Version 1; it just uses different derivative notation. With Version 2, we identify the outer function $y = f(u)$ and the inner function $u = g(x)$. Then, $\dfrac{d}{dx}[f(g(x))]$ is the product of $f'(u)$ evaluated at $u = g(x)$ and $g'(x)$.

EXAMPLE 2 Version 2 of the Chain Rule Use Version 2 of the Chain Rule to calculate the derivatives of each of the following functions.

a. $(6x^3 + 3x + 1)^{10}$ **b.** $\sqrt{5x^2 + 1}$ **c.** $\left(\dfrac{5t^2}{3t^2 + 2}\right)^3$

SOLUTION

a. The inner function of $(6x^3 + 3x + 1)^{10}$ is $g(x) = 6x^3 + 3x + 1$, and the outer function is $f(u) = u^{10}$. The derivative of the outer function is $f'(u) = 10u^9$, which, when evaluated at $g(x)$ is $10(6x^3 + 3x + 1)^9$. The derivative of the inner function is $g'(x) = 18x^2 + 3$. Multiplying the derivatives of the outer and inner functions, we have

$$\frac{d}{dx}[(6x^3 + 3x + 1)^{10}] = \underbrace{10(6x^3 + 3x + 1)^9}_{f'(u)\text{ evaluated at }g(x)} \cdot \underbrace{(18x^2 + 3)}_{g'(x)}$$

$$= 30(6x^2 + 1)(6x^3 + 3x + 1)^9 \qquad \text{Factor and simplify.}$$

b. The inner function of $\sqrt{5x^2 + 1}$ is $g(x) = 5x^2 + 1$, and the outer function is $f(u) = \sqrt{u}$. The derivatives of these functions are $f'(u) = \dfrac{1}{2\sqrt{u}}$ and $g'(x) = 10x$.

Therefore,

$$\frac{d}{dx}\sqrt{5x^2 + 1} = \underbrace{\frac{1}{2\sqrt{5x^2 + 1}}}_{\substack{f'(u)\text{ evaluated}\\\text{at }g(x)}} \cdot \underbrace{10x}_{g'(x)} = \frac{5x}{\sqrt{5x^2 + 1}}.$$

c. The inner function of $\left(\dfrac{5t^2}{3t^2 + 2}\right)^3$ is $g(t) = \dfrac{5t^2}{3t^2 + 2}$. The outer function is $f(u) = u^3$, whose derivative is $f'(u) = 3u^2$. The derivative of the inner function requires the Quotient Rule. Applying the Chain Rule, we have

$$\frac{d}{dt}\left(\frac{5t^2}{3t^2 + 2}\right)^3 = \underbrace{3\left(\frac{5t^2}{3t^2 + 2}\right)^2}_{\substack{f'(u)\text{ evaluated}\\\text{at }g(t)}} \cdot \underbrace{\frac{(3t^2 + 2)10t - 5t^2(6t)}{(3t^2 + 2)^2}}_{g'(t)\text{ by the Quotient Rule}} = \frac{1500t^5}{(3t^2 + 2)^4}$$

Related Exercises 17–30 ◄

The Chain Rule is also used to calculate the derivative of a composite function for a specific value of the variable. If $h(x) = f(g(x))$ and a is a real number, then $h'(a) = f'(g(a))g'(a)$. Therefore, $h'(a)$ is the derivative of f evaluated at $g(a)$ multiplied by the derivative of g evaluated at a.

EXAMPLE 3 Calculating derivatives at a point Let $h(x) = f(g(x))$. Use the values in Table 3.3 to calculate $h'(1)$ and $h'(2)$.

SOLUTION We use $h'(a) = f'(g(a))g'(a)$ with $a = 1$:

$$h'(1) = f'(g(1))g'(1) = f'(2)g'(1) = 7 \cdot 3 = 21$$

With $a = 2$, we have

$$h'(2) = f'(g(2))g'(2) = f'(1)g'(2) = 5 \cdot 4 = 20.$$

Related Exercises 31–32 ◄

Table 3.3

x	$f'(x)$	$g(x)$	$g'(x)$
1	5	2	3
2	7	1	4

Chain Rule for Powers

The Chain Rule leads to a general derivative rule that works for powers of differentiable functions. In fact, we have already used it in several examples. Consider the function $f(x) = (g(x))^n$, where n is an integer. Letting $f(u) = u^n$ be the outer function and $u = g(x)$ be the inner function, we obtain the Chain Rule for powers of functions.

> **THEOREM 3.15   Chain Rule for Powers**
> If g is differentiable for all x in the domain and n is an integer, then
>
> $$\frac{d}{dx}[(g(x))^n] = n(g(x))^{n-1}g'(x).$$

EXAMPLE 4   Chain Rule for powers Find $\dfrac{d}{dx}(\tan x + 10)^{21}$.

SOLUTION With $g(x) = \tan x + 10$, the Chain Rule gives

$$\frac{d}{dx}(\tan x + 10)^{21} = 21(\tan x + 10)^{20}\frac{d}{dx}(\tan x + 10)$$

$$= 21(\tan x + 10)^{20}\sec^2 x.$$

Related Exercises 33–36 ◄

The Composition of Three or More Functions

We can differentiate the composition of three or more functions by applying the Chain Rule repeatedly, as shown in the following example.

EXAMPLE 5   Composition of three functions Calculate the derivative of $\sin(e^{\cos x})$.

SOLUTION The inner function of $\sin(e^{\cos x})$ is $e^{\cos x}$. Because $e^{\cos x}$ is also a composition of two functions, the Chain Rule is used again to calculate $\dfrac{d}{dx}(e^{\cos x})$, where $\cos x$ is the inner function:

$$\frac{d}{dx}\Big[\underbrace{\sin(\underbrace{e^{\cos x}}_{\text{inner}})}_{\text{outer}}\Big] = \cos(e^{\cos x})\frac{d}{dx}(e^{\cos x}) \qquad \text{Chain Rule}$$

$$= \cos(e^{\cos x})\,e^{\cos x}\cdot\underbrace{\frac{d}{dx}(\cos x)}_{\frac{d}{dx}(e^{\cos x})} \qquad \text{Chain Rule}$$

$$= \cos(e^{\cos x})\cdot e^{\cos x}(-\sin x) \qquad \text{Differentiate } \cos x.$$

$$= -\sin x \cdot e^{\cos x}\cdot\cos(e^{\cos x}) \qquad \text{Simplify.}$$

Related Exercises 37–46 ◄

> Before dismissing the function in Example 5 as merely a tool to teach the Chain Rule, consider the graph of a related function, $y = \sin(e^{1.3\cos x}) + 1$ (Figure 3.46). This periodic function has two peaks per cycle and could be used as a simple model of traffic flow (two rush hours followed by light traffic in the middle of the night), tides (high tide, medium tide, high tide, low tide, …), or the presence of certain allergens in the air (peaks in the spring and fall).

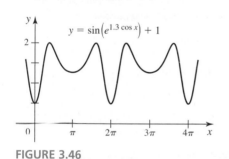

FIGURE 3.46

QUICK CHECK 4 Let $y = \tan^{10}(x^5)$. Find f, g, and h such that $y = f(u)$, where $u = g(v)$ and $v = h(x)$. ◄

Proof of the Chain Rule

Suppose f and g are differentiable functions and $h(x) = f(g(x))$. By the definition of the derivative of h,

$$h'(a) = \lim_{x \to a}\frac{h(x) - h(a)}{x - a} = \lim_{x \to a}\frac{f(g(x)) - f(g(a))}{x - a}. \qquad (1)$$

We assume that $g(a) \neq g(x)$ for values of x near a but not equal to a. This assumption holds for most, but not all, functions encountered in this text. For a proof of the Chain Rule without this assumption, see Exercise 81.

We multiply the right side of equation (1) by $\dfrac{g(x) - g(a)}{g(x) - g(a)}$, which equals 1, and let $v = g(x)$ and $u = g(a)$. The result is

$$h'(a) = \lim_{x \to a} \frac{f(g(x)) - f(g(a))}{g(x) - g(a)} \cdot \frac{g(x) - g(a)}{x - a}$$

$$= \lim_{x \to a} \frac{f(v) - f(u)}{v - u} \cdot \frac{g(x) - g(a)}{x - a}$$

By assumption, g is a differentiable function; therefore, it is continuous. This means that $\lim_{x \to a} g(x) = g(a)$, so $v \to u$ as $x \to a$. Consequently,

$$h'(a) = \underbrace{\lim_{v \to u} \frac{f(v) - f(u)}{v - u}}_{f'(u)} \cdot \underbrace{\lim_{x \to a} \frac{g(x) - g(a)}{x - a}}_{g'(a)} = f'(u)g'(a).$$

Because f and g are differentiable, the two limits in this expression exist; therefore $h'(a)$ exists. Noting that $u = g(a)$, we have $h'(a) = f'(g(a))g'(a)$. Replacing a with the variable x gives the Chain Rule: $h'(x) = f'(g(x))g'(x)$. ◄

SECTION 3.6 EXERCISES

Review Questions

1. Two equivalent forms of the Chain Rule for calculating the derivative of $y = f(g(x))$ are presented in this section. State both forms.

2. Let $h(x) = f(g(x))$, where f and g are differentiable on their domains. If $g(1) = 3$ and $g'(1) = 5$, what else do you need to know to calculate $h'(1)$?

3. Fill in the blanks. The derivative of $f(g(x))$ equals f' evaluated at _____ multiplied by g' evaluated at _____.

4. Identify the inner and outer functions in the composition $\cos^4 x$.

5. Identify the inner and outer functions in the composition $(x^2 + 10)^{-5}$.

6. Express $Q(x) = \cos^4(x^2 + 1)$ as the composition of three functions; that is, identify f, g, and h so that $Q(x) = f(g(h(x)))$.

Basic Skills

7–16. Version 1 of the Chain Rule *Use Version 1 of the Chain Rule to calculate* $\dfrac{dy}{dx}$.

7. $y = (3x + 7)^{10}$ 8. $y = (5x^2 + 11x)^{20}$ 9. $y = \sqrt{x^2 + 1}$

10. $y = e^{\sqrt{x}}$ 11. $y = \tan(5x^2)$ 12. $y = \sin\left(\dfrac{x}{4}\right)$

13. $y = \sec(e^x)$ 14. $y = \left(\dfrac{3x}{4x + 2}\right)^5$ 15. $y = e^{-x^2}$

16. $y = [(x + 2)(3x^3 + 3x)]^4$

17–28. Version 2 of the Chain Rule *Use Version 2 of the Chain Rule to calculate the derivatives of the following composite functions.*

17. $y = (3x^2 + 7x)^{10}$ 18. $y = \sqrt{x^2 + 9}$

19. $y = 5(7x^3 + 1)^{-3}$ 20. $y = \cos(5t + 1)$

21. $y = \tan(e^x)$ 22. $y = e^{\tan t}$

23. $y = \sin(4x^3 + 3x + 1)$ 24. $y = \csc(t^2 + t)$

25. $y = \theta^2 \sec(5\theta)$ 26. $y = \cos^4 \theta + \sin^4 \theta$

27. $y = [\sec x + \tan x]^5$ 28. $y = \sin(4 \cos z)$

29–30. Similar-looking composite functions *Two composite functions are given that look similar, but in fact are quite different. Identify the inner function $u = g(x)$ and the outer function $y = f(u)$; then evaluate $\dfrac{dy}{dx}$ using the Chain Rule.*

29. **a.** $y = \cos^3 x$ **b.** $y = \cos(x^3)$

30. **a.** $y = (e^x)^3$ **b.** $y = e^{(x^3)}$

31. **Chain Rule using a table** Let $h(x) = f(g(x))$ and $p(x) = g(f(x))$. Use the table to compute the following derivatives.

   a. $h'(3)$ **b.** $h'(2)$ **c.** $p'(4)$ **d.** $p'(2)$ **e.** $h'(5)$

x	1	2	3	4	5
$f(x)$	0	3	5	1	0
$f'(x)$	5	2	−5	−8	−10
$g(x)$	4	5	1	3	2
$g'(x)$	2	10	20	15	20

32. **Chain Rule using a table** Let $h(x) = f(g(x))$ and $k(x) = g(g(x))$. Use the table to compute the following derivatives.

   a. $h'(1)$ **b.** $h'(2)$ **c.** $h'(3)$ **d.** $k'(3)$ **e.** $k'(1)$ **f.** $k'(5)$

x	1	2	3	4	5
$f'(x)$	−6	−3	8	7	2
$g(x)$	4	1	5	2	3
$g'(x)$	9	7	3	−1	−5

33–36. Chain Rule for powers *Use the Chain Rule to find the derivative of the following functions.*

33. $y = (2x^6 - 3x^3 + 3)^{25}$

34. $y = (\cos x + 2\sin x)^8$

35. $y = (1 + 2\tan x)^{15}$

36. $y = (1 - e^x)^4$

37–46. Repeated use of the Chain Rule *Calculate the derivative of the following functions.*

37. $\sqrt{1 + \cot^2 x}$

38. $\sqrt{(3x - 4)^2 + 3x}$

39. $\sin^5(\cos(3x))$

40. $\cos^4(7x^3)$

41. $\tan(e^{\sqrt{3x}})$

42. $(1 - e^{-0.05x})^{-1}$

43. $\sqrt{x + \sqrt{x}}$

44. $\sqrt{x + \sqrt{x + \sqrt{x}}}$

45. $f(g(x^2))$, where f and g are differentiable for all real numbers

46. $f(\sqrt{g(x^2)})$, where f and g are differentiable for all real numbers and g is nonnegative

Further Explorations

47. Explain why or why not Determine whether the following statements are true and give an explanation or counterexample.

   a. The function $x\sin x$ can be differentiated without using the Chain Rule.

   b. The function $(x^2 + 10)^{-12}$ must be differentiated using the Chain Rule.

   c. The derivative of a product is *not* the product of the derivatives, but the derivative of a composition is a product of derivatives.

   d. $\dfrac{d}{dx}P(Q(x)) = P'(x)Q'(x)$

48–51. Second derivatives *Find $\dfrac{d^2y}{dx^2}$ for the following functions.*

48. $y = x\cos x^2$

49. $y = \sin(x^2)$

50. $y = \sqrt{3x^3 + 4x + 1}$

51. $y = e^{-2x^2}$

52. Derivatives by different methods

   a. Calculate $\dfrac{d}{dx}(x^2 + x)^2$ using the Chain Rule. Simplify your answer.

   b. Expand $(x^2 + x)^2$ first and then calculate the derivative. Verify that your answer agrees with part (a).

53–54. Square root derivatives *Find the derivative of the following functions.*

53. $y = \sqrt{f(x)}$, where f is differentiable and nonnegative

54. $y = \sqrt{f(x)g(x)}$, where f and g are differentiable and nonnegative

55. Tangent lines Determine an equation of the line tangent to the graph of $y = \dfrac{(x^2 - 1)^2}{x^3 - 6x - 1}$ at the point $(3, 8)$. Graph the function and the tangent line.

56. Tangent lines Determine equations of the lines tangent to the graph of $y = x\sqrt{5 - x^2}$ at the points $(1, 2)$ and $(-2, -2)$. Graph the function and the tangent lines.

57. Tangent lines Assume f and g are differentiable functions with $h(x) = f(g(x))$. Suppose the equation of the line tangent to the graph of g at the point $(4, 7)$ is $y = 3x - 5$ and the equation of the line tangent to the graph of f at $(7, 9)$ is $y = -2x + 23$.

   a. Calculate $h(4)$ and $h'(4)$.

   b. Determine an equation of the line tangent to the graph of h at the point on the graph where $x = 4$.

58. Tangent lines Assume f is a differentiable function whose graph passes through the point $(1, 4)$. If $g(x) = f(x^2)$ and the line tangent to the graph of f at $(1, 4)$ is $y = 3x - 1$, determine each of the following.

   a. $g(1)$ **b.** $g'(x)$ **c.** $g'(1)$

   d. Find an equation of the line tangent to the graph of g when $x = 1$.

59. Tangent lines Find the equation of the line tangent to $y = e^{2x}$ at $x = \frac{1}{2}\ln 3$. Graph the function and the tangent line.

60. Composition containing $\sin x$ Suppose f is differentiable on $[-2, 2]$ with $f'(0) = 3$ and $f'(1) = 5$. Let $g(x) = f(\sin x)$. Evaluate the following expressions.

   a. $g'(0)$ **b.** $g'\left(\dfrac{\pi}{2}\right)$ **c.** $g'(\pi)$

61. Composition containing $\sin x$ Suppose f is differentiable for all real numbers with $f(0) = -3$, $f(1) = 3$, $f'(0) = 3$, and $f'(1) = 5$. Let $g(x) = \sin(\pi f(x))$. Evaluate the following expressions.

   a. $g'(0)$ **b.** $g'(1)$

Applications

62–64. Vibrations of a spring *Suppose an object of mass m is attached to the end of a spring hanging from the ceiling. We say that the mass is at its equilibrium position $y = 0$ when the spring hangs at rest. Suppose you push the mass to a position y_0 units above its equilibrium position and release it. As the mass oscillates up and down (neglecting any friction in the system), the position y of the mass after t seconds is given by the equation*

$$y = y_0\cos\left(t\sqrt{\dfrac{k}{m}}\right) \qquad (2)$$

where k is a constant measuring the stiffness of the spring (the larger the value of k, the stiffer the spring) and y is positive in the upward direction.

62. Use equation (2) to answer the following questions.

   a. Find $\dfrac{dy}{dt}$, the velocity of the mass. (Assume k and m are constant.)

   b. How would the velocity be affected if the experiment were repeated with four times the mass on the end of the spring?

   c. How would the velocity be affected if the experiment were repeated with a spring having four times the stiffness (k is increased by a factor of 4)?

   d. Assume that y has units of meters, t has units of seconds, m has units of kg, and k has units of kg/s^2. Show that the units of the velocity in part (a) are consistent.

63. Use equation (2) to answer the following questions.

   a. Find the second derivative $\dfrac{d^2y}{dt^2}$.

   b. Verify that $\dfrac{d^2y}{dt^2} = -\dfrac{k}{m}y$.

64. Use equation (2) to answer the following questions.

a. The *period T* is the time required by the mass to complete one oscillation. Show that $T = 2\pi\sqrt{\dfrac{m}{k}}$.

b. Assume k is constant and calculate $\dfrac{dT}{dm}$.

c. Give a physical explanation of why $\dfrac{dT}{dm}$ is positive.

☐ 65. A damped oscillator The displacement of a mass on a spring suspended from the ceiling is given by $y = 10e^{-t/2}\cos\left(\dfrac{\pi t}{8}\right)$.

a. Graph the displacement function.
b. Compute and graph the velocity of the mass, $v(t) = y'(t)$.
c. Verify that the velocity is zero when the mass reaches the high and low points of its oscillation.

66. Oscillator equation A mechanical oscillator (such as a mass on a spring or a pendulum) subject to frictional forces satisfies the equation (called a differential equation)

$$y''(t) + 2y'(t) + 5y(t) = 0,$$

where y is the displacement of the oscillator from its equilibrium position. Verify by substitution that the function $y(t) = e^{-t}(\sin 2t - 2\cos 2t)$ satisfies this equation.

☐ 67. Hours of daylight The number of hours of daylight at any point on Earth fluctuates throughout the year. In the northern hemisphere, the shortest day is on the winter solstice and the longest day is on the summer solstice. At 40° north latitude, the length of a day is approximated by

$$D(t) = 12 - 3\cos\left[\frac{2\pi(t + 10)}{365}\right],$$

where D is measured in hours and $0 \le t \le 365$ is measured in days, with $t = 0$ corresponding to January 1.

a. Approximately how much daylight is there on March 1 ($t = 59$)?
b. Find the rate at which the daylight function changes.
c. Find the rate at which the daylight function changes on March 1. Convert your answer to units of min/day and explain what this result means.
d. Graph the function $y = D'(t)$ using a graphing utility.
e. At what times of year is the length of day changing most rapidly? Least rapidly?

☐ 68. A mixing tank A 500-L tank is filled with pure water. At time $t = 0$, a salt solution begins flowing into the tank at a rate of 5 L/min. At the same time, the (fully mixed) solution flows out of the tank at a rate of 5.5 L/min. The mass of salt in grams in the tank at any time $t \ge 0$ is given by

$$M(t) = 250(1000 - t)[1 - 10^{-30}(1000 - t)^{10}]$$

and the volume of solution in the tank is given by $V(t) = 500 - 0.5t$ L.

a. Graph the mass function and verify that $M(0) = 0$.
b. Graph the volume function and verify that the tank is empty when $t = 1000$ min.

c. The concentration of the salt solution in the tank is given by $C(t) = M(t)/V(t)$ g/L. Graph the concentration function and comment on its properties. Specifically, what are $C(0)$ and $C(1000)$?
d. Find the rate of change of the mass $M'(t)$ for $0 \le t \le 1000$.
e. Find the rate of change of the concentration $C'(t)$ for $0 \le t \le 1000$.
f. For what times is the concentration of the solution increasing? Decreasing?

☐ 69. Power and Energy The total energy in megawatt-hr (MWh) used by a town is given by

$$E(t) = 400t + \frac{2400}{\pi}\sin\left(\frac{\pi t}{12}\right)$$

where $t \ge 0$ is measured in hours, with $t = 0$ corresponding to noon.

a. Find the power, or rate of energy consumption, $P(t) = E'(t)$ in units of megawatts (MW).
b. At what time of day is the rate of energy consumption a maximum? What is the power at that time of day?
c. At what time of day is the rate of energy consumption a minimum? What is the power at that time of day?
d. Sketch a graph of the power function reflecting the actual times when energy use is a minimum or maximum.

Additional Exercises

70. Deriving Trigonometric Identities

a. Recall that $\cos 2t = \cos^2 t - \sin^2 t$. Use differentiation to find a trigonometric identity for $\sin 2t$.
b. Verify that you obtain the same identity for $\sin 2t$ as in part (a) if you use the identity $\cos 2t = 2\cos^2 t - 1$.
c. Verify that you obtain the same identity for $\sin 2t$ as in part (a) if you use the identity $\cos 2t = 1 - 2\sin^2 t$.

71. Proof of $\cos^2 x + \sin^2 x = 1$ Let $f(x) = \cos^2 x + \sin^2 x$.

a. Use the Chain Rule to show that $f'(x) = 0$.
b. Assume that if $f' = 0$, then f is a constant function. Calculate $f(0)$ and use it with part (a) to explain why $\cos^2 x + \sin^2 x = 1$.

72. Using the Chain Rule to prove that $\dfrac{d}{dx}(e^{kx}) = ke^{kx}$

a. Identify the inner function $g(x)$ and the outer function $f(u)$ for the composition $f(g(x)) = e^{kx}$, where k is a real number.
b. Use the Chain Rule to show that $\dfrac{d}{dx}(e^{kx}) = ke^{kx}$.

73. Deriving the Quotient Rule using the Product Rule and Chain Rule Suppose you forgot the Quotient Rule for calculating $\dfrac{d}{dx}\left[\dfrac{f(x)}{g(x)}\right]$. Use the Chain Rule and Product Rule with the identity $\dfrac{f(x)}{g(x)} = f(x)[g(x)]^{-1}$ to derive the Quotient Rule.

74. The Chain Rule for second derivatives

a. Derive a formula for the second derivative, $\dfrac{d^2}{dx^2}[f(g(x))]$.

b. Use the formula in part (a) to calculate

$$\frac{d^2}{dx^2}\sin(3x^4 + 5x^2 + 2).$$

75–78. Calculating limits *The following limits are the derivatives of a composite function h at some point x = a.*

a. *Find a composite function h and the value of a.*

b. *Use the Chain Rule to find each limit. Verify your answer by using the limit command on a calculator.*

75. $\displaystyle\lim_{x\to 2}\frac{(x^2 - 3)^5 - 1}{x - 2}$

76. $\displaystyle\lim_{x\to 0}\frac{\sqrt{4 + \sin x} - 2}{x}$

77. $\displaystyle\lim_{h\to 0}\frac{\sin(\pi/2 + h)^2 - \sin(\pi^2/4)}{h}$

78. $\displaystyle\lim_{h\to 0}\frac{\dfrac{1}{3[(1+h)^5 + 7]^{10}} - \dfrac{1}{3(8)^{10}}}{h}$

79. Assuming that f is differentiable for all x, simplify

$$\lim_{x\to 5}\frac{f(x^2) - f(25)}{x - 5}.$$

80. Derivatives of even and odd functions Recall that f is even if $f(x) = f(-x)$ for all x in the domain of f, and f is odd if $f(x) = -f(-x)$ for all x in the domain of f.

a. If f is a differentiable, even function on its domain, determine whether f' is even, odd, or neither.

b. If f is a differentiable, odd function on its domain, determine whether f' is even, odd, or neither.

81. A general proof of the Chain Rule Let f and g be differentiable functions with $h(x) = f(g(x))$. For a given constant a, let $u = g(a)$ and $v = g(x)$, and define

$$H(v) = \begin{cases} \dfrac{f(v) - f(u)}{v - u} - f'(u) & \text{if } v \ne u \\[2mm] 0 & \text{if } v = u \end{cases}$$

a. Show that $\displaystyle\lim_{v\to u} H(v) = 0$.

b. For any value of u show that

$$f(v) - f(u) = (H(v) + f'(u))(v - u).$$

c. Show that

$$h'(a) = \lim_{x\to a}\left[(H(g(x)) + f'(g(a)))\cdot\frac{g(x) - g(a)}{x - a}\right].$$

d. Show that $h'(a) = f'(g(a))g'(a)$.

QUICK CHECK ANSWERS

1. The expansion of $(5x + 4)^{100}$ contains 101 terms. It would take too much time to calculate both the expansion and the derivative. **2.** The inner function is $u = 5x + 4$, and the outer function is $y = u^3$. **4.** $f(u) = u^{10}$; $u = g(v) = \tan v$; $v = h(x) = x^5$ ◄

3.7 Implicit Differentiation

This chapter has been devoted to calculating derivatives of functions of the form $y = f(x)$, where y is defined *explicitly* as a function of x. However, relationships between variables are often expressed *implicitly*. For example, the equation of the unit circle $x^2 + y^2 = 1$, when written $x^2 + y^2 - 1 = 0$, has the *implicit* form $F(x, y) = 0$. This equation does not represent a single function because its graph fails the vertical line test (Figure 3.47a). If, however, the equation $x^2 + y^2 = 1$ is solved for y, then *two* functions, $y = -\sqrt{1 - x^2}$ and $y = \sqrt{1 - x^2}$, emerge (Figure 3.47b). Having identified two explicit functions that form the circle, their derivatives are found using the Chain Rule:

$$\text{If } y = \sqrt{1 - x^2}, \text{ then } \frac{dy}{dx} = -\frac{x}{\sqrt{1 - x^2}}. \tag{1}$$

$$\text{If } y = -\sqrt{1 - x^2}, \text{ then } \frac{dy}{dx} = \frac{x}{\sqrt{1 - x^2}}. \tag{2}$$

We use equation (1) to find the slope of the curve at any point on the upper half of the unit circle and equation (2) to find the slope of the curve at any point on the lower half of the circle.

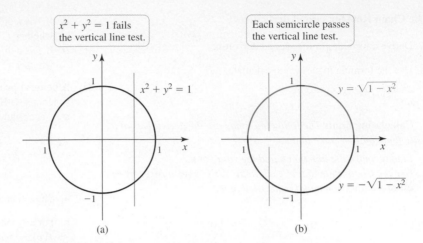

FIGURE 3.47   (a)   (b)

QUICK CHECK 1 The equation $x - y^2 = 0$ implicitly defines what two functions? ◄

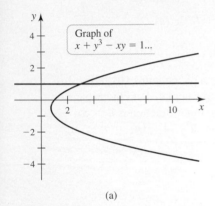

(a)

FIGURE 3.48

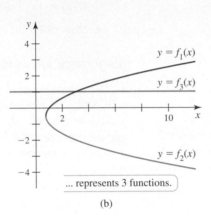

(b)

While it is straightforward to solve some implicit equations for y (such as $x^2 + y^2 = 1$ or $x - y^2 = 0$), it is difficult or impossible to solve other equations for y. For example, the graph of $x + y^3 - xy = 1$ (Figure 3.48a) represents three functions: the upper half of a parabola $y = f_1(x)$, the lower half of a parabola $y = f_2(x)$, and the horizontal line $y = f_3(x)$ (Figure 3.48b). Solving for y to obtain these three functions is challenging (Exercise 55), and even after solving for y, derivatives for each of the three functions must be calculated separately. The goal of this section is to find a *single* expression for the derivative *directly* from an equation $F(x, y) = 0$ without first solving for y. This technique, called **implicit differentiation**, is demonstrated through examples.

EXAMPLE 1   Implicit differentiation

a. Calculate $\dfrac{dy}{dx}$ directly from the equation for the unit circle $x^2 + y^2 = 1$.

b. Find the slope of the unit circle at $\left(\dfrac{1}{2}, \dfrac{\sqrt{3}}{2}\right)$ and $\left(\dfrac{1}{2}, -\dfrac{\sqrt{3}}{2}\right)$.

SOLUTION

a. To indicate the choice of x as the independent variable, it is helpful to replace the variable y with $y(x)$:

$$x^2 + (y(x))^2 = 1 \quad \text{Replace } y \text{ by } y(x).$$

We now take the derivative of each term in the equation *with respect to x*:

$$\underbrace{\frac{d}{dx}(x^2)}_{2x} + \underbrace{\frac{d}{dx}[y(x)]^2}_{\text{Use the Chain Rule}} = \underbrace{\frac{d}{dx}(1)}_{0}$$

By the Chain Rule, $\dfrac{d}{dx}[y(x)]^2 = 2y(x)y'(x)$, or $\dfrac{d}{dx}(y^2) = 2y\dfrac{dy}{dx}$. Using this result, we have

$$2x + 2y\frac{dy}{dx} = 0.$$

The last step is to solve for $\dfrac{dy}{dx}$:

$$2y\frac{dy}{dx} = -2x \quad \text{Subtract } 2x \text{ from both sides.}$$

$$\frac{dy}{dx} = -\frac{x}{y} \quad \text{Divide by } 2y \text{ and simplify.}$$

This result holds provided $y \neq 0$. At the points $(1, 0)$ and $(-1, 0)$, the circle has vertical tangent lines.

b. Notice that the derivative $\dfrac{dy}{dx} = -\dfrac{x}{y}$ depends on *both* x and y. Therefore, to find the slope of the circle at $\left(\dfrac{1}{2}, \dfrac{\sqrt{3}}{2}\right)$, we substitute both $x = 1/2$ and $y = \sqrt{3}/2$ into the derivative formula. The result is

$$\left.\frac{dy}{dx}\right|_{\left(\frac{1}{2}, \frac{\sqrt{3}}{2}\right)} = -\frac{1/2}{\sqrt{3}/2} = -\frac{1}{\sqrt{3}}.$$

The slope of the curve at $\left(\dfrac{1}{2}, -\dfrac{\sqrt{3}}{2}\right)$ is

$$\left.\frac{dy}{dx}\right|_{\left(\frac{1}{2}, -\frac{\sqrt{3}}{2}\right)} = -\frac{1/2}{-\sqrt{3}/2} = \frac{1}{\sqrt{3}}.$$

The curve and tangent lines are shown in Figure 3.49.   *Related Exercises 5–20* ◄

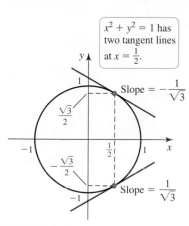

FIGURE 3.49

Example 1 illustrates the technique of implicit differentiation. It is done without solving for y, and it produces $\dfrac{dy}{dx}$ in terms of *both* x and y. The derivative obtained in Example 1 is consistent with the derivatives calculated explicitly in equations (1) and (2). For the upper half of the circle, substituting $y = \sqrt{1 - x^2}$ into the implicit derivative $\dfrac{dy}{dx} = -\dfrac{x}{y}$ gives

$$\frac{dy}{dx} = -\frac{x}{y} = -\frac{x}{\sqrt{1 - x^2}},$$

which agrees with equation (1). For the lower half of the circle, substituting $y = -\sqrt{1 - x^2}$ into $\dfrac{dy}{dx} = -\dfrac{x}{y}$ gives

$$\frac{dy}{dx} = -\frac{x}{y} = \frac{x}{\sqrt{1 - x^2}},$$

which is consistent with equation (2). Therefore, implicit differentiation gives a single unified derivative $\dfrac{dy}{dx} = -\dfrac{x}{y}$.

QUICK CHECK 2 Use implicit differentiation to find $\dfrac{dy}{dx}$ for $x - y^2 = 3$. ◄

Slopes of Tangent Lines

Derivatives obtained by implicit differentiation typically depend on *x and y*. Therefore, the slope of a curve at a particular point (x, y) requires both the x- and y-coordinates of the point. These coordinates are also needed to find an equation of the tangent line at that point.

QUICK CHECK 3 If a function is defined explicitly in the form $y = f(x)$, which coordinates are needed to find the slope of a tangent line—the x-coordinate, the y-coordinate, or both? ◄

EXAMPLE 2 **Finding tangent lines with implicit functions** Find an equation of the line tangent to the curve $x^2 + xy - y^3 = 7$ at $(3, 2)$.

SOLUTION We calculate the derivative of each term of the equation $x^2 + xy - y^3 = 7$:

$$\frac{d}{dx}(x^2) + \frac{d}{dx}(xy) - \frac{d}{dx}(y^3) = \frac{d}{dx}(7) \qquad \text{Differentiate each term.}$$

$$2x + \underbrace{y + xy'}_{\text{Product Rule}} - \underbrace{3y^2 y'}_{\text{Chain Rule}} = 0 \qquad \text{Calculate the derivatives.}$$

$$3y^2 y' - xy' = 2x + y \qquad \text{Group the terms containing } y'.$$

$$y' = \frac{2x + y}{3y^2 - x} \qquad \text{Factor and solve for } y'.$$

> Because y is a function of x, we have
> $$\frac{d}{dx}(x) = 1 \quad \text{and}$$
> $$\frac{d}{dx}(y) = y'$$
> To differentiate y^3 with respect to x, we need the Chain Rule.

To find the slope of the tangent line at $(3, 2)$, we substitute $x = 3$ and $y = 2$ into the derivative formula:

$$\left.\frac{dy}{dx}\right|_{(3,2)} = \left.\frac{2x + y}{3y^2 - x}\right|_{(3,2)} = \frac{8}{9}$$

An equation of the line passing through $(3, 2)$ with slope $\frac{8}{9}$ is

$$y - 2 = \frac{8}{9}(x - 3) \quad \text{or} \quad y = \frac{8}{9}x - \frac{2}{3}.$$

Figure 3.50 shows the graphs of the curve $x^2 + xy - y^3 = 7$ and the tangent line.

Related Exercises 21–26 ◄

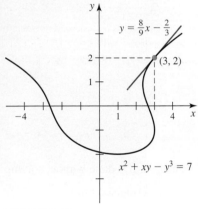

FIGURE 3.50

Higher-Order Derivatives of Implicit Functions

In previous sections of this chapter, we found higher-order derivatives $\dfrac{d^n y}{dx^n}$ by first calculating $\dfrac{dy}{dx}, \dfrac{d^2 y}{dx^2}, \ldots,$ and $\dfrac{d^{n-1} y}{dx^{n-1}}$. The same approach is used with implicit differentiation.

EXAMPLE 3 **A second derivative** Find $\dfrac{d^2 y}{dx^2}$ if $x^2 + y^2 = 1$.

SOLUTION The first derivative $\dfrac{dy}{dx} = -\dfrac{x}{y}$ was computed in Example 1.

We now calculate the derivative of each side of this equation and solve for the second derivative:

$$\frac{d}{dx}\left(\frac{dy}{dx}\right) = \frac{d}{dx}\left(-\frac{x}{y}\right) \qquad \text{Take derivatives with respect to } x.$$

$$\frac{d^2y}{dx^2} = -\frac{y \cdot 1 - x\dfrac{dy}{dx}}{y^2} \qquad \text{Quotient Rule}$$

$$= -\frac{y - x\left(-\dfrac{x}{y}\right)}{y^2} \qquad \text{Substitute for } \frac{dy}{dx}.$$

$$= -\frac{x^2 + y^2}{y^3} \qquad \text{Simplify.}$$

$$= -\frac{1}{y^3} \qquad x^2 + y^2 = 1$$

Related Exercises 27–32 ◄

The Power Rule for Rational Exponents

The Extended Power Rule states that $\dfrac{d}{dx}(x^n) = nx^{n-1}$ if n is an integer. Using implicit differentiation this rule can be extended to rational values of n such as $\frac{1}{2}$ or $-\frac{5}{3}$. Assume p and q are integers with $q \neq 0$ and let $y = x^{p/q}$, where $x \geq 0$ when q is even. By raising each side of $y = x^{p/q}$ to the power q, we obtain $y^q = x^p$. Assuming that y is a differentiable function of x on its domain, both sides of $y^q = x^p$ are differentiated with respect to x:

$$qy^{q-1}\frac{dy}{dx} = px^{p-1}$$

We divide both sides by qy^{q-1} and simplify:

$$\frac{dy}{dx} = \frac{p}{q} \cdot \frac{x^{p-1}}{y^{q-1}} = \frac{p}{q} \cdot \frac{x^{p-1}}{(x^{p/q})^{q-1}} \qquad \text{Substitute } x^{p/q} \text{ for } y.$$

$$= \frac{p}{q} \cdot \frac{x^{p-1}}{x^{p-p/q}} \qquad \text{Multiply exponents in the denominator.}$$

$$= \frac{p}{q} \cdot x^{p/q-1} \qquad \text{Simplify by combining exponents.}$$

If we let $n = \dfrac{p}{q}$, then $\dfrac{d}{dx}(x^n) = nx^{n-1}$. So, the power rule for rational exponents is the same as the power rule for integers.

> ➤ The assumption that $y = x^{p/q}$ is differentiable on its domain is proved in Section 3.8, where the Power Rule is proved for all real powers; that is, we prove that $\dfrac{d}{dx}(x^n) = nx^{n-1}$ holds for any real number n.

THEOREM 3.16   Power Rule for Rational Exponents
Assume p and q are integers with $q \neq 0$. Then,

$$\frac{d}{dx}(x^{p/q}) = \frac{p}{q}x^{p/q-1}$$

provided that $x \geq 0$ when q is even.

EXAMPLE 4   **Rational exponent** Calculate $\dfrac{dy}{dx}$ for the following functions.

a. $y = \sqrt{x}$ 　　　　　　　　　　　　　　**b.** $y = (x^6 + 3x)^{2/3}$

> The derivative of $\sqrt{x}$ (Example 4a) was determined using the limit definition of the derivative in Section 3.1.

SOLUTION

a. $\dfrac{dy}{dx} = \dfrac{d}{dx}(x^{1/2}) = \dfrac{1}{2}x^{-1/2} = \dfrac{1}{2\sqrt{x}}$

b. We apply the Chain Rule, where the outer function is $u^{2/3}$ and the inner function is $x^6 + 3x$:

$$\dfrac{dy}{dx} = \dfrac{d}{dx}[(x^6 + 3x)^{2/3}] = \underbrace{\dfrac{2}{3}(x^6 + 3x)^{-1/3}}_{\substack{\text{derivative of}\\\text{outer function}}}\underbrace{(6x^5 + 3)}_{\substack{\text{derivative of}\\\text{inner function}}}$$

$$= \dfrac{2(2x^5 + 1)}{(x^6 + 3x)^{1/3}}$$

Related Exercises 33–40 ◀

EXAMPLE 5   **Implicit differentiation with rational exponents** Find the slope of the curve $2(x + y)^{1/3} = y$ at the point $(4, 4)$.

SOLUTION We begin by differentiating both sides of the given equation:

$$\dfrac{2}{3}(x + y)^{-2/3}\left(1 + \dfrac{dy}{dx}\right) = \dfrac{dy}{dx} \qquad \text{\small Implicit differentiation, Chain Rule, Theorem 3.16}$$

$$\dfrac{2}{3}(x + y)^{-2/3} = \dfrac{dy}{dx} - \dfrac{2}{3}(x + y)^{-2/3}\dfrac{dy}{dx} \qquad \text{\small Expand and collect like terms.}$$

$$\dfrac{2}{3}(x + y)^{-2/3} = \dfrac{dy}{dx}\left(1 - \dfrac{2}{3}(x + y)^{-2/3}\right) \qquad \text{\small Factor out } \dfrac{dy}{dx}.$$

We now solve for dy/dx:

$$\dfrac{dy}{dx} = \dfrac{\dfrac{2}{3}(x + y)^{-2/3}}{1 - \dfrac{2}{3}(x + y)^{-2/3}} \qquad \text{\small Divide by } 1 - \dfrac{2}{3}(x + y)^{-2/3}.$$

$$\dfrac{dy}{dx} = \dfrac{2}{3(x + y)^{2/3} - 2} \qquad \text{\small Multiply by } 3(x + y)^{2/3} \text{ and simplify.}$$

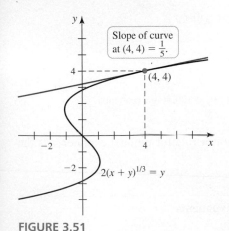

Slope of curve at $(4, 4) = \frac{1}{5}$.

$(4, 4)$

$2(x + y)^{1/3} = y$

FIGURE 3.51

Note that the point $(4, 4)$ *does* lie on the curve (Figure 3.51). The slope of the curve at $(4, 4)$ is found by substituting $x = 4$ and $y = 4$ into the formula for $\dfrac{dy}{dx}$:

$$\dfrac{dy}{dx}\bigg|_{(4, 4)} = \dfrac{2}{3(8)^{2/3} - 2} = \dfrac{1}{5}$$

Related Exercises 41–46 ◀

SECTION 3.7 EXERCISES

Review Questions

1. For some equations, such as $x^2 + y^2 = 1$ or $x - y^2 = 0$, it is possible to solve for y and then calculate $\dfrac{dy}{dx}$. Even in these cases, explain why implicit differentiation is usually a more efficient method for calculating the derivative.

2. Explain the differences between computing the derivatives of functions that are defined implicitly and explicitly.

3. Why are both the x-coordinate and the y-coordinate generally needed to find the slope of the tangent line at a point for an implicitly defined function?

4. In this section, for what values of n did we prove that
$$\frac{d}{dx}(x^n) = nx^{n-1}?$$

Basic Skills

5–10. Implicit differentiation *Carry out the following steps.*

   a. *Use implicit differentiation to find the derivative $\dfrac{dy}{dx}$.*

   b. *Find the slope of the curve at the given point.*

5. $y^2 = 4x$; $(1, 2)$ **6.** $y^2 + 3x = 2$; $(-1, \sqrt{5})$

7. $\sin(y) = 5x^4 - 5$; $(1, \pi)$ **8.** $5\sqrt{x} - 10\sqrt{y} = \sin x$; $(4\pi, \pi)$

9. $\cos(y) = x$; $\left(0, \dfrac{\pi}{2}\right)$ **10.** $\tan(xy) = x + y$; $(0, 0)$

11–20. Implicit differentiation *Use implicit differentiation to find $\dfrac{dy}{dx}$.*

11. $\sin(xy) = x + y$ **12.** $e^{xy} = 2y$

13. $\cos(y^2) + x = e^y$ **14.** $y = \dfrac{x+1}{y-1}$

15. $x^3 = \dfrac{x+y}{x-y}$ **16.** $(xy + 1)^3 = x - y^2 + 8$

17. $6x^3 + 7y^3 = 13xy$ **18.** $(x^2 + y^2)(x^2 + y^2 + x) = 8xy^2$

19. $\sqrt{x^4 + y^2} = 5x + 2y^3$ **20.** $\sqrt{3x^7 + y^2} = \sin^2 y + 100xy$

21–26. Tangent lines *Carry out the following steps.*

   a. *Verify that the given point lies on the curve.*

   b. *Determine an equation of the line tangent to the curve at the given point.*

21. $x^2 + xy + y^2 = 7$; $(2, 1)$ **22.** $x^4 - x^2y + y^4 = 1$; $(-1, 1)$

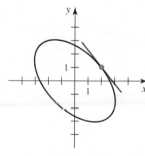

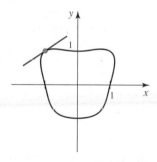

23. $\sin y + 5x = y^2$; $\left(\dfrac{\pi^2}{5}, \pi\right)$ **24.** $x^3 + y^3 = 2xy$; $(1, 1)$

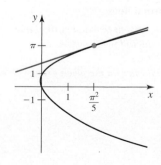

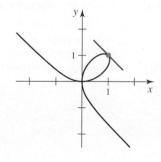

25. $\cos(x - y) + \sin y = \sqrt{2}$;
$\left(\dfrac{\pi}{2}, \dfrac{\pi}{4}\right)$

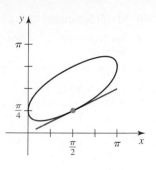

26. $(x^2 + y^2)^2 = \dfrac{25}{4}xy^2$;

$(1, 2)$

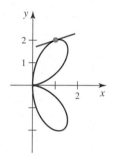

27–32. Second derivatives *Find $\dfrac{d^2y}{dx^2}$.*

27. $x + y^2 = 1$ **28.** $2x^2 + y^2 = 4$ **29.** $\sqrt{y} + xy = 1$

30. $x^4 + y^4 = 64$ **31.** $e^{2y} + x = y$ **32.** $\sin x + x^2y = 10$

33–40. Derivatives of functions with rational exponents *Find $\dfrac{dy}{dx}$.*

33. $y = x^{5/4}$ **34.** $y = \sqrt[3]{x^2 - x + 1}$

35. $y = (5x + 1)^{2/3}$ **36.** $y = e^x\sqrt{x^3}$

37. $y = \sqrt[4]{\dfrac{2x}{4x - 3}}$ **38.** $y = (2x + 3)^2(4x + 6)^{1/4}$

39. $y = x\sqrt[3]{x^2 + 5x + 1}$ **40.** $y = \dfrac{x}{\sqrt[5]{x} + x}$

41–46. Implicit differentiation with rational exponents *Determine the slope of the following curves at the given point.*

41. $\sqrt[3]{x} + \sqrt[3]{y^4} = 2$; $(1, 1)$ **42.** $x^{2/3} + y^{2/3} = 2$; $(1, 1)$

43. $xy^{1/3} + y = 10$; $(1, 8)$ **44.** $(x + y)^{2/3} = y$; $(4, 4)$

45. $xy + x^{3/2}y^{-1/2} = 2$; $(1, 1)$ **46.** $xy^{5/2} + x^{3/2}y = 12$; $(4, 1)$

Further Explorations

47. Explain why or why not Determine whether the following statements are true and give an explanation or counterexample.

   a. For any equation containing the variables x and y, the derivative dy/dx can be found by first using algebra to rewrite the equation in the form $y = f(x)$.

   b. For the equation of a circle of radius r, $x^2 + y^2 = r^2$, we have
   $$\frac{dy}{dx} = -\frac{x}{y} \text{ for } y \neq 0 \text{ and any real number } r > 0.$$

   c. If $x = 1$, then by implicit differentiation, $1 = 0$.

   d. If $xy = 1$, then $y' = 1/x$.

48–50. Multiple tangent lines *Complete the following steps.*

a. *Find equations of all lines tangent to the curve at the given value of x.*

b. *Graph the tangent lines on the given graph.*

48. $x + y^3 - y = 1$; $x = 1$

49. $x + y^2 - y = 1$; $x = 1$

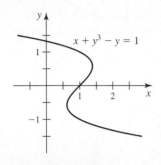

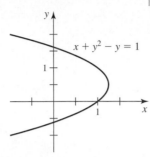

50. $4x^3 = y^2(4 - x)$; $x = 2$ (cissoid of Diocles)

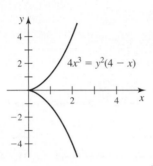

51. Multiple tangent lines Let $y(x^2 + 4) = 8$ (witch of Agnesi).

a. Use implicit differentiation to find $\dfrac{dy}{dx}$.

b. Find equations of all lines tangent to the curve $y(x^2 + 4) = 8$ when $y = 1$.

c. Solve the equation $y(x^2 + 4) = 8$ for y to find an explicit expression for y and then calculate $\dfrac{dy}{dx}$.

d. Verify that the results of parts (a) and (c) are consistent.

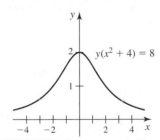

52. Vertical tangent lines

a. Determine the points where the curve $x + y^3 - y = 1$ has a vertical tangent line (see Exercise 48).

b. Does the curve have any horizontal tangent lines? Explain.

53. Vertical tangent lines

a. Determine the points where the curve $x + y^2 - y = 1$ has a vertical tangent line (see Exercise 49).

b. Does the curve have any horizontal tangent lines? Explain.

54–58. Identifying functions from an equation *The following equations implicitly define one or more functions.*

a. *Find $\dfrac{dy}{dx}$ using implicit differentiation.*

b. *Solve the given equation for y to identify the implicitly defined functions $y = f_1(x), y = f_2(x), \ldots$.*

c. *Use the functions found in part (b) to graph the given equation.*

d. *Find the derivative of each function in part (b) and verify that your results are consistent with part (a).*

54. $y^3 = ax^2$ (Neile's semicubical parabola)

55. $x + y^3 - xy = 1$ (*Hint:* Rewrite as $y^3 - 1 = xy - x$ and then factor both sides.)

56. $y^2 = \dfrac{x^2(4 - x)}{4 + x}$ (right strophoid)

57. $x^4 = 2(x^2 - y^2)$ (eight curve)

58. $y^2(x + 2) = x^2(6 - x)$ (trisectrix)

59–64. Normal lines *A **normal line** to a curve passes through a point P on the curve perpendicular to the line tangent to the curve at P (see figure). Use the following equations and graphs to determine an equation of the normal line at the given point and illustrate your work by graphing the curve with the normal line.*

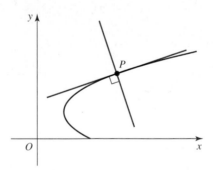

59. Exercise 21   **60.** Exercise 22   **61.** Exercise 23

62. Exercise 24   **63.** Exercise 25   **64.** Exercise 26

65–68. Visualizing tangent and normal lines

a. *Determine an equation of the tangent line and normal line at the given point (x_0, y_0) on the following curves. (See instructions for Exercises 59–64.)*

b. *Graph the tangent and normal lines on the given graph.*

65. $3x^3 + 7y^3 = 10y$;
$(x_0, y_0) = (1, 1)$

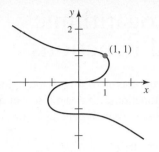

66. $x^4 = 2x^2 + 2y^2$;
$(x_0, y_0) = (2, 2)$
(kampyle of Eudoxus)

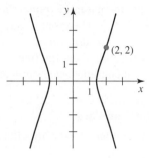

67. $(x^2 + y^2 - 2x)^2 = 2(x^2 + y^2)$;
$(x_0, y_0) = (2, 2)$
(limaçon of Pascal)

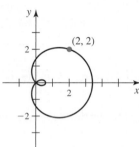

68. $(x^2 + y^2)^2 = \dfrac{25}{3}(x^2 - y^2)$;

$(x_0, y_0) = (2, -1)$
(lemniscate of Bernoulli)

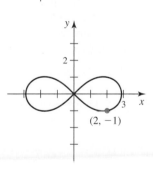

Applications

69. Cobb-Douglas production function The output of an economic system Q, subject to two inputs, such as labor L and capital K, is often modeled by the Cobb-Douglas production function $Q = cL^aK^b$. When $a + b = 1$ the case is called *constant returns to scale*. Suppose $Q = 1280$, $a = \frac{1}{3}$, $b = \frac{2}{3}$, and $c = 40$.

a. Find the rate of change of capital with respect to labor, dK/dL.
b. Evaluate the derivative in part (a) with $L = 8$ and $K = 64$.

70. Surface area of a cone The lateral surface area of a cone of radius r and height h (the surface area excluding the base) is

$A = \pi r\sqrt{r^2 + h^2}$. Find dr/dh for a cone with a lateral surface area of $A = 1500\pi$ cm². Evaluate this derivative when $r = 30$ cm and $h = 40$ cm.

71. Volume of a spherical cap Imagine slicing through a sphere with a plane (sheet of paper). The smaller piece produced is called a spherical cap. Its volume is $V = \pi h^2(3r - h)/3$, where r is the radius of the sphere and h is the thickness of the cap. Find dr/dh for a sphere with a volume of $5\pi/3$ m³. Evaluate this derivative when $r = 2$ m and $h = 1$ m.

72. Volume of a torus The volume of a torus (doughnut or bagel) with an inner radius of a and an outer radius of b is $V = \pi^2(b + a)(b - a)^2/4$. Find db/da for a torus with a volume of $64\pi^2$ in³. Evaluate this derivative when $a = 6$ in and $b = 10$ in.

Additional Exercises

73–75. Orthogonal trajectories *Two curves are* orthogonal *to each other if their tangent lines are perpendicular at each point of intersection (recall that two lines are perpendicular to each other if their slopes are negative reciprocals). A family of curves forms* **orthogonal trajectories** *with another family of curves if each curve in one family is orthogonal to each curve in the other family. For example, the parabolas $y = cx^2$ form orthogonal trajectories with the family of ellipses $x^2 + 2y^2 = k$, where c and k are constants (see figure).*

    Use implicit differentiation if needed to find dy/dx for each equation of the following pairs. Use the derivatives to explain why the families of curves form orthogonal trajectories.

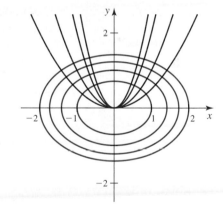

73. $y = mx$; $x^2 + y^2 = a^2$, where m and a are constants

74. $y = cx^2$; $x^2 + 2y^2 = k$, where c and k are constants

75. $xy = a$; $x^2 - y^2 = b$, where a and b are constants

QUICK CHECK ANSWERS

1. $y = \sqrt{x}$ and $y = -\sqrt{x}$ **2.** $\dfrac{dy}{dx} = \dfrac{1}{2y}$ **3.** Only the x-coordinate is needed. ◄

3.8 Derivatives of Logarithmic and Exponential Functions

We return now to the major theme of this chapter: developing rules of differentiation for the standard families of functions. First, we discover how to differentiate the natural logarithmic function. From there, we treat general exponential and logarithmic functions.

The derivative of $y = \ln x$

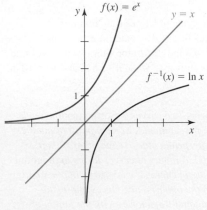

FIGURE 3.52

Recall from Section 1.3 that the natural exponential function $f(x) = e^x$ is a one-to-one function on the interval $(-\infty, \infty)$. Therefore, it has an inverse, which is the natural logarithmic function $f^{-1}(x) = \ln x$. The domain of f^{-1} is the range of f, which is $(0, \infty)$. The graphs of f and f^{-1} are symmetric about the line $y = x$ (Figure 3.52). This inverse relationship has several important consequences, summarized as follows.

> **Inverse Properties for e^x and $\ln x$**
>
> 1. $e^{\ln x} = x$ for $x > 0$, and $\ln(e^x) = x$ for all x.
>
> 2. $y = \ln x$ if and only if $x = e^y$.
>
> 3. For real numbers x and $b > 0$, $b^x = e^{(\ln b^x)} = e^{x \ln b}$.

QUICK CHECK 1 Simplify $e^{2 \ln x}$. Express 5^x using the base e. ◄

➤ Figure 3.52 also provides evidence that $\ln x$ is differentiable for $x > 0$: Its graph is smooth with no jumps or cusps.

With these preliminary observations, we now determine the derivative of $\ln x$. A theorem we prove in Section 3.9 says that because e^x is differentiable on its domain, its inverse $\ln x$ is also differentiable on its domain.

To find the derivative of $y = \ln x$, we begin with inverse property 2 and write $x = e^y$, where $x > 0$. The key step is to compute dy/dx using implicit differentiation. Using the Chain Rule to differentiate both sides of $x = e^y$ with respect to x, we have

$$x = e^y \qquad y = \ln x \Rightarrow x = e^y$$

$$1 = e^y \cdot \frac{dy}{dx} \qquad \text{Differentiate both sides with respect to } x.$$

$$\frac{dy}{dx} = \frac{1}{e^y} = \frac{1}{x} \qquad \text{Solve for } dy/dx \text{ and use } x = e^y.$$

Therefore,

$$\frac{d}{dx}(\ln x) = \frac{1}{x}.$$

Because the domain of the natural logarithm is $(0, \infty)$, this rule is limited to positive values of x (Figure 3.53a).

An important extension is obtained by considering the function $\ln |x|$, which is defined for all $x \neq 0$. By the definition of the absolute value,

➤ Recall that
$$|x| = \begin{cases} x & \text{if } x \geq 0 \\ -x & \text{if } x < 0 \end{cases}$$

$$\ln |x| = \begin{cases} \ln x & \text{if } x > 0 \\ \ln (-x) & \text{if } x < 0 \end{cases}$$

For $x > 0$, it follows immediately that

$$\frac{d}{dx}\left(\ln |x|\right) = \frac{d}{dx}(\ln x) = \frac{1}{x}.$$

When $x < 0$, a similar calculation using the Chain Rule reveals that

$$\frac{d}{dx}\left(\ln|x|\right) = \frac{d}{dx}\left(\ln(-x)\right) = \frac{1}{(-x)}(-1) = \frac{1}{x}.$$

Therefore, we have the result that the derivative of $\ln|x|$ is $\frac{1}{x}$ (Figure 3.53b).

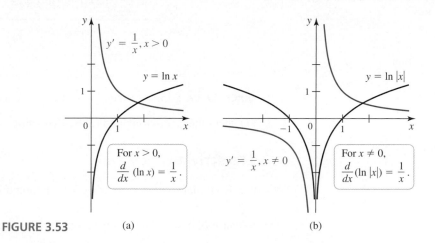

FIGURE 3.53   (a)   (b)

THEOREM 3.17   Derivative of ln x

$$\frac{d}{dx}(\ln x) = \frac{1}{x} \quad \text{for } x > 0 \qquad \frac{d}{dx}\left(\ln|x|\right) = \frac{1}{x} \quad \text{for } x \neq 0$$

EXAMPLE 1   Derivatives involving ln x Find $\dfrac{dy}{dx}$ for the following functions.

a. $y = \ln(4x)$ **b.** $y = x\ln x$ **c.** $y = \ln|\sec x|$ **d.** $y = \dfrac{\ln(x^2)}{x^2}$

SOLUTION

a. Using the Chain Rule,

$$\frac{dy}{dx} = \frac{d}{dx}\left[\ln(4x)\right] - \frac{1}{4x}\cdot 4 = \frac{1}{x}.$$

> Because $\ln x$ and $\ln 4x$ differ by a constant ($\ln 4x = \ln x + \ln 4$), the derivatives of $\ln x$ and $\ln 4x$ are equal.

An alternative method uses a property of logarithms before differentiating:

$$\frac{d}{dx}(\ln 4x) = \frac{d}{dx}(\ln 4 + \ln x) \quad \ln(ab) = \ln a + \ln b$$

$$= 0 + \frac{1}{x} = \frac{1}{x} \qquad \ln 4 \text{ is a constant.}$$

b. By the Product Rule,

$$\frac{dy}{dx} = \frac{d}{dx}\left[x\ln x\right] = 1\cdot\ln x + x\cdot\frac{1}{x} = \ln x + 1.$$

c. Using the Chain Rule,

$$\frac{dy}{dx} = \frac{1}{\sec x}\left[\frac{d}{dx}(\sec x)\right] = \frac{1}{\sec x}(\sec x \tan x) = \tan x.$$

> The fact that $\ln x^2 = 2 \ln x$ was used to simplify the result. It could have been used prior to differentiation to avoid the Chain Rule.

d. The Quotient Rule and Chain Rule give

$$\frac{dy}{dx} = \frac{x^2\left(\dfrac{1}{x^2} \cdot 2x\right) - (\ln x^2)\, 2x}{\left(x^2\right)^2} = \frac{2x - 4x \ln x}{x^4} = \frac{2 - 4 \ln x}{x^3}.$$

Related Exercises 9–16 ◄

QUICK CHECK 2 Find $\dfrac{d}{dx}(\ln x^p)$, where $x > 0$ and p is a rational number, in two ways:
(1) using the Chain Rule and (2) by first using a property of logarithms. ◄

The derivative of b^x

A rule similar to $\dfrac{d}{dx}(e^x) = e^x$ exists for computing the derivative of b^x, where $b > 0$.

Because $b^x = e^{x \ln b}$ by inverse property 3, its derivative is

$$\frac{d}{dx}(b^x) = \frac{d}{dx}(e^{x \ln b}) = \underbrace{e^{x \ln b}}_{b^x} \cdot \ln b \quad \text{Chain Rule with } \frac{d}{dx}(x \ln b) = \ln b$$

Noting that $e^{x \ln b} = b^x$ results in the following theorem.

> Check that when $b = e$, Theorem 3.18 becomes
> $$\frac{d}{dx}(e^x) = e^x.$$

THEOREM 3.18   Derivative of b^x
If $b > 0$, then for all x,

$$\frac{d}{dx}(b^x) = b^x \ln b.$$

Notice that when $b > 1$, $\ln b > 0$ and the graph of $y = b^x$ has tangent lines with positive slopes for all x. When $0 < b < 1$, $\ln b < 0$ and the graph of $y = b^x$ has tangent lines with negative slopes for all x. In either case, the tangent line at $(0, 1)$ has slope $\ln b$ (Figure 3.54).

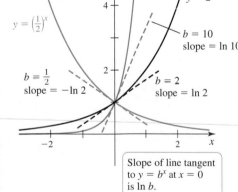

$y = 10^x$

$y = 2^x$

$y = \left(\dfrac{1}{2}\right)^x$

$b = 10$
slope $= \ln 10$

$b = \dfrac{1}{2}$
slope $= -\ln 2$

$b = 2$
slope $= \ln 2$

Slope of line tangent to $y = b^x$ at $x = 0$ is $\ln b$.

FIGURE 3.54

EXAMPLE 2   Derivatives with b^x Find the derivative of the following functions.

a. $f(x) = 3^x$ **b.** $g(t) = 108(2)^{t/12}$

SOLUTION

a. Using Theorem 3.18, $f'(x) = 3^x \cdot \ln 3$.

b.

$$g'(t) = 108\frac{d}{dt}\left(2^{t/12}\right) \qquad \text{Constant Multiple Rule}$$

$$= 108 \cdot \ln 2 \cdot 2^{t/12} \underbrace{\frac{d}{dt}\left(\frac{t}{12}\right)}_{1/12} \qquad \text{Chain Rule}$$

$$= 9 \ln 2 \cdot 2^{t/12} \qquad \text{Simplify.}$$

Related Exercises 17–22 ◄

Table 3.4

Mother's Age	Incidence of Down Syndrome	Decimal Equivalent
30	1 in 900	0.00111
35	1 in 400	0.00250
36	1 in 300	0.00333
37	1 in 230	0.00435
38	1 in 180	0.00556
39	1 in 135	0.00741
40	1 in 105	0.00952
42	1 in 60	0.01667
44	1 in 35	0.02875
46	1 in 20	0.05000
48	1 in 16	0.06250
49	1 in 12	0.08333

Source: E.G. Hook and A. Lindsjo, *Down Syndrome in Live Births by Single Year Maternal Age.*

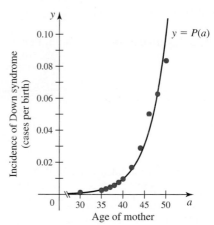

FIGURE 3.55

> The model in Example 3 was created using a method called *exponential regression.* The parameters A and B are chosen so that the function $P(a) = A B^a$ fits the data as closely as possible.

EXAMPLE 3   An exponential model Table 3.4 and Figure 3.55 show how the incidence of Down syndrome in newborn infants increases with the age of the mother. The data can be modeled with the exponential function $P(a) = \dfrac{1}{1,613,000} 1.2733^a$, where a is the age of the mother (in years) and $P(a)$ is the incidence (number of Down syndrome children per total births).

a. According to the model, at what age is the incidence of Down syndrome equal to 0.01 (that is, 1 in 100)?

b. Compute $P'(a)$.

c. Find $P'(35)$ and $P'(46)$, and interpret each.

SOLUTION

a. We let $P(a) = 0.01$ and solve for a:

$$0.01 = \frac{1}{1,613,000} 1.2733^a$$

$$\ln 16,130 = \ln\left(1.2733^a\right) \qquad \text{Multiply both sides by 1,613,000, and take logarithms of both sides.}$$

$$\ln 16,130 = a \ln 1.2733 \qquad \text{Property of logarithms}$$

$$a = \frac{\ln 16,130}{\ln 1.2733} \approx 40 \text{ years old} \qquad \text{Solve for } a.$$

b. $$P'(a) = \frac{1}{1,613,000} \frac{d}{da}\left(1.2733^a\right)$$

$$= \frac{1}{1,613,000} 1.2733^a \cdot \ln 1.2733$$

$$\approx \frac{1}{6,676,000} 1.2733^a$$

c. The derivative measures the rate of change of the incidence with respect to age. For a 35-year-old woman,

$$P'(35) = \frac{1}{6,676,000} 1.2733^{35} \approx 0.0007,$$

which means the incidence increases at a rate of about 0.0007/year. By age 46, the rate of change is

$$P'(46) = \frac{1}{6,676,000} 1.2733^{46} \approx 0.01,$$

which is a significant increase over the rate of change of the incidence at age 35.

Related Exercises 23–25 ◄

QUICK CHECK 3 Suppose $A = 500(1.045)^t$. Compute $\dfrac{dA}{dt}$. ◄

The General Power Rule

As it stands now, the Power Rule for derivatives says that $\dfrac{d}{dx}(x^p) = px^{p-1}$ for rational powers p. The rule is now extended to all real powers.

THEOREM 3.19   General Power Rule
For real numbers p and for $x > 0$,

$$\frac{d}{dx}(x^p) = px^{p-1}.$$

Furthermore, if u is a positive differentiable function on its domain, then

$$\frac{d}{dx}(u(x)^p) = p(u(x))^{p-1} \cdot u'(x).$$

Proof For $x > 0$ and real numbers p, we have $x^p = e^{p \ln x}$ by inverse property (3). Therefore, the derivative of x^p is computed as follows:

$$\frac{d}{dx}(x^p) = \frac{d}{dx}(e^{p \ln x}) \quad \text{Inverse property (3)}$$

$$= e^{p \ln x} \cdot \frac{p}{x} \quad \text{Chain Rule, } \frac{d}{dx}(p \ln x) = \frac{p}{x}$$

$$= x^p \cdot \frac{p}{x} \quad e^{p \ln x} = x^p$$

$$= px^{p-1} \quad \text{Simplify.}$$

We see that $\dfrac{d}{dx}(x^p) = px^{p-1}$ for all real powers p. The second part of the General Power Rule follows from the Chain Rule. ◄

EXAMPLE 4   Computing derivatives Find the derivative of the following functions.

a. $y = x^\pi$ **b.** $y = \pi^x$ **c.** $y = (x^2 + 4)^e$

SOLUTION

> Recall that power functions have the variable in the base, while exponential functions have the variable in the exponent.

a. With $y = x^\pi$, we have a power function with an irrational exponent; by the General Power Rule,

$$\frac{dy}{dx} = \pi x^{\pi - 1} \quad \text{for } x > 0.$$

b. Here we have an exponential function with base $b = \pi$. By Theorem 3.18,

$$\frac{dy}{dx} = \pi^x \cdot \ln \pi.$$

c. The Chain Rule and General Power Rule are required:

$$\frac{dy}{dx} = e(x^2 + 4)^{e-1} \cdot 2x = 2ex(x^2 + 4)^{e-1}$$

Because $x^2 + 4 > 0$ for all x, the result is valid for all x.   *Related Exercises 26–32* ◄

Functions of the form $f(x) = [g(x)]^{h(x)}$, where both g and h are nonconstant functions, are neither exponential functions nor power functions (they are sometimes called *tower functions*). In order to compute their derivatives, we use the identity $b^x = e^{x \ln b}$ to rewrite f with base e:

$$f(x) = [g(x)]^{h(x)} = e^{h(x) \ln g(x)}.$$

This function carries the restriction $g(x) > 0$. The derivative of f is then computed using the methods developed in this section.

EXAMPLE 5   **Finding a horizontal tangent line**   Determine whether the graph of $f(x) = x^x$, for $x > 0$, has any horizontal tangent lines.

SOLUTION   A horizontal tangent occurs when $f'(x) = 0$. In order to find the derivative, we first write $f(x) = x^x = e^{x \ln x}$. Then,

$$\frac{d}{dx}(x^x) = \frac{d}{dx}(e^{x \ln x})$$

$$= e^{x \ln x} \cdot \frac{d}{dx}(x \ln x) \qquad \text{Chain Rule}$$

$$= e^{x \ln x}\left(1 \cdot \ln x + x \cdot \frac{1}{x}\right) \qquad \text{Product Rule}$$

$$= x^x (\ln x + 1) \qquad \text{Simplify; } e^{x \ln x} = x^x.$$

The equation $f'(x) = 0$ implies that $x^x = 0$ or $\ln x + 1 = 0$. The first equation has no solution because $x^x = e^{x \ln x} > 0$ for all $x > 0$. We solve the second equation, $\ln x + 1 = 0$, as follows:

$$\ln x = -1$$

$$e^{\ln x} = e^{-1} \qquad \text{Exponentiate both sides.}$$

$$x = \frac{1}{e} \qquad e^{\ln x} = x$$

Therefore, the graph of $f(x) = x^x$ (Figure 3.56) has a single horizontal tangent at $(e^{-1}, f(e^{-1})) \approx (0.368, 0.692)$.

Related Exercises 33–36 ◄

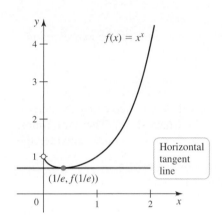

FIGURE 3.56

Derivatives of General Logarithmic Functions

The general exponential function $f(x) = b^x$ is one-to-one when $b > 0$. The inverse function is $f^{-1}(x) = \log_b x$, the logarithmic function with base b. The technique used to differentiate the natural logarithm applies to the general logarithmic function. We begin with the inverse relationship

$$y = \log_b x \iff x = b^y.$$

Applying implicit differentiation to both sides of $x = b^y$, we obtain

$$1 = b^y \cdot \ln b \cdot \frac{dy}{dx} \qquad \text{Implicit differentiation}$$

$$\frac{dy}{dx} = \frac{1}{b^y \ln b} \qquad \text{Solve for } \frac{dy}{dx}.$$

$$\frac{dy}{dx} = \frac{1}{x \ln b} \qquad b^y = x.$$

> An alternative proof of Theorem 3.20 uses the change-of-base formula
>
> $\log_b x = \dfrac{\ln x}{\ln b}$ (Section 1.3).
>
> Differentiating both sides of this equation gives the same result.

QUICK CHECK 4 Compute dy/dx for $y = \log_3 x$. ◄

THEOREM 3.20   Derivative of $\log_b x$

If $b > 0$ with $b \neq 1$, then

$$\frac{d}{dx}(\log_b x) = \frac{1}{x \ln b} \quad \text{for } x > 0 \qquad \frac{d}{dx}(\log_b |x|) = \frac{1}{x \ln b} \quad \text{for } x \neq 0$$

EXAMPLE 6 **Derivatives with general logarithms** Compute the derivative of each of the following functions.

a. $f(x) = \log_5(2x + 1)$ **b.** $T(n) = n \log_2 n$

SOLUTION

a. We use Theorem 3.20 with the Chain Rule assuming $2x + 1 > 0$:

$$f'(x) = \frac{1}{(2x + 1)\ln 5} \cdot 2 = \frac{2}{\ln 5} \cdot \frac{1}{2x + 1}.$$

> The function in Example 6b is used in computer science as an estimate on the computing time needed to carry out a *sorting algorithm* on a list of *n* items.

b.
$$T'(n) = n \cdot \frac{1}{n \ln 2} + \log_2 n = \frac{1}{\ln 2} + \log_2 n \quad \text{Product Rule}$$

We can change bases and write the result in base e:

$$T'(n) = \frac{1}{\ln 2} + \frac{\ln n}{\ln 2} = \frac{1 + \ln n}{\ln 2}$$

Related Exercises 37–42 ◄

QUICK CHECK 5 Show that the derivative computed in Example 6b can be expressed in base 2 as $T'(n) = \log_2(en)$. ◄

Logarithmic Differentiation

Products, quotients, and powers of functions are usually differentiated using the derivative rules of the same name (perhaps combined with the Chain Rule). There are times, however, when the direct computation of a derivative is very tedious. Consider the function

$$f(x) = \frac{(x^3 - 1)^4 \sqrt{3x - 1}}{x^2 + 4}.$$

> The properties of logarithms needed for logarithmic differentiation are:
>
> **1.** $\ln(xy) = \ln x + \ln y$
>
> **2.** $\ln(x/y) = \ln x - \ln y$
>
> **3.** $\ln(x^y) = y \ln x$
>
> All three properties are used in Example 7.

We would need the Quotient, Product, and Chain Rules just to compute $f'(x)$, and simplifying the result would require additional work. The properties of logarithms reviewed in Section 1.3 are useful for differentiating such functions.

EXAMPLE 7 **Logarithmic differentiation** Let $f(x) = \dfrac{(x^3 - 1)^4 \sqrt{3x - 1}}{x^2 + 4}$ and compute $f'(x)$.

SOLUTION We begin by taking the natural logarithm of both sides and simplifying the result:

> In the event that $f \leq 0$ for some values of x, $\ln[f(x)]$ is not defined. In that case, we generally find the derivative of $|y| = |f(x)|$.

$$\ln[f(x)] = \ln\left[\frac{(x^3 - 1)^4 \sqrt{3x - 1}}{x^2 + 4}\right]$$

$$= \ln(x^3 - 1)^4 + \ln\sqrt{3x - 1} - \ln(x^2 + 4) \quad \log(xy) = \log x + \log y$$

$$= 4\ln(x^3 - 1) + \tfrac{1}{2}\ln(3x - 1) - \ln(x^2 + 4) \quad \log x^y = y \log x$$

We now differentiate both sides using the Chain Rule; specifically the derivative of the left side is $\dfrac{d}{dx}[\ln f(x)] = \dfrac{f'(x)}{f(x)}$. Therefore,

$$\frac{f'(x)}{f(x)} = 4 \cdot \frac{1}{x^3 - 1} \cdot 3x^2 + \frac{1}{2} \cdot \frac{1}{3x - 1} \cdot 3 - \frac{1}{x^2 + 4} \cdot 2x.$$

Solving for $f'(x)$, we have

$$f'(x) = f(x)\left[\frac{12x^2}{x^3 - 1} + \frac{3}{2(3x - 1)} - \frac{2x}{x^2 + 4}\right].$$

Finally, we replace $f(x)$ with the original function:

$$f'(x) = \frac{(x^3 - 1)^4\sqrt{3x - 1}}{x^2 + 4}\left[\frac{12x^2}{x^3 - 1} + \frac{3}{2(3x - 1)} - \frac{2x}{x^2 + 4}\right]$$

Related Exercises 43–50 ◄

Logarithmic differentiation also provides an alternative method for finding derivatives of functions of the form $g(x)^{h(x)}$. The derivative of $f(x) = x^x$ (Example 5) is computed as follows, assuming $x > 0$:

$$f(x) = x^x$$

$\ln(f(x)) = \ln(x^x) = x \ln x$ Take logarithms of both sides; use properties.

$\dfrac{1}{f(x)}f'(x) = \left(1 \cdot \ln x + x \cdot \dfrac{1}{x}\right)$ Differentiate both sides.

$f'(x) = f(x)(\ln x + 1)$ Solve for $f'(x)$ and simplify.

$f'(x) = x^x(\ln x + 1)$ Replace $f(x)$ with x^x.

This result agrees with Example 5. The decision about which method to use is largely one of preference.

SECTION 3.8 EXERCISES

Review Questions

1. Use $x = e^y$ to explain why $\dfrac{d}{dx}(\ln x) = \dfrac{1}{x}$ for $x > 0$.

2. Sketch the graph of $f(x) = \ln|x|$ and explain how the graph shows that $f'(x) = 1/x$.

3. Show that $\dfrac{d}{dx}(\ln kx) = \dfrac{d}{dx}(\ln x)$, where $x > 0$ and $k > 0$ is a real number.

4. State the rule of differentiation for the exponential function $f(x) = b^x$. How does it differ from the derivative formula for e^x?

5. State the rule of differentiation for the logarithmic function $f(x) = \log_b x$. How does it differ from the derivative formula for $\ln x$?

6. Explain why $b^x = e^{x \ln b}$.

7. Express the function $f(x) = g(x)^{h(x)}$ in terms of the natural logarithmic and natural exponential functions (base e).

8. Explain the general procedure of logarithmic differentiation.

Basic Skills

9–16. Derivatives involving ln x *Find the following derivatives. Give the intervals on which the results are valid.*

9. $\dfrac{d}{dx}(\ln x^2)$

10. $\dfrac{d}{dx}(\ln 2x^8)$

11. $\dfrac{d}{dx}\left[\ln\left(\dfrac{x + 1}{x - 1}\right)\right]$

12. $\dfrac{d}{dx}(e^x \ln x)$

13. $\dfrac{d}{dx}[(x^2 + 1) \ln x]$

14. $\dfrac{d}{dx}(\ln|x^2 - 1|)$

15. $\dfrac{d}{dx}[\ln(\ln x)]$

16. $\dfrac{d}{dx}[\ln(\cos^2 x)]$

17–22. Derivatives of b^x *Find the derivatives of the following functions.*

17. $y = 5 \cdot 4^x$

18. $y = 4^{-x} \sin x$

19. $y = x^3 \cdot 3^x$

20. $P = \dfrac{40}{1 + 2^{-t}}$

21. $A = 250(1.045)^{4t}$

22. $y = \ln(10^x)$

23. **Exponential model** The following table shows the *time of useful consciousness* at various altitudes in the situation where a pressurized airplane suddenly loses pressure. The change in pressure drastically reduces available oxygen, and hypoxia sets in. The upper value of each time interval is roughly modeled by $T = 10 \cdot 2^{-0.274a}$, where T measures time in minutes and a is the altitude over 22,000 in thousands of feet ($a = 0$ corresponds to 22,000 ft).

Altitude (in feet)	Time of useful consciousness
22,000	5 to 10 min
25,000	3 to 5 min
28,000	2.5 to 3 min
30,000	1 to 2 min
35,000	30 to 60 s
40,000	15 to 20 s
45,000	9 to 15 s

a. A Learjet flying at 38,000 ft ($a = 16$) suddenly loses pressure when the seal on a window fails. How long do the pilot and passengers have to deploy oxygen masks before they become incapacitated, according to the model?

b. What is the average rate of change of T with respect to a over the interval from 24,000 to 30,000 ft (include units)?

c. Find the instantaneous rate of change dT/da, compute it at 30,000 ft, and interpret its meaning.

24. **Magnitude of an earthquake** The energy (in joules) released by an earthquake of magnitude M is given by the equation $E = 25,000 \cdot 10^{1.5M}$. (This equation can be solved for M to define the magnitude of a given earthquake; it is a refinement of the original Richter scale created by Charles Richter in 1935.)

a. Compute the energy released by earthquakes of magnitude 1, 2, 3, 4, and 5. Plot the points on a graph and join them with a smooth curve.

b. Compute dE/dM and evaluate it for $M = 3$. What does this derivative mean? (M has no units, so the units of the derivative are J per change in magnitude.)

25. **Diagnostic scanning** Iodine-123 is a radioactive isotope used in medicine to test the function of the thyroid gland. Its *half-life* is about 13.1 hr. If a 350-microcurie (μCi) dose of iodine-123 is administered to a patient, the quantity Q left in the body after t hours is approximately $Q = 350\left(\frac{1}{2}\right)^{t/13.1}$.

a. How long does it take for the level of iodine-123 to drop to 10 μCi?

b. Find the rate of change of the quantity of iodine-123 at 12 hr, 1 day, and 2 days. What do your answers say about the rate at which iodine decreases as time increases?

26–32. General Power Rule *Use the General Power Rule where appropriate to find the derivative of the following functions.*

26. $f(x) = 2x^{\sqrt{2}}$ 27. $g(y) = e^y \cdot y^e$ 28. $s(t) = \cos(2^t)$

29. $r = e^{2\theta}$ 30. $y = \ln(x^3 + 1)^{\pi}$

31. $f(x) = (2x - 3)x^{3/2}$ 32. $y = \tan(x^{0.74})$

33–36. Tangent lines and general exponential functions

33. Find an equation of the line tangent to $y = x^{\sin x}$ at the point $x = 1$.

34. Determine whether the graph of $y = x^{\sqrt{x}}$ has any horizontal tangent lines.

35. The graph of $y = (x^2)^x$ has two horizontal tangent lines. Find equations for both of them.

36. The graph of $y = x^{\ln x}$ has one horizontal tangent line. Find an equation for it.

37–42. Derivatives of logarithmic functions *Calculate the derivative of the following functions.*

37. $y = 4\log_3(x^2 - 1)$ 38. $y = \log_{10} x$

39. $y = \cos x \ln(\cos^2 x)$ 40. $y = \log_8\left(|\tan x|\right)$

41. $y = \dfrac{1}{\log_4 x}$ 42. $y = \log_2(\log_2 x)$

43–50. Logarithmic differentiation *Use logarithmic differentiation to evaluate $f'(x)$.*

43. $f(x) = \dfrac{(x + 1)^{10}}{(2x - 4)^8}$ 44. $f(x) = x^2 \cos x$

45. $f(x) = x^{\ln x}$ 46. $f(x) = \dfrac{\tan^{10} x}{(5x + 3)^6}$

47. $f(x) = \dfrac{(x + 1)^{3/2}(x - 4)^{5/2}}{(5x + 3)^{2/3}}$ 48. $f(x) = \dfrac{x^8 \cos^3 x}{\sqrt{x - 1}}$

49. $f(x) = (\sin x)^{\tan x}$, $0 < x < \pi, x \neq \pi/2$

50. $f(x) = \left(1 + \dfrac{1}{x}\right)^{2x}$

Further Explorations

51. **Explain why or why not** Determine whether the following statements are true and give an explanation or counterexample.

a. The derivative of $\log_2 9 = 1/(9 \ln 2)$.

b. $\ln(x + 1) + \ln(x - 1) = \ln(x^2 - 1)$.

c. The exponential function 2^{x+1} can be written in base e as $e^{2\ln(x+1)}$.

d. $\dfrac{d}{dx}\left(\sqrt{2}^{\,x}\right) = x\sqrt{2}^{\,x-1}$

e. $\dfrac{d}{dx}\left(x^{\sqrt{2}}\right) = \sqrt{2}x^{\sqrt{2}-1}$

52–55. Higher-order derivatives *Find the following higher-order derivatives.*

52. $\dfrac{d^3}{dx^3}\left(x^{4.2}\right)\Big|_{x=1}$ 53. $\dfrac{d^2}{dx^2}(\log x)$

54. $\dfrac{d^n}{dx^n}(2^x)$ 55. $\dfrac{d^3}{dx^3}(x^2 \ln x)$

56–58. Derivatives by different methods *Calculate the derivative of the following functions (i) using the fact that $b^x = e^{x \ln b}$ and (ii) by using logarithmic differentiation. Verify that both answers are the same.*

56. $y = (x^2 + 1)^x$ 57. $y = 3^x$ 58. $y = (g(x))^{h(x)}$

59–64. Derivatives of logarithmic functions *Use the properties of logarithms to simplify the following functions before computing $f'(x)$.*

59. $f(x) = \ln(3x + 1)^4$ 60. $f(x) = \ln\dfrac{2x}{(x^2 + 1)^3}$

61. $f(x) = \log\sqrt{10x}$ 62. $f(x) = \log_2\dfrac{8}{\sqrt{x + 1}}$

63. $f(x) = \ln\dfrac{(2x - 1)(x + 2)^3}{(1 - 4x)^2}$ 64. $f(x) = \ln(\sec^4 x \tan^2 x)$

65. **Tangent lines** Find the equation of the line tangent to $y = 2^{\sin x}$ at $x = \pi/2$. Graph the function and the tangent line.

66. **Horizontal tangents** The graph of $y = \cos x \cdot \ln\cos^2 x$ has seven horizontal tangent lines on the interval $[0, 2\pi]$. Find the x-coordinates of all points at which these tangent lines occur.

67–74. General logarithmic and exponential derivatives *Compute the following derivatives. Use logarithmic differentiation where appropriate.*

67. $\dfrac{d}{dx}(x^{10x})$ **68.** $\dfrac{d}{dx}(2x)^{2x}$ **69.** $\dfrac{d}{dx}(x^{\cos x})$

70. $\dfrac{d}{dx}(x^{\pi} + \pi^x)$ **71.** $\dfrac{d}{dx}\left(1 + \dfrac{1}{x}\right)^x$ **72.** $\dfrac{d}{dx}(1 + x^2)^{\sin x}$

73. $\dfrac{d}{dx}\left[x^{(x^{10})}\right]$ **74.** $\dfrac{d}{dx}(\ln x)^{x^2}$

Applications

75–78. Logistic growth *Scientists often use the* logistic growth *function* $P(t) = \dfrac{P_0 K}{P_0 + (K - P_0)e^{-r_0 t}}$ *to model population growth, where P_0 is the population at some reference time $t = 0$, K is the* **carrying capacity**, *and r_0 is the* base growth rate. *The carrying capacity is a theoretical upper bound on the total population that the surrounding environment can support. The figure shows the* sigmoid *(S-shaped) curve associated with a typical logistic model.*

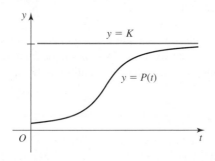

75. Gone fishing When a reservoir is created by a new dam, 50 fish are introduced into the reservoir, which has an estimated carrying capacity of 8000 fish. A logistic model of the fish population is $P(t) = \dfrac{400{,}000}{50 + 7950e^{-0.5t}}$, where t is measured in years.

  a. Graph P using a graphing utility. Experiment with different windows until you produce an S-shaped curve characteristic of the logistic model. What window works well for this function?

  b. How long does it take for the population to reach 5000 fish? How long does it take for the population to reach 90% of the carrying capacity?

  c. How fast (in fish per year) is the population growing at $t = 0$? At $t = 5$?

  d. Graph P' and use the graph to estimate the year in which the population is growing fastest.

76. World population (part 1) The population of the world reached 6 billion in 1999 ($t = 0$). Assume the carrying capacity is 15 billion and the base growth rate is $r_0 = 0.025$ per year.

  a. Write a logistic growth function for the world's population (in billions), and graph your equation on the interval $0 \le t \le 200$ using a graphing utility.

  b. What will the population be in the year 2020? When will it reach 12 billion?

77. World population (part 2) The *relative growth rate r* of a function f measures the change in the function compared to its value at a particular point. It is computed as $r(t) = f'(t)/f(t)$.

  a. Confirm that the relative growth rate in 1999 ($t = 0$) for the logistic model in Exercise 76 is $r(0) = P'(0)/P(0) = 0.015$. This means the world's population was growing at 1.5% per year in 1999.

  b. Compute the relative growth rate of the world's population in 2010 and 2020. What appears to be happening to the relative growth rate as time increases?

  c. Evaluate $\lim\limits_{t \to \infty} r(t) = \lim\limits_{t \to \infty} \dfrac{P'(t)}{P(t)}$, where $P(t)$ is the logistic growth function from Exercise 76. What does your answer say about populations that follow a logistic growth pattern?

78. Population crash The logistic model can be used for situations in which the initial population P_0 is above the carrying capacity K. For example, consider a deer population of 1500 on an island where a large fire has reduced the carrying capacity to 1000 deer.

  a. Assuming a base growth rate of $r_0 = 0.1$ and an initial population of $P(0) = 1500$, write a logistic growth function for the deer population and graph it. Based on the graph, what happens to the deer population in the long run?

  b. How fast (in deer per year) is the population declining immediately after the fire at $t = 0$?

  c. How long does it take for the deer population to decline to 1200 deer?

79. Savings Plan Beginning at age 30, a self-employed plumber saves $250 per month in a retirement account until he reaches age 65. The account offers 6% interest, compounded monthly. The balance in the account after t years is given by $A(t) = 50{,}000(1.005^{12t} - 1)$.

  a. Compute the balance in the account after 5, 15, 25, and 35 years. What is the average rate of change in the value of the account over the intervals $[5, 15]$, $[15, 25]$, and $[25, 35]$?

  b. Suppose the plumber started saving at age 25 instead of age 30. Find the balance at age 65 (after 40 years of investing).

  c. Use the derivative dA/dt to explain the surprising result in part (b) and to explain the advice: Start saving for retirement as early as possible.

Additional Exercises

80. Tangency question It is easily verified that the graphs of $y = x^2$ and $y = e^x$ have no points of intersection (for $x > 0$), while the graphs of $y = x^3$ and $y = e^x$ have two points of intersection. It follows that for some real number $2 < p < 3$, the graphs of $y = x^p$ and $y = e^x$ have exactly one point of intersection (for $x > 0$). Using analytical and/or graphical methods, determine p and the coordinates of the single point of intersection.

81. Tangency question It is easily verified that the graphs of $y = 1.1^x$ and $y = x$ have two points of intersection, while the graphs of $y = 2^x$ and $y = x$ have no points of intersection. It follows that for some real number $1 < p < 2$, the graphs of $y = p^x$ and $y = x$ have exactly one point of intersection. Using analytical and/or graphical methods, determine p and the coordinates of the single point of intersection.

82. Triple intersection Graph the functions $f(x) = x^3$, $g(x) = 3^x$, and $h(x) = x^x$ and find their common intersection point (exactly).

83–86. Calculating limits exactly *Use the definition of the derivative to evaluate the following limits.*

83. $\displaystyle\lim_{x \to e} \frac{\ln x - 1}{x - e}$

84. $\displaystyle\lim_{h \to 0} \frac{\ln(e^8 + h) - 8}{h}$

85. $\displaystyle\lim_{h \to 0} \frac{(3 + h)^{3+h} - 27}{h}$

86. $\displaystyle\lim_{x \to 2} \frac{5^x - 25}{x - 2}$

87. Derivative of $u(x)^{v(x)}$ Use logarithmic differentiation to prove that

$$\frac{d}{dx}[u(x)^{v(x)}] = u(x)^{v(x)}\left[\frac{v(x)}{u(x)}\frac{du}{dx} + \ln u(x)\frac{dv}{dx}\right].$$

QUICK CHECK ANSWERS

1. x^2; $e^{x\ln 5}$ **2.** Either way, $\dfrac{d}{dx}(\ln x^p) = \dfrac{p}{x}$.

3. $\dfrac{dA}{dt} = 500(1.045)^t \cdot \ln 1.045 \approx 22(1.045)^t$ **4.** $\dfrac{dy}{dx} = \dfrac{1}{x \ln 3}$

5. $T'(n) = \log_2 n + \dfrac{1}{\ln 2} = \log_2 n + \dfrac{1}{\dfrac{\log_2 2}{\log_2 e}} = \log_2 n + \log_2 e$

$= \log_2(en)$ ◄

3.9 Derivatives of Inverse Trigonometric Functions

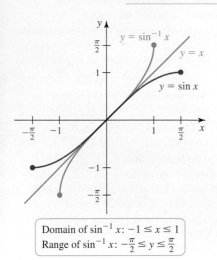

Domain of $\sin^{-1} x$: $-1 \le x \le 1$
Range of $\sin^{-1} x$: $-\frac{\pi}{2} \le y \le \frac{\pi}{2}$

FIGURE 3.57

The inverse trigonometric functions, introduced in Section 1.4, are major players in calculus. In this section, we develop the derivatives of the six inverse trigonometric functions and begin an exploration of their many applications. A method for differentiating the inverses of more general functions is also presented.

Inverse Sine and Its Derivative

Recall from Section 1.4 that $y = \sin^{-1} x$ is the value of y such that $x = \sin y$, where $-\pi/2 \le y \le \pi/2$. The domain of $\sin^{-1} x$ is $\{x: -1 \le x \le 1\}$ (Figure 3.57). The derivative of $y = \sin^{-1} x$ follows by differentiating both sides of $x = \sin y$ with respect to x, simplifying, and solving for dy/dx:

$$x = \sin y \qquad y = \sin^{-1} x \iff x = \sin y$$

$$\frac{d}{dx}(x) = \frac{d}{dx}(\sin y) \qquad \text{Differentiate with respect to } x.$$

$$1 = (\cos y)\frac{dy}{dx} \qquad \text{Chain Rule on the right side}$$

$$\frac{dy}{dx} = \frac{1}{\cos y} \qquad \text{Solve for } \frac{dy}{dx}.$$

The identity $\sin^2 y + \cos^2 y = 1$ is used to express this derivative in terms of x. Solving for $\cos y$ yields

$$\cos y = \pm\sqrt{1 - \underbrace{\sin^2 y}_{x^2}} \qquad x = \sin y \Rightarrow x^2 = \sin^2 y$$

$$= \pm\sqrt{1 - x^2}$$

Because y is restricted to the interval $-\pi/2 \le y \le \pi/2$, we have $\cos y \ge 0$. Therefore, we choose the positive branch of the square root, and it follows that

$$\frac{dy}{dx} = \frac{d}{dx}(\sin^{-1} x) = \frac{1}{\sqrt{1 - x^2}}.$$

This result is consistent with the graph of $f(x) = \sin^{-1} x$ (Figure 3.58).

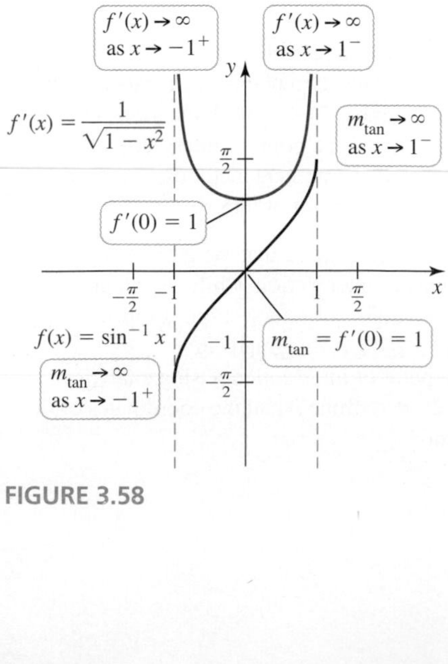

$f'(x) \to \infty$ as $x \to -1^+$

$f'(x) \to \infty$ as $x \to 1^-$

$f'(x) = \dfrac{1}{\sqrt{1 - x^2}}$

$m_{\tan} \to \infty$ as $x \to 1^-$

$f'(0) = 1$

$f(x) = \sin^{-1} x$

$m_{\tan} = f'(0) = 1$

$m_{\tan} \to \infty$ as $x \to -1^+$

FIGURE 3.58

THEOREM 3.21 Derivative of Inverse Sine

$$\frac{d}{dx}(\sin^{-1} x) = \frac{1}{\sqrt{1 - x^2}} \quad \text{for } -1 < x < 1$$

QUICK CHECK 1 Is $f(x) = \sin^{-1} x$ an even or odd function? Is $f'(x)$ an even or odd function? ◄

EXAMPLE 1 Derivatives involving the inverse sine Compute the following derivatives.

a. $\dfrac{d}{dx}[\sin^{-1}(x^2 - 1)]$ **b.** $\dfrac{d}{dx}[\cos(\sin^{-1} x)]$

SOLUTION We apply the Chain Rule for both derivatives.

a. $\dfrac{d}{dx}[\sin^{-1}(\underbrace{x^2 - 1}_{u})] = \underbrace{\frac{1}{\sqrt{1 - (x^2 - 1)^2}}}_{\substack{\text{derivative of } \sin^{-1} u \\ \text{evaluated at } u = x^2 - 1}} \cdot \underbrace{2x}_{u'(x)} = \frac{2x}{\sqrt{2x^2 - x^4}}$

> The result in Example 1b could have been obtained by noting that $\cos(\sin^{-1} x) = \sqrt{1 - x^2}$ and differentiating this expression (Exercise 65).

b. $\dfrac{d}{dx}[\cos(\underbrace{\sin^{-1} x}_{u})] = \underbrace{-\sin(\sin^{-1} x)}_{\substack{\text{the derivative of the} \\ \text{outer function } \cos u \\ \text{evaluated at } u = \sin^{-1} x}} \cdot \underbrace{\frac{1}{\sqrt{1 - x^2}}}_{\substack{\text{the derivative of the} \\ \text{inner function } \sin^{-1} x}} = \frac{-x}{\sqrt{1 - x^2}}$

This result is valid for $-1 < x < 1$, where $\sin(\sin^{-1} x) = x$. *Related Exercises 7–12* ◄

Derivative of Inverse Tangent and Secant

The derivatives of the inverse tangent and inverse secant are derived using a method similar to that used for the inverse sine. Once these three derivative results are known, the derivatives of the inverse cosine, cotangent, and cosecant follow immediately.

Inverse Tangent Recall from Section 1.4 that $y = \tan^{-1} x$ is the value of y such that $x = \tan y$, where $-\pi/2 < y < \pi/2$. The domain of $y = \tan^{-1} x$ is $\{x: -\infty < x < \infty\}$ (Figure 3.59). To find $\dfrac{dy}{dx}$, we differentiate both sides of $x = \tan y$ with respect to x and simplify:

$$x = \tan y \qquad y = \tan^{-1} x \iff x = \tan y$$

$$\frac{d}{dx}(x) = \frac{d}{dx}(\tan y) \qquad \text{Differentiate with respect to } x.$$

$$1 = \sec^2 y \cdot \frac{dy}{dx} \qquad \text{Chain Rule}$$

$$\frac{dy}{dx} = \frac{1}{\sec^2 y} \qquad \text{Solve for } \frac{dy}{dx}.$$

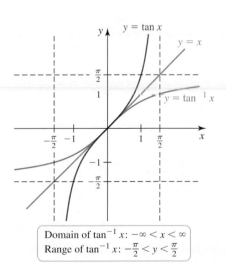

Domain of $\tan^{-1} x$: $-\infty < x < \infty$
Range of $\tan^{-1} x$: $-\frac{\pi}{2} < y < \frac{\pi}{2}$

FIGURE 3.59

To express this derivative in terms of x, we combine the trigonometric identity $\sec^2 y = 1 + \tan^2 y$ with $x = \tan y$ to obtain $\sec^2 y = 1 + x^2$. Substituting this result into the expression for dy/dx, it follows that

$$\frac{dy}{dx} = \frac{d}{dx}(\tan^{-1} x) = \frac{1}{1 + x^2}.$$

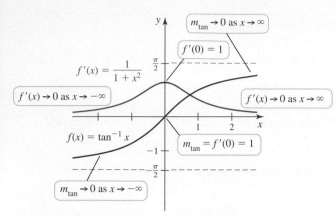

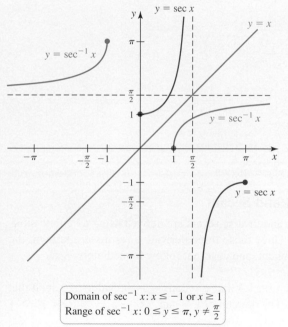

FIGURE 3.60

Domain of $\sec^{-1} x$: $x \leq -1$ or $x \geq 1$
Range of $\sec^{-1} x$: $0 \leq y \leq \pi$, $y \neq \dfrac{\pi}{2}$

FIGURE 3.61

The graphs of the inverse tangent and its derivative (Figure 3.60) are informative. Letting $f(x) = \tan^{-1} x$ and $f'(x) = \dfrac{1}{1 + x^2}$, we see that $f'(0) = 1$, which is the maximum value of the derivative; that is, $\tan^{-1} x$ has its maximum slope at $x = 0$. As $x \to \infty$, $f'(x)$ approaches zero; likewise, as $x \to -\infty$, $f'(x)$ approaches zero.

QUICK CHECK 2 How do the slopes of the lines tangent to the graph of $y = \tan^{-1} x$ behave as $x \to \infty$? ◄

Inverse Secant   Recall from Section 1.4 that $y = \sec^{-1} x$ is the value of y such that $x = \sec y$, where $0 \leq y \leq \pi$, with $y \neq \pi/2$. The domain of $y = \sec^{-1} x$ is $\{x: |x| \geq 1\}$ (Figure 3.61).

The derivative of the inverse secant presents a new twist. Let $y = \sec^{-1} x$, or $x = \sec y$, and then differentiate both sides of $x = \sec y$ with respect to x:

$$1 = \sec y \tan y \frac{dy}{dx}$$

Solving for $\dfrac{dy}{dx}$ produces

$$\frac{dy}{dx} = \frac{d}{dx}(\sec^{-1} x) = \frac{1}{\sec y \tan y}.$$

The final step is to express $\sec y \tan y$ in terms of x by using the identity $\sec^2 y = 1 + \tan^2 y$. Solving this equation for $\tan y$, we have

$$\tan y = \pm\sqrt{\underbrace{\sec^2 y}_{x^2} - 1} = \pm\sqrt{x^2 - 1}.$$

Two cases must be examined to resolve the sign on the square root:

- By the definition of $y = \sec^{-1} x$, if $x \geq 1$, then $0 \leq y < \pi/2$ and $\tan y > 0$. In this case we choose the positive branch and take $\tan y = \sqrt{x^2 - 1}$.
- However, if $x \leq -1$, then $\pi/2 < y \leq \pi$ and $\tan y < 0$. Now we choose the negative branch.

This argument accounts for the $\tan y$ factor in the derivative. For the $\sec y$ factor, we have $\sec y = x$. Therefore, the derivative of the inverse secant is

$$\frac{d}{dx}(\sec^{-1} x) = \begin{cases} \dfrac{1}{x\sqrt{x^2 - 1}} & \text{if } x > 1 \\[2ex] -\dfrac{1}{x\sqrt{x^2 - 1}} & \text{if } x < -1 \end{cases}$$

which is an awkward result. The absolute value helps here: Recall that $|x| = x$ if $x > 0$ and $|x| = -x$ if $x < 0$. It follows that

$$\frac{d}{dx}(\sec^{-1} x) = \frac{1}{|x|\sqrt{x^2 - 1}} \qquad \text{for } |x| > 1.$$

Geometrically, we see that the slope of the inverse secant function is always positive, which is consistent with this derivative result (Figure 3.61).

Derivatives of Other Inverse Trigonometric Functions

The hard work is complete. The derivative of the inverse cosine results from the identity

▶ This identity was proved in Example 5 of
Section 1.4.

$$\cos^{-1} x + \sin^{-1} x = \frac{\pi}{2}.$$

Differentiating both sides of this equation with respect to x, we find that

$$\frac{d}{dx}(\cos^{-1} x) + \underbrace{\frac{d}{dx}(\sin^{-1} x)}_{1/\sqrt{1-x^2}} = \underbrace{\frac{d}{dx}\left(\frac{\pi}{2}\right)}_{0}.$$

Solving for $\frac{d}{dx}(\cos^{-1} x)$, the required derivative is

$$\frac{d}{dx}(\cos^{-1} x) = -\frac{1}{\sqrt{1-x^2}}.$$

In a similar manner, the analogous identities

$$\cot^{-1} x + \tan^{-1} x = \frac{\pi}{2} \quad \text{and} \quad \csc^{-1} x + \sec^{-1} x = \frac{\pi}{2}$$

are used to show that the derivatives of $\cot^{-1} x$ and $\csc^{-1} x$ are the negative of the derivatives of $\tan^{-1} x$ and $\sec^{-1} x$, respectively (Exercise 63).

THEOREM 3.22   Derivatives of Inverse Trigonometric Functions

$$\frac{d}{dx}(\sin^{-1} x) = \frac{1}{\sqrt{1-x^2}}, \qquad \frac{d}{dx}(\cos^{-1} x) = -\frac{1}{\sqrt{1-x^2}} \quad \text{for } -1 < x < 1$$

$$\frac{d}{dx}(\tan^{-1} x) = \frac{1}{1+x^2}, \qquad \frac{d}{dx}(\cot^{-1} x) = -\frac{1}{1+x^2} \quad \text{for } -\infty < x < \infty$$

$$\frac{d}{dx}(\sec^{-1} x) = \frac{1}{|x|\sqrt{x^2-1}}, \qquad \frac{d}{dx}(\csc^{-1} x) = -\frac{1}{|x|\sqrt{x^2-1}} \quad \text{for } |x| > 1$$

QUICK CHECK 3 Summarize how the derivatives of inverse trigonometric functions are related to the derivatives of the corresponding inverse cofunctions (for example, inverse tangent and inverse cotangent). ◀

EXAMPLE 2   Derivatives of inverse trigonometric functions

a. Evaluate $f'(2\sqrt{3})$, where $f(x) = x \tan^{-1}(x/2)$.

b. Find an equation of the line tangent to the graph of $g(x) = \sec^{-1}(2x)$ at the point $(1, \pi/3)$.

SOLUTION

a. $f'(x) = 1 \cdot \tan^{-1}\left(\frac{x}{2}\right) + x \underbrace{\frac{1}{1+(x/2)^2} \cdot \frac{1}{2}}_{\frac{d}{dx}(\tan^{-1}(x/2))}$ Product Rule and Chain Rule

$$= \tan^{-1}\left(\frac{x}{2}\right) + \frac{2x}{4+x^2} \qquad \text{Simplify.}$$

We evaluate f' at $x = 2\sqrt{3}$ and note that $\tan^{-1}(\sqrt{3}) = \pi/3$:

$$f'(2\sqrt{3}) = \tan^{-1}(\sqrt{3}) + \frac{2(2\sqrt{3})}{4+(2\sqrt{3})^2} = \frac{\pi}{3} + \frac{\sqrt{3}}{4}$$

b. The slope of the tangent line at $(1, \pi/3)$ is $g'(1)$. Using the Chain Rule, we have

$$g'(x) = \frac{d}{dx}(\sec^{-1} 2x) = \frac{2}{|2x|\sqrt{4x^2 - 1}} = \frac{1}{|x|\sqrt{4x^2 - 1}}.$$

It follows that $g'(1) = 1/\sqrt{3}$. An equation of the tangent line is

$$\left(y - \frac{\pi}{3}\right) = \frac{1}{\sqrt{3}}(x - 1) \quad \text{or} \quad y = \frac{1}{\sqrt{3}}x + \frac{\pi}{3} - \frac{1}{\sqrt{3}}.$$

Related Exercises 13–28 ◄

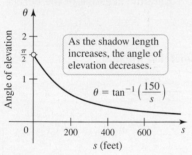

FIGURE 3.62

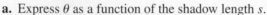

As the shadow length increases, the angle of elevation decreases.

$$\theta = \tan^{-1}\left(\frac{150}{s}\right)$$

FIGURE 3.63

EXAMPLE 3 Shadows in a ballpark As the sun descends behind the 150-ft wall of a baseball stadium, the shadow of the wall moves across the field (Figure 3.62). Let ℓ be the line segment between the edge of the shadow and the sun, and let θ be the angle of elevation of the sun—the angle between ℓ and the horizontal. The length of the shadow s is the distance between the edge of the shadow and the base of the wall.

a. Express θ as a function of the shadow length s.

b. Compute $d\theta/ds$ when $s = 200$ ft and explain what this rate of change measures.

SOLUTION

a. The tangent of θ $[$(opposite side)/(adjacent side)$]$ is

$$\tan \theta = \frac{150}{s},$$

where $s > 0$. Taking the inverse tangent of both sides of this equation, we find that

$$\theta = \tan^{-1}\left(\frac{150}{s}\right).$$

As shown in Figure 3.63, as the shadow length approaches zero, the sun's angle of elevation θ approaches $\pi/2$, which means the sun is overhead. As the shadow length increases, θ decreases and approaches zero.

b. Using the Chain Rule, we have

$$\frac{d\theta}{ds} = \frac{1}{1 + (150/s)^2} \frac{d}{ds}\left(\frac{150}{s}\right) \qquad \text{Chain Rule; } \frac{d}{du}(\tan^{-1} u) = \frac{1}{1 + u^2}$$

$$= \frac{1}{1 + (150/s)^2}\left(-\frac{150}{s^2}\right) \qquad \text{Evaluate the derivative.}$$

$$= -\frac{150}{s^2 + 22{,}500} \qquad \text{Simplify.}$$

Notice that $d\theta/ds$ is negative for all values of s, which means longer shadows are associated with smaller angles of elevation (Figure 3.63). At $s = 200$ ft, we have

$$\left.\frac{d\theta}{ds}\right|_{s=200} = -\frac{150}{200^2 + 150^2} = -0.0024 \frac{\text{rad}}{\text{ft}}.$$

When the length of the shadow is $s = 200$ ft, the angle of elevation is changing by -0.0024 rad/ft, or -0.138 degrees/ft.

Related Exercises 29–30 ◄

QUICK CHECK 4 Example 3 makes the claim that $d\theta/ds = -0.0024$ rad/ft is equivalent to -0.138 degrees/ft. Verify this claim. ◄

Derivatives of Inverse Functions in General

We found the derivatives of the inverse trigonometric functions using implicit differentiation. However, this approach does not always work. For example, suppose we know only f and its derivative f' and wish to evaluate the derivative of f^{-1}. The key to finding the derivative of the inverse function lies in the symmetry of the graphs of f and f^{-1}.

EXAMPLE 4 Linear functions, inverses, and derivatives Consider the general linear function $y = f(x) = mx + b$, where m and b are constants.

a. Write the inverse of f in the form $y = f^{-1}(x)$.

b. Find the derivative of the inverse $\dfrac{d}{dx}[f^{-1}(x)]$.

c. Consider the specific case $f(x) = 2x - 6$. Graph f and f^{-1}, and find the slope of each line.

SOLUTION

a. Solving $y = mx + b$ for x, we find that $mx = y - b$, or

$$x = \frac{y}{m} - \frac{b}{m}.$$

Writing this function in the form $y = f^{-1}(x)$ (by reversing the roles of x and y), we have

$$y = f^{-1}(x) = \frac{x}{m} - \frac{b}{m}$$

which describes a line with slope $1/m$.

b. The derivative of f^{-1} is

$$(f^{-1})'(x) = \frac{1}{m} = \frac{1}{f'(x)}.$$

Notice that $f'(x) = m$, so the derivative of f^{-1} is the reciprocal of $f'(x)$.

c. In the case that $f(x) = 2x - 6$, we have $f^{-1}(x) = x/2 + 3$. The graphs of these two lines are symmetric about the line $y = x$ (Figure 3.64). Furthermore, the slope of the line $y = f(x)$ is 2 and the slope of $y = f^{-1}(x)$ is $\frac{1}{2}$; that is, the slopes (and, therefore, the derivatives) are reciprocals of each other.   *Related Exercise 31* ◀

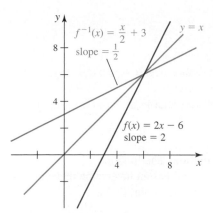

$f^{-1}(x) = \dfrac{x}{2} + 3$

slope $= \dfrac{1}{2}$

$y = x$

$f(x) = 2x - 6$

slope $= 2$

FIGURE 3.64

The reciprocal property obeyed by f' and $(f^{-1})'$ in Example 4 holds for all functions. Figure 3.65 shows the graphs of a typical one-to-one function and its inverse. It also shows a pair of symmetric points—(x_0, y_0) on the graph of f and (y_0, x_0) on the graph of f^{-1}— along with the tangent lines at these points. Notice that as the lines tangent to the graph of f get steeper (as x increases), the corresponding lines tangent to the graph of f^{-1} get less steep. The next theorem makes this relationship precise.

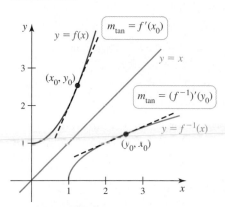

$y = f(x)$

$m_{\tan} = f'(x_0)$

(x_0, y_0)

$y = x$

$m_{\tan} = (f^{-1})'(y_0)$

$y = f^{-1}(x)$

(y_0, x_0)

FIGURE 3.65

▷ The result of Theorem 3.23 is also written in the form

$$(f^{-1})'(f(x_0)) = \frac{1}{f'(x_0)}$$

or

$$(f^{-1})'(y_0) = \frac{1}{f'(f^{-1}(y_0))}$$

THEOREM 3.23 Derivative of the Inverse Function

Let f be differentiable and have an inverse on an interval I. Let x_0 be a point of I at which $f'(x_0) \neq 0$. Then, f^{-1} is differentiable at $y_0 = f(x_0)$ and

$$(f^{-1})'(y_0) = \frac{1}{f'(x_0)} \quad \text{where} \quad y_0 = f(x_0).$$

To understand this theorem, suppose that (x_0, y_0) is a point on the graph of f, which means that (y_0, x_0) is the corresponding point on the graph of f^{-1}. Then the slope of the line tangent to the graph of f^{-1} at the point (y_0, x_0) is the reciprocal of the slope of the line tangent to the graph of f at the point (x_0, y_0). Importantly, the theorem says that we can evaluate the derivative of the inverse function without finding the inverse function itself.

Proof Before doing a short calculation, we note two facts:

- At a point x_0 where f is differentiable, $y_0 = f(x_0)$ and $x_0 = f^{-1}(y_0)$.

- As a differentiable function, f is continuous at x_0 (Theorem 3.1), which implies that f^{-1} is also continuous at y_0 (Theorem 2.13). Therefore, as $y \to y_0$, $x \to x_0$.

Using the definition of the derivative, we have

$$(f^{-1})'(y_0) = \lim_{y \to y_0} \frac{f^{-1}(y) - f^{-1}(y_0)}{y - y_0} \qquad \text{Definition of derivative of } f^{-1}$$

$$= \lim_{y \to y_0} \frac{x - x_0}{f(x) - f(x_0)} \qquad y = f(x) \text{ and } x = f^{-1}(y)$$

$$= \lim_{y \to y_0} \frac{1}{\dfrac{f(x) - f(x_0)}{x - x_0}} \qquad \frac{a}{b} = \frac{1}{b/a}$$

$$= \frac{1}{f'(x_0)}. \qquad \text{Definition of derivative of } f$$

We have shown that $(f^{-1})'(y_0)$ exists (f^{-1} is differentiable at y_0) and it equals the reciprocal of $f'(x_0)$. ◄

QUICK CHECK 5 Sketch the graphs of $y = \sin x$ and $y = \sin^{-1} x$. Then, verify that Theorem 3.23 holds at the point $(0, 0)$. ◄

EXAMPLE 5 Derivative of an inverse function The function $f(x) = \sqrt{x} + x^2 + 1$ is one-to-one for $x \geq 0$ and has an inverse on that interval. Find the slope of the curve $y = f^{-1}(x)$ at the point $(3, 1)$.

SOLUTION The point $(1, 3)$ is on the graph of f; therefore, $(3, 1)$ is on the graph of f^{-1}. In this case, the slope of the curve $y = f^{-1}(x)$ at the point $(3, 1)$ is the reciprocal of the slope of the curve $y = f(x)$ at $(1, 3)$ (Figure 3.66). Note that $f'(x) = \dfrac{1}{2\sqrt{x}} + 2x$, which means that $f'(1) = \dfrac{1}{2} + 2 = \dfrac{5}{2}$. Therefore,

$$(f^{-1})'(3) = \frac{1}{f'(1)} = \frac{1}{5/2} = \frac{2}{5}.$$

Observe that it is not necessary to find a formula for f^{-1} in order to evaluate its derivative at a point. *Related Exercises 32–42* ◄

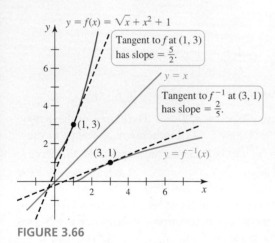

FIGURE 3.66

SECTION 3.9 EXERCISES

Review Questions

1. State the derivative formulas for $\sin^{-1} x$, $\tan^{-1} x$, and $\sec^{-1} x$.

2. What is the slope of the line tangent to the graph of $y = \sin^{-1} x$ at $x = 0$?

3. What is the slope of the line tangent to the graph of $y = \tan^{-1} x$ at $x = -2$?

4. How are the derivatives of $\sin^{-1} x$ and $\cos^{-1} x$ related?

5. If f is a one-to-one function with $f(2) = 8$ and $f'(2) = 4$, what is the value of $(f^{-1})'(8)$?

6. Explain how to find $(f^{-1})'(y_0)$, given that $y_0 = f(x_0)$.

Basic Skills

7–12. Derivatives of inverse sine *Evaluate the derivatives of the following functions.*

7. $f(x) = \sin^{-1}(2x)$

8. $f(x) = x \sin^{-1} x$

9. $f(w) = \cos[\sin^{-1}(2w)]$

10. $f(x) = \sin^{-1}(\ln x)$

11. $f(x) = \sin^{-1}(e^{-2x})$

12. $f(x) = \sin^{-1}(e^{\sin x})$

13–28. Derivatives *Evaluate the derivatives of the following functions.*

13. $f(y) = \tan^{-1}(2y^2 - 4)$

14. $g(z) = \tan^{-1}(1/z)$

15. $f(z) = \cot^{-1}\sqrt{z}$

16. $f(x) = \sec^{-1}\sqrt{x}$

17. $f(x) = \cos^{-1}(1/x)$

18. $f(t) = (\cos^{-1} t)^2$

19. $f(u) = \csc^{-1}(2u + 1)$

20. $f(t) = \ln(\tan^{-1} t)$

21. $f(y) = \cot^{-1}[1/(y^2 + 1)]$

22. $f(w) = \sin[\sec^{-1}(2w)]$

23. $f(x) = \sec^{-1}(\ln x)$

24. $f(x) = \tan^{-1}(e^{4x})$

25. $f(x) = \csc^{-1}(\tan e^x)$

26. $f(x) = \sin[\tan^{-1}(\ln x)]$

27. $f(s) = \cot^{-1}(e^s)$

28. $f(x) = 1/(\tan^{-1}(x^2 + 4))$

⊤ 29. Angular size A boat sails directly toward a 150-m skyscraper that stands on the edge of a harbor. The angular size θ of the building is the angle formed by lines from the top and bottom of the building to the observer (see figure).

    a. What is the rate of change of the angular size, $d\theta/dx$, when the boat is $x = 500$ m from the building?

    b. Graph $d\theta/dx$ as a function of x and determine the point at which the angular size changes most rapidly.

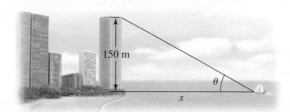

⊤ 30. Angle of elevation A small plane flies horizontally on a line 400 m directly above an observer with a speed of 70 m/s. Let θ be the angle of elevation of the plane (see figure).

    a. What is the rate of change of the angle of elevation, $d\theta/dx$, when the plane is $x = 500$ m past the observer?

    b. Graph $d\theta/dx$ as a function of x and determine the point at which θ changes most rapidly.

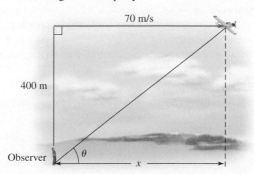

31–34. Derivatives of inverse functions at a point *Find the derivative of the inverse of the following functions at the specified point on the graph of the inverse function. You do not need to find f^{-1}.*

31. $f(x) = 3x + 4$; $(16, 4)$

32. $f(x) = x^2 + 1$ for $x \geq 0$; $(5, 2)$

33. $f(x) = \tan x$; $(1, \pi/4)$

34. $f(x) = x^2 - 2x - 3$ for $x \geq 1$; $(12, -3)$

35–38. Slopes of tangent lines *Given the function f, find the slope of the line tangent to the graph of f^{-1} at the specified point on the graph of f^{-1}.*

35. $f(x) = \sqrt{x}$; $(2, 4)$

36. $f(x) = x^3$; $(8, 2)$

37. $f(x) = (x + 2)^2$ for $x \geq -2$; $(36, 4)$

38. $f(x) = -x^2 + 8$ for $x \geq 0$; $(7, 1)$

39–42. Derivatives and inverse functions

39. Find $(f^{-1})'(3)$ if $f(x) = x^3 + x + 1$.

40. Find the slope of the curve $y = f^{-1}(x)$ at $(4, 7)$ if the slope of the curve $y = f(x)$ at $(7, 4)$ is $\frac{2}{3}$.

41. If the slope of the curve $y = f^{-1}(x)$ at $(4, 7)$ is $\frac{4}{5}$, find $f'(7)$.

42. If the slope of the curve $y = f(x)$ at $(4, 7)$ is $\frac{1}{5}$, find $(f^{-1})'(7)$.

Further Explorations

43. Explain why or why not Determine whether the following statements are true and give an explanation or counterexample.

    a. $\dfrac{d}{dx}(\sin^{-1} x + \cos^{-1} x) = 0$ **b.** $\dfrac{d}{dx}(\tan^{-1} x) = \sec^2 x$

    c. The lines tangent to the graph of $y = \sin^{-1} x$ on the interval $[-1, 1]$ have a minimum slope of 1.

    d. The lines tangent to the graph of $y = \sin x$ on the interval $[-\pi/2, \pi/2]$ have a maximum slope of 1.

    e. If $f(x) = 1/x$, then $[f^{-1}(x)]' = -\dfrac{1}{x^2}$

⊤ 44–47. Graphing f and f′

    a. *Graph f with a graphing utility.*

    b. *Compute and graph f′.*

    c. *Verify that the zeros of f′ correspond to points at which f has a horizontal tangent line.*

44. $f(x) = (x - 1)\sin^{-1} x$ on $[-1, 1]$

45. $f(x) = (x^2 - 1)\sin^{-1} x$ on $[-1, 1]$

46. $f(x) = (\sec^{-1} x)/x$ on $[1, \infty)$

47. $f(x) = e^{-x}\tan^{-1} x$ on $[0, \infty)$

⊤ 48. Graphing with inverse trigonometric functions

    a. Graph the function $f(x) = \dfrac{\tan^{-1} x}{x^2 + 1}$.

    b. Compute and graph f' and determine (perhaps approximately) the points at which $f'(x) = 0$.

    c. Verify that the zeros of f' correspond to points at which f has a horizontal tangent line.

49–56. Derivatives of inverse functions *Consider the following functions (on the given interval, if specified). Find the inverse function, express it as a function of x, and find the derivative of the inverse function.*

49. $f(x) = 3x - 4$

50. $f(x) = |x + 2|, \ x \le -2$

51. $f(x) = x^2 - 4, \ x > 0$

52. $f(x) = \dfrac{x}{x + 5}$

53. $f(x) = \sqrt{x + 2}, \ x \ge -2$

54. $f(x) = x^{2/3}, \ x > 0$

55. $f(x) = x^{-1/2}, \ x > 0$

56. $f(x) = x^3 + 3$

Applications

57. Towing a boat A boat is towed toward a dock by a cable attached to a winch that stands 10 ft above the water level (see figure). Let θ be the angle of elevation of the winch and let ℓ be the length of the cable as the boat is towed toward the dock.

  a. Show that the rate of change of θ with respect to ℓ is
  $$\frac{d\theta}{d\ell} = \frac{-10}{\ell\sqrt{\ell^2 - 100}}.$$

  b. Compute $\dfrac{d\theta}{d\ell}$ when $\ell = 50, 20,$ and 11 ft.

  c. Find $\displaystyle\lim_{\ell \to 10^+} \frac{d\theta}{d\ell}$, and explain what is happening as the last foot of cable is reeled in (note that the boat is at the dock when $\ell = 10$).

  d. It is evident from the figure that θ increases as the boat is towed to the dock. Why, then, is $d\theta/d\ell$ negative?

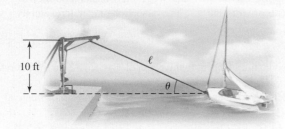

58. Tracking a dive A biologist standing at the bottom of an 80-ft vertical cliff watches a peregrine falcon dive from the top of the cliff at a 45° angle from the horizontal (see figure).

  a. Express the angle of elevation θ from the biologist to the falcon as a function of the height h of the bird above the ground. (*Hint:* The vertical distance between the top of the cliff and the falcon is $80 - h$.)

  b. What is the rate of change of θ with respect to the bird's height when it is 60 ft above the ground?

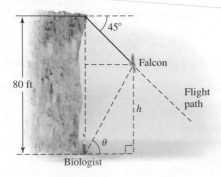

59. Angle to a particle A particle travels clockwise on a circular path of diameter R, monitored by a sensor on the circle at point P; the other endpoint of the diameter on which the sensor lies is Q (see figure). Let θ be the angle between the diameter PQ and the line from the sensor to the particle. Let c be the length of the chord from the particle's position to Q.

  a. Calculate $d\theta/dc$.

  b. Evaluate $\dfrac{d\theta}{dc}\bigg|_{c=0}$.

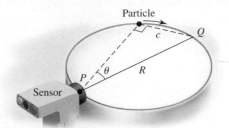

60. Angle to a particle, part II The drawing in Exercise 59 shows the particle traveling away from the sensor, which may have influenced your solution (we expect you used the inverse sine function). Suppose instead that the particle approaches the sensor (see figure). How would this change the solution? Explain the differences in the two answers.

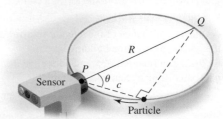

Additional Exercises

61. Derivative of the inverse sine Find the derivative of the inverse sine using Theorem 3.23.

62. Derivative of the inverse cosine Find the derivative of the inverse cosine in the following two ways.

  a. Using Theorem 3.23
  b. Using the identity $\sin^{-1} x + \cos^{-1} x = \pi/2$

63. Derivative of $\cot^{-1} x$ and $\csc^{-1} x$ Use a trigonometric identity to show that the derivatives of the inverse cotangent and inverse cosecant differ from the derivatives of the inverse tangent and inverse secant, respectively, by a multiplicative factor of -1.

64. Tangents and inverses Suppose $y = L(x) = ax + b$ (with $a \ne 0$) is the equation of the line tangent to the graph of a one-to-one function f at (x_0, y_0). Also, suppose that $y = M(x) = cx + d$ is the equation of the line tangent to the graph of f^{-1} at (y_0, x_0).

  a. Express b in terms of a, x_0, and y_0.
  b. Express c in terms of a, and d in terms of a, x_0, and y_0.
  c. Prove that $L^{-1}(x) = M(x)$.

65–68. Identity proofs *Prove the following identities and give the values of x for which they are true.*

65. $\cos(\sin^{-1} x) = \sqrt{1 - x^2}$

66. $\cos(2\sin^{-1} x) = 1 - 2x^2$

67. $\tan(2\tan^{-1} x) = \dfrac{2x}{1 - x^2}$

68. $\sin(2\sin^{-1} x) = 2x\sqrt{1 - x^2}$

3.10 Related Rates

We now return to the theme of derivatives as rates of change in problems in which the variables change with respect to *time*. The essential feature of these problems is that two or more variables, which are related in a known way, are themselves changing in time. Here are two examples illustrating this type of problem.

- An oil rig springs a leak and the oil spreads in a (roughly) circular patch around the rig. If the radius of the oil patch increases at a known rate, how fast is the area of the patch changing (Example 1)?

- Two airliners approach an airport with known speeds, one flying west and one flying north. How fast is the distance between the airliners changing (Example 2)?

In the first problem, the two related variables are the radius and the area of the oil patch. Both are changing in time. The second problem has three related variables: the positions of the two airliners and the distance between them. Again, the three variables change in time. The goal in both problems is to determine the rate of change of one of the variables at a specific moment of time—hence the name *related rates*.

We present a progression of examples in this section. After the first example, a general procedure is given for solving related-rate problems.

EXAMPLE 1   Spreading oil An oil rig springs a leak in calm seas and the oil spreads in a circular patch around the rig. If the radius of the oil patch increases at a rate of 30 m/hr, how fast is the area of the patch increasing when the patch has a radius of 100 m (Figure 3.67)?

SOLUTION Two variables change simultaneously: the radius of the circle and its area. The key relationship between the radius and area is $A = \pi r^2$. It helps to rewrite the basic relationship showing explicitly which quantities vary in time. In this case, we rewrite A and r as $A(t)$ and $r(t)$ to emphasize that they change with respect to t (time). The general expression relating the radius and area at any time t is $A(t) = \pi r(t)^2$.

The goal is to find the rate of change of the area of the circle, which is $A'(t)$. In order to introduce derivatives into the problem, we differentiate the area relation $A(t) = \pi r(t)^2$ with respect to t:

$$A'(t) = \frac{d}{dt}\left(\pi r(t)^2\right)$$

$$= \pi \frac{d}{dt}\left[r(t)^2\right]$$

$$= \pi[2r(t)]r'(t) \quad \text{Chain Rule}$$

$$= 2\pi r(t)\,r'(t) \quad \text{Simplify.}$$

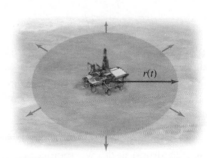

FIGURE 3.67

Substituting the given values $r(t) = 100$ m and $r'(t) = 30$ m/hr, we have (including units)

$$A'(t) = 2\pi r(t)\, r'(t)$$

$$= 2\pi(100 \text{ m})\left(30\, \frac{\text{m}}{\text{hr}}\right)$$

$$= 6000\,\pi\, \frac{\text{m}^2}{\text{hr}}$$

We see that the area of the oil spill increases at a rate of $6000\pi \approx 18{,}850$ m²/hr. Including units is a simple and valuable way to check your work. In this case, we expect an answer with units of area per unit time, so m²/hr makes sense.

Notice that the rate of change of the area depends on the radius of the spill. As the radius increases, the rate of change of the area also increases—assuming the radius increases at a constant rate.   *Related Exercises 5–13* ◄

> It is important to remember that substitution of specific values of the variables occurs *after* differentiating.

QUICK CHECK 1 In Example 1, what is the rate of change of the area when the radius is 200 m? 300 m? ◄

Using Example 1 as a template, we offer a set of guidelines for solving related-rate problems. There are always variations that arise for individual problems, but here is a general procedure.

PROCEDURE   Steps for Related-Rate Problems

1. Read the problem carefully, making a sketch to organize the given information. Identify the rates that are given and the rate that is to be determined.

2. Write one or more equations that express the basic relationships among the variables.

3. Introduce rates of change by differentiating the appropriate equation(s) with respect to time t.

4. Substitute known values and solve for the desired quantity.

5. Check that units are consistent and the answer is reasonable. (For example, does it have the correct sign?)

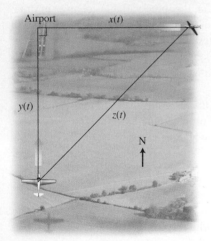

Airport $x(t)$

$y(t)$

$z(t)$

N

FIGURE 3.68

> In Example 1, we replaced A and r by $A(t)$ and $r(t)$, respectively, to remind us of the independent variable. After some practice, this replacement is not necessary.

EXAMPLE 2   Converging airplanes Two small planes approach an airport, one flying due west at 120 mi/hr and the other flying due north at 150 mi/hr. Assuming they fly at the same constant elevation, how fast is the distance between the planes changing when the westbound plane is 180 mi from the airport and the northbound plane is 225 mi from the airport?

SOLUTION A sketch such as Figure 3.68 helps us visualize the problem and organize the information. Let $x(t)$ and $y(t)$ denote the distance from the airport to the westbound and northbound planes, respectively. The paths of the two planes form the legs of a right triangle and the distance between them, denoted $z(t)$, is the hypotenuse. By the Pythagorean theorem, $z^2 = x^2 + y^2$.

Our aim is to find dz/dt, the rate of change of the distance between the planes. We first differentiate both sides of $z^2 = x^2 + y^2$ with respect to t:

$$\frac{d}{dt}(z^2) = \frac{d}{dt}(x^2 + y^2) \;\;\Rightarrow\;\; 2z\frac{dz}{dt} = 2x\frac{dx}{dt} + 2y\frac{dy}{dt}$$

Notice that the Chain Rule is needed because x, y, and z are functions of t. Solving for dz/dt results in

$$\frac{dz}{dt} = \frac{2x\frac{dx}{dt} + 2y\frac{dy}{dt}}{2z} = \frac{x\frac{dx}{dt} + y\frac{dy}{dt}}{z}.$$

> One could solve the equation
> $z^2 = x^2 + y^2$ for z, with the result
>
> $$z = \sqrt{x^2 + y^2},$$
>
> and then differentiate.
> However, it is much easier to differentiate
> implicitly as shown in the example.

QUICK CHECK 2 Assuming the same plane speeds as in Example 2, how fast is the distance between the planes changing if $x = 60$ mi and $y = 75$ mi? ◄

This equation relates the unknown rate dz/dt to the known quantities $x, y, z, dx/dt$, and dy/dt. For the westbound plane, $dx/dt = -120$ mi/hr (negative because the distance is decreasing), and for the northbound plane, $dy/dt = -150$ mi/hr. At the moment of interest, when $x = 180$ mi and $y = 225$ mi, the distance between the planes is

$$z = \sqrt{x^2 + y^2} = \sqrt{180^2 + 225^2} \approx 288 \text{ mi}.$$

Substituting these values gives

$$\frac{dz}{dt} = \frac{x\frac{dx}{dt} + y\frac{dy}{dt}}{z} \approx \frac{(180 \text{ mi})(-120 \text{ mi/hr}) + (225 \text{ mi})(-150 \text{ mi/hr})}{288 \text{ mi}}$$

$$\approx -192 \text{ mi/hr}$$

Notice that $dz/dt < 0$, which means the distance between the planes is *decreasing* at a rate of about 192 mi/hr.   *Related Exercises 14–20* ◄

EXAMPLE 3 **Sandpile** Sand falls from an overhead bin, accumulating in a conical pile with a radius that is always three times its height. If the sand falls from the bin at a rate of $12 \text{ ft}^3/\text{min}$, how fast is the height of the sandpile changing when the pile is 10 ft high?

SOLUTION A sketch of the problem (Figure 3.69) shows the three relevant variables: the volume V, the radius r, and the height h of the sandpile. The aim is to find the rate of change of the height dh/dt at the instant that $h = 10$ ft. The basic relationship among the variables is the formula for the volume of a cone, $V = \frac{1}{3}\pi r^2 h$. We now use the given fact that the radius is always three times the height. Substituting $r = 3h$ into the volume relationship gives V in terms of h:

$$V = \frac{1}{3}\pi r^2 h = \frac{1}{3}\pi (3h)^2 h = 3\pi h^3$$

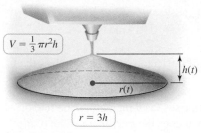

$V = \frac{1}{3}\pi r^2 h$

$h(t)$

$r(t)$

$r = 3h$

FIGURE 3.69

Rates of change are introduced by differentiating both sides of $V = 3\pi h^3$ with respect to t. Using the Chain Rule, we have

$$\frac{dV}{dt} = 9\pi h^2 \frac{dh}{dt}.$$

Now we find dh/dt at the instant that $h = 10$ ft, given that $dV/dt = 12 \text{ ft}^3/\text{min}$. Solving for dh/dt and substituting these values, we have

$$\frac{dh}{dt} = \frac{dV/dt}{9\pi h^2} \qquad \text{Solve for } \frac{dh}{dt}.$$

$$= \frac{12 \text{ ft}^3/\text{min}}{9\pi (10 \text{ ft})^2} \approx 0.0042 \frac{\text{ft}}{\text{min}} \qquad \text{Substitute for } \frac{dV}{dt} \text{ and } h.$$

At the instant that the sandpile is 10 ft high, the height is changing at a rate of 0.0042 ft/min. Notice how the units work out consistently.   *Related Exercises 21–25* ◄

QUICK CHECK 3 In Example 3, what is the rate of change of the height when $h = 2$ ft? Does the rate of change of the height increase or decrease with increasing height? ◄

EXAMPLE 4 **Observing a launch** An observer stands 200 m from the launch site of a hot-air balloon. The balloon rises vertically at a constant rate of 4 m/s. How fast is the angle of elevation of the balloon increasing 30 s after the launch? (The angle of elevation is the angle between the ground and the observer's line of sight to the balloon.)

SOLUTION Figure 3.70 shows the geometry of the launch. As the balloon rises, its distance from the ground y and its angle of elevation θ change simultaneously. An equation expressing the relationship between these variables is $\tan\theta = y/200$.

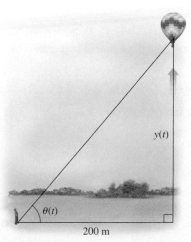

$y(t)$

$\theta(t)$

200 m

FIGURE 3.70

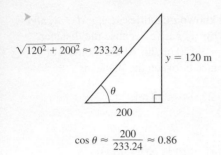

$$\sqrt{120^2 + 200^2} \approx 233.24$$

$y = 120\,\text{m}$

θ

200

$$\cos\theta \approx \frac{200}{233.24} \approx 0.86$$

In order to find $d\theta/dt$, we differentiate both sides of this relationship using the Chain Rule:

$$\sec^2\theta\,\frac{d\theta}{dt} = \frac{1}{200}\frac{dy}{dt}$$

Next we solve for $\dfrac{d\theta}{dt}$:

$$\frac{d\theta}{dt} = \frac{dy/dt}{200\sec^2\theta} = \frac{(dy/dt)\cdot\cos^2\theta}{200}$$

The rate of change of the angle of elevation depends on the angle of elevation and the speed of the balloon. Thirty seconds after the launch, the balloon has risen $y = (4\,\text{m/s})(30\,\text{s}) = 120\,\text{m}$. To complete the problem, we need the value of $\cos\theta$. Note that when $y = 120\,\text{m}$, the distance between the observer and the balloon is

$$d = \sqrt{120^2 + 200^2} \approx 233.24\,\text{m}.$$

Therefore, $\cos\theta \approx 200/233.24 \approx 0.86$ (see margin figure), and the rate of change of the angle of elevation is

$$\frac{d\theta}{dt} = \frac{(dy/dt)\cdot\cos^2\theta}{200} \approx \frac{(4\,\text{m/s})(0.86^2)}{200\,\text{m}} = 0.015\,\text{rad/s}.$$

> The solution to Example 4 is reported in units of rad/s. Where did the radians come from? Because a radian has no physical dimensions (it is the ratio of an arc length and a radius), no unit appears. We write rad/s for clarity because $d\theta/dt$ is the rate of change of an angle.

At this instant, the balloon is rising at an angular rate of 0.015 rad/s, or slightly less than $1°/\text{s}$, as seen by the observer.   *Related Exercises 26–31* ◄

> Recall that to convert radians to degrees, we use
>
> $$\text{degrees} = \frac{180}{\pi}\cdot\text{radians}.$$

QUICK CHECK 4 In Example 4, notice that as the balloon rises (as θ increases), the rate of change of the angle of elevation decreases to zero. When does the maximum value of $\theta'(t)$ occur and what is it? ◄

SECTION 3.10 EXERCISES

Review Questions

1. Give an example in which one dimension of a geometric figure changes and produces a corresponding change in the area or volume of the figure.

2. Explain how implicit differentiation can simplify the work in a related-rates problem.

3. If two opposite sides of a rectangle increase in length, how must the other two opposite sides change if the area of the rectangle is to remain constant?

4. Explain why the term *related rates* describes the problems of this section.

Basic Skills

5. **Expanding square** The sides of a square increase in length at a rate of 2 m/s.

   a. At what rate is the area of the square changing when the sides are 10 m long?
   b. At what rate is the area of the square changing when the sides are 20 m long?
   c. Draw a graph of how the rate of change of the area varies with the side length.

6. **Expanding cube** The edges of a cube increase at a rate of 2 cm/s. How fast is the volume changing when the length of each edge is 50 cm?

7. **Shrinking circle** A circle has an initial radius of 50 ft when the radius begins decreasing at a rate of 2 ft/min. What is the rate of change of the area at the instant the radius is 10 ft?

8. **Shrinking cube** The volume of a cube decreases at a rate of $0.5\,\text{ft}^3/\text{min}$. What is the rate of change of the side length when the side lengths are 12 ft?

9. **Balloons** A spherical balloon is inflated and its volume increases at a rate of $15\,\text{in}^3/\text{min}$. What is the rate of change of its radius when the radius is 10 in?

10. **Piston compression** A piston is seated at the top of a cylindrical chamber with radius 5 cm when it starts moving into the chamber at a constant speed of 3 cm/s (see figure). What is the rate of change of the volume of the cylinder when the piston is 2 cm from the base of the chamber?

5 cm — Piston

11. **Melting snowball** A spherical snowball melts at a rate proportional to its surface area. Show that the rate of change of the radius is constant. (*Hint:* Surface area = $4\pi r^2$.)

12. **Expanding rectangle** A rectangle initially has dimensions 2 cm by 4 cm. All sides begin increasing in length at a rate of 1 cm/s. At what rate is the area of the rectangle increasing after 20 s?

13. **Filling a pool** A swimming pool is 50 m long and 20 m wide. Its depth decreases linearly along the length from 3 m to 1 m (see figure). It is initially empty and is filled at a rate of 1 m³/min. How fast is the water level rising 250 min after the filling begins? How long will it take to fill the pool?

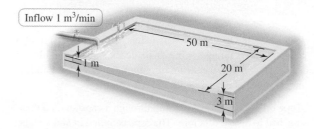

14. **Altitude of a jet** A jet ascends at a 10° angle from the horizontal with an airspeed of 550 mi/hr (its speed along its line of flight is 550 mi/hr). How fast is the altitude of the jet increasing? If the sun is directly overhead, how fast is the shadow of the jet moving on the ground?

15. **Rate of dive of a submarine** A surface ship is moving in a straight line at 10 km/hr. At the same time, an enemy submarine maintains a position directly below the ship while diving at an angle that is 20° below the horizontal. How fast is the submarine's altitude decreasing?

16. **Divergent paths** Two boats leave a port at the same time, one traveling west at 20 mi/hr and the other traveling southwest at 15 mi/hr. At what rate is the distance between them changing 30 min after they leave the port?

17. **Ladder against the wall** A 13-ft ladder is leaning against a vertical wall (see figure) when Jack begins pulling the foot of the ladder away from the wall at a rate of 0.5 ft/s. How fast is the top of the ladder sliding down the wall when the foot of the ladder is 5 ft from the wall?

18. **Ladder against the wall again** A 12-ft ladder is leaning against a vertical wall when Jack begins pulling the foot of the ladder away from the wall at a rate of 0.2 ft/s. What is the configuration of the ladder at the instant that the vertical speed of the top of the ladder equals the horizontal speed of the foot of the ladder?

19. **Moving shadow** A 5-ft-tall woman walks at 8 ft/s toward a street light that is 20 ft above the ground. What is the rate of change of the length of her shadow when she is 15 ft from the street light? At what rate is the tip of her shadow moving?

20. **Baseball runners** Runners stand at first and second base in a baseball game. At the moment a ball is hit, the runner at first base runs to second base at 18 ft/s; simultaneously the runner on second runs to third base at 20 ft/s. How fast is the distance between the runners changing 1 s after the ball is hit (see figure)? (*Hint:* The distance between consecutive bases is 90 ft and the bases lie at the corners of a square.)

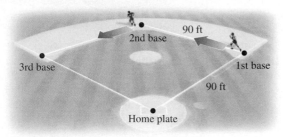

21. **Growing sandpile** Sand falls from an overhead bin and accumulates in a conical pile with a radius that is always three times its height. Suppose the height of the pile increases at a rate of 2 cm/s when the pile is 12 cm high. At what rate is the sand leaving the bin at that instant?

22. **Drinking a soda** At what rate is soda being sucked out of a cylindrical glass that is 6 in tall and has a radius of 2 in? The depth of the soda decreases at a constant rate of 0.25 in/s.

23. **Draining a tank** An inverted conical water tank with a height of 12 ft and a radius of 6 ft is drained through a hole in the vertex at a rate of 2 ft³/s (see figure). What is the rate of change of the water depth when the water depth is 3 ft? (*Hint:* Use similar triangles.)

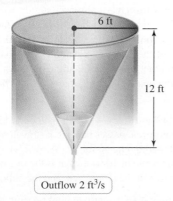

24. Filling a hemispherical tank A hemispherical tank with a radius of 10 m is filled from an input pipe at a rate of 3 m³/min (see figure). (*Hint:* The volume of a cap of thickness h sliced from a sphere of radius r is $\pi h^2(3r - h)/3$.)

a. How fast is the water level rising when the water level is 5 m from the bottom of the tank?

b. What is the rate of change of the surface area of the water when the water is 5 m deep?

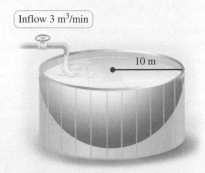

Inflow 3 m³/min

10 m

25. Draining a half-cylinder tank A trough is shaped like a half cylinder with length 5 m and radius 1 m. The tank is full of water when a valve is opened and water flows out of the bottom of the tank at a rate of 1.5 m³/hr (see figure). (*Hint:* The area of a sector of a circle of radius r subtended by an angle θ is $r^2\theta/2$.)

a. How fast is the water level changing when the water level is 0.5 m from the bottom of the trough?

b. What is the rate of change of the surface area of the water when the water is 0.5 m deep.

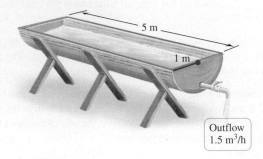

5 m

1 m

Outflow
1.5 m³/h

26. Observing a launch An observer stands 300 ft from the launch site of a hot-air balloon. The balloon is launched vertically and maintains a constant upward velocity of 20 ft/s. What is the rate of change of the angle of elevation of the balloon when it is 400 ft from the ground? The angle of elevation is the angle θ between the observer's line of sight to the balloon and the ground.

27. Another balloon story A hot-air balloon is 150 ft above the ground when a motorcycle passes beneath it (traveling in a straight line on a horizontal road) going 40 mi/hr (58.67 ft/s). If the balloon is rising vertically at a rate of 10 ft/s, what is the rate of change of the distance between the motorcycle and the balloon 10 s later?

28. Fishing story A fly fisherman hooks a trout and begins turning his circular reel at 1.5 rev/s. If the radius of the reel (and the fishing line on it) is 2 in, then how fast is he reeling in his fishing line?

29. Another fishing story A fisherman hooks a trout and reels in his line at 4 in/s. Assume the tip of the fishing rod is 12 ft above the water and the fish is pulled horizontally directly towards the fisherman (see figure). Find the horizontal speed of the fish when it is 20 ft from the fisherman.

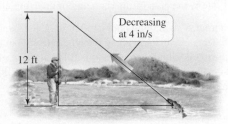

Decreasing at 4 in/s

12 ft

30. Flying a kite Once Kate's kite reaches a height of 50 ft (above her hands), it rises no higher but drifts due east in a wind blowing 5 ft/s. How fast is the string running through Kate's hands at the moment that she has released 120 ft of string?

31. Rope on a boat A rope passing through a capstan on a dock is attached to a boat offshore. The rope is pulled in at a constant rate of 3 ft/s and the capstan is 5 ft vertically above the water. How fast is the boat traveling when it is 10 ft from the dock?

Further Explorations

32. Parabolic motion An arrow is shot into the air and moves along the parabolic path $y = x(50 - x)$ (see figure). The horizontal component of velocity is always 30 ft/s. What is the vertical component of velocity when (i) $x = 10$ and (ii) $x = 40$?

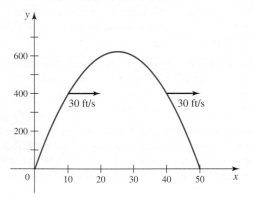

30 ft/s   30 ft/s

33. Time-lagged flights An airliner passes over an airport at noon traveling 500 mi/hr due west. At 1:00 P.M., another airliner passes over the same airport at the same elevation traveling due north at 550 mi/hr. Assuming both airliners maintain their (equal) elevations, how fast is the distance between them changing at 2:30 P.M.?

34. Disappearing triangle An equilateral triangle initially has sides of length 20 ft when each vertex moves toward the midpoint of the opposite side at a rate of 1.5 ft/min. Assuming the triangle remains equilateral, what is the rate of change of the area of the triangle at the instant the triangle disappears?

35. Clock hands The hands of the clock in the tower of the Houses of Parliament in London are approximately 3 m and 2.5 m in length. How fast is the distance between the tips of the hands changing at 9:00? (*Hint:* Use the Law of Cosines.)

36. Filling two pools Two cylindrical swimming pools are being filled simultaneously at the same rate (m³/min; see figure). The smaller pool has a radius of 5 m, and the water level rises at a rate of 0.5 m/min. The larger pool has a radius of 8 m. How fast is the water level rising in the larger pool?

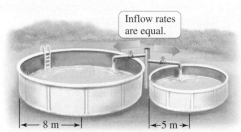

Inflow rates are equal.

←— 8 m —→   ←5 m→

37. Filming a race A camera is set up at the starting line of a drag race 50 ft from a dragster at the starting line (camera 1 in the figure). Two seconds after the start of the race, the dragster has traveled 100 ft and the camera is turning at 0.75 rad/s while filming the dragster.

a. What is the speed of the dragster at this point?

b. A second camera (camera 2 in the figure) filming the dragster is located on the starting line 100 ft away from the dragster at the start of the race. How fast is this camera turning 2 s after the start of the race?

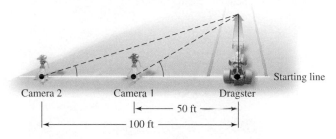

Camera 2   Camera 1   Dragster   Starting line

|←—— 50 ft ——→|

|←———— 100 ft ————→|

38. Two tanks A conical tank with an upper radius of 4 m and a height of 5 m drains into a cylindrical tank with a radius of 4 m and a height of 5 m (see figure). If the water level in the conical tank drops at a rate of 0.5 m/min, at what rate does the water level in the cylindrical tank rise when the water level in the conical tank is 3 m? 1 m?

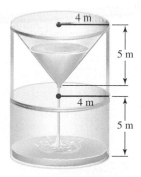

4 m

5 m

4 m

5 m

39. Oblique tracking A port and a radar station are 2 mi apart on a straight shore running east and west. A ship leaves the port at

noon traveling northeast at a rate of 15 mi/hr. If the ship maintains its speed and course, what is the rate of change of the tracking angle θ between the shore and the line between the radar station and the ship at 12:30 P.M.? (*Hint:* Use the Law of Sines.)

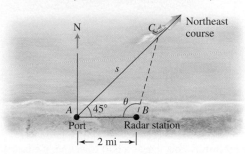

N   C   Northeast course

s

θ

A 45°   B

Port   Radar station

|←— 2 mi —→|

40. Oblique tracking A ship leaves port traveling southwest at a rate of 12 mi/hr. At noon, the ship reaches its closest approach to a radar station, which is on the shore 1.5 mi from the port. If the ship maintains its speed and course, what is the rate of change of the tracking angle θ between the radar station and the ship at 1:30 P.M. (see figure)? (*Hint:* Use the Law of Sines.)

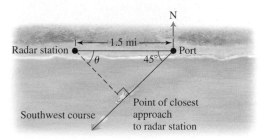

N

Radar station • |←—1.5 mi—→| • Port

θ   45°

Point of closest approach to radar station

Southwest course

41. Watching an elevator An observer is 20 m above the ground floor of a large hotel atrium looking at a glass-enclosed elevator shaft that is 20 m horizontally from the observer (see figure). The angle of elevation of the elevator is the angle that the observer's line of sight makes with the horizontal (it may be positive or negative). Assuming that the elevator rises at a rate of 5 m/s, what is the rate of change of the angle of elevation when the elevator is 10 m above the ground? When the elevator is 40 m above the ground?

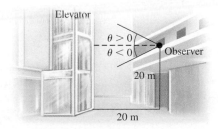

Elevator

θ > 0
θ < 0   Observer

20 m

20 m

42. A lighthouse problem A lighthouse stands 500 m off of a straight shore, the focused beam of its light revolving four times each minute. As shown in the figure on the next page, P is the point on shore closest to the lighthouse and Q is a point on the shore 200 m from P. What is the speed of the beam along the shore when it strikes the point Q? Describe how the speed of the beam along the shore varies with the distance between P and Q. Neglect the height of the lighthouse.

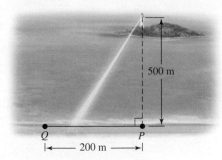

43. Navigation A boat leaves a port traveling due east at 12 mi/hr. At the same time, another boat leaves the same port traveling northeast at 15 mi/hr. The angle θ of the line between the boats is measured relative to due north (see figure). What is the rate of change of this angle 30 min after the boats leave the port? 2 hr after the boats leave the port?

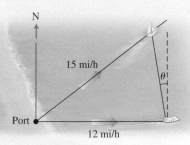

44. Watching a Ferris wheel An observer stands 20 m from the bottom of a 10-m-tall Ferris wheel on a line that is perpendicular to the face of the Ferris wheel. The wheel revolves at a rate of π rad/min and the observer's line of sight with a specific seat on the wheel makes an angle θ with the ground (see figure). Forty seconds after that seat leaves the lowest point on the wheel, what is the rate of change of θ? Assume the observer's eyes are level with the bottom of the wheel.

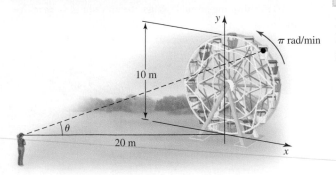

45. Viewing angle The bottom of a large theater screen is 3 ft above your eye level and the top of the screen is 10 ft above your eye level. Assume you walk away from the screen (perpendicular to the screen) at a rate of 3 ft/s while looking at the screen. What is the rate of change of the viewing angle θ when you are 30 ft from the wall on which the screen hangs, assuming the floor is flat (see figure)?

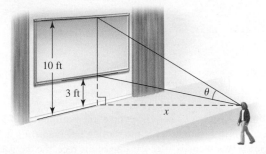

46. Searchlight—wide beam A revolving searchlight, which is 100 m from the nearest point on a straight highway, casts a horizontal beam along a highway (see figure). The beam leaves the spotlight at an angle of $\pi/16$ rad and revolves at a rate of $\pi/6$ rad/s. Let w be the width of the beam as it sweeps along the highway and θ be the angle that the center of the beam makes with the perpendicular to the highway. What is the rate of change of w when $\theta = \pi/3$? Neglect the height of the lighthouse.

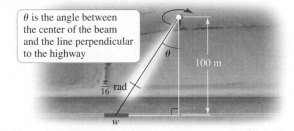

θ is the angle between the center of the beam and the line perpendicular to the highway

QUICK CHECK ANSWERS

1. $12{,}000\pi$ m²/hr, $18{,}000\pi$ m²/hr   2. -192 mi/hr
3. 0.11 ft/min; decreases with height   4. $t = 0, \theta = 0$,
$\theta'(0) = 0.02$ rad/s ◄

CHAPTER 3 REVIEW EXERCISES

1. **Explain why or why not** Determine whether the following statements are true and give an explanation or counterexample.

   a. The function $f(x) = |2x + 1|$ is continuous for all x; therefore, it is differentiable for all x.

   b. If $\dfrac{d}{dx}[f(x)] = \dfrac{d}{dx}[g(x)]$, then $f = g$.

   c. For any function f, $\dfrac{d}{dx}|f(x)| = |f'(x)|$.

   d. The value of $f'(a)$ fails to exist only if the curve $y = f(x)$ has a vertical tangent line at $x = a$.

   e. An object can have negative acceleration and increasing speed.

2–5. Tangent lines

   a. Use either definition of the derivative to determine the slope of the curve $y = f(x)$ at the given point P.

   b. Find an equation of the line tangent to the curve $y = f(x)$ at point P; then, graph the curve and the tangent line.

2. $f(x) = 4x^2 - 7x + 5;\ P(2, 7)$

3. $f(x) = 5x^3 + x;\ P(1, 6)$

4. $y = f(x) = \dfrac{x + 3}{2x + 1};\ P(0, 3)$

5. $f(x) = \dfrac{1}{2\sqrt{3x + 1}};\ P\left(0, \dfrac{1}{2}\right)$

6. **Calculating average and instantaneous velocities** Suppose the height s of an object (in m) above the ground after t seconds is approximated by the function $s = f(t) = -4.9t^2 + 25t + 1$.

   a. Make a table showing the average velocities of the object from time $t = 1$ to $t = 1 + h$ for $h = 0.01, 0.001, 0.0001$, and 0.00001.

   b. Use the table in part (a) to estimate the instantaneous velocity of the object at $t = 1$ s.

   c. Use limits to verify your estimate in part (b).

7. **Population of the United States in the 20th century** The population of the United States (in millions) by decade is given in the table, where t is the number of years after 1900. These data are plotted and connected with a smooth curve $y = p(t)$ in the figure.

   a. Compute the average rate of population growth from 1950 to 1960.

   b. Explain why the average rate of growth from 1950 to 1960 is a good approximation to the (instantaneous) rate of growth in 1955.

   c. Estimate the instantaneous rate of growth in 1985.

Year	1900	1910	1920	1930	1940	1950
t	0	10	20	30	40	50
$p(t)$	76.21	92.23	106.02	123.2	132.16	152.32

Year	1960	1970	1980	1990	2000	2010
t	60	70	80	90	100	110
$p(t)$	179.32	203.30	226.54	248.71	281.42	308.94

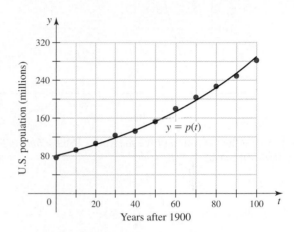

8. **Growth rate of bacteria** Suppose the following graph represents the number of bacteria in a culture t hours after the start of an experiment.

   a. At approximately what time is the instantaneous growth rate the greatest? Estimate the growth rate at this time.

   b. At approximately what time in the interval $0 \le t \le 36$ is the instantaneous growth rate the least? Estimate the instantaneous growth rate at this time.

   c. What is the average growth rate over the interval $0 \le t \le 36$?

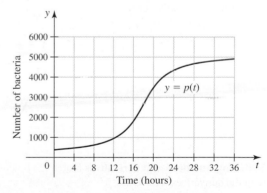

9. **Velocity of a skydiver** Assume the graph on the next page represents the distance (in m) fallen by a skydiver t seconds after jumping out of a plane.

   a. Estimate the velocity of the skydiver at $t = 15$ s.

   b. Estimate the velocity of the skydiver at $t = 70$ s.

   c. Estimate the average velocity of the skydiver between $t = 20$ s and $t = 90$ s.

d. Sketch a graph of the velocity function for $0 \le t \le 120$.

e. What significant event do you think occurred at $t = 30$ s?

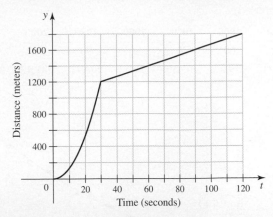

10–11. Using the definition of the derivative *Use the limit definition of the derivative to do the following.*

10. Verify that $f'(x) = 4x - 3$ if $f(x) = 2x^2 - 3x + 1$.

11. Verify that $g'(x) = \dfrac{1}{\sqrt{2x - 3}}$ if $g(x) = \sqrt{2x - 3}$.

12. Sketching a derivative graph Sketch a graph of f' for the function f shown in the figure.

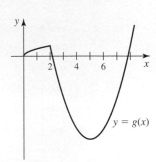

13. Sketching a derivative graph Sketch a graph of g' for the function g shown in the figure.

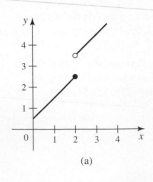

14. Matching functions and derivatives Match the functions (a)–(d) with the derivatives (A)–(D).

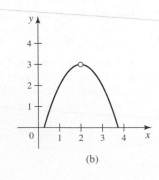

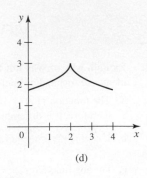

(c)

(d)

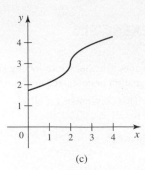

(A)

(B)

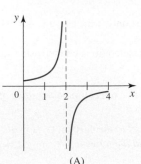

(C)

(D)

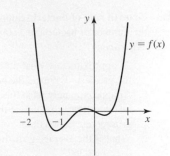

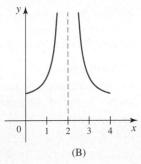

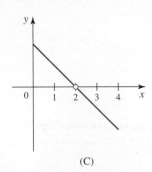

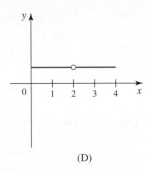

15–36. Evaluating derivatives *Evaluate and simplify the following derivatives.*

15. $\dfrac{d}{dx}\left(\dfrac{2}{3}x^3 + \pi x^2 + 7x + 1\right)$

16. $\dfrac{d}{dx}\left(2x\sqrt{x^2 - 2x + 2}\right)$

17. $\dfrac{d}{dt}\left(5t^2 \sin t\right)$

18. $\dfrac{d}{dx}\left(5x + \sin^3 x + \sin x^3\right)$

19. $\dfrac{d}{d\theta}\left[4\tan\left(\theta^2 + 3\theta + 2\right)\right]$

20. $\dfrac{d}{dx}\left(\csc^5 3x\right)$

21. $\dfrac{d}{du}\left(\dfrac{4u^2 + u}{8u + 1}\right)$

22. $\dfrac{d}{dt}\left(\dfrac{3t^2 - 1}{3t^2 + 1}\right)^{-3}$

23. $\dfrac{d}{d\theta}\left[\tan\left(\sin\theta\right)\right]$

24. $\dfrac{d}{dv}\left(\dfrac{v}{3v^2 + 2v + 1}\right)^{1/3}$

25. $\dfrac{d}{dx}\left[2x\left(\sin x\right)\sqrt{3x - 1}\right]$

26. $\dfrac{d}{dx}\left(xe^{-10x}\right)$

27. $\dfrac{d}{dx}\left(x\ln^2 x\right)$

28. $\dfrac{d}{dw}\left(e^{-w}\ln w\right)$

29. $\dfrac{d}{dx}\left(2^{x^2 - x}\right)$

30. $\dfrac{d}{dx}\left(\log_3\left(x + 8\right)\right)$

31. $\dfrac{d}{dx}\left[\sin^{-1}\left(\dfrac{1}{x}\right)\right]$ **32.** $\dfrac{d}{dx}(x^{\sin x})$

33. $f'(1)$ when $f(x) = x^{1/x}$ **34.** $f'(1)$ when $f(x) = \tan^{-1}(4x^2)$

35. $\dfrac{d}{dx}\left(x\sec^{-1}x\right)\Big|_{x=\frac{2}{\sqrt3}}$ **36.** $\dfrac{d}{dx}\left[\tan^{-1}(e^{-x})\right]\Big|_{x=0}$

37–39. Implicit differentiation *Calculate $y'(x)$ for the following relations.*

37. $y = \dfrac{e^y}{1 + \sin x}$ **38.** $\sin(x)\cos(y-1) = \dfrac{1}{2}$

39. $y\sqrt{x^2 + y^2} = 15$

40. Quadratic functions

 a. Show that if $(a, f(a))$ is any point on the graph of $f(x) = x^2$, then the slope of the tangent line at that point is $m = 2a$.
 b. Show that if $(a, f(a))$ is any point on the graph of $f(x) = bx^2 + cx + d$, then the slope of the tangent line at that point is $m = 2ab + c$.

41–44. Tangent lines *Find an equation of the line tangent to the following curves at the given point.*

41. $y = 3x^3 + \sin x;\ x = 0$

42. $y = \dfrac{4x}{x^2 + 3};\ x = 3$

43. $y + \sqrt{xy} = 6;\ (x, y) = (1, 4)$

44. $x^2y + y^3 = 75;\ (x, y) = (4, 3)$

45. Horizontal tangent line For what value(s) of x is the line tangent to the curve $y = x\sqrt{6 - x}$ horizontal?

46. A parabola property Let $f(x) = x^2$.

 a. Show that $\dfrac{f(x) - f(y)}{x - y} = f'\left(\dfrac{x + y}{2}\right)$ for all $x \neq y$.
 b. Is this property true for $f(x) = ax^2$, where a is a real number?
 c. Give a geometrical interpretation of this property.
 d. Is this property true for $f(x) = ax^3$?

47–48. Higher-order derivatives *Find $y', y'', $ and y''' for the following functions.*

47. $y = \sin\sqrt{x}$ **48.** $y = \sqrt{x + 2}\,(x - 3)$

49–52. Derivative formulas *Evaluate the following derivatives. Express your answers in terms of f, g, f' and g'.*

49. $\dfrac{d}{dx}[x^2 f(x)]$ **50.** $\dfrac{d}{dx}\sqrt{\dfrac{f(x)}{g(x)}}$

51. $\dfrac{d}{dx}\left(\dfrac{x \cdot f(x)}{g(x)}\right)$ **52.** $\dfrac{d}{dx}f\left(\sqrt{g(x)}\right)$

53. Finding derivatives from a table Find the values of the following derivatives using the table.

x	1	3	5	7	9
$f(x)$	3	1	9	7	5
$f'(x)$	7	9	5	1	3
$g(x)$	9	7	5	3	1
$g'(x)$	5	9	3	1	7

 a. $\dfrac{d}{dx}[f(x) + 2g(x)]\Big|_{x=3}$ b. $\dfrac{d}{dx}\left[\dfrac{x \cdot f(x)}{g(x)}\right]\Big|_{x=1}$ c. $\dfrac{d}{dx}f[g(x^2)]\Big|_{x=3}$

54–55. Limits *The following limits represent the derivative of a function f at a point a. Find a possible f and a, and then evaluate the limit.*

54. $\lim\limits_{h\to 0}\dfrac{\sin^2\left(\dfrac{\pi}{4} + h\right) - \dfrac{1}{2}}{h}$ **55.** $\lim\limits_{x\to 5}\dfrac{\tan\left(\pi\sqrt{3x - 11}\right)}{x - 5}$

56–57. Derivative of the inverse at a point *Consider the following functions. In each case, without finding the inverse, evaluate the derivative of the inverse at the given point.*

56. $f(x) = 1/(x + 1)$ at $f(0)$

57. $f(x) = x^4 - 2x^2 - x$ at $f(0)$

58–59. Derivative of the inverse *Find the derivative of the inverse of the following functions. Express the result with x as the independent variable.*

58. $f(x) = 12x - 16$ **59.** $f(x) = x^{-1/3}$

60. A function and its inverse function The function $f(x) = \dfrac{x}{x + 1}$ is one-to-one for $x > -1$ and has an inverse on that interval.

 a. Graph f for $x > -1$.
 b. Find the inverse function f^{-1} corresponding to the function graphed in part (a). Graph f^{-1} on the same set of axes as in part (a).
 c. Evaluate the derivative of f^{-1} at the point $\left(\dfrac{1}{2}, 1\right)$.
 d. Sketch the tangent lines on the graphs of f and f^{-1} at $\left(1, \dfrac{1}{2}\right)$ and $\left(\dfrac{1}{2}, 1\right)$, respectively.

61. Derivative of the inverse in two ways Let $f(x) = \sin x$, $f^{-1}(x) = \sin^{-1}(x)$, and $(x_0, y_0) = (\pi/4, 1/\sqrt{2})$.

 a. Evaluate $(f^{-1})'(1/\sqrt{2})$ using Theorem 3.23 $[(f^{-1})'(y_0) = 1/f'(x_0)]$.
 b. Evaluate $(f^{-1})'(1/\sqrt{2})$ directly by differentiating f^{-1}. Check for agreement with part (a).

62. Velocity of a rocket The height in feet of a rocket above the ground is given by $s(t) = \dfrac{200t^2}{t^2 + 1}$ for $t \geq 0$.

 a. Graph the height function and describe the motion of the rocket.
 b. Find the velocity of the rocket, $v(t) = s'(t)$.
 c. Graph the velocity function and determine the approximate time at which the velocity is a maximum.

63. Marginal and average cost Suppose the cost of producing x lawnmowers is $C(x) = -0.02x^2 + 400x + 5000$.

   a. Determine the average and marginal costs for $x = 3000$ lawnmowers.

   b. Interpret the meaning of your results in part (a).

64. Marginal and average cost Suppose a company produces fly rods. Assume $C(x) = -0.0001x^3 + 0.05x^2 + 60x + 800$ represents the cost of making x fly rods.

   a. Determine the average and marginal costs for $x = 400$ fly rods.

   b. Interpret the meaning of your results in part (a).

⊤ 65. Population growth Suppose $p(t) = -1.7t^3 + 72t^2 + 7200t + 80{,}000$ is the population of a city t years after 1950.

   a. Determine the average rate of growth of the city from 1950 to 2000.

   b. What was the rate of growth of the city in 1990?

⊤ 66. Position of a piston The distance between the head of a piston and the end of a cylindrical chamber is given by $x(t) = \dfrac{8t}{t+1}$ cm, for $t \geq 0$ s. The radius of the cylinder is 4 cm.

   a. Find the volume of the chamber for $t \geq 0$.

   b. Find the rate of change of the volume $V'(t)$ for $t \geq 0$.

   c. Graph the derivative of the volume function. On what intervals is the volume increasing? Decreasing?

67. Boat rates Two boats leave a dock at the same time. One boat travels south at 30 mi/hr and the other travels east at 40 mi/hr. After half an hour, how fast is the distance between the boats increasing?

68. Rate of inflation of a balloon A spherical balloon is inflated at a rate of 10 cm^3/min. At what rate is the diameter of the balloon increasing when the balloon has a diameter of 5 cm?

69. Rate of descent of a hot-air balloon A rope is attached to the bottom of a hot-air balloon that is floating above a flat field. If the angle of the rope to the ground remains at 65° and the rope is pulled in at 5 ft/s, how quickly is the elevation of the balloon changing?

70. Filling a tank Water flows into a conical tank at a rate of 2 ft^3/min. If the radius of the top of the tank is 4 ft and the height is 6 ft, determine how quickly the water level is rising when the water is 2 ft deep in the tank.

71. Angle of elevation A jet flies horizontally 500 ft directly above a spectator at an air show at 450 mi/hr. Determine how quickly the angle of elevation (between the ground and the line from the spectator to the jet) is changing 2 s later.

72. Viewing angle A man whose eye level is 6 ft above the ground walks toward a billboard at a rate of 2 ft/s. The bottom of the billboard is 10 ft above the ground and it is 15 ft high. The man's viewing angle is the angle formed by the lines between the man's eyes and the top and bottom of the billboard. At what rate is the viewing angle changing when the man is 30 ft from the billboard?

Chapter 3 Guided Projects

Applications of the material in this chapter and related topics can be found in the following Guided Projects. For additional information, see the Preface.

• Numerical differentiation

• Elasticity in economics

• Enzyme kinetics

• Pharmacokinetics—drug metabolism

4

Applications of the Derivative

Chapter Preview Much of the previous chapter was devoted to the basic mechanics of derivatives: evaluating them and interpreting them as rates of change. We now apply derivatives to a variety of mathematical questions about the properties of functions and their graphs. One outcome of this work is a set of analytical curve-sketching methods that produce accurate graphs of functions. Equally important, derivatives allow us to formulate and solve a wealth of practical problems. For example, a weather probe dropped from an airplane accelerates until it reaches its terminal velocity: When is the acceleration the greatest? An economist has a mathematical model that relates the demand for a product to its price: What price maximizes the revenue? In this chapter, we develop the tools needed to answer such questions. In addition, we begin an ongoing discussion about approximating functions, we present an important result called the Mean Value Theorem, and we work with a powerful method that enables us to evaluate a new kind of limit.

4.1 Maxima and Minima

With a working understanding of derivatives, we now undertake one of the fundamental tasks of calculus: analyzing the behavior and producing accurate graphs of functions. An important question associated with any function concerns its maximum and minimum values: On a given interval (perhaps the entire domain), where does the function assume its largest and smallest values? Questions about maximum and minimum values take on added significance when a function represents a practical quantity, such as the profits of a company, the surface area of a container, or the speed of a space vehicle.

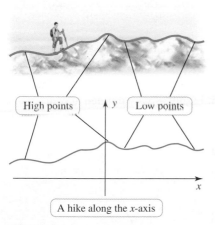

A hike along the x-axis

FIGURE 4.1

Absolute Maxima and Minima

Imagine taking a long hike through varying terrain from west to east. Your elevation changes as you walk over hills, through valleys, and across plains, and you reach several high and low points along the journey. Analogously, when we examine a function over an interval on the x-axis, its values increase and decrease, reaching high points and low points (Figure 4.1). You can view our study of functions in this chapter as an exploratory hike along the x-axis.

> ➤ Absolute maximum and minimum values are also called *global* maximum and minimum values. The plural of maximum is maxima; the plural of minimum is minima. *Extrema* (plural) and *extremum* (singular) refer to either maxima or minima.

DEFINITION   Absolute Maximum and Minimum

Let f be defined on an interval I containing c. Then, f has an **absolute maximum** value on I at c if $f(c) \geq f(x)$ for every x in I. Similarly, f has an **absolute minimum** value on I at c if $f(c) \leq f(x)$ for every x in I.

The existence and location of absolute extreme values depend on both the function and the interval of interest. Figure 4.2 shows various cases for the function $f(x) = x^2$. Notice that if the interval of interest is not closed, a function may not attain absolute extreme values (Figure 4.2a, c, and d).

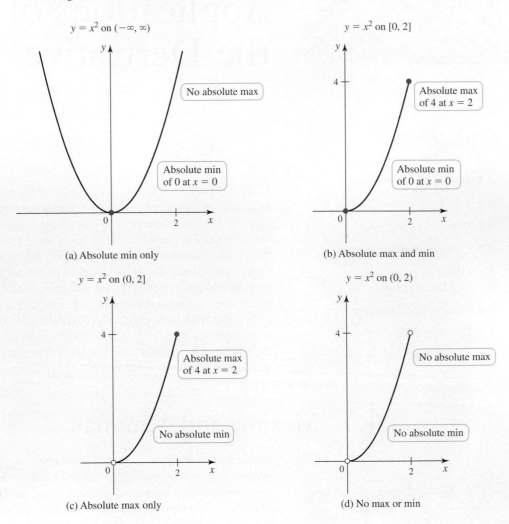

FIGURE 4.2. The function $f(x) = x^2$ has different absolute extrema depending on the interval of interest.

However, defining a function on a closed interval is not enough to guarantee the existence of absolute extreme values. Both functions in Figure 4.3 are defined at every point of a closed interval, but neither function attains an absolute maximum—the discontinuity in each function prevents it from happening.

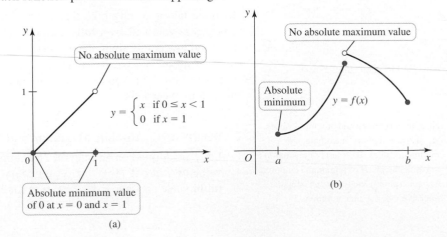

FIGURE 4.3

It turns out that *two* conditions ensure the existence of absolute minimum and maximum values on an interval: The function must be continuous on an interval and the interval must be closed and bounded.

> The proof of the Extreme Value Theorem relies on some deep properties of the real numbers, found in advanced books.

THEOREM 4.1   Extreme Value Theorem

A function that is continuous on a closed interval $[a, b]$ has an absolute maximum value and an absolute minimum value on that interval.

QUICK CHECK 1 Sketch the graph of a function that is continuous on an interval but does not have an absolute minimum value. Sketch the graph of a function that is defined on a closed interval but does not have an absolute minimum value. ◄

EXAMPLE 1   Locating absolute maximum and minimum values For the functions in Figure 4.4, identify the location of the absolute maximum value and the absolute minimum value on the interval $[a, b]$. Do the functions meet the conditions of the Extreme Value Theorem?

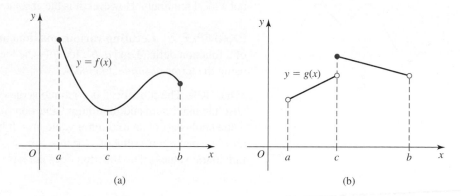

FIGURE 4.4   (a)   (b)

SOLUTION

a. The function f is continuous on the closed interval $[a, b]$, so the Extreme Value Theorem guarantees an absolute maximum (which occurs at a) and an absolute minimum (which occurs at c).

b. The function g does not satisfy the conditions of the Extreme Value Theorem because it is not continuous, and it is defined only on the open interval (a, b). It does not have an absolute minimum value. It does, however, have an absolute maximum at c. Therefore, a function may violate the conditions of the Extreme Value Theorem and still have an absolute minimum or maximum (or both). *Related Exercises 11–14* ◄

Local Maxima and Minima

Figure 4.5 shows a function defined on the interval $[a, b]$. It has an absolute minimum at the endpoint a and an absolute maximum at the interior point e. In addition, the function

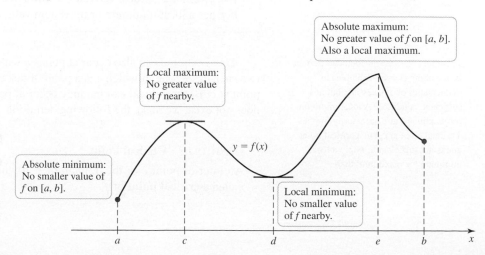

Absolute maximum:
No greater value of f on $[a, b]$.
Also a local maximum.

Local maximum:
No greater value of f nearby.

Absolute minimum:
No smaller value of f on $[a, b]$.

Local minimum:
No smaller value of f nearby.

FIGURE 4.5

has special behavior at c, where its value is greatest *among nearby points*, and at d, where its value is least *among nearby points*. A point at which a function takes on the maximum or minimum value among nearby points is important.

> Local maximum and minimum values are also called *relative maximum and minimum values*. *Local extrema* (plural) and *local extremum* (singular) refer to either local maxima or local minima.

DEFINITION   Local Maximum and Minimum Values

Suppose I is an interval on which f is defined and c is an interior point of I. If $f(c) \geq f(x)$ for all x in some open interval containing c, then $f(c)$ is a **local maximum** value of f. If $f(c) \leq f(x)$ for all x in some open interval containing c, then $f(c)$ is a **local minimum** value of f.

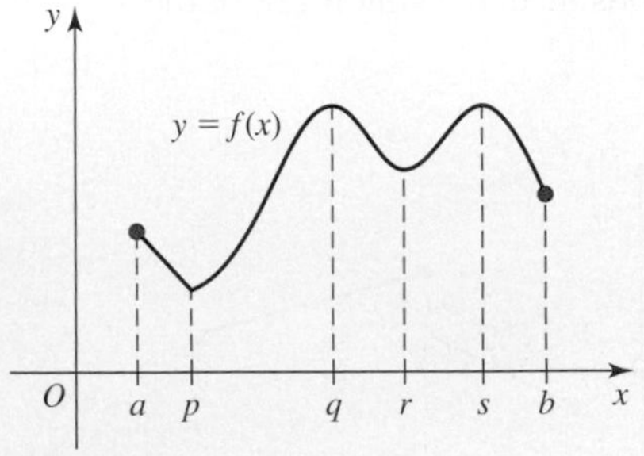

Note that local maxima and minima occur at interior points of the interval of interest, not at endpoints. For example, in Figure 4.5, the minimum value that occurs at the endpoint a is not a local minimum. However, it is the absolute minimum of the function on $[a, b]$.

EXAMPLE 2   Locating various maxima and minima   Figure 4.6 shows the graph of a function defined on $[a, b]$. Identify the location of the various maxima and minima using the terms *absolute* and *local*.

SOLUTION   The function f is continuous on a closed interval; by Theorem 4.1, it has absolute maximum and minimum values on $[a, b]$. The function has a local minimum value and its absolute minimum value at p. It has another local minimum value at r. The absolute maximum value of f occurs at both q and s (which also correspond to local maximum values). The function does not have extrema at the endpoints a and b.

Related Exercises 15–22 ◄

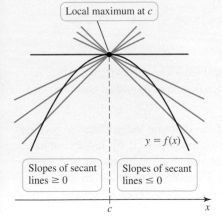

FIGURE 4.6

Critical Points   Another look at Figure 4.6 shows that local maxima and minima occur at points in the open interval (a, b) where the derivative is zero ($x = q, r$, and s) and at points where the derivative fails to exist ($x = p$). We now make this observation precise.

Figure 4.7 illustrates a function that is differentiable at c with a local maximum at c. For x near c with $x < c$, the secant lines between the points $(x, f(x))$ and $(c, f(c))$ have nonnegative slopes. For x near c with $x > c$, the secant lines between the points $(x, f(x))$ and $(c, f(c))$ have nonpositive slopes. As $x \to c$, the slopes of these secant lines approach the slope of the tangent line at $(c, f(c))$. These observations imply that the slope of the tangent line must be both nonnegative and nonpositive, which happens only if $f'(c) = 0$. Similar reasoning leads to the same conclusion for a function with a local minimum at c: $f'(c)$ must be zero. This argument is an outline of the proof (Exercise 75) of the following theorem.

Local maximum at c

Slopes of secant lines ≥ 0

Slopes of secant lines ≤ 0

$y = f(x)$

c

FIGURE 4.7

THEOREM 4.2   Local Extreme Point Theorem

If f has a local minimum or maximum value at c and $f'(c)$ exists, then $f'(c) = 0$.

> Theorem 4.2, often attributed to Fermat, is one of the clearest examples in mathematics of a necessary, but not sufficient, condition. A local minimum (or maximum) at c necessarily implies a critical point at c, but a critical point at c is not sufficient to imply a local minimum (or maximum) there.

Local extrema can also occur at points c where $f'(c)$ does not exist. Figure 4.8 shows two such cases, one in which c is a point of discontinuity and one in which f has a corner point at c. Because local extema may occur at points c where $f'(c) = 0$ *and* where $f'(c)$ does not exist, we make the following definition.

DEFINITION   Critical Point

An interior point c of the domain of f at which $f'(c) = 0$ or $f'(c)$ fails to exist is called a **critical point** of f.

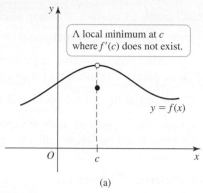

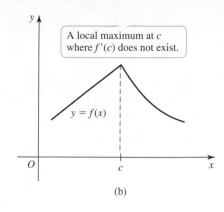

FIGURE 4.8

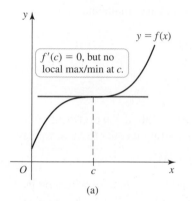

(a)

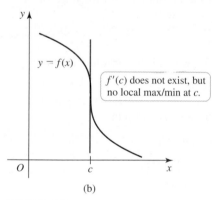

(b)

FIGURE 4.9

Note that the converse of Theorem 2 is not necessarily true. It is possible that $f'(c) = 0$ at a point without a local maximum or local minimum value occurring there (Figure 4.9a). It is also possible that $f'(c)$ fails to exist, with no local extreme value occurring at c (Figure 4.9b). Therefore, critical points are *candidates* for local extreme points, but you must determine whether they actually correspond to local maxima or minima. This procedure is discussed in Section 4.2.

EXAMPLE 3 **Locating critical points** Find the critical points of $f(x) = x^2 \ln x$.

SOLUTION Note that f is differentiable on its domain, which is $(0, \infty)$. By the Product Rule,

$$f'(x) = 2x \cdot \ln x + x^2 \cdot \frac{1}{x} = x(2 \ln x + 1).$$

Setting $f'(x) = 0$ gives $x(2 \ln x + 1) = 0$, which has the solution $x = e^{-1/2} = 1/\sqrt{e}$. Because $x = 0$ is not in the domain of f, it is not a critical point. Therefore, the only critical point is $x = 1/\sqrt{e} \approx 0.61$. A graph of f (Figure 4.10) reveals that a local (and, indeed, absolute) minimum value occurs at $(1/\sqrt{e}, -1/(2e))$.

Related Exercises 23–30 ◀

QUICK CHECK 2 Consider the function $f(x) = x^3$. Where is the critical point of f? Does f have a local maximum or minimum at the critical point? ◀

Locating Absolute Maxima and Minima

Theorem 4.1 guarantees the existence of absolute extreme values of a continuous function on a closed interval $[a, b]$, but it doesn't say where these values are located. Two observations lead to a procedure for locating absolute extreme values.

- An absolute extreme value in the interior of an interval is also a local extreme value, and we know that local extreme values occur at the critical points of f.
- Absolute extreme values may also occur at the endpoints of the interval of interest.

These two facts suggest the following procedure for locating the absolute extreme values of a continuous function on a closed interval.

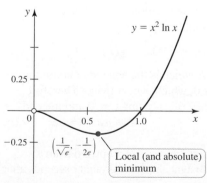

FIGURE 4.10

> **PROCEDURE   Locating Absolute Maximum and Minimum Values**
>
> Assume the function f is continuous on the closed interval $[a, b]$.
>
> 1. Locate the critical points c in (a, b), where $f'(c) = 0$ or $f'(c)$ does not exist. These points are candidates for absolute maxima and minima.
>
> 2. Evaluate f at the critical points and at the endpoints of $[a, b]$.
>
> 3. Choose the largest and smallest values of f from Step 2 for the absolute maximum and minimum values, respectively.

If the interval of interest is an open interval, then absolute extreme values—if they exist—occur at interior points.

EXAMPLE 4   Absolute extreme values Find the absolute maximum and minimum values of the following functions.

a. $f(x) = x^4 - 2x^3$ on the interval $[-2, 2]$

b. $g(x) = x^{2/3}(2 - x)$ on the interval $[-1, 2]$

SOLUTION

a. Because f is a polynomial, its derivative exists everywhere. So, if f has critical points, they are points at which $f'(x) = 0$. Computing f' and setting it equal to zero, we have

$$f'(x) = 4x^3 - 6x^2 = 2x^2(2x - 3) = 0.$$

Solving this equation gives the critical points $x = 0$ and $x = \frac{3}{2}$, both of which lie in the interval $[-2, 2]$; these points and the endpoints are candidates for the location of absolute extrema. Evaluating f at each of these points, we have

$$f(-2) = 32 \qquad f(0) = 0 \qquad f\left(\tfrac{3}{2}\right) = -\tfrac{27}{16} \qquad f(2) = 0.$$

The largest of these function values is $f(-2) = 32$, which is the absolute maximum of f on $[-2, 2]$. The smallest of these values is $f\left(\tfrac{3}{2}\right) = -\tfrac{27}{16}$, which is the absolute minimum of f on $[-2, 2]$. The graph of f (Figure 4.11) shows that the critical point $x = 0$ corresponds to neither a local maximum nor a local minimum.

b. Differentiating $g(x) = x^{2/3}(2 - x) = 2x^{2/3} - x^{5/3}$, we have

$$g'(x) = \frac{4}{3}x^{-1/3} - \frac{5}{3}x^{2/3} = \frac{4 - 5x}{3\sqrt[3]{x}}.$$

Because $g'(0)$ is undefined and 0 is in the domain of g, $x = 0$ is a critical point. In addition, $g'(x) = 0$ when $4 - 5x = 0$, so $x = \frac{4}{5}$ is also a critical point. These two critical points and the endpoints are *candidates* for the location of absolute extrema. The next step is to evaluate f at the critical points and endpoints:

$$g(-1) = 3 \qquad g(0) = 0 \qquad g(4/5) \approx 1.03 \qquad g(2) = 0$$

The largest of these function values is $g(-1) = 3$, which is the absolute maximum value of g on $[-1, 2]$. The least of these values is 0, which occurs twice. Therefore, g has its absolute minimum value on $[-1, 2]$ at the critical point $x = 0$ and the endpoint $x = 2$ (Figure 4.12).   *Related Exercises 31–42* ◄

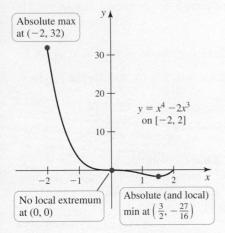

Absolute max at $(-2, 32)$

$y = x^4 - 2x^3$ on $[-2, 2]$

No local extremum at $(0, 0)$

Absolute (and local) min at $\left(\tfrac{3}{2}, -\tfrac{27}{16}\right)$

FIGURE 4.11

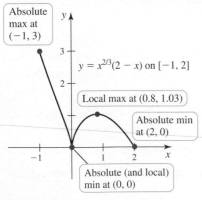

Absolute max at $(-1, 3)$

$y = x^{2/3}(2 - x)$ on $[-1, 2]$

Local max at $(0.8, 1.03)$

Absolute min at $(2, 0)$

Absolute (and local) min at $(0, 0)$

FIGURE 4.12

➤ The derivation of the position function for an object moving in a gravitational field is given in Section 6.1.

EXAMPLE 5   Trajectory high point A stone is launched vertically upward from a bridge 80 ft above the ground at a speed of 64 ft/s. Its height above the ground t seconds after the launch is given by

$$f(t) = -16t^2 + 64t + 80 \quad \text{for } 0 \le t \le 5.$$

When does the stone reach its maximum height?

SOLUTION We must evaluate the height function at the critical points and at the endpoints. The critical points satisfy the equation

$$f'(t) = -32t + 64 = -32(t - 2) = 0,$$

so the only critical point is $t = 2$. We now evaluate f at the endpoints and at the critical point:

$$f(0) = 80 \qquad f(2) = 144 \qquad f(5) = 0$$

On the interval $[0, 5]$, the absolute maximum occurs at $t = 2$ s, at which time the stone reaches a height of 144 ft. Because $f'(t)$ is the velocity of the stone, the maximum height occurs at the instant the velocity is zero. *Related Exercises 43–46* ◄

SECTION 4.1 EXERCISES

Review Questions

1. What does it mean for a function to have an absolute extreme value at a point c of an interval $[a, b]$?

2. What are local maximum and minimum values of a function?

3. What conditions must be met to ensure that a function has an absolute maximum value and an absolute minimum value on an interval?

4. Sketch the graph of a function that is continuous on an open interval (a, b) but has neither an absolute maximum nor an absolute minimum value on (a, b).

5. Sketch the graph of a function that has an absolute maximum, a local minimum, but no absolute minimum on $[0, 3]$.

6. What is a critical point of a function?

7. Sketch the graph of a function f that has a local maximum value at a point c where $f'(c) = 0$.

8. Sketch the graph of a function f that has a local minimum value at a point c where $f'(x)$ is undefined.

9. How do you determine the absolute maximum and minimum values of a continuous function on a closed interval?

10. Explain how a function can have an absolute minimum value at an endpoint of an interval.

Basic Skills

11–14. Absolute maximum/minimum values from graphs *Use the following graphs to identify the points (if any) on the interval $[a, b]$ at which the function has an absolute maximum value or an absolute minimum value.*

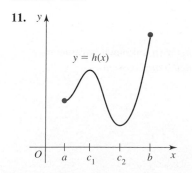

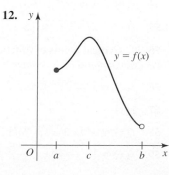

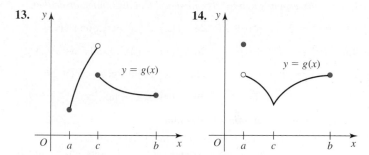

15–18. Local and absolute extreme values *Use the following graphs to identify the points on the interval $[a, b]$ at which local and absolute extreme values occur.*

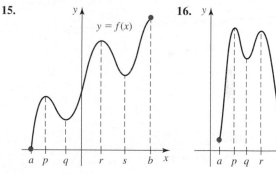

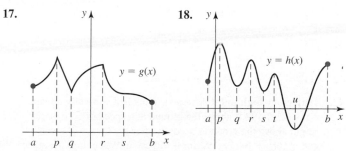

19–22. Designing a function *Sketch the graph of a continuous function f on $[0, 4]$ satisfying the given properties.*

19. $f'(x) = 0$ for $x = 1$ and 2; f has an absolute maximum at $x = 4$; f has an absolute minimum at $x = 0$; and f has a local minimum at $x = 2$.

20. $f'(x) = 0$ for $x = 1, 2,$ and 3; f has an absolute minimum at $x = 1$; f has no local extremum at $x = 2$; and f has an absolute maximum at $x = 3$.

21. f' is undefined at $x = 1$ and 3; $f'(2) = 0$; f has a local maximum at $x = 1$; f has a local minimum at $x = 2$; f has an absolute maximum at $x = 3$; and f has an absolute minimum at $x = 4$.

22. $f'(x) = 0$ at $x = 1$ and 3; $f'(2)$ is undefined; f has an absolute maximum at $x = 2$; f has neither a local maximum nor a local minimum at $x = 1$; and f has an absolute minimum at $x = 3$.

23–30. Locating critical points

　　a. Find the critical points of the following functions on the domain or on the given interval.
　　b. Use a graphing utility to determine whether each critical point corresponds to a local minimum, local maximum, or neither.

23. $f(x) = 3x^2 - 4x + 2$　　**24.** $f(x) = \frac{1}{8}x^3 - \frac{1}{2}x$ on $[-1, 3]$

25. $f(x) = x/(x^2 + 1)$　　**26.** $f(x) = 12x^5 - 20x^3$ on $[-0.5, 2]$

27. $f(x) = (e^x + e^{-x})/2$　　**28.** $f(x) = \sin x \cos x$ on $[0, 2\pi]$

29. $f(x) = 1/x - \ln x$　　**30.** $f(x) = x - \tan^{-1} x$

31–42. Absolute maxima and minima

　　a. Find the critical points of f on the given interval.
　　b. Determine the absolute extreme values of f on the given interval.
　　c. Use a graphing utility to confirm your conclusions.

31. $f(x) = x^2 - 10$ on $[-2, 3]$

32. $f(x) = (x + 1)^{4/3}$ on $[-8, 8]$

33. $f(x) = \cos^2 x$ on $[0, \pi]$

34. $f(x) = x/(x^2 + 1)^2$ on $[-2, 2]$

35. $f(x) = \sin 3x$ on $[-\pi/4, \pi/3]$

36. $f(x) = x^{2/3}$ on $[-8, 8]$

37. $f(x) = (2x)^x$ on $[0.1, 1]$

38. $f(x) = xe^{-x/2}$ on $[0, 5]$

39. $f(x) = x^2 + \cos^{-1} x$ on $[-1, 1]$

40. $f(x) = x\sqrt{2 - x^2}$ on $[-\sqrt{2}, \sqrt{2}]$

41. $f(x) = \dfrac{x}{\sqrt{4 - x^2}}$ on $(-2, 2)$

42. $f(x) = x^3 - 2x^2 - 5x + 6$ on $[4, 8]$

43. Trajectory high point A stone is launched vertically upward from a cliff 192 ft above the ground at a speed of 64 ft/s. Its height above the ground t seconds after the launch is given by $s = -16t^2 + 64t + 192$ for $0 \le t \le 6$. When does the stone reach its maximum height?

44. Maximizing revenue A sales analyst determines that the revenue from sales of fruit smoothies is given by $R(x) = -60x^2 + 300x$, where x is the price in dollars charged per item, with $0 \le x \le 5$.

　　a. Find the critical points of the revenue function.
　　b. Determine the absolute maximum value of the revenue function and give the price that maximizes the revenue.

45. Maximizing profit Suppose a tour guide has a bus that holds a maximum of 100 people. Assume his profit (in dollars) for taking n people on a city tour is $P(n) = n(50 - 0.5n) - 100$. (Although P is defined only for positive integers, treat it as a continuous function.)

　　a. How many people should the guide take on a tour to maximize the profit?
　　b. Suppose the bus holds a maximum of 45 people. How many people should be taken on a tour to maximize the profit?

46. Maximizing rectangle perimeters All rectangles with an area of 64 m² have a perimeter given by $P(x) = 2x + 128/x$, where x is the length of one side of the rectangle. Find the absolute minimum value of the perimeter function. What are the dimensions of the rectangle with minimum perimeter?

Further Explorations

47. Explain why or why not Determine whether the following statements are true and give an explanation or counterexample.

　　a. The function $f(x) = \sqrt{x}$ has a local maximum on the interval $[0, 1]$.
　　b. If a function has an absolute maximum, then the function must be continuous on a closed interval.
　　c. A function f has the property that $f'(2) = 0$. Therefore, f has a local maximum or minimum at $x = 2$.
　　d. Absolute extreme values on an interval always occur at a critical point or an endpoint of the interval.
　　e. A function f has the property that $f'(3)$ does not exist. Therefore, $x = 3$ is a critical point of f.

48–55. Absolute maxima and minima

　　a. Find the critical points of f on the given interval.
　　b. Determine the absolute extreme values of f on the given interval.
　　c. Use a graphing utility to confirm your conclusions.

48. $f(x) = (x - 2)^{1/2}$; $[2, 6]$

49. $f(x) = 2^x \sin x$; $[-2, 6]$

50. $f(x) = x^{1/2}(x^2/5 - 4)$; $[0, 4]$

51. $f(x) = \sec x$; $[-\pi/4, \pi/4]$

52. $f(x) = x^{1/3}(x + 4)$; $[-27, 27]$

53. $f(x) = x^3 e^{-x}$; $[-1, 5]$

54. $f(x) = x \ln(x/5)$; $[0.1, 5]$

55. $f(x) = x/\sqrt{x - 4}$; $[6, 12]$

56–59. Critical points of functions with unknown parameters *Find the critical points of f. Assume a and b are constants.*

56. $f(x) = x/\sqrt{x - a}$

57. $f(x) = x\sqrt{x - a}$

58. $f(x) = x^3 - 3ax^2 + 3a^2x - a^3 + b$

59. $f(x) = \frac{1}{5}x^5 - a^4 x$

T 60–65. Critical points and extreme values

a. *Find the critical points of the following functions on the given interval.*

b. *Use a graphing device to determine whether the critical points correspond to local maxima, local minima, or neither.*

c. *Find the absolute maximum and minimum values on the given interval.*

60. $f(x) = 6x^4 - 16x^3 - 45x^2 + 54x + 23$; $[-5, 5]$

61. $f(\theta) = 2 \sin \theta + \cos \theta$; $[-2\pi, 2\pi]$

62. $f(x) = x^{2/3}(4 - x^2)$; $[-3, 4]$

63. $g(x) = (x - 3)^{5/3}(x + 2)$; $[-4, 4]$

64. $f(t) = 3t/(t^2 + 1)$; $[-2, 2]$

65. $h(x) = (5 - x)/(x^2 + 2x - 3)$; $[-10, 10]$

T 66–67. Absolute value functions *Graph the following functions and determine the local and absolute extreme values on the given interval.*

66. $f(x) = |x - 3| + |x + 2|$; $[-4, 4]$

67. $g(x) = |x - 3| - 2|x + 1|$; $[-2, 3]$

Applications

68. **Minimum surface area box** All boxes with a square base and a volume of 50 ft³ have a surface area given by $S(x) = 2x^2 + 200/x$, where x is the length of the sides of the base. Find the absolute minimum of the surface area function. What are the dimensions of the box with minimum surface area?

T 69. **Every second counts** You must get from a point P on the straight shore of a lake to a stranded swimmer who is 50 m from a point Q on the shore that is 50 m from you (see figure). If you can swim at a speed of 2 m/s and run at a speed of 4 m/s, at what point along the shore, x meters from Q, should you stop running and start swimming if you want to reach the swimmer in the minimum time?

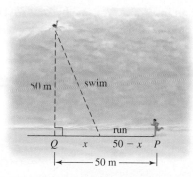

a. Find the function T that gives the travel time as a function of x, where $0 \leq x \leq 50$.

b. Find the critical point of T on $(0, 50)$.

c. Evaluate T at the critical point and the endpoints ($x = 0$ and $x = 50$) to verify that the critical point corresponds to an absolute minimum. What is the minimum travel time?

d. Graph the function T to check your work.

T 70. **Dancing on a parabola** Suppose that two people, A and B, walk along the parabola $y = x^2$ in such a way that the line segment L between them is always perpendicular to the line tangent to the parabola at A's position. What are the positions of A and B when L has minimum length?

a. Assume that A's position is (a, a^2), where $a > 0$. Find the slope of the line tangent to the parabola at A and find the slope of the line that is perpendicular to the tangent line at A.

b. Find the equation of the line joining A and B when A is at (a, a^2).

c. Find the position of B on the parabola when A is at (a, a^2).

d. Write the function $F(a)$ that gives the *square* of the distance between A and B as it varies with a. (The square of the distance is minimized at the same point that the distance is minimized; it is easier to work with the square of the distance.)

e. Find the critical point of F on the interval $a > 0$.

f. Evaluate F at the critical point and verify that it corresponds to an absolute minimum. What are the positions of A and B that minimize the length of L? What is the minimum length?

g. Graph the function F to check your work.

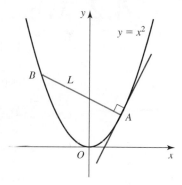

Additional Exercises

71. **Values of related functions** Suppose f is differentiable on $(-\infty, \infty)$ and assume it has a local extreme value at the point $x = 2$ where $f(2) = 0$. Let $g(x) = xf(x) + 1$ and let $h(x) = xf(x) + x + 1$ for all values of x.

a. Evaluate $g(2), h(2), g'(2)$, and $h'(2)$.

b. Does either g or h have a local extreme value at $x = 2$? Explain.

72. **Extreme values of parabolas** Consider the function $f(x) = ax^2 + bx + c$, with $a \neq 0$. Explain geometrically why f has exactly one absolute extreme value on $(-\infty, \infty)$. Find the critical points to determine the value of x at which f has an extreme value.

73. **Even and odd functions**

a. Suppose an even function f has a local minimum at c. Does f have a local maximum or minimum at $-c$? Explain. (An even function satisfies $f(x) = f(-x)$.)

b. Suppose an odd function f has a local minimum at c. Does f have a local maximum or minimum at $-c$? Explain. (An odd function satisfies $f(x) = -f(-x)$.)

74. A family of double-humped functions Consider the functions $f(x) = x/(x^2 + 1)^n$, where n is a positive integer.

   a. Show that these functions are odd for all positive integers n.

   b. Show that the critical points of these functions are $x = \pm\sqrt{\dfrac{1}{2n-1}}$ for all positive integers n. (Start with the special cases $n = 1$ and $n = 2$.)

   c. Show that as n increases the absolute maximum values of these functions decrease.

   d. Use a graphing utility to verify your conclusions.

75. Proof of the Local Extreme Point Theorem Prove Theorem 4.2 for a local maximum: If f has a local maximum at the point c and $f'(c)$ exists, then $f'(c) = 0$. Use the following steps.

   a. If f has a local maximum at c, then what is the sign of $f(x) - f(c)$ if x is near c and $x > c$? What is the sign of $f(x) - f(c)$ if x is near c and $x < c$?

   b. If $f'(c)$ exists, then it is defined by $\displaystyle\lim_{x \to c} \dfrac{f(x) - f(c)}{x - c}$. Examine this limit as $x \to c^+$ and conclude that $f'(c) \le 0$.

   c. Examine the limit in part (b) as $x \to c^-$ and conclude that $f'(c) \ge 0$.

   d. Combine parts (b) and (c) to conclude that $f'(c) = 0$.

QUICK CHECK ANSWERS

1. The continuous function $f(x) = x$ does not have an absolute minimum on the open interval $(0, 1)$. The function $f(x) = -x$ on $\left[0, \frac{1}{2}\right)$ and $f(x) = 0$ on $\left[\frac{1}{2}, 1\right]$ does not have an absolute minimum on $[0, 1]$.

2. The critical point is $x = 0$. Although $f'(0) = 0$, the function has neither a local minimum nor maximum at $x = 0$. ◄

4.2 What Derivatives Tell Us

In the previous section, we saw that the derivative is a tool for finding critical points, which are related to local maxima and minima. As we show in this section, derivatives (first *and* second derivatives) tell us much more about the behavior of functions.

Increasing and Decreasing Functions

We have used the terms *increasing* and *decreasing* informally in earlier sections to describe a function or its graph. For example, the graph in Figure 4.13a rises as x increases, so the corresponding function is increasing. In Figure 4.13b, the graph falls as x increases, so the corresponding function is decreasing. The following definition makes these ideas precise.

> ➤ A function is called **monotonic** if it is either increasing or decreasing. Some books make a further distinction by defining **nondecreasing** ($f(x_2) \ge f(x_1)$ whenever $x_2 > x_1$) and **nonincreasing** ($f(x_2) \le f(x_1)$ whenever $x_2 > x_1$).

DEFINITION Increasing and Decreasing Functions

Suppose a function f is defined on an interval I. We say that f is **increasing** on I if $f(x_2) > f(x_1)$ whenever x_1 and x_2 are in I and $x_2 > x_1$. We say that f is **decreasing** on I if $f(x_2) < f(x_1)$ whenever x_1 and x_2 are in I and $x_2 > x_1$.

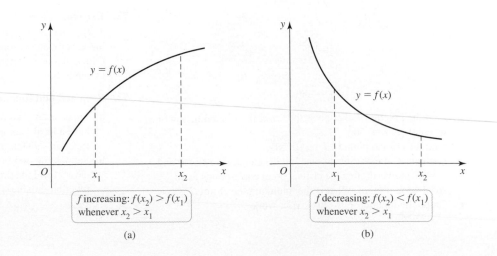

FIGURE 4.13   (a)   (b)

Intervals of Increase and Decrease The graph of a function f gives us an idea of the intervals on which f is increasing and decreasing. But how do we determine those intervals precisely? This question is answered by making a connection to the derivative.

Recall that the derivative of a function gives the slopes of tangent lines. If the derivative is positive on an interval, the tangent lines on that interval have positive slopes, and the function is increasing on the interval (Figure 4.14a). Said differently, positive derivatives on an interval imply positive rates of change on the interval, which, in turn, indicate an increase in function values.

Similarly, if the derivative is negative on an interval, the tangent lines on that interval have negative slopes, and the function is decreasing on that interval (Figure 4.14b). These observations are proved in Section 4.6 using a result called the Mean Value Theorem.

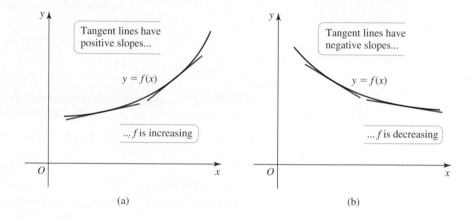

FIGURE 4.14

(a)   (b)

> The converse of Theorem 4.3 may not be true. According to the definition, $f(x) = x^3$ is increasing on $(-\infty, \infty)$, but it is not true that $f'(x) > 0$ on $(-\infty, \infty)$ (because $f'(0) = 0$).

THEOREM 4.3 Test for Intervals of Increase and Decrease

Suppose f is continuous on an interval I and differentiable at every interior point of I. If $f'(x) > 0$ at all interior points of I, then f is increasing on I. If $f'(x) < 0$ at all interior points of I, then f is decreasing on I.

QUICK CHECK 1 Explain why a positive derivative on an interval implies that the function is increasing on the interval. ◄

EXAMPLE 1 Sketching a function Sketch a function f continuous on its domain $(-\infty, \infty)$ satisfying the following conditions:

1. $f' > 0$ on $(-\infty, 0)$, $(4, 6)$, and $(6, \infty)$
2. $f' < 0$ on $(0, 4)$
3. $f'(0)$ is undefined
4. $f'(4) - f'(6) = 0$

SOLUTION By condition (1), f is increasing on the intervals $(-\infty, 0)$, $(4, 6)$, and $(6, \infty)$. By condition (2), f is decreasing on $(0, 4)$. Condition (3) implies f has a cusp or corner at $x = 0$, and by condition (4), the graph has a horizontal tangent line at $x = 4$ and $x = 6$. It is useful to summarize these results (Figure 4.15) before sketching a graph. One of many possible graphs satisfying these conditions is shown in Figure 4.16.

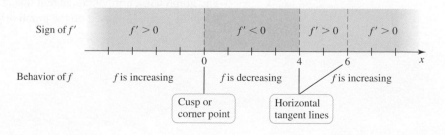

FIGURE 4.15

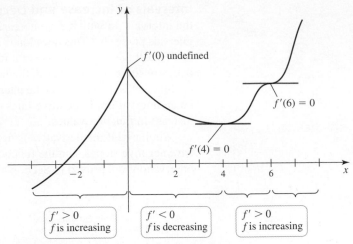

FIGURE 4.16

Related Exercises 11–16 ◀

EXAMPLE 2   Intervals of increase and decrease Find the intervals on which the following functions are increasing and decreasing.

a. $f(x) = xe^{-x}$ **b.** $f(x) = 2x^3 + 3x^2 + 1$

SOLUTION

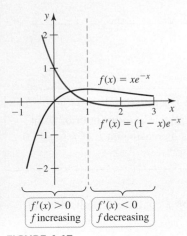

FIGURE 4.17

a. By the Product Rule, $f'(x) = e^{-x} + x(-e^{-x}) = (1 - x)e^{-x}$. Solving $f'(x) = 0$ and noting that $e^{-x} \neq 0$ for all x, the sole critical point is $x = 1$. Therefore, if f' changes sign, then it does so at $x = 1$ and nowhere else. By evaluating f' at a selected point in $(-\infty, 1)$ and $(1, \infty)$, we can determine the sign of f' on the entire interval:

- At $x = 0, f'(0) = 1 > 0$. So, $f' > 0$ on $(-\infty, 1)$, which means that f is increasing on $(-\infty, 1)$. (In fact, f is increasing on $(-\infty, 1]$.)
- At $x = 2, f'(2) = -e^{-2} < 0$. So $f' < 0$ on $(1, \infty)$, which means that f is decreasing on $(1, \infty)$. (In fact, f is decreasing on $[1, \infty)$.)

Note also that the graph has a horizontal tangent line at $x = 1$. We verify these conclusions by plotting f and f' (Figure 4.17).

b. In this case, $f'(x) = 6x^2 + 6x = 6x(x + 1)$. To find the intervals of increase, we first solve $6x(x + 1) = 0$ and determine that the critical points are $x = 0$ and $x = -1$. If f' changes sign, then it does so at these points and nowhere else; that is, f' has the same sign throughout each of the intervals $(-\infty, -1)$, $(-1, 0)$, and $(0, \infty)$. Evaluating f' at selected points of each interval determines the sign of f' on that interval.

- At $x = -2, f'(-2) = 12 > 0$, so $f' > 0$ and f is increasing on $(-\infty, -1)$.
- At $x = -\frac{1}{2}, f'\left(-\frac{1}{2}\right) = -\frac{3}{2} < 0$, so $f' < 0$ and f is decreasing on $(-1, 0)$.
- At $x = 1, f'(1) = 12 > 0$, so $f' > 0$ and f is increasing on $(0, \infty)$.

The graph has a horizontal tangent line at $x = -1$ and $x = 0$. Figure 4.18 shows the graph of f superimposed on the graph of f', confirming our conclusions.

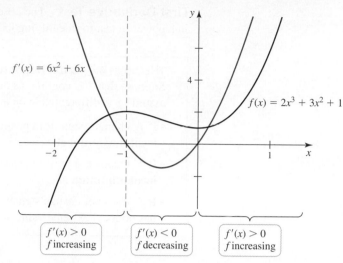

FIGURE 4.18

$$f'(x) > 0 \quad\quad f'(x) < 0 \quad\quad f'(x) > 0$$
$$f \text{ increasing} \quad f \text{ decreasing} \quad f \text{ increasing}$$

Related Exercises 17–30 ◄

Identifying Local Maxima and Minima

Using what we know about increasing and decreasing functions, we can now identify local extrema. Suppose $x = c$ is a critical point of f, where $f'(c) = 0$. Suppose also that f' changes sign at c with $f'(x) < 0$ on an interval (a, c) to the left of c and $f'(x) > 0$ on an interval (c, b) to the right of c. In this case f' is decreasing to the left of c and increasing to the right of c, which means that f has a local minimum at c, as shown in Figure 4.19a.

Similarly, suppose f' changes sign at c with $f'(x) > 0$ on an interval (a, c) to the left of c and $f'(x) < 0$ on an interval (c, b) to the right of c. Then, f is increasing to the left of c and decreasing to the right of c, so f has a local maximum at c, as shown in Figure 4.19b.

Figure 4.20 shows typical features of a function on an interval $[a, b]$. At local maxima or minima (c_2, c_3, and c_4), f' changes sign. Although c_1 and c_5 are critical points, f' does not change sign at these points, so there is no local maximum or minimum at these points. As emphasized earlier, *critical points do not always correspond to local extreme values.*

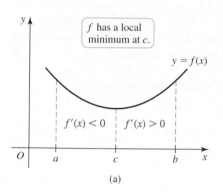

(a)

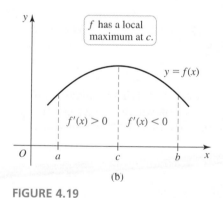

(b)

FIGURE 4.19

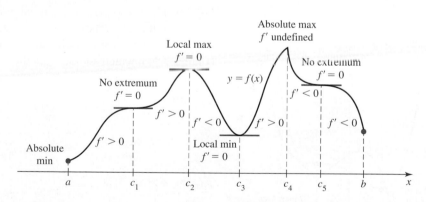

FIGURE 4.20

QUICK CHECK 2 Sketch a function f that is differentiable on $(-\infty, \infty)$ with the following properties: (i) $x = 0$ and $x = 2$ are critical points; (ii) f is increasing on $(-\infty, 2)$; (iii) f is decreasing on $(2, \infty)$. ◄

First Derivative Test   The observations used to interpret Figure 4.20 are summarized in a powerful test for identifying local maxima and minima.

THEOREM 4.4   First Derivative Test
Suppose that f is continuous on an interval that contains a critical point c and assume f is differentiable on an interval containing c, except perhaps at c itself.

- If f' changes sign from positive to negative as x increases through c, then f has a **local maximum** at c.

- If f' changes sign from negative to positive as x increases through c, then f has a **local minimum** at c.

- If f' does not change sign at c (from positive to negative or vice versa), then f has no local extreme value at c.

Proof   Suppose that $f'(x) > 0$ on an interval (a, c), which means that f is increasing on (a, c), which, in turn, implies that $f(x) < f(c)$ for all x in (a, c). Similarly, suppose that $f'(x) < 0$ on an interval (c, b), which means that f is decreasing on (c, b), which, in turn, implies that $f(x) < f(c)$ for all x in (c, b). Therefore, $f(x) \leq f(c)$ for all x in (a, b) and f has a local maximum at c. The proofs of the other two cases are similar.   ◄

EXAMPLE 3   Using the First Derivative Test   Consider the function
$$f(x) = 3x^4 - 4x^3 - 6x^2 + 12x + 1.$$

a. Find the intervals on which f is increasing and decreasing.

b. Identify the local extrema of f.

SOLUTION

a. Differentiating f, we find that
$$\begin{aligned} f'(x) &= 12x^3 - 12x^2 - 12x + 12 \\ &= 12(x^3 - x^2 - x + 1) \\ &= 12(x + 1)(x - 1)^2 \end{aligned}$$

Solving $f'(x) = 0$ gives the critical points $x = -1$ and $x = 1$. The critical points determine the intervals $(-\infty, -1)$, $(-1, 1)$, and $(1, \infty)$ on which f' does not change sign. Choosing a test point in each interval, a sign graph of f' is constructed (Figure 4.21), which summarizes the behavior of f.

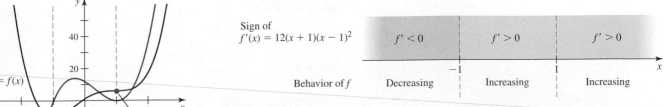

Sign of
$f'(x) = 12(x + 1)(x - 1)^2$

| $f' < 0$ | $f' > 0$ | $f' > 0$ |

Behavior of f Decreasing   Increasing   Increasing

FIGURE 4.21

b. Because f' changes sign from negative to positive as x passes through the critical point $x = -1$, it follows by the First Derivative Test that f has a local minimum value of $f(-1) = -10$ at $x = -1$. As x passes through $x = 1$, f does not change sign, so f does not have a local extreme value at the critical point $x = 1$ (Figure 4.22).

Related Exercises 31–38 ◄

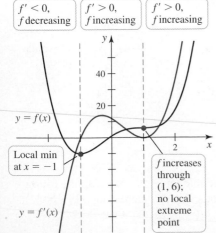

$f' < 0,$ $f' > 0,$ $f' > 0,$
f decreasing f increasing f increasing

$y = f(x)$

Local min
at $x = -1$

$y = f'(x)$

f increases through $(1, 6)$; no local extreme point

FIGURE 4.22

EXAMPLE 4 **Extreme points** Find the local extrema of the function $f(x) = x^{2/3}(2 - x)$.

SOLUTION In Example 4b of Section 4.1, we found that

$$f'(x) = \frac{4}{3}x^{-1/3} - \frac{5}{3}x^{2/3} = \frac{4 - 5x}{3x^{1/3}}$$

and that the critical points of f are $x = 0$ and $x = \frac{4}{5}$. These two critical points are *candidates* for local extrema, and Theorem 4.4 is used to classify each as a local minimum, local maximum, or neither.

Using Figure 4.23, we see that f has a local minimum at $x = 0$ and a local maximum at $x = \frac{4}{5}$. These observations are confirmed by the graphs of f and f' (Figure 4.24).

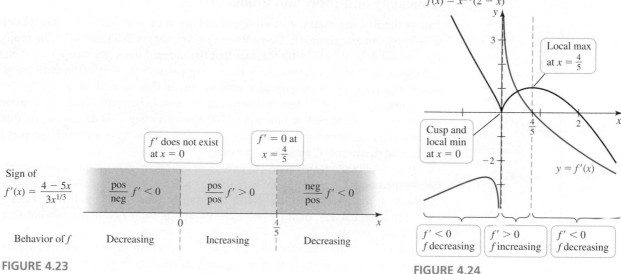

Sign of
$f'(x) = \dfrac{4 - 5x}{3x^{1/3}}$

$\dfrac{\text{pos}}{\text{neg}}\ f' < 0$ f' does not exist at $x = 0$ $\dfrac{\text{pos}}{\text{pos}}\ f' > 0$ $f' = 0$ at $x = \frac{4}{5}$ $\dfrac{\text{neg}}{\text{pos}}\ f' < 0$

Behavior of f Decreasing   Increasing   Decreasing

FIGURE 4.23

$f(x) = x^{2/3}(2 - x)$

Local max at $x = \frac{4}{5}$

Cusp and local min at $x = 0$

$y = f'(x)$

$f' < 0$ f decreasing $f' > 0$ f increasing $f' < 0$ f decreasing

FIGURE 4.24

Related Exercises 31–38 ◀

QUICK CHECK 3 Explain how the First Derivative Test determines whether $f(x) = x^2$ has a local maximum or local minimum at $x = 0$. ◀

Absolute Extreme Values on Any Interval Theorem 4.1 guarantees the existence of absolute extreme values only on closed intervals. What can be said about absolute extrema on intervals that are not closed? The following theorem provides a valuable test.

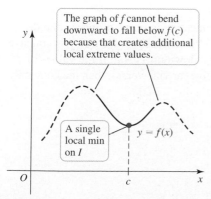

The graph of f cannot bend downward to fall below $f(c)$ because that creates additional local extreme values.

A single local min on I

$y = f(x)$

FIGURE 4.25

THEOREM 4.5 **One Local Extremum Implies Absolute Extremum**
Suppose f is continuous on an interval I that contains one local extremum at c.

- If a local minimum occurs at c, then $f(c)$ is the absolute minimum of f on I.
- If a local maximum occurs at c, then $f(c)$ is the absolute maximum of f on I.

The proof of Theorem 4.5 is beyond the scope of this text, although Figure 4.25 illustrates why the theorem is plausible. Assume f has exactly one local minimum on I at c. Notice that there is no other point on the graph at which f has a value less than $f(c)$. If such a point did exist, the graph would have to bend downward to drop below $f(c)$. Because f is continuous, this cannot happen as it implies additional local extreme values on I. A similar argument applies to a solitary local maximum.

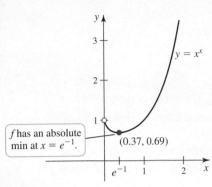

f has an absolute min at $x = e^{-1}$.

(0.37, 0.69)

FIGURE 4.26

EXAMPLE 5   Finding an absolute extremum   Verify that $f(x) = x^x$ has an absolute extreme value on its domain.

SOLUTION   First note that f is continuous on its domain $(0, \infty)$. Because $f(x) = x^x = e^{x \ln x}$, it follows that

$$f'(x) = e^{x \ln x}(1 + \ln x) = x^x(1 + \ln x).$$

Solving $f'(x) = 0$ gives a single critical point $x = e^{-1}$; there is no point in the domain at which $f'(x)$ does not exist. The critical point splits the domain of f into the intervals $(0, e^{-1})$ and (e^{-1}, ∞). Evaluating the sign of f' on each interval gives $f'(x) < 0$ on $(0, e^{-1})$ and $f'(x) > 0$ on (e^{-1}, ∞); therefore, by Theorem 4.4, a local minimum occurs at $x = e^{-1}$. Because it is the only local extremum on $(0, \infty)$, it follows from Theorem 4.5 that the absolute minimum of f occurs at $x = e^{-1}$ (Figure 4.26). Its value is $f(e^{-1}) \approx 0.69$.   *Related Exercises 39–42* ◄

Concavity and Inflection Points

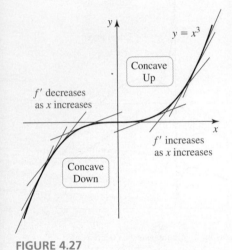

FIGURE 4.27

Just as the first derivative is related to the slope of tangent lines, the second derivative also has geometric meaning. Consider $f(x) = x^3$, shown in Figure 4.27. Its graph bends upward for $x > 0$, reflecting the fact that the tangent lines get steeper as x increases. It follows that the first derivative is increasing for $x > 0$. A function with the property that f' is increasing on an interval is **concave up** on that interval.

Similarly, $f(x) = x^3$ bends downward for $x < 0$ because it has a decreasing first derivative on that interval. A function with the property that f' is decreasing as x increases on an interval is **concave down** on that interval. We now have a useful interpretation of the second derivative: It measures *concavity*.

Here is another useful characterization of concavity. If a function is concave up at a point (any point $x > 0$ in Figure 4.27), then its graph near that point lies *above* the tangent line at that point. Similarly, if a function is concave down at a point (any point $x < 0$ in Figure 4.27), then its graph near that point lies *below* the tangent line at that point (Exercise 84).

Finally, imagine a function that changes concavity (from up to down, or vice versa) at a point c. For example, $f(x) = x^3$ in Figure 4.27 changes from concave down to concave up as x passes through $x = 0$. A point on the graph of f at which f changes concavity is called an **inflection point**.

DEFINITION   Concavity and Inflection Point

Let f be differentiable on an open interval I. If f' is increasing on I, then f is **concave up** on I. If f' is decreasing on I, then f is **concave down** on I.

If f is continuous at c and f changes concavity at c (from up to down, or vice versa), then f has an **inflection point** at c.

Applying Theorem 4.3 to f' leads to a test for concavity in terms of the second derivative. Specifically, if $f'' > 0$ on an interval I, then f' is increasing on I and f is concave up on I. Similarly, if $f'' < 0$ on I, then f is concave down on I. And if the values of f'' pass through zero at a point c (from positive to negative, or vice versa), then the concavity of f changes at c and f has an inflection point at c (Figure 4.28a).

> **THEOREM 4.6   Test for Concavity**
> Suppose that f'' exists on an interval I.
>
> - If $f'' > 0$ on I, then f is concave up on I.
> - If $f'' < 0$ on I, then f is concave down on I.
> - If c is a point of I at which $f''(c) = 0$ and f'' changes sign at c, then f has an inflection point at c.

There are a few important but subtle points here. The fact that $f''(c) = 0$ does not necessarily imply that f has an inflection point at c. A good example is $f(x) = x^4$. Although $f''(0) = 0$, the concavity does not change at $x = 0$ (a similar function is shown in Figure 4.28b).

Typically, if f has an inflection point at c, then $f''(c) = 0$, reflecting the smooth change in concavity. However, an inflection point may also occur at a point where f'' does not exist. For example, the function $f(x) = x^{1/3}$ has a vertical tangent line and an inflection point at $x = 0$ (a similar function is shown in Figure 4.28c).

▶ The function shown in Figure 4.28d, with behavior similar to that of $f(x) = x^{2/3}$, does not have an inflection point at c and $f''(c)$ does not exist.

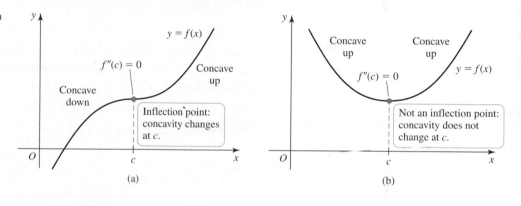

(a)

(b)

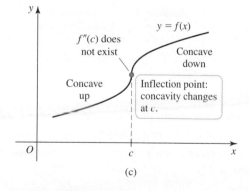

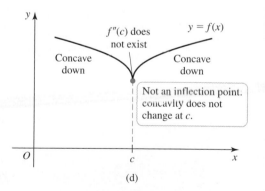

FIGURE 4.28

(c)

(d)

QUICK CHECK 4 Verify that the function $f(x) = x^4$ is concave up for $x > 0$ and for $x < 0$. Is $x = 0$ an inflection point? Explain. ◀

EXAMPLE 6   Interpreting concavity Sketch a function satisfying each set of conditions for $-\infty < t < \infty$.

a. $f'(t) > 0$ and $f''(t) > 0$ **b.** $g'(t) > 0$ and $g''(t) < 0$

c. Would you rather have f or g as a function representing the market value of a house that you own?

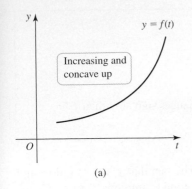

Increasing and concave up

$y = f(t)$

(a)

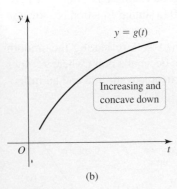

Increasing and concave down

$y = g(t)$

(b)

FIGURE 4.29

SOLUTION

a. Figure 4.29a shows the graph of a function that is increasing ($f'(t) > 0$) and concave up ($f''(t) > 0$).

b. Figure 4.29b shows the graph of a function that is increasing ($g'(t) > 0$) and concave down ($g''(t) < 0$).

c. Because f increases at an *increasing* rate and g increases at a *decreasing* rate, f would be a better function for the value of your house.   *Related Exercises 43–46* ◄

EXAMPLE 7   Detecting concavity Identify the intervals on which the following functions are concave up or concave down. Then locate the inflection points.

a. $f(x) = 3x^4 - 4x^3 - 6x^2 + 12x + 1$ **b.** $f(x) = \sin^{-1} x$ on $(-1, 1)$

SOLUTION

a. This function was considered in Example 3, where we found that

$$f'(x) = 12(x + 1)(x - 1)^2.$$

It follows that

$$f''(x) = 12(x - 1)(3x + 1).$$

We see that $f''(x) = 0$ at $x = 1$ and $x = -\frac{1}{3}$. These points are *candidates* for inflection points, and it must be determined whether the concavity changes at these points. The sign graph in Figure 4.30 shows the following:

- $f''(x) > 0$ and f is concave up on $\left(-\infty, -\frac{1}{3}\right)$ and $(1, \infty)$.
- $f''(x) < 0$ and f is concave down on $\left(-\frac{1}{3}, 1\right)$.

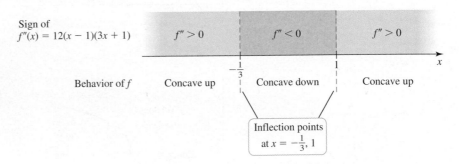

Sign of
$f''(x) = 12(x - 1)(3x + 1)$

$f'' > 0$ $f'' < 0$ $f'' > 0$

$-\frac{1}{3}$ 1 x

Behavior of f Concave up   Concave down   Concave up

Inflection points
at $x = -\frac{1}{3}, 1$

FIGURE 4.30

We see that the sign of f'' changes at $x = -\frac{1}{3}$ and at $x = 1$, so the concavity of f also changes at these points. Therefore, inflection points occur at $x = -\frac{1}{3}$ and $x = 1$. The graphs of f and f'' (Figure 4.31) show that the concavity of f changes at the zeros of f''.

b. The first derivative of $f(x) = \sin^{-1} x$ is $f'(x) = 1/\sqrt{1 - x^2}$. We use the Chain Rule to compute the second derivative:

$$f''(x) = -\frac{1}{2}(1 - x^2)^{-3/2} \cdot (-2x) = \frac{x}{(1 - x^2)^{3/2}}$$

The only zero of f'' is $x = 0$, and because its denominator is positive on $(-1, 1)$, f'' changes sign at $x = 0$ from negative to positive. Therefore, f is concave down on $(-1, 0)$ and concave up on $(0, 1)$, with an inflection point at $x = 0$ (Figure 4.32).

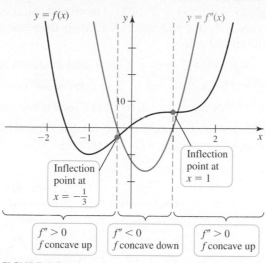

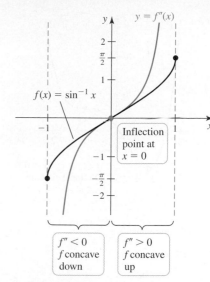

FIGURE 4.31

FIGURE 4.32

Related Exercises 47–56 ◀

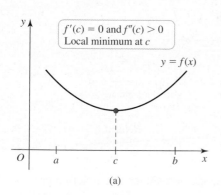

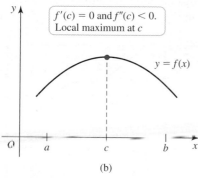

FIGURE 4.33

> In the inconclusive case of Theorem 4.7 where $f''(c) = 0$, it is usually best to use the First Derivative Test.

Second Derivative Test It is now a short step to a test that uses the second derivative to identify local maxima and minima (Figure 4.33).

THEOREM 4.7   Second Derivative Test for Local Extrema
Suppose that f'' is continuous on an open interval containing c with $f'(c) = 0$.

- If $f''(c) > 0$, then f has a local minimum at c.
- If $f''(c) < 0$, then f has a local maximum at c.
- If $f''(c) = 0$, then the test is inconclusive.

Proof Because $f''(c) > 0$ and f'' is continuous on an interval I containing c, it follows that $f'' > 0$ on I and f' is increasing on I. Because $f'(c) = 0$, it follows that f' changes sign at c from negative to positive, which, by the First Derivative Test, implies that f has a local minimum at c. The proofs of the other two cases are similar.   ◀

QUICK CHECK 5 Make a sketch of a function with $f'(x) > 0$ and $f''(x) > 0$ on an interval. Make a sketch of a function with $f'(x) < 0$ and $f''(x) < 0$ on an interval. ◀

EXAMPLE 8   The Second Derivative Test Use the Second Derivative Test to locate the local extrema of the following functions.

a. $f(x) = 3x^4 - 4x^3 - 6x^2 + 12x + 1$ on $[-2, 2]$ **b.** $f(x) = \sin^2 x$

SOLUTION

a. This function was considered in Examples 3 and 7, where we found that

$$f'(x) = 12(x + 1)(x - 1)^2 \quad \text{and} \quad f''(x) = 12(x - 1)(3x + 1).$$

Therefore, the critical points of f are $x = -1$ and $x = 1$. Evaluating f'' at the critical points, we find that $f''(-1) = 48 > 0$. By the Second Derivative Test, f has a local minimum at $x = -1$. At the other critical point, $f''(1) = 0$, so the test is inconclusive. You can check that the first derivative does not change sign at $x = 1$, which means f does not have a local maximum or minimum at $x = 1$ (Figure 4.34).

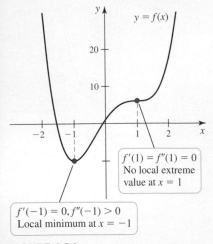

$f'(1) = f''(1) = 0$
No local extreme value at $x = 1$

$f'(-1) = 0, f''(-1) > 0$
Local minimum at $x = -1$

FIGURE 4.34

b. Using the Chain Rule and a trigonometric identity, we have $f'(x) = 2 \sin x \cos x = \sin 2x$ and $f''(x) = 2 \cos 2x$. The critical points occur when $f'(x) = \sin 2x = 0$, or when $x = 0, \pm\pi/2, \pm\pi, \ldots$. To apply the Second Derivative Test, we evaluate f'' at the critical points:

- $f''(0) = 2 > 0$, so f has a local minimum at $x = 0$.
- $f''(\pm\pi/2) = -2 < 0$, so f has a local maximum at $x = \pm\pi/2$.
- $f''(\pm\pi) = 2 > 0$, so f has a local minimum at $x = \pm\pi$.

This pattern continues, and we see that f has alternating local maxima and minima, evenly spaced every $\pi/2$ units (Figure 4.35).

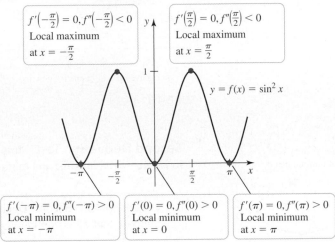

$f'\left(-\frac{\pi}{2}\right) = 0, f''\left(-\frac{\pi}{2}\right) < 0$
Local maximum at $x = -\frac{\pi}{2}$

$f'\left(\frac{\pi}{2}\right) = 0, f''\left(\frac{\pi}{2}\right) < 0$
Local maximum at $x = \frac{\pi}{2}$

$y = f(x) = \sin^2 x$

$f'(-\pi) = 0, f''(-\pi) > 0$
Local minimum at $x = -\pi$

$f'(0) = 0, f''(0) > 0$
Local minimum at $x = 0$

$f'(\pi) = 0, f''(\pi) > 0$
Local minimum at $x = \pi$

FIGURE 4.35

Related Exercises 57–62 ◄

Recap of Derivative Properties

This section has demonstrated that the first and second derivatives of a function provide valuable information about its graph. The relationships among a function's derivatives, and its extreme points and concavity are summarized in Figure 4.36.

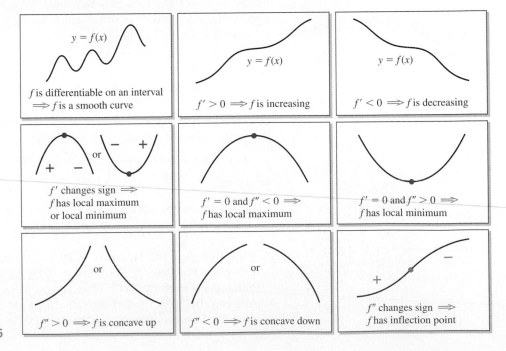

$y = f(x)$

f is differentiable on an interval $\Longrightarrow f$ is a smooth curve

$y = f(x)$

$f' > 0 \Longrightarrow f$ is increasing

$y = f(x)$

$f' < 0 \Longrightarrow f$ is decreasing

or

f' changes sign $\Longrightarrow$ f has local maximum or local minimum

$f' = 0$ and $f'' < 0 \Longrightarrow$ f has local maximum

$f' = 0$ and $f'' > 0 \Longrightarrow$ f has local minimum

or

$f'' > 0 \Longrightarrow f$ is concave up

or

$f'' < 0 \Longrightarrow f$ is concave down

f'' changes sign $\Longrightarrow$ f has inflection point

FIGURE 4.36

SECTION 4.2 EXERCISES

Review Questions

1. Explain how the first derivative of a function determines where the function is increasing and decreasing.

2. Explain how to apply the First Derivative Test.

3. Sketch the graph of a function that has neither a local maximum nor a local minimum at a point where $f'(x) = 0$.

4. Explain how to apply the Second Derivative Test.

5. Assume that f is twice differentiable at c and that f has a local maximum at c. Explain why $f''(c) < 0$.

6. Sketch a function that changes from concave up to concave down as x increases. Describe how the second derivative of this function changes.

7. What is an inflection point?

8. Sketch the graph of a function that does not have an inflection point at a point where $f''(x) = 0$.

9. Is it possible for a function to satisfy $f(x) > 0$, $f'(x) > 0$, and $f''(x) > 0$ on an interval? Explain.

10. Suppose f is continuous on an interval containing a critical point c and $f''(c) = 0$. How do you determine if f has a local extreme value at $x = c$?

Basic Skills

11–14. Sketches from properties *Sketch a function that is continuous on* $(-\infty, \infty)$ *and has the following properties. Use a number line to summarize information about the function.*

11. $f'(x) < 0$ on $(-\infty, 2)$; $f'(x) > 0$ on $(2, 5)$; $f'(x) < 0$ on $(5, \infty)$

12. $f'(-1)$ is undefined; $f'(x) > 0$ on $(-\infty, -1)$; $f'(x) < 0$ on $(-1, \infty)$

13. $f(0) = f(4) = f'(0) = f'(2) = f'(4) = 0$; $f(x) \geq 0$ on $(-\infty, \infty)$

14. $f'(-2) = f'(2) = f'(6) = 0$; $f'(x) \geq 0$ on $(-\infty, \infty)$

15–16. Functions from derivatives *The following figures give the graph of the derivative of a continuous function f that passes through the origin. Sketch a graph of f on the same set of axes.*

15.

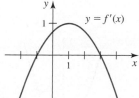

16.

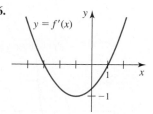

17–22. Increasing and decreasing functions *Find the intervals on which f is increasing and decreasing. Superimpose the graphs of f and f' to verify your work.*

17. $f(x) = 4 - x^2$ 18. $f(x) = x^2 - 16$ 19. $f(x) = (x - 1)^2$

20. $f(x) = x^3 + 4x$

21. $f(x) = 12 + x - x^2$

22. $f(x) = x^4 - 4x^3 + 4x^2$

23–30. Increasing and decreasing functions *Find the intervals on which f is increasing and decreasing.*

23. $f(x) = 3 \cos 3x$ on $[-\pi, \pi]$ 24. $f(x) = \cos^2 x$ on $[-\pi, \pi]$

25. $f(x) = x^{4/3}$ 26. $f(x) = x^2\sqrt{x + 5}$

27. $f(x) = \tan^{-1} x$ 28. $f(x) = \ln|x|$

29. $f(x) = -12x^5 + 75x^4 - 80x^3$

30. $f(x) = x^2 - 2 \ln x$

31–38. First Derivative Test

   a. Locate the critical points of the given function.

   b. Use the First Derivative Test to locate the local maximum and minimum values.

   c. Identify the absolute minimum and maximum values of the function on the given interval (when they exist).

31. $f(x) = x^2 + 3$; $[-3, 2]$

32. $f(x) = -x^2 - x + 2$; $[-4, 4]$

33. $f(x) = x\sqrt{9 - x^2}$; $[-3, 3]$

34. $f(x) = 2x^3 + 3x^2 - 12x + 1$; $[-2, 4]$

35. $f(x) = x^{2/3}(x - 4)$; $[-5, 5]$

36. $f(x) = \dfrac{x^2}{x^2 - 1}$; $[-4, 4]$

37. $f(x) = \sqrt{x} \ln x$; $(0, \infty)$

38. $f(x) = \tan^{-1} x - x^3$; $[-1, 1]$

39–42. Absolute extreme values *Verify that the following functions satisfy the conditions of Theorem 4.5 on their domains. Then, find the location and value of the absolute extrema guaranteed by the theorem.*

39. $f(x) = xe^{-x}$ 40. $f(x) = 4x + 1/\sqrt{x}$

41. $A(r) = 24/r + 2\pi r^2$, $r > 0$ 42. $f(x) = x\sqrt{3 - x}$

43–46. Sketching curves *Sketch a graph of the function f that is continuous on $(-\infty, \infty)$ and has the following properties.*

43. $f'(x) > 0, f''(x) > 0$

44. $f'(x) < 0$ and $f''(x) > 0$ on $(-\infty, 0)$; $f'(x) > 0$ and $f''(x) > 0$ on $(0, \infty)$

45. $f'(x) < 0$ and $f''(x) < 0$ on $(-\infty, 0)$; $f'(x) < 0$ and $f''(x) > 0$ on $(0, \infty)$

46. $f'(x) < 0$ and $f''(x) > 0$ on $(-\infty, 0)$; $f'(x) < 0$ and $f''(x) < 0$ on $(0, \infty)$

47–56. Concavity *Determine the intervals on which the following functions are concave up or concave down. Identify any inflection points.*

47. $f(x) = 5x^4 - 20x^3 + 10$ 48. $f(x) = -\dfrac{1}{1 + x^2}$

49. $g(t) = \ln(3t^2 + 1)$

50. $g(x) = \sqrt[3]{x - 4}$

51. $f(x) = e^{-x^2/2}$

52. $f(x) = \tan^{-1} x$

53. $f(x) = \sqrt{x} \ln x$

54. $h(t) = 2 + \cos 2t$ for $-\pi \le t \le \pi$

55. $g(t) = 3t^5 - 30t^4 + 80t^3 + 100$

56. $f(x) = 2x^4 + 8x^3 + 12x^2 - x - 2$

57–62. Second Derivative Test *Locate the critical points of the following functions. Then, use the Second Derivative Test to determine whether they correspond to local minima or local maxima or whether the test is inconclusive.*

57. $f(x) = 4 - x^2$

58. $g(x) = x^3 - 6$

59. $f(x) = 2x^3 - 3x^2 + 12$

60. $p(x) = (x - 4)/(x^2 + 20)$

61. $f(x) = x^2 e^{-x}$

62. $g(x) = x^4/2 - 12x^2$

Further Explorations

63. Explain why or why not Determine whether the following statements are true and give an explanation or counterexample.

   a. If $f'(x) > 0$ and $f''(x) < 0$ on an interval, then f is increasing at a decreasing rate.

   b. If $f'(c) > 0$ and $f''(c) = 0$, then f has a local maximum at c.

   c. Two functions that differ by a constant increase and decrease on the same intervals.

   d. If f and g increase on an interval, then the product fg also increases on that interval.

   e. There exists a function f that is continuous on $(-\infty, \infty)$ with exactly three critical points, all of which correspond to local maxima.

64–65. Functions from derivatives *Consider the following graphs of f' and f''. On the same set of axes, sketch the graph of a possible function f. The graphs of f are not unique.*

64.

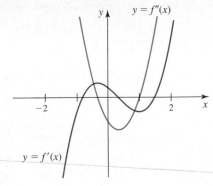

65.

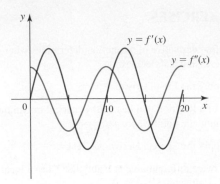

66. Is it possible? Determine whether the following properties can be satisfied by a function that is continuous on $(-\infty, \infty)$. If such a function is possible, provide an example or a sketch of the function. If such a function is not possible, explain why.

   a. A function f is concave down and positive everywhere.

   b. A function f is increasing and concave down everywhere.

   c. A function f has exactly two local extrema and three inflection points.

   d. A function f has exactly four zeros and two local extrema.

67. Matching derivatives and functions The following figures show the graphs of three functions (graphs a–c). Match each function with its first derivative (graphs d–f) *and* its second derivative (graphs g–i).

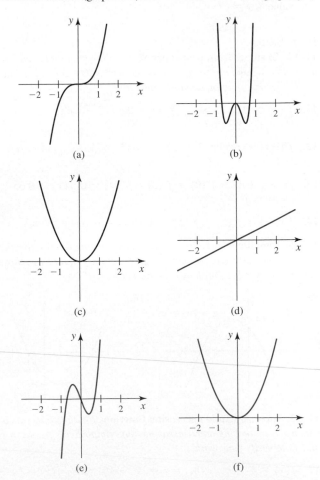

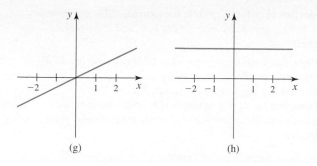

(g)   (h)

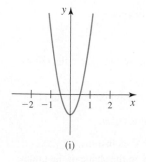

(i)

68. Graphical analysis The accompanying figure shows the graphs of f, f', and f''. Which curve is which?

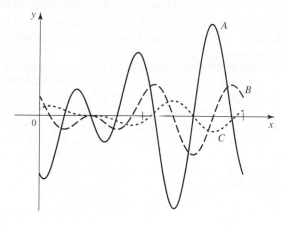

69. Sketching graphs Sketch the graph of a function f continuous on $[a, b]$ such that f, f', and f'' have the signs indicated in the following table on $[a, b]$. There are eight different cases lettered A–H and eight different graphs.

Case	A	B	C	D	E	F	G	H
f	+	+	+	+	−	−	−	−
f'	+	+	−	−	+	+	−	−
f''	+	−	+	−	+	−	+	−

70–73. Designer functions *Sketch the graph of a function that is continuous on $(-\infty, \infty)$ and satisfies the following sets of conditions.*

70. $f''(x) > 0$ on $(-\infty, -2)$; $f''(-2) = 0$; $f'(-1) = f'(1) = 0$; $f''(2) = 0$; $f'(3) = 0$; $f''(x) > 0$ on $(4, \infty)$

71. $f(-2) = f''(-1) = 0$; $f'\left(-\frac{3}{2}\right) = 0$; $f(0) = f'(0) = 0$; $f(1) = f'(1) = 0$

72. $f(x) > f'(x) > 0$ for all x; $f''(1) = 0$

73. $f''(x) > 0$ on $(-\infty, -2)$; $f''(x) < 0$ on $(-2, 1)$; $f''(x) > 0$ on $(1, 3)$; $f''(x) < 0$ on $(3, \infty)$

74. Strength of concavity The functions $f(x) = ax^2$, where $a > 0$, are concave up for all x. Graph these functions for $a = 1, 5,$ and 10, and discuss how the concavity varies with a. How does a change the appearance of the graph?

75. Interpreting the derivative The graph of f' on the interval $[-3, 2]$ is shown in the figure.

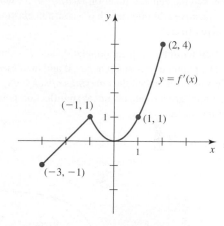

a. On what interval(s) is f increasing? Decreasing?
b. Find the critical points of f. Which critical points correspond to local maxima? Local minima? Neither?
c. At what points does f have an inflection point?
d. On what intervals is f concave up? Concave down?
e. Sketch the graph of f''.
f. Sketch one possible graph of f.

76–79. Second Derivative Test *Locate the critical points of the following functions and use the Second Derivative Test to determine whether they correspond to local minima, local maxima, or neither.*

76. $p(t) = 2t^3 + 3t^2 - 36t$

77. $f(x) = \dfrac{x^4}{4} - \dfrac{5x^3}{3} - 4x^2 + 48x$

78. $h(x) = (x + a)^4$, a constant

79. $f(x) = x^3 + 2x^2 + 4x - 1$

80. Concavity of parabolas Consider the general parabola described by the function $f(x) = ax^2 + bx + c$. For what values of $a, b,$ and c is f concave up? For what values of $a, b,$ and c is f concave down?

Applications

81. Demand functions and elasticity Economists use *demand functions* to describe how much of a commodity can be sold at varying prices. For example, the demand function $D(p) = 500 - 10p$ says that at a price of $p = 10$, a quantity of $D(10) = 400$ units of the commodity can be sold. The elasticity $E = \dfrac{dD}{dp} \dfrac{p}{D}$ of the demand gives the approximate percent change in the demand for every 1% change in the price. (See the Guided Projects for more on demand functions and elasticity.)

a. Compute the elasticity of the demand function $D(p) = 500 - 10p$.

b. If the price is $12 and increases by 4.5%, what is the approximate percent change in the demand?

c. Show that for the linear demand function $D(p) = a - bp$, where a and b are positive real numbers, the elasticity is a decreasing function for $p \geq 0$ and $p \neq a/b$.

d. Show that the demand function $D(p) = a/p^b$, where a and b are positive real numbers, has a constant elasticity for all positive prices.

82. Population models A typical population curve is shown in the figure. The population is small at $t = 0$ and increases toward a steady-state level called the *carrying capacity*. Explain why the maximum growth rate occurs at an inflection point of the population curve.

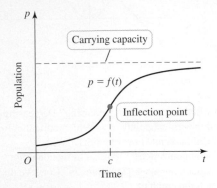

83. Population models The population of a species is given by the function $P(t) = \dfrac{Kt^2}{t^2 + b}$, where $t \geq 0$ is measured in years and K and b are positive real numbers.

a. With $K = 300$ and $b = 30$, what is $\lim_{t \to \infty} P(t)$, the carrying capacity of the population?

b. With $K = 300$ and $b = 30$, when does the maximum growth rate occur?

c. For arbitrary positive values of K and b, when does the maximum growth rate occur (in terms of K and b)?

Additional Exercises

84. Tangent lines and concavity Give an argument to support the claim that if a function is concave up at a point, then the tangent line lies below the curve near that point.

85. Symmetry of cubics Consider the general cubic polynomial $f(x) = x^3 + ax^2 + bx + c$, where a, b, and c are real numbers.

a. Show that f has exactly one inflection point and it occurs at $x^* = -a/3$.

b. Show that f is an odd function with respect to the inflection point $(x^*, f(x^*))$. This means that $f(x^*) - f(x^* + x) = f(x^* - x) - f(x^*)$ for all x.

86. Properties of cubics Consider the general cubic polynomial $f(x) = x^3 + ax^2 + bx + c$, where a, b, and c are real numbers.

a. Prove that f has exactly one local minimum and one local maximum provided that $a^2 > 3b$.

b. Prove that f has no extreme values if $a^2 < 3b$.

c. Prove that for all real values of a, b, and c, the function has exactly one inflection point. Where is the inflection point located?

■ 87. A family of single-humped functions Consider the functions $f(x) = \dfrac{1}{x^{2n} + 1}$, where n is a positive integer.

a. Show that these functions are even.

b. Show that the graphs of these functions intersect at the points $\left(\pm 1, \frac{1}{2}\right)$ for all positive values of n.

c. Show that the inflection points of these functions occur at $x = \pm \sqrt[2n]{\dfrac{2n - 1}{2n + 1}}$ for all positive values of n.

d. Use a graphing utility to verify your conclusions.

e. Describe how the inflection points and the shape of the graph change as n increases.

88. Even quartics Consider the quartic (fourth-degree) polynomial $f(x) = x^4 + bx^2 + d$ consisting only of even-powered terms.

a. Show that the graph of f is symmetric about the y-axis.

b. Show that if $b \geq 0$, then f has one critical point and no inflection points.

c. Show that if $b < 0$, then f has three critical points and two inflection points. Find the critical points and inflection points, and show that they alternate along the x-axis. Explain why one critical point is always $x = 0$.

d. Prove that the number of distinct real roots of f depends on the values of the coefficients b and d, as shown in the figure. The curve that divides the plane is the parabola $d = b^2/4$.

e. Find the number of real roots when $b = 0$ or $d = 0$ or $d = b^2/4$.

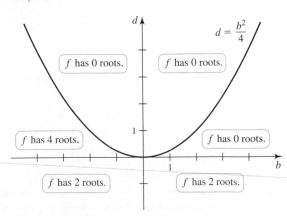

89. General quartic Show that the general quartic (fourth-degree) polynomial $f(x) = x^4 + ax^3 + bx^2 + cx + d$ has either zero or two inflection points, and the latter case occurs provided that $b < 3a^2/8$.

1. Positive derivatives on an interval mean the curve is rising on the interval, which means the function is increasing on the interval.   **2.** The graph of f rises for $x < 0$. At $x = 0$, the graph flattens out momentarily, then continues to rise for $0 < x < 2$. There is a local maximum at $x = 2$ and f is decreasing for $x > 2$.

3. $f'(x) < 0$ on $(-\infty, 0)$ and $f'(x) > 0$ on $(0, \infty)$. Therefore, f has a local minimum at $x = 0$ by the First Derivative Test.   **4.** $f''(x) = 12x^2$, so $f''(x) > 0$ for $x < 0$ and for $x > 0$. There is no inflection point at $x = 0$ because the second derivative does not change sign.
5. The first curve should be rising and concave up. The second curve should be falling and concave down.

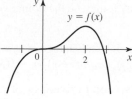

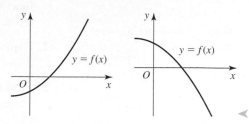

4.3 Graphing Functions

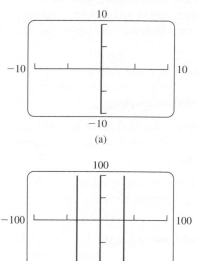

FIGURE 4.37

Over the span of three chapters, we have collected the tools required for a comprehensive approach to graphing functions. These *analytical methods* are indispensable, even with the availability of powerful graphing utilities, as illustrated by the following example.

Calculators and Analysis

Suppose you want to graph the harmless-looking function $f(x) = x^3/3 - 400x$. If you plot f using a typical graphing utility with a default window of $[-10, 10] \times [-10, 10]$, the resulting graph is shown in Figure 4.37a; one vertical line appears on the screen. Zooming out to the window $[-100, 100] \times [-100, 100]$ produces three vertical lines (Figure 4.37b) which is not an accurate graph of the function. Expanding the window even more to $[-1000, 1000] \times [-1000, 1000]$ is no better. So, what do we do?

QUICK CHECK 1 Try to graph $f(x) = x^3/3 - 400x$ by using various windows on a graphing utility. Can you find a window that gives a better graph of f than those in Figure 4.37? ◄

Like most functions, $f(x) = x^3/3 - 400x$ has a reasonable graph, but it cannot be found automatically by letting technology do all the work. Here is the message of this section: Graphing utilities are valuable for exploring functions, producing preliminary graphs, and checking your work. But they should not be relied on exclusively because they cannot explain *why* a graph has its shape. Rather, graphing utilities should be used in an interactive way with the analytical methods presented in this chapter.

Graphing Guidelines

The following set of guidelines need not be followed exactly for every function, and you will find that several steps can often be done at once. Depending on the specific problem, some of the steps are best done analytically, while other steps can be done with a graphing utility. Experiment with both approaches and try to find a good balance. We also present a schematic record-keeping procedure to keep track of discoveries as they are made.

> The precise order of these steps may vary from one problem to another.

Graphing Guidelines for $y = f(x)$

1. **Identify the domain or interval of interest.** On what interval should the function be graphed? It may be the domain of the function or some subset of the domain.

2. **Exploit symmetry.** Take advantage of symmetry. For example, is the function *even* $(f(-x) = f(x))$, *odd* $(f(-x) = -f(x))$, or neither?

3. **Find the first and second derivatives.** They are needed to determine extreme values, concavity, inflection points, and intervals of increase and decrease. Computing derivatives—particularly second derivatives—may not be practical, so some functions may need to be graphed without complete derivative information.

4. **Find critical points and possible inflection points.** Determine points at which $f'(x) = 0$ or f' is undefined. Determine points at which $f''(x) = 0$ or f'' is undefined.

5. **Find intervals on which the function is increasing/decreasing and concave up/down.** The first derivative determines the intervals of increase and decrease. The second derivative determines the intervals on which the function is concave up and concave down.

6. **Identify extreme values and inflection points.** Use either the First or the Second Derivative Test to classify the critical points. Both *x*- and *y*-coordinates of maxima, minima, and inflection points are needed for graphing.

7. **Locate vertical/horizontal asymptotes and determine end behavior.** Vertical asymptotes often occur at zeros of denominators. Horizontal asymptotes require examining limits as $x \to \pm\infty$ and determine end behavior.

8. **Find the intercepts.** The *y*-intercept of the graph is found by setting $x = 0$. The *x*-intercepts are the real zeros (or roots) of a function: those values of *x* that satisfy $f(x) = 0$.

9. **Choose an appropriate graphing window and make a graph.** Use the results of the above steps to graph the function. If you use graphing software, check for consistency with your analytical work. Is your graph *complete*—that is, does it show all the essential details of the function?

EXAMPLE 1   A warm-up Given the following information about the first and second derivatives of a function f continuous on $(-\infty, \infty)$, summarize the information using a number line, and then sketch a possible graph of f.

$$f' < 0, f'' > 0 \text{ on } (-\infty, 0) \qquad f' > 0, f'' > 0 \text{ on } (0, 1) \qquad f' > 0, f'' < 0 \text{ on } (1, 2)$$
$$f' < 0, f'' < 0 \text{ on } (2, 3) \qquad f' < 0, f'' > 0 \text{ on } (3, 4) \qquad f' > 0, f'' > 0 \text{ on } (4, \infty)$$

SOLUTION We illustrate the given information on a number line. For example, on the interval $(-\infty, 0)$, f is decreasing and concave up; so we sketch a segment of a curve with these properties on this interval (Figure 4.38). Continuing in this manner, we obtain a useful summary of the properties of f.

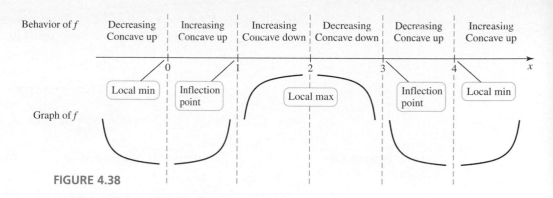

Behavior of f | Decreasing Concave up | Increasing Concave up | Increasing Concave down | Decreasing Concave down | Decreasing Concave up | Increasing Concave up

Local min | Inflection point | Local max | Inflection point | Local min

Graph of f

FIGURE 4.38

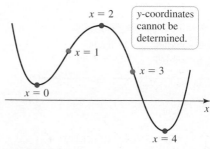

FIGURE 4.39

Assembling the information shown in Figure 4.38, a rough graph of f is produced (Figure 4.39). Notice that derivative information is not sufficient to determine the y-coordinates of points on the curve.   *Related Exercises 7–8* ◀

> **QUICK CHECK 2** Explain why the function f and $f + C$, where C is a constant, have the same derivative properties. ◀

EXAMPLE 2   A deceptive polynomial Use the graphing guidelines to graph $f(x) = \dfrac{x^3}{3} - 400x$ on its domain.

SOLUTION

1. **Domain** The domain of any polynomial is $(-\infty, \infty)$.

2. **Symmetry** Because f consists of odd powers of the variable, it is an odd function. Its graph is symmetric about the origin.

3. **Derivatives** The derivatives of f are

$$f'(x) = x^2 - 400 \quad \text{and} \quad f''(x) = 2x.$$

> Notice that the first derivative of an odd polynomial is an even polynomial and the second derivative is an odd polynomial.

4. **Critical points and possible inflection points** Solving $f'(x) = 0$, we find that the critical points are $x = \pm 20$. Solving $f''(x) = 0$, we see that a possible inflection point occurs at $x = 0$.

> See Appendix A for solving inequalities using test values.

5. **Increasing/decreasing and concavity** Note that

$$f'(x) = x^2 - 400 = (x - 20)(x + 20).$$

Solving the inequality $f'(x) < 0$, we find that f is decreasing on the interval $(-20, 20)$. Solving the inequality $f'(x) > 0$ reveals that f is increasing on the intervals $(-\infty, -20)$ and $(20, \infty)$ (Figure 4.40). By the First Derivative Test, we have enough information to conclude that f has a local maximum at $x = -20$ and a local minimum at $x = 20$.

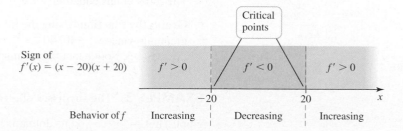

Critical points

Sign of $f'(x) = (x - 20)(x + 20)$ $f' > 0$ | $f' < 0$ | $f' > 0$

-20 20 x

Behavior of f Increasing | Decreasing | Increasing

FIGURE 4.40

Furthermore, $f''(x) = 2x < 0$ on $(-\infty, 0)$, so f is concave down on this interval. Also, $f''(x) > 0$ on $(0, \infty)$, so f is concave up on $(0, \infty)$ (Figure 4.41).

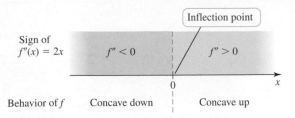

FIGURE 4.41

The evidence obtained so far is summarized in Figure 4.42.

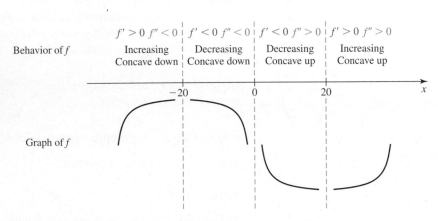

FIGURE 4.42

6. **Extreme values and inflection points** In this case, the Second Derivative Test is easily applied and it confirms what we have already learned. Because $f''(-20) < 0$ and $f''(20) > 0$, f has a local maximum at $x = -20$ and a local minimum at $x = 20$. The corresponding function values are $f(-20) = 16{,}000/3 = 5333\frac{1}{3}$ and $f(20) = -f(-20) = -5333\frac{1}{3}$. Finally, we see that f'' changes sign at $x = 0$, making $(0, 0)$ an inflection point.

7. **Asymptotes and end behavior** Polynomials have neither vertical nor horizontal asymptotes. Because the highest-power term in the polynomial is x^3 (an odd power) and the leading coefficient is positive, we have the end behavior

$$\lim_{x \to \infty} f(x) = \infty \quad \text{and} \quad \lim_{x \to -\infty} f(x) = -\infty.$$

8. **Intercepts** The y-intercept is $(0, 0)$. We solve the equation $f(x) = 0$ to find the x-intercepts:

$$\frac{x^3}{3} - 400x = x\left(\frac{x^2}{3} - 400\right) = 0$$

The roots of this equation are $x = 0$ and $x = \pm\sqrt{1200} \approx \pm 34.6$.

9. **Graph the function** Using the information found in Steps 1–8, we choose the graphing window $[-40, 40] \times [-6000, 6000]$ and produce the graph shown in Figure 4.43. Notice that the symmetry detected in Step 2 is evident in this graph.

Related Exercises 9–14 ◀

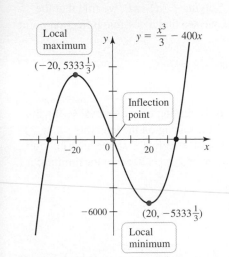

FIGURE 4.43

EXAMPLE 3 **The surprises of a rational function** Use the graphing guidelines to graph $f(x) = \dfrac{10x^3}{x^2 - 1}$ on its domain.

SOLUTION

1. **Domain** The zeros of the denominator are $x = \pm 1$, so the domain is $\{x : x \neq \pm 1\}$.

2. **Symmetry** This function consists of an odd function divided by an even function. The product or quotient of an even function and an odd function is odd. Therefore, the graph is symmetric about the origin.

3. **Derivatives** The Quotient Rule is used to find the first and second derivatives:

$$f'(x) = \frac{10x^2(x^2 - 3)}{(x^2 - 1)^2} \quad \text{and} \quad f''(x) = \frac{20x(x^2 + 3)}{(x^2 - 1)^3}.$$

4. **Critical points and possible inflection points** The solutions of $f'(x) = 0$ occur where the numerator equals 0, provided the denominator is nonzero at those points. Solving $10x^2(x^2 - 3) = 0$ gives the critical points $x = 0$ and $x = \pm\sqrt{3}$. The solutions of $f''(x) = 0$ are found by solving $20x(x^2 + 3) = 0$; we see that the only possible inflection point occurs at $x = 0$.

5. **Increasing/decreasing and concavity** To find the sign of f', first note that the denominator of f' is nonnegative, as is the factor $10x^2$ in the numerator. So, the sign of f' is determined by the factor $x^2 - 3$, which is negative on $(-\sqrt{3}, \sqrt{3})$ and positive on $(-\infty, -\sqrt{3})$ and $(\sqrt{3}, \infty)$. Therefore, f is decreasing on $(-\sqrt{3}, \sqrt{3})$ and increasing on $(-\infty, -\sqrt{3})$ and $(\sqrt{3}, \infty)$.

    The sign of f'' is a bit trickier. Because $x^2 + 3$ is positive, the sign of f'' is determined by the sign of x in the numerator and $(x^2 - 1)^3$ in the denominator. When x and $(x^2 - 1)^3$ have the same sign, $f''(x) > 0$; when x and $(x^2 - 1)^3$ have opposite signs, $f''(x) < 0$ (Table 4.1). The results of this analysis are shown in Figure 4.44.

Table 4.1

	$20x$	$x^2 + 3$	$(x^2 - 1)^3$	**Sign of f''**
$(-\infty, -1)$	$-$	$+$	$+$	$-$
$(-1, 0)$	$-$	$+$	$-$	$+$
$(0, 1)$	$+$	$+$	$-$	$-$
$(1, \infty)$	$+$	$+$	$+$	$+$

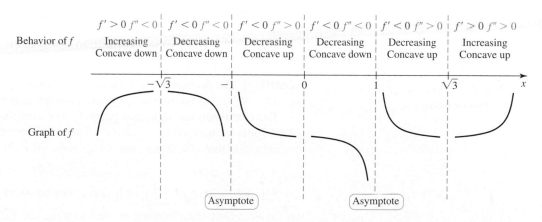

FIGURE 4.44

6. **Extreme values and inflection points** The First Derivative Test is easily applied by looking at Figure 4.44. The function is increasing on $(-\infty, -\sqrt{3})$ and decreasing on $(-\sqrt{3}, \sqrt{3})$; therefore, f has a local maximum at $x = -\sqrt{3}$, where $f(-\sqrt{3}) = -15\sqrt{3}$. Similarly, f has a local minimum at $x = \sqrt{3}$, where $f(\sqrt{3}) = 15\sqrt{3}$. (These results could also be obtained with the Second Derivative Test.) There is no local extreme value at the critical point $x = 0$, only a horizontal tangent line.

   Using the calculations of Step 5, we see that f'' changes sign at $x = \pm 1$ and at $x = 0$. The points $x = \pm 1$ are not in the domain of f, so they cannot correspond to inflection points. However, there is an inflection point at $(0, 0)$.

7. **Asymptotes and end behavior** Recall from Section 2.4 that zeros of the denominator, which in this case are $x = \pm 1$, are candidates for vertical asymptotes. Checking the sign of f on either side of $x = \pm 1$, we find

$$\lim_{x \to -1^-} f(x) = -\infty \qquad \lim_{x \to -1^+} f(x) = +\infty$$

$$\lim_{x \to 1^-} f(x) = -\infty \qquad \lim_{x \to 1^+} f(x) = +\infty$$

It follows that f has vertical asymptotes at $x = \pm 1$. The degree of the numerator is greater than the degree of the denominator, so there are no horizontal asymptotes. It can be shown that f has the slant asymptote (see Section 2.5) $y = 10x$.

8. **Intercepts** The zeros of a rational function coincide with the zeros of the numerator, provided that those points are not also zeros of the denominator. In this case the zeros of f satisfy $10x^3 = 0$, or $x = 0$ (which is not a zero of the denominator). Therefore, $(0, 0)$ is both the x- and y-intercept.

9. **Graphing** We now assemble an accurate graph of f, as shown in Figure 4.45. A window of $[-3, 3] \times [-40, 40]$ gives a complete graph of the function. Notice that the symmetry about the origin deduced in Step 2 is apparent in the graph. *Related Exercises 15–20*◄

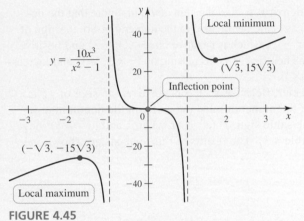

$y = \dfrac{10x^3}{x^2 - 1}$

Local minimum

$(\sqrt{3}, 15\sqrt{3})$

Inflection point

$(-\sqrt{3}, -15\sqrt{3})$

Local maximum

FIGURE 4.45

QUICK CHECK 3 Verify that the function f in Example 3 is symmetric about the origin by showing that $f(-x) = -f(x)$.◄

In the next two examples, we show how the guidelines may be streamlined to some extent.

> The function $f(x) = e^{-x^2}$ and the family of functions $f(x) = ce^{-ax^2}$ are central to the study of statistics. They have bell-shaped curves and describe Gaussian or normal distributions.

EXAMPLE 4 The normal distribution Analyze the function $f(x) = e^{-x^2}$ and draw its graph.

SOLUTION The domain of f is all real numbers and $f(x) > 0$ for all x. Because $f(-x) = f(x)$, it is an even function and its graph is symmetric about the y-axis.

Extreme points and inflection points follow from the derivatives of f. Using the Chain Rule, we have $f'(x) = -2xe^{-x^2}$. The critical points satisfy $f'(x) = 0$, which has the single root $x = 0$ (because $e^{-x^2} > 0$ for all x). It now follows that

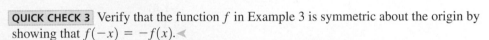

• $f'(x) > 0$ for $x < 0$, so f is increasing on $(-\infty, 0)$.

• $f'(x) < 0$ for $x > 0$, so f is decreasing on $(0, \infty)$.

By the First Derivative Test, we see that f has a local maximum (and an absolute maximum by Theorem 4.5) at $x = 0$ where $f(0) = 1$.

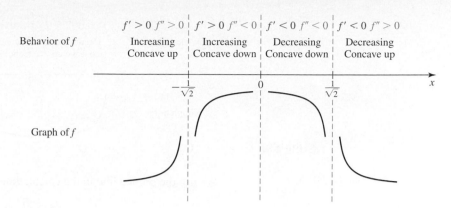

FIGURE 4.46

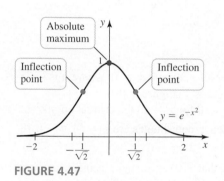

FIGURE 4.47

Differentiating $f'(x) = -2xe^{-x^2}$ with the Product Rule yields

$$f''(x) = e^{-x^2}(-2) + (-2x)(-2xe^{-x^2}) \quad \text{Product Rule}$$
$$= 2e^{-x^2}(2x^2 - 1) \quad \text{Simplify.}$$

Again using the fact that $e^{-x^2} > 0$ for all x, we see that $f''(x) = 0$ when $2x^2 - 1 = 0$ or when $x = \pm 1/\sqrt{2}$; these values are candidates for inflection points. Observe that $f''(x) > 0$ and f is concave up on $(-\infty, -1/\sqrt{2})$ and $(1/\sqrt{2}, \infty)$, while $f''(x) < 0$ and f is concave down on $(-1/\sqrt{2}, 1/\sqrt{2})$. Because f'' changes sign at $x = \pm 1/\sqrt{2}$, we have inflection points at $(\pm 1/\sqrt{2}, 1/\sqrt{e})$ (Figure 4.46).

To determine the end behavior, notice that $\lim\limits_{x \to \pm\infty} e^{-x^2} = 0$, so $y = 0$ is a horizontal asymptote of f. Assembling all of these facts, an accurate graph can now be drawn (Figure 4.47). *Related Exercises 21–36*◄

EXAMPLE 5 **Roots and cusps** Graph $f(x) = \frac{1}{8}x^{2/3}(9x^2 - 8x - 16)$ on its domain.

SOLUTION The domain of f is $(-\infty, \infty)$. The polynomial factor in f consists of both even and odd powers, so f has no special symmetry. Computing the first derivative is straightforward if you first expand f as a sum of three terms:

$$f'(x) = \frac{d}{dx}\left(\frac{9x^{8/3}}{8} - x^{5/3} - 2x^{2/3}\right) \quad \text{Expand } f.$$

$$= 3x^{5/3} - \frac{5}{3}x^{2/3} - \frac{4}{3}x^{-1/3} \quad \text{Differentiate.}$$

$$= \frac{(x-1)(9x+4)}{3x^{1/3}} \quad \text{Simplify.}$$

The critical points are now identified: f' is undefined at $x = 0$ (because $x^{-1/3}$ is undefined there) and $f'(x) = 0$ at $x = 1$ and $x = -\frac{4}{9}$. So we have three critical points to analyze. Table 4.2 tracks the signs of the three factors in f' and shows the sign of f' on the relevant intervals; this information is recorded in Figure 4.48.

Table 4.2

	$\dfrac{x^{-1/3}}{3}$	$9x + 4$	$x - 1$	Sign of f'
$\left(-\infty, -\frac{4}{9}\right)$	−	−	−	−
$\left(-\frac{4}{9}, 0\right)$	−	+	−	+
$(0, 1)$	+	+	−	−
$(1, \infty)$	+	+	+	+

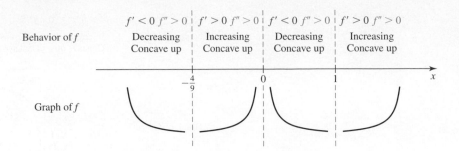

Behavior of f

$f' < 0 \; f'' > 0$	$f' > 0 \; f'' > 0$	$f' < 0 \; f'' > 0$	$f' > 0 \; f'' > 0$
Decreasing Concave up	Increasing Concave up	Decreasing Concave up	Increasing Concave up

Graph of f

FIGURE 4.48

We use the second line in the calculation of f' to compute the second derivative:

$$f''(x) = \frac{d}{dx}\left(3x^{5/3} - \frac{5}{3}x^{2/3} - \frac{4}{3}x^{-1/3}\right)$$

$$= 5x^{2/3} - \frac{10}{9}x^{-1/3} + \frac{4}{9}x^{-4/3} \qquad \text{Differentiate.}$$

$$= \frac{45x^2 - 10x + 4}{9x^{4/3}} \qquad \text{Simplify.}$$

Solving $f''(x) = 0$, we discover that $f''(x) > 0$ for all x, except $x = 0$, where it is undefined. Therefore, f is concave up on $(-\infty, 0)$ and $(0, \infty)$ (Figure 4.48).

By the Second Derivative Test, because $f''(x) > 0$ for $x \neq 0$, the critical points $x = -\frac{4}{9}$ and $x = 1$ correspond to local minima; their y-coordinates are $f\left(-\frac{4}{9}\right) \approx -0.78$ and $f(1) = -\frac{15}{8} = -1.875$.

What about the third critical point $x = 0$? Note that $f(0) = 0$, and f is increasing just to the left of 0 and decreasing just to the right. By the First Derivative Test, f has a local maximum at $x = 0$. Furthermore, $f'(x) \to \infty$ as $x \to 0^-$ and $f'(x) \to -\infty$ as $x \to 0^+$, so the graph of f has a cusp at $x = 0$.

As $x \to \pm\infty$, f is dominated by its highest-power term, which is $9x^{8/3}/8$. This term becomes large and positive as $x \to \pm\infty$; therefore, f has no absolute maximum. Its absolute minimum occurs at $x = 1$ because, comparing the two local minima,

$$f(1) < f\left(-\frac{4}{9}\right).$$

The roots of f satisfy $\frac{1}{8}x^{2/3}(9x^2 - 8x - 16) = 0$, which gives $x = 0$ and

$$x = \frac{4}{9}(1 \pm \sqrt{10}) \approx -0.96 \quad \text{or} \quad 1.85 \quad \text{Use the quadratic formula.}$$

With the information gathered in this analysis, we obtain the graph shown in Figure 4.49.

Related Exercises 21–36 ◀

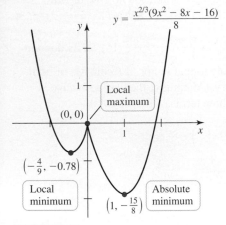

$$y = \frac{x^{2/3}(9x^2 - 8x - 16)}{8}$$

Local maximum

$(0, 0)$

$\left(-\frac{4}{9}, -0.78\right)$

Local minimum

Absolute minimum

$\left(1, -\frac{15}{8}\right)$

FIGURE 4.49

SECTION 4.3 EXERCISES

Review Questions

1. Why is it important to determine the domain of f before graphing f?

2. Explain why it is useful to know about symmetry in a function.

3. Can the graph of a polynomial have vertical or horizontal asymptotes? Explain.

4. Where are the vertical asymptotes of a rational function located?

5. How do you find the absolute maximum and minimum values of a function that is continuous on a closed interval?

6. Describe the possible end behavior of a polynomial.

Basic Skills

7–8. Shape of the curve *Sketch a curve with the following properties.*

7. $x < 3 \qquad f' < 0$ and $f'' < 0$

   $x > 3 \qquad f' < 0$ and $f'' > 0$

8. $x < -1 \qquad\quad f' < 0$ and $f'' < 0$

   $-1 < x < 2 \qquad f' < 0$ and $f'' > 0$

   $2 < x < 8 \qquad\; f' > 0$ and $f'' > 0$

   $8 < x < 10 \qquad f' > 0$ and $f'' < 0$

   $x > 10 \qquad\quad\; f' > 0$ and $f'' > 0$

9–14. Graphing polynomials *Sketch a graph of the following polynomials. Identify local extrema, inflection points, and x- and y-intercepts when they exist.*

9. $f(x) = \frac{1}{3}x^3 - 2x^2 - 5x + 2$ **10.** $f(x) = \frac{1}{15}x^3 - x + 1$

11. $f(x) = x^4 - 6x^2$ **12.** $f(x) = 2x^6 - 3x^4$

13. $f(x) = 3x^4 + 4x^3 - 12x^2$

14. $f(x) = x^3 - 33x^2 + 216x - 2$

15–20. Graphing rational functions *Use the guidelines of this section to make a complete graph of f.*

15. $f(x) = \dfrac{x^2}{x - 2}$ **16.** $f(x) = \dfrac{x^2}{x^2 - 4}$

17. $f(x) = \dfrac{3x - 5}{x^2 - 1}$ **18.** $f(x) = \dfrac{2x - 3}{2x - 8}$

19. $f(x) = \dfrac{x^2 + 12}{2x + 1}$ **20.** $f(x) = \dfrac{4x + 4}{x^2 + 3}$

21–32. More graphing *Make a complete graph of the following functions. If an interval is not specified, graph the function on its domain. Use a graphing utility to check your work.*

21. $f(x) = x + 2\cos x$ on $[-2\pi, 2\pi]$ **22.** $f(x) = x^{1/3}(x - 2)^2$

23. $f(x) = \sin x - x$ on $[0, 2\pi]$ **24.** $f(x) = x\sqrt{x + 4}$

25. $g(t) = e^{-t}\sin t$ on $[-\pi, \pi]$ **26.** $g(x) = x^2 \ln x$

27. $f(x) = x + \tan x$ on $\left(-\dfrac{3\pi}{2}, \dfrac{3\pi}{2}\right)$ **28.** $f(x) = (\ln x)/x^2$

29. $f(x) = x \ln x$ **30.** $g(x) = e^{-x^2/2}$

31. $p(x) = xe^{-x^2}$ **32.** $g(x) = 1/(e^{-x} - 1)$

33–36. Graphing with technology *Make a complete graph of the following functions. A graphing utility is useful in locating local extreme values and inflection points.*

33. $f(x) = \dfrac{\tan^{-1} x}{x^2 + 1}$ **34.** $f(x) = \dfrac{\sqrt{4x^2 + 1}}{x^2 + 1}$

35. $f(x) = \dfrac{x \sin x}{x^2 + 1}$ on $[-2\pi, 2\pi]$ **36.** $f(x) = x/\ln x$

Further Explorations

37. Explain why or why not Determine whether the following statements are true and give an explanation or counterexample.

  a. The zeros of f' are $x = -3, 1$, and 4, so the local extrema are located at these points.

  b. The zeros of f'' are $x = -2$ and 4, so the inflection points are also located at these points.

  c. The zeros of the denominator of f are $x = -3$ and 4, so f has vertical asymptotes at these points.

  d. If a rational function has a finite limit as $x \to \infty$, it must have a finite limit as $x \to -\infty$.

38–41. Functions from derivatives *Use the derivative f' to determine the local minima and maxima of f and the intervals of increase and decrease. Sketch a possible graph of f (f is not unique).*

38. $f'(x) = (x - 1)(x + 2)(x + 4)$

39. $f'(x) = 10 \sin 2x$ on $[-2\pi, 2\pi]$

40. $f'(x) = \dfrac{x - 1}{(x - 2)^2(x - 3)}$

41. $f'(x) = \dfrac{x + 2}{x^2(x - 6)}$

42–43. Functions from graphs *Use the graphs of f' and f'' to find the critical points and inflection points of f, the intervals on which f is increasing and decreasing, and the intervals of concavity. Then, graph f assuming $f(0) = 0$.*

42.

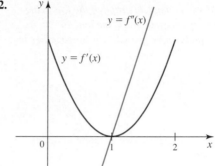

43.

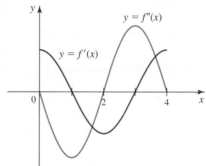

44–47. Nice cubics and quartics *The following third- and fourth-degree polynomials have a property that makes them relatively easy to graph. Make a complete graph and describe the property.*

44. $f(x) = x^4 + 8x^3 - 270x^2 + 1$

45. $f(x) = x^3 - 6x^2 - 135x$

46. $f(x) = x^3 - 147x + 286$

47. $f(x) = x^3 - 3x^2 - 144x - 140$

48. Oscillations Consider the function $f(x) = \cos(\ln x)$ for $x > 0$. Use analytical techniques and a graphing utility.

  a. Locate all local extrema on the interval $(0, 4]$.

  b. Identify the inflection points on the interval $(0, 4]$.

  c. Locate the three smallest zeros of f on the interval $(0.1, \infty)$.

  d. Sketch the graph of f.

49. Local max/min of $x^{1/x}$ Use analytical methods to find all local extreme points of the function $f(x) = x^{1/x}$ for $x > 0$. Verify your work using a graphing utility.

50. Local max/min of x^x Use analytical methods to find all local extreme points of the function $f(x) = x^x$ for $x > 0$. Verify your work using a graphing utility.

51–54. Designer functions *Sketch a continuous function f on some interval that has the properties described.*

51. The function f has one inflection point but no local minima or maxima.

52. The function f has three real zeros and exactly two local minima.

53. The function f satisfies $f'(-2) = 2$, $f'(0) = 0$, $f'(1) = -3$, $f'(4) = 1$.

54. The function f has the same finite limit as $x \to \pm\infty$ and has exactly one local minimum and one local maximum.

55–62. More graphing *Make a complete graph of the following functions. If an interval is not specified, graph the function on its domain. Use analytical methods and a graphing utility together in a complementary way.*

55. $f(x) = \dfrac{-x\sqrt{x^2 - 4}}{x - 2}$

56. $f(x) = 3\sqrt[4]{x} - \sqrt{x} - 2$

57. $f(x) = 3x^4 - 44x^3 + 60x^2$ (*Hint:* Two different graphing windows may be needed.)

58. $f(x) = \dfrac{1}{1 + \cos(\pi x)}$ on $(1, 3)$

59. $f(x) = 10x^6 - 36x^5 - 75x^4 + 300x^3 + 120x^2 - 720x$

60. $f(x) = \dfrac{\sin(\pi x)}{1 + \sin(\pi x)}$ on $[0, 2]$ (*Hint:* Two different graphing windows may be needed.)

61. $f(x) = \dfrac{x\sqrt{|x^2 - 1|}}{x^4 + 1}$

62. $f(x) = \sin(3\pi \cos x)$ on $[-\pi/2, \pi/2]$

63. Hidden oscillations Use analytical methods together with a graphing utility to graph the following functions on the interval $[-2\pi, 2\pi]$. Define f at $x = 0$ so that it is continuous there. Be sure to uncover all relevant features of the graph.

a. $f(x) = \dfrac{1 - \cos^3 x}{x^2}$ **b.** $f(x) = \dfrac{1 - \cos^5 x}{x^2}$

64. Cubic with parameters Locate all local maxima and minima of $f(x) = x^3 - 3bx^2 + 3a^2x + 23$, where a and b are constants, in the following cases.

a. $|a| < |b|$ **b.** $|a| > |b|$ **c.** $|a| = |b|$

Applications

65. Height vs. volume The figure shows six containers, each of which is filled from the top. Assume that water is poured into the containers at a constant rate and each container is filled in 10 s. Assume also that the horizontal cross sections of the containers are always circles. Let $h(t)$ be the depth of water in the container at time t for $0 \le t \le 10$.

a. For each container, sketch a graph of the function $y = h(t)$ for $0 \le t \le 10$.

b. Explain why h is an increasing function.

c. Describe the concavity of the function. Identify inflection points when they occur.

d. For each container, where does h' (the derivative of h) have an absolute maximum on $[0, 10]$?

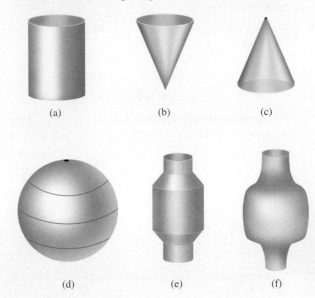

(a)   (b)   (c)

(d)   (e)   (f)

66. A pursuit curve Imagine a man standing 1 mi east of a crossroads. At noon, a dog starts walking north from the crossroads at 1 mi/hr (see figure). At the same instant, the man starts walking and at all times walks directly toward the dog at $s > 1$ mi/hr. The path in the xy-plane followed by the man as he pursues his dog is given by the function

$$y = f(x) = \frac{s}{2}\left(\frac{x^{(s+1)/s}}{s + 1} - \frac{x^{(s-1)/s}}{s - 1}\right) + \frac{s}{s^2 - 1}.$$

Select various values of $s > 1$ and graph this pursuit curve. Comment on the changes in the curve as s increases.

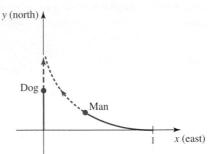

Additional Exercises

67. Derivative information Suppose f is concave up on $(-\infty, 0)$ and $(0, \infty)$. Assume f has a local maximum at $x = 0$. What, if anything, do you know about $f'(0)$? Explain with an illustration.

68. $e^\pi > \pi^e$ Prove that $e^\pi > \pi^e$ by first finding the maximum value of $f(x) = \ln x/x$.

69–75. Special curves *The following classical curves have been studied by generations of mathematicians. Use analytical methods (including implicit differentiation) and a graphing utility to graph the curves described by the following equations. Include as much detail as possible.*

69. $x^{2/3} + y^{2/3} = 1$ Astroid or hypocycloid with four cusps

70. $y = \dfrac{8}{x^2 + 4}$ Witch of Agnesi

71. $x^3 + y^3 = 3xy$ Folium of Descartes

72. $y^2 = \dfrac{x^3}{2 - x}$ Cissoid of Diocles

73. $y^4 - x^4 - 4y^2 + 5x^2 = 0$ Devil's curve

74. $y^2 = x^3(1 - x)$ Pear curve

75. $x^4 - x^2 + y^2 = 0$ Figure-8 curve

76. Elliptic curves The equation $y^2 = x^3 - ax + 3$, where a is a parameter, defines a well-known family of *elliptic curves*.

a. Verify that if $a = 3$, the graph consists of a single curve.
b. Verify that if $a = 4$, the graph consists of two distinct curves.
c. By experimentation, determine the value of a ($3 < a < 4$) at which the graph separates into two curves.

77. Lamé Curves The equation $|y/a|^n + |x/a|^n = 1$, where n and a are positive real numbers, defines the family of Lamé curves. Make a complete graph of this function with $a = 1$ for $n = \dfrac{2}{3}, 1, 2, 3$. Describe the progression that you observe as n increases.

78. An exotic curve (Putnam Exam 1942) Find the coordinates of four local maxima of the function $f(x) = \dfrac{x}{1 + x^6 \sin^2 x}$ and graph the function for $0 \le x \le 10$.

79. A family of superexponential functions Let $f(x) = (a - x)^x$, where $a > 0$.

a. What is the domain of f (in terms of a)?
b. Describe the end behavior of f (near the boundary of its domain).
c. Compute f'. Then graph f and f' for $a = 0.5, 1, 2,$ and 3.
d. Show that f has a single local maximum at the point z that satisfies $z = (a - z) \ln (a - z)$.
e. Describe how z [found in part (d)] varies as a increases. Describe how $f(z)$ varies as a increases.

80. x^y **versus** y^x Consider positive real numbers x and y. Notice that $4^3 < 3^4$, while $3^2 > 2^3$ and $4^2 = 2^4$. Describe the regions in the first quadrant of the xy-plane in which $x^y > y^x$ and $x^y < y^x$.

QUICK CHECK ANSWERS

1. Make the window larger in the y-direction.
2. Notice that f and $f + C$ have the same derivatives.
3. $f(-x) = \dfrac{10(-x)^3}{(-x)^2 - 1} = -\dfrac{10x^3}{x^2 - 1} = -f(x)$ ◄

4.4 Optimization Problems

The theme of this section is *optimization*, a topic arising in many disciplines that rely on mathematics. A structural engineer may seek the dimensions of a beam that maximizes strength for a specified cost. A packaging designer may seek the dimensions of a container that maximizes the capacity of the container for a given surface area. Airline strategists need to find the best allocation of airliners among several hubs in order to minimize fuel costs and maximize passenger miles. In all these examples, the challenge is to find an *efficient* way to carry out a task, where "efficient" could mean least expensive, most profitable, least time consuming, or, as you will see, many other measures.

To introduce the ideas behind optimization problems, think about pairs of nonnegative real numbers x and y between 0 and 20 with the property that their sum is 20, that is, $x + y = 20$. Of all possible pairs, which has the greatest product?

Table 4.3 displays a few cases showing how the product of two nonnegative numbers varies while their sum remains constant. The condition that $x + y = 20$ is called a **constraint**: It tells us to consider only (nonnegative) values of x and y satisfying this equation.

The quantity that we wish to maximize (or minimize in other cases) is called the **objective function**; in this case, the objective function is the product $P = xy$. From Table 4.3 it appears that the product is greatest if both x and y are near the middle of the interval $[0, 20]$.

This simple problem has all the essential features of optimization problems. At their heart, all optimization problems take the following form:

What is the maximum (minimum) value of an objective function subject to the given constraint(s)?

Table 4.3

x	y	$x + y$	$P = xy$
1	19	20	19
5.5	14.5	20	79.75
9	11	20	99
13	7	20	91
18	2	20	36

▷ In this problem it is just as easy to eliminate x as y. In other problems, eliminating one variable may result in less work than eliminating other variables.

For the problem at hand, this question would be stated as "What pair of nonnegative numbers maximizes $P = xy$ subject to the constraint $x + y = 20$?" The first step is to use the constraint to express the objective function $P = xy$ in terms of a single variable. In this case, the constraint is

$$x + y = 20, \quad \text{or} \quad y = 20 - x.$$

Substituting for y, the objective function becomes

$$P = xy = x(20 - x) = 20x - x^2,$$

which is a function of the single variable x. Notice that the values of x lie in the interval $0 \le x \le 20$ with $P(0) = P(20) = 0$.

To maximize P, we first find the critical points by solving

$$P'(x) = 20 - 2x = 0$$

to obtain the solution $x = 10$. To find the absolute maximum value of P on the interval $[0, 20]$, we check the endpoints and the critical points. Because $P(0) = P(20) = 0$ and $P(10) = 100$, we conclude that P has its absolute maximum value at $x = 10$. By the constraint $x + y = 20$, the numbers with the greatest product are $x = y = 10$, and their product is $P = 100$.

Figure 4.50 summarizes this problem. We see the constraint line $x + y = 20$ in the xy-plane. Above the line is the objective function $P = xy$. As x and y vary along the constraint line, the objective function changes, reaching a maximum value of 100 when $x = y = 10$.

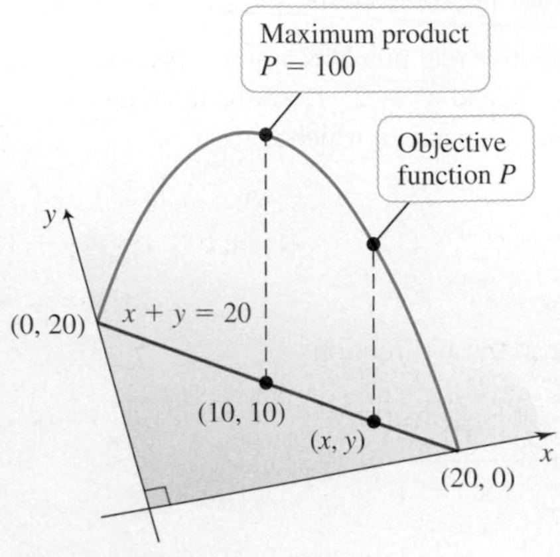

FIGURE 4.50

QUICK CHECK 1 Verify that in the previous example the same result is obtained if the constraint $x + y = 20$ is used to eliminate x rather than y. ◁

Most optimization problems have the same basic structure as the preceding example: There is an objective function, which may involve several variables, and one or more constraints. The methods of calculus (Sections 4.1 and 4.2) are used to find the minimum or maximum values of the objective function.

EXAMPLE 1   Rancher's dilemma A rancher has 400 ft of fence for constructing a rectangular corral. One side of the corral will be formed by a barn and requires no fence. Three exterior fences and two interior fences partition the corral into three rectangular regions. What dimensions of the corral maximize the enclosed area? What is the area of that corral?

SOLUTION We first sketch the corral (Figure 4.51), where x is the width and y is the length of the corral. The amount of fence required is $4x + y$, so the constraint is $4x + y = 400$, or $y = 400 - 4x$.

The objective function to be maximized is the area of the corral, $A = xy$. Using $y = 400 - 4x$, we eliminate y and express A as a function of x:

$$A = xy = x(400 - 4x) = 400x - 4x^2$$

Notice that the width of the corral must be at least $x = 0$, and it cannot exceed $x = 100$ (because 400 ft of fence are available). Therefore, we maximize $A(x) = 400x - 4x^2$ for $0 \le x \le 100$. The critical points of the objective function satisfy

$$A'(x) = 400 - 8x = 0,$$

▷ Recall from Section 4.1 that the absolute extreme points occur at critical points or endpoints.

which has the solution $x = 50$. To find the absolute maximum value of A, we check the endpoints of $[0, 100]$ and the critical point $x = 50$. Because $A(0) = A(100) = 0$ and $A(50) = 10,000$, the absolute maximum value of A occurs when $x = 50$. Using the constraint, the optimal length of the corral is $y = 400 - 4(50) = 200$ ft. Therefore, the

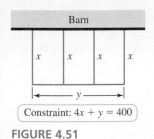

FIGURE 4.51

Constraint: $4x + y = 400$

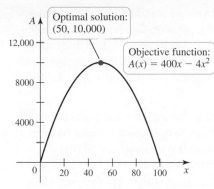

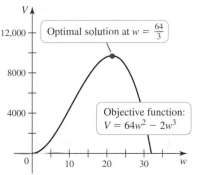

FIGURE 4.52

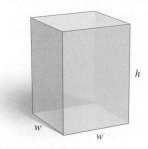

Objective function: $V = w^2h$
Constraint: $2w + h = 64$

FIGURE 4.53

maximum area of 10,000 ft^2 is achieved with dimensions $x = 50$ ft and $y - 200$ ft. The objective function A is shown in Figure 4.52.   *Related Exercises 5–10* ◄

QUICK CHECK 2 Find the objective function in Example 1 (in terms of x) if there is no interior fence and if there is one interior fence. ◄

EXAMPLE 2 **Airline regulations** Suppose an airline policy states that all baggage must be box-shaped with a sum of length, width, and height not exceeding 64 in. What are the dimensions and volume of a square-based box with the greatest volume under these conditions?

SOLUTION We sketch a square-based box whose length and width are w and whose height is h (Figure 4.53). By the airline policy, the box with greatest volume satisfies $2w + h = 64$. The objective function is the volume, $V = w^2h$. Either w or h may be eliminated from the objective function; substituting $h = 64 - 2w$, the volume is

$$V = w^2h = w^2(64 - 2w) = 64w^2 - 2w^3.$$

The objective function has now been expressed in terms of a single variable. Notice that w is nonnegative and cannot exceed 32, so the domain of V is $0 \le w \le 32$. The critical points satisfy

$$V'(w) = 128w - 6w^2 = 2w(64 - 3w) = 0,$$

which has roots $w = 0$ and $w = \frac{64}{3} \approx 21.3$. By the First (or Second) Derivative Test, $w = \frac{64}{3}$ corresponds to a local maximum. At the endpoints, $V(0) = V(32) = 0$. Therefore, the volume function has an absolute maximum of $V(64/3) \approx 9709$ in^3. The dimensions of the optimal box are $w = \ell = 64/3$ in and $h = 64 - 2w = 64/3$ in, so the optimal box is a cube. A graph of the volume function is shown in Figure 4.54.
            Related Exercises 11–13 ◄

QUICK CHECK 3 Find the objective function in Example 2 (in terms of w) if the constraint is that the sum of length and width and height cannot exceed 108 in. ◄

Optimization Guidelines With two examples providing some insight, we present a procedure for solving optimization problems. These guidelines provide a general framework, but the details may vary depending upon the problem.

FIGURE 4.54

Guidelines for Optimization Problems

1. Read the problem carefully, identify the variables, and organize the given information with a picture.

2. Identify the objective function (the function to be optimized). Write it in terms of the variables of the problem.

3. Identify the constraint(s). Write them in terms of the variables of the problem.

4. Use the constraint(s) to eliminate all but one independent variable of the objective function.

5. With the objective function expressed in terms of a single variable, find the interval of interest for that variable.

6. Use methods of calculus to find the absolute maximum or minimum value of the objective function on the interval of interest. If necessary, check the endpoints.

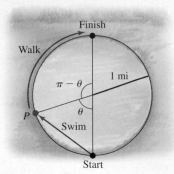

FIGURE 4.55

> You can check two special cases: If the
> entire trip is done walking, the travel time
> is $(\pi \text{ mi})/(3 \text{ mi/hr}) \approx 1.05 \text{ hr}$. If the
> entire trip is done swimming, the travel
> time is $(2 \text{ mi})/(2 \text{ mi/hr}) \approx 1 \text{ hr}$.

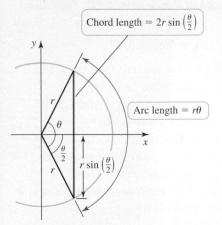

FIGURE 4.56

> To show that the chord length of a circle
> is $2r \sin(\theta/2)$, draw a line from the
> center of the circle to the midpoint of the
> chord. This line bisects the angle θ. Using
> a right triangle, half the length of the
> chord is $r \sin(\theta/2)$.

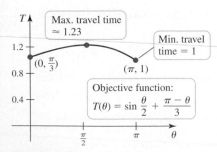

FIGURE 4.57

EXAMPLE 3 Walking and swimming Suppose you are standing on the shore of a
circular pond with radius 1 mi and you want to get to a point on the shore directly opposite your position (on the other end of a diameter). You plan to swim at 2 mi/hr from your
current position to another point P on the shore and then walk at 3 mi/hr along the shore
to the terminal point (Figure 4.55). How should you choose P to minimize the total time
for the trip?

SOLUTION As shown in Figure 4.55, the initial point is chosen arbitrarily, and the terminal point is at the other end of a diameter. The easiest way to describe the transition point P
is to refer to the central angle θ. If $\theta = 0$, then the entire trip is done by walking; if
$\theta = \pi$, the entire trip is done by swimming. So the interval of interest is $0 \le \theta \le \pi$.

   The objective function is the total travel time as it varies with θ. For each leg of the
trip (swim and walk), the travel time is the distance traveled divided by the speed. We
need a few facts from circular geometry. The length of the swimming leg is the length of
the chord of the circle corresponding to the angle θ. For a circle of radius r, this chord
length is given by $2r \sin(\theta/2)$ (Figure 4.56). So, the time for the swimming leg (with
$r = 1$ and speed 2 mi/hr) is

$$\frac{\text{distance}}{\text{rate}} = \frac{2 \sin(\theta/2)}{2} = \sin\frac{\theta}{2}.$$

The length of the walking leg is the length of the arc of the circle corresponding to the
angle $\pi - \theta$. For a circle of radius r, the arc length corresponding to an angle θ is $r\theta$
(Figure 4.56). Therefore, the time for the walking leg (with an angle $\pi - \theta$, $r = 1$, and
speed 3 mi/hr) is

$$\frac{\text{distance}}{\text{rate}} = \frac{\pi - \theta}{3}.$$

The total travel time for the trip is the objective function

$$T(\theta) = \sin\frac{\theta}{2} + \frac{\pi - \theta}{3}, \qquad 0 \le \theta \le \pi.$$

   We now analyze the objective function. The critical points of T satisfy

$$\frac{dT}{d\theta} = \frac{1}{2}\cos\frac{\theta}{2} - \frac{1}{3} = 0 \quad \text{or} \quad \cos\frac{\theta}{2} = \frac{2}{3}.$$

Using a calculator, the only solution in the interval $[0, \pi]$ is $\theta \approx 1.68 \text{ rad} \approx 96°$, which is
the critical point.

   Evaluating the objective function at the critical point and at the endpoints, we find
that $T(1.68) \approx 1.23 \text{ hr}$, $T(0) = \pi/3 = 1.05 \text{ hr}$, and $T(\pi) = 1 \text{ hr}$. We conclude that the
minimum travel time is $T(\pi) = 1 \text{ hr}$ when the entire trip is done swimming. The
maximum travel time, corresponding to $\theta \approx 96°$, is $T \approx 1.23 \text{ hr}$.

   The objective function is shown in Figure 4.57. In general, the maximum and minimum
travel times depend on the walking and swimming speeds (Exercise 14).

Related Exercises 14–15 ◄

EXAMPLE 4 Ladder over the fence An 8-foot-tall fence runs parallel to the side of
a house 3 feet away (Figure 4.58a). What is the length of the shortest ladder that clears
the fence and reaches the house? Assume that the vertical wall of the house and the horizontal ground have infinite extent (see Exercise 17 for more realistic assumptions).

SOLUTION Let's first ask why we expect a minimum ladder length. You could put the foot
of the ladder far from the fence, making it clear the fence at a shallow angle; but the ladder
would be very long. Or you could put the foot of the ladder close to the fence, making it
clear the fence at a steep angle; but again, the ladder would be long. Somewhere between
these extremes, there is a ladder position that minimizes the ladder length.

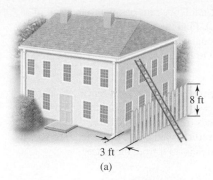

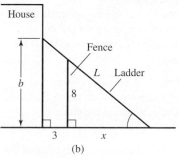

FIGURE 4.58

The objective function in this problem is the ladder length L. The position of the ladder is specified by x, the distance between the foot of the ladder and the fence (Figure 4.58b). The goal is to express L as a function of x, where $x > 0$.

The Pythagorean theorem gives the relationship

$$L^2 = (x + 3)^2 + b^2,$$

where b is the height of the top of the ladder above the ground. Similar triangles give the constraint $8/x = b/(3 + x)$. We now solve the constraint equation for b and substitute to express L^2 in terms of x:

$$L^2 = (x + 3)^2 + \underbrace{\left(\frac{8(x + 3)}{x}\right)^2}_{b} = (x + 3)^2 \left(1 + \frac{64}{x^2}\right)$$

At this juncture, we could find the critical points of L by first solving the preceding equation for L, and then solving $L' = 0$. However, the solution is simplified considerably if we note that L is a nonnegative function. Therefore, L and L^2 have local extrema at the same points, and we choose to minimize L^2. The derivative of L^2 is

$$\frac{d}{dx}\left[(x + 3)^2 \left(1 + \frac{64}{x^2}\right)\right] = (x + 3)^2 \left(-\frac{128}{x^3}\right) + 2(x + 3)\left(1 + \frac{64}{x^2}\right) \quad \text{Chain Rule and Product Rule}$$

$$= \frac{2(x + 3)(x^3 - 192)}{x^3} \quad \text{Simplify.}$$

Because $x > 0$, we have $x + 3 \neq 0$; so the condition $\dfrac{d}{dx}(L^2) = 0$ becomes $x^3 - 192 = 0$, or $x = 4\sqrt[3]{3} \approx 5.77$. By the First Derivative Test, this critical point corresponds to a local minimum. By Theorem 4.5, this solitary local minimum is also the absolute minimum on the interval $(0, \infty)$. Therefore, the minimum ladder length occurs when the foot of the ladder is approximately 5.77 ft from the fence. The minimum ladder length is $\sqrt{L^2(5.77)} \approx 15$ ft.

Related Exercises 16–17 ◄

SECTION 4.4 EXERCISES

Review Questions

1. Fill in the blanks: The goal of an optimization problem is to find the maximum or minimum value of the _____ function subject to the _____.

2. If the objective function involves more than one independent variable, how are the extra variables eliminated?

3. If the objective function is $Q = x^2 y$ and you know that $x + y = 10$, write the objective function first in terms of x and then in terms of y.

4. Suppose you wish to minimize the objective function on a closed interval, but you find that it has only a single local maximum. Where should you look for the solution to the problem?

Basic Skills

5. **Maximum area rectangles** Of all rectangles with a perimeter of 10 m, which one has the maximum area? (Give the dimensions.)

6. **Minimum perimeter rectangles** Of all rectangles with a fixed area A, which one has the minimum perimeter? (Give the dimensions in terms of A.)

7. **Maximum product** What two nonnegative real numbers with a sum of 23 have the largest possible product?

8. **Maximum length** What two nonnegative real numbers a and b whose sum is 23 maximize $a^2 + b^2$? Minimize $a^2 + b^2$?

9. **Minimum sum** What two positive real numbers whose product is 50 have the smallest possible sum?

10. **Pen problems**
    a. A rectangular pen is built with one side against a barn. Two hundred meters of fencing are used for the other three sides of the pen. What dimensions maximize the area of the pen?
    b. A rancher plans to make four identical and adjacent rectangular pens against a barn, each with an area of 100 m² (see figure). What are the dimensions of each pen that minimize the amount of fence that must be used?

Barn			
100	100	100	100

c. Two rectangular pens are built against a barn. Two hundred meters of fencing are to be used for the three sides and the diagonal dividing fence (see figure). What dimensions maximize the area of the pen?

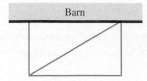

11. **Minimum surface area box** Of all boxes with a square base and a volume of 100 m^3, which one has the minimum surface area? (Give its dimensions.)

12. **Maximum volume box** Suppose an airline policy states that all baggage must be box-shaped with a sum of length, width, and height not exceeding 108 in. What are the dimensions and volume of a square-based box with the greatest volume under these conditions?

13. **Shipping crates** A square-based, box-shaped shipping crate is designed to have a volume of 16 ft^3. The material used to make the base costs twice as much (per ft^2) as the material in the sides, and the material used to make the top costs half as much (per ft^2) as the material in the sides. What are the dimensions of the crate that minimize the cost of materials?

14. **Walking and swimming** A man wishes to get from an initial point on the shore of a circular pond with radius 1 mi to a point on the shore directly opposite (on the other end of the diameter). He plans to swim from the initial point to another point on the shore and then walk along the shore to the terminal point.

   a. If he swims at 2 mi/hr and walks at 4 mi/hr, what are the minimum and maximum times for the trip?
   b. If he swims at 2 mi/hr and walks at 1.5 mi/hr, what are the minimum and maximum times for the trip?
   c. If he swims at 2 mi/hr, what is the minimum walking speed for which it is quickest to walk the entire distance?

15. **Walking and rowing** A boat on the ocean is 4 mi from the nearest point on a straight shoreline; that point is 6 mi from a restaurant on the shore. A woman plans to row the boat straight to a point on the shore and then walk along the shore to the restaurant.

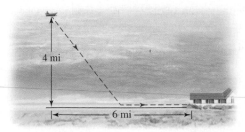

   a. If she walks at 3 mi/hr and rows at 2 mi/hr, at which point on the shore should she land to minimize the total travel time?
   b. If she walks at 3 mi/hr, what is the minimum speed at which she must row so that the quickest way to the restaurant is to row directly (with no walking)?

16. **Shortest ladder** A 10-ft-tall fence runs parallel to the wall of a house at a distance of 4 ft. Find the length of the shortest ladder that extends from the ground, over the fence, to the house. Assume the vertical wall of the house and the horizontal ground have infinite extent.

17. **Shortest ladder—more realistic** An 8-ft-tall fence runs parallel to the wall of a house at a distance of 5 ft. Find the length of the shortest ladder that extends from the ground, over the fence, to the house. Assume that the vertical wall of the house is 20 ft high and the horizontal ground extends 20 ft from the fence.

Further Explorations and Applications

18. **Rectangles beneath a parabola** A rectangle is constructed with its base on the x-axis and two of its vertices on the parabola $y = 16 - x^2$. What are the dimensions of the rectangle with the maximum area? What is the area?

19. **Rectangles beneath a semicircle** A rectangle is constructed with its base on the diameter of a semicircle with radius 5 cm and with two vertices on the semicircle. What are the dimensions of the rectangle with maximum area?

20. **Circle and square** A piece of wire 60 cm in length is cut, and the resulting two pieces are formed to make a circle and a square. Where should the wire be cut to (a) minimize and (b) maximize the combined area of the circle and the square?

21. **Maximum volume cone** A cone is constructed by cutting a sector of angle θ from a circular sheet of metal with radius 20 cm. The cut sheet is then folded up and welded (see figure). What angle θ maximizes the volume of the cone?

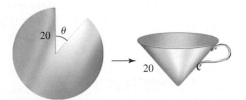

22. **Covering a marble** Imagine a flat-bottomed cylindrical pot with a circular cross section of radius 4 in. A marble with radius $0 < r < 4$ in is placed in the bottom of the pot. What is the radius of the marble that requires the most water to cover it completely?

23. **Optimal garden** A rectangular flower garden with an area of 30 m^2 is surrounded by a grass border 1 m wide on two sides and 2 m wide on the other two sides (see figure). What dimensions of the garden minimize the combined area of the garden and borders?

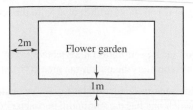

24. **Rectangles beneath a line**

  a. A rectangle is constructed with one side on the positive x-axis, one side on the positive y-axis, and one vertex on the line $y = 10 - 2x$. What dimensions maximize the area of the rectangle? What is the maximum area?

  b. Is it possible to construct a rectangle with a greater area than that found in part (a) by placing one side of the rectangle on the line $y = 10 - 2x$, one vertex on the positive x-axis, and one vertex on the positive y-axis? Find the dimensions of the rectangle of maximum area that can be constructed in this way.

25. **Kepler's wine barrel** Several mathematical stories originated with the second wedding of the mathematician and astronomer Johannes Kepler. Here is one: While shopping for wine for his wedding, Kepler noticed that the price of a barrel of wine (here assumed to be a cylinder) was determined solely by the length d of a dipstick that was inserted diagonally through a hole in the top of the barrel to the edge of the base of the barrel (see figure). Kepler realized that this measurement does not determine the volume of the barrel and that for a fixed value of d, the volume varies with the radius r and height h of the barrel. For a fixed value of d, what is the ratio r/h that maximizes the volume of the barrel?

26. **Folded boxes**

  a. Squares with sides of length x are cut out of each corner of a rectangular piece of cardboard measuring 3 ft by 4 ft. The resulting piece of cardboard is then folded into a box without a lid. Find the volume of the largest box that can be formed in this way.

  b. Suppose that in part (a) the original piece of cardboard is a square with sides of length ℓ. Find the volume of the largest box that can be formed in this way.

  c. Suppose that in part (a) the original piece of cardboard is a rectangle with sides of length ℓ and L. Holding ℓ fixed, find the size of the corner squares x that maximizes the volume of the box as $L \to \infty$. (*Source: Mathematics Teacher*, November 2002)

27. **Making silos** A grain silo consists of a cylindrical concrete tower surmounted by a metal hemispherical dome. The metal in the dome costs 1.5 times as much as the concrete (per unit of surface area). If the volume of the silo is 750 m³, what are the dimensions of the silo (radius and height of the cylindrical tower) that minimize the cost of the materials? Assume the silo has no floor and no flat ceiling under the dome.

28. **Suspension system** A load must be suspended 6 m below a high ceiling using cables attached to two supports that are 2 m apart (see figure). How far below the ceiling (x in the figure) should the cables bc joined to minimize the total length of cable used?

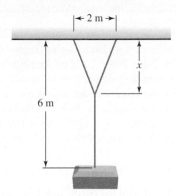

29. **Light sources** The intensity of a light source at a distance is directly proportional to the strength of the source and inversely proportional to the square of the distance from the source. Two light sources, one twice as strong as the other, are 12 m apart. At what point on the line segment joining the sources is the intensity the weakest?

30. **Crease-length problem** A rectangular sheet of paper of width a and length b, where $0 < a < b$, is folded by taking one corner of the sheet and placing it at a point P on the opposite long side of the sheet (see figure). The fold is then flattened to form a crease across the sheet. Assuming that the fold is made so that there is no flap extending beyond the original sheet, find the point P that produces the crease of minimum length. What is the length of that crease?

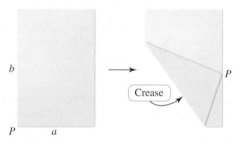

31. **Laying cable** An island is 3.5 mi from the nearest point on a straight shoreline; that point is 8 mi from a power station (see figure). A utility company plans to lay electrical cable underwater from the island to the shore and then underground along the shore to the power station. Assume that it costs $2400/mi to lay underwater cable and $1200/mi to lay underground cable. At what point should the underwater cable meet the shore in order to minimize the cost of the project?

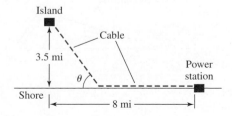

32. Laying cable again Solve the problem in Exercise 31, but this time minimize the cost with respect to the smaller angle θ between the underwater cable and the shore. (You should get the same answer.)

33. Sum of isosceles distances

a. An isosceles triangle has a base of length 4 and two sides of length $2\sqrt{2}$. Let P be a point on the perpendicular bisector of the base. Find the location P that minimizes the sum of the distances between P and the three vertices.

b. Assume in part (a) that the height of the isosceles triangle is $h > 0$ and its base has length 4. Show that the location of P that gives a minimum solution is independent of h for

$$h \geq \frac{2}{\sqrt{3}}.$$

34. Circle in a triangle What are the radius and area of the circle of maximum area that can be inscribed in an isosceles triangle whose two equal sides have length 1?

35. Slant height and cones Among all right circular cones with a slant height of 3, what are the dimensions (radius and height) that maximize the volume of the cone? The slant height of a cone is the distance from the outer edge of the base to the vertex.

36. Blood testing Suppose that a blood test for a disease must be given to a population of N people, where N is large. At most N individual blood tests must be done. The following strategy reduces the number of tests. Suppose 100 people are selected from the population and their blood samples are pooled. One test determines whether any of the 100 people test positive. If the test is positive, those 100 people are tested individually, making 101 tests necessary. However, if the pooled sample tests negative, then 100 people have been tested with one test. This procedure is then repeated. Probability theory shows that if the group size is x (for example, $x = 100$, as described here), then the average number of blood tests required to test N people is $N(1 - q^x + 1/x)$, where q is the probability that any one person tests negative. What group size x minimizes the average number of tests in the case that $N = 10{,}000$ and $q = 0.95$? Assume that x is a nonnegative real number.

37. Crankshaft A crank of radius r rotates with an angular frequency ω. It is connected to a piston by a connecting rod of length L (see figure). The acceleration of the piston varies with the position of the crank according to the function

$$a(\theta) = \omega^2 r\left(\cos\theta + \frac{r\cos 2\theta}{L}\right).$$

For fixed ω and r, find the values of θ, with $0 \leq \theta \leq 2\pi$, for which the acceleration of the piston is a maximum and minimum.

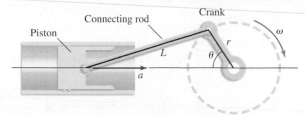

38. Metal rain gutters A rain gutter is made from sheets of metal 9 in wide. The gutters have a 3-in base and two 3-in sides, folded up at an angle θ (see figure). What angle θ maximizes the cross-sectional area of the gutter?

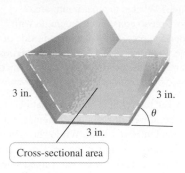

Cross-sectional area

39. Optimal soda can

a. **Classical problem** Find the radius and height of a cylindrical soda can with a volume of 354 cm³ that minimize the surface area.

b. **Real problem** Compare your answer in part (a) to a real soda can, which has a volume of 354 cm³, a radius of 3.1 cm, and a height of 12.0 cm, to conclude that real soda cans do not seem to have an optimal design. Then use the fact that real soda cans have a double thickness in their top and bottom surfaces to find the radius and height that minimizes the surface area of a real can (the surface areas of the top and bottom are now twice their values in part (a)). Are these dimensions closer to the dimensions of a real soda can?

40. Cylinder and cones (Putnam Exam 1938) Right circular cones of height h and radius r are attached to each end of a right circular cylinder of height h and radius r, forming a double-pointed object. For a given surface area A, what are the dimensions r and h that maximize the volume of the object?

41. Viewing angles An auditorium with a flat floor has a large screen on one wall. The lower edge of the screen is 3 ft above eye level and the upper edge of the screen is 10 ft above eye level (see figure). How far from the screen should you stand to maximize your viewing angle?

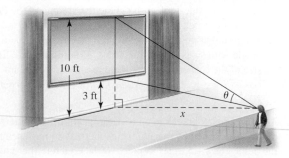

42. Searchlight problem—narrow beam A searchlight is 100 m from the nearest point on a straight highway (see figure). As it rotates, the searchlight casts a horizontal beam that intersects the highway in a point. If the light revolves at a rate of $\pi/6$ rad/s, find the rate at which the beam sweeps along the highway as a function of θ. For what value of θ is this rate minimized?

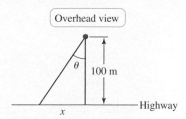

Overhead view

100 m

x

Highway

43. Watching a Ferris wheel An observer stands 20 m from the bottom of a Ferris wheel on a line that is perpendicular to the face of the wheel, with her eyes at the level of the bottom of the wheel. The wheel revolves at a rate of π rad/min and the observer's line of sight with a specific seat on the Ferris wheel makes an angle θ with the horizontal (see figure). At what time during a full revolution is θ changing most rapidly?

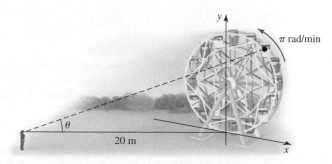

y

π rad/min

θ

20 m

x

44. Maximum angle Find the value of x that maximizes θ in the figure.

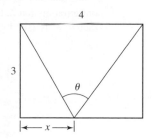

4

3

θ

x

45. Maximum volume cylinder in a sphere Find the dimensions of the right circular cylinder of maximum volume that can be placed inside of a sphere of radius R.

46. Rectangles in triangles Find the dimensions and area of the rectangle of maximum area that can be inscribed in the following figures.

    a. A right triangle with a given hypotenuse length L
    b. An equilateral triangle with a given side length L
    c. A right triangle with a given area A
    d. An arbitrary triangle with a given area A (The result applies to any triangle, but first consider triangles for which all the angles are less than or equal to 90°.)

47. Cylinder in a cone A right circular cylinder is placed inside a cone of radius R and height H so that the base of the cylinder lies on the base of the cone.

    a. Find the dimensions of the cylinder with maximum volume. Specifically, show that the volume of the maximum-volume cylinder is $\frac{4}{9}$ the volume of the cone.

    b. Find the dimensions of the cylinder with maximum lateral surface area (area of the curved surface).

48. Maximizing profit Suppose you own a tour bus and you book groups of 20 to 70 people for a day tour. The cost per person is $30 minus $0.25 for every ticket sold. If gas and other miscellaneous costs are $200, how many tickets should you sell to maximize your profit? Treat the number of tickets as a nonnegative real number.

49. Cone in a cone A right circular cone is inscribed inside a larger right circular cone with a volume of 150 cm³. The axes of the cones coincide and the vertex of the inner cone touches the center of the base of the outer cone. Find the ratio of the heights of the cones that maximizes the volume of the inner cone.

50. Another pen problem A rancher is building a horse pen on the corner of her property using 1000 ft of fencing. Because of the unusual shape of her property, the pen must be built in the shape of a trapezoid (see figure).

    a. Determine the lengths of the sides that maximize the area of the pen.

    b. Suppose there is already a fence along the side of the property opposite the side of length y. Find the lengths of the sides that maximize the area of the pen, using 1000 ft of fencing.

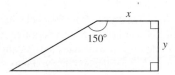

x

150°

y

51. Minimum-length roads A house is located at each corner of a square with side lengths of 1 mi. What is the length of the shortest road system with straight roads that connects all of the houses by roads (that is, a road system that allows one to drive from any house to any other house)? (*Hint:* Place two points inside the square at which roads meet.) (*Source:* Halmos, *Problems for Mathematicians Young and Old.*)

52. Light transmission A window consists of a rectangular pane of clear glass surmounted by a semicircular pane of tinted glass. The clear glass transmits twice as much light per unit of surface area as the tinted glass. Of all such windows with a fixed perimeter P, what are the dimensions of the window that transmits the most light?

53. Slowest shortcut Suppose you are standing in a field near a straight section of railroad tracks just as the locomotive of a train passes the point nearest to you, which is $\frac{1}{4}$ mi away. The train, with length $\frac{1}{3}$ mi, is traveling at 20 mi/hr. If you start running in a straight line across the field, how slowly can you run and still catch the train? In which direction should you run?

54. The arbelos An arbelos is the region enclosed by three mutually tangent semicircles; it is the region inside the larger semicircle and outside the two smaller semicircles (see figure).

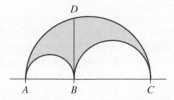

D

A B C

a. Given an arbelos in which the diameter of the largest circle is 1, what positions of point B maximize the area of the arbelos?

b. Show that the area of the arbelos is the area of a circle whose diameter is the distance BD in the figure.

55. Proximity questions

a. What point on the line $y = 3x + 4$ is closest to the origin?

b. What point on the parabola $y = 1 - x^2$ is closest to the point $(1, 1)$?

c. Find the point on the graph of $y = \sqrt{x}$ that is nearest the point $(p, 0)$ if (i) $p > \frac{1}{2}$; and (ii) $0 < p < \frac{1}{2}$.

56. Turning a corner with a pole

a. What is the length of the longest pole that can be carried horizontally around a corner at which a 3-ft corridor and a 4-ft corridor meet at right angles?

b. What is the length of the longest pole that can be carried horizontally around a corner at which a corridor that is a ft wide and a corridor that is b ft wide meet at right angles?

c. What is the length of the longest pole that can be carried horizontally around a corner at which a corridor that is $a = 5$ ft wide and a corridor that is $b = 5$ ft wide meet at an angle of $120°$?

d. What is the length of the longest pole that can be carried around a corner at which a corridor that is a ft wide and a corridor that is b ft wide meet at right angles, assuming there is an 8-ft ceiling and that you may tilt the pole at any angle?

57. Travel costs A simple model for travel costs involves the cost of gasoline and the cost of a driver. Specifically, assume that gasoline costs $\$p$/gallon and the vehicle gets g miles per gallon. Also, assume that the driver earns $\$w$/hour.

a. A plausible function to describe how gas mileage (in mi/gal) varies with speed is $g(v) = v(85 - v)/60$. Evaluate $g(0)$, $g(40)$, and $g(60)$ and explain why these values are reasonable.

b. At what speed does the gas mileage function have its maximum?

c. Explain why the cost of a trip of length L miles is $C(v) = Lp/g(v) + Lw/v$.

d. Let $L = 400$ mi, $p = \$4$/gal, and $w = \$20$/hr. At what (constant) speed should the vehicle be driven to minimize the cost of the trip?

e. Should the optimal speed be increased or decreased [compared with part (d)] if L is increased from 400 mi to 500 mi? Explain.

f. Should the optimal speed be increased or decreased [compared with part (d)] if p is increased from $\$4$/gal to $\$4.20$/gal? Explain.

g. Should the optimal speed be increased or decreased [compared with part (d)] if w is decreased from $\$20$/hr to $\$15$/hr? Explain.

58. Do dogs know calculus? A mathematician stands on a beach with his dog at point A. He throws a tennis ball so that it hits the water at point B. The dog, wanting to get to the tennis ball as quickly as possible, runs along the straight beach line to point D and then swims from point D to point B to retrieve his ball. Assume point C is the closest point on the edge of the beach from the tennis ball (see figure).

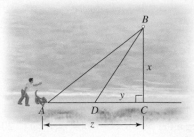

a. Assume the dog runs at r m/s and swims at s m/s, where $r > s$. Also assume the lengths of BC, CD, and AC are x, y, and z, respectively. Find a function $T(y)$ representing the total time it takes for the dog to get to the ball.

b. Verify that the value of y that minimizes the time it takes to retrieve the ball is $y = \dfrac{x}{\sqrt{r/s + 1}\sqrt{r/s - 1}}$.

c. If the dog runs at 8 m/s and swims at 1 m/s, what ratio y/x produces the fastest retrieving time?

d. A dog named Elvis who runs at 6.4 m/s and swims at 0.910 m/s was found to use an average ratio y/x of 0.144 to retrieve his ball. Does Elvis appear to know calculus? (*Source: College Mathematics Journal*, May 2003)

59. Fermat's Principle

a. Two poles of heights m and n are separated by a horizontal distance d. A rope is stretched from the top of one pole to the ground and then to the top of the other pole. Show that the configuration that requires the least amount of rope occurs when $\theta_1 = \theta_2$ (see figure).

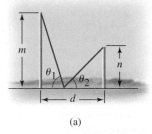

(a)

b. Fermat's Principle states that when light travels between two points in the same medium (at a constant speed), it travels on the path that minimizes the travel time. Show that when light from a source A reflects off of a surface and is received at point B, the angle of incidence equals the angle of reflection, or $\theta_1 = \theta_2$ (see figure).

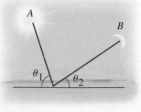

(b)

60. Snell's Law Suppose that a light source at A is in a medium in which light travels at a speed v_1 and the point B is in a medium in which light travels at a speed v_2 (see figure). Using Fermat's Principle, which states that light travels along the path that requires the minimum travel time (Exercise 59), show that the path taken between points A and B satisfies $(\sin \theta_1)/v_1 = (\sin \theta_2)/v_2$.

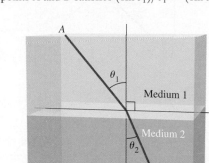

61. Tree notch (Putnam Exam 1938, rephrased) A notch is cut in a cylindrical vertical tree trunk. The notch penetrates to the axis of the cylinder and is bounded by two half-planes that intersect on a diameter D of the tree. The angle between the two half planes is θ. Prove that for a given tree and fixed angle θ, the volume of the notch is minimized by taking the bounding planes at equal angles to the horizontal plane that also passes through D.

62. Gliding mammals Many species of small mammals (such as flying squirrels and marsupial gliders) have the ability to walk and glide. Recent research suggests that these animals choose the most energy-efficient means of travel. According to one empirical model, the energy required for a glider with body mass m to walk a horizontal distance D is 8.46 $Dm^{2/3}$ (where m is measured in grams,

D is measured in meters, and energy is measured in microliters of oxygen consumed in respiration). The energy cost of climbing to a height $D \tan \theta$ and gliding at an angle of θ (below the horizontal, with $\theta = 0$ representing perfectly horizontal flight and $\theta > 45°$ representing controlled falling) a horizontal distance D is modeled by 1.36 $mD \tan \theta$. Therefore, the function

$$S(m, \theta) = 8.46m^{2/3} - 1.36m \tan \theta$$

gives the energy difference per horizontal meter traveled between walking and gliding: If $S > 0$ for given values of m and θ, then it is more costly to walk than glide.

a. For what glide angles is it more efficient for a 200 gram animal to glide rather than walk?
b. Find the threshold function $\theta = g(m)$ that gives the curve along which walking and gliding are equally efficient. Is it an increasing or decreasing function of body mass?
c. In order to make gliding more efficient than walking, do larger gliders have a larger or smaller selection of glide angles that they can use?
d. Let $\theta = 25°$ (a typical glide angle) and graph S as a function of m for $0 \le m \le 3000$ (in grams). For what values of m is gliding more efficient?
e. For $\theta = 25°$, what value of m (call it m^*) maximizes S?
f. Does m^*, as defined in part (e), increase or decrease with increasing θ? That is, as a glider reduces its glide angle, does its optimal size become larger or smaller?
g. Assuming Dumbo is a gliding elephant whose weight is one metric ton (10^6 g), what glide angle would Dumbo use to be more efficient at gliding than walking?

(*Source: Energetic savings and the body size distribution of gliding mammals, Roman Dial, Evolutionary Ecology Research,* **5** (2003): 1151–1162.)

QUICK CHECK ANSWERS

2. $A = 400x - 2x^2, A = 400x - 3x^2$
3. $V = 108w^2 - 2w^3$ ◀

4.5 Linear Approximation and Differentials

Imagine plotting a smooth curve with a graphing utility. Now pick a point P on the curve, draw the line tangent to the curve at P, and zoom in on it several times. As you successively enlarge the curve near P, it looks more and more like the tangent line (Figure 4.59a). This fundamental observation—that smooth curves appear straighter on smaller scales—is the basis of many important mathematical ideas, one of which is *linear approximation*.

Now, consider a curve with a corner or cusp at a point Q (Figure 4.59b). No amount of magnification "straightens out" the curve or removes the corner at Q. The different behavior at P and Q is related to the idea of differentiability: The function in Figure 4.59a is differentiable at P, whereas the function in Figure 4.59b is not differentiable at Q. One of the requirements for the techniques presented in this section is that the function be differentiable at the point in question.

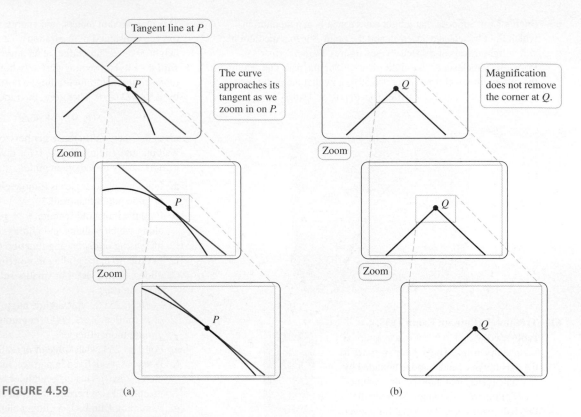

FIGURE 4.59   (a)   (b)

Linear Approximation

Figure 4.59a suggests that when we zoom in on the graph of a smooth function at a point P, the curve approaches its tangent line at P. This fact is the key to understanding linear approximation. The idea is to use the line tangent to the curve at P to approximate the value of the function at points near P. Here's how it works.

Assume f is differentiable on an interval containing the point a. The slope of the line tangent to the curve at the point $(a, f(a))$ is $f'(a)$. Therefore, the equation of the tangent line is

$$y - f(a) = f'(a)(x - a) \quad \text{or} \quad y = \underbrace{f(a) + f'(a)(x - a)}_{L(x)}.$$

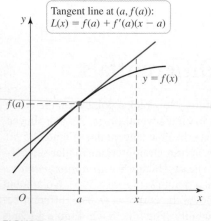

Tangent line at $(a, f(a))$:
$L(x) = f(a) + f'(a)(x - a)$

This tangent line represents a new function L that we call the *linear approximation* to f at the point a (Figure 4.60). If f and f' are easy to evaluate at a, then the value of f at points near a is easily approximated using the linear approximation L. That is,

$$f(x) \approx L(x) = f(a) + f'(a)(x - a).$$

This approximation improves as x approaches a.

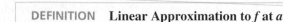

FIGURE 4.60

DEFINITION   **Linear Approximation to f at a**

Suppose f is differentiable on an interval I containing the point a. The **linear approximation** to f at a is the linear function

$$L(x) = f(a) + f'(a)(x - a) \quad \text{for } x \text{ in } I.$$

QUICK CHECK 1 Sketch the graph of a function f that is concave up at a point $(a, f(a))$. Sketch the linear approximation to f at a. Is the graph of the linear approximation above or below the graph of f? ◄

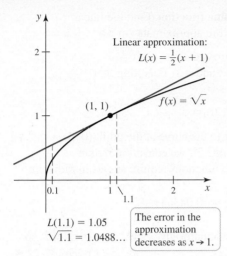

Linear approximation:
$$L(x) = \tfrac{1}{2}(x + 1)$$

$(1, 1)$

$f(x) = \sqrt{x}$

$L(1.1) = 1.05$
$\sqrt{1.1} = 1.0488\ldots$

The error in the approximation decreases as $x \to 1$.

FIGURE 4.61

Table 4.4

x	$L(x)$	Exact $\sqrt{x}$	Percent error
1.2	1.1	1.0954...	0.4%
1.1	1.05	1.0488...	0.1%
1.01	1.005	1.0049...	0.001%
1.001	1.0005	1.0005...	0.00001%

> We choose $a = \frac{9}{100}$ because it is close to 0.1 and its square root is easy to evaluate.

EXAMPLE 1 Linear approximations and errors

a. Find the linear approximation to $f(x) = \sqrt{x}$ at $x = 1$ and use it to approximate $\sqrt{1.1}$.

b. Use linear approximation to estimate the value of $\sqrt{0.1}$.

SOLUTION

a. We construct the linear approximation

$$L(x) = f(a) + f'(a)(x - a),$$

where $f(x) = \sqrt{x}$, $f'(x) = 1/(2\sqrt{x})$, and $a = 1$. Noting that $f(a) = f(1) = 1$ and $f'(a) = f'(1) = \frac{1}{2}$, we have

$$L(x) = 1 + \frac{1}{2}(x - 1) = \frac{1}{2}(x + 1),$$

which is an equation of the line tangent to the curve at the point $(1, 1)$ (Figure 4.61). Because $x = 1.1$ is near $x = 1$, we approximate $\sqrt{1.1}$ by $L(1.1)$:

$$\sqrt{1.1} \approx L(1.1) = \frac{1}{2}(1.1 + 1) = 1.05.$$

The exact value is $f(1.1) = \sqrt{1.1} = 1.0488\ldots$; therefore, the linear approximation has an error of about 0.1%. Furthermore, our approximation is an *overestimate* because the tangent line lies above the graph of f. In Table 4.4 we see several approximations to $\sqrt{x}$ for x near 1 and the associated errors. Clearly, the errors decrease as x approaches 1.

b. If the linear approximation $L(x) = \frac{1}{2}(x + 1)$ obtained in part (a) is used to approximate $\sqrt{0.1}$, we have

$$\sqrt{0.1} \approx L(0.1) = \frac{1}{2}(0.1 + 1) = 0.55.$$

A calculator gives $\sqrt{0.1} = 0.3162\ldots$, which shows that the approximation is well off the mark. The error arises because the tangent line through $(1, 1)$ is not close to the curve at $x = 0.1$ (Figure 4.61). For this reason, we seek a different value of a, with the requirement that it is near $x = 0.1$, and both $f(a)$ and $f'(a)$ are easily computed. It is tempting to try $a = 0$, but $f'(0)$ is undefined. One choice that works well is $a = \frac{9}{100} = 0.09$. Using the linear approximation $L(x) = f(a) + f'(a)(x - a)$, we have

$$\sqrt{0.1} \approx L(0.1) = \overbrace{\sqrt{\frac{9}{100}}}^{f(a)} + \overbrace{\frac{1}{2\sqrt{9/100}}}^{f'(a)} \overbrace{\left(\frac{1}{10} - \frac{9}{100}\right)}^{(x - a)}$$

$$= \frac{3}{10} + \frac{10}{6}\left(\frac{1}{100}\right)$$

$$= \frac{19}{60} \approx 0.3167$$

This approximation agrees with the exact value to three decimal places.

Related Exercises 7–22 ◄

QUICK CHECK 2 Suppose you want to use linear approximation to estimate $\sqrt{0.18}$. What is a good choice for a? ◄

EXAMPLE 2   Linear approximation for the sine function Find the linear approximation to $f(x) = \sin x$ at $x = 0$, and use it to approximate $\sin 2.5°$.

SOLUTION We begin by constructing a linear approximation $L(x) = f(a) + f'(a)(x - a)$, where $f(x) = \sin x$ and $a = 0$. Noting that $f(0) = 0$ and $f'(0) = \cos(0) = 1$, we have

$$L(x) = 0 + 1(x - 0) = x.$$

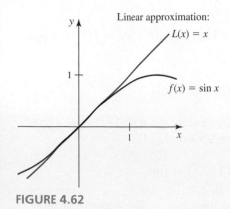

Linear approximation:

$L(x) = x$

$f(x) = \sin x$

FIGURE 4.62

Again, the linear approximation is the line tangent to the curve at the point $(0, 0)$ (Figure 4.62). Before using $L(x)$ to approximate $\sin 2.5°$, we convert to radian measure (the derivative formulas for trigonometric functions require angles in radians):

$$2.5° = 2.5°\left(\frac{\pi}{180°}\right) = \frac{\pi}{72} \approx 0.04363 \text{ rad}$$

Therefore, $\sin 2.5° \approx L(0.04363) = 0.04363$. A calculator gives $\sin 2.5° \approx 0.04362$, so the approximation is accurate to four decimal places.   *Related Exercises 7–22* ◄

In Examples 1 and 2, we used a calculator to check the accuracy of our approximations. This begs the question: Why bother with linear approximation when a calculator does a better job? There are some good answers to that question.

Linear approximation is actually just the first step in the larger process of *polynomial approximation*. While linear approximation does a decent job of estimating function values when x is near a, we can generally do better with higher-degree polynomials. These ideas are explored further in Chapter 9.

Linear approximation also allows us to approximate a complicated function with a simple linear function. In Example 2, we found the *small-angle approximation to the sine function*; $\sin x \approx x$ for x near 0. Finally, as will be shown in Chapter 9, linear approximation allows us to estimate errors in approximations.

QUICK CHECK 3 Explain why the linear approximation to $f(x) = \cos x$ at $x = 0$ is $L(x) = 1$. ◄

A Variation on Linear Approximation

Linear approximation says that a function f can be approximated as

$$f(x) \approx f(a) + f'(a)(x - a)$$

where a is fixed and x is a nearby point. We first rewrite this expression as

$$\underbrace{f(x) - f(a)}_{\Delta y} \approx f'(a)\underbrace{(x - a)}_{\Delta x}.$$

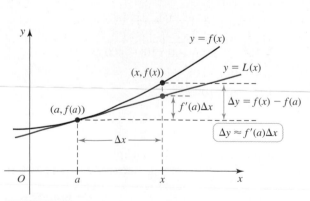

$y = f(x)$

$(x, f(x))$

$y = L(x)$

$(a, f(a))$

$f'(a)\Delta x$

$\Delta y = f(x) - f(a)$

Δx

$\boxed{\Delta y \approx f'(a)\Delta x}$

FIGURE 4.63

Recall that it is customary to use the notation Δ (capital Greek delta) to denote a change. The factor $x - a$ is the change in the x-coordinate between a and a nearby point x. Similarly, $f(x) - f(a)$ is the corresponding change in the y-coordinate (Figure 4.63). So, we write this approximation as

$$\Delta y \approx f'(a)\,\Delta x.$$

In other words, a change in y (the function value) can be approximated by the corresponding change in x magnified or diminished by a factor of $f'(a)$. This interpretation states the familiar fact that $f'(a)$ is the rate of change of y with respect to x.

> **Relationship Between Δx and Δy**
>
> Suppose f is differentiable on an interval I containing the point a. The change in the value of f between two points a and $a + \Delta x$ is approximately
>
> $$\Delta y \approx f'(a)\,\Delta x$$
>
> where $a + \Delta x$ is in I.

EXAMPLE 3 Estimating changes with linear approximations

a. Approximate the change in $y = f(x) = x^9 - 2x + 1$ when x changes from 1.00 to 1.05.

b. Approximate the change in the surface area of a spherical hot-air balloon when the radius decreases from 4 m to 3.9 m.

SOLUTION

a. The change in y is $\Delta y \approx f'(a)\,\Delta x$, where $a = 1$, $\Delta x = 0.05$, and $f'(x) = 9x^8 - 2$. Substituting these values, we find that

$$\Delta y \approx f'(a)\,\Delta x = f'(1) \cdot 0.05 = 7 \cdot 0.05 = 0.35.$$

If x increases from 1.00 to 1.05, then y increases by approximately 0.35.

> ➤ Notice that the units in these calculations are consistent. If r has units of meters (m), S' has units of m^2/m = m, so ΔS has units of m^2 as it should.

b. The surface area of a sphere is $S = 4\pi r^2$, so the change in the surface area when the radius changes by Δr is $\Delta S \approx S'(a)\,\Delta r$. Substituting $S'(r) = 8\pi r$, $a = 4$, and $\Delta r = -0.1$, the approximate change in the surface area is

$$\Delta S \approx S'(a)\,\Delta r = S'(4) \cdot (-0.1) = 32\pi \cdot (-0.1) \approx -10.05.$$

The change in surface area is approximately $-10.05\ m^2$; it is negative, reflecting a decrease.   *Related Exercises 23–28* ◄

> **QUICK CHECK 4** Given that the volume of a sphere is $V = 4\pi r^3/3$, find an expression for the approximate change in the volume when the radius changes from a to $a + \Delta r$. ◄

> **SUMMARY   Uses of Linear Approximation**
>
> • To approximate f near $x = a$, use
>
> $$f(x) \approx L(x) = f(a) + f'(a)(x - a).$$
>
> • To approximate the change Δy in the dependent variable given a change Δx in the independent variable, use
>
> $$\Delta y \approx f'(a)\,\Delta x.$$

Differentials

We now introduce an important concept that allows us to distinguish two related quantities:

- The change in the function $y = f(x)$ as x changes from a to $a + \Delta x$ (which we call Δy, as before)

- The change in the linear approximation $y = L(x)$ as x changes from a to $a + \Delta x$ (which we will call the *differential* dy)

Consider a function $y = f(x)$ differentiable on an interval containing a. If the x-coordinate changes from a to $a + \Delta x$, the corresponding change in the function is *exactly*

$$\Delta y = f(a + \Delta x) - f(a).$$

Using the linear approximation $L(x) = f(a) + f'(a)(x - a)$, the change in L as x changes from a to $a + \Delta x$ is

$$\Delta L = L(a + \Delta x) - L(a)$$
$$= \underbrace{[f(a) + f'(a)(a + \Delta x - a)]}_{L(a + \Delta x)} - \underbrace{[f(a) + f'(a)(a - a)]}_{L(a)}$$
$$= f'(a)\,\Delta x$$

In order to distinguish Δy and ΔL, we define two new variables called **differentials**. The differential dx is simply Δx; the differential dy is the change in the linear approximation, which is $\Delta L = f'(a)\,\Delta x$. Using this notation,

$$\Delta L = \underbrace{dy}_{\substack{\text{same} \\ \text{as } \Delta L}} = f'(a)\,\Delta x = f'(a)\,\underbrace{dx}_{\substack{\text{same} \\ \text{as } \Delta x}}.$$

Therefore, at the point a, we have $dy = f'(a)\,dx$. More generally, we replace the fixed point a by a variable point x and write

$$dy = f'(x)\,dx.$$

DEFINITION   Differentials

Let f be differentiable on an interval containing x. A small change in x is denoted by the **differential** dx. The corresponding change in $y = f(x)$ is approximated by the **differential** $dy = f'(x)\,dx$; that is,

$$\Delta y = f(x + dx) - f(x) \approx dy = f'(x)\,dx.$$

▷ Of the two coinventors of calculus, Gottfried Leibnitz relied on the idea of differentials in his development of calculus. Leibnitz's notation for differentials is essentially the same as the notation we use today. An Irish philosopher of the day, Bishop Berkeley, called differentials "the ghost of departed quantities."

Figure 4.64 shows that if $\Delta x = dx$ is small, then the change in f, which is Δy, is well approximated by the change in the linear approximation, which is dy. Furthermore, the approximation $\Delta y \approx dy$ improves as dx approaches 0. The notation for differentials is consistent with the notation for the derivative: If we divide both sides of $dy = f'(x)\,dx$ by dx, we have

$$\frac{dy}{dx} = \frac{f'(x)\,dx}{dx} = f'(x).$$

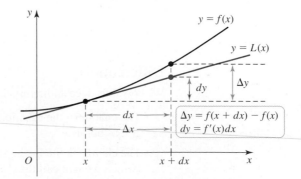

FIGURE 4.64

EXAMPLE 4   Differentials as change Use the notation of differentials to write the approximate change in $f(x) = 3\cos^2 x$ given a small change dx.

▷ Recall that $\sin 2x = 2\sin x \cos x$.

SOLUTION With $f(x) = 3\cos^2 x$, we have $f'(x) = -6\cos x \sin x = -3\sin 2x$. Therefore,

$$dy = f'(x)\,dx = -3\sin 2x\,dx.$$

The interpretation is that a small change dx in the variable x produces an approximate change of $dy = -3 \sin 2x \, dx$ in y. For example, if x increases from $x = \pi/4$ to $x = \pi/4 + 0.1$, then $dx = 0.1$ and

$$dy = -3 \sin (\pi/2)(0.1) = -0.3.$$

The approximate change in the function is -0.3, which means a decrease of approximately 0.3.

Related Exercises 29–38 ◄

SECTION 4.5 EXERCISES

Review Questions

1. Sketch the graph of a smooth function f and label a point $P(a, (f(a))$ on the curve. Draw the line that represents the linear approximation to f at P.

2. Suppose you find the linear approximation to a differentiable function at a local maximum of that function. Describe the graph of the linear approximation.

3. How can linear approximation be used to approximate the value of a function f near a point at which f and f' are easily evaluated?

4. How can linear approximation be used to approximate the change in $y = f(x)$ given a change in x?

5. Given a function f differentiable on its domain, write and explain the relationship between the differentials dx and dy.

6. Does the differential dy represent the change in f or the change in the linear approximation to f? Explain.

Basic Skills

7–12. Linear approximation

    a. Write the equation of the line that represents the linear approximation to the following functions at the given point a.
    b. Graph the function and the linear approximation near $x = a$.
    c. Use the linear approximation to estimate the given function value.
    d. Compute the percent error in your approximation: $100 \cdot |\text{approx} - \text{exact}|/|\text{exact}|$

7. $f(x) = 12 - x^2$; $a = 2$; $f(2.1)$

8. $f(x) = \sin x$; $a = \pi/4$; $f(0.75)$

9. $f(x) = \ln (1 + x)$; $a = 0$; $f(0.9)$

10. $f(x) = x/(x + 1)$; $a = 1$; $f(1.1)$

11. $f(x) = \cos x$; $a = 0$; $f(-0.01)$

12. $f(x) = e^x$; $a = 0$; $f(0.05)$

13–22. Estimations with linear approximation *Use linear approximations to estimate the following quantities. Choose a value of a to produce a small error.*

13. $1/203$	14. $\tan 3°$	15. $\sqrt{146}$	16. $\sqrt[3]{65}$
17. $\ln (1.05)$	18. $\sqrt{5/29}$	19. $e^{0.06}$	20. $1/\sqrt{119}$
21. $1/\sqrt[3]{510}$	22. $\cos 31°$		

23–28. Approximating changes

23. Approximate the change in the volume of a sphere when its radius changes from $r = 5$ ft to $r = 5.1$ ft $\left(V(r) = \frac{4}{3}\pi r^3\right)$.

24. Approximate the change in the atmospheric pressure when the altitude increases from $z = 2$ km to $z = 2.01$ km $(P(z) = 1000 \, e^{-z/10})$.

25. Approximate the change in the volume of a right circular cylinder of fixed radius $r = 20$ cm when its height decreases from $h = 12$ cm to $h = 11.9$ cm $(V(h) = \pi r^2 \, h)$.

26. Approximate the change in the volume of a right circular cone of fixed height $h = 4$ m when its radius increases from $r = 3$ m to $r = 3.05$ m $(V(r) = \pi r^2 \, h/3)$.

27. Approximate the change in the lateral surface area (excluding the area of the base) of a right circular cone of fixed height of $h = 6$ m when its radius decreases from $r = 10$ m to $r = 9.9$ m $(S = \pi r \sqrt{r^2 + h^2})$.

28. Approximate the change in the magnitude of the electrostatic force between two charges when the distance between them increases from $r = 20$ m to $r = 21$ m $(F(r) = 0.01/r^2)$.

29–38. Differentials *Consider the following functions and express the relationship between a small change in x and the corresponding change in y in the form $dy = f'(x) \, dx$.*

29. $f(x) = 2x + 1$	30. $\sin^2 x$
31. $f(x) = 1/x^3$	32. $f(x) = e^{2x}$
33. $f(x) = 2 - a \cos x$	34. $f(x) = (4 + x)/(4 - x)$
35. $f(x) = 3x^3 - 4x$	36. $f(x) = \sin^{-1} x$
37. $f(x) = \tan x$	38. $f(x) = \ln (1 - x)$

Further Explorations

39. **Explain why or why not** Determine whether the following statements are true and give an explanation or counterexample.

    a. The linear approximation to $f(x) = x^2$ at the point $(0, 0)$ is $L(x) = 0$.
    b. Linear approximation provides a good approximation to $f(x) = |x|$ at $(0, 0)$.
    c. If $f(x) = mx + b$, then at any point $x = a$, the linear approximation to f is $L(x) = f(x)$.

40–43. Linear approximation *Write an equation of the line that represents the linear approximation to the following functions at a. Then (a) graph the function and the linear approximation near a; (b) use the linear approximation to estimate the given quantity; and (c) compute the percent error in your approximation.*

40. $f(x) = \tan x$; $a = 0$; $\tan 1.5°$

41. $f(x) = 1/(x + 1)$; $a = 0$; $1/1.1$

42. $f(x) = \cos x$; $a = \pi/4$; $\cos(0.8)$

43. $f(x) = e^{-x}$; $a = 0$; $e^{-0.03}$

Applications

44. Ideal Gas Law The pressure P, temperature T, and volume V of an ideal gas are related by $PV = nRT$, where n is the number of moles of the gas and R is the universal gas constant. For the purposes of this exercise, let $nR = 1$; thus, $P = T/V$.

   a. Suppose that the volume is held constant and the temperature increases by $\Delta T = 0.05$. What is the approximate change in the pressure? Does the pressure increase or decrease?

   b. Suppose that the temperature is held constant and the volume increases by $\Delta V = 0.1$. What is the approximate change in the pressure? Does the pressure increase or decrease?

   c. Suppose that the pressure is held constant and the volume increases by $\Delta V = 0.1$. What is the approximate change in the temperature? Does the temperature increase or decrease?

45. Errors in approximations Suppose $f(x) = \sqrt[3]{x}$ is to be approximated near $x = 8$. Find the linear approximation to f at $x = 8$. Then, complete the following table, showing the errors in various approximations. Use a calculator to obtain the exact values. The percent error is $100 \cdot (\text{approximation} - \text{exact})/\text{exact}$. Comment on the behavior of the errors as x approaches 8.

x	Linear approx	Exact value	Percent error
8.1			
8.01			
8.001			
8.0001			
7.9999			
7.999			
7.99			
7.9			

46. Errors in approximations Suppose $f(x) = 1/(1 + x)$ is to be approximated near $x = 0$. Find the linear approximation to f at

$x = 0$. Then, complete the following table showing the errors in various approximations. Use a calculator to obtain the exact values. The percent error is $100 \cdot (\text{approximation} - \text{exact})/\text{exact}$. Comment on the behavior of the errors as x approaches 0.

x	Linear approx	Exact value	Percent error
0.1			
0.01			
0.001			
0.0001			
−0.0001			
−0.001			
−0.01			
−0.1			

Additional Exercises

47. Linear approximation and the second derivative Draw the graph of a function f such that $f(1) = f'(1) = f''(1) = 1$. Draw the linear approximation to the function at the point $(1, 1)$. Now draw the graph of another function g such that $g(1) = g'(1) = 1$ and $g''(1) = 10$. (It is not possible to represent the second derivative exactly, but your graphs should reflect the fact that $f''(1)$ is relatively small and $g''(1)$ is relatively large.) Now, suppose that linear approximations are used to approximate $f(1.1)$ and $g(1.1)$.

   a. Which function value has the more accurate linear approximation near $x = 1$ and why?

   b. Explain why the error in the linear approximation to f near a point a is proportional to the magnitude of $f''(a)$.

4.6 Mean Value Theorem

The *Mean Value Theorem* is a cornerstone in the theoretical framework of calculus. Several critical theorems (some stated in previous sections) rely on the Mean Value Theorem; the theorem also appears in practical applications. We begin with a preliminary result known as Rolle's Theorem.

Rolle's Theorem

Consider a function f that is continuous on a closed interval $[a, b]$ and differentiable on the open interval (a, b). Furthermore, assume f has the special property that $f(a) = f(b)$ (Figure 4.65). The statement of Rolle's Theorem is not surprising: It says that somewhere between a and b, there is at least one point at which f has a horizontal tangent line.

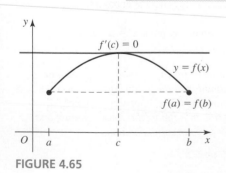

FIGURE 4.65

Michel Rolle (1652–1719) is one of the less celebrated mathematicians, whose name is nevertheless attached to a theorem. He worked in Paris most of his life as a scribe and published his theorem in 1691.

The Extreme Value Theorem, discussed in Section 4.1, states that a function that is continuous on a closed bounded interval attains its absolute maximum and minimum values on that interval.

> **THEOREM 4.8   Rolle's Theorem**
> Let f be continuous on a closed interval $[a, b]$ and differentiable on (a, b) with $f(a) = f(b)$. There is at least one point c in (a, b) such that $f'(c) = 0$.

Proof The function f satisfies the conditions of Theorem 4.1 (Extreme Value Theorem) and thus attains its absolute maximum and minimum values on $[a, b]$. Those values are attained either at an endpoint or at an interior point c.

Case 1: First suppose that f attains both its absolute maximum and minimum values at the endpoints. Because $f(a) = f(b)$, the maximum and minimum values are equal, and it follows that f is a constant function on $[a, b]$. Therefore, $f'(x) = 0$ for all x in (a, b), and the conclusion of the theorem holds.

Case 2: Assume at least one of the absolute extreme values of f does not occur at an endpoint. Then, f must attain an absolute extreme value at an interior point of $[a, b]$; therefore, f must have either a local maximum or a local minimum at a point c in (a, b). Because f is differentiable on (a, b), we know from Theorem 4.2 that at a local extremum the derivative is zero. Thus, $f'(c) = 0$ for at least one point c of (a, b), and again the conclusion of the theorem holds.   ◄

Why does Rolle's Theorem require continuity? A function that is not continuous on $[a, b]$ may have identical values at both endpoints and still not have a horizontal tangent line at any point on the interval (Figure 4.66a). Similarly, a function that is continuous on $[a, b]$ but not differentiable at a point of (a, b) may also fail to have a horizontal tangent line (Figure 4.66b).

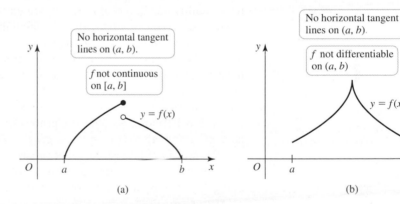

FIGURE 4.66

(a)   (b)

QUICK CHECK 1 Where on the interval $[0, 4]$ does $f(x) = 4x - x^2$ have a horizontal tangent line? ◄

EXAMPLE 1   Verifying Rolle's Theorem Find an interval on which Rolle's Theorem applies to $f(x) = x^3 - 7x^2 + 10x$. Then find all the points on that interval at which $f'(c) = 0$.

SOLUTION Because f is a polynomial, it is everywhere continuous and differentiable. We need an interval $[a, b]$ with the property that $f(a) = f(b)$. Noting that $f(x) = x(x - 2)(x - 5)$, we choose the interval $[0, 5]$, because $f(0) = f(5) = 0$ (other intervals are possible). The goal is to find points c in the interval $(0, 5)$ at which $f'(c) = 0$, which amounts to the familiar task of finding the critical points of f. The critical points satisfy

$$f'(x) = 3x^2 - 14x + 10 = 0.$$

Using the quadratic formula, the roots are

$$x = \frac{7 \pm \sqrt{19}}{3}, \quad \text{or} \quad x \approx 0.88 \quad \text{and} \quad x \approx 3.79.$$

As shown in Figure 4.67, the graph of f has two points at which the tangent line is horizontal.

Related Exercises 7–12 ◄

Mean Value Theorem

The Mean Value Theorem is easily understood with the aid of a picture. Figure 4.68 shows a function f differentiable on (a, b) with a secant line passing through $(a, f(a))$ and $(b, f(b))$; its slope is the average rate of change of f over $[a, b]$. The Mean Value Theorem claims that there exists a point c in (a, b) at which the slope of the tangent line at c is equal to the slope of the secant line. In other words, we can find a point on the graph of f where the tangent line is parallel to the secant line.

> **THEOREM 4.9 Mean Value Theorem**
> If f is continuous on the closed interval $[a, b]$ and differentiable on the open interval (a, b), then there is at least one point c in (a, b) such that
> $$\frac{f(b) - f(a)}{b - a} = f'(c).$$

Proof The strategy of the proof is to use the function f of the Mean Value Theorem to form a new function g that satisfies Rolle's Theorem. Notice that the continuity and differentiability conditions of Rolle's Theorem and the Mean Value Theorem are the same. We devise g so that it satisfies the condition that $g(a) = g(b) = 0$.

As shown in Figure 4.69, the chord between $(a, f(a))$ and $(b, f(b))$ is a segment of the straight line described by a function ℓ. We now define a new function g that measures the vertical distance between the given function f and the line ℓ. This function is simply $g(x) = f(x) - \ell(x)$. Because f and ℓ are continuous on $[a, b]$ and differentiable on (a, b), it follows that g is also continuous on $[a, b]$ and differentiable on (a, b). Furthermore, because the graphs of f and ℓ intersect at $x = a$ and $x = b$, we have $g(a) = f(a) - \ell(a) = 0$ and $g(b) = f(b) - \ell(b) = 0$.

We now have a function g that satisfies the conditions of Rolle's Theorem. By that theorem, we are guaranteed the existence of at least one point c in the interval (a, b) such that $g'(c) = 0$. By the definition of g, this condition implies that $f'(c) - \ell'(c) = 0$, or $f'(c) = \ell'(c)$.

We are almost finished. What is $\ell'(c)$? It is just the slope of the chord, which is

$$\frac{f(b) - f(a)}{b - a}.$$

Therefore, $f'(c) = \ell'(c)$ implies that

$$\frac{f(b) - f(a)}{b - a} = f'(c).$$
◄

QUICK CHECK 2 Sketch the graph of a function that illustrates why the continuity condition of the Mean Value Theorem is needed. Sketch the graph of a function that illustrates why the differentiability condition of the Mean Value Theorem is needed. ◄

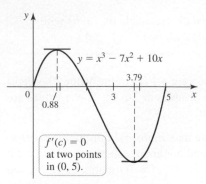

FIGURE 4.67

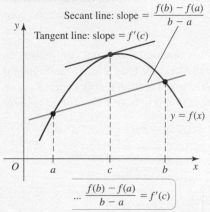

FIGURE 4.68

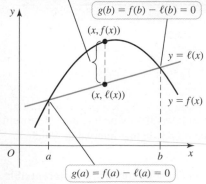

FIGURE 4.69

> The proofs of Rolle's Theorem and the Mean Value Theorem are nonconstructive: The theorems claim that a certain point exists, but their proofs do not say how to find it.

The following situation offers an interpretation of the Mean Value Theorem. Imagine taking 2 hours to drive to a town 100 miles away. While your average speed is 100 mi/2 hr = 50 mi/hr, your instantaneous speed (measured by the speedometer) almost certainly varies. The Mean Value Theorem says that at some point during the trip, your instantaneous speed equals your average speed, which is 50 mi/hr.

> Meteorologists look for "steep" lapse rates in the layer of the atmosphere where the pressure is between 700 and 500 hPa (hectopascals). This range of pressure typically corresponds to altitudes between 3 km and 5.5 km. The data in Example 2 were recorded in Denver at nearly the same time a tornado struck 50 mi to the north.

EXAMPLE 2   **Mean Value Theorem in action**   The *lapse rate* is the rate at which the temperature T decreases in the atmosphere with respect to increasing altitude z. It is typically reported in units of °C/km and is defined by $\gamma = -dT/dz$. When the lapse rate rises above 7°C/km in a certain layer of the atmosphere, it indicates favorable conditions for thunderstorm and tornado formation, provided other atmospheric conditions are also present.

Suppose the temperature at $z = 2.9$ km is $T = 7.6$°C and the temperature at $z = 5.6$ km is $T = -14.3$°C. Assume also that the temperature function is continuous and differentiable at all altitudes of interest. What can a meteorologist conclude from these data?

SOLUTION   Figure 4.70 shows the two data points plotted on a graph of altitude and temperature. The slope of the line joining these points is

$$\frac{-14.3°C - 7.6°C}{5.6\,km - 2.9\,km} = -8.1°C/km,$$

which means, on average, the temperature is decreasing at 8.1°C per km in the layer of air between 2.9 km and 5.6 km. With only two data points, we cannot know the entire temperature profile. The Mean Value Theorem, however, guarantees that there is at least one altitude at which $dT/dz = -8.1$°C/km. At each such altitude, the lapse rate is $\gamma = -dT/dz = 8.1$°C/km. Because this lapse rate is above the 7°C/km threshold associated with unstable weather, the meteorologist might expect an increased likelihood of severe storms.

Related Exercises 13–14 ◄

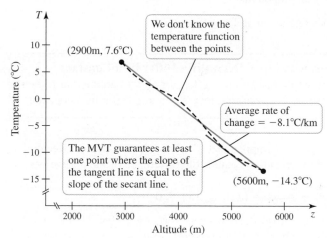

FIGURE 4.70

EXAMPLE 3   **Verifying the Mean Value Theorem**   Determine whether the function $f(x) = 2x^3 - 3x + 1$ satisfies the conditions of the Mean Value Theorem on the interval $[-2, 2]$. If so, find the point(s) guaranteed to exist by the theorem.

SOLUTION   The polynomial f is everywhere continuous and differentiable, so it satisfies the conditions of the Mean Value Theorem. The average rate of change of the function on the interval $[-2, 2]$ is

$$\frac{f(2) - f(-2)}{2 - (-2)} = \frac{11 - (-9)}{4} = 5.$$

The goal is to find points in $(-2, 2)$ at which the line tangent to the curve has a slope of 5—that is, to find points at which $f'(x) = 5$. Differentiating f, this condition becomes

$$f'(x) = 6x^2 - 3 = 5 \quad \text{or} \quad x^2 = \frac{4}{3}.$$

Therefore, the points guaranteed to exist by the Mean Value Theorem are $x = \pm2/\sqrt{3} \approx \pm1.15$. The tangent lines have slope 5 at the points $(\pm2/\sqrt{3}, f(\pm2/\sqrt{3}))$ (Figure 4.71).

Related Exercises 15–22 ◄

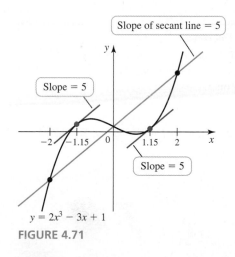

FIGURE 4.71

Consequences of the Mean Value Theorem

We close with several results—some postponed from previous sections—that follow from the Mean Value Theorem.

We already know that the derivative of a constant function is zero; that is, if $f(x) = C$, then $f'(x) = 0$ (Theorem 3.2). Theorem 4.10 states the converse of this result.

> **THEOREM 4.10   Zero Derivative Implies Constant Function**
> If f is differentiable and $f'(x) = 0$ at all points of an interval I, then f is a constant function on I.

Proof Suppose $f'(x) = 0$ on $[a, b]$, where a and b are distinct points of I. By the Mean Value Theorem, there exists a point c in (a, b) such that

$$\frac{f(b) - f(a)}{b - a} = \underbrace{f'(c) = 0.}_{\substack{f'(x) = 0 \text{ for} \\ \text{all } x \text{ in } I}}$$

Multiplying both sides of this equation by $b - a \neq 0$, it follows that $f(b) = f(a)$, and this is true for every pair of points a and b in I. If $f(b) = f(a)$ for every pair of points in an interval, then f is a constant function on that interval.   ◄

Theorem 4.11 builds on the conclusion of Theorem 4.10.

> **THEOREM 4.11   Functions with Equal Derivatives Differ by a Constant**
> If two functions have the property that $f'(x) = g'(x)$ for all x of an interval I, then $f(x) - g(x) = C$ on I, where C is a constant; that is, f and g differ by a constant.

QUICK CHECK 3 Give two linear functions f and g that satisfy $f'(x) = g'(x)$; that is, the lines have equal slopes. Show that f and g differ by a constant. ◄

Proof The fact that $f'(x) = g'(x)$ on I implies that $f'(x) - g'(x) = 0$ on I. Recall that the derivative of a difference of two functions equals the difference of the derivatives, so we can write

$$f'(x) - g'(x) = (f - g)'(x) = 0.$$

Now we have a function $f - g$ whose derivative is zero on I. By Theorem 4.10, $f(x) - g(x) = C$ for all x in I, where C is a constant; that is, f and g differ by a constant.   ◄

In Section 4.2, we stated and gave an argument to support the test for intervals of increase and decrease. With the Mean Value Theorem, we can prove this important result.

> **THEOREM 4.12   Intervals of Increase and Decrease**
> Suppose f is continuous on an interval I and differentiable at all interior points of I. If $f'(x) > 0$ at all interior points of I, then f is increasing on I. If $f'(x) < 0$ at all interior points of I, then f is decreasing on I.

Proof Let a and b be any two distinct points in the interval I with $b > a$. By the Mean Value Theorem,

$$\frac{f(b) - f(a)}{b - a} = f'(c)$$

for some c between a and b. Equivalently,

$$f(b) - f(a) = f'(c)(b - a).$$

Notice that $b - a > 0$ by assumption. So, if $f'(c) > 0$, then $f(b) - f(a) > 0$. Therefore, for all a and b in I with $b > a$, we have $f(b) > f(a)$, which implies that f is increasing on I. Similarly if $f'(c) < 0$, then $f(b) - f(a) < 0$ or $f(b) < f(a)$. It follows that f is decreasing on I.   ◄

SECTION 4.6 EXERCISES

Review Questions

1. Explain Rolle's Theorem with a sketch.

2. Draw the graph of a function for which the conclusion of Rolle's Theorem does not hold.

3. Explain why Rolle's Theorem cannot be applied to the function $f(x) = |x|$ on the interval $[-a, a]$ for any $a > 0$.

4. Explain the Mean Value Theorem with a sketch.

5. Draw the graph of a function for which the conclusion of the Mean Value Theorem does not hold.

6. At what points c does the conclusion of the Mean Value Theorem hold for $f(x) = x^3$ on the interval $[-10, 10]$?

Basic Skills

7–12. Rolle's Theorem *Determine whether Rolle's Theorem applies to the following functions on the given interval. If so, find the point(s) that are guaranteed to exist by Rolle's Theorem.*

7. $f(x) = x(x - 1)^2$; $[0, 1]$

8. $f(x) = \sin 2x$; $[0, \pi/2]$

9. $f(x) = \cos 4x$; $[\pi/8, 3\pi/8]$

10. $f(x) = 1 - |x|$; $[-1, 1]$

11. $f(x) = 1 - x^{2/3}$; $[-1, 1]$

12. $f(x) = x^3 - 2x^2 - 8x$; $[-4, 2]$

13. **Lapse rates in the atmosphere** Concurrent measurements indicate that at an elevation of 6.1 km, the temperature is $-10.3°C$, and at an elevation of 3.2 km, the temperature is 8.0°C. Based on the Mean Value Theorem, can you conclude that the lapse rate exceeds the threshold value of 7°C/km at some intermediate elevation? Explain.

14. **Drag racer acceleration** The fastest drag racers can reach a speed of 330 mi/hr over a quarter-mile strip in 4.45 s (from a standing start). Complete the following sentence about such a drag racer: At some point during the race, the maximum acceleration of the drag race is at least _____ mi/hr/s.

15–22. Mean Value Theorem

a. *Determine whether the Mean Value Theorem applies to the following functions on the given interval $[a, b]$.*

b. *If so, find or approximate the point(s) that are guaranteed to exist by the Mean Value Theorem.*

c. *Make a sketch of the function and the line that passes through $(a, f(a))$ and $(b, f(b))$. Then, mark the points (if they exist) at which the slope of the function equals the slope of the secant line.*

15. $f(x) = 7 - x^2$; $[-1, 2]$

16. $f(x) = 3 \sin 2x$; $[0, \pi/4]$

17. $f(x) = e^x$; $[0, \ln 4]$

18. $f(x) = \ln 2x$; $[1, e]$

19. $f(x) = \sin^{-1} x$; $[0, 1/2]$

20. $f(x) = x + 1/x$; $[1, 3]$

21. $f(x) = 2x^{1/3}$; $[-8, 8]$

22. $f(x) = x/(x + 2)$; $[-1, 2]$

Further Explorations

23. **Explain why or why not** Determine whether the following statements are true and give an explanation or counterexample.

a. The continuous function $f(x) = 1 - |x|$ satisfies the conditions of the Mean Value Theorem on the interval $[-1, 1]$.

b. Two differentiable functions that differ by a constant always have the same derivative.

c. If $f'(x) = 0$, then $f(x) = 10$.

24–26. Questions about derivatives

24. Without evaluating derivatives, which of the following functions have the same derivative: $f(x) = \ln x$, $g(x) = \ln 2x$, $h(x) = \ln x^2$, $p(x) = \ln 10x^2$?

25. Without evaluating derivatives, which of the functions $g(x) = 2x^{10}$, $h(x) = x^{10} + 2$, or $p(x) = x^{10} - \ln 2$ have the same derivative as $f(x) = x^{10}$?

26. Find all functions f whose derivative is $f'(x) = x + 1$.

27. **Mean Value Theorem and graphs** By visual inspection, locate all points on the graph at which the slope of the tangent line equals the average rate of change of the function over the interval $[-4, 4]$.

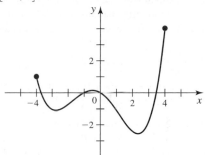

Applications

28. **Avalanche forecasting** Avalanche forecasters measure the *temperature gradient dT/dh*, which is the rate at which the temperature in a snowpack T changes with respect to its depth h. If the temperature gradient is large, it may lead to a weak layer of snow in the snowpack. When these weak layers collapse, avalanches occur. Avalanche forecasters use the following rule of thumb: If dT/dh exceeds 10°C/m anywhere in the snowpack, conditions are favorable for weak layer formation and the risk of avalanche increases. Assume the temperature function is continuous and differentiable.

a. An avalanche forecaster digs a snow pit and takes two temperature measurements. At the surface ($h = 0$) the temperature is $-12°C$. At a depth of 1.1 m, the temperature is 2°C. Using the Mean Value Theorem, what can he conclude about the temperature gradient? Is the formation of a weak layer likely?

b. One mile away, a skier finds that the temperature at a depth of 1.4 m is $-1°C$, and at the surface it is $-12°C$. What can be concluded about the temperature gradient? Is the formation of a weak layer in her location likely?

c. Because snow is an excellent insulator, the temperature of snow-covered ground is near 0°C. Furthermore, the surface temperature of snow in a particular area does not vary much from one location to the next. Explain why a weak layer is more likely to form in places where the snowpack is not too deep.

d. The term *isothermal* is used to describe the situation where all layers of the snowpack are at the same temperature (typically near the freezing point). Is a weak layer likely to form in isothermal snow? Explain.

29. Mean Value Theorem and the police A state patrol officer saw a car start from rest at a highway on-ramp. She radioed ahead to another patrol officer 30 mi along the highway. When the car reached the location of the second officer 28 min later, it was clocked going 60 mi/hr. The driver of the car was given a ticket for exceeding the 60-mi/hr speed limit. Why can the officer conclude that the driver exceeded the speed limit?

30. Mean Value Theorem and the police again Compare carefully to Exercise 29. A state patrol officer saw a car start from rest at a highway on-ramp. She radioed ahead to another officer 30 mi along the highway. When the car reached the location of the second officer 30 min later, it was clocked going 60 mi/hr. Can the patrol officer conclude that the driver exceeded the speed limit?

31. Running pace Explain why if a runner completes a 6.2-mi (10-km) race in 32 min, then he must have been running at exactly 11 mi/hr at least twice in the race. Assume the runners speed is 0 at the finish line.

Additional Exercises

32. Mean Value Theorem for linear functions Interpret the Mean Value Theorem when it is applied to any linear function.

33. Mean Value Theorem for quadratic functions Consider the quadratic function $f(x) = Ax^2 + Bx + C$, where A, B, and C are real numbers with $A \neq 0$. Show that when the Mean Value Theorem is applied to f on the interval $[a, b]$, the number c guaranteed by the theorem is the midpoint of the interval.

34. Means

a. Show that the point c guaranteed to exist by the Mean Value Theorem for $f(x) = x^2$ on $[a, b]$ is the arithmetic mean of a and b; that is, $c = (a + b)/2$.

b. Show that the point c guaranteed to exist by the Mean Value Theorem for $f(x) = 1/x$ on $[a, b]$, where $0 < a < b$, is the geometric mean of a and b; that is, $c = \sqrt{ab}$.

35. Equal derivatives Verify that the functions $f(x) = \tan^2 x$ and $g(x) = \sec^2 x$ have the same derivative. What can you say about the difference $f - g$? Explain.

36. Equal derivatives Verify that the functions $f(x) = \sin^2 x$ and $g(x) = -\cos^2 x$ have the same derivative. What can you say about the difference $f - g$? Explain.

37. 100-m speed The Jamaican sprinter Usain Bolt set a world record of 9.58 s in the 100-meter dash in the summer of 2009. Did his speed ever exceed 37 km/hr during the race? Explain.

38. Condition for nondifferentiabilty Suppose $f'(x) < 0 < f''(x)$ for $x < a$ and $f'(x) > 0 > f''(x)$ for $x > a$. Prove that f is not differentiable at $x = a$. (*Hint:* Assume f is differentiable at a, and apply the Mean Value Theorem to f'.) More generally, show that if f' and f'' change sign at the same point, then f is not differentiable at that point.

39. Generalized Mean Value Theorem Suppose f and g are functions that are continuous on $[a, b]$ and differentiable on (a, b), where $g(a) \neq g(b)$. Then, there is a point c in (a, b) at which

$$\frac{f(b) - f(a)}{g(b) - g(a)} = \frac{f'(c)}{g'(c)}.$$

This result is known as the **Generalized (or Cauchy's) Mean Value Theorem**.

a. If $g(x) = x$, then show that the Generalized Mean Value Theorem reduces to the Mean Value Theorem.

b. Suppose $f(x) = x^2 - 1$, $g(x) = 4x + 2$, and $[a, b] = [0, 1]$. Find a value of c satisfying the Generalized Mean Value Theorem.

QUICK CHECK ANSWERS

1. $x = 2$ **2.** The functions shown in Figure 4.66 provide examples.   **3.** The graphs of $f(x) = 3x$ and $g(x) = 3x + 2$ have the same slope. Note that $f(x) - g(x) = -2$, a constant. ◄

4.7 L'Hôpital's Rule

The study of limits in Chapter 2 was thorough but not exhaustive. Some limits, called *indeterminate forms*, cannot generally be evaluated using the techniques presented in Chapter 2. These limits tend to be the more interesting limits that arise in practice. A powerful result called *l'Hôpital's Rule* enables us to evaluate such limits with relative ease.

Here is how indeterminate forms arise. If f is a *continuous* function at a point a, then we know that $\lim_{x \to a} f(x) = f(a)$, allowing the limit to be evaluated by computing $f(a)$. But there are many limits that cannot be evaluated by substitution. In fact, we encountered such a limit in Section 3.4:

$$\lim_{x \to 0} \frac{\sin x}{x} = 1.$$

If we attempt to substitute $x = 0$ into $(\sin x)/x$, we get $0/0$, which has no meaning. Yet we proved that $(\sin x)/x$ has limit 1 at $x = 0$ (Theorem 3.11). This limit is an example of an *indeterminate form*.

The meaning of an *indeterminate form* is further illustrated by $\displaystyle\lim_{x \to \infty} \frac{ax}{x}$, where $a \neq 0$. This limit has the indeterminate form ∞/∞ (meaning that the numerator and denominator of ax/x become arbitrarily large in magnitude as $x \to \infty$), but the actual value of the limit is $\displaystyle\lim_{x \to \infty} \frac{ax}{x} = \lim_{x \to \infty} a = a$. In general, a limit with the form ∞/∞ or $0/0$ can have *any* value—which is why these limits must be handled carefully.

L'Hôpital's Rule for the Form 0/0

Consider a function of the form $f(x)/g(x)$ and assume that $\displaystyle\lim_{x \to a} f(x) = \lim_{x \to a} g(x) = 0$.

Then, the limit $\displaystyle\lim_{x \to a} \frac{f(x)}{g(x)}$ has the indeterminate form $0/0$. We first state l'Hôpital's Rule and then prove a special case.

> **THEOREM 4.13 L'Hôpital's Rule**
>
> Suppose f and g are differentiable on an open interval I containing a with $g'(x) \neq 0$ on I when $x \neq a$. If $\displaystyle\lim_{x \to a} f(x) = \lim_{x \to a} g(x) = 0$, then
>
> $$\lim_{x \to a} \frac{f(x)}{g(x)} = \lim_{x \to a} \frac{f'(x)}{g'(x)},$$
>
> provided the limit on the right side exists (or is $\pm\infty$). The rule also applies if $x \to a$ is replaced by $x \to \pm\infty$, $x \to a^+$, or $x \to a^-$.

Proof (special case) The proof of this theorem relies on the Generalized Mean Value Theorem (Exercise 39 of Section 4.6). We prove a special case of the theorem in which we assume that f' and g' are continuous at a, $f(a) = g(a) = 0$, and $g'(a) \neq 0$. We have

$$\lim_{x \to a} \frac{f'(x)}{g'(x)} = \frac{f'(a)}{g'(a)} \qquad \text{Continuity of } f' \text{ and } g'$$

$$= \frac{\displaystyle\lim_{x \to a} \frac{f(x) - f(a)}{x - a}}{\displaystyle\lim_{x \to a} \frac{g(x) - g(a)}{x - a}} \qquad \text{Definition of } f'(a) \text{ and } g'(a)$$

$$= \lim_{x \to a} \frac{\dfrac{f(x) - f(a)}{x - a}}{\dfrac{g(x) - g(a)}{x - a}} \qquad \text{Limit of a quotient, } g'(a) \neq 0$$

$$= \lim_{x \to a} \frac{f(x) - f(a)}{g(x) - g(a)} \qquad \text{Cancel } x - a.$$

$$= \lim_{x \to a} \frac{f(x)}{g(x)} \qquad f(a) = g(a) = 0$$

◄

▶ The notations $0/0$ and ∞/∞ are merely symbols used to describe various types of indeterminate forms. The notation $0/0$ does not imply division by 0.

▶ Guillaume François l'Hôpital (lo-pee-tal) (1661–1704) is credited with writing the first calculus textbook. Much of the material in the book, including l'Hôpital's Rule, was provided by the Swiss mathematician Johann Bernoulli (1667–1748).

▶ The definition of the derivative provides an example of an indeterminate form:

$$f'(x) = \lim_{h \to 0} \frac{f(x + h) - f(x)}{h}$$

has the form $0/0$.

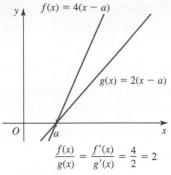

$f(x) = 4(x - a)$

$g(x) = 2(x - a)$

$$\frac{f(x)}{g(x)} = \frac{f'(x)}{g'(x)} = \frac{4}{2} = 2$$

FIGURE 4.72

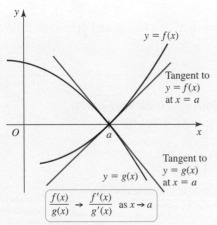

$y = f(x)$

Tangent to
$y = f(x)$
at $x = a$

Tangent to
$y = g(x)$
at $x = a$

$y = g(x)$

$$\frac{f(x)}{g(x)} \to \frac{f'(x)}{g'(x)} \text{ as } x \to a$$

FIGURE 4.73

QUICK CHECK 1 Which of the following functions lead to an indeterminate form as $x \to 0$: $f(x) = x^2/(x + 2)$, $g(x) = (\tan 3x)/x$, or $h(x) = (1 - \cos x)/x^2$? ◄

▶ This limit in part (a) can also be evaluated by factoring the numerator and canceling $(x - 1)$:

$$\lim_{x \to 1} \frac{x^3 + x^2 - 2x}{x - 1}$$

$$= \lim_{x \to 1} \frac{x(x - 1)(x + 2)}{x - 1}$$

$$= \lim_{x \to 1} x(x + 2) = 3$$

The geometry of l'Hôpital's Rule offers some insight. First consider two *linear* functions, f and g, whose graphs both pass through the point $(a, 0)$ with slopes 4 and 2 respectively; this means that

$$f(x) = 4(x - a) \quad \text{and} \quad g(x) = 2(x - a).$$

Furthermore $f(a) = g(a) = 0$, $f'(x) = 4$, and $g'(x) = 2$ (Figure 4.72).

Looking at the quotient f/g, we see that

$$\frac{f(x)}{g(x)} = \frac{4(x - a)}{2(x - a)} = \frac{4}{2} = \frac{f'(x)}{g'(x)} \qquad \text{Exactly}$$

This argument may be generalized, and we find that for any linear functions f and g with $f(a) = g(a) = 0$,

$$\lim_{x \to a} \frac{f(x)}{g(x)} = \lim_{x \to a} \frac{f'(x)}{g'(x)}$$

provided $g'(a) \ne 0$.

If f and g are not linear functions, we replace them by their linear approximations at $(a, 0)$ (Figure 4.73). Zooming in on the point a, the curves are close to their respective tangent lines $y = f'(a)(x - a)$ and $y = g'(a)(x - a)$, which have slopes $f'(a)$ and $g'(a) \ne 0$, respectively. Therefore, near $x = a$ we have

$$\frac{f(x)}{g(x)} \approx \frac{f'(a)(x - a)}{g'(a)(x - a)} = \frac{f'(a)}{g'(a)}.$$

Therefore, the ratio of the functions is well approximated by the ratio of the derivatives. And in the limit as $x \to a$, we again have

$$\lim_{x \to a} \frac{f(x)}{g(x)} = \lim_{x \to a} \frac{f'(x)}{g'(x)}.$$

EXAMPLE 1   Using l'Hôpital's Rule Evaluate the following limits.

a. $\displaystyle\lim_{x \to 1} \frac{x^3 + x^2 - 2x}{x - 1}$ **b.** $\displaystyle\lim_{x \to 0} \frac{\sqrt{9 + 3x} - 3}{x}$

SOLUTION

a. Direct substitution of $x = 1$ into $\dfrac{x^3 + x^2 - 2x}{x - 1}$ produces the indeterminate form $0/0$. Applying l'Hôpital's Rule with $f(x) = x^3 + x^2 - 2x$ and $g(x) = x - 1$ gives

$$\lim_{x \to 1} \frac{x^3 + x^2 - 2x}{x - 1} = \lim_{x \to 1} \frac{f'(x)}{g'(x)} = \lim_{x \to 1} \frac{3x^2 + 2x - 2}{1} = 3.$$

b. Substituting $x = 0$ into this function produces the indeterminate form $0/0$. Let $f(x) = \sqrt{9 + 3x} - 3$ and $g(x) = x$, and note that $f'(x) = \dfrac{3}{2\sqrt{9 + 3x}}$ and $g'(x) = 1$. Applying l'Hôpital's Rule, we have

$$\underbrace{\lim_{x \to 0} \frac{\sqrt{9 + 3x} - 3}{x}}_{f/g} = \underbrace{\lim_{x \to 0} \frac{\dfrac{3}{2\sqrt{9 + 3x}}}{1}}_{f'/g'} = \frac{1}{2}.$$

Related Exercises 13–18 ◄

L'Hôpital's Rule requires evaluating $\lim_{x \to a} f'(x)/g'(x)$. It may happen that this second limit is another indeterminate form to which l'Hôpital's Rule may be applied again.

EXAMPLE 2 L'Hôpital's Rule repeated Evaluate the following limits.

a. $\lim_{x \to 0} \dfrac{e^x - x - 1}{x^2}$ **b.** $\lim_{x \to 2} \dfrac{x^3 - 3x^2 + 4}{x^4 - 4x^3 + 7x^2 - 12x + 12}$.

SOLUTION

a. This limit has the indeterminate form 0/0. Applying l'Hôpital's Rule, we have

$$\lim_{x \to 0} \frac{e^x - x - 1}{x^2} = \lim_{x \to 0} \frac{e^x - 1}{2x},$$

which is another limit of the form 0/0. Therefore, we apply l'Hôpital's Rule again:

$$\lim_{x \to 0} \frac{e^x - x - 1}{x^2} = \lim_{x \to 0} \frac{e^x - 1}{2x} \quad \text{L'Hôpital's Rule}$$

$$= \lim_{x \to 0} \frac{e^x}{2} \quad \text{L'Hôpital's Rule again}$$

$$= \frac{1}{2} \quad \text{Evaluate limit.}$$

b. Evaluating the numerator and denominator at $x = 2$, we see that this limit has the form 0/0. Applying l'Hôpital's Rule twice, we have

$$\lim_{x \to 2} \frac{x^3 - 3x^2 + 4}{x^4 - 4x^3 + 7x^2 - 12x + 12} = \underbrace{\lim_{x \to 2} \frac{3x^2 - 6x}{4x^3 - 12x^2 + 14x - 12}}_{\text{limit of the form 0/0}} \quad \text{L'Hôpital's Rule}$$

$$= \lim_{x \to 2} \frac{6x - 6}{12x^2 - 24x + 14} \quad \text{L'Hôpital's Rule again}$$

$$= \frac{3}{7} \quad \text{Evaluate limit.}$$

It is easy to overlook a crucial step in this computation: After applying l'Hôpital's Rule the first time, you *must* establish that the new limit is an indeterminate form before applying l'Hôpital's Rule a second time. *Related Exercises 19–26* ◄

Indeterminate Form ∞/∞

L'Hôpital's Rule also applies directly to limits of the form $\lim_{x \to a} f(x)/g(x)$, where $\lim_{x \to a} f(x) = \pm\infty$ and $\lim_{x \to a} g(x) = \pm\infty$; this indeterminate form is denoted ∞/∞. The proof of this result is found in advanced books.

THEOREM 4.14 L'Hôpital's Rule (∞/∞)
Suppose that f and g are differentiable on an open interval I containing a, with $g'(x) \neq 0$ on I when $x \neq a$. If $\lim_{x \to a} f(x) = \pm\infty$ and $\lim_{x \to a} g(x) = \pm\infty$, then

$$\lim_{x \to a} \frac{f(x)}{g(x)} = \lim_{x \to a} \frac{f'(x)}{g'(x)}$$

provided the limit on the right side exists (or is $\pm\infty$). The rule also applies for $x \to \pm\infty$, $x \to a^+$, or $x \to a^-$.

QUICK CHECK 2 Which of the following functions lead to an indeterminate form as $x \to \infty$: $f(x) = \sin x/x$, $g(x) = 2^x/x^2$, or $h(x) = (3x^2 + 4)/x^2$? ◄

EXAMPLE 3 L'Hôpital's Rule for ∞/∞ Evaluate the following limits.

a. $\displaystyle\lim_{x \to \infty} \frac{4x^3 - 6x^2 + 1}{2x^3 - 10x + 3}$ **b.** $\displaystyle\lim_{x \to \pi/2^-} \frac{1 + \tan x}{\sec x}$

SOLUTION

> As shown in Section 2.5, this limit could also be evaluated by first dividing the numerator and denominator by x^3.

a. This limit has the indeterminate form ∞/∞ because both the numerator and the denominator approach $+\infty$ as $x \to \infty$. Applying l'Hôpital's Rule three times, we have

$$\underbrace{\lim_{x \to \infty} \frac{4x^3 - 6x^2 + 1}{2x^3 - 10x + 3}}_{\infty/\infty} = \underbrace{\lim_{x \to \infty} \frac{12x^2 - 12x}{6x^2 - 10}}_{\infty/\infty} = \underbrace{\lim_{x \to \infty} \frac{24x - 12}{12x}}_{\infty/\infty} = \lim_{x \to \infty} \frac{24}{12} = 2.$$

b. In this limit both the numerator and the denominator approach $+\infty$ as $x \to \pi/2^-$. L'Hôpital's Rule gives us

$$
\begin{aligned}
\lim_{x \to \pi/2^-} \frac{1 + \tan x}{\sec x} &= \lim_{x \to \pi/2^-} \frac{\sec^2 x}{\sec x \tan x} &&\text{L'Hôpital's Rule} \\
&= \lim_{x \to \pi/2^-} \frac{1}{\sin x} &&\text{Simplify.} \\
&= 1 &&\text{Evaluate limit.}
\end{aligned}
$$

Related Exercises 27–30 ◄

Related Indeterminate Forms: $0 \cdot \infty$ and $\infty - \infty$

The limit $\displaystyle\lim_{x \to a} f(x)g(x)$, where $\displaystyle\lim_{x \to a} f(x) = 0$ and $\displaystyle\lim_{x \to a} g(x) = \pm\infty$, is an indeterminate form that we denote $0 \cdot \infty$. *L'Hôpital's Rule cannot be directly applied to this limit.* The following examples illustrate how this indeterminate form can be recast in the form $0/0$ or ∞/∞.

EXAMPLE 4 L'Hôpital's Rule for $0 \cdot \infty$ Evaluate $\displaystyle\lim_{x \to \infty} x^2 \sin\left(\frac{1}{4x^2}\right)$.

SOLUTION This limit has the form $0 \cdot \infty$. A common technique that converts this form to either $0/0$ or ∞/∞ is to *divide by the reciprocal*. We rewrite the limit and apply l'Hôpital's Rule:

$$\underbrace{\lim_{x \to \infty} x^2 \sin\left(\frac{1}{4x^2}\right)}_{0 \cdot \infty \text{ form}} = \underbrace{\lim_{x \to \infty} \frac{\sin\left(\dfrac{1}{4x^2}\right)}{\left(\dfrac{1}{x^2}\right)}}_{\text{recast in } 0/0 \text{ form}} \qquad x^2 = \frac{1}{1/x^2}$$

$$
\begin{aligned}
&= \lim_{x \to \infty} \frac{\cos\left(\dfrac{1}{4x^2}\right)\dfrac{1}{4}(-2x^{-3})}{-2x^{-3}} &&\text{L'Hôpital's Rule} \\
&= \frac{1}{4} \lim_{x \to \infty} \cos\left(\frac{1}{4x^2}\right) &&\text{Simplify.} \\
&= \frac{1}{4} &&\frac{1}{4x^2} \to 0, \cos 0 = 1
\end{aligned}
$$

QUICK CHECK 3 What is the form of the limit $\displaystyle\lim_{x \to \pi/2^-} (x - \pi/2)(\tan x)$? Write it in the form $0/0$. ◄

Related Exercises 31–34 ◄

Indeterminate Form $\infty - \infty$ Limits of the form $\lim\limits_{x \to a} (f(x) - g(x))$, where $\lim\limits_{x \to a} f(x) = \infty$ and $\lim\limits_{x \to a} g(x) = \infty$, are indeterminate forms that we denote $\infty - \infty$. L'Hôpital's Rule cannot be applied directly to an $\infty - \infty$ form. It must first be expressed in the form $0/0$ or ∞/∞. With the $\infty - \infty$ form, it is easy to reach erroneous conclusions. For example, if $f(x) = 3x + 5$ and $g(x) = 3x$, then the $\infty - \infty$ form has the limit

$$\lim_{x \to \infty} [(3x + 5) - (3x)] = 5.$$

However, if $f(x) = 3x$ and $g(x) = 2x$, then the $\infty - \infty$ form has the limit

$$\lim_{x \to \infty} (3x - 2x) = \lim_{x \to \infty} x = \infty.$$

These examples show again why indeterminate forms are deceptive. Before proceeding, we introduce another useful technique.

Occasionally, it helps to convert a limit as $x \to \infty$ to a limit as $t \to 0^+$ (or vice versa) by a *change of variables*. To evaluate $\lim\limits_{x \to \infty} f(x)$, we define $t = 1/x$ and note that as $x \to \infty, t \to 0^+$. Then,

$$\lim_{x \to \infty} f(x) = \lim_{t \to 0^+} f\left(\frac{1}{t}\right).$$

This idea is illustrated in the next example.

EXAMPLE 5 **L'Hôpital's Rule for $\infty - \infty$** Evaluate $\lim\limits_{x \to \infty} (x - \sqrt{x^2 - 3x})$.

SOLUTION As $x \to \infty$, both terms in the difference $x - \sqrt{x^2 - 3x}$ approach ∞ and this limit has the form $\infty - \infty$. We first factor x from the expression and form a quotient:

$$
\begin{aligned}
\lim_{x \to \infty} (x - \sqrt{x^2 - 3x}) &= \lim_{x \to \infty} (x - \sqrt{x^2(1 - 3/x)}) && \text{Factor } x^2 \text{ under square root.} \\
&= \lim_{x \to \infty} x(1 - \sqrt{1 - 3/x}) && x > 0, \text{ so } \sqrt{x^2} = x \\
&= \lim_{x \to \infty} \frac{1 - \sqrt{1 - 3/x}}{1/x} && \begin{array}{l}\text{Write } 0 \cdot \infty \text{ form as } 0/0 \\ \text{form}; x = \dfrac{1}{1/x}\end{array}
\end{aligned}
$$

This new limit has the form $0/0$, and l'Hôpital's Rule may be applied.

One way to proceed is to use the change of variables $t = 1/x$:

$$
\begin{aligned}
\lim_{x \to \infty} \frac{1 - \sqrt{1 - 3/x}}{1/x} &= \lim_{t \to 0^+} \frac{1 - \sqrt{1 - 3t}}{t} && \text{Let } t = 1/x; \text{ replace } \lim_{x \to \infty} \text{ by } \lim_{t \to 0^+}. \\
&= \lim_{t \to 0^+} \frac{\dfrac{3}{2\sqrt{1 - 3t}}}{1} && \text{L'Hôpital's Rule} \\
&= \frac{3}{2} && \text{Evaluate limit.}
\end{aligned}
$$

Related Exercises 35–38 ◄

Indeterminate Forms $1^\infty, 0^0,$ and ∞^0

The indeterminate forms $1^\infty, 0^0,$ and ∞^0 all arise in limits of the form $\lim\limits_{x \to a} f(x)^{g(x)}$, where $x \to a$ could be replaced by $x \to a^\pm$ or $x \to \pm\infty$. L'Hôpital's Rule cannot be applied directly to the indeterminate forms $1^\infty, 0^0,$ and ∞^0. They must first be expressed in the form $0/0$ or ∞/∞. Here is how we proceed.

The inverse relationship between $\ln x$ and e^x says that $f^g = e^{g \ln f}$, so we first write

$$\lim_{x \to a} f(x)^{g(x)} = \lim_{x \to a} e^{g(x) \ln f(x)}.$$

By the continuity of the exponential function, we switch the order of the limit and the exponential function; therefore,

$$\lim_{x \to a} f(x)^{g(x)} = \lim_{x \to a} e^{g(x) \ln f(x)} = e^{\lim_{x \to a} g(x) \ln f(x)}$$

provided $\lim_{x \to a} g(x) \ln f(x)$ exists. Therefore, $\lim_{x \to a} f(x)^{g(x)}$ is evaluated using the following two steps.

PROCEDURE   **Indeterminate forms 1^∞, 0^0, and ∞^0**

Assume $\lim_{x \to a} f(x)^{g(x)}$ has the indeterminate form 1^∞, 0^0, or ∞^0.

1. Evaluate $L = \lim_{x \to a} g(x) \ln f(x)$. This limit can often be put in the form $0/0$ or ∞/∞, both of which are handled by l'Hôpital's Rule.

2. Then $\lim_{x \to a} f(x)^{g(x)} = e^L$.

QUICK CHECK 4 Explain why a limit of the form 0^∞ is not an indeterminate form. ◄

EXAMPLE 6   **Indeterminate forms 0^0 and 1^∞**   Evaluate the following limits.

a. $\lim_{x \to 0^+} x^x$ **b.** $\lim_{x \to \infty} \left(1 + \dfrac{1}{x} \right)^x$

SOLUTION

a. This limit has the form 0^0. Using the given two-step procedure, we note that $x^x = e^{x \ln x}$ and first evaluate

$$L = \lim_{x \to 0^+} x \ln x.$$

This limit has the form $0 \cdot \infty$, which may be put in the form ∞/∞ so that l'Hôpital's Rule can be applied:

$$L = \lim_{x \to 0^+} x \ln x = \lim_{x \to 0^+} \frac{\ln x}{1/x} \qquad x = \frac{1}{1/x}$$

$$= \lim_{x \to 0^+} \frac{1/x}{-1/x^2} \qquad \text{L'Hôpital's Rule for } \infty/\infty \text{ form}$$

$$= \lim_{x \to 0^+} (-x) = 0 \qquad \text{Simplify and evaluate the limit.}$$

The second step is to exponentiate:

$$\lim_{x \to 0^+} x^x = e^L = e^0 = 1$$

We conclude that $\lim_{x \to 0^+} x^x = 1$ (Figure 4.74).

b. This limit has the form 1^∞. Noting that $(1 + 1/x)^x = e^{x \ln (1 + 1/x)}$, the first step is to evaluate

$$L = \lim_{x \to \infty} x \ln \left(1 + \frac{1}{x} \right),$$

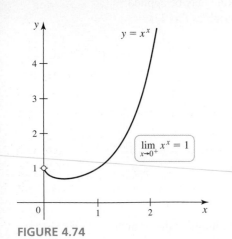

$y = x^x$

$$\lim_{x \to 0^+} x^x = 1$$

FIGURE 4.74

The limit in Example 6b is often given as a definition of e. It is a special case of the more general limit

$$\lim_{x \to \infty} \left(1 + \frac{a}{x}\right)^x = e^a.$$

See Exercise 93.

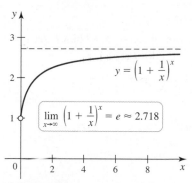

FIGURE 4.75

which has the form $0 \cdot \infty$. Proceeding as in part (a), we have

$$L = \lim_{x \to \infty} x \ln\left(1 + \frac{1}{x}\right) = \lim_{x \to \infty} \frac{\ln(1 + 1/x)}{1/x} \qquad x = \frac{1}{1/x}$$

$$= \lim_{x \to \infty} \frac{\frac{1}{1 + 1/x} \cdot \left(-\frac{1}{x^2}\right)}{\left(-\frac{1}{x^2}\right)} \qquad \text{L'Hôpital's Rule for 0/0 form}$$

$$= \lim_{x \to \infty} \frac{1}{1 + 1/x} = 1 \qquad \text{Simplify and evaluate.}$$

The second step is to exponentiate:

$$\lim_{x \to \infty} \left(1 + \frac{1}{x}\right)^x = e^L = e^1 = e$$

The function $y = (1 + 1/x)^x$ (Figure 4.75) has a horizontal asymptote at $e \approx 2.71828$.

Related Exercises 39–48 ◀

Growth Rates of Functions

An important use of l'Hôpital's Rule is to compare the growth rates of functions. Here are two questions—one practical and one theoretical—that demonstrate the importance of comparative growth rates of functions.

- Models of epidemics produce more complicated functions than the one given here, but they have the same general features.

- A particular theory for modeling the spread of an epidemic predicts that the number of infected people t days after the start of the epidemic is given by the function

$$N(t) = 2.5t^2 e^{-0.01t} = 2.5 \frac{t^2}{e^{0.01t}}.$$

Question: In the long run (as $t \to \infty$), does the epidemic spread or does it die out?

- The Prime Number Theorem was proved simultaneously (two different proofs) in 1896 by Jacques Hadamard and Charles de la Vallée Poussin, relying on fundamental ideas contributed by Riemann.

- A prime number is an integer $p \geq 2$ that has only two divisors, 1 and itself. The first few prime numbers are 2, 3, 5, 7, and 11. A celebrated theorem states that the number of prime numbers less than x is approximately

$$P(x) = \frac{x}{\ln x}, \qquad \text{for large values of } x.$$

Question: According to this function, is the number of prime numbers infinite?

These two questions involve a comparison of two functions. In the first question, if t^2 grows faster than $e^{0.01t}$ as $t \to \infty$, then $\lim_{t \to \infty} N(t) = \infty$ and the epidemic grows. If $e^{0.01t}$ grows faster than t^2 as $t \to \infty$, then $\lim_{t \to \infty} N(t) = 0$ and the epidemic dies out. We will explain what is meant by *grows faster than* in a moment.

In the second example, the comparison is between x and $\ln x$. If x grows faster than $\ln x$ as $x \to \infty$, then $\lim_{x \to \infty} P(x) = \infty$ and the number of prime numbers is infinite.

Our goal is to obtain a ranking of the following families of functions based on their growth rates:

- Another function with a large growth rate is the factorial function, defined for integers as $f(n) = n! = n(n-1) \cdots 2 \cdot 1$.

- mx, where $m > 0$ (represents linear functions)
- x^p, where $p > 0$ (represents polynomials and algebraic functions)
- x^x (sometimes called a *superexponential* or *tower function*)
- $\ln x$

QUICK CHECK 5 Before proceeding, use your intuition and rank these classes of functions in order of their growth rates. ◄

• $\ln^q x$, where $q > 0$

• $x^p \ln x$, where $p > 0$ (a combination of powers and logarithms)

• e^x

We need to be precise about growth rates and what it means for f to grow faster than g as $x \to \infty$. We work with the following definitions.

DEFINITION   Growth Rates of Functions (as $x \to \infty$)

Suppose f and g are functions with $\lim\limits_{x \to \infty} f(x) = \lim\limits_{x \to \infty} g(x) = \infty$. Then **$f$ grows faster than g** as $x \to \infty$ if

$$\lim_{x \to \infty} \frac{g(x)}{f(x)} = 0 \quad \text{or, equivalently, if} \quad \lim_{x \to \infty} \frac{f(x)}{g(x)} = \infty.$$

The functions f and g have **comparable growth rates** if

$$\lim_{x \to \infty} \frac{f(x)}{g(x)} = M$$

where $0 < M < \infty$ (M is nonzero and is finite).

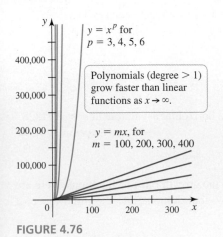

FIGURE 4.76

The idea of growth rates is illustrated nicely with graphs. Figure 4.76 shows a family of linear functions of the form $y = mx$, where $m > 0$, and a family of polynomials of the form $y = x^p$, where $p > 1$. We see that the polynomials grow faster (their curves rise at a greater rate) than the linear functions as $x \to \infty$.

Figure 4.77 shows that exponential functions of the form $y = b^x$, where $b > 1$, grow faster than power functions of the form $y = x^p$, where $p > 0$, as $x \to \infty$.

QUICK CHECK 6 Compare the growth rates of $f(x) = x^2$ and $g(x) = x^3$ as $x \to \infty$. Compare the growth rates of $f(x) = x^2$ and $g(x) = 10x^2$ as $x \to \infty$. ◄

We now begin a systematic comparison of growth rates. Note that a growth rate limit involves an indeterminate form ∞ / ∞, so l'Hôpital's Rule is always in the picture.

EXAMPLE 7   Powers of x vs. powers of $\ln x$ Compare the growth rates as $x \to \infty$ of the following pairs of functions.

a. $f(x) = \ln x$ and $g(x) = x^p$, where $p > 0$

b. $f(x) = \ln^q x$ and $g(x) = x^p$, where $p > 0$ and $q > 0$

SOLUTION

a. The limit of the ratio of the two functions is

$$\lim_{x \to \infty} \frac{\ln x}{x^p} = \lim_{x \to \infty} \frac{1/x}{px^{p-1}} \quad \text{L'Hôpital's Rule}$$

$$= \lim_{x \to \infty} \frac{1}{px^p} \quad \text{Simplify.}$$

$$= 0 \quad \text{Evaluate the limit.}$$

We see that any positive power of x grows faster than $\ln x$.

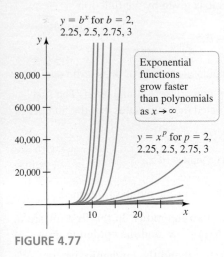

FIGURE 4.77

b. We compare $\ln^q x$ and x^p by observing that

$$\lim_{x\to\infty}\frac{\ln^q x}{x^p} = \lim_{x\to\infty}\left(\frac{\ln x}{x^{p/q}}\right)^q = \left(\underbrace{\lim_{x\to\infty}\frac{\ln x}{x^{p/q}}}_{0}\right)^q.$$

By part (a), $\displaystyle\lim_{x\to\infty}\frac{\ln x}{x^{p/q}} = 0$ (because $p/q > 0$). Therefore, $\displaystyle\lim_{x\to\infty}\frac{\ln^q x}{x^p} = 0$ (because $q > 0$). We conclude that positive powers of x grow faster than positive powers of $\ln x$.

Related Exercises 49–60 ◄

EXAMPLE 8   Powers of x vs. exponentials Compare the rates of growth of $f(x) = x^p$ and $g(x) = e^x$ as $x \to \infty$, where p is a positive real number.

SOLUTION The goal is to evaluate $\displaystyle\lim_{x\to\infty}\frac{x^p}{e^x}$ for $p > 0$. This comparison is most easily done using Example 7 and a change of variables. We let $x = \ln t$ and note that as $x \to \infty$, we also have $t \to \infty$. With this substitution, $x^p = \ln^p t$ and $e^x = e^{\ln t} = t$. Therefore,

$$\lim_{x\to\infty}\frac{x^p}{e^x} = \lim_{t\to\infty}\frac{\ln^p t}{t} = 0. \quad \text{Example 7}$$

We see that increasing exponential functions grow faster than positive powers of x.

Related Exercises 49–60 ◄

These examples, together with the comparison of exponential functions b^x and the superexponential x^x (see Exercise 94), establish a ranking of growth rates.

THEOREM 4.15   Ranking Growth Rates as $x \to \infty$
Let $f \ll g$ mean that g grows faster than f as $x \to \infty$. With positive real numbers $p, q, r,$ and s and $b > 1$,

$$\ln^q x \ll x^p \ll x^p \ln^r x \ll x^{p+s} \ll b^x \ll x^x.$$

You should try to build these relative growth rates into your intuition. They are useful in future chapters (Chapter 8 on sequences, in particular), and they can be used to evaluate limits at infinity quickly.

Pitfalls in Using l'Hôpital's Rule

We close with a short list of common pitfalls of l'Hôpital's Rule.

1. L'Hôpital's Rule says $\displaystyle\lim_{x\to a}\frac{f(x)}{g(x)} = \lim_{x\to a}\frac{f'(x)}{g'(x)}$, not

$$\lim_{x\to a}\frac{f(x)}{g(x)} = \lim_{x\to a}\left[\frac{f(x)}{g(x)}\right]' \quad \text{or} \quad \lim_{x\to a}\frac{f(x)}{g(x)} = \lim_{x\to a}\left[\frac{1}{g(x)}\right]'f'(x).$$

In other words, you should evaluate $f'(x)$ and $g'(x)$, form their quotient, and then take the limit. Don't confuse l'Hôpital's Rule with the Quotient Rule.

2. Be sure that the given limit involves the indeterminate form $0/0$ or ∞/∞ before applying l'Hôpital's Rule. For example, consider the following erroneous use of l'Hôpital's Rule:

$$\lim_{x\to 0}\frac{1-\sin x}{\cos x} = \lim_{x\to 0}\frac{-\cos x}{\sin x},$$

which does not exist. The limit is not an indeterminate form in the first place. This limit should be evaluated by direct substitution:

$$\lim_{x \to 0} \frac{1 - \sin x}{\cos x} = \frac{1 - \sin 0}{1} = 1$$

3. When using l'Hôpital's Rule repeatedly, be sure to simplify expressions as much as possible at each step and evaluate the limit as soon as the new limit is no longer an indeterminate form.

4. Repeated use of l'Hôpital's Rule occasionally leads to unending cycles, in which case other methods must be used. Limits of the form $\lim\limits_{x \to \infty} \dfrac{\sqrt{ax + 1}}{\sqrt{bx + 1}}$, where a and b are real numbers, lead to such behavior (see Exercise 85).

SECTION 4.7 EXERCISES

Review Questions

1. Explain with examples what is meant by the indeterminate form 0/0.

2. Why are special methods, such as l'Hôpital's Rule, needed to evaluate indeterminate forms (as opposed to substitution)?

3. Explain the steps used to apply l'Hôpital's Rule to a limit of the form 0/0.

4. To which indeterminate forms does l'Hôpital's Rule apply *directly*?

5. Explain how to convert a limit of the form $0 \cdot \infty$ to a limit of the form 0/0 or ∞/∞.

6. Give an example of a limit of the form ∞/∞ as $x \to 0$.

7. Explain why the form 1^∞ is indeterminate and cannot be evaluated by substitution. Explain how the competing functions behave.

8. Give the two-step method for attacking a limit of the form $\lim\limits_{x \to a} f(x)^{g(x)}$.

9. In terms of limits, what does it mean for f to grow faster than g as $x \to \infty$?

10. In terms of limits, what does it mean for the rates of growth of f and g to be comparable as $x \to \infty$?

11. Rank the functions x^3, $\ln x$, x^x, and 2^x in order of increasing growth rates as $x \to \infty$.

12. Rank the functions x^{100}, $\ln x^{10}$, x^x, and 10^x in order of increasing growth rates as $x \to \infty$.

Basic Skills

13–18. 0/0 form *Evaluate the following limits using l'Hôpital's Rule.*

13. $\lim\limits_{x \to 2} \dfrac{x^2 - 2x}{8 - 6x + x^2}$

14. $\lim\limits_{x \to -1} \dfrac{x^4 + x^3 + 2x + 2}{x + 1}$

15. $\lim\limits_{x \to 0} \dfrac{3 \sin 4x}{5x}$

16. $\lim\limits_{x \to 2\pi} \dfrac{x \sin x + x^2 - 4\pi^2}{x - 2\pi}$

17. $\lim\limits_{u \to \pi/4} \dfrac{\tan u - \cot u}{u - \pi/4}$

18. $\lim\limits_{z \to 0} \dfrac{\tan 4z}{\tan 7z}$

19–26. 0/0 form *Evaluate the following limits.*

19. $\lim\limits_{x \to 0} \dfrac{1 - \cos 3x}{8x^2}$

20. $\lim\limits_{x \to 0} \dfrac{\sin^2 3x}{x^2}$

21. $\lim\limits_{x \to -1} \dfrac{x^3 - x^2 - 5x - 3}{x^4 + 2x^3 - x^2 - 4x - 2}$

22. $\lim\limits_{x \to 1} \dfrac{x^n - 1}{x - 1}$, n is a positive integer

23. $\lim\limits_{v \to 3} \dfrac{v - 1 - \sqrt{v^2 - 5}}{v - 3}$

24. $\lim\limits_{y \to 2} \dfrac{y^2 + y - 6}{\sqrt{8 - y^2} - y}$

25. $\lim\limits_{h \to 0} \dfrac{\sin(x + h) - \sin x}{h}$, x is a real number

26. $\lim\limits_{x \to 2} \dfrac{\sqrt[3]{3x + 2} - 2}{x - 2}$

27–30. ∞/∞ form *Evaluate the following limits.*

27. $\lim\limits_{x \to \infty} \dfrac{3x^4 - x^2}{6x^4 + 12}$

28. $\lim\limits_{x \to \infty} \dfrac{4x^3 - 2x^2 + 6}{\pi x^3 + 4}$

29. $\lim\limits_{x \to \infty} \dfrac{\sqrt{8 - 4x^2}}{3x^3 + x - 1}$

30. $\lim\limits_{x \to \pi/2} \dfrac{2 \tan x}{\sec^2 x}$

31–34. $0 \cdot \infty$ form *Evaluate the following limits.*

31. $\lim\limits_{x \to 0} x \csc x$

32. $\lim\limits_{x \to 1^-} (1 - x) \tan\left(\dfrac{\pi x}{2}\right)$

33. $\lim\limits_{x \to (\pi/2)^-} \left(\dfrac{\pi}{2} - x\right) \sec x$

34. $\lim\limits_{x \to 0^+} (\sin x)\sqrt{\dfrac{1 - x}{x}}$

35–38. $\infty - \infty$ form *Evaluate the following limits.*

35. $\lim\limits_{x \to 0^+} \left(\cot x - \dfrac{1}{x}\right)$

36. $\lim\limits_{x \to \infty} \left(x - \sqrt{x^2 + 1}\right)$

37. $\lim\limits_{\theta \to \pi/2^-} (\tan \theta - \sec \theta)$

38. $\lim\limits_{x \to \infty} \left(x - \sqrt{x^2 + 4x}\right)$

39–48. $1^\infty, 0^0, \infty^0$ forms *Evaluate the following limits or explain why they do not exist. Check your results by graphing.*

39. $\lim\limits_{x \to 0^+} x^{2x}$

40. $\lim\limits_{x \to 0} (1 + 4x)^{3/x}$

41. $\lim\limits_{\theta \to \pi/2^-} (\tan \theta)^{\cos \theta}$

42. $\lim\limits_{\theta \to 0^+} (\sin \theta)^{\tan \theta}$

43. $\lim\limits_{x \to 0^+} (1 + x)^{\cot x}$

44. $\lim\limits_{x \to \infty} \left(1 + \dfrac{1}{x}\right)^{\ln x}$

45. $\lim\limits_{x \to 0^+} (\tan x)^x$

46. $\lim\limits_{z \to \infty} \left(1 + \dfrac{10}{z^2}\right)^{z^2}$

47. $\displaystyle\lim_{x\to 0} (x + \cos x)^{1/x}$

48. $\displaystyle\lim_{x\to 0^+} \left(\tfrac{1}{3}\cdot 3^x + \tfrac{2}{3}\cdot 2^x\right)^{1/x}$

49–60. Comparing growth rates *Use limit methods to determine which of the two given functions grows faster, or state that they have comparable growth rates.*

49. $x^{10};\ e^{0.01x}$

50. $x^2 \ln x;\ \ln^2 x$

51. $\ln x^{20};\ \ln x$

52. $\ln x;\ \ln(\ln x)$

53. $100^x;\ x^x$

54. $x^2 \ln x;\ x^3$

55. $x^{20};\ 1.00001^x$

56. $x^{10} \ln^{10} x;\ x^{11}$

57. $x^x;\ (x/2)^x$

58. $\ln \sqrt{x};\ \ln^2 x$

59. $e^{x^2};\ e^{10x}$

60. $e^{x^2};\ x^{x/10}$

Further Explorations

61. Explain why or why not Determine whether the following statements are true and give an explanation or counterexample.

   a. By l'Hôpital's Rule, $\displaystyle\lim_{x\to 2}\frac{x-2}{x^2-1} = \lim_{x\to 2}\frac{1}{2x} = \frac{1}{4}$.

   b. $\displaystyle\lim_{x\to 0} (x \sin x) = \lim_{x\to 0} f(x)g(x) = \lim_{x\to 0} f'(x)\lim_{x\to 0} g'(x) =$
$\left(\displaystyle\lim_{x\to 0} 1\right)\left(\displaystyle\lim_{x\to 0} \cos x\right) = 1$.

   c. $\displaystyle\lim_{x\to 0^+} x^{1/x}$ is an indeterminate form.

   d. The number 1 raised to any fixed power is 1. Therefore, because $(1 + x) \to 1$ as $x \to 0$, $(1 + x)^{1/x} \to 1$ as $x \to 0$.

   e. The functions $\ln x^{100}$ and $\ln x$ have comparable growth rates as $x \to \infty$.

   f. The function e^x grows faster than 2^x as $x \to \infty$.

62–63. Two methods *Evaluate the following limits in two different ways: Use the methods of Chapter 2 and use l'Hôpital's Rule.*

62. $\displaystyle\lim_{x\to\infty}\frac{100x^3 - 3}{x^4 - 2}$

63. $\displaystyle\lim_{x\to\infty}\frac{2x^3 - x^2 + 1}{5x^3 + 2x}$

64. L'Hôpital's example Evaluate one of the limits l'Hôpital used in his own textbook in about 1700:

$$\lim_{x\to a}\frac{\sqrt{2a^3x - x^4} - a\sqrt[3]{a^2x}}{a - \sqrt[4]{ax^3}}, \text{ where } a \text{ is a real number}$$

65–76. Miscellaneous limits by any means *Use analytical methods to evaluate the following limits.*

65. $\displaystyle\lim_{x\to 6}\frac{\sqrt[3]{5x+2} - 2}{1/x - 1/6}$

66. $\displaystyle\lim_{t\to\pi/2^+}\frac{\tan 3t}{\sec 5t}$

67. $\displaystyle\lim_{x\to\infty} (\sqrt{x-2} - \sqrt{x-4})$

68. $\displaystyle\lim_{x\to\pi/2} (\pi - 2x)\tan x$

69. $\displaystyle\lim_{x\to\infty} x^3\left(\frac{1}{x} - \sin\frac{1}{x}\right)$

70. $\displaystyle\lim_{x\to\infty} (x^2 e^{1/x} - x^2 - x)$

71. $\displaystyle\lim_{x\to 1^+}\left(\frac{1}{x-1} - \frac{1}{\sqrt{x-1}}\right)$

72. $\displaystyle\lim_{x\to 0^+} x^{\ln x}$

73. $\displaystyle\lim_{x\to\infty}\frac{\log_2 x}{\log_3 x}$

74. $\displaystyle\lim_{x\to\infty} (\log_2 x - \log_3 x)$

75. $\displaystyle\lim_{n\to\infty}\frac{1 + 2 + \cdots + n}{n^2}$

76. $\displaystyle\lim_{x\to 0}\left(\frac{\sin x}{x}\right)^{1/x^2}$

77. It may take time The ranking of growth rates given in the text applies for $x \to \infty$. However, these rates may not be evident for small values of x. For example, an exponential grows faster than any power of x. However, for $1 < x < 19{,}800$, x^2 is greater than $e^{x/1000}$. For the following pairs of functions, estimate the point at which the faster growing function overtakes the slower growing function (for the last time).

   a. $\ln^3 x$ and $x^{0.3}$

   b. $2^{x/100}$ and x^3

   c. $x^{x/100}$ and e^x

   d. $\ln^{10} x$ and $e^{x/10}$

78–81. Limits with parameters *Evaluate the following limits and give the limit in terms of the parameters a and b, which are positive real numbers. In each case, graph the function for specific values of the parameters to check your results.*

78. $\displaystyle\lim_{x\to 0} (1 + ax)^{b/x}$

79. $\displaystyle\lim_{x\to 0^+} (a^x - b^x)^x, a > b > 0$

80. $\displaystyle\lim_{x\to 0^+} (a^x - b^x)^{1/x}, a > b > 0$

81. $\displaystyle\lim_{x\to 0}\frac{a^x - b^x}{x}$

Applications

82. An optics limit The theory of interference of coherent oscillators requires the limit $\displaystyle\lim_{\delta\to 2m\pi}\frac{\sin^2(N\delta/2)}{\sin^2(\delta/2)}$, where N is a positive integer and m is any integer. Show that the value of this limit is N^2.

83. Compound interest Suppose you make a deposit of $\$P$ into a savings account that earns interest at a rate of $100r\%$ per year.

   a. Show that if interest is compounded once per year, then the balance after t years is $B(t) = P(1 + r)^t$.

   b. If interest is compounded m times per year, then the balance after t years is $B(t) = P(1 + r/m)^{mt}$. For example, $m = 12$ corresponds to monthly compounding, and the interest rate for each month is $r/12$. In the limit $m \to \infty$, the compounding is said to be *continuous*. Show that with continuous compounding, the balance after t years is $B(t) = Pe^{rt}$.

84. Algorithm complexity The complexity of a computer algorithm is the number of operations or steps the algorithm needs to complete its task assuming there are n pieces of input (for example, the number of steps needed to put n numbers in ascending order). Four algorithms for doing the same task have complexities of A: $n^{3/2}$, B: $n \log_2 n$, C: $n(\log_2 n)^2$, and D: $\sqrt{n}\log_2 n$. Rank the algorithms in order of increasing efficiency for large values of n. Graph the complexities as they vary with n and comment on your observations.

Additional Exercises

85. L'Hôpital loops Consider the limit $\displaystyle\lim_{x\to\infty}\frac{\sqrt{ax+b}}{\sqrt{cx+d}}$, where $a, b, c,$ and d are positive real numbers. Show that l'Hôpital's Rule fails for this limit. Find the limit using another method.

86. General $\infty - \infty$ result Let a and b be positive real numbers. Evaluate $\displaystyle\lim_{x\to\infty} (ax - \sqrt{a^2x^2 - bx})$ in terms of a and b.

87. Exponential functions and powers Show that any exponential, b^x for $b > 1$, grows faster than x^p for $p > 0$.

88. Exponentials with different bases Show that $f(x) = a^x$ grows faster than $g(x) = b^x$ as $x \to \infty$ if $1 < b < a$.

89. Logs with different bases Show that $f(x) = \log_a x$ and $g(x) = \log_b x$, where $a > 1$ and $b > 1$, grow at a comparable rate as $x \to \infty$.

90. Factorial growth rate The factorial function is usually defined for positive integers as $n! = n(n-1)(n-2)\cdots 3\cdot 2\cdot 1$. For example, $5! = 5\cdot 4\cdot 3\cdot 2\cdot 1 = 120$. A valuable result that gives good approximations to $n!$ for large values of n is Stirling's formula, $n! \approx \sqrt{2\pi n}\, n^n e^{-n}$ (see Guided Projects for more on Stirling's formula). Use this formula and a calculator to determine where the factorial function appears in the ranking of growth rates.

91. A geometric limit Let $f(\theta)$ be the area of the triangle ABP (see figure) and let $g(\theta)$ be the area of the region between the chord PB and the arc PB. Evaluate $\lim_{\theta \to 0} g(\theta)/f(\theta)$.

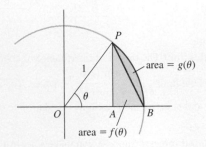

92. A fascinating function Consider the function $f(x) = (ab^x + (1-a)c^x)^{1/x}$, where a, b, and c are positive real numbers, $0 < a < 1$.

　a. Graph f for several sets of (a, b, c). Verify that in all cases that f is an increasing function for all x with a single inflection point.

　b. Use analytical methods to determine $\lim_{x \to 0} f(x)$ in terms of a, b, and c.

　c. Use analytical methods to determine $\lim_{x\to\infty} f(x)$ and $\lim_{x\to -\infty} f(x)$.

　d. Estimate the location of the inflection point (in terms of a, b, and c).

93. Exponential limit Prove that $\lim_{x\to\infty}\left(1 + \dfrac{a}{x}\right)^x = e^a$ for $a \neq 0$.

94. Exponentials vs. super exponentials Show that x^x grows faster than b^x as $x \to \infty$ for $b > 1$.

95. Exponential growth rates

　a. For what values of $b > 0$ does b^x grow faster than e^x as $x \to \infty$?

　b. Compare the growth rates of e^x and e^{ax} as $x \to \infty$ for $a > 0$.

96. A max/min detector Consider the function $f(t) = (ax^t + by^t)^{1/t}$, where a, b, x, and y are positive real numbers with $a + b = 1$.

　a. Show that $\lim_{t\to 0} f(t) = x^a y^b$.

　b. Show that $\lim_{t\to\infty} f(t) = \max\{x, y\}$ and

$$\lim_{t\to -\infty} f(t) = \min\{x, y\}.$$

QUICK CHECK ANSWERS

1. g and h　**2.** g and h　**3.** $0\cdot\infty; (x - \pi/2)/\cot x$　**4.** The form 0^∞ (for example, $\lim_{x\to 0^+} x^{1/x}$) is not indeterminate, because as the base goes to zero, raising it to larger and larger powers drives the entire function to zero.　**6.** x^3 grows faster than x^2 as $x \to \infty$, whereas x^2 and $10x^2$ have comparable growth rates as $x \to \infty$. ◄

4.8　Antiderivatives

The goal of differentiation is to find the derivative f' of a given function f. The reverse process, called *antidifferentiation*, is equally important: Given a function f, we look for an *antiderivative* function F whose derivative is f; that is, a function F such that $F' = f$.

> **DEFINITION　Antiderivative**
>
> A function F is an **antiderivative** of f on an interval I provided $F'(x) = f(x)$ for all x in I.

In this section, we revisit derivative formulas developed in previous chapters to discover corresponding antiderivative formulas.

Thinking Backward

Consider the function $f(x) = 1$ and the derivative formula $\dfrac{d}{dx}(x) = 1$. We see that an antiderivative of f is $F(x) = x$ because $F'(x) = 1 = f(x)$. Using the same logic, we can write

$$\frac{d}{dx}(x^2) = 2x \quad \Rightarrow \quad \text{an antiderivative of } f(x) = 2x \text{ is } F(x) = x^2$$

QUICK CHECK 1 Verify by differentiation that x^3 is an antiderivative of $3x^2$ and $-\cos x$ is an antiderivative of $\sin x$. ◄

$$\frac{d}{dx}(\sin x) = \cos x \quad \Rightarrow \quad \text{an antiderivative of } f(x) = \cos x \text{ is } F(x) = \sin x.$$

Each of these proposed antiderivative formulas is easily checked by showing that $F' = f$.

An immediate question arises: Does a function have more than one antiderivative? To answer this question, let's focus on $f(x) = 1$ and the antiderivative $F(x) = x$. Because the derivative of a constant C is zero, we see that $F(x) = x + C$ is also an antiderivative of $f(x) = 1$, which is easy to check:

$$F'(x) = \frac{d}{dx}(x + C) = 1 = f(x)$$

Therefore, $f(x) = 1$ actually has an infinite number of antiderivatives. For the same reason, any function of the form $F(x) = x^2 + C$ is an antiderivative of $f(x) = 2x$, and any function of the form $F(x) = \sin x + C$ is an antiderivative of $f(x) = \cos x$, where C is an arbitrary constant.

We might ask whether there are still *more* antiderivatives of a given function. The following theorem provides the answer.

THEOREM 4.16 The Family of Antiderivatives
Let F be any antiderivative of f. Then *all* the antiderivatives of f have the form $F + C$, where C is an arbitrary constant.

Proof Suppose that F and G are antiderivatives of f on an interval I. Then $F' = f$ and $G' = f$, which implies that $F' = G'$ on I. From Theorem 4.11, which states that functions with equal derivatives differ by a constant, it follows that $G = F + C$. Therefore, all antiderivatives of f have the form $F + C$, where C is an arbitrary constant. ◄

Theorem 4.16 says that while there are infinitely many antiderivatives of a function, they are all of one family, namely, those functions of the form $F + C$. Because the antiderivatives of a particular function differ by a constant, the antiderivatives are vertical translations of one another (Figure 4.78).

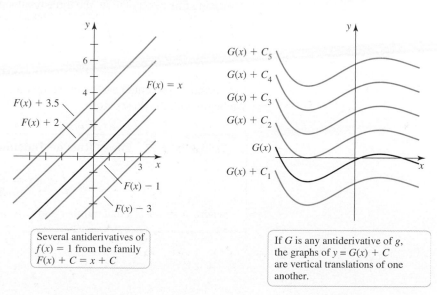

Several antiderivatives of $f(x) = 1$ from the family $F(x) + C = x + C$

If G is any antiderivative of g, the graphs of $y = G(x) + C$ are vertical translations of one another.

FIGURE 4.78

EXAMPLE 1   **Finding antiderivatives**   Use what you know about derivatives to find all antiderivatives of the following functions.

a. $f(x) = 3x^2$ **b.** $f(x) = \dfrac{1}{1 + x^2}$ **c.** $f(x) = \sin x$

SOLUTION

a. Note that $\dfrac{d}{dx}(x^3) = 3x^2$. Reversing this derivative formula says that an antiderivative of $f(x) = 3x^2$ is x^3. By Theorem 4.16, the complete family of antiderivatives is $F(x) = x^3 + C$, where C is an arbitrary constant.

b. Because $\dfrac{d}{dx}(\tan^{-1} x) = \dfrac{1}{1 + x^2}$, all antiderivatives of f are of the form $F(x) = \tan^{-1} x + C$, where C is an arbitrary constant.

c. Recall that $\dfrac{d}{dx}(\cos x) = -\sin x$. We seek a function whose derivative is $\sin x$, not $-\sin x$. Observing that $\dfrac{d}{dx}(-\cos x) = \sin x$, it follows that the antiderivatives are $F(x) = -\cos x + C$, where C is an arbitrary constant.   *Related Exercises 11–18* ◄

QUICK CHECK 2 Find the family of antiderivatives for each of $f(x) = e^x$, $g(x) = 4x^3$, and $h(x) = \sec^2 x$. ◄

Indefinite Integrals

The notation $\dfrac{d}{dx}(f)$ means *take the derivative of f*. We need analogous notation for antiderivatives. For historical reasons that become apparent in the next chapter, the notation that means *find the antiderivatives of f* is the **indefinite integral** $\int f(x)\, dx$. Every time an indefinite integral sign $\int$ appears, it is followed by a function called the **integrand**, which in turn is followed by the differential dx. For now dx simply means that x is the independent variable, or the **variable of integration**. The notation $\int f(x)\, dx$ represents *all* of the antiderivatives of f.

Using this new notation, the three results of Example 1 are written as

$$\int 3x^2\, dx = x^3 + C, \quad \int \frac{1}{1 + x^2}\, dx = \tan^{-1} x + C, \text{ and } \int \sin x\, dx = -\cos x + C,$$

where C is an arbitrary constant called a **constant of integration**. Virtually all the derivative formulas presented earlier in the text may be written in terms of indefinite integrals. We begin with the Power Rule.

THEOREM 4.17   Power Rule for Indefinite Integrals

$$\int x^p\, dx = \frac{x^{p+1}}{p + 1} + C,$$

where $p \neq -1$ is a real number and C is an arbitrary constant.

> Notice that if $p = -1$ in this antiderivative formula, then $F(x)$ is undefined. The antiderivative of $f(x) = x^{-1}$ is discussed shortly.

> Any indefinite integral calculation can be checked by differentiation: The derivative of the alleged indefinite integral must equal the integrand.

Proof The theorem says that the antiderivatives of $f(x) = x^p$ are of the form $F(x) = \dfrac{x^{p+1}}{p+1} + C$. Differentiating F, we verify that $F'(x) = f(x)$:

$$F'(x) = \frac{d}{dx}\left(\frac{x^{p+1}}{p+1} + C\right)$$

$$= \frac{d}{dx}\left(\frac{x^{p+1}}{p+1}\right) + \underbrace{\frac{d}{dx}(C)}_{0}$$

$$= \frac{(p+1)x^{(p+1)-1}}{p+1} + 0 = x^p$$

Theorems 3.4 and 3.5 (Section 3.2) state the Constant Multiple and Sum Rules for derivatives. Here are the corresponding antiderivative rules, which are proved by differentiation.

THEOREM 4.18   Constant Multiple and Sum Rules

Constant Multiple Rule: $\displaystyle\int cf(x)\,dx = c\int f(x)\,dx$

Sum Rule: $\displaystyle\int (f(x) + g(x))\,dx = \int f(x)\,dx + \int g(x)\,dx$

EXAMPLE 2   Indefinite integrals Determine the following indefinite integrals.

a. $\displaystyle\int (3x^5 + 2 - 5x^{-3/2})\,dx$ **b.** $\displaystyle\int \left(\frac{4x^{19} - 5x^{-8}}{x^2}\right)dx$

SOLUTION

> $\int dx$ means $\int 1\,dx$, which is the indefinite integral of the constant function $f(x) = 1$, so $\int dx = x + C$.

> Each indefinite integral produces an arbitrary constant, all of which may be combined in one arbitrary constant called C.

a. $\displaystyle\int (3x^5 + 2 - 5x^{-3/2})\,dx = \int 3x^5\,dx + \int 2\,dx - \int 5x^{-3/2}\,dx$ Sum Rule

$$= 3\int x^5\,dx + 2\int dx - 5\int x^{-3/2}\,dx \quad \text{Constant Multiple Rule}$$

$$= 3\cdot\frac{x^6}{6} + 2\cdot x - 5\cdot\frac{x^{-1/2}}{(-1/2)} + C \quad \text{Power Rule}$$

$$= \frac{x^6}{2} + 2x + 10x^{-1/2} + C \quad \text{Simplify.}$$

b. $\displaystyle\int \left(\frac{4x^{19} - 5x^{-8}}{x^2}\right)dx = \int (4x^{17} - 5x^{-10})\,dx$ Simplify the integrand.

$$= 4\int x^{17}\,dx - 5\int x^{-10}\,dx \quad \text{Sum and Constant Multiple Rules}$$

$$= 4\cdot\frac{x^{18}}{18} - 5\cdot\frac{x^{-9}}{(-9)} + C \quad \text{Power Rule}$$

$$= \frac{2x^{18}}{9} + \frac{5x^{-9}}{9} + C \quad \text{Simplify.}$$

Both of these results should be checked by differentiation.   *Related Exercises 19–26*◄

Indefinite Integrals of Trigonometric Functions

Any derivative formula can be restated in terms of an indefinite integral formula. For example, by the Chain Rule we know that

$$\frac{d}{dx}(\cos 3x) = -3\sin 3x.$$

Therefore, we can immediately write

$$\int -3\sin 3x \, dx = \cos 3x + C.$$

Factoring -3 from the left side and dividing through by -3, we have

$$\int \sin 3x \, dx = -\frac{1}{3}\cos 3x + C.$$

This argument works if we replace 3 by any constant $a \neq 0$. Similar reasoning leads to the results in Table 4.5, where $a \neq 0$ and C is an arbitrary constant.

Table 4.5   Indefinite Integrals of Trigonometric Functions

1. $\dfrac{d}{dx}(\sin ax) = a\cos ax \quad \rightarrow \quad \displaystyle\int \cos ax \, dx = \frac{1}{a}\sin ax + C$

2. $\dfrac{d}{dx}(\cos ax) = -a\sin ax \quad \rightarrow \quad \displaystyle\int \sin ax \, dx = -\frac{1}{a}\cos ax + C$

3. $\dfrac{d}{dx}(\tan ax) = a\sec^2 ax \quad \rightarrow \quad \displaystyle\int \sec^2 ax \, dx = \frac{1}{a}\tan ax + C$

4. $\dfrac{d}{dx}(\cot ax) = -a\csc^2 ax \quad \rightarrow \quad \displaystyle\int \csc^2 ax \, dx = -\frac{1}{a}\cot ax + C$

5. $\dfrac{d}{dx}(\sec ax) = a\sec ax \tan ax \quad \rightarrow \quad \displaystyle\int \sec ax \tan ax \, dx = \frac{1}{a}\sec ax + C$

6. $\dfrac{d}{dx}(\csc ax) = -a\csc ax \cot ax \quad \rightarrow \quad \displaystyle\int \csc ax \cot ax \, dx = -\frac{1}{a}\csc ax + C$

QUICK CHECK 3 Use differentiation to verify that $\displaystyle\int \sin 2x \, dx = -\frac{1}{2}\cos 2x + C.$ ◄

EXAMPLE 3   Indefinite integrals of trigonometric functions Determine the following indefinite integrals.

a. $\displaystyle\int \sec^2 3x \, dx$ **b.** $\displaystyle\int \cos\left(\frac{x}{2}\right) dx$

SOLUTION These integrals follow directly from Table 4.5 and can be verified by differentiation.

a. Letting $a = 3$ in result (3) of Table 4.5, we have

$$\int \sec^2 3x \, dx = \frac{\tan 3x}{3} + C.$$

b. We let $a = \frac{1}{2}$ in result (1) of Table 4.5, which says that

$$\int \cos\left(\frac{x}{2}\right) dx = \frac{\sin (x/2)}{\frac{1}{2}} + C = 2\sin\left(\frac{x}{2}\right) + C.$$

Related Exercises 27–32 ◄

Other Indefinite Integrals

We now complete the process of rewriting familiar derivative results in terms of indefinite integrals. For example, because $\dfrac{d}{dx}(e^{ax}) = ae^{ax}$, where $a \neq 0$, we can divide both sides of this equation by a and write

$$\int e^{ax}\,dx = \frac{1}{a}e^{ax} + C.$$

Similarly, because $\dfrac{d}{dx}\left(\ln|x|\right) = \dfrac{1}{x}$ for $x \neq 0$, it follows that $\displaystyle\int \dfrac{dx}{x} = \ln|x| + C$. Notice that this result fills the gap in the Power Rule for the case $p = -1$. The same reasoning leads to the indefinite integrals in Table 4.6, where $a \neq 0$ and C is an arbitrary constant.

▷ Tables 4.5 and 4.6 are subsets of the table of integrals at the end of the book.

Table 4.6   Other Indefinite Integrals

7. $\dfrac{d}{dx}(e^{ax}) = ae^{ax} \;\rightarrow\; \displaystyle\int e^{ax}\,dx = \dfrac{1}{a}e^{ax} + C$

8. $\dfrac{d}{dx}\left(\ln|x|\right) = \dfrac{1}{x} \;\rightarrow\; \displaystyle\int \dfrac{dx}{x} = \ln|x| + C$

9. $\dfrac{d}{dx}\left[\sin^{-1}\left(\dfrac{x}{a}\right)\right] = \dfrac{1}{\sqrt{a^2 - x^2}} \;\rightarrow\; \displaystyle\int \dfrac{dx}{\sqrt{a^2 - x^2}} = \sin^{-1}\left(\dfrac{x}{a}\right) + C$

10. $\dfrac{d}{dx}\left[\tan^{-1}\left(\dfrac{x}{a}\right)\right] = \dfrac{a}{a^2 + x^2} \;\rightarrow\; \displaystyle\int \dfrac{dx}{a^2 + x^2} = \dfrac{1}{a}\tan^{-1}\left(\dfrac{x}{a}\right) + C$

11. $\dfrac{d}{dx}\left(\sec^{-1}\left|\dfrac{x}{a}\right|\right) = \dfrac{a}{x\sqrt{x^2 - a^2}} \;\rightarrow\; \displaystyle\int \dfrac{dx}{x\sqrt{x^2 - a^2}} = \dfrac{1}{a}\sec^{-1}\left|\dfrac{x}{a}\right| + C$

EXAMPLE 4   Indefinite integrals Determine the following indefinite integrals.

a. $\displaystyle\int e^{-10x}\,dx$ **b.** $\displaystyle\int \dfrac{4}{\sqrt{9 - x^2}}\,dx$ **c.** $\displaystyle\int \dfrac{dx}{16x^2 + 1}$

SOLUTION

a. Setting $a = -10$ in result (7) of Table 4.6, we find that

$$\int e^{-10x}\,dx = -\frac{1}{10}e^{-10x} + C,$$

which should be verified by differentiation.

b. Setting $a = 3$ in result (9) of Table 4.6, we have

$$\int \frac{4}{\sqrt{9 - x^2}}\,dx = 4\int \frac{dx}{\sqrt{3^2 - x^2}} = 4\sin^{-1}\left(\frac{x}{3}\right) + C.$$

c. An algebra step is needed to put this integral in a form that matches Table 4.6. We first write

$$\int \frac{dx}{16x^2 + 1} = \frac{1}{16}\int \frac{dx}{x^2 + \left(\frac{1}{16}\right)} = \frac{1}{16}\int \frac{dx}{x^2 + \left(\frac{1}{4}\right)^2}.$$

Setting $a = \frac{1}{4}$ in result (10) of Table 4.6 gives

$$\int \frac{dx}{16x^2 + 1} = \frac{1}{16}\int \frac{dx}{x^2 + \left(\frac{1}{4}\right)^2} = \left(\frac{1}{16}\right)4\tan^{-1}4x + C = \frac{1}{4}\tan^{-1}4x + C.$$

Related Exercises 33–38 ◀

Introduction to Differential Equations

Suppose you know that the derivative of a function f satisfies the equation

$$f'(x) = 2x + 10.$$

QUICK CHECK 4 Explain why an antiderivative of f' is f. ◄

To solve this *differential equation* for the function f, we note that the solutions are antiderivatives of $2x + 10$, which are $x^2 + 10x + C$, where C is an arbitrary constant. So we have found an infinite number of solutions, all of the form $f(x) = x^2 + 10x + C$.

Now consider a more general differential equation of the form $f'(x) = G(x)$, where G is given and f is unknown. The solution consists of antiderivatives of G, which involve an arbitrary constant. In most practical cases, the differential equation is accompanied by an **initial condition** that allows us to determine the arbitrary constant. Therefore, we consider problems of the form

$$f'(x) = G(x), \quad \text{where } G \text{ is given} \qquad \text{Differential equation}$$
$$f(a) = b, \qquad \text{where } a, b \text{ are given} \qquad \text{Initial condition}$$

A differential equation coupled with an initial condition is called an **initial value problem**.

EXAMPLE 5   An initial value problem Solve the initial value problem $f'(x) = x^2 - 2x$ with $f(1) = \frac{1}{3}$.

SOLUTION The solution is an antiderivative of $x^2 - 2x$. Therefore,

$$f(x) = \frac{x^3}{3} - x^2 + C,$$

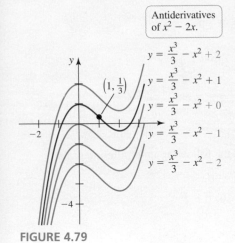

Antiderivatives of $x^2 - 2x$.

$y = \dfrac{x^3}{3} - x^2 + 2$

$y = \dfrac{x^3}{3} - x^2 + 1$

$y = \dfrac{x^3}{3} - x^2 + 0$

$y = \dfrac{x^3}{3} - x^2 - 1$

$y = \dfrac{x^3}{3} - x^2 - 2$

FIGURE 4.79

where C is an arbitrary constant. We have determined that the solution is a member of a family of functions, all of which differ by a constant. This family of functions, called the **general solution**, is shown in Figure 4.79, where we see curves for various choices of C.

Using the initial condition $f(1) = \frac{1}{3}$, we must find the particular function in the general solution whose graph passes through the point $\left(1, \frac{1}{3}\right)$. Imposing the condition $f(1) = \frac{1}{3}$, we reason as follows:

$$f(x) = \frac{x^3}{3} - x^2 + C \qquad \text{General solution}$$

$$f(1) = \frac{1}{3} - 1 + C \qquad \text{Substitute } x = 1.$$

$$\frac{1}{3} = \frac{1}{3} - 1 + C \qquad f(1) = \frac{1}{3}$$

$$C = 1 \qquad \text{Solve for } C.$$

Therefore, the solution to the initial value problem is

> It is advisable to check that the solution satisfies the original problem: $f'(x) = x^2 - 2x$ and $f(1) = \frac{1}{3} - 1 + 1 = \frac{1}{3}$.

$$f(x) = \frac{x^3}{3} - x^2 + 1$$

which is just one of the curves in the family shown in Figure 4.79.

Related Exercises 39–54 ◄

Motion Problems Revisited

QUICK CHECK 5 Position is an antiderivative of velocity. But there are infinitely many antiderivatives that differ by a constant. Explain how two objects can have the same velocity function but two different position functions. ◄

> The convention with motion problems is to assume that motion begins at $t = 0$. This means that initial conditions are specified at $t = 0$.

Antiderivatives allow us to revisit the topic of one-dimensional motion introduced in Section 3.5. Suppose the position of an object that moves along a line relative to an origin is $s(t)$, where $t \geq 0$ measures elapsed time. The velocity of the object is $v(t) = s'(t)$, which may now be read in terms of antiderivatives: *The position function is an antiderivative of the velocity.* If we are given the velocity function of an object and its position at a particular time, we can determine its position at all future times by solving an initial value problem.

We also know that the acceleration $a(t)$ of an object moving in one dimension is the rate of change of the velocity, which means $a(t) = v'(t)$. In antiderivative terms, this says that the velocity is an antiderivative of the acceleration. Thus, if we are given the acceleration of an object and its velocity at a particular time, we can determine its velocity at all times. These ideas lie at the heart of modeling the motion of objects.

Initial Value Problems for Velocity and Position

Suppose an object moves along a line with a (known) velocity $v(t)$ for $t \geq 0$. Then its position is found by solving the initial value problem

$$s'(t) = v(t), \quad s(0) = s_0, \quad \text{where } s_0 \text{ is the initial position.}$$

If the acceleration of the object $a(t)$ is given, then its velocity is found by solving the initial value problem

$$v'(t) = a(t), \quad v(0) = v_0, \quad \text{where } v_0 \text{ is the initial velocity.}$$

EXAMPLE 6 A race Runner A begins at the point $s(0) = 0$ and runs with velocity $v(t) = 2t$. Runner B begins with a head start at the point $S(0) = 8$ and runs with velocity $V(t) = 2$. Find the positions of the runners for $t \geq 0$ and determine who is ahead at $t = 6$ time units.

SOLUTION Let the position of Runner A be $s(t)$, with an initial position $s(0) = 0$. Then, the position function satisfies the initial value problem

$$s'(t) = 2t, \quad s(0) = 0.$$

The solution is an antiderivative of $s'(t) = 2t$, which has the form $s(t) = t^2 + C$. Substituting $s(0) = 0$, we find that $C = 0$. Therefore, the position of Runner A is given by $s(t) = t^2$ for $t \geq 0$.

Let the position of Runner B be $S(t)$, with an initial position $S(0) = 8$. This position function satisfies the initial value problem

$$S'(t) = 2, \quad S(0) = 8.$$

The antiderivatives of $S'(t) = 2$ are $S(t) = 2t + C$. Substituting $S(0) = 8$ implies that $C = 8$. Therefore, the position of Runner B is given by $S(t) = 2t + 8$ for $t \geq 0$.

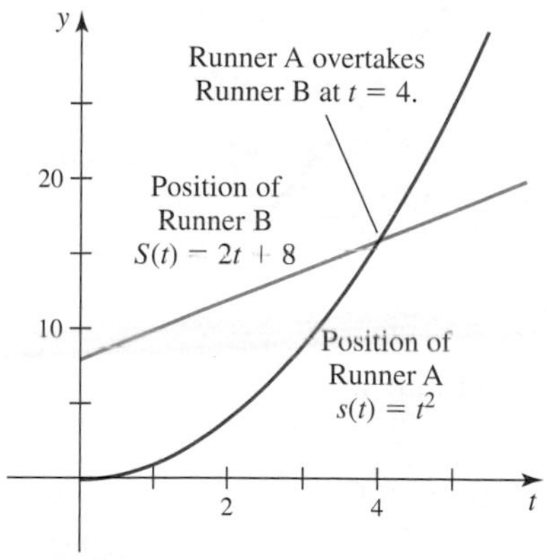

The graphs of the position functions are shown in Figure 4.80. Runner B begins with a head start but is overtaken when $s(t) = S(t)$, or when $t^2 = 2t + 8$. The solutions of this equation are $t = 4$ and $t = -2$. Only the positive solution is relevant because the race takes place for $t \geq 0$, so Runner A overtakes Runner B at $t = 4$, when $s = S = 16$. When $t = 6$, Runner A has the lead. *Related Exercises 55–62* ◄

FIGURE 4.80

EXAMPLE 7 Motion with gravity Neglecting air resistance, the motion of an object moving vertically near Earth's surface is determined by the acceleration due to gravity, which is approximately 9.8 m/s^2. Suppose a stone is thrown vertically upward at $t = 0$ with a velocity of 40 m/s from the edge of a cliff that is 100 m above a river.

a. Find the velocity $v(t)$ of the object for $t \geq 0$.

b. Find the position $s(t)$ of the object for $t \geq 0$.

c. Find the maximum height of the object above the river.

d. With what speed does the object strike the river?

SOLUTION We establish a coordinate system in which the positive s-axis points vertically upward with $s = 0$ corresponding to the river (Figure 4.81). Let $s(t)$ be the position of the stone measured relative to the river for $t \geq 0$. The initial velocity of the stone is $v(0) = 40$ m/s and the initial position of the stone is $s(0) = 100$ m.

> The acceleration due to gravity at Earth's surface is approximately $g = 9.8$ m/s², or $g = 32$ ft/s². It varies even at sea level from about 9.8640 at the poles to 9.7982 at the equator. The equation $v'(t) = -g$ is an instance of Newton's Second Law of Motion, assuming no other forces (such as air resistance) are present.

a. The acceleration due to gravity points in the *negative s-direction*. Therefore, the initial value problem governing the motion of the object is

$$\text{acceleration} = v'(t) = -9.8, \; v(0) = 40.$$

The antiderivatives of -9.8 are $v(t) = -9.8t + C$. The initial condition $v(0) = 40$ gives $C = 40$. Therefore, the velocity of the stone is

$$v(t) = -9.8t + 40.$$

As shown in Figure 4.82, the velocity decreases from its initial value $v(0) = 40$ until it reaches zero at the high point of the trajectory. This point is reached when

$$v(t) = -9.8t + 40 = 0$$

or when $t \approx 4.1$ s. For $t > 4.1$, the velocity becomes increasingly negative as the stone falls to Earth.

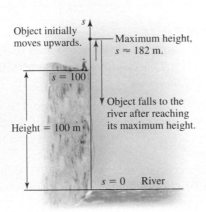

FIGURE 4.81

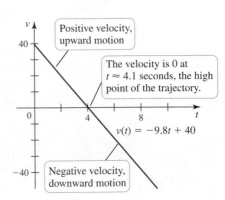

FIGURE 4.82

b. Knowing the velocity function of the stone, we can determine its position. The position function satisfies the initial value problem

$$v(t) = s'(t) = -9.8t + 40, \; s(0) = 100.$$

The antiderivatives of $-9.8t + 40$ are

$$s(t) = -4.9t^2 + 40t + C.$$

The initial condition $s(0) = 100$ implies $C = 100$, so the position function of the stone is

$$s(t) = -4.9t^2 + 40t + 100$$

as shown in Figure 4.83. The parabolic graph of the position function is not the actual trajectory of the stone; the stone moves vertically along the s-axis.

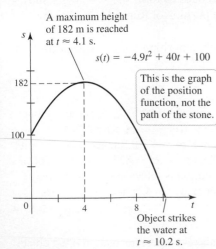

FIGURE 4.83

c. The position function of the stone increases for $0 < t < 4.1$. At $t \approx 4.1$, the stone reaches a high point of $s(4.1) \approx 182$ m.

d. For $t > 4.1$, the position function decreases, and the stone strikes the river when $s(t) = 0$. The roots of this equation are $t \approx 10.2$ and $t \approx -2.0$. Only the first root is relevant because the motion takes place for $t \geq 0$. Therefore, the stone strikes the ground at $t \approx 10.2$ s. Its speed at this instant is $|v(10.2)| \approx |-60| = 60$ m/s.

Related Exercises 63–66 ◄

SECTION 4.8 EXERCISES

Review Questions

1. Fill in the blank with the words *derivative* or *antiderivative*: If $F'(x) = f(x)$, then f is the _____ of F and F is the _____ of f.

2. Describe the set of antiderivatives of $f(x) = 0$.

3. Describe the set of antiderivatives of $f(x) = 1$.

4. Why do two different antiderivatives of a function differ by a constant?

5. Give the antiderivatives of x^p. For what values of p does your answer apply?

6. Give the antiderivatives of e^{-x}.

7. Give the antiderivatives of $1/x$ for $x > 0$.

8. Evaluate $\int \cos ax \, dx$ and $\int \sin ax \, dx$.

9. If $F(x) = x^2 - 3x + C$ and $F(-1) = 4$, what is the value of C?

10. For a given function f, explain the steps used to solve the initial value problem $F'(t) = f(t)$, $F(0) = 10$.

Basic Skills

11–18. Finding antiderivatives *Find all the antiderivatives of the following functions. Check your work by taking derivatives.*

11. $f(x) = 5x^4$

12. $g(x) = 11x^{10}$

13. $f(x) = \sin 2x$

14. $g(x) = -4 \cos 4x$

15. $P(x) = 3 \sec^2 x$

16. $Q(s) = \csc^2 s$

17. $f(y) = -2/y^3$

18. $H(z) = -6z^{-7}$

19–26. Indefinite integrals *Determine the following indefinite integrals. Check your work by differentiation.*

19. $\int (3x^5 - 5x^9) \, dx$

20. $\int (3u^{-2} - 4u^2 + 1) \, du$

21. $\int \left(4\sqrt{x} - \dfrac{4}{\sqrt{x}}\right) dx$

22. $\int \left(\dfrac{5}{t^2} + 4t^2\right) dt$

23. $\int (5s + 3)^2 \, ds$

24. $\int 5m(12m^3 - 10m) \, dm$

25. $\int (3x^{1/3} + 4x^{-1/3} + 6) \, dx$

26. $\int 6\sqrt[3]{x} \, dx$

27–32. Indefinite integrals involving trigonometric functions *Determine the following indefinite integrals. Check your work by differentiation.*

27. $\int (\sin 2y + \cos 3y) \, dy$

28. $\int \left[\sin 4t - \sin\left(\dfrac{t}{4}\right)\right] dt$

29. $\int (\sec^2 x - 1) \, dx$

30. $\int 2 \sec^2 2v \, dv$

31. $\int (\sec^2 \theta + \sec \theta \tan \theta) \, d\theta$

32. $\int \dfrac{\sin \theta - 1}{\cos^2 \theta} \, d\theta$

33–38. Other indefinite integrals *Determine the following indefinite integrals. Check your work by differentiation.*

33. $\int \dfrac{1}{2y} \, dy$

34. $\int (e^{2t} + 2\sqrt{t}) \, dt$

35. $\int \dfrac{6}{\sqrt{25 - x^2}} \, dx$

36. $\int \dfrac{3}{4 + v^2} \, dv$

37. $\int \dfrac{1}{x\sqrt{x^2 - 100}} \, dx$

38. $\int \dfrac{2}{16z^2 + 25} \, dz$

39–42. Particular antiderivatives *For the following functions f, find the antiderivative F that satisfies the given condition.*

39. $f(x) = x^5 - 2x^{-2} + 1$; $F(1) - 0$

40. $f(t) = \sec^2 t$; $F(\pi/4) = 1$

41. $f(v) = \sec v \tan v$; $F(0) = 2$

42. $f(x) = (4\sqrt{x} + 6/\sqrt{x})/x^2$; $F(1) = 4$

43–48. Solving initial value problems *Find the solution of the following initial value problems.*

43. $f'(x) = 2x - 3$; $f(0) = 4$

44. $g'(x) = 7x^6 - 4x^3 + 12$; $g(1) = 24$

45. $g'(x) = 7x\left(x^6 - \dfrac{1}{7}\right)$; $g(1) = 2$

46. $h'(t) = 6 \sin 3t$; $h(\pi/6) = 6$

47. $f'(u) = 4(\cos u - \sin 2u)$; $f(\pi/6) = 0$

48. $p'(t) = 10e^{-t}$; $p(0) = 100$

49–54. Graphing general solutions *Graph several functions that satisfy each of the following differential equations. Then, find and graph the particular function that satisfies the given initial condition.*

49. $f'(x) = 2x - 5$, $f(0) = 4$ **50.** $f'(x) = 3x^2 - 1$, $f(1) = 2$

51. $f'(x) = 3x + \sin \pi x$, $f(2) = 3$

52. $f'(s) = 4 \sec s \tan s$, $f(\pi/4) = 1$

53. $f'(t) = 1/t$, $f(1) = 4$ **54.** $f'(x) = 2 \cos 2x$, $f(0) = 1$

55–60. Velocity to position *Given the following velocity functions of an object moving along a line, find the position function with the given initial position. Then, graph both the velocity and position functions.*

55. $v(t) = 2t + 4$; $s(0) = 0$ **56.** $v(t) = e^{-2t} + 4$; $s(0) = 2$

57. $v(t) = 2\sqrt{t}$; $s(0) = 1$ **58.** $v(t) = 2 \cos t$; $s(0) = 0$

59. $v(t) = 6t^2 + 4t - 10$; $s(0) = 0$

60. $v(t) = 2 \sin 2t$; $s(0) = 0$

61–62. Races *The velocity function and initial position of Runners A and B are given. Analyze the race that results by graphing the position functions of the runners and finding the time and positions (if any) at which they first pass each other.*

61. A: $v(t) = \sin t$, $s(0) = 0$;   B: $V(t) = \cos t$, $S(0) = 0$

62. A: $v(t) = 2e^{-t}$, $s(0) = 0$;   B: $V(t) = 4e^{-4t}$, $S(0) = 10$

63–66. Motion with gravity *Consider the following descriptions of the vertical motion of an object subject only to the acceleration due to gravity.*

   a. *Find the velocity of the object for all relevant times.*
   b. *Find the position of the object for all relevant times.*
   c. *Find the time when the object reaches its highest point (What is the height?)*
   d. *Find the time when the object strikes the ground*

63. A softball is popped up vertically (from the ground) with a velocity of 30 m/s.

64. A stone is thrown vertically upward with a velocity of 30 m/s from the edge of a cliff 200 m above a river.

65. A payload is released at an elevation of 400 m from a hot-air balloon that is rising at a rate of 10 m/s.

66. A payload is dropped at an elevation of 400 m from a hot-air balloon that is descending at a rate of 10 m/s.

Further Explorations

67. Explain why or why not Determine whether the following statements are true and give an explanation or counterexample.

   a. $F(x) = x^3 - 4x + 100$ and $G(x) = x^3 - 4x - 100$ are antiderivatives of the same function.
   b. If $F'(x) = f(x)$, then f is an antiderivative of F.
   c. If $F'(x) = f(x)$, then $\int f(x)\, dx = F(x) + C$.
   d. $f(x) = x^3 + 3$ and $g(x) = x^3 - 4$ are derivatives of the same function.
   e. If $F'(x) = G'(x)$, then $F(x) = G(x)$.

68–75. Miscellaneous indefinite integrals *Determine the following indefinite integrals. Check your work by differentiation.*

68. $\displaystyle \int \left(\sqrt[3]{x^2} + \sqrt{x^3} \right) dx$ **69.** $\displaystyle \int \frac{e^{2x} - e^{-2x}}{2}\, dx$

70. $\displaystyle \int (4 \cos 4w - 3 \sin 3w)\, dw$ **71.** $\displaystyle \int (\csc^2 \theta + 2\theta^2 - 3\theta)\, d\theta$

72. $\displaystyle \int (\csc^2 \theta + 1)\, d\theta$ **73.** $\displaystyle \int \frac{1 + \sqrt{x}}{x}\, dx$

74. $\displaystyle \int \frac{2 + x^2}{1 + x^2}\, dx$ **75.** $\displaystyle \int \sqrt{x}\, (2x^6 - 4\sqrt[3]{x})\, dx$

76–79. Functions from higher derivatives *Find the function F that satisfies the following differential equations and initial conditions.*

76. $F''(x) = 1$, $F'(0) = 3$, $F(0) = 4$

77. $F''(x) = \cos x$, $F'(0) = 3$, $F(\pi) = 4$

78. $F'''(x) = 4x$, $F''(0) = 0$, $F'(0) = 1$, $F(0) = 3$

79. $F'''(x) = 672x^5 + 24x$, $F''(0) = 0$, $F'(0) = 2$, $F(0) = 1$

Applications

80. Mass on a spring A mass oscillates up and down on the end of a spring. Find its position s relative to the equilibrium position if its acceleration is $a(t) = \sin (\pi t)$, and its initial velocity and position are $v(0) = 3$ and $s(0) = 0$, respectively.

81. Flow rate A large tank is filled with water when an outflow valve is opened at $t = 0$. Water flows out at a rate given by $Q'(t) = 0.1(100 - t^2)$ gal/min for $0 \le t \le 10$ min.

   a. Find the amount of water $Q(t)$ that has flowed out of the tank after t minutes, given the initial condition $Q(0) = 0$.
   b. Graph the flow function Q for $0 \le t \le 10$.
   c. How much water flows out of the tank in 10 min?

82. General headstart problem Suppose that object A is located at $s = 0$ at time $t = 0$ and starts moving along the s-axis with a velocity given by $v(t) = 2at$, where $a > 0$. Object B is located at $s = c > 0$ at $t = 0$ and starts moving along the s-axis with a constant velocity given by $V(t) = b > 0$. Show that A always overtakes B at time

$$t = \frac{b + \sqrt{b^2 + 4ac}}{2a}.$$

Additional Exercises

83. Using identities Use the identities $\sin^2 x = (1 - \cos 2x)/2$ and $\cos^2 x = (1 + \cos 2x)/2$ to find $\int \sin^2 x\, dx$ and $\int \cos^2 x\, dx$.

84–87. Verifying indefinite integrals *Verify the following indefinite integrals by differentiation. These integrals are derived in later chapters.*

84. $\displaystyle \int \frac{\cos \sqrt{x}}{\sqrt{x}}\, dx = 2 \sin \sqrt{x} + C$

85. $\displaystyle \int \frac{x}{\sqrt{x^2 + 1}}\, dx = \sqrt{x^2 + 1} + C$

86. $\int x^2 \cos x^3 \, dx = \dfrac{1}{3} \sin x^3 + C$

87. $\int \dfrac{x}{(x^2 - 1)^2} \, dx = -\dfrac{1}{2(x^2 - 1)} + C$

CHAPTER 4 REVIEW EXERCISES

1. **Explain why or why not** Determine whether the following statements are true and give an explanation or counterexample.

   a. If $f'(c) = 0$, then f has a local minimum or maximum at c.
   b. If $f''(c) = 0$, then f has an inflection point at $(c, f(c))$.
   c. $F(x) = x^2 + 10$ and $G(x) = x^2 - 100$ are antiderivatives of the same function.
   d. Between two local minima of a continuous function on $(-\infty, \infty)$, there must be a local maximum.

2. **Locating extrema** Consider the graph of a function f on the interval $[-3, 3]$.

   a. Identify the local minima and maxima of f.
   b. Identify the absolute minimum and absolute maximum of f (if they exist).
   c. Give the approximate co-ordinates of the inflection point(s) of f.
   d. Give the approximate co-ordinates of the zero(s) of f.
   e. On what intervals (approximately) is f concave up?
   f. On what intervals (approximately) is f concave down?

3–4. Designer functions *Sketch the graph of a continuous function that satisfies the following conditions.*

3. f is continuous on the interval $[-4, 4]$, $f'(x) = 0$ for $x = -2, 0$, and 3; f has an absolute minimum at $x = 3$; f has a local minimum at $x = -2$; f has a local maximum at $x = 0$; f has an absolute maximum at $x = -4$.

4. f is continuous on $(-\infty, \infty)$; $f'(x) < 0$ and $f''(x) < 0$ on $(-\infty, 0)$; $f'(x) > 0$ and $f''(x) > 0$ on $(0, \infty)$

5. **Functions from derivatives** Given the graphs of f' and f'', sketch a possible graph of f.

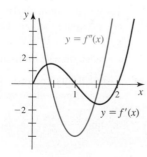

6–10. Critical points *Find the critical points of the following functions on the given intervals. Identify the absolute minimum and absolute maximum values (if possible). Graph the function to confirm your conclusions.*

6. $f(x) = \sin 2x + 3$; $[-\pi, \pi]$

7. $f(x) = 2x^3 - 3x^2 - 36x + 12$; $(-\infty, \infty)$

8. $f(x) = 4x^{1/2} - x^{5/2}$; $[0, 4]$

9. $f(x) = 2x \ln x + 10$; $(0, 4)$

10. $g(x) = x^{1/3}(9 - x^2)$; $[-4, 4]$

11. **Absolute values** Consider the function $f(x) = |x - 2| + |x + 3|$ on $[-4, 4]$. Graph f, identify the critical points, and give the coordinates of the local and absolute extreme values.

12. **Inflection points** Does $f(x) = 2x^5 - 10x^4 + 20x^3 + x + 1$ have any inflection points? If so, identify them.

13–20. Curve sketching *Use the guidelines of this chapter to make a complete graph of the following functions on their domains or on the given interval. Use a graphing utility to check your work.*

13. $f(x) = x^4/2 - 3x^2 + 4x + 1$

14. $f(x) = \dfrac{3x}{x^2 + 3}$

15. $f(x) = 4 \cos[\pi(x - 1)]$ on $[0, 2]$

16. $f(x) = \dfrac{x^2 + x}{4 - x^2}$

17. $f(x) = \sqrt[3]{x} - \sqrt{x} + 2$

18. $f(x) = \dfrac{\cos \pi x}{1 + x^2}$ on $[-2, 2]$

19. $f(x) = x^{2/3} + (x + 2)^{1/3}$

20. $f(x) = x(x - 1)e^{-x}$

21. **Optimization** A right triangle has legs of length h and r and a hypotenuse of length 4 (see figure). It is revolved about the leg of length h to sweep out a right circular cone. What values of h and r maximize the volume of the cone?

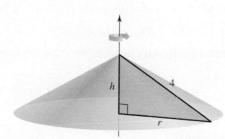

▣ 22. Rectangles beneath a curve A rectangle is constructed with one side on the positive x-axis, one side on the positive y-axis, and one vertex on the curve $y = \cos x$ for $0 < x < \pi/2$. Approximate the dimensions of the rectangle that maximize the area of the rectangle. What is the area?

23. Maximum length What two nonnegative real numbers a and b whose sum is 23 (a) minimize $a^2 + b^2$? (b) Maximize $a^2 + b^2$?

24. Nearest point What point of the graph of $f(x) = \frac{5}{2} - x^2$ is closest to the origin? (*Hint:* You can minimize the square of the distance.)

25. Mean Value Theorem The population of a culture of cells grows according to the function $P(t) = \dfrac{100t}{t + 1}$, where $t \geq 0$ is measured in weeks.

a. What is the average rate of change in the population over the interval $[0, 8]$?

b. At what point of the interval $[0, 8]$ is the instantaneous rate of change equal to the average rate of change?

26–35. Limits *Evaluate the following limits. Use l'Hôpital's Rule when needed.*

26. $\displaystyle \lim_{t \to 2} \frac{t^3 - t^2 - 2t}{t^2 - 4}$

27. $\displaystyle \lim_{t \to 0} \frac{1 - \cos 6t}{2t}$

28. $\displaystyle \lim_{x \to \infty} \frac{5x^2 + 2x - 5}{\sqrt{x^4 - 1}}$

29. $\displaystyle \lim_{\theta \to 0} \frac{3 \sin^2 2\theta}{\theta^2}$

30. $\displaystyle \lim_{x \to \infty} \left(\sqrt{x^2 + x + 1} - \sqrt{x^2 - x} \right)$

31. $\displaystyle \lim_{\theta \to 0} 2\theta \cot 3\theta$

32. $\displaystyle \lim_{x \to 0} \frac{e^{-2x} - 1 + 2x}{x^2}$

33. $\displaystyle \lim_{y \to 0^+} \frac{\ln^{10} y}{\sqrt{y}}$

34. $\displaystyle \lim_{\theta \to 0} \frac{3 \sin 8\theta}{8 \sin 3\theta}$

35. $\displaystyle \lim_{x \to 1} \frac{x^4 - x^3 - 3x^2 + 5x - 2}{x^3 + x^2 - 5x + 3}$

36–41. $1^\infty, 0^0, \infty^0$ forms *Evaluate the following limits. Check your results by graphing.*

36. $\displaystyle \lim_{x \to \infty} \frac{\ln x^{100}}{\sqrt{x}}$

37. $\displaystyle \lim_{x \to \pi/2^-} (\sin x)^{\tan x}$

38. $\displaystyle \lim_{x \to \infty} \frac{\ln^3 x}{\sqrt{x}}$

39. $\displaystyle \lim_{x \to \infty} \ln \left(\frac{x + 1}{x - 1} \right)$

40. $\displaystyle \lim_{x \to \infty} x^{1/x}$

41. $\displaystyle \lim_{x \to \infty} \left(1 - \frac{3}{x} \right)^x$

42–49. Comparing growth rates *Determine which of the two functions grows faster, or state that they have comparable growth rates.*

42. $10x$ and $\ln x$

43. $x^{1/2}$ and $x^{1/3}$

44. $\ln x$ and $\log_{10} x$

45. $\sqrt{x}$ and $\ln^{10} x$

46. $10x$ and $\ln x^2$

47. e^x and 3^x

48. $\sqrt{x^6 + 10}$ and x^3

49. 2^x and $4^{x/2}$

50–61. Indefinite integrals *Determine the following indefinite integrals.*

50. $\displaystyle \int (x^8 - 3x^3 + 1) \, dx$

51. $\displaystyle \int \left(\frac{1}{x^2} - \frac{2}{x^{5/2}} \right) dx$

52. $\displaystyle \int \frac{x^4 - 2\sqrt{x} + 2}{x^2} \, dx$

53. $\displaystyle \int (1 + \cos 3\theta) \, d\theta$

54. $\displaystyle \int 2 \sec^2 x \, dx$

55. $\displaystyle \int \sec 2x \tan 2x \, dx$

56. $\displaystyle \int 2e^{2x} \, dx$

57. $\displaystyle \int \frac{12}{x} \, dx$

58. $\displaystyle \int \frac{dx}{\sqrt{1 - x^2}}$

59. $\displaystyle \int \frac{dx}{x^2 + 1}$

60. $\displaystyle \int \frac{1 + \tan \theta}{\sec \theta} \, d\theta$

61. $\displaystyle \int \left(\sqrt[4]{x^3} + \sqrt{x^5} \right) dx$

62–65. Functions from derivatives *Find the function with the following properties.*

62. $f'(x) = 3x^2 - 1$ and $f(0) = 10$

63. $f'(t) = \sin t + 2t$ and $f(0) = 5$

64. $g'(t) = t^2 + t^{-2}$ and $g(1) = 1$

65. $h'(x) = \sin^2 x$ and $h(1) = 1$ (*Hint:* $\sin^2 x = (1 - \cos 2x)/2$.)

66. Motion along a line Two objects move along the x-axis with position functions $x_1(t) = 2 \sin t$ and $x_2(t) = \sin (t - \pi/2)$. At what times on the interval $[0, 2\pi]$ are the objects closest to each other and farthest from each other?

67. Vertical motion with gravity A rocket is launched vertically upward with an initial velocity of 120 m/s from a platform that is 125 m above the ground. Assume that the only force at work is gravity. Determine and graph the velocity and position functions of the rocket for $t \geq 0$. Then describe the motion in words.

68. Logs of logs Compare the growth rates of $\ln x, \ln (\ln x)$, and $\ln [\ln (\ln x)]$.

69. Two limits with exponentials Evaluate $\displaystyle \lim_{x \to 0^+} \frac{x}{\sqrt{1 - e^{-x^2}}}$ and $\displaystyle \lim_{x \to 0^+} \frac{x^2}{1 - e^{-x^2}}$ and confirm your result by graphing.

70. Geometric mean Prove that $\displaystyle \lim_{r \to 0} \left(\frac{a^r + b^r + c^r}{3} \right)^{1/r} = \sqrt[3]{abc}$, where a, b, and c are positive real numbers.

71–72. Two methods *Evaluate the following limits in two different ways: Use the methods of Chapter 2 and use l'Hôpital's Rule.*

71. $\displaystyle \lim_{x \to \infty} \frac{2x^5 - x + 1}{5x^6 + x}$

72. $\displaystyle \lim_{x \to \infty} \frac{4x^4 - \sqrt{x}}{2x^4 + x^{-1}}$

73. Towers of exponents The functions $f(x) = (x^x)^x$ and $g(x) = x^{(x^x)}$ are different functions. For example, $f(3) = 19{,}683$ and $g(3) \approx 7.6 \times 10^{12}$. Determine whether $\displaystyle \lim_{x \to 0^+} f(x)$ and $\displaystyle \lim_{x \to 0^+} g(x)$ are indeterminate forms and evaluate the limits.

74. Cosine limits Let n be a positive integer. Use graphical and/or analytical methods to verify the following limits:

a. $\displaystyle \lim_{x \to 0} \frac{1 - \cos x^n}{x^{2n}} = \frac{1}{2}$

b. $\displaystyle \lim_{x \to 0} \frac{1 - \cos^n x}{x^2} = \frac{n}{2}$

75. Limits for e Consider the function $g(x) = (1 + 1/x)^{x+a}$. Show that if $0 \le a < \frac{1}{2}$, then $g(x) \to e$ from *below* as $x \to \infty$; if $\frac{1}{2} \le a < 1$, then $g(x) \to e$ from *above* as $x \to \infty$.

76. A family of super-exponential functions Let $f(x) = (a + x)^x$, where $a > 0$.

   a. What is the domain of f (in terms of a)?

   b. Describe the end behavior of f (near the boundary of its domain or as $|x| \to \infty$).

   c. Compute f'. Then graph f and f' for $a = 0.5, 1, 2, 3$.

   d. Show that f has a single local minimum at the point z that satisfies $(z + a)\ln(z + a) + z = 0$.

   e. Describe how z [found in part (d)] varies as a increases. Describe how $f(z)$ varies as a increases.

Chapter 4 Guided Projects

Applications of the material in this chapter and related topics can be found in the following Guided Projects. For additional information, see the Preface.

• Oscillators

• Newton's method

• Ice cream, geometry, and calculus

5

Integration

Chapter Preview   We are now at a critical point in the calculus story. Many would argue that this chapter is the cornerstone of calculus because it explains the relationship between the two processes of calculus: differentiation and integration. We begin by explaining why finding the area of regions bounded by the graphs of functions is such an important problem in calculus. Then you will see how antiderivatives lead to definite integrals, which are used to solve this problem. But there is more to the story. You will also see the remarkable connection between derivatives and integrals, which is expressed in the Fundamental Theorem of Calculus. Once we know how—in principle—to evaluate definite integrals, a long agenda unfolds that will be addressed over the next three chapters. In this chapter, we develop key properties of definite integrals, investigate a few of their many applications, and present the first of several powerful techniques for evaluating definite integrals.

5.1 Approximating Areas under Curves

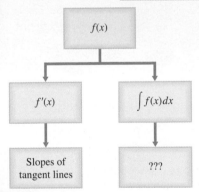

FIGURE 5.1

The derivative of a function is associated with rates of change and slopes of tangent lines. We also know that antiderivatives (or indefinite integrals) reverse the derivative operation. Figure 5.1 summarizes our current understanding and raises the question: What is the geometric meaning of the integral? The following example reveals a clue.

Area under a Velocity Curve

Consider an object moving along a line. You learned in previous chapters that the slope of the line tangent to the graph of the position function at a certain time gives the velocity at that time. We now turn the situation around. If we know the velocity function of a moving object, what can we learn about its position function?

Imagine a car traveling at a constant velocity of 60 mi/hr along a straight highway over a two-hour period. The graph of the velocity function $v = 60$ on the interval $0 \le t \le 2$ is a horizontal line (Figure 5.2). The displacement of the car between $t = 0$ and $t = 2$ hr is found by a familiar formula:

$$\text{displacement} = \text{rate} \cdot \text{time}$$

$$= 60 \text{ mi/hr} \cdot 2 \text{ hr} = 120 \text{ mi}$$

▷ Recall from Section 3.5 that the *displacement* of an object moving along a line is the difference between its initial and final position. If the velocity of an object is positive, its displacement equals the distance traveled.

This product is the area of the rectangle formed by the velocity curve and the t-axis between $t = 0$ and $t = 2$ hr (Figure 5.3). In the case of constant positive velocity, we see that the area between the velocity curve and the t-axis is the displacement of the moving object.

▶ The side lengths of the rectangle in Figure 5.3 have units mi/hr and hr. Therefore, the units of the area are mi/hr · hr = mi, which is the unit of displacement.

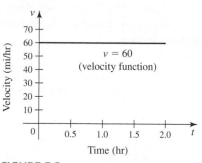

FIGURE 5.2

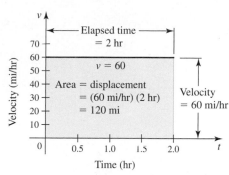

FIGURE 5.3

QUICK CHECK 1 What is the displacement of an object that travels at a constant velocity of 10 mi/hr for a half hour, 20 mi/hr for the next half hour, and 30 mi/hr for the next hour? ◀

Because objects do not necessarily move at a constant velocity, we must extend these ideas to positive velocities that *change* over an interval of time. One strategy is to divide the time interval into many subintervals and approximate the velocity on each subinterval by a constant velocity. Then the displacements on each subinterval are calculated and summed. This strategy produces only an approximation to the displacement; however, this approximation generally improves as the number of subintervals increases.

EXAMPLE 1 Approximating the displacement Suppose the velocity in m/s of an object moving along a line is given by the function $v = t^2$, where $0 \le t \le 8$. Approximate the displacement of the object by dividing the time interval $[0, 8]$ into n subintervals of equal length. On each subinterval, approximate the velocity by a constant equal to the value of v evaluated at the midpoint of the subinterval.

a. Begin by dividing $[0, 8]$ into $n = 2$ subintervals: $[0, 4]$ and $[4, 8]$.

b. Divide $[0, 8]$ into $n = 4$ subintervals: $[0, 2]$, $[2, 4]$, $[4, 6]$, and $[6, 8]$.

c. Divide $[0, 8]$ into $n = 8$ subintervals of equal length.

SOLUTION

a. We divide the interval $[0, 8]$ into $n = 2$ subintervals, $[0, 4]$ and $[4, 8]$, each with length 4. The velocity on each subinterval is approximated using the value of v evaluated at the midpoint of that subinterval (Figure 5.4a).

- We approximate the velocity on $[0, 4]$ by $v(2) = 2^2 = 4$ m/s. Traveling at 4 m/s for 4 s results in a displacement of 4 m/s · 4 s = 16 m.

- We approximate the velocity on $[4, 8]$ by $v(6) = 6^2 = 36$ m/s. Traveling at 36 m/s for 4 s results in a displacement of 36 m/s · 4 s = 144 m.

Therefore, an approximation to the displacement over the entire interval $[0, 8]$ is

$$(v(2) \cdot 4 \, \text{s}) + (v(6) \cdot 4 \, \text{s}) = (4 \, \text{m/s} \cdot 4 \, \text{s}) + (36 \, \text{m/s} \cdot 4 \, \text{s}) = 160 \, \text{m}.$$

b. With $n = 4$ (Figure 5.4b), each subinterval has length 2. The approximate displacement over the entire interval is

$$\underbrace{(1\,\text{m/s} \cdot 2\,\text{s})}_{v(1)} + \underbrace{(9\,\text{m/s} \cdot 2\,\text{s})}_{v(3)} + \underbrace{(25\,\text{m/s} \cdot 2\,\text{s})}_{v(5)} + \underbrace{(49\,\text{m/s} \cdot 2\,\text{s})}_{v(7)} = 168\,\text{m}.$$

c. With $n = 8$ subintervals (Figure 5.4c), the approximation to the displacement is 170 m. In each case, the approximate displacement is the sum of the areas of the rectangles under the velocity curve.

> The midpoint of each subinterval is used to approximate the velocity over that subinterval.

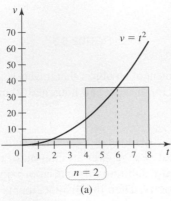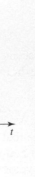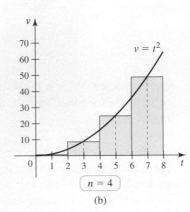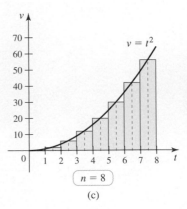

FIGURE 5.4

Related Exercises 9–14 ◄

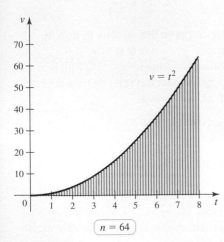

FIGURE 5.5

QUICK CHECK 2 In Example 1, if we used $n = 32$ subintervals of equal length, what would be the length of each subinterval? Find the midpoint of the first and last subinterval. ◄

The progression in Example 1 may be continued. Larger values of n mean more rectangles; in general, more rectangles give a better fit to the region under the curve (Figure 5.5). With the help of a calculator, we can generate the approximations in Table 5.1 using $n = 1$, 2, 4, 8, 16, 32, and 64 subintervals. Observe that as n increases, the approximations appear to approach a limit of approximately 170.7 m. The limit is the exact displacement, which is represented by the area of the region under the velocity curve. This strategy of taking limits of sums is developed fully in Section 5.2.

Table 5.1 Approximations to the area under the velocity curve $v = t^2$ on $[0, 8]$

Number of subintervals	Length of each subinterval	Approximate displacement (area under curve)
1	8 s	128.0 m
2	4 s	160.0 m
4	2 s	168.0 m
8	1 s	170.0 m
16	0.5 s	170.5 m
32	0.25 s	170.625 m
64	0.125 s	170.65625 m

> The language "the area of the region bounded by the graph of a function" is often abbreviated as "the area under the curve."

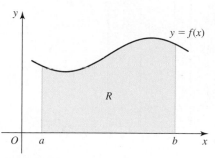

FIGURE 5.6

Approximating Areas by Riemann Sums

We now develop a method for approximating areas under curves. Consider a continuous function f that is nonnegative on an interval $[a, b]$. The goal is to approximate the area of the region R bounded by the graph of f and the x-axis from $x = a$ to $x = b$ (Figure 5.6). We begin by dividing the interval $[a, b]$ into n subintervals of equal length,

$$[x_0, x_1], [x_1, x_2], \ldots, [x_{n-1}, x_n],$$

where $a = x_0$ and $b = x_n$ (Figure 5.7). The length of each subinterval, denoted Δx, is found by dividing the length of the interval by n:

$$\Delta x = \frac{b - a}{n}$$

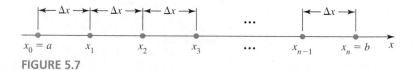

FIGURE 5.7

DEFINITION   Regular Partition

Suppose $[a, b]$ is a closed interval containing n subintervals

$$[x_0, x_1], [x_1, x_2], \ldots, [x_{n-1}, x_n]$$

of equal length $\Delta x = \dfrac{b - a}{n}$ with $a = x_0$ and $b = x_n$. The endpoints $x_0, x_1, x_2, \ldots,$ x_{n-1}, x_n of the subintervals are called **grid points** and they create a **regular partition** of the interval $[a, b]$. In general, the kth grid point is

$$x_k = a + k\,\Delta x \qquad \text{for } k = 0, 1, 2, \ldots, n.$$

QUICK CHECK 3 If the interval $[1, 9]$ is partitioned into 4 subintervals of equal length, what is Δx? List the grid points x_0, x_1, x_2, x_3, and x_4. ◄

> Although the idea of integration was developed in the 17th century, it was almost 200 years later that the German mathematician Bernhard Riemann (1826–1866) worked on the mathematical theory underlying integration.

In the kth subinterval $[x_{k-1}, x_k]$, we choose any point $\overline{x}_k$ and build a rectangle whose height is $f(\overline{x}_k)$, the value of f at $\overline{x}_k$ (Figure 5.8). The area of the rectangle on the kth subinterval is

$$\text{height} \cdot \text{base} = f(\overline{x}_k)\,\Delta x, \qquad \text{where } k = 1, 2, \ldots, n.$$

Summing the areas of the rectangles in Figure 5.8, we obtain an approximation to the area of R, which is called a **Riemann sum**:

$$f(\overline{x}_1)\,\Delta x + f(\overline{x}_2)\,\Delta x + \cdots + f(\overline{x}_n)\,\Delta x$$

Three notable Riemann sums are the left, right, and midpoint Riemann sums.

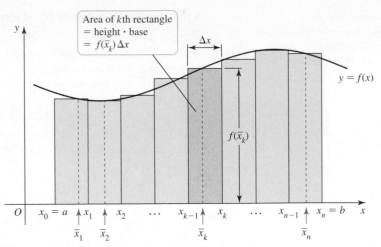

FIGURE 5.8

DEFINITION   Riemann Sum

Suppose f is defined on a closed interval $[a, b]$, which is divided into n subintervals of equal length Δx. If $\overline{x}_k$ is any point in the kth subinterval $[x_{k-1}, x_k]$, for $k = 1, 2, \ldots, n$, then

$$f(\overline{x}_1) \Delta x + f(\overline{x}_2) \Delta x + \cdots + f(\overline{x}_n) \Delta x$$

is called a **Riemann sum** for f on $[a, b]$. This sum is a **left Riemann sum** if $\overline{x}_k$ is the left endpoint of $[x_{k-1}, x_k]$ (Figure 5.9); a **right Riemann sum** if $\overline{x}_k$ is the right endpoint of $[x_{k-1}, x_k]$ (Figure 5.10); and a **midpoint Riemann sum** if $\overline{x}_k$ is the midpoint of $[x_{k-1}, x_k]$ (Figure 5.11), for $k = 1, 2, \ldots, n$.

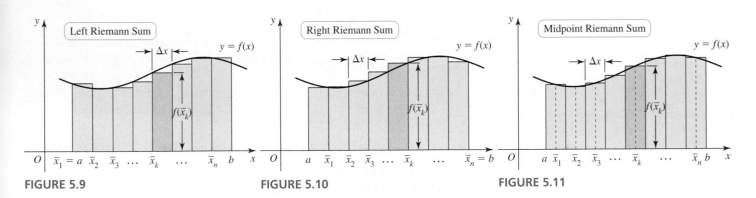

FIGURE 5.9   FIGURE 5.10   FIGURE 5.11

EXAMPLE 2   Area under the sine curve Let R be the region bounded by the graph of $f(x) = \sin x$ and the x-axis between $x = 0$ and $x = \pi/2$.

a. Approximate the area of R using a left Riemann sum with $n = 6$ subintervals. Illustrate the sum with the appropriate rectangles.

b. Approximate the area of R using a right Riemann sum with $n = 6$ subintervals. Illustrate the sum with the appropriate rectangles.

c. How do the area approximations in parts (a) and (b) compare to the actual area under the curve?

SOLUTION Dividing the interval $[a, b] = [0, \pi/2]$ into $n = 6$ subintervals means the length of each subinterval is

$$\Delta x = \frac{b - a}{n} = \frac{\pi/2 - 0}{6} = \frac{\pi}{12}.$$

a. To find the left Riemann sum, we set $\overline{x}_1, \overline{x}_2, \ldots, \overline{x}_6$ equal to the left endpoints of the six subintervals. The heights of the rectangles are $f(\overline{x}_k)$, for $k = 1, \ldots, 6$.

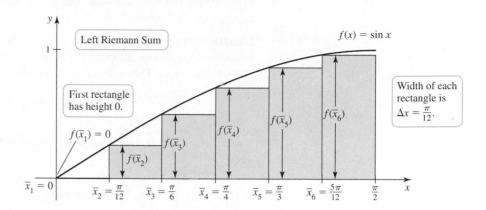

FIGURE 5.12

The resulting left Riemann sum (Figure 5.12) is

$$f(\overline{x}_1)\Delta x + f(\overline{x}_2)\Delta x + \cdots + f(\overline{x}_6)\Delta x$$

$$= \left[\sin(0)\cdot\frac{\pi}{12}\right] + \left[\sin\left(\frac{\pi}{12}\right)\cdot\frac{\pi}{12}\right] + \left[\sin\left(\frac{\pi}{6}\right)\cdot\frac{\pi}{12}\right]$$

$$+ \left[\sin\left(\frac{\pi}{4}\right)\cdot\frac{\pi}{12}\right] + \left[\sin\left(\frac{\pi}{3}\right)\cdot\frac{\pi}{12}\right] + \left[\sin\left(\frac{5\pi}{12}\right)\cdot\frac{\pi}{12}\right]$$

$$\approx 0.863.$$

b. In a right Riemann sum, the right endpoints are used for $\overline{x}_1, \overline{x}_2, \ldots, \overline{x}_6$, and the heights of the rectangles are $f(\overline{x}_k)$, for $k = 1, \ldots, 6$.

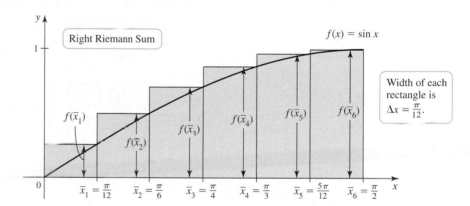

FIGURE 5.13

The resulting right Riemann sum (Figure 5.13) is

$$f(\overline{x}_1)\Delta x + f(\overline{x}_2)\Delta x + \cdots + f(\overline{x}_6)\Delta x$$

$$= \left[\sin\left(\frac{\pi}{12}\right)\cdot\frac{\pi}{12}\right] + \left[\sin\left(\frac{\pi}{6}\right)\cdot\frac{\pi}{12}\right] + \left[\sin\left(\frac{\pi}{4}\right)\cdot\frac{\pi}{12}\right]$$

$$+ \left[\sin\left(\frac{\pi}{3}\right)\cdot\frac{\pi}{12}\right] + \left[\sin\left(\frac{5\pi}{12}\right)\cdot\frac{\pi}{12}\right] + \left[\sin\left(\frac{\pi}{2}\right)\cdot\frac{\pi}{12}\right]$$

$$\approx 1.125.$$

QUICK CHECK 4 If the function in Example 2 is $f(x) = \cos x$, does the left Riemann sum or the right Riemann sum overestimate the area under the curve? ◄

c. Looking at the graphs, we see that the left Riemann sum in part (a) underestimates the actual area of R, whereas the right Riemann sum in part (b) overestimates the area of R. Therefore, the area of R is between 0.863 and 1.125. As the number of rectangles increases, these approximations improve.

Related Exercises 15–20 ◄

EXAMPLE 3   A midpoint Riemann sum Let R be the region bounded by the graph of $f(x) = \sin x$ and the x-axis between $x = 0$ and $x = \pi/2$. Approximate the area of R using a midpoint Riemann sum with $n = 6$ subintervals. Illustrate the sum with the appropriate rectangles.

SOLUTION The grid points and the length of the subintervals $\Delta x = \pi/12$ are the same as in Example 2. To find the midpoint Riemann sum, we set $\overline{x}_1, \overline{x}_2, \ldots, \overline{x}_6$ equal to the midpoints of the subintervals. The midpoint of the first subinterval is the average of x_0 and x_1, which is

$$\overline{x}_1 = \frac{x_1 + x_0}{2} = \frac{\pi/12 + 0}{2} = \frac{\pi}{24}.$$

The remaining midpoints are also computed by averaging the two nearest grid points.

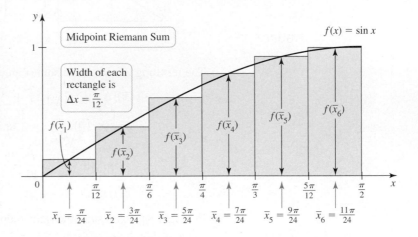

FIGURE 5.14

The resulting midpoint Riemann sum (Figure 5.14) is

$$f(\overline{x}_1)\,\Delta x + f(\overline{x}_2)\,\Delta x + \cdots + f(\overline{x}_6)\,\Delta x$$

$$= \left[\sin\left(\frac{\pi}{24}\right) \cdot \frac{\pi}{12}\right] + \left[\sin\left(\frac{3\pi}{24}\right) \cdot \frac{\pi}{12}\right] + \left[\sin\left(\frac{5\pi}{24}\right) \cdot \frac{\pi}{12}\right]$$

$$+ \left[\sin\left(\frac{7\pi}{24}\right) \cdot \frac{\pi}{12}\right] + \left[\sin\left(\frac{9\pi}{24}\right) \cdot \frac{\pi}{12}\right] + \left[\sin\left(\frac{11\pi}{24}\right) \cdot \frac{\pi}{12}\right]$$

$$\approx 1.003.$$

Comparing the midpoint Riemann sum (Figure 5.14) with the left (Figure 5.12) and right (Figure 5.13) Riemann sums suggests that the midpoint sum is a more accurate estimate of the area under the curve.   *Related Exercises 21–26* ◀

Table 5.2

x	$f(x)$
0	1
0.5	3
1.0	4.5
1.5	5.5
2.0	6.0

EXAMPLE 4   Riemann sums from tables Estimate the area A under the graph of f on the interval $[0, 2]$ using left and right Riemann sums with $n = 4$, when f is continuous but known only at the points in Table 5.2.

SOLUTION With $n = 4$ subintervals on the interval $[0, 2]$, $\Delta x = 2/4 = 0.5$. Using the left endpoint of each subinterval, the left Riemann sum is

$$A \approx (f(0) + f(0.5) + f(1.0) + f(1.5))\,\Delta x = (1 + 3 + 4.5 + 5.5)0.5 = 7.0.$$

Using the right endpoint of each subinterval, the right Riemann sum is

$$A \approx (f(0.5) + f(1.0) + f(1.5) + f(2.0))\,\Delta x = (3 + 4.5 + 5.5 + 6.0)0.5 = 9.5.$$

With only five function values, these estimates of the area are necessarily crude. Better estimates are obtained by using more subintervals and more function values.

Related Exercises 27–30 ◀

Sigma (Summation) Notation

Working with Riemann sums is cumbersome with large numbers of subintervals. Therefore, we pause for a moment to introduce some notation that simplifies our work.

Sigma (or **summation**), **notation** is used to express sums in a compact way. For example, the sum $1 + 2 + 3 + \cdots + 10$ is represented in sigma notation as $\sum_{k=1}^{10} k$. Here is how the notation works. The symbol Σ (*sigma*, the Greek capital S) stands for *sum*. The **index** k takes on all integer values from the lower limit ($k = 1$) to the upper limit ($k = 10$). The expression that immediately follows Σ (the **summand**) is evaluated for each value of k, and the resulting values are summed. Here are some examples:

$$\sum_{k=1}^{99} k = 1 + 2 + 3 + \cdots + 99 = 4950 \qquad \sum_{k=1}^{n} k = 1 + 2 + \cdots + n$$

$$\sum_{k=0}^{3} k^2 = 0^2 + 1^2 + 2^2 + 3^2 = 14 \qquad \sum_{k=1}^{4} (2k + 1) = 3 + 5 + 7 + 9 = 24$$

$$\sum_{k=-1}^{2} (k^2 + k) = [(-1)^2 + (-1)] + (0^2 + 0) + (1^2 + 1) + (2^2 + 2) = 8$$

The index in a sum is a *dummy variable*. It is internal to the sum, so it does not matter what symbol you choose as an index. For example,

$$\sum_{k=1}^{99} k = \sum_{n=1}^{99} n = \sum_{p=1}^{99} p.$$

Two properties of sums and sigma notation are useful in upcoming work. Suppose that $\{a_1, a_2, \ldots, a_n\}$ and $\{b_1, b_2, \ldots, b_n\}$ are two sets of real numbers, and suppose that c is a real number. Then we can factor constants out of a sum:

$$\textit{Constant Multiple Rule} \qquad \sum_{k=1}^{n} c a_k = c \sum_{k=1}^{n} a_k$$

We can also split a sum into two sums:

$$\textit{Addition Rule} \qquad \sum_{k=1}^{n} (a_k + b_k) = \sum_{k=1}^{n} a_k + \sum_{k=1}^{n} b_k$$

In the coming examples and exercises, the following summation formulas are essential.

> ➤ Formulas for the sums of powers of positive integers have been known for centuries. The formulas for powers $p = 0, 1, 2,$ and 3 are relatively simple. The formulas become complicated as p increases.

THEOREM 5.1   Sums of Positive Integers

Let n be a positive integer.

Sum of a constant c	$\displaystyle\sum_{k=1}^{n} c = cn$
Sum of the first n integers	$\displaystyle\sum_{k=1}^{n} k = \frac{n(n + 1)}{2}$
Sum of squares of the first n integers	$\displaystyle\sum_{k=1}^{n} k^2 = \frac{n(n + 1)(2n + 1)}{6}$
Sum of cubes of the first n integers	$\displaystyle\sum_{k=1}^{n} k^3 = \frac{n^2(n + 1)^2}{4}$

Related Exercises 31–34 ◄

Riemann Sums Using Sigma Notation

With sigma notation, a Riemann sum has the convenient compact form

$$f(\overline{x}_1)\,\Delta x + f(\overline{x}_2)\,\Delta x + \cdots + f(\overline{x}_n)\,\Delta x = \sum_{k=1}^{n} f(\overline{x}_k)\,\Delta x.$$

To express left, right, and midpoint Riemann sums in sigma notation, we must identify the points $\overline{x}_k$.

- For left Riemann sums, the left endpoints of the subintervals are $\overline{x}_k = a + (k-1)\,\Delta x$, for $k = 1, \ldots, n$.
- For right Riemann sums, the right endpoints of the subintervals are $\overline{x}_k = a + k\,\Delta x$, for $k = 1, \ldots, n$.
- For midpoint Riemann sums, the midpoints of the subintervals are $\overline{x}_k = a + \left(k - \frac{1}{2}\right)\Delta x$, for $k = 1, \ldots, n$.

The three Riemann sums are written compactly as follows.

DEFINITION   **Left, Right, and Midpoint Riemann Sums in Sigma Notation**

Suppose f is defined on a closed interval $[a, b]$, which is divided into n subintervals of equal length Δx. If $\overline{x}_k$ is a point in the kth subinterval $[x_{k-1}, x_k]$, for $k = 1, 2, \ldots, n$, then the **Riemann sum** of f on $[a, b]$ is $\displaystyle\sum_{k=1}^{n} f(\overline{x}_k)\,\Delta x$. Three cases arise in practice:

- **left Riemann sum** if $\overline{x}_k = a + (k-1)\,\Delta x$
- **right Riemann sum** if $\overline{x}_k = a + k\,\Delta x$
- **midpoint Riemann sum** if $\overline{x}_k = a + \left(k - \frac{1}{2}\right)\Delta x$,   for $k = 1, 2, \ldots, n$

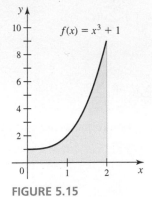

FIGURE 5.15

$f(x) = x^3 + 1$

EXAMPLE 5   Calculating Riemann sums Evaluate the left, right, and midpoint Riemann sums of $f(x) = x^3 + 1$ between $a = 0$ and $b = 2$ using $n = 50$ subintervals. Make a conjecture about the exact area of the region under the curve (Figure 5.15).

SOLUTION With $n = 50$, the length of each subinterval is

$$\Delta x = \frac{b - a}{n} = \frac{2 - 0}{50} = \frac{1}{25} = 0.04.$$

The value of $\overline{x}_k$ for the left Riemann sum is

$$\overline{x}_k = a + (k-1)\,\Delta x = 0 + 0.04(k-1) = 0.04k - 0.04$$

for $k = 1, 2, \ldots, 50$. Therefore, the left Riemann sum, evaluated with a calculator, is

$$\sum_{k=1}^{n} f(\overline{x}_k)\,\Delta x = \sum_{k=1}^{50} f(0.04k - 0.04)0.04 \approx 5.8416.$$

To evaluate the right Riemann sum, we let $\overline{x}_k = a + k\,\Delta x = 0.04k$ and find that

$$\sum_{k=1}^{n} f(\overline{x}_k)\,\Delta x = \sum_{k=1}^{50} f(0.04k)0.04 \approx 6.1616.$$

For the midpoint Riemann sum, we let

$$\overline{x}_k = a + \left(k - \frac{1}{2}\right)\Delta x = 0 + 0.04\left(k - \frac{1}{2}\right) = 0.04k - 0.02.$$

The value of the sum is

$$\sum_{k=1}^{n} f(\overline{x}_k)\Delta x = \sum_{k=1}^{50} f(0.04k - 0.02)0.04 \approx 5.9992.$$

Because f is increasing on $[0, 2]$, the left Riemann sum underestimates the area of the shaded region in Figure 5.15, while the right Riemann sum overestimates the area. Therefore, the exact area lies between 5.8416 and 6.1616. The midpoint Riemann sum usually gives the best estimate for increasing or decreasing functions; a reasonable estimate of the area under the curve is 6.

ALTERNATIVE SOLUTION It is worth examining another approach to Example 5 that reappears in Section 5.2. Consider the right Riemann sum given previously:

$$\sum_{k=1}^{n} f(\overline{x}_k)\Delta x = \sum_{k=1}^{50} f(0.04k)0.04$$

Rather than evaluating this sum with a calculator, we note that $f(0.04k) = (0.04k)^3 + 1$ and then use the properties of sums:

$$\sum_{k=1}^{n} f(\overline{x}_k)\Delta x = \sum_{k=1}^{50} \underbrace{[(0.04k)^3 + 1]}_{f(\overline{x}_k)}\underbrace{0.04}_{\Delta x}$$

$$= \sum_{k=1}^{50} (0.04k)^3\, 0.04 + \sum_{k=1}^{50} 1 \cdot 0.04 \qquad \sum(a_k + b_k) = \sum a_k + \sum b_k$$

$$= (0.04)^4 \sum_{k=1}^{50} k^3 + 0.04 \sum_{k=1}^{50} 1 \qquad \sum ca_k = c \sum a_k$$

Using the summation formulas for powers of integers in Theorem 5.1, we find that

$$\sum_{k=1}^{50} 1 = 50 \quad \text{and} \quad \sum_{k=1}^{50} k^3 = \frac{50^2 \cdot 51^2}{4}.$$

Substituting the values of these sums into the right Riemann sum, its value is

$$\sum_{k=1}^{50} f(\overline{x}_k)\Delta x = \frac{3851}{625} = 6.1616,$$

confirming the result given in the first solution. The idea of evaluating Riemann sums for arbitrary values of n is used in Section 5.2, where we evaluate the limit of the Riemann sum as $n \to \infty$.

Related Exercises 35–44 ◄

SECTION 5.1 EXERCISES

Review Questions

1. Suppose an object moves along a line at 15 m/s for $0 \le t < 2$ and at 25 m/s for $2 \le t \le 5$, where t is measured in seconds. Sketch the graph of the velocity function and find the displacement of the object for $0 \le t \le 5$.

2. Given the graph of the positive velocity of an object moving along a line, what is the geometrical representation of its displacement over a time interval $[a, b]$?

3. Suppose you want to approximate the area of the region bounded by the graph of $f(x) = \cos x$ and the x-axis between $x = 0$ and $x = \pi/2$. Explain a possible strategy.

4. Explain how Riemann sum approximations to the area of a region under a curve change as the number of subintervals increases.

5. Suppose the interval $[1, 3]$ is partitioned into $n = 4$ subintervals. What is the subinterval length Δx? List the grid points x_0, x_1, x_2, x_3, x_4. Which points are used for the left, right, and midpoint Riemann sums?

6. Suppose the interval $[2, 6]$ is partitioned into $n = 4$ subintervals with grid points $x_0 = 2, x_1 = 3, x_2 = 4, x_3 = 5$, and $x_4 = 6$. Write but do not evaluate the left, right, and midpoint Riemann sums for $f(x) = x^2$.

7. Does the right Riemann sum underestimate or overestimate the area of the region under the graph of a positive decreasing function? Explain.

8. Does the left Riemann sum underestimate or overestimate the area of the region under the graph of a positive increasing function? Explain.

Basic Skills

9. Approximating displacement The velocity in ft/s of an object moving along a line is given by $v = 3t^2 + 1$ on the interval $0 \le t \le 4$.

    a. Divide the interval $[0, 4]$ into $n = 4$ subintervals, $[0, 1]$, $[1, 2]$, $[2, 3]$, and $[3, 4]$. On each subinterval, assume the object moves at a constant velocity equal to the value of v evaluated at the midpoint of the subinterval and use these approximations to estimate the displacement of the object on $[0, 4]$ (see part (a) of the figure).

    b. Repeat part (a) for $n = 8$ subintervals (see part (b) of the figure).

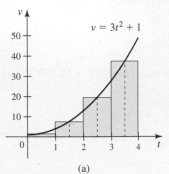

      (a)   (b)

10. Approximating displacement The velocity in ft/s of an object moving along a line is given by $v = \sqrt{10t}$ on the interval $1 \le t \le 7$.

    a. Divide the time interval $[1, 7]$ into $n = 3$ subintervals, $[1, 3]$, $[3, 5]$, and $[5, 7]$. On each subinterval, assume the object moves at a constant velocity equal to the value of v evaluated at the midpoint of the subinterval and use these approximations to estimate the displacement of the object on $[1, 7]$ (see part (a) of the figure).

    b. Repeat part (a) for $n = 6$ subintervals (see part (b) of the figure).

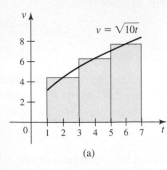

      (a)   (b)

11–14. Approximating displacement *The velocity of an object is given by the following functions on a specified interval. Approximate the displacement of the object on this interval by subdividing the interval into the indicated number of subintervals. Use the left endpoint of each subinterval to compute the height of the rectangles.*

11. $v = 1/(2t + 1)$ (m/s)   for $0 \le t \le 8$; $n = 4$

12. $v = t^2/2 + 4$ (ft/s)   for $0 \le t \le 12$; $n = 6$

13. $v = 4\sqrt{t + 1}$ (mi/hr)   for $0 \le t \le 15$; $n = 5$

14. $v = (t + 3)/6$ (m/s)   for $0 \le t \le 4$; $n = 4$

15–16. Left and right Riemann sums *Use the figures to calculate the left and right Riemann sums for f on the given interval and the given value of n.*

15. $f(x) = x + 1$ on $[1, 6]$; $n = 5$

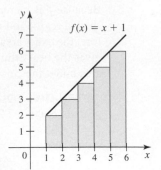
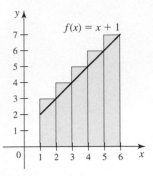

16. $f(x) = \dfrac{1}{x}$ on $[1, 5]$; $n = 4$

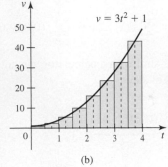
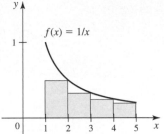

17–20. Left and right Riemann sums *Complete the following steps for the given function, interval, and value of n.*

    a. *Sketch the graph of the function on the given interval.*

    b. *Calculate Δx and the grid points $x_0, x_1, \ldots, x_n$.*

    c. *Illustrate the left and right Riemann sums, and determine which Riemann sum underestimates and which sum overestimates the area under the curve.*

    d. *Calculate the left and right Riemann sums.*

17. $f(x) = x^2 - 1$ on $[2, 4]$; $n = 4$

18. $f(x) = 2x^2$ on $[1, 6]$; $n = 5$

19. $f(x) = \cos x$ on $[0, \pi/2]$; $n = 4$

20. $f(x) = \cos x$ on $[-\pi/2, \pi/2]$; $n = 6$

21. A midpoint Riemann sum Approximate the area of the region bounded by the graph of $f(x) = 100 - x^2$ and the x-axis on $[0, 10]$ with $n = 5$ subintervals. Use the midpoint of each subinterval to determine the height of each rectangle (see figure).

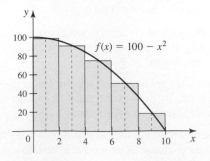

22. A midpoint Riemann sum Approximate the area of the region bounded by the graph of $f(t) = \cos(t/2)$ and the t-axis on $[0, \pi]$ with $n = 4$ subintervals. Use the midpoint of each subinterval to determine the height of each rectangle (see figure).

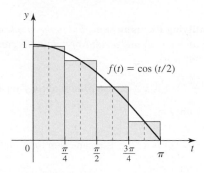

$f(t) = \cos(t/2)$

23–26. Midpoint Riemann sums *Complete the following steps for the given function, interval, and value of n.*

 a. *Sketch the graph of the function on the given interval.*
 b. *Calculate Δx and the grid points $x_0, x_1, \ldots, x_n$.*
 c. *Illustrate the midpoint Riemann sum by sketching the appropriate rectangles.*
 d. *Calculate the midpoint Riemann sum.*

23. $f(x) = \sqrt{x}$ on $[1, 3]$; $n = 4$

24. $f(x) = x^2$ on $[0, 4]$; $n = 4$

25. $f(x) = \dfrac{1}{x}$ on $[1, 6]$; $n = 5$

26. $f(x) = 4 - x$ on $[-1, 4]$; $n = 5$

27–28. Riemann sums from tables *Use the tabulated values of f to evaluate the left and right Riemann sums for the given value of n.*

27. $n = 4$; $[0, 2]$

x	0	0.5	1	1.5	2
$f(x)$	5	3	2	1	1

28. $n = 8$; $[1, 5]$

x	1	1.5	2	2.5	3	3.5	4	4.5	5
$f(x)$	0	2	3	2	2	1	0	2	3

29. Displacement from a table of velocities The velocities (in mi/hr) of an automobile moving along a straight highway over a 2-hr period $0 \le t \le 2$ are given in the following table.

t (hr)	0	0.25	0.5	0.75	1	1.25	1.5	1.75	2
v (mi/hr)	50	50	60	60	55	65	50	60	70

 a. Sketch a smooth curve passing through the data points.
 b. Find the midpoint Riemann sum approximation to the displacement on $[0, 2]$ with $n = 2$ and $n = 4$.

30. Displacement from a table of velocities The velocities (in m/s) of an automobile moving along a straight freeway over a 4-s period $0 \le t \le 4$ are given in the following table.

t (s)	0	0.5	1	1.5	2	2.5	3	3.5	4
v (m/s)	20	25	30	35	30	30	35	40	40

 a. Sketch a smooth curve passing through the data points.
 b. Find the midpoint Riemann sum approximation to the displacement on $[0, 4]$ with $n = 2$ and $n = 4$ subintervals.

31. Sigma notation Express the following sums using sigma notation. (Answers are not unique.)

 a. $1 + 2 + 3 + 4 + 5$ **b.** $4 + 5 + 6 + 7 + 8 + 9$
 c. $1^2 + 2^2 + 3^2 + 4^2$ **d.** $1 + \frac{1}{2} + \frac{1}{3} + \frac{1}{4}$

32. Sigma notation Express the following sums using sigma notation. (Answers are not unique.)

 a. $1 + 3 + 5 + 7 + \cdots + 99$
 b. $4 + 9 + 14 + \cdots + 44$
 c. $3 + 8 + 13 + \cdots + 63$
 d. $\dfrac{1}{1 \cdot 2} + \dfrac{1}{2 \cdot 3} + \dfrac{1}{3 \cdot 4} + \cdots + \dfrac{1}{49 \cdot 50}$

33. Sigma notation Evaluate the following expressions.

 a. $\displaystyle\sum_{k=1}^{10} k$ **b.** $\displaystyle\sum_{k=1}^{6} (2k + 1)$

 c. $\displaystyle\sum_{k=1}^{4} k^2$ **d.** $\displaystyle\sum_{n=1}^{5} (1 + n^2)$

 e. $\displaystyle\sum_{m=1}^{3} \frac{2m + 2}{3}$ **f.** $\displaystyle\sum_{j=1}^{3} (3j - 4)$

 g. $\displaystyle\sum_{p=1}^{5} (2p + p^2)$ **h.** $\displaystyle\sum_{n=0}^{4} \sin \frac{n\pi}{2}$

34. Evaluating sums Evaluate the following expressions by two methods.

 (i) Use Theorem 5.1.   **(ii)** Use a calculator.

 a. $\displaystyle\sum_{k=1}^{45} k$ **b.** $\displaystyle\sum_{k=1}^{45} (5k - 1)$ **c.** $\displaystyle\sum_{k=1}^{75} 2k^2$

 d. $\displaystyle\sum_{n=1}^{50} (1 + n^2)$ **e.** $\displaystyle\sum_{m=1}^{75} \frac{2m + 2}{3}$ **f.** $\displaystyle\sum_{j=1}^{20} (3j - 4)$

 g. $\displaystyle\sum_{p=1}^{35} (2p + p^2)$ **h.** $\displaystyle\sum_{n=0}^{40} (n^2 + 3n - 1)$

35–38. Riemann sums for larger values of n *Complete the following steps for the given function f and interval.*

 a. *For the given value of n, use sigma notation to write the left, right, and midpoint Riemann sums. Then evaluate each sum using a calculator.*
 b. *Based on the approximations found in part (a), estimate the area of the region bounded by the graph of f on the interval.*

35. $f(x) = \sqrt{x}$ for $[0, 4]$; $n = 40$

36. $f(x) = x^2 + 1$ for $[-1, 1]$; $n = 50$

37. $f(x) = x^2 - 1$ for $[2, 7]$; $n = 75$

38. $f(x) = \cos 2x$ for $[0, \pi/4]$; $n = 60$

39–44. Approximating areas with a calculator *Use a calculator and right Riemann sums to approximate the area of the region described. Present your calculations in a table showing the approximations for* $n = 10, 30, 60,$ *and* 80 *subintervals. Comment on whether your approximations appear to approach a limit.*

39. The region bounded by the graph of $f(x) = 4 - x^2$ and the x-axis on the interval $[-2, 2]$

40. The region bounded by the graph of $f(x) = x^2 + 1$ and the x-axis on the interval $[0, 2]$

41. The region bounded by the graph of $f(x) = 2 - 2\sin x$ and the x-axis on the interval $[-\pi/2, \pi/2]$

42. The region bounded by the graph of $f(x) = 2^x$ and the x-axis on the interval $[1, 2]$

43. The region bounded by the graph of $f(x) = \ln x$ and the x-axis on the interval $[1, e]$

44. The region bounded by the graph of $f(x) = \sqrt{x + 1}$ and the x-axis on the interval $[0, 3]$

Further Explorations

45. **Explain why or why not** State whether the following statements are true and give an explanation or counterexample.

   a. Consider the linear function $f(x) = 2x + 5$ and the region bounded by its graph and the x-axis on the interval $[3, 6]$. Suppose the area of this region is approximated using midpoint Riemann sums. Then the approximations give the exact area of the region for any number of subintervals.

   b. A left Riemann sum always overestimates the area of a region bounded by a positive increasing function and the x-axis on an interval $[a, b]$.

   c. For an increasing or decreasing nonconstant function and a given value of n on an interval $[a, b]$, the value of the midpoint Riemann sum always lies between the values of the left and right Riemann sums.

46–47. Riemann sums *Evaluate the Riemann sum for f on the given interval for the given values of n and $\overline{x}_k$. Sketch the graph of f and the rectangles used in the Riemann sum.*

46. $f(x) = x^2 + 2$ for $[0, 2]$; $n = 2$; $\overline{x}_1 = 0.25$ and $\overline{x}_2 = 1.75$

47. $f(x) = 1/x$ for $[1, 3]$; $n = 5$; $\overline{x}_1 = 1.1, \overline{x}_2 = 1.5, \overline{x}_3 = 2,$ $\overline{x}_4 = 2.3,$ and $\overline{x}_5 = 3$

48. **Riemann sums for a semicircle** Let $f(x) = \sqrt{1 - x^2}$.

   a. Show that the graph of f is the upper half of a circle of radius 1 centered at the origin.

   b. Estimate the area between the graph of f and the x-axis on the interval $[-1, 1]$ using a midpoint Riemann sum with $n = 25$.

   c. Repeat part (b) using $n = 75$ rectangles.

   d. What happens to the midpoint Riemann sums on $[-1, 1]$ as $n \to \infty$?

49–52. Sigma notation for Riemann sums *Use sigma notation to write the following Riemann sums. Then, evaluate each Riemann sum using Theorem 5.1 or a calculator.*

49. The right Riemann sum for $f(x) = x + 1$ on $[0, 4]$ with $n = 50$

50. The left Riemann sum for $f(x) = e^x$ on $[0, \ln 2]$ with $n = 40$

51. The midpoint Riemann sum for $f(x) = x^3$ on $[3, 11]$ with $n = 32$

52. The midpoint Riemann sum for $f(x) = 1 + \cos(\pi x)$ on $[0, 2]$ with $n = 50$

53–56. Identifying Riemann sums *Fill in the blanks with right, left, or midpoint; an interval; and a value of n. In some cases, more than one answer may work.*

53. $\displaystyle\sum_{k=1}^{4} f(1 + k) \cdot 1$ is a ____ Riemann sum for f on the interval $[\_\_, \_\_]$ with $n = \_\_$

54. $\displaystyle\sum_{k=1}^{4} f(2 + k) \cdot 1$ is a ____ Riemann sum for f on the interval $[\_\_, \_\_]$ with $n = \_\_$

55. $\displaystyle\sum_{k=1}^{4} f(1.5 + k) \cdot 1$ is a ____ Riemann sum for f on the interval $[\_\_, \_\_]$ with $n = \_\_$

56. $\displaystyle\sum_{k=1}^{8} f\left(1.5 + \frac{k}{2}\right) \cdot \frac{1}{2}$ is a ____ Riemann sum for f on the interval $[\_\_, \_\_]$ with $n = \_\_$

57. **Approximating areas** Estimate the area of the region bounded by the graph of $f(x) = x^2 + 2$ and the x-axis on $[0, 2]$ in the following ways.

   a. Divide $[0, 2]$ into $n = 4$ subintervals and approximate the area of the region using a left Riemann sum. Illustrate the solution geometrically.

   b. Divide $[0, 2]$ into $n = 4$ subintervals and approximate the area of the region using a midpoint Riemann sum. Illustrate the solution geometrically.

   c. Divide $[0, 2]$ into $n = 4$ subintervals and approximate the area of the region using a right Riemann sum. Illustrate the solution geometrically.

58. **Approximating area from a graph** Approximate the area of the region bounded by the graph (see figure) and the x-axis by dividing the interval $[0, 6]$ into $n = 3$ subintervals. Then use left and right Riemann sums to obtain two different approximations.

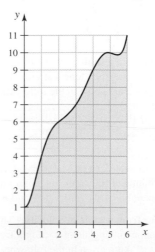

59. Approximating area from a graph Approximate the area of the region under the graph (see figure) by dividing the interval $[1, 7]$ into $n = 6$ subintervals. Then use left and right Riemann sums to obtain two different approximations.

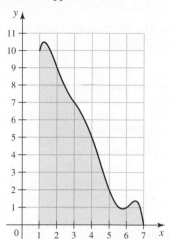

Applications

60. Displacement from a velocity graph Consider the velocity function for an object moving along a line shown in the figure.

   a. Describe the motion of the object over the interval $[0, 6]$.

   b. Use geometry to find the displacement of the object between $t = 0$ and $t = 3$.

   c. Use geometry to find the displacement of the object between $t = 3$ and $t = 5$.

   d. Assuming that the velocity remains 30 m/s for $t \geq 4$, find the function that gives the displacement between $t = 0$ and any time $t \geq 5$.

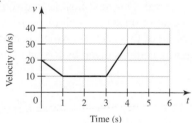

61. Displacement from a velocity graph Consider the velocity function for an object moving along a line shown in the figure.

   a. Describe the motion of the object over the interval $[0, 6]$.

   b. Use geometry to find the displacement of the object between $t = 0$ and $t = 2$.

   c. Use geometry to find the displacement of the object between $t = 2$ and $t = 5$.

   d. Assuming that the velocity remains 10 m/s for $t \geq 5$, find the function that gives the displacement between $t = 0$ and any time $t \geq 5$.

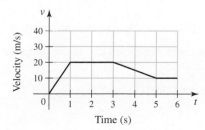

62. Flow rates Suppose a gauge at the outflow of a reservoir measures the flow rate of water in units of ft³/hr. In Chapter 6 we show that the total amount of water that flows out of the reservoir is the area under the flow rate curve. Consider the flow rate function shown in the figure.

   a. Find the amount of water (in units of ft³) that flows out of the reservoir over the interval $[0, 4]$.

   b. Find the amount of water that flows out of the reservoir over the interval $[8, 10]$.

   c. Does more water flow out of the reservoir over the interval $[0, 4]$ or $[4, 6]$?

   d. Show that the units of your answer are consistent with the units of the variables on the axes.

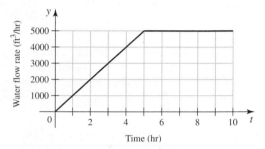

63. Mass from density A thin 10-cm rod is made of an alloy whose density varies along its length according to the function shown in the figure. Assume density is measured in units of g/cm. In Chapter 6, we show that the mass of the rod is the area under the density curve.

   a. Find the mass of the left half of the rod ($0 \leq x \leq 5$).

   b. Find the mass of the right half of the rod ($5 \leq x \leq 10$).

   c. Find the mass of the entire rod ($0 \leq x \leq 10$).

   d. Estimate the point along the rod at which it will balance (called the center of mass).

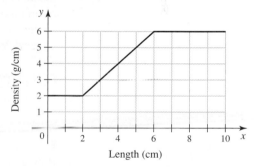

64–65. Displacement from velocity *The following functions describe the velocity of a car (in mi/hr) moving along a straight highway for a 3-hr interval. In each case, find the function that gives the displacement of the car over the interval $[0, t]$, where $0 \leq t \leq 3$.*

64. $v(t) = \begin{cases} 40 & \text{if } 0 \leq t \leq 1.5 \\ 50 & \text{if } 1.5 < t \leq 3 \end{cases}$

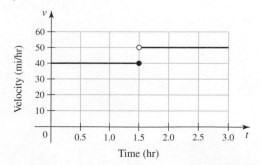

65. $v(t) = \begin{cases} 30 & \text{if } 0 \leq t \leq 2 \\ 50 & \text{if } 2 < t \leq 2.5 \\ 44 & \text{if } 2.5 < t \leq 3 \end{cases}$

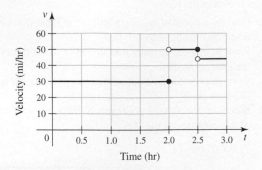

Time (hr)

66–69. Functions with absolute value *Use a calculator and the method of your choice to approximate the area of the following regions. Present your calculations in a table, showing approximations using n = 16, 32, and 64 subintervals. Comment on whether your approximations appear to approach a limit.*

66. The region bounded by the graph of $f(x) = |25 - x^2|$ and the x-axis on the interval $[0, 10]$

67. The region bounded by the graph of $f(x) = |x(x^2 - 1)|$ and the x-axis on the interval $[-1, 1]$

68. The region bounded by the graph of $f(x) = |\cos 2x|$ and the x-axis on the interval $[0, \pi]$

69. The region bounded by the graph of $f(x) = |1 - x^3|$ and the x-axis on the interval $[-1, 2]$

Additional Exercises

70. Riemann sums for constant functions Let $f(x) = c$, where $c > 0$, be a constant function on $[a, b]$. Prove that any Riemann sum for any value of n gives the exact area of the region between the graph of f and the x-axis on $[a, b]$.

71. Riemann sums for linear functions Assume that the linear function $f(x) = mx + c$ is positive on the interval $[a, b]$. Prove that the midpoint Riemann sum with any value of n gives the exact area of the region between the graph of f and the x-axis on $[a, b]$.

QUICK CHECK ANSWERS

1. 45 mi   **2.** 0.25, 0.125, 7.875   **3.** $\Delta x = 2$; $\{1, 3, 5, 7, 9\}$
4. The left sum overestimates the area. ◄

5.2 Definite Integrals

We introduced Riemann sums in Section 5.1 as a way to approximate the area of a region bounded by a curve $y = f(x)$ and the x-axis on an interval $[a, b]$. In that discussion, we assumed f to be nonnegative on the interval. Our next task is to discover the geometric meaning of Riemann sums when f is negative on some or all of $[a, b]$. Once this matter is settled, we can proceed to the main event of this section, which is to define the *definite integral*. With definite integrals, the approximations given by Riemann sums become exact.

Net Area

How do we interpret Riemann sums when f is negative at some or all points of $[a, b]$? The answer follows directly from the Riemann sum definition.

EXAMPLE 1 Interpreting Riemann Sums Evaluate and interpret the following Riemann sums for $f(x) = 1 - x^2$ on the interval $[a, b]$ with n equally spaced subintervals.

a. A midpoint Riemann sum with $[a, b] = [1, 3]$ and $n = 4$

b. A left Riemann sum with $[a, b] = [0, 3]$ and $n = 6$

SOLUTION

a. The length of each subinterval is $\Delta x = \dfrac{b - a}{n} = \dfrac{3 - 1}{4} = 0.5$. So the grid points are

$$x_0 = 1, \quad x_1 = 1.5, \quad x_2 = 2, \quad x_3 = 2.5, \quad x_4 = 3.$$

To compute the midpoint Riemann sum, we evaluate f at the midpoints of the sub-intervals, which are

$$\bar{x}_1 = 1.25, \quad \bar{x}_2 = 1.75, \quad \bar{x}_3 = 2.25, \quad \bar{x}_4 = 2.75.$$

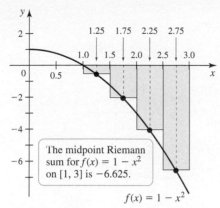

The midpoint Riemann sum for $f(x) = 1 - x^2$ on [1, 3] is -6.625.

$f(x) = 1 - x^2$

FIGURE 5.16

The resulting midpoint Riemann sum is

$$\sum_{k=1}^{n} f(\overline{x}_k)\,\Delta x = \sum_{k=1}^{4} f(\overline{x}_k)(0.5)$$

$$= f(1.25)(0.5) + f(1.75)(0.5) + f(2.25)(0.5) + f(2.75)(0.5)$$

$$= (-0.5625 - 2.0625 - 4.0625 - 6.5625)0.5$$

$$= -6.625.$$

All values of $f(\overline{x}_k)$ are negative, so the Riemann sum is also negative. Because area is always a nonnegative quantity, this Riemann sum does not approximate an area. Notice, however, that the values of $f(\overline{x}_k)$ are the *negative* of the heights of the corresponding rectangles (Figure 5.16). Therefore, the Riemann sum is an approximation to the *negative* of the area of the region bounded by the curve.

b. The length of each subinterval is $\Delta x = \dfrac{b - a}{n} = \dfrac{3 - 0}{6} = 0.5$ and the grid points are

$$x_0 = 0, \quad x_1 = 0.5, \quad x_2 = 1, \quad x_3 = 1.5, \quad x_4 = 2, \quad x_5 = 2.5, \quad x_6 = 3.$$

To calculate the left Riemann sum, we set $\overline{x}_1, \overline{x}_2, \dots, \overline{x}_6$ equal to the left endpoints of the subintervals:

$$\overline{x}_1 = 0, \quad \overline{x}_2 = 0.5, \quad \overline{x}_3 = 1, \quad \overline{x}_4 = 1.5, \quad \overline{x}_5 = 2, \quad \overline{x}_6 = 2.5$$

The resulting left Riemann sum is

$$\sum_{k=1}^{n} f(\overline{x}_k)\,\Delta x = \sum_{k=1}^{6} f(\overline{x}_k)(0.5)$$

$$= \underbrace{(f(0) + f(0.5) + f(1)}_{\text{nonnegative contribution}} + \underbrace{f(1.5) + f(2) + f(2.5))}_{\text{negative contribution}} 0.5$$

$$= (1 + 0.75 + 0 - 1.25 - 3 - 5.25)0.5$$

$$= -3.875.$$

In this case the values of $f(\overline{x}_k)$ are nonnegative for $k = 1, 2,$ and 3 and negative for $k = 4, 5,$ and 6 (Figure 5.17). Where f is positive, we get positive contributions to the Riemann sum and where f is negative, we get negative contributions to the sum.

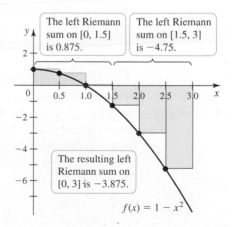

The left Riemann sum on [0, 1.5] is 0.875.

The left Riemann sum on [1.5, 3] is -4.75.

The resulting left Riemann sum on [0, 3] is -3.875.

$f(x) = 1 - x^2$

FIGURE 5.17

Related Exercises 11–18 ◄

Let's recap what was learned in Example 1. On intervals where $f(x) < 0$, Riemann sums approximate the *negative* of the area of the region bounded by the curve (Figure 5.18).

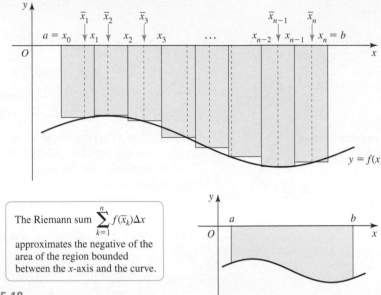

The Riemann sum $\displaystyle\sum_{k=1}^{n} f(\overline{x}_k)\Delta x$ approximates the negative of the area of the region bounded between the x-axis and the curve.

FIGURE 5.18

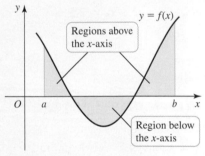

FIGURE 5.19

> Net area suggests the difference between positive and negative contributions much like net change or net profit. Some texts use the term **signed area** for net area.

QUICK CHECK 2 Sketch a continuous function f that is positive over the interval $[0, 1]$, negative over the interval $[1, 2]$, such that the net area of the region bounded by the graph of f and the x-axis on $[0, 2]$ is zero. ◄

In the more general case that f is positive on only part of $[a, b]$, we get positive contributions to the sum where f is positive and negative contributions to the sum where f is negative. In this case, Riemann sums approximate the area of the regions that lie above the x-axis *minus* the area of the regions that lie *below* the x-axis (Figure 5.19). This difference between the positive and negative contributions is called the *net area*; it can be positive, negative, or zero.

QUICK CHECK 1 Suppose $f(x) = -5$. What is the net area of the region bounded by the graph of f and the x-axis on the interval $[1, 5]$? Make a sketch of the function and the region. ◄

DEFINITION   Net Area

Consider the region R bounded by the graph of a continuous function f and the x-axis between $x = a$ and $x = b$. The **net area** of R is the sum of the areas of the parts of R that lie above the x-axis *minus* the sum of the areas of the parts of R that lie below the x-axis on $[a, b]$.

The Definite Integral

Riemann sums for f on $[a, b]$ give *approximations* to the net area of the region bounded by the graph of f and the x-axis between $x = a$ and $x = b$. How can we make these approximations exact? If f is continuous on $[a, b]$, it is reasonable to expect the Riemann sum approximations to approach the exact value of the net area as the number of subintervals $n \to \infty$ and as the length of the subintervals $\Delta x \to 0$ (Figure 5.20). In terms of limits, we write

$$\text{net area} = \lim_{n \to \infty} \sum_{k=1}^{n} f(\overline{x}_k)\,\Delta x.$$

The Riemann sums we have used so far involve partitions in which the subintervals have the same length Δx.

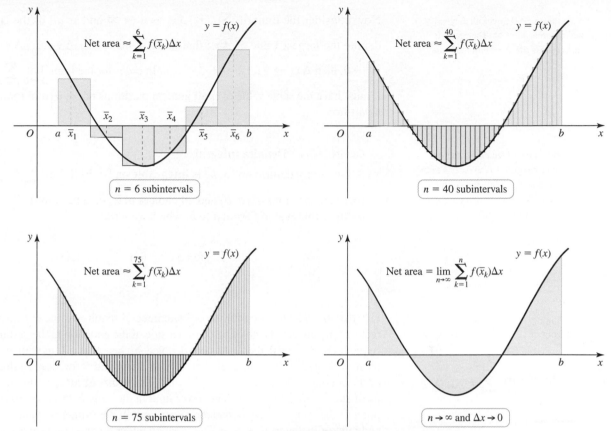

FIGURE 5.20. As the number of subintervals n increases, the Riemann sum approaches the net area of the region between the curve $y = f(x)$ and the x-axis on $[a, b]$.

We now introduce partitions of $[a, b]$ in which the lengths of the subintervals are not necessarily equal. A **general partition** of $[a, b]$ consists of the n subintervals

$$[x_0, x_1], [x_1, x_2], \ldots, [x_{n-1}, x_n],$$

where $x_0 = a$ and $x_n = b$. The length of the kth subinterval is $\Delta x_k = x_k - x_{k-1}$, for $k = 1, \ldots, n$. We let $\overline{x}_k$ be any point in the subinterval $[x_{k-1}, x_k]$. This general partition is used to define the *general Riemann sum*.

DEFINITION **General Riemann Sum**

Suppose $[x_0, x_1], [x_1, x_2], \ldots, [x_{n-1}, x_n]$ are subintervals of $[a, b]$ with

$$a = x_0 < x_1 < x_2 < \cdots < x_{n-1} < x_n = b.$$

Let Δx_k be the length of the subinterval $[x_{k-1}, x_k]$ and let $\overline{x}_k$ be any point in $[x_{k-1}, x_k]$ for $k = 1, 2, \ldots, n$.

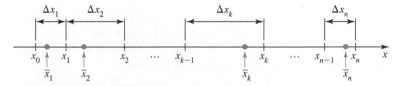

If f is defined on $[a, b]$, the sum

$$\sum_{k=1}^{n} f(\overline{x}_k) \Delta x_k = f(\overline{x}_1) \Delta x_1 + f(\overline{x}_2) \Delta x_2 + \cdots + f(\overline{x}_n) \Delta x_n$$

is called a **general Riemann sum for f on $[a, b]$.**

> Note that $\Delta \to 0$ forces all $\Delta x_k \to 0$, which forces $n \to \infty$. Therefore, it suffices to write $\Delta \to 0$ in the limit.

Now consider the limit of $\sum_{k=1}^{n} f(\overline{x}_k)\Delta x_k$ as $n \to \infty$ and as *all* of the $\Delta x_k \to 0$. We let Δ denote the largest value of Δx_k; that is, $\Delta = \max\{\Delta x_1, \Delta x_2, \ldots, \Delta x_n\}$. Observe that if $\Delta \to 0$, then $\Delta x_k \to 0$ for $k = 1, 2, \ldots, n$. In order for the limit $\lim_{\Delta \to 0} \sum_{k=1}^{n} f(\overline{x}_k)\Delta x_k$ to exist, it must have the same value over all general partitions of $[a, b]$ and for all choices of $\overline{x}_k$ on a partition.

> It is imperative to remember that the indefinite integral $\int f(x)\,dx$ is a family of functions of x, while the definite integral $\int_a^b f(x)\,dx$ is a real number (the net area of a region).

DEFINITION   Definite Integral

A function f defined on $[a, b]$ is **integrable** on $[a, b]$ if $\lim_{\Delta \to 0} \sum_{k=1}^{n} f(\overline{x}_k)\Delta x_k$ exists (over all partitions of $[a, b]$ and all choices of $\overline{x}_k$ on a partition). This limit is the **definite integral of f from a to b**, which we write

$$\int_a^b f(x)\,dx = \lim_{\Delta \to 0} \sum_{k=1}^{n} f(\overline{x}_k)\Delta x_k.$$

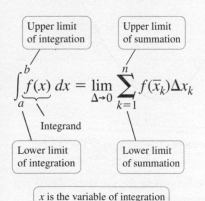

FIGURE 5.21

Notation   The notation for the definite integral requires some explanation. There is a direct match between the notation on either side of the equation in the definition (Figure 5.21). In the limit as $\Delta \to 0$, the finite sum, denoted $\sum$, becomes a sum with an infinite number of terms, denoted $\int$. The integral sign $\int$ is an elongated *S* for sum. In this limit, the lengths of the subintervals Δx_k are replaced by dx. The **limits of integration**, a and b, and the limits of summation also match: The lower limit in the sum, $k = 1$, corresponds to the left endpoint of the interval, $x = a$, and the upper limit in the sum, $k = n$, corresponds to the right endpoint of the interval, $x = b$. The function under the integral sign is called the **integrand**. Finally, the differential dx in the integral is an essential part of the notation; it tells us that the **variable of integration** is x.

The variable of integration is a dummy variable that is completely internal to the integral. It does not matter what the variable of integration is called, as long as it does not conflict with other variables that are in use. Therefore, the integrals in Figure 5.22 all have the same meaning.

> For Leibniz, who introduced this notation in 1675, dx represented the width of an infinitesimally thin rectangle and $f(x)\,dx$ represented the area of such a rectangle. He used $\int_a^b f(x)\,dx$ to denote the sum of all these areas from a to b.

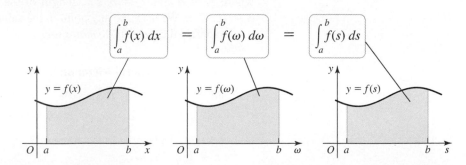

FIGURE 5.22

The strategy of slicing a region into smaller parts, summing the results from the parts, and taking a limit is used repeatedly in calculus and its applications. We call this strategy the **slice-and-sum method**. It often results in a Riemann sum whose limit is a definite integral.

Evaluating Definite Integrals

> A function f is bounded on an interval I if there is a number M such that $|f(x)| < M$ for all x in I.

Most of the functions encountered in this text are integrable (see Exercise 79 for an exception). In fact, if f is continuous on $[a, b]$ or if f is bounded on $[a, b]$ with a finite number of discontinuities, then f is integrable on $[a, b]$. The proof of this result goes beyond the scope of this text.

> **THEOREM 5.2   Integrable Functions**
> If f is continuous on $[a, b]$ or bounded on $[a, b]$ with a finite number of discontinuities, then f is integrable on $[a, b]$.

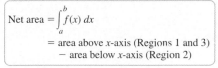

Net area $= \displaystyle\int_a^b f(x)\, dx$

$=$ area above x-axis (Regions 1 and 3)
$-$ area below x-axis (Region 2)

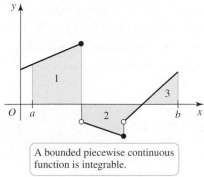

A bounded piecewise continuous function is integrable.

FIGURE 5.23

When f is continuous, we have seen that the definite integral $\int_a^b f(x)\, dx$ is the net area bounded by the graph of f and the x-axis on $[a, b]$. Figure 5.23 illustrates how the idea of net area carries over to piecewise continuous functions.

QUICK CHECK 3 Graph $f(x) = x$ and use geometry to evaluate $\int_{-1}^1 x\, dx$. ◄

EXAMPLE 2   Identifying the limit of a sum Assume that

$$\lim_{\Delta \to 0} \sum_{k=1}^n (3\overline{x}_k^2 + 2\overline{x}_k + 1)\,\Delta x_k$$

is the limit of a Riemann sum for a function f on $[1, 3]$. Identify the function f and express the limit as a definite integral. What does the definite integral represent geometrically?

SOLUTION By comparing the sum $\sum_{k=1}^n (3\overline{x}_k^2 + 2\overline{x}_k + 1)\,\Delta x_k$ to the general Riemann sum $\sum_{k=1}^n f(\overline{x}_k)\,\Delta x_k$, we see that $f(x) = 3x^2 + 2x + 1$. Because f is a polynomial, it is continuous on $[1, 3]$ and is, therefore, integrable on $[1, 3]$. It follows that

$$\lim_{\Delta \to 0} \sum_{k=1}^n (3\overline{x}_k^2 + 2\overline{x}_k + 1)\,\Delta x_k = \int_1^3 (3x^2 + 2x + 1)\, dx.$$

Because f is positive on $[1, 3]$, the definite integral $\int_1^3 (3x^2 + 2x + 1)\, dx$ is the area of the region bounded by the curve $y = 3x^2 + 2x + 1$ and the x-axis on $[1, 3]$ (Figure 5.24).

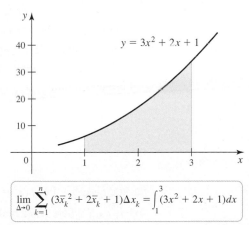

FIGURE 5.24   *Related Exercises 19–22* ◄

EXAMPLE 3   Evaluating definite integrals using geometry Use familiar area formulas to evaluate the following definite integrals.

a. $\displaystyle\int_2^4 (2x + 3)\, dx$ **b.** $\displaystyle\int_1^6 (2x - 6)\, dx$ **c.** $\displaystyle\int_3^4 \sqrt{1 - (x - 3)^2}\, dx$

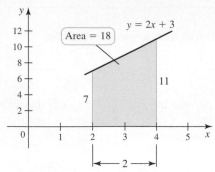

FIGURE 5.25

SOLUTION To evaluate these definite integrals geometrically, a sketch of the corresponding region is essential.

a. The definite integral $\int_2^4 (2x + 3)\,dx$ is the area of the trapezoid bounded by the x-axis
and the line $y = 2x + 3$ from $x = 2$ to $x = 4$ (Figure 5.25). The width of its base is
2 and the lengths of its two parallel sides are $f(2) = 7$ and $f(4) = 11$. Using the area
formula for a trapezoid we have

$$\int_2^4 (2x + 3)\,dx = \frac{1}{2}\cdot 2(11 + 7) = 18.$$

➤ A trapezoid and its area. When $a = 0$,
we get the area of a triangle. When
$a = b$, we get the area of a rectangle.

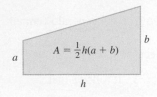

b. A sketch shows that the regions bounded by the line $y = 2x - 6$ and the x-axis are
triangles (Figure 5.26). The area of the triangle on the interval $[1, 3]$ is $\frac{1}{2}\cdot 2\cdot 4 = 4$.
Similarly, the area of the triangle on $[3, 6]$ is $\frac{1}{2}\cdot 3\cdot 6 = 9$. The definite integral is the
net area of the entire region, which is the area of the triangle above the x-axis minus
the area of the triangle below the x-axis:

$$\int_1^6 (2x - 6)\,dx = \text{net area} = 9 - 4 = 5$$

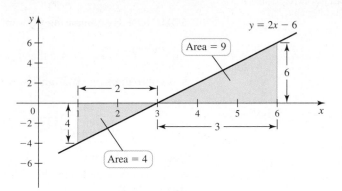

FIGURE 5.26

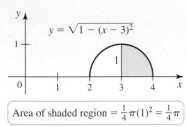

FIGURE 5.27

c. We first let $y = \sqrt{1 - (x - 3)^2}$ and observe that $y \geq 0$ when $2 \leq x \leq 4$. Squaring
both sides leads to the equation $(x - 3)^2 + y^2 = 1$, whose graph is a circle of radius 1
centered at $(3, 0)$. Because $y \geq 0$, the graph of $y = \sqrt{1 - (x - 3)^2}$ is the upper half
of the circle. It follows that the integral $\int_3^4 \sqrt{1 - (x - 3)^2}\,dx$ is the area of a quarter
circle of radius 1 (Figure 5.27). Therefore,

$$\int_3^4 \sqrt{1 - (x - 3)^2}\,dx = \frac{1}{4}\pi(1)^2 = \frac{\pi}{4}.$$

Related Exercises 23–30 ◄

QUICK CHECK 4 Let $f(x) = 5$ and use geometry to evaluate $\int_1^3 f(x)\,dx$. What is the value
of $\int_a^b c\,dx$ where c is a real number? ◄

EXAMPLE 4 Definite integrals from graphs Figure 5.28 shows the graph of a function f with the areas of the regions bounded by its graph and the x-axis given. Find the
values of the following definite integrals.

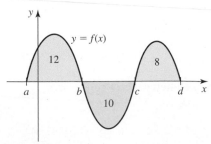

FIGURE 5.28

a. $\displaystyle\int_a^b f(x)\,dx$ **b.** $\displaystyle\int_b^c f(x)\,dx$ **c.** $\displaystyle\int_a^c f(x)\,dx$ **d.** $\displaystyle\int_b^d f(x)\,dx$

SOLUTION

a. Because f is positive on $[a, b]$, the value of the definite integral is the area of the region between the graph and the x-axis on $[a, b]$; that is, $\int_a^b f(x)\, dx = 12$.

b. Because f is negative on $[b, c]$, the value of the definite integral is the negative of the area of the corresponding region; that is, $\int_b^c f(x)\, dx = -10$.

c. The value of the definite integral is the area of the region on $[a, b]$ (where f is positive) minus the area of the region on $[b, c]$ (where f is negative). Therefore, $\int_a^c f(x)\, dx = 12 - 10 = 2$.

d. Reasoning as in part (c), we have $\int_b^d f(x)\, dx = -10 + 8 = -2$.

Related Exercises 31–38 ◀

Properties of Definite Integrals

Recall that the definite integral $\int_a^b f(x)\, dx$ was defined assuming that $a < b$. There are, however, occasions when it is necessary to allow the limits of integration to be reversed. If f is integrable on $[a, b]$, we define

$$\int_b^a f(x)\, dx = -\int_a^b f(x)\, dx.$$

In other words, reversing the limits of integration changes the sign of the integral.

Another fundamental property of integrals is that if we integrate from a point to itself, then the length of the interval of integration is zero, which means the definite integral is also zero.

QUICK CHECK 5 Evaluate $\int_a^b f(x)\, dx + \int_b^a f(x)\, dx$ if f is integrable on $[a, b]$. ◀

> **DEFINITION   Reversing Limits and Identical Limits**
>
> Suppose f is integrable on $[a, b]$.
>
> **1.** $\int_b^a f(x)\, dx = -\int_a^b f(x)\, dx$ **2.** $\int_a^a f(x)\, dx = 0$

Integral of a Sum   Definite integrals possess other properties that often simplify their evaluation. Assume f and g are integrable on $[a, b]$. The first property states that their sum $f + g$ is integrable on $[a, b]$ and the integral of their sum is the sum of their integrals:

$$\int_a^b (f(x) + g(x))\, dx = \int_a^b f(x)\, dx + \int_a^b g(x)\, dx$$

We prove this property, assuming that f and g are continuous. In this case, $f + g$ is continuous and, therefore, integrable. We then have

$$\int_a^b (f(x) + g(x))\, dx = \lim_{\Delta \to 0} \sum_{k=1}^n [f(\overline{x}_k) + g(\overline{x}_k)]\, \Delta x_k \qquad \text{Definition of definite integral}$$

$$= \lim_{\Delta \to 0} \left[\sum_{k=1}^n f(\overline{x}_k)\, \Delta x_k + \sum_{k=1}^n g(\overline{x}_k)\, \Delta x_k \right] \qquad \text{Split into two finite sums.}$$

$$= \lim_{\Delta \to 0} \sum_{k=1}^n f(\overline{x}_k)\, \Delta x_k + \lim_{\Delta \to 0} \sum_{k=1}^n g(\overline{x}_k)\, \Delta x_k \qquad \text{Split into two limits.}$$

$$= \int_a^b f(x)\, dx + \int_a^b g(x)\, dx. \qquad \text{Definition of definite integral}$$

Constants in Integrals   Another property of definite integrals is that constants can be factored out of definite integrals. If f is integrable on $[a, b]$ and c is a constant, then cf is integrable on $[a, b]$ and

$$\int_a^b cf(x)\,dx = c\int_a^b f(x)\,dx.$$

The justification (Exercise 77) is based on the fact that for finite sums,

$$\sum_{k=1}^n cf(\overline{x}_k)\,\Delta x_k = c\sum_{k=1}^n f(\overline{x}_k)\,\Delta x_k.$$

Integrals over Subintervals   If c lies between a and b, then the integral on $[a, b]$ may be split into two integrals. As shown in Figure 5.29, we have the property

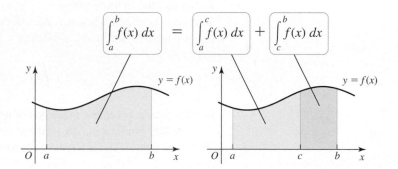

FIGURE 5.29

It is surprising that this same property also holds when c lies outside the interval $[a, b]$. For example, if $a < b < c$ and f is integrable on $[a, c]$, then it follows (Figure 5.30) that

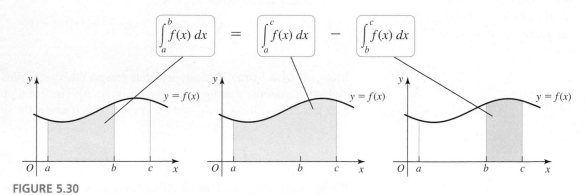

FIGURE 5.30

Because $\int_c^b f(x)\,dx = -\int_b^c f(x)\,dx$, we have the original property $\int_a^b f(x)\,dx = \int_a^c f(x)\,dx + \int_c^b f(x)\,dx$.

Integrals of Absolute Values   Finally, how do we interpret $\int_a^b |f(x)|\,dx$, the integral of the absolute value of a function? The graphs f and $|f|$ are shown in Figure 5.31. The integral $\int_a^b |f(x)|\,dx$ gives the area of regions R_1^* and R_2. But R_1 and R_1^* have the same area; therefore, $\int_a^b |f(x)|\,dx$ also gives the area of R_1 and R_2. The conclusion is that $\int_a^b |f(x)|\,dx$ is the area of the entire region (above and below the x-axis) that lies between the graph of f and the x-axis on $[a, b]$.

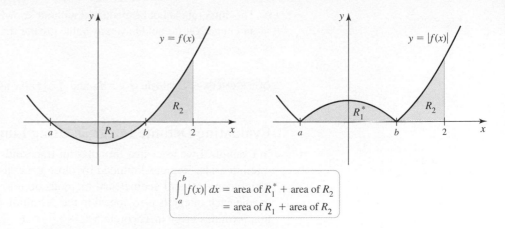

FIGURE 5.31

$$\int_a^b |f(x)|\, dx = \text{area of } R_1^* + \text{area of } R_2$$
$$= \text{area of } R_1 + \text{area of } R_2$$

Table 5.3   Properties of Definite Integrals

Let f and g be integrable functions on an interval that contains a, b, and c.

1. $\displaystyle\int_a^a f(x)\, dx = 0$ Definition

2. $\displaystyle\int_b^a f(x)\, dx = -\int_a^b f(x)\, dx$ Definition

3. $\displaystyle\int_a^b (f(x) + g(x))\, dx = \int_a^b f(x)\, dx + \int_a^b g(x)\, dx$

4. $\displaystyle\int_a^b cf(x)\, dx = c\int_a^b f(x)\, dx$ For any constant c

5. $\displaystyle\int_a^b f(x)\, dx = \int_a^c f(x)\, dx + \int_c^b f(x)\, dx$

6. The function $|f|$ is integrable on $[a, b]$ and $\int_a^b |f(x)|\, dx$ is the sum of the areas of the regions bounded by the graph of f and the x-axis on $[a, b]$.

EXAMPLE 5   Properties of integrals   Assume that $\int_0^5 f(x)\, dx = 3$ and $\int_0^7 f(x)\, dx = -10$. Evaluate the following integrals, if possible.

a. $\displaystyle\int_0^7 2f(x)\, dx$ **b.** $\displaystyle\int_5^7 f(x)\, dx$ **c.** $\displaystyle\int_5^0 f(x)\, dx$ **d.** $\displaystyle\int_7^0 6f(x)\, dx$ **e.** $\displaystyle\int_0^7 |f(x)|\, dx$

SOLUTION

a. By Property 4 of Table 5.3, $\int_0^7 2f(x)\, dx = 2\int_0^7 f(x)\, dx = 2 \cdot (-10) = -20$.

b. By Property 5 of Table 5.3, $\int_0^7 f(x)\, dx = \int_0^5 f(x)\, dx + \int_5^7 f(x)\, dx$.

   Therefore, $\int_5^7 f(x)\, dx = \int_0^7 f(x)\, dx - \int_0^5 f(x)\, dx = -10 - 3 = -13$.

c. By Property 2 of Table 5.3,

$$\int_5^0 f(x)\, dx = -\int_0^5 f(x)\, dx = -3.$$

d. Reversing limits and using Properties 2 and 4 of Table 5.3, we have

$$\int_7^0 6f(x)\, dx = -\int_0^7 6f(x)\, dx = -6\int_0^7 f(x)\, dx = (-6)(-10) = 60.$$

e. This integral cannot be evaluated without knowing the intervals on which f is positive and negative. It could have any value greater than or equal to 10.

Related Exercises 39–44 ◄

QUICK CHECK 6 Evaluate $\int_{-1}^{2} x\, dx$ and $\int_{-1}^{2} |x|\, dx$ using geometry. ◄

Evaluating Definite Integrals Using Limits

In Example 3 we used area formulas for trapezoids, triangles, and circles to evaluate definite integrals. Regions bounded by more general functions have curved boundaries for which conventional geometrical methods do not work. At the moment the only way to handle such integrals is to appeal to the definition of the definite integral and the summation formulas given in Theorem 5.1.

We know that if f is integrable on $[a, b]$, then $\int_a^b f(x)\, dx = \lim\limits_{\Delta \to 0} \sum\limits_{k=1}^{n} f(\bar{x}_k)\Delta x_k$ for any partition of $[a, b]$ and any points $\bar{x}_k$. To simplify these calculations, we use equally spaced grid points and right Riemann sums. That is, for each value of n we let

$$\Delta x_k = \Delta x = \frac{b - a}{n} \text{ and } \bar{x}_k = a + k\,\Delta x, \text{ for } k = 1, 2, \ldots, n. \text{ Then, as } n \to \infty \text{ and}$$
$$\Delta \to 0,$$

$$\int_a^b f(x)\, dx = \lim_{\Delta \to 0} \sum_{k=1}^{n} f(\bar{x}_k)\,\Delta x_k = \lim_{n \to \infty} \sum_{k=1}^{n} f(a + k\Delta x)\,\Delta x.$$

EXAMPLE 6 Evaluating definite integrals Find the value of $\int_0^2 (x^3 + 1)\, dx$ by evaluating a right Riemann sum and letting $n \to \infty$.

SOLUTION Based on approximations found in Example 5, Section 5.1, we conjectured that the value of this integral is 6. To verify this conjecture, we now evaluate the integral exactly.

The interval $[a, b] = [0, 2]$ is divided into n subintervals of length $\Delta x = \dfrac{b - a}{n} = \dfrac{2}{n}$, which produces the grid points

$$\bar{x}_k = a + k\,\Delta x = 0 + k \cdot \frac{2}{n} = \frac{2k}{n}, \qquad \text{for } k = 1, 2, \ldots, n.$$

Letting $f(x) = x^3 + 1$, the right Riemann sum is

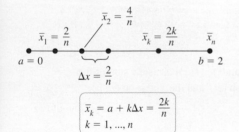

$$\bar{x}_2 = \frac{4}{n}$$
$$\bar{x}_1 = \frac{2}{n} \qquad \bar{x}_k = \frac{2k}{n} \qquad \bar{x}_n$$
$$a = 0 \qquad\qquad\qquad b = 2$$
$$\Delta x = \frac{2}{n}$$

$$\boxed{\begin{array}{l} \bar{x}_k = a + k\Delta x = \dfrac{2k}{n} \\ k = 1, \ldots, n \end{array}}$$

$$\sum_{k=1}^{n} f(\bar{x}_k)\,\Delta x = \sum_{k=1}^{n} \left[\left(\frac{2k}{n} \right)^3 + 1 \right] \frac{2}{n}$$

$$= \frac{2}{n} \sum_{k=1}^{n} \left(\frac{8k^3}{n^3} + 1 \right) \qquad\qquad \sum_{k=1}^{n} c a_k = c \sum_{k=1}^{n} a_k$$

$$= \frac{2}{n} \left(\frac{8}{n^3} \sum_{k=1}^{n} k^3 + \sum_{k=1}^{n} 1 \right) \qquad \sum_{k=1}^{n}(a_k + b_k) = \sum_{k=1}^{n} a_k + \sum_{k=1}^{n} b_k$$

> An analogous calculation could be done using left Riemann sums or midpoint Riemann sums.

$$= \frac{2}{n} \left[\frac{8}{n^3} \left(\frac{n^2(n + 1)^2}{4} \right) + n \right] \qquad \sum_{k=1}^{n} k^3 = \frac{n^2(n + 1)^2}{4} \text{ and } \sum_{k=1}^{n} 1 = n; \text{ Theorem 5.1}$$

$$= \frac{4(n^2 + 2n + 1)}{n^2} + 2. \qquad \text{Simplify.}$$

Now we evaluate $\int_0^2 (x^3 + 1)\, dx$ by letting $n \to \infty$ in the Riemann sum:

$$\int_0^2 (x^3 + 1)\, dx = \lim_{n \to \infty} \sum_{k=1}^{n} f(\bar{x}_k)\, \Delta x$$

$$= \lim_{n \to \infty} \left[\frac{4(n^2 + 2n + 1)}{n^2} + 2 \right]$$

$$= 4 \lim_{n \to \infty} \left(\frac{n^2 + 2n + 1}{n^2} \right) + \lim_{n \to \infty} 2$$

$$= 4(1) + 2 = 6$$

Therefore, $\int_0^2 (x^3 + 1)\, dx = 6$, confirming our conjecture in Example 5, Section 5.1.

Related Exercises 45–50 ◄

The Riemann sum calculations in Example 6 are tedious even if f is a simple function. For polynomials of degree 4 and higher, the calculations are much more challenging, and for rational and transcendental functions, advanced mathematical results are needed. The next section introduces more efficient methods for evaluating definite integrals.

SECTION 5.2 EXERCISES

Review Questions

1. Explain what net area means.

2. How do you interpret geometrically the definite integral of a function that changes sign on the interval of integration?

3. When does the net area of a region equal the area of a region? When does the net area of a region differ from the area of a region?

4. Suppose that $f(x) < 0$ on the interval $[a, b]$. Using Riemann sums, explain why the definite integral $\int_a^b f(x)\, dx$ is negative.

5. Use graphs to evaluate $\int_0^{2\pi} \sin x\, dx$ and $\int_0^{2\pi} \cos x\, dx$.

6. Explain how the notation for Riemann sums, $\sum_{k=1}^{n} f(\bar{x}_k)\, \Delta x$, corresponds to the notation for the definite integral, $\int_a^b f(x)\, dx$.

7. Give a geometrical explanation of why $\int_a^a f(x)\, dx = 0$.

8. Use Table 5.3 to rewrite $\int_1^6 (2x^3 - 4x)\, dx$ as the sum of two integrals.

9. Use geometry to find a formula for $\int_0^a x\, dx$, in terms of a.

10. If f is integrable and $\int_a^b |f(x)|\, dx = 0$, what can you conclude about f?

Basic Skills

11–14. Approximating net area *The following functions are negative on the given interval.*

    a. *Sketch the function on the given interval.*
    b. *Approximate the net area bounded by the graph of f and the x-axis on the interval using a left, right, and midpoint Riemann sum with n = 4.*

11. $f(x) = -2x - 1$; $[0, 4]$
12. $f(x) = -4 - x^3$; $[3, 7]$
13. $f(x) = \sin 2x$; $[\pi/2, \pi]$
14. $f(x) = x^3 - 1$; $[-2, 0]$

15–18. Approximating net area *The following functions are positive and negative on the given interval.*

    a. *Sketch the function on the given interval.*
    b. *Approximate the net area bounded by the graph of f and the x-axis on the interval using a left, right, and midpoint Riemann sum with n = 4.*
    c. *Use the sketch in part (a) to show which intervals of $[a, b]$ make positive and negative contributions to the net area.*

15. $f(x) = 4 - 2x$; $[0, 4]$ 16. $f(x) = 8 - 2x^2$; $[0, 4]$
17. $f(x) = \sin 2x$; $[0, 3\pi/4]$ 18. $f(x) = x^3$; $[-1, 2]$

19–22. Identifying definite integrals as limits of sums *Consider the following limits of Riemann sums of a function f on $[a, b]$. Identify f and express the limit as a definite integral.*

19. $\displaystyle\lim_{\Delta \to 0} \sum_{k=1}^{n} (\bar{x}_k^2 + 1)\, \Delta x_k$; $[0, 2]$

20. $\displaystyle\lim_{\Delta \to 0} \sum_{k=1}^{n} (4 - \bar{x}_k^2)\, \Delta x_k$; $[-2, 2]$

21. $\displaystyle\lim_{\Delta \to 0} \sum_{k=1}^{n} \bar{x}_k \ln \bar{x}_k\, \Delta x_k$; $[1, 2]$

22. $\displaystyle\lim_{\Delta \to 0} \sum_{k=1}^{n} |\bar{x}_k^2 - 1|\, \Delta x_k$; $[-2, 2]$

23–30. Net area and definite integrals *Use geometry (not Riemann sums) to evaluate the following definite integrals. Sketch a graph of the integrand, show the region in question, and interpret your result.*

23. $\displaystyle\int_0^4 (8 - 2x)\, dx$ 24. $\displaystyle\int_{-4}^2 (2x + 4)\, dx$

25. $\displaystyle\int_{-1}^{2} (-|x|)\, dx$

26. $\displaystyle\int_{0}^{2} (1 - |x|)\, dx$

27. $\displaystyle\int_{0}^{4} \sqrt{16 - x^2}\, dx$

28. $\displaystyle\int_{-1}^{3} \sqrt{4 - (x-1)^2}\, dx$

29. $\displaystyle\int_{0}^{4} f(x)\, dx$ where $f(x) = \begin{cases} 5 & \text{if } x \le 2 \\ 3x - 1 & \text{if } x > 2 \end{cases}$

30. $\displaystyle\int_{1}^{10} g(x)\, dx$ where $g(x) = \begin{cases} 4x & \text{if } 0 \le x \le 2 \\ -8x + 16 & \text{if } 2 < x \le 3 \\ -8 & \text{if } x > 3 \end{cases}$

31–34. Net area from graphs *The figure shows the areas of regions bounded by the graph of f and the x-axis. Evaluate the following integrals.*

31. $\displaystyle\int_{0}^{a} f(x)\, dx$

32. $\displaystyle\int_{0}^{b} f(x)\, dx$

33. $\displaystyle\int_{a}^{c} f(x)\, dx$

34. $\displaystyle\int_{0}^{c} f(x)\, dx$

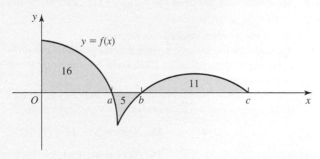

35–38. Net area from graphs *The accompanying figure shows four regions bounded by the graph of $y = x \sin x$: R_1, R_2, R_3, and R_4, whose areas are 1, $\pi - 1$, $\pi + 1$, and $2\pi - 1$, respectively. (We verify these results later in the text.) Use this information to evaluate the following integrals.*

35. $\displaystyle\int_{0}^{\pi} x \sin x\, dx$

36. $\displaystyle\int_{0}^{3\pi/2} x \sin x\, dx$

37. $\displaystyle\int_{0}^{2\pi} x \sin x\, dx$

38. $\displaystyle\int_{\pi/2}^{2\pi} x \sin x\, dx$

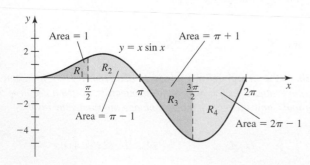

39. Properties of integrals Use only the fact that $\int_{0}^{4} 3x(4 - x)\, dx = 32$ and the definitions and properties of integrals to evaluate the following integrals, if possible.

a. $\displaystyle\int_{4}^{0} 3x(4 - x)\, dx$

b. $\displaystyle\int_{0}^{4} x(x - 4)\, dx$

c. $\displaystyle\int_{4}^{0} 6x(4 - x)\, dx$

d. $\displaystyle\int_{0}^{8} 3x(4 - x)\, dx$

40. Properties of integrals Suppose $\int_{1}^{4} f(x)\, dx = 8$ and $\int_{1}^{6} f(x)\, dx = 5$. Evaluate the following integrals.

a. $\displaystyle\int_{1}^{4} (-3 f(x))\, dx$

b. $\displaystyle\int_{1}^{4} 3f(x)\, dx$

c. $\displaystyle\int_{6}^{4} 12 f(x)\, dx$

d. $\displaystyle\int_{4}^{6} 3 f(x)\, dx$

41. Properties of integrals Suppose $\int_{0}^{3} f(x)\, dx = 2$, $\int_{3}^{6} f(x)\, dx = -5$, and $\int_{3}^{6} g(x)\, dx = 1$. Evaluate the following integrals.

a. $\displaystyle\int_{0}^{3} 5f(x)\, dx$

b. $\displaystyle\int_{3}^{6} [-3g(x)]\, dx$

c. $\displaystyle\int_{3}^{6} (3f(x) - g(x))\, dx$

d. $\displaystyle\int_{6}^{3} [f(x) + 2g(x)]\, dx$

42. Properties of integrals Suppose that $f(x) \ge 0$ on $[0, 2]$, $f(x) \le 0$ on $[2, 5]$, $\int_{0}^{2} f(x)\, dx = 6$, and $\int_{2}^{5} f(x)\, dx = -8$. Evaluate the following integrals.

a. $\displaystyle\int_{0}^{5} f(x)\, dx$

b. $\displaystyle\int_{0}^{5} |f(x)|\, dx$

c. $\displaystyle\int_{2}^{5} 4|f(x)|\, dx$

d. $\displaystyle\int_{0}^{5} (f(x) + |f(x)|)\, dx$

43–44. Using properties of integrals *Use the value of the first integral I to evaluate the two given integrals.*

43. $I = \displaystyle\int_{0}^{1} (x^3 - 2x)\, dx = -\dfrac{3}{4}$

a. $\displaystyle\int_{0}^{1} (4x - 2x^3)\, dx$

b. $\displaystyle\int_{1}^{0} (2x - x^3)\, dx$

44. $I = \displaystyle\int_{0}^{\pi/2} (\cos \theta - 2 \sin \theta)\, d\theta = -1$

a. $\displaystyle\int_{0}^{\pi/2} (2 \sin \theta - \cos \theta)\, d\theta$

b. $\displaystyle\int_{\pi/2}^{0} (4 \cos \theta - 8 \sin \theta)\, d\theta$

45–50. Limits of sums *Use the definition of the definite integral to evaluate the following definite integrals. Use right Riemann sums and Theorem 5.1.*

45. $\int_0^2 (2x + 1)\, dx$

46. $\int_1^5 (1 - x)\, dx$

47. $\int_3^7 (4x + 6)\, dx$

48. $\int_0^2 (x^2 - 1)\, dx$

49. $\int_1^4 (x^2 - 1)\, dx$

50. $\int_0^2 4x^3\, dx$

Further Explorations

51. Explain why or why not Determine whether the following statements are true and give an explanation or counterexample.

    a. If f is a constant function on the interval $[a, b]$, then the right and left Riemann sums give the exact value of $\int_a^b f(x)\, dx$ for any n.

    b. If f is a linear function on the interval $[a, b]$, then a midpoint Riemann sum gives the exact value of $\int_a^b f(x)\, dx$ for any n.

    c. $\int_0^{2\pi/a} \sin ax\, dx = \int_0^{2\pi/a} \cos ax\, dx = 0$ (*Hint:* Graph the functions and use properties of trigonometric functions).

    d. If $\int_a^b f(x)\, dx = \int_b^a f(x)\, dx$, then f is a constant function.

    e. Property 4 of Table 5.3 implies that $\int_a^b x f(x)\, dx = x \int_a^b f(x)\, dx$.

52–55. Approximating definite integrals *Complete the following steps for the given integral and the given value of n.*

    a. *Sketch the graph of the integrand on the interval of integration.*

    b. *Calculate Δx and the grid points $x_0, x_1, \ldots, x_n$, assuming a regular partition.*

    c. *Calculate the left and right Riemann sums for the given value of n.*

    d. *Determine which Riemann sum (left or right) underestimates the value of the definite integral and which overestimates the value of the definite integral.*

52. $\int_0^2 (x^2 - 2)\, dx; \ n = 4$

53. $\int_3^6 (1 - 2x)\, dx; \ n = 6$

54. $\int_0^{\pi/2} \cos x\, dx; \ n = 4$

55. $\int_1^7 \frac{1}{x}\, dx; \ n = 6$

56–60. Approximating definite integrals with a calculator *Consider the following definite integrals.*

    a. *Write the left and right Riemann sums in sigma notation for $n = 20, 50$, and 100. Then evaluate the sums using a calculator.*

    b. *Based upon your answers to part (a), make a conjecture about the value of the definite integral.*

56. $\int_4^9 3\sqrt{x}\, dx$

57. $\int_0^1 (x^2 + 1)\, dx$

58. $\int_0^1 \tan\left(\frac{\pi x}{4}\right) dx$

59. $\int_0^1 e^x\, dx$

60. $\int_{-1}^1 \cos\left(\frac{x\pi}{2}\right) dx$

61–64. Riemann sums with midpoints and a calculator *Consider the following definite integrals.*

    a. *Write the midpoint Riemann sum in sigma notation for an arbitrary value of n.*

    b. *Evaluate each sum using a calculator with $n = 20, 50$, and 100. Use these values to estimate the value of the integral.*

61. $\int_1^4 2\sqrt{x}\, dx$

62. $\int_{-1}^2 \sin\left(\frac{x\pi}{4}\right) dx$

63. $\int_0^4 (4x - x^2)\, dx$

64. $\int_0^{1/2} \sin^{-1} x\, dx$

65. More properties of integrals Consider two functions f and g on $[1, 6]$ such that $\int_1^6 f(x)\, dx = 10$, $\int_1^6 g(x)\, dx = 5$, $\int_4^6 f(x)\, dx = 5$, and $\int_1^4 g(x)\, dx = 2$. Evaluate the following integrals.

    a. $\int_1^4 3 f(x)\, dx$

    b. $\int_1^6 (f(x) - g(x))\, dx$

    c. $\int_1^4 (f(x) - g(x))\, dx$

    d. $\int_4^6 (g(x) - f(x))\, dx$

    e. $\int_4^6 8 g(x)\, dx$

    f. $\int_4^1 2 f(x)\, dx$

66–69. Area versus net area *Graph the following functions. Then use geometry (not Riemann sums) to find the area and the net area of the region described.*

66. The region between the graph of $y = 4x - 8$ and the x-axis for $-4 \le x \le 8$

67. The region between the graph of $y = -3x$ and the x-axis for $-2 \le x \le 2$

68. The region between the graph of $y = 3x - 6$ and the x-axis for $0 \le x \le 6$

69. The region between the graph of $y = 1 - |x|$ and the x-axis for $-2 \le x \le 2$

70–73. Area by geometry *Use geometry to evaluate the following integrals.*

70. $\int_{-2}^3 |x + 1|\, dx$

71. $\int_1^6 |2x - 4|\, dx$

72. $\int_1^6 (3x - 6)\, dx$

73. $\int_{-6}^4 \sqrt{24 - 2x - x^2}\, dx$

Additional Exercises

74. Integrating piecewise continuous functions Suppose f is continuous on the interval $[a, c]$ and on the interval $(c, b]$, where $a < c < b$, with a finite jump at $x = c$. Form a uniform partition on the interval $[a, c]$ with n grid points and another uniform partition on the interval $[c, b]$ with m grid points, where $x = c$ is a grid point of both partitions. Write a Riemann sum for $\int_a^b f(x)\,dx$ and separate it into two pieces for $[a, c]$ and $[c, b]$. Explain why $\int_a^b f(x)\,dx = \int_a^c f(x)\,dx + \int_c^b f(x)\,dx$.

75–76. Piecewise continuous functions *Use geometry and the result of Exercise 74 to evaluate the following integrals.*

75. $\displaystyle\int_0^{10} f(x)\,dx$ where $f(x) = \begin{cases} 2 & \text{if } 0 \le x < 5 \\ 3 & \text{if } 5 \le x \le 10 \end{cases}$

76. $\displaystyle\int_1^6 f(x)\,dx$ where $f(x) = \begin{cases} 2x & \text{if } 1 \le x < 4 \\ 10 - 2x & \text{if } 4 \le x \le 6 \end{cases}$

77. Constants in integrals Use the definition of the definite integral to justify the property $\int_a^b cf(x)\,dx = c\int_a^b f(x)\,dx$, where f is continuous and c is a real number.

78. Exact area Consider the linear function $f(x) = 2px + q$ on the interval $[a, b]$, where a, b, p, and q are positive constants.

a. Show that the midpoint Riemann sum with n subintervals equals $\int_a^b f(x)\,dx$ for any value of n.
b. Show that $\int_a^b f(x)\,dx = (b - a)[p(b + a) + q]$.

79. A nonintegrable function Consider the function defined on $[0, 1]$ such that $f(x) = 1$ if x is a rational number and $f(x) = 0$ if x is irrational. This function has an infinite number of discontinuities, and the integral $\int_0^1 f(x)\,dx$ does not exist. Show that if you consider only right, left, and midpoint Riemann sums on *regular* partitions with n subintervals, then they equal 1 for all n.

80. Powers of x by Riemann sums Consider the integral $I(p) = \int_0^1 x^p\,dx$ where p is a positive integer.

a. Write the left Riemann sum for the integral with n subintervals.
b. It is a fact (proved by the 17th-century mathematicians Fermat and Pascal) that $\displaystyle\lim_{n \to \infty} \frac{1}{n} \sum_{k=0}^{n-1} \left(\frac{k}{n}\right)^p = \frac{1}{p+1}$.
Use this fact to evaluate $I(p)$.

QUICK CHECK ANSWERS

1. -20 **2.** $f(x) = 1 - x$ is one possibility. **3.** 0 **4.** 10; $c(b - a)$ **5.** 0 **6.** $\frac{3}{2}; \frac{5}{2}$ ◀

5.3 Fundamental Theorem of Calculus

Evaluating definite integrals using limits of Riemann sums, as described in Section 5.2, is usually not possible or practical. Fortunately, there is a powerful and practical method for evaluating definite integrals, which is developed in this section. Along the way, we discover the inverse relationship between differentiation and integration, expressed in the most important result of calculus, the Fundamental Theorem of Calculus.

Area Functions

The concept of an area function is crucial to the discussion about the connection between derivatives and integrals. We start with a continuous function $y = f(t)$ defined for $t \ge a$, where a is a fixed number. The *area function* for f with left endpoint a is denoted $A(x)$; it gives the net area of the region bounded by the graph of f and the t-axis between $t = a$ and $t = x$ (Figure 5.32). The net area of this region is also given by the definite integral

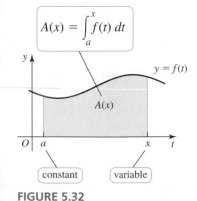

$$A(x) = \int_a^x f(t)\,dt$$

FIGURE 5.32

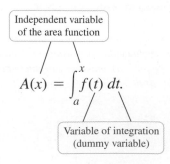

Independent variable of the area function

$$A(x) = \int_a^x f(t)\,dt.$$

Variable of integration (dummy variable)

▶ A dummy variable is a placeholder; its role can be played by any symbol that does not conflict with other variables in the problem.

Notice that x is the upper limit of the integral *and* the independent variable of the area function: As x changes, so does the net area under the curve. Because the symbol x is already in use as the independent variable for A, we must choose another symbol for the variable of integration. Any symbol—except x—can be used because it is a *dummy variable*; we have chosen t as the integration variable.

Figure 5.33 gives a general view of how an area function is generated. Suppose that f is a continuous function and a is a fixed number. Now choose a point $b > a$. The net area of the region between the graph of f and the t-axis on the interval $[a, b]$ is $A(b)$. Moving the right endpoint to $(c, 0)$ or $(d, 0)$ produces different regions with net areas $A(c)$ and $A(d)$, respectively. In general, if $x > a$ is a variable point, then $A(x) = \int_a^x f(t)\, dt$ is the net area of the region between the graph of f and the t-axis on the interval $[a, x]$.

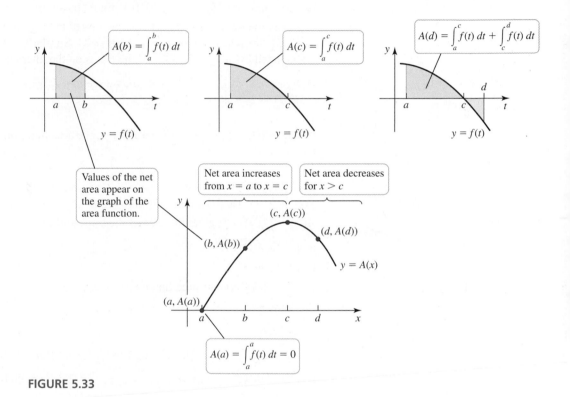

FIGURE 5.33

Figure 5.33 shows how $A(x)$ varies with respect to x. Notice that $A(a) = \int_a^a f(t)\, dt = 0$. Then, for $x > a$ the net area increases until $x = c$, at which point $f(c) = 0$. For $x > c$, the function f is negative, which produces a negative contribution to the area function. As a result, the area function decreases for $x > c$.

DEFINITION   Area Function

Let f be a continuous function for $t \geq a$. The **area function for f with left endpoint a** is

$$A(x) = \int_a^x f(t)\, dt,$$

where $x \geq a$. The area function gives the net area of the region bounded by the graph of f and the t-axis on the interval $[a, x]$.

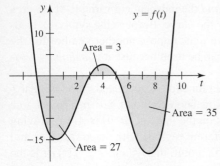

FIGURE 5.34

EXAMPLE 1 **Area of regions** The graph of f is shown in Figure 5.34 with areas of various regions marked. Let $A(x) = \int_{-1}^{x} f(t)\,dt$ and $F(x) = \int_{3}^{x} f(t)\,dt$ be two area functions for f (note the different left endpoints). Evaluate the following area functions.

a. $A(3)$ and $F(3)$ **b.** $A(5)$ and $F(5)$ **c.** $A(9)$ and $F(9)$

SOLUTION

a. The value of $A(3) = \int_{-1}^{3} f(t)\,dt$ is the net area of the region bounded by the graph of f and the t-axis on the interval $[-1, 3]$. Using the graph of f, we see that $A(3) = -27$ (because this region has an area of 27 and lies below the t-axis). On the other hand, $F(3) = \int_{3}^{3} f(t)\,dt = 0$ by Property 1 of Table 5.3.

b. The value of $A(5) = \int_{-1}^{5} f(t)\,dt$ is found by subtracting the area of the region that lies below the t-axis on $[-1, 3]$ from the area of the region that lies above the t-axis on $[3, 5]$. Therefore, $A(5) = 3 - 27 = -24$. Similarly, $F(5)$ is the net area of the region bounded by the graph of f and the t-axis on the interval $[3, 5]$; therefore, $F(5) = 3$.

c. Reasoning as in parts (a) and (b), we see that $A(9) = -27 + 3 - 35 = -59$ and $F(9) = 3 - 35 = -32$.   *Related Exercises 11–12* ◄

QUICK CHECK 1 In Example 1, let $B(x)$ be the area function for f with left endpoint 5. Evaluate $B(5)$ and $B(9)$. ◄

EXAMPLE 2 **Area of a trapezoid** Consider the trapezoid bounded by the line $f(t) = 2t + 3$ and the t-axis from $t = 2$ to $t = x$ (Figure 5.35). The area function $A(x) = \int_{2}^{x} f(t)\,dt$ gives the area of the trapezoid for $x \geq 2$.

a. Evaluate $A(2)$.

b. Evaluate $A(5)$.

c. Find and graph the area function $y = A(x)$ for $x \geq 2$.

d. Compare the derivative of A to f.

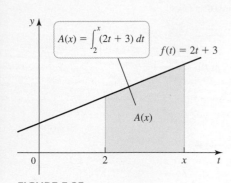

FIGURE 5.35

SOLUTION

a. By Property 1 of Table 5.3, $A(2) = \int_{2}^{2} (2t + 3)\,dt = 0$.

b. Notice that $A(5)$ is the area of the trapezoid (Figure 5.35) bounded by the line $y = 2t + 3$ and the t-axis on the interval $[2, 5]$. Using the area formula for a trapezoid (Figure 5.36), we find that

$$A(5) = \int_{2}^{5} (2t + 3)\,dt = \frac{1}{2} \underbrace{(5 - 2)}_{\substack{\text{distance between} \\ \text{parallel sides}}} \underbrace{(f(2) + f(5))}_{\substack{\text{sum of parallel} \\ \text{side lengths}}} = \frac{1}{2} \cdot 3(7 + 13) = 30.$$

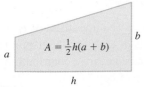

FIGURE 5.36

c. Now the right endpoint of the base is a variable $x \geq 2$ (Figure 5.37). The distance between the parallel sides of the trapezoid is $x - 2$. By the area formula for a trapezoid, the area of this trapezoid for any $x \geq 2$ is

$$A(x) = \frac{1}{2} \underbrace{(x - 2)}_{\substack{\text{distance between} \\ \text{parallel sides}}} \underbrace{(f(2) + f(x))}_{\substack{\text{sum of parallel} \\ \text{side lengths}}}$$

$$= \frac{1}{2}(x - 2)(7 + 2x + 3)$$

$$= (x - 2)(x + 5)$$

$$= x^2 + 3x - 10.$$

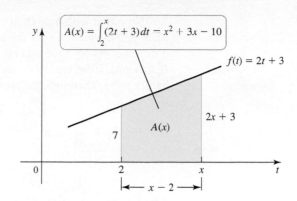

FIGURE 5.37

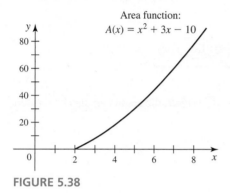

FIGURE 5.38

> Recall that if $A'(x) = f(x)$, then f is the derivative of A; equivalently, A is an antiderivative of f.

Expressing the area function in terms of an integral with a variable upper limit we have

$$A(x) = \int_2^x (2t + 3) \, dt = x^2 + 3x - 10.$$

Because the line $f(t) = 2t + 3$ is above the t-axis for $t \geq 2$, the area function $A(x) = x^2 + 3x - 10$ is an increasing function of x with $A(2) = 0$ (Figure 5.38).

d. Differentiating the area function, we find that

$$A'(x) = \frac{d}{dx}(x^2 + 3x - 10) = 2x + 3 = f(x).$$

Therefore, $A'(x) = f(x)$, or equivalently, the area function A is an antiderivative of f. We soon show this relationship is not an accident; it is one part of the Fundamental Theorem of Calculus.

Related Exercises 13–22 ◄

QUICK CHECK 2 Verify that the area function in Example 2 gives the correct area when $x = 6$ and $x = 10$. ◄

Fundamental Theorem of Calculus

Example 2 suggests that the area function A for a linear function f is an antiderivative of f; that is, $A'(x) = f(x)$. Our goal is to show that this conjecture holds for more general functions. Let's start with an intuitive argument.

Assume that f is a continuous function defined on an interval $[a, b]$. As before, $A(x) = \int_a^x f(t) \, dt$ is the area function for f with a left endpoint a: It gives the net area of the region bounded by the graph of f and the t-axis on the interval $[a, x]$ for $x \geq a$. Figure 5.39 is the key to the argument.

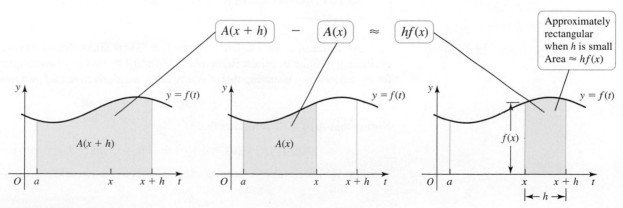

FIGURE 5.39

Note that with $h > 0$, $A(x + h)$ is the area of the region whose base is the interval $[a, x + h]$, while $A(x)$ is the area of the region whose base is the interval $[a, x]$. So the difference $A(x + h) - A(x)$ is the area of the region whose base is the interval $[x, x + h]$. If h is small, the region in question is nearly rectangular with a base of length h and a height $f(x)$. Therefore, the area of this region is approximately

$$A(x + h) - A(x) \approx h\,f(x).$$

Dividing by h, we have

$$\frac{A(x + h) - A(x)}{h} \approx f(x).$$

> Recall that
>
> $$f'(x) = \lim_{h \to 0} \frac{f(x + h) - f(x)}{h}.$$
>
> If the function f is replaced by A, then
>
> $$A'(x) = \lim_{h \to 0} \frac{A(x + h) - A(x)}{h}.$$

An analogous argument can be made with $h < 0$. Now observe that as h tends to zero, this approximation improves. In the limit as $h \to 0$, we have

$$\underbrace{\lim_{h \to 0} \frac{A(x + h) - A(x)}{h}}_{A'(x)} = \underbrace{\lim_{h \to 0} f(x)}_{f(x)}.$$

We see that indeed $A'(x) = f(x)$. Because $A(x) = \int_a^x f(t)\,dt$, the result can also be written

$$A'(x) = \frac{d}{dx}\underbrace{\int_a^x f(t)\,dt}_{A(x)} = f(x),$$

which says that the derivative of the integral of f is f. A formal proof that $A'(x) = f(x)$ is given at the end of the section; but for the moment, we have a plausible argument. This conclusion is the first part of the Fundamental Theorem of Calculus.

THEOREM 5.3 (PART 1)   Fundamental Theorem of Calculus
If f is continuous on $[a, b]$, then the area function

$$A(x) = \int_a^x f(t)\,dt \qquad \text{for } a \le x \le b$$

is continuous on $[a, b]$ and differentiable on (a, b). The area function satisfies $A'(x) = f(x)$; or, equivalently,

$$A'(x) = \frac{d}{dx}\int_a^x f(t)\,dt = f(x),$$

which means that the area function of f is an antiderivative of f.

Given that A is an antiderivative of f, it is one short step to a powerful method for evaluating definite integrals. Remember (Section 4.8) that any two antiderivatives of f differ by a constant. Assuming that F is any other antiderivative of f, we have

$$F(x) = A(x) + C \qquad \text{for all } x.$$

Noting that $A(a) = 0$, it follows that

$$F(b) - F(a) = (A(b) + C) - (A(a) + C) = A(b).$$

Writing $A(b)$ in terms of a definite integral leads to the remarkable result

$$A(b) = \int_a^b f(x)\,dx = F(b) - F(a).$$

We have shown that to evaluate a definite integral of f, we

- find any antiderivative of f, call it F
- compute $F(b) - F(a)$, the difference in the values of F between the upper and lower limits of integration.

This process is the essence of the second part of the Fundamental Theorem of Calculus.

THEOREM 5.3 (PART 2)   Fundamental Theorem of Calculus
If f is continuous on $[a, b]$ and F is any antiderivative of f, then

$$\int_a^b f(x)\,dx = F(b) - F(a).$$

It is customary and convenient to denote the difference $F(b) - F(a)$ by $F(x)\big|_a^b$. Using this shorthand, the Fundamental Theorem is summarized in Figure 5.40.

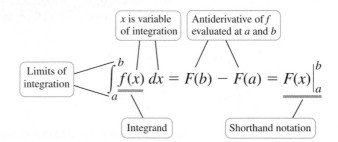

FIGURE 5.40

QUICK CHECK 3 Evaluate $\left(\dfrac{x}{x+1}\right)\bigg|_1^2$. ◄

The Inverse Relationship between Differentiation and Integration   It is worth pausing to observe that the two parts of the Fundamental Theorem express the inverse relationship between differentiation and integration. Part 1 of the Fundamental Theorem says

$$\frac{d}{dx}\int_a^x f(t)\,dt = f(x)$$

or the derivative of the integral of f is f itself.

Noting that f is an antiderivative of f', Part 2 of the Fundamental Theorem says

$$\int_a^b f'(x)\,dx = f(b) - f(a),$$

QUICK CHECK 4 Explain why f is an antiderivative of f'. ◄

or the definite integral of the derivative of f is given in terms of f evaluated at two points. In other words, the integral "undoes" the derivative.

EXAMPLE 3   **Evaluating definite integrals** Evaluate the following definite integrals using the Fundamental Theorem of Calculus, Part 2. Interpret each result geometrically.

a. $\displaystyle\int_0^{10} (60x - 6x^2)\,dx$ **b.** $\displaystyle\int_0^{2\pi} 3\sin x\,dx$ **c.** $\displaystyle\int_{1/16}^{1/4} \frac{\sqrt{t} - 1}{t}\,dt$

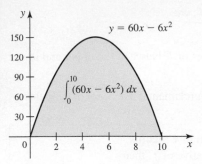

FIGURE 5.41

▶ The arbitrary constant C may always be omitted when evaluating definite integrals. It is added in when evaluating at the upper limit and then subtracted out when evaluating at the lower limit.

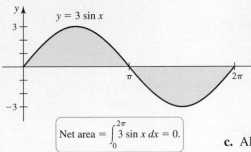

Net area $= \displaystyle\int_0^{2\pi} 3 \sin x \, dx = 0.$

FIGURE 5.42

▶ We know that

$$\frac{d}{dt}(t^{1/2}) = \frac{1}{2}t^{-1/2}.$$

Therefore,

$$\int \frac{1}{2}t^{-1/2}\, dt = t^{1/2} + C$$

and

$$\int \frac{dt}{\sqrt{t}} = \int t^{-1/2}\, dt = 2t^{1/2} + C.$$

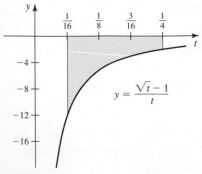

FIGURE 5.43

SOLUTION

a. Using the antiderivative rules of Section 4.8, an antiderivative of $60x - 6x^2$ is $30x^2 - 2x^3$. By the Fundamental Theorem, the value of the definite integral is

$$\int_0^{10} (60x - 6x^2)\, dx = (30x^2 - 2x^3)\Big|_0^{10} \qquad \text{Fundamental Theorem}$$

$$= (30\cdot 10^2 - 2\cdot 10^3) - (30\cdot 0^2 - 2\cdot 0^3) \qquad \begin{array}{l}\text{Evaluate at } x = 10 \\ \text{and } x = 0.\end{array}$$

$$= (3000 - 2000) - 0$$

$$= 1000. \qquad \text{Simplify.}$$

Because f is positive on $[0, 10]$, the definite integral $\int_0^{10}(60x - 6x^2)\, dx$ is the area of the region between the graph of f and the x-axis on the interval $[0, 10]$ (Figure 5.41).

b. As shown in Figure 5.42, the region bounded by the graph of $f(x) = 3 \sin x$ and the x-axis on $[0, 2\pi]$ consists of two parts, one above the x-axis and one below the x-axis. By the symmetry of f, these two regions have the same area, so the definite integral over $[0, 2\pi]$ is zero. Let's confirm this fact. An antiderivative of $f(x) = 3 \sin x$ is $-3 \cos x$. Therefore, the value of the definite integral is

$$\int_0^{2\pi} 3 \sin x \, dx = -3 \cos x \Big|_0^{2\pi} \qquad \text{Fundamental Theorem}$$

$$= [-3 \cos(2\pi)] - [-3 \cos(0)] \qquad \text{Substitute.}$$

$$= -3 - (-3) = 0. \qquad \text{Simplify.}$$

c. Although the variable of integration is t, rather than x, we proceed as before after simplifying the integrand:

$$\frac{\sqrt{t} - 1}{t} = \frac{1}{\sqrt{t}} - \frac{1}{t}$$

Finding antiderivatives with respect to t and applying the Fundamental Theorem, we have

$$\int_{1/16}^{1/4} \frac{\sqrt{t} - 1}{t}\, dt = \int_{1/16}^{1/4} \left(t^{-1/2} - \frac{1}{t}\right) dt \qquad \begin{array}{l}\text{Simplify the} \\ \text{integrand.}\end{array}$$

$$= 2t^{1/2} - \ln|t|\, \Big|_{1/16}^{1/4} \qquad \begin{array}{l}\text{Fundamental} \\ \text{Theorem}\end{array}$$

$$= \left[2\left(\frac{1}{4}\right)^{1/2} - \ln\left(\frac{1}{4}\right)\right] - \left[2\left(\frac{1}{16}\right)^{1/2} - \ln\left(\frac{1}{16}\right)\right] \qquad \text{Evaluate.}$$

$$= 1 - \ln\left(\frac{1}{4}\right) - \frac{1}{2} + \ln\left(\frac{1}{16}\right) \qquad \text{Simplify.}$$

$$= \frac{1}{2} - \ln 4 \approx -0.8863.$$

The definite integral is negative because the graph of f lies below the t-axis (Figure 5.43). *Related Exercises 23–40* ◀

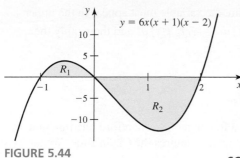

FIGURE 5.44

EXAMPLE 4 **Net areas and definite integrals** The graph of $f(x) = 6x(x + 1)(x - 2)$ is shown in Figure 5.44. The region R_1 is bounded by the curve and the x-axis on the interval $[-1, 0]$, and R_2 is bounded by the curve and the x-axis on the interval $[0, 2]$.

a. Find the *net area* of the region between the curve and the x-axis on the interval $[-1, 2]$.

b. Find the *area* of the region between the curve and the x-axis on the interval $[-1, 2]$.

SOLUTION

a. The net area of the region is given by a definite integral. The integrand f is first expanded in order to find an antiderivative:

$$\int_{-1}^{2} f(x)\, dx = \int_{-1}^{2} (6x^3 - 6x^2 - 12x)\, dx \quad \text{Expanding } f$$

$$= \left(\frac{3}{2}x^4 - 2x^3 - 6x^2\right)\Big|_{-1}^{2} \quad \text{Fundamental Theorem}$$

$$= -\frac{27}{2} \quad \text{Simplify.}$$

The net area of the region between the curve and the x-axis on $[-1, 2]$ is $-\frac{27}{2}$, which is the area of R_1 *minus* the area of R_2 (Figure 5.44). Because R_2 has a larger area than R_1, the net area is negative.

b. The region R_1 lies above the x-axis, so its area is

$$\int_{-1}^{0} (6x^3 - 6x^2 - 12x)\, dx = \left(\frac{3}{2}x^4 - 2x^3 - 6x^2\right)\Big|_{-1}^{0} = \frac{5}{2}.$$

The region R_2 lies below the x-axis, so its net area is negative:

$$\int_{0}^{2} (6x^3 - 6x^2 - 12x)\, dx = \left(\frac{3}{2}x^4 - 2x^3 - 6x^2\right)\Big|_{0}^{2} = -16$$

Therefore, the *area* of R_2 is $-(-16) = 16$. The combined area of R_1 and R_2 is $\frac{5}{2} + 16 = \frac{37}{2}$. We could also find the area of this region directly by evaluating $\int_{-1}^{2} |f(x)|\, dx$.

Related Exercises 41–50 ◄

Examples 3 and 4 make use of Part 2 of the Fundamental Theorem, which is the most potent tool for evaluating definite integrals. The remaining examples illustrate the use of the equally important Part 1 of the Fundamental Theorem.

EXAMPLE 5 **Derivatives of integrals** Use Part 1 of the Fundamental Theorem to simplify the following expressions.

a. $\dfrac{d}{dx} \displaystyle\int_{1}^{x} \sin^2 t\, dt$ **b.** $\dfrac{d}{dx} \displaystyle\int_{x}^{5} \sqrt{t^2 + 1}\, dt$ **c.** $\dfrac{d}{dx} \displaystyle\int_{0}^{x^2} \cos t^2\, dt$

SOLUTION

a. Using Part 1 of the Fundamental Theorem, we see that

$$\frac{d}{dx} \int_{1}^{x} \sin^2 t\, dt = \sin^2 x.$$

b. To apply Part 1 of the Fundamental Theorem, the variable must appear in the upper limit. Therefore, we use the fact that $\int_a^b f(t)\,dt = -\int_b^a f(t)\,dt$ and then apply the Fundamental Theorem:

$$\frac{d}{dx}\int_x^5 \sqrt{t^2 + 1}\,dt = -\frac{d}{dx}\int_5^x \sqrt{t^2 + 1}\,dt = -\sqrt{x^2 + 1}$$

c. The upper limit of the integral is not x, but a function of x. Therefore, the function to be differentiated is a composite function, which requires the Chain Rule. We let $u = x^2$ to produce

$$y = g(u) = \int_0^u \cos t^2\,dt.$$

By the Chain Rule,

$$\frac{d}{dx}\int_0^{x^2} \cos t^2\,dt = \frac{dy}{dx}$$

> Example 5c illustrates one case of Leibniz's Rule:
>
> $$\frac{d}{dx}\int_a^{g(x)} f(t)\,dt = f(g(x))g'(x).$$

$$= \frac{dy}{du}\frac{du}{dx} \qquad \text{Chain Rule}$$

$$= \left[\frac{d}{du}\int_0^u \cos t^2\,dt\right](2x) \qquad \text{Substitute for } g; \text{ note that } u'(x) = 2x.$$

$$= (\cos u^2)(2x) \qquad \text{Fundamental Theorem}$$

$$= 2x\cos x^4. \qquad \text{Substitute } u = x^2.$$

Related Exercises 51–56 ◄

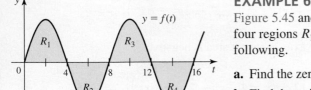

FIGURE 5.45

EXAMPLE 6 Working with area functions Consider the function f shown in Figure 5.45 and its area function $A(x) = \int_0^x f(t)\,dt$, for $0 \le x \le 17$. Assume that the four regions R_1, R_2, R_3, and R_4 have the same area. Based on the graph of f, do the following.

a. Find the zeros of A on $[0, 17]$.

b. Find the points on $[0, 17]$ at which A has local maxima or local minima.

c. Sketch a graph of A for $0 \le x \le 17$.

SOLUTION

a. The area function $A(x) = \int_0^x f(t)\,dt$ gives the net area bounded by the graph of f and the t-axis on the interval $[0, x]$ (Figure 5.46a). Therefore, $A(0) = \int_0^0 f(t)\,dt = 0$. Because R_1 and R_2 have the same area but lie on opposite sides of the t-axis, it follows that $A(8) = \int_0^8 f(t)\,dt = 0$. Similarly, $A(16) = \int_0^{16} f(t)\,dt = 0$. Therefore, the zeros of A are $x = 0$, 8, and 16.

b. Observe that the function f is positive for $0 < t < 4$, which implies that $A(x)$ increases as x increases from 0 to 4 (Figure 5.46b). Then, as x increases from 4 to 8, $A(x)$ decreases because f is negative for $4 < t < 8$ (Figure 5.46c). Similarly, $A(x)$ increases as x increases from $x = 8$ to $x = 12$ (Figure 5.46d) and decreases from $x = 12$ to $x = 16$. By the First Derivative Test, A has local minima at $x = 8$ and $x = 16$ and local maxima at $x = 4$ and $x = 12$ (Figure 5.46e).

c. Combining the observations in parts (a) and (b) leads to a qualitative sketch of A (Figure 5.46e). Note that $A(x) \ge 0$ for all $x \ge 0$. It is not possible to determine function values (y-coordinates) on the graph of A.

> Recall that local extrema occur only at interior points of the domain.

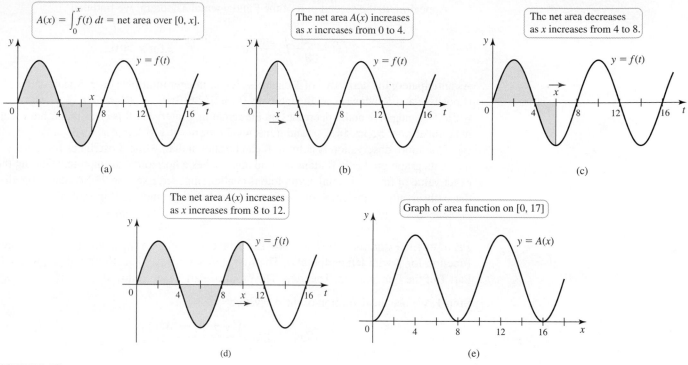

FIGURE 5.46

Related Exercises 57–68 ◀

EXAMPLE 7 **The sine integral function** Let

$$g(t) = \begin{cases} \dfrac{\sin t}{t} & \text{if } t > 0 \\ 1 & \text{if } t = 0 \end{cases}$$

Graph the *sine integral function* $S(x) = \int_0^x g(t)\, dt$, for $x \geq 0$.

SOLUTION Notice that S is an area function for g. The independent variable of S is x, while t has been chosen as the (dummy) variable of integration. A good way to start is by graphing the integrand g (Figure 5.47a). The function oscillates with a decreasing amplitude with $g(0) = 1$. Beginning with $S(0) = 0$, the area function S increases until $x = \pi$ because g is positive on $(0, \pi)$. However, on $(\pi, 2\pi)$, g is negative and the net area decreases. Then, on $(2\pi, 3\pi)$, g is positive again, so S again increases. Therefore, the graph of S has alternating local maxima and minima. Because the amplitude of g decreases, each maximum is less than the previous maximum and each minimum is greater than the previous minimum (Figure 5.47b). Determining the exact value of S at these maxima and minima is difficult.

FIGURE 5.47

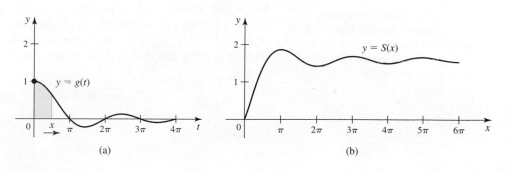

Appealing to the second part of the Fundamental Theorem, we find that

$$S'(x) = \frac{d}{dx} \int_0^x \frac{\sin t}{t} \, dt = \frac{\sin x}{x} \qquad \text{for } x > 0.$$

As anticipated, the derivative of S changes sign at integer multiples of π. Specifically, S' is positive and S increases on the intervals $(0, \pi), (2\pi, 3\pi), \ldots, (2n\pi, (2n + 1)\pi), \ldots$, while S' is negative and S decreases on the remaining intervals. It is clear that S has local maxima at $x = \pi, 3\pi, 5\pi, \ldots$, and it has local minima at $x = 2\pi, 4\pi, 6\pi, \ldots$.

One more observation is helpful. It can be shown that, while S oscillates for increasing x, its graph gradually flattens out and approaches a horizontal asymptote. (Finding the exact value of this horizontal asymptote is challenging; see Exercise 97.) Assembling all these observations, the graph of the sine integral function emerges (Figure 5.47b).

Related Exercises 69–72 ◄

Proof of the Fundamental Theorem Let f be continuous on $[a, b]$ and let A be the area function for f with left endpoint a. The first step is to prove that $A'(x) = f(x)$, which is Part 1 of the Fundamental Theorem. The proof of Part 2 then follows.

Step 1. We use the definition of the derivative,

$$A'(x) = \lim_{h \to 0} \frac{A(x + h) - A(x)}{h}.$$

First assume that $h > 0$. Using Figure 5.48 and Property 5 of Table 5.3, we have

$$A(x + h) - A(x) = \int_a^{x+h} f(t) \, dt - \int_a^x f(t) \, dt = \int_x^{x+h} f(t) \, dt.$$

That is, $A(x + h) - A(x)$ is the net area of the region bounded by the curve on the interval $[x, x + h]$.

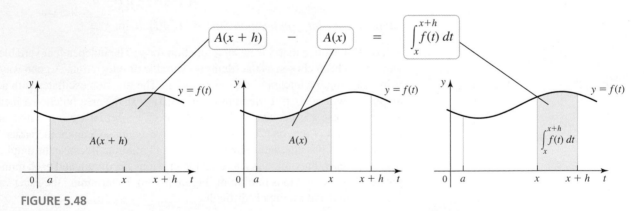

FIGURE 5.48

▶ The quantities m and M exist for any $h > 0$; however, they also depend on h. Figure 5.48 illustrates the case $0 \le m \le M$. The argument that follows holds for the general case.

Let m and M be the minimum and maximum values of f on $[x, x + h]$, respectively, which exist by the continuity of f. In the case that $0 \le m \le M$ (Figure 5.49), $A(x + h) - A(x)$ is greater than or equal to the area of a rectangle with height m and width h and it is less than or equal to the area of a rectangle with height M and width h; that is,

$$mh \le A(x + h) - A(x) \le Mh.$$

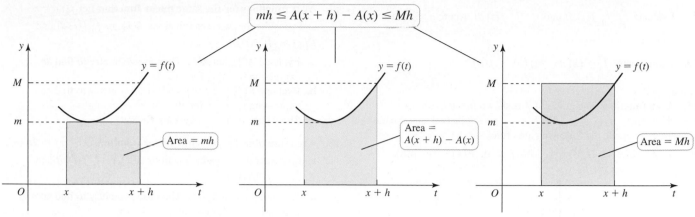

FIGURE 5.49

Dividing these inequalities by h, we have

$$m \leq \frac{A(x+h) - A(x)}{h} \leq M.$$

The case $h < 0$ is handled similarly and leads to the same conclusion.

We now take the limit as $h \to 0$ across these inequalities. As $h \to 0$, m and M squeeze together toward the value of $f(x)$, because f is continuous at x. At the same time, as $h \to 0$, the quotient that is sandwiched between m and M approaches $A'(x)$:

$$\underbrace{\lim_{h \to 0} m}_{f(x)} = \underbrace{\lim_{h \to 0} \frac{A(x+h) - A(x)}{h}}_{A'(x)} = \underbrace{\lim_{h \to 0} M}_{f(x)}$$

By the Squeeze Theorem (Theorem 2.5), we conclude that $A'(x) = f(x)$.

> Once again we use an important fact: Two antiderivatives of the same function differ by a constant.

Step 2. Having established that the area function A is an antiderivative of f, we know that $F(x) = A(x) + C$, where F is any antiderivative of f and C is a constant. Noting that $A(a) = 0$, it follows that

$$F(b) - F(a) = (A(b) + C) - (A(a) + C) = A(b).$$

Writing $A(b)$ in terms of a definite integral, we have

$$A(b) = \int_a^b f(x)\,dx = F(b) - F(a),$$

which is part 2 of the Fundamental Theorem.   ◄

SECTION 5.3 EXERCISES

Review Questions

1. Suppose F is an antiderivative of f and A is an area function of f. What is the relationship between f and A?

2. Suppose F is an antiderivative of f and A is an area function of f. What is the relationship between F and A?

3. Explain in words and write mathematically how the Fundamental Theorem of Calculus is used to evaluate definite integrals.

4. Let $f(x) = c$, where c is a positive constant. Explain why an area function of f is an increasing function.

5. The linear function $f(x) = 3 - x$ is decreasing on the interval $[0, 3]$. Is its area function on the interval $[0, 3]$ increasing or decreasing? Draw a picture and explain.

6. Evaluate $\int_0^2 3x^2\,dx$ and $\int_{-2}^2 3x^2\,dx$.

7. Explain in words and express mathematically the inverse relationship between differentiation and integration as given by the Fundamental Theorem of Calculus.

8. Why can the constant of integration be omitted from the anti-derivative when evaluating a definite integral?

9. Evaluate $\dfrac{d}{dx}\int_a^x f(t)\,dt$ and $\dfrac{d}{dx}\int_a^b f(t)\,dt$, where a and b are constants.

10. Explain why $\int_a^b f'(x)\,dx = f(b) - f(a)$.

Basic Skills

11. Area functions The graph of f is shown in the figure. Let $A(x) = \int_{-2}^x f(t)\,dt$ and $F(x) = \int_4^x f(t)\,dt$ be two area functions for f. Evaluate the following area functions.

   a. $A(-2)$ **b.** $F(8)$ **c.** $A(4)$ **d.** $F(4)$ **e.** $A(8)$

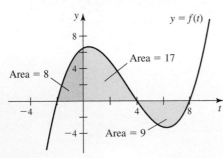

12. Area functions The graph of f is shown in the figure. Let $A(x) = \int_0^x f(t)\,dt$ and $F(x) = \int_2^x f(t)\,dt$ be two area functions for f. Evaluate the following area functions.

   a. $A(2)$ **b.** $F(5)$ **c.** $A(0)$ **d.** $F(8)$ **e.** $A(8)$

   f. $A(5)$ **g.** $F(2)$

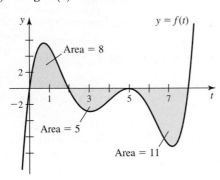

13–16. Area functions for constant functions *Consider the following functions f and real numbers a (see figure).*

   a. *Find and graph the area function $A(x) = \int_a^x f(t)\,dt$ for f.*

   b. *Verify that $A'(x) = f(x)$.*

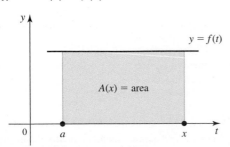

13. $f(t) = 5,\ a = 0$ **14.** $f(t) = 10,\ a = 4$

15. $f(t) = 5,\ a = -5$ **16.** $f(t) = 2,\ a = -3$

17. Area functions for the same linear function Let $f(t) = t$ and consider the two area functions $A(x) = \int_0^x f(t)\,dt$ and $F(x) = \int_2^x f(t)\,dt$.

   a. Evaluate $A(2)$ and $A(4)$. Then use geometry to find an expression for $A(x)$ for all $x \geq 0$.

   b. Evaluate $F(4)$ and $F(6)$. Then use geometry to find an expression for $F(x)$ for all $x \geq 2$.

   c. Show that $A(x) - F(x)$ is a constant.

18. Area functions for the same linear function Let $f(t) = 2t - 2$ and consider the two area functions $A(x) = \int_1^x f(t)\,dt$ and $F(x) = \int_4^x f(t)\,dt$.

   a. Evaluate $A(2)$ and $A(3)$. Then use geometry to find an expression for $A(x)$ for all $x \geq 1$.

   b. Evaluate $F(5)$ and $F(6)$. Then use geometry to find an expression for $F(x)$ for all $x \geq 4$.

   c. Show that $A(x) - F(x)$ is a constant.

19–22. Area functions for linear functions *Consider the following functions f and real numbers a (see figure).*

   a. *Find and graph the area function $A(x) = \int_a^x f(t)\,dt$.*

   b. *Verify that $A'(x) = f(x)$.*

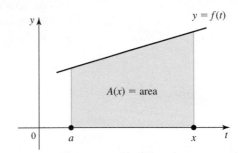

19. $f(t) = t + 5,\ a = -5$ **20.** $f(t) = 2t + 5,\ a = 0$

21. $f(t) = 3t + 1,\ a = 2$ **22.** $f(t) = 4t + 2,\ a = 0$

23–24. Definite integrals *Evaluate the following integrals using the Fundamental Theorem of Calculus. Discuss whether your result is consistent with the figure.*

23. $\displaystyle\int_0^1 (x^2 - 2x + 3)\,dx$ **24.** $\displaystyle\int_{-\pi/4}^{7\pi/4} (\sin x + \cos x)\,dx$

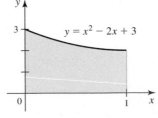

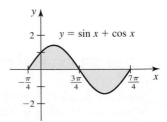

25–30. Definite integrals *Evaluate the following integrals using the Fundamental Theorem of Calculus. Sketch the graph of the integrand and shade the region whose net area you have found.*

25. $\displaystyle\int_1^4 (1 - x)(x - 4)\,dx$ **26.** $\displaystyle\int_0^\pi (1 - \sin x)\,dx$

27. $\int_{2}^{3} (x^2 - x - 6) \, dx$ **28.** $\int_{0}^{1} (x - \sqrt{x}) \, dx$

29. $\int_{0}^{5} (x^2 - 9) \, dx$ **30.** $\int_{1/2}^{2} \left(1 - \frac{1}{x^2}\right) dx$

31–40. Definite integrals *Evaluate the following integrals using the Fundamental Theorem of Calculus.*

31. $\int_{-2}^{2} (x^2 - 4) \, dx$ **32.** $\int_{0}^{\ln 8} e^x \, dx$

33. $\int_{1/2}^{1} (x^{-3} - 8) \, dx$ **34.** $\int_{0}^{4} x(x - 2)(x - 4) \, dx$

35. $\int_{0}^{\pi/4} \sec^2 \theta \, d\theta$ **36.** $\int_{0}^{1/2} \frac{dx}{\sqrt{1 - x^2}}$

37. $\int_{-2}^{-1} x^{-3} \, dx$ **38.** $\int_{-\pi/2}^{\pi/2} (\cos x - 1) \, dx$

39. $\int_{1}^{2} \frac{3}{t} \, dt$ **40.** $\int_{4}^{9} \frac{x - \sqrt{x}}{x^3} \, dx$

41–44. Areas *Find (i) the net area and (ii) the area of the following regions. Graph the function and indicate the region in question.*

41. The region bounded by $y = x^{1/2}$ and the x-axis between $x = 1$ and $x = 4$

42. The region above the x-axis bounded by $y = 4 - x^2$

43. The region below the x-axis bounded by $y = x^4 - 16$

44. The region bounded by $y = 6 \cos x$ and the x-axis between $x = -\pi/2$ and $x = \pi$

45–50. Areas of regions *Find the area of the region R bounded by the graph of f and the x-axis on the given interval. Graph f and the region R.*

45. $f(x) = x^2 - 25$; $[2, 4]$ **46.** $f(x) = x^3 - 1$; $[-1, 2]$

47. $f(x) = \frac{1}{x}$; $[-2, -1]$

48. $f(x) = x(x + 1)(x - 2)$; $[-1, 2]$

49. $f(x) = \sin x$; $[-\pi/4, 3\pi/4]$

50. $f(x) = \cos x$; $[\pi/2, \pi]$

51–56. Derivatives of integrals *Simplify the following expressions.*

51. $\frac{d}{dx} \int_{3}^{x} (t^2 + t + 1) \, dt$ **52.** $\frac{d}{dx} \int_{0}^{x} e^t \, dt$

53. $\frac{d}{dx} \int_{2}^{x^3} \frac{dp}{p^2}$ **54.** $\frac{d}{dx} \int_{x^2}^{10} \frac{dz}{z^2 + 1}$

55. $\frac{d}{dx} \int_{x}^{1} \sqrt{t^4 + 1} \, dt$ **56.** $\frac{d}{dx} \int_{x}^{0} \frac{dp}{p^2 + 1}$

57. Matching functions with area functions Match the functions f whose graphs are given in (a)–(d) with the area functions $A(x) = \int_{0}^{x} f(t) \, dt$, whose graphs are given in (A)–(D).

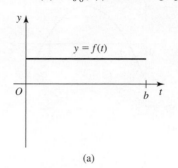

(a)

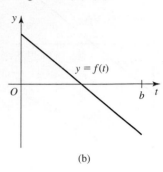

(b)

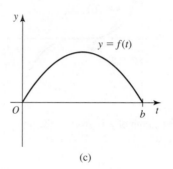

(c)

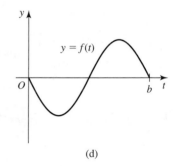

(d)

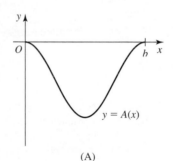

(A)

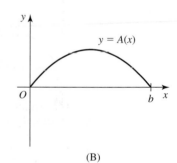

(B)

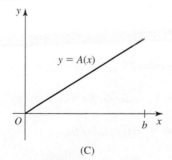

(C)

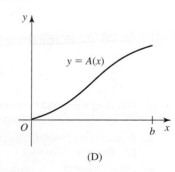

(D)

58–61. Working with area functions *Consider the following graphs of functions f.*

   a. Estimate the zeros of the area function $A(x) = \int_{0}^{x} f(t) \, dt$ for $0 \le x \le 10$.

   b. Estimate the points (if any) at which A has a local maximum or minimum.

   c. Sketch a rough graph of A for $0 \le x \le 10$ without a scale on the y-axis.

58.

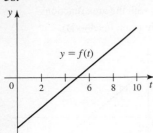

59.

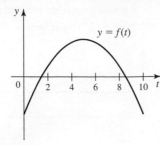

60.

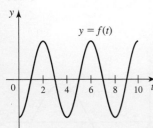

61.

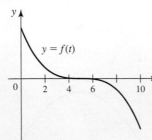

62. Area functions from graphs The graph of f is given in the figure. Let $A(x) = \int_0^x f(t)\, dt$ and evaluate $A(1)$, $A(2)$, $A(4)$, and $A(6)$.

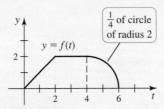

63. Area functions from graphs The graph of f is given in the figure. Let $A(x) = \int_0^x f(t)\, dt$ and evaluate $A(2)$, $A(5)$, $A(8)$, and $A(12)$.

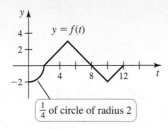

64–68. Working with area functions *Consider the function f and the points a, b, and c.*

a. Find the area function $A(x) = \int_a^x f(t)\, dt$ using the Fundamental Theorem.
b. Graph f and A.
c. Evaluate $A(b)$ and $A(c)$ and interpret the results using the graphs of part (b).

64. $f(x) = \sin x$; $a = 0, b = \pi/2, c = \pi$

65. $f(x) = e^x$; $a = 0, b = \ln 2, c = \ln 4$

66. $f(x) = x^3 + 1$; $a = 0, b = 2, c = 3$

67. $f(x) = x^{1/2}$; $a = 1, b = 4, c = 9$

68. $f(x) = 1/x$; $a = 1, b = 4, c = 6$

69–72. Functions defined by integrals *Consider the function g, which is given in terms of a definite integral with a variable upper limit.*

a. Graph the integrand.
b. Calculate $g'(x)$.
c. Graph g, showing all of your work and reasoning.

69. $g(x) = \displaystyle\int_0^x \sin^2 t\, dt$ **70.** $g(x) = \displaystyle\int_0^x (t^2 + 1)\, dt$

71. $g(x) = \displaystyle\int_0^x \sin(\pi t^2)\, dt$ (a Fresnel integral)

72. $g(x) = \displaystyle\int_0^x \cos(\pi \sqrt{t})\, dt$

Further Explorations

73. Explain why or why not Determine whether the following statements are true and give an explanation or counterexample.

a. Suppose that f is a positive decreasing function for $x > 0$. Then the area function $A(x) = \int_0^x f(t)\, dt$ is an increasing function of x.
b. Suppose that f is a negative increasing function for $x > 0$. Then the area function $A(x) = \int_0^x f(t)\, dt$ is a decreasing function of x.
c. The functions $p(x) = \sin 3x$ and $q(x) = 4\sin 3x$ are antiderivatives of the same function.
d. If $A(x) = 3x^2 - x + 2$ is an area function for f, then $B(x) = 3x^2 - x$ is also an area function for f.

74–82. Definite integrals *Evaluate the following definite integrals using the Fundamental Theorem of Calculus.*

74. $\dfrac{1}{2}\displaystyle\int_0^{\ln 2} e^x\, dx$ **75.** $\displaystyle\int_1^4 \dfrac{x-2}{\sqrt{x}}\, dx$ **76.** $\displaystyle\int_1^2 \left(\dfrac{2}{s} - \dfrac{4}{s^3}\right) ds$

77. $\displaystyle\int_0^{\pi/3} \sec x \tan x\, dx$ **78.** $\displaystyle\int_{\pi/4}^{\pi/2} \csc^2 \theta\, d\theta$ **79.** $\displaystyle\int_1^8 \sqrt[3]{y}\, dy$

80. $\displaystyle\int_{\sqrt{2}}^2 \dfrac{dx}{x\sqrt{x^2-1}}$ **81.** $\displaystyle\int_1^2 \dfrac{z^2+4}{z}\, dz$ **82.** $\displaystyle\int_0^{\sqrt{3}} \dfrac{dx}{1+x^2}$

83–86. Areas of regions *Find the area of the region R bounded by the graph of f and the x-axis on the given interval. Graph f and show the region R.*

83. $f(x) = 2 - |x|$; $[-2, 4]$

84. $f(x) = (1 - x^2)^{-1/2}$; $[-1/2, \sqrt{3}/2]$

85. $f(x) = x^4 - 4$; $[1, 4]$ **86.** $f(x) = x^2(x - 2)$; $[-1, 3]$

87–90. Derivatives and integrals *Simplify the given expressions. Assume that derivatives are continuous on the interval of integration.*

87. $\displaystyle\int_3^8 f'(t)\, dt$ **88.** $\dfrac{d}{dx} \displaystyle\int_0^{x^2} \dfrac{1}{t^2+4}\, dt$

89. $\dfrac{d}{dx} \displaystyle\int_0^{\cos x} (t^4 + 6)\, dt$ **90.** $\dfrac{d}{dx} \displaystyle\int_x^1 e^{t^2}\, dt$

Additional Exercises

91. Zero net area Consider the function $f(x) = x^2 - 4x$.

    a. Graph f on the interval $x \geq 0$.

    b. For what value of $b > 0$ is $\int_0^b f(x)\,dx = 0$?

    c. In general, for the function $f(x) = x^2 - ax$, where $a > 0$, for what value of $b > 0$ (as a function of a) is $\int_0^b f(x)\,dx = 0$?

92. Cubic zero net area Consider the graph of the cubic $y = x(x - a)(x - b)$ where $0 < a < b$. Verify that the graph bounds a region above the x-axis for $0 < x < a$ and bounds a region below the x-axis for $a < x < b$. What is the relationship between a and b if the areas of these two regions are equal?

93. Maximum net area What value of $b > -1$ maximizes the integral

$$\int_{-1}^{b} x^2(3 - x)\,dx?$$

94. Maximum net area Graph the function $f(x) = 8 + 2x - x^2$ and determine the values of a and b that maximize the value of the integral

$$\int_{a}^{b} (8 + 2x - x^2)\,dx.$$

95. An integral equation Use the Fundamental Theorem of Calculus, Part 1, to find the function f that satisfies the equation

$$\int_{0}^{x} f(t)\,dt = 2\cos x + 3x + 2.$$

96. Max/min of area functions Suppose f is continuous on $[0, \infty)$ and $A(x)$ is the net area bounded by the graph of f and the t-axis on $[0, x]$. Show that the maxima and minima of A occur at the zeros of f. Verify this fact with the function $f(x) = x^2 - 10x$.

97. Asymptote of sine integral Use a calculator to approximate

$$\lim_{x \to \infty} S(x) = \lim_{x \to \infty} \int_{0}^{x} \frac{\sin t}{t}\,dt,$$

where S is the sine integral function (see Example 7). Show your work and describe your reasoning.

98. Sine integral Show that the sine integral $S(x) = \int_{0}^{x} \frac{\sin t}{t}\,dt$ satisfies the (differential) equation $xS'(x) + 2S''(x) + xS'''(x) = 0$.

99. Fresnel integral Show that the Fresnel integral $S(x) = \int_{0}^{x} \sin(t^2)\,dt$ satisfies the (differential) equation

$$(S'(x))^2 + \left(\frac{S''(x)}{2x}\right)^2 = 1.$$

100. Variable integration limits Evaluate $\dfrac{d}{dx} \int_{-x}^{x} (t^2 + t)\,dt$.

    (*Hint:* Separate the integral into two pieces.)

QUICK CHECK ANSWERS

1. $0, -35$ **2.** $A(6) = 44$; $A(10) = 120$ **3.** $\frac{2}{3} - \frac{1}{2} = \frac{1}{6}$
4. If f is differentiated, we get f'. Thus f is an antiderivative of f'. ◄

5.4 Working with Integrals

With the Fundamental Theorem of Calculus in hand, we may begin an investigation of integration and its applications. In this section we discuss the role of symmetry in integrals, use the slice-and-sum strategy to define the average value of a function, and then explore a theoretical result called the Mean Value Theorem for integrals.

Integrating Even and Odd Functions

Symmetry appears throughout mathematics in many different forms, and its use often leads to insights and efficiencies. Here we use the symmetry of a function to simplify integral calculations.

    Section 1.1 introduced the symmetry of even and odd functions. An **even function** satisfies the property that $f(-x) = f(x)$, which means that its graph is symmetric about the y-axis (Figure 5.50a). Examples of even functions are $f(x) = \cos x$ and $f(x) = x^n$, where n is an even integer. An **odd function** satisfies the property that $f(-x) = -f(x)$, which means that its graph is symmetric about the origin (Figure 5.50b). Examples of odd functions are $f(x) = \sin x$ and $f(x) = x^n$, where n is an odd integer.

    Special things happen when we integrate even and odd functions on intervals centered at the origin. First, suppose f is an even function and consider $\int_{-a}^{a} f(x)\,dx$. From

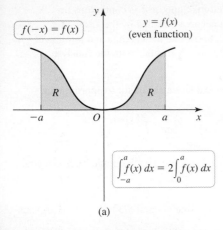

$$f(-x) = f(x)$$

$y = f(x)$
(even function)

R R

$-a$ O a x

$$\int_{-a}^{a} f(x)\, dx = 2\int_{0}^{a} f(x)\, dx$$

(a)

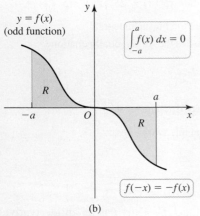

$y = f(x)$
(odd function)

$$\int_{-a}^{a} f(x)\, dx = 0$$

R

$-a$ O a x

R

$$f(-x) = -f(x)$$

(b)

FIGURE 5.50

Figure 5.50a, we see that the integral of f on $[-a, 0]$ equals the integral of f on $[0, a]$. Therefore, the integral on $[-a, a]$ is twice the integral on $[0, a]$, or

$$\int_{-a}^{a} f(x)\, dx = 2\int_{0}^{a} f(x)\, dx.$$

On the other hand, suppose f is an odd function and consider $\int_{-a}^{a} f(x)\, dx$. As shown in Figure 5.50b, the integral on the interval $[-a, 0]$ is the negative of the integral on $[0, a]$. Therefore, the integral on $[-a, a]$ is zero, or

$$\int_{-a}^{a} f(x)\, dx = 0.$$

We summarize these results in the following theorem.

THEOREM 5.4   Integrals of Even and Odd Functions
Let a be a positive real number and let f be an integrable function on the interval $[-a, a]$.

- If f is even, $\int_{-a}^{a} f(x)\, dx = 2\int_{0}^{a} f(x)\, dx$.
- If f is odd, $\int_{-a}^{a} f(x)\, dx = 0$.

QUICK CHECK 1 If f and g are both even functions, is the product fg even or odd? Use the facts that $f(-x) = f(x)$ and $g(-x) = g(x)$. ◄

EXAMPLE 1   Integrating symmetric functions   Evaluate the following integrals using symmetry arguments.

a. $\displaystyle\int_{-2}^{2} (x^4 - 3x^3)\, dx$ **b.** $\displaystyle\int_{-\pi/2}^{\pi/2} (\cos x - 4\sin^3 x)\, dx$

SOLUTION

a. Using Properties 3 and 4 of Table 5.3, we split the integral into two integrals and use symmetry:

$$\int_{-2}^{2} (x^4 - 3x^3)\, dx = \int_{-2}^{2} x^4\, dx - 3\int_{-2}^{2} x^3\, dx$$

$$= 2\int_{0}^{2} x^4\, dx - 0 \qquad x^4 \text{ is even, } x^3 \text{ is odd.}$$

$$= 2\left(\frac{x^5}{5}\right)\Bigg|_{0}^{2} \qquad \text{Fundamental Theorem}$$

$$= 2\left(\frac{32}{5}\right) = \frac{64}{5} \qquad \text{Simplify.}$$

Notice how the odd-powered term of the integrand is eliminated by symmetry. Integration of the even-powered term is simplified because the lower limit is zero.

There are a couple of ways to see that $\sin^3 x$ is an odd function. Its graph is symmetric about the origin. Or by analogy, take an odd power of x and raise it to an odd power. For example, $(x^5)^3 = x^{15}$, which is odd. See Exercises 47–50 for direct proofs of symmetry in composite functions.

b. The $\cos x$ term is an even function, so it can be integrated on the interval $[0, \pi/2]$. What about $\sin^3 x$? It is an odd function raised to an odd power, which results in an odd function; its integral on $[-\pi/2, \pi/2]$ is zero. Therefore,

$$\int_{-\pi/2}^{\pi/2} (\cos x - 4 \sin^3 x)\, dx = 2 \int_0^{\pi/2} \cos x \, dx - 0 \quad \text{Symmetry}$$

$$= 2 \sin x \Big|_0^{\pi/2} \qquad \text{Fundamental Theorem}$$

$$= 2(1 - 0) = 2. \qquad \text{Simplify.}$$

Related Exercises 7–18 ◀

Average Value of a Function

If five people weigh 155, 143, 180, 105, and 123 lb, their average (mean) weight is

$$\frac{155 + 143 + 180 + 105 + 123}{5} = 141.2 \text{ lb.}$$

This idea generalizes quite naturally to functions. Consider a function f that is continuous on $[a, b]$. Let the grid points $x_0 = a, x_1, x_2, \ldots, x_n = b$ form a regular partition of $[a, b]$ with $\Delta x = \dfrac{b - a}{n}$. We now select a point $\overline{x}_k$ in each subinterval and compute $f(\overline{x}_k)$ for $k = 1, \ldots, n$. The values of $f(\overline{x}_k)$ may be viewed as a sampling of f on $[a, b]$. The average of these function values is

$$\frac{f(\overline{x}_1) + f(\overline{x}_2) + \cdots + f(\overline{x}_n)}{n}.$$

Noting that $n = \dfrac{b - a}{\Delta x}$, we write the average of the n sample values as the Riemann sum

$$\frac{f(\overline{x}_1) + f(\overline{x}_2) + \cdots + f(\overline{x}_n)}{(b - a)/\Delta x} = \frac{1}{b - a} \sum_{k=1}^{n} f(\overline{x}_k) \Delta x.$$

Now suppose we increase n, taking more and more samples of f, while Δx decreases to zero. The limit of this sum is a definite integral that gives the average value $\overline{f}$ on $[a, b]$:

$$\overline{f} = \frac{1}{b - a} \lim_{n \to \infty} \sum_{k=1}^{n} f(\overline{x}_k) \Delta x$$

$$= \frac{1}{b - a} \int_a^b f(x)\, dx$$

This definition of the average value of a function is analogous to the definition of the average of a finite set of numbers.

> **DEFINITION   Average Value of a Function**
>
> The average value of an integrable function f on the interval $[a, b]$ is
>
> $$\overline{f} = \frac{1}{b - a} \int_a^b f(x)\, dx.$$

The average value of a function f on an interval $[a, b]$ has a clear geometrical interpretation. Multiplying both sides of the definition of average value by $(b - a)$, we have

$$\underbrace{(b - a)\overline{f}}_{\substack{\text{net area of} \\ \text{rectangle}}} = \underbrace{\int_a^b f(x)\,dx.}_{\substack{\text{net area of region} \\ \text{bounded by curve}}}$$

We see that the average value is the height of the rectangle with base $[a, b]$ that has the same net area as the region bounded by the graph of f on the interval $[a, b]$ (Figure 5.51). (We need to use net area in case f is negative on part of $[a, b]$, which could make $\overline{f}$ negative.)

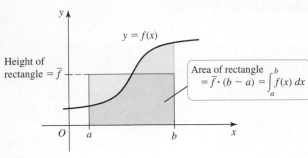

FIGURE 5.51

QUICK CHECK 2 What is the average value of a constant function on an interval? What is the average value of an odd function on an interval $[-a, a]$? ◄

EXAMPLE 2 Average elevation A hiking trail has an elevation given by

$$f(x) = 60x^3 - 650x^2 + 1200x + 4500,$$

where f is measured in feet above sea level and x represents horizontal distance along the trail in mi, with $0 \le x \le 5$. What is the average elevation of the trail?

SOLUTION The trail ranges between elevations of about 2000 and 5000 ft (Figure 5.52). If we let the endpoints of the trail correspond to the horizontal distances $a = 0$ and $b = 5$ mi, the average elevation of the trail is

$$\overline{f} = \frac{1}{5}\int_0^5 (60x^3 - 650x^2 + 1200x + 4500)\,dx$$

$$= \frac{1}{5}\left(60\frac{x^4}{4} - 650\frac{x^3}{3} + 1200\frac{x^2}{2} + 4500x\right)\Big|_0^5 \quad \text{Fundamental Theorem}$$

$$= 3958\tfrac{1}{3}\ \text{ft.} \qquad\qquad\qquad\qquad\qquad \text{Simplify.}$$

The average elevation of the trail is slightly less than 3960 ft.

Related Exercises 19–28 ◄

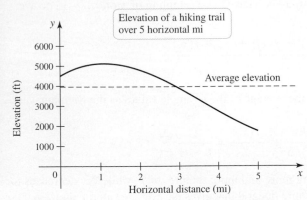

FIGURE 5.52

EXAMPLE 3 Average distance Suppose you walk at constant speed along a semicircle with a radius of 1 km. What is your average distance from the *base* of the semicircle over your entire trip?

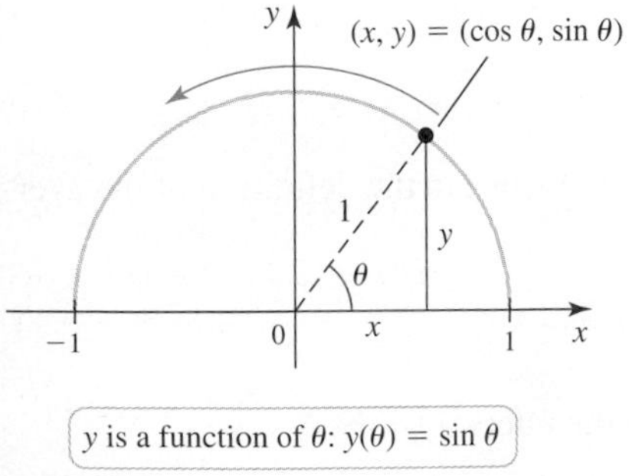

SOLUTION As you walk along the semicircle, your x- and y-coordinates change continuously (Figure 5.53). The goal is to find the average value of the y-coordinate. The easiest way to describe the walk is in terms of the angle θ, which varies from 0 to π. For each value of θ corresponding to points along the path, the y-coordinate is $y = \sin \theta$. Therefore, the average value of the y-coordinate as you walk along the path is

$$\overline{y} = \frac{1}{\pi - 0}\int_0^\pi y(\theta)\,d\theta = \frac{1}{\pi}\int_0^\pi \sin\theta\,d\theta \quad \text{Substitute for } y.$$

$$= \frac{1}{\pi}(-\cos\theta)\Big|_0^\pi \qquad\qquad\qquad \text{Fundamental Theorem}$$

$$= -\frac{1}{\pi}[(-1)-1] = \frac{2}{\pi}. \qquad\qquad \text{Simplify.}$$

FIGURE 5.53

> In Example 3, the average distance to the x-axis is computed relative to a person walking on the semicircle. One could also compute the average height of the semicircle above the x-axis, which is $\pi/4$.

Your average distance from the base of the semicircle is $2/\pi \approx 0.64$ km.

Related Exercises 29–30 ◄

Mean Value Theorem for Integrals

The average value of a function brings us close to an important theoretical result. The **Mean Value Theorem for Integrals** says that if f is continuous on $[a, b]$ then there is at least one point c in the interval $[a, b]$ such that $f(c)$ equals the average value of f on $[a, b]$. In other words, the horizontal line $y = \overline{f} = f(c)$ intersects the graph of f for some point c in $[a, b]$ (Figure 5.54). If f were not continuous, such a point might not exist.

> Compare this statement to that of the Mean Value Theorem for derivatives: There is at least one point c in (a, b) such that $f'(c)$ equals the average slope of f.

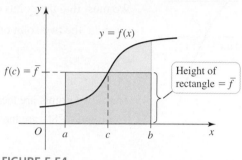

FIGURE 5.54

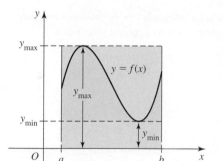

Area of smallest rectangle (red)
$= (b - a) y_{min}$
Area under curve (red + green)
$= \int_a^b f(x)\, dx$
Area of largest rectangle (red + green + blue)
$= (b - a) y_{max}$

FIGURE 5.55

THEOREM 5.5   Mean Value Theorem for Integrals
Let f be continuous on the interval $[a, b]$. There exists a point c in $[a, b]$ such that

$$f(c) = \overline{f} = \frac{1}{b - a} \int_a^b f(x)\, dx.$$

Proof The proof offers a nice application of the Extreme Value Theorem (Theorem 4.1) and the Intermediate Value Theorem (Theorem 2.15). Because f is continuous on a closed interval $[a, b]$, it attains a minimum value y_{min} and a maximum value y_{max} on $[a, b]$. Also note that

$$(b - a)y_{min} \leq \int_a^b f(x)\, dx \leq (b - a)y_{max} \qquad \text{(Figure 5.55)}.$$

These inequalities are true because if every function value in the Riemann sum for the integral is replaced by y_{min}, we obtain a lower bound on the integral. If every function value in the Riemann sum for the integral is replaced by y_{max}, we obtain an upper bound on the integral. Dividing through these inequalities by $(b - a)$, we have

$$y_{min} \leq \underbrace{\frac{1}{b - a} \int_a^b f(x)\, dx}_{\overline{f}} \leq y_{max}.$$

Because f is continuous on $[a, b]$, f assumes all values between y_{min} and y_{max} by the Intermediate Value Theorem. In particular, $y_{min} \leq \overline{f} \leq y_{max}$, so there must be a point c in $[a, b]$ for which $f(c) = \overline{f} = \frac{1}{b - a} \int_a^b f(x)\, dx.$ ◄

QUICK CHECK 3 Explain why $f(x) = 0$ for at least one point of $[a, b]$ if f is continuous and $\int_a^b f(x)\, dx = 0$. ◄

EXAMPLE 4   Average value equals function value Find the point(s) on the interval $[0, 1]$ at which the value of $f(x) = 2x(1 - x)$ equals its average value on $[0, 1]$.

SOLUTION The average value of f on $[0, 1]$ is

$$\bar{f} = \frac{1}{1 - 0}\int_0^1 2x(1 - x)\, dx = \left(x^2 - \frac{2}{3}x^3\right)\Big|_0^1 = \frac{1}{3}.$$

We must find the points on $[0, 1]$ at which $f(x) = \frac{1}{3}$ (Figure 5.56). Using the quadratic formula, the two solutions of $f(x) = 2x(1 - x) = \frac{1}{3}$ are

$$\frac{1 - \sqrt{1/3}}{2} \approx 0.211 \quad \text{and} \quad \frac{1 + \sqrt{1/3}}{2} \approx 0.789.$$

These two points are located symmetrically on either side of $x = \frac{1}{2}$. The two solutions, 0.211 and 0.789, are the same for $f(x) = ax(1 - x)$ for any value of a (Exercise 51).

Related Exercises 31–36 ◄

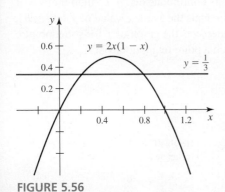

FIGURE 5.56

SECTION 5.4 EXERCISES

Review Questions

1. If f is an odd function, why is $\int_{-a}^{a} f(x)\, dx = 0$?

2. If f is an even function, why is $\int_{-a}^{a} f(x)\, dx = 2\int_0^a f(x)\, dx$?

3. Is x^{12} an even or odd function? Is $\sin(x^2)$ an even or odd function?

4. Explain how to find the average value of a function on an interval $[a, b]$ and why this definition is analogous to the definition of the average of a finite set of numbers.

5. Explain the statement that a continuous function on an interval $[a, b]$ equals its average value at some point on $[a, b]$.

6. Sketch the function $y = x$ on the interval $[0, 2]$ and let R be the region bounded by $y = x$ and the x-axis on $[0, 2]$. Now sketch a rectangle in the first quadrant whose base is $[0, 2]$ and whose area equals the area of R.

Basic Skills

7–14. Symmetry in integrals *Use symmetry to evaluate the following integrals.*

7. $\displaystyle\int_{-2}^{2} (3x^8 - 2)\, dx$

8. $\displaystyle\int_{-\pi/4}^{\pi/4} \cos x\, dx$

9. $\displaystyle\int_{-2}^{2} (x^9 - 3x^5 + 2x^2 - 10)\, dx$

10. $\displaystyle\int_{-\pi/2}^{\pi/2} 5\sin x\, dx$

11. $\displaystyle\int_{-10}^{10} \frac{x}{\sqrt{200 - x^2}}\, dx$

12. $\displaystyle\int_{-\pi/2}^{\pi/2} (\cos 2x + \cos x \sin x - 3\sin x^5)\, dx$

13. $\displaystyle\int_{-\pi/4}^{\pi/4} \sin^5 x\, dx$

14. $\displaystyle\int_{-1}^{1} (1 - |x|)\, dx$

15–18. Symmetry and definite integrals *Use symmetry to evaluate the following integrals. Draw a figure to interpret your result.*

15. $\displaystyle\int_{-\pi}^{\pi} \sin x\, dx$

16. $\displaystyle\int_{0}^{2\pi} \cos x\, dx$

17. $\displaystyle\int_{0}^{\pi} \cos x\, dx$

18. $\displaystyle\int_{0}^{2\pi} \sin x\, dx$

19–24. Average values *Find the average value of the following functions on the given interval. Draw a graph of the function and indicate the average value.*

19. $f(x) = 1/x$; $[1, e]$

20. $f(x) = e^{2x}$; $[0, \ln 2]$

21. $f(x) = \cos x$; $[-\pi/2, \pi/2]$

22. $f(x) = x(1 - x)$; $[0, 1]$

23. $f(x) = x^n$; $[0, 1]$ for any positive integer n

24. $f(x) = x^{1/n}$; $[0, 1]$ for any positive integer n

25. **Average distance on a parabola** What is the average distance between the parabola $y = 10x(20 - x)$ and the x-axis on the interval $[0, 20]$?

26. **Average elevation** The elevation of a path is given by $f(x) = x^3 - 5x^2 + 10$, where x measures horizontal distances. Draw a graph of the elevation function and find its average value for $0 \le x \le 4$.

27. **Average height of an arch** The height of an arch above the ground is given by the function $y = 10\sin x$ for $0 \le x \le \pi$. What is the average height of the arch above the ground?

28. **Average height of a wave** The surface of a water wave is described by $y = 5(1 + \cos x)$ for $-\pi \le x \le \pi$, where $y = 0$ corresponds to a trough of the wave. Find the average height of the wave above the trough on $[-\pi, \pi]$.

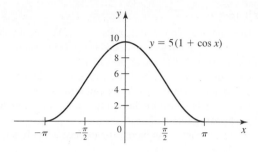

$y = 5(1 + \cos x)$

29. Average distance on a semicircle Suppose you walk the length of a semicircle with radius 2 km at a constant speed. What is your average distance from the *base* of the semicircle?

30. Average distance on a semicircle Suppose you walk the length of a semicircle with radius 2 km at a constant speed. What is your average distance from the *center* of the semicircle?

31–36. Mean Value Theorem for Integrals *Find or approximate the point(s) at which the given function equals its average value on the given interval.*

31. $f(x) = 8 - 2x$; $[0,4]$ **32.** $f(x) = e^x$; $[0,2]$

33. $f(x) = 1 - x^2/a^2$; $[0,a]$, where a is a real number

34. $f(x) = \dfrac{\pi}{4}\sin x$; $[0,\pi]$ **35.** $f(x) = 1 - |x|$; $[-1,1]$

36. $f(x) = 1/x$; $[1,4]$

Further Explorations

37. Explain why or why not Determine whether the following statements are true and give an explanation or counterexample.

a. If f is symmetric about the line $x = 2$, then $\int_0^4 f(x)\,dx = 2\int_0^2 f(x)\,dx.$

b. If f has the property $f(a + x) = -f(a - x)$ for all x, where a is a constant, then $\int_{a-2}^{a+2} f(x)\,dx = 0.$

c. The average value of a linear function on an interval $[a, b]$ is the function value at the midpoint of the interval.

d. Consider the function $f(x) = x(a - x)$ on the interval $[0, a]$ for $a > 0$. Its average value on $[0, a]$ is $\frac{1}{2}$ of its maximum value.

38–41. Symmetry in integrals *Use symmetry to evaluate the following integrals.*

38. $\displaystyle\int_{-\pi/4}^{\pi/4} \tan x\,dx$ **39.** $\displaystyle\int_{-\pi/4}^{\pi/4} \sec^2 x\,dx$

40. $\displaystyle\int_{-2}^{2}(1 - |x|^3)\,dx$ **41.** $\displaystyle\int_{-2}^{2}\frac{x^3 - 4x}{x^2 + 1}\,dx$

Applications

42. Root mean square The root mean square (or RMS) is used to measure the average value of oscillating functions (for example, sine and cosine functions that describe the current, voltage, or power in an alternating circuit). The RMS of a function f on the interval $[0, T]$ is

$$\bar{f}_{RMS} = \sqrt{\frac{1}{T}\int_0^T f(t)^2\,dt}.$$

Compute the RMS of $f(t) = A\sin(\omega t)$, where A and ω are positive constants and T is any integer multiple of the period of f, which is $2\pi/\omega$.

43. Gateway Arch The Gateway Arch in St. Louis is 630 ft high and has a 630-ft base. It can be modeled by the parabola

$$y = 630\left[1 - \left(\frac{x}{315}\right)^2\right].$$

Find the average height of the arch above the ground.

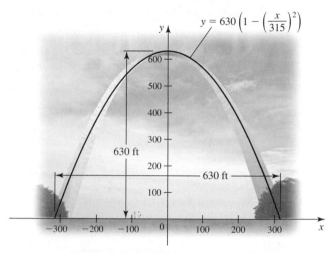

$y = 630\left(1 - \left(\frac{x}{315}\right)^2\right)$

44. Another Gateway Arch Another description of the Gateway Arch is

$$y = 1260 - 315(e^{0.00418x} + e^{-0.00418x}),$$

where the base of the arch is $[-315, 315]$ and x and y are measured in ft. Find the average height of the arch above the ground.

45. Planetary orbits The planets orbit the Sun in elliptical orbits with the Sun at one focus (see Section 10.4 for more on ellipses). The equation of an ellipse whose dimensions are $2a$ in the x-direction and $2b$ in the y-direction is $\dfrac{x^2}{a^2} + \dfrac{y^2}{b^2} = 1.$

a. Let d^2 denote the square of the distance from a planet to the center of the ellipse at $(0, 0)$. Integrate over the interval $[-a, a]$ to show that the average value of d^2 is $(a^2 + 2b^2)/3$.

b. Show that in the case of a circle ($a = b = R$), the average value in part (a) is R^2.

c. Assuming $0 < b < a$, the coordinates of the Sun are $\left(\sqrt{a^2 - b^2}, 0\right)$. Let D^2 denote the square of the distance from the planet to the Sun. Integrate over the interval $[-a, a]$ to show that the average value of D^2 is $(4a^2 - b^2)/3$.

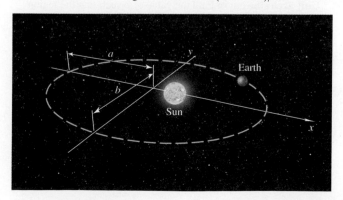

Additional Exercises

▣ 46. Comparing a sine and a quadratic function Consider the functions $f(x) = \sin x$ and $g(x) = \dfrac{4}{\pi^2} x(\pi - x)$.

    a. Carefully graph f and g on the same set of axes. Verify that both functions have a single local maximum on the interval $[0, \pi]$ and they have the same maximum value on $[0, \pi]$.

    b. On the interval $[0, \pi]$, which is true: $f(x) \geq g(x)$, $g(x) \geq f(x)$, or neither?

    c. Compute and compare the average values of f and g on $[0, \pi]$.

47–50. Symmetry of composite functions *Prove that the integrand is either even or odd. Then give the value of the integral or show how it can be simplified. Assume that f and g are even functions and p and q are odd functions.*

47. $\displaystyle\int_{-a}^{a} f(g(x)) \, dx$ **48.** $\displaystyle\int_{-a}^{a} f(p(x)) \, dx$

49. $\displaystyle\int_{-a}^{a} p(g(x)) \, dx$ **50.** $\displaystyle\int_{-a}^{a} p(q(x)) \, dx$

51. Average value with a parameter Consider the function $f(x) = ax(1 - x)$ on the interval $[0, 1]$, where a is a positive real number.

    a. Find the average value of f as a function of a.

    b. Find the points at which the value of f equals its average value and prove that they are independent of a.

52. Square of the average For what functions f is it true that the square of the average value of f equals the average value of the square of f over all intervals $[a, b]$?

53. Problems of antiquity Several calculus problems were solved by Greek mathematicians long before the discovery of calculus. The following problems were solved by Archimedes using methods that predated calculus by 2000 years.

    a. Show that the area of a segment of a parabola is $4/3$ that of its inscribed triangle of greatest area. In other words, the area bounded by the parabola $y = a^2 - x^2$ and the x-axis is $4/3$ the area of the triangle with vertices $(\pm a, 0)$ and $(0, a^2)$. Assume that $a > 0$, but is unspecified.

    b. Show that the area bounded by the parabola $y = a^2 - x^2$ and the x-axis is $2/3$ the area of the rectangle with vertices $(\pm a, 0)$ and $(\pm a, a^2)$. Assume that $a > 0$, but is unspecified.

54. Unit area sine curve Find the value of c such that the region bounded by $y = c \sin x$ and the x-axis on the interval $[0, \pi]$ has area 1.

55. Unit area cubic Find the value of $c > 0$ such that the region bounded by the cubic $y = x(x - c)^2$ and the x-axis on the interval $[0, c]$ has area 1.

56. Unit area

    a. Consider the curve $y = 1/x$ for $x \geq 1$. For what value of $b > 0$ does the region bounded by this curve and the x-axis on the interval $[1, b]$ have an area of 1?

    b. Consider the curve $y = 1/x^p$, where $x \geq 1$ and $p > 1$ is a rational number. For what value of b (as a function of p) does the region bounded by this curve and the x-axis on the interval $[1, b]$ have unit area?

    c. Is $b(p)$ in part (b) an increasing or decreasing function of p? Explain.

57. A sine integral by Riemann sums Consider the integral $I = \int_0^{\pi/2} \sin x \, dx$.

    a. Write the left Riemann sum for I with n subintervals.

    b. Show that $\displaystyle\lim_{\theta \to 0} \theta \left(\frac{\cos \theta + \sin \theta - 1}{2(1 - \cos \theta)} \right) = 1$.

    c. It is a fact that $\displaystyle\sum_{k=0}^{n-1} \sin\left(\frac{\pi k}{2n}\right) = \frac{\cos\left(\dfrac{\pi}{2n}\right) + \sin\left(\dfrac{\pi}{2n}\right) - 1}{2\left[1 - \cos\left(\dfrac{\pi}{2n}\right)\right]}$.

    Use this fact and part (b) to evaluate I by taking the limit of the Riemann sum as $n \to \infty$.

58. Alternate definitions of means Consider the function

$$f(t) = \frac{\int_a^b x^{t+1} \, dx}{\int_a^b x^t \, dx}.$$

Show that the following means can be defined in terms of f.

    a. Arithmetic mean: $f(0) = \dfrac{a + b}{2}$

    b. Geometric mean: $f\left(-\dfrac{3}{2}\right) = \sqrt{ab}$

    c. Harmonic mean: $f(-3) = \dfrac{2ab}{a + b}$

    d. Logarithmic mean: $f(-1) = \dfrac{b - a}{\ln b - \ln a}$

(Source: Mathematics Magazine 78, No. 5 (December 2005))

59. Fill in the following table with either **even** or **odd** and prove each result. Assume n is an integer and f^n means the nth power of f.

	f is even	f is odd
n is even	f^n is _____	f^n is _____
n is odd	f^n is _____	f^n is _____

60. Average value of the derivative Suppose that f' is a continuous function for all real numbers. Show that the average value of the derivative on an interval $[a, b]$ is $\overline{f}' = \dfrac{f(b) - f(a)}{b - a}$. Interpret this result in terms of secant lines.

61. **Symmetry about a point** A function f is symmetric about a point (c, d) if whenever $(c - x, d - y)$ is on the graph, then so is $(c + x, d + y)$. Functions that are symmetric about a point (c, d) are easily integrated on an interval with midpoint c.

    a. Show that if f is symmetric about (c, d) and $a > 0$, then
    $$\int_{c-a}^{c+a} f(x)\, dx = 2af(c) = 2ad.$$

    b. Graph the function $f(x) = \sin^2 x$ on the interval $[0, \pi/2]$ and show that the function is symmetric about the point $(\pi/4, 1/2)$.

    c. Using only the graph of f (and no integration), show that
    $$\int_0^{\pi/2} \sin^2 x\, dx = \frac{\pi}{4}.$$

    (See the Guided Projects for more on symmetry in integrals.)

62. **Bounds on an integral** Suppose f is continuous on $[a, b]$ with $f''(x) > 0$ on the interval. It can be shown that

$$(b - a)f\left(\frac{a + b}{2}\right) \le \int_a^b f(x)\, dx \le (b - a)\frac{f(a) + f(b)}{2}.$$

a. Assuming f is nonnegative on $[a, b]$, draw a figure to illustrate the geometric meaning of these inequalities. Discuss your conclusions.

b. Divide these inequalities by $(b - a)$ and interpret the resulting inequalities in terms of the average value of f on $[a, b]$.

QUICK CHECK ANSWERS

1. $f(-x)g(-x) = f(x)g(x)$; therefore, fg is even.
2. The average value is the constant; the average value is 0.
3. The average value is zero on the interval; by the Mean Value Theorem for Integrals, $f(x) = 0$ at some point on the interval. ◄

5.5 Substitution Rule

Given just about any differentiable function, with enough know-how and persistence, you can compute its derivative. But the same cannot be said of antiderivatives. Many functions, even relatively simple ones, do not have antiderivatives that can be expressed in terms of familiar functions. Examples are $\sin(x^2)$, $(\sin x)/x$, and x^x. At the moment, the number of functions for which we can find antiderivatives is extremely limited. The immediate goal of this section is to enlarge the family of functions for which we can find antiderivatives. This campaign resumes in Chapter 7, where additional integration methods are developed.

Indefinite Integrals

One way to find new antiderivative rules is to start with familiar derivative rules and work backward. When applied to the Chain Rule, this strategy leads to the Substitution Rule. A few examples illustrate the technique.

EXAMPLE 1   Antiderivatives by trial and error Find $\int \cos 2x\, dx$.

SOLUTION The closest familiar indefinite integral related to this problem is

> We assume C is an arbitrary constant without stating so each time it appears.

$$\int \cos x\, dx = \sin x + C,$$

which is true because

$$\frac{d}{dx}(\sin x + C) = \cos x.$$

Therefore, we might *incorrectly* conclude that the indefinite integral of $\cos 2x$ is $\sin 2x + C$. However, by the Chain Rule,

$$\frac{d}{dx}(\sin 2x + C) = 2\cos 2x \neq \cos 2x.$$

Note that $\sin 2x$ fails to be an antiderivative of $\cos 2x$ by a multiplicative factor of 2. A small adjustment corrects this problem. Let's try $\frac{1}{2}\sin 2x$:

$$\frac{d}{dx}\left(\frac{1}{2}\sin 2x\right) = \frac{1}{2}\cdot 2\cos 2x = \cos 2x.$$

It works! So we have

$$\int \cos 2x \, dx = \frac{1}{2}\sin 2x + C.$$

Related Exercises 9–12 ◀

The trial-and-error approach of Example 1 does not work for complicated integrals. To develop a systematic method, consider a composite function $F(g(x))$, where F is an antiderivative of f; that is, $F' = f$. Using the Chain Rule to differentiate the composite function $F(g(x))$, we find that

$$\frac{d}{dx}[F(g(x))] = \underbrace{F'(g(x))}_{f(g(x))}g'(x) = f(g(x))g'(x).$$

This equation says that $F(g(x))$ is an antiderivative of $f(g(x))g'(x)$, which is written

$$\int f(g(x))g'(x)\,dx = F(g(x)) + C, \tag{1}$$

where F is any antiderivative of f.

Why is this approach called the *Substitution Rule* (or *Change of Variables Rule*)? In the composite function $f(g(x))$ in equation (1), we identify the "inner function" as $u = g(x)$, which implies that $du = g'(x)\,dx$. Making this identification, the integral in equation (1) is written

$$\int \underbrace{f(g(x))}_{f(u)}\underbrace{g'(x)dx}_{du} = \int f(u)\,du = F(u) + C.$$

> You can call the new variable anything you want because it is just another variable of integration. Typically, u is a standard choice for the new variable.

We see that the integral $\int f(g(x))g'(x)\,dx$ with respect to x is replaced by a new integral $\int f(u)\,du$ with respect to the new variable u. In other words, we have substituted the new variable u for the old variable x. Of course, if the new integral with respect to u is no easier to find than the original integral, then the change of variables has not helped. The Substitution Rule requires some practice until certain patterns become familiar.

THEOREM 5.6   Substitution Rule for Indefinite Integrals
Let $u = g(x)$, where g' is continuous on an interval, and let f be continuous on the corresponding range of g. On that interval,

$$\int f(g(x))g'(x)\,dx = \int f(u)\,du.$$

PROCEDURE   **Substitution Rule (Change of Variables)**

1. Given an indefinite integral involving a composite function $f(g(x))$, identify an inner function $u = g(x)$ such that a constant multiple of $g'(x)$ (equivalently, $u'(x)$) appears in the integrand.

2. Substitute $u = g(x)$ and $du = g'(x)\,dx$ in the integral.

3. Evaluate the new indefinite integral with respect to u.

4. Write the result in terms of x using $u = g(x)$.

Disclaimer: Not all integrals yield to the Substitution Rule.

EXAMPLE 2   **Perfect substitutions** Use the Substitution Rule to find the following indefinite integrals. Check your work by differentiating.

a. $\displaystyle\int 2(2x + 1)^3 \, dx$ **b.** $\displaystyle\int 10e^{10x} \, dx$

SOLUTION

a. We identify $u = 2x + 1$ as the inner function of the composite function $(2x + 1)^3$. Therefore, we choose the new variable $u = 2x + 1$, which implies that $du = 2\,dx$. Notice that $du = 2\,dx$ appears as a factor in the integrand. The change of variables looks like this:

$$\int \underbrace{(2x + 1)^3}_{u^3} \cdot \underbrace{2\,dx}_{du} = \int u^3 \, du \qquad \text{Substitute } u = 2x + 1, du = 2\,dx.$$

$$= \frac{u^4}{4} + C \qquad \text{Antiderivative}$$

$$= \frac{(2x + 1)^4}{4} + C \quad \text{Replace } u \text{ by } 2x + 1.$$

Notice that the final step uses $u = 2x + 1$ to return to the original variable.

> It is a good idea to check the result. By the Chain Rule, we have
> $$\frac{d}{dx}\left[\frac{(2x + 1)^4}{4} + C\right] = 2(2x + 1)^3.$$

b. The composite function e^{10x} has the inner function $u = 10x$, which implies that $du = 10\,dx$. The change of variables appears as

$$\int \underbrace{e^{10x}}_{e^u} \underbrace{10\,dx}_{du} = \int e^u \, du \qquad \text{Substitute } u = 10x, du = 10\,dx.$$

$$= e^u + C \qquad \text{Antiderivative}$$

$$= e^{10x} + C. \quad \text{Replace } u \text{ by } 10x.$$

In checking, we see that $\dfrac{d}{dx}(e^{10x} + C) = e^{10x} \cdot 10 = 10e^{10x}$.

Related Exercises 13–16 ◄

QUICK CHECK 1 Find a new variable u so that $\int 4x^3(x^4 + 5)^{10}\,dx = \int u^{10}\,du.$ ◄

EXAMPLE 3   **Introducing a constant** Find the following indefinite integrals.

a. $\int x^4(x^5 + 6)^9 \, dx$ **b.** $\int \cos^3 x \sin x \, dx$

SOLUTION

a. The inner function of the composite function $(x^5 + 6)^9$ is $x^5 + 6$ and its derivative $5x^4$ also appears in the integrand (up to a multiplicative factor). Therefore, we use the substitution $u = x^5 + 6$, which implies that $du = 5x^4 \, dx$ or $x^4 \, dx = 1/5 \, du$. By the Substitution Rule,

$$\int \underbrace{(x^5 + 6)^9}_{u^9} \underbrace{x^4 \, dx}_{\frac{1}{5} du} = \int u^9 \frac{1}{5} \, du \qquad \begin{array}{l} \text{Substitute } u = x^5 + 6, \\[4pt] du = 5x^4 \, dx \Rightarrow x^4 \, dx = \dfrac{1}{5} \, du \end{array}$$

$$= \frac{1}{5} \int u^9 \, du \qquad \int c \, f(x) \, dx = c \int f(x) \, dx$$

$$= \frac{1}{5} \cdot \frac{u^{10}}{10} + C \qquad \text{Antiderivative}$$

$$= \frac{1}{50}(x^5 + 6)^{10} + C. \quad \text{Replace } u \text{ by } x^5 + 6.$$

b. The integrand can be written as $(\cos x)^3 \sin x$. The inner function in the composition is $\cos x$, which suggests the substitution $u = \cos x$. Note that $du = -\sin x \, dx$ or $\sin x \, dx = -du$. The change of variables appears as

$$\int \underbrace{\cos^3 x}_{u^3} \underbrace{\sin x \, dx}_{-du} = -\int u^3 \, du \qquad \text{Substitute } u = \cos x, du = -\sin x \, dx.$$

$$= -\frac{u^4}{4} + C \qquad \text{Antiderivative}$$

$$= -\frac{\cos^4 x}{4} + C. \quad \text{Replace } u \text{ by } \cos x.$$

Related Exercises 17–28 ◄

QUICK CHECK 2 In Example 3a, explain why the same substitution would not work as well for the integral $\int x^3 (x^5 + 6)^9 \, dx$. ◄

Sometimes the choice for a u-substitution is not so obvious *or* more than one u-substitution works. The following example illustrates both of these points.

EXAMPLE 4 Variations on the substitution method Find $\displaystyle\int \frac{x}{\sqrt{x + 1}} \, dx$.

SOLUTION

Substitution 1 The composite function $\sqrt{x + 1}$ suggests the new variable $u = x + 1$. You might doubt whether this choice will work because $du = dx$ and the x in the numerator of the integrand is unaccounted for. But let's proceed. Letting $u = x + 1$, we have $x = u - 1, du = dx$, and

$$\int \frac{x}{\sqrt{x + 1}} \, dx = \int \frac{u - 1}{\sqrt{u}} \, du \qquad \text{Substitute } u = x + 1, du = dx.$$

$$= \int \left(\sqrt{u} - \frac{1}{\sqrt{u}} \right) du \quad \text{Rewrite integrand.}$$

$$= \int (u^{1/2} - u^{-1/2}) \, du. \quad \text{Fractional powers}$$

We integrate each term individually and then return to the original variable x:

$$\int (u^{1/2} - u^{-1/2})\, du = \frac{2}{3} u^{3/2} - 2u^{1/2} + C \qquad \text{Antiderivatives}$$

$$= \frac{2}{3}(x+1)^{3/2} - 2(x+1)^{1/2} + C \quad \text{Replace } u \text{ by } x + 1.$$

$$= \frac{2}{3}(x+1)^{1/2}(x-2) + C \qquad \begin{array}{l}\text{Factor out } (x+1)^{1/2} \text{ and}\\ \text{simplify.}\end{array}$$

Substitution 2 Another possible substitution is $u = \sqrt{x+1}$. Now $u^2 = x + 1$, $x = u^2 - 1$, and $dx = 2u\, du$. Making these substitutions leads to

$$\int \frac{x}{\sqrt{x+1}}\, dx = \int \frac{u^2-1}{u} 2u\, du \qquad \text{Substitute } u = \sqrt{x+1}, x = u^2 - 1.$$

$$= 2\int (u^2 - 1)\, du \qquad \text{Simplify the integrand.}$$

$$= 2\left(\frac{u^3}{3} - u\right) + C \qquad \text{Antiderivatives}$$

$$= \frac{2}{3}(x+1)^{3/2} - 2(x+1)^{1/2} + C \quad \text{Replace } u \text{ by } \sqrt{x+1}.$$

$$= \frac{2}{3}(x+1)^{1/2}(x-2) + C. \qquad \text{Factor out } (x+1)^{1/2} \text{ and simplify.}$$

The same indefinite integral is found using either substitution.   *Related Exercises 29–34* ◄

> In Substitution 2, you could also use the fact that
> $$u'(x) = \frac{1}{2\sqrt{x+1}},$$
> which implies
> $$du = \frac{1}{2\sqrt{x+1}}\, dx.$$

Definite Integrals

The Substitution Rule is also used for definite integrals; in fact, there are two ways to proceed.

- You may use the Substitution Rule to find an antiderivative F, and then use the Fundamental Theorem to evaluate $F(b) - F(a)$.

- Alternatively, once you have changed variables from x to u, you may also change the limits of integration and complete the integration with respect to u. Specifically, if $u = g(x)$, the lower limit $x = a$ is replaced by $u = g(a)$ and the upper limit $x = b$ is replaced by $u = g(b)$.

The second option tends to be more efficient, and we use it whenever possible. A few examples illustrate this idea.

THEOREM 5.7   Substitution Rule for Definite Integrals
Let $u = g(x)$, where g' is continuous on $[a, b]$, and let f be continuous on the range of g. Then

$$\int_a^b f(g(x))g'(x)\, dx = \int_{g(a)}^{g(b)} f(u)\, du.$$

EXAMPLE 5   Definite integrals Evaluate the following integrals.

a. $\displaystyle\int_0^2 \frac{dx}{(x+3)^3}$
b. $\displaystyle\int_0^4 \frac{x}{x^2+1}\, dx$
c. $\displaystyle\int_0^{\pi/2} \sin^4 x \cos x\, dx$

SOLUTION

> When the integrand has the form $f(ax + b)$, the substitution $u = ax + b$ is often effective.

a. Let the new variable be $u = x + 3$ and then $du = dx$. Because we have changed the variable of integration from x to u, the limits of integration must also be expressed in terms of u. In this case,

$$x = 0 \text{ implies } u = 0 + 3 = 3 \quad \text{Lower limit}$$
$$x = 2 \text{ implies } u = 2 + 3 = 5 \quad \text{Upper limit}$$

The entire integration is carried out as follows:

$$\int_0^2 \frac{dx}{(x + 3)^3} = \int_3^5 u^{-3}\, du \qquad \text{Substitute } u = x + 3, du = dx.$$

$$= -\frac{u^{-2}}{2}\bigg|_3^5 \qquad \text{Fundamental Theorem}$$

$$= -\frac{1}{2}(5^{-2} - 3^{-2}) = \frac{8}{225} \quad \text{Simplify.}$$

b. Notice that a multiple of the derivative of the denominator appears in the numerator; therefore, we let $u = x^2 + 1$. Then $du = 2x\, dx$, or $x\, dx = \frac{1}{2}\, du$. Changing limits of integration,

$$x = 0 \text{ implies } u = 0 + 1 = 1 \qquad \text{Lower limit}$$
$$x = 4 \text{ implies } u = 4^2 + 1 = 17 \quad \text{Upper limit}$$

Changing variables, we have

$$\int_0^4 \frac{x}{x^2 + 1}\, dx = \frac{1}{2}\int_1^{17} u^{-1}\, du \qquad \text{Substitute } u = x^2 + 1, du = 2x\, dx.$$

$$= \frac{1}{2}(\ln|u|)\bigg|_1^{17} \qquad \text{Fundamental Theorem}$$

$$= \frac{1}{2}(\ln 17 - \ln 1) \quad \text{Simplify.}$$

$$= \frac{1}{2}\ln 17 \approx 1.417. \quad \ln 1 = 0$$

c. Let $u = \sin x$, which implies that $du = \cos x\, dx$. The lower limit of integration becomes $u = 0$ and the upper limit becomes $u = 1$. Changing variables, we have

$$\int_0^{\pi/2} \sin^4 x \cos x\, dx = \int_0^1 u^4\, du \qquad u = \sin x, du = \cos x\, dx$$

$$= \left(\frac{u^5}{5}\right)\bigg|_0^1 = \frac{1}{5}. \quad \text{Fundamental Theorem}$$

Related Exercises 35–44 ◄

The Substitution Rule enables us to find two standard integrals that appear frequently in practice, $\int \sin^2 x\, dx$ and $\int \cos^2 x\, dx$. These integrals are handled using the identities

$$\sin^2 x = \frac{1 - \cos 2x}{2} \quad \text{and} \quad \cos^2 x = \frac{1 + \cos 2x}{2}.$$

EXAMPLE 6   Integral of $\cos^2 \theta$ Evaluate $\int_0^{\pi/2} \cos^2 \theta\, d\theta$.

SOLUTION Working with the indefinite integral first, we use the identity for $\cos^2 \theta$:

$$\int \cos^2 \theta\, d\theta = \int \frac{1 + \cos 2\theta}{2}\, d\theta = \frac{1}{2}\int d\theta + \frac{1}{2}\int \cos 2\theta\, d\theta.$$

The change of variables $u = 2\theta$ is now used for the second integral, and we have

$$\int \cos^2 \theta \, d\theta = \frac{1}{2} \int d\theta + \frac{1}{2} \int \cos 2\theta \, d\theta$$

$$= \frac{1}{2} \int d\theta + \frac{1}{2} \cdot \frac{1}{2} \int \cos u \, du \quad u = 2\theta, du = 2 \, d\theta$$

$$= \frac{\theta}{2} + \frac{1}{4} \sin 2\theta + C. \qquad \text{Evaluate integrals; } u = 2\theta.$$

> See Exercise 86 for a generalization of Example 6. Trigonometric integrals involving powers of $\sin x$ and $\cos x$ are explored in greater detail in Section 7.2.

Using the Fundamental Theorem of Calculus, the value of the definite integral is

$$\int_0^{\pi/2} \cos^2 \theta \, d\theta = \left(\frac{\theta}{2} + \frac{1}{4} \sin 2\theta \right) \Big|_0^{\pi/2}$$

$$= \left(\frac{\pi}{4} + \frac{1}{4} \sin \pi \right) - \left(0 + \frac{1}{4} \sin 0 \right) = \frac{\pi}{4}.$$

Related Exercises 45–50 ◄

Geometry of Substitution

The Substitution Rule may be interpreted graphically. To keep matters simple, consider the integral $\int_0^2 2(2x + 1) \, dx$. The graph of the integrand $y = 2(2x + 1)$ on the interval $[0, 2]$ is shown in Figure 5.57, along with the region R whose area is given by the integral. The change of variables $u = 2x + 1, du = 2 \, dx$, $u(0) = 1$, and $u(2) = 5$ leads to the new integral

$$\int_0^2 2(2x + 1) \, dx = \int_1^5 u \, du.$$

Figure 5.57 also shows the graph of the new integrand $y = u$ on the interval $[1, 5]$ and the region R' whose area is given by the new integral. You can check that the areas of R and R' are equal. An analogous interpretation may be given to more complicated integrands and substitutions.

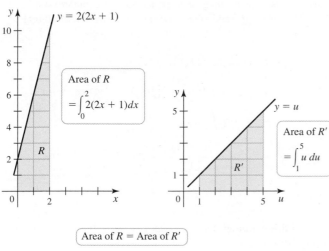

FIGURE 5.57

QUICK CHECK 3 Changes of variables occur frequently in mathematics. For example, suppose you want to solve the equation $x^4 - 13x^2 + 36 = 0$. If you use the substitution $u = x^2$, what is the new equation that must be solved for u? What are the roots of the original equation? ◄

SECTION 5.5 EXERCISES

Review Questions

1. On which derivative rule is the Substitution Rule based?

2. Explain why the Substitution Rule is referred to as a change of variables.

3. The composite function $f(g(x))$ consists of an inner function g and an outer function f. When doing a change of variables, which function is often a likely choice for a new variable u?

4. Find a suitable substitution for evaluating $\int \tan x \sec^2 x \, dx$, and explain your choice.

5. When using a change of variables $u = g(x)$ to evaluate the definite integral $\int_a^b f(g(x))g'(x) \, dx$, how are the limits of integration transformed?

6. If the change of variables $u = x^2 - 4$ is used to evaluate the definite integral $\int_2^4 f(x) \, dx$, what are the new limits of integration?

7. Find $\int \cos^2 x \, dx$.

8. What identity is needed to find $\int \sin^2 x \, dx$?

Basic Skills

9–12. Trial and error *Find an antiderivative of the following functions by trial and error. Check your answer by differentiation.*

9. $f(x) = (x + 1)^{12}$

10. $f(x) = e^{3x+1}$

11. $f(x) = \sqrt{2x + 1}$

12. $f(x) = \cos(2x + 5)$

13–16. Substitution given *Use the given substitution to find the following indefinite integrals. Check your answer by differentiation.*

13. $\displaystyle\int 2x(x^2 + 1)^4 \, dx, \; u = x^2 + 1$

14. $\displaystyle\int 8x \cos(4x^2 + 3) \, dx, \; u = 4x^2 + 3$

15. $\displaystyle\int \sin^3 x \cos x \, dx, \; u = \sin x$

16. $\displaystyle\int (6x + 1)\sqrt{3x^2 + x} \, dx, \; u = 3x^2 + x$

17–28. Indefinite integrals *Use a change of variables to find the following indefinite integrals. Check your work by differentiation.*

17. $\displaystyle\int 2x(x^2 - 1)^{99} \, dx$

18. $\displaystyle\int xe^{x^2} \, dx$

19. $\displaystyle\int \frac{2x^2}{\sqrt{1 - 4x^3}} \, dx$

20. $\displaystyle\int \frac{(\sqrt{x} + 1)^4}{2\sqrt{x}} \, dx$

21. $\displaystyle\int (x^2 + x)^{10} (2x + 1) \, dx$

22. $\displaystyle\int \frac{1}{10x - 3} \, dx$

23. $\displaystyle\int x^3(x^4 + 16)^6 \, dx$

24. $\displaystyle\int \sin^{10} \theta \cos \theta \, d\theta$

25. $\displaystyle\int \frac{1}{\sqrt{1 - 9x^2}} \, dx$

26. $\displaystyle\int x^9 \sin x^{10} \, dx$

27. $\displaystyle\int (x^6 - 3x^2)^4 (x^5 - x) \, dx$

28. $\displaystyle\int \frac{x}{x - 2} \, dx$ (*Hint: Let $u = x - 2$*)

29–34. Variations on the substitution method *Find the following integrals.*

29. $\displaystyle\int \frac{x}{\sqrt{x - 4}} \, dx$

30. $\displaystyle\int \frac{y^2}{(y + 1)^4} \, dy$

31. $\displaystyle\int \frac{x}{\sqrt[3]{x + 4}} \, dx$

32. $\displaystyle\int \frac{e^x - e^{-x}}{e^x + e^{-x}} \, dx$

33. $\displaystyle\int x\sqrt[3]{2x + 1} \, dx$

34. $\displaystyle\int (x + 1)\sqrt{3x + 2} \, dx$

35–44. Definite integrals *Use a change of variables to evaluate the following definite integrals.*

35. $\displaystyle\int_0^1 2x(4 - x^2) \, dx$

36. $\displaystyle\int_0^2 \frac{2x}{(x^2 + 1)^2} \, dx$

37. $\displaystyle\int_0^{\pi/2} \sin^2 \theta \cos \theta \, d\theta$

38. $\displaystyle\int_0^{\pi/4} \frac{\sin x}{\cos^2 x} \, dx$

39. $\displaystyle\int_{-1}^2 x^2 e^{x^3 + 1} \, dx$

40. $\displaystyle\int_0^4 \frac{p}{\sqrt{9 + p^2}} \, dp$

41. $\displaystyle\int_{\pi/4}^{\pi/2} \frac{\cos x}{\sin^2 x} \, dx$

42. $\displaystyle\int_0^{\pi/4} \frac{\sin x}{\cos^3 x} \, dx$

43. $\displaystyle\int_{2/(5\sqrt{3})}^{2/5} \frac{dx}{x\sqrt{25x^2 - 1}}$

44. $\displaystyle\int_0^3 \frac{v^2 + 1}{\sqrt{v^3 + 3v + 4}} \, dv$

45–50. Integrals with $\sin^2 x$ and $\cos^2 x$ *Evaluate the following integrals.*

45. $\displaystyle\int_{-\pi}^{\pi} \cos^2 x \, dx$

46. $\displaystyle\int \sin^2 x \, dx$

47. $\displaystyle\int \sin^2\left(\theta + \frac{\pi}{6}\right) d\theta$

48. $\displaystyle\int_0^{\pi/4} \cos^2 8\theta \, d\theta$

49. $\displaystyle\int_{-\pi/4}^{\pi/4} \sin^2 2\theta \, d\theta$

50. $\displaystyle\int x \cos^2(x^2) \, dx$

Further Explorations

51. Explain why or why not Determine whether the following statements are true and give an explanation or counterexample. Assume that f, f', and f'' are continuous functions for all real numbers.

a. $\displaystyle\int f(x)f'(x) \, dx = \frac{1}{2}(f(x))^2 + C$

b. $\displaystyle\int (f(x))^n \, f'(x) \, dx = \frac{1}{n + 1}(f(x))^{n+1} + C, \; n \neq -1$

c. $\displaystyle\int \sin 2x \, dx = 2 \int \sin x \, dx$

d. $\displaystyle\int (x^2 + 1)^9 \, dx = \frac{(x^2 + 1)^{10}}{10} + C$

e. $\displaystyle\int_a^b f'(x)f''(x) \, dx = f'(b) - f'(a)$

52–64. Additional integrals *Use a change of variables to evaluate the following integrals.*

52. $\displaystyle\int \sec 4w \tan 4w \, dw$

53. $\displaystyle\int \sec^2 10x \, dx$

54. $\displaystyle\int (\sin^5 x + 3\sin^3 x - \sin x) \cos x \, dx$

55. $\displaystyle\int \frac{\csc^2 x}{\cot^3 x} \, dx$

56. $\displaystyle\int (x^{3/2} + 8)^5 \sqrt{x} \, dx$

57. $\displaystyle\int \sin x \sec^8 x \, dx$

58. $\displaystyle\int \frac{e^{2x}}{e^{2x} + 1} \, dx$

59. $\displaystyle\int_0^1 x\sqrt{1 - x^2} \, dx$

60. $\displaystyle\int_1^{e^2} \frac{\ln x}{x} \, dx$

61. $\displaystyle\int_2^3 \frac{x}{\sqrt[3]{x^2 - 1}} \, dx$

62. $\displaystyle\int_0^6 \frac{dx}{x^2 + 36}$

63. $\displaystyle\int_0^2 x^3 \sqrt{16 - x^4} \, dx$

64. $\displaystyle\int_{\sqrt{2}}^{\sqrt{3}} (x - 1)(x^2 - 2x)^{11} \, dx$

65–68. Areas of regions *Find the area of the following regions.*

65. The region bounded by the graph of $f(x) = x \sin(x^2)$ and the x-axis between $x = 0$ and $x = \sqrt{\pi}$

66. The region bounded by the graph of $f(\theta) = \cos\theta \sin\theta$ and the θ-axis between $\theta = 0$ and $\theta = \pi/2$

67. The region bounded by the graph of $f(x) = (x - 4)^4$ and the x-axis between $x = 2$ and $x = 6$

68. The region bounded by the graph of $f(x) = \dfrac{x}{\sqrt{x^2 - 9}}$ and the x-axis between $x = 4$ and $x = 5$

69. Morphing parabolas The family of parabolas $y = (1/a) - x^2/a^3$, where $a > 0$, has the property that for $x \geq 0$, the x-intercept is $(a, 0)$ and the y-intercept is $(0, 1/a)$. Let $A(a)$ be the area of the region in the first quadrant bounded by the parabola and the x-axis. Find $A(a)$ and determine whether it is an increasing, decreasing, or constant function of a.

Applications

▣ 70. Periodic motion An object moves in one dimension with a velocity in m/s given by $v(t) = 8\cos(\pi t/6)$.

    a. Graph the velocity function.

    b. As will be discussed in Chapter 6, the position of the object is given by $s(t) = \int_0^t v(y)\, dy$ for $t \geq 0$. Find the position function for all $t \geq 0$.

    c. What is the period of the motion—that is, starting at any point, how long does it take the object to return to that position?

71. Population models The population of a culture of bacteria has a growth rate given by $p'(t) = \dfrac{200}{(t + 1)^r}$ bacteria per hour, for $t \geq 0$, where $r > 1$ is a real number. In Chapter 6 it will be shown that the increase in the population over the time interval $[0, t]$ is given by $\int_0^t p'(s)\, ds$. (Note that the growth rate decreases in time, reflecting competition for space and food.)

    a. Using the population model with $r = 2$, what is the increase in the population over the time interval $0 \leq t \leq 4$?

    b. Using the population model with $r = 3$, what is the increase in the population over the time interval $0 \leq t \leq 6$?

    c. Let ΔP be the increase in the population over a fixed time interval $[0, T]$. For fixed T, does ΔP increase or decrease with the parameter r? Explain.

    d. A lab technician measures an increase in the population of 350 bacteria over the 10-hr period $[0, 10]$. Estimate the value of r that best fits this data point.

    e. Looking ahead: Work with the population model using $r = 3$ (part (b)) and find the increase in population over the time interval $[0, T]$ for any $T > 0$. If the culture is allowed to grow indefinitely $(T \to \infty)$, does the bacteria population increase without bound? Or does it approach a finite limit?

72. Consider the right triangle with vertices $(0, 0)$, $(0, b)$, and $(a, 0)$, where $a > 0$ and $b > 0$. Show that the average vertical distance from points on the x-axis to the hypotenuse is $b/2$ for all $a > 0$.

▣ 73. Average value of sine functions Use a graphing utility to verify that the functions $f(x) = \sin kx$ have a period of $2\pi/k$, where $k = 1, 2, 3, \ldots$. Equivalently, the first "hump" of $f(x) = \sin kx$ occurs on the interval $[0, \pi/k]$. Verify that the average value of the first hump of $f(x) = \sin kx$ is independent of k. What is the average value? (See Section 5.4 for average value.)

Additional Exercises

74. Looking ahead Integrals of $\tan x$ and $\cot x$

    a. Use a change of variables to show that

$$\int \tan x\, dx = -\ln|\cos x| + C = \ln|\sec x| + C.$$

    b. Show that

$$\int \cot x\, dx = \ln|\sin x| + C.$$

75. Looking ahead Integrals of $\sec x$ and $\csc x$

    a. Multiply the numerator and denominator of $\sec x$ by $\sec x + \tan x$; then use a change of variables to show that

$$\int \sec x\, dx = \ln|\sec x + \tan x| + C.$$

    b. Show that

$$\int \csc x\, dx = -\ln|\csc x + \cot x| + C.$$

76. Equal areas The area of the shaded region under the curve $y = 2\sin 2x$ in (a) equals the area of the shaded region under the curve $y = \sin x$ in (b). Explain why this is true without computing areas.

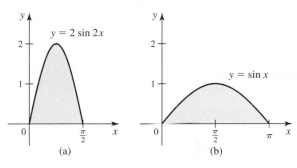

(a)   (b)

77. Equal areas The area of the shaded region under the curve $y = \dfrac{(\sqrt{x} - 1)^2}{2\sqrt{x}}$ in (a) equals the area of the shaded region under the curve $y = x^2$ in (b). Without computing areas, explain why.

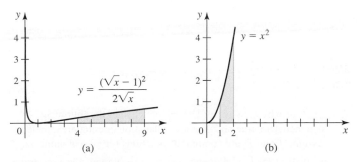

(a)   (b)

78–82. General results *Evaluate the following integrals in which the function f is unspecified. Note $f^{(p)}$ is the pth derivative of f and f^p is the pth power of f. Assume f and its derivatives are continuous for all real numbers.*

78. $\displaystyle \int (5f^3(x) + 7f^2(x) + f(x))f'(x)\, dx$

79. $\displaystyle \int_1^2 (5f^3(x) + 7f^2(x) + f(x))f'(x)\, dx,$

    where $f(1) = 4$, $f(2) = 5$

80. $\int_0^1 f'(x)f''(x)\,dx$, where $f'(0) = 3$ and $f'(1) = 2$

81. $\int (f^{(p)}(x))^n f^{(p+1)}(x)\,dx$, where p is a positive integer, $n \neq -1$

82. $\int 2(f^2(x) + 2f(x))f(x)f'(x)\,dx$

83–85. More than one way *Occasionally, two different substitutions do the job. Use both of the given substitutions to evaluate the following integrals.*

83. $\int_0^1 x\sqrt{x + a}\,dx;\ a > 0 \qquad (u = \sqrt{x + a} \text{ and } u = x + a)$

84. $\int_0^1 x\sqrt[p]{x + a}\,dx;\ a > 0 \qquad (u = \sqrt[p]{x + a} \text{ and } u = x + a)$

85. $\int \sec^3 \theta \tan \theta\,d\theta \qquad (u = \cos \theta \text{ and } u = \sec \theta)$

86. $\sin^2 ax$ and $\cos^2 ax$ integrals Use the Substitution Rule to prove that

$$\int \sin^2 ax\,dx = \frac{x}{2} - \frac{\sin(2ax)}{4a} + C \quad \text{and}$$

$$\int \cos^2 ax\,dx = \frac{x}{2} + \frac{\sin(2ax)}{4a} + C$$

87. Integral of $\sin^2 x \cos^2 x$ Consider the integral

$$I = \int \sin^2 x \cos^2 x\,dx.$$

    a. Find I using the identity $\sin 2x = 2 \sin x \cos x$.
    b. Find I using the identity $\cos^2 x = 1 - \sin^2 x$.
    c. Confirm that the results in parts (a) and (b) are consistent and compare the work involved in each method.

88. Substitution: shift Perhaps the simplest change of variables is the shift or translation given by $u = x + c$, where c is a real number.

a. Prove that shifting a function does not change the net area under the curve, in the sense that

$$\int_a^b f(x + c)\,dx = \int_{a+c}^{b+c} f(u)\,du.$$

b. Draw a picture to illustrate this change of variables in the case that $f(x) = \sin x, a = 0, b = \pi, c = \pi/2$.

89. Substitution: scaling Another change of variables that can be interpreted geometrically is the scaling $u = cx$, where c is a real number. Prove and interpret the fact that

$$\int_a^b f(cx)\,dx = \frac{1}{c}\int_{ac}^{bc} f(u)\,du.$$

Draw a picture to illustrate this change of variables in the case that $f(x) = \sin x, a = 0, b = \pi, c = \frac{1}{2}$.

90–93. Multiple substitutions *Use two or more substitutions to find the following integrals.*

90. $\int x \sin^4(x^2) \cos(x^2)\,dx$

    (*Hint:* Begin with $u = x^2$, then use $v = \sin u$.)

91. $\int \dfrac{dx}{\sqrt{1 + \sqrt{1 + x}}} \qquad \left(\text{\textit{Hint:} Begin with } u = \sqrt{1 + x}.\right)$

92. $\int \tan^{10}(4x) \sec^2(4x)\,dx \quad$ (*Hint:* Begin with $u = 4x$.)

93. $\int_0^{\pi/2} \dfrac{\cos \theta \sin \theta}{\sqrt{\cos^2 \theta + 16}}\,d\theta \quad$ (*Hint:* Begin with $u = \cos \theta$.)

QUICK CHECK ANSWERS

1. $u = x^4 + 5$ **2.** With $u = x^5 + 6$, we have $du = 5x^4$, and x^4 does not appear in the integrand. **3.** New equation: $u^2 - 13u + 36 = 0$; roots: $x = \pm 2, \pm 3$ ◄

CHAPTER 5 REVIEW EXERCISES

1. **Explain why or why not** Determine whether the following statements are true and give an explanation or counterexample. Assume f and f' are continuous functions for all real numbers.

    a. If $A(x) = \int_a^x f(t)\,dt$ and $f(t) = 2t - 3$, then A is a quadratic function.

    b. Given an area function $A(x) = \int_a^x f(t)\,dt$ and an antiderivative F of f, it follows that $A'(x) = F(x)$.

    c. $\int_a^b f'(x)\,dx = f(b) - f(a)$

    d. If $\int_a^b |f(x)|\,dx = 0$, then $f(x) = 0$ on $[a, b]$.

    e. If the average value of f on $[a, b]$ is zero, then $f(x) = 0$ on $[a, b]$.

    f. $\int_a^b (2f(x) - 3g(x))\,dx = 2\int_a^b f(x)\,dx + 3\int_b^a g(x)\,dx$

    g. $\int f'(g(x))g'(x)\,dx = f(g(x)) + C$

2. **Velocity to displacement** An object travels on the x-axis with a velocity given by $v(t) = 2t + 5$, for $0 \le t \le 4$.

    a. How far does the object travel for $0 \le t \le 4$?
    b. What is the average value of v on the interval $[0, 4]$?
    c. True or false: The object would travel as far as in part (a) if it traveled at its average velocity (a constant) for $0 \le t \le 4$.

3. Area by geometry Use geometry to evaluate $\int_0^7 f(x)\,dx$, where the graph of f is given in the figure.

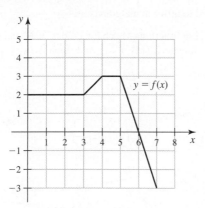

4. Displacement by geometry Use geometry to find the displacement of an object moving along a line for $0 \le t \le 8$, where the graph of its velocity $v = g(t)$ is given in the figure.

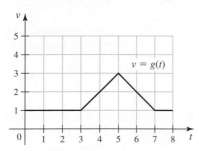

5. Area by geometry Use geometry to evaluate $\int_0^4 \sqrt{8x - x^2}\,dx$ (*Hint:* Complete the square of $8x - x^2$ first).

6. Bagel output The manager of a bagel bakery collects the following production rate data (in bagels per minute) at six different times during the morning. Estimate the total number of bagels produced between 6:00 and 7:30 a.m.

Time of day (a.m.)	Production rate (bagels/min)
6:00	45
6:15	60
6:30	75
6:45	60
7:00	50
7:15	40

7. Integration by Riemann sums Consider the integral $\int_1^4 (3x - 2)\,dx$.
   a. Give the right Riemann sum for the integral with $n = 3$.
   b. Use summation notation to write the right Riemann sum for an arbitrary positive integer n.
   c. Evaluate the definite integral by taking the limit as $n \to \infty$ of the Riemann sum in part (b).

8. Evaluating Riemann sums Consider the function $f(x) = 3x + 4$ on the interval $[3, 7]$. Show that the midpoint Riemann sum with $n = 4$ gives the exact area of the region bounded by the graph.

9. Sum to integral Evaluate the following limit by identifying the integral that it represents:

$$\lim_{n \to \infty} \sum_{k=1}^{n} \left[\left(\frac{4k}{n} \right)^8 + 1 \right] \left(\frac{4}{n} \right).$$

10. Area function by geometry Use geometry to find the area $A(x)$ that is bounded by the graph of $f(t) = 2t - 4$ and the t-axis between the point $(2, 0)$ and the variable point $(x, 0)$, where $x \ge 2$. Verify that $A'(x) = f(x)$.

11–26. Evaluating integrals *Evaluate the following integrals.*

11. $\displaystyle\int_{-2}^{2} (3x^4 - 2x + 1)\,dx$

12. $\displaystyle\int \cos 3x\,dx$

13. $\displaystyle\int_0^2 (x + 1)^3\,dx$

14. $\displaystyle\int_0^1 (4x^{21} - 2x^{16} + 1)\,dx$

15. $\displaystyle\int (9x^8 - 7x^6)\,dx$

16. $\displaystyle\int_{-2}^{2} e^{4x+8}\,dx$

17. $\displaystyle\int_0^1 \sqrt{x}(\sqrt{x} + 1)\,dx$

18. $\displaystyle\int \frac{x^2}{x^3 + 27}\,dx$

19. $\displaystyle\int_0^1 \frac{dx}{\sqrt{4 - x^2}}$

20. $\displaystyle\int y^2 (3y^3 + 1)^4\,dy$

21. $\displaystyle\int_0^3 \frac{x}{\sqrt{25 - x^2}}\,dx$

22. $\displaystyle\int x \sin x^2 \cos^8 x^2\,dx$

23. $\displaystyle\int \sin^2 5\theta\,d\theta$

24. $\displaystyle\int_0^\pi (1 - \cos^2 3\theta)\,d\theta$

25. $\displaystyle\int \frac{x^2 + 2x - 2}{x^3 + 3x^2 - 6x}\,dx$

26. $\displaystyle\int_0^{\ln 2} \frac{e^x}{1 + e^{2x}}\,dx$

27. Symmetry properties Suppose that $\int_0^4 f(x)\,dx = 10$ and $\int_0^4 g(x)\,dx = 20$. Furthermore, suppose that f is an even function and g is an odd function. Evaluate the following integrals.
   a. $\displaystyle\int_{-4}^{4} f(x)\,dx$ **b.** $\displaystyle\int_{-4}^{4} 3g(x)\,dx$ **c.** $\displaystyle\int_{-4}^{4} (4f(x) - 3g(x))\,dx$

28. Properties of integrals The figure shows the areas of regions bounded by the graph of f and the x-axis. Evaluate the following integrals.
   a. $\displaystyle\int_a^c f(x)\,dx$ **b.** $\displaystyle\int_b^d f(x)\,dx$ **c.** $2\displaystyle\int_c^b f(x)\,dx$
   d. $4\displaystyle\int_a^d f(x)\,dx$ **e.** $3\displaystyle\int_a^b f(x)\,dx$ **f.** $2\displaystyle\int_b^d f(x)\,dx$

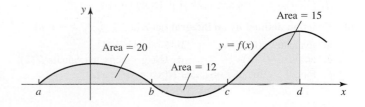

29–34. Properties of integrals *Suppose that* $\int_1^4 f(x)\,dx = 6$, $\int_1^4 g(x)\,dx = 4$, *and* $\int_3^4 f(x)\,dx = 2$. *Evaluate the following integrals or state that there is not enough information.*

29. $\displaystyle\int_1^4 3f(x)\,dx$

30. $\displaystyle -\int_4^1 2f(x)\,dx$

31. $\displaystyle\int_1^4 (3f(x) - 2g(x))\,dx$

32. $\displaystyle\int_1^4 f(x)g(x)\,dx$

33. $\displaystyle\int_1^3 \frac{3f(x)}{g(x)}\,dx$

34. $\displaystyle\int_3^1 (f(x) - g(x))\,dx$

35. Displacement from velocity A particle moves along a line with a velocity given by $v(t) = 5\sin(\pi t)$ starting with an initial position $s(0) = 0$. Find the displacement of the particle between $t = 0$ and $t = 2$, which is given by $s(t) = \int_0^2 v(t)\,dt$. Find the distance traveled by the particle during this interval, which is $\int_0^2 |v(t)|\,dt$.

36. Average height A baseball is launched into the outfield on a parabolic trajectory given by $y = 0.01x(200 - x)$. Find the average height of the baseball during its flight.

37. Average values Find the average value of the functions shown in (a) and (b). Integration is not needed.

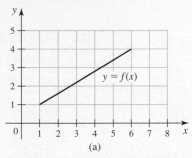

(a)

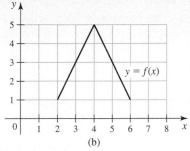

(b)

38. An unknown function The function f satisfies the equation $3x^4 - 2 = \int_a^x f(t)\,dt$. Find f and check your answer by substitution.

39. An unknown function Assume f' is a continuous function, $\int_1^2 f'(2x)\,dx = 10$, and $f(2) = 4$. Evaluate $f(4)$.

40. Function defined by an integral Let $H(x) = \int_0^x \sqrt{4 - t^2}\,dt$.
 a. Evaluate $H(0)$.
 b. Evaluate $H'(1)$.
 c. Evaluate $H'(2)$.
 d. Use geometry to evaluate $H(2)$.
 e. Find the value of s such that $H(x) = sH(-x)$.

41. Function defined by an integral Make a graph of the function $f(x) = \int_1^x \frac{dt}{t}$ for $x \geq 1$. Be sure to include all of the evidence you used to arrive at the graph.

42. Identifying functions Match the graphs A, B, and C in the figure with the functions $f(x)$, $f'(x)$, and $\int_0^x f(t)\,dt$.

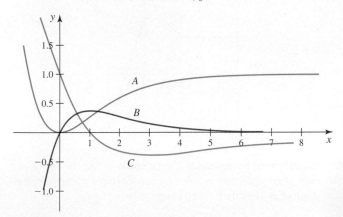

43. Geometry of integrals Without evaluating the integrals, explain why the following statement is true for positive integers n:

$$\int_0^1 x^n\,dx + \int_0^1 \sqrt[n]{x}\,dx = 1.$$

44. Change of variables Use the change of variables $u^3 = x^2 - 1$ to evaluate the integral $\int_1^3 x\,\sqrt[3]{x^2 - 1}\,dx$.

45. Inverse tangent integral Prove that, for nonzero constants a and b,

$$\int \frac{dx}{a^2 x^2 + b^2} = \frac{1}{ab}\tan^{-1}\left(\frac{ax}{b}\right).$$

46–51. Additional integrals *Evaluate the following integrals.*

46. $\displaystyle\int \frac{\sin 2x}{1 + \cos^2 x}\,dx \quad (\sin 2x = 2\sin x \cos x)$

47. $\displaystyle\int \frac{1}{x^2}\sin\left(\frac{1}{x}\right)dx$

48. $\displaystyle\int \frac{(\tan^{-1} x)^5}{1 + x^2}\,dx$

49. $\displaystyle\int \frac{dx}{(\tan^{-1} x)(1 + x^2)}$

50. $\displaystyle\int \frac{\sin^{-1} x}{\sqrt{1 - x^2}}\,dx$

51. $\displaystyle\int \frac{e^x - e^{-x}}{e^x + e^{-x}}\,dx$

52. Area with a parameter Let $a > 0$ and consider the family of functions $f(x) = \sin ax$ on the interval $[0, \pi/a]$.
 a. Graph f for $a = 1, 2, 3$.
 b. Let $g(a)$ be the area of the region bounded by the graph of f and the x-axis on the interval $[0, \pi/a]$. Graph g for $0 < a < \infty$. Is g an increasing function, a decreasing function, or neither?

53. Equivalent equations Explain why a function that satisfies the equation $u(x) + 2\int_0^x u(t)\,dt = 10$ also satisfies the equation $u'(x) + 2u(x) = 0$.

54. Area function properties Consider the function $f(x) = x^2 - 5x + 4$ and the area function $A(x) = \int_0^x f(t)\, dt$.

a. Graph f on the interval $[0, 6]$.
b. Compute and graph A on the interval $[0, 6]$.
c. Show that the local extrema of A occur at the zeros of f.
d. Give a geometrical and analytical explanation for the observation in part (c).
e. Find the approximate zeros of A, other than 0, and call them x_1 and x_2.
f. Find b such that the area bounded by the graph of f and the x-axis on the interval $[0, x_1]$ equals the area bounded by the graph of f and the x-axis on the interval $[x_1, b]$.
g. If f is an integrable function and $A(x) = \int_a^x f(t)\, dt$, is it always true that the local extrema of A occur at the zeros of f? Explain.

55. Function defined by an integral Let
$$f(x) = \int_0^x (t - 1)^{15}(t - 2)^9\, dt.$$

a. Find the intervals on which f is increasing and the intervals on which f is decreasing.
b. Find the intervals on which f is concave up and the intervals on which f is concave down.
c. For what values of x does f have local minima? Local maxima?
d. Where are the inflection points of f?

56. Exponential inequalities Sketch a graph of $f(t) = e^t$ on an arbitrary interval $[a, b]$. Use the graph and compare areas of regions to prove that

$$e^{(a+b)/2} < \frac{e^b - e^a}{b - a} < \frac{e^a + e^b}{2}.$$

(*Source: Mathematics Magazine* **81,** no. 5 (December 2008): 374)

Chapter 5 Guided Projects

Applications of the material in this chapter and related topics can be found in the following Guided Projects. For additional information, see the Preface.

• Limits of sums
• Symmetry in integrals

• Distribution of wealth

6 Applications of Integration

Chapter Preview Now that we have some basic techniques for evaluating integrals, we turn our attention to the uses of integration, which are virtually endless. Some uses of integration are theoretical and some are practical. We first illustrate the general rule that if the rate of change of a quantity is known, then integration can be used to determine the net change or future value of that quantity over a certain time interval. Next, we explore some rich geometric applications of integration: computing the area of regions bounded by several curves, the volume of three-dimensional solids, and the length of curves. A variety of physical applications of integration include finding the work done by a variable force and computing the total force exerted by water behind a dam. All of these applications are unified by their use of the *slice-and-sum* strategy. We end this chapter by revisiting the logarithmic function and exploring the many applications of the exponential function.

6.1 Velocity and Net Change

In previous chapters we established the relationship between the position and velocity of an object moving along a line. With integration, we can now say much more about this relationship. Once we relate velocity and position through integration, we can make analogous observations about a variety of other practical problems, which include fluid flow, population growth, manufacturing costs, and production and consumption of natural resources. The ideas in this section come directly from the Fundamental Theorem of Calculus, and they are among the most powerful applications of calculus.

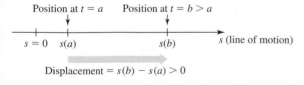

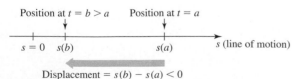

FIGURE 6.1

Velocity, Position, and Displacement

Suppose you are driving along a straight highway and your position relative to a reference point or origin is $s(t)$ for times $t \geq 0$ (Figure 6.1). Your **displacement** over a time interval $[a, b]$ is the *change in the position* between those times, or $s(b) - s(a)$. If $s(b) > s(a)$, then your displacement is positive; when $s(b) < s(a)$, your displacement is negative.

Now assume that $v(t)$ is the velocity of the object at a particular time t. Recall from Chapter 3 that $v(t) = s'(t)$, which means that s is an antiderivative of v. From the Fundamental Theorem of Calculus, it follows that

$$\int_a^b v(t)\, dt = \int_a^b s'(t)\, dt = s(b) - s(a) = \text{displacement}.$$

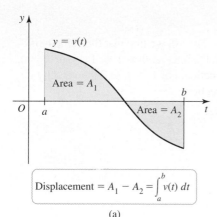

$$\text{Displacement} = A_1 - A_2 = \int_a^b v(t)\, dt$$

(a)

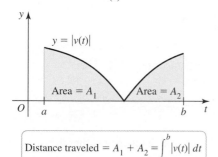

$$\text{Distance traveled} = A_1 + A_2 = \int_a^b |v(t)|\, dt$$

(b)

FIGURE 6.2

We see that the definite integral $\int_a^b v(t)\, dt$ is the displacement (change in position) between times $t = a$ and $t = b$. Equivalently, the displacement over the time interval $[a, b]$ is the net area under the velocity curve over $[a, b]$ (Figure 6.2a).

Not to be confused with the displacement is the **distance traveled** over a time interval, which is the *total distance traveled* by the object, independent of the direction of motion. If the velocity is positive, the object moves in the positive direction and the displacement equals the distance traveled. However, if the velocity changes sign, then the displacement and the distance traveled are not generally equal.

QUICK CHECK 1 A policeman leaves his station at 9 a.m. on a north-south freeway, traveling north (the positive direction) for 40 mi between 9 a.m. and 10 a.m. From 10 a.m. to 11 a.m., he travels south to a point 20 mi south of the station. What is the distance traveled and the displacement between 9:00 a.m. and 11:00 a.m.? ◄

To compute the distance traveled, we need the *magnitude*, but not the *sign* of the velocity. The magnitude of the velocity $|v(t)|$ is called the **speed**. The distance traveled over a small time interval dt is $|v(t)|\, dt$ (speed multiplied by elapsed time). Summing these distances, the distance traveled over the time interval $[a, b]$ is the integral of the speed; that is,

$$\text{distance traveled} = \int_a^b |v(t)|\, dt.$$

As shown in Figure 6.2b, integrating the speed produces the area (not net area) bounded by the velocity curve and the t-axis, which corresponds to the distance traveled. The distance traveled is always nonnegative.

DEFINITIONS   Position, Velocity, Displacement, and Distance

1. The **position** of an object at time t, denoted $s(t)$, is the location of the object relative to the origin.

2. The **velocity** of an object at time t is $v(t) = s'(t)$.

3. The **displacement** of the object between $t = a$ and $t = b > a$ is

$$s(b) - s(a) = \int_a^b v(t)\, dt.$$

4. The **distance traveled** by the object between $t = a$ and $t = b > a$ is

$$\int_a^b |v(t)|\, dt,$$

where $|v(t)|$ is the speed of the object at time t.

QUICK CHECK 2 Describe a possible motion of an object along a line for $0 \le t \le 5$ for which the displacement and the distance traveled are different. ◄

EXAMPLE 1   Displacement from velocity A cyclist pedals along a straight road with velocity $v(t) = 2t^2 - 8t + 6$ mi/hr for $0 \le t \le 3$, where t is measured in hours.

a. Graph the velocity function over the interval $[0, 3]$. Determine when the cyclist moves in the positive direction and when she moves in the negative direction.

b. Find the displacement of the cyclist (in miles) on the time intervals $[0, 1]$, $[1, 3]$, and $[0, 3]$. Interpret these results.

c. Find the distance traveled over the interval $[0, 3]$.

SOLUTION

a. By solving $v(t) = 2t^2 - 8t + 6 = 2(t - 1)(t - 3) = 0$, we find that the velocity is zero at $t = 1$ and $t = 3$. The velocity is positive on the interval $0 \le t < 1$ (Figure 6.3a), which means the cyclist moves in the positive s direction. For $1 < t < 3$ the velocity is negative and the cyclist moves in the negative s direction.

b. The displacement (in miles) over the interval $[0, 1]$ is

$$s(1) - s(0) = \int_0^1 v(t)\, dt$$

$$= \int_0^1 (2t^2 - 8t + 6)\, dt \qquad \text{Substitute for } v.$$

$$= \left(\frac{2}{3}t^3 - 4t^2 + 6t\right)\Big|_0^1 = \frac{8}{3}. \qquad \text{Evaluate integral.}$$

A similar calculation shows that the displacement over the interval $[1, 3]$ is

$$s(3) - s(1) = \int_1^3 v(t)\, dt = -\frac{8}{3}.$$

Over the interval $[0, 3]$, the displacement is $\frac{8}{3} + \left(-\frac{8}{3}\right) = 0$. This means that the cyclist returns to the starting point after three hours.

c. From part (b), we can deduce the total distance traveled by the cyclist. On the interval $[0, 1]$ the distance traveled is $\frac{8}{3}$ mi; on the interval $[1, 3]$, the distance traveled is also $\frac{8}{3}$ mi. Therefore, the distance traveled on $[0, 3]$ is $\frac{16}{3}$ mi. Alternatively (Figure 6.3b), we can integrate the speed and get the same result:

$$\int_0^3 |v(t)|\, dt = \int_0^1 (2t^2 - 8t + 6)\, dt + \int_1^3 (-(2t^2 - 8t + 6))\, dt \quad \text{Definition of } |v(t)|$$

$$= \left(\frac{2}{3}t^3 - 4t^2 + 6t\right)\Big|_0^1 + \left(-\frac{2}{3}t^3 + 4t^2 - 6t\right)\Big|_1^3 \qquad \text{Evaluate integrals.}$$

$$= \frac{16}{3} \qquad \text{Simplify.}$$

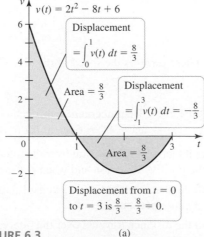

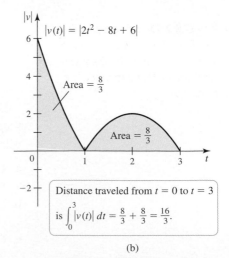

FIGURE 6.3   (a)   (b)

Related Exercises 7–10 ◄

Future Value of the Position Function

To find the displacement of an object, we do not need to know its initial position. For example, whether an object moves from $s = -20$ to $s = -10$ or from $s = 50$ to $s = 60$, its displacement is 10 units. What happens if we are interested in the actual *position* of the object at some future time?

Suppose we know the velocity of an object and its initial position $s(0)$. The goal is to find the position $s(t)$ at some future time $t \geq 0$. The Fundamental Theorem of Calculus gives us the answer directly. Because the velocity v is an antiderivative of the position s, we have

$$\int_0^t v(x)\, dx = \int_0^t s'(x)\, dx = s(x)\Big|_0^t = s(t) - s(0).$$

> Note that t is the independent variable of the position function. Therefore, another (dummy) variable, in this case x, must be used as the variable of integration.

Rearranging this expression, we have the following result.

> Theorem 6.1 is a consequence (actually a restatement) of the Fundamental Theorem of Calculus.

THEOREM 6.1   Position from Velocity

Given the velocity $v(t)$ of an object moving along a line and its initial position $s(0)$, the position function of the object for future times $t \geq 0$ is

$$s(t) = \underbrace{s(0)}_{\substack{\text{position at} \\ \text{time } t}} + \underbrace{\int_0^t v(x)\, dx}_{\substack{\text{displacement} \\ \text{over } [0, t]}}.$$

position at   initial
time t position

Theorem 6.1 says that to find the position $s(t)$, we add the displacement over the interval $[0, t]$ to the initial position $s(0)$.

QUICK CHECK 3 Is the position $s(t)$ a number or a function? For fixed times $t = a$ and $t = b$, is the displacement $s(b) - s(a)$ a number or a function? ◄

There are two *equivalent* ways to determine the position function:

• Using antiderivatives (Section 4.8)

• Using Theorem 6.1

The latter method is usually more efficient, but either method produces the same result. The following example illustrates both approaches.

EXAMPLE 2   Position from velocity A block hangs at rest from a massless spring at the origin ($s = 0$). At $t = 0$, the block is pulled downward $\frac{1}{4}$ m to its initial position $s(0) = -\frac{1}{4}$ and released (Figure 6.4). Its velocity is given by $v(t) = \frac{1}{4}\sin t$ (in m/s) for $t \geq 0$. Assume that the upward direction is positive.

a. Find the position of the block for $t \geq 0$.

b. Graph the position function for $0 \leq t \leq 3\pi$.

c. When does the block move through the origin for the first time?

d. When does the block reach its highest point for the first time and what is its position at that time? When does the block return to its lowest point?

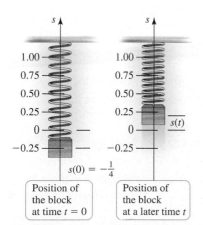

FIGURE 6.4

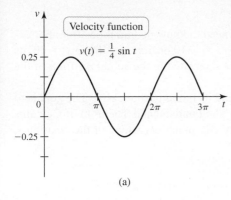

Velocity function

$v(t) = \frac{1}{4}\sin t$

(a)

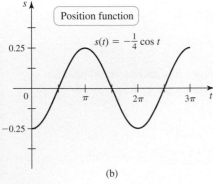

Position function

$s(t) = -\frac{1}{4}\cos t$

(b)

FIGURE 6.5

▷ It is worth repeating that to find the displacement, we need to know only the velocity. To find the position, we must know both the velocity and the initial position $s(0)$.

SOLUTION

a. The velocity function (Figure 6.5a) is positive for $0 < t < \pi$, which means the block moves in the positive (upward) direction. At $t = \pi$, the block comes to rest momentarily; for $\pi < t < 2\pi$, the block moves in the negative (downward) direction. We let $s(t)$ be the position at time $t \geq 0$ with the initial position $s(0) = -\frac{1}{4}$ m.

Method 1: Using antiderivatives Because the position is an antiderivative of the velocity, we have

$$s(t) = \int v(t)\, dt = \int \frac{1}{4}\sin t\, dt = -\frac{1}{4}\cos t + C.$$

To determine the arbitrary constant C, we substitute the initial condition $s(0) = -\frac{1}{4}$ into the expression for $s(t)$:

$$-\frac{1}{4} = -\frac{1}{4}\cos 0 + C$$

Solving for C, we find that $C = 0$. Therefore, the position for any time $t \geq 0$ is

$$s(t) = -\frac{1}{4}\cos t$$

Method 2: Using Theorem 6.1 Alternatively, we may use the relationship

$$s(t) = s(0) + \int_0^t v(x)\, dx.$$

Substituting $v(x) = \frac{1}{4}\sin x$ and $s(0) = -\frac{1}{4}$, the position function is

$$s(t) = \underbrace{-\frac{1}{4}}_{s(0)} + \int_0^t \underbrace{\frac{1}{4}\sin x}_{v(x)}\, dx$$

$$= -\frac{1}{4} - \left(\frac{1}{4}\cos x\right)\Big|_0^t \quad \text{Evaluate integral.}$$

$$= -\frac{1}{4} - \frac{1}{4}(\cos t - 1) \quad \text{Simplify.}$$

$$= -\frac{1}{4}\cos t. \quad \text{Simplify.}$$

b. The graph of the position function is shown in Figure 6.5b. We see that $s(0) = -\frac{1}{4}$ m, as prescribed.

c. The block initially moves in the positive s direction (upward), reaching the origin $(s = 0)$ when $s(t) = -\frac{1}{4}\cos t = 0$. So the block arrives at the origin for the first time when $t = \pi/2$.

d. The block moves in the positive direction and reaches its high point for the first time when $t = \pi$; the position at that moment is $s(\pi) = \frac{1}{4}$ m. The block then reverses direction and moves in the negative (downward) direction, reaching its low point at $t = 2\pi$. This motion repeats every 2π seconds. *Related Exercises 11–18* ◄

QUICK CHECK 4 Without doing further calculations, what are the displacement and distance traveled by the block in Example 2 over the interval $[0, 2\pi]$? ◄

▷ The terminal velocity of an object depends on its density, shape, size, and the medium through which it falls. Estimates for human beings in free fall vary from 120 mi/hr (54 m/s) to 180 mi/hr (80 m/s).

EXAMPLE 3   Skydiving Suppose a skydiver leaps from a hovering helicopter and falls in a straight line. He falls at a terminal velocity of 80 m/s for 19 seconds, at which time he opens his parachute. The velocity decreases linearly to 6 m/s over a two-second period and then remains constant until he reaches the ground at $t = 40$ s. The motion is described by the velocity function

$$v(t) = \begin{cases} 80 & \text{if } 0 \le t < 19 \\ 783 - 37t & \text{if } 19 \le t < 21 \\ 6 & \text{if } 21 \le t \le 40 \end{cases}$$

Determine the altitude from which the skydiver jumped.

SOLUTION We let the position of the skydiver increase *downward* with the origin ($s = 0$) corresponding to the position of the helicopter. The velocity (Figure 6.6) is positive, so the distance traveled by the skydiver equals the displacement, which is

$$\int_0^{40} |v(t)|\, dt = \int_0^{19} 80\, dt + \int_{19}^{21} (783 - 37t)\, dt + \int_{21}^{40} 6\, dt$$

$$= 80t \Big|_0^{19} + \left(783t - \frac{37t^2}{2}\right)\Bigg|_{19}^{21} + 6t \Big|_{21}^{40} \qquad \text{Fundamental Theorem}$$

$$= 1720. \qquad\qquad\qquad\qquad\qquad\qquad\qquad \text{Evaluate and simplify.}$$

The skydiver jumped from 1720 m above the ground. Notice that the displacement of the skydiver is the area under the velocity curve.

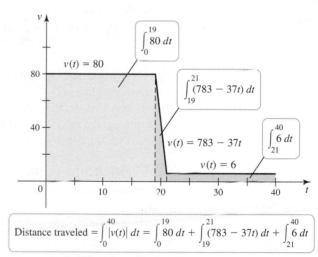

$$\text{Distance traveled} = \int_0^{40} |v(t)|\, dt = \int_0^{19} 80\, dt + \int_{19}^{21} (783 - 37t)\, dt + \int_{21}^{40} 6\, dt$$

FIGURE 6.6

Related Exercises 19–20 ◀

QUICK CHECK 5 Suppose (unrealistically) in Example 3 that the velocity of the skydiver is 80 m/s for $0 < t < 20$ and then it changes instantaneously to 6 m/s for $20 < t < 40$. Sketch the velocity function and, without integrating, find the distance the skydiver falls in 40 s. ◀

Acceleration

Because the acceleration of an object moving along a line is given by $a(t) = v'(t)$, the relationship between velocity and acceleration is the same as the relationship between position and velocity. Given the acceleration of an object, the change in velocity over an interval $[a, b]$ is

$$\text{change in velocity} = v(b) - v(a) = \int_a^b v'(t)\, dt = \int_a^b a(t)\, dt.$$

Furthermore, if we know the acceleration and initial velocity $v(0)$, then the velocity at future times can also be found.

> Theorem 6.2 is a consequence of the Fundamental Theorem of Calculus.

THEOREM 6.2   Velocity from Acceleration

Given the acceleration $a(t)$ of an object moving along a line and its initial velocity $v(0)$, the velocity of the object for future times $t \geq 0$ is

$$v(t) = v(0) + \int_0^t a(x)\, dx.$$

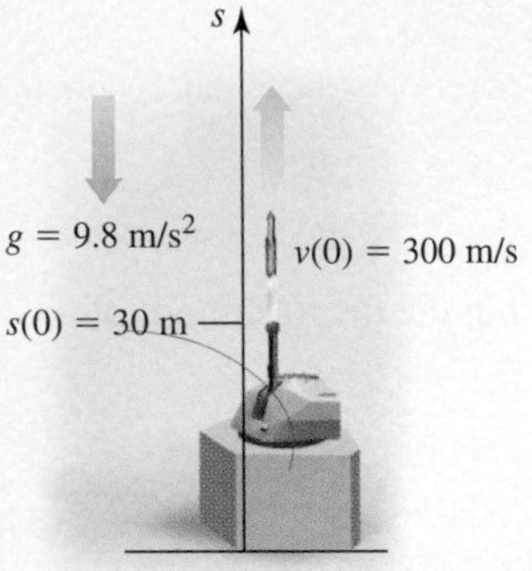

EXAMPLE 4   Motion in a gravitational field An artillery shell is fired directly upward with an initial velocity of 300 m/s from a point 30 m above the ground (Figure 6.7). Assume that only the force of gravity acts on the shell and it produces an acceleration of 9.8 m/s². Find the velocity of the shell for $t \geq 0$.

SOLUTION We let the positive direction be upward with the origin ($s = 0$) corresponding to the ground. The initial velocity of the shell is $v(0) = 300$ m/s. The acceleration due to gravity is downward; therefore, $a(t) = -9.8$ m/s². The velocity for $t \geq 0$ is

$$v(t) = \underbrace{v(0)}_{300\ m/s} + \int_0^t \underbrace{a(x)}_{-9.8\ m/s^2}\, dx = 300 + \int_0^t (-9.8)\, dx = 300 - 9.8t.$$

The velocity decreases from its initial value of 300 m/s, reaching zero at the high point of the trajectory when $v(t) = 300 - 9.8t = 0$, or at $t \approx 30.6$ s (Figure 6.8). At this point the velocity becomes negative, and the shell begins its descent to Earth.

Knowing the velocity function, you could now find the position function using the methods of Example 3.   *Related Exercises 21–27* ◄

Net Change and Future Value

Everything we have said about velocity, position, and displacement carries over to more general situations. Suppose you are interested in some quantity Q that changes over *time*; Q may represent the amount of water in a reservoir, the population of a cell culture, or the amount of a resource that is consumed or produced. If you are given the rate Q' at which Q changes, then integration allows you to calculate either the net change in the quantity Q or the future value of Q.

We argue just as we did for velocity and position: Because $Q(t)$ is an antiderivative of $Q'(t)$, the Fundamental Theorem of Calculus tells us that

$$\int_a^b Q'(t)\, dt = Q(b) - Q(a) = \text{net change in } Q \text{ over } [a, b].$$

Geometrically, the net change in Q over the time interval $[a, b]$ is the net area under the graph of Q' over $[a, b]$.

Alternatively, suppose we are given both the rate of change Q' and the initial value $Q(0)$. Integrating over the interval $[0, t]$, where $t \geq 0$, we have

$$\int_0^t Q'(x)\, dx = Q(t) - Q(0).$$

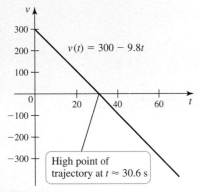

FIGURE 6.7

FIGURE 6.8

For Figure 6.7: $g = 9.8$ m/s² ; $v(0) = 300$ m/s ; $s(0) = 30$ m

For Figure 6.8: $v(t) = 300 - 9.8t$; High point of trajectory at $t \approx 30.6$ s

> Note that the units in the integral are consistent. For example, if Q' has units of gallons/second, and t and x have units of seconds, then $Q'(x)\, dx$ has units of (gallon/second)(seconds) = gal, which are the units of Q.

Rearranging this equation, we write the value of Q at any future time $t \geq 0$ as

$$Q(t) = \underbrace{Q(0)}_{\substack{\text{future} \\ \text{value}}} + \underbrace{Q(0)}_{\substack{\text{initial} \\ \text{value}}} + \underbrace{\int_0^t Q'(x)\,dx.}_{\substack{\text{net change} \\ \text{over }[0,t]}}$$

> At the risk of being repetitious, Theorem 6.3 is also a consequence of the Fundamental Theorem of Calculus. We assume that Q' is an integrable function.

THEOREM 6.3   Net Change and Future Value

Suppose a quantity Q changes over time at a known rate Q'. Then the **net change** in Q between $t = a$ and $t = b$ is

$$\underbrace{Q(b) - Q(a)}_{\text{net change in }Q} = \int_a^b Q'(t)\,dt.$$

Given the initial value $Q(0)$, the **future value** of Q at future times $t \geq 0$ is

$$Q(t) = Q(0) + \int_0^t Q'(x)\,dx.$$

The correspondences between velocity-displacement problems and more general problems are shown in Table 6.1.

Table 6.1

Velocity-Displacement Problems	General Problems
Position $s(t)$	Quantity $Q(t)$ (such as volume or population size)
Velocity: $s'(t) = v(t)$	Rate of change: $Q'(t)$
Displacement: $s(b) - s(a) = \displaystyle\int_a^b v(t)\,dt$	Net change: $Q(b) - Q(a) = \displaystyle\int_a^b Q'(t)\,dt$
Future position: $s(t) = s(0) + \displaystyle\int_0^t v(x)\,dx$	Future value of Q: $Q(t) = Q(0) + \displaystyle\int_0^t Q'(x)\,dx$

EXAMPLE 5   Cell growth A culture of cells in a lab has a population of 100 cells when nutrients are added at time $t = 0$. Suppose the population $N(t)$ increases at a rate given by

$$N'(t) = 90e^{-0.1t} \quad \text{cells/hr.}$$

> Although N is a positive integer (the number of cells), we treat it as a continuous variable in this example.

Find $N(t)$ for $t \geq 0$.

SOLUTION As shown in Figure 6.9, the growth rate is large when t is small (plenty of food and space) and decreases as t increases. Knowing that the initial population is $N(0) = 100$ cells, we can find the population $N(t)$ at any future time $t \geq 0$ using Theorem 6.3:

$$N(t) = N(0) + \int_0^t N'(x)\,dx$$

$$= \underbrace{100}_{N(0)} + \int_0^t \underbrace{90e^{-0.1x}}_{N'(x)}\,dx$$

$$= 100 + \left[\left(\frac{90}{-0.1}\right)e^{-0.1x}\right]\Big|_0^t \quad \text{Fundamental Theorem}$$

$$= 1000 - 900e^{-0.1t} \quad \text{Simplify.}$$

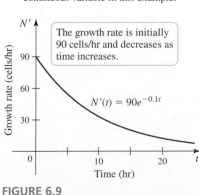

The growth rate is initially 90 cells/hr and decreases as time increases.

$N'(t) = 90e^{-0.1t}$

FIGURE 6.9

The graph of the population function (Figure 6.10) shows that the population increases, but at a decreasing rate. Note that the initial condition $N(0) = 100$ cells is satisfied and that the population size approaches 1000 cells as $t \to \infty$.

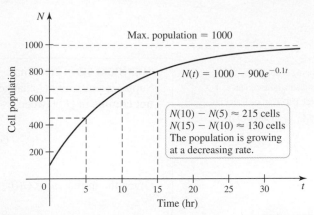

FIGURE 6.10

Related Exercises 28–34 ◄

EXAMPLE 6   Production costs A book publisher estimates that the marginal cost of a particular title (in dollars/book) is given by

$$C'(x) = 12 - 0.0002x,$$

where $0 \le x \le 50{,}000$ is the number of books printed. What is the cost of producing the 12,001st through the 15,000th book?

SOLUTION Recall from Section 3.5 that the cost function $C(x)$ is the cost required to produce x units of a product. The marginal cost $C'(x)$ is the approximate cost of producing one additional unit after x units have already been produced. The cost of producing books $x = 12{,}001$ through $x = 15{,}000$ is the cost of producing 15,000 books minus the cost of producing the first 12,000 books. Therefore, the cost of producing books 12,001 through 15,000 is

> ➤ Although x is a positive integer (the number of books produced), we treat it as a continuous variable in this example.

$$
\begin{aligned}
C(15{,}000) - C(12{,}000) &= \int_{12{,}000}^{15{,}000} C'(x)\,dx \\[2mm]
&= \int_{12{,}000}^{15{,}000} (12 - 0.0002x)\,dx \quad &\text{Substitute for } C'(x). \\[2mm]
&= (12x - 0.0001x^2)\Big|_{12{,}000}^{15{,}000} \quad &\text{Fundamental Theorem} \\[2mm]
&= \$27{,}900 \quad &\text{Simplify.}
\end{aligned}
$$

Related Exercises 35–38 ◄

> **QUICK CHECK 6** Would the cost of increasing the production from 9000 books to 12,000 books be more or less than the cost of increasing the production from 12,000 books to 15,000 books? Explain. ◄

SECTION 6.1 EXERCISES

Review Questions

1. Explain the meaning of position, displacement, and distance traveled as they apply to an object moving along a line.

2. Suppose the velocity of an object moving along a line is positive. Are position, displacement, and distance traveled equal? Explain.

3. Given the velocity function v of an object moving along a line, explain how definite integrals can be used to find the displacement of the object.

4. Explain how to use definite integrals to find the net change in a quantity, given the rate of change of that quantity.

5. Given the rate of change of a quantity Q and its initial value $Q(0)$, explain how to find the value of Q at a future time $t \geq 0$.

6. What is the result of integrating a population growth rate between two times $t = a$ and $t = b$?

Basic Skills

⊤ **7–10. Displacement from velocity** *Assume t is time measured in seconds and velocities have units of m/s.*

    a. Graph the velocity function over the given interval. Then determine when the motion is in the positive direction and when it is in the negative direction.

    b. Find the displacement over the given interval.

    c. Find the distance traveled over the given interval.

7. $v(t) = 6 - 2t;\ 0 \leq t \leq 6$

8. $v(t) = 10 \sin 2t;\ 0 \leq t \leq 2\pi$

9. $v(t) = t^3 - 5t^2 + 6t;\ 0 \leq t \leq 5$

10. $v(t) = 50e^{-2t};\ 0 \leq t \leq 4$

⊤ **11–14. Position from velocity** *Consider an object moving along a line with the following velocities and initial positions.*

    a. Graph the velocity function on the given interval and determine when the object is moving in the positive direction and when it is moving in the negative direction.

    b. Determine the position function for $t \geq 0$ using both the antiderivative method and the Fundamental Theorem of Calculus (Theorem 6.1). Check for agreement between the two methods.

    c. Graph the position function on the given interval.

11. $v(t) = 6 - 2t$ on $[0, 5];\ s(0) = 0$

12. $v(t) = 3 \sin \pi t$ on $[0, 4];\ s(0) = 1$

13. $v(t) = 9 - t^2$ on $[0, 4];\ s(0) = -2$

14. $v(t) = 1/(t + 1)$ on $[0, 8];\ s(0) = -4$

⊤ **15. Oscillating motion** A mass hanging from a spring is set in motion and its ensuing velocity is given by $v(t) = 2\pi \cos \pi t$ for $t \geq 0$. Assume that the positive direction is upward and $s(0) = 0$.

    a. Determine the position function for $t \geq 0$.

    b. Graph the position function on the interval $[0, 3]$.

    c. At what times does the mass reach its lowest point the first three times?

    d. At what times does the mass reach its highest point the first three times?

16. **Cycling distance** A cyclist rides down a long straight road at a velocity (in m/min) given by $v(t) = 400 - 20t$, for $0 \leq t \leq 10$ min.

    a. How far does the cyclist travel in the first 5 min?

    b. How far does the cyclist travel in the first 10 min?

    c. How far has the cyclist traveled when her velocity is 250 m/min?

17. **Flying into a headwind** The velocity of an airplane flying into a headwind is given by $v(t) = 30(16 - t^2)$ mi/hr for $0 \leq t \leq 3$ hr. Assume that $s(0) = 0$.

    a. Determine and graph the position function for $0 \leq t \leq 3$.

    b. How far does the airplane travel in the first 2 hr?

    c. How far has the airplane traveled at the instant its velocity reaches 400 mi/hr?

18. **Day hike** The velocity (in mi/hr) of a hiker walking along a straight trail is given by $v(t) = 3 \sin^2 (\pi t/2)$, for $0 \leq t \leq 4$ hr. Assume that $s(0) = 0$.

    a. Determine and graph the position function for $0 \leq t \leq 4$.

    b. What is the distance traveled by the hiker in the first 15 minutes of the hike? (*Hint*: $\sin^2 t = \frac{1}{2}(1 - \cos 2t)$.)

    c. What is the hiker's position at $t = 3$?

19. **Piecewise velocity** The velocity of a (fast) automobile on a straight highway is given by the function

$$v(t) = \begin{cases} 3t & \text{if } 0 \leq t < 20 \\ 60 & \text{if } 20 \leq t < 45 \\ 240 - 4t & \text{if } t \geq 45 \end{cases}$$

where t is measured in seconds and v has units of m/s.

    a. Graph the velocity function for $0 \leq t \leq 100$. When is the velocity a maximum? When is the velocity zero?

    b. What is the distance traveled by the automobile in the first 30 s?

    c. What is the distance traveled by the automobile in the first 60 s?

    d. What is the position of the automobile when $t = 75$?

20. **Probe speed** A data collection probe is dropped from a stationary balloon and it falls with a velocity (in m/s) given by $v(t) = 9.8t$, neglecting air resistance. After 10 s, a chute deploys and the probe immediately slows to a constant speed of 10 m/s, which it maintains until it enters the ocean.

    a. Graph the velocity function.

    b. How far does the probe fall in the first 30 s after it is released?

    c. If the probe was released from an altitude of 3 km, when does it enter the ocean?

21–24. Position and velocity from acceleration *Find the position and velocity of an object moving along a straight line with the given acceleration, initial velocity, and initial position. Assume units of meters and seconds.*

21. $a(t) = -9.8,\ v(0) = 20,\ s(0) = 0$

22. $a(t) = e^{-t},\ v(0) = 60,\ s(0) = 40$

23. $a(t) = -0.01t,\ v(0) = 10,\ s(0) = 0$

24. $a(t) = 20/(t + 2)^2,\ v(0) = 20,\ s(0) = 10$

25. **Acceleration** A drag racer accelerates at $a(t) = 88$ ft/s^2. Assume that $v(0) = 0$ and $s(0) = 0$.

    a. Determine and graph the position function for $t \geq 0$.

    b. How far does the racer travel in 4 s?

    c. At this rate, how long will it take the racer to travel $\frac{1}{4}$ mi?

    d. How long does it take the racer to travel 300 ft?

    e. How far has the racer traveled when it reaches a speed of 178 ft/s?

26. **Deceleration** A car slows down with an acceleration of $a(t) = -15$ ft/s^2. Assume that $v(0) = 60$ ft/s and $s(0) = 0$.

    a. Determine and graph the position function for $t \geq 0$.

    b. How far does the car travel in the time it takes to come to rest?

27. **Approaching a station** At $t = 0$, a train approaching a station begins decelerating from a speed of 80 mi/hr according to the acceleration function $a(t) = -1280(1 + 8t)^{-3}$ mi/hr^2, where $t \geq 0$. How far does the train travel between $t = 0$ and $t = 0.2$? Between $t = 0.2$ and $t = 0.4$?

28. **Peak oil extraction** The owners of an oil reserve begin extracting oil at $t = 0$. Based on estimates of the reserves, suppose the projected extraction rate is given by $Q'(t) = 3t^2 (40 - t)^2$, where $0 \leq t \leq 40$, Q is measured in millions of barrels, and t is measured in years.

   a. When does the peak extraction rate occur?
   b. How much oil is extracted in the first 10, 20, and 30 years?
   c. What is the total amount of oil extracted in 40 years?
   d. Is one-fourth of the total oil extracted in the first one-fourth of the extraction period? Explain.

29. **Oil production** An oil refinery produces oil at a variable rate given by

$$Q'(t) = \begin{cases} 800 & \text{if } 0 \leq t < 30 \\ 2600 - 60t & \text{if } 30 \leq t < 40 \\ 200 & \text{if } t \geq 40 \end{cases}$$

   where t is measured in days and Q is measured in barrels.

   a. How many barrels are produced in the first 35 days?
   b. How many barrels are produced in the first 50 days?
   c. Without using calculus, determine the number of barrels produced over the interval $[60, 80]$.

30–33. Population growth

30. Starting with an initial value of $P(0) = 55$, the population of a prairie dog community grows at a rate of $P'(t) = 20 - t/5$ (in units of prairie dogs/month), for $0 \leq t \leq 200$.

   a. What is the population 6 months later?
   b. Find the population $P(t)$ for $0 \leq t \leq 200$.

31. When records were first kept ($t = 0$), the population of a rural town was 250 people. During the following years, the population grew at a rate of $P'(t) = 30(1 + \sqrt{t})$.

   a. What is the population after 20 years?
   b. Find the population $P(t)$ at any time $t \geq 0$.

32. The population of a community of foxes is observed to fluctuate on a 10-year cycle due to variations in the availability of prey. When population measurements began ($t = 0$ years), the population was 35 foxes. The growth rate in units of foxes/yr was observed to be

$$P'(t) = 5 + 10 \sin\left(\frac{\pi t}{5}\right).$$

   a. What is the population 15 years later? 35 years later?
   b. Find the population $P(t)$ at any time $t \geq 0$.

33. A culture of bacteria in a petri dish has an initial population of 1500 cells and grows at a rate of $N'(t) = 100e^{-0.25t}$ cells/day.

   a. What is the population after 20 days? after 40 days?
   b. Find the population $N(t)$ at any time $t \geq 0$.

34. **Endangered species** The population of an endangered species changes at a rate given by $P'(t) = 30 - 20t$ (individuals/year). Assume the initial population of the species is 300 individuals.

   a. What is the population after 5 years?
   b. When will the species become extinct?
   c. How does the extinction time change if the initial population is 100 individuals? 400 individuals?

35–38. Marginal cost *Consider the following marginal cost functions.*

   a. *Find the additional cost incurred in dollars when production is increased from 100 units to 150 units.*
   b. *Find the additional cost incurred in dollars when production is increased from 500 units to 550 units.*

35. $C'(x) = 2000 - 0.5x$ 36. $C'(x) = 200 - 0.05x$

37. $C'(x) = 300 + 10x - 0.01x^2$

38. $C'(x) = 3000 - x - 0.001x^2$

Further Explorations

39. **Explain why or why not** Determine whether the following statements are true and give an explanation or counterexample.

   a. The distance traveled by an object moving along a line is the same as the displacement of the object.
   b. When the velocity is positive on an interval, the displacement and the distance traveled on that interval are equal.
   c. Consider a tank that is filled and drained at a flow rate of $R(t) = 1 - t^2/100$ (gal/min), for $t \geq 0$. It follows that the volume of water in the tank increases for 10 min and then decreases until the tank is empty.
   d. A particular marginal cost function has the property that it is positive and decreasing. The cost of increasing production from A units to $2A$ units is greater than the cost of increasing production from $2A$ units to $3A$ units.

40–41. Velocity graphs *The figures show velocity functions for motion along a straight line. Assume the motion begins with an initial position of $s(0) = 0$. Determine the following:*

   a. *The displacement between $t = 0$ and $t = 5$*
   b. *The distance traveled between $t = 0$ and $t = 5$*
   c. *The position at $t = 5$*
   d. *A piecewise function for $s(t)$*

40. 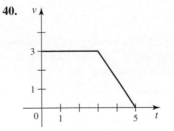   41.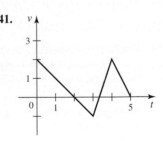

42–45. Equivalent constant velocity *Consider the following velocity functions. In each case, complete the sentence: The same distance could have been traveled over the given time period at a constant velocity of _____.*

42. $v(t) = 2t + 6$ for $0 \leq t \leq 8$

43. $v(t) = 1 - t^2/16$ for $0 \leq t \leq 4$

44. $v(t) = 2\sin t$ for $0 \leq t \leq \pi$

45. $v(t) = t(25 - t^2)^{1/2}$ for $0 \leq t \leq 5$

46. **Where do they meet?** Kelly started at noon ($t = 0$) riding a bike from Niwot to Berthoud, a distance of 20 km, with velocity $v(t) = 15/(t + 1)^2$ (decreasing because of fatigue). Sandy started at noon ($t = 0$) riding a bike in the opposite direction from

Berthoud to Niwot with velocity $u(t) = 20/(t + 1)^2$ (also decreasing because of fatigue). Assume distance is measured in kilometers and time is measured in hours.

a. Make a graph of Kelly's distance from Niwot as a function of time.

b. Make a graph of Sandy's distance from Berthoud as a function of time.

c. How far has each person traveled when they meet? When do they meet?

d. If the riders' speeds are $v(t) = A/(t + 1)^2$ and $u(t) = B/(t + 1)^2$ and the distance between the towns is D, what conditions on A, B, and D must be met to ensure that the riders will pass each other?

e. Looking ahead: With the velocity functions given in part (d), make a conjecture about the maximum distance each person can ride (given unlimited time).

47. Bike race Theo and Sasha start at the same place on a straight road riding bikes with the following velocities (measure in mi/hr):

Theo: $v_T(t) = 10$ for $t \geq 0$
Sasha: $v_S(t) = 15t$ for $0 \leq t \leq 1$ and
$v_S(t) = 15$ for $t > 1$

a. Graph the velocity functions for both riders.

b. If the riders ride for 1 hr, who rides farther? Interpret your answer geometrically using the graphs of part (a).

c. If the riders ride for 2 hr, who rides farther? Interpret your answer geometrically using the graphs of part (a).

d. Which rider arrives first at the 10, 15, and 20 mile markers of the race? Interpret your answer geometrically using the graphs of part (a).

e. Suppose Sasha gives Theo a head start of 0.2 mi and the riders ride for 20 mi. Who wins the race?

f. Suppose Sasha gives Theo a head start of 0.2 hr and the riders ride for 20 mi. Who wins the race?

48. Two runners At noon ($t = 0$), Alicia starts running along a long straight road at 4 mi/hr. Her velocity decreases according to the function $v(t) = 4/(t + 1)$ for $t \geq 0$. At noon, Boris also starts running along the same road with a 2-mi head start on Alicia; his velocity is given by $u(t) = 2/(t + 1)$ for $t \geq 0$.

a. Find the position functions for Alicia and Boris, where $s = 0$ corresponds to Alice's starting point.

b. When, if ever, does Alicia overtake Boris?

49. Running in a wind A strong west wind blows across a circular running track. Abe and Bess start at the south end of the track and at the same time, Abe starts running clockwise and Bess starts running counterclockwise. Abe runs with a speed (in units of mi/hr) given by $u(\varphi) = 3 - 2 \cos \varphi$ and Bess runs with a speed given by $v(\theta) = 3 + 2 \cos \theta$, where φ and θ are the central angles of the runners.

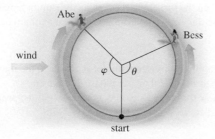

a. Graph the speed functions u and v, and explain why they describe the runners' speeds (in light of the wind).

b. Which runner has the greater average speed for one lap?

c. Challenge: If the track has a radius of $\frac{1}{10}$ mi, how long does it take each runner to complete one lap and who wins the race?

Applications

50. Filling a tank A 200-L cistern is empty when water begins flowing into it (at $t = 0$) at a rate in L/min given by $Q'(t) = 3\sqrt{t}$.

a. How much water flows into the cistern in 1 hr?

b. Find and graph the function that gives the amount of water in the tank at any time $t \geq 0$.

c. When will the tank be full?

51. Depletion of natural resources Suppose that $r(t) = r_0 e^{-kt}$ is the rate at which a nation extracts oil, where $r_0 = 10^7$ barrels/yr is the current rate of extraction. Suppose also that the estimate of the total oil reserve is 2×10^9 barrels.

a. Find the minimum decay constant k for which the total oil reserves will last forever.

b. Suppose $r_0 = 2 \times 10^7$ barrels/yr and the decay constant k is the minimum value found in part (a). How long will the total oil reserves last?

52. Snow plow problem With snow on the ground and falling at a constant rate, a snow plow began plowing down a long straight road at noon. The plow traveled twice as far in the first hour as it did in the second hour. At what time did the snow start falling? Assume the plowing rate is inversely proportional to the depth of the snow.

53. Filling a reservoir A reservoir with a capacity of 2500 m³ is filled with a single inflow pipe. The reservoir is empty and the inflow pipe is opened at $t = 0$. Letting $Q(t)$ be the amount of water in the reservoir at time t, the flow rate of water into the reservoir (in m³/hr) oscillates on a 24-hr cycle (see figure) and is given by

$$Q'(t) = 20\left[1 + \cos\left(\frac{\pi t}{12}\right)\right].$$

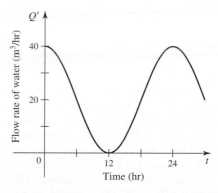

a. How much water flows into the reservoir in the first 2 hr?

b. Find and graph the function that gives the amount of water in the reservoir over the interval $[0, t]$ where $t \geq 0$.

c. When is the reservoir full?

54. Blood flow A typical human heart pumps 20 mL of blood with each stroke (stroke volume). Assuming a heart rate of 60 beats/min, a reasonable model for the outflow rate of the heart is $V'(t) = 20(1 + \sin(2\pi t))$, where $V(t)$ is the total volume of blood pumped at time t measured in milliliters. Assume that $V(0) = 0$.

a. Graph the outflow rate function.
b. Verify that the amount of blood pumped over a one-second interval is 20 mL.
c. Find the function that gives the total blood pumped between $t = 0$ and a future time $t > 0$.
d. What is the cardiac output over a period of 1 min? (Use calculus, then check your answer with algebra.)

55. Air flow in the lungs A reasonable model (with different parameters for different people) for the flow of air in and out of the lungs is

$$V'(t) = -\frac{\pi V_0}{10} \sin\left(\frac{\pi t}{5}\right),$$

where $V(t)$ is the volume of air in the lungs at time $t \geq 0$, measured in liters, t is measured in seconds, and V_0 is the capacity of the lungs. The time $t = 0$ corresponds to a time at which the lungs are full and exhalation begins.

a. Graph the flow rate function with $V_0 = 10\,\text{L}$.
b. Find and graph the function V, assuming that $V(0) = V_0 = 10\,\text{L}$.
c. What is the breathing rate in breaths/min?

56. Oscillating growth rates Some species have growth rates that oscillate with an (approximately) constant period P. Consider the growth rate function

$$N'(t) = A \sin\left(\frac{2\pi t}{P}\right) + r,$$

where A and r are constants with units of individuals/yr. A species becomes extinct if its population ever reaches 0 after $t = 0$.

a. Suppose $P = 10$, $A = 20$, and $r = 0$. If the initial population is $N(0) = 10$, does the population ever become extinct? Explain.
b. Suppose $P = 10$, $A = 20$, and $r = 0$. If the initial population is $N(0) = 100$, does the population ever become extinct? Explain.
c. Suppose $P = 10$, $A = 50$, and $r = 5$. If the initial population is $N(0) = 10$, does the population ever become extinct? Explain.
d. Suppose $P = 10$, $A = 50$, and $r = -5$. Find the initial population $N(0)$ needed to ensure that the population never becomes extinct.

57. Power and energy Power and energy are often used interchangeably, but they are quite different. **Energy** is what makes matter move or heat up and is measured in units of **joules** (J) or **Calories** (Cal), where 1 Cal = 4184 J. One hour of walking consumes roughly 10^6 J, or 250 Cal. On the other hand, **power** is the rate at which energy is used and is measured in **watts** (W; 1 W = 1 J/s). Other useful units of power are **kilowatts** (1 kW = 10^3 W) and **megawatts** (1 MW = 10^6 W). If energy is used at a rate of 1 kW for 1 hr, the total amount of energy used is 1 **kilowatt-hour** (kWh), which is 3.6×10^6 J.

Suppose the power function of a large city over a 24-hr period is given by

$$P(t) = E'(t) = 300 - 200 \sin\left(\frac{\pi t}{12}\right),$$

where P is measured in MW and $t = 0$ corresponds to 6:00 p.m. (see figure).

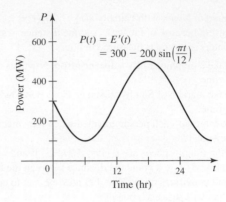

a. How much energy is consumed by this city in a typical 24-hr period? Express the answer in MWh and in J.
b. Burning 1 kg of coal produces about 450 kWh of energy. How many kg of coal are required to meet the energy needs of the city for 1 day? For 1 yr?
c. Fission of 1 gram of uranium-235 (U-235) produces about 16,000 kWh of energy. How many grams of uranium are needed to meet the energy needs of the city for 1 day? For 1 yr?
d. A typical wind turbine can generate electricity at a rate of about 200 kW. Approximately how many wind turbines are needed to meet the average energy needs of the city?

58. Variable gravity At Earth's surface the acceleration due to gravity is approximately $g = 9.8\,\text{m/s}^2$ (with local variations). However, the acceleration decreases with distance from the surface according to Newton's law of gravitation. At a distance of y meters from Earth's surface, the acceleration is given by

$$a(y) = -\frac{g}{(1 + y/R)^2}$$

where $R = 6.4 \times 10^6$ m is the radius of Earth.

a. Suppose a projectile is launched upward with an initial velocity of v_0 m/s. Let $v(t)$ be its velocity and $y(t)$ its height (in meters) above the surface t seconds after the launch. Neglecting forces such as air resistance, explain why $\dfrac{dv}{dt} = a(y)$ and $\dfrac{dy}{dt} = v(t)$.
b. Use the Chain Rule to show that $\dfrac{dv}{dt} = \dfrac{1}{2}\dfrac{d}{dy}(v^2)$.
c. Show that the equation of motion for the projectile is $\dfrac{1}{2}\dfrac{d}{dy}(v^2) = a(y)$, where $a(y)$ is given previously.
d. Integrate both sides of the equation in part (c) with respect to y using the fact that when $y = 0$, $v = v_0$. Show that

$$\frac{1}{2}(v^2 - v_0^2) = gR\left(\frac{1}{1 + y/R} - 1\right).$$

e. When the projectile reaches its maximum height, $v = 0$. Use this fact to determine that the maximum height is

$$y_{\max} = \frac{Rv_0^2}{2gR - v_0^2}.$$

f. Graph $y_{\max}$ as a function of v_0. What is the maximum height when $v_0 = 500\,\text{m/s}$, 1500 m/s, and 5 km/s?
g. Show that the value of v_0 needed to put the projectile into orbit (called the escape velocity) is $\sqrt{2gR}$.

1. Displacement $= -20$ mi (20 mi south); distance traveled $= 100$ mi.   **2.** Suppose the object moves in the positive direction for $0 \le t \le 3$ and then moves in the negative direction for $3 < t \le 5$.   **3.** A function; a number

4. Displacement $= 0$; distance traveled $= 1$ **5.** 1720 m
6. The production cost would increase more between 9000 and 12,000 books than between 12,000 and 15,000 books. Graph C' and look at the area under the curve. ◄

6.2 Regions Between Curves

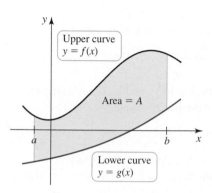

FIGURE 6.11

In this section, the method for finding the area of a region bounded by a single curve is generalized to regions bounded by two or more curves. Consider two continuous functions f and g defined on an interval $[a, b]$, where $f(x) \ge g(x)$ (Figure 6.11). The goal is to find the area A of the region bounded by the two curves and the vertical lines $x = a$ and $x = b$.

Once again we rely on the *slice-and-sum* strategy (Section 5.2) for finding areas by Riemann sums. The interval $[a, b]$ is partitioned into n subintervals using uniformly spaced grid points separated by a distance $\Delta x = (b - a)/n$ (Figure 6.12). On each subinterval, we build a rectangle extending from the lower curve to the upper curve. On the kth subinterval, a point $\overline{x}_k$ is chosen, and the height of the corresponding rectangle is taken to be $f(\overline{x}_k) - g(\overline{x}_k)$. Therefore, the area of the kth rectangle is $(f(\overline{x}_k) - g(\overline{x}_k)) \Delta x$ (Figure 6.13). Summing the areas of the n rectangles gives an approximation to the area of the region between the curves:

$$A \approx \sum_{k=1}^{n} (f(\overline{x}_k) - g(\overline{x}_k)) \Delta x$$

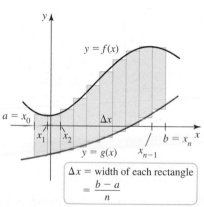

FIGURE 6.12

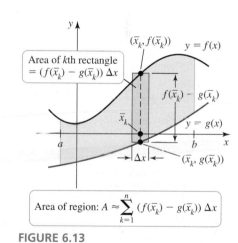

FIGURE 6.13

As the number of grid points increases, Δx approaches zero and these sums approach the area between the curves; that is,

$$A = \lim_{n \to \infty} \sum_{k=1}^{n} (f(\overline{x}_k) - g(\overline{x}_k)) \Delta x.$$

The limit of these Riemann sums is a definite integral of the function $f - g$.

> It is helpful to interpret the area formula: $f(x) - g(x)$ is the length of a rectangle and dx is its width. We sum (integrate) the areas of the rectangles $(f(x) - g(x))\, dx$ to obtain the area of the region.

DEFINITION   Area of a Region Between Two Curves

Suppose that f and g are continuous functions with $f(x) \geq g(x)$ on the interval $[a, b]$. The area of the region bounded by the graphs of f and g on $[a, b]$ is

$$A = \int_a^b (f(x) - g(x))\, dx.$$

QUICK CHECK 1 In the area formula for a region between two curves, verify that if the lower curve is $g(x) = 0$, the formula becomes the usual formula for the area of the region bounded by $y = f(x)$ and the x-axis. ◄

EXAMPLE 1   Area between curves Find the area of the region bounded by the graphs of $f(x) = \dfrac{1}{1 + x^2}$, $g(x) = x - \dfrac{1}{2}$, and the y-axis (Figure 6.14).

> The intersection point satisfies the equation $\dfrac{1}{1 + x^2} = x - \dfrac{1}{2}$, which has the same roots as the cubic equation $2x^3 - x^2 + 2x - 3 = 0$. Using synthetic division or a root finder, we find that $x = 1$ is the only real root.

SOLUTION A key step in the solution of many area problems is finding the intersection points of the boundary curves, which often determine the limits of integration. The intersection point of these two curves satisfies the equation $\dfrac{1}{1 + x^2} = x - \dfrac{1}{2}$, whose only real solution is $x = 1$. Because the intersection point is the rightmost boundary point of the region, its x-coordinate becomes the upper limit of integration. The line $x = 0$ (the y-axis) bounds the region on the left, which gives the lower limit of integration. The graph of f is the upper curve and the graph of g is the lower curve on the interval $[0, 1]$, so the area of the region is

$$
\begin{aligned}
A &= \int_0^1 \left[\frac{1}{1 + x^2} - \left(x - \frac{1}{2} \right) \right] dx && \text{Substitute for } f \text{ and } g.\\[2mm]
&= \left(\tan^{-1} x - \frac{x^2}{2} + \frac{x}{2} \right) \Bigg|_0^1 && \text{Fundamental Theorem}\\[2mm]
&= \left(\tan^{-1} 1 - \frac{1}{2} + \frac{1}{2} \right) - 0 = \frac{\pi}{4}. && \text{Evaluate and simplify.}
\end{aligned}
$$

Related Exercises 5–14 ◄

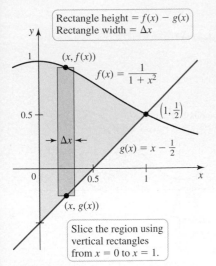

Rectangle height $= f(x) - g(x)$
Rectangle width $= \Delta x$

$f(x) = \dfrac{1}{1 + x^2}$

$(x, f(x))$

$\left(1, \frac{1}{2}\right)$

$g(x) = x - \dfrac{1}{2}$

$(x, g(x))$

Slice the region using vertical rectangles from $x = 0$ to $x = 1$.

FIGURE 6.14

QUICK CHECK 2 Interpret the area formula in the form $A = \int_a^b f(x)\, dx - \int_a^b g(x)\, dx$. ◄

EXAMPLE 2   Compound region Find the area of the region between the graphs of $f(x) = x + 3$ and $g(x) = |2x|$ (Figure 6.15a).

SOLUTION The lower boundary of the region in question is bounded by two different branches of the absolute value function. In situations like this, the region is divided into two (or more) subregions, whose areas are found independently and then summed; these regions are labeled R_1 and R_2 (Figure 6.15b). By the definition of absolute value,

$$
g(x) = |2x| =
\begin{cases}
2x & \text{if } x \geq 0\\
-2x & \text{if } x < 0
\end{cases}
$$

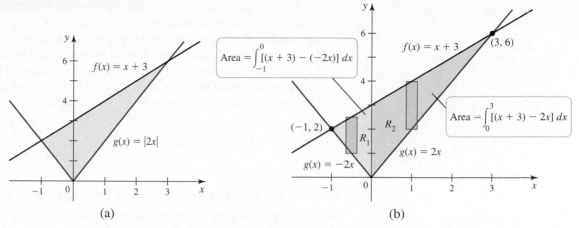

FIGURE 6.15   (a)   (b)

The left intersection point of f and g satisfies $-2x = x + 3$, or $x = -1$. The right intersection point satisfies $2x = x + 3$, or $x = 3$. We see that the region R_1 is bounded by the lines $y = x + 3$ and $y = -2x$ from $x = -1$ to $x = 0$. Similarly, region R_2 is bounded by the lines $y = x + 3$ and $y = 2x$ from $x = 0$ to $x = 3$ (Figure 6.15b). Therefore,

$$A = \underbrace{\int_{-1}^{0} [(x + 3) - (-2x)] \, dx}_{\text{area of region } R_1} + \underbrace{\int_{0}^{3} [(x + 3) - 2x] \, dx}_{\text{area of region } R_2}$$

$$= \int_{-1}^{0} (3x + 3) \, dx + \int_{0}^{3} (-x + 3) \, dx.$$

Evaluating the integrals, the area of the shaded region is

$$A = \int_{-1}^{0} (3x + 3) \, dx + \int_{0}^{3} (-x + 3) \, dx \qquad \text{Simplify integrands.}$$

$$= \left(\frac{3}{2} x^2 + 3x \right) \Big|_{-1}^{0} + \left(-\frac{x^2}{2} + 3x \right) \Big|_{0}^{3} \qquad \text{Fundamental Theorem}$$

$$= 0 - \left(\frac{3}{2} - 3 \right) + \left(-\frac{9}{2} + 9 \right) - 0 = 6. \qquad \text{Simplify.}$$

Related Exercises 15–22 ◄

Integrating with Respect to y

There are occasions when it is convenient to reverse the roles of x and y. Consider the regions shown in Figure 6.16 that are bounded by the graphs of $x = f(y)$ and $x = g(y)$, where $f(y) \geq g(y)$ for $c \leq y \leq d$ (which implies that the graph of f lies to the right of the graph of g). The lower and upper boundaries of the regions are $y = c$ and $y = d$, respectively.

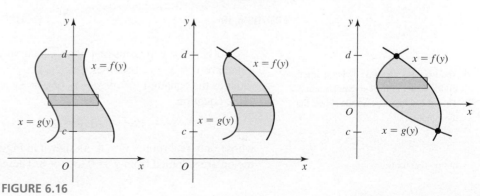

FIGURE 6.16

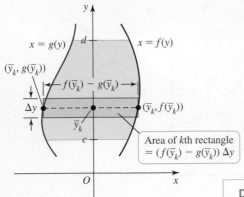

Area of region: $A \approx \sum_{k=1}^{n} (f(\bar{y}_k) - g(\bar{y}_k))\,\Delta y$

FIGURE 6.17

▶ This area formula is identical to the one given on page 384; it is now expressed with respect to the y-axis. In this case, $f(y) - g(y)$ is the length of a rectangle and dy is its width. We sum (integrate) the areas of the rectangles $(f(y) - g(y))\,dy$ to obtain the area of the region.

In cases such as these, we treat y as the independent variable and divide the interval $[c, d]$ into n subintervals of width $\Delta y = (d - c)/n$ (Figure 6.17). On the kth subinterval, a point $\bar{y}_k$ is selected and we construct a rectangle that extends from the left curve to the right curve. The kth rectangle has length $f(\bar{y}_k) - g(\bar{y}_k)$, and so the area of the kth rectangle is $(f(\bar{y}_k) - g(\bar{y}_k))\,\Delta y$. The area of the region is approximated by the sum of the areas of the rectangles. In the limit as $n \to \infty$ and $\Delta y \to 0$, the area of the region is given as the definite integral

$$A = \lim_{n \to \infty} \sum_{k=1}^{n} (f(\bar{y}_k) - g(\bar{y}_k))\,\Delta y = \int_{c}^{d} (f(y) - g(y))\,dy.$$

DEFINITION   Area of a Region Between Two Curves with Respect to y

Suppose that f and g are continuous functions with $f(y) \geq g(y)$ on the interval $[c, d]$. The area of the region bounded by the graphs $x = f(y)$ and $x = g(y)$ on $[c, d]$ is

$$A = \int_{c}^{d} (f(y) - g(y))\,dy.$$

EXAMPLE 3   Integrating with respect to y Find the area of the region R bounded by the graphs of $y = x^3$, $y = x + 6$, and the x-axis.

SOLUTION The area of this region could be found by integrating with respect to x. But this approach requires splitting the region into two pieces (Figure 6.18). Alternatively, we can view y as the independent variable, express the bounding curves as functions of y, and make horizontal slices parallel to the x-axis (Figure 6.19).

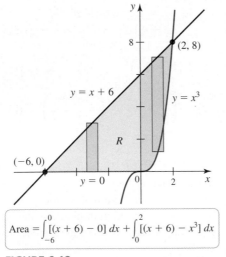

Area $= \int_{-6}^{0} [(x + 6) - 0]\,dx + \int_{0}^{2} [(x + 6) - x^3]\,dx$

FIGURE 6.18

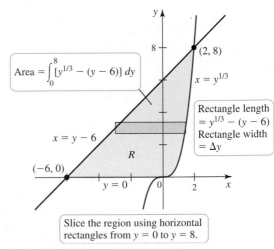

Slice the region using horizontal rectangles from $y = 0$ to $y = 8$.

FIGURE 6.19

▶ You may use synthetic division or a root finder to factor this cubic polynomial. Then the quadratic formula shows that the equation

$$y^2 - 10y + 27 = 0$$

has no real roots.

Solving for x in terms of y, the right curve $y = x^3$ becomes $x = f(y) = y^{1/3}$. The left curve $y = x + 6$ becomes $x = g(y) = y - 6$. The intersection point of the curves satisfies the equation $y^{1/3} = y - 6$, or $y = (y - 6)^3$. Expanding this equation gives the cubic equation

$$y^3 - 18y^2 + 107y - 216 = (y - 8)(y^2 - 10y + 27) = 0,$$

whose only real root is $y = 8$. As shown in Figure 6.19, the areas of the slices through the region are summed from $y = 0$ to $y = 8$. Therefore, the area of the region is given by

QUICK CHECK 3 The region R is bounded by the curve $y = \sqrt{x}$, the line $y = x - 2$, and the x-axis. Express the area of R in terms of (a) integral(s) with respect to x and (b) integral(s) with respect to y. ◄

$$\int_0^8 \left[y^{1/3} - (y - 6) \right] dy = \left(\frac{3}{4} y^{4/3} - \frac{y^2}{2} + 6y \right) \Big|_0^8 \quad \text{Fundamental Theorem}$$

$$= \left(\frac{3}{4} \cdot 16 - 32 + 48 \right) - 0 = 28. \quad \text{Simplify.}$$

Related Exercises 23–32 ◄

EXAMPLE 4 Calculus and geometry Find the area of the region R in the first quadrant bounded by the curves $y = x^{2/3}$ and $y = x - 4$ (Figure 6.20).

SOLUTION Slicing the region vertically and integrating with respect to x requires two integrals. Slicing the region horizontally requires a single integral with respect to y. The second approach appears to involve less work.

Slicing horizontally, the right bounding curve is $x = y + 4$ and the left bounding curve is $x = y^{3/2}$. The two curves intersect at $(8, 4)$, so the limits of integration are $y = 0$ and $y = 4$. The area of R is

$$\int_0^4 \underbrace{(y + 4}_{\text{right curve}} - \underbrace{y^{3/2})}_{\text{left curve}} \, dy = \left(\frac{y^2}{2} + 4y - \frac{2}{5} y^{5/2} \right) \Big|_0^4 = \frac{56}{5}.$$

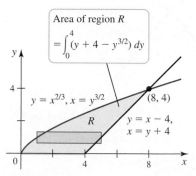

Area of region R
$$= \int_0^4 (y + 4 - y^{3/2}) \, dy$$

$y = x^{2/3}, \, x = y^{3/2}$

R

$(8, 4)$

$y = x - 4,$
$x = y + 4$

FIGURE 6.20

Can this area be found using a different approach? Sometimes it helps to use geometry. Notice that the region R can be formed by taking the entire region under the curve $y = x^{2/3}$ on the interval $[0, 8]$ and then removing a triangle whose base is the interval $[4, 8]$ (Figure 6.21). The area of the region R_1 under the curve $y = x^{2/3}$ is

$$\int_0^8 x^{2/3} \, dx = \frac{3}{5} x^{5/3} \Big|_0^8 = \frac{96}{5}.$$

The triangle R_2 has a base of length 4 and a height of 4, so its area is $\frac{1}{2} \cdot 4 \cdot 4 = 8$. Therefore, the area of R is $\frac{96}{5} - 8 = \frac{56}{5}$, which agrees with the first calculation.

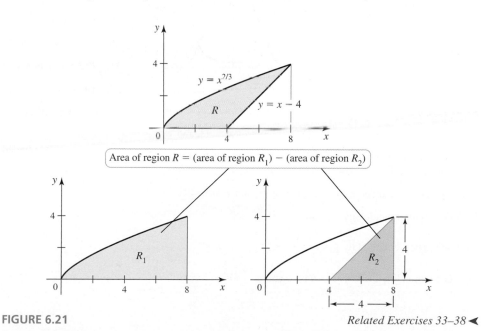

$y = x^{2/3}$

R

$y = x - 4$

Area of region R = (area of region R_1) − (area of region R_2)

R_1

R_2

FIGURE 6.21

Related Exercises 33–38 ◄

QUICK CHECK 4 An alternative way to determine the area of the region in Example 3 (Figure 6.18) is to compute $18 + \int_0^2 (x + 6 - x^3) \, dx$. Why? ◄

SECTION 6.2 EXERCISES

Review Questions

1. Draw the graphs of two continuous functions f and g that intersect exactly twice. Explain how to use integration to find the area of the region bounded by the two curves.

2. Draw the graphs of two continuous functions f and g that intersect exactly three times. Explain how to use integration to find the area of the region bounded by the two curves.

3. Make a sketch to show a case in which the area bounded by two curves is most easily found by integrating with respect to x.

4. Make a sketch to show a case in which the area bounded by two curves is most easily found by integrating with respect to y.

Basic Skills

5–8. Finding area *Determine the area of the shaded region in the following figures.*

5.
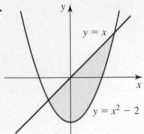

6.
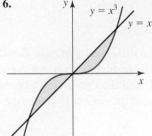

7.
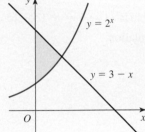

8.
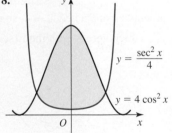

9–14. Regions between curves *Sketch the region and find its area.*

9. The region bounded by $y = 2(x + 1)$, $y = 3(x + 1)$, and $x = 4$

10. The region bounded by $y = \cos x$ and $y = \sin x$ between $x = \pi/4$ and $x = 5\pi/4$

11. The region bounded by $y = e^x$, $y = e^{-2x}$, and $x = \ln 4$

12. The region bounded by $y = x$ and $y = x^2 - 2$

13. The region bounded by $y = \dfrac{4}{1 + x^2}$ and $y = 1$

14. The region bounded by $y = 64\sqrt{x}$ and $y = 8x^2$

15–22. Compound regions *Sketch the following regions (if figure is not given) and then find the area.*

15. The region bounded by $y = \sin x$, $y = \cos x$, and the x-axis between $x = 0$ and $x = \pi/2$

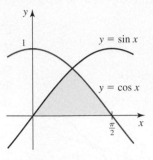

16. The regions between $y = \sin x$ and $y = \sin 2x$, for $0 \le x \le \pi$

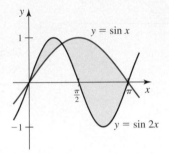

17. The region bounded by $y = x$, $y = 1/x$, $y = 0$, and $x = 2$

18. The region in the first quadrant bounded by $y = (x - 1)^3$ and $y = x - 1$

19. The region bounded by $y = 1 - |x|$ and the x-axis

20. The regions bounded by $y = x^3$ and $y = 9x$

21. The region bounded by $y = |x - 3|$ and $y = x/2$

22. The regions bounded by $y = x^2(3 - x)$ and $y = 12 - 4x$

23–26. Integrating with respect to y *Sketch the following regions (if a figure is not given) and find the area.*

23. The region bounded by $y = 8 - 2x$, $y = x + 8$, and $y = 0$

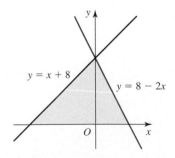

24. The region bounded by $y = \ln x$, $y = 2$, $y = 0$, and $x = 0$

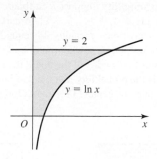

25. The region bounded by $y = 4 - x^2$ and $y - x - 2 = 0$

26. The region bounded by $y = \ln x^2$, $y = \ln x$, and $x = e^2$

27–30. Two approaches *Express the area of the following shaded regions in terms of (a) one or more integrals with respect to x, and (b) one or more integrals with respect to y. You do not need to evaluate the integrals.*

27.

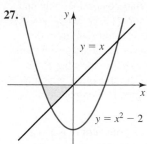

$y = x$

$y = x^2 - 2$

28.

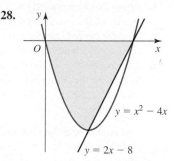

$y = x^2 - 4x$

$y = 2x - 8$

29.

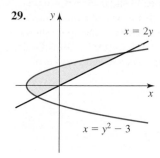

$x = 2y$

$x = y^2 - 3$

30.

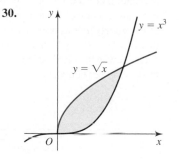

$y = x^3$

$y = \sqrt{x}$

31–32. Two approaches *Find the area of the following regions by (a) integrating with respect to x, and then (b) integrating with respect to y. Be sure your results agree. In each case, sketch the bounding curves and the region in question.*

31. The region bounded by $y = \sqrt{x}$, $y = 2x - 15$, and $y = 0$

32. The region in the first quadrant bounded by $y = x^{1/3}$, $18y - x - 27 = 0$, and $y = 0$

33–38. Any method *Use any method (including geometry) to find the area of the following regions. In each case, sketch the bounding curves and the region in question.*

33. The region in the first quadrant bounded by $y = x^{2/3}$ and $y = 4$

34. The region in the first quadrant bounded by $y = 2$ and $y = 2 \sin x$ on the interval $[0, \pi/2]$

35. The region bounded by $y = e^x$, $y = 2e^{-x} + 1$, and $x = 0$

36. The region below the line $y = 2$ and above the curve $y = \sec^2 x$ on the interval $[0, \pi/4]$

37. The region between the line $y = x$ and the curve $y = 2x\sqrt{1 - x^2}$ in the first quadrant

38. The region bounded by $x = y^2 - 4$, $y = x/3$.

Further Explorations

39. Explain why or why not Determine whether the following statements are true and give an explanation or counterexample.

    a. The area of the region bounded by $y = x$ and $x = y^2$ can be found only by integrating with respect to x.

    b. The area of the region between $y = \sin x$ and $y = \cos x$ on the interval $[0, \pi/2]$ is $\int_0^{\pi/2}(\cos x - \sin x)\, dx$.

    c. Without evaluating integrals, $\int_0^1 (x - x^2)\, dx = \int_0^1 (\sqrt{y} - y)\, dy$

40–43. Regions between curves *Sketch the region and find its area.*

40. The region bounded by $y = \sin x$ and $y = x(x - \pi)$ for $0 \le x \le \pi$

41. The region bounded by $y = (x - 1)^2$ and $y = 7x - 19$

42. The region bounded by $y = 5/4$ and $y = \dfrac{1}{\sqrt{1 - x^2}}$

43. The region bounded by $y = x^2 - 2x + 1$ and $y = 5x - 9$

44–50. *Either* method *Use the most efficient strategy for computing the area of the following regions.*

44. The region bounded by $x = y(y - 1)$ and $x = -y(y - 1)$

45. The region bounded by $x = y(y - 1)$ and $y = x/3$

46. The region bounded by $y = x^3$, $y = -x^3$, and $3y - 7x - 10 = 0$

47. The region bounded by $y = \sqrt{x}$, $y = 2x - 15$, and $y = 0$

48. The region bounded by $y = x^2 - 4$, $4y - 5x - 5 = 0$, and $y = 0$

49. The region in the first quadrant bounded by $y = \dfrac{5}{2} - \dfrac{1}{x}$ and $y = x$.

50. The region in the first quadrant bounded by $y = x^{-1}$, $y = 4x$, and $y = x/4$

51. Comparing areas Let $f(x) = x^p$ and $g(x) = x^{1/q}$, where $p > 1$ and $q > 1$ are positive integers. Let R_1 be the region in the first quadrant between $y = f(x)$ and $y = x$ and let R_2 be the region in the first quadrant between $y = g(x)$ and $y = x$.

    a. Find the area of R_1 and R_2 when $p = q$, and determine which region has the greater area.

    b. Find the area of R_1 and R_2 when $p > q$, and determine which region has the greater area.

    c. Find the area of R_1 and R_2 when $p < q$, and determine which region has the greater area.

52–55. Complicated regions *Find the area of the regions shown in the following figures.*

52.

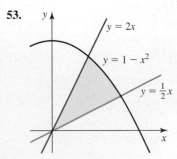

$y = 4\sqrt{2x}$

$y = 2x^2$

$y = -4x + 6$

53.

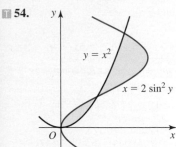

$y = 2x$

$y = 1 - x^2$

$y = \frac{1}{2}x$

54.

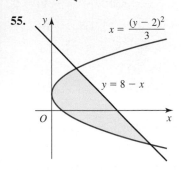

$y = x^2$

$x = 2\sin^2 y$

55.

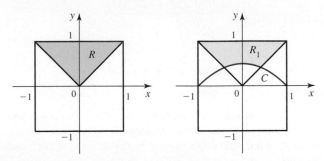

$x = \dfrac{(y-2)^2}{3}$

$y = 8 - x$

56–59. Roots and powers *Find the area of the following regions, expressing your results in terms of the positive integer $n \geq 2$.*

56. The region bounded by $f(x) = x$ and $g(x) = x^n$, for $x \geq 0$

57. The region bounded by $f(x) = x$ and $g(x) = x^{1/n}$, for $x \geq 0$

58. The region bounded by $f(x) = x^{1/n}$ and $g(x) = x^n$, for $x \geq 0$

59. Let A_n be the area of the region bounded by $f(x) = x^{1/n}$ and $g(x) = x^n$ on the interval $[0, 1]$, where n is a positive integer. Evaluate $\lim_{n \to \infty} A_n$ and interpret the result.

Applications

60. Geometric probability Suppose a dartboard occupies the square $\{(x, y): 0 \leq |x| \leq 1, 0 \leq |y| \leq 1\}$. A dart is thrown randomly at the board many times (meaning it is equally likely to land at any point in the square). In what fraction of the throws does the dart land closer to the edge of the board than the center? Equivalently, what is the probability that the dart lands closer to the edge of the board than the center? Proceed as follows.

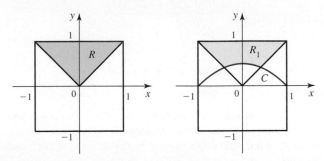

a. Argue that by symmetry it is necessary to consider only one quarter of the board, say the region R: $\{(x, y): |x| \leq y \leq 1\}$.

b. Find the curve C in this region that is equidistant from the center of the board and the top edge of the board (see figure).

c. The probability that the dart lands closer to the edge of the board than the center is the ratio of the area of the region R_1 above C to the area of the entire region R. Compute this probability.

61. Lorenz curves and the Gini index A **Lorenz curve** is given by $y = L(x)$, where $0 \leq x \leq 1$ represents the lowest fraction of the population of a society in terms of wealth and $0 \leq y \leq 1$ represents the fraction of the total wealth that is owned by that fraction of the society. For example, the Lorenz curve in the figure shows that $L(0.5) = 0.2$, which means that the lowest 0.5 (50%) of the society owns 0.2 (20%) of the wealth. (See Guided Projects for more on Lorenz curves.)

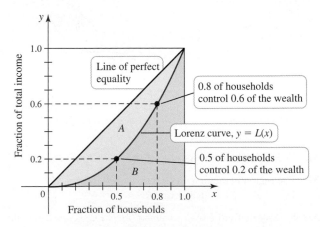

a. A Lorenz curve $y = L(x)$ is accompanied by the line $y = x$, called the **line of perfect equality**. Explain why this line is given this name.

b. Explain why a Lorenz curve satisfies the conditions $L(0) = 0, L(1) = 1,$ and $L'(x) \geq 0$ on $[0, 1]$.

c. Graph the Lorenz curves $L(x) = x^p$ corresponding to $p = 1.1, 1.5, 2, 3, 4$. Which value of p corresponds to the *most* equitable distribution of wealth (closest to the line of perfect equality)? Which value of p corresponds to the *least* equitable distribution of wealth? Explain.

d. The information in the Lorenz curve is often summarized in a single measure called the **Gini index**, which is defined as follows. Let A be the area of the region between $y = x$ and $y = L(x)$ (see figure) and let B be the area of the region between $y = L(x)$ and the x-axis. Then the Gini index is $G = \dfrac{A}{A + B}$. Show that $G = 2A = 1 - 2\int_0^1 L(x)\,dx$.

e. Compute the Gini index for the cases $L(x) = x^p$ and $p = 1.1, 1.5, 2, 3, 4$.

f. What is the smallest interval $[a, b]$ on which values of the Gini index lie, for $L(x) = x^p$ with $p \geq 1$? Which endpoints of $[a, b]$ correspond to the least and most equitable distribution of wealth?

g. Consider the Lorenz curve described by $L(x) = 5x^2/6 + x/6$. Show that it satisfies the conditions $L(0) = 0, L(1) = 1,$ and $L'(x) \geq 0$ on $[0, 1]$. Find the Gini index for this function.

Additional Exercises

62. **Equal area properties for parabolas** Consider the parabola $y = x^2$. Let $P, Q,$ and R be points on the parabola with R between P and Q on the curve. Let $\ell_P, \ell_Q,$ and ℓ_R be the lines tangent to the parabola at $P, Q,$ and R, respectively (see figure). Let P' be the intersection point of $\ell_Q,$ and ℓ_R; let Q' be the intersection point of ℓ_P and ℓ_R; and let R' be the intersection point of ℓ_P and ℓ_Q. Prove that Area $\triangle PQR = 2 \cdot$ Area $\triangle P'Q'R'$ in the following cases. (In fact, the property holds for any three points on any parabola.) (*Mathematics Magazine* 81, No. 2 (April 2008): 83–95.)

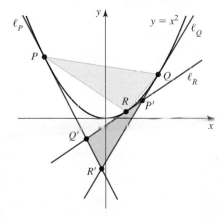

a. $P(-a, a^2), Q(a, a^2),$ and $R(0, 0),$ where a is a positive real number

b. $P(-a, a^2), Q(b, b^2),$ and $R(0, 0),$ where a and b are positive real numbers

c. $P(-a, a^2), Q(b, b^2),$ and R is any point between P and Q on the curve

63. **Minimum area** Graph the curves $y = (x + 1)(x - 2)$ and $y = ax + 1$ for various values of a. For what value of a is the area of the region between the two curves a minimum?

64. **An area function** Graph the curves $y = a^2x^3$ and $y = \sqrt{x}$ for various values of $a > 0$. Note how the area $A(a)$ between the curves varies with a. Find and graph the area function $A(a)$. For what value of a is $A(a) = 16$?

65. **Area of a curve defined implicitly** Determine the area of the shaded region bounded by the curve $x^2 = y^4(1 - y^3)$ (see figure).

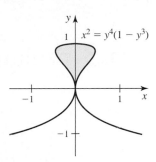

66. **Rewrite first** Find the area of the region bounded by the curve $x = \dfrac{1}{2y} - \sqrt{\dfrac{1}{4y^2} - 1}$ and the line $x = 1$ in the first quadrant. (*Hint:* Express y in terms of x.)

67. **Area function for a cubic** Consider the cubic polynomial $f(x) = x(x - a)(x - b),$ where $0 \leq a \leq b$.

a. For a fixed value of b, find the function $F(a) = \int_0^b f(x)\,dx$. For what value of a (which depends on b) is $F(a) = 0$?

b. For a fixed value of b, find the function $A(a)$ that gives the area of the region bounded by the graph of f and the x-axis between $x = 0$ and $x = b$. Graph this function and show that it has a minimum at $a = b/2$. What is the maximum value of $A(a)$, and where does it occur (in terms of b)?

68. **Differences of even functions** Assume f and g are even, integrable functions on $[-a, a],$ where $a > 1$. Suppose $f(x) > g(x) > 0$ on $[-a, a]$ and that the area bounded by the graphs of f and g on $[-a, a]$ is 10. What is the value of $\int_0^{\sqrt{a}} x[f(x^2) - g(x^2)]\,dx$?

69. **Roots and powers** Consider the functions $f(x) = x^n$ and $g(x) = x^{1/n},$ where $n \geq 2$ is a positive integer.

a. Graph f and g for $n = 2, 3,$ and 4 for $x \geq 0$.

b. Give a geometric interpretation of the area function $A_n(x) = \int_0^x (f(s) - g(s))\,ds$ for $n = 2, 3, 4, \ldots$ and $x > 0$.

c. Find the positive root of $A_n(x) = 0$ in terms of n. Does the root increase or decrease with n?

70. **Shifting sines** Consider the functions $f(x) = a \sin 2x$ and $g(x) = (\sin x)/a,$ where $a > 0$ is a real number.

a. Graph the two functions on the interval $[0, \pi/2]$ for $a = \frac{1}{2},$ 1, and 2.

b. Show that the curves have an intersection point x^* (other than $x = 0$) on $[0, \pi/2]$ that satisfies $\cos x^* = 1/(2a^2),$ provided $a \geq 1/\sqrt{2}$.

c. Find the area of the region between the two curves on $[0, x^*]$ when $a = 1$.

d. Show that as $a \to 1/\sqrt{2},$ the area of the region between the two curves on $[0, x^*]$ approaches zero.

6.3 Volume by Slicing

We have seen that integration is used to compute the area of two-dimensional regions bounded by curves. Integrals are also used to find the volume of three-dimensional regions (or solids). Once again, the slice-and-sum method is the key to solving these problems.

General Slicing Method

Consider a solid object that extends in the x-direction from $x = a$ to $x = b$. Imagine making a cut through the solid, perpendicular to the x-axis at a particular point x, and suppose the area of the cross section through the solid created by the cut is a known integrable function $A(x)$ (Figure 6.22).

To find the volume of this solid, we first divide $[a, b]$ into n subintervals of length $\Delta x = (b - a)/n$. The endpoints of the subintervals are the grid points $x_0 = a, x_1, x_2, \ldots, x_n = b$. We now make cuts through the solid perpendicular to the x-axis at each grid point, which produces n slices of thickness Δx. (Imagine cutting a loaf of bread to create n slices of equal width.) On each subinterval, an arbitrary point $\overline{x}_k$ is identified. The kth slice through the solid has a thickness Δx, and we take $A(\overline{x}_k)$ as a representative cross-sectional area of the slice. Therefore, the volume of the kth slice is approximately $A(\overline{x}_k)\,\Delta x$ (Figure 6.23). Summing the volumes of the slices, the approximate volume of the solid is

$$V \approx \sum_{k=1}^{n} A(\overline{x}_k)\,\Delta x.$$

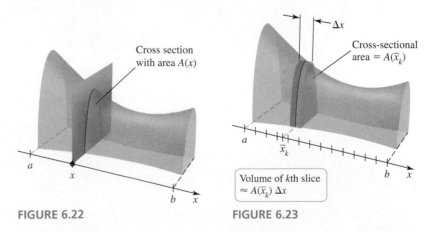

FIGURE 6.22   **FIGURE 6.23**

As the number of slices increases ($n \to \infty$) and the thickness of each slice goes to zero ($\Delta x \to 0$), the exact volume V is obtained in terms of a definite integral (Figure 6.24):

$$V = \lim_{n \to \infty} \sum_{k=1}^{n} A(\overline{x}_k)\,\Delta x = \int_a^b A(x)\,dx.$$

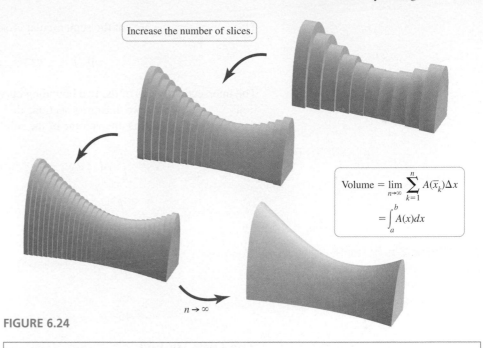

Increase the number of slices.

$$\text{Volume} = \lim_{n \to \infty} \sum_{k=1}^{n} A(\overline{x}_k)\Delta x$$

$$= \int_a^b A(x)dx$$

$n \to \infty$

FIGURE 6.24

> The factors in this volume integral have meaning: $A(x)$ is the cross-sectional area of a slice and dx is its thickness. Summing (integrating) the volumes of the slices $A(x)\,dx$ gives the volume of the solid.

General Slicing Method

Suppose a solid object extends from $x = a$ to $x = b$ and a cross section of the solid perpendicular to the x-axis at a particular point x has an area given by an integrable function $A(x)$. The volume of the solid is

$$V = \int_a^b A(x)\,dx.$$

QUICK CHECK 1 Explain why the volume, as given by the general slicing method, is equal to the average value of $A(x)$ on $[a, b]$ multiplied by $b - a$. ◄

EXAMPLE 1 **Volume of a "parabolic hemisphere"** A solid has a base that is bounded by the curves $y = x^2$ and $y = 2 - x^2$ in the xy-plane. Cross sections through the solid perpendicular to the x-axis are semicircular disks. Find the volume of the solid.

SOLUTION Because a typical cross section perpendicular to the x-axis is a semicircular disk (Figure 6.25), the area of a cross section is $\frac{1}{2}\pi r^2$, where r is the radius of the cross section. The key observation is that this radius is one-half of the distance between the upper bounding curve $y = 2 - x^2$ and the lower bounding curve $y = x^2$. So the radius at the point x is

$$r = \frac{1}{2}[(2 - x^2) - x^2] = 1 - x^2.$$

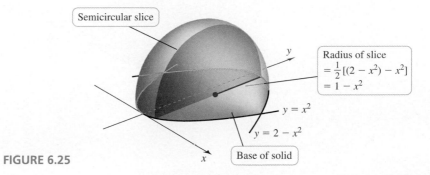

Semicircular slice

Radius of slice
$= \frac{1}{2}[(2 - x^2) - x^2]$
$= 1 - x^2$

y

$y = x^2$

$y = 2 - x^2$

x

Base of solid

FIGURE 6.25

This means that the area of the semicircular cross section at the point x is

$$A(x) = \frac{1}{2}\pi r^2 = \frac{\pi}{2}(1 - x^2)^2.$$

The intersection points of the two bounding curves satisfy $2 - x^2 = x^2$, which has solutions $x = \pm 1$. Therefore, the cross sections lie between $x = -1$ and $x = 1$. Integrating the cross-sectional areas, the volume of the solid is

$$
\begin{aligned}
V &= \int_{-1}^{1} A(x)\,dx & \text{General slicing method} \\[2mm]
&= \int_{-1}^{1} \frac{\pi}{2}(1 - x^2)^2\,dx & \text{Substitute for } A(x). \\[2mm]
&= \frac{\pi}{2}\int_{-1}^{1}(1 - 2x^2 + x^4)\,dx & \text{Expand integand.} \\[2mm]
&= \frac{8\pi}{15} & \text{Evaluate.}
\end{aligned}
$$

Related Exercises 7–14 ◄

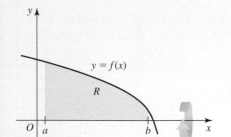

QUICK CHECK 2 In Example 1, what is the cross-sectional area function $A(x)$ if cross sections perpendicular to the base are squares rather than semicircles? ◄

The Disk Method

We now consider a specific type of solid known as a *solid of revolution*. Suppose f is a continuous function with $f(x) \geq 0$ on an interval $[a, b]$. Let R be the region bounded by the graph of f, the x-axis, and the lines $x = a$ and $x = b$ (Figure 6.26). Now revolve R around the x-axis. As R revolves once around the x-axis, it sweeps out a three-dimensional **solid of revolution** (Figure 6.27). The goal is to find the volume of this solid, and it may be done using the general slicing method.

$y = f(x)$

R

FIGURE 6.26

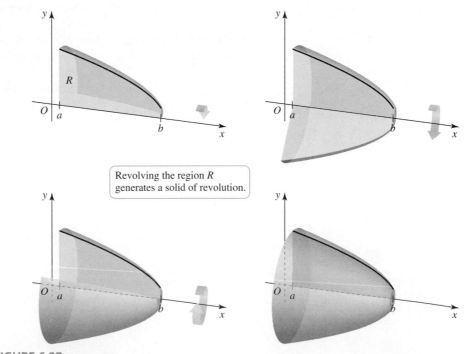

Revolving the region R generates a solid of revolution.

FIGURE 6.27

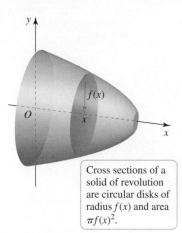

Cross sections of a solid of revolution are circular disks of radius $f(x)$ and area $\pi f(x)^2$.

FIGURE 6.28

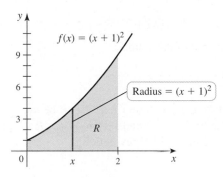

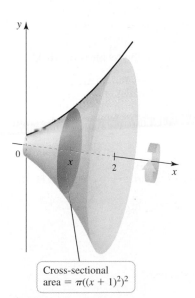

Cross-sectional area $= \pi((x+1)^2)^2$

FIGURE 6.29

QUICK CHECK 3 What solid results when the region R is revolved about the x-axis if (a) R is a square with vertices $(0, 0)$, $(0, 2)$, $(2, 0)$, and $(2, 2)$ and (b) R is a triangle with vertices $(0, 0)$, $(0, 2)$, and $(2, 0)$? ◄

With a solid of revolution, the cross-sectional area function has a special form because all cross sections perpendicular to the x-axis are *circular disks* with radius $f(x)$ (Figure 6.28). Therefore, the cross section at the point x, where $a \le x \le b$, has area

$$A(x) = \pi(\text{radius})^2 = \pi f(x)^2.$$

By the general slicing method, the volume of the solid is

$$V = \int_a^b A(x)\, dx = \int_a^b \pi f(x)^2\, dx.$$

Because each slice through the solid is a circular disk, the resulting method is called the **disk method**.

Disk Method About the x-Axis

Let f be continuous with $f(x) \ge 0$ on the interval $[a, b]$. If the region R bounded by the graph of f, the x-axis, and the lines $x = a$ and $x = b$ is revolved about the x-axis, the volume of the resulting solid of revolution is

$$V = \int_a^b \pi f(x)^2\, dx.$$

EXAMPLE 2   Disk method at work Let R be the region bounded by the curve $f(x) = (x + 1)^2$, the x-axis, and the lines $x = 0$ and $x = 2$. Find the volume of the solid of revolution obtained by revolving R about the x-axis.

SOLUTION When the region R is revolved about the x-axis, it generates a solid of revolution (Figure 6.29). A cross section perpendicular to the x-axis at the point $0 \le x \le 2$ is a circular disk of radius $f(x)$. Therefore, a typical cross section has area

$$A(x) = \pi f(x)^2 = \pi((x + 1)^2)^2.$$

Integrating these cross-sectional areas between $x = 0$ and $x = 2$ gives the volume of the solid:

$$V = \int_0^2 A(x)\, dx = \int_0^2 \pi\big((x + 1)^2\big)^2 dx \quad \text{Substitute for } A(x).$$

$$= \int_0^2 \pi(x + 1)^4\, dx \qquad \text{Simplify.}$$

$$= \pi \frac{u^5}{5}\bigg|_1^3 = \frac{242\,\pi}{5} \qquad \text{Let } u = x + 1 \text{ and evaluate.}$$

Related Exercises 15–22 ◄

Washer Method A slight variation on the disk method enables us to compute the volume of more exotic solids of revolution. Suppose that R is the region bounded by the graphs of f and g between $x = a$ and $x = b$, where $f(x) \ge g(x) \ge 0$ (Figure 6.30). If R is revolved about the x-axis to generate a solid of revolution, the resulting solid generally has a hole through it.

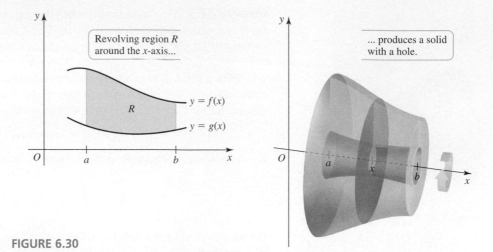

FIGURE 6.30

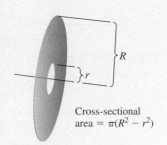

Cross-sectional
area = $\pi(R^2 - r^2)$

Once again we apply the general slicing method. In this case, a cross section through the solid perpendicular to the x-axis is a circular *washer* with an outer radius of $R = f(x)$ and a hole with a radius of $r = g(x)$, where $a \leq x \leq b$. The area of the cross section is the area of the entire disk minus the area of the hole, or

$$A(x) = \pi(R^2 - r^2) = \pi(f(x)^2 - g(x)^2)$$

(Figure 6.31). The general slicing method gives the area of the solid.

Washer Method About the x-Axis

Let f and g be continuous functions with $f(x) \geq g(x) \geq 0$ on $[a, b]$. Let R be the region bounded by $y = f(x)$, $y = g(x)$, and the lines $x = a$ and $x = b$. When R is revolved about the x-axis, the volume of the resulting solid of revolution is

$$V = \int_a^b \pi(f(x)^2 - g(x)^2)\, dx.$$

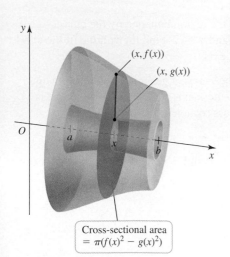

$(x, f(x))$

$(x, g(x))$

Cross-sectional area
$= \pi(f(x)^2 - g(x)^2)$

FIGURE 6.31

▶ The washer method is really two applications of the disk method. We compute the volume of the entire solid without the hole (by the disk method) and then subtract the volume of the hole (also computed by the disk method).

QUICK CHECK 4 Show that when $g(x) = 0$ in the washer method, the result is the disk method. ◀

EXAMPLE 3 Volume by the washer method The region R is bounded by the graphs of $f(x) = \sqrt{x}$ and $g(x) = x^2$ between $x = 0$ and $x = 1$. What is the volume of the solid that results when R is revolved about the x-axis?

SOLUTION The region R is bounded by the graphs of f and g with $f(x) \geq g(x)$ on $[0, 1]$, so the washer method is applicable (Figure 6.32). The area of a typical cross section at the point x is

$$A(x) = \pi(f(x)^2 - g(x)^2) = \pi((\sqrt{x})^2 - (x^2)^2) = \pi(x - x^4).$$

Therefore, the volume of the solid is

$$V = \int_0^1 \pi(x - x^4)\, dx \qquad \text{Washer method}$$

$$= \pi\left(\frac{x^2}{2} - \frac{x^5}{5}\right)\Big|_0^1 = \frac{3\pi}{10}. \qquad \text{Fundamental Theorem}$$

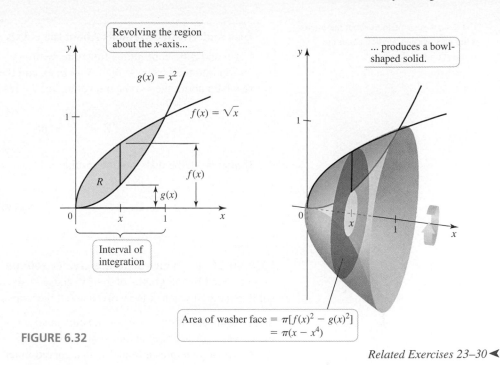

FIGURE 6.32

Related Exercises 23–30 ◄

➤ See Exercises 56–60 for problems in which regions are revolved about lines other than coordinate axes.

QUICK CHECK 5 Suppose the region in Example 3 is revolved about the line $y = -1$ instead of the x-axis. (a) What is the inner radius of a typical washer? (b) What is the outer radius of a typical washer? ◄

Revolving About the y-Axis

Everything you learned about revolving regions about the x-axis applies to revolving regions about the y-axis. Consider a region R bounded by the curve $x = p(y)$ on the right, the curve $x = q(y)$ on the left, and the horizontal lines $y = c$ and $y = d$ (Figure 6.33).

To find the volume of the solid generated when R is revolved about the y-axis, we use the general slicing method—now with respect to the y-axis. The area of a typical cross section is $A(y) = \pi(p(y)^2 - q(y)^2)$, where $c \leq y \leq d$. As before, integrating these cross-sectional areas of the solid gives the volume.

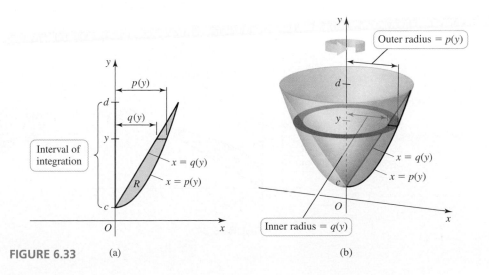

FIGURE 6.33   (a)   (b)

> The disk/washer method about the y-axis is the disk/washer method about the x-axis with x replaced by y.

Disk and Washer Methods About the y-Axis

Let p and q be continuous functions with $p(y) \geq q(y) \geq 0$ on $[c, d]$. Let R be the region bounded by $x = p(y)$, $x = q(y)$, and the lines $y = c$ and $y = d$. When R is revolved about the y-axis, the volume of the resulting solid of revolution is given by

$$V = \int_c^d \pi (p(y)^2 - q(y)^2)\, dy$$

If $q(y) = 0$, the disk method results:

$$V = \int_c^d \pi p(y)^2\, dy$$

EXAMPLE 4   Which solid has greater volume? Let R be the region in the first quadrant bounded by the graphs of $x = y^3$ and $x = 4y$. Which is greater, the volume of the solid generated when R is revolved about the x-axis or the y-axis?

SOLUTION Solving $y^3 = 4y$—or, equivalently, $y(y^2 - 4) = 0$—we find that the bounding curves of R intersect at the points $(0, 0)$ and $(8, 2)$. When the region R is revolved about the y-axis, it generates a funnel with a curved inner surface (Figure 6.34). Washer-shaped cross sections perpendicular to the y-axis extend from $y = 0$ to $y = 2$. The outer radius of the cross section at the point y is determined by the line $x = p(y) = 4y$. The inner radius of the cross section at the point y is determined by the curve $x = q(y) = y^3$. Applying the washer method, the volume of this solid is

$$V = \int_0^2 \pi(p(y)^2 - q(y)^2)\, dy \qquad \text{Washer method}$$

$$= \int_0^2 \pi(16y^2 - y^6)\, dy \qquad \text{Substitute for } p \text{ and } q.$$

$$= \pi\left(\frac{16}{3}y^3 - \frac{y^7}{7}\right)\Bigg|_0^2 \qquad \text{Fundamental Theorem}$$

$$= \frac{512\pi}{21} \approx 76.60. \qquad \text{Evaluate.}$$

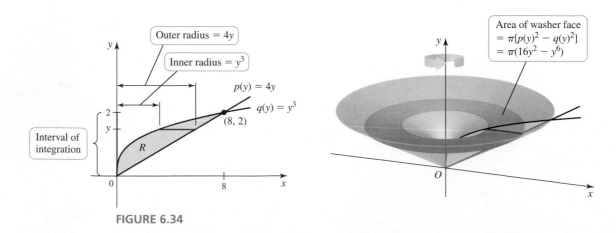

FIGURE 6.34

When the region R is revolved about the x-axis, it generates a funnel with a straight inner surface (Figure 6.35). Vertical slices through the solid between $x = 0$ and $x = 8$ produce washers. The outer radius of the washer at the point x is determined by the curve $x = y^3$, or $y = f(x) = x^{1/3}$. The inner radius is determined by $x = 4y$, or $y = g(x) = x/4$. The volume of the resulting solid is

$$V = \int_0^8 \pi(f(x)^2 - g(x)^2)\, dx \quad \text{Washer method}$$

$$= \int_0^8 \pi\left(x^{2/3} - \frac{x^2}{16}\right) dx \quad \text{Substitute for } f \text{ and } g.$$

$$= \pi\left(\frac{3}{5}x^{5/3} - \frac{x^3}{48}\right)\Bigg|_0^8 \quad \text{Fundamental Theorem}$$

$$= \frac{128\pi}{15} \approx 26.81. \quad \text{Evaluate.}$$

We see that revolving the region about the y-axis produces a solid of greater volume.

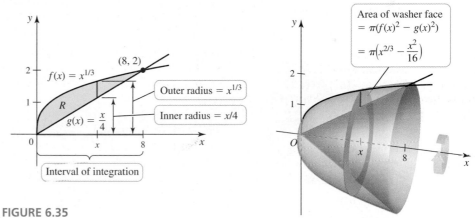

FIGURE 6.35

Related Exercises 31–40 ◄

QUICK CHECK 6 The region in the first quadrant bounded by $y = x$ and $y = x^3$ is revolved about the y-axis. Give the integral for the volume of the solid that is generated. ◄

SECTION 6.3 EXERCISES

Review Questions

1. Suppose a cut is made through a solid object perpendicular to the x-axis at a particular point x. Explain the meaning of $A(x)$.

2. Describe how a solid of revolution is generated.

3. The region bounded by the curves $y = 2x$ and $y = x^2$ is revolved about the x-axis. Give an integral for the volume of the solid that is generated.

4. The region bounded by the curves $y = 2x$ and $y = x^2$ is revolved about the y-axis. Give an integral for the volume of the solid that is generated.

5. Why is the disk method a special case of the general slicing method?

6. A solid has a circular base and cross sections perpendicular to the base are squares. What method should be used to find the volume of the solid?

Basic Skills

7–14. General slicing method *Use the general slicing method to find the volume of the following solids.*

7. A triangular wedge whose perpendicular sides have lengths 3, 4, and 5 (use calculus)

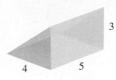

8. The solid with a circular base of radius 5 whose cross sections perpendicular to the base and parallel to the x-axis are equilateral triangles

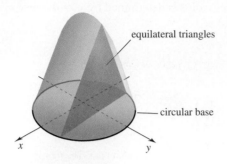

equilateral triangles

circular base

x y

9. The solid with a semicircular base of radius 5 whose cross sections perpendicular to the base and parallel to the diameter are squares

10. The solid whose base is the region bounded by $y = x^2$ and the line $y = 1$ and whose cross sections perpendicular to the base and parallel to the x-axis are squares

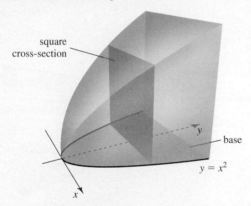

square cross-section

y

base

$y = x^2$

x

11. The solid whose base is the triangle with vertices $(0, 0)$, $(2, 0)$, and $(0, 2)$ and whose cross sections perpendicular to the base and parallel to the y-axis are semicircles

12. The pyramid with a square base 4 m on a side and a height of 2 m (Do not use the volume formula for a pyramid. Use slices that produce square cross sections.)

13. The tetrahedron (pyramid with four triangular faces), all of whose edges have length 4

14. A circular cylinder of radius r and height h whose curved surface is at an angle of $\pi/4$ rad (45°) to the base

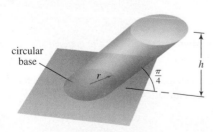

circular base

r

$\dfrac{\pi}{4}$

h

15–22. Disk method *Let R be the region bounded by the following curves. Use the disk method to find the volume of the solid generated when R is revolved about the x-axis.*

15. $y = 2x, y = 0, x = 3$ (Verify that your answer agrees with the volume formula for a cone.)

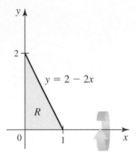

y

(3, 6)

$y = 2x$

R

0   3 x

16. $y = 2 - 2x, y = 0, x = 0$ (Verify that your answer agrees with the volume formula for a cone.)

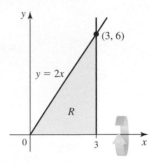

y

2

$y = 2 - 2x$

R

0   1 x

17. $y = e^{-x}, y = 0, x = 0, x = \ln 4$

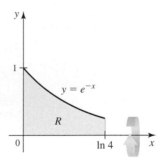

y

1

$y = e^{-x}$

R

0          $\ln 4$ x

18. $y = \cos x, y = 0, x = 0$ (Recall that $\cos^2 x = \frac{1}{2}(1 + \cos 2x)$.)

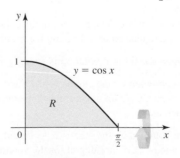

y

1

$y = \cos x$

R

0          $\dfrac{\pi}{2}$ x

19. $y = \sin x, y = 0$, for $0 \le x \le \pi$ (Recall that $\sin^2 x = \frac{1}{2}(1 - \cos 2x)$.)

20. $y = \sqrt{25 - x^2}, y = 0$ (Verify that your answer agrees with the volume formula for a sphere.)

21. $y = \dfrac{1}{\sqrt[4]{1 - x^2}}, y = 0, x = 0$ and $x = \frac{1}{2}$

22. $y = \sec x, y = 0, x = 0$ and $x = \pi/4$

23–30. Washer method *Let R be the region bounded by the following curves. Use the washer method to find the volume of the solid generated when R is revolved about the x-axis.*

23. $y = x, y = 2\sqrt{x}$ 24. $y = 2x, y = 16x^{1/4}$

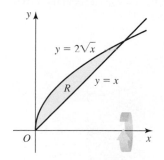

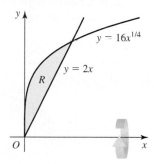

25. $y = e^{x/2}, y = e^{-x}, x = \ln 2, x = \ln 8$

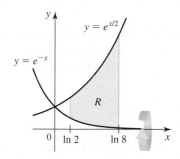

26. $y = x, y = x + 2, x = 0, x = 4$

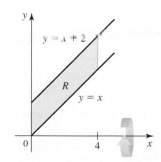

27. $y = 4x, y = 4x^2 - x^3$ 28. $y = \sqrt{\sin x}, y = 1, \; x = 0$

29. $y = \sin x, y = \sqrt{\sin x}$ for $0 \le x \le \pi/2$

30. $y = |x|, y = 12 - x^2$

31–36. Disks/washers about the y-axis *Let R be the region bounded by the following lines and curves. Use the disk or washer method to find the volume of the solid generated when R is revolved about the y-axis.*

31. $y = x, y = 2x, y = 6$ 32. $y = 0, y = \ln x, y = 2, x = 0$

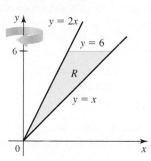

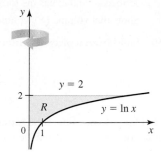

33. $y = x^3, y = 0, x = 2$ 34. $y = \sqrt{x}, y = 0, x = 4$

35. $x = \sqrt{16 - y^2}, x = 0$ 36. $y = \sin^{-1} x, x = 0, y = \pi/4$

37–40. Which is greater? *For the following regions R, determine which is greater—the volume of the solid generated when R is revolved about the x-axis or about the y-axis.*

37. R is bounded by $y = 2x$, the x-axis, and $x = 5$

38. R is bounded by $y = 4 - 2x$, the x-axis, and the y-axis

39. R is bounded by $y = 1 - x^3$, the x-axis, and the y-axis

40. R is bounded by $y = x^2$ and $y = \sqrt{8x}$

Further Explorations

41. **Explain why or why not** Determine whether the following statements are true and give an explanation or counterexample.

    a. A pyramid is a solid of revolution.

    b. The volume of a hemisphere can be computed using the disk method.

    c. Let R_1 be the region bounded by $y = \cos x$ and the x-axis on $[-\pi/2, \pi/2]$. Let R_2 be the region bounded by $y = \sin x$ and the x-axis on $[0, \pi]$. The volumes of the solids generated when R_1 and R_2 are revolved about the x-axis are equal.

42–48. Solids of revolution *Find the volume of the solid of revolution. Sketch the region in question.*

42. The region bounded by $y = (\ln x)/\sqrt{x}, y = 0$, and $x = 2$ revolved about the x-axis

43. The region bounded by $y = 1/\sqrt{x}, y = 0, x = 2$, and $x = 6$ revolved about the x-axis

44. The region bounded by $y = \sqrt{\tan x}, y = 1$, and $x = 0$ revolved about the x-axis

45. The region bounded by $y = e^x, y = 0, x = 0$, and $x = 2$ revolved about the x-axis

46. The region bounded by $y = e^{-x}, y = e^x, x = 0$, and $x = \ln 4$ revolved about the x-axis

47. The region bounded by $y = \ln x, y = \ln x^2$, and $y = \ln 8$ revolved about the y-axis

48. The region bounded by $y = e^{-x}$, $y = 0$, $x = 0$, and $x = p > 0$ revolved about the x-axis (Is the volume bounded as $p \to \infty$?)

49. Fermat's volume calculation (1636) Let R be the region bounded by $y = \sqrt{x + a}$ (with $a > 0$), the y-axis, and the x-axis. Let S be the solid generated by rotating R about the y-axis. Let T be the inscribed cone that has the same circular base as S and height $\sqrt{a}$. Show that volume $(S)/\text{volume}(T) = \frac{8}{5}$.

50. Solid from a piecewise function Let

$$f(x) = \begin{cases} x & \text{if } 0 \le x \le 2 \\ 2x - 2 & \text{if } 2 < x \le 5 \\ -2x + 18 & \text{if } 5 < x \le 6 \end{cases}$$

Find the volume of the solid formed when the region bounded by the graph of f, the x-axis, and the line $x = 6$ is revolved about the x-axis.

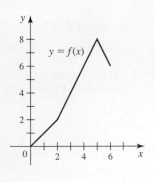

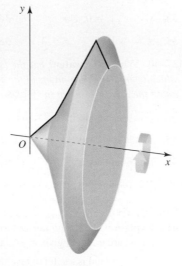

51. Solids from integrals Sketch the solid of revolution whose volume by the disk method is given by the following integrals. Indicate the function that generates the solid.

a. $\displaystyle\int_0^{\pi} \pi \sin^2 x \, dx$ **b.** $\displaystyle\int_0^2 \pi(x^2 + 2x + 1) \, dx$

Applications

52. Volume of a wooden object A solid wooden object turned on a lathe has a length of 50 cm and diameters (measured in cm) shown in the figure. (A lathe is a tool that spins and cuts a block of wood so that it has circular cross sections.) Use left Riemann sums to estimate the volume of the object.

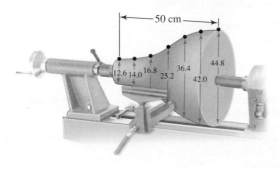

53. Cylinder, cone, hemisphere A right circular cylinder with height R and radius R has a volume of $V_C = \pi R^3$ (height = radius).

   a. Find the volume of the cone that is inscribed in the cylinder with the same base as the cylinder and height R. Express the volume in terms of V_C.

   b. Find the volume of the hemisphere that is inscribed in the cylinder with the same base as the cylinder. Express the volume in terms of V_C.

54. Water in a bowl A hemispherical bowl of radius 8 inches is filled to a depth of h in, where $0 \le h \le 8$. Find the volume of water in the bowl as a function of h. (Check the special cases $h = 0$ and $h = 8$.)

55. A torus (doughnut) Find the volume of the torus formed when the circle of radius 2 centered at $(3, 0)$ is revolved about the y-axis. Use geometry to evaluate the integral.

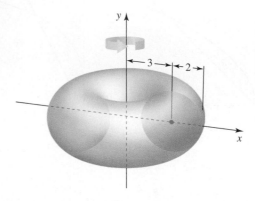

56–60. Different axes of revolution *Find the volume of the solid that is generated when the region in the first quadrant bounded by $y = x^2$, $y = 4$, and $x = 0$ is revolved about the following lines.*

56. y-axis   **57.** $y = -2$ **58.** $x = -1$

59. $y = 6$ **60.** $x = 2$

61. Different axes of revolution Suppose R is the region bounded by $y = f(x)$ and $y = g(x)$ on the interval $[a, b]$, where $f(x) \ge g(x) \ge 0$.

   a. Show that if R is revolved about the horizontal line $y = y_0$ that lies below R, then by the washer method, the volume of the resulting solid is

$$V = \int_a^b \pi[(f(x) - y_0)^2 - (g(x) - y_0)^2] \, dx.$$

   b. How is this formula changed if the line $y = y_0$ lies above R?

62. Which is greater? Let R be the region bounded by $y = x^2$ and $y = \sqrt{x}$. Which is greater, the volume of the solid generated when R is revolved about the x-axis or about the line $y = 1$?

Additional Exercises

63. Cavalieri's principle *Cavalieri's principle* states that if two solids of equal altitudes have the same cross-sectional areas at every height, then they have equal volumes (see figure).

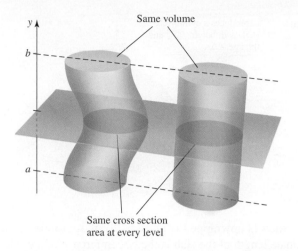

Same volume

Same cross section
area at every level

a. Use the theory of this section to justify Cavalieri's principle.
b. Find the radius of a circular cylinder of height 10 m that has the same volume as a 2-m by 2-m by 10-m box.

64. Limiting volume Consider the region R in the first quadrant bounded by $y = x^{1/n}$ and $y = x^n$, where n is a positive number.

  a. Find the volume $V(n)$ of the solid generated when R is revolved about the x-axis. Express your answer in terms of n.
  b. Evaluate $\lim_{n \to \infty} V(n)$. Interpret this limit geometrically.

QUICK CHECK ANSWERS

1. The average value of A on $[a, b]$ is $\overline{A} = \dfrac{1}{b - a} \displaystyle\int_a^b A(x)\,dx$.
Therefore, $V = (b - a)\overline{A}$.   **2.** $A(x) = (2 - 2x^2)^2$
3. (a) A cylinder with height 2 and radius 2; (b) a cone with height 2 and base radius 2   **4.** When $g(x) = 0$, the washer method $V = \int_a^b \pi(f(x)^2 - g(x)^2)\,dx$ reduces to the disk method $V = \int_a^b \pi(f(x)^2)\,dx$.   **5.** (a) Inner radius $= \sqrt{x} + 1$; (b) outer radius $= x^2 + 1$ **6.** $\int_0^1 \pi(y^{2/3} - y^2)\,dy$ ◄

6.4 Volume by Shells

You can solve a lot of challenging volume problems using the disk/washer method. There are, however, some volume problems that are difficult to solve with this method. For this reason, we extend our discussion of volume problems to the *shell method*, which—like the disk/washer method—is used to compute the volume of solids of revolution.

Cylindrical Shells

Let R be a region bounded by the graph of f, the x-axis, and the lines $x = a$ and $x = b$, where $0 \le a < b$ and $f(x) \ge 0$ on $[a, b]$. When R is revolved about the y-axis, a solid is generated (Figure 6.36) whose volume is computed with the slice-and-sum strategy.

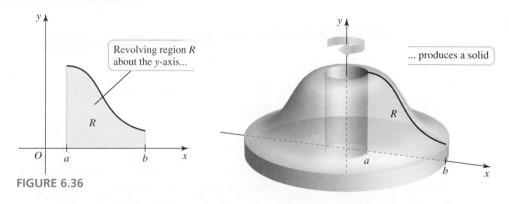

Revolving region R about the y-axis...

... produces a solid

FIGURE 6.36

We divide $[a, b]$ into n subintervals of length $\Delta x = (b - a)/n$, and identify an arbitrary point $\bar{x}_k$ on the kth subinterval for $k = 1, \ldots, n$. Now observe the rectangle built on the kth subinterval with a height of $f(\bar{x}_k)$ and a width Δx (Figure 6.37). As it revolves about the y-axis, this rectangle sweeps out a thin *cylindrical shell*.

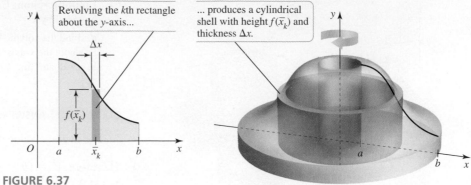

FIGURE 6.37

When the kth cylindrical shell is unwrapped (Figure 6.38), it approximates a thin rectangular slab. The approximate length of the slab is the circumference of a circle with radius $\overline{x}_k$, which is $2\pi\overline{x}_k$. The height of the slab is the height of the original rectangle $f(\overline{x}_k)$ and its thickness is Δx; therefore, the volume of the kth shell is approximately

$$\underbrace{2\pi\overline{x}_k}_{\text{length}} \cdot \underbrace{f(\overline{x}_k)}_{\text{height}} \cdot \underbrace{\Delta x}_{\text{thickness}} = 2\pi\overline{x}_k f(\overline{x}_k)\,\Delta x.$$

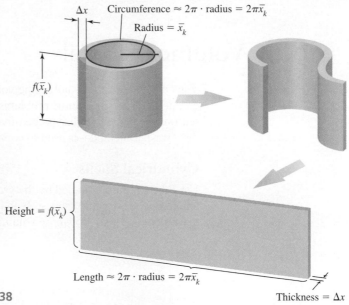

FIGURE 6.38

Summing the volumes of the n cylindrical shells gives an approximation to the volume of the entire solid:

$$V \approx \sum_{k=1}^{n} 2\pi\overline{x}_k f(\overline{x}_k)\,\Delta x$$

As n increases and as Δx approaches 0 (Figure 6.39), we obtain the exact volume of the solid as a definite integral:

$$V = \lim_{n\to\infty} \sum_{k=1}^{n} 2\pi\,\underbrace{\overline{x}_k}_{\substack{\text{shell}\\\text{circumference}}}\,\overbrace{f(\overline{x}_k)}^{\substack{\text{shell}\\\text{height}}}\,\underbrace{\Delta x}_{\substack{\text{shell}\\\text{thickness}}} = \int_a^b 2\pi x f(x)\,dx$$

> Rather than memorizing, think of the meaning of the factors in this formula: $f(x)$ is the height of a single cylindrical shell, $2\pi x$ is the circumference of the shell, and dx corresponds to the thickness of a shell. Therefore, $2\pi x f(x)\, dx$ represents the volume of a single shell, and we sum the volumes from $x = a$ to $x = b$. Notice that the integrand for the shell method is the function $A(x)$ that gives the surface area of the shell of radius x for $a \le x \le b$.

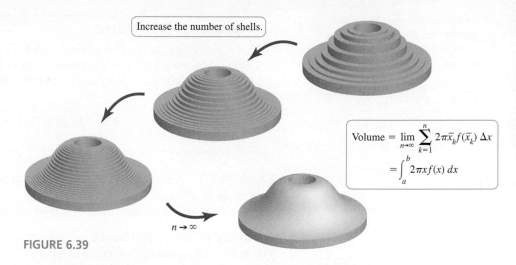

Increase the number of shells.

$$\text{Volume} = \lim_{n\to\infty} \sum_{k=1}^{n} 2\pi \overline{x}_k f(\overline{x}_k)\, \Delta x$$
$$= \int_a^b 2\pi x f(x)\, dx$$

$n \to \infty$

FIGURE 6.39

Before doing examples, we generalize this method as we did for the disk method. Suppose that the region R is bounded by two curves, $y = f(x)$ and $y = g(x)$, where $f(x) \ge g(x)$ on $[a, b]$ (Figure 6.40). What is the volume of the solid generated when R is revolved about the y-axis?

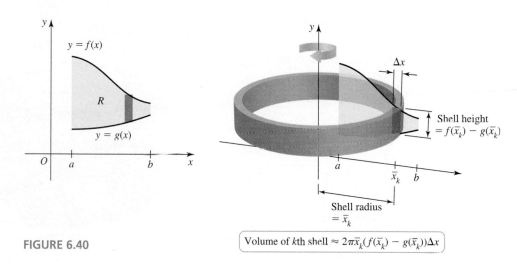

FIGURE 6.40

Volume of kth shell $\approx 2\pi \overline{x}_k(f(\overline{x}_k) - g(\overline{x}_k))\Delta x$

The situation is similar to the case we just considered. A typical rectangle in R sweeps out a cylindrical shell, but now the height of the kth shell is $f(\overline{x}_k) - g(\overline{x}_k)$, for $k = 1, \dots n$. As before, we take the radius of the kth shell to be $\overline{x}_k$, which means the volume of the kth shell is approximated by $2\pi \overline{x}_k(f(\overline{x}_k) - g(\overline{x}_k))\, \Delta x$ (Figure 6.40). Summing the volumes of all the shells gives an approximation to the volume of the entire solid:

$$V \approx \sum_{k=1}^{n} \underbrace{2\pi \overline{x}_k}_{\substack{\text{circumference} \\ \text{of shell}}} \underbrace{(f(\overline{x}_k) - g(\overline{x}_k))}_{\text{height of shell}}\Delta x$$

Taking the limit as $n \to \infty$ (which implies that $\Delta x \to 0$), the exact volume is the definite integral

$$V = \lim_{n\to\infty} \sum_{k=1}^{n} 2\pi\, \overline{x}_k(f(\overline{x}_k) - g(\overline{x}_k))\, \Delta x = \int_a^b 2\pi x\, (f(x) - g(x))\, dx.$$

> An analogous formula for the shell method when R is revolved about the x-axis is obtained by reversing the roles of x and y:

$$V = \int_c^d 2\pi y(f(y) - g(y)) \, dy$$

Volume by the Shell Method

Let f and g be continuous functions with $f(x) \geq g(x)$ on $[a, b]$. If R is the region bounded by the curves $y = f(x)$ and $y = g(x)$ between the lines $x = a$ and $x = b$, the volume of the solid generated when R is revolved about the y-axis is

$$V = \int_a^b 2\pi x(f(x) - g(x)) \, dx.$$

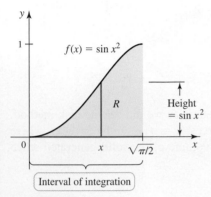

FIGURE 6.41

> When computing volumes using the shell method, it is best to sketch the region R in the xy-plane and draw a slice through the region that generates a typical shell.

EXAMPLE 1   A sine bowl Let R be the region bounded by the graph of $f(x) = \sin x^2$, the x-axis, and the vertical line $x = \sqrt{\pi/2}$ (Figure 6.41). Find the volume of the solid generated when R is revolved about the y-axis.

SOLUTION Revolving R about the y-axis produces a bowl-shaped region (Figure 6.42). The radius of a typical cylindrical shell is x and its height is $f(x) = \sin x^2$. Therefore, the volume by the shell method is

$$V = \int_a^b 2\pi x f(x) \, dx = \int_0^{\sqrt{\pi/2}} 2\pi x \sin x^2 \, dx.$$

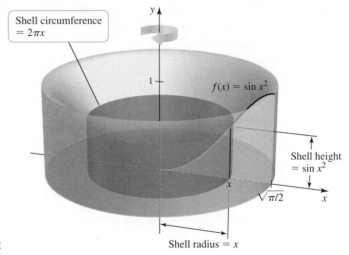

FIGURE 6.42

Now we make the change of variables $u = x^2$, which means that $du = 2x \, dx$. The lower limit $x = 0$ becomes $u = 0$ and the upper limit $x = \sqrt{\pi/2}$ becomes $u = \pi/2$. The volume of the solid is

$$V = \int_0^{\sqrt{\pi/2}} 2\pi x \sin x^2 \, dx = \pi \int_0^{\pi/2} \sin u \, du \qquad u = x^2, du = 2x \, dx$$

$$= \pi \left(-\cos u \right) \Big|_0^{\pi/2} \qquad \text{Fundamental Theorem}$$

$$= \pi[0 - (-1)] = \pi. \quad \text{Simplify.}$$

Related Exercises 5–10 ◄

QUICK CHECK 1 The triangle bounded by the x-axis, the line $y = 2x$, and the line $x = 1$ is revolved about the y-axis. Give an integral that equals the volume of the resulting solid using the shell method? ◄

> We could use the disk/washer method to compute the volume, but notice that this approach requires splitting the region into two subregions. A better approach is to use the shell method and integrate along the y-axis.

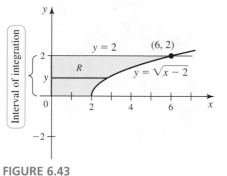

FIGURE 6.43

EXAMPLE 2 Shells about the x-axis

Let R be the region in the first quadrant bounded by the graph of $y = \sqrt{x - 2}$ and the line $y = 2$.

a. Find the volume of the solid generated when R is revolved about the x-axis.

b. Find the volume of the solid generated when R is revolved about the line $y = -2$.

SOLUTION

a. The revolution is about the x-axis, so the integration in the shell method is with respect to y. A typical shell runs parallel to the x-axis and has radius y, where $0 \le y \le 2$; the shells extend from the y-axis to the curve $y = \sqrt{x - 2}$ (Figure 6.43). Solving $y = \sqrt{x - 2}$ for x, we have $x = y^2 + 2$, which is the height of the shell at the point y (Figure 6.44a). Integrating with respect to y, the volume of the region is

$$V = \int_0^2 \underbrace{2\pi y}_{\substack{\text{shell} \\ \text{circumference}}} \underbrace{(y^2 + 2)}_{\substack{\text{shell} \\ \text{height}}} dy = 2\pi \int_0^2 (y^3 + 2y)\, dy = 16\pi.$$

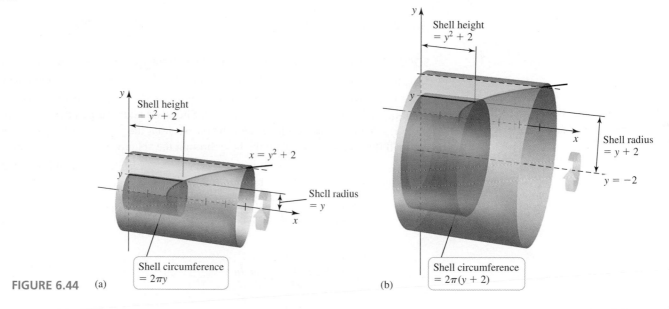

FIGURE 6.44   (a)   (b)

b. Revolving R about the line $y = -2$ produces a solid with a cylindrical hole through it (Figure 6.44b). To find the volume of this solid, we carry out the calculation in part (a) with a single change: The radius of a typical shell at a point y is now $y + 2$ (the distance to the x-axis plus 2 units from the x-axis to the axis of revolution). With this change, the volume of this solid is

$$V = \int_0^2 \underbrace{2\pi(y + 2)}_{\substack{\text{shell} \\ \text{circumference}}} \underbrace{(y^2 + 2)}_{\substack{\text{shell} \\ \text{height}}} dy = 2\pi \int_0^2 (y^3 + 2y^2 + 2y + 4)\, dy = \frac{128\pi}{3}.$$

Related Exercises 11–20 ◀

QUICK CHECK 2 Write the volume integral in Example 2 in the case that R is revolved about the line $y = -5$. ◀

EXAMPLE 3   Volume of a drilled sphere A cylindrical hole with radius r is drilled symmetrically through the center of a sphere with radius R, where $r \leq R$. What is the volume of the remaining material?

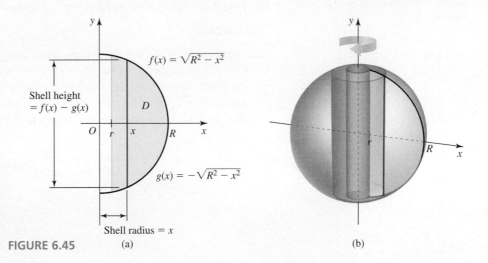

FIGURE 6.45   (a)   (b)

SOLUTION The y-axis is chosen to coincide with the axis of the cylindrical hole. We let D be the region in the xy-plane bounded above by $f(x) = \sqrt{R^2 - x^2}$, the upper half of a circle of radius R, and bounded below by $g(x) = -\sqrt{R^2 - x^2}$, the lower half of a circle of radius R, for $r \leq x \leq R$ (Figure 6.45a). Slices are taken perpendicular to the x-axis from $x = r$ to $x = R$. When a slice is revolved about the y-axis, it sweeps out a cylindrical shell that is concentric with the hole through the sphere (Figure 6.45b). The radius of a typical shell is x and its height is $f(x) - g(x) = 2\sqrt{R^2 - x^2}$. Therefore, the volume of the material that remains in the sphere is

$$V = \int_r^R 2\pi x \left(2\sqrt{R^2 - x^2}\right) dx$$

$$= -2\pi \int_{R^2-r^2}^0 \sqrt{u} \, du \qquad u = R^2 - x^2, du = -2x \, dx$$

$$= 2\pi \left(\frac{2}{3}u^{3/2}\right)\Big|_0^{R^2-r^2} \qquad \text{Fundamental Theorem}$$

$$= \frac{4\pi}{3}\left(R^2 - r^2\right)^{3/2}. \qquad \text{Simplify.}$$

It is important to check the result by looking at special cases. In the case that $r = R$ (the radius of the hole equals the radius of the sphere), our calculation gives a volume of 0, which is correct. In the case that $r = 0$ (no hole in the sphere), our calculation gives the correct volume of a sphere, $\frac{4}{3}\pi R^3$.   *Related Exercises 21–26* ◄

Restoring Order

After working with slices, disks, washers, and shells, you may feel somewhat overwhelmed. How do you choose a method and which method is best?

First, notice that the disk method is just a special case of the washer method. So, for solids of revolution, the choice is between the washer method and the shell method. In *principle*, either method can be used. In *practice*, one method usually produces an integral that is easier to evaluate than the other method. The following table summarizes these methods.

SUMMARY Disk/washer and Shell Methods

Integration with respect to x	**Disk/washer method about the x-axis** Disks/washers are *perpendicular* to the x-axis.
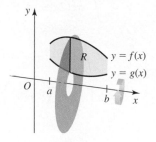	$$\int_a^b \pi(f(x)^2 - g(x)^2)\,dx$$
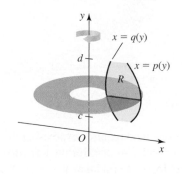	**Shell method about the y-axis** Shells are *parallel* to the y-axis. $$\int_a^b 2\pi x(f(x) - g(x))\,dx$$
Integration with respect to y	**Disk/washer method about the y-axis** Disks/washers are *perpendicular* to the y-axis.
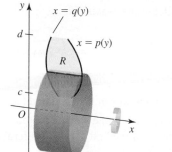	$$\int_c^d \pi(p(y)^2 - q(y)^2)\,dy$$
	Shell method about the x-axis Shells are *parallel* to the x-axis. $$\int_c^d 2\pi y(p(y) - q(y))\,dy$$

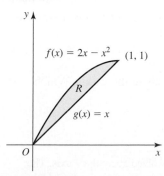

FIGURE 6.46

EXAMPLE 4 Volume by which method? The region R is bounded by the graphs of $f(x) = 2x - x^2$ and $g(x) = x$ on the interval $[0, 1]$ (Figure 6.46). Use the washer method and the shell method to find the volume of the solid formed when R is revolved about the x-axis.

SOLUTION Solving $f(x) = g(x)$, we find that the curves intersect at the points $(0, 0)$ and $(1, 1)$. Using the washer method, the upper bounding curve is the graph of f, the lower bounding curve is the graph of g, and a typical washer is perpendicular to the x-axis (Figure 6.47). Therefore, the volume is

$$V = \int_0^1 \pi\big[(2x - x^2)^2 - x^2\big]\, dx \quad \text{Washer method}$$

$$= \pi \int_0^1 (x^4 - 4x^3 + 3x^2)\, dx \quad \text{Expand integrand.}$$

$$= \pi \left(\frac{x^5}{5} - x^4 + x^3\right)\bigg|_0^1 = \frac{\pi}{5}. \quad \text{Evaluate integral.}$$

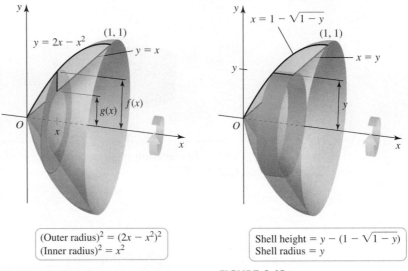

▶ To solve $y = 2x - x^2$ for x, write the equation as $x^2 - 2x + y = 0$ and complete the square or use the quadratic formula.

(Outer radius)^2 = $(2x - x^2)^2$
(Inner radius)^2 = x^2

FIGURE 6.47

Shell height = $y - (1 - \sqrt{1 - y})$
Shell radius = y

FIGURE 6.48

The shell method requires expressing the bounding curves in the form $x = p(y)$ for the right curve and $x = q(y)$ for the left curve. The right curve is $x = y$. Solving $y = 2x - x^2$ for x, we find that $x = 1 - \sqrt{1 - y}$ describes the left curve. A typical shell is parallel to the x-axis (Figure 6.48). Therefore, the volume is

$$V = \int_0^1 2\pi y \big[\underbrace{y}_{p(y)} - \underbrace{(1 - \sqrt{1 - y})}_{q(y)}\big]\, dy.$$

Although this integral can be evaluated $\left(\text{and equals } \frac{\pi}{5}\right)$, it is decidedly more difficult than the integral required by the washer method. In this case, the washer method is preferable. Of course, the shell method may be preferable for other problems.

Related Exercises 27–32 ◀

QUICK CHECK 3 Suppose the region in Example 4 is revolved about the y-axis. Which method (washer or shell) leads to an easier integral? ◀

SECTION 6.4 EXERCISES

Review Questions

1. Assume $f(x) \geq g(x) \geq 0$ on $[a, b]$. The region bounded by the graphs of f and g and the lines $x = a$ and $x = b$ is revolved about the y-axis. Write the integral given by the shell method that equals the volume of the resulting solid.

2. Fill in the blanks: A region R is revolved about the y-axis. The volume of the resulting solid could (in principle) be found by using the disk/washer method and integrating with respect to _____ or using the shell method and integrating with respect to _____.

3. Fill in the blanks: A region R is revolved about the x-axis. The volume of the resulting solid could (in principle) be found by using the disk/washer method and integrating with respect to _____ or using the shell method and integrating with respect to _____ .

4. Are shell method integrals easier to evaluate than washer method integrals? Explain.

Basic Skills

5–10. Shell method *Let R be the region bounded by the following curves. Use the shell method to find the volume of the solid generated when R is revolved about the y-axis.*

5. $y = (1 + x^2)^{-1}, y = 0, x = 0,$ and $x = 2$

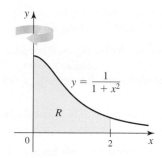

6. $y = 6 - x, y = 0, x = 2,$ and $x = 4$

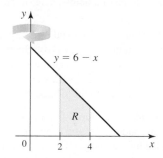

7. $y = 3x, y = 3,$ and $x = 0$ (Do not use the volume formula for a cone.)

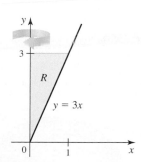

8. $y = \sqrt{x}, y = 0,$ and $x = 4$

9. $y = \cos x^2, y = 0,$ for $0 \le x \le \sqrt{\pi/2}$

10. $y = \sqrt{4 - 2x^2}, y = 0,$ and $x = 0,$ in the first quadrant

11–16. Shell method *Let R be the region bounded by the following curves. Use the shell method to find the volume of the solid generated when R is revolved about the x-axis.*

11. $y = \sqrt{x}, y = 0,$ and $x = 4$

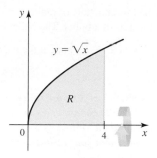

12. $y = 8, y = 2x + 2, x = 0,$ and $x = 2$

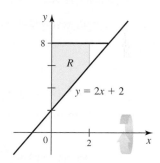

13. $y = 4 - x, y = 2,$ and $x = 0$

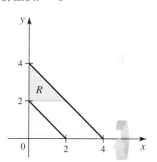

14. $y = x^3, y = 8,$ and $x = 0$

15. $y = 2x^{-3/2}, y = 2, y = 16,$ and $x = 0$

16. $y = \sqrt{50 - 2x^2},$ in the first quadrant

17–20. Shell method about other lines *Let R be the region bounded by $y = x^2, x = 1,$ and $y = 0$. Use the shell method to find the volume of the solid generated when R is revolved about the following lines.*

17. $x = -2$ **18.** $x = 2$ **19.** $y = -2$ **20.** $y = 2$

21–26. Shell method *Use the shell method to find the volume of the following solids.*

21. A right circular cone of radius 3 and height 8

22. The solid formed when a hole of radius 2 is drilled symmetrically along the axis of a right circular cylinder of height 6 and radius 4

23. The solid formed when a hole of radius 3 is drilled symmetrically along the axis of a right circular cone of radius 6 and height 9

24. The solid formed when a hole of radius 3 is drilled symmetrically through the center of a sphere of radius 6

25. The *ellipsoid* formed when the ellipse $x^2 + 2y^2 = 4$ is revolved about the y-axis

26. The solid formed when a hole of radius $r \le R$ is drilled symmetrically along the axis of a bullet. The bullet is formed by revolving the parabola $y = 6\left(1 - \dfrac{x^2}{R^2}\right)$ about the y-axis, where $0 \le x \le R$.

27–32. Washers vs. shells *Let R be the region bounded by the following curves. Let S be the solid generated when R is revolved about the given axis. If possible, find the volume of S by both the disk/washer and shell methods. Check that your results agree and state which method is easiest to apply.*

27. $y = x, y = x^{1/3}$; in the first quadrant; revolved about the x-axis

28. $y = x^2/8, y = 2 - x, x = 0$; revolved about the y-axis

29. $y = 1/(x + 1), y = 1 - x/3$; revolved about the x-axis

30. $y = (x - 2)^3 - 2, x = 0, y = 25$; revolved about the y-axis

31. $y = \sqrt{\ln x}, y = \sqrt{\ln x^2}, y = 1$; revolved about the x-axis

32. $y = 6/(x + 3), y = 2 - x$; revolved about the x-axis

Further Explorations

33. Explain why or why not Determine whether the following statements are true and give an explanation or counterexample.

   a. When using the shell method, the axis of the cylindrical shells is parallel to the axis of revolution.

   b. If a region is revolved about the y-axis, then the shell method must be used.

   c. If a region is revolved about the x-axis, then in principle it is possible to use the disk/washer method and integrate with respect to x or the shell method and integrate with respect to y.

34–38. Solids of revolution *Find the volume of the following solids of revolution. Sketch the region in question.*

34. The region bounded by $y = (\ln x)/x^2, y = 0, x = 1$, and $x = 3$ revolved about the y-axis

35. The region bounded by $y = 1/x^2, y = 0, x = 2$, and $x = 8$ revolved about the y-axis

36. The region bounded by $y = 1/(x^2 + 1), y = 0, x = 1$, and $x = 4$ revolved about the y-axis

37. The region bounded by $y = e^x/x, y = 0, x = 1$, and $x = 2$ revolved about the y-axis

38. The region bounded by $y^2 = \ln x, y^2 = \ln x^3$, and $y = 2$ revolved about the x-axis

39–46. Choose your method *Find the volume of the following solids using the method of your choice.*

39. The solid formed when the region bounded by $y = x^2$ and $y = 2 - x^2$ is revolved about the x-axis

40. The solid formed when the region bounded by $y = \sin x$ and $y = 1 - \sin x$ between $x = \pi/6$ and $x = 5\pi/6$ is revolved about the x-axis

41. The solid formed when the region bounded by $y = x, y = 2x + 2$, $x = 2$, and $x = 6$ is revolved about the y-axis

42. The solid formed when the region bounded by $y = x^3$, the x-axis, and $x = 2$ is revolved about the x-axis

43. The solid whose base is the region bounded by $y = x^2$ and the line $y = 1$ and whose cross sections perpendicular to the base and parallel to the x-axis are semicircles

44. The solid formed when the region bounded by $y = 2, y = 2x + 2$, and $x = 6$ is revolved about the y-axis

45. The solid whose base is the square with vertices $(1, 0), (0, 1)$, $(-1, 0)$, and $(0, -1)$ and whose cross sections perpendicular to the base and perpendicular to the x-axis are semicircles

46. The solid formed when the region bounded by $y = \sqrt{x}$, the x-axis, and $x = 4$ is revolved about the x-axis

47. Equal volumes Consider the region R bounded by the curves $y = ax^2 + 1, y = 0, x = 0$, and $x = 1$, for $a \ge -1$. Let S_1 and S_2 be solids generated when R is revolved about the x- and y-axes, respectively.

   a. Find V_1 and V_2, the volumes of S_1 and S_2, as functions of a.

   b. Are there values of $a \ge -1$ for which $V_1(a) = V_2(a)$?

48. A hemisphere by several methods Let R be the region in the first quadrant bounded by the circle $x^2 + y^2 = r^2$ and the coordinate axes. Find the volume of a hemisphere of radius r in the following ways.

   a. Revolve R about the x-axis and use the disk method.

   b. Revolve R about the x-axis and use the shell method.

   c. Assume the base of the hemisphere is in the xy-plane and use the general slicing method with slices perpendicular to the xy-plane and parallel to the x-axis.

49. A cone by two methods Verify that the volume of a right circular cone with a base radius of r and a height of h is $\pi r^2 h/3$. Use the region bounded by the line $y = rx/h$, the x-axis, and the line $x = h$, where the region is rotated around the x-axis. Then (a) use the disk method and integrate with respect to x, and (b) use the shell method and integrate with respect to y.

50. A spherical cap Consider the cap of thickness h that has been sliced from a sphere of radius r (see figure). Verify that the volume of the cap is $\pi h^2 (3r - h)/3$ using (a) the washer method, (b) the shell method, and (c) the general slicing method. Check for consistency among the three methods and check the special cases $h = r$ and $h = 0$.

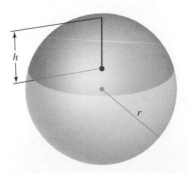

Applications

51. **Water in a bowl** A hemispherical bowl of radius 8 inches is filled to a depth of h inches, where $0 \le h \le 8$ ($h = 0$ corresponds to an empty bowl). Use the shell method to find the volume of water in the bowl as a function of h. (Check the special cases $h = 0$ and $h = 8$.)

52. **Wedge from a tree** Imagine a cylindrical tree of radius a. A wedge is cut from the tree by making two cuts: one in a horizontal plane P perpendicular to the axis of the cylinder, and one that makes an angle θ with P, intersecting P along a diameter of the cylinder (see figure). What is the volume of the wedge?

T 53. **A torus (doughnut)** Find the volume of the torus formed when a circle of radius 2 centered at $(3, 0)$ is revolved about the y-axis. Use the shell method. You may need a computer algebra system or table of integrals to evaluate the integral.

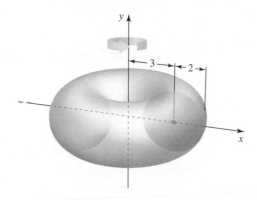

54. **Different axes of revolution** Suppose R is the region bounded by $y = f(x)$ and $y = g(x)$ on the interval $[a, b]$, where $f(x) \ge g(x)$.
   a. Show that if R is revolved about the vertical line $x = x_0$, where $x_0 < a$, then by the shell method, the volume of the resulting solid is $V = \int_a^b 2\pi(x - x_0)(f(x) - g(x))\,dx$.
   b. How is this formula changed if $x_0 > b$?

Additional Exercises

55. **Ellipsoids** An ellipse centered at the origin is described by the equation $x^2/a^2 + y^2/b^2 = 1$. If an ellipse R is revolved about either axis, the resulting solid is an *ellipsoid*.
   a. Find the volume of the ellipsoid generated when R is revolved about the x-axis (in terms of a and b).
   b. Find the volume of the ellipsoid generated when R is revolved about the y-axis (in terms of a and b).
   c. Should the results of parts (a) and (b) agree? Explain.

56. **Change of variables** Suppose $f(x) > 0$ for all x and $\int_0^4 f(x)\,dx = 10$. Let R be the region in the first quadrant bounded by the coordinate axes, $y = f(x^2)$, and $x = 2$. Find the volume of the solid generated by revolving R around the y-axis.

57. **Equal integrals** Without evaluating integrals, explain why the following equalities are true. (*Hint:* Draw pictures.)

   a. $\pi \displaystyle\int_0^4 (8 - 2x)^2\,dx = 2\pi \int_0^8 y\left(4 - \frac{y}{2}\right)dy$

   b. $\displaystyle\int_0^2 (25 - (x^2 + 1)^2)\,dx = 2\int_1^5 y\sqrt{y - 1}\,dy$

58. **Volumes without calculus** Solve the following problems with *and* without calculus. A good picture helps!
   a. A cube with side length r is inscribed in a sphere, which is inscribed in a right circular cone, which is inscribed in a right circular cylinder. The side length (slant height) of the cone is equal to its diameter. What is the volume of the cylinder?
   b. A cube is inscribed in a right circular cone with a radius of 1 and a height of 3. What is the volume of the cube?
   c. A cylindrical hole 10 in long is drilled symmetrically through the center of a sphere. How much material is left in the sphere? (There *is* enough information given.)

1. $\int_0^1 2\pi x(2x)\,dx$ 2. $V = \int_0^2 2\pi(y + 5)(y^2 + 2)\,dy$
3. The shell method is easier. ◄

6.5 Length of Curves

A space shuttle orbits Earth in an elliptical path. How far does it travel in one orbit? A baseball slugger launches a home run into the upper deck and the sportscaster claims it landed 480 feet from home plate. But how far did the ball actually travel along its flight path? These questions deal with the length of trajectories or, more generally, with *arc length*. As you will see, their answers can be found by integration.

There are two common ways to formulate problems about arc length: The curve may be given explicitly in the form $y = f(x)$ or it may be defined *parametrically*. In this section we deal with the first case. Parametric curves are introduced in Section 10.1 and the associated arc length problem is discussed in Section 11.8.

Arc Length for $y = f(x)$

Suppose a curve is given by $y = f(x)$, where f is a function with a continuous first derivative on the interval $[a, b]$. The goal is to determine how far you would travel if you walked along the curve from $(a, f(a))$ to $(b, f(b))$. This distance is the arc length, which we denote L.

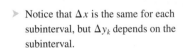 *More generally, we may choose any point in the kth subinterval and Δx may vary from one subinterval to the next. Using right endpoints, as we do here, simplifies the discussion and leads to the same result.*

As shown in Figure 6.49, we divide $[a, b]$ into n subintervals of length $\Delta x = (b - a)/n$, where x_k is the right endpoint of the kth subinterval, for $k = 1, \dots, n$. Joining the corresponding points on the curve by line segments, we obtain a polygonal line with n line segments. If n is large and Δx is small, the length of the polygonal line is a good approximation to the length of the actual curve. The strategy is to find the length of the polygonal line and then let n increase, while Δx goes to zero, to get the exact length of the curve.

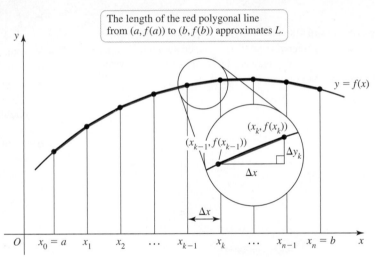

The length of the red polygonal line from $(a, f(a))$ to $(b, f(b))$ approximates L.

FIGURE 6.49

 Notice that Δx is the same for each subinterval, but Δy_k depends on the subinterval.

The kth segment of the polygonal line is the hypotenuse of a right triangle with sides of length Δx and $|\Delta y_k| = |f(x_k) - f(x_{k-1})|$. The length of each line segment is

$$\sqrt{(\Delta x)^2 + (\Delta y_k)^2} \quad \text{for} \quad k = 1, 2, \dots, n.$$

Summing these lengths, we obtain the length of the polygonal line, which approximates the length L of the curve:

$$L \approx \sum_{k=1}^{n} \sqrt{(\Delta x)^2 + (\Delta y_k)^2}$$

In previous applications of the integral, we would, at this point, take the limit as $n \to \infty$ and $\Delta x \to 0$ to obtain a definite integral. However, because of the presence of the Δy_k term, we must complete one additional step before taking a limit. Notice that the slope of the line segment on the kth subinterval is $\Delta y_k / \Delta x$ (rise over run). By the

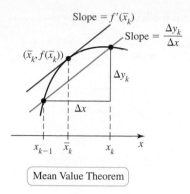

Slope = $f'(\bar{x}_k)$

Slope = $\dfrac{\Delta y_k}{\Delta x}$

$(\bar{x}_k, f(\bar{x}_k))$

Δy_k

Δx

x_{k-1} $\bar{x}_k$ x_k

Mean Value Theorem

Mean Value Theorem (see margin figure and Section 4.6), this slope equals $f'(\bar{x}_k)$ for some point $\bar{x}_k$ on the kth subinterval. Therefore,

$$L \approx \sum_{k=1}^{n} \sqrt{(\Delta x)^2 + (\Delta y_k)^2}$$

$$= \sum_{k=1}^{n} \sqrt{(\Delta x)^2\left[1 + \left(\frac{\Delta y_k}{\Delta x}\right)^2\right]} \qquad \text{Factor out } (\Delta x)^2.$$

$$= \sum_{k=1}^{n} \sqrt{1 + \left(\frac{\Delta y_k}{\Delta x}\right)^2}\,\Delta x \qquad \text{Bring } \Delta x \text{ out of the square root.}$$

$$= \sum_{k=1}^{n} \sqrt{1 + f'(\bar{x}_k)^2}\,\Delta x. \qquad \text{Mean Value Theorem}$$

Now we have a Riemann sum. As n increases and as Δx tends to zero, the sum approaches a definite integral, which is also the exact length of the curve. We have

$$L = \lim_{n\to\infty} \sum_{k=1}^{n} \sqrt{1 + f'(\bar{x}_k)^2}\,\Delta x = \int_a^b \sqrt{1 + f'(x)^2}\,dx.$$

> Note that $1 + f'(x)^2$ is positive, so the square root in the integrand is defined whenever f' exists. To ensure that $\sqrt{1 + f'(x)^2}$ is integrable, we require that f' be continuous.

DEFINITION   Arc Length for $y = f(x)$

Let f have a continuous first derivative on the interval $[a, b]$. The length of the curve from $(a, f(a))$ to $(b, f(b))$ is

$$L = \int_a^b \sqrt{1 + f'(x)^2}\,dx.$$

QUICK CHECK 1 What does the arc length formula give for the length of the line $y = x$ between $x = 0$ and $x = a$, where $a \geq 0$? ◄

EXAMPLE 1   Arc length Find the length of the curve $f(x) = x^{3/2}$ between $x = 0$ and $x = 4$ (Figure 6.50).

SOLUTION Notice that $f'(x) = \frac{3}{2}x^{1/2}$, which is continuous on the interval $[0, 4]$. Using the arc length formula, we have

$$L = \int_a^b \sqrt{1 + f'(x)^2}\,dx = \int_0^4 \sqrt{1 + \left(\frac{3}{2}x^{1/2}\right)^2}\,dx \quad \text{Substitute for } f'(x).$$

$$= \int_0^4 \sqrt{1 + \frac{9}{4}x}\,dx \qquad \text{Simplify.}$$

$$= \frac{4}{9}\int_1^{10} \sqrt{u}\,du \qquad u = 1 + \frac{9x}{4}, du = \frac{9}{4}dx$$

$$= \frac{4}{9}\left(\frac{2}{3}u^{3/2}\right)\Big|_1^{10} \qquad \text{Fundamental Theorem}$$

$$= \frac{8}{27}(10^{3/2} - 1). \qquad \text{Simplify.}$$

The length of the curve is $\frac{8}{27}(10^{3/2} - 1) \approx 9.1$ units. *Related Exercises 3–10* ◄

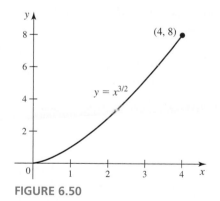

y

8

(4, 8)

6

4

$y = x^{3/2}$

2

0   1   2   3   4 x

FIGURE 6.50

EXAMPLE 2 Arc length of an exponential curve Find the length of the curve $f(x) = 2e^x + \frac{1}{8}e^{-x}$ on the interval $[0, \ln 2]$.

SOLUTION We first calculate $f'(x) = 2e^x - \frac{1}{8}e^{-x}$ and $f'(x)^2 = 4e^{2x} - \frac{1}{2} + \frac{1}{64}e^{-2x}$. The length of the curve on the interval $[0, \ln 2]$ is

$$
\begin{aligned}
L = \int_0^{\ln 2} \sqrt{1 + f'(x)^2}\, dx &= \int_0^{\ln 2} \sqrt{1 + \left(4e^{2x} - \frac{1}{2} + \frac{1}{64}e^{-2x}\right)}\, dx \\[2mm]
&= \int_0^{\ln 2} \sqrt{4e^{2x} + \frac{1}{2} + \frac{1}{64}e^{-2x}}\, dx && \text{Simplify.} \\[2mm]
&= \int_0^{\ln 2} \sqrt{\left(2e^x + \frac{1}{8}e^{-x}\right)^2}\, dx && \text{Factor.} \\[2mm]
&= \int_0^{\ln 2} \left(2e^x + \frac{1}{8}e^{-x}\right) dx && \text{Simplify.} \\[2mm]
&= \left(2e^x - \frac{1}{8}e^{-x}\right)\Bigg|_0^{\ln 2} = \frac{33}{16}. && \begin{array}{l}\text{Evaluate the}\\ \text{integral.}\end{array}
\end{aligned}
$$

Related Exercises 3–10 ◀

EXAMPLE 3 Circumference of a circle Confirm that the circumference of a circle of radius a is $2\pi a$.

SOLUTION The upper half of a circle of radius a centered at $(0, 0)$ is given by the function $f(x) = \sqrt{a^2 - x^2}$ for $|x| \le a$ (Figure 6.51). So we might consider using the arc length formula on the interval $[-a, a]$ to find the length of a semicircle. However, the circle has vertical tangent lines at $x = \pm a$ and $f'(\pm a)$ is undefined, which prevents us from using the arc length formula. An alternative approach is to use symmetry and avoid the points $x = \pm a$. For example, let's compute the length of one-eighth of the circle on the interval $[0, a/\sqrt{2}]$ (Figure 6.51).

We first determine that $f'(x) = -\dfrac{x}{\sqrt{a^2 - x^2}}$, which is continuous on $[0, a/\sqrt{2}]$. The length of one-eighth of the circle is

$$
\begin{aligned}
\int_0^{a/\sqrt{2}} \sqrt{1 + f'(x)^2}\, dx &= \int_0^{a/\sqrt{2}} \sqrt{1 + \left(-\frac{x}{\sqrt{a^2 - x^2}}\right)^2}\, dx \\[2mm]
&= a \int_0^{a/\sqrt{2}} \frac{dx}{\sqrt{a^2 - x^2}} && \text{Simplify; } a > 0. \\[2mm]
&= a \sin^{-1}\left(\frac{x}{a}\right)\Bigg|_0^{a/\sqrt{2}} && \text{Integrate.} \\[2mm]
&= a\left[\sin^{-1}\left(\frac{1}{\sqrt{2}}\right) - 0\right] && \text{Evaluate.} \\[2mm]
&= \frac{\pi a}{4} && \text{Simplify.}
\end{aligned}
$$

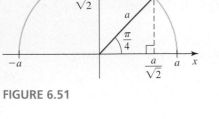

FIGURE 6.51

> The arc length integral for the semicircle on $[-a, a]$ is an example of an *improper integral*, a topic considered in Section 7.6.

It follows that the circumference of the full circle is $8(\pi a/4) = 2\pi a$ units.

Related Exercises 3–10 ◀

EXAMPLE 4 Looking ahead Consider the segment of the parabola $f(x) = x^2$ on the interval $[0, 2]$.

a. Write the integral for the length of the curve.

b. Use a calculator to evaluate the integral.

SOLUTION

a. Noting that $f'(x) = 2x$, the arc length integral is

$$\int_0^2 \sqrt{1 + f'(x)^2}\, dx = \int_0^2 \sqrt{1 + 4x^2}\, dx.$$

> When relying on technology, it is a good idea to check whether an answer is plausible. In Example 4, we found the arc length of $y = x^2$ on $[0, 2]$ is approximately 4.647. The straight-line distance between $(0, 0)$ and $(2, 4)$ is $\sqrt{20} \approx 4.472$, so our answer is reasonable.

b. Even simple functions can lead to arc length integrals that are difficult, if not impossible, to evaluate analytically. Using integration techniques presented so far, this integral cannot be evaluated (the required method is given in Section 7.3). Without an analytical method, we may use numerical integration to *approximate* the value of a definite integral (Section 7.5). Many calculators have built-in functions for this purpose. For this integral, the approximate arc length is

$$\int_0^2 \sqrt{1 + 4x^2}\, dx \approx 4.647. \qquad \textit{Related Exercises 11–20} \blacktriangleleft$$

Arc Length for $x = g(y)$

Sometimes it is advantageous to describe a curve as a function of y—that is, $x = g(y)$. The arc length formula in this case is derived exactly as in the case of $y = f(x)$, switching the roles of x and y. The result is the following arc length formula.

DEFINITION Arc Length for $x = g(y)$

Let $x = g(y)$ have a continuous first derivative on the interval $[c, d]$. The length of the curve from $(g(c), c)$ to $(g(d), d)$ is

$$L = \int_c^d \sqrt{1 + g'(y)^2}\, dy.$$

QUICK CHECK 2 What does the arc length formula give for the length of the line $x = y$ between $y = c$ and $y = d$, where $d \geq c$? Is the result correct? ◄

EXAMPLE 5 Arc length Find the length of the curve $y = f(x) = x^{2/3}$ between $x = 0$ and $x = 8$ (Figure 6.52).

SOLUTION The derivative of $f(x) = x^{2/3}$ is $f'(x) = \frac{2}{3}x^{-1/3}$, which is undefined at $x = 0$. Therefore, the arc length formula with respect to x cannot be used, yet the curve certainly appears to have a well-defined length.

The key is to describe the curve with y as the independent variable. Solving $y = x^{2/3}$ for x, we have $x = g(y) = \pm y^{3/2}$. Notice that when $x = 8$, $y = 8^{2/3} = 4$, which says that we should use the positive branch of $y^{3/2}$. Therefore, finding the length of the curve $y = f(x) = x^{2/3}$ from $x = 0$ to $x = 8$ is equivalent to finding the length of the curve $x = g(y) = y^{3/2}$ from $y = 0$ to $y = 4$. This is precisely the problem solved in Example 1. The arc length is $\frac{8}{27}(10^{3/2} - 1) \approx 9.1$ units. *Related Exercises 21–24* ◄

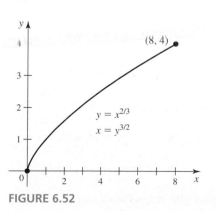

FIGURE 6.52

QUICK CHECK 3 Write the integral for the length of the curve $x = \sin y$ on the interval $0 \leq y \leq \pi$. ◄

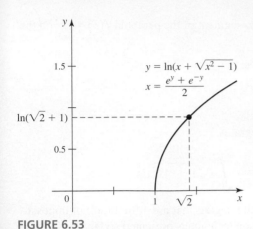

FIGURE 6.53

▸ The function $\frac{1}{2}(e^y + e^{-y})$ is the **hyperbolic cosine**, denoted cosh y. The function $\frac{1}{2}(e^y - e^{-y})$ is the **hyperbolic sine**, denoted sinh y.

EXAMPLE 6   Ingenuity required Find the length of the curve
$y = f(x) = \ln(x + \sqrt{x^2 - 1})$ on the interval $[1, \sqrt{2}]$ (Figure 6.53).

SOLUTION Calculating f' shows that the graph of f has a vertical tangent line at $x = 1$. Therefore, the integrand in the arc length integral is undefined at $x = 1$. An alternative strategy is to express the function in the form $x = g(y)$ and evaluate the arc length integral with respect to y. Noting that $x \geq 1$ and $y \geq 0$, we solve $y = \ln(x + \sqrt{x^2 - 1})$ for x in the following steps:

$$e^y = x + \sqrt{x^2 - 1} \qquad \text{Exponentiate both sides.}$$

$$e^y - x = \sqrt{x^2 - 1} \qquad \text{Subtract } x \text{ from both sides.}$$

$$e^{2y} - 2e^y x = -1 \qquad \text{Square both sides and cancel } x^2.$$

$$x = \frac{e^{2y} + 1}{2e^y} = \frac{e^y + e^{-y}}{2} \qquad \text{Solve for } x.$$

We conclude that the given curve is also described by the function $x = g(y) = \dfrac{e^y + e^{-y}}{2}$. The interval $1 \leq x \leq \sqrt{2}$ corresponds to the interval $0 \leq y \leq \ln(\sqrt{2} + 1)$ (Figure 6.53). Note that $g'(y) = \dfrac{e^y - e^{-y}}{2}$ is continuous on $[0, \ln(\sqrt{2} + 1)]$. The arc length is

$$\int_0^{\ln(\sqrt{2}+1)} \sqrt{1 + g'(y)^2} \, dy = \int_0^{\ln(\sqrt{2}+1)} \sqrt{1 + \left(\frac{e^y - e^{-y}}{2}\right)^2} \, dy \quad \text{Substitute for } g'(y).$$

$$= \frac{1}{2}\int_0^{\ln(\sqrt{2}+1)} (e^y + e^{-y}) \, dy \qquad \text{Simplify.}$$

$$= \frac{1}{2}(e^y - e^{-y})\Big|_0^{\ln(\sqrt{2}+1)} = 1. \qquad \begin{array}{l}\text{Fundamental}\\\text{Theorem}\end{array}$$

Related Exercises 21–24 ◀

SECTION 6.5 EXERCISES

Review Questions

1. Explain the steps required to find the length of a curve $y = f(x)$ between $x = a$ and $x = b$.

2. Explain the steps required to find the length of a curve $x = g(y)$ between $y = c$ and $y = d$.

Basic Skills

3–10. Arc length calculations *Find the arc length of the following curves on the given interval by integrating with respect to x.*

3. $y = 2x + 1$; $[1, 5]$ (use calculus)

4. $y = \frac{1}{2}(e^x + e^{-x})$; $[-\ln 2, \ln 2]$

5. $y = \frac{1}{3}x^{3/2}$; $[0, 60]$

6. $y = 3\ln x - \dfrac{x^2}{24}$; $[1, 6]$

7. $y = \dfrac{(x^2 + 2)^{3/2}}{3}$; $[0, 1]$

8. $y = \dfrac{x^{3/2}}{3} - x^{1/2}$; $[4, 16]$

9. $y = \dfrac{x^4}{4} + \dfrac{1}{8x^2}$; $[1, 2]$

10. $y = \frac{2}{3}x^{3/2} - \frac{1}{2}x^{1/2}$; $[1, 9]$

⊡ 11–20. Arc length by calculator

a. *Write and simplify the integral that gives the arc length of the following curves on the given interval.*

b. *If necessary, use a calculator to evaluate or approximate the integral.*

11. $y = x^2$ on $[-1, 1]$

12. $y = \sin x$ on $[0, \pi]$

13. $y = \ln x$ on $[1, 4]$

14. $y = \dfrac{x^3}{3}$ on $[-1, 1]$

15. $y = \sqrt{x - 2}$ on $[3, 4]$

16. $y = \dfrac{8}{x^2}$ on $[1, 4]$

17. $y = \cos 2x$ on $[0, \pi]$

18. $y = 4x - x^2$ on $[0, 4]$

19. $y = \dfrac{1}{x}$ on $[1, 10]$

20. $y = \dfrac{1}{x^2 + 1}$ on $[-5, 5]$

⊡ 21–24. Arc length calculations with respect to y *Find the arc length of the following curves by integrating with respect to y.*

21. $x = 2y - 4$ for $-3 \leq y \leq 4$

22. $y = \ln(x - \sqrt{x^2 - 1})$ for $1 \leq x \leq \sqrt{2}$

23. $x = \dfrac{y^4}{4} + \dfrac{1}{8y^2}$ for $1 \leq y \leq 2$

24. $x = 2e^{\sqrt{2}y} + \dfrac{1}{16}e^{-\sqrt{2}y}$ for $0 \leq y \leq \dfrac{\ln 2}{\sqrt{2}}$

Further Explorations

25. Explain why or why not Determine whether the following statements are true and give an explanation or counterexample.

    a. $\displaystyle\int_a^b \sqrt{1 + f'(x)^2}\,dx = \int_a^b (1 + f'(x))\,dx$

    b. Assuming f' is continuous on the interval $[a, b]$, the length of the curve $y = f(x)$ between $x = a$ and $x = b$ is the area under the curve $y = \sqrt{1 + f'(x)^2}$ between $x = a$ and $x = b$.

    c. Arc length may be negative if $f(x) < 0$ on part of the interval in question.

26. Arc length for a line Consider the line $y = mx + c$ on the interval $[a, b]$. Use the arc length formula to show that the length of the line is $(b - a)\sqrt{1 + m^2}$. Verify this result by computing the length of the segment of the line extending from $x = a$ to $x = b$ using the distance formula.

27. Functions from arc length What differentiable functions have an arc length on the interval $[a, b]$ given by the following integrals? Note that the answers are not unique. Give all functions that satisfy the conditions.

    a. $\displaystyle\int_a^b \sqrt{1 + 16x^4}\,dx$ **b.** $\displaystyle\int_a^b \sqrt{1 + 36\cos^2(2x)}\,dx$

28. Function from arc length What curve passes through the point $(1, 5)$ and has an arc length on the interval $[2, 6]$ given by $\displaystyle\int_2^6 \sqrt{1 + 16x^{-6}}\,dx$?

29. Cosine vs. parabola Which curve has the greater length on the interval $[-1, 1]$, $y = 1 - x^2$ or $y = \cos(\pi x/2)$?

30. Function defined as an integral Write the integral that gives the length of the curve $y = f(x) = \displaystyle\int_0^x \sin t\,dt$ on the interval $[0, \pi]$.

Applications

31. Golden Gate cables The profile of the cables on a suspension bridge may be modeled by a parabola. The central span of the Golden Gate Bridge (see figure) is 1280 m long and 152 m high. The parabola $y = 0.00037x^2$ gives a good fit to the shape of the cables, where $|x| \leq 640$, and x and y are measured in meters. Approximate the length of the cables that stretch between the tops of the two towers.

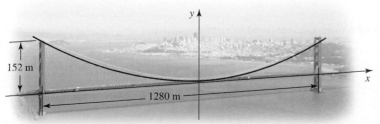

32. Gateway Arch The shape of the Gateway Arch in St. Louis (with a height and a base length of 630 ft) is modeled by the function $y = -630\cosh(x/239.2) + 1260$, where $|x| \leq 315$, and x and y are measured in feet (see figure). The function $\cosh x$ is the **hyperbolic cosine**, defined by $\cosh x = \dfrac{e^x + e^{-x}}{2}$ (see the Guided Projects for more on hyperbolic functions). Estimate the length of the Gateway Arch.

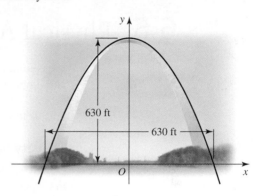

Additional Exercises

33. Lengths of related curves Suppose the graph of f on the interval $[a, b]$ has length L, where f' is continuous on $[a, b]$. Evaluate the following integrals in terms of L.

    a. $\displaystyle\int_{a/2}^{b/2} \sqrt{1 + f'(2x)^2}\,dx$ **b.** $\displaystyle\int_{a/c}^{b/c} \sqrt{1 + f'(cx)^2}\,dx$ if $c \neq 0$

34. Lengths of symmetric curves Suppose a curve is described by $y = f(x)$ on the interval $[-b, b]$, where f' is continuous on $[-b, b]$. Show that if f is symmetric about the origin (f is odd) *or* f is symmetric about the y-axis (f is even), then the length of the curve $y = f(x)$ from $x = -b$ to $x = b$ is twice the length of the curve from $x = 0$ to $x = b$. Use a geometric argument and then prove it analytically using calculus.

35. A family of exponential functions

    a. Show that the arc length integral for the function $f(x) = Ae^{ax} + \dfrac{1}{4Aa^2}e^{-ax}$, where $a > 0$ and $A > 0$, may be integrated using methods you already know.

    b. Verify that the arc length of the curve $y = f(x)$ on the interval $[0, \ln 2]$ is
$$A(2^a - 1) - \dfrac{1}{4a^2 A}(2^{-a} - 1).$$

36. Bernoulli's "parabolas" Johann Bernoulli (1667–1748) evaluated the arc length of curves of the form $y = x^{(2n+1)/2n}$, where n is a positive integer, on the interval $[0, a]$.

    a. Write the arc length integral.

    b. Make the change of variables $u^2 = 1 + \left(\dfrac{2n + 1}{2n}\right)^2 x^{1/n}$ to obtain a new integral with respect to u.

    c. Use the Binomial Theorem to expand this integrand and evaluate the integral.

d. The case $n = 1$ ($y = x^{3/2}$) was done in Example 1. With $a = 1$, evaluate the arc length in the cases $n = 2$ and $n = 3$. Does the arc length increase or decrease with n?

e. Graph the arc length of the parabolas for $a = 1$ as a function of n.

6.6 Physical Applications

We continue this chapter on applications of integration with several problems from physics and engineering. The physical themes in these problems are mass, work, pressure, and force. The common mathematical theme is the use of the slice-and-sum strategy, which always leads to a definite integral.

Density and Mass

Density is the concentration of mass in an object and is usually measured in units of mass per volume (for example, g/cm³). An object with *uniform* density satisfies the basic relationship

$$\text{mass} = \text{density} \cdot \text{volume}.$$

> In Chapter 13, we return to mass calculations for two- and three-dimensional objects (plates and solids).

When the density of an object *varies*, this formula no longer holds, and we must appeal to calculus.

In this section we introduce mass calculations for thin objects that can be viewed as line segments (such as wires or thin bars). The bar shown in Figure 6.54 has a density ρ that varies along its length. For one-dimensional objects we use *linear density* with units of mass per length (for example, g/cm). What is the mass of such an object?

$x = a$ $x = b$

FIGURE 6.54

QUICK CHECK 1 In Figure 6.54, suppose $a = 0$, $b = 3$, and the density of the rod in g/cm is $\rho(x) = (4 - x)$. (a) Where is the rod lightest and heaviest? (b) What is the density at the middle of the bar? ◄

We begin by dividing the bar, represented by the interval $a \le x \le b$, into n subintervals of equal length $\Delta x = (b - a)/n$ (Figure 6.55). Let $\overline{x}_k$ be any point in the kth subinterval for $k = 1, \ldots, n$. The mass of the kth segment of the bar, denoted m_k, is approximately the density at $\overline{x}_k$ multiplied by the length of the interval, or $m_k \approx \rho(\overline{x}_k)\Delta x$. So the approximate mass of the entire bar is

Mass of kth subinterval:
$m_k \approx \rho(\overline{x}_k)\Delta x$

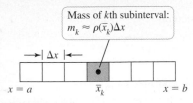

$x = a$ $\overline{x}_k$ $x = b$

FIGURE 6.55

$$\sum_{k=1}^{n} m_k \approx \sum_{k=1}^{n} \underbrace{\rho(\overline{x}_k)\Delta x}_{m_k}.$$

> Note that the units of the integral work out as they should: ρ has units of mass per length and dx has units of length; so $\rho(x)\, dx$ has units of mass.

The exact mass is obtained by taking the limit as $n \to \infty$ and as $\Delta x \to 0$, which produces a definite integral.

> Another interpretation of the mass integral is that mass equals the average value of the density multiplied by the length of the bar $b - a$.

DEFINITION Mass of a One-Dimensional Object

Suppose a thin bar or wire can be represented as a line segment on the interval $a \le x \le b$ with a density function ρ (with units of mass per length). The **mass** of the object is

$$m = \int_a^b \rho(x)\, dx.$$

EXAMPLE 1   **Mass from variable density**   A thin 2-m bar, represented by the interval $0 \le x \le 2$, is made of an alloy whose density in units of kg/m is given by $\rho(x) = (1 + x^2)$. What is the mass of the bar?

SOLUTION   The mass of the bar in kilograms is

$$m = \int_a^b \rho(x)\, dx = \int_0^2 (1 + x^2)\, dx = \left(x + \frac{x^3}{3} \right)\Big|_0^2 = \frac{14}{3}.$$

Related Exercises 9–16 ◀

QUICK CHECK 2   A thin bar occupies the interval $0 \le x \le 2$ and it has a density in kg/m of $\rho(x) = (1 + x^2)$. Using the minimum value of the density, what is a lower bound for the mass of the object? Using the maximum value of the density, what is an upper bound for the mass of the object? ◀

Work

Work can be described as the change in energy when a force causes a displacement of an object. When you carry a refrigerator up a flight of stairs or push a stalled car, you apply a force that results in the displacement of an object, and work is done. If a *constant* force F displaces an object a distance d in the direction of the force, the work done is the force multiplied by the distance:

$$\text{Work} = \text{force} \cdot \text{distance}$$

It is easiest to use metric units for force and work. A newton (N) is the force required to give a 1-kg mass an acceleration of 1 m/s^2. A joule (J) is 1 newton-meter (N-m), the work done by a 1-N force over a distance of 1 m.

    Calculus enters the picture with *variable* forces. Suppose an object is moved along the x-axis by a variable force F that is directed along the x-axis (Figure 6.56). How much work is done in moving the object between $x = a$ and $x = b$? Once again, we use the slice-and-sum strategy.

    The interval $[a, b]$ is divided into n subintervals of equal length $\Delta x = (b - a)/n$. We let $\overline{x}_k$ be any point in the kth subinterval, for $k = 1, \dots, n$. On that subinterval the force is approximately constant with a value of $F(\overline{x}_k)$. Therefore, the work done in moving the object across the kth subinterval is approximately $F(\overline{x}_k)\,\Delta x$ (force · distance). Summing the work done over each of the n subintervals, the total work over the interval $[a, b]$ is approximately

$$W \approx \sum_{k=1}^n F(\overline{x}_k)\, \Delta x.$$

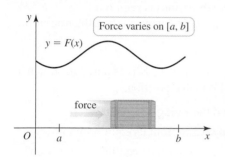

Force varies on $[a, b]$

$y = F(x)$

force

FIGURE 6.56

This approximation becomes exact when we take the limit as $n \to \infty$ and $\Delta x \to 0$. The total work done is the integral of the force over the interval $[a, b]$ (or, equivalently, the net area under the force curve in Figure 6.56).

QUICK CHECK 3   Explain why the sum of the work over n subintervals is only an approximation to the total work. ◀

DEFINITION   **Work**

The work done by a variable force F in moving an object along a line from $x = a$ to $x = b$ in the direction of the force is

$$W = \int_a^b F(x)\, dx.$$

    An application of force and work that is easy to visualize is the stretching and compression of a spring. Suppose an object is attached to a spring on a frictionless horizontal surface; the object can slide back and forth only under the influence of the spring. We say that the spring is at *equilibrium* when it is neither compressed nor stretched. It is convenient to let x be the position of the object, where $x = 0$ is the equilibrium position (Figure 6.57).

> Hooke's law was proposed by the English scientist Robert Hooke (1635–1703), who also coined the biological term *cell*.
>
> Larger values of the spring constant k correspond to stiffer springs. Hooke's law works well for springs made of many common materials. However, some springs obey more complicated spring laws (see Exercise 41).

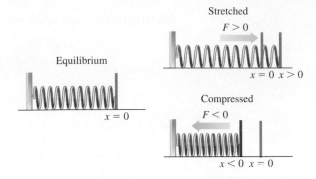

FIGURE 6.57

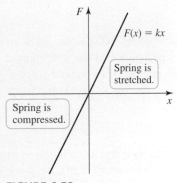

FIGURE 6.58

According to **Hooke's law**, the force required to keep the spring in a compressed or stretched position x units from the equilibrium position is $F(x) = kx$, where the spring constant k measures the stiffness of the spring. Note that to stretch the spring to a position $x > 0$, a force $F > 0$ (in the positive direction) is required. To compress the spring to a position $x < 0$, a force $F < 0$ (in the negative direction) is required (Figure 6.58). In other words, the force required to displace the spring is always in the direction of the displacement.

EXAMPLE 2   Compressing a spring Suppose a force of 10 N is required to stretch a spring 0.1 m from its equilibrium position and hold it in that position.

a. Assuming that the spring obeys Hooke's law, find the spring constant k.

b. How much work is needed to *compress* the spring 0.5 m from its equilibrium position?

c. How much work is needed to *stretch* the spring 0.25 m from its equilibrium position?

d. How much additional work is required to stretch the spring 0.25 m if it has already been stretched 0.1 m from its equilibrium position?

SOLUTION

a. The fact that a force of 10 N is required to keep the spring stretched at $x = 0.1$ m means (by Hooke's law) that $F(0.1) = k(0.1\text{ m}) = 10$ N. Solving for the spring constant, we find that $k = 100$ N/m. Therefore, Hooke's law for this spring is $F(x) = 100x$.

b. The work required to compress the spring from $x = 0$ to $x = -0.5$ is

$$W = \int_a^b F(x)\,dx = \int_0^{-0.5} 100x\,dx = 50x^2\Big|_0^{-0.5} = 12.5\text{ J}.$$

> Notice again that the units in the integral are consistent. If F has units of N and x has units of m, then W has units of $F\,dx$, or N-m, which are the units of work (1 N-m $= 1$ J).

c. The work required to stretch the spring from $x = 0$ to $x = 0.25$ is

$$W = \int_a^b F(x)\,dx = \int_0^{0.25} 100x\,dx = 50x^2\Big|_0^{0.25} = 3.125\text{ J}.$$

d. The work required to stretch the spring from $x = 0.1$ to $x = 0.35$ is

$$W = \int_a^b F(x)\,dx = \int_{0.1}^{0.35} 100x\,dx = 50x^2\Big|_{0.1}^{0.35} = 5.625\text{ J}.$$

QUICK CHECK 4 In Example 2, explain why more work is needed in part (d) than in part (c), even though the displacement is the same. ◄

Comparing parts (c) and (d), we see that more work is required to stretch the spring 0.25 m starting at $x = 0.1$ than starting at $x = 0$.   *Related Exercises 17–22* ◄

Lifting Problems Another common work problem arises when the motion is vertical and the force is the gravitational force. The gravitational force exerted on an object with a mass of m is $F = mg$, where $g \approx 9.8 \text{ m/s}^2$ is the acceleration due to gravity near the surface of Earth. The work in joules required to lift an object of mass m a vertical distance of y meters is

$$\text{work} = \text{force} \cdot \text{distance} = mgy.$$

This type of problem becomes interesting when the object being lifted is a body of water, a rope, or chain. In these situations, different parts of the object are lifted different distances—so integration is necessary. Here is a typical situation and the strategy used.

Suppose a fluid such as water is pumped out of a tank to a height h above the bottom of the tank. How much work is required, assuming the tank is full of water? Three key observations lead to the solution:

- Water from different levels of the tank is lifted different vertical distances, requiring different amounts of work.

- Water from the same horizontal plane is lifted the same distance, requiring the same amount of work.

- A volume V of water has mass ρV, where $\rho = 1 \text{ g/cm}^3 = 1000 \text{ kg/m}^3$ is the density of water.

> The choice of a coordinate system is somewhat arbitrary and may depend on the geometry of the problem. You can let the y-axis point upward or downward, and there are usually several logical choices for the location of $y = 0$. You should experiment with different coordinate systems.

To solve this problem, we let the y-axis point upward with $y = 0$ at the bottom of the tank. The body of water that must be lifted extends from $y = 0$ to $y = b$ (which *may* be the top of the tank). The level to which the water must be raised is $y = h$, where $h \geq b$ (Figure 6.59). We now slice the water into n horizontal layers, each having thickness Δy. The kth layer occupying the interval $[y_{k-1}, y_k]$, for $k = 1, \ldots, n$, is approximately $\bar{y}_k$ units above the bottom of the tank, where $\bar{y}_k$ is any point in $[y_{k-1}, y_k]$.

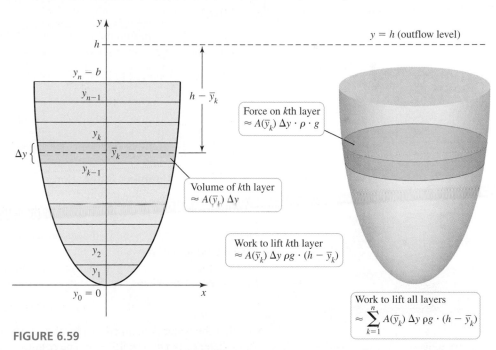

FIGURE 6.59

The cross-sectional area of the kth layer at $\bar{y}_k$, denoted $A(\bar{y}_k)$, is determined by the shape of the tank; the solution depends on being able to find A for all values of y. Because the volume of the kth layer is approximately $A(\bar{y}_k)\Delta y$, the force on the kth layer (its weight) is

$$F_k = mg \approx \underbrace{A(\bar{y}_k)\Delta y}_{\text{volume}} \cdot \underbrace{\rho}_{\text{density}} \cdot g.$$

To reach the level $y = h$, the kth layer is lifted an approximate distance of $(h - \bar{y}_k)$ (Figure 6.59). So the work in lifting the kth layer to a height h is approximately

$$W_k = \underbrace{A(\bar{y}_k)\Delta y \rho g}_{\text{force}} \cdot \underbrace{(h - \bar{y}_k)}_{\text{distance}}.$$

Summing the work required to lift all the layers to a height h, the total work is

$$W \approx \sum_{k=1}^{n} W_k = \sum_{k=1}^{n} A(\bar{y}_k)\rho g(h - \bar{y}_k)\Delta y.$$

This approximation becomes more accurate as the width of the layers Δy tends to zero and the number of layers tends to infinity. In this limit, we obtain a definite integral from $y = 0$ to $y = b$. The total work required to empty the tank is

$$W = \lim_{n \to \infty} \sum_{k=1}^{n} A(\bar{y}_k)\rho g(h - \bar{y}_k)\Delta y = \int_0^b \rho g A(y)(h - y)\,dy.$$

This derivation assumes that the *bottom* of the tank is at $y = 0$, in which case the distance that the slice at level y must be lifted is $D(y) = h - y$. If you choose a different location for the origin, the function D will be different. Here is a general procedure for any choice of origin.

Solving Lifting Problems

1. Draw a y-axis in the vertical direction (parallel to gravity) and choose a convenient origin. Assume the interval $[a, b]$ corresponds to the vertical extent of the fluid.

2. For $a \le y \le b$, find the cross-sectional area $A(y)$ of the horizontal slices and the distance $D(y)$ the slices must be lifted.

3. The work required to lift the water is

$$W = \int_a^b \rho g A(y) D(y)\,dy.$$

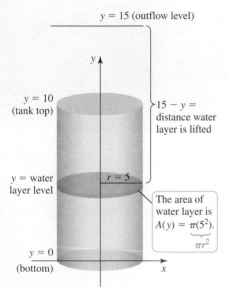

$y = 15$ (outflow level)

$y = 10$
(tank top)

$15 - y =$
distance water
layer is lifted

$y = $ water
layer level

$r = 5$

The area of
water layer is
$A(y) = \pi(5^2)$.
$\underbrace{\qquad}_{\pi r^2}$

$y = 0$
(bottom)

x

FIGURE 6.60

EXAMPLE 3   Pumping water   How much work is needed to pump all the water out of a cylindrical tank with a height of 10 m and a radius of 5 m? The water is pumped to an outflow pipe 15 m above the bottom of the tank.

SOLUTION   Figure 6.60 shows the cylindrical tank filled to capacity and the outflow 15 m above the bottom of the tank. We let $y = 0$ represent the bottom of the tank and $y = 10$ represent the top of the tank. In this case, all horizontal slices have a circular cross section of radius $r = 5$ m. Therefore, for $0 \le y \le 10$, the cross-sectional area is

$$A(y) = \pi r^2 = \pi 5^2 = 25\pi.$$

Note that the water is pumped to a level $h = 15$ m above the bottom of the tank, so the lifting distance is $D(y) = 15 - y$. The resulting work integral is

$$W = \int_0^{10} \rho g \underbrace{A(y)}_{25\pi} \underbrace{D(y)}_{15-y}\,dy = 25\pi\rho g \int_0^{10} (15 - y)\,dy.$$

▶ Recall that $g \approx 9.8 \text{ m/s}^2$. You should verify that the units are consistent in this calculation: The units of ρ, g, $A(y)$, $D(y)$, and dy are kg/m^3, m/s^2, m^2, m, and m, respectively. The resulting units of W are $\text{kg} \cdot \text{m}^2/\text{s}^2$, or J. A more convenient unit for large amounts of work and energy is the kilowatt-hr, which is 3.6 million joules.

Substituting $\rho = 1000 \text{ kg/m}^3$ and $g = 9.8 \text{ m/s}^2$, the total work is

$$W = 25\pi\rho g \int_0^{10} (15 - y) \, dy$$

$$= 25\pi \underbrace{(1000)}_{\rho} \underbrace{(9.8)}_{g} \left(15y - \frac{1}{2}y^2 \right) \Big|_0^{10}$$

$$\approx 7.7 \times 10^7.$$

The work required to pump the water out of the tank is approximately 77 million J.

Related Exercises 23–28 ◀

QUICK CHECK 5 In the previous example, how would the integral change if the outflow pipe were at the top of the tank? ◀

EXAMPLE 4 **Pumping gasoline** A cylindrical tank with a length of 10 m and a radius of 5 m is on its side and half-full of gasoline (Figure 6.61). How much work is required to empty the tank through an outlet pipe at the top of the tank? (The density of gasoline is $\rho \approx 737 \text{ kg/m}^3$.)

The equation of the right side of the circle is $x = \sqrt{25 - y^2}$

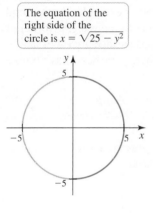

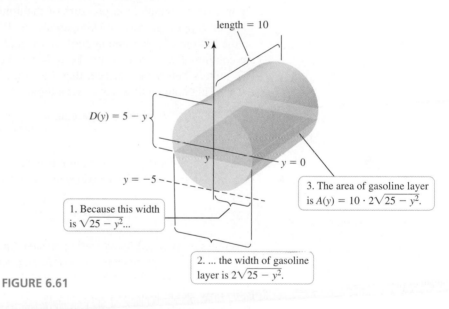

FIGURE 6.61

1. Because this width is $\sqrt{25 - y^2}$...

2. ... the width of gasoline layer is $2\sqrt{25 - y^2}$.

3. The area of gasoline layer is $A(y) = 10 \cdot 2\sqrt{25 - y^2}$.

$D(y) = 5 - y$

▶ Again, there are several choices for the location of the origin. The location in this example makes $A(y)$ easy to compute.

SOLUTION In this problem we choose a different origin by letting $y = 0$ and $y = -5$ correspond to the center and the bottom of the tank, respectively. For $-5 \le y \le 0$, a horizontal layer of gasoline located at a depth y is a rectangle with a length of 10 and width of $2\sqrt{25 - y^2}$ (Figure 6.61). Therefore, the cross-sectional area of the layer at depth y is

$$A(y) = 20\sqrt{25 - y^2}.$$

The distance the layer at level y must be lifted to reach the top of the tank is $D(y) = 5 - y$, where $5 \le D(y) \le 10$. The resulting work integral is

$$W = \underbrace{737}_{\rho}\underbrace{(9.8)}_{g} \int_{-5}^{0} \underbrace{20\sqrt{25 - y^2}}_{A(y)} \underbrace{(5 - y)}_{D(y)} \, dy = 144{,}452 \int_{-5}^{0} \sqrt{25 - y^2} \, (5 - y) \, dy.$$

This integral is evaluated by splitting it into two pieces and recognizing that one piece is the area of a quarter circle of radius 5:

$$\int_{-5}^{0} \sqrt{25 - y^2}\,(5 - y)\,dy = 5\underbrace{\int_{-5}^{0} \sqrt{25 - y^2}\,dy}_{\text{area of quarter circle}} - \underbrace{\int_{-5}^{0} y\sqrt{25 - y^2}\,dy}_{\text{let } u = 25 - y^2;\ du = -2y\,dy}$$

$$= 5 \cdot \frac{25\pi}{4} + \frac{1}{2}\int_{0}^{25} \sqrt{u}\,du$$

$$= \frac{125\pi}{4} + \frac{1}{3}u^{3/2}\Big|_{0}^{25} = \frac{375\pi + 500}{12}$$

Multiplying this result by 144,452, we find that the work required is approximately 20.2 million joules.   *Related Exercises 23–29* ◄

Force and Pressure

Another application of integration deals with the force exerted on a surface by a body of water. Again, we need a few physical principles.

Pressure is a force per unit area, measured in units such as newtons per square meter (N/m^2). For example, the pressure of the atmosphere on the surface of Earth is about $14\ lb/in^2$ (approximately 100 kilopascals, or $10^5\ N/m^2$). As another example, if you stood on the bottom of a swimming pool, you would feel pressure due to the weight (force) of the column of water above your head. If your head is flat and has surface area A m^2 and it is h meters below the surface, then the column of water above your head has volume Ah m^3. That column of water exerts a force:

$$F = \text{mass} \cdot \text{acceleration} = \underbrace{\text{volume} \cdot \text{density}}_{\text{mass}} \cdot g = Ah\rho g,$$

where ρ is the density of water and g is the acceleration due to gravity. Therefore, the pressure on your head is the force divided by the surface area of your head:

$$\text{pressure} = \frac{\text{force}}{A} = \frac{Ah\rho g}{A} = \rho g h.$$

> ➤ We have chosen $y = 0$ to be the base of the dam. Depending on the geometry of the problem, it may be more convenient (less computation) to let $y = 0$ be at the top of the dam. Experiment with different choices.

This pressure is called **hydrostatic pressure** (meaning the pressure of *water at rest*), and it has the following important property: *It has the same magnitude in all directions.* Specifically, the hydrostatic pressure on a vertical wall of the swimming pool at a depth h is also $\rho g h$. This is the only fact needed to find the total force on vertical walls such as dams. We assume that the water completely covers the face of the dam.

The first step in finding the force on the face of the dam is to introduce a coordinate system. We choose a y-axis pointing upward with $y = 0$ corresponding to the base of the dam and $y = a$ corresponding to the top of the dam (Figure 6.62). Because the pressure varies with depth (y-direction), the dam is sliced horizontally into n strips of equal thickness Δy. The kth strip corresponds to the interval $[y_{k-1}, y_k]$, and we let $\bar{y}_k$ be any point in that interval. The depth of that strip is approximately $h = a - \bar{y}_k$, so the hydrostatic pressure on that strip is approximately $\rho g(a - \bar{y}_k)$.

The crux of any dam problem is finding the width of the strips as a function of y, which we denote $w(y)$. Each dam has its own width function; however, once the width function is known, the solution follows directly. The approximate area of the kth strip is its width multiplied by its thickness, or $w(\bar{y}_k)\,\Delta y$. The force on the kth strip (which is the area of the strip multiplied by the pressure) is approximately

$$F_k = \underbrace{w(\bar{y}_k)\,\Delta y}_{\text{area of strip}}\ \underbrace{\rho g(a - \bar{y}_k)}_{\text{pressure}}.$$

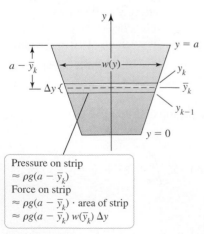

Pressure on strip
$\approx \rho g(a - \bar{y}_k)$
Force on strip
$\approx \rho g(a - \bar{y}_k) \cdot$ area of strip
$\approx \rho g(a - \bar{y}_k)\,w(\bar{y}_k)\,\Delta y$

FIGURE 6.62

Summing the forces over the n strips, the total force is approximately

$$F \approx \sum_{k=1}^{n} F_k = \sum_{k=1}^{n} \rho g(a - \bar{y}_k)w(\bar{y}_k)\Delta y.$$

To find the exact force, we let the thickness of the strips tend to zero and the number of strips tend to infinity, which produces a definite integral. The limits of integration correspond to the base ($y = 0$) and top ($y = a$) of the dam. Therefore, the total force on the dam is

$$F = \lim_{n \to \infty} \sum_{k=1}^{n} \rho g(a - \bar{y}_k)w(\bar{y}_k)\Delta y = \int_0^a \rho g(a - y)w(y)\,dy.$$

Solving Force/Pressure Problems

1. Draw a y-axis on the face of the dam in the vertical direction and choose a convenient origin (often taken to be the base of the dam).

2. Find the width function $w(y)$ for each value of y on the face of the dam.

3. If the base of the dam is at $y = 0$ and the top of the dam is at $y = a$, then the total force on the dam is

$$F = \int_0^a \rho g \underbrace{(a - y)}_{\text{depth}}\underbrace{w(y)}_{\text{width}}\,dy.$$

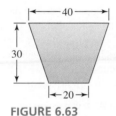

FIGURE 6.63

EXAMPLE 5 **Pressure on a dam** A large vertical dam in the shape of a symmetric trapezoid has a height of 30 m, a width of 20 m at its base, and a width of 40 m at the top (Figure 6.63). What is the total force on the face of the dam when the reservoir is full?

SOLUTION We place the origin at the center of the base of the dam (Figure 6.64). The right slanted edge of the dam is a segment of the line that passes through the points $(10, 0)$ and $(20, 30)$. An equation of that line is

$$y - 0 = \frac{30}{10}(x - 10) \quad \text{or} \quad y = 3x - 30 \quad \text{or} \quad x = \frac{1}{3}(y + 30).$$

> You should check the width function:
> $w(0) = 20$ (the width of the dam at its base) and $w(30) = 40$ (the width of the dam at its top).

Notice that at a depth of y, where $0 \le y \le 30$, the width of the dam is

$$w(y) = 2x = \frac{2}{3}(y + 30).$$

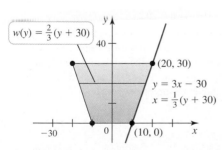

FIGURE 6.64

Using $\rho = 1000 \text{ kg/m}^3$ and $g = 9.8 \text{ m/s}^2$, the total force on the dam is

$$F = \int_0^a \rho g(a - y)w(y)\,dy \qquad \text{Force integral}$$

$$= \rho g \int_0^{30} \underbrace{(30 - y)}_{a - y}\underbrace{\frac{2}{3}(y + 30)}_{w(y)}\,dy \qquad \text{Substitute.}$$

$$= \frac{2}{3}\rho g \int_0^{30} (900 - y^2)\,dy \qquad \text{Simplify.}$$

$$= \frac{2}{3}\rho g\left(900y - \frac{y^3}{3}\right)\Big|_0^{30} \qquad \text{Fundamental Theorem}$$

$$\approx 1.18 \times 10^8 \text{ N}.$$

The force of 1.18×10^8 N on the dam amounts to about 26 million pounds, or 13,000 tons.

Related Exercises 30–38 ◄

SECTION 6.6 EXERCISES

Review Questions

1. If a 1-m cylindrical bar has a constant density of 1 g/cm for its left half and a constant density 2 g/cm for its right half, what is its mass?

2. Explain how to find the mass of a one-dimensional object with a variable density ρ.

3. Explain how to find the work done in moving an object along a line in the direction of a constant force.

4. Why must integration be used to find the work done by a variable force?

5. Why must integration be used to find the work required to pump water out of a tank?

6. Why must integration be used to find the total force on the face of a dam?

7. What is the pressure on a horizontal surface with an area of 2 m² that is 4 m underwater?

8. Explain why you integrate in the vertical direction (parallel to the acceleration due to gravity) rather than the horizontal direction to find the force on the face of a dam.

Basic Skills

9–16. Mass of one-dimensional objects *Find the mass of the following thin bars with the given density function.*

9. $\rho(x) = 1 + \sin x$; for $0 \le x \le \pi$

10. $\rho(x) = 1 + x^3$; for $0 \le x \le 1$

11. $\rho(x) = 2 - x/2$; for $0 \le x \le 2$

12. $\rho(x) = 5e^{-2x}$; for $0 \le x \le 4$

13. $\rho(x) = x\sqrt{2 - x^2}$; for $0 \le x \le 1$

14. $\rho(x) = \begin{cases} 1 & \text{if } 0 \le x \le 2 \\ 2 & \text{if } 2 < x \le 3 \end{cases}$

15. $\rho(x) = \begin{cases} 1 & \text{if } 0 \le x \le 2 \\ 1 + x & \text{if } 2 < x \le 4 \end{cases}$

16. $\rho(x) = \begin{cases} x^2 & \text{if } 0 \le x \le 1 \\ x(2 - x) & \text{if } 1 < x \le 2 \end{cases}$

17. **Work from force** How much work is required to move an object from $x = 1$ to $x = 5$ (measured in meters) in the presence of a constant force of 5 N acting along the x-axis?

18. **Work from force** How much work is required to move an object from $x = 0$ to $x = 3$ (measured in meters) with a force (in N) given by $F(x) = 2/x^2$ acting along the x-axis?

19. **Working a spring** A spring on a horizontal surface can be stretched and held 0.5 m from its equilibrium position with a force of 50 N.
    a. How much work is done in stretching the spring 1.5 m from its equilibrium position?
    b. How much work is done in compressing the spring 0.5 m from its equilibrium position?

20. **Shock absorber** A heavy-duty shock absorber is compressed 2 cm from its equilibrium position by a mass of 500 kg. How much work is required to compress the shock absorber 4 cm from its equilibrium position? (A mass of 500 kg exerts a force (in newtons) of 500g, where $g \approx 9.8$ m/s².)

21. **Additional stretch** It takes 100 J of work to stretch a spring 0.5 m from its equilibrium position. How much work is needed to stretch it an additional 0.75 m?

22. **Work function** A spring has a restoring force given by $F(x) = 25x$. Let $W(x)$ be the work required to stretch the spring from its equilibrium position $(x = 0)$ to a variable distance x. Graph the work function. Compare the work required to stretch the spring x units from equilibrium to the work required to compress the spring x units from equilibrium.

23. **Emptying a swimming pool** A swimming pool has the shape of a box with a base that measures 25 m by 15 m and a depth of 2.5 m. How much work is required to pump the water out of the pool when it is full?

24. **Emptying a cylindrical tank** A cylindrical water tank has height 8 m and radius 2 m (see figure).
    a. If the tank is full of water, how much work is required to pump the water to the level of the top of the tank and out of the tank?
    b. Is it true that it takes half as much work to pump the water out of the tank when it is half full as when it is full? Explain.

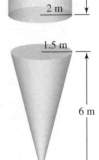

8 m

2 m

25. **Emptying a conical tank** A water tank is shaped like an inverted cone with height 6 m and base radius 1.5 m (see figure).
    a. If the tank is full, how much work is required to pump the water to the level of the top of the tank and out of the tank?
    b. Is it true that it takes half as much work to pump the water out of the tank when it is filled to half its depth as when it is full? Explain.

1.5 m

6 m

26. **Emptying a real swimming pool** A swimming pool is 20 m long and 10 m wide, with a bottom that slopes uniformly from a depth of 1 m at one end to a depth of 2 m at the other end (see figure). Assuming the pool is full, how much work is required to pump the water to a level 0.2 m above the top of the pool?

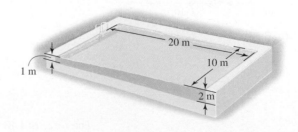

20 m

10 m

1 m

2 m

27. **Filling a spherical tank** A spherical water tank with an inner radius of 8 m has its lowest point 2 m above the ground. It is filled by a pipe that feeds the tank at its lowest point (see figure).

    a. Neglecting the volume of the inflow pipe, how much work is required to fill the tank if it is initially empty?
    b. Now assume that the inflow pipe feeds the tank at the top of the tank. Neglecting the volume of the inflow pipe, how much work is required to fill the tank if it is initially empty?

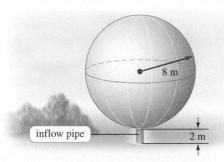

28. **Emptying a water trough** A water trough has a semicircular cross section with a radius of 0.25 m and a length of 3 m (see figure).

    a. How much work is required to pump the water out of the trough when it is full?
    b. If the length is doubled, is the required work doubled? Explain.
    c. If the radius is doubled, is the required work doubled? Explain.

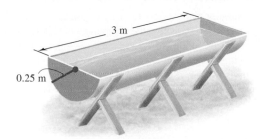

29. **Emptying a water trough** A cattle trough has a trapezoidal cross section with a height of 1 m and horizontal sides of width $\frac{1}{2}$ m and 1 m. Assume the length of the trough is 10 m (see figure).

    a. How much work is required to pump the water out of the trough (to the level of the top of the trough) when it is full?
    b. If the length is doubled, is the required work doubled? Explain.

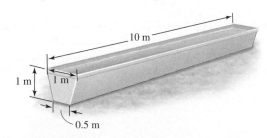

30–36. Force on dams *The following figures show the shape and dimensions of small dams. Assuming the water level is at the top of the dam, find the total force on the face of the dam.*

30.

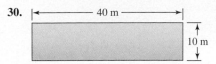

31.

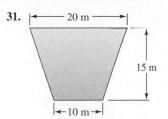

32.

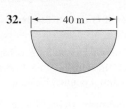

33.

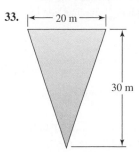

34. **Parabolic dam** The lower edge of a dam is defined by the parabola $y = x^2/16$ (see figure). Use a coordinate system with $y = 0$ at the bottom of the dam to determine the total force on the dam. Lengths are measured in meters.

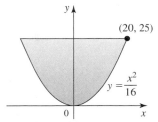

35. **Force on a building** A large building shaped like a box is 50 m high with a face that is 80 m wide. A strong wind blows directly at the face of the building, exerting a pressure of 150 N/m² at the ground and increasing with height according to $P(y) = 150 + 2y$, where y is the height above the ground. Calculate the total force on the building, which is a measure of the resistance that must be included in the design of the building.

36–38. Force on a window *A diving pool that is 4 m deep and full of water has a viewing window on one of its vertical walls. Find the force on the following windows.*

36. The window is a square, 0.5 m on a side, with the lower edge of the window on the bottom of the pool.

37. The window is a square, 0.5 m on a side, with the lower edge of the window 1 m from the bottom of the pool.

38. The window is a circle, with a radius of 0.5 m, tangent to the bottom of the pool.

Further Explorations

39. **Explain why or why not** Determine whether the following statements are true and give an explanation or counterexample.

    a. The mass of a thin wire is the length of the wire times its average density over its length.
    b. The work required to stretch a linear spring (that obeys Hooke's law) 100 cm from equilibrium is the same as the work required to compress it 100 cm from equilibrium.
    c. The work required to lift a 10-kg object vertically 10 m is the same as the work required to lift a 20-kg object vertically 5 m.
    d. The total force on a 10-ft² region on the (horizontal) floor of a pool is the same as the total force on a 10-ft² region on a (vertical) wall of the pool.

40. Mass of two bars Two bars of length L have densities of $\rho_1(x) = 4e^{-x}$ and $\rho_2(x) = 6e^{-2x}$, for $0 \le x \le L$.

a. For what values of L is bar 1 heavier than bar 2?

b. As the lengths of the bars increase, do their masses increase without bound? Explain.

41. A nonlinear spring Hooke's law is applicable to idealized (linear) springs that are not stretched or compressed too far. Consider a nonlinear spring whose restoring force is given by $F(x) = 16x - 0.1x^3$, for $|x| \le 7$.

a. Graph the restoring force and interpret it.

b. How much work is done in stretching the spring from its equilibrium position ($x = 0$) to $x = 1.5$?

c. How much work is done in compressing the spring from its equilibrium position ($x = 0$) to $x = -2$?

42. A vertical spring A 10-kg mass is attached to a spring that hangs vertically and is stretched 2 m from the equilibrium position of the spring. Assume a linear spring with $F(x) = kx$.

a. How much work is required to compress the spring and lift the mass 0.5 m?

b. How much work is required to stretch the spring and lower the mass 0.5 m?

43. Drinking juice A glass has circular cross sections that taper (linearly) from a radius of 5 cm at the top of the glass to a radius of 4 cm at the bottom. The glass is 15 cm high and full of orange juice. How much work is required to drink all the juice through a straw if your mouth is 5 cm above the top of the glass? Assume the density of orange juice equals the density of water.

44. Upper and lower half A cylinder with height 8 m and radius 3 m is filled with water and must be emptied through an outlet pipe 2 m above the top of the cylinder.

a. Compute the work required to empty the water in the top half of the tank.

b. Compute the work required to empty the (equal amount of) water in the lower half of the tank.

c. Interpret the results of parts (a) and (b).

Applications

45. Work in a gravitational field For large distances from the surface of Earth, the gravitational force is given by $F(x) = GMm/(x + R)^2$, where $G = 6.7 \times 10^{-11}\,\text{N} \cdot \text{m}^2/\text{kg}^2$ is the gravitational constant, $M = 6 \times 10^{24}$ kg is the mass of Earth, m is the mass of the object in the gravitational field, $R = 6.378 \times 10^6$ m is the radius of Earth, and $x \ge 0$ is the distance above the surface of Earth (in meters).

a. How much work is required to launch a rocket with a mass of 500 kg in a vertical flight path to a height of 2500 km (from Earth's surface)?

b. Find the work required to launch the rocket to a height of x kilometers for $x > 0$.

c. How much work is required to reach outer space ($x \to \infty$)?

d. Equate the work in part (b) to the initial kinetic energy of the rocket, $\frac{1}{2}mv^2$, to compute the escape velocity of the rocket.

46. Work by two different integrals A rigid body with a mass of 2 kg moves along a line due to a force that produces a position function $x(t) = 4t^2$, where x is measured in meters and t is measured in seconds. Find the work done during the first 5 s in two ways.

a. Note that $x''(t) = 8$; then use Newton's second law, ($F = ma = mx''(t)$) to evaluate the work integral $W = \int_{x_0}^{x_f} F(x)\,dx$, where x_0 and x_f are the initial and final positions, respectively.

b. Change variables in the work integral and integrate with respect to t. Be sure your answer agrees with part (a).

47. Winding a chain A 30-m-long chain hangs vertically from a cylinder attached to a winch. Assume there is no friction in the system and that the chain has a density of 5 kg/m.

a. How much work is required to wind the entire chain onto the cylinder using the winch?

b. How much work is required to wind the chain onto the cylinder if a 50-kg block is attached to the end of the chain?

48. Coiling a rope A 60-m-long, 9.4-mm-diameter rope hangs free from a ledge. The density of the rope is 55 g/m. How much work is needed to pull the entire rope to the ledge?

49. Lifting a pendulum A body of mass m is suspended by a rod of length L that pivots without friction (see figure). The mass is slowly lifted along a circular arc to a height h.

a. Assuming that the only force acting on the mass is the gravitational force, show that the component of this force acting along the arc of motion is $F = mg \sin\theta$.

b. Noting that an element of length along the path of the pendulum is $ds = L\,d\theta$, evaluate an integral in θ to show that the work done in lifting the mass to a height h is mgh.

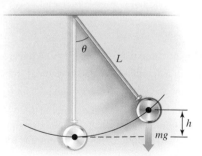

50. Orientation and force A plate shaped like an equilateral triangle 1 m on a side is placed on a vertical wall 1 m below the surface of a pool filled with water. On which plate in the figure is the force greater? Try to anticipate the answer and then compute the force on each plate.

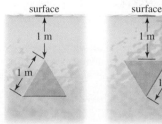

51. Orientation and force A square plate 1 m on a side is placed on a vertical wall 1 m below the surface of a pool filled with water. On which plate in the figure is the force greater? Try to anticipate the answer and then compute the force on each plate.

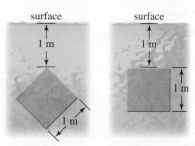

52. A calorie-free milkshake? Suppose a cylindrical glass with a diameter of $\frac{1}{12}$ m and a height of $\frac{1}{10}$ m is filled to the brim with a 400-Cal milkshake. If you have a straw that is 1.1 m long (so the top of the straw is 1 m above the top of the glass), do you burn off all the calories in the milkshake in drinking it? Assume that the density of the milkshake is 1 g/cm^3 ($1 \text{ Cal} = 4184 \text{ J}$).

53. Critical depth A large tank has a plastic window on one wall that is designed to withstand a force of 90,000 N. The square window is 2 m on a side, and its lower edge is 1 m from the bottom of the tank.

   a. If the tank is filled to a depth of 4 m, will the window withstand the resulting force?

   b. What is the maximum depth to which the tank can be filled without the window failing?

54. Buoyancy Archimedes' principle says that the buoyant force exerted on an object that is (partially or totally) submerged in water is equal to the weight of the water displaced by the object (see figure). Let $\rho_w = 1 \text{ g/cm}^3 = 1000 \text{ kg/m}^3$ be the density of water

and let ρ be the density of an object in water. Let $f = \rho/\rho_w$. If $0 < f \le 1$, then the object floats with a fraction f of its volume submerged; if $f > 1$, then the object sinks.

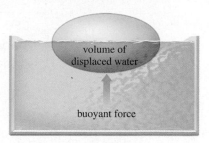

Consider a cubical box with sides 2 m long floating in water with one-half of its volume submerged ($\rho = \rho_w/2$). Find the force required to fully submerge the box (so its top surface is at the water level).

(See Guided Projects for further explorations of buoyancy problems.)

QUICK CHECK ANSWERS

1. (a) The bar is heaviest at the left end and lightest at the right end. (b) $\rho = 2.5 \text{ g/cm}$.   **2.** Minimum mass $= 2 \text{ kg}$; maximum mass $= 10 \text{ kg}$ **3.** We assume that the force is constant over each subinterval, when, in fact, it varies over each subinterval.   **4.** The restoring force of the spring increases as the spring is stretched ($F(x) = 100x$). Greater restoring forces are encountered on the interval $[0.1, 0.35]$ than on the interval $[0, 0.25]$.   **5.** The factor $(15 - y)$ in the integral is replaced by $(10 - y)$. ◄

6.7 Logarithmic and Exponential Functions Revisited

In previous chapters, we worked extensively with logarithmic and exponential functions. However, we skipped some of the theoretical issues surrounding these functions, relying instead on numerical and graphical evidence to derive their fundamental properties. For example, in Section 2.6 it was stated, without formal proof, that $f(x) = e^x$ is a continuous function defined for *all* real numbers. And in Section 3.2, we appealed to numerical approximations to a limit to claim that $\dfrac{d}{dx}(e^x) = e^x$. Assuming the truth of these results, we then determined properties of logarithmic and exponential functions. We now place logarithmic and exponential functions on a solid foundation by presenting a more rigorous development of these functions.

The Natural Logarithm

Our aim is to develop the properties of the natural logarithm using definite integrals.

DEFINITION   The Natural Logarithm

The **natural logarithm** of a number $x > 0$, denoted $\ln x$, is defined as

$$\ln x = \int_1^x \frac{1}{t} \, dt.$$

All the familiar geometric and algebraic properties of the natural logarithmic function follow directly from this new integral definition.

Properties of the Natural Logarithm

Domain, range, and sign Because the natural logarithm is defined as a definite integral, its value is the net area under the curve $y = 1/t$ between $t = 1$ and $t = x$. The integrand is undefined at $t = 0$, so the domain of $\ln x$ is $(0, \infty)$. On the interval $(1, \infty)$, $\ln x$ is positive because the net area of the region under the curve is positive (Figure 6.65a). On $(0, 1)$, we have $\int_1^x \frac{1}{t} \, dt = -\int_x^1 \frac{1}{t} \, dt$, which implies $\ln x$ is negative (Figure 6.65b). As expected, when $x = 1$, we have $\ln 1 = \int_1^1 \frac{1}{t} \, dt = 0$. The net area interpretation of $\ln x$ also implies that the range of $\ln x$ is $(-\infty, \infty)$ (see Exercise 58 for an outline of a proof).

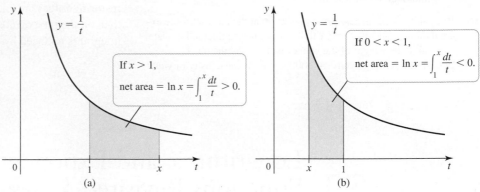

FIGURE 6.65

Derivative The derivative of the natural logarithm follows immediately from its definition and the Fundamental Theorem of Calculus:

> Recall that by the Fundamental Theorem of Calculus
>
> $$\frac{d}{dx} \int_a^x f(t) \, dt = f(x).$$

$$\frac{d}{dx}(\ln x) = \frac{d}{dx} \int_1^x \frac{dt}{t} = \frac{1}{x}, \quad \text{for } x > 0$$

We have two important consequences:

• Because its derivative is defined for $x > 0$, $\ln x$ is a differentiable function for $x > 0$, which means it is continuous on its domain (Theorem 3.1).

• Because $1/x > 0$ for $x > 0$, $\ln x$ is strictly increasing and one-to-one on its domain; therefore, it has a well-defined inverse.

The Chain Rule allows us to extend the derivative property to all nonzero real numbers (Exercise 56). By differentiating $\ln(-x)$ for $x < 0$, we find that

$$\frac{d}{dx}(\ln|x|) = \frac{1}{x}.$$

More generally, by the Chain Rule,

$$\frac{d}{dx}(\ln|u(x)|) = \frac{d}{du}(\ln|u|)u'(x) = \frac{u'(x)}{u(x)}.$$

Graph of $\ln x$ As noted before, $\ln x$ is continuous and strictly increasing for $x > 0$. The second derivative, $\frac{d^2}{dx^2}(\ln x) = -\frac{1}{x^2}$, is negative for all x, which implies the graph of $\ln x$ is concave down for $x > 0$. As demonstrated in Exercise 58,

$$\lim_{x\to\infty} \ln x = \infty, \quad \text{and} \quad \lim_{x\to 0^+} \ln x = -\infty.$$

This information, coupled with the fact that $\ln 1 = 0$, gives the graph of $y = \ln x$ (Figure 6.66).

Logarithm of a product The familiar logarithm property

$$\ln xy = \ln x + \ln y \qquad \text{for } x > 0, \quad y > 0$$

may be proved using the integral definition:

$$\ln xy = \int_1^{xy} \frac{dt}{t} \qquad\qquad \text{Definition of } \ln xy$$

$$= \int_1^x \frac{dt}{t} + \int_x^{xy} \frac{dt}{t} \qquad \text{Additive property of integrals}$$

$$= \int_1^x \frac{dt}{t} + \int_1^y \frac{du}{u} \qquad \text{Substitute } u = t/x \text{ in second integral.}$$

$$= \ln x + \ln y \qquad\qquad \text{Definition of } \ln$$

Logarithm of a quotient Assuming $x > 0$ and $y > 0$, the product property and a bit of algebra give

$$\ln x = \ln\left(y \cdot \frac{x}{y}\right) = \ln y + \ln\left(\frac{x}{y}\right).$$

Solving for $\ln(x/y)$, we have

$$\ln\left(\frac{x}{y}\right) = \ln x - \ln y,$$

which is the quotient property for logarithms.

Logarithm of a power By the product rule for logarithms, if $x > 0$ and p is a positive integer, then

$$\ln x^p = \ln\underbrace{(x \cdot x \cdots x)}_{p \text{ factors}} = \underbrace{\ln x + \cdots + \ln x}_{p \text{ terms}} = p \ln x.$$

Later in this section, we prove that $\ln x^p = p \ln x$ for $x > 0$ and for all real numbers p.

Integrals Because $\frac{d}{dx}(\ln|x|) = \frac{1}{x}$, we have

$$\int \frac{1}{x}\,dx = \ln|x| + C.$$

QUICK CHECK 1 What is the domain of $\ln|x|$?◄

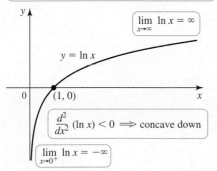

$\frac{d}{dx}(\ln x) > 0 \implies \ln x$ is increasing for $x > 0$

$\lim_{x\to\infty} \ln x = \infty$

$y = \ln x$

0 $(1, 0)$

$\frac{d^2}{dx^2}(\ln x) < 0 \implies$ concave down

$\lim_{x\to 0^+} \ln x = -\infty$

FIGURE 6.66

We have shown that the familiar properties of $\ln x$ follow from its new integral definition.

THEOREM 6.4   Properties of the Natural Logarithm

1. The domain and range of $y = \ln x$ are $(0, \infty)$ and $(-\infty, \infty)$, respectively.

2. $\dfrac{d}{dx}(\ln |u(x)|) = \dfrac{u'(x)}{u(x)},\ u(x) \neq 0$

3. $\ln (xy) = \ln x + \ln y \quad (x > 0, y > 0)$

4. $\ln (x/y) = \ln x - \ln y \quad (x > 0, y > 0)$

5. $\ln x^p = p \ln x \quad (x > 0, p \text{ a real number})$

6. $\displaystyle\int \frac{1}{x}\, dx = \ln |x| + C$

EXAMPLE 1   Integrals with $\ln x$ Evaluate $\displaystyle\int_0^4 \frac{x}{x^2 + 9}\, dx.$

SOLUTION

$$\int_0^4 \frac{x}{x^2 + 9}\, dx = \frac{1}{2}\int_9^{25} \frac{du}{u} \qquad \text{Let } u = x^2 + 9;\ du = 2x\, dx.$$

$$= \frac{1}{2} \ln |u| \Big|_9^{25} \qquad \text{Fundamental Theorem}$$

$$= \frac{1}{2}(\ln 25 - \ln 9) \qquad \text{Evaluate.}$$

$$= \ln \frac{5}{3} \qquad \text{Properties of logarithms}$$

Related Exercises 7–14 ◄

The Question of Base

The natural logarithm *is* a logarithm, but what is its base? We now determine the base b such that $\ln x = \log_b x$. Two steps are needed: We show that b exists; then we identify the value of b.

Recall that $\log_b b = 1$ for any base $b > 0$ (Section 1.3). Therefore, the number b that we seek has the property $\ln b = 1$, or

$$\ln b = \int_1^b \frac{dt}{t} = 1.$$

We see that b is the number that makes the area of the region under the curve $y = 1/t$ on the interval $[1, b]$ exactly 1 (Figure 6.67).

Computations with Riemann sums show that $\ln 2 = \displaystyle\int_1^2 \frac{dt}{t} < 1$ and that $\ln 3 = \displaystyle\int_1^3 \frac{dt}{t} > 1$ (Exercise 59). Because $\ln x$ is a continuous function, the Intermediate Value Theorem says that there is a number b with $2 < b < 3$ such that $\ln b = 1$. That completes the first step: We know that b exists and lies between 2 and 3. To estimate b, we use the fact that the derivative of $\ln x$ at $x = 1$ is 1. By the definition of the derivative, it follows that

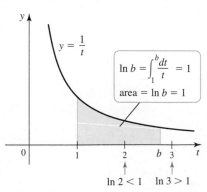

FIGURE 6.67

$$1 = \left.\frac{d}{dx}(\ln x)\right|_{x=1} = \lim_{h \to 0} \frac{\ln(1+h) - \ln 1}{h} \qquad \text{Derivative of } \ln x \text{ at } x = 1$$

$$= \lim_{h \to 0} \frac{\ln(1+h)}{h} \qquad \ln 1 = 0$$

$$= \lim_{h \to 0} \ln(1+h)^{1/h} \qquad p \ln x = \ln x^p$$

> We rely on Theorem 2.11 of Section 2.6 here. If f is continuous at $g(a)$ and g is continuous at a, then $\lim_{x \to a} f(g(x)) = f(\lim_{x \to a} g(x))$.

The natural logarithm is continuous for $x > 0$, so it is permissible to interchange the order of $\lim_{h \to 0}$ and the evaluation of $\ln(1+h)^{1/h}$. The result is that

$$\ln\left(\underbrace{\lim_{h \to 0}(1+h)^{1/h}}_{b}\right) = 1.$$

Observe that the limit within the brackets is b because $\ln b = 1$ and only one number satisfies this equation. Therefore, we have isolated b as a limit:

$$b = \lim_{h \to 0}(1+h)^{1/h}.$$

It is evident from the values in Table 6.2 that $(1+h)^{1/h} \to 2.71828\ldots$ as $h \to 0$. We first encountered the exact value of this limit in Section 3.2: It is the mathematical constant e, and it has been computed to millions of digits. A better approximation is

$$e \approx 2.718281828459045.$$

With this argument, we have identified the base of the natural logarithm; it is $b = e$.

Table 6.2

h	$(1+h)^{1/h}$	h	$(1+h)^{1/h}$
10^{-1}	2.593742	-10^{-1}	2.867972
10^{-2}	2.704814	-10^{-2}	2.731999
10^{-3}	2.716924	-10^{-3}	2.719642
10^{-4}	2.718146	-10^{-4}	2.718418
10^{-5}	2.718268	-10^{-5}	2.718295
10^{-6}	2.718280	-10^{-6}	2.718283
10^{-7}	2.718282	-10^{-7}	2.718282

> The number e was defined in Section 3.2 as the number such that $\lim_{h \to 0} \dfrac{e^h - 1}{h} = 1$.

DEFINITION Base of the Natural Logarithm

The **natural logarithm** is the logarithm with a base of $e = \lim_{h \to 0}(1+h)^{1/h} \approx 2.71828$. It follows that $\ln e = 1$.

The Exponential Function

We have established that $f(x) = \ln x$ is a continuous, increasing function on the interval $(0, \infty)$. Therefore, it is one-to-one on this interval and its inverse function exists. We denote the inverse function $f^{-1}(x) = \exp(x)$. Its graph is obtained by reflecting the graph of $f(x) = \ln x$ about the line $y = x$ (Figure 6.68). The domain of $\exp(x)$ is $(-\infty, \infty)$ because the range of $\ln x$ is $(-\infty, \infty)$, and the range of $\exp(x)$ is $(0, \infty)$ because the domain of $\ln x$ is $(0, \infty)$.

The usual relationships between a function and its inverse also hold:

- $y = \exp(x)$ if and only if $x = \ln y$
- $\exp(\ln x) = x$ for $x > 0$ and $\ln(\exp(x)) = x$ for all x

We now appeal to the properties of $\ln x$ and use the inverse relations between $\ln x$ and $\exp(x)$ to show that $\exp(x)$ satisfies the required properties of any exponential function. For example, if $x_1 = \ln y_1$ (which implies that $y_1 = \exp(x_1)$) and $x_2 = \ln y_2$ (which implies that $y_2 = \exp(x_2)$), then

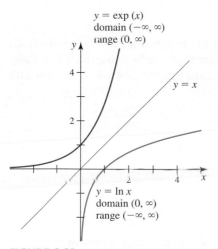

$y = \exp(x)$
domain $(-\infty, \infty)$
range $(0, \infty)$

$y = x$

$y = \ln x$
domain $(0, \infty)$
range $(-\infty, \infty)$

FIGURE 6.68

> Note that we already know two important values of $\exp(x)$. Because $\ln e = 1$, we have $\exp(1) = e$. Because $\ln 1 = 0$, $\exp(0) = 1$.

$$\exp(x_1 + x_2) = \exp(\underbrace{\ln y_1 + \ln y_2}_{\ln y_1 y_2}) \qquad \text{Substitute } x_1 = \ln y_1, x_2 = \ln y_2.$$

$$= \exp(\ln y_1 y_2) \qquad \text{Properties of logarithms}$$

$$= y_1 y_2 \qquad \text{Inverse property of } \exp x \text{ and } \ln x$$

$$= \exp(x_1)\exp(x_2). \qquad y_1 = \exp(x_1), y_2 = \exp(x_2)$$

Therefore, $\exp(x)$ satisfies the property of exponential functions $b^{x_1 + x_2} = b^{x_1} b^{x_2}$. Similar arguments show that $\exp(x)$ satisfies the defining properties of all exponential functions (Exercise 57).

We have shown that $\exp(x)$ is an exponential function and it is the inverse function of $\ln x$. We have also shown that $\ln x$ is the logarithmic function base e. Therefore, the base for $\exp(x)$ is also the number e, and we have $\exp(x) = e^x$ for all real numbers x.

DEFINITION   The Exponential Function

The **(natural) exponential function** is the exponential function with the base $e \approx 2.71828$. It is the inverse function of the natural logarithm $\ln x$.

The essential properties of the exponential function are summarized in the following theorem.

THEOREM 6.5   Properties of e^x

The exponential function $f(x) = e^x$ satisfies the following properties, all of which follow from the integral definition of $\ln x$. Let x and y be real numbers.

1. $e^{x+y} = e^x e^y$

2. $e^{x-y} = e^x / e^y$

3. $(e^x)^y = e^{xy}$

4. $\ln(e^x) = x$

5. $e^{\ln x} = x$, for $x > 0$

QUICK CHECK 2 Simplify $e^{\ln 2x}$, $\ln(e^{2x})$, $e^{2\ln x}$, $\ln(2e^x)$. ◄

Derivatives and Integrals   The derivative of the exponential function follows directly from Theorem 3.23 (derivatives of inverse functions) or by using the Chain Rule. Taking the latter course, we observe that $\ln(e^x) = x$ and then differentiate both sides with respect to x:

$$\frac{d}{dx}(\ln e^x) = \frac{d}{dx}(x)$$

$$\underbrace{\qquad}_{1}$$

$$\frac{1}{e^x}\frac{d}{dx}(e^x) = 1 \qquad\qquad \frac{d}{dx}(\ln u) = \frac{u'(x)}{u(x)}$$

$$\frac{d}{dx}(e^x) = e^x \qquad\qquad \text{Solve for } \frac{d}{dx}(e^x).$$

QUICK CHECK 3 What is the slope of the curve $y = e^x$ at $x = \ln 2$? What is the area of the region bounded by the graph of $y = e^x$ and the x-axis between $x = 0$ and $x = \ln 2$? ◄

Once again, we obtain the remarkable result that the exponential function is its own derivative. Of course, it immediately follows that e^x is its own antiderivative up to a constant; that is,

$$\int e^x \, dx = e^x + C.$$

Extending these results using the Chain Rule, we have the following theorem.

THEOREM 6.6   Derivative and Integral of the Exponential Function

For real numbers x,

$$\frac{d}{dx}(e^{u(x)}) = e^{u(x)} u'(x) \quad \text{and} \quad \int e^x \, dx = e^x + C.$$

EXAMPLE 2 **Integrals with e^x** Evaluate $\displaystyle\int \frac{e^x}{1+e^x}\,dx$.

SOLUTION The change of variables $u = 1 + e^x$ implies $du = e^x\,dx$:

$$\int \underbrace{\frac{1}{1+e^x}}_{u}\underbrace{e^x\,dx}_{du} = \int \frac{1}{u}\,du \qquad u = 1 + e^x,\ du = e^x\,dx$$

$$= \ln|u| + C \qquad \text{Antiderivative of } u^{-1}$$

$$= \ln(1 + e^x) + C \quad \text{Replace } u \text{ by } 1 + e^x.$$

Note that the absolute value may be removed from $\ln|u|$ because $1 + e^x > 0$ for all x.

Related Exercises 15–18 ◄

General Logarithmic and Exponential Functions

The goal of this section has been accomplished. We have developed the properties of the natural logarithmic and exponential functions beginning with the integral definition of the natural logarithm. With these two functions on firm ground, the next step is to establish the properties of logarithmic and exponential functions with a general positive base other than e. However, for the most part, this work has already been done. If you look in Section 3.8, you will find the basic derivative results involving exponential functions b^x and logarithmic functions $\log_b x$, where $b > 0$. We close this section by summarizing those results and filling in a missing integral.

First, it is important to note that the function b^x is defined for all bases $b > 0$ and for all real numbers x. The reason is that

> ► The change-of-base formula relating $\log_b x$ to the natural logarithm is
>
> $$\log_b x = \frac{\ln x}{\ln b}.$$

$$b^x = (e^{\ln b})^x = e^{x \ln b},$$

where we now know that e^x is defined for all real numbers x. Just as b^x is defined in terms of e^x, $\log_b x$ is evaluated in terms of $\ln x$ via the change of base formula.

Theorems 3.18 and 3.20 give us the derivative results for exponential and logarithmic functions with a general base $b > 0$. Extending those results with the Chain Rule, we have the following derivatives.

SUMMARY **Derivatives with Other Bases**

Let $b > 0$ and $b \neq 1$. Then,

$$\frac{d}{dx}(\log_b u(x)) = \frac{u'(x)}{u(x)\ln b}, \text{ for } u(x) > 0 \qquad \frac{d}{dx}(b^{u(x)}) = (\ln b)b^{u(x)}u'(x)$$

Writing the derivative result for b^x in terms of an indefinite integral, we have another useful integral.

THEOREM 6.7 **Integral of b^x**

For $b > 0$ and $b \neq 1$, $\displaystyle\int b^x\,dx = \frac{1}{\ln b}b^x + C$.

QUICK CHECK 4 Verify that the derivative and integral results for a general base b reduce to the expected results when $b = e$. ◄

EXAMPLE 3 **Integrals involving exponentials with other bases** Evaluate the following integrals.

a. $\displaystyle\int x\,3^{x^2}\,dx$ **b.** $\displaystyle\int_1^4 \frac{6^{-\sqrt{x}}}{\sqrt{x}}\,dx$

SOLUTION

a. $\displaystyle\int x\,3^{x^2}\,dx = \frac{1}{2}\int 3^u\,du$ $u = x^2, du = 2x\,dx$

$\qquad\qquad\quad = \frac{1}{2}\frac{1}{\ln 3}\,3^u + C$ Integrate.

$\qquad\qquad\quad = \frac{1}{2\ln 3}\,3^{x^2} + C$ Substitute $u = x^2$.

b. $\displaystyle\int_1^4 \frac{6^{-\sqrt{x}}}{\sqrt{x}}\,dx = -2\int_{-1}^{-2} 6^u\,du$ $u = -\sqrt{x}, du = -\frac{1}{2\sqrt{x}}\,dx$

$\qquad\qquad\qquad = -\frac{2}{\ln 6}\,6^u\Big|_{-1}^{-2}$ Fundamental Theorem

$\qquad\qquad\qquad = \frac{5}{18\ln 6}$ Simplify.

Related Exercises 19–22 ◄

General Power Rule

The definition of e^x for all real numbers x has several significant consequences. First, it fills a gap: Until now, we did not have a definition of expressions such as $x^{\sqrt{2}}$ and x^x (for real numbers $x > 0$) and 2^x (for real numbers x). With the identity $x^p = e^{p\ln x}$, the expression x^p now has meaning for real numbers p and $x > 0$. For example,

$$x^{\sqrt{2}} = e^{\sqrt{2}\ln x}, \quad x^x = e^{x\ln x}, \quad \text{and } 2^x = e^{x\ln 2}.$$

The definition of e^x also enables us to confirm an important property of logarithms. Taking logarithms of both sides of $x^p = e^{p\ln x}$, we see that $\ln x^p = p\ln x$, for real numbers p and $x > 0$, as stated earlier in the section.

Finally, we can fill another gap in our derivative knowledge. It has been shown that the power rule

$$\frac{d}{dx}(x^p) = px^{p-1}$$

applies when p is a rational number. This result is extended to all real values of p by differentiating $x^p = e^{p\ln x}$:

$$\frac{d}{dx}(x^p) = \frac{d}{dx}(e^{p\ln x})\quad x^p = e^{p\ln x}$$

$$= \underbrace{e^{p\ln x}}_{x^p}\frac{p}{x}\qquad \text{Chain Rule}$$

$$= x^p\frac{p}{x}\qquad e^{p\ln x} = x^p$$

$$= px^{p-1}\qquad \text{Simplify.}$$

THEOREM 6.8 General Power Rule

For any real number p,

$$\frac{d}{dx}(x^p) = px^{p-1} \quad \text{and} \quad \frac{d}{dx}[u(x)^p] = pu(x)^{p-1}u'(x).$$

EXAMPLE 4 **Derivative of a tower function** Evaluate the derivative of $f(x) = x^{2x}$.

SOLUTION We use the inverse relationship $e^{\ln x} = x$ to write $x^{2x} = e^{\ln(x^{2x})} = e^{2x \ln x}$. It follows that

$$\frac{d}{dx}(x^{2x}) = \frac{d}{dx}(e^{2x \ln x})$$

$$= \underbrace{e^{2x \ln x}}_{x^{2x}} \frac{d}{dx}(2x \ln x) \qquad \frac{d}{dx}(e^{u(x)}) = e^{u(x)}u'(x)$$

$$= x^{2x}\left(2 \ln x + 2x \cdot \frac{1}{x}\right) \qquad \text{Product Rule}$$

$$= 2x^{2x}(1 + \ln x) \qquad \text{Simplify.} \qquad \text{Related Exercises 23–28} \blacktriangleleft$$

SECTION 6.7 EXERCISES

Review Questions

1. What are the domain and range of $\ln x$?

2. Give a geometrical interpretation of the function $\ln x = \int_1^x \frac{dt}{t}$.

3. Evaluate $\int 4^x \, dx$.

4. What is the inverse function of $\ln x$, and what are its domain and range?

5. Express $3^x, x^\pi, x^{\sin x}$ using the base e.

6. Evaluate $\dfrac{d}{dx}(3^x)$.

Basic Skills

7–10. Derivatives with ln x *Evaluate the following derivatives.*

7. $\dfrac{d}{dx}(x \ln(x^3))\Big|_{x=1}$

8. $\dfrac{d}{dx}(\ln(\ln x))$

9. $\dfrac{d}{dx}(\sin(\ln x))$

10. $\dfrac{d}{dx}(\ln(\cos^2 x))$

11–14. Integrals with ln x *Evaluate the following integrals. Include absolute values only when needed.*

11. $\displaystyle\int_0^3 \frac{2x - 1}{x + 1} \, dx$

12. $\displaystyle\int \tan 10x \, dx$

13. $\displaystyle\int_3^4 \frac{dx}{2x \ln x \ln^3(\ln x)}$

14. $\displaystyle\int_0^{\pi/2} \frac{\sin x}{1 + \cos x} \, dx$

15–18. Integrals with e^x *Evaluate the following integrals.*

15. $\displaystyle\int \frac{e^x + e^{-x}}{e^x - e^{-x}} \, dx$

16. $\displaystyle\int \frac{e^{\sin x}}{\sec x} \, dx$

17. $\displaystyle\int \frac{e^{\sqrt{x}}}{\sqrt{x}} \, dx$

18. $\displaystyle\int_{-2}^2 \frac{e^{x/2}}{e^{x/2} + 1} \, dx$

19–22. Integrals with general bases *Evaluate the following integrals.*

19. $\displaystyle\int_{-1}^1 10^x \, dx$

20. $\displaystyle\int_0^{\pi/2} 4^{\sin x} \cos x \, dx$

21. $\displaystyle\int_1^2 (1 + \ln x)x^x \, dx$

22. $\displaystyle\int_{1/3}^{1/2} \frac{10^{1/x}}{x^2} \, dx$

23–28. Derivatives *Evaluate the derivatives of the following functions.*

23. $f(x) = (2x)^{4x}$

24. $f(x) = x^\pi$

25. $h(x) = 2^{(x^2)}$

26. $h(t) = (\sin t)^{\sqrt{t}}$

27. $H(x) = (x + 1)^{2x}$

28. $p(x) = x^{-\ln x}$

Further Explorations

29. **Explain why or why not** Determine whether the following statements are true and give an explanation or counterexample. Assume $x > 0$ and $y > 0$.

    a. $\ln(xy) = \ln x + \ln y$ **b.** $\ln 0 = 1$

    c. $\ln(x + y) = \ln x + \ln y$ **d.** $2^x = e^{2 \ln x}$

    e. The area under the curve $y = 1/x$ and the x-axis on the interval $[1, e]$ is 1.

30. **Logarithm properties** Use the integral definition of the natural logarithm to prove that $\ln(x/y) = \ln x - \ln y$.

31–34. Calculator limits *Use a calculator to make a table similar to Table 6.2 to approximate the following limits. Confirm your result with l'Hôpital's Rule.*

31. $\displaystyle\lim_{h \to 0} (1 + 2h)^{1/h}$

32. $\displaystyle\lim_{h \to 0} (1 + 3h)^{2/h}$

33. $\displaystyle\lim_{x \to 0} \frac{2^x - 1}{x}$

34. $\displaystyle\lim_{x \to 0} \frac{\ln(1 + x)}{x}$

35. **Zero net area** Consider the function $f(x) = \dfrac{1 - x}{x}$.

    a. Are there numbers $0 < a < 1$ such that $\displaystyle\int_{1-a}^{1+a} f(x) \, dx = 0$?

    b. Are there numbers $a > 1$ such that $\displaystyle\int_{1/a}^a f(x) \, dx = 0$?

36. Behavior at the origin Using calculus and accurate sketches, explain how the graphs of $f(x) = x^p \ln x$ differ as $x \to 0$ for $p = \frac{1}{2}, 1, 2$.

37. Average value What is the average value of $f(x) = 1/x$ on the interval $[1, p]$ for $p > 1$? What is the average value of f as $p \to \infty$?

38–45. Miscellaneous derivatives *Compute the following derivatives using the method of your choice.*

38. $\dfrac{d}{dx}(x^{2x})$

39. $\dfrac{d}{dx}(e^{-10x^2})$

40. $\dfrac{d}{dx}(x^{\tan x})$

41. $\dfrac{d}{dx}\left[\left(\dfrac{1}{x}\right)^x\right]$

42. $\dfrac{d}{dx}(x^e + e^x)$

43. $\dfrac{d}{dx}\left(1 + \dfrac{4}{x}\right)^x$

44. $\dfrac{d}{dx}\left(x^{(x^{10})}\right)$

45. $\dfrac{d}{dx}(\cos(x^{2\sin x}))$

46–54. Miscellaneous integrals *Evaluate the following integrals.*

46. $\displaystyle\int 7^{2x}\,dx$

47. $\displaystyle\int 3^{-2x}\,dx$

48. $\displaystyle\int_0^5 5^{5x}\,dx$

49. $\displaystyle\int x^2 10^{x^3}\,dx$

50. $\displaystyle\int_0^\pi \cos x \cdot 2^{\sin x}\,dx$

51. $\displaystyle\int_1^{2e} \dfrac{3^{\ln x}}{x}\,dx$

52. $\displaystyle\int \dfrac{\sin(\ln x)}{4x}\,dx$

53. $\displaystyle\int_1^{e^2} \dfrac{(\ln x)^5}{x}\,dx$

54. $\displaystyle\int \dfrac{\ln^2 x + 2\ln x - 1}{x}\,dx$

Applications

55. Probability as an integral Two points P and Q are chosen randomly, one on each of two adjacent sides of a unit square (see figure). What is the probability that the area of the triangle formed by the sides of the square and the line segment PQ is less than one-fourth the area of the square? Begin by showing that x and y must satisfy $xy < \frac{1}{2}$ in order for the area condition to be met. Then argue that the required probability is $\dfrac{1}{2} + \displaystyle\int_{1/2}^1 \dfrac{dx}{2x}$ and evaluate the integral.

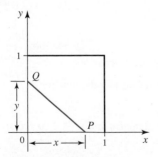

Additional Exercises

56. Derivative of $\ln |x|$ Differentiate $\ln x$ for $x > 0$ and differentiate $\ln(-x)$ for $x < 0$ to conclude that $\dfrac{d}{dx}(\ln|x|) = \dfrac{1}{x}$.

57. Properties of e^x Use the inverse relations between $\ln x$ and e^x and the properties of $\ln x$ to prove the following properties.

   a. $e^{x-y} = \dfrac{e^x}{e^y}$ **b.** $(e^x)^y = e^{xy}$

58. $\ln x$ is unbounded Use the following argument to show that $\lim\limits_{x\to\infty} \ln x = \infty$ and $\lim\limits_{x\to 0^+} \ln x = -\infty$.

   a. Make a sketch of the function $f(x) = 1/x$ on the interval $[1, 2]$. Explain why the area of the region bounded by $y = f(x)$ and the x-axis on $[1, 2]$ is $\ln 2$.

   b. Construct a rectangle over the interval $[1, 2]$ with height $\frac{1}{2}$. Explain why $\ln 2 > \frac{1}{2}$.

   c. Show that $\ln 2^n > n/2$ and $\ln 2^{-n} < -n/2$.

   d. Conclude that $\lim\limits_{x\to\infty} \ln x = \infty$ and $\lim\limits_{x\to 0^+} \ln x = -\infty$.

59. Bounds on e Use a left Riemann sum with at least $n = 2$ subintervals of equal length to approximate $\ln 2 = \displaystyle\int_1^2 \dfrac{dt}{t}$ and show that $\ln 2 < 1$. Use a right Riemann sum with at least $n = 7$ subintervals of equal length to approximate $\ln 3 = \displaystyle\int_1^3 \dfrac{dt}{t}$ and show that $\ln 3 > 1$.

60. Alternate proof of product property Assume that $y > 0$ is fixed and that $x > 0$. Show that $\dfrac{d}{dx}(\ln xy) = \dfrac{d}{dx}(\ln x)$. Recall that if two functions have the same derivative, they differ by a constant. Set $x = 1$ to evaluate the constant and prove that $\ln xy = \ln x + \ln y$.

61. Harmonic sum In Chapter 8, we will encounter the harmonic sum $1 + \dfrac{1}{2} + \dfrac{1}{3} + \cdots + \dfrac{1}{n}$. Use a right Riemann sum to approximate $\displaystyle\int_1^n \dfrac{dx}{x}$ (with unit spacing between the grid points) to show that $1 + \dfrac{1}{2} + \dfrac{1}{3} + \cdots + \dfrac{1}{n} > \ln(n + 1)$. Use this fact to conclude that $\lim\limits_{n\to\infty}\left(1 + \dfrac{1}{2} + \dfrac{1}{3} + \cdots + \dfrac{1}{n}\right)$ does not exist.

QUICK CHECK ANSWERS

1. $\{x: x \ne 0\}$ **2.** $2x, 2x, x^2, \ln 2 + x$ **3.** Slope $= 2$; area $= 1$ **4.** Note that when $b = e$, we have $\ln b = 1$. ◄

6.8 Exponential Models

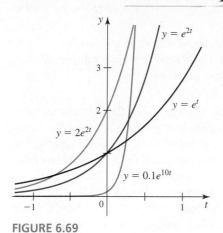

FIGURE 6.69

▸ The derivative $\dfrac{dy}{dt}$ is the *absolute* growth rate but is usually more simply called the *growth rate*.

▸ A consumer price index that increases at a constant rate of 4% per year increases exponentially. A currency that is devalued at a constant rate of 3% per month decreases exponentially. By contrast, linear growth is characterized by constant absolute growth rates, such as 500 people per year or $400 per month.

The uses of exponential functions are wide-ranging. In this section, you will see them applied to problems in finance, medicine, ecology, biology, economics, pharmacokinetics, anthropology, and physics.

Exponential Growth

Exponential growth models use functions of the form $y(t) = Ce^{kt}$, where C is a constant and the **rate constant** k is positive (Figure 6.69).

If we start with the exponential growth function $y(t) = Ce^{kt}$ and take its derivative, we find that

$$\frac{dy}{dt} = \frac{d}{dt}(Ce^{kt}) = C \cdot ke^{kt} = k(\underbrace{Ce^{kt}}_{y})$$

that is, $\dfrac{dy}{dt} = ky$. Here is the first insight about exponential functions: *Their rate of change is proportional to their value*. If y represents a population, then $\dfrac{dy}{dt}$ is the **growth rate** with units such as people/month or cells/hr. And if y is an exponential function, then the more people present, the faster the population grows.

Another way to talk about growth rates is to use the **relative growth rate**, which is the growth rate divided by the current value of that quantity—that is, $\dfrac{1}{y}\dfrac{dy}{dt}$. For example, if y is a population, the relative growth rate is the fraction or percentage by which the population grows each unit of time. Examples of relative growth rates are *5% per year* or a *factor of 1.2 per month*. Therefore, when the equation $\dfrac{dy}{dt} = ky$ is written in the form $\dfrac{1}{y}\dfrac{dy}{dt} = k$, it has another interpretation. It says that *a quantity that grows exponentially has a constant relative growth rate*. Constant relative or percentage change is the hallmark of exponential growth.

EXAMPLE 1   Linear vs. exponential growth Suppose the population of the town of Pine is given by $P(t) = 1500 + 125t$, while the population of the town of Spruce is given by $S(t) = 1500e^{0.1t}$, where $t \geq 0$ is measured in years. Find the growth rates and the relative growth rates of the two towns.

SOLUTION Note that Pine grows according to a linear function, while Spruce grows exponentially (Figure 6.70). The growth rate of Pine is $\dfrac{dP}{dt} = 125$ people/yr, which is constant for all times. The growth rate of Spruce is

$$\frac{dS}{dt} = 0.1(\underbrace{1500e^{0.1t}}_{S(t)}) = 0.1S(t),$$

showing that the growth rate is proportional to the population. The relative growth rate of Pine is $\dfrac{1}{P}\dfrac{dP}{dt} = \dfrac{125}{1500 + 125t}$, which decreases in time. The relative growth rate of Spruce is

$$\frac{1}{S}\frac{dS}{dt} = \frac{0.1 \cdot 1500e^{0.1t}}{1500e^{0.1t}} = 0.1,$$

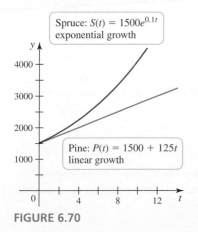

FIGURE 6.70

which is constant for all times. In summary, the linear population function has a *constant absolute growth rate*, while the exponential population function has a *constant relative growth rate*.   *Related Exercises 9–10* ◄

QUICK CHECK 1 Population A increases at a constant rate of 4%/yr. Population B increases at a constant rate of 500 people/yr. Which population exhibits exponential growth? What kind of growth is exhibited by the other population? ◄

The rate constant k in $y(t) = Ce^{kt}$ determines the growth rate of the exponential function. We adopt the convention that $k > 0$; then it is clear that $y(t) = Ce^{kt}$ describes exponential growth and $y(t) = Ce^{-kt}$ describes exponential decay, to be discussed shortly. For problems that involve time, the units of k are time^{-1}; for example, if t is measured in months, the units of k are month^{-1}. In this way, the exponent kt is dimensionless (without units).

> The unit time^{-1} is read *per unit time*. For example, month^{-1} is read *per month*.

Unless there is good reason to do otherwise, it is customary to take $t = 0$ as the reference point for time. Notice that with $y(t) = Ce^{kt}$, we have $y(0) = C$. Therefore, C has a simple meaning: It is the **initial value** of the quantity of interest, which we denote y_0. In the examples that follow, two pieces of information are typically given: the initial condition and clues for determining the rate constant k. The initial condition and the rate constant determine an exponential growth function completely.

Exponential Growth Functions

Exponential growth is described by functions of the form $y(t) = y_0 e^{kt}$. The initial value of y at $t = 0$ is $y(0) = y_0$ and the **rate constant $k > 0$** determines the rate of growth. Exponential growth is characterized by a constant relative growth rate.

Because exponential growth is characterized by a constant relative growth rate, the time required for a quantity to double (a 100% increase) is constant. Therefore, one way to describe an exponentially growing quantity is to give its **doubling time**. To compute the time it takes for the function $y(t) = y_0 e^{kt}$ to double in value, say from y_0 to $2y_0$, we find the value of t that satisfies

$$y(t) = 2y_0 \quad \text{or} \quad y_0 e^{kt} = 2y_0.$$

> Note that the initial value y_0 appears on both sides of this equation. It may be canceled, meaning that the doubling time is independent of the initial condition: *The doubling time is constant for all t.*

Canceling y_0 from the equation $y_0 e^{kt} = 2y_0$ leaves the equation $e^{kt} = 2$. Taking logarithms of both sides, we have $\ln e^{kt} = \ln 2$, or $kt = \ln 2$, which has the solution $t = \dfrac{\ln 2}{k}$.

We denote this doubling time T_2 so that $T_2 = \dfrac{\ln 2}{k}$. If y increases exponentially, the time it takes to double from 100 to 200 is the same as the time it takes to double from 1000 to 2000.

QUICK CHECK 2 Verify that the time needed for $y(t) = y_0 e^{kt}$ to double from y_0 to $2y_0$ is the same as the time needed to double from $2y_0$ to $4y_0$. ◄

DEFINITION   Doubling Time

The quantity described by the function $y(t) = y_0 e^{kt}$ for $k > 0$ has a constant doubling time of $T_2 = \dfrac{\ln 2}{k}$, with the same units as t.

EXAMPLE 2   World population Human population growth rates vary geographically and fluctuate over time. The overall growth rate for world population peaked at an annual

World population

1804	1 billion
1927	2 billion
1960	3 billion
1974	4 billion
1987	5 billion
1999	6 billion
2011	7 billion (proj.)

➤ It is a common mistake to assume that if the annual growth rate is 1.4% per year, then $k = 1.4\% = 0.014 \text{ yr}^{-1}$. The rate constant k must be calculated, as it is in Example 2 to give $k = 0.013976$. For larger growth rates, the difference between k and the growth rate is greater.

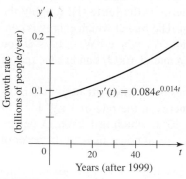

FIGURE 6.71

➤ Converted to a daily rate (dividing by 365), the world population in 2010 increased at a rate of roughly 268,000 people per day.

rate of 2.1% per year in the 1960s. Assume a world population of 6.0 billion in 1999 ($t = 0$) and 6.9 billion in 2009 ($t = 10$).

a. Find an exponential growth function for the world population that fits the two data points.

b. Find the doubling time for the world population using the model in part (a).

c. Find the (absolute) growth rate $y'(t)$ and graph it for $0 \le t \le 50$.

d. How fast was the population growing in 2010 ($t = 11$)?

SOLUTION

a. Let $y(t)$ be world population measured in billions of people t years after 1999. We use the growth function $y(t) = y_0 e^{kt}$, where y_0 and k must be determined. The initial value is $y_0 = 6$ (billion). To determine the rate constant k, we use the fact that $y(10) = 6.9$. Substituting $t = 10$ into the growth function with $y_0 = 6$ implies

$$y(10) = 6e^{10k} = 6.9.$$

Solving for k yields the rate constant $k = \dfrac{\ln(6.9/6)}{10} \approx 0.013976 \approx 0.014 \text{ yr}^{-1}$.

Therefore, the growth function is

$$y(t) = 6e^{0.014t}.$$

b. The doubling time of the population is

$$T_2 = \frac{\ln 2}{k} \approx \frac{\ln 2}{0.014} \approx 50 \text{ years.}$$

c. Working with the growth function $y(t) = 6e^{0.014t}$, we find that

$$y'(t) = 6(0.014)e^{0.014t} = 0.084e^{0.014t}$$

which has units of *billions of people/yr*. As shown in Figure 6.71 the growth rate itself increases exponentially.

d. In 2010 ($t = 11$), the growth rate was

$$y'(11) = 0.084e^{(0.014)(11)} \approx 0.098 \text{ billion people/yr}$$

or roughly 98 million people/yr. *Related Exercises 11–16* ◄

QUICK CHECK 3 Assume $y(t) = 100e^{0.05t}$. By what percentage does y increase when t increases by 1 unit? ◄

A Financial Model Exponential functions are used in many financial applications, several of which are explored in the exercises. For now, consider a simple savings account in which an initial deposit earns interest that is reinvested in the account. Interest payments are made on a regular basis (for example, annually, monthly, daily) or interest may be compounded continuously. With continuous compounding, the balance in the account increases exponentially at a rate that can be determined from the advertised **annual percentage yield** (or **APY**) of the account. Assuming that no additional deposits are made, the balance in the account is given by the exponential growth function $y(t) = y_0 e^{kt}$, where y_0 is the initial deposit, t is measured in years, and k is determined by the annual percentage yield.

EXAMPLE 3 **Continuous compounding** The APY of a savings account is the percentage increase in the balance over the course of a year. Suppose you deposit $500 in a savings account that has an APY of 6.18% per year with continuous compounding. Assume that the interest rate remains constant and that no additional deposits or withdrawals are made. How long will it take for the balance to reach $2500?

▶ The rate constant k, in this case $0.06 = 6\%$, is the factor by which the balance increases if interest is compounded once at the end of the year. It is often advertised by banks as the *annual percentage rate* (or APR). If the balance increases by 6.18% in one year, it increases by a factor of 1.0618 in one year.

SOLUTION Because the balance grows by a fixed percentage every year, it grows exponentially. Letting $y(t)$ be the balance t years after the initial deposit of $y_0 = \$500$, we have $y(t) = y_0 e^{kt}$, where the rate constant k must be determined. Note that if the initial balance is y_0, one year later the balance is 6.18% more, or

$$y(1) = 1.0618\, y_0 = y_0 e^k.$$

Solving for k, we find that the rate constant is

$$k = \ln 1.0618 \approx 0.060 \text{ yr}^{-1}.$$

Therefore, the balance at any time $t \geq 0$ is $y(t) = 500 e^{0.060t}$. To determine the time required for the balance to reach \$2500, we solve the equation

$$y(t) = 500 e^{0.060t} = 2500.$$

Dividing by 500 and taking the natural logarithm of both sides yields

$$0.060t = \ln 5.$$

The balance reaches \$2500 in $t = (\ln 5)/0.060 \approx 26.8$ yr.   *Related Exercises 11–16* ◀

Resource Consumption   Among the many resources that people use, energy is certainly one of the most important. The basic unit of energy is the **joule** (J), roughly the energy needed to lift a 0.1-kg object (say an orange) 1 m. The *rate* at which energy is consumed is called **power**. The basic unit of power is the **watt** (W), where $1 \text{ W} = 1 \text{ J/s}$. If you turn on a 100-W lightbulb for 1 min, you use energy at a rate of 100 J/s and use a total of $100 \text{ J/s} \cdot 60 \text{ s} = 6000 \text{ J}$ of energy.

A more useful measure of energy for large quantities is the **kilowatt-hour** (kWh). A kilowatt is 1000 W or 1000 J/s. So if you consume energy at the rate of 1 kW for 1 hr (3600 s), you use a total of $1000 \text{ J/s} \cdot 3600 \text{ s} = 3.6 \times 10^6 \text{ J}$, which is 1 kWh. A person running for 1 hr consumes roughly 1 kWh of energy. A typical house uses on the order of 1000 kWh of energy in a month.

Assume that the total energy used (by a person, machine, or city) is given by the function $E(t)$. Because the power $P(t)$ is the rate at which energy is used, we have $P(t) = E'(t)$. Using the ideas of Section 6.1, the total amount of energy used between the times $t = a$ and $t = b$ is

$$\text{total energy used} = \int_a^b E'(t)\, dt = \int_a^b P(t)\, dt.$$

We see that energy is the area under the power curve. With this background, we can investigate a situation in which the rate of energy consumption increases exponentially.

EXAMPLE 4   Energy consumption   At the beginning of 2006, the rate of energy consumption for the city of Denver was 7000 megawatts (MW), where $1 \text{ MW} = 10^6 \text{ W}$. That rate was expected to increase at an annual growth rate of 2% per year.

a. Find the function that gives the power or rate of energy consumption for all times after the beginning of 2006.

b. Find the total amount of energy used during the year 2010.

c. Find the function that gives the total (cumulative) amount of energy used by the city between 2006 and any time $t \geq 0$.

SOLUTION

▶ In one year, the power function increases by 2% or by a factor of 1.02.

a. Let $t \geq 0$ be the number of years after the beginning of 2006, and let $P(t)$ be the power function that gives the rate of energy consumption at time t. Because P increases at a constant rate of 2% per year, it increases exponentially. Therefore, $P(t) = P_0 e^{kt}$, where $P_0 = 7000 \text{ MW}$. We determine k as before by setting $t = 1$; after one year the power is

$$P(1) = P_0 e^k = 1.02 P_0.$$

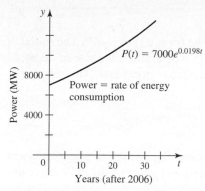

FIGURE 6.72

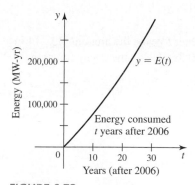

FIGURE 6.73

Canceling P_0 and solving for k, we find that $k = \ln(1.02) \approx 0.0198$. Therefore, the power function (Figure 6.72) is

$$P(t) = 7000e^{0.0198t} \quad \text{for } t \geq 0.$$

b. The entire year 2010 corresponds to the interval $4 \leq t \leq 5$. Substituting $P(t) = 7000e^{0.0198t}$, the total energy used in 2010 was

$$\int_4^5 P(t)\, dt = \int_4^5 7000e^{0.0198t}\, dt \quad \text{Substitute for } P(t).$$

$$= \frac{7000}{0.0198} e^{0.0198t} \Big|_4^5 \quad \text{Fundamental Theorem}$$

$$\approx 7652 \text{ MW-years} \quad \text{Evaluate.}$$

Because the units of P are MW and t is measured in yr, the units of energy are MW-years. To convert to MWh, we multiply by 8760 hr/yr to get the total energy of about 6.7×10^7 MWh (or 6.7×10^{10} kWh).

c. The total energy used between $t = 0$ and at any future time t is given by the future value formula (Section 6.1):

$$E(t) = E(0) + \int_0^t E'(s)\, ds = E(0) + \int_0^t P(s)\, ds.$$

Assuming $t = 0$ corresponds to the beginning of 2006, we take $E(0) = 0$. Substituting again for the power function P, the total energy at time t is

$$E(t) = E(0) + \int_0^t P(s)\, ds$$

$$= 0 + \int_0^t 7000e^{0.0198s}\, ds \quad \text{Substitute for } P(s) \text{ and } E(0).$$

$$= \frac{7000}{0.0198} e^{0.0198s} \Big|_0^t \quad \text{Fundamental Theorem}$$

$$\approx 353{,}535(e^{0.0198t} - 1) \text{ MW-yr} \quad \text{Evaluate.}$$

As shown in Figure 6.73, when the rate of energy consumption increases exponentially, the total amount of energy consumed also increases exponentially.

Related Exercises 11–16 ◄

Exponential Decay

Everything you have learned about exponential growth carries over directly to exponential decay. A function that decreases exponentially has the form $y(t) = y_0 e^{-kt}$, where $y_0 = y(0)$ is the initial value and $k > 0$ is the rate constant.

Exponential decay is characterized by a constant relative decay rate and by a constant half-life. For example, radioactive plutonium has a half-life of 24,000 years. An initial sample of 1 mg decays to 0.5 mg after 24,000 years and to 0.25 mg after 48,000 years. To compute the half-life, we determine the time required for the quantity $y(t) = y_0 e^{-kt}$ to reach one half of its current value; that is, we solve $y_0 e^{-kt} = y_0/2$ for t. Canceling y_0 and taking logarithms of both sides, we find that

QUICK CHECK 4 If a quantity decreases by a factor of 8 every 30 yr, what is its half-life? ◄

$$e^{-kt} = \frac{1}{2} \quad \Rightarrow \quad -kt = \ln\left(\frac{1}{2}\right) = -\ln 2 \quad \Rightarrow \quad t = \frac{\ln 2}{k}.$$

The half-life is given by the same formula as the doubling time.

> **Exponential Decay Functions**
>
> Exponential decay is described by functions of the form $y(t) = y_0 e^{-kt}$. The initial value of y is $y(0) = y_0$, and the rate constant $k > 0$ determines the rate of decay. Exponential decay is characterized by a constant relative decay rate. The constant half-life is $T_{1/2} = \dfrac{\ln 2}{k}$, with the same units as t.

Radiometric Dating   A powerful method for estimating the age of ancient objects (for example, fossils, bones, meteorites, and cave paintings) relies on the radioactive decay of certain elements. A common version of radiometric dating uses the carbon isotope C-14, which is present in all living matter. When a living organism dies, it ceases to replace C-14, and the C-14 that is present decays with a half-life of about $T_{1/2} = 5730$ yr. Comparing the C-14 in a living organism to the amount in a dead sample provides an estimate of its age.

EXAMPLE 5   Radiometric dating   Researchers determine that a fossilized bone has 30% of the C-14 of a live bone. Estimate the age of the bone. Assume a half-life for C-14 of 5730 yr.

SOLUTION   The exponential decay function $y(t) = y_0 e^{-kt}$ may be invoked as it applies to all decay processes with a constant half-life. By the half-life formula, $T_{1/2} = (\ln 2)/k$. Substituting $T_{1/2} = 5730$ yr, the rate constant is

$$k = \frac{\ln 2}{T_{1/2}} = \frac{\ln 2}{5730 \text{ yr}} \approx 0.000121 \text{ yr}^{-1}.$$

Assume that the amount of C-14 in a living bone is y_0. Over t years, the amount of C-14 in the fossilized bone decays to 30% of its initial value, or $0.3 y_0$. Using the decay function, we have

$$0.3 y_0 = y_0 e^{-0.000121t}.$$

Solving for t, the age of the bone in years is:

$$t = \frac{\ln 0.3}{-0.000121} \approx 9950.$$

Related Exercises 17–24 ◄

Pharmacokinetics   Pharmacokinetics describes the processes by which drugs are assimilated by the body. The elimination of most drugs from the body may be modeled by an exponential decay function with a known half-life (alcohol is a notable exception). The simplest models assume that an entire drug dose is immediately absorbed into the blood. This assumption is a bit of an idealization; more refined mathematical models can account for the absorption process.

> **Half-lives of common drugs**
>
> | | |
> |---|---|
> | Penicillin | 1 hr |
> | Amoxicillin | 1 hr |
> | Nicotine | 2 hr |
> | Morphine | 3 hr |
> | Tetracycline | 9 hr |
> | Digitalis | 33 hr |
> | Phenobarbitol | 2–6 days |

EXAMPLE 6   Pharmacokinetics   An exponential decay function $y(t) = y_0 e^{-kt}$ models the amount of drug in the blood t hr after an initial dose of $y_0 = 100$ mg is administered. Assume the half-life of the drug is 16 hours.

a. Find the exponential decay function that governs the amount of drug in the blood.

b. How much time is required for the drug to reach 1% of the initial dose (1 mg)?

c. If a second 100-mg dose is given 12 hr after the first dose, how much time is required for the drug level to reach 1 mg?

SOLUTION

a. Knowing that the half-life is 16 hr, the rate constant is

$$k = \frac{\ln 2}{T_{1/2}} = \frac{\ln 2}{16 \text{ hr}} \approx 0.0433 \text{ hr}^{-1}.$$

Therefore, the decay function is $y(t) = 100e^{-0.0433t}$.

b. The time required for the drug to reach 1 mg is the solution of

$$100e^{-0.0433t} = 1.$$

Solving for t, we have

$$t = \frac{\ln 0.01}{-0.0433 \text{ hr}^{-1}} \approx 106 \text{ hr}.$$

It takes more than 4 days for the drug to be reduced to 1% of the initial dose.

c. Using the exponential decay function of part (a), the amount of drug in the blood after 12 hr is

$$y(12) = 100e^{-0.0433 \cdot 12} \approx 59.5 \text{ mg}.$$

The second 100-mg dose given after 12 hr increases the amount of drug (assuming instantaneous absorption) to 159.5 mg. This amount becomes the new initial condition for another exponential decay process (Figure 6.74). Measuring t from the time of the second dose, the amount of drug in the blood is

$$y(t) = 159.5e^{-0.0433t}.$$

The amount of drug reaches 1 mg when

$$y(t) = 159.5e^{-0.0433t} = 1$$

which implies that

$$t = \frac{-\ln 159.5}{-0.0433 \text{ hr}^{-1}} = 117.1 \text{ hr}.$$

Approximately 117 hr after the second dose (or 129 hr after the first dose), the drug reaches 1% of the initial dose.

Related Exercises 17–24 ◄

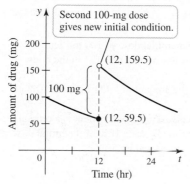

Second 100-mg dose gives new initial condition.

(12, 159.5)

100 mg

(12, 59.5)

FIGURE 6.74

SECTION 6.8 EXERCISES

Review Questions

1. In terms of relative growth rate, what is the defining property of exponential growth?

2. Give two pieces of information that may be used to formulate an exponential growth or decay function.

3. Explain the meaning of doubling time.

4. Explain the meaning of half-life.

5. How are the rate constant and the doubling time related?

6. How are the rate constant and the half-life related?

7. Give two examples of processes that are modeled by exponential growth.

8. Give two examples of processes that are modeled by exponential decay.

Basic Skills

9–10. Absolute and relative growth rates *Two functions f and g are given. Show that the growth rate of the linear function is constant and the relative growth rate of the exponential function is constant.*

9. $f(t) = 100 + 10.5t$, $g(t) = 100e^{t/10}$

10. $f(t) = 2200 + 400t$, $g(t) = 400 \cdot 2^{t/20}$

11–14. Designing exponential growth functions *Devise the exponential growth function that fits the given data, then answer the accompanying questions. Be sure to identify the reference point ($t = 0$) and units of time.*

11. **Population** The population of a town with a 2010 population of 90,000 grows at a rate of 2.4%/yr. In what year will the population double its initial value (to 180,000)?

12. **Population** The population of Clark County, Nevada, was 1.9 million in 2008. Assuming an annual growth rate of 4.5%/yr, what will the county population be in 2020?

13. **Rising costs** Between 2005 and 2010, the average rate of inflation was about 3%/yr (as measured by the Consumer Price Index). If a cart of groceries cost $100 in 2005, what will it cost in 2015 assuming the rate of inflation remains constant?

14. **Cell growth** The number of cells in a tumor doubles every 6 weeks starting with 8 cells. After how many weeks does the tumor have 1500 cells?

15. **Projection sensitivity** According to the 2000 census, the U.S. population was 281 million with an estimated growth rate of 0.7%/yr.

   a. Based on these figures, find the doubling time and project the population in 2100.
   b. Suppose the actual growth rates are just 0.2 percentage points lower and higher than 0.7%/yr (0.5% and 0.9%). What are the resulting doubling times and projected 2100 population?
   c. Comment on the sensitivity of these projections to the growth rate.

16. **Oil consumption** Starting in 2010 ($t = 0$), the rate at which oil is consumed by a small country increases at a rate of 1.5%/yr, starting with an initial rate of 1.2 million barrels per year.

   a. How much oil is consumed over the course of the year 2010 (between $t = 0$ and $t = 1$)?
   b. Find the function that gives the amount of oil consumed between $t = 0$ and any future time t.
   c. After how many years will the amount of oil consumed reach 10 million barrels?

17–20. Designing exponential decay functions *Devise an exponential decay function that fits the following data; then answer the accompanying questions. Be sure to identify the reference point ($t = 0$) and units of time.*

17. **Crime rate** The homicide rate decreases at a rate of 3%/yr in a city that had 800 homicides/yr in 2010. At this rate, when will the homicide rate reach 600 homicides/yr?

18. **Drug metabolism** A drug is eliminated from the body at a rate of 15%/hr. After how many hours does the amount of drug reach 10% of the initial dose?

19. **Atmospheric pressure** The pressure of Earth's atmosphere at sea level is approximately 1000 millibars and decreases exponentially with elevation. At an elevation of 30,000 ft (approximately the altitude of Mt. Everest), the pressure is one-third of the sea-level pressure. At what elevation is the pressure half of the sea-level pressure? At what elevation is it 1% of the sea-level pressure?

20. **China's population** China's one-child policy was implemented with a goal of reducing China's population to 700 million by 2050 (from 1.2 billion in 2000). Suppose China's population declines at a rate of 0.5%/yr. Will this rate of decline be sufficient to meet the goal?

21. **Valium metabolism** The drug Valium is eliminated from the bloodstream with a half-life of 36 hr. Suppose that a patient receives an initial dose of 20 mg of Valium at midnight.

   a. How much Valium is in the patient's blood at noon the next day?
   b. When will the Valium concentration reach 10% of its initial level?

22. **Carbon dating** The half-life of C-14 is about 5730 yr.

   a. Archaeologists find a piece of cloth painted with organic dyes. Analysis of the dye in the cloth shows that only 77% of the C-14 originally in the dye remains. When was the cloth painted?
   b. A well-preserved piece of wood found at an archaeological site has 6.2% of the C-14 that it had when it was alive. Estimate when the wood was cut.

23. **Uranium dating** Uranium-238 (U-238) has a half-life of 4.5 billion yr. Geologists find a rock containing a mixture of U-238 and lead, and determine that 85% of the original U-238 remains; the other 15% has decayed into lead. How old is the rock?

24. **Radioiodine treatment** Roughly 12,000 Americans are diagnosed with thyroid cancer every year, which accounts for 1% of all cancer cases. It occurs in women three times as frequently as in men. Fortunately, thyroid cancer can be treated successfully in many cases with radioactive iodine, or I-131. This unstable form of iodine has a half-life of 8 days and is given in small doses measured in millicuries.

   a. Suppose a patient is given an initial dose of 100 millicuries. Find the function that gives the amount of I-131 in the body after $t \geq 0$ days.
   b. How long does it take for the amount of I-131 to reach 10% of the initial dose?
   c. Finding the initial dose to give a particular patient is a critical calculation. How does the time to reach 10% of the initial dose change if the initial dose is increased by 5%?

Further Explorations

25. **Explain why or why not** Determine whether the following statements are true and give an explanation or counterexample.

   a. A quantity that increases at 6%/yr obeys the growth function $y(t) = y_0 e^{0.06t}$.
   b. If a quantity increases by 10%/yr, it increases by 30% over 3 yr.
   c. A quantity decreases by one-third every month. Therefore, it decreases exponentially.
   d. If the rate constant of an exponential growth function is increased, its doubling time is decreased.
   e. If a quantity increases exponentially, the time required to increase by a factor of 10 remains constant for all time.

26. **Tripling time** A quantity increases according to the exponential function $y(t) = y_0 e^{kt}$. What is the tripling time for the quantity? What is the time required for the quantity to increase p-fold?

27. **Constant doubling time** Prove that the doubling time for an exponentially increasing quantity is constant for all time.

28. **Overtaking** City A has a current population of 500,000 people and grows at a rate of 3%/yr. City B has a current population of 300,000 and grows at a rate of 5%/yr.

   a. When will the cities have the same population?
   b. Suppose City C has a current population of $y_0 < 500,000$ and a growth rate of $p > 3\%$/yr. What is the relationship between y_0 and p such that the Cities A and C have the same population in 10 yr?

29. A slowing race Starting at the same time and place, Abe and Bob race, running at velocities $u(t) = 4/(t + 1)$ mi/hr and $v(t) = 4e^{-t/2}$ mi/hr, respectively, for $t \geq 0$.

   a. Who is ahead after $t = 5$ hr? After $t = 10$ hr?

   b. Find and graph the position functions of both runners. Which runner can run only a finite distance in an unlimited amount of time?

Applications

30. Law of 70 Bankers use the law of 70, which says that if an account increases at a fixed rate of $p\%$/yr, its doubling time is approximately $70/p$. Explain why and when this statement is true.

31. Compounded inflation The U.S. government reports the rate of inflation (as measured by the Consumer Price Index) both monthly and annually. Suppose that, for a particular month, the *monthly* rate of inflation is reported as 0.8%. Assuming that this rate remains constant, what is the corresponding *annual* rate of inflation? Is the annual rate 12 times the monthly rate? Explain.

32. Acceleration, velocity, position Suppose the acceleration of an object moving along a line is given by $a(t) = -kv(t)$, where k is a positive constant and v is the object's velocity. Assume that the initial velocity and position are given by $v(0) = 10$ and $s(0) = 0$, respectively.

   a. Use $a(t) = v'(t)$ to find the velocity of the object as a function of time.

   b. Use $v(t) = s'(t)$ to find the position of the object as a function of time.

   c. Use the fact that $dv/dt = (dv/ds)(ds/dt)$ by the Chain Rule to find the velocity as a function of position.

33. Free fall (adapted from Putnam Exam, 1939) An object falls freely in a straight line and experiences air resistance proportional to its speed; this means its acceleration is $a(t) = -kv(t)$. The speed of the object decreases from 1000 ft/s to 900 ft/s over a distance of 1200 ft. Approximate the time required for this deceleration to occur. (Exercise 32 may be useful.)

34. A running model A model for the startup of a runner in a short race results in the velocity function $v(t) = a(1 - e^{-t/c})$, where a and c are physical constants and v has units of m/s. Source: *A Theory of Competitive Running*, Joe Keller, *Physics Today*, 26 (Sept 1973).

   a. Graph the velocity function for $a = 12$ and $c = 2$. What is the runner's maximum velocity?

   b. Using the velocity in part (a) and assuming $s(0) = 0$, find the position function $s(t)$ for $t \geq 0$.

   c. Graph the position function and estimate the time required to run 100 m.

35. Tumor growth Suppose the cells of a tumor are idealized as spheres each with a radius of 5 μm (micron). The number of cells has a doubling time of 35 days. Approximately how long will it take a single cell to grow into a multi-celled spherical tumor with a volume of 0.5 cm^3 (1 cm = 10,000 μm)? Assume that the tumor spheres are tightly packed.

36. Carbon emissions from China and the United States The burning of fossil fuels releases greenhouse gases into the atmosphere. In 1995, the United States emitted about 1.4 billion tons of carbon into the atmosphere, nearly one-fourth of the world total. China was the second largest contributor, emitting about 850 million tons of carbon. However, emissions from China were rising at a rate of about 4%/yr, while U.S. emissions were rising at about 1.3%/yr. Using these growth rates, project greenhouse gas emissions from the United States and China in 2020. Graph the projected emissions for both countries. Comment on your observations.

37. A revenue model The owner of a clothing store understands that the demand for shirts decreases with the price. In fact, she has developed a model that predicts that at a price of \$$x$, she can sell $D(x) = 40e^{-x/50}$ shirts in a day. It follows that the revenue (total money taken in) in a day is $R(x) = xD(x)$ (\$$x$/shirt · $D(x)$ shirts). What price should the owner charge to maximize revenue?

Additional Exercises

38. Geometric means A quantity grows exponentially according to $y(t) = y_0 e^{kt}$. What is the relationship between m, n, and p such that $y(p) = \sqrt{y(m)y(n)}$?

39. Equivalent growth functions The same exponential growth function can be written in the forms $y(t) = y_0 e^{kt}$, $y(t) = y_0(1 + r)^t$, and $y(t) = y_0 2^{t/T_2}$. Derive the relationships among k, r, and T_2.

40. General relative growth rates Define the relative growth rate of the function f over the time interval T to be the relative change in f over an interval of length T:

$$R_T = \frac{f(t + T) - f(t)}{f(t)}.$$

Show that for the exponential function $y(t) = y_0 e^{kt}$, the relative growth rate R_T is constant for any T; that is, choose any T and show that R_T is constant for all t.

QUICK CHECK **ANSWERS**

1. Population A grows exponentially; population B grows linearly. **3.** The function $100e^{0.05t}$ increases by a factor of 1.0513, or by 5.13%, in 1 unit of time. **4.** 10 yr. ◄

CHAPTER 6   REVIEW EXERCISES

1. **Explain why or why not** Determine whether the following statements are true and give an explanation or counterexample.

   a. A region R is revolved about the y-axis to generate a solid S. To find the volume of S, you could use either the disk/washer method and integrate with respect to y or the shell method and integrate with respect to x.

   b. Given only the velocity of an object moving on a line, it is possible to find its displacement, but not its position.

   c. If water flows into a tank at a constant rate (for example 6 gal/min), the volume of water in the tank increases according to a linear function of time.

   d. The variable $y = t + 1$ doubles in value whenever t increases by 1 unit.

   e. $\ln xy = (\ln x)(\ln y)$

   f. The function $y = Ae^{0.1t}$ increases by 10% when t increases by 1 unit.

2. **Displacement from velocity** The velocity of an object moving along a line is given by $v(t) = 20 \cos \pi t$ (in ft/s). What is the displacement of the object after 1.5 s?

3. **Position, displacement, and distance** A projectile is launched vertically from the ground at $t = 0$ and its velocity in flight is given by $v(t) = 20 - 10t$ m/s. Find the position, displacement, and distance traveled after t seconds, for $0 \le t \le 4$.

4. **Deceleration** At $t = 0$, a car begins decelerating from a velocity of 80 ft/s at a constant rate of 5 ft/s². Find its position function assuming $s(0) = 0$.

5. **An oscillator** The acceleration of an object moving along a line is given by $a(t) = 2 \sin\left(\dfrac{\pi t}{4}\right)$. The initial velocity and position are $v(0) = -\dfrac{8}{\pi}$ and $s(0) = 0$.

   a. Find the velocity and position for $t \ge 0$.

   b. What are the minimum and maximum values of s?

   c. Find the average velocity and average position over the interval $[0, 8]$.

6. **A race** Starting at the same point on a straight road, Anna and Benny begin running with velocities given by $v_A(t) = 2t + 1$ and $v_B(t) = 4 - t$ (in mi/hr), respectively.

   a. Graph the velocity functions for $0 \le t \le 4$.

   b. If the runners run for 1 hr, who runs farther? Interpret your conclusion geometrically using the graph in part (a).

   c. If the runners run for 6 mi, who wins the race? Interpret your conclusion geometrically using the graph in part (a).

7. **Fuel consumption** A small plane in flight consumes fuel at a rate in gal/min given by

   $$R'(t) = \begin{cases} 4t^{1/3} & \text{if } 0 \le t \le 8 \text{ (take-off)} \\ 2 & \text{if } t > 8 \text{ (cruising)} \end{cases}$$

   a. Find a function R that gives the total fuel consumed for $0 \le t \le 8$.

   b. Find a function R that gives the total fuel consumed for $t \ge 0$.

   c. If the fuel tank capacity is 150 gal, when does the fuel run out?

8. **Variable flow rate** Water flows out of a large tank at a rate (in m³/hr) given by $V'(t) = 10/(t + 1)$. If the tank initially holds 750 m³ of water, when will the tank be empty?

T 9. **Decreasing velocity** A projectile is fired upward and its velocity in m/s is given by $v(t) = 200e^{-t/10}$ for $t \ge 0$.

   a. Graph the velocity function for $t \ge 0$.

   b. When does the velocity reach 50 m/s?

   c. Find and graph the position function for the projectile for $t \ge 0$ assuming $s(0) = 0$.

   d. Given unlimited time, can the projectile travel 2500 m? If so, at what time does the distance traveled equal 2500 m?

T 10. **Another decreasing velocity** A projectile is fired upward and its velocity in m/s is given by $v(t) = \dfrac{200}{\sqrt{t + 1}}$ for $t \ge 0$.

   a. Graph the velocity function for $t \ge 0$.

   b. Find and graph the position function for the projectile for $t \ge 0$ assuming $s(0) = 0$.

   c. Given unlimited time, can the projectile travel 2500 m? If so, at what time does the distance traveled equal 2500 m?

T 11. **An exponential bike ride** Tom and Sue took a bike ride, both starting at the same time and position. Tom started riding at 20 mi/hr, and his velocity decreased according to the function $v(t) = 20e^{-2t}$ for $t \ge 0$. Sue started riding at 15 mi/hr, and her velocity decreased according to the function $u(t) = 15e^{-t}$ for $t \ge 0$.

   a. Find and graph the position functions of Tom and Sue.

   b. Find the times at which the riders had the same position at the same time.

   c. Who ultimately took the lead and remained in the lead?

T 12–17. **Areas of regions** *Use any method to find the area of the region described. Make a graph of the region in question.*

12. The region in the first quadrant bounded by $y = x^p$ and $y = \sqrt[p]{x}$, where $p = 100$ and $p = 1000$.

13. The region in the first quadrant bounded by $y = \sqrt{4 - x^2}$ and $y = \sqrt{25 - x^2}$

14. The regions R_1 and R_2 (separately) shown in the figure, which are formed by the graphs of $y = 16 - x^2$ and $y = 5x - 8$

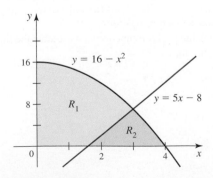

15. The region bounded by $y = x^2$, $y = 2x^2 - 4x$, and $y = 0$

16. The region in the first quadrant bounded by the curve
$\sqrt{x} + \sqrt{y} = 1$

17. The region in the first quadrant bounded by $y = x/6$ and
$y = 1 - |x/2 - 1|$

18. **An area function** Let $R(x)$ be the area of the shaded region
between the graphs of $y = f(t)$ and $y = g(t)$ in the figure.

   a. Sketch a plausible graph of R for $a \le x \le c$.
   b. Give expressions for $R(x)$ and $R'(x)$ for $a \le x \le c$.

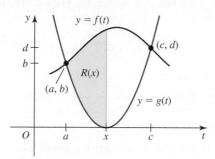

19. **An area function** Consider the functions $y = \dfrac{x^2}{a}$ and $y = \sqrt{\dfrac{x}{a}}$,
where $a > 0$. Find $A(a)$, the area of the region between the
curves.

20. **Three methods** The quarter circle of radius R in the first quadrant
$(x^2 + y^2 = R^2$ for $x \ge 0$ and $y \ge 0)$ is revolved about the x-axis
to produce a hemisphere. Find the volume of the hemisphere in
the following ways:

   a. Apply the disk method and integrate with respect to x.
   b. Apply the shell method and integrate with respect to y.
   c. Apply the general slicing method and integrate with respect to y.

21–25. Volumes of solids *Choose the general slicing method, the
disk/washer method, or the shell method to find the volume of the
following solids.*

21. A pyramid has a square base in the xy-plane with vertices at
$(1, 1), (1, -1), (-1, 1)$, and $(-1, -1)$. All cross sections of the
pyramid parallel to the xy-plane are squares and the height of the
pyramid is 12 units. What is the volume of the pyramid?

22. The region bounded by the curves $y = -x^2 + 2x + 2$ and
$y = 2x^2 - 4x + 2$ is revolved about the x-axis. What is the
volume of the solid that is generated?

23. The region bounded by the curves $y = 1 + \sqrt{x}$ and
$y = 1 - \sqrt{x}$, and the line $x = 1$ is revolved about the y-axis.
Find the volume of the resulting solid by (a) integrating with
respect to x and (b) integrating with respect to y. Be sure your
answers agree.

24. The region bounded by the curves $y = 2e^{-x}, y = e^x$, and the
y-axis is revolved about the x-axis. What is the volume of the
solid that is generated?

25. Find the volume of a right circular cone with radius r and height h
by treating it as a solid of revolution.

26. **Area and volume** The region R is bounded by the curves
$x = y^2 + 2, y = x - 4$, and $y = 0$ (see figure).

   a. Write a single integral that gives the area of R.

b. Write a single integral that gives the volume of the solid
generated when R is revolved about the x-axis.
c. Write a single integral that gives the volume of the solid
generated when R is revolved about the y-axis.
d. Suppose S is a solid whose base is R and whose cross sections
perpendicular to R and parallel to the x-axis are semicircles.
Write a single integral that gives the volume of S.

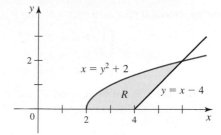

27. **Comparing volumes** Let R be the region bounded by
$y = 1/x^p$ and the x-axis on the interval $[1, a]$, where $p > 0$
and $a > 1$ (see figure). Let V_x and V_y be the volumes of the
solids generated when R is revolved about the x- and y-axes,
respectively.

   a. With $a = 2$ and $p = 1$, which is greater, V_x or V_y?
   b. With $a = 4$ and $p = 3$, which is greater, V_x or V_y?
   c. Find a general expression for V_x in terms of a and p. Note
   that $p = \frac{1}{2}$ is a special case. What is V_x when $p = \frac{1}{2}$?
   d. Find a general expression for V_y in terms of a and p. Note
   that $p = 2$ is a special case. What is V_y when $p = 2$?
   e. Explain how parts (c) and (d) demonstrate that
   $$\lim_{h \to 0} \frac{a^h - 1}{h} = \ln a.$$
   f. Can you find any values of a and p for which $V_x > V_y$?

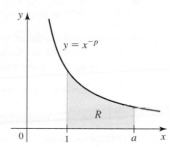

28–31. Arc length *Find the length of the following curves.*

28. $y = x^3/6 + 1/(2x)$ on the interval $[1, 2]$.

29. $y = x^{1/2} - x^{3/2}/3$ on the interval $[1, 3]$.

30. $y = x^3/3 + x^2 + x + 1/(4x + 4)$ on the interval $[0, 4]$.

31. Find the length of the curve $y = \ln x$ between $x = 1$ and
$x = b > 1$ given that

$$\int \frac{\sqrt{x^2 + a^2}}{x} \, dx = \sqrt{x^2 + a^2} - a \ln\left(\frac{a + \sqrt{x^2 + a^2}}{x}\right) + C.$$

Use any means to approximate the value of b for which the curve
has length 2.

32–34. Variable density in one dimension *Find the mass of the following thin bars.*

32. A bar on the interval $0 \le x \le 9$ with a density (in g/cm) given by $\rho(x) = 3 + 2\sqrt{x}$

33. A 3-m bar with a density (in g/m) of $\rho(x) = 150e^{-x/3}$ for $0 \le x \le 3$.

34. A bar on the interval $0 \le x \le 6$ with a density

$$\rho(x) = \begin{cases} 1 & \text{if } 0 \le x < 2 \\ 2 & \text{if } 2 \le x < 4 \\ 4 & \text{if } 4 \le x \le 6 \end{cases}$$

35. Spring work It takes 50 J of work to stretch a spring 0.2 m from its equilibrium position. How much work is needed to stretch it an additional 0.5 m?

36. Pumping water A cylindrical water tank has a height of 6 m and a radius of 4 m. How much work is required to empty the full tank by pumping the water to an outflow pipe at the top of the tank?

37. Force on a dam Find the total force on the face of a semicircular dam with a radius of 20 m when its reservoir is full of water. The diameter of the semicircle is the top of the dam.

38–41. Integrals *Evaluate the following integrals.*

38. $\displaystyle\int \frac{e^x}{4e^x + 6}\, dx$

39. $\displaystyle\int_{e^2}^{e^8} \frac{dx}{x \ln x}$

40. $\displaystyle\int_{1}^{4} \frac{10^{\sqrt{x}}}{\sqrt{x}}\, dx$

41. $\displaystyle\int \frac{x + 4}{x^2 + 8x + 25}\, dx$

42. Radioactive decay The mass of radioactive material in a sample has decreased by 30% since the decay began. Assuming a half-life of 1500 years, how long ago did the decay begin?

43. Population growth Growing from an initial population of 150,000 at a constant annual growth rate of 4%/yr, how long will it take a city to reach a population of 1 million?

44. Savings account A savings account advertises an annual percentage yield (APY) of 5.4%, which means that the balance in the account increases at an annual growth rate of 5.4%/yr.

   a. Find the balance in the account for $t \ge 0$ with an initial deposit of $1500, assuming the APY remains fixed and no additional deposits or withdrawals are made.

   b. What is the doubling time of the balance?

   c. After how many years does the balance reach $5000?

45–46. Curve sketching *Use the graphing techniques of Section 4.3 to graph the following functions on their domains. Identify local extreme points, inflection points, concavity, and end behavior. Use a graphing utility only to check your work.*

45. $f(x) = e^x(x^2 - x)$ **46.** $f(x) = \ln x - \ln^2 x$

47. Log-normal probability distribution A commonly used distribution in probability and statistics is the log-normal distribution. (If the logarithm of a variable has a normal distribution, then the variable itself has a log-normal distribution.) The distribution function is

$$f(x) = \frac{1}{x\sigma\sqrt{2\pi}} e^{-\ln^2 x/(2\sigma^2)}, \quad \text{for } x \ge 0$$

where $\ln x$ has zero mean and standard deviation $\sigma > 0$.

   a. Graph f for $\sigma = \frac{1}{2}, 1, 2$. Based on your graphs, does $\lim_{x \to 0} f(x)$ appear to exist?

   b. Evaluate $\lim_{x \to 0} f(x)$. (*Hint:* Let $x = e^y$.)

   c. Show that f has a single local maximum at $x^* = e^{-\sigma^2}$.

   d. Evaluate $f(x^*)$ and express the result as a function of σ.

   e. For what value of $\sigma > 0$ in part (d) does $f(x^*)$ have a minimum?

48. Equal area property for parabolas Let $f(x) = ax^2 + bx + c$ be an arbitrary quadratic function and choose two points $x = p$ and $x = q$. Let L_1 be the line tangent to the graph of f at the point $(p, f(p))$, and let L_2 be the line tangent to the graph at the point $(q, f(q))$. Let $x = s$ be the vertical line through the intersection point of L_1 and L_2. Finally, let R_1 be the region bounded by $y = f(x)$, L_1, and the vertical line $x = s$, and let R_2 be the region bounded by $y = f(x)$, L_2, and the vertical line $x = s$. Prove that the area of R_1 equals the area of R_2.

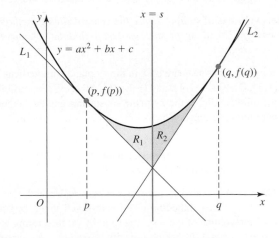

Chapter 6 Guided Projects

Applications of the material in this chapter and related topics can be found in the following Guided Projects. For additional information, see the Preface.

- Means and tangent lines
- Geometric probability
- Designing a water clock
- Dipstick problems
- Optimizing fuel use

- Mathematics of the CD player
- Buoyancy and Archimedes' principle
- Hyperbolic functions
- Inverse sine from geometry
- Landing an airliner

7

Integration Techniques

Chapter Preview The Substitution Rule introduced in Chapter 5 is a powerful method for evaluating a wide range of integrals. However, there are many commonly occurring integrals that *cannot* be handled by the Substitution Rule. Therefore, this chapter has a critical purpose: to develop additional integration techniques that enable us to evaluate a far greater number of integrals. The new *analytical methods* (pencil-and-paper methods) we introduce here are integration by parts, trigonometric substitution, and partial fractions. And yet, even with these new methods, it is important to recognize that there are still many integrals that do not yield to them. For this reason, we also introduce alternative strategies for evaluating indefinite integrals and computer-based methods for approximating definite integrals. The discussion then turns to integrals that have either infinite integrands or infinite intervals of integration. These *improper integrals* offer surprising results and have many practical applications. The chapter closes with an introductory survey of differential equations, a vast topic that has a central place in both the theory and applications of mathematics.

7.1 Integration by Parts

The Substitution Rule (Section 5.5) arises when we reverse the Chain Rule for derivatives. In this section, we employ a similar strategy and reverse the Product Rule for derivatives. The result is an integration technique called *integration by parts*. To illustrate the importance of integration by parts, consider the indefinite integrals

$$\int e^x \, dx = e^x + C \quad \text{and} \quad \int xe^x \, dx = ?$$

The first integral is an elementary integral that we have already encountered. The second integral is only slightly different—and yet, the appearance of the product xe^x in the integrand makes this integral (at the moment) impossible to evaluate. Integration by parts is ideally suited for evaluating integrals of *products* of functions. Such integrals arise frequently.

Integration by Parts for Indefinite Integrals

Given two differentiable functions u and v, the Product Rule states that

$$\frac{d}{dx}[u(x)v(x)] = u'(x)v(x) + u(x)v'(x).$$

By integrating both sides, we can write this rule in terms of an indefinite integral:

$$u(x)v(x) = \int [u'(x)v(x) + u(x)v'(x)] \, dx$$

Rearranging this expression in the form

$$\int u(x)\underbrace{v'(x)\,dx}_{dv} = u(x)v(x) - \int v(x)\underbrace{u'(x)\,dx}_{du}$$

leads to the basic relationship for **integration by parts**. It is expressed more compactly by letting $du = u'(x)\,dx$ and $dv = v'(x)\,dx$. Suppressing the independent variable x, we have

$$\int u\,dv = uv - \int v\,du.$$

The integral $\int u\,dv$ is viewed as the given integral, and we use integration by parts to express it in terms of a new integral $\int v\,du$. The technique is successful if the new integral can be evaluated.

Integration by Parts

Suppose that u and v are differentiable functions. Then,

$$\int u\,dv = uv - \int v\,du.$$

EXAMPLE 1 Integration by parts Evaluate $\int xe^x\,dx$.

SOLUTION The presence of *products* in the integrand often suggests integration by parts. We split the product into two factors, one of which must be identified as u and the other as dv (the latter always includes the differential dx). Powers of x are *often* good choices for u. The choice for dv should be easy to integrate because it produces the function v on the right side of the integration by parts formula ($v = \int dv$). In this case, the choices $u = x$ and $dv = e^x\,dx$ are advisable. It follows that $du = dx$. The relationship $dv = e^x\,dx$ means that v is an antiderivative of e^x, which implies $v = e^x$. A table is helpful for organizing these calculations.

▶ The integration by parts calculation may be done without including the constant of integration—as long as it is included in the final result.

Functions in original integral	$u = x$	$dv = e^x\,dx$
Functions in new integral	$du = dx$	$v = e^x$

The integration by parts rule is now applied:

$$\int \underbrace{x}_{u}\ \underbrace{e^x\,dx}_{dv} = \underbrace{x}_{u}\ \underbrace{e^x}_{v} - \int \underbrace{e^x}_{v}\ \underbrace{dx}_{du}$$

The original integral $\int xe^x\,dx$ has been replaced by the integral of e^x, which is easier to evaluate: $\int e^x\,dx = e^x + C$. The entire procedure looks like this:

$$\int xe^x\,dx = xe^x - \int e^x\,dx \qquad \text{Integration by parts}$$

$$= xe^x - e^x + C \qquad \text{Evaluate the new integral.}$$

Related Exercises 7–22 ◀

▶ To make the table, first write the functions in the original integral:

$$u = \underline{\quad}, dv = \underline{\quad}.$$

Then find the functions in the new integral by differentiating u and integrating dv:

$$du = \underline{\quad}, v = \underline{\quad}.$$

EXAMPLE 2 Integration by parts Evaluate $\int x\sin x\,dx$.

SOLUTION Remembering that powers of x are often a good choice for u, we form the following table.

$u = x$	$dv = \sin x\,dx$
$du = dx$	$v = -\cos x$

Applying integration by parts, we have

$$\int \underset{u}{\underbrace{x}} \underset{dv}{\underbrace{\sin x \, dx}} = \underset{u}{\underbrace{x}} \underset{v}{\underbrace{(-\cos x)}} - \int \underset{v}{\underbrace{(-\cos x)}} \underset{du}{\underbrace{dx}} \qquad \text{Integration by parts}$$

$$= -x \cos x + \sin x + C \qquad \text{Evaluate } \int \cos x \, dx = \sin x.$$

Related Exercises 7–22 ◄

QUICK CHECK 1 What is the best choice for u and dv in evaluating $\int x \cos x \, dx$? ◄

In general, integration by parts works when we can easily integrate the choice for dv and when the new integral is easier to evaluate than the original. Integration by parts is often used for integrals of the form $\int x^n f(x) \, dx$, where n is a positive integer. Such integrals generally require the repeated use of integration by parts, as shown in the following example.

EXAMPLE 3   Repeated use of integration by parts

a. Evaluate $\int x^2 e^x \, dx$.

b. How would you evaluate $\int x^n e^x \, dx$, where n is a positive integer?

SOLUTION

a. The factor x^2 is a good choice for u, leaving $dv = e^x \, dx$. We then have

$u = x^2$	$dv = e^x \, dx$
$du = 2x \, dx$	$v = e^x$

$$\int \underset{u}{\underbrace{x^2}} \underset{dv}{\underbrace{e^x \, dx}} = \underset{u}{\underbrace{x^2}} \underset{v}{\underbrace{e^x}} - \int \underset{v}{\underbrace{e^x}} \underset{du}{\underbrace{2x \, dx}}.$$

Notice that the new integral on the right side is simpler than the original integral because the power of x has been reduced by one. In fact, the new integral was evaluated in Example 1. Therefore, after using integration by parts twice, we have

$$\int x^2 e^x \, dx = x^2 e^x - 2 \int x e^x \, dx \qquad \text{Integration by parts}$$

$$= x^2 e^x - 2(x e^x - e^x) + C \qquad \text{Result of Example 1}$$

$$= e^x (x^2 - 2x + 2) + C. \qquad \text{Simplify.}$$

$u = x^n$	$dv = e^x \, dx$
$du = n x^{n-1} \, dx$	$v = e^x$

b. We now let $u = x^n$ and $dv = e^x \, dx$. The integration takes the form

$$\int x^n e^x \, dx = x^n e^x - n \int x^{n-1} e^x \, dx.$$

➤ An integral identity in which the power of the variable is reduced is called a **reduction formula**. Other examples of reduction formulas are explored in Exercises 42–49.

We see that integration by parts reduces the power of the variable in the integrand. The integral in part (a) with $n = 2$ requires two uses of integration by parts. You can probably anticipate that evaluating the integral $\int x^n e^x \, dx$ requires n applications of integration by parts to reach the integral $\int e^x \, dx$, which is easily evaluated.

Related Exercises 7–22 ◄

EXAMPLE 4   Repeated use of integration by parts   Evaluate $\int e^{2x} \sin x \, dx$.

➤ In Example 4, we could also use $u = \sin x$ and $dv = e^{2x} \, dx$. In general, some trial and error may be required when using integration by parts. Effective choices come with practice.

SOLUTION The integrand consists of a product, which suggests integration by parts. In this case there is no obvious choice for u and dv, so let's try the following choices:

$u = e^{2x}$	$dv = \sin x \, dx$
$du = 2e^{2x} \, dx$	$v = -\cos x$

The integral then becomes

$$\int e^{2x} \sin x \, dx = -e^{2x} \cos x + 2 \int e^{2x} \cos x \, dx. \tag{1}$$

The original integral has been expressed in terms of a new integral, $\int e^{2x} \cos x \, dx$, which appears no easier to evaluate than the original integral. It is tempting to start over with a new choice of u and dv, but a little persistence pays off. Suppose we evaluate $\int e^{2x} \cos x \, dx$ using integration by parts with the following choices:

$u = e^{2x}$	$dv = \cos x \, dx$
$du = 2e^{2x} \, dx$	$v = \sin x$

Integrating by parts, we have

$$\int e^{2x} \cos x \, dx = e^{2x} \sin x - 2 \int e^{2x} \sin x \, dx. \tag{2}$$

Now observe that equation (2) contains the original integral, $\int e^{2x} \sin x \, dx$. Substituting the result of equation (2) into equation (1), we find that

$$\int e^{2x} \sin x \, dx = -e^{2x} \cos x + 2 \int e^{2x} \cos x \, dx$$

$$= -e^{2x} \cos x + 2 \left(e^{2x} \sin x - 2 \int e^{2x} \sin x \, dx \right) \quad \text{Substitute for } \int e^{2x} \cos x \, dx.$$

$$= -e^{2x} \cos x + 2e^{2x} \sin x - 4 \int e^{2x} \sin x \, dx. \quad \text{Simplify.}$$

Now it is a matter of solving for $\int e^{2x} \sin x \, dx$ and including the constant of integration. We find that

$$\int e^{2x} \sin x \, dx = \frac{1}{5} e^{2x} (2 \sin x - \cos x) + C.$$

Related Exercises 23–28 ◄

Integration by Parts for Definite Integrals

Integration by parts with definite integrals presents two options. You can use the method outlined in Examples 1–4 to find an antiderivative and then evaluate it at the upper and lower limits of integration. Alternatively, the limits of integration can be incorporated directly into the integration by parts process. With the second approach, integration by parts for definite integrals has the following form.

▷ Integration by parts for definite integrals still has the form

$$\int u \, dv = uv - \int v \, du.$$

However, both definite integrals must be written with respect to x.

Integration by Parts for Definite Integrals

Let u and v be differentiable. Then,

$$\int_a^b u(x) v'(x) \, dx = u(x) v(x) \Big|_a^b - \int_a^b v(x) u'(x) \, dx.$$

EXAMPLE 5 **A definite integral** Evaluate $\int_1^2 \ln x \, dx$.

SOLUTION This example is instructive because the integrand does not appear to be a product. The key is to view the integrand as the product $(\ln x)(1 \, dx)$. Then, the following choices are plausible:

$u = \ln x$	$dv = dx$
$du = \dfrac{1}{x} dx$	$v = x$

Using integration by parts, we have

$$\int_1^2 \underbrace{\ln x}_{u} \underbrace{dx}_{dv} = ((\ln x)\, x)\Big|_1^2 - \int_1^2 \underbrace{x}_{v} \underbrace{\frac{1}{x}\, dx}_{du} \qquad \text{Integration by parts}$$

$$= x \ln x \Big|_1^2 - \int_1^2 dx \qquad\qquad \text{Simplify.}$$

$$= (2 \ln 2 - 0) - (2 - 1) \qquad \text{Evaluate.}$$

$$= 2 \ln 2 - 1 \approx 0.386. \qquad\qquad \text{Simplify.}$$

Related Exercises 29–36 ◄

In Example 5 we evaluated a definite integral of $\ln x$. The corresponding indefinite integral can be added to our list of integration formulas.

> **Integral of ln x**
> $$\int \ln x\, dx = x \ln x - x + C$$

QUICK CHECK 2 Verify by differentiation that $\int \ln x\, dx = x \ln x - x + C.$ ◄

EXAMPLE 6   Solids of revolution Let R be the region bounded by $y = \ln x$, the x-axis, and the line $x = a$, where $a > 1$ (Figure 7.1). Find the volume of the solid that is generated when the region R is revolved about the x-axis.

SOLUTION Revolving R about the x-axis generates a solid whose volume is computed with the disk method (Section 6.3). Its volume is

$$V = \pi \int_1^a (\ln x)^2\, dx.$$

We integrate by parts with the following assignments:

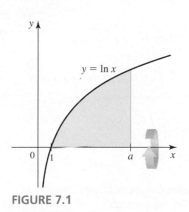

$y = \ln x$

FIGURE 7.1

$u = (\ln x)^2$	$dv = dx$
$du = \dfrac{2 \ln x}{x}\, dx$	$v = x$

> Recall that if $f(x) \geq 0$ on $[a, b]$ and the region bounded by the graph of f and the x-axis on $[a, b]$ is revolved about the x-axis, then the volume of the solid generated is
> $$V = \pi \int_a^b f(x)^2\, dx.$$

The integration is carried out as follows, using the indefinite integral of $\ln x$ given above:

$$V = \pi \int_1^a (\ln x)^2\, dx \qquad\qquad \text{Disk method}$$

$$= \pi \left[\underbrace{(\ln x)^2}_{u} \underbrace{x}_{v}\Big|_1^a - \int_1^a \underbrace{x}_{v} \underbrace{\frac{2 \ln x}{x}\, dx}_{du} \right] \qquad \text{Integration by parts}$$

$$= \pi \left[x(\ln x)^2 \Big|_1^a - 2 \int_1^a \ln x\, dx \right] \qquad \text{Simplify.}$$

$$= \pi \left(x(\ln x)^2 \Big|_1^a - 2(x \ln x - x)\Big|_1^a \right) \qquad \int \ln x\, dx = x \ln x - x + C$$

$$= \pi(a(\ln a)^2 - 2a \ln a + 2a - 2) \qquad \text{Evaluate and simplify.}$$

Related Exercises 37–40 ◄

QUICK CHECK 3 How many times do you need to integrate by parts to reduce $\int_1^a (\ln x)^6\, dx$ to an integral of $\ln x$? ◄

SECTION 7.1 EXERCISES

Review Questions

1. On which derivative rule is integration by parts based?

2. How would you choose the term dv when evaluating $\int x^n e^{ax}\, dx$ using integration by parts?

3. How would you choose the term u when evaluting $\int x^n \cos ax\, dx$ using integration by parts?

4. Explain how integration by parts is used to evaluate a definite integral.

5. For what type of integrand is integration by parts useful?

6. How would you choose u and dv to simplify $\int x^4 e^{-2x}\, dx$?

Basic Skills

7–22. Integration by parts *Evaluate the following integrals.*

7. $\displaystyle\int x \cos x\, dx$

8. $\displaystyle\int x \sin 2x\, dx$

9. $\displaystyle\int t e^t\, dt$

10. $\displaystyle\int 2x e^{3x}\, dx$

11. $\displaystyle\int x^2 \sin 2x\, dx$

12. $\displaystyle\int s e^{-2s}\, ds$

13. $\displaystyle\int x^2 e^{4x}\, dx$

14. $\displaystyle\int \theta \sec^2 \theta\, d\theta$

15. $\displaystyle\int x^2 \ln x\, dx$

16. $\displaystyle\int x \ln x\, dx$

17. $\displaystyle\int \frac{\ln x}{x^{10}}\, dx$

18. $\displaystyle\int \sin^{-1} x\, dx$

19. $\displaystyle\int \tan^{-1} x\, dx$

20. $\displaystyle\int x \sec^{-1} x\, dx,\ x \geq 1$

21. $\displaystyle\int x \sin x \cos x\, dx$

22. $\displaystyle\int x \tan^{-1}(x^2)\, dx$

23–28. Repeated integration by parts *Evaluate the following integrals.*

23. $\displaystyle\int e^x \cos x\, dx$

24. $\displaystyle\int e^{3x} \cos 2x\, dx$

25. $\displaystyle\int e^{-x} \sin 4x\, dx$

26. $\displaystyle\int x^2 \ln^2 x\, dx$

27. $\displaystyle\int t^3 e^{-t}\, dt$

28. $\displaystyle\int e^{-2\theta} \sin 6\theta\, d\theta$

29–36. Definite integrals *Evaluate the following definite integrals.*

29. $\displaystyle\int_0^\pi x \sin x\, dx$

30. $\displaystyle\int_1^e \ln 2x\, dx$

31. $\displaystyle\int_0^{\pi/2} x \cos 2x\, dx$

32. $\displaystyle\int_0^{\ln 2} x e^x\, dx$

33. $\displaystyle\int_1^{e^2} x^2 \ln x\, dx$

34. $\displaystyle\int_0^{1/\sqrt{2}} y \tan^{-1} y^2\, dy$

35. $\displaystyle\int_{1/2}^{\sqrt{3}/2} \sin^{-1} y\, dy$

36. $\displaystyle\int_{2/\sqrt{3}}^{2} z \sec^{-1} z\, dz$

37–40. Volumes of solids *Find the volume of the solid that is generated when the region is revolved as described.*

37. The region bounded by $f(x) = e^{-x}$, $x = \ln 2$, and the coordinate axes is revolved about the y-axis.

38. The region bounded by $f(x) = \sin x$ and the x-axis on $[0, \pi]$ is revolved about the y-axis.

39. The region bounded by $f(x) = x \ln x$ and the x-axis on $[1, e^2]$ is revolved about the x-axis.

40. The region bounded by $f(x) = e^{-x}$ and the x-axis on $[0, \ln 2]$ is revolved about the line $x = \ln 2$.

Further Explorations

41. **Explain why or why not** Determine whether the following statements are true and give an explanation or counterexample.

    a. $\displaystyle\int uv'\, dx = \left(\int u\, dx \right)\left(\int v'\, dx \right)$

    b. $\displaystyle\int uv'\, dx = uv - \int vu'\, dx$

42–45. Reduction formulas *Use integration by parts to derive the following reduction formulas.*

42. $\displaystyle\int x^n e^{ax}\, dx = \frac{x^n e^{ax}}{a} - \frac{n}{a} \int x^{n-1} e^{ax}\, dx$ for $a \neq 0$

43. $\displaystyle\int x^n \cos ax\, dx = \frac{x^n \sin ax}{a} - \frac{n}{a} \int x^{n-1} \sin ax\, dx$ for $a \neq 0$

44. $\displaystyle\int x^n \sin ax\, dx = -\frac{x^n \cos ax}{a} + \frac{n}{a} \int x^{n-1} \cos ax\, dx$ for $a \neq 0$

45. $\displaystyle\int \ln^n x\, dx = x \ln^n x - n \int \ln^{n-1} x\, dx$

46–49. Applying reduction formulas *Use the reduction formulas in Exercises 42–45 to evaluate the following integrals.*

46. $\displaystyle\int x^2 e^{3x}\, dx$

47. $\displaystyle\int x^2 \cos 5x\, dx$

48. $\displaystyle\int x^3 \sin x\, dx$

49. $\displaystyle\int \ln^4 x\, dx$

50–51. Integrals involving $\int \ln x\, dx$ *Use a substitution to reduce the following integrals to $\int \ln u\, du$. Then, evaluate the resulting integral.*

50. $\displaystyle\int \cos x \ln (\sin x)\, dx$

51. $\displaystyle\int \sec^2 x \ln (\tan x + 2)\, dx$

52. **Two methods**

    a. Evaluate $\int x \ln x^2\, dx$ using the substitution $u = x^2$ and evaluating $\int \ln u\, du$.

    b. Evaluate $\int x \ln x^2\, dx$ using integration by parts.

    c. Verify that your answers to parts (a) and (b) are consistent.

53. **Logarithm base b** Prove that $\int \log_b x\, dx = \dfrac{1}{\ln b}(x \ln x - x) + C$.

54. Two integration methods Evaluate $\int \sin x \cos x \, dx$ using integration by parts. Then evaluate the integral using a substitution. Reconcile your answers.

55. Combining two integration methods Evaluate $\int \cos (\sqrt{x}) \, dx$ using a substitution followed by integration by parts.

56. Combining two integration methods Evaluate $\int_0^{\pi^2/4} \sin (\sqrt{x}) \, dx$ using a substitution followed by integration by parts.

57. Function defined as an integral Find the arc length of the function $f(x) = \int_e^x \sqrt{\ln^2 t - 1} \, dt$ on $[e, e^3]$.

58. A family of exponentials The curves $y = xe^{-ax}$ are shown in the figure for $a = 1, 2$, and 3.

    a. Find the area of the region bounded by $y = xe^{-x}$ and the x-axis on the interval $[0, 4]$.

    b. Find the area of the region bounded by $y = xe^{-ax}$ and the x-axis on the interval $[0, 4]$, where $a > 0$.

    c. Find the area of the region bounded by $y = xe^{-ax}$ and the x-axis on the interval $[0, b]$. Because this area depends on a and b, we call it $A(a, b)$, where $a > 0$ and $b > 0$.

    d. Use part (c) to show that $A(1, \ln b) = 4A(2, (\ln b)/2)$.

    e. Does this pattern continue? Is it true that $A(1, \ln b) = a^2 A(a, (\ln b)/a)$?

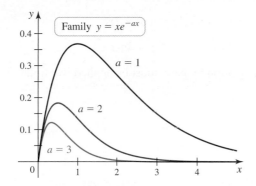

59. Solid of revolution Find the volume of the solid generated when the region bounded by $y = \cos x$ and the x-axis on the interval $[0, \pi/2]$ is revolved about the y-axis.

60. Between the sine and inverse sine Find the area of the region bounded by the curves $y = \sin x$ and $y = \sin^{-1} x$ on the interval $[0, 1]$.

61. Comparing volumes Let R be the region bounded by $y = \sin x$ and the x-axis on the interval $[0, \pi]$. Which is greater, the volume of the solid generated when R is revolved about the x-axis or the volume of the solid generated when R is revolved about the y-axis?

62. Log integrals Use integration by parts to show that for $m \neq -1$,

$$\int x^m \ln x \, dx = \frac{x^{m+1}}{m + 1} \left(\ln x - \frac{1}{m + 1} \right) + C$$

and for $m = -1$,

$$\int \frac{\ln x}{x} \, dx = \frac{1}{2} \ln^2 x + C.$$

63. A useful integral

    a. Use integration by parts to show that if f' is continuous

$$\int x f'(x) \, dx = x f(x) - \int f(x) \, dx.$$

    b. Use part (a) to evaluate $\int xe^{3x} \, dx$.

64. Integrating inverse functions

    a. Let $y = f^{-1}(x)$, which means $x = f(y)$ and $dx = f'(y) \, dy$. Show that

$$\int f^{-1}(x) \, dx = \int y f'(y) \, dy.$$

    b. Use the result of Exercise 63 to show that

$$\int f^{-1}(x) \, dx = y f(y) - \int f(y) \, dy.$$

    c. Use the result of part (b) to evaluate $\int \ln x \, dx$ (express the result in terms of x).

    d. Use the result of part (b) to evaluate $\int \sin^{-1} x \, dx$.

    e. Use the result of part (b) to evaluate $\int \tan^{-1} x \, dx$.

65. Integral of $\sec^3 x$ Use integration by parts to show that

$$\int \sec^3 x \, dx = \frac{1}{2} \sec x \tan x + \frac{1}{2} \int \sec x \, dx.$$

66. Two useful exponential integrals Use integration by parts to derive the following formulas for real numbers a and b.

$$\int e^{ax} \sin bx \, dx = \frac{e^{ax} (a \sin bx - b \cos bx)}{a^2 + b^2} + C$$

$$\int e^{ax} \cos bx \, dx = \frac{e^{ax} (a \cos bx + b \sin bx)}{a^2 + b^2} + C$$

Applications

67. Oscillator displacements Suppose an oscillator (such as a pendulum or a mass on a spring) that is slowed by friction has the position function $s(t) = e^{-t} \sin t$.

    a. Graph the position function. At what times does the oscillator pass through the position $s = 0$?

    b. Find the average value of the position on the interval $[0, \pi]$.

    c. Generalize part (b) and find the average value of the position on the interval $[n\pi, (n + 1)\pi]$ for $n = 0, 1, 2, \ldots$.

    d. Let a_n be the absolute value of the average position on the intervals $[n\pi, (n + 1)\pi]$ for $n = 0, 1, 2, \ldots$. Describe the pattern in the numbers $a_0, a_1, a_2, \ldots$.

Additional Exercises

68. Find the error Suppose you evaluate $\int \frac{dx}{x}$ using integration by parts. With $u = 1/x$ and $dv = dx$, you find that $du = -1/x^2 \, dx$, $v = x$, and

$$\int \frac{dx}{x} = \left(\frac{1}{x} \right) x - \int x \left(-\frac{1}{x^2} \right) dx = 1 + \int \frac{dx}{x}.$$

You conclude that $0 = 1$. Explain the problem with the calculation.

69. Proof without words Explain how the diagram in the figure illustrates integration by parts for definite integrals.

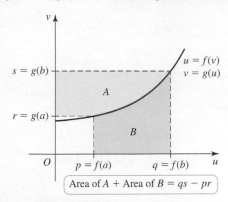

Area of A + Area of B = $qs - pr$

70. An identity Show that if f and g have continuous second derivatives and $f(0) = f(1) = g(0) = g(1) = 0$, then

$$\int_0^1 f''(x)g(x)\,dx = \int_0^1 f(x)g''(x)\,dx.$$

71. Possible and impossible integrals Let $I_n = \int x^n e^{-x^2}\,dx$, where n is a nonnegative integer.

a. $I_0 = \int e^{-x^2}\,dx$ cannot be expressed in terms of elementary functions. Evaluate I_1.
b. Use integration by parts to evaluate I_3.
c. Use integration by parts and the result of part (b) to evaluate I_5.
d. Show that, in general, if n is odd, then $I_n = -\dfrac{1}{2}e^{-x^2}p_{n-1}(x)$, where p_{n-1} is an even polynomial of degree $n - 1$.
e. Argue that if n is even, then I_n cannot be expressed in terms of elementary functions.

72. Looking ahead (to Chapter 9) Suppose that a function f has derivatives of all orders near $x = 0$. By the Fundamental Theorem of Calculus,

$$f(x) - f(0) = \int_0^x f'(t)\,dt.$$

a. Evaluate the integral using integration by parts to show that

$$f(x) = f(0) + xf'(0) + \int_0^x f''(t)(x - t)\,dt.$$

b. Show (by observing a pattern or using induction) that by integrating by parts n times,

$$f(x) = f(0) + xf'(0) + \frac{1}{2!}x^2 f''(0) + \cdots + \frac{1}{n!}x^n f^{(n)}(0)$$

$$+ \frac{1}{n!}\int_0^x f^{(n+1)}(t)(x - t)^n\,dt + \cdots$$

This expression, called the *Taylor series* for f at $x = 0$, is revisited in Chapter 9.

QUICK CHECK ANSWERS

1. Let $u = x$ and $dv = \cos x\,dx$.

2. $\dfrac{d}{dx}(x \ln x - x + C) = \ln x$

3. Integration by parts must be applied five times. ◄

7.2 Trigonometric Integrals

At the moment, our inventory of integrals involving trigonometric functions is rather limited. For example, we can integrate $\sin ax$ and $\cos ax$, where a is a constant, but missing from the list are integrals of $\tan ax$, $\cot ax$, $\sec ax$, and $\csc ax$. It turns out that integrals of powers of trigonometric functions, such as $\int \cos^5 x\,dx$ and $\int \cos^2 x \sin^4 x\,dx$, are also important. The goal of this section is to develop techniques for integrating integrals involving trigonometric functions. These techniques are indispensable when we use *trigonometric substitutions* in the next section.

Integrating Powers of $\sin x$ and $\cos x$

Two strategies are employed when evaluating integrals of the form $\int \sin^m x\,dx$ or $\int \cos^n x\,dx$, where m and n are positive integers. Both use trigonometric identities to recast the integrand, as shown in the first example.

EXAMPLE 1 **Powers of sine or cosine** Evaluate the following integrals.

a. $\int \cos^5 x \, dx$ **b.** $\int \sin^4 x \, dx$

SOLUTION

> Pythagorean identities:
>
> $$\cos^2 x + \sin^2 x = 1$$
> $$1 + \tan^2 x = \sec^2 x$$
> $$\cot^2 x + 1 = \csc^2 x$$

a. Integrals involving odd powers of $\cos x$ (or $\sin x$) are most easily evaluated by splitting off a single factor of $\cos x$ (or $\sin x$). In this case, we rewrite $\cos^5 x$ as $\cos^4 x \cdot \cos x$. Now, $\cos^4 x$ can be written in terms of $\sin x$ using the identity $\cos^2 x = 1 - \sin^2 x$. The result is an integrand that readily yields to the substitution $u = \sin x$:

$$\int \cos^5 x \, dx = \int \cos^4 x \cdot \cos x \, dx \qquad \text{Split off } \cos x.$$

$$= \int (1 - \sin^2 x)^2 \cdot \cos x \, dx \qquad \text{Pythagorean identity}$$

$$= \int (1 - u^2)^2 \, du \qquad \text{Let } u = \sin x; \, du = \cos x \, dx.$$

$$= \int (1 - 2u^2 + u^4) \, du \qquad \text{Expand.}$$

$$= u - \frac{2}{3}u^3 + \frac{1}{5}u^5 + C \qquad \text{Integrate.}$$

$$= \sin x - \frac{2}{3}\sin^3 x + \frac{1}{5}\sin^5 x + C \quad \text{Replace } u \text{ with } \sin x.$$

> Use the phrase "sine is minus" to remember that a minus sign is associated with the half-angle formula for $\sin^2 x$, while a positive sign is used for $\cos^2 x$.

b. With even powers of $\sin x$ or $\cos x$, we use the half-angle formulas

$$\sin^2 x = \frac{1 - \cos 2x}{2} \quad \text{and} \quad \cos^2 x = \frac{1 + \cos 2x}{2}$$

to reduce the powers in the integrand:

$$\int \sin^4 x \, dx = \int \underbrace{\left(\frac{1 - \cos 2x}{2}\right)^2}_{\sin^2 x} dx \qquad \text{Half-angle formula}$$

$$= \frac{1}{4}\int (1 - 2\cos 2x + \cos^2 2x) \, dx \quad \text{Expand the integrand.}$$

Using the half-angle formula again for $\cos^2 2x$, the evaluation may be completed:

$$\int \sin^4 x \, dx = \frac{1}{4}\int \left(1 - 2\cos 2x + \underbrace{\frac{1 + \cos 4x}{2}}_{\cos^2 2x}\right) dx \quad \text{Half-angle formula}$$

$$= \frac{1}{4}\int \left(\frac{3}{2} - 2\cos 2x + \frac{1}{2}\cos 4x\right) dx \qquad \text{Simplify.}$$

$$= \frac{3x}{8} - \frac{1}{4}\sin 2x + \frac{1}{32}\sin 4x + C \qquad \text{Evaluate the integrals.}$$

Related Exercises 9–12 ◄

QUICK CHECK 1 Evaluate $\int \sin^3 x \, dx$ by splitting off a factor of $\sin x$, rewriting $\sin^2 x$ in terms of $\cos x$, and using an appropriate u-substitution. ◄

Integrating Products of sin x and cos x

We now consider integrals of the form $\int \sin^m x \cos^n x \, dx$. If m is an odd, positive integer, we split off a factor of $\sin x$ and write the remaining even power of $\sin x$ in terms of cosine functions. This step prepares the integrand for the substitution $u = \cos x$, and the resulting integral is readily solved. A similar strategy is used when n is an odd, positive integer.

If both m and n are even, positive integers, the half-angle formulas are used to transform the integrand into a polynomial in $\cos 2x$, each of whose terms can be integrated, as shown in Example 2.

EXAMPLE 2   **Products of sine and cosine**   Evaluate the following integrals.

a. $\int \sin^4 x \cos^2 x \, dx$ **b.** $\int \sin^3 x \cos^{-2} x \, dx$

SOLUTION

a. When both powers are even, the half-angle formulas are used:

$$\int \sin^4 x \cos^2 x \, dx = \int \underbrace{\left(\frac{1 - \cos 2x}{2}\right)^2}_{\sin^2 x} \underbrace{\left(\frac{1 + \cos 2x}{2}\right)}_{\cos^2 x} dx \qquad \text{Half-angle formulas}$$

$$= \frac{1}{8} \int (1 - \cos 2x - \cos^2 2x + \cos^3 2x) \, dx \quad \text{Expand.}$$

The third term in the integrand is rewritten with a half-angle formula. For the last term, a factor of $\cos 2x$ is split off, and the resulting even power of $\cos 2x$ is written in terms of $\sin 2x$ to prepare for a u-substitution:

$$\int \sin^4 x \cos^2 x \, dx =$$

$$\frac{1}{8} \int \left[1 - \cos 2x - \overbrace{\left(\frac{1 + \cos 4x}{2}\right)}^{\cos^2 2x} \right] dx + \frac{1}{8} \int \overbrace{(1 - \sin^2 2x)}^{\cos^2 2x} \cdot \cos 2x \, dx$$

Finally, the integrals are evaluated, using the substitution $u = \sin 2x$ for the second integral. After simplification, we find that

$$\int \sin^4 x \cos^2 x \, dx = \frac{1}{16} x - \frac{1}{64} \sin 4x - \frac{1}{48} \sin^3 2x + C.$$

b. When at least one power is odd, the following approach works:

$$\int \sin^3 x \cos^{-2} x \, dx = \int \sin^2 x \cos^{-2} x \cdot \sin x \, dx \qquad \text{Split off } \sin x.$$

$$= \int (1 - \cos^2 x) \cos^{-2} x \cdot \sin x \, dx \quad \text{Pythagorean identity}$$

$$= -\int (1 - u^2) u^{-2} \, du \qquad u = \cos x; \, du = -\sin x \, dx$$

$$= \int (1 - u^{-2}) \, du = u + \frac{1}{u} + C \quad \text{Evaluate the integral.}$$

$$= \cos x + \sec x + C \qquad \text{Replace } u \text{ with } \cos x.$$

Related Exercises 13–18 ◄

QUICK CHECK 2 What strategy would you use to evaluate $\int \sin^3 x \cos^3 x \, dx$? ◄

Table 7.1 summarizes the techniques used to evaluate integrals of the form $\int \sin^m x \cos^n x \, dx$.

Table 7.1

$\int \sin^m x \cos^n x \, dx$	Strategy
m odd, n real	Split off $\sin x$, rewrite the resulting even power of $\sin x$ in terms of $\cos x$, and then use $u = \cos x$.
n odd, m real	Split off $\cos x$, rewrite the resulting even power of $\cos x$ in terms of $\sin x$, and then use $u = \sin x$.
m and n both even, nonnegative integers	Use half-angle identities to transform the integrand into a polynomial in $\cos 2x$, and apply the preceding strategies once again to powers greater than 1 $(\cos^2 2x, \cos^3 2x, \text{ etc.})$.

Reduction Formulas

Evaluating an integral such as $\int \sin^8 x \, dx$ using the method of Example 1b would be tedious, at best. For this reason, *reduction formulas* have been developed to ease the workload. A reduction formula equates an integral involving a power of a function with another integral in which the power is reduced; several reduction formulas were encountered in Exercises 42–45 of Section 7.1. Here are some frequently used reduction formulas for trigonometric integrals.

Reduction Formulas

Assume n is a positive integer.

1. $\displaystyle \int \sin^n x \, dx = -\frac{\sin^{n-1} x \cos x}{n} + \frac{n-1}{n} \int \sin^{n-2} x \, dx$

2. $\displaystyle \int \cos^n x \, dx = \frac{\cos^{n-1} x \sin x}{n} + \frac{n-1}{n} \int \cos^{n-2} x \, dx$

3. $\displaystyle \int \tan^n x \, dx = \frac{\tan^{n-1} x}{n-1} - \int \tan^{n-2} x \, dx, \; n \neq 1$

4. $\displaystyle \int \sec^n x \, dx = \frac{\sec^{n-2} x \tan x}{n-1} + \frac{n-2}{n-1} \int \sec^{n-2} x \, dx, \; n \neq 1$

Formulas 1, 3, and 4 are derived in Exercises 52–54. The derivation of formula 2 is similar to that of formula 1.

EXAMPLE 3 Powers of $\tan x$ Evaluate $\int \tan^4 x \, dx$.

SOLUTION Reduction formula 3 gives

$$\int \tan^4 x \, dx = \frac{1}{3}\tan^3 x - \underbrace{\int \tan^2 x \, dx}_{\text{use (3) again}}$$

$$= \frac{1}{3}\tan^3 x - \left(\tan x - \int \underbrace{\tan^0 x \, dx}_{=1}\right)$$

$$= \frac{1}{3}\tan^3 x - \tan x + x + C$$

An alternative solution uses the identity $\tan^2 x = \sec^2 x - 1$:

$$\int \tan^4 x \, dx = \int \tan^2 x \, (\sec^2 x - 1) \, dx$$

$$= \int \tan^2 x \sec^2 x \, dx - \int \tan^2 x \, dx$$

The substitution $u = \tan x$, $du = \sec^2 x \, dx$ is used in first integral, while the identity $\tan^2 x = \sec^2 x - 1$ is used again in the second integral:

$$\int \tan^4 x \, dx = \int \underbrace{\tan^2 x}_{u^2} \underbrace{\sec^2 x \, dx}_{du} - \int \tan^2 x \, dx$$

$$= \int u^2 \, du - \int (\sec^2 x - 1) \, dx \qquad \text{Substitution and identity}$$

$$= \frac{u^3}{3} - \tan x + x + C \qquad\qquad \text{Evaluate integrals.}$$

$$= \frac{1}{3} \tan^3 x - \tan x + x + C \qquad\qquad u = \tan x$$

Related Exercises 19–24 ◀

Note that for odd powers of $\tan x$ and $\sec x$, the use of reduction formula 3 or 4 will eventually lead to $\int \tan x \, dx$ or $\int \sec x \, dx$. Theorem 7.1 gives these integrals, along with the integrals of $\cot x$ and $\csc x$.

THEOREM 7.1   Integrals of $\tan x$, $\cot x$, $\sec x$, and $\csc x$

$$\int \tan x \, dx = -\ln|\cos x| + C = \ln|\sec x| + C \qquad \int \cot x \, dx = \ln|\sin x| + C$$

$$\int \sec x \, dx = \ln|\sec x + \tan x| + C \qquad \int \csc x \, dx = -\ln|\csc x + \cot x| + C$$

Proof In the first integral, $\tan x$ is expressed as the ratio of $\sin x$ and $\cos x$ to prepare for a standard substitution:

$$\int \tan x \, dx = \int \frac{\sin x}{\cos x} \, dx$$

$$= -\int \frac{1}{u} \, du \qquad\qquad u = \cos x;\ du = -\sin x \, dx$$

$$= -\ln|u| + C = -\ln|\cos x| + C$$

Using properties of logarithms, the integral can also be written

$$\int \tan x \, dx = -\ln|\cos x| + C = \ln|(\cos x)^{-1}| + C = \ln|\sec x| + C.$$

Derivations of the remaining integrals are left to Exercises 34–37.   ◀

Integrating Products of $\tan x$ and $\sec x$

Integrals of the form $\int \tan^m x \sec^n x \, dx$ are evaluated using methods analogous to those used for $\int \sin^m x \cos^n x \, dx$. For example, if n is even, we split off a factor of $\sec^2 x$ and write the remaining even power of $\sec x$ in terms of $\tan x$. This step prepares the integral for the substitution $u = \tan x$. If m is odd, we split off a factor of $\sec x \tan x$ (the derivative of $\sec x$), which prepares the integral for the substitution $u = \sec x$. If m is even and n is odd, the integrand is expressed as a polynomial in $\sec x$, each of whose terms is handled by a reduction formula. Example 4 illustrates these techniques.

EXAMPLE 4   Products of $\tan x$ and $\sec x$ Evaluate the integrals.

a. $\int \tan^3 x \sec^4 x \, dx$ **b.** $\int \tan^2 x \sec x \, dx$

SOLUTION

a. With an even power of $\sec x$, we split off a factor of $\sec^2 x$, and prepare the integral for the substitution $u = \tan x$:

$$\int \tan^3 x \sec^4 x \, dx = \int \tan^3 x \sec^2 x \cdot \sec^2 x \, dx$$

$$= \int \tan^3 x \left(\tan^2 x + 1\right) \cdot \sec^2 x \, dx \quad \sec^2 x = \tan^2 x + 1$$

$$= \int u^3(u^2 + 1) \, du \qquad\qquad u = \tan x; \, du = \sec^2 x \, dx$$

$$= \frac{1}{6} \tan^6 x + \frac{1}{4} \tan^4 x + C \qquad\qquad \text{Evaluate; } u = \tan x.$$

Because the integrand also has an odd power of $\tan x$, an alternative solution is to split off a factor of $\sec x \tan x$, and prepare the integral for the substitution $u = \sec x$.

> In Example 4a, the two methods produce results that look different, but are equivalent. This is common when evaluating trigonometric integrals. For instance, try $\int \sin^4 x \, dx$ using reduction formula 1, and compare your answer to
>
> $$\frac{3x}{8} - \frac{1}{4}\sin 2x + \frac{1}{32}\sin 4x + C,$$
>
> the solution found in Example 1b.

$$\int \tan^3 x \sec^4 x \, dx = \int \underbrace{\tan^2 x}_{\sec^2 x - 1} \sec^3 x \cdot \sec x \tan x \, dx$$

$$= \int (\sec^2 x - 1) \sec^3 x \cdot \sec x \tan x \, dx$$

$$= \int (u^2 - 1) u^3 \, du \qquad\qquad \begin{array}{l} u = \sec x; \\ du = \sec x \tan x \, dx \end{array}$$

$$= \frac{1}{6} \sec^6 x - \frac{1}{4} \sec^4 x + C \qquad\qquad \text{Evaluate; } u = \sec x.$$

The apparent difference in the two solutions given here is reconciled by using the identity $1 + \tan^2 x = \sec^2 x$ to transform the second result into the first, the only difference being an additive constant, which is part of C.

b. In this case, we write the even power of $\tan x$ in terms of $\sec x$:

$$\int \tan^2 x \sec x \, dx = \int (\sec^2 x - 1) \sec x \, dx \qquad\qquad \tan^2 x = \sec^2 x - 1$$

$$= \int \sec^3 x \, dx - \int \sec x \, dx$$

$$\overbrace{}^{\text{reduction formula 4}}$$

$$= \frac{1}{2} \sec x \tan x + \frac{1}{2} \int \sec x \, dx - \int \sec x \, dx$$

$$= \frac{1}{2} \sec x \tan x - \frac{1}{2} \ln |\sec x + \tan x| + C \qquad \begin{array}{l} \text{Add secant integrals;} \\ \text{use Theorem 7.1.} \end{array}$$

Related Exercises 25–32 ◄

Table 7.2 summarizes the methods used to integrate $\int \tan^m x \sec^n x \, dx$. Analogous techniques are used for $\int \cot^m x \csc^n x \, dx$.

Table 7.2

$\int \tan^m x \sec^n x \, dx$	Strategy
n even	Split off $\sec^2 x$, rewrite the remaining even power of $\sec x$ in terms of $\tan x$, and use $u = \tan x$.
m odd	Split off $\sec x \tan x$, rewrite the remaining even power of $\tan x$ in terms of $\sec x$, and use $u = \sec x$.
m even and n odd	Rewrite the even power of $\tan x$ in terms of $\sec x$ to produce a polynomial in $\sec x$; apply reduction formula 4 to each term.

SECTION 7.2 EXERCISES

Review Questions

1. State the half-angle identities used to integrate $\sin^2 x$ and $\cos^2 x$.

2. State the three Pythagorean identities.

3. Describe the method used to integrate $\sin^3 x$.

4. Describe the method used to integrate $\sin^m x \cos^n x$ for m even and n odd.

5. What is a reduction formula?

6. How would you evaluate $\int \cos^2 x \sin^3 x \, dx$?

7. How would you evaluate $\int \tan^{10} x \sec^2 x \, dx$?

8. How would you evaluate $\int \sec^{12} x \tan x \, dx$?

Basic Skills

9–12. Integrals of $\sin x$ or $\cos x$ *Evaluate the following integrals.*

9. $\displaystyle\int \sin^2 x \, dx$

10. $\displaystyle\int \cos^4 2x \, dx$

11. $\displaystyle\int \sin^5 x \, dx$

12. $\displaystyle\int \cos^3 20x \, dx$

13–18. Integrals of $\sin x$ and $\cos x$ *Evaluate the following integrals.*

13. $\displaystyle\int \sin^2 x \cos^2 x \, dx$

14. $\displaystyle\int \sin^3 x \cos^5 x \, dx$

15. $\displaystyle\int \sin^5 x \cos^{-2} x \, dx$

16. $\displaystyle\int \sin^{-3/2} x \cos^3 x \, dx$

17. $\displaystyle\int \sin^2 x \cos^4 x \, dx$

18. $\displaystyle\int \sin^3 x \cos^{3/2} x \, dx$

19–24. Integrals of $\tan x$ or $\cot x$ *Evaluate the following integrals.*

19. $\displaystyle\int \tan^2 x \, dx$

20. $\displaystyle\int 6 \sec^4 x \, dx$

21. $\displaystyle\int \tan^3 4x \, dx$

22. $\displaystyle\int \sec^5 \theta \, d\theta$

23. $\displaystyle\int 20 \tan^6 x \, dx$

24. $\displaystyle\int \cot^5 3x \, dx$

25–32. Integrals of $\tan x$ and $\sec x$ *Evaluate the following integrals.*

25. $\displaystyle\int \sec^2 x \tan^{1/2} x \, dx$

26. $\displaystyle\int \sec^{-2} x \tan^3 x \, dx$

27. $\displaystyle\int \frac{\csc^4 x}{\cot^2 x} \, dx$

28. $\displaystyle\int \csc^{10} x \cot^3 x \, dx$

29. $\displaystyle\int_0^{\pi/4} \sec^4 \theta \, d\theta$

30. $\displaystyle\int \tan^5 \theta \sec^4 \theta \, d\theta$

31. $\displaystyle\int_{\pi/6}^{\pi/3} \cot^3 \theta \, d\theta$

32. $\displaystyle\int_0^{\pi/4} \tan^3 \theta \sec^2 \theta \, d\theta$

Further Explorations

33. **Explain why or why not** Determine whether the following statements are true and give an explanation or counterexample.

    a. If m is a positive integer, then $\int_0^\pi \cos^{2m+1} x \, dx = 0$.

    b. If m is a positive integer, then $\int_0^\pi \sin^m x \, dx = 0$.

34–37. Integrals of $\cot x$, $\sec x$, and $\csc x$

34. Use a change of variables to prove that $\int \cot x \, dx = \ln|\sin x| + C$.

35. Prove that $\int \sec x \, dx = \ln|\sec x + \tan x| + C$. (*Hint:* Multiply numerator and denominator of the integrand by $\sec x + \tan x$; then make a change of variables with $u = \sec x + \tan x$.)

36. Prove that $\int \csc x \, dx = -\ln|\csc x + \cot x| + C$. (*Hint:* Use a method analogous to that used in Exercise 35.)

37. Use the results of Theorem 7.1 to find the indefinite integral of $\tan ax$ and $\sec ax$, where a is a nonzero real number.

38. **Comparing areas** The region R_1 is bounded by the graph of $y = \tan x$ and the x-axis on the interval $[0, \pi/3]$. The region R_2 is bounded by the graph of $y = \sec x$ and the x-axis on the interval $[0, \pi/6]$. Which region has the greater area?

39. **Region between curves** Find the area of the region bounded by the graphs of $y = \tan x$ and $y = \sec x$ on the interval $[0, \pi/4]$.

40–45. Additional integrals *Evaluate the following integrals.*

40. $\displaystyle\int_0^{\sqrt{\pi/2}} x \sin^3(x^2)\, dx$

41. $\displaystyle\int \frac{\sec^4(\ln\theta)}{\theta}\, d\theta$

42. $\displaystyle\int_{\pi/6}^{\pi/2} \frac{dy}{\sin y}$

43. $\displaystyle\int_{-\pi/3}^{\pi/3} \sqrt{\sec^2\theta - 1}\, d\theta$

44. $\displaystyle\int_{-\pi/4}^{\pi/4} \tan^3 x \sec^2 x\, dx$

45. $\displaystyle\int_0^{\pi} (1 - \cos 2x)^{3/2}\, dx$

46–49. Square roots *Evaluate the following integrals.*

46. $\displaystyle\int_{-\pi/4}^{\pi/4} \sqrt{1 + \cos 4x}\, dx$

47. $\displaystyle\int_0^{\pi/2} \sqrt{1 - \cos 2x}\, dx$

48. $\displaystyle\int_0^{\pi/8} \sqrt{1 - \cos 8x}\, dx$

49. $\displaystyle\int_0^{\pi/4} (1 + \cos 4x)^{3/2}\, dx$

50. Sine football Find the volume of the solid generated when the region bounded by the graph of $y = \sin x$ and the x-axis on the interval $[0, \pi]$ is revolved about the x-axis.

51. Arc length Find the length of the curve $y = \ln(\cos x)$ for $0 \le x \le \pi/4$.

52. A sine reduction formula Use integration by parts to obtain the following reduction formula for positive integers n:

$$\int \sin^n x\, dx = -\sin^{n-1} x \cos x + (n-1)\int \sin^{n-2} x \cos^2 x\, dx.$$

Then use an identity to obtain the reduction formula

$$\int \sin^n x\, dx = -\frac{\sin^{n-1} x \cos x}{n} + \frac{n-1}{n}\int \sin^{n-2} x\, dx.$$

Use this reduction formula to evaluate $\int \sin^6 x\, dx$.

53. A tangent reduction formula Prove that for positive integers $n \neq 1$,

$$\int \tan^n x\, dx = \frac{\tan^{n-1} x}{n-1} - \int \tan^{n-2} x\, dx.$$

Use the formula to evaluate $\int_0^{\pi/4} \tan^3 x\, dx$.

54. A secant reduction formula Prove that for positive integers $n \neq 1$,

$$\int \sec^n x\, dx = \frac{\sec^{n-2} x \tan x}{n-1} + \frac{n-2}{n-1}\int \sec^{n-2} x\, dx.$$

(*Hint:* Integrate by parts with $u = \sec^{n-2} x$ and $dv = \sec^2 x\, dx$.)

Applications

55–59. Integrals of the form $\int \sin mx \cos nx\, dx$ *Use the following three identities to evaluate the given integrals.*

$$\sin mx \sin nx = \frac{1}{2}[\cos((m-n)x) - \cos((m+n)x)]$$

$$\sin mx \cos nx = \frac{1}{2}[\sin((m-n)x) + \sin((m+n)x)]$$

$$\cos mx \cos nx = \frac{1}{2}[\cos((m-n)x) + \cos((m+n)x)]$$

55. $\displaystyle\int \sin 3x \cos 7x\, dx$

56. $\displaystyle\int \sin 5x \sin 7x\, dx$

57. $\displaystyle\int \sin 3x \sin 2x\, dx$

58. $\displaystyle\int \cos x \cos 2x\, dx$

59. Prove the following **orthogonality relations** (which are used to generate *Fourier series*). Assume m and n are integers with $m \neq n$.

a. $\displaystyle\int_0^{\pi} \sin mx \sin nx\, dx = 0$

b. $\displaystyle\int_0^{\pi} \cos mx \cos nx\, dx = 0$

c. $\displaystyle\int_0^{\pi} \sin mx \cos nx\, dx = 0$

60. Mercator map projection The Mercator map projection was proposed by the Flemish geographer Gerardus Mercator (1512–1594). The stretching of the Mercator map as a function of the latitude θ is given by the function

$$G(\theta) = \int_0^{\theta} \sec x\, dx.$$

Graph G for $0 \le \theta < \pi/2$. (See the Guided Projects for a derivation of this integral.)

Additional Exercises

61. Exploring powers of sine and cosine

a. Graph the functions $f_1(x) = \sin^2 x$ and $f_2(x) = \sin^2 2x$ on the interval $[0, \pi]$. Find the area under these curves on $[0, \pi]$.

b. Graph a few more of the functions $f_n(x) = \sin^2 nx$ on the interval $[0, \pi]$, where n is a positive integer. Find the area under these curves on $[0, \pi]$. Comment on your observations.

c. Prove that $\int_0^{\pi} \sin^2(nx)\, dx$ has the same value for all positive integers n.

d. Does the conclusion of part (c) hold if sine is replaced by cosine?

e. Repeat parts (a), (b), and (c) with $\sin^2 x$ replaced by $\sin^4 x$. Comment on your observations.

f. Challenge problem: Show that for $m = 1, 2, 3, \ldots$,

$$\int_0^{\pi} \sin^{2m} x\, dx = \int_0^{\pi} \cos^{2m} x\, dx = \pi \cdot \frac{1 \cdot 3 \cdot 5 \cdots (2m-1)}{2 \cdot 4 \cdot 6 \cdots 2m}.$$

QUICK CHECK ANSWERS

1. $\frac{1}{3}\cos^3 x - \cos x + C$ **2.** Write $\int \sin^3 x \cos^3 x\, dx = \int \sin^2 x \cos^3 x \sin x\, dx = \int(1 - \cos^2 x)\cos^3 x \sin x\, dx$. Then, use the substitution $u = \cos x$. Or, begin by writing $\int \sin^3 x \cos^3 x\, dx = \int \sin^3 x \cos^2 x \cos x\, dx$. ◄

7.3 Trigonometric Substitutions

In Section 6.5, we wrote the arc length integral for the segment of the parabola $y = x^2$ on the interval $[0, 2]$ as

$$\int_0^2 \sqrt{1 + 4x^2}\, dx = \int_0^2 2\sqrt{\tfrac{1}{4} + x^2}\, dx.$$

At the time, we did not have the analytical methods needed to evaluate this integral. The goal of this section is to develop the tools needed to evaluate such integrals.

Integrals similar to the arc length integral for the parabola arise in many different situations. For example, electrostatic, magnetic, and gravitational forces obey an inverse square law (their strength is proportional to $1/r^2$, where r is a distance). Computing these force fields in two dimensions leads to integrals such as $\displaystyle\int \frac{dx}{\sqrt{x^2 + a^2}}$ or $\displaystyle\int \frac{dx}{(x^2 + a^2)^{3/2}}$.

It turns out that integrals containing the terms $a^2 \pm x^2$ or $x^2 - a^2$, where a is a constant, can be simplified using somewhat unexpected substitutions involving trigonometric functions. The new integrals produced by these substitutions are often trigonometric integrals of the variety studied in the preceding section.

Integrals Involving $a^2 - x^2$

▷ The following thinking might lead you to the substitution $x = a \sin \theta$. The term $\sqrt{a^2 - x^2}$ looks like the length of one side of a right triangle whose hypotenuse has length a and whose other side has length x. Labeling one acute angle θ, we see that $x = a \sin \theta$.

Suppose you are faced with an integral whose integrand contains the term $a^2 - x^2$, where a is a positive constant. Observe what happens when x is replaced with $a \sin \theta$:

$$
\begin{aligned}
a^2 - x^2 &= a^2 - (a \sin \theta)^2 && \text{Replace } x \text{ with } a \sin \theta. \\
&= a^2 - a^2 \sin^2 \theta && \text{Simplify.} \\
&= a^2(1 - \sin^2 \theta) && \text{Factor.} \\
&= a^2 \cos^2 \theta && 1 - \sin^2 \theta = \cos^2 \theta
\end{aligned}
$$

This calculation shows that the substitution $x = a \sin \theta$ turns the difference $a^2 - x^2$ into the product $a^2 \cos^2 \theta$. The resulting integral—now with respect to θ—is often easier to evaluate than the original integral. The details of this procedure are spelled out in the following examples.

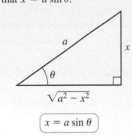

$$x = a \sin \theta$$

QUICK CHECK 1 Use a substitution of the form $x = a \sin \theta$ to transform $9 - x^2$ into a product. ◀

EXAMPLE 1 Area of a circle Verify that the area of a circle of radius a is πa^2.

SOLUTION The function $f(x) = \sqrt{a^2 - x^2}$ describes the upper half of a circle centered at the origin with radius a (Figure 7.2). The region under this curve on the interval $[0, a]$ is a quarter-circle. Therefore, the area of the full circle is $4\int_0^a \sqrt{a^2 - x^2}\, dx$.

Because the integrand contains the expression $a^2 - x^2$, we use the trigonometric substitution $x = a \sin \theta$. As with all substitutions, the differential associated with the substitution must be computed:

$$x = a \sin \theta \quad \text{implies that} \quad dx = a \cos \theta\, d\theta$$

The substitution $x = a \sin \theta$ can also be written $\theta = \sin^{-1}(x/a)$, where $-\pi/2 \le \theta \le \pi/2$ (Figure 7.3). Notice that the new variable θ plays the role of an angle. The substitution works nicely, because when x is replaced by $a \sin \theta$ in the integrand, we have

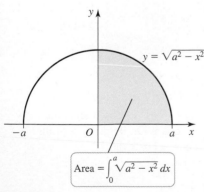

$$y = \sqrt{a^2 - x^2}$$

$$\text{Area} = \int_0^a \sqrt{a^2 - x^2}\, dx$$

FIGURE 7.2

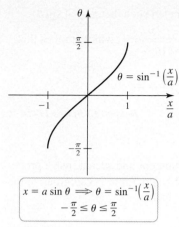

$$x = a \sin \theta \implies \theta = \sin^{-1}\left(\frac{x}{a}\right)$$
$$-\frac{\pi}{2} \le \theta \le \frac{\pi}{2}$$

FIGURE 7.3

➤ The key identities for integrating $\sin^2 \theta$ and $\cos^2 \theta$ are

$$\sin^2 \theta = \frac{1 - \cos 2\theta}{2}$$

$$\cos^2 \theta = \frac{1 + \cos 2\theta}{2}$$

$$\sqrt{a^2 - x^2} = \sqrt{a^2 - (a \sin \theta)^2} \quad \text{Replace } x \text{ with } a \sin \theta.$$
$$= \sqrt{a^2(1 - \sin^2 \theta)} \quad \text{Factor.}$$
$$= \sqrt{a^2 \cos^2 \theta} \quad 1 - \sin^2 \theta = \cos^2 \theta$$
$$= |a \cos \theta| \quad \sqrt{x^2} = |x|$$
$$= a \cos \theta \quad a > 0, \cos \theta > 0 \text{ for } -\frac{\pi}{2} \le \theta \le \frac{\pi}{2}$$

We also change the limits of integration: When $x = 0$, $\theta = \sin^{-1} 0 = 0$; when $x = a$, $\theta = \sin^{-1}(a/a) = \sin^{-1} 1 = \pi/2$. Making these substitutions, the integral is evaluated as follows:

$$4 \int_0^a \sqrt{a^2 - x^2}\, dx = 4 \int_0^{\pi/2} \underbrace{a \cos \theta}_{\substack{\text{integrand} \\ \text{simplified}}} \cdot \underbrace{a \cos \theta\, d\theta}_{dx} \qquad x = a \sin \theta, dx = a \cos \theta\, d\theta$$

$$= 4a^2 \int_0^{\pi/2} \cos^2 \theta\, d\theta \qquad \text{Simplify.}$$

$$= 4a^2 \left(\frac{\theta}{2} + \frac{\sin 2\theta}{4} \right)\Big|_0^{\pi/2} \qquad \cos^2 \theta = \frac{1 + \cos 2\theta}{2}$$

$$= 4a^2 \left[\left(\frac{\pi}{4} + 0 \right) - (0 + 0) \right] = \pi a^2 \quad \text{Simplify.}$$

A similar calculation (Exercise 56) gives the area of an ellipse.   *Related Exercises 7–10* ◄

EXAMPLE 2   Sine substitution Evaluate $\displaystyle\int \frac{dx}{(16 - x^2)^{3/2}}$.

SOLUTION The factor $16 - x^2$ has the form $a^2 - x^2$ with $a = 4$, so we use the substitution $x = 4 \sin \theta$. It follows that $dx = 4 \cos \theta\, d\theta$. We now simplify $(16 - x^2)^{3/2}$:

$$(16 - x^2)^{3/2} = \left(16 - (4 \sin \theta)^2 \right)^{3/2} \quad \text{Substitute } x = 4 \sin \theta.$$
$$= \left(16(1 - \sin^2 \theta) \right)^{3/2} \quad \text{Factor.}$$
$$= (16 \cos^2 \theta)^{3/2} \quad 1 - \sin^2 \theta = \cos^2 \theta$$
$$= 64 \cos^3 \theta \quad \text{Simplify.}$$

Replacing the factors $(16 - x^2)^{3/2}$ and dx of the original integral with appropriate expressions in θ, we have

$$\int \frac{\overbrace{dx}^{4 \cos \theta\, d\theta}}{\underbrace{(16 - x^2)^{3/2}}_{64 \cos^3 \theta}} = \int \frac{4 \cos \theta}{64 \cos^3 \theta}\, d\theta$$

$$= \frac{1}{16} \int \frac{d\theta}{\cos^2 \theta}$$

$$= \frac{1}{16} \int \sec^2 \theta\, d\theta \quad \text{Simplify.}$$

$$= \frac{1}{16} \tan \theta + C \quad \text{Evaluate the integral.}$$

The final step is to express this result in terms of x. In many integrals, this step is most easily done with a reference triangle showing the relationship between x and θ. Figure 7.4

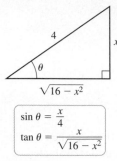

$$\sin \theta = \frac{x}{4}$$

$$\tan \theta = \frac{x}{\sqrt{16 - x^2}}$$

FIGURE 7.4

shows a right triangle with an angle θ and with the sides labeled such that $x = 4 \sin \theta$ (or $\sin \theta = x/4$). Using this triangle, we see that $\tan \theta = \dfrac{x}{\sqrt{16 - x^2}}$, which implies that

$$\int \frac{dx}{(16 - x^2)^{3/2}} = \frac{1}{16} \tan \theta + C = \frac{x}{16\sqrt{16 - x^2}} + C.$$

Related Exercises 11–14 ◄

Integrals Involving $a^2 + x^2$ or $x^2 - a^2$

The other standard trigonometric substitutions, involving tangent and secant, use a procedure similar to that used for the sine substitution. Figure 7.5 and Table 7.3 summarize the three basic trigonometric substitutions for real numbers $a > 0$.

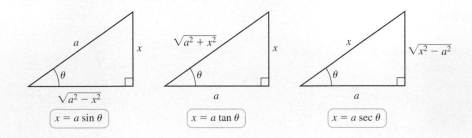

FIGURE 7.5

$x = a \sin \theta$ $x = a \tan \theta$ $x = a \sec \theta$

Table 7.3

The Integral Contains...	Corresponding Substitution	Useful Identity
$a^2 - x^2$	$x = a \sin \theta, \ -\dfrac{\pi}{2} \le \theta \le \dfrac{\pi}{2}$	$a^2 - a^2 \sin^2 \theta = a^2 \cos^2 \theta$
$a^2 + x^2$	$x = a \tan \theta, \ -\dfrac{\pi}{2} < \theta < \dfrac{\pi}{2}$	$a^2 + a^2 \tan^2 \theta = a^2 \sec^2 \theta$
$x^2 - a^2$	$x = a \sec \theta, \ \begin{cases} 0 \le \theta < \dfrac{\pi}{2} & \text{for } x \ge a \\[2mm] \dfrac{\pi}{2} < \theta \le \pi & \text{for } x \le -a \end{cases}$	$a^2 \sec^2 \theta - a^2 = a^2 \tan^2 \theta$

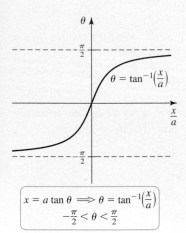

$$\theta = \tan^{-1}\left(\frac{x}{a}\right)$$

$$x = a \tan \theta \implies \theta = \tan^{-1}\left(\frac{x}{a}\right)$$
$$-\frac{\pi}{2} < \theta < \frac{\pi}{2}$$

FIGURE 7.6

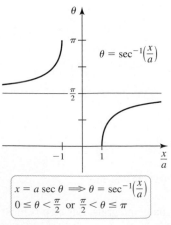

$$\theta = \sec^{-1}\left(\frac{x}{a}\right)$$

$$x = a \sec \theta \implies \theta = \sec^{-1}\left(\frac{x}{a}\right)$$
$$0 \le \theta < \frac{\pi}{2} \text{ or } \frac{\pi}{2} < \theta \le \pi$$

FIGURE 7.7

In order for the tangent substitution $x = a \tan \theta$ to be well defined, the angle θ must be restricted to the interval $-\pi/2 < \theta < \pi/2$, which is consistent with the definition of $\tan^{-1}(x/a)$ (Figure 7.6). On this interval, $\sec \theta > 0$ and with $a > 0$, it is valid to write

$$\sqrt{a^2 + x^2} = \sqrt{a^2 + (a \tan \theta)^2} = \sqrt{a^2 \underbrace{(1 + \tan^2 \theta)}_{\sec^2 \theta}} = a \sec \theta.$$

With the secant substitution, there is a technicality. As discussed in Section 1.4, $\theta = \sec^{-1}(x/a)$ is defined for $x \ge a$, in which case $0 \le \theta < \pi/2$, *and* for $x \le -a$, in which case $\pi/2 < \theta \le \pi$ (Figure 7.7). These restrictions on θ must be treated carefully when simplifying integrands with a factor of $\sqrt{x^2 - a^2}$. Because $\tan \theta$ is positive in the first quadrant but negative in the second, we have

$$\sqrt{x^2 - a^2} = \sqrt{a^2 \underbrace{(\sec^2 \theta - 1)}_{\tan^2 \theta}} = |a \tan \theta| = \begin{cases} a \tan \theta & \text{if } 0 \le \theta < \dfrac{\pi}{2} \\[3mm] -a \tan \theta & \text{if } \dfrac{\pi}{2} < \theta \le \pi \end{cases}$$

QUICK CHECK 2 What change of variables would you try on the integrals

(a) $\displaystyle\int \frac{x^2}{\sqrt{x^2 + 9}}\,dx$ and

(b) $\displaystyle\int \frac{3}{x\sqrt{16 - x^2}}\,dx?$ ◄

When evaluating a definite integral, you should check the limits of integration to see which of these two cases applies. For indefinite integrals, a piecewise formula is often needed, unless a restriction on the variable is given in the problem (see Exercises 75–78).

EXAMPLE 3 **Arc length of a parabola** Evaluate $\int_0^2 \sqrt{1 + 4x^2}\,dx$, the arc length of the segment of the parabola $y = x^2$ on $[0, 2]$.

SOLUTION Removing a factor of 4 from the square root, we have

$$\int_0^2 \sqrt{1 + 4x^2}\,dx = 2 \int_0^2 \sqrt{\tfrac{1}{4} + x^2}\,dx = 2 \int_0^2 \sqrt{\left(\tfrac{1}{2}\right)^2 + x^2}\,dx.$$

The integrand contains the expression $a^2 + x^2$, with $a = \tfrac{1}{2}$, which suggests the substitution $x = \tfrac{1}{2}\tan\theta$. It follows that $dx = \tfrac{1}{2}\sec^2\theta\,d\theta$, and

$$\sqrt{\left(\tfrac{1}{2}\right)^2 + x^2} = \sqrt{\left(\tfrac{1}{2}\right)^2 + \left(\tfrac{1}{2}\tan\theta\right)^2} = \frac{1}{2}\sqrt{\underbrace{1 + \tan^2\theta}_{\sec^2\theta}} = \frac{1}{2}\sec\theta.$$

> Because we are evaluating a definite integral, we could change the limits of integration to $\theta = 0$ and $\theta = \tan^{-1} 4$. However, $\tan^{-1} 4$ is not a standard angle, so it is easier to express the antiderivative in terms of x and use the original limits of integration.

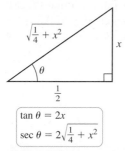

$$\tan\theta = 2x$$
$$\sec\theta = 2\sqrt{\tfrac{1}{4} + x^2}$$

FIGURE 7.8

Setting aside the limits of integration for the moment, we compute the antiderivative

$$2\int \sqrt{\left(\tfrac{1}{2}\right)^2 + x^2}\,dx = 2\int \frac{1}{2}\sec\theta \underbrace{\frac{1}{2}\sec^2\theta\,d\theta}_{dx} \qquad x = \frac{1}{2}\tan\theta,\, dx = \frac{1}{2}\sec^2\theta\,d\theta$$

$$= \frac{1}{2}\int \sec^3\theta\,d\theta \qquad\qquad \text{Simplify.}$$

$$= \frac{1}{4}(\sec\theta\tan\theta + \ln|\sec\theta + \tan\theta|) \qquad \begin{array}{l}\text{Reduction formula 4,}\\ \text{Section 7.2}\end{array}$$

Using a reference triangle (Figure 7.8), we express the antiderivative in terms of the original variable x and evaluate the definite integral:

$$2\int_0^2 \sqrt{\left(\tfrac{1}{2}\right)^2 + x^2}\,dx = \frac{1}{4}\left(\underbrace{2\sqrt{\tfrac{1}{4} + x^2}}_{\sec\theta}\underbrace{2x}_{\tan\theta} + \ln\left|\underbrace{2\sqrt{\tfrac{1}{4} + x^2}}_{\sec\theta} + \underbrace{2x}_{\tan\theta}\right|\right)\Bigg|_0^2$$

$$\tan\theta = 2x,\, \sec\theta = 2\sqrt{\tfrac{1}{4} + x^2}$$

$$= \frac{1}{4}\left(4\sqrt{17} + \ln\left(\sqrt{17} + 4\right)\right) \approx 4.65$$

Related Exercises 15–46 ◄

QUICK CHECK 3 The integral
$$\int \frac{dx}{a^2 + x^2} = \frac{1}{a}\tan^{-1}\frac{x}{a} + C \text{ was}$$
given in Section 4.8. Verify this result with the appropriate trigonometric substitution. ◄

EXAMPLE 4 **Another tangent substitution** Evaluate $\displaystyle\int \frac{dx}{(1 + x^2)^2}$.

SOLUTION The factor $1 + x^2$ suggests the substitution $x = \tan\theta$. It follows that $dx = \sec^2\theta\,d\theta$ and

$$(1 + x^2)^2 = (1 + \tan^2\theta)^2 = \sec^4\theta.$$

Substituting these factors leads to

$$\int \frac{dx}{(1 + x^2)^2} = \int \frac{\sec^2\theta}{\sec^4\theta}\,d\theta \qquad x = \tan\theta,\, dx = \sec^2\theta\,d\theta$$

$$= \int \cos^2\theta\,d\theta \qquad\qquad \text{Simplify.}$$

$$= \left(\frac{\theta}{2} + \frac{\sin 2\theta}{4}\right) + C \qquad \text{Integrate } \cos^2\theta = \frac{1 + \cos 2\theta}{2}.$$

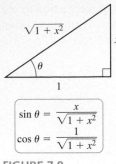

$$\sin \theta = \frac{x}{\sqrt{1 + x^2}}$$
$$\cos \theta = \frac{1}{\sqrt{1 + x^2}}$$

FIGURE 7.9

The final step is to return to the original variable x. The first term $\theta/2$ is replaced by $\frac{1}{2}\tan^{-1} x$. The second term involving $\sin 2\theta$ requires the identity $\sin 2\theta = 2 \sin \theta \cos \theta$. The reference triangle (Figure 7.9) tells us that

$$\frac{1}{4} \sin 2\theta = \frac{1}{2} \sin \theta \cos \theta = \frac{1}{2} \cdot \frac{x}{\sqrt{1 + x^2}} \cdot \frac{1}{\sqrt{1 + x^2}} = \frac{1}{2} \cdot \frac{x}{1 + x^2}.$$

The integration can now be completed:

$$\int \frac{dx}{(1 + x^2)^2} = \left(\frac{\theta}{2} + \frac{\sin 2\theta}{4} \right) + C$$

$$= \frac{1}{2} \tan^{-1} x + \frac{x}{2(1 + x^2)} + C$$

Related Exercises 15–46 ◀

EXAMPLE 5   **A secant substitution**   Evaluate $\displaystyle\int_1^4 \frac{\sqrt{x^2 + 4x - 5}}{x + 2} \, dx.$

> Recall that to complete the square with $x^2 + bx + c$, you add and subtract $(b/2)^2$ to the expression, and then factor to form a perfect square. You could also make the single substitution $x + 2 = 3 \sec \theta$ in Example 5.

SOLUTION   This example illustrates a useful preliminary step before making a trigonometric substitution. The integrand does not contain any of the patterns in Table 7.3 that suggest a trigonometric substitution. Completing the square does, however, lead to one of those patterns. Noting that $x^2 + 4x - 5 = (x + 2)^2 - 9$, we change variables with $u = x + 2$ and write the integral as

$$\int_1^4 \frac{\sqrt{x^2 + 4x - 5}}{x + 2} \, dx = \int_1^4 \frac{\sqrt{(x + 2)^2 - 9}}{x + 2} \, dx \qquad \text{Complete the square.}$$

$$= \int_3^6 \frac{\sqrt{u^2 - 9}}{u} \, du \qquad \begin{aligned} &u = x + 2, du = dx \\ &\text{Change limits of integration.} \end{aligned}$$

> The substitution $u = 3 \sec \theta$ can be rewritten as $\theta = \sec^{-1}(u/3)$. Because $u \geq 3$ in the integral $\displaystyle\int_3^6 \frac{\sqrt{u^2 - 9}}{u} \, du,$ we have $0 \leq \theta < \dfrac{\pi}{2}.$

This new integral calls for the secant substitution $u = 3 \sec \theta$ (where $0 \leq \theta < \pi/2$), which implies that $du = 3 \sec \theta \tan \theta \, d\theta$ and $\sqrt{u^2 - 9} = 3 \tan \theta$. We also change the limits of integration: When $u = 3, \theta = 0$, and when $u = 6, \theta = \pi/3$. The complete integration can now be done:

$$\int_1^4 \frac{\sqrt{x^2 + 4x - 5}}{x + 2} \, dx = \int_3^6 \frac{\sqrt{u^2 - 9}}{u} \, du \qquad u = x + 2, du = dx$$

$$= \int_0^{\pi/3} \frac{3 \tan \theta}{3 \sec \theta} \, 3 \sec \theta \tan \theta \, d\theta \qquad u = 3 \sec \theta, du = 3 \sec \theta \tan \theta \, d\theta$$

$$= 3 \int_0^{\pi/3} \tan^2 \theta \, d\theta \qquad \text{Simplify.}$$

$$= 3 \int_0^{\pi/3} (\sec^2 \theta - 1) \, d\theta \qquad \tan^2 \theta = \sec^2 \theta - 1$$

$$= 3 (\tan \theta - \theta) \Big|_0^{\pi/3} \qquad \text{Evaluate integrals.}$$

$$= 3\sqrt{3} - \pi \qquad \text{Simplify.}$$

Related Exercises 15–46 ◀

SECTION 7.3 EXERCISES

Review Questions

1. What change of variables is suggested by an integral containing $\sqrt{x^2 - 9}$?

2. What change of variables is suggested by an integral containing $\sqrt{x^2 + 36}$?

3. What change of variables is suggested by an integral containing $\sqrt{100 - x^2}$?

4. If $x = 4 \tan \theta$, express $\sin \theta$ in terms of x.

5. If $x = 2 \sin \theta$, express $\cot \theta$ in terms of x.

6. If $x = 8 \sec \theta$, express $\tan \theta$ in terms of x.

Basic Skills

7–10. *Evaluate the following integrals.*

7. $\displaystyle\int_0^{5/2} \frac{dx}{\sqrt{25 - x^2}}$

8. $\displaystyle\int_0^{3/2} \frac{dx}{(9 - x^2)^{3/2}}$

9. $\displaystyle\int_5^{10} \sqrt{100 - x^2}\, dx$

10. $\displaystyle\int_0^{\sqrt{2}} \frac{x^2}{\sqrt{4 - x^2}}\, dx$

11–14. *Evaluate the following integrals.*

11. $\displaystyle\int \frac{dx}{(16 - x^2)^{1/2}}$

12. $\displaystyle\int \sqrt{36 - x^2}\, dx$

13. $\displaystyle\int \frac{\sqrt{9 - x^2}}{x}\, dx$

14. $\displaystyle\int (36 - 9x^2)^{-3/2}\, dx$

15–40. *Evaluate the following integrals.*

15. $\displaystyle\int \sqrt{64 - x^2}\, dx$

16. $\displaystyle\int \frac{dx}{\sqrt{x^2 - 49}},\ x > 7$

17. $\displaystyle\int \frac{dx}{\sqrt{36 - x^2}}$

18. $\displaystyle\int \frac{dx}{\sqrt{16 + 4x^2}}$

19. $\displaystyle\int \frac{dx}{\sqrt{x^2 - 81}},\ x > 9$

20. $\displaystyle\int \frac{dx}{\sqrt{1 - 2x^2}}$

21. $\displaystyle\int \frac{dx}{(1 + 4x^2)^{3/2}}$

22. $\displaystyle\int \frac{dx}{(x^2 - 36)^{3/2}},\ x > 6$

23. $\displaystyle\int \frac{x^2}{\sqrt{16 - x^2}}\, dx$

24. $\displaystyle\int \frac{dx}{(81 + x^2)^2}$

25. $\displaystyle\int \frac{\sqrt{x^2 - 9}}{x}\, dx,\ x > 3$

26. $\displaystyle\int \sqrt{9 - 4x^2}\, dx$

27. $\displaystyle\int \frac{x^2}{\sqrt{4 + x^2}}\, dx$

28. $\displaystyle\int \frac{\sqrt{4x^2 - 1}}{x^2}\, dx,\ x > \frac{1}{2}$

29. $\displaystyle\int \frac{dx}{\sqrt{3 - 2x - x^2}}$

30. $\displaystyle\int \frac{x^4}{1 + x^2}\, dx$

31. $\displaystyle\int \frac{\sqrt{9x^2 - 25}}{x^3}\, dx,\ x > \frac{5}{3}$

32. $\displaystyle\int \frac{\sqrt{9 - x^2}}{x^2}\, dx$

33. $\displaystyle\int \frac{x^2}{(25 + x^2)^2}\, dx$

34. $\displaystyle\int \frac{dx}{x^2 \sqrt{9x^2 - 1}},\ x > \frac{1}{3}$

35. $\displaystyle\int \frac{x^2}{(100 - x^2)^{3/2}}\, dx$

36. $\displaystyle\int \frac{dx}{x^3 \sqrt{x^2 - 100}},\ x > 10$

37. $\displaystyle\int \frac{x^3}{(81 - x^2)^2}\, dx$

38. $\displaystyle\int \frac{dx}{x^3 \sqrt{x^2 - 1}},\ x > 1$

39. $\displaystyle\int \frac{dx}{x(x^2 - 1)^{3/2}},\ x > 1$

40. $\displaystyle\int \frac{x^3}{(x^2 - 16)^{3/2}}\, dx,\ x < -4$

41–46. Evaluating definite integrals *Evaluate the following definite integrals.*

41. $\displaystyle\int_0^1 \frac{dx}{\sqrt{x^2 + 16}}$

42. $\displaystyle\int_{8\sqrt{2}}^{16} \frac{dx}{\sqrt{x^2 - 64}}$

43. $\displaystyle\int_0^{1/3} \frac{dx}{(9x^2 + 1)^{3/2}}$

44. $\displaystyle\int_{10/\sqrt{3}}^{10} \frac{dx}{\sqrt{x^2 - 25}}$

45. $\displaystyle\int_{4/\sqrt{3}}^4 \frac{dx}{x^2(x^2 - 4)}$

46. $\displaystyle\int_6^{6\sqrt{3}} \frac{x^2}{(x^2 + 36)^2}\, dx$

Further Explorations

47. **Explain why or why not** Determine whether the following statements are true and give an explanation or counterexample.

   a. If $x = 4 \tan \theta$, then $\csc \theta = 4/x$.

   b. The integral $\int_1^2 \sqrt{1 - x^2}\, dx$ does not have a finite real value.

   c. The integral $\int_1^2 \sqrt{x^2 - 1}\, dx$ does not have a finite real value.

   d. The integral $\displaystyle\int \frac{dx}{x^2 + 4x + 9}$ cannot be evaluated using a trigonometric substitution.

48–55. Completing the square *Evaluate the following integrals.*

48. $\displaystyle\int \frac{dx}{x^2 - 2x + 10}$

49. $\displaystyle\int \frac{dx}{x^2 + 6x + 18}$

50. $\displaystyle\int \frac{dx}{2x^2 - 12x + 36}$

51. $\displaystyle\int \frac{x^2 - 2x + 1}{\sqrt{x^2 - 2x + 10}}\, dx$

52. $\displaystyle\int \frac{x^2 + 2x + 4}{\sqrt{x^2 - 4x}}\, dx,\ x > 4$

53. $\displaystyle\int \frac{x^2 - 8x + 16}{(9 + 8x - x^2)^{3/2}}\, dx$

54. $\displaystyle\int_1^4 \frac{dx}{x^2 - 2x + 10}$

55. $\displaystyle\int_{1/2}^{(\sqrt{2}+3)/(2\sqrt{2})} \frac{dx}{8x^2 - 8x + 11}$

56. Area of an ellipse The upper half of the ellipse centered at the origin with axes of length $2a$ and $2b$ is described by $y = \dfrac{b}{a}\sqrt{a^2 - x^2}$ (see figure). Find the area of the ellipse in terms of a and b.

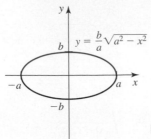

57. Area of a segment of a circle Use two approaches to show that the area of a cap (or segment) of a circle of radius r subtended by an angle θ (see figure) is given by

$$A_{\text{seg}} = \frac{1}{2}r^2(\theta - \sin\theta).$$

a. Find the area using geometry (no calculus).
b. Find the area using calculus.

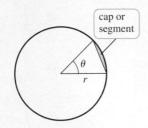

58. Area of a lune A lune is a crescent-shaped region bounded by the arcs of two circles. Let C_1 be a circle of radius 4 centered at the origin. Let C_2 be a circle of radius 3 centered at the point $(2, 0)$. Find the area of the lune (shaded in the figure) that lies inside C_1 and outside C_2.

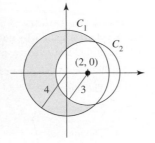

59. Area and volume Consider the function $f(x) = (9 + x^2)^{-1/2}$ and the region R on the interval $[0, 4]$ (see figure).

a. Find the area of R.
b. Find the volume of the solid generated when R is revolved about the x-axis.
c. Find the volume of the solid generated when R is revolved about the y-axis.

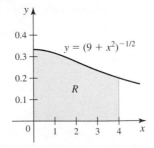

T 60. Area of a region Graph the function $f(x) = (16 + x^2)^{-3/2}$ and find the area of the region bounded by the curve and the x-axis on the interval $[0, 3]$.

61. Arc length of a parabola Find the length of the curve $y = ax^2$ from $x = 0$ to $x = 10$, where $a > 0$ is a real number.

62. Comparing areas On the interval $[0, 2]$, the graphs of $f(x) = x^2/3$ and $g(x) = x^2(9 - x^2)^{-1/2}$ have similar shapes.

a. Find the area of the region bounded by the graph of f and the x-axis on the interval $[0, 2]$.
b. Find the area of the region bounded by the graph of g and the x-axis on the interval $[0, 2]$.
c. Which region has the greater area?

T 63–65. Using the integral of $\sec^3 u$ By reduction formula 4 in Section 7.2,

$$\int \sec^3 u\, du = \frac{1}{2}\left(\sec u \tan u + \ln|\sec u + \tan u|\right) + C.$$

Graph the following functions and find the area under the curve on the given interval.

63. $f(x) = (9 - x^2)^{-2},\ \left[0, \frac{3}{2}\right]$ **64.** $f(x) = (4 + x^2)^{1/2},\ [0, 2]$

65. $f(x) = (x^2 - 25)^{1/2},\ [5, 10]$

66–67. Asymmetric integrands *Evaluate the following integrals. Consider completing the square.*

66. $\displaystyle\int \frac{dx}{\sqrt{(x - 1)(3 - x)}}$ **67.** $\displaystyle\int_{2+\sqrt{2}}^{4} \frac{dx}{\sqrt{(x - 1)(x - 3)}}$

68. Clever substitution Evaluate $\displaystyle\int \frac{dx}{1 + \sin x + \cos x}$ using the substitution $x = 2\tan^{-1}\theta$. The identities $\sin x = 2\sin\dfrac{x}{2}\cos\dfrac{x}{2}$ and $\cos x = \cos^2\dfrac{x}{2} - \sin^2\dfrac{x}{2}$ are helpful.

Applications

69. A torus (doughnut) Find the volume of the solid torus formed when the circle of radius 4 centered at $(0, 6)$ is revolved about the x-axis.

70. Bagel wars Bob and Bruce bake bagels (shaped like tori). They both make standard bagels that have an inner radius of 0.5 in and an outer radius of 2.5 in. Bob plans to increase the volume of his bagels by decreasing the inner radius by 20% (leaving the outer radius unchanged). Bruce plans to increase the volume of his bagels by increasing the outer radius by 20% (leaving the inner radius unchanged). Whose new bagels will have the greater volume? Does this result depend on the size of the original bagels? Explain.

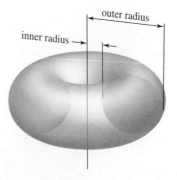

71. Electric field due to a line of charge A total charge of Q is distributed uniformly on a line segment of length $2L$ along the y-axis (see figure). The x-component of the electric field at a point $(a, 0)$ on the x-axis is given by

$$E_x(a) = \frac{kQa}{2L} \int_{-L}^{L} \frac{dy}{(a^2 + y^2)^{3/2}}$$

where k is a physical constant and $a > 0$.

a. Confirm that $E_x(a) = \dfrac{kQ}{a\sqrt{a^2 + L^2}}$.

b. Letting $\rho = Q/2L$ be the charge density on the line segment, show that if $L \to \infty$, then $E_x(a) = 2k\rho/a$.

(See the Guided Projects for a derivation of this and other similar integrals.)

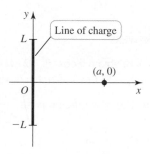

72. Magnetic field due to current in a straight wire A long straight wire of length $2L$ on the y-axis carries a current I. According to the Biot-Savart Law, the magnitude of the magnetic field due the current at a point $(a, 0)$ is given by

$$B(a) = \frac{\mu_0 I}{4\pi} \int_{-L}^{L} \frac{\sin\theta}{r^2} \, dy$$

where μ_0 is a physical constant, $a > 0$, and θ, r, and y are related as shown in the figure.

a. Show that the magnitude of the magnetic field at $(a, 0)$ is

$$B(a) = \frac{\mu_0 I L}{2\pi a\sqrt{a^2 + L^2}}.$$

b. What is the magnitude of the magnetic field at $(a, 0)$ due to an infinitely long wire $(L \to \infty)$?

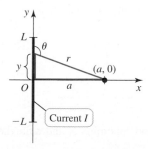

73. Fastest descent time The cycloid is the curve traced by a point on the rim of a rolling wheel. Imagine a wire shaped like an inverted cycloid (see figure). A bead sliding down this wire without friction has some remarkable properties. Among all wire shapes, the cycloid is the shape that produces the fastest descent time (see the Guided Projects for more about this *brachistochrone property*). It can be shown that the descent time between any two points $0 \le a \le b \le \pi$ on the curve is

$$\text{descent time} = \int_{a}^{b} \sqrt{\frac{1 - \cos t}{g(\cos a - \cos t)}} \, dt.$$

where g is the acceleration due to gravity, $t = 0$ corresponds to the top of the wire, and $t = \pi$ corresponds to the lowest point on the wire.

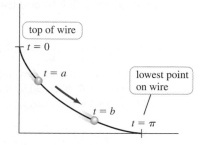

a. Find the descent time on the interval $[a, b]$ by making the substitution $u = \cos t$.

b. Show that when $b = \pi$, the descent time is the same for all values of a; that is, the descent time to the bottom of the wire is the same for all starting points.

74. Maximum path length of a projectile (Adapted from Putnam Exam 1940) A projectile is launched from the ground with an initial speed V at an angle θ from the horizontal. Assume that the x-axis is the horizontal ground and y is the height above the ground. Neglecting air resistance and letting g be the acceleration due to gravity, it can be shown that the trajectory of the projectile is given by

$$y = -\frac{1}{2}kx^2 + y_{max}, \quad \text{where} \quad k = \frac{g}{(V\cos\theta)^2}$$

$$\text{and} \quad y_{max} = \frac{(V\sin\theta)^2}{2g}.$$

a. Note that the high point of the trajectory occurs at $(0, y_{max})$. If the projectile is on the ground at $(-a, 0)$ and $(a, 0)$, what is a?

b. Show that the length of the trajectory (arc length) is $2\int_0^a \sqrt{1 + k^2x^2} \, dx$.

c. Evaluate the arc length integral and express your result in terms of V, g, and θ.

d. For a fixed value of V and g, show that the launch angle θ that maximizes the length of the trajectory satisfies $(\sin\theta)\ln(\sec\theta + \tan\theta) = 1$.

e. Use a graphing utility to approximate the optimal launch angle.

Additional Exercises

75–78. Care with the secant substitution *Recall that the substitution $x = a\sec\theta$ implies that $x \ge a$ (in which case $0 \le \theta < \pi/2$ and $\tan\theta \ge 0$) or $x \le -a$ (in which case $\pi/2 < \theta \le \pi$ and $\tan\theta \le 0$)*

75. Show that

$$\int \frac{dx}{x\sqrt{x^2-1}} = \begin{cases} \sec^{-1}x + C = \tan^{-1}\sqrt{x^2-1} + C & \text{if } x > 1 \\ -\sec^{-1}x + C = -\tan^{-1}\sqrt{x^2-1} + C & \text{if } x < -1 \end{cases}$$

76. Evaluate for $\int \frac{\sqrt{x^2-1}}{x^3}\,dx$ for $x > 1$ and for $x < -1$.

77. Graph the function $f(x) = \frac{\sqrt{x^2-9}}{x}$ and consider the region bounded by the curve and the x-axis on $[-6, -3]$. Then, evaluate $\int_{-6}^{-3} \frac{\sqrt{x^2-9}}{x}\,dx$. Be sure the result is consistent with the graph.

78. Graph the function $f(x) = \frac{1}{x\sqrt{x^2-36}}$ on its domain. Then, find the area of the region R_1 bounded by the curve and the x-axis on $[-12, -12/\sqrt{3}]$ and the region R_2 bounded by the curve and the x-axis on $[12/\sqrt{3}, 12]$. Be sure your results are consistent with the graph.

79. Visual Proof Let $F(x) = \int_0^x \sqrt{a^2-t^2}\,dt$. The figure shows that $F(x) = $ area of sector OAB + area of triangle OBC.

a. Use the figure to prove that $F(x) = $
$$\frac{a^2\sin^{-1}(x/a)}{2} + \frac{x\sqrt{a^2-x^2}}{2}.$$

b. Conclude that $\int \sqrt{a^2-x^2}\,dx = $
$$\frac{a^2\sin^{-1}(x/a)}{2} + \frac{x\sqrt{a^2-x^2}}{2} + C.$$

[*Source: The College Mathematics Journal* 34, no. 3 (May 2003)]

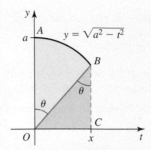

QUICK CHECK ANSWERS

1. Use $x = 3\sin\theta$ to obtain $9\cos^2\theta$. **2.** (a) Use $x = 3\tan\theta$. (b) Use $x = 4\sin\theta$. **3.** Let $x = a\tan\theta$, so that $dx = a\sec^2\theta\,d\theta$. The new integral is $\int \frac{a\sec^2\theta\,d\theta}{a^2(1+\tan^2\theta)} = \frac{1}{a}\int d\theta = \frac{1}{a}\theta + C = \frac{1}{a}\tan^{-1}\frac{x}{a} + C.$ ◄

7.4 Partial Fractions

> Recall that a rational function has the form p/q, where p and q are polynomials.

Later in this chapter, we will see that finding the velocity of a skydiver requires evaluating an integral of the form $\int \frac{dv}{a - bv^2}$, where a and b are constants. Similarly, finding the population of a species that is limited in size involves an integral of the form $\int \frac{dP}{aP(1 - bP)}$, where a and b are constants. These integrals have the common feature that their integrands are rational functions. Similar integrals result from modeling mechanical and electrical networks. The goal of this section is to introduce the *method of partial fractions* for integrating rational functions. When combined with standard and trigonometric substitutions, this method allows us (in principle) to integrate any rational function.

Method of Partial Fractions

Given a function such as

$$f(x) = \frac{1}{x-2} + \frac{2}{x+4},$$

it is a straightforward task to find a common denominator and write the equivalent expression

$$f(x) = \frac{(x+4) + 2(x-2)}{(x-2)(x+4)} = \frac{3x}{(x-2)(x+4)} = \frac{3x}{x^2 + 2x - 8}.$$

The purpose of partial fractions is to reverse this process. Given a rational function that is difficult to integrate, the method of partial fractions produces an equivalent function that is much easier to integrate.

$$\underset{\text{rational function}}{\frac{3x}{(x - 2)(x + 4)}} = \frac{3x}{x^2 + 2x - 8} \quad \xrightarrow[\text{partial fractions}]{\text{method of}} \quad \underset{\text{partial fraction decomposition}}{\frac{1}{x - 2} + \frac{2}{x + 4}}$$

Difficult to integrate: **Easy to integrate:**

$$\int \frac{3x}{x^2 + 2x - 8}\, dx \qquad\qquad \int \left(\frac{1}{x - 2} + \frac{2}{x + 4} \right) dx$$

QUICK CHECK 1 Find an antiderivative of $f(x) = \dfrac{1}{x - 2} + \dfrac{2}{x + 4}$. ◄

The Key Idea Working with the same function, $f(x) = \dfrac{3x}{(x - 2)(x + 4)}$, our objective is to write it in the form

$$\frac{A}{x - 2} + \frac{B}{x + 4}$$

> Notice that the numerator of the original rational function does not affect the form of the partial fraction decomposition. The constants A and B are called *undetermined coefficients*.

where A and B are constants to be determined. This expression is called the **partial fraction decomposition** of the original function; in this case, it has two terms, one for each factor in the denominator of the original function.

The constants A and B are determined using the condition that the original function f and its partial fraction decomposition must be equal for all values of x; that is,

$$\frac{3x}{(x - 2)(x + 4)} = \frac{A}{x - 2} + \frac{B}{x + 4}. \tag{1}$$

> This step requires that $x \neq 2$ and $x \neq -4$; both values are outside the domain of f.

Multiplying both sides of equation (1) by $(x - 2)(x + 4)$ gives

$$3x = A(x + 4) + B(x - 2).$$

Collecting like powers of x results in

$$3x = (A + B)x + (4A - 2B). \tag{2}$$

If equation (2) is to hold for all values of x, then

- the coefficients of x^1 on both sides of the equation must match;
- the coefficients of x^0 (that is, the constants) on both sides of the equation must match.

$$3x + 0 = \underbrace{(A + B)}x + \overbrace{(4A - 2B)}$$

This observation leads to two equations for A and B.

$$\text{Match coefficients of } x^1: \quad 3 = A + B$$
$$\text{Match coefficients of } x^0: \quad 0 = 4A - 2B$$

The first equation says that $A = 3 - B$. Substituting $A = 3 - B$ into the second equation gives the equation $0 = 4(3 - B) - 2B$. Solving for B, we find that $6B = 12$, or $B = 2$. The value of A now follows; we have $A = 3 - B = 1$.

Substituting these values of A and B into equation (1), the partial fraction decomposition is

$$\frac{3x}{(x - 2)(x + 4)} = \frac{1}{x - 2} + \frac{2}{x + 4}.$$

Simple Linear Factors

The previous example illustrates the case of **simple linear factors**; this means the denominator of the original function consists only of linear factors of the form $(x - r)$, which appear to the first power and no higher power. Here is the general procedure for this case.

> Like a fraction, a rational function is said to be in **reduced form** if the numerator and denominator have no common factors and it is said to be **proper** if the degree of the numerator is less than the degree of the denominator.

PROCEDURE   Partial Fractions with Simple Linear Factors

Suppose $f(x) = p(x)/q(x)$, where p and q are polynomials with no common factors and with the degree of p less than the degree of q. Assume that q is the product of simple linear factors. The partial fraction decomposition is obtained as follows.

Step 1. Factor the denominator q in the form $(x - r_1)(x - r_2)\cdots(x - r_n)$, where $r_1, \ldots, r_n$ are real numbers.

Step 2. Partial fraction decomposition Form the partial fraction decomposition by writing

$$\frac{p(x)}{q(x)} = \frac{A_1}{(x - r_1)} + \frac{A_2}{(x - r_2)} + \cdots + \frac{A_n}{(x - r_n)}.$$

> **QUICK CHECK 2** If the denominator of a rational function is $(x - 1)(x + 5)(x - 10)$, what is the general form of its partial fraction decomposition? ◄

Step 3. Clear denominators Multiply both sides of the equation in Step 2 by $q(x) = (x - r_1)(x - r_2)\cdots(x - r_n)$, which produces conditions for $A_1, \ldots, A_n$.

Step 4. Solve for coefficients Match like powers of x in Step 3 to solve for the undetermined coefficients $A_1, \ldots, A_n$.

EXAMPLE 1   Integrating with partial fractions

a. Find the partial fraction decomposition for $f(x) = \dfrac{3x^2 + 7x - 2}{x^3 - x^2 - 2x}$.

b. Evaluate $\int f(x)\, dx$.

SOLUTION

a. The partial fraction decomposition is done in four steps.

Step 1: Factoring the denominator, we find that

$$x^3 - x^2 - 2x = x(x + 1)(x - 2)$$

in which only simple linear factors appear.

Step 2: The partial fraction decomposition has one term for each factor in the denominator:

$$\frac{3x^2 + 7x - 2}{x(x + 1)(x - 2)} = \frac{A}{x} + \frac{B}{x + 1} + \frac{C}{x - 2} \qquad (3)$$

> You can call the undetermined coefficients $A_1, A_2, A_3, \ldots$ or $A, B, C, \ldots$. The latter may be preferable because it avoids subscripts.

The goal is to find the undetermined coefficients A, B, and C.

Step 3: We multiply both sides of equation (3) by $x(x + 1)(x - 2)$:

$$3x^2 + 7x - 2 = A(x + 1)(x - 2) + Bx(x - 2) + Cx(x + 1)$$
$$= (A + B + C)x^2 + (-A - 2B + C)x - 2A$$

Step 4: We now match coefficients of x^2, x^1, and x^0 on both sides of the equation in Step 3.

$$\text{Match coefficients of } x^2: \qquad A + B + C = 3$$
$$\text{Match coefficients of } x^1: \qquad -A - 2B + C = 7$$
$$\text{Match coefficients of } x^0: \qquad -2A = -2$$

The third equation implies that $A = 1$, which is substituted into the first two equations to give

$$B + C = 2 \quad \text{and} \quad -2B + C = 8$$

Solving for B and C, we conclude that $A = 1$, $B = -2$, and $C = 4$. Substituting the values of A, B, and C into equation (3), the partial fraction decomposition is

$$f(x) = \frac{1}{x} - \frac{2}{x+1} + \frac{4}{x-2}.$$

b. Integration is now straightforward:

$$\int \frac{3x^2 + 7x - 2}{x^3 - x^2 - 2x}\, dx = \int \left(\frac{1}{x} - \frac{2}{x+1} + \frac{4}{x-2} \right) dx \qquad \text{Partial fractions}$$

$$= \ln|x| - 2\ln|x+1| + 4\ln|x-2| + K \qquad \text{Integrate; arbitrary constant } K.$$

$$= \ln \frac{|x|(x-2)^4}{(x+1)^2} + K \qquad \text{Properties of logarithms}$$

Related Exercises 5–18 ◄

A Shortcut   Solving for more than three unknown coefficients in a partial fraction decomposition may be difficult. In the case of simple linear factors, a shortcut saves work. In Example 1, Step 3 led to the equation

$$3x^2 + 7x - 2 = A(x+1)(x-2) + Bx(x-2) + Cx(x+1).$$

> In cases other than simple linear factors, the shortcut can be used to determine some, but not all, of the coefficients, which reduces the work required to find the remaining coefficients.

Because this equation holds for *all* values of x, it must hold for any particular value of x. By choosing values of x judiciously, it is easy to solve for A, B, and C. For example, setting $x = 0$ in this equation results in $-2 = -2A$, or $A = 1$. Setting $x = -1$ results in $-6 = 3B$, or $B = -2$, and setting $x = 2$ results in $24 = 6C$, or $C = 4$. In each case, we choose a value of x that eliminates all but one term on the right side of the equation.

Repeated Linear Factors

> *Simple* means the factor is raised to the first power; *repeated* means the factor is raised to a power higher than the first power.

The preceding discussion relies on the assumption that the denominator of the rational function can be factored into simple linear factors of the form $(x - r)$. But what about denominators such as $x^2(x - 3)$, or $(x + 2)^2(x - 4)^3$, in which linear factors are raised to integer powers greater than 1? In these cases we have **repeated linear factors**, and a modification to the previous procedure must be made.

Here is the modification: Suppose the factor $(x - r)^m$ appears in the denominator, where $m > 1$ is an integer. Then there must be a partial fraction for each power of $(x - r)$ up to and including the mth power. For example, if $x^2(x - 3)^4$ appears in the denominator, then the partial fraction decomposition includes the terms

> Think of x^2 as the repeated linear factor $(x - 0)^2$.

$$\frac{A}{x} + \frac{B}{x^2} + \frac{C}{(x-3)} + \frac{D}{(x-3)^2} + \frac{E}{(x-3)^3} + \frac{F}{(x-3)^4}.$$

The rest of the partial fraction procedure remains the same, although the amount of work increases as the number of coefficients increases.

PROCEDURE   **Partial Fractions for Repeated Linear Factors**

Suppose the repeated linear factor $(x - r)^m$ appears in the denominator of a proper rational function in reduced form. The partial fraction decomposition has a partial fraction for each power of $(x - r)$ up to and including the mth power; that is, the partial fraction decomposition contains the sum

$$\frac{A_1}{(x-r)} + \frac{A_2}{(x-r)^2} + \frac{A_3}{(x-r)^3} + \cdots + \frac{A_m}{(x-r)^m}$$

where $A_1, \ldots, A_m$ are constants to be determined.

EXAMPLE 2 Integrating with repeated linear factors Evaluate $\int f(x)\, dx$, where
$$f(x) = \frac{5x^2 - 3x + 2}{x^3 - 2x^2}.$$

QUICK CHECK 3 State the form of the partial fraction decomposition of the rational function $p(x)/q(x)$ if $q(x) = x^2(x - 3)^2(x - 1)$. ◄

SOLUTION The denominator factors as $x^3 - 2x^2 = x^2(x - 2)$, so it has one simple linear factor $(x - 2)$ and one repeated linear factor x^2. The partial fraction decomposition has the form

$$\frac{5x^2 - 3x + 2}{x^2(x - 2)} = \frac{A}{x} + \frac{B}{x^2} + \frac{C}{(x - 2)}.$$

Multiplying both sides of the partial fraction decomposition by $x^2(x - 2)$, we find

$$5x^2 - 3x + 2 = Ax(x - 2) + B(x - 2) + Cx^2$$
$$= (A + C)x^2 + (-2A + B)x - 2B$$

The coefficients A, B, and C are determined by matching the coefficients of x^2, x^1, and x^0:

> The shortcut can be used to obtain two of the three coefficients easily. Choosing $x = 0$ allows B to be determined. Choosing $x = 2$ determines C. To find A, any other value of x may be substituted.

Match coefficients of x^2: $A + C = 5$

Match coefficients of x^1: $-2A + B = -3$

Match coefficients of x^0: $-2B = 2$

Solving these three equations in three unknowns results in the solution $A = 1$, $B = -1$, and $C = 4$. When A, B, and C are substituted, the partial fraction decomposition is

$$f(x) = \frac{1}{x} - \frac{1}{x^2} + \frac{4}{x - 2}.$$

Integration is now straightforward:

$$\int \frac{5x^2 - 3x + 2}{x^3 - 2x^2}\, dx = \int \left(\frac{1}{x} - \frac{1}{x^2} + \frac{4}{x - 2} \right) dx \qquad \text{Partial fractions}$$

$$= \ln|x| + \frac{1}{x} + 4\ln|x - 2| + K \qquad \text{Integrate; arbitrary constant } K.$$

$$= \frac{1}{x} + \ln\left(|x|(x - 2)^4\right) + K. \qquad \text{Properties of logarithms}$$

Related Exercises 19–25 ◄

Irreducible Quadratic Factors

It is a fact that a polynomial with real-valued coefficients can be written as the product of linear factors of the form $x - r$ and *irreducible quadratic factors* of the form $ax^2 + bx + c$, where $r, a, b,$ and c are real numbers. By irreducible, we mean that $ax^2 + bx + c$ cannot be factored further over the real numbers. For example, the polynomial

> The quadratic $ax^2 + bx + c$ has no real roots and cannot be factored over the real numbers if $b^2 - 4ac < 0$.

$$x^9 + 4x^8 + 6x^7 + 34x^6 + 64x^5 - 84x^4 - 287x^3 - 500x^2 - 354x - 180$$

factors as

$$\underbrace{(x - 2)}_{\substack{\text{linear} \\ \text{factor}}} \underbrace{(x + 3)^2}_{\substack{\text{repeated} \\ \text{linear} \\ \text{factor}}} \underbrace{(x^2 - 2x + 10)}_{\substack{\text{irreducible} \\ \text{quadratic} \\ \text{factor}}} \underbrace{(x^2 + x + 1)^2}_{\substack{\text{repeated} \\ \text{irreducible} \\ \text{quadratic factor}}}.$$

In this factored form, we see linear factors (simple and repeated) and irreducible quadratic factors (simple and repeated).

With irreducible quadratic factors, two cases must be considered: simple and repeated factors. Simple quadratic factors are examined in the following examples, and repeated quadratic factors (which generally involve long computations) are explored in the exercises.

PROCEDURE   **Partial Fractions with Simple Irreducible Quadratic Factors**

Suppose a simple irreducible factor $ax^2 + bx + c$ appears in the denominator of a proper rational function in reduced form. The partial fraction decomposition contains a term of the form

$$\frac{Ax + B}{ax^2 + bx + c}$$

where A and B are unknown coefficients to be determined.

EXAMPLE 3   Setting up partial fractions Give the appropriate form of the partial fraction decomposition for the following functions.

a. $\dfrac{x^2 + 1}{x^4 - 4x^3 - 32x^2}$

b. $\dfrac{10}{(x - 2)^2(x^2 + 2x + 2)}$

SOLUTION

a. The denominator factors as $x^2(x^2 - 4x - 32) = x^2(x - 8)(x + 4)$. Therefore, x is a repeated linear factor, and $(x - 8)$ and $(x + 4)$ are simple linear factors. The required form of the decomposition is

$$\frac{A}{x} + \frac{B}{x^2} + \frac{C}{x - 8} + \frac{D}{x + 4}.$$

We see that the factor $x^2 - 4x - 32$ is quadratic, but it can be further factored, so it is not irreducible.

b. The denominator is already fully factored. The quadratic factor $x^2 + 2x + 2$ cannot be factored further using real numbers; therefore, it is irreducible. The form of the decomposition is

$$\frac{A}{x - 2} + \frac{B}{(x - 2)^2} + \frac{Cx + D}{x^2 + 2x + 2}.$$

Related Exercises 26–29 ◄

EXAMPLE 4   Integrating with partial fractions Evaluate

$$\int \frac{7x^2 - 13x + 13}{(x - 2)(x^2 - 2x + 3)}\, dx.$$

SOLUTION The appropriate form of the partial fraction decomposition is

$$\frac{7x^2 - 13x + 13}{(x - 2)(x^2 - 2x + 3)} = \frac{A}{x - 2} + \frac{Bx + C}{x^2 - 2x + 3}.$$

Note that the irreducible quadratic factor requires $Bx + C$ in the numerator of the second fraction. Multiplying both sides of this equation by $(x - 2)(x^2 - 2x + 3)$ leads to

$$7x^2 - 13x + 13 = A(x^2 - 2x + 3) + (Bx + C)(x - 2)$$
$$= (A + B)x^2 + (-2A - 2B + C)x + (3A - 2C)$$

Matching coefficients of equal powers of x results in the equations

$$A + B = 7 \qquad -2A - 2B + C = -13 \qquad 3A - 2C = 13.$$

Solving this system of equations gives $A = 5$, $B = 2$, and $C = 1$; therefore, the original integral can be written as

$$\int \frac{7x^2 - 13x + 13}{(x - 2)(x^2 - 2x + 3)}\, dx = \int \frac{5}{x - 2}\, dx + \int \frac{2x + 1}{x^2 - 2x + 3}\, dx.$$

Let's work on the second (more difficult) integral. The substitution $u = x^2 - 2x + 3$ would work if $du = (2x - 2)\, dx$ appeared in the numerator. For this reason, we write the numerator as $2x + 1 = (2x - 2) + 3$ and split the integral:

$$\int \frac{2x + 1}{x^2 - 2x + 3}\, dx = \int \frac{2x - 2}{x^2 - 2x + 3}\, dx + \int \frac{3}{x^2 - 2x + 3}\, dx$$

Assembling all the pieces, we have

$$\int \frac{7x^2 - 13x + 13}{(x - 2)(x^2 - 2x + 3)}\, dx$$

$$= \int \frac{5}{x - 2}\, dx + \underbrace{\int \frac{2x - 2}{x^2 - 2x + 3}\, dx}_{\text{let } u = x^2 - 2x + 3} + \underbrace{\int \frac{3}{x^2 - 2x + 3}\, dx}_{(x - 1)^2 + 2}$$

$$= 5\ln|x - 2| + \ln|x^2 - 2x + 3| + \frac{3}{\sqrt{2}}\tan^{-1}\left(\frac{x - 1}{\sqrt{2}}\right) + C \quad \text{Integrate.}$$

$$= \ln|(x - 2)^5(x^2 - 2x + 3)| + \frac{3}{\sqrt{2}}\tan^{-1}\left(\frac{x - 1}{\sqrt{2}}\right) + C \quad \text{Property of logarithms}$$

To evaluate the last integral $\displaystyle\int \frac{3\, dx}{x^2 - 2x + 3}$, we completed the square in the denominator and used the substitution $u = x - 1$ to produce $\displaystyle\int \frac{3\, du}{u^2 + 2}$, which is a standard form.

Related Exercises 30–36 ◄

Final Note   The preceding discussion of partial fraction decomposition assumes that $f(x) = p(x)/q(x)$ is a proper rational function. If this is not the case and we are faced with an improper rational function f, we divide the denominator into the numerator and express f in two parts. One part will be a polynomial, and the other will be a proper rational function. For example, given the function

$$f(x) = \frac{2x^3 + 11x^2 + 28x + 33}{x^2 - x + 6}$$

we perform long division:

$$
\begin{array}{r}
2x + 13 \\
x^2 - x + 6 \overline{)\,2x^3 + 11x^2 + 28x + 33} \\
\underline{2x^3 - 2x^2 + 12x } \\
13x^2 + 16x + 33 \\
\underline{13x^2 - 13x + 78} \\
29x - 45
\end{array}
$$

It follows that

$$f(x) = \underbrace{2x + 13}_{\substack{\text{polynomial} \\ \text{easy to} \\ \text{integrate}}} + \underbrace{\frac{29x - 45}{x^2 - x + 6}}_{\substack{\text{apply partial fraction} \\ \text{decomposition}}}.$$

The first piece is easily integrated, and the second piece now qualifies for the methods described in this section.

SUMMARY Partial Fraction Decompositions

Let $f(x) = p(x)/q(x)$ be a proper rational function in reduced form. Assume the denominator q has been factored completely over the real numbers and m is a positive integer.

1. **Simple linear factor** A factor $x - r$ in the denominator requires the partial fraction $\dfrac{A}{x - r}$.

2. **Repeated linear factor** A factor $(x - r)^m$ with $m > 1$ in the denominator requires the partial fractions

$$\frac{A_1}{(x - r)} + \frac{A_2}{(x - r)^2} + \frac{A_3}{(x - r)^3} + \cdots + \frac{A_m}{(x - r)^m}.$$

3. **Simple irreducible quadratic factor** An irreducible factor $ax^2 + bx + c$ in the denominator requires the partial fraction

$$\frac{Ax + B}{ax^2 + bx + c}.$$

4. **Repeated irreducible quadratic factor** (See Exercises 67–70.) An irreducible factor $(ax^2 + bx + c)^m$ with $m > 1$ in the denominator requires the partial fractions

$$\frac{A_1x + B_1}{ax^2 + bx + c} + \frac{A_2x + B_2}{(ax^2 + bx + c)^2} + \cdots + \frac{A_mx + B_m}{(ax^2 + bx + c)^m}.$$

SECTION 7.4 EXERCISES

Review Questions

1. What kinds of functions can be integrated using partial fraction decomposition?

2. Give an example of each of the following.
   a. A simple linear factor
   b. A repeated linear factor
   c. A simple irreducible quadratic factor
   d. A repeated irreducible quadratic factor

3. What term(s) should appear in the partial fraction decomposition of a proper rational function with each of the following?
   a. A factor of $x - 3$ in the denominator
   b. A factor of $(x - 4)^3$ in the denominator
   c. A factor of $x^2 + 2x + 6$ in the denominator

4. What is the first step in integrating $\dfrac{x^2 + 2x - 3}{x + 1}$?

Basic Skills

5–8. Setting up partial fraction decomposition *Give the appropriate form of the partial fraction decomposition for the following functions.*

5. $\dfrac{2}{x^2 - 2x - 8}$

6. $\dfrac{x - 9}{x^2 - 3x - 18}$

7. $\dfrac{x^2}{x^3 - 16x}$

8. $\dfrac{x^2 - 3x}{x^3 - 3x^2 - 4x}$

9–18. Simple linear factors *Evaluate the following integrals.*

9. $\displaystyle\int \frac{dx}{(x - 1)(x + 2)}$

10. $\displaystyle\int \frac{8}{(x - 2)(x + 6)}\,dx$

11. $\displaystyle\int \frac{3}{x^2 - 1}\,dx$

12. $\displaystyle\int \frac{dt}{t^2 - 9}$

13. $\displaystyle\int \frac{2}{x^2 - x - 6}\,dx$

14. $\displaystyle\int \frac{3}{x^3 - x^2 - 12x}\,dx$

15. $\displaystyle\int \frac{dx}{x^2 - 2x - 24}$

16. $\displaystyle\int \frac{y + 1}{y^3 + 3y^2 - 18y}\,dy$

17. $\displaystyle\int \frac{1}{x^4 - 10x^2 + 9}\,dx$

18. $\displaystyle\int \frac{2}{x^2 - 4x - 32}\,dx$

19–25. Repeated linear factors *Evaluate the following integrals.*

19. $\displaystyle\int \frac{3}{x^3 - 9x^2}\,dx$

20. $\displaystyle\int \frac{x}{(x - 6)(x + 2)^2}\,dx$

21. $\displaystyle\int \frac{x}{(x + 3)^2}\,dx$

22. $\displaystyle\int \frac{dx}{x^3 - 2x^2 - 4x + 8}$

23. $\displaystyle\int \frac{2}{x^3 + x^2}\, dx$

24. $\displaystyle\int \frac{2}{t^3(t + 1)}\, dt$

25. $\displaystyle\int \frac{x - 5}{x^2(x + 1)}\, dx$

26–29. Setting up partial fraction decompositions *Give the appropriate form of the partial fraction decomposition for the following functions.*

26. $\displaystyle\frac{2}{x(x^2 - 6x + 9)}$

27. $\displaystyle\frac{20x}{(x - 1)^2(x^2 + 1)}$

28. $\displaystyle\frac{x^2}{x^3(x^2 + 1)}$

29. $\displaystyle\frac{2x^2 + 3}{(x^2 - 8x + 16)(x^2 + 3x + 4)}$

30–36. Simple irreducible quadratic factors *Evaluate the following integrals.*

30. $\displaystyle\int \frac{x^2 + 2}{x(x^2 + 5x + 8)}\, dx$

31. $\displaystyle\int \frac{2}{(x - 4)(x^2 + 2x + 6)}\, dx$

32. $\displaystyle\int \frac{z + 1}{z(z^2 + 4)}\, dz$

33. $\displaystyle\int \frac{x^2}{(x - 1)(x^2 + 4x + 5)}\, dx$

34. $\displaystyle\int \frac{2x + 1}{x^2 + 4}\, dx$

35. $\displaystyle\int \frac{x^2}{x^3 - x^2 + 4x - 4}\, dx$

36. $\displaystyle\int \frac{1}{(y^2 + 1)(y^2 + 2)}\, dy$

Further Explorations

37. Explain why or why not Determine whether the following statements are true and give an explanation or counterexample.

  a. To evaluate $\displaystyle\int \frac{4x^6}{x^4 + 3x^2}\, dx$, the first step is to find the partial fraction decomposition of the integrand.

  b. The easiest way to evaluate $\displaystyle\int \frac{6x + 1}{3x^2 + x}\, dx$ is with a partial fraction decomposition of the integrand.

  c. The rational function $f(x) = \dfrac{1}{x^2 - 13x + 42}$ has an irreducible quadratic denominator.

  d. The rational function $f(x) = \dfrac{1}{x^2 - 13x + 43}$ has an irreducible quadratic denominator.

38–41. Areas of regions *Find the area of the following regions. In each case, graph the relevant functions and show the region in question.*

38. The region bounded by the curve $y = x/(1 + x)$, the x-axis, and the line $x = 4$.

39. The region bounded by the curve $y = 10/(x^2 - 2x - 24)$, the x-axis, and the lines $x = -2$ and $x = 2$.

40. The region bounded by the curves $y = 1/x$, $y = x/(3x + 4)$, and the line $x = 10$.

41. The region bounded entirely by the curve $y = \dfrac{x^2 - 4x - 4}{x^2 - 4x - 5}$ and the x-axis.

42–47. Volumes of solids *Find the volume of the following solids.*

42. The region bounded by $y = 1/(x + 1)$, $y = 0$, $x = 0$, and $x = 2$ is revolved about the y-axis.

43. The region bounded by $y = x/(x + 1)$, the x-axis, and $x = 4$ is revolved about the x-axis.

44. The region bounded by $y = (1 - x^2)^{-1/2}$ and $y = 4$ is revolved about the x-axis.

45. The region bounded by $y = \dfrac{1}{\sqrt{x(3 - x)}}$, $y = 0$, $x = 1$, and $x = 2$ is revolved about the x-axis.

46. The region bounded by $y = \dfrac{1}{\sqrt{4 - x^2}}$, $y = 0$, $x = -1$, and $x = 1$ is revolved about the x-axis.

47. The region bounded by $y = 1/(x + 2)$, $y = 0$, $x = 0$, and $x = 3$ is revolved about the line $x = -1$.

48. What's wrong? Explain why the coefficients A and B cannot be found if we set

$$\frac{x^2}{(x - 4)(x + 5)} = \frac{A}{x - 4} + \frac{B}{x + 5}.$$

49–59. Preliminary steps *The following integrals require a preliminary step such as long division or a change of variables before using partial fractions. Evaluate these integrals.*

49. $\displaystyle\int \frac{dx}{1 + e^x}$

50. $\displaystyle\int \frac{x^4 + 1}{x^3 + 9x}\, dx$

51. $\displaystyle\int \frac{3x^2 + 4x - 6}{x^2 - 3x + 2}\, dx$

52. $\displaystyle\int \frac{2x^3 + x^2 - 6x + 7}{x^2 + x - 6}\, dx$

53. $\displaystyle\int \frac{dt}{2 + e^{-t}}$

54. $\displaystyle\int \frac{dx}{e^x + e^{2x}}$

55. $\displaystyle\int \frac{\sec \theta}{1 + \sin \theta}\, d\theta$

56. $\displaystyle\int \sqrt{e^x + 1}\, dx$

57. $\displaystyle\int \frac{e^x}{(e^x - 1)(e^x + 2)}\, dx$

58. $\displaystyle\int \frac{\cos x}{(\sin^3 x - 4 \sin x)}\, dx$

59. $\displaystyle\int \frac{dx}{(e^x + e^{-x})^2}$

60–65. Fractional powers *Use the indicated substitution to convert the given integral to an integral of a rational function. Evaluate the resulting integral.*

60. $\displaystyle\int \frac{dx}{x - \sqrt[3]{x}};\ x = u^3$

61. $\displaystyle\int \frac{dx}{\sqrt[4]{x + 2} + 1};\ x + 2 = u^4$

62. $\displaystyle\int \frac{dx}{x\sqrt{1 + 2x}};\ 1 + 2x = u^2$

63. $\displaystyle\int \frac{dx}{\sqrt{x} + \sqrt[3]{x}}; \ x = u^6$

64. $\displaystyle\int \frac{dx}{x - \sqrt[4]{x}}; \ x = u^4$

65. $\displaystyle\int \frac{dx}{\sqrt{1 + \sqrt{x}}}; \ x = (u^2 - 1)^2$

66. Arc length of the natural logarithm Consider the curve $y = \ln x$.

    a. Find the length of the curve from $x = 1$ to $x = a$ and call it $L(a)$. (*Hint:* The change of variables $u = \sqrt{x^2 + 1}$ allows evaluation by partial fractions.)

    b. Graph $L(a)$.

    c. As a increases, $L(a)$ increases as what power of a?

67–70. Repeated quadratic factors *Refer to the summary box on p. 483 and evaluate the following integrals.*

67. $\displaystyle\int \frac{2}{x(x^2 + 1)^2}\, dx$

68. $\displaystyle\int \frac{dx}{(x + 1)(x^2 + 2x + 2)^2}$

69. $\displaystyle\int \frac{x}{(x - 1)(x^2 + 2x + 2)^2}\, dx$

70. $\displaystyle\int \frac{x^3 + 1}{x(x^2 + x + 1)^2}\, dx$

71. Two methods Evaluate $\displaystyle\int \frac{dx}{x^2 - 1}$ for $x > 1$ in two ways: using partial fractions and a trigonometric substitution. Reconcile your two answers.

72–78. Rational functions of trigonometric functions *An integrand with trigonometric functions in the numerator and denominator can often be converted to a rational integrand using the substitution $u = \tan(x/2)$ or $x = 2\tan^{-1} u$. The following relations are used in making this change of variables.*

A: $\ dx = \dfrac{2}{1 + u^2}\, du$ B: $\ \sin x = \dfrac{2u}{1 + u^2}$ C: $\ \cos x = \dfrac{1 - u^2}{1 + u^2}$

72. Verify relation A by differentiating $x = 2\tan^{-1} u$. Verify relations B and C using a right-triangle diagram and the double-angle formulas

$$\sin x = 2\sin\left(\frac{x}{2}\right)\cos\left(\frac{x}{2}\right) \text{ and } \cos x = 2\cos^2\left(\frac{x}{2}\right) - 1.$$

73. Evaluate $\displaystyle\int \frac{dx}{1 + \sin x}$.

74. Evaluate $\displaystyle\int \frac{dx}{2 + \cos x}$.

75. Evaluate $\displaystyle\int \frac{dx}{1 - \cos x}$.

76. Evaluate $\displaystyle\int \frac{dx}{1 + \sin x + \cos x}$.

77. Evaluate $\displaystyle\int \frac{d\theta}{\cos \theta - \sin \theta}$.

78. Evaluate $\displaystyle\int \sec t \, dt$.

Applications

79. Three start-ups Three cars, A, B, and C, start from rest and accelerate along a line according to the following velocity functions:

$$v_A(t) = \frac{88t}{t + 1} \qquad v_B(t) = \frac{88t^2}{(t + 1)^2} \qquad v_C(t) = \frac{88t^2}{t^2 + 1}$$

    a. After $t = 1$ s, which car has traveled farthest?

    b. After $t = 5$ s, which car has traveled farthest?

    c. Find the position functions for the three cars assuming that all cars start at the origin.

    d. Which car ultimately gains the lead and remains in front?

80. Skydiving A skydiver has a downward velocity given by

$$v(t) = V\left(\frac{1 - e^{-2gt/V}}{1 + e^{-2gt/V}}\right),$$

where $t = 0$ is the instant the skydiver starts falling, $g \approx 9.8\,\text{m/s}^2$ is the acceleration due to gravity, and V is the terminal velocity of the skydiver.

    a. Evaluate $v(0)$ and $\displaystyle\lim_{t \to \infty} v(t)$ and interpret these results.

    b. Graph the velocity function.

    c. Verify by integration that the position function is given by

$$s(t) = Vt + \frac{V^2}{g}\ln\left(\frac{1 + e^{-2gt/V}}{2}\right)$$

    where $s'(t) = v(t)$ and $s(0) = 0$.

    d. Graph the position function.

(See the Guided Projects for more details on free fall and terminal velocity.)

Additional Exercises

81. $\pi < \dfrac{22}{7}$ One of the earliest approximations to π is $\dfrac{22}{7}$. Verify that $0 < \displaystyle\int_0^1 \frac{x^4(1 - x)^4}{1 + x^2}\, dx = \frac{22}{7} - \pi$. Why can you conclude that $\pi < \dfrac{22}{7}$?

82. Challenge Show that with the change of variables $u = \sqrt{\tan x}$, the integral $\int \sqrt{\tan x}\, dx$ can be converted to an integral amenable to partial fractions. Evaluate $\int_0^{\pi/4} \sqrt{\tan x}\, dx$.

QUICK CHECK ANSWERS

1. $\ln|x - 2| + 2\ln|x + 4| = \ln|(x - 2)(x + 4)^2|$
2. $A/(x - 1) + B/(x + 5) + C/(x - 10)$
3. $A/x + B/x^2 + C/(x - 3) + D/(x - 3)^2 + E/(x - 1)$ ◄

7.5 Other Integration Strategies

The integration methods studied so far—various substitutions, integration by parts, and partial fractions—are examples of *analytical methods*; they are done with pencil and paper and they give exact results. While many important integrals can be evaluated with analytical methods, many more integrals lie beyond their reach. For example, the following integrals cannot be evaluated in terms of familiar functions:

$$\int e^{x^2}\,dx \qquad \int \sin(x^2)\,dx \qquad \int \frac{\sin x}{x}\,dx \qquad \int \frac{e^{-x}}{x}\,dx \qquad \int \ln(\ln x)\,dx$$

The next two sections survey alternative strategies for evaluating integrals when standard analytical methods do not work. These strategies fall into three categories.

1. **Tables of integrals** The endpapers of this text contain a table of many standard integrals. Because these integrals were evaluated analytically, using tables is considered an analytical method. Tables of integrals also contain reduction formulas like those discussed in Sections 7.1 and 7.2.

2. **Computer algebra systems** Computer algebra systems have elaborate sets of rules to evaluate difficult integrals. Many definite and indefinite integrals can be evaluated exactly with such systems.

3. **Numerical methods** The value of a definite integral can be approximated accurately using numerical methods introduced in the next section. *Numerical* means that these methods compute numbers rather than manipulate symbols. Computers and calculators often have built-in functions to carry out these calculations.

Figure 7.10 is a chart of the various integration strategies and how they are related.

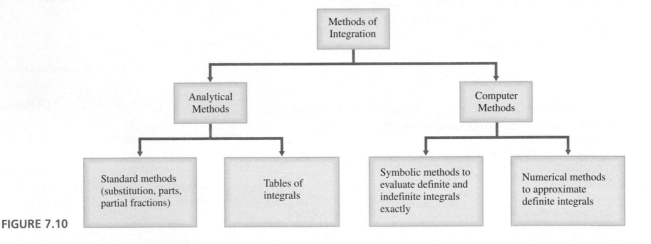

FIGURE 7.10

> A short table of integrals can be found at the end of the book. Longer tables of integrals are found online and in venerable collections such as the *CRC Mathematical Tables* and *Handbook of Mathematical Functions,* by Abramowitz and Stegun.

Using Tables of Integrals

Given a specific integral, you *may* be able to find the identical integral in a table of integrals. More likely, some preliminary work is needed to convert the given integral into one that appears in a table. Most tables give only indefinite integrals, although some tables include special definite integrals. The following examples illustrate various ways in which tables of integrals are used.

EXAMPLE 1 **Using tables of integrals** Evaluate the integral $\int \dfrac{dx}{x\sqrt{2x-9}}$.

SOLUTION It is worth noting that this integral may be evaluated with the change of variables $u^2 = 2x - 9$. Alternatively, a table of integrals includes the integral

$$\int \frac{dx}{x\sqrt{ax-b}} = \frac{2}{\sqrt{b}}\tan^{-1}\sqrt{\frac{ax-b}{b}} + C, \quad \text{where} \quad b > 0,$$

which matches the given integral. Letting $a = 2$ and $b = 9$, we find that

$$\int \frac{dx}{x\sqrt{2x-9}} = \frac{2}{\sqrt{9}}\tan^{-1}\sqrt{\frac{2x-9}{9}} + C = \frac{2}{3}\tan^{-1}\frac{\sqrt{2x-9}}{3} + C.$$

Related Exercises 5–20 ◄

> Letting $u^2 = 2x - 9$, we have $u\, du = dx$ and $x = \frac{1}{2}(u^2 + 9)$. Therefore,
> $$\int \frac{dx}{x\sqrt{2x-9}} = 2\int \frac{du}{u^2 + 9}.$$

EXAMPLE 2 **Preliminary work** Evaluate $\int \sqrt{x^2 + 6x}\, dx$.

SOLUTION Most tables of integrals do not include this integral. The nearest integral you are likely to find is $\int \sqrt{x^2 \pm a^2}\, dx$. The given integral can be put into this form by completing the square and using a substitution:

$$x^2 + 6x = x^2 + 6x + 9 - 9 = (x+3)^2 - 9$$

With the change of variables $u = x + 3$, the evaluation appears as follows:

$$
\begin{aligned}
\int \sqrt{x^2 + 6x}\, dx &= \int \sqrt{(x+3)^2 - 9}\, dx && \text{Complete the square.} \\[2mm]
&= \int \sqrt{u^2 - 9}\, du && u = x + 3,\, du = dx \\[2mm]
&= \frac{u}{2}\sqrt{u^2 - 9} - \frac{9}{2}\ln\left|u + \sqrt{u^2 - 9}\right| + C && \text{Table of integrals} \\[2mm]
&= \frac{x+3}{2}\sqrt{(x+3)^2 - 9} - \frac{9}{2}\ln\left|x + 3 + \sqrt{(x+3)^2 - 9}\right| + C \\[2mm]
&= \frac{x+3}{2}\sqrt{x^2 + 6x} - \frac{9}{2}\ln\left|x + 3 + \sqrt{x^2 + 6x}\right| + C.
\end{aligned}
$$

Related Exercises 21–32 ◄

EXAMPLE 3 **Using tables of integrals for area** Find the area of the region bounded by the curve $y = \dfrac{1}{1 + \sin x}$ and the x-axis between $x = 0$ and $x = \pi$.

SOLUTION The region in question (Figure 7.11) lies entirely above the x-axis, so its area is $\int_0^\pi \dfrac{dx}{1 + \sin x}$. A matching integral in a table of integrals is

$$\int \frac{dx}{1 + \sin ax} = -\frac{1}{a}\tan\left(\frac{\pi}{4} - \frac{ax}{2}\right) + C.$$

Evaluating the definite integral with $a = 1$, we have

$$\int_0^\pi \frac{dx}{1 + \sin x} = -\tan\left(\frac{\pi}{4} - \frac{x}{2}\right)\Big|_0^\pi = -\tan\left(-\frac{\pi}{4}\right) - \left(-\tan\frac{\pi}{4}\right) = 2.$$

Related Exercises 33–40 ◄

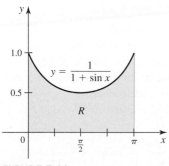

FIGURE 7.11

QUICK CHECK 1 Use the result of Example 3 to evaluate $\displaystyle\int_0^{\pi/2} \dfrac{dx}{1 + \sin x}$. ◄

Using a Computer Algebra System

Computer algebra systems evaluate many integrals exactly using symbolic methods, and they approximate many definite integrals using numerical methods. Different software packages may produce different results for the same indefinite integral; but, ultimately, they must agree. The discussion that follows does not rely on one particular computer algebra system. Rather, it illustrates results from different systems and shows some of the idiosyncrasies of using a computer algebra system.

> **QUICK CHECK 2** Using one computer algebra system, it was found that $\int \sin x \cos x \, dx = \frac{1}{2} \sin^2 x + C$; using another computer algebra system, it was found that $\int \sin x \cos x \, dx = -\frac{1}{2} \cos^2 x + C$. Reconcile the two answers. ◄

▶ Most computer algebra systems do not include the constant of integration after evaluating an indefinite integral. But, it should always be included when reporting the result.

EXAMPLE 4   Apparent discrepancies Evaluate $\int \dfrac{dx}{\sqrt{e^x + 1}}$ using tables and a computer algebra system.

SOLUTION Using one particular computer algebra system, we find that

$$\int \frac{dx}{\sqrt{e^x + 1}} = -2 \tanh^{-1}\left(\sqrt{e^x + 1}\right) + C,$$

▶ The *hyperbolic tangent* is defined as

$$\tanh x = \frac{e^x - e^{-x}}{e^x + e^{-x}}.$$

Its inverse is the *inverse hyperbolic tangent*, written $\tanh^{-1} x$.

where $\tanh^{-1}$ is the *inverse hyperbolic tangent* function. However, we can obtain a result in terms of more familiar functions by first using the substitution $u = e^x$, which implies that $du = e^x \, dx$ or $dx = du/e^x = du/u$. The integral becomes

$$\int \frac{dx}{\sqrt{e^x + 1}} = \int \frac{du}{u\sqrt{u + 1}}.$$

Using a computer algebra system again, we obtain

$$\int \frac{dx}{\sqrt{e^x + 1}} = \int \frac{du}{u\sqrt{u + 1}} = \ln\left(\sqrt{1 + u} - 1\right) - \ln\left(\sqrt{1 + u} + 1\right)$$

$$= \ln\left(\sqrt{1 + e^x} - 1\right) - \ln\left(\sqrt{1 + e^x} + 1\right).$$

A table of integrals leads to a third equivalent form of the integral:

$$\int \frac{dx}{\sqrt{e^x + 1}} = \int \frac{du}{u\sqrt{u + 1}} = \ln\left(\frac{\sqrt{u + 1} - 1}{\sqrt{u + 1} + 1}\right) + C$$

$$= \ln\left(\frac{\sqrt{e^x + 1} - 1}{\sqrt{e^x + 1} + 1}\right) + C$$

Often, the difference between two results is a few steps of algebra or a trigonometric identity. In this case, the final two results are reconciled using logarithm properties. This example illustrates that computer algebra systems generally do not include constants of integration and may omit absolute values when logarithms appear. It is important for the user to determine whether integration constants and absolute values are needed.

Related Exercises 41–56 ◄

> **QUICK CHECK 3** Using partial fractions, we know that $\int \dfrac{dx}{x(x + 1)} = \ln\left|\dfrac{x}{x + 1}\right| + C$.
> Using a computer algebra system, we find that $\int \dfrac{dx}{x(x + 1)} = \ln x - \ln(x + 1)$. What is wrong with the result from the computer algebra system? ◄

EXAMPLE 5 Symbolic vs. numerical integration Use a computer algebra system to evaluate $\int_0^1 \sin(x^2)\, dx$.

SOLUTION Sometimes a computer algebra system gives the exact value of an integral in terms of an unfamiliar function, or it may not be able to evaluate the integral exactly. For example, one particular computer algebra system returns the result

$$\int_0^1 \sin(x^2)\, dx = \sqrt{\frac{\pi}{2}}\, S\left(\sqrt{\frac{2}{\pi}}\right)$$

where S is a function called the *Fresnel integral function*

$\left(S(x) = \int_0^x \sin\left(\frac{\pi t^2}{2}\right) dt \right)$. However, if the computer algebra system is instructed to compute an approximate solution, the result is

$$\int_0^1 \sin(x^2)\, dx \approx 0.3102683017,$$

which is an excellent approximation.   *Related Exercises 41–56* ◄

SECTION 7.5 EXERCISES

Review Questions

1. Give some examples of analytical methods for evaluating integrals.

2. Does a computer algebra system give an exact result for an indefinite integral? Explain.

3. Why might an integral found in a table differ from the same integral evaluated by a computer algebra system?

4. Is a reduction formula an analytical method or a numerical method? Explain.

Basic Skills

5–20. Table lookup integrals *Use a table of integrals to evaluate the following indefinite integrals.*

5. $\int \dfrac{dx}{\sqrt{x^2 + 16}}$

6. $\int \dfrac{dx}{\sqrt{x^2 - 25}}$

7. $\int \dfrac{3u}{2u + 7}\, du$

8. $\int \dfrac{dy}{y(2y + 9)}$

9. $\int \dfrac{dx}{1 - \cos 4x}$

10. $\int \dfrac{dx}{x\sqrt{81 - x^2}}$

11. $\int \dfrac{dx}{\sqrt{4x + 1}}$

12. $\int \sqrt{4x + 12}\, dx$

13. $\int \dfrac{dx}{\sqrt{9x^2 - 100}}$

14. $\int \dfrac{dx}{225 - 16x^2}$

15. $\int \dfrac{dx}{(16 + 9x^2)^{3/2}}$

16. $\int \sqrt{4x^2 - 9}\, dx$

17. $\int \dfrac{dx}{x\sqrt{144 - x^2}}$

18. $\int \dfrac{dx}{x(x^3 + 8)}$

19. $\int \dfrac{dx}{x(x^{10} + 1)}$

20. $\int \dfrac{dx}{x(x^8 - 256)}$

21–32. Preliminary work *Use a table of integrals to evaluate the following indefinite integrals. These integrals require preliminary work, such as completing the square or changing variables, before they can be found in a table.*

21. $\int \dfrac{dx}{x^2 + 2x + 10}$

22. $\int \sqrt{x^2 - 4x + 8}\, dx$

23. $\int \dfrac{dx}{\sqrt{x^2 - 6x}}$

24. $\int \dfrac{dx}{\sqrt{x^2 + 10x}}$

25. $\int \dfrac{e^x}{\sqrt{e^{2x} + 4}}\, dx$

26. $\int \dfrac{\sqrt{\ln^2 x + 4}}{x}\, dx$

27. $\int \dfrac{\cos x}{\sin^2 x + 2 \sin x}\, dx$

28. $\int \dfrac{\cos^{-1}\sqrt{x}}{\sqrt{x}}\, dx$

29. $\int \dfrac{\tan^{-1} x^3}{x^4}\, dx$

30. $\int \dfrac{e^t}{\sqrt{3 + 4e^t}}\, dt$

31. $\int \dfrac{\ln x \sin^{-1}(\ln x)}{x}\, dx$

32. $\int \dfrac{dt}{\sqrt{1 + 4e^t}}$

33–40. Geometry problems *Use a table of integrals to solve the following problems.*

33. Find the length of the curve $y = x^2/4$ on the interval $[0, 8]$.

34. Find the length of the curve $y = x^{3/2} + 8$ on the interval $[0, 2]$.

35. Find the length of the curve $y = e^x$ on the interval $[0, \ln 2]$.

36. The region bounded by the graph of $y = 1/(x + 10)$ and the x-axis on the interval $[0, 3]$ is revolved about the x-axis. What is the volume of the solid that is formed?

37. The region bounded by the graph of $y = \dfrac{1}{\sqrt{x+4}}$ and the x-axis on the interval $[0, 12]$ is revolved about the y-axis. What is the volume of the solid that is formed?

38. Find the area of the region bounded by the graph of $y = \dfrac{1}{\sqrt{x^2 - 2x + 2}}$ and the x-axis between $x = 0$ and $x = 3$.

39. The region bounded by the graphs of $y = \pi/2$, $y = \sin^{-1} x$, and the y-axis is revolved about the y-axis. What is the volume of the solid that is formed?

40. The graphs of $f(x) = \dfrac{2}{x^2 + 1}$ and $g(x) = \dfrac{7}{4\sqrt{x^2 + 1}}$ are shown in the accompanying figure. Which is greater, the average value of f or that of g on the interval $[-1, 1]$?

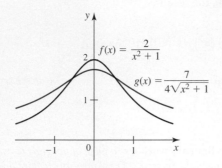

41–48. Indefinite integrals *Use a computer algebra system to evaluate the following indefinite integrals. Assume that a is a positive real number.*

41. $\displaystyle\int \dfrac{x}{\sqrt{2x+3}}\, dx$ **42.** $\displaystyle\int \sqrt{4x^2 + 36}\, dx$ **43.** $\displaystyle\int \tan^2 3x\, dx$

44. $\displaystyle\int (a^2 - x^2)^{-2}\, dx$ **45.** $\displaystyle\int \dfrac{(x^2 - a^2)^{3/2}}{x}\, dx$ **46.** $\displaystyle\int \dfrac{dx}{x(a^2 - x^2)^2}$

47. $\displaystyle\int (a^2 - x^2)^{3/2}\, dx$ **48.** $\displaystyle\int (x^2 + a^2)^{-5/2}\, dx$

49–56. Definite integrals *Use a computer algebra system to evaluate the following definite integrals. In each case, find an exact value of the integral (obtained by a symbolic method) and find an approximate value (obtained by a numerical method). Compare the results.*

49. $\displaystyle\int_{2/3}^{4/5} x^8\, dx$ **50.** $\displaystyle\int_0^{\pi/2} \cos^6 x\, dx$

51. $\displaystyle\int_0^4 (9 + x^2)^{3/2}\, dx$ **52.** $\displaystyle\int_{1/2}^1 \dfrac{\sin^{-1} x}{x}\, dx$

53. $\displaystyle\int_0^{\pi/2} \dfrac{dx}{1 + \tan^2 x}$ **54.** $\displaystyle\int_0^{2\pi} \dfrac{dx}{(4 + 2\sin x)^2}$

55. $\displaystyle\int_0^1 \ln x \ln (1 + x)\, dx$ **56.** $\displaystyle\int_0^{\pi/4} \ln (1 + \tan x)\, dx$

Further Explorations

57. Explain why or why not Determine whether the following statements are true and give an explanation or counterexample.

  a. It is possible that a computer algebra system says
  $$\int \dfrac{dx}{x(x-1)} = \ln(x-1) - \ln x$$ and a table of integrals says
  $$\int \dfrac{dx}{x(x-1)} = \ln\left|\dfrac{x-1}{x}\right| + C.$$

  b. A computer algebra system working in symbolic mode could give the result $\int_0^1 x^8\, dx = \frac{1}{9}$, and a computer algebra system working in approximate (numerical) mode could give the result $\int_0^1 x^8\, dx = 0.11111111$.

58. Apparent discrepancy Three different computer algebra systems give the following results:
$$\int \dfrac{dx}{x\sqrt{x^4 - 1}} = \dfrac{1}{2}\cos^{-1}\sqrt{x^{-4}} = \dfrac{1}{2}\cos^{-1} x^{-2} = \dfrac{1}{2}\tan^{-1}\sqrt{x^4 - 1}.$$

Explain how they can all be correct.

59. Reconciling results Using one computer algebra system, it was found that $\displaystyle\int \dfrac{dx}{1 + \sin x} = \dfrac{\sin x - 1}{\cos x}$ and using another computer algebra system, it was found that $\displaystyle\int \dfrac{dx}{1 + \sin x} = \dfrac{2\sin(x/2)}{\cos(x/2) + \sin(x/2)}$. Reconcile the two answers.

60. Apparent discrepancy Resolve the apparent discrepancy between
$$\int \dfrac{dx}{x(x-1)(x+2)} = \dfrac{1}{6}\ln\dfrac{(x-1)^2|x+2|}{|x|^3} + C \quad \text{and}$$
$$\int \dfrac{dx}{x(x-1)(x+2)} = \dfrac{\ln|x-1|}{3} + \dfrac{\ln|x+2|}{6} - \dfrac{\ln|x|}{2} + C$$

61–64. Reduction formulas *Use the reduction formulas in a table of integrals to evaluate the following integrals.*

61. $\displaystyle\int x^3 e^{2x}\, dx$ **62.** $\displaystyle\int x^2 e^{-3x}\, dx$

63. $\displaystyle\int \tan^4 3y\, dy$ **64.** $\displaystyle\int \sec^4 4x\, dx$

65–70. Double table lookup *The following integrals may require more than one table lookup. Evaluate the integrals using a table of integrals, then check your answer with a computer algebra system. When the parameter a appears, assume a > 0.*

65. $\displaystyle\int x\sin^{-1} 2x\, dx$ **66.** $\displaystyle\int 4x\cos^{-1} 10x\, dx$ **67.** $\displaystyle\int \dfrac{\tan^{-1} x}{x^2}\, dx$

68. $\displaystyle\int \dfrac{\sin^{-1} ax}{x^2}\, dx$ **69.** $\displaystyle\int \dfrac{dx}{\sqrt{2ax - x^2}}$ **70.** $\int \sqrt{2ax - x^2}\, dx$

Applications

71. Period of a pendulum Consider a pendulum with a length of L meters swinging only under the influence of gravity. Suppose the pendulum starts swinging with an initial displacement of

θ_0 radians (see figure). The period (time to complete one full cycle) is given by

$$T = \frac{4}{\omega} \int_0^{\pi/2} \frac{d\varphi}{\sqrt{1 - k^2 \sin^2 \varphi}}$$

where $\omega^2 = g/L$, $g \approx 9.8 \text{ m/s}^2$ is the acceleration due to gravity, and $k^2 = \sin^2(\theta_0/2)$. Assume $L = 9.8 \text{ m}$, which means $\omega = 1 \text{ s}^{-1}$.

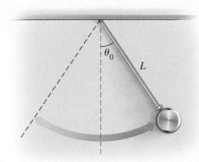

a. Use a computer algebra system to find the period of the pendulum for $\theta_0 = 0.1, 0.2, \ldots, 0.9, 1.0$ rad.
b. For small values of θ_0, the period should be approximately 2π s. For what values of θ_0 are your computed values within 10% of 2π (relative error less than 0.1)?

Additional Exercises

▮ 72. **Arc length of a parabola** Let $L(c)$ be the length of the parabola $f(x) = x^2$ from $x = 0$ to $x = c$, where $c \geq 0$ is a constant.

a. Find an expression for L and graph the function.
b. Is L concave up or concave down on $[0, \infty)$?
c. Show that as c becomes large and positive, the arc length function increases as c^2; that is, $L(c) \approx kc^2$, where k is a constant.

73–76. Deriving formulas *Evaluate the following integrals. Assume a and b are real numbers and n is an integer.*

73. $\displaystyle\int \frac{x}{ax + b}\, dx;$ Use $u = ax + b$.

74. $\displaystyle\int \frac{x}{\sqrt{ax + b}}\, dx;$ Use $u^2 = ax + b$.

75. $\int x(ax + b)^n\, dx;$ Use $u = ax + b$.

76. $\int x^n \sin^{-1} x\, dx;$ Use integration by parts.

▮ 77. **Powers of sine and cosine** It can be shown that

$$\int_0^{\pi/2} \sin^n x\, dx = \int_0^{\pi/2} \cos^n x\, dx =$$

$$\begin{cases} \dfrac{1 \cdot 3 \cdot 5 \cdots (n - 1)}{2 \cdot 4 \cdot 6 \cdots n} \cdot \dfrac{\pi}{2} & \text{if } n \geq 2 \text{ is an even integer} \\[2mm] \dfrac{2 \cdot 4 \cdot 6 \cdots (n - 1)}{3 \cdot 5 \cdot 7 \cdots n} & \text{if } n \geq 3 \text{ is an odd integer} \end{cases}$$

a. Use a computer algebra system to confirm this result for $n = 2, 3, 4$, and 5.
b. Evaluate the integrals with $n = 10$ and confirm the result.
c. Using graphing and/or symbolic computation, determine whether the values of the integrals increase or decrease as n increases.

▮ 78. **A remarkable integral** It is a fact that $\displaystyle\int_0^{\pi/2} \frac{dx}{1 + \tan^m x} = \frac{\pi}{4}$ for *all* real numbers m.

a. Graph the integrand for $m = -2, -3/2, -1, -1/2, 0, 1/2, 1, 3/2$, and 2, and explain geometrically how the area under the curve on the interval $[0, \pi/2]$ remains constant as m varies.
b. Use a computer algebra system to confirm that the integral is constant for all m.

QUICK CHECK ANSWERS

1. 1   **2.** Because $\sin^2 x = 1 - \cos^2 x$, the two results differ by a constant, which can be absorbed in the arbitrary constant C.   **3.** The second result agrees with the first for $x > 0$ after using $\ln a - \ln b = \ln(a/b)$. The second result should have absolute values and an arbitrary constant. ◄

7.6 Numerical Integration

Situations arise in which the analytical methods we have developed so far cannot be used to evaluate a definite integral. For example, an integrand may not have an obvious antiderivative (such as $\cos(x^2)$ and $1/\ln x$), or perhaps the integrand is represented by individual data points, which makes finding an antiderivative impossible.

When analytical methods fail, we often turn to *numerical methods*, which are typically done on a calculator or computer. These methods do not produce exact values of definite integrals, but they provide approximations that are generally quite accurate. Many calculators, software packages, and computer algebra systems have built-in numerical integration methods. In this section, we explore some of these methods.

Absolute and Relative Error

Because numerical methods do not typically produce exact results, we should be concerned about the accuracy of approximations, which leads to the ideas of *absolute* and *relative error.*

DEFINITIONS   Absolute and Relative Error

Suppose c is a computed numerical solution to a problem having an exact solution x. There are two common measures of the error in c as an approximation to x:

$$\textbf{absolute error} = |c - x|$$

and

$$\textbf{relative error} = \frac{|c - x|}{|x|} \quad (\text{if } x \neq 0)$$

▷ Because the exact solution is usually not known, the goal in practice is to estimate the maximum size of the error.

EXAMPLE 1   Absolute and relative error The ancient Greeks used $\frac{22}{7}$ to approximate the value of π. Determine the absolute and relative error in this approximation to π.

SOLUTION Letting $c = \frac{22}{7}$ be the approximate value of $x = \pi$, we find that

$$\text{absolute error} = \left| \frac{22}{7} - \pi \right| \approx 0.00126$$

and

$$\text{relative error} = \frac{|22/7 - \pi|}{|\pi|} \approx 0.000402 \approx 0.04\%$$

Related Exercises 7–10 ◀

Midpoint Rule

Many numerical integration methods are based on the ideas that underlie Riemann sums; these methods approximate the net area of regions bounded by curves. A typical problem is shown in Figure 7.12, where we see a function f defined on an interval $[a, b]$. The goal is to approximate the value of $\int_a^b f(x)\, dx$. As with Riemann sums, we first partition the

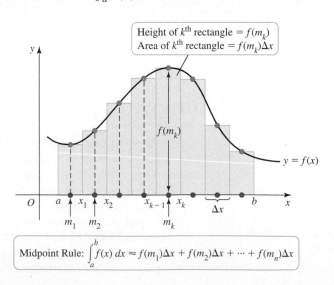

FIGURE 7.12

interval $[a, b]$ into n subintervals of equal length $\Delta x = (b - a)/n$. This partition establishes $n + 1$ grid points

$$x_0 = a, \quad x_1 = a + \Delta x, \quad x_2 = a + 2\Delta x, \ldots, \quad x_k = a + k\Delta x, \ldots, \quad x_n = b.$$

The kth subinterval is $[x_{k-1}, x_k]$, for $k = 1, 2, \ldots, n$.

The Midpoint Rule approximates the region under the curve using rectangles. The bases of the rectangles have width Δx. The height of the kth rectangle is $f(m_k)$, where $m_k = (x_{k-1} + x_k)/2$ is the midpoint of the kth subinterval (Figure 7.12). Therefore, the net area of the kth rectangle is $f(m_k)\,\Delta x$.

Let $M(n)$ be the Midpoint Rule approximation to the integral using n rectangles. Summing the net areas of the rectangles, we have

> ▶ The Midpoint Rule is a Riemann sum that uses midpoints of subintervals for $\overline{x}_k$.

> ▶ Recall that if $f(m_k) < 0$ for some k, then the net area of that rectangle is negative, which makes a negative contribution to the approximation (Section 5.2).

$$\int_a^b f(x)\,dx \approx M(n)$$

$$= f(m_1)\Delta x + f(m_2)\Delta x + \cdots + f(m_n)\Delta x$$

$$= \sum_{k=1}^{n} f\left(\frac{x_{k-1} + x_k}{2}\right)\Delta x$$

Just as with Riemann sums, the Midpoint Rule approximations to $\int_a^b f(x)\,dx$ generally improve as n increases.

DEFINITION Midpoint Rule

Suppose f is defined and integrable on $[a, b]$. The Midpoint Rule approximation to $\int_a^b f(x)\,dx$ using n equally spaced subintervals on $[a, b]$ is

$$M(n) = f(m_1)\Delta x + f(m_2)\Delta x + \cdots + f(m_n)\Delta x$$

$$= \sum_{k=1}^{n} f\left(\frac{x_{k-1} + x_k}{2}\right)\Delta x,$$

where $\Delta x = (b - a)/n$, $x_k = a + k\Delta x$, and m_k is the midpoint of $[x_{k-1}, x_k]$, for $k = 1, \ldots, n$.

QUICK CHECK 1 To apply the Midpoint Rule on the interval $[3, 11]$ with $n = 4$, at what points must the integrand be evaluated? ◄

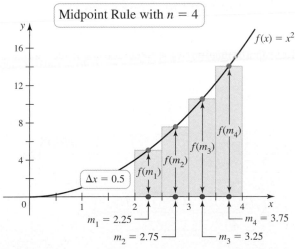

Midpoint Rule with $n = 4$

$f(x) = x^2$

$f(m_4)$
$f(m_3)$
$f(m_2)$
$f(m_1)$
$\Delta x = 0.5$

$m_1 = 2.25$
$m_2 = 2.75$
$m_3 = 3.25$
$m_4 = 3.75$

FIGURE 7.13

EXAMPLE 2 Applying the Midpoint Rule Approximate $\int_2^4 x^2\,dx$ using the Midpoint Rule with $n = 4$ and $n = 8$ subintervals.

SOLUTION With $a = 2, b = 4$, and $n = 4$ subintervals, the length of each subinterval is $\Delta x = (b - a)/n = 2/4 = 0.5$. The grid points are

$$x_0 = 2, \quad x_1 = 2.5, \quad x_2 = 3, \quad x_3 = 3.5, \quad \text{and} \quad x_4 = 4.$$

The integrand must be evaluated at the midpoints (Figure 7.13):

$$m_1 = 2.25, \quad m_2 = 2.75, \quad m_3 = 3.25, \quad m_4 = 3.75.$$

With $f(x) = x^2$ and $n = 4$, the Midpoint Rule approximation is

$$M(4) = f(m_1)\Delta x + f(m_2)\Delta x + f(m_3)\Delta x + f(m_4)\Delta x$$
$$= (m_1^2 + m_2^2 + m_3^2 + m_4^2)\Delta x$$
$$= (2.25^2 + 2.75^2 + 3.25^2 + 3.75^2) \cdot 0.5$$
$$= 18.625.$$

The exact area of the region is $\frac{56}{3}$, so the Midpoint Rule has an absolute error of

$$|18.625 - 56/3| \approx 0.0417$$

and a relative error of

$$\left| \frac{18.625 - 56/3}{56/3} \right| \approx 0.00223 = 0.223\%.$$

Using $n = 8$ subintervals, the midpoint approximation is

$$M(8) = \sum_{k=1}^{8} f(m_k)\Delta x = 18.65625,$$

which has an absolute error of about 0.0104 and a relative error of about 0.0558%. We see that increasing n and using more rectangles decreases the error in the approximations.

Related Exercises 11–14 ◄

The Trapezoid Rule

Area of a trapezoid

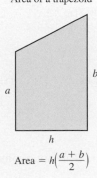

$$\text{Area} = h\left(\frac{a+b}{2}\right)$$

Another method for estimating $\int_a^b f(x)\,dx$ is the Trapezoid Rule, which uses the same partition of the interval $[a, b]$ described for the Midpoint Rule. Instead of approximating the region under the curve by rectangles, the Trapezoid Rule uses (what else?) trapezoids. The bases of the trapezoids have length Δx. The sides of the kth trapezoid have lengths $f(x_{k-1})$ and $f(x_k)$, for $k = 1, 2, \ldots, n$ (Figure 7.14). Therefore, the area of the kth trapezoid is $\left(\dfrac{f(x_{k-1}) + f(x_k)}{2}\right)\Delta x.$

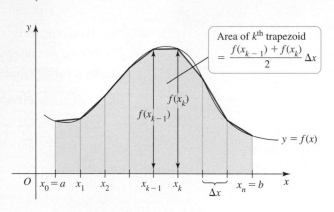

Trapezoid Rule: $\int_a^b f(x)dx \approx \left[\frac{1}{2}f(x_0) + f(x_1) + \cdots + f(x_{n-1}) + \frac{1}{2}f(x_n)\right]\Delta x$

FIGURE 7.14

Letting $T(n)$ be the Trapezoid Rule approximation to the integral using n subintervals, we have

$$\int_a^b f(x)\,dx \approx T(n)$$

$$= \underbrace{\left(\frac{f(x_0) + f(x_1)}{2}\right)\Delta x}_{\text{first trapezoid}} + \underbrace{\left(\frac{f(x_1) + f(x_2)}{2}\right)\Delta x}_{\text{second trapezoid}} + \cdots + \underbrace{\left(\frac{f(x_{n-1}) + f(x_n)}{2}\right)\Delta x}_{n\text{th trapezoid}}$$

$$= \left(\frac{f(x_0)}{2} + \underbrace{\frac{f(x_1)}{2} + \frac{f(x_1)}{2}}_{f(x_1)} + \cdots + \underbrace{\frac{f(x_{n-1})}{2} + \frac{f(x_{n-1})}{2}}_{f(x_{n-1})} + \frac{f(x_n)}{2}\right)\Delta x$$

$$= \left(\frac{f(x_0)}{2} + \underbrace{f(x_1) + \cdots + f(x_{n-1})}_{\sum_{k=1}^{n-1} f(x_k)} + \frac{f(x_n)}{2}\right)\Delta x$$

As with the Midpoint Rule, the Trapezoid Rule approximations generally improve as n increases.

> **DEFINITION Trapezoid Rule**
>
> Suppose f is defined and integrable on $[a, b]$. The Trapezoid Rule approximation to $\int_a^b f(x)\, dx$ using n equally spaced subintervals on $[a, b]$ is
>
> $$T(n) = \left(\frac{1}{2} f(x_0) + \sum_{k=1}^{n-1} f(x_k) + \frac{1}{2} f(x_n) \right) \Delta x,$$
>
> where $\Delta x = (b - a)/n$ and $x_k = a + k\Delta x$, for $k = 0, 1, \ldots, n$.

QUICK CHECK 2 Does the Trapezoid Rule underestimate or overestimate the value of $\int_0^4 x^2\, dx$? ◄

EXAMPLE 3 Applying the Trapezoid Rule Approximate $\int_2^4 x^2\, dx$ using the Trapezoid Rule with $n = 4$ subintervals.

SOLUTION As in Example 2, the grid points are

$$x_0 = 2, \quad x_1 = 2.5, \quad x_2 = 3, \quad x_3 = 3.5, \quad \text{and} \quad x_4 = 4.$$

With $f(x) = x^2$ and $n = 4$, the Trapezoid Rule approximation is

$$
\begin{aligned}
T(4) &= \tfrac{1}{2} f(x_0)\Delta x + f(x_1)\Delta x + f(x_2)\Delta x + f(x_3)\Delta x + \tfrac{1}{2} f(x_4)\Delta x \\
&= \left(\tfrac{1}{2} x_0^2 + x_1^2 + x_2^2 + x_3^2 + \tfrac{1}{2} x_4^2 \right) \Delta x \\
&= \left(\tfrac{1}{2} \cdot 2^2 + 2.5^2 + 3^2 + 3.5^2 + \tfrac{1}{2} \cdot 4^2 \right) \cdot 0.5 \\
&= 18.75.
\end{aligned}
$$

Figure 7.15 shows the approximation with $n = 4$ trapezoids. The exact area of the region is $56/3$, so the Trapezoid Rule approximation has an absolute error of about 0.0833 and a relative error of approximately 0.00446, or 0.446%. Increasing n decreases this error.   *Related Exercises 15–18* ◄

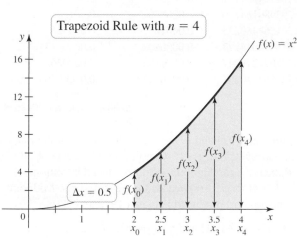

Trapezoid Rule with $n = 4$

$f(x) = x^2$

$f(x_4)$
$f(x_3)$
$f(x_2)$
$f(x_1)$
$\Delta x = 0.5$ $f(x_0)$

FIGURE 7.15

EXAMPLE 4 Errors in the Midpoint and Trapezoid Rules Given that

$$\int_0^1 xe^{-x}\, dx = 1 - 2e^{-1},$$

find the absolute errors in the Midpoint Rule and Trapezoid Rule approximations to the integral with $n = 4, 8, 16, 32, 64$, and 128 subintervals.

SOLUTION Because the exact value of the integral is known (which often does *not* happen in practice), we can compute the error in various approximations. For example, if $n = 16$, then

$$\Delta x = \frac{1}{16} \quad \text{and} \quad x_k = \frac{k}{16}, \quad \text{for } k = 0, 1, \ldots, n.$$

Using sigma notation and a computer algebra system, we have

$$M(16) = \sum_{k=1}^{\overset{n}{\overset{\frown}{16}}} f\left(\overbrace{\frac{(k-1)/16}{}}^{x_{k-1}} + \overbrace{k/16}^{x_k} \over 2 \right) \overbrace{\frac{1}{16}}^{\Delta x} = \sum_{k=1}^{16} f\left(\frac{2k - 1}{32} \right) \frac{1}{16} \approx 0.26440383609318$$

and

$$T(16) = \left(\frac{1}{2} \underbrace{f(0)}_{x_0 = a} + \sum_{k=1}^{\overset{n-1}{\overset{\frown}{15}}} \underbrace{f(k/16)}_{x_k} + \frac{1}{2} \underbrace{f(1)}_{x_{16} = b} \right) \frac{1}{16} \approx 0.26391564480235.$$

The absolute error in the Midpoint Rule approximation with $n = 16$ is $|M(16) - (1 - 2e^{-1})| \approx 0.000163$. The absolute error in the Trapezoid Rule approximation with $n = 16$ is $|T(16) - (1 - 2e^{-1})| \approx 0.000325$.

The Midpoint Rule and Trapezoid Rule approximations to the integral, together with the associated absolute errors, are shown in Table 7.4 for various values of n. Notice that as n increases, the errors in both methods decrease, as expected. With $n = 128$ subintervals, the approximations $M(128)$ and $T(128)$ agree to four decimal places. Based on these approximations, a good approximation to the integral is 0.2642. The way in which the errors decrease is also worth noting. If you look carefully at both error columns in Table 7.4, you will see that each time n is doubled (or Δx is halved), the error decreases by a factor of approximately 4.

Table 7.4

n	$M(n)$	$T(n)$	Error $M(n)$	Error $T(n)$
4	0.26683456310319	0.25904504019141	0.00259	0.00520
8	0.26489148795740	0.26293980164730	0.000650	0.00130
16	0.26440383609318	0.26391564480235	0.000163	0.000325
32	0.26428180513718	0.26415974044777	0.0000407	0.0000814
64	0.26425129001915	0.26422077279247	0.0000102	0.0000203
128	0.26424366077837	0.26423603140581	0.00000254	0.00000509

QUICK CHECK 3 Compute the approximate factor by which the error decreases in Table 7.4 between $T(16)$ and $T(32)$; between $T(32)$ and $T(64)$. ◄

Related Exercises 19–26 ◄

Table 7.5

Year	World Oil Production (billions barrels/yr)
1992	22.3
1993	21.9
1994	21.5
1995	21.9
1996	22.3
1997	23.0
1998	23.7
1999	24.5
2000	23.7
2001	25.2
2002	24.8
2003	24.5
2004	25.2
2005	25.9
2006	26.3
2007	27.0
2008	27.5

EXAMPLE 5 **World oil production** Table 7.5 and Figure 7.16 show data for the rate of world oil production (in billions of barrels/yr) over a 16-year period. If the rate of oil production is given by the function R, then the total amount of oil produced in billions of barrels over the time period $a \leq t \leq b$ is $Q = \int_a^b R(t)\, dt$ (Section 6.1). Use the Midpoint and Trapezoid Rules to approximate the total oil produced between 1992 and 2008.

SOLUTION For convenience, let $t = 0$ represent 1992 and $t = 16$ represent 2008. We let $R(t)$ be the rate of oil production in the year corresponding to t (for example, $R(6) = 23.7$ is the rate in 1998). The goal is to approximate $Q = \int_0^{16} R(t)\, dt$. If we use $n = 4$ subintervals, then $\Delta t = 4$ yr. The resulting Midpoint and Trapezoid Rule approximations (in billions of barrels) are

$$Q \approx M(4) = [R(2) + R(6) + R(10) + R(14)]\Delta t$$
$$= (21.5 + 23.7 + 24.8 + 26.3)4$$
$$= 385.2$$

and

$$Q \approx T(4) = \left[\frac{1}{2}R(0) + R(4) + R(8) + R(12) + \frac{1}{2}R(16)\right]\Delta t$$
$$= \left(\frac{1}{2} \cdot 22.3 + 22.3 + 23.7 + 25.2 + \frac{1}{2} \cdot 27.5\right)4$$
$$= 384.4.$$

The two methods give reasonable agreement. Using $n = 8$ subintervals, with $\Delta t = 2$ yr, similar calculations give the approximations

$$Q \approx M(8) = 387.8 \quad \text{and} \quad Q \approx T(8) = 384.8.$$

The given data do not allow us to compute the next Midpoint Rule approximation $M(16)$. However, we can compute the next Trapezoid Rule approximation $T(16)$ and here is a good way to do it. If $T(n)$ and $M(n)$ are known, then the next Trapezoid Rule approximation is (Exercise 58)

$$T(2n) = \frac{T(n) + M(n)}{2}.$$

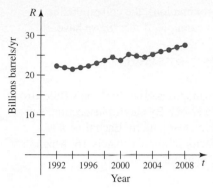

FIGURE 7.16

Source: U.S. Energy Information Administration

Using this trick, we find that

$$T(16) = \frac{T(8) + M(8)}{2} = \frac{384.8 + 387.8}{2} = 386.3.$$

Based on these calculations, the best approximation to the total oil produced between 1992 and 2008 is 386.3 billion barrels.   *Related Exercises 27–30* ◄

Simpson's Rule

The Midpoint Rule and the Trapezoid Rule can be improved by approximating the graph of f with curves, rather than line segments. Let's return to the partition used by the Midpoint and Trapezoid Rules, but now suppose we work with three neighboring points on the curve $y = f(x)$, say $(x_0, f(x_0))$, $(x_1, f(x_1))$, and $(x_2, f(x_2))$. These three points determine a *parabola*, and it is easy to find the net area bounded by the parabola on the interval $[x_0, x_2]$. When this idea is applied to every group of three consecutive points along the interval of integration, the result is **Simpson's Rule**. With n subintervals, Simpson's Rule is denoted $S(n)$ and is given by

$$\int_a^b f(x)\,dx \approx S(n)$$

$$= (f(x_0) + 4f(x_1) + 2f(x_2) + 4f(x_3) + \cdots + 2f(x_{n-2}) + 4f(x_{n-1}) + f(x_n))\frac{\Delta x}{3}$$

Notice that apart from the first and last terms, the coefficients alternate between 4 and 2; **n must be an even integer** for this rule to work.

You can use the formula for Simpson's Rule given above; but here is an easier way. If you already have the Trapezoid Rule approximations $T(n)$ and $T(2n)$, the next Simpson's Rule approximation follows immediately with a simple calculation (Exercise 60):

$$S(2n) = \frac{4T(2n) - T(n)}{3}.$$

DEFINITION   Simpson's Rule

Suppose f is defined and integrable on $[a, b]$. The Simpson's Rule approximation to $\int_a^b f(x)\,dx$ using n equally spaced subintervals on $[a, b]$ is

$$S(n) = [f(x_0) + 4f(x_1) + 2f(x_2) + 4f(x_3) + \cdots + 4f(x_{n-1}) + f(x_n)]\frac{\Delta x}{3},$$

where n is an even integer, $\Delta x = (b - a)/n$, and $x_k = a + k\Delta x$, for $k = 0, 1, \ldots, n$. Alternatively, if the Trapezoid Rule approximations $T(2n)$ and $T(n)$ are known, then

$$S(2n) = \frac{4T(2n) - T(n)}{3}.$$

EXAMPLE 6   Errors in the Trapezoid Rule and Simpson's Rule Given that $\int_0^1 xe^{-x}\,dx = 1 - 2e^{-1}$, find the absolute errors in the Trapezoid Rule and Simpson's Rule approximations to the integral with $n = 8, 16, 32, 64$, and 128 subintervals.

SOLUTION Because the shortcut formula for Simpson's Rule is based on values generated by the Trapezoid Rule, it is best to calculate the Trapezoid Rule approximations first. The second column of Table 7.6 shows the Trapezoid Rule approximations computed in

Example 4. Having a column of Trapezoid Rule approximations, the corresponding Simpson's Rule approximations are easily found. For example, if $n = 8$, we have

$$S(8) = \frac{4T(8) - T(4)}{3} \approx 0.26423805546593.$$

The table also shows the absolute errors in the approximations. The Simpson's Rule errors decrease much more quickly than the Trapezoid Rule errors. By careful inspection, you will see that the Simpson's Rule errors decrease with a clear pattern: Each time n is doubled (or Δx is halved), the errors decrease by a factor of approximately 16, which makes Simpson's Rule a more efficient and accurate method.

Table 7.6

n	$T(n)$	$S(n)$	Error $T(n)$	Error $S(n)$
4	0.25904504019141		0.00520	
8	0.26293980164730	0.26423805546593	0.00130	0.00000306
16	0.26391564480235	0.26424092585404	0.000325	0.000000192
32	0.26415974044777	0.26424110566291	0.0000814	0.0000000120
64	0.26422077279247	0.26424111690738	0.0000203	0.000000000750
128	0.26423603140581	0.26424111761026	0.00000509	0.0000000000469

QUICK CHECK 4 Compute the approximate factor by which the error decreases in Table 7.6 between $S(16)$ and $S(32)$ and between $S(32)$ and $S(64)$. ◄

Related Exercises 31–38 ◄

Errors in Numerical Integration

A detailed analysis of the errors in the three methods we have discussed goes beyond the scope of the book. We state without proof the standard error theorems for the methods and note that Examples 3, 4, and 6 are consistent with these results.

> **THEOREM 7.2   Errors in Numerical Integration**
> Assume that f'' is continuous on the interval $[a, b]$ and that k is a real number such that $|f''(x)| < k$ for all x in $[a, b]$. The absolute errors in approximating the integral $\int_a^b f(x)\, dx$ by the Midpoint Rule and Trapezoid Rule with n subintervals satisfy the inequalities
>
> $$E_M \leq \frac{k(b - a)}{24}(\Delta x)^2 \quad \text{and} \quad E_T \leq \frac{k(b - a)}{12}(\Delta x)^2$$
>
> respectively, where $\Delta x = (b - a)/n$.
>   Assume that $f^{(4)}$ is continuous on the interval $[a, b]$ and that K is a real number such that $|f^{(4)}(x)| < K$ on $[a, b]$. The error in approximating the integral $\int_a^b f(x)\, dx$ by Simpson's Rule with n subintervals satisfies the inequality
>
> $$E_S \leq \frac{K(b - a)}{180}(\Delta x)^4.$$

The absolute errors associated with the Midpoint Rule and Trapezoid Rule are proportional to $(\Delta x)^2$. So, if Δx is reduced by a factor of 2, the errors decrease roughly by a factor of 4, as seen in Example 4. Simpson's Rule is a more accurate method; its error is proportional to $(\Delta x)^4$, which means that if Δx is reduced by a factor of 2, the errors decrease roughly by a factor of 16, as seen in Example 6. Computing both the Trapezoid Rule and Simpson's Rule together, as shown in Example 6, is a powerful method that produces accurate approximations with relatively little work.

SECTION 7.6 EXERCISES

Review Questions

1. If the interval $[4, 18]$ is partitioned into $n = 28$ subintervals of equal width, what is Δx?

2. Explain geometrically how the Midpoint Rule is used to approximate a definite integral.

3. Explain geometrically how the Trapezoid Rule is used to approximate a definite integral.

4. If the Midpoint Rule is used on the interval $[-1, 11]$ with $n = 3$ subintervals, at what x-coordinates is the integrand evaluated?

5. If the Trapezoid Rule is used on the interval $[-1, 9]$ with $n = 5$ subintervals, at what x-coordinates is the integrand evaluated?

6. State how to compute the Simpson's Rule approximation $S(2n)$ if the Trapezoid Rule approximations $T(2n)$ and $T(n)$ are known.

Basic Skills

7–10. Absolute and relative error *Compute the absolute and relative errors in using c to approximate x.*

7. $x = \pi$; $c = 3.14$

8. $x = \sqrt{2}$; $c = 1.414$

9. $x = e$; $c = 2.72$

10. $x = e$; $c = 2.718$

11–14. Midpoint Rule approximations *Find the indicated Midpoint Rule approximations to the following integrals.*

11. $\displaystyle\int_2^{10} 2x^2 \, dx$ using $n = 1, 2$, and 4 subintervals

12. $\displaystyle\int_1^9 x^3 \, dx$ using $n = 1, 2$, and 4 subintervals

13. $\displaystyle\int_0^1 \sin \pi x \, dx$ using $n = 6$ subintervals

14. $\displaystyle\int_0^1 e^{-x} \, dx$ using $n = 8$ subintervals

15–18. Trapezoid Rule approximations *Find the indicated Trapezoid Rule approximations to the following integrals.*

15. $\displaystyle\int_2^{10} 2x^2 \, dx$ using $n = 2, 4$, and 8 subintervals

16. $\displaystyle\int_1^9 x^3 \, dx$ using $n = 2, 4$, and 8 subintervals

17. $\displaystyle\int_0^1 \sin \pi x \, dx$ using $n = 6$ subintervals

18. $\displaystyle\int_0^1 e^{-x} \, dx$ using $n = 8$ subintervals

19. **Midpoint Rule, Trapezoid Rule and relative error** Find the Midpoint and Trapezoid Rule approximations to $\int_0^1 \sin \pi x \, dx$ using $n = 25$ subintervals. Compute the relative error of each approximation.

20. **Midpoint Rule, Trapezoid Rule and relative error** Find the Midpoint and Trapezoid Rule approximations to $\int_0^1 e^{-x} \, dx$ using $n = 50$ subintervals. Compute the relative error of each approximation.

21–26. Comparing the Midpoint and Trapezoid Rules *Apply the Midpoint and Trapezoid Rules to the following integrals. Make a table similar to Table 7.4 showing the approximations and errors for $n = 4, 8, 16,$ and 32. The exact values of the integrals are given for computing the error.*

21. $\displaystyle\int_1^5 (3x^2 - 2x) \, dx = 100$

22. $\displaystyle\int_{-2}^6 \left(\frac{x^3}{16} - x\right) dx = 4$

23. $\displaystyle\int_0^{\pi/4} 3 \sin 2x \, dx = \frac{3}{2}$

24. $\displaystyle\int_1^e \ln x \, dx = 1$

25. $\displaystyle\int_0^\pi \sin x \cos 3x \, dx = 0$

26. $\displaystyle\int_0^8 e^{-2x} \, dx = \frac{1 - e^{-16}}{2} \approx 0.4999999$

27–30. Temperature data *Hourly temperature data for Boulder, CO, San Francisco, CA, Nantucket, MA, and Duluth, MN, over a 12-hr period on the same day of January are shown in the figure. Assume that these data are taken from a continuous temperature function $T(t)$. The average temperature over the 12-hr period is $\overline{T} = \dfrac{1}{12}\displaystyle\int_0^{12} T(t) \, dt$.*

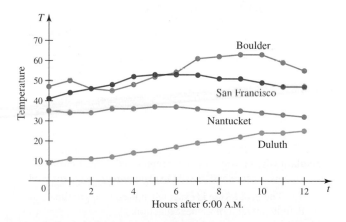

Hours after 6:00 A.M.

t	0	1	2	3	4	5	6	7	8	9	10	11	12
B	47	50	46	45	48	52	54	61	62	63	63	59	55
SF	41	44	46	48	52	53	53	53	51	51	49	47	47
N	35	34	34	36	36	37	37	36	35	35	34	33	32
D	9	11	11	12	14	15	17	19	20	22	24	24	25

27. Find an accurate approximation to the average temperature over the 12-hr period for Boulder. State your method.

28. Find an accurate approximation to the average temperature over the 12-hour period for San Francisco. State your method.

29. Find an accurate approximation to the average temperature over the 12-hr period for Nantucket. State your method.

30. Find an accurate approximation to the average temperature over the 12-hr period for Duluth. State your method.

31–34. Trapezoid Rule and Simpson's Rule *Consider the following integrals and the given values of n.*

   a. *Find the Trapezoid Rule approximations to the integral using n and 2n subintervals.*
   b. *Find the Simpson's Rule approximation to the integral using 2n subintervals. It is easiest to obtain Simpson's Rule approximations from the Trapezoid Rule approximations, as in Example 6.*
   c. *Compute the absolute errors in the Trapezoid Rule and Simpson's Rule with 2n subintervals.*

31. $\int_0^1 e^{2x}\, dx;\ n = 25$

32. $\int_0^2 x^4\, dx;\ n = 30$

33. $\int_1^e \frac{1}{x}\, dx;\ n = 50$

34. $\int_0^{\pi/4} \frac{1}{1 + x^2}\, dx;\ n = 64$

35–38. Simpson's Rule *Apply Simpson's Rule to the following integrals. It is easiest to obtain the Simpson's Rule approximations from the Trapezoid Rule approximations, as in Example 6. Make a table similar to Table 7.6 showing the approximations and errors for n = 4, 8, 16, and 32. The exact values of the integrals are given for computing the error.*

35. $\int_0^4 (3x^5 - 8x^3)\, dx = 1536$

36. $\int_1^e \ln x\, dx = 1$

37. $\int_0^\pi e^{-t} \sin t\, dt = \frac{1}{2}(e^{-\pi} + 1)$

38. $\int_0^6 3e^{-3x}\, dx = 1 - e^{-18} \approx 1.000000$

Further Explorations

39. **Explain why or why not** Determine whether the following statements are true and give an explanation or counterexample.

   a. The Trapezoid Rule is exact when used to approximate the definite integral of a linear function.
   b. If the number of subintervals used in the Midpoint Rule is increased by a factor of 3, the error is expected to decrease by a factor of 8.
   c. If the number of subintervals used in the Trapezoid Rule is increased by a factor of 4, the error is expected to decrease by a factor of 16.

40–43. Comparing the Midpoint and Trapezoid Rules *Compare the errors in the Midpoint and Trapezoid Rules with n = 4, 8, 16, and 32 subintervals when they are applied to the following integrals (with their exact values given).*

40. $\int_0^{\pi/2} \sin^6 x\, dx = \frac{5\pi}{32}$

41. $\int_0^{\pi/2} \cos^9 x\, dx = \frac{128}{315}$

42. $\int_0^1 (8x^7 - 7x^8)\, dx = \frac{2}{9}$

43. $\int_0^\pi \ln(5 + 3\cos x)\, dx = \pi \ln \frac{9}{2}$

44–47. Using Simpson's Rule *Approximate the following integrals using Simpson's Rule. Experiment with values of n to ensure that the error is less than 10^{-3}.*

44. $\int_0^{2\pi} \frac{dx}{(5 + 3\sin x)^2} = \frac{5\pi}{32}$

45. $\int_0^\pi \frac{\cos x}{\frac{5}{4} - \cos x}\, dx = \frac{2\pi}{3}$

46. $\int_0^\pi \ln(2 + \cos x)\, dx = \pi \ln\left(\frac{2 + \sqrt{3}}{2}\right)$

47. $\int_0^\pi \sin 6x \cos 3x\, dx = \frac{4}{9}$

Applications

48. **Period of a pendulum** A standard pendulum of length L swinging under only the influence of gravity (no resistance) has a period of

$$T = \frac{4}{\omega} \int_0^{\pi/2} \frac{d\varphi}{\sqrt{1 - k^2 \sin^2 \varphi}}$$

where $\omega^2 = g/L$, $k^2 = \sin^2(\theta_0/2)$, $g \approx 9.8\ \text{m/s}^2$ is the acceleration due to gravity, and θ_0 is the initial angle from which the pendulum is released (in radians). Use numerical integration to approximate the period of a pendulum with $L = 1$ m that is released from an angle of $\theta_0 = \pi/4$ rad.

49. **Arc length of an ellipse** The length of an ellipse with axes of length $2a$ and $2b$ is

$$\int_0^{2\pi} \sqrt{a^2 \cos^2 t + b^2 \sin^2 t}\, dt.$$

Use numerical integration and experiment with different values of n to approximate the length of an ellipse with $a = 4$ and $b = 8$.

50. **Sine Integral** The theory of diffraction produces the sine integral function $\text{Si}(x) = \int_0^x \frac{\sin t}{t}\, dt$. Use the Midpoint Rule to approximate $\text{Si}(1)$ and $\text{Si}(10)$. (Recall that $\lim_{x \to 0} (\sin x)/x = 1$.) Experiment with the number of subintervals until you obtain approximations that have an error less than 10^{-3}. A rule of thumb is that if two successive approximations differ by less than 10^{-3}, then the error is usually less than 10^{-3}.

51. Normal distribution of heights The heights of U.S. men are normally distributed with a mean of 69 in and a standard deviation of 3 in. This means that the fraction of men with a height between a and b (with $a < b$) inches is given by the integral

$$\frac{1}{3\sqrt{2\pi}} \int_a^b e^{-[(x-69)/3]^2/2}\, dx.$$

What percentage of American men are between 66 and 72 inches in height? Use the method of your choice and experiment with the number of subintervals until you obtain successive approximations that differ by less than 10^{-3}.

52. Normal distribution of movie lengths A recent study revealed that the lengths of U.S. movies are normally distributed with a mean of 110 min and a standard deviation of 22 min. This means that the fraction of movies with lengths between a and b minutes (with $a < b$) is given by the integral

$$\frac{1}{22\sqrt{2\pi}} \int_a^b e^{-[(x-110)/22]^2/2}\, dx.$$

What percentage of U.S. movies are between 1 hr and 1.5 hr long (60–90 min)?

53. U.S. oil produced and imported The figure shows the rate at which U.S. oil was produced and imported between 1920 and 2005 in units of millions of barrels per day. The total amount of oil produced or imported is given by the area of the region under the corresponding curve. Be careful with units because both days and years are used in this data set.

   a. Use numerical integration to estimate the amount of U.S. oil produced between 1940 and 2000. Use the method of your choice and experiment with values of n.

   b. Use numerical integration to estimate the amount of oil imported between 1940 and 2000. Use the method of your choice and experiment with values of n.

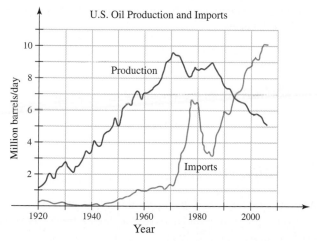

Source: U.S. Energy Information Administration

Additional Exercises

54. Estimating error Refer to Theorem 7.2 and let $f(x) = e^{x^2}$.

   a. Find a Trapezoid Rule approximation to $\int_0^1 e^{x^2}\, dx$ using $n = 50$ subintervals.

   b. Calculate $f''(x)$.

   c. Explain why $|f''(x)| < 18$ on $[0, 1]$, given that $e < 3$.

   d. Use Theorem 7.2 to find an upper bound on the absolute error in the estimate found in part (a).

55. Estimating error Refer to Theorem 7.2 and let $f(x) = \sin e^x$.

   a. Find a Trapezoid Rule approximation to $\int_0^1 \sin(e^x)\, dx$ using $n = 40$ subintervals.

   b. Calculate $f''(x)$.

   c. Explain why $|f''(x)| < 6$ on $[0, 1]$, given that $e < 3$. (*Hint:* Graph f''.)

   d. Find an upper bound on the absolute error in the estimate found in part (a) using Theorem 7.2.

56. Exact Trapezoid Rule Prove that the Trapezoid Rule is exact (no error) when approximating the definite integral of a linear function.

57. Exact Simpson's Rule Prove that Simpson's Rule is exact (no error) when approximating the definite integral of a linear function *and* a quadratic function.

58. Shortcut for the Trapezoid Rule Prove that if you have $M(n)$ and $T(n)$ (a Midpoint Rule approximation and a Trapezoid Rule approximation with n subintervals), then $T(2n) = (T(n) + M(n))/2$.

59. Trapezoid Rule and concavity Suppose f is positive and its first two derivatives are continuous on $[a, b]$. If f'' is positive on $[a, b]$, then is a Trapezoid Rule estimate of $\int_a^b f(x)\, dx$ an underestimate or overestimate of the integral? Justify your answer using Theorem 7.2 and an illustration.

60. Shortcut for Simpson's Rule Using the notation of the text, prove that $S(2n) = \dfrac{4T(2n) - T(n)}{3}$ for $n \geq 1$.

61. Another Simpson's Rule formula Another Simpson's Rule formula is $S(2n) = \dfrac{2M(n) + T(n)}{3}$ for $n \geq 1$. Use this rule to estimate $\int_1^e 1/x\, dx$ using $n = 10$ subintervals.

QUICK CHECK ANSWERS

1. 4, 6, 8, 10   **2.** Overestimates   **3.** 4 and 4
4. 16 and 16 ◄

7.7 Improper Integrals

The definite integrals we have encountered so far involve finite-valued functions and finite intervals of integration. In this section, you will see that definite integrals can sometimes be evaluated when these conditions are not met. Here is an example. The energy required to launch a rocket from the surface of Earth ($R = 6370$ km from the center of Earth) to an altitude H is given by an integral of the form $\int_R^{R+H} k/x^2\, dx$, where k is a constant that includes the mass of the rocket, the mass of Earth, and the gravitational constant. This integral may be evaluated for any finite altitude $H > 0$. Now suppose that the aim is to launch the rocket to an arbitrarily large altitude H so that it escapes Earth's gravitational field. The energy required is given by the preceding integral as $H \to \infty$, which we write $\int_R^{\infty} k/x^2\, dx$. This integral is an example of an *improper integral*, and it has a finite value (which explains why it is possible to launch rockets to outer space). For historical reasons, the term *improper integral* is used for cases in which

- the interval of integration is infinite, or
- the integrand is unbounded on the interval of integration.

In this section, we explore improper integrals and their many uses.

Infinite Intervals

A simple example illustrates what can happen when integrating a function over an infinite interval. Consider the integral $\displaystyle\int_1^b \frac{1}{x^2}\, dx$, for any real number $b > 1$. As shown in Figure 7.17, this integral gives the area of the region bounded by the curve $y = x^{-2}$ and the x-axis between $x = 1$ and $x = b$. In fact, the value of the integral is

$$\int_1^b \frac{1}{x^2}\, dx = -\frac{1}{x}\Big|_1^b = 1 - \frac{1}{b}.$$

For example, if $b = 2$, the area under the curve is $\frac{1}{2}$; if $b = 3$, the area under the curve is $\frac{2}{3}$. In general, as b increases, the area under the curve increases.

Now let's ask what happens to the area as b becomes arbitrarily large. Letting $b \to \infty$, the area of the region under the curve is

$$\lim_{b \to \infty}\left(1 - \frac{1}{b}\right) = 1.$$

We have discovered, surprising as it may seem, a curve of *infinite* length that bounds a region with *finite* area (1 square unit).

We express this result as

$$\int_1^{\infty} \frac{1}{x^2}\, dx = 1,$$

which is an improper integral because ∞ appears in the upper limit. In general, to evaluate $\int_a^{\infty} f(x)\, dx$, we first integrate over a finite interval $[a, b]$ and then let $b \to \infty$. A similar procedure is used to evaluate $\int_{-\infty}^b f(x)\, dx$ and $\int_{-\infty}^{\infty} f(x)\, dx$.

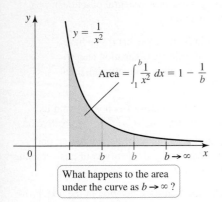

Area $= \displaystyle\int_1^b \frac{1}{x^2}\, dx = 1 - \frac{1}{b}$

$y = \dfrac{1}{x^2}$

What happens to the area under the curve as $b \to \infty$?

FIGURE 7.17

DEFINITIONS **Improper Integrals over Infinite Intervals**

1. If f is continuous on $[a, \infty)$, then

$$\int_a^\infty f(x)\,dx = \lim_{b \to \infty} \int_a^b f(x)\,dx,$$

provided the limit exists.

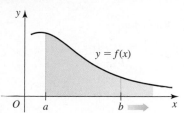

2. If f is continuous on $(-\infty, b]$, then

$$\int_{-\infty}^b f(x)\,dx = \lim_{a \to -\infty} \int_a^b f(x)\,dx,$$

provided the limit exists.

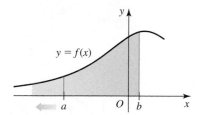

> Doubly infinite integrals (Case 3 in the definition) must be evaluated as two independent limits and not as
>
> $$\int_{-\infty}^\infty f(x)\,dx = \lim_{b \to \infty} \int_{-b}^b f(x)\,dx.$$

3. If f is continuous on $(-\infty, \infty)$, then

$$\int_{-\infty}^\infty f(x)\,dx = \lim_{a \to -\infty} \int_a^c f(x)\,dx$$

$$+ \lim_{b \to \infty} \int_c^b f(x)\,dx$$

provided both limits exist, where c is any real number.

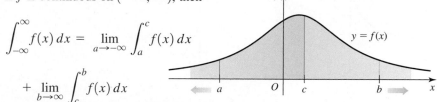

In each case, if the limit exists, the improper integral is said to **converge**; if it does not exist, the improper integral is said to **diverge**.

EXAMPLE 1 **Infinite intervals** Evaluate each integral.

a. $\displaystyle\int_0^\infty e^{-3x}\,dx$ **b.** $\displaystyle\int_0^\infty \frac{1}{1 + x^2}\,dx$

SOLUTION

a. Using the definition of the improper integral, we have

$$\int_0^\infty e^{-3x}\,dx = \lim_{b \to \infty} \int_0^b e^{-3x}\,dx \qquad \text{Definition of improper integral}$$

$$= \lim_{b \to \infty} \left(-\frac{1}{3} e^{-3x}\right)\Big|_0^b \qquad \text{Evaluate the integral.}$$

$$= \lim_{b \to \infty} \frac{1}{3}(1 - e^{-3b}) \qquad \text{Simplify.}$$

$$= \frac{1}{3}\left(1 - \underbrace{\lim_{b \to \infty} \frac{1}{e^{3b}}}_{\text{equals } 0}\right) = \frac{1}{3}. \quad e^{-3b} = \frac{1}{e^{3b}}$$

In this case the limit exists, so the integral converges and the region under the curve has a finite area of $\frac{1}{3}$ (Figure 7.18).

> Recall that
> $$\int \frac{dx}{a^2 + x^2}$$
> $$= \frac{1}{a}\tan^{-1}\left(\frac{x}{a}\right) + C$$
>
> The graph of $y = \tan^{-1} x$ shows that
> $$\lim_{x \to \infty} \tan^{-1} x = \frac{\pi}{2}.$$

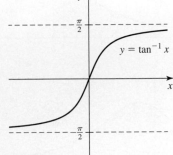

b. Using the definition of the improper integral, we have

$$\int_0^\infty \frac{dx}{1 + x^2} = \lim_{b \to \infty} \int_0^b \frac{dx}{1 + x^2} \qquad \text{Definition of improper integral}$$

$$= \lim_{b \to \infty} \left. (\tan^{-1} x)\right|_0^b \qquad \text{Evaluate the integral.}$$

$$= \lim_{b \to \infty} (\tan^{-1} b - \tan^{-1} 0) \quad \text{Simplify.}$$

$$= \frac{\pi}{2} - 0 = \frac{\pi}{2}. \qquad \lim_{b \to \infty} \tan^{-1} b = \frac{\pi}{2}, \tan^{-1}(0) = 0$$

Figure 7.19 shows the region whose finite area is given by this integral.

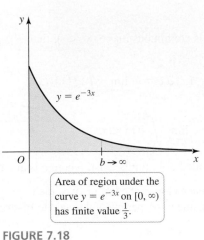

Area of region under the curve $y = e^{-3x}$ on $[0, \infty)$ has finite value $\frac{1}{3}$.

FIGURE 7.18

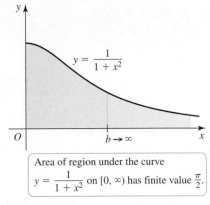

Area of region under the curve $y = \dfrac{1}{1 + x^2}$ on $[0, \infty)$ has finite value $\frac{\pi}{2}$.

FIGURE 7.19

Related Exercises 5–20 ◄

QUICK CHECK 1 The function $f(x) = 1 + x^{-1}$ decreases to 1 as $x \to \infty$. Does $\int_1^\infty f(x)\,dx$ exist? ◄

EXAMPLE 2   The family $f(x) = 1/x^p$ Consider the family of functions $f(x) = 1/x^p$, where p is a real number. For what values of p does $\int_1^\infty f(x)\,dx$ converge?

SOLUTION For $p > 0$, the functions $f(x) = 1/x^p$ approach zero as $x \to \infty$, with larger values of p giving greater rates of decrease (Figure 7.20). Assuming $p \neq 1$, the integral is evaluated as follows:

> Recall for $p \neq -1$,
> $$\int \frac{1}{x^p}\,dx = \int x^{-p}\,dx$$
> $$= \frac{x^{-p+1}}{-p + 1} + C$$
> $$= \frac{x^{1-p}}{1 - p} + C$$

$$\int_1^\infty \frac{1}{x^p}\,dx = \lim_{b \to \infty} \int_1^b x^{-p}\,dx \qquad \text{Definition of improper integral}$$

$$= \frac{1}{1 - p} \lim_{b \to \infty} \left(\left. x^{1-p} \right|_1^b \right) \qquad \text{Evaluate the integral on a finite interval.}$$

$$= \frac{1}{1 - p} \lim_{b \to \infty} (b^{1-p} - 1) \quad \text{Simplify.}$$

It is easiest to consider three cases.

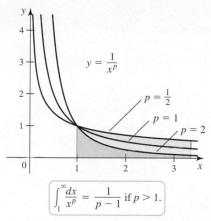

$$\int_1^\infty \frac{dx}{x^p} = \frac{1}{p-1} \text{ if } p > 1.$$

FIGURE 7.20

▶ Example 2 is important in the study of infinite series in Chapter 8. It shows that a continuous function f must do more than simply decrease to zero for its integral on $[a, \infty)$ to converge; it must decrease to zero *sufficiently fast*.

▶ The solid in Example 3a, called *Gabriel's horn* or *Torricelli's trumpet*, has finite volume and infinite surface area (see Section 14.6).

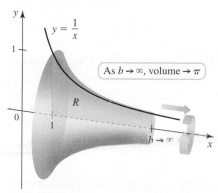

FIGURE 7.21

▶ Recall that if $f(x) > 0$ on $[a, b]$ and the region bounded by the graph of f and the x-axis on $[a, b]$ is revolved about the y-axis, the volume of the solid generated is

$$V = 2\pi \int_a^b x f(x)\, dx.$$

Case 1: If $p > 1$, then $p - 1 > 0$, and $b^{1-p} = 1/b^{p-1}$ approaches 0 as $b \to \infty$. Therefore,

$$\int_1^\infty \frac{1}{x^p}\, dx = \frac{1}{1-p} \underbrace{\lim_{b \to \infty} (b^{1-p} - 1)}_{\substack{\text{approaches} \\ 0}} = \frac{1}{p-1}.$$

Case 2: If $p < 1$, then $1 - p > 0$, and

$$\int_1^\infty \frac{1}{x^p}\, dx = \lim_{b \to \infty} \frac{1}{1-p} \underbrace{(b^{1-p} - 1)}_{\substack{\text{arbitrarily} \\ \text{large}}} = \infty.$$

Case 3: If $p = 1$, then $\int_1^\infty \frac{1}{x}\, dx = \lim_{b \to \infty} (\ln b) = \infty$; so the integral diverges.

In summary, $\int_1^\infty \frac{1}{x^p}\, dx = \frac{1}{p-1}$ if $p > 1$, and the integral diverges if $p \le 1$.

Related Exercises 5–20 ◀

QUICK CHECK 2 Use the result of Example 2 to evaluate $\int_1^\infty \frac{1}{x^4}\, dx.$ ◀

EXAMPLE 3 Solids of revolution Let R be the region bounded by the graph of $y = x^{-1}$ and the x-axis for $x \ge 1$.

a. What is the volume of the solid that is generated when R is revolved about the x-axis?

b. What is the volume of the solid that is generated when R is revolved about the y-axis?

SOLUTION

a. The region in question and the corresponding solid of revolution are shown in Figure 7.21. We use the disk method (Section 6.3) over the interval $[1, b]$ and then let $b \to \infty$:

$$\text{Volume} = \pi \int_1^\infty \big(f(x)\big)^2\, dx \qquad \text{Disk method}$$

$$= \pi \lim_{b \to \infty} \int_1^b \frac{1}{x^2}\, dx \qquad \text{Definition of improper integral}$$

$$= \pi \lim_{b \to \infty} \left(1 - \frac{1}{b}\right) = \pi \qquad \text{Evaluate the integral.}$$

The improper integral exists and the solid has a volume of π cubic units.

b. The region in question and the corresponding solid of revolution are shown in Figure 7.22. Using the shell method (Section 6.4) on the interval $[1, b)$ and letting $b \to \infty$, the volume is given by

$$\text{volume} = 2\pi \int_1^\infty x f(x)\, dx \quad \text{Shell method}$$

$$= 2\pi \int_1^\infty 1\, dx \qquad f(x) = x^{-1}$$

$$= 2\pi \lim_{b \to \infty} \int_1^b 1 \, dx \quad \text{Definition of improper integral}$$

$$= 2\pi \lim_{b \to \infty} (b - 1) \quad \text{Evaluate the integral over a finite interval.}$$

$$= \infty$$

In this case, the volume of the solid is infinite.

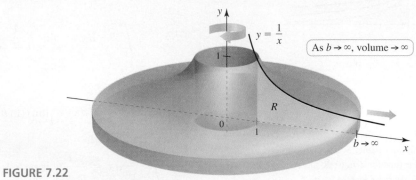

$$y = \frac{1}{x}$$

As $b \to \infty$, volume $\to \infty$

R

$b \to \infty$

FIGURE 7.22

Related Exercises 21–26 ◄

Unbounded Integrands

Improper integrals also occur when the integrand becomes infinite somewhere on the interval of integration. Consider the function $f(x) = 1/\sqrt{x}$ (Figure 7.23). Let's examine the area of the region bounded by the graph of f between $x = 0$ and $x = 1$. Notice that f is not even defined at $x = 0$, and it increases without bound as $x \to 0^+$.

The idea here is to replace the lower limit 0 with a nearby positive number c and then consider the integral $\int_c^1 \frac{1}{\sqrt{x}} \, dx$, where $0 < c < 1$. We find that

$$\int_c^1 \frac{1}{\sqrt{x}} \, dx = 2\sqrt{x} \Big|_c^1 = 2(1 - \sqrt{c}).$$

To find the area of the region under the curve over the entire interval $[0, 1]$, we let $c \to 0^+$. The resulting area, which we denote $\int_0^1 \frac{dx}{\sqrt{x}}$, is

$$\lim_{c \to 0^+} \int_c^1 \frac{1}{\sqrt{x}} \, dx = \lim_{c \to 0^+} 2(1 - \sqrt{c}) = 2.$$

Once again we have a surprising result: Although the region in question has a boundary curve with infinite length, the area of the region is finite.

QUICK CHECK 3 Explain why the one-sided limit $c \to 0^+$ (instead of a two-sided limit) must be used in this example. ◄

The preceding example shows that if a function is unbounded at a point c, it may be possible to integrate that function over an interval that contains c. The point c may occur at either endpoint or at an interior point of the interval of integration.

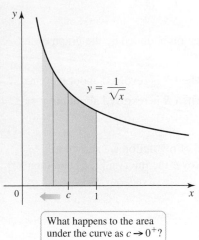

$$y = \frac{1}{\sqrt{x}}$$

What happens to the area under the curve as $c \to 0^+$?

FIGURE 7.23

▶ The functions $f(x) = 1/x^p$ are unbounded at $x = 0$ for $p > 0$. It can be shown (Exercise 60) that

$$\int_0^1 \frac{dx}{x^p} = \frac{1}{1 - p}$$

provided $p < 1$. Otherwise, the integral diverges.

DEFINITIONS   Improper Integrals with an Unbounded Integrand

1. Suppose f is continuous on $(a, b]$ with $\lim\limits_{x \to a^+} f(x) = \pm\infty$. Then,

$$\int_a^b f(x)\, dx = \lim_{c \to a^+} \int_c^b f(x)\, dx,$$

provided the limit exists.

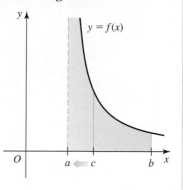

2. Suppose f is continuous on $[a, b)$ with $\lim\limits_{x \to b^-} f(x) = \pm\infty$. Then,

$$\int_a^b f(x)\, dx = \lim_{c \to b^-} \int_a^c f(x)\, dx,$$

provided the limit exists.

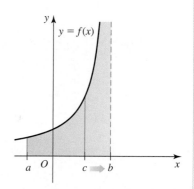

3. Suppose f is continuous on $[a, b]$ except at the interior point p where f is unbounded. Then,

$$\int_a^b f(x)\, dx = \int_a^p f(x)\, dx + \int_p^b f(x)\, dx$$

provided the improper integrals on the right side exist.

In each case, if the limit exists, the improper integral is said to **converge**; if it does not exist, the improper integral is said to **diverge**.

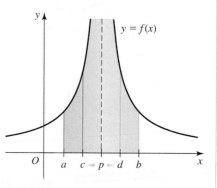

EXAMPLE 4   Infinite integrand Find the area of the region R between the graph of $f(x) = \dfrac{1}{\sqrt{9 - x^2}}$ and the x-axis on the interval $[-3, 3]$ (if it exists).

SOLUTION The integrand is even and has vertical asymptotes at $x = \pm 3$ (Figure 7.24). By symmetry, the area of R is given by

$$\int_{-3}^{3} \frac{1}{\sqrt{9 - x^2}}\, dx = 2 \int_{0}^{3} \frac{1}{\sqrt{9 - x^2}}\, dx,$$

▷ Recall that
$$\int \frac{dx}{\sqrt{a^2 - x^2}} = \sin^{-1}\left(\frac{x}{a}\right) + C.$$

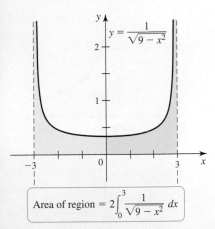

$$\text{Area of region} = 2\int_0^3 \frac{1}{\sqrt{9 - x^2}}\, dx$$

FIGURE 7.24

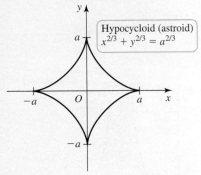

FIGURE 7.25

assuming these improper integrals exist. Because the integrand is unbounded at $x = 3$, we replace the upper limit with $x = c$, evaluate the resulting integral, and then let $c \to 3^-$:

$$2\int_0^3 \frac{dx}{\sqrt{9 - x^2}} = 2\lim_{c \to 3^-} \int_0^c \frac{dx}{\sqrt{9 - x^2}} \qquad \text{Definition of improper integral}$$

$$= 2\lim_{c \to 3^-} \sin^{-1}\left(\frac{x}{3}\right)\Big|_0^c \qquad \text{Evaluate the integral.}$$

$$= 2\lim_{c \to 3^-} \left(\underbrace{\sin^{-1}\left(\frac{c}{3}\right)}_{\text{approaches } \pi/2} - \underbrace{\sin^{-1} 0}_{\text{equals } 0}\right) \qquad \text{Simplify.}$$

Note that as $c \to 3^-$, $\sin^{-1}(c/3) \to \sin^{-1} 1 = \pi/2$. Therefore, the area of R is

$$2\int_0^3 \frac{1}{\sqrt{9 - x^2}}\, dx = 2\left(\frac{\pi}{2} - 0\right) = \pi.$$

Related Exercises 27–40 ◀

EXAMPLE 5 Length of a hypocycloid Find the length L of the complete hypocycloid (or astroid; Figure 7.25) given by $x^{2/3} + y^{2/3} = a^{2/3}$, where $a > 0$.

SOLUTION Solving the equation $x^{2/3} + y^{2/3} = a^{2/3}$ for y, we find that the curve is described by the functions $f(x) = \pm(a^{2/3} - x^{2/3})^{3/2}$ (corresponding to the upper and lower halves of the curve). By symmetry, the length of the entire curve is four times the length of the curve in the first quadrant, which is given by $f(x) = (a^{2/3} - x^{2/3})^{3/2}$ for $0 \le x \le a$. We need the derivative f' for the arc length integral:

$$f'(x) = \frac{3}{2}(a^{2/3} - x^{2/3})^{1/2}\left(-\frac{2}{3}x^{-1/3}\right) = -x^{-1/3}(a^{2/3} - x^{2/3})^{1/2}$$

Now the arc length can be computed:

$$L = 4\int_0^a \sqrt{1 + f'(x)^2}\, dx$$

$$= 4\int_0^a \sqrt{1 + \left(-x^{-1/3}(a^{2/3} - x^{2/3})^{1/2}\right)^2}\, dx \qquad \text{Substitute for } f'.$$

$$= 4\int_0^a \sqrt{a^{2/3} x^{-2/3}}\, dx \qquad \text{Simplify.}$$

$$= 4a^{1/3}\int_0^a x^{-1/3}\, dx \qquad \text{Simplify.}$$

Because $x^{-1/3} \to \infty$ as $x \to 0^+$, the resulting integral is an improper integral, which is handled in the usual manner:

$$L = 4a^{1/3} \lim_{c \to 0^+} \int_c^a x^{-1/3}\, dx \qquad \text{Improper integral}$$

$$= 4a^{1/3} \lim_{c \to 0^+} \left(\frac{3}{2} x^{2/3}\right)\Big|_c^a \qquad \text{Integrate.}$$

$$= 6a^{1/3} \lim_{c \to 0^+} \underbrace{(a^{2/3} - c^{2/3})}_{\to 0} \qquad \text{Simplify.}$$

$$= 6a \qquad \text{Evaluate limit.}$$

The length of the entire hypocycloid is $6a$ units.

Related Exercises 41–42 ◀

EXAMPLE 6 **Bioavailability** The most efficient way to deliver a drug to its intended target site is to administer it intravenously (directly into the blood). If a drug is administered in any other way (for example, orally, nasal inhalant, or skin patch), then some of the drug is typically lost due to absorption before it gets to the blood. By definition, the bioavailability of a drug measures the effectiveness of a nonintravenous method compared to the intravenous method. The bioavailability of intravenous dosing is 100%.

Let the functions $C_i(t)$ and $C_o(t)$ give the concentration of a drug in the blood, for times $t \geq 0$, using intravenous and oral dosing, respectively. (These functions can be determined through clinical experiments.) Assuming the same amount of drug is initially administered by both methods, the bioavailability for an oral dose is defined to be

$$F = \frac{\mathrm{AUC}_o}{\mathrm{AUC}_i} = \frac{\displaystyle\int_0^\infty C_o(t)\,dt}{\displaystyle\int_0^\infty C_i(t)\,dt}$$

where AUC is used in the pharmacology literature to mean *area under the curve*.

Suppose the concentration of a certain drug in the blood in mg/L when given intravenously is $C_i(t) = 100e^{-0.3t}$, where $t \geq 0$ is measured in hours. Suppose also that concentration of the same drug when delivered orally is $C_o(t) = 90(e^{-0.3t} - e^{-2.5t})$ (Figure 7.26). Find the bioavailability of the drug.

SOLUTION Evaluating the integrals of the concentration functions, we find that

$$\mathrm{AUC}_i = \int_0^\infty C_i(t)\,dt = \int_0^\infty 100e^{-0.3t}\,dt$$

$$= \lim_{b\to\infty} \int_0^b 100e^{-0.3t}\,dt \qquad \text{Improper integral}$$

$$= \lim_{b\to\infty} \frac{1000}{3}(1 - \underbrace{e^{-0.3b}}_{\substack{\text{approaches}\\\text{zero}}}) \qquad \text{Evaluate the integral.}$$

$$= \frac{1000}{3} \qquad \text{Evaluate the limit.}$$

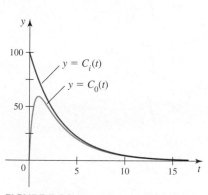

FIGURE 7.26

Similarly,

$$\mathrm{AUC}_o = \int_0^\infty C_o(t)\,dt = \int_0^\infty 90(e^{-0.3t} - e^{-2.5t})\,dt$$

$$= \lim_{b\to\infty} \int_0^b 90(e^{-0.3t} - e^{-2.5t})\,dt \qquad \text{Improper integral}$$

$$= \lim_{b\to\infty} \left[300(1 - \underbrace{e^{-0.3b}}_{\substack{\text{approaches}\\\text{zero}}}) - 36(1 - \underbrace{e^{-2.5b}}_{\substack{\text{approaches}\\\text{zero}}})\right] \qquad \text{Evaluate the integral.}$$

$$= 264 \qquad \text{Evaluate the limt.}$$

Therefore, the bioavailability is $F = 264/(1000/3) = 0.792$, which means oral administration of the drug is roughly 80% as effective as intravenous dosing. Notice that F is the ratio of the areas under the two curves on the interval $[0, \infty)$.   *Related Exercises 43–46* ◄

SECTION 7.7 EXERCISES

Review Questions

1. What are the two general ways in which an improper integral may occur?

2. Explain how to evaluate $\int_a^\infty f(x)\,dx$.

3. Explain how to evaluate $\int_0^1 x^{-1/2}\,dx$.

4. For what values of p does $\int_1^\infty x^{-p}\,dx$ converge?

Basic Skills

5–20. Infinite intervals of integration *Evaluate the following integrals or state that they diverge.*

5. $\displaystyle\int_1^\infty x^{-2}\,dx$

6. $\displaystyle\int_0^\infty \frac{dx}{(x+1)^3}$

7. $\displaystyle\int_2^\infty \frac{dx}{\sqrt{x}}$

8. $\displaystyle\int_0^\infty \frac{dx}{\sqrt[3]{x+2}}$

9. $\displaystyle\int_0^\infty e^{-2x}\,dx$

10. $\displaystyle\int_2^\infty \frac{dx}{x\ln x}$

11. $\displaystyle\int_{e^2}^\infty \frac{dx}{x\ln^p x},\ p>1$

12. $\displaystyle\int_0^\infty \frac{x}{\sqrt[5]{x^2+1}}\,dx$

13. $\displaystyle\int_0^\infty xe^{-x^2}\,dx$

14. $\displaystyle\int_0^\infty \cos x\,dx$

15. $\displaystyle\int_2^\infty \frac{\cos(\pi/x)}{x^2}\,dx$

16. $\displaystyle\int_0^\infty \frac{dx}{1+x^2}$

17. $\displaystyle\int_0^\infty \frac{x}{\sqrt{x^4+1}}\,dx$

18. $\displaystyle\int_a^\infty \sqrt{e^{-x}}\,dx,$ for any finite constant a

19. $\displaystyle\int_2^\infty \frac{x}{(x+2)^2}\,dx$

20. $\displaystyle\int_1^\infty \frac{\tan^{-1} x}{x^2+1}\,dx$

21–26. Volumes on infinite intervals *Find the volume of the described solid of revolution or state that it does not exist.*

21. The region bounded by $f(x) = x^{-2}$ and the x-axis on the interval $[1,\infty)$ is revolved about the x-axis.

22. The region bounded by $f(x) = (x^2+1)^{-1/2}$ and the x-axis on the interval $[2,\infty)$ is revolved about the x-axis.

23. The region bounded by $f(x) = \sqrt{\dfrac{x+1}{x^3}}$ and the x-axis on the interval $[1,\infty)$ is revolved about the x-axis.

24. The region bounded by $f(x) = (x+1)^{-3}$ and the x-axis on the interval $[0,\infty)$ is revolved about the y-axis.

25. The region bounded by $f(x) = \dfrac{1}{\sqrt{x}\,\ln x}$ and the x-axis on the interval $[2,\infty)$ is revolved about the x-axis.

26. The region bounded by $f(x) = \dfrac{\sqrt{x}}{\sqrt[3]{x^2+1}}$ and the x-axis on the interval $[0,\infty)$ is revolved about the x-axis.

27–36. Integrals with unbounded integrands *Evaluate the following integrals or state that they diverge.*

27. $\displaystyle\int_0^8 \frac{dx}{\sqrt[3]{x}}$

28. $\displaystyle\int_0^{\pi/2} \tan\theta\,d\theta$

29. $\displaystyle\int_0^1 \frac{x^3}{x^4-1}\,dx$

30. $\displaystyle\int_1^\infty \frac{dx}{\sqrt[3]{x-1}}$

31. $\displaystyle\int_0^{10} \frac{dx}{\sqrt[4]{10-x}}$

32. $\displaystyle\int_1^{11} \frac{dx}{(x-3)^{2/3}}$

33. $\displaystyle\int_0^1 \ln x^2\,dx$

34. $\displaystyle\int_{-1}^1 \frac{x}{x^2+2x+1}\,dx$

35. $\displaystyle\int_{-2}^2 \frac{dx}{\sqrt{4-x^2}}$

36. $\displaystyle\int_0^{\pi/2} \sec\theta\,d\theta$

37–40. Volumes with infinite integrands *Find the volume of the described solid of revolution or state that it does not exist.*

37. The region bounded by $f(x) = (x-1)^{-1/4}$ and the x-axis on the interval $(1,2]$ is revolved about the x-axis.

38. The region bounded by $f(x) = (x^2-1)^{-1/4}$ and the x-axis on the interval $(1,2]$ is revolved about the y-axis.

39. The region bounded by $f(x) = (4-x)^{-1/3}$ and the x-axis on the interval $[0,4)$ is revolved about the y-axis.

40. The region bounded by $f(x) = (x+1)^{-3/2}$ and the y-axis on the interval $(-1,1]$ is revolved about the line $x=-1$.

41. **Arc length** Find the length of the hypocycloid (or astroid) $x^{2/3} + y^{2/3} = 4$.

42. **Circumference of a circle** Use calculus to find the circumference of a circle with radius a.

43. **Bioavailability** When a drug is given intravenously, the concentration of the drug in the blood is $C_i(t) = 250e^{-0.08t}$, for $t \geq 0$. When the same drug is given orally, the concentration of the drug in the blood is $C_0(t) = 200(e^{-0.08t} - e^{-1.8t})$, for $t \geq 0$. Compute the bioavailability of the drug.

44. **Draining a pool** Water is drained from a swimming pool at a rate given by $R(t) = 100e^{-0.05t}$ gal/hr. If the drain is left open indefinitely, how much water is drained from the pool?

45. **Maximum distance** An object moves on a line with velocity $v(t) = 10/(t+1)^2$ mi/hr for $t \geq 0$. What is the maximum distance the object can travel?

46. Depletion of oil reserves Suppose that the rate at which a company extracts oil is given by $r(t) = r_0 e^{-kt}$, where $r_0 = 10^7$ barrels/yr and $k = 0.005 \text{ yr}^{-1}$. Suppose also the estimate of the total oil reserve is 2×10^9 barrels. If the extraction continues indefinitely, will the reserve be exhausted?

Further Explorations

47. Explain why or why not Determine whether the following statements are true and give an explanation or counterexample.

a. If f is continuous and $0 < f(x) < g(x)$ on the interval $[0, \infty)$ and $\int_0^\infty g(x)\,dx = M < \infty$, then $\int_0^\infty f(x)\,dx$ exists.

b. If $\lim_{x \to \infty} f(x) = 1$, then $\int_0^\infty f(x)\,dx$ exists.

c. If $\int_0^1 x^{-p}\,dx$ exists, then $\int_0^1 x^{-q}\,dx$ exists, where $q > p$.

d. If $\int_1^\infty x^{-p}\,dx$ exists, then $\int_1^\infty x^{-q}\,dx$ exists, where $q > p$.

e. $\displaystyle\int_1^\infty \frac{dx}{x^{3p+2}}$ exists for $p > -\frac{1}{3}$.

48. Incorrect calculation What is wrong with this calculation?

$$\int_{-1}^1 \frac{dx}{x} = \ln|x|\Big|_{-1}^1 = \ln 1 - \ln 1 = 0$$

49. Using symmetry Use symmetry to evaluate the following integrals.

a. $\displaystyle\int_{-\infty}^\infty e^{-|x|}\,dx$

b. $\displaystyle\int_{-\infty}^\infty \frac{x^3}{1 + x^8}\,dx$

50. Integral with a parameter For what values of p does the integral

$$\int_2^\infty \frac{dx}{x \ln^p x}$$

exist and what is its value (in terms of p)?

51. Improper integrals by numerical methods Use the Trapezoid Rule (Section 7.6) to approximate $\int_0^R e^{-x^2}\,dx$ with $R = 2, 4$, and 8. For each value of R, take $n = 4, 8, 16$, and 32, and compare approximations with successive values of n. Use these approximations to approximate $I = \int_0^\infty e^{-x^2}\,dx$.

52–54. Integration by parts *Use integration by parts to evaluate the following improper integrals.*

52. $\displaystyle\int_0^\infty xe^{-x}\,dx$

53. $\displaystyle\int_0^1 x \ln x\,dx$

54. $\displaystyle\int_1^\infty \frac{\ln x}{x^2}\,dx$

55. A close comparison Graph the integrands; then, evaluate and compare the values of $\int_0^\infty xe^{-x^2}\,dx$ and $\int_0^\infty x^2 e^{-x^2}\,dx$.

56. Area between curves Let R be the region bounded by the graphs of $y = x^{-p}$ and $y = x^{-q}$ for $x \geq 1$, where $q > p > 1$. Find the area of R.

57. Area between curves Let R be the region bounded by the graphs of $y = e^{-ax}$ and $y = e^{-bx}$ for $x \geq 0$, where $a > b > 0$. Find the area of R.

58. An area function Let $A(a)$ denote the area of the region bounded by $y = e^{-ax}$ and the x-axis on the interval $[0, \infty)$. Graph the function $A(a)$ for $0 < a < \infty$. Describe how the area of the region decreases as the parameter a increases.

59. Regions bounded by exponentials Let $a > 0$ and let R be the region bounded by the graph of $y = e^{-ax}$ and the x-axis on the interval $[b, \infty)$.

a. Find $A(a, b)$, the area of R as a function of a and b.

b. Find the relationship $b = g(a)$ such that $A(a, b) = 2$.

c. What is the minimum value of b (call it b^*) such that when $b > b^*$, $A(a, b) = 2$ for some value of $a > 0$?

60. The family $f(x) = 1/x^p$ revisited Consider the family of functions $f(x) = 1/x^p$, where p is a real number. For what values of p does the integral $\int_0^1 f(x)\,dx$ exist? What is its value?

61. When is the volume finite? Let R be the region bounded by the graph of $f(x) = x^{-p}$ and the x-axis for $0 < x < 1$.

a. Let S be the solid generated when R is revolved about the x-axis. For what values of p is the volume of S finite?

b. Let S be the solid generated when R is revolved about the y-axis. For what values of p is the volume of S finite?

62. When is the volume finite? Let R be the region bounded by the graph of $f(x) = x^{-p}$ and the x-axis for $x \geq 1$.

a. Let S be the solid generated when R is revolved about the x-axis. For what values of p is the volume of S finite?

b. Let S be the solid generated when R is revolved about the y-axis. For what values of p is the volume of S finite?

63–66. By all means *Use any means to verify (or approximate as closely as possible) the following integrals.*

63. $\displaystyle\int_0^{\pi/2} \ln(\sin x)\,dx = \int_0^{\pi/2} \ln(\cos x)\,dx = -\frac{\pi \ln 2}{2}$

64. $\displaystyle\int_0^\infty \frac{\sin^2 x}{x^2}\,dx = \frac{\pi}{2}$

65. $\displaystyle\int_0^\infty \ln\left(\frac{e^x + 1}{e^x - 1}\right)dx = \frac{\pi^2}{4}$

66. $\displaystyle\int_0^1 \frac{\ln x}{1 + x}\,dx = -\frac{\pi^2}{12}$

Applications

67. Perpetual annuity Imagine that today you deposit $\$B$ in a savings account that earns interest at a rate of $p\%$ per year compounded continuously (see Section 6.8). The goal is to draw an income of $\$I$ per year from the account forever. The amount of money that must be deposited is $B = I \int_0^\infty e^{-rt}\,dt$, where $r = p/100$. Suppose you find an account that earns 12% interest annually and you wish to have an income from the account of \$5000 per year. How much must you deposit today?

68. Draining a tank Water is drained from a 3000-gal tank at a rate that starts at 100 gal/hr and decreases continuously by 5%/hr. If the drain is left open indefinitely, how much water is drained from the tank? Can a full tank be emptied at this rate?

69. Decaying oscillations Let $a > 0$ and b be real numbers. Use integration to confirm the following identities.

a. $\displaystyle\int_0^\infty e^{-ax} \cos bx\,dx = \frac{a}{a^2 + b^2}$

b. $\displaystyle\int_0^\infty e^{-ax} \sin bx\,dx = \frac{b}{a^2 + b^2}$

70. Electronic chips Suppose the probability that a particular computer chip fails after $t = a$ hours of operation is $0.00005 \int_a^\infty e^{-0.00005t} \, dt$.

  a. Find the probability that the computer chip fails after 15,000 hr of operation.
  b. Of the chips that are still operating after 15,000 hr, what fraction of these will operate for at least another 15,000 hr?
  c. Evaluate $0.00005 \int_0^\infty e^{-0.00005t} \, dt$ and interpret its meaning.

71. Average lifetime The average time until a computer chip fails (see Exercise 70) is $0.00005 \int_0^\infty t e^{-0.00005t} \, dt$. Find this value.

72. The Eiffel Tower property Let R be the region between the curves $y = e^{-cx}$ and $y = -e^{-cx}$ on the interval $[a, \infty)$, where $a \geq 0$ and $c > 0$. The center of mass of R is located at $(\bar{x}, 0)$,

where $\bar{x} = \dfrac{\int_a^\infty x e^{-cx} \, dx}{\int_a^\infty e^{-cx} \, dx}$. (The profile of the Eiffel Tower is modeled by the two exponential curves.)

  a. For $a = 0$ and $c = 2$, sketch the curves that define R and find the center of mass of R. Indicate the location of the center of mass.
  b. With $a = 0$ and $c = 2$, find equations of the tangent lines to the curves at the points corresponding to $x = 0$.
  c. Show that the tangent lines intersect at the center of mass.
  d. Show that this same property holds for any $a \geq 0$ and any $c > 0$; that is, the tangent lines to the curves $y = \pm e^{-cx}$ at $x = a$ intersect at the center of mass of R.

(*Source:* P. Weidman and I. Pinelis, *Comptes Rendu, Mechanique* 332 (2004): 571–584. Also see the Guided Projects.)

73. Escape velocity and black holes The work required to launch an object from the surface of Earth to outer space is given by $W = \int_R^\infty F(x) \, dx$, where $R = 6370$ km is the approximate radius of Earth, $F(x) = GMm/x^2$ is the gravitational force between Earth and the object, G is the gravitational constant, M is the mass of Earth, m is the mass of the object, and $GM = 4 \times 10^{14} \, \text{m}^3/\text{s}^2$.

  a. Find the work required to launch an object in terms of m.
  b. What escape velocity v_e is required to give the object a kinetic energy $\frac{1}{2} m v_e^2$ equal to W?
  c. The French scientist Laplace anticipated the existence of black holes in the 18th century with the following argument: If a body has an escape velocity that equals or exceeds the speed of light, $c = 300,000$ km/s, then light cannot escape the body and it cannot be seen. Show that such a body has a radius $R \leq 2GM/c^2$. For Earth to be a black hole, what would its radius need to be?

74. Adding a proton to a nucleus The nucleus of an atom is positively charged because it consists of positively charged protons and uncharged neutrons. To bring a free proton toward a nucleus, a repulsive force $F(r) = kqQ/r^2$ must be overcome, where $q = 1.6 \times 10^{-19}$ C is the charge on the proton, $k = 9 \times 10^9$ N-m^2/C^2, Q is the charge on the nucleus, and r is the distance between the center of the nucleus and the proton. Find the work required to bring a free proton (assumed to be a point mass) from a large distance $(r \to \infty)$ to the edge of a nucleus that has a charge $Q = 50q$ and a radius of 6×10^{-11} m.

75. Gaussians An important function in statistics is the Gaussian (or normal distribution, or bell-shaped curve), $f(x) = e^{-ax^2}$.

  a. Graph the Gaussian for $a = 0.5, 1$, and 2.
  b. Given that $\int_{-\infty}^\infty e^{-ax^2} \, dx = \sqrt{\dfrac{\pi}{a}}$, compute the area under the curves in part (a).
  c. Complete the square to evaluate $\int_{-\infty}^\infty e^{-(ax^2+bx+c)} \, dx$, where $a > 0$, b, and c are real numbers.

76–80. Laplace transforms *A powerful tool in solving problems in engineering and physics is the Laplace transform. Given a function* $f(t)$, *the Laplace transform is a new function* $F(s)$ *defined by*

$$F(s) = \int_0^\infty e^{-st} f(t) \, dt,$$

where we assume that s is a positive real number. For example, to find the Laplace transform of $f(t) = e^{-t}$, *the following improper integral is evaluated using integration by parts:*

$$F(s) = \int_0^\infty e^{-st} e^{-t} \, dt = \int_0^\infty e^{-(s+1)t} \, dt = \frac{1}{s+1}$$

Verify the following Laplace transforms, where a is a real number.

76. $f(t) = 1 \quad \rightarrow \quad F(s) = \dfrac{1}{s}$

77. $f(t) = e^{at} \quad \rightarrow \quad F(s) = \dfrac{1}{s-a}$

78. $f(t) = t \quad \rightarrow \quad F(s) = \dfrac{1}{s^2}$

79. $f(t) = \sin at \quad \rightarrow \quad F(s) = \dfrac{a}{s^2+a^2}$

80. $f(t) = \cos at \quad \rightarrow \quad F(s) = \dfrac{s}{s^2+a^2}$

Additional Exercises

81. Improper integrals Evaluate the following improper integrals (Putnam Exam, 1939).

  a. $\displaystyle \int_1^3 \frac{dx}{\sqrt{(x-1)(3-x)}}$
  b. $\displaystyle \int_1^\infty \frac{dx}{e^{x+1} + e^{3-x}}$

82. A better way Compute $\int_0^1 \ln x \, dx$ using integration by parts. Then explain why $-\int_0^\infty e^{-x} \, dx$ (an easier integral) gives the same result.

83. Competing powers For what values of p is $\displaystyle \int_0^\infty \frac{dx}{x^p + x^{-p}} < \infty$?

84. Gamma function The gamma function is defined by $\Gamma(p) = \int_0^\infty x^{p-1} e^{-x} \, dx$, for p not equal to zero or a negative integer.

  a. Use the reduction formula

  $$\int_0^\infty x^p e^{-x} \, dx = p \int_0^\infty x^{p-1} e^{-x} \, dx \quad \text{for } p = 1, 2, 3, \ldots$$

  to show that $\Gamma(p+1) = p!$ (p factorial).

b. Use the substitution $x = u^2$ and the fact that $\int_0^\infty e^{-u^2}\, du = \dfrac{\sqrt{\pi}}{2}$

to show that $\Gamma\left(\dfrac{1}{2}\right) = \sqrt{\pi}$.

85. Many methods needed Show that $\displaystyle\int_0^\infty \dfrac{\sqrt{x}\,\ln x}{(1+x)^2}\, dx = \pi$ in the following steps.

a. Integrate by parts with $u = \sqrt{x}\,\ln x$.
b. Change variables by letting $y = 1/x$.

c. Show that $\displaystyle\int_0^1 \dfrac{\ln x}{\sqrt{x}(1+x)}\, dx = -\int_1^\infty \dfrac{\ln x}{\sqrt{x}(1+x)}\, dx$

and conclude that $\displaystyle\int_0^\infty \dfrac{\ln x}{\sqrt{x}(1+x)}\, dx = 0$.

d. Evaluate the remaining integral using the change of variables $z = \sqrt{x}$.

(*Source: Mathematics Magazine* 59, no. 1, February 1986: 49)

86. Riemann sums to integrals Show that

$$L = \lim_{n\to\infty}\left(\frac{1}{n}\ln n! - \ln n\right) = -1 \text{ in the following steps.}$$

a. Note that $n! = n(n-1)(n-2)\cdots 1$ and use $\ln(ab) = \ln a + \ln b$ to show that

$$L = \lim_{n\to\infty}\left[\left(\frac{1}{n}\sum_{k=1}^n \ln k\right) - \ln n\right]$$

$$= \lim_{n\to\infty}\frac{1}{n}\sum_{k=1}^n \ln\left(\frac{k}{n}\right)$$

b. Identify the limit of this sum as a Riemann sum for $\int_0^1 \ln x\, dx$. Integrate this improper integral by parts and reach the desired conclusion.

QUICK CHECK ANSWERS

1. The integral diverges. $\displaystyle\lim_{b\to\infty}\int_1^b(1+x^{-1})\, dx =$ $\displaystyle\lim_{b\to\infty}(x + \ln x)\big|_1^b$ does not exist. **2.** $\frac{1}{3}$ **3.** c must approach 0 through values in the interval of integration $(0, 1)$. Therefore, $c \to 0^+$. ◄

7.8 Introduction to Differential Equations

> If you read Sections 4.8 and 6.1, then you have already encountered a preview of differential equations. Given the derivative of a function, these two sections show how to find the function itself by integration. This process amounts to solving a differential equation.

If you had to demonstrate the utility of mathematics to a skeptic, a convincing way would be to cite *differential equations*. This vast subject lies at the heart of mathematical modeling and is used in engineering, the natural and biological sciences, economics, management, and finance. Differential equations rely heavily on calculus, and they are usually studied in advanced courses that follow calculus. Nevertheless, you have now seen enough calculus to understand a brief survey of differential equations and appreciate their power.

An Overview

A differential equation involves an unknown function y and its derivatives. The unknown in a differential equation is not a number (as in an algebraic equation), but rather a function or a relationship. Here are some examples of differential equations:

$$\text{(A) } \frac{dy}{dx} + 4y = \cos x \qquad \text{(B) } \frac{d^2 y}{dx^2} + 16y = 0 \qquad \text{(C) } y'(t) = 0.1y(100 - y)$$

In each case, the goal is to find a function y that satisfies the equation.

> Common choices for the independent variable in a differentiable equation are x and t, with t being used for time-dependent problems.

The **order** of a differential equation is the order of the highest-order derivative that appears in the equation. Of the three differential equations just given, (A) and (C) are first order, and (B) is second order. A differential equation is **linear** if the unknown function and its derivatives appear only to the first power and are not composed with other functions. Of these equations, (A) and (B) are linear, but (C) is nonlinear (because the right side contains y^2).

> A *linear* differential equation cannot have terms such as y^2, yy', or $\sin y$, where y is the unknown function. The most general first-order linear equation has the form $y' + py = q$, where p and q are functions of the independent variable.

Solving a first-order differential equation requires integration—you must "undo" the derivative $y'(t)$ in order to find $y(t)$. Integration introduces an arbitrary constant, so the **general solution** of a first-order differential equation involves one arbitrary constant. Similarly, the general solution of a second-order differential equation involves two arbitrary constants; with an nth-order differential equation, the general solution involves n arbitrary constants.

As discussed in Section 4.8, if a differential equation is accompanied by *initial conditions* then it is possible to determine specific values of the arbitrary constants. A differential equation together with the appropriate number of initial conditions is called an **initial value problem**. The solution of an initial value problem must satisfy both the differential equation and the initial conditions.

EXAMPLE 1 **An initial value problem** Solve the initial value problem

$$y'(t) = 10e^{-t/2}, \qquad y(0) = 4.$$

SOLUTION The solution is found by taking an indefinite integral of both sides of the differential equation with respect to t:

$$\underbrace{\int y'(t)\, dt}_{y(t)} = \int 10e^{-t/2}\, dt \qquad \text{Integrate both sides with respect to } t.$$

> The two integrals in this calculation both produce an arbitrary constant of integration. These two constants may be combined as one arbitrary constant.

$$y(t) = -20e^{-t/2} + C \quad \text{Evaluate integrals; } y(t) \text{ is an antiderivative of } y'(t).$$

We have found the general solution, which involves one arbitrary constant. To determine its value, we use the initial condition by substituting $t = 0$ and $y = 4$ into the general solution:

$$y(0) = (-20e^{-t/2} + C)|_{t=0} = -20 + C = 4 \Rightarrow C = 24$$

Therefore, the solution of the initial value problem is $y(t) = -20e^{-t/2} + 24$ (Figure 7.27). You should check that this function satisfies both the differential equation and the initial condition.

Related Exercises 9–12 ◄

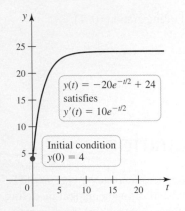

FIGURE 7.27

> The solution of the equation $y'(t) = ky(t)$ is $y(t) = Ce^{kt}$, so it models exponential growth when $k > 0$ and exponential decay when $k < 0$.

QUICK CHECK 1 What is the order of the equation in Example 1? Is it linear or nonlinear? ◄

A First-Order Linear Differential Equation

In Section 6.8 we studied functions that exhibit exponential growth or decay. Such functions have the property that their rate of change at a particular point is proportional to the function value at that point. In other words, these functions satisfy a first-order differential equation of the form $y'(t) = ky(t)$, where k is a real number. You should verify by substitution that the function $y(t) = Ce^{kt}$ is the general solution of this equation, where C is an arbitrary constant.

Now, let's generalize and consider the first-order linear equation $y'(t) = ky(t) + b$, where k and b are real numbers. Solutions of this equation have a wide range of behavior (depending on the values of k and b), and the equation itself has many modeling applications. Specifically, the terms of the equation have the following meaning:

$$\underbrace{y'(t)}_{\substack{\text{rate of change} \\ \text{of } y}} = \underbrace{ky(t)}_{\substack{\text{natural growth or} \\ \text{decay rate of } y}} + \underbrace{b}_{\substack{\text{growth or decay} \\ \text{rate due to external} \\ \text{effects}}}$$

For example, if y represents the number of fish in a hatchery, then $ky(t)$ (with $k > 0$) models exponential growth in the fish population, in the absence of other factors, and $b < 0$ is the harvesting rate at which the population is depleted. As another example, if y represents the amount of a drug in the blood, then $ky(t)$ (with $k < 0$) models exponential decay of the drug through the kidneys, and $b > 0$ is the rate at which the drug is added to the blood intravenously. We can give an explicit solution for the equation $y'(t) = ky(t) + b$.

We begin by dividing both sides of the equation $y'(t) = ky + b$ by $ky + b$, which gives

$$\frac{y'(t)}{ky + b} = 1.$$

Because the goal is to determine $y(t)$ from $y'(t)$, we integrate both sides of this equation with respect to t:

$$\int \frac{y'(t)}{ky + b}\, dt = \int dt.$$

The factor $y'(t)\, dt$ on the left side is simply dy. Making this substitution and evaluating the integrals, we have

> The arbitrary constant of integration needs to be included in only one of the integrals.

$$\int \frac{dy}{ky + b} = \int dt \quad \text{or} \quad \frac{1}{k} \ln|ky + b| = t + C.$$

For the moment, we assume that $ky + b \geq 0$, or $y \geq -b/k$, so the absolute value may be removed. Multiplying through by k, exponentiating both sides of the equation, and solving for y gives the solution $y(t) = Ce^{kt} - b/k$. In the process of solving for y, we have successively redefined C; for example, if C is arbitrary, then kC and e^C are also arbitrary. You can also show that if $ky + b < 0$, or $y < -b/k$, then the same solution results.

> The equation $y'(t) = ky(t) + b$ is one of many first-order linear differential equations. If k and b are functions of t, the equation is still first-order linear.

Solution of a First-Order Linear Differential Equation

The general solution of the first-order linear equation $y'(t) = ky(t) + b$, where k and b are real numbers, is $y(t) = Ce^{kt} - b/k$, where C is an arbitrary constant. Given an initial condition, the value of C may be determined.

QUICK CHECK 2 Verify by substitution that $y(t) = Ce^{kt} - b/k$ is a solution of $y'(t) = ky(t) + b$. ◄

EXAMPLE 2   An initial value problem for drug dosing A drug is administered to a patient through an intravenous line at a rate of 6 mg/hr. The drug has a half-life that corresponds to a rate constant of 0.03/hr (Section 6.8). Let $y(t)$ be the amount of drug in the blood for $t \geq 0$. Solve the following initial value problem and interpret the solution.

$$\text{Differential equation:} \quad y'(t) = -0.03y(t) + 6$$
$$\text{Initial condition:} \qquad\quad y(0) = 0$$

SOLUTION The equation has the form $y'(t) = ky(t) + b$, where $k = -0.03$ and $b = 6$. Therefore, the general solution is $y(t) = Ce^{-0.03t} + 200$. To determine the value of C for this particular problem, we substitute $y(0) = 0$ into the general solution. We find that $y(0) = C + 200 = 0$, which implies that $C = -200$. Therefore, the solution of the initial value problem is

$$y(t) = -200e^{-0.03t} + 200 = 200(1 - e^{-0.03t}).$$

The graph of the solution (Figure 7.28) reveals an important fact: The amount of drug in the blood increases monotonically, but it approaches a steady-state level of

$$\lim_{t \to \infty} y(t) = \lim_{t \to \infty} \left[200(1 - e^{-0.03t}) \right] = 200 \text{ mg}.$$

Related Exercises 13–22 ◄

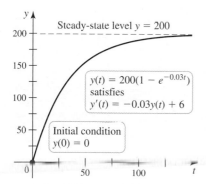

FIGURE 7.28

QUICK CHECK 3 What is the solution of $y'(t) = 3y(t) + 6$ with the initial condition $y(0) = 14$? ◄

Separable First-Order Differential Equations

The most general first-order differential equation has the form $y'(t) = F(t, y)$, where F is a given function that may involve both variables. We have a *chance* of solving such an equation if it can be written in the form

$$g(y)\, y'(t) = h(t),$$

in which the terms that involve y appear on one side of the equation *separated* from the terms that involve t. An equation that can be written in this form is said to be **separable**.

> The result of this change of variables is that the left side of the equation is integrated with respect to y and the right side is integrated with respect to t. With this justification, this shortcut is permissible and is often taken.

QUICK CHECK 4 Write
$y'(t) = (t^2 + 1)/y^3$
in separated form. ◄

The solution of the linear equation $y'(t) = ky(t) + b$ given above is a specific example of the method for solving separable differential equations. In general, we solve the separable equation $g(y)\, y'(t) = h(t)$ by integrating both sides of the equation with respect to t:

$$\int g(y) \underbrace{y'(t)\, dt}_{dy} = \int h(t)\, dt \quad \text{Integrate both sides.}$$

$$\int g(y)\, dy = \int h(t)\, dt \quad \text{Change of variables on the left side}$$

A change of variables on the left side of the equation leaves us with two integrals to evaluate, one with respect to y and one with respect to t. Finding a solution depends on evaluating these integrals.

EXAMPLE 3 A separable equation Find the function that satisfies the following initial value problem.

$$\frac{dy}{dx} = y^2 e^{-x}, \qquad y(0) = \frac{1}{2}$$

SOLUTION The equation can be written in separable form when we divide both sides of the equation by y^2 to give $y'(x)/y^2 = e^{-x}$. We now integrate both sides of the equation with respect to x and evaluate the resulting integrals:

$$\int \frac{1}{y^2} \underbrace{y'(x)\, dx}_{dy} = \int e^{-x}\, dx$$

$$\int \frac{dy}{y^2} = \int e^{-x}\, dx \quad \text{Change of variables on the left side}$$

$$-\frac{1}{y} = -e^{-x} + C \quad \text{Evaluate the integrals.}$$

Solving for y gives the general solution

$$y(x) = \frac{1}{e^{-x} - C}.$$

The initial condition $y(0) = \frac{1}{2}$ implies that

$$y(0) = \frac{1}{e^0 - C} = \frac{1}{1 - C} = \frac{1}{2}.$$

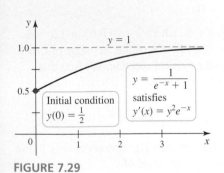

FIGURE 7.29

Solving for C gives $C = -1$, so the solution to the initial value problem is $y(x) = \dfrac{1}{e^{-x} + 1}$. The solution (Figure 7.29) has a graph that begins at $\left(0, \frac{1}{2}\right)$ and increases to approach the asymptote $y = 1$ because $\displaystyle\lim_{x \to \infty} \frac{1}{e^{-x} + 1} = 1$.

Related Exercises 23–32 ◄

> The logistic equation is used to describe the population of many different species as well as the spread of rumors and epidemics (Exercises 33–34).

EXAMPLE 4 Logistic population growth Fifty fruit flies are in a large container at the beginning of an experiment. Let $P(t)$ be the number of fruit flies in the container t days later. At first, the population grows exponentially, but due to limited space and food supply, the growth rate decreases and the population is prevented from growing without bound. This experiment can be modeled by the *logistic equation*

$$\frac{dP}{dt} = 0.1P\left(1 - \frac{P}{300}\right)$$

together with the initial condition $P(0) = 50$. Solve this initial value problem.

SOLUTION We see that the equation is separable by writing it in the form

$$\frac{1}{P\left(1 - \dfrac{P}{300}\right)} \cdot \frac{dP}{dt} = 0.1.$$

Integrating both sides with respect to t leads to the equation

$$\int \frac{1}{P\left(1 - \dfrac{P}{300}\right)} \, dP = \underbrace{\int 0.1 \, dt}_{0.1\,t \,+\, C} \tag{1}$$

The integral on the right side of equation (1) is $\int 0.1 \, dt = 0.1t + C$. Because the integrand on the left side is a rational function in P, we use partial fractions. You should verify that

$$\frac{1}{P\left(1 - \dfrac{P}{300}\right)} = \frac{300}{P(300 - P)} = \frac{1}{P} + \frac{1}{300 - P},$$

and therefore,

$$\int \frac{1}{P\left(1 - \dfrac{P}{300}\right)} \, dP = \int \left(\frac{1}{P} + \frac{1}{300 - P}\right) dP = \ln \left|\frac{P}{300 - P}\right| + C.$$

> Notice again that two constants of integration have been combined into one.

Equation (1) now becomes

$$\ln \left|\frac{P}{300 - P}\right| = 0.1t + C. \tag{2}$$

The final step is to solve for P, which is tangled up inside the logarithm. To simplify matters, we assume that if the initial population $P(0)$ is between 0 and 300, then $0 < P(t) < 300$ for all $t > 0$. This assumption (which can be verified independently) allows us to remove the absolute value on the left side of equation (2).

> It is always a good idea to check that the final solution satisfies the initial condition. In this case, $P(0) = 50$.

Using the initial condition $P(0) = 50$ and solving for C (Exercise 60), we find that $C = \ln \frac{1}{5}$. It follows that the solution of the initial value problem is

$$P(t) = \frac{300}{1 + 5e^{-0.1t}}.$$

The graph of the solution shows that the population increases, but not without bound (Figure 7.30). Instead, it approaches a steady state value of

$$\lim_{t \to \infty} P(t) = \lim_{t \to \infty} \frac{300}{1 + 5e^{-0.1t}} = 300,$$

which is the maximum population that the environment (space and food supply) can sustain. This steady-state population is called the **carrying capacity**.

Related Exercises 33–34 ◄

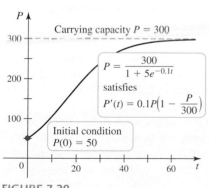

FIGURE 7.30

Direction Fields

The geometry of first-order differential equations is beautifully displayed using *direction fields*. Consider the general first-order differential equation $y'(t) = F(t, y)$, where F is a given function involving t and/or y. A solution of this equation has the property that at each point (t, y) of the solution curve, the slope of the curve is $F(t, y)$. A **direction field** is simply a picture that shows the slope of the solution at selected points of the ty-plane.

➤ Drawing direction fields by hand can be tedious. It's best to use a calculator or software.

For example, consider the equation $y'(t) = y^2 e^{-t}$. We choose a regular grid of points in the ty-plane, and at each point (t, y) we make a small line segment with slope $y^2 e^{-t}$. The line segment at a point P gives the slope of the solution curve that passes through P (Figure 7.31). For example, along the t-axis ($y = 0$), the slopes of the line segments are $F(t, 0) = 0$. And along the y-axis ($t = 0$), the slopes of the line segments are $F(0, y) = y^2$.

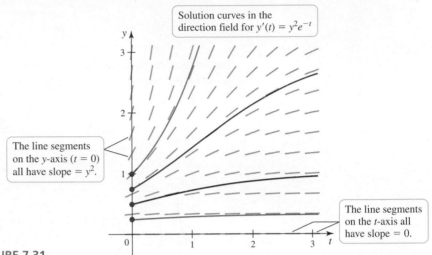

FIGURE 7.31

Now suppose an initial condition $y(a) = A$ is given. We start at the point (a, A) in the direction field and sketch a curve in the positive t-direction that follows the flow of the direction field. At each point of the solution curve, the slope matches the direction field. A different initial condition gives a different solution curve (Figure 7.31). The collection of solution curves for several different initial conditions is a representation of the general solution of the equation.

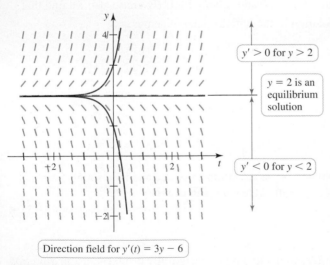

Direction field for $y'(t) = 3y - 6$

FIGURE 7.32

EXAMPLE 5 Direction field for a linear equation Sketch the direction field for the first-order linear equation $y'(t) = 3y - 6$. For what initial conditions at $t = 0$ are the solutions increasing?

SOLUTION Notice that $y'(t) = 0$ when $y = 2$. Therefore, the direction field has a horizontal line at $y = 2$. This line corresponds to an *equilibrium solution*, a solution that is constant in time: If the initial condition is $y(0) = 2$, then the solution is $y = 2$ for $t \geq 0$.

We also see that $y'(t) > 0$ when $y > 2$. Therefore, the direction field has small line segments with positive slopes above the line $y = 2$. When $y < 2$, $y'(t) < 0$, which means the direction field has small line segments with negative slopes below the line $y = 2$ (Figure 7.32).

Using the direction field, it now follows that if the initial condition satisfies $y(0) > 2$, the resulting solution increases for $t \geq 0$. If the initial condition satisfies $y(0) < 2$, the resulting solution decreases for $t \geq 0$.   *Related Exercises 35–40* ◀

QUICK CHECK 5 In Example 5, describe the behavior of the solution that results from the initial condition (a) $y(-1) = 3$ and (b) $y(-2) = 0$. ◀

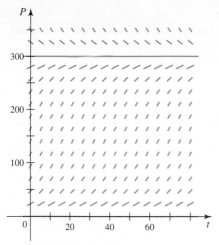

FIGURE 7.33

▷ The constant solutions $P = 0$ and $P = 300$ are equilibrium solutions. The solution $P = 0$ is an *unstable* equilibrium because nearby solution curves move away from $P = 0$. By contrast, the solution $P = 300$ is a stable equilibrium because nearby solution curves are attracted to $P = 300$.

EXAMPLE 6 Direction field for the logistic equation Consider the logistic equation of Example 4,

$$\frac{dP}{dt} = 0.1P\left(1 - \frac{P}{300}\right) \qquad \text{for } t \geq 0$$

and its direction field (Figure 7.33). Sketch the solution curves corresponding to the initial conditions $y(0) = 50$, $y(0) = 150$, and $y(0) = 350$.

SOLUTION A few preliminary observations are useful. Because P represents a population, we assume that $P \geq 0$.

- Notice that $\frac{dP}{dt} = 0$ when $P = 0$ or $P = 300$. Therefore, if the initial population is $P = 0$ or $P = 300$, then $\frac{dP}{dt} = 0$ for all $t \geq 0$, and the solution is constant. For this reason, the direction field has horizontal line segments at $P = 0$ and $P = 300$.
- The equation implies that $dP/dt > 0$ provided $0 < P < 300$. Therefore, the direction field has positive slopes, and the solutions are increasing for $t \geq 0$ and $0 < P < 300$.
- The equation also implies that $dP/dt < 0$ provided $P > 300$. Therefore, the direction field has negative slopes, and the solutions are decreasing for $t \geq 0$ and $P > 300$.

Figure 7.34 shows the direction field with three solution curves corresponding to three different initial conditions. The horizontal line $P = 300$ corresponds to the carrying capacity of the population. We see that if the initial population is less than 300, the resulting solution increases to the carrying capacity from below. If the initial population is greater than 300, the resulting solution decreases to the carrying capacity from above.

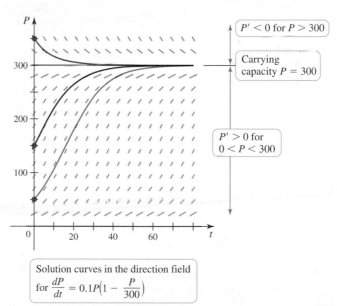

Solution curves in the direction field for $\frac{dP}{dt} = 0.1P\left(1 - \frac{P}{300}\right)$

FIGURE 7.34

Related Exercises 35–40 ◀

Direction fields are useful for at least two reasons. As shown in Example 5, a direction field provides valuable qualitative information about the solutions of a differential equation *without solving the equation*. In addition, it turns out that direction fields are the basis for many computer-based methods for approximating solutions of a differential equation. The computer begins with the initial condition and advances the solution in small steps, always following the direction field at each time step.

SECTION 7.8 EXERCISES

Review Questions

1. What is the order of $y''(t) + 9y(t) = 10$?

2. Is $y''(t) + 9y(t) = 10$ linear or nonlinear?

3. How many arbitrary constants appear in the general solution of $y''(t) + 9y(t) = 10$?

4. If the general solution of a differential equation is $y(t) = Ce^{-3t} + 10$, what is the solution that satisfies the initial condition $y(0) = 5$?

5. What is a separable first-order differential equation?

6. Is the equation $t^2 y'(t) = (t + 4)/y^2$ separable?

7. Explain how to solve a separable differential equation of the form $g(y) y'(t) = h(t)$.

8. Explain how to sketch the direction field of the equation $y'(t) = F(t, y)$, where F is given.

Basic Skills

9–12. Warm-up initial value problems *Solve the following problems.*

9. $y'(t) = 3t^2 - 4t + 10$, $y(0) = 20$

10. $\dfrac{dy}{dt} = 8e^{-4t} + 1$, $y(0) = 5$

11. $y'(t) = (2t^2 + 4)/t$, $y(1) = 2$

12. $\dfrac{dy}{dx} = 3\cos 2x + 2\sin 3x$, $y(\pi/2) = 8$

13–16. First-order linear equations *Find the general solution of the following equations.*

13. $y'(t) = 3y - 4$

14. $\dfrac{dy}{dx} = -y + 2$

15. $y'(x) = -2y - 4$

16. $\dfrac{dy}{dt} = 2y + 6$

17–20. Initial value problems *Solve the following problems.*

17. $y'(t) = 3y - 6$, $y(0) = 9$

18. $\dfrac{dy}{dx} = -y + 2$, $y(0) = -2$

19. $y'(t) = -2y - 4$, $y(0) = 0$

20. $\dfrac{du}{dx} = 2u + 6$, $u(1) = 6$

21. **Intravenous drug dosing** The amount of drug in the blood of a patient (in mg) due to an intravenous line is governed by the initial value problem

$$y'(t) = -0.02y + 3, \quad y(0) = 0 \qquad \text{for } t \geq 0$$

where t is measured in hours.

a. Find and graph the solution of the initial value problem.
b. What is the steady-state level of the drug?
c. When does the drug level reach 90% of the steady-state value?

22. **Fish harvesting** A fish hatchery has 500 fish at time $t = 0$, when harvesting begins at a rate of b fish/yr, where $b > 0$. The fish population is modeled by the initial value problem

$$y'(t) = 0.1y - b, \quad y(0) = 500 \qquad \text{for } t \geq 0$$

where t is measured in years.

a. Find the fish population for $t \geq 0$ in terms of the harvesting rate b.
b. Graph the solution in the case that $b = 40$ fish/yr. Describe the solution.
c. Graph the solution in the case that $b = 60$ fish/yr. Describe the solution.

23–26. Separable differential equations *Find the general solution of the following equations.*

23. $\dfrac{dy}{dt} = \dfrac{3t^2}{y}$

24. $\dfrac{dy}{dx} = y(x^2 + 1)$, where $y > 0$

25. $y'(t) = e^{y/2}\sin t$

26. $x^2 \dfrac{dw}{dx} = \sqrt{w}(3x + 1)$

27–32. Separable differential equations *Determine whether the following equations are separable. If so, solve the given initial value problem.*

27. $\dfrac{dy}{dt} = ty + 2$, $y(1) = 2$

28. $y'(t) = y(4t^3 + 1)$, $y(0) = 4$

29. $y'(t) = \dfrac{e^t}{2y}$, $y(\ln 2) = 1$

30. $(\sec x) y'(x) = y^3$, $y(0) = 3$

31. $\dfrac{dy}{dx} = e^{x-y}$, $y(0) = \ln 3$

32. $y'(t) = 2e^{3y-t}$, $y(0) = 0$

33. **Logistic equation for a population** A community of hares on an island has a population of 50 when observations begin at $t = 0$. The population for $t \geq 0$ is modeled by the initial value problem

$$\dfrac{dP}{dt} = 0.08P\left(1 - \dfrac{P}{200}\right), \qquad P(0) = 50$$

a. Find and graph the solution of the initial value problem.
b. What is the steady-state population?

34. **Logistic equation for an epidemic** When an infected person is introduced into a closed and otherwise healthy community, the number of people who become infected with the disease (in the absence of any intervention) may be modeled by the logistic equation

$$\dfrac{dP}{dt} = kP\left(1 - \dfrac{P}{A}\right), \qquad P(0) = P_0$$

where k is a positive infection rate, A is the number of people in the community, and P_0 is the number of infected people at $t = 0$. The model assumes no recovery or intervention.

a. Find the solution of the initial value problem in terms of k, A, and P_0.

b. Graph the solution in the case that $k = 0.025$, $A = 300$, and $P_0 = 1$.

c. For fixed values of k and A, describe the long-term behavior of the solutions for any P_0 with $0 < P_0 < A$.

35–36. Direction fields *A differential equation and its direction field are given. Sketch a graph of the solution that results with each initial condition.*

35. $y'(t) = \dfrac{t^2}{y^2 + 1}$,
$y(0) = -2$ and
$y(-2) = 0$

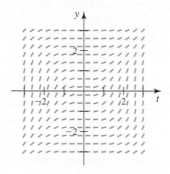

36. $y'(t) = \dfrac{\sin t}{y}$,
$y(-2) = -2$ and
$y(-2) = 2$

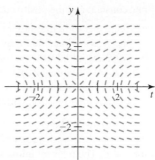

37. Matching direction fields Match equations (a)–(d) with the direction fields A–D.

(a) $y'(t) = t/2$ **(b)** $y'(t) = y/2$
(c) $y'(t) = (t^2 + y^2)/2$ **(d)** $y'(t) = y/t$

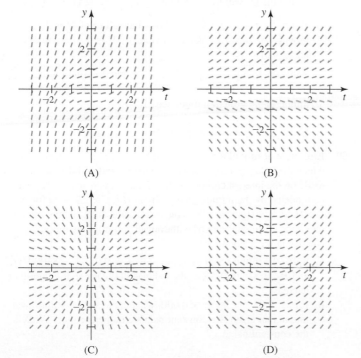

38–40. Sketching direction fields *Use the window $[-2, 2] \times [-2, 2]$ to sketch a direction field for the following equations. Then sketch the solution curve that corresponds to the given initial condition.*

38. $y'(t) = y - 3$, $y(0) = 1$

39. $y'(x) = \sin x$, $y(-2) = 2$

40. $y'(t) = \sin y$, $y(-2) = \frac{1}{2}$

Further Explorations

41. Explain why or why not Determine whether the following statements are true and give an explanation or counter-example.

    a. The general solution of $y'(t) = 20y$ is $y = e^{20t}$.
    b. The functions $y = 2e^{-2t}$ and $y = 10e^{-2t}$ do not both satisfy the differential equation $y' + 2y = 0$.
    c. The equation $y'(t) = ty + 2y + 2t + 4$ is not separable.
    d. A solution of $y'(t) = 2\sqrt{y}$ is $y = (t + 1)^2$.

42–47. Equilibrium solutions *A differential equation of the form $y'(t) = F(y)$ is said to be **autonomous** (the function F depends only on y). The constant function $y = y_0$ is an equilibrium solution of the equation provided $F(y_0) = 0$ (because then $y'(t) = 0$, and the solution remains constant for all t). Note that equilibrium solutions correspond to horizontal line segments in the direction field. Note also that for autonomous equations, the direction field is independent of t. Consider the following equations.*

    a. *Find all equilibrium solutions.*
    b. *Sketch the direction field on either side of the equilibrium solutions for $t \geq 0$.*
    c. *Sketch the solution curve that corresponds to the initial condition $y(0) = 1$.*

42. $y'(t) = 2y + 4$ **43.** $y'(t) = y^2$

44. $y'(t) = y(2 - y)$ **45.** $y'(t) = y(y - 3)$

46. $y'(t) = \sin y$ **47.** $y'(t) = y(y - 3)(y + 2)$

48–51. Solving initial value problems *Solve the following problems using the method of your choice.*

48. $u'(t) = 4u - 2$, $u(0) = 4$ **49.** $\dfrac{dp}{dt} = \dfrac{p + 1}{t^2}$, $p(1) = 3$

50. $\dfrac{dz}{dx} = \dfrac{z^2}{1 + x^2}$, $z(0) = \dfrac{1}{6}$

51. $w'(t) = 2t \cos^2 w$, $w(0) = \pi/4$

52. Optimal harvesting rate Let $y(t)$ be the population of a species that is being harvested. Consider the harvesting model $y'(t) = 0.008y - h$, $y(0) = y_0$, where $h > 0$ is the annual harvesting rate and y_0 is the initial population of the species.

    a. If $y_0 = 2000$, what harvesting rate should be used to maintain a constant population of $y = 2000$ for $t \geq 0$?
    b. If the harvesting rate is $h = 200$/year, what initial population ensures a constant population for $t \geq 0$?

Applications

53. Logistic equation for spread of rumors Sociologists model the spread of rumors using logistic equations. The key assumption is that at any given time a fraction y of the population, where $0 \le y \le 1$, knows the rumor, while the remaining fraction $1 - y$ does not. Furthermore, the rumor spreads by interactions between those who know the rumor and those who do not. The number of such interactions is proportional to $y(1 - y)$. Therefore, the equation that describes the spread of the rumor is $y'(t) = ky(1 - y)$, where k is a positive real number. The fraction of people who initially know the rumor is $y(0) = y_0$, where $0 < y_0 < 1$.

    a. Solve this initial value problem and give the solution in terms of k and y_0.

    b. Assume $k = 0.3 \text{ weeks}^{-1}$ and graph the solution for $y_0 = 0.1$ and $y_0 = 0.7$.

    c. Describe and interpret the long-term behavior of the rumor function for any $0 < y_0 < 1$.

54. Free fall An object in free fall may be modeled by assuming that the only forces at work are the gravitational force and resistance (friction due to the medium in which the object falls). By Newton's second law (mass × acceleration = the sum of the external forces), the velocity of the object satisfies the differential equation

$$\underbrace{m}_{\text{mass}} \cdot \underbrace{v'(t)}_{\text{acceleration}} = \underbrace{mg + f(v)}_{\substack{\text{external} \\ \text{forces}}},$$

where f is a function that models the resistance and the positive direction is downward. One common assumption (often used for motion in air) is that $f(v) = -kv^2$, where $k > 0$ is a drag coefficient.

    a. Show that the equation can be written in the form $v'(t) = g - av^2$, where $a = k/m$.

    b. For what (positive) value of v is $v'(t) = 0$? (This equilibrium solution is called the **terminal velocity**.)

    c. Find the solution of this separable equation assuming $v(0) = 0$ and $0 < v(t)^2 < g/a$, for $t \ge 0$.

    d. Graph the solution found in part (c) with $g = 9.8 \text{ m/s}^2$, $m = 1 \text{ kg}$, and $k = 0.1 \text{ kg/m}$, and verify that the terminal velocity agrees with the value found in part (b).

55. Free fall Using the background given in Exercise 54, assume the resistance is given by $f(v) = -Rv$, where $R > 0$ is a drag coefficient (an assumption often made for a heavy medium such as water or oil).

    a. Show that the equation can be written in the form $v'(t) = g - bv$, where $b = R/m$.

    b. For what (positive) value of v is $v'(t) = 0$? (This equilibrium solution is called the **terminal velocity**.)

    c. Find the solution of this separable equation assuming $v(0) = 0$ and $0 < v < g/b$.

    d. Graph the solution found in part (c) with $g = 9.8 \text{ m/s}^2$, $m = 1 \text{ kg}$, and $R = 0.1 \text{ kg/s}$, and verify that the terminal velocity agrees with the value found in part (b).

56. Torricelli's Law An open cylindrical tank initially filled with water drains through a hole in the bottom of the tank according to Torricelli's Law (see figure). If $h(t)$ is the depth of water in the tank for $t \ge 0$, then Torricelli's Law implies $h'(t) = -2k\sqrt{h}$, where $k > 0$ is a constant that includes the acceleration due to gravity,

the radius of the tank, and the radius of the drain. Assume that the initial depth of the water is $h(0) = H$.

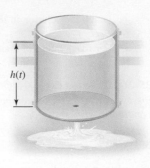

    a. Find the general solution of the equation.

    b. Find the solution in the case that $k = 0.1$ and $H = 0.5 \text{ m}$.

    c. In general, how long does it take for the tank to drain in terms of k and H?

57. Chemical rate equations The reaction of chemical compounds can often be modeled by differential equations. Let $y(t)$ be the concentration of a substance in reaction for $t \ge 0$ (typical units of y are moles/L). The change in the concentration of the substance, under appropriate conditions, is $\dfrac{dy}{dt} = -ky^n$, where $k > 0$ is a rate constant and the positive integer n is the order of the reaction.

    a. Show that for a first-order reaction ($n = 1$), the concentration obeys an exponential decay law.

    b. Solve the initial value problem for a second-order reaction ($n = 2$) assuming $y(0) = y_0$.

    c. Graph and compare the concentration for a first-order and second-order reaction with $k = 0.1$ and $y_0 = 1$.

58. Tumor growth The growth of cancer tumors may be modeled by the Gompertz growth equation. Let $M(t)$ be the mass of the tumor for $t \ge 0$. The relevant initial value problem is

$$\frac{dM}{dt} = -aM \ln\left(\frac{M}{K}\right), \qquad M(0) = M_0,$$

where a and K are positive constants and $0 < M_0 < K$.

    a. Graph the growth rate function $R(M) = -aM \ln\left(\dfrac{M}{K}\right)$ assuming $a = 1$ and $K = 4$. For what values of M is the growth rate positive? For what value of M is the growth rate a maximum?

    b. Solve the initial value problem and graph the solution for $a = 1$, $K = 4$, and $M_0 = 1$. Describe the growth pattern of the tumor. Is the growth unbounded? If not, what is the limiting size of the tumor?

    c. In the general equation, what is the meaning of K?

59. Endowment model An endowment is an investment account in which the balance ideally remains constant and withdrawals are made on the interest earned by the account. Such an account may be modeled by the initial value problem $B'(t) = aB - m$ for $t \ge 0$, with $B(0) = B_0$. The constant a reflects the annual interest rate, m is the annual rate of withdrawal, and B_0 is the initial balance in the account.

    a. Solve the initial value problem with $a = 0.05$, $m = \$1000/\text{yr}$, and $B_0 = \$15,000$. Does the balance in the account increase or decrease?

    b. If $a = 0.05$ and $B_0 = \$50,000$, what is the annual withdrawal rate m that ensures a constant balance in the account? What is the constant balance?

Additional Exercises

60. Solution of the logistic equation Consider the solution of the logistic equation in Example 4.

a. From the general solution $\ln\left|\dfrac{P}{300-P}\right| = 0.1t + C$, show that the initial condition $P(0) = 50$ implies that $C = \ln\frac{1}{5}$.

b. Solve for P and show that $P(t) = \dfrac{300}{1 + 5e^{-0.1t}}$.

61. Direction field analysis Consider the general first-order initial value problem $y'(t) = ay + b, y(0) = y_0$, for $t \geq 0$, where $a, b,$ and y_0 are real numbers.

a. Explain why $y = -b/a$ is an equilibrium solution and corresponds to a horizontal line in the direction field.

b. Draw a representative direction field in the case that $a > 0$. Show that if $y_0 > -b/a$, then the solution increases for $t \geq 0$, and if $y_0 < -b/a$, then the solution decreases for $t \geq 0$.

c. Draw a representative direction field in the case that $a < 0$. Show that if $y_0 > -b/a$, then the solution decreases for $t \geq 0$, and if $y_0 < -b/a$, then the solution increases for $t \geq 0$.

1. The equation is first order and linear. **3.** The solution is $y(t) = 16e^{3t} - 2$. **4.** $y^3 y'(t) = t^2 + 1$ **5. a.** Solution increases for $t \geq -1$. **b.** Solution decreases for $t \geq -2$. ◄

CHAPTER 7 REVIEW EXERCISES

1. Explain why or why not Determine whether the following statements are true and give an explanation or counterexample.

a. The integral $\int x^2 e^{2x}\, dx$ can be evaluated analytically using integration by parts.

b. To evaluate the integral $\displaystyle\int \frac{dx}{\sqrt{x^2 - 100}}$ analytically, it is best to use partial fractions.

c. One computer algebra system produces $\int 2\sin x \cos x\, dx = \sin^2 x$. Another computer algebra system produces $\int 2\sin x \cos x\, dx = -\cos^2 x$. One computer algebra system is wrong (apart from a missing constant of integration).

2–19. Integrals Evaluate the following integrals analytically.

2. $\displaystyle\int x^2 \cos x\, dx$

3. $\displaystyle\int e^x \sin x\, dx$

4. $\displaystyle\int_1^e x^7 \ln x\, dx$

5. $\displaystyle\int \cos^2 4\theta\, d\theta$

6. $\displaystyle\int \sin 3x \cos^6 3x\, dx$

7. $\displaystyle\int \sec^5 z \tan z\, dz$

8. $\displaystyle\int_0^{\pi/2} \cos^4 x\, dx$

9. $\displaystyle\int_0^{\pi/6} \sin^5 \theta\, d\theta$

10. $\displaystyle\int \tan^4 u\, du$

11. $\displaystyle\int \frac{dx}{\sqrt{4 - x^2}}$

12. $\displaystyle\int \frac{dx}{\sqrt{9x^2 - 25}}$ for $x > \frac{5}{3}$

13. $\displaystyle\int \frac{dy}{y^2\sqrt{9 - y^2}}$

14. $\displaystyle\int_0^{\sqrt{3}/2} \frac{x^2}{(1 - x^2)^{3/2}}\, dx$

15. $\displaystyle\int_0^{\sqrt{3}/2} \frac{4}{9 + 4x^2}\, dx$

16. $\displaystyle\int \frac{(1 - u^2)^{5/2}}{u^8}\, du$

17. $\displaystyle\int \frac{dx}{x^2 - 2x - 15}$

18. $\displaystyle\int \frac{dx}{x^3 - 2x^2}$

19. $\displaystyle\int_0^1 \frac{dy}{(y + 1)(y^2 + 1)}$

20–22. Table of integrals Use a table of integrals to evaluate the following integrals.

20. $\displaystyle\int x(2x + 3)^5\, dx$

21. $\displaystyle\int \frac{dx}{x\sqrt{4x - 6}}$

22. $\displaystyle\int_0^{\pi/2} \frac{d\theta}{1 + \sin 2\theta}$

23–24. Approximations Use a computer algebra system to approximate the value of the following integrals.

23. $\displaystyle\int_{-1}^1 e^{-2x^2}\, dx$

24. $\displaystyle\int_1^{\sqrt{e}} x^3 (\ln x)^3\, dx$

25. Numerical integration methods Let $I = \int_0^3 x^2\, dx = 9$ and consider the Trapezoid Rule (T_n) and the Midpoint Rule (M_n) approximations to I.

a. Compute T_6 and M_6.

b. Compute T_{12} and M_{12}.

26. Errors in numerical integration Let $I = \int_{-1}^{2}\left(x^7 - 3x^5 - x^2 + \frac{7}{8}\right) dx$ and note that $I = 0$.

  a. Complete the following table with Trapezoid Rule (T_n) and Midpoint Rule (M_n) approximations to I for various values of n.

  b. Fill in the error columns with the absolute errors in the approximations in part (a).

n	T_n	M_n	Abs error in T_n	Abs error in M_n
4				
8				
16				
32				
64				

  c. How do the errors in T_n decrease as n doubles in size?

  d. How do the errors in M_n decrease as n doubles in size?

27. Best approximation Let $I = \int_{0}^{1}\frac{x^2 - x}{\ln x} dx$. Use any method you choose to find a good approximation to I. You may use the facts that $\lim_{x\to 0^+}\frac{x^2 - x}{\ln x} = 0$ and $\lim_{x\to 1}\frac{x^2 - x}{\ln x} = 1$.

28–31. Improper integrals *Evaluate the following integrals.*

28. $\int_{1}^{\infty}\frac{dx}{(x + 1)^9}$

29. $\int_{0}^{\infty} xe^{-x}\, dx$

30. $\int_{0}^{8}\frac{dx}{\sqrt{2x}}$

31. $\int_{0}^{3}\frac{dx}{\sqrt{9 - x^2}}$

32–37. Preliminary work *Make a change of variables or use an algebra step before evaluating the following integrals.*

32. $\int_{-1}^{1}\frac{dx}{x^2 + 2x + 5}$

33. $\int\frac{dx}{x^2 - x - 2}$

34. $\int\frac{3x^2 + x - 3}{x^2 - 1}\, dx$

35. $\int\frac{2x^2 - 4x}{x^2 - 4}\, dx$

36. $\int_{1/12}^{1/4}\frac{dx}{\sqrt{x}(1 + 4x)}$

37. $\int\frac{e^{2t}}{(1 + e^{4t})^{3/2}}\, dt$

38. Two ways Evaluate $\int\frac{dx}{4 - x^2}$ using partial fractions and a trigonometric substitution, and show that the results are consistent.

39–42. Volumes *The region R is bounded by the curve $y = \ln x$ and the x-axis on the interval $[1, e]$. Find the volume of the solid that is generated when R is revolved in the following ways.*

39. About the x-axis

40. About the y-axis

41. About the line $x = 1$

42. About the line $y = 1$

43. Comparing volumes Let R be the region bounded by the graph of $y = \sin x$ and the x-axis on the interval $[0, \pi]$. Which is greater, the volume of the solid generated when R is revolved about the x-axis or the y-axis?

44. Comparing areas Show that the area of the region bounded by the graph of $y = ae^{-ax}$ and the x-axis on the interval $[0, \infty)$ is the same for all values of $a > 0$.

45. Zero log integral It is evident from the graph of $y = \ln x$ that for every real number a with $0 < a < 1$, there is a unique real number $b = g(a)$ with $b > 1$, such that $\int_{a}^{b}\ln x\, dx = 0$ (the net area bounded by the graph of $y = \ln x$ on $[a, b]$ is 0).

  a. Approximate $b = g\left(\frac{1}{2}\right)$.

  b. Approximate $b = g\left(\frac{1}{3}\right)$.

  c. Find the equation satisfied by all pairs of numbers (a, b) such that $b = g(a)$.

  d. Is g an increasing or decreasing function of a? Explain.

46. Arc length Find the length of the curve $y = \ln x$ from $x = 1$ to $x = e^2$.

47. Average velocity Find the average velocity of a projectile whose velocity over the interval $0 \le t \le \pi$ is given by $v(t) = 10\sin 3t$.

48. Comparing distances Starting at the same time and place ($t = 0$ and $s = 0$), the velocity of car A is given by $u(t) = 40/(t + 1)$ (mi/hr) and the velocity of car B is given by $v(t) = 40e^{-t/2}$ (mi/hr).

  a. After $t = 2$ hr, which car has traveled the greater distance?

  b. After $t = 3$ hr, which car has traveled the greater distance?

  c. If allowed to travel indefinitely ($t \to \infty$), which car will travel a finite distance?

49. Traffic flow When data from a traffic study are fitted to a curve, the flow rate of cars past a point on a highway is approximated by $R(t) = 800te^{-t/2}$ cars/hr. How many cars pass the measuring site during the time interval $0 \le t \le 4$ hr?

50. Comparing integrals Graph the functions $f(x) = \pm 1/x^2$, $g(x) = (\cos x)/x^2$, and $h(x) = (\cos^2 x)/x^2$. Without evaluating integrals and knowing that $\int_{1}^{\infty} f(x)\, dx$ has a finite value, determine whether $\int_{1}^{\infty} g(x)\, dx$ and $\int_{1}^{\infty} h(x)\, dx$ have finite values.

51. A family of logarithm integrals Let $I(p) = \int_{1}^{e}\frac{\ln x}{x^p}\, dx$, where p is a real number.

  a. Find an expression for $I(p)$ for all real values of p.

  b. Evaluate $\lim_{p\to\infty} I(p)$ and $\lim_{p\to-\infty} I(p)$.

  c. For what value of p is $I(p) = 1$?

52. CAS approximation Use a computer algebra system to determine the integer n that satisfies $\int_{0}^{1/2}\frac{\ln(1 + 2x)}{x}\, dx = \frac{\pi^2}{n}$.

53. CAS approximation Use a computer algebra system to determine the integer n that satisfies $\int_{0}^{1}\frac{\sin^{-1}x}{x}\, dx = \frac{\pi\ln 2}{n}$.

54. Two worthy integrals

a. Let $I(a) = \int_0^\infty \dfrac{dx}{(1 + x^a)(1 + x^2)}$, where a is a real number. Evaluate $I(a)$ and show that its value is independent of a. (*Hint:* Split the integral into two integrals over $[0, 1]$ and $[1, \infty)$; then, use a change of variables to convert the second integral into an integral over $[0, 1]$.)

b. Let f be any positive continuous function on $[0, \pi/2]$ and evaluate $\int_0^{\pi/2} \dfrac{f(\cos x)}{f(\cos x) + f(\sin x)} dx.$

(*Hint:* Use the identity $\cos(\pi/2 - x) = \sin x$.)
(*Source: Mathematics Magazine*, 81, no. 2, April 2008: 152–154)

55–58. Initial value problems *Use the method of your choice to solve the following initial value problems.*

55. $y'(t) = 2y + 4, \; y(0) = 8$

56. $\dfrac{dy}{dt} = \dfrac{2ty}{\ln y}, \; y(2) = e$

57. $y'(t) = \dfrac{t + 1}{2ty}, \; y(1) = 4$

58. $\dfrac{dy}{dt} = \sqrt{y} \sin t, \; y(0) = 4$

59. Limit of a solution Evaluate $\lim_{t \to \infty} y(t)$, where y is the solution of the initial value problem $y'(t) = \dfrac{\sec y}{t^2}, y(1) = 0.$

60–62. Sketching direction fields *Use the window $[-2, 2] \times [-2, 2]$ to sketch a direction field for the given differential equation. Then, sketch the solution curve that corresponds to the given initial condition.*

60. $y'(t) = 3y - 6, \; y(0) = 1$

61. $y'(t) = t^2, \; y(-1) = -1$

62. $y'(t) - y - t, \; y(-2) = \frac{1}{2}$

63. Enzyme kinetics The consumption of a substrate in a reaction involving an enzyme is often modeled using Michaelis-Menton kinetics, which involves the initial value problem $\dfrac{ds}{dt} = -\dfrac{Qs}{K + s}$,

$s(0) = s_0$, where $s(t)$ is the amount of substrate present at time $t > 0$, and Q and K are positive constants. Solve the initial value problem with $Q = 10, K = 5$, and $s_0 = 50$. Notice that the solution can be expressed explicitly only with t as a function of s. Graph the solution and describe how s behaves as $t \to \infty$. (See the Guided Projects for more on enzyme kinetics.)

64. Investment model An investment account that earns interest and also has regular deposits can be modeled by the initial value problem $B'(t) = aB + m$ for $t \geq 0$, with $B(0) = B_0$. The constant a reflects the monthly interest rate, m is the rate of monthly deposits, and B_0 is the initial balance in the account. Solve the initial value problem with $a = 0.005, m = \$100/\text{month}$, and $B_0 = \$100$. After how many months does the account have a balance of $\$7500$?

65. Comparing volumes Let R be the region bounded by $y = \ln x$, the x-axis, and the line $x = a$, where $a > 1$.

a. Find the volume $V_1(a)$ of the solid generated when R is revolved about the x-axis (as a function of a).

b. Find the volume $V_2(a)$ of the solid generated when R is revolved about the y-axis (as a function of a).

c. Graph V_1 and V_2. For what values of $a > 1$ is $V_1(a) > V_2(a)$?

66. Equal volumes

a. Let R be the region bounded by the graph of $f(x) = x^{-p}$ and the x-axis for $x \geq 1$. Let V_1 and V_2 be the volumes of the solids generated when R is revolved about the x-axis and the y-axis, respectively, if they exist. For what values of p (if any) is $V_1 = V_2$?

b. Repeat part (a) on the interval $(0, 1]$.

67. Equal volumes Let R_1 be the region bounded by the graph of $y = e^{-ax}$ and the x-axis on the interval $[0, b]$ where $a > 0$ and $b > 0$. Let R_2 be the region bounded by the graph of $y = e^{-ax}$ and the x-axis on the interval $[b, \infty)$. Let V_1 and V_2 be the volumes of the solids generated when R_1 and R_2 are revolved about the x-axis. Find and graph the relationship between a and b for which $V_1 = V_2$.

Chapter 7 Guided Projects

Applications of the material in this chapter and related topics can be found in the following Guided Projects. For additional information, see the Preface.

- Cooling coffee
- Euler's method for differential equations
- Terminal velocity
- A pursuit problem
- How long will your iPod last?

- Simpson's rule
- Predator-prey models
- Period of the pendulum
- Logistic growth
- Mercator projections

8

Sequences and Infinite Series

Chapter Preview This chapter covers topics that lie at the foundation of calculus—indeed, at the foundation of mathematics. The first task is to make a clear distinction between a *sequence* and an *infinite series*. A sequence is an ordered *list* of numbers, $a_1, a_2, \ldots$, while an infinite series is a *sum* of numbers, $a_1 + a_2 + \cdots$. The idea of convergence to a limit is important for both sequences and series, but convergence is analyzed differently in the two cases. To determine limits of sequences, we use the same tools used for limits at infinity of functions. Convergence of infinite series is a different matter, and we develop the required methods in this chapter. The study of infinite series begins with the ubiquitous *geometric series*; it has theoretical importance and it is used to answer many practical questions (When is your auto loan paid off? How much antibiotic do you have in your blood if you take three pills a day?). We then present several tests that are used to determine whether series with positive terms converge. Finally, alternating series, whose terms alternate in sign, are discussed in anticipation of power series in the next chapter.

8.1 An Overview

> Keeping with common practice, the terms *series* and *infinite series* are used interchangeably throughout this chapter.

> The dots (. . .) after the last number (called an *ellipsis*) mean that the list goes on indefinitely.

To understand sequences and series, you must understand how they differ and how they are related. The purposes of this opening section are to introduce sequences and series in concrete terms and to illustrate their differences and their crucial relationships with each other.

Examples of Sequences

Consider the following *list* of numbers:

$$\{1, 4, 7, 10, 13, 16, \ldots\}$$

Each number in the list is obtained by adding 3 to the previous number. With this rule, we could extend the list indefinitely.

This list is an example of a **sequence**, where each number in the sequence is called a **term** of the sequence. We denote sequences in any of the following forms:

$$\{a_1, a_2, a_3, \ldots, a_n, \ldots\} \qquad \{a_n\}_{n=1}^{\infty} \qquad \{a_n\}$$

The subscript n that appears in a_n is called an **index**, and it indicates the order of terms in the sequence. The choice of a starting index is arbitrary, but sequences usually begin with $n = 0$ or $n = 1$.

The sequence $\{1, 4, 7, 10, \dots\}$ can be defined in two ways. First, we have the rule that each term of the sequence is 3 more than the previous term; that is, $a_2 = a_1 + 3$, $a_3 = a_2 + 3$, $a_4 = a_3 + 3$, and so forth. In general, we see that

$$a_1 = 1 \quad \text{and} \quad a_{n+1} = a_n + 3, \qquad \text{for } n = 1, 2, 3, \dots .$$

This way of defining a sequence is called a **recurrence relation** (or an **implicit formula**). It specifies the initial term of the sequence (in this case, $a_1 = 1$) and gives a general rule for computing the next term of the sequence from previous terms. For example, if you know a_{100}, the recurrence relation can be used to find a_{101}.

Suppose instead you want to find a_{147} directly without computing the first 146 terms of the sequence. The first four terms of the sequence can be written

$$a_1 = 1 + (3 \cdot 0), \qquad a_2 = 1 + (3 \cdot 1), \qquad a_3 = 1 + (3 \cdot 2), \qquad a_4 = 1 + (3 \cdot 3).$$

Observe the pattern: The nth term of the sequence is 1 plus 3 multiplied by $n - 1$, or

$$a_n = 1 + 3(n - 1) = 3n - 2, \qquad \text{for } n = 1, 2, 3, \dots$$

With this **explicit formula**, the nth term of the sequence is determined directly from the value of n. For example, with $n = 147$,

$$a_{147} = 3 \cdot \underset{n}{147} - 2 = 439.$$

QUICK CHECK 1 Find a_{10} for the sequence $\{1, 4, 7, 10, \dots\}$ using the recurrence relation and then again using the explicit formula for the nth term. ◄

> When defined by an explicit formula $a_n = f(n)$, it is evident that sequences are functions. The domain is the set of positive, or nonnegative, integers, and one real number a_n is assigned to each integer in the domain.

DEFINITION   Sequence

A **sequence** $\{a_n\}$ is an ordered list of numbers of the form

$$\{a_1, a_2, a_3, \dots, a_n, \dots\}.$$

A sequence may be generated by a **recurrence relation** of the form $a_{n+1} = f(a_n)$, for $n = 1, 2, 3, \dots$, where a_1 is given. A sequence may also be defined with an **explicit formula** for the nth term in the form $a_n = f(n)$, for $n = 1, 2, 3, \dots .$

EXAMPLE 1   Explicit formulas Use the explicit formula for $\{a_n\}_{n=1}^{\infty}$ to write the first four terms of each sequence. Sketch a graph of the sequence.

a. $a_n = \dfrac{1}{2^n}$ **b.** $a_n = \dfrac{(-1)^n n}{n^2 + 1}$

SOLUTION

a. Substituting $n = 1, 2, 3, 4, \dots$ into the explicit formula $a_n = \dfrac{1}{2^n}$, we find that the terms of the sequence are

$$\left\{ \frac{1}{2}, \frac{1}{2^2}, \frac{1}{2^3}, \frac{1}{2^4}, \dots \right\} = \left\{ \frac{1}{2}, \frac{1}{4}, \frac{1}{8}, \frac{1}{16}, \dots \right\}.$$

The graph of a sequence is like the graph of a function that is defined only on a set of integers. In this case, we plot the coordinate pairs (n, a_n) for $n = 1, 2, 3, \dots$, resulting in a graph consisting of individual points. The graph of the sequence $a_n = \dfrac{1}{2^n}$ suggests that the terms of this sequence approach 0 as n increases (Figure 8.1).

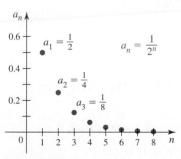

FIGURE 8.1

> The "switch" $(-1)^n$ is used frequently to alternate the signs of the terms of sequences and series.

b. Substituting $n = 1, 2, 3, 4, \dots$ into the explicit formula, the terms of the sequence are

$$\left\{ \frac{(-1)^1(1)}{1^2 + 1}, \frac{(-1)^2 2}{2^2 + 1}, \frac{(-1)^3 3}{3^2 + 1}, \frac{(-1)^4 4}{4^2 + 1}, \dots \right\} = \left\{ -\frac{1}{2}, \frac{2}{5}, -\frac{3}{10}, \frac{4}{17}, \dots \right\}.$$

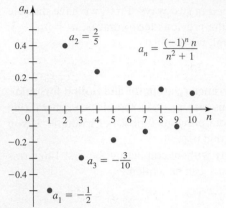

FIGURE 8.2

From the graph (Figure 8.2), we see that the terms of the sequence alternate in sign and appear to approach 0 as n increases.   *Related Exercises 9–12* ◄

EXAMPLE 2   Recurrence relations Use the recurrence relation for $\{a_n\}_{n=1}^{\infty}$ to write the first four terms of the sequences

$$a_{n+1} = 2a_n + 1, a_1 = 1 \quad \text{and} \quad a_{n+1} = 2a_n + 1, a_1 = -1.$$

SOLUTION Notice that the recurrence relation is the same for the two sequences; only the first term differs. The first four terms of the two sequences are as follows.

n	a_n with $a_1 = 1$	a_n with $a_1 = -1$
1	$a_1 = 1$ (given)	$a_1 = -1$ (given)
2	$a_2 = 2a_1 + 1 = 2 \cdot 1 + 1 = 3$	$a_2 = 2a_1 + 1 = 2(-1) + 1 = -1$
3	$a_3 = 2a_2 + 1 = 2 \cdot 3 + 1 = 7$	$a_3 = 2a_2 + 1 = 2(-1) + 1 = -1$
4	$a_4 = 2a_3 + 1 = 2 \cdot 7 + 1 = 15$	$a_4 = 2a_3 + 1 = 2(-1) + 1 = -1$

We see that the terms of the first sequence increase without bound, while all terms of the second sequence are -1. Clearly, the initial term of the sequence has a lot to say about the behavior of the entire sequence.   *Related Exercises 13–16* ◄

QUICK CHECK 2 Find an explicit formula for the sequence $\{1, 3, 7, 15, \dots\}$ (Example 2). ◄

EXAMPLE 3   Working with sequences Consider the following sequences.

a. $\{a_n\} = \{-2, 5, 12, 19, \dots\}$ **b.** $\{b_n\} = \{3, 6, 12, 24, 48, \dots\}$

(i)   Find the next two terms of the sequence.

(ii) Find a recurrence relation that generates the sequence.

(iii) Find an explicit formula for the nth term of the sequence.

SOLUTION

a. (i)   Each term is obtained by adding 7 to its predecessor. The next two terms are $19 + 7 = 26$ and $26 + 7 = 33$.

> In Example 3, we chose the starting index to be $n = 0$. Other choices are possible.

(ii) Because each term is seven more than its predecessor, the recurrence relation is

$$a_{n+1} = a_n + 7, a_0 = -2, \quad \text{for } n = 0, 1, 2, \dots$$

(iii) Notice that $a_0 = -2, a_1 = -2 + (1 \cdot 7)$, and $a_2 = -2 + (2 \cdot 7)$, so the explicit formula is

$$a_n = 7n - 2, \quad \text{for } n = 0, 1, 2, \dots.$$

b. (i)   Each term is obtained by multiplying its predecessor by 2. The next two terms are $48 \cdot 2 = 96$ and $96 \cdot 2 = 192$.

(ii) Because each term is two times its predecessor, the recurrence relation is

$$a_{n+1} = 2a_n, a_0 = 3, \quad \text{for } n = 0, 1, 2, \dots$$

(iii) To obtain the explicit formula, note that $a_0 = 3, a_1 = 3(2^1)$, and $a_2 = 3(2^2)$. In general,

$$a_n = 3(2^n), \quad \text{for } n = 0, 1, 2, \dots.$$

Related Exercises 17–22 ◄

Limit of a Sequence

Perhaps the most important question about a sequence is this: If you go farther and farther out in the sequence, $a_{100}, \ldots, a_{10,000}, \ldots, a_{100,000}, \ldots$, how do the terms of the sequence behave? Do they approach a specific number, and if so, what is that number? Or do they grow in magnitude without bound? Or do they wander around with or without a pattern?

The long-term behavior of a sequence is described by its **limit**. The limit of a sequence is defined rigorously in the next section. For now, we work with an informal definition.

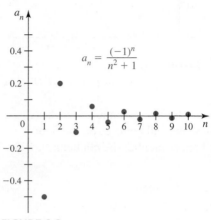

FIGURE 8.3

> **DEFINITION**   **Limit of a Sequence**
>
> If the terms of a sequence $\{a_n\}$ approach a unique number L as n increases, then we say $\lim\limits_{n \to \infty} a_n = L$ exists, and the sequence **converges** to L. If the terms of the sequence do not approach a single number as n increases, the sequence has no limit, and the sequence **diverges**.

EXAMPLE 4   **Limit of a sequence**   Write the first four terms of each sequence. If you believe the sequence converges, make a conjecture about its limit. If the sequence appears to diverge, explain why.

a. $\left\{ \dfrac{(-1)^n}{n^2 + 1} \right\}_{n=1}^{\infty}$ Explicit formula

b. $\{\cos(n\pi)\}_{n=1}^{\infty}$ Explicit formula

c. $\{a_n\}_{n=1}^{\infty}$, where $a_{n+1} = -2a_n, a_1 = 1$ Recurrence relation

SOLUTION

a. Beginning with $n = 1$, the first four terms of the sequence are

$$\left\{ \frac{(-1)^1}{1^2 + 1}, \frac{(-1)^2}{2^2 + 1}, \frac{(-1)^3}{3^2 + 1}, \frac{(-1)^4}{4^2 + 1}, \ldots \right\} = \left\{ -\frac{1}{2}, \frac{1}{5}, -\frac{1}{10}, \frac{1}{17}, \ldots \right\}.$$

The terms decrease in magnitude and approach zero with alternating signs. The limit appears to be 0 (Figure 8.3).

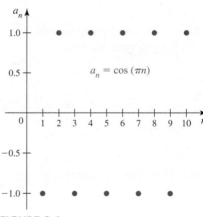

FIGURE 8.4

b. The first four terms of the sequence are

$$\{\cos \pi, \cos 2\pi, \cos 3\pi, \cos 4\pi, \ldots \} = \{-1, 1, -1, 1, \ldots \}.$$

In this case, the terms of the sequence alternate between -1 and $+1$, and never approach a single value. Thus, the sequence diverges (Figure 8.4).

c. The first four terms of the sequence are

$$\{1, -2a_1, -2a_2, -2a_3, \ldots \} = \{1, -2, 4, -8, \ldots \}.$$

Because the magnitudes of the terms increase without bound, the sequence diverges (Figure 8.5).   *Related Exercises 23–30* ◀

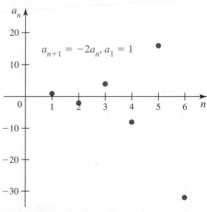

FIGURE 8.5

EXAMPLE 5   **Limit of a sequence**   Enumerate and graph the terms of the following sequence and make a conjecture about its limit.

$$a_n = \frac{4n^3}{n^3 + 1}, \qquad \text{for } n = 1, 2, 3, \ldots \quad \text{Explicit formula}$$

SOLUTION The first 14 terms of the sequence $\{a_n\}$ are tabulated in Table 8.1 and graphed in Figure 8.6. The terms appear to approach 4.

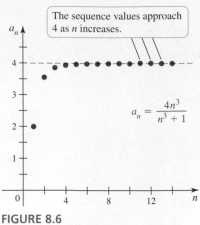

The sequence values approach 4 as n increases.

$$a_n = \frac{4n^3}{n^3 + 1}$$

FIGURE 8.6

Related Exercises 31–44 ◄

Table 8.1

n	a_n	n	a_n
1	2.000	8	3.992
2	3.556	9	3.995
3	3.857	10	3.996
4	3.938	11	3.997
5	3.968	12	3.998
6	3.982	13	3.998
7	3.988	14	3.999

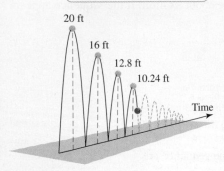

The height of each bounce of the basketball is 0.8 of the height of the previous bounce.

20 ft

16 ft

12.8 ft

10.24 ft

Time

FIGURE 8.7

EXAMPLE 6   A bouncing ball A basketball tossed straight up in the air reaches a high point and falls to the floor. Assume that each time the ball bounces on the floor it rebounds to 0.8 of its previous height. Let h_n be the high point after the nth bounce, with the initial height being $h_0 = 20$ ft.

a. Find a recurrence relation and an explicit formula for the sequence $\{h_n\}$.

b. What is the high point after the 10th bounce? after the 20th bounce?

c. Speculate on the limit of the sequence $\{h_n\}$.

SOLUTION

a. We first write and graph the heights of the ball for several bounces using the rule that each height is 0.8 of the previous height (Figure 8.7). For example, we have

$$h_0 = 20 \text{ ft}$$
$$h_1 = 0.8\, h_0 = 16 \text{ ft}$$
$$h_2 = 0.8\, h_1 = 0.8^2\, h_0 = 12.80 \text{ ft}$$
$$h_3 = 0.8\, h_2 = 0.8^3\, h_0 = 10.24 \text{ ft}$$
$$h_4 = 0.8\, h_3 = 0.8^4\, h_0 \approx 8.19 \text{ ft}.$$

Each number in the list is 0.8 of the previous number. Therefore, the recurrence relation for the sequence of heights is

$$h_{n+1} = 0.8\, h_n, \qquad \text{for } n = 0, 1, 2, 3, \ldots, h_0 = 20 \text{ ft}.$$

To find an explicit formula for the nth term, note that

$$h_1 = h_0 \cdot 0.8, \qquad h_2 = h_0 \cdot 0.8^2, \qquad h_3 = h_0 \cdot 0.8^3, \qquad \text{and} \qquad h_4 = h_0 \cdot 0.8^4.$$

In general, we have

$$h_n = h_0 \cdot 0.8^n = 20 \cdot 0.8^n, \qquad \text{for } n = 0, 1, 2, 3, \ldots,$$

which is an explicit formula for the terms of the sequence.

b. Using the explicit formula for the sequence, we see that after $n = 10$ bounces, the next height is

$$h_{10} = 20 \cdot 0.8^{10} \approx 2.15 \text{ ft}.$$

After $n = 20$ bounces, the next height is

$$h_{20} = 20 \cdot 0.8^{20} \approx 0.23 \text{ ft}.$$

c. The terms of the sequence (Figure 8.8) appear to decrease and approach 0. A reasonable conjecture is that $\lim\limits_{n \to \infty} h_n = 0$.

Related Exercises 45–48 ◄

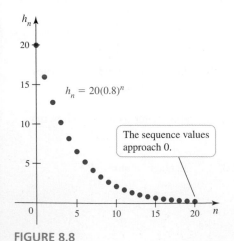

$h_n = 20(0.8)^n$

The sequence values approach 0.

FIGURE 8.8

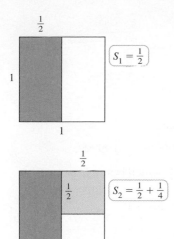

$\left(S_1 = \frac{1}{2}\right)$

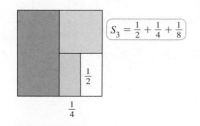

$\left(S_2 = \frac{1}{2} + \frac{1}{4}\right)$

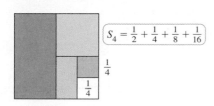

$\left(S_3 = \frac{1}{2} + \frac{1}{4} + \frac{1}{8}\right)$

$\left(S_4 = \frac{1}{2} + \frac{1}{4} + \frac{1}{8} + \frac{1}{16}\right)$

⋮

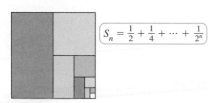

$\left(S_n = \frac{1}{2} + \frac{1}{4} + \cdots + \frac{1}{2^n}\right)$

FIGURE 8.9

Infinite Series and the Sequence of Partial Sums

An infinite series can be viewed as a sum of an infinite set of numbers; it has the form

$$a_1 + a_2 + \cdots + a_n + \cdots = \sum_{k=1}^{\infty} a_k,$$

where the terms of the series, $a_1, a_2, \ldots$, are real numbers. *An infinite series is quite distinct from a sequence.* We first answer the question: How is it possible to sum an infinite set of numbers and produce a finite number? Here is an informative example.

Consider a unit square (sides of length 1) that is subdivided as shown in Figure 8.9. We let S_n be the area of the colored region in the nth figure of the progression. The area of the colored region in the first figure is

$$S_1 = 1 \cdot \frac{1}{2} = \frac{1}{2}.$$

The area of the colored region in the second figure is S_1 plus the area of the smaller blue square, which is $\frac{1}{2} \cdot \frac{1}{2} = \frac{1}{4}$. Therefore,

$$S_2 = \frac{1}{2} + \frac{1}{4}.$$

The area of the colored region in the third figure is S_2 plus the area of the smaller green rectangle, which is $\frac{1}{2} \cdot \frac{1}{4} = \frac{1}{8}$. Therefore,

$$S_3 = \frac{1}{2} + \frac{1}{4} + \frac{1}{8}.$$

Continuing in this manner, we find that

$$S_n = \frac{1}{2} + \frac{1}{4} + \frac{1}{8} + \cdots + \frac{1}{2^n}.$$

If this process is continued indefinitely, the area of the colored region S_n approaches the area of the square, which is 1. So, it is plausible that

$$\lim_{n \to \infty} S_n = \underbrace{\frac{1}{2} + \frac{1}{4} + \frac{1}{8} + \cdots}_{\text{sum continues indefinitely}} = 1.$$

This example shows that it is possible to sum an infinite set of numbers and obtain a finite number—in this case, the sum is 1. The sequence $\{S_n\}$ generated in this example is extremely important. It is called a *sequence of partial sums*, and its limit is the value of the infinite series $\frac{1}{2} + \frac{1}{4} + \frac{1}{8} + \cdots$.

EXAMPLE 7   **Working with series** Consider the infinite series

$$0.9 + 0.09 + 0.009 + 0.0009 + \cdots,$$

where each term of the sum is $1/10$ of the previous term.

a. Find the sum of the first one, two, three, four, and five terms of the series.

b. What value would you assign to the infinite series $0.9 + 0.09 + 0.009 + \cdots$?

SOLUTION

a. Let S_n denote the sum of the first n terms of the given series. Then,

$$S_1 = 0.9$$
$$S_2 = 0.9 + 0.09 = 0.99$$
$$S_3 = 0.9 + 0.09 + 0.009 = 0.999$$
$$S_4 = 0.9 + 0.09 + 0.009 + 0.0009 = 0.9999$$
$$S_5 = 0.9 + 0.09 + 0.009 + 0.0009 + 0.00009 = 0.99999.$$

b. Notice that the sums $S_1, S_2, \ldots, S_n$ form a sequence $\{S_n\}$, which is a *sequence of partial sums*. As more and more terms are included, the values of S_n approach 1. Therefore, a reasonable conjecture for the value of the series is 1:

$$\underbrace{0.9}_{S_1 = 0.9} + 0.09 + 0.009 + 0.0009 + \cdots = 1$$

$$\underbrace{}_{S_2 = 0.99}$$

$$\underbrace{}_{S_3 = 0.999}$$

> **QUICK CHECK 3** Reasoning as in Example 7, what is the value of $0.3 + 0.03 + 0.003 + \cdots$? ◄

Related Exercises 49–52 ◄

The general nth term of the sequence in Example 7 can be written as

$$S_n = \underbrace{0.9 + 0.09 + 0.009 + \cdots + 0.0\ldots9}_{n \text{ terms}} = \sum_{k=1}^{n} 9 \cdot 0.1^k.$$

> ➤ Recall the summation notation introduced in Chapter 5: $\sum_{k=1}^{n} a_k$ means $a_1 + a_2 + \cdots + a_n$.

We observed that $\lim\limits_{n \to \infty} S_n = 1$. For this reason, we write

$$\lim_{n \to \infty} S_n = \lim_{n \to \infty} \underbrace{\sum_{k=1}^{n} 9 \cdot 0.1^k}_{S_n} = \underbrace{\sum_{k=1}^{\infty} 9 \cdot 0.1^k}_{\text{new object}} = 1.$$

By letting $n \to \infty$ a new mathematical object $\sum\limits_{k=1}^{\infty} 9 \cdot 0.1^k$ is created. It is an infinite series and it is the *limit of the sequence of partial sums*.

> ➤ The term *series* is used for historical reasons. When you see *series*, you should think *sum*.

DEFINITION   Infinite Series

Given a set of numbers $\{a_1, a_2, a_3, \ldots\}$, the sum

$$a_1 + a_2 + a_3 + \cdots = \sum_{k=1}^{\infty} a_k$$

is called an **infinite series**. Its **sequence of partial sums** $\{S_n\}$ has the terms

$$S_1 = a_1$$
$$S_2 = a_1 + a_2$$
$$S_3 = a_1 + a_2 + a_3$$
$$\vdots$$
$$S_n = a_1 + a_2 + a_3 + \cdots + a_n = \sum_{k=1}^{n} a_k, \quad \text{for } n = 1, 2, 3, \ldots$$

> **QUICK CHECK 4** Do the series $\sum\limits_{k=1}^{\infty} 1$ and $\sum\limits_{k=1}^{\infty} k$ converge or diverge? ◄

If the sequence of partial sums $\{S_n\}$ has a limit L, the infinite series **converges** to that limit, and we write

$$\sum_{k=1}^{\infty} a_k = \lim_{n \to \infty} \underbrace{\sum_{k=1}^{n} a_k}_{S_n} = \lim_{n \to \infty} S_n = L.$$

If the sequence of partial sums diverges, the infinite series also **diverges**.

EXAMPLE 8 **Sequence of partial sums** Consider the infinite series

$$\sum_{k=1}^{\infty} \frac{1}{k(k+1)}.$$

a. Find the first four terms of the sequence of partial sums.

b. Find an expression for S_n and make a conjecture about the value of the series.

SOLUTION

a. The sequence of partial sums can be evaluated explicitly:

$$S_1 = \sum_{k=1}^{1} \frac{1}{k(k+1)} = \frac{1}{2}$$

$$S_2 = \sum_{k=1}^{2} \frac{1}{k(k+1)} = \frac{1}{2} + \frac{1}{6} = \frac{2}{3}$$

$$S_3 = \sum_{k=1}^{3} \frac{1}{k(k+1)} = \frac{1}{2} + \frac{1}{6} + \frac{1}{12} = \frac{3}{4}$$

$$S_4 = \sum_{k=1}^{4} \frac{1}{k(k+1)} = \frac{1}{2} + \frac{1}{6} + \frac{1}{12} + \frac{1}{20} = \frac{4}{5}$$

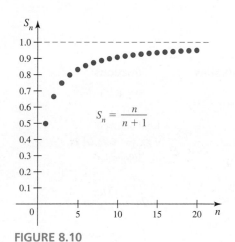

FIGURE 8.10

b. Based on the pattern in the sequence of partial sums, a reasonable conjecture is that $S_n = \dfrac{n}{n+1}$, for $n = 1, 2, 3, \ldots$, which produces the sequence $\left\{ \dfrac{1}{2}, \dfrac{2}{3}, \dfrac{3}{4}, \dfrac{4}{5}, \dfrac{5}{6}, \ldots \right\}$ (Figure 8.10). Because $\displaystyle\lim_{n\to\infty} \frac{n}{n+1} = 1$, we conclude that

$$\lim_{n\to\infty} S_n = \sum_{k=1}^{\infty} \frac{1}{k(k+1)} = 1.$$

Related Exercises 53–56 ◄

QUICK CHECK 5 Find the first four terms of the sequence of partial sums for the series $\displaystyle\sum_{k=1}^{\infty} (-1)^k k$. Does the series converge or diverge? ◄

Summary

This section has shown that there are three key ideas to keep in mind.

- A *sequence* $\{a_1, a_2, \ldots, a_n \ldots\}$ is an ordered *list* of numbers.

- An *infinite series* $\displaystyle\sum_{k=1}^{\infty} a_k = a_1 + a_2 + a_3 + \cdots$ is a *sum* of numbers.

- The *sequence of partial sums* $S_n = a_1 + a_2 + \cdots + a_n$ is used to evaluate the series $\displaystyle\sum_{k=1}^{\infty} a_k$.

For sequences, we ask about the behavior of the individual terms as we go out farther and farther in the list; that is, we ask about $\displaystyle\lim_{n\to\infty} a_n$. For infinite series, we examine the sequence of partial sums related to the series. If the sequence of partial sums $\{S_n\}$ has a limit, then the infinite series $\sum_{k=1}^{\infty} a_k$ converges to that limit. If the sequence of partial sums does not have a limit, the infinite series diverges.

Table 8.2 shows the correspondences between sequences/series and functions, and between summing and integration. For a sequence, the index n plays the role of the independent variable and takes on integer values; the terms of the sequence $\{a_n\}$ correspond to the dependent variable.

With sequences $\{a_n\}$, the idea of accumulation corresponds to summation, whereas with functions, accumulation corresponds to integration. A finite sum is analogous to integrating a function over a finite interval. An infinite series is analogous to integrating a function over an infinite interval.

Table 8.2

	Sequences/Series	Functions
Independent variable	n	x
Dependent variable	a_n	$f(x)$
Domain	Integers	Real numbers
	e.g., $n = 0, 1, 2, 3, \ldots$	e.g., $\{x : x \geq 0\}$
Accumulation	Sums	Integrals
Accumulation over a finite interval	$\sum_{k=0}^{n} a_k$	$\int_{0}^{n} f(x)\,dx$
Accumulation over an infinite interval	$\sum_{k=0}^{\infty} a_k$	$\int_{0}^{\infty} f(x)\,dx$

SECTION 8.1 EXERCISES

Review Questions

1. Define *sequence* and give an example.

2. Suppose the sequence $\{a_n\}$ is defined by the explicit formula $a_n = 1/n$, for $n = 1, 2, 3, \ldots$. Write out the first five terms of the sequence.

3. Suppose the sequence $\{a_n\}$ is defined by the recurrence relation $a_{n+1} = na_n$, for $n = 1, 2, 3, \ldots$, where $a_1 = 1$. Write out the first five terms of the sequence.

4. Define *finite sum* and give an example.

5. Define *infinite series* and give an example.

6. Given the series $\sum_{k=1}^{\infty} k$, evaluate the first four terms of its sequence of partial sums $S_n = \sum_{k=1}^{n} k$.

7. The terms of a sequence of partial sums are defined by
$S_n = \sum_{k=1}^{n} k^2$, for $n = 1, 2, 3, \ldots$. Evaluate the first four terms of the sequence.

8. Consider the infinite series $\sum_{k=1}^{\infty} \dfrac{1}{k}$. Evaluate the first four terms of the sequence of partial sums.

Basic Skills

9–12. Explicit formulas *Write the first four terms of the sequence* $\{a_n\}_{n=1}^{\infty}$.

9. $a_n = 1/10^n$

10. $a_n = n + 1/n$

11. $a_n = 1 + \sin(\pi n/2)$

12. $a_n = 2n^2 - 3n + 1$

13–16. Recurrence relations *Write the first four terms of the sequence* $\{a_n\}$ *defined by the following recurrence relations.*

13. $a_{n+1} = 3a_n - 12; \quad a_1 = 10$

14. $a_{n+1} = a_n^2 - 1; \quad a_1 = 1$

15. $a_{n+1} = 3a_n^2 + n + 1; \quad a_1 = 0$

16. $a_{n+1} = a_n + a_{n-1}; \quad a_1 = 1, a_0 = 1$

17–22. Enumerated sequences *Several terms of a sequence* $\{a_n\}_{n=1}^{\infty}$ *are given.*

    a. *Find the next two terms of the sequence.*

    b. *Find a recurrence relation that generates the sequence (supply the initial value of the index and the first term of the sequence).*

    c. *Find an explicit formula for the general nth term of the sequence.*

17. $\left\{1, \frac{1}{2}, \frac{1}{4}, \frac{1}{8}, \frac{1}{16}, \dots\right\}$ **18.** $\{1, -2, 3, -4, 5, \dots\}$

19. $\{1, 2, 4, 8, 16, \dots\}$ **20.** $\{1, 4, 9, 16, 25, \dots\}$

21. $\{1, 3, 9, 27, 81, \dots\}$ **22.** $\{64, 32, 16, 8, 4, \dots\}$

23–30. Limits of sequences *Write the terms* $a_1, a_2, a_3,$ *and* a_4 *of the following sequences. If the sequence appears to converge, make a conjecture about its limit. If the sequence diverges, explain why.*

23. $a_n = 10^n - 1; \quad n = 1, 2, 3, \dots$

24. $a_n = n^8 + 1; \quad n = 1, 2, 3, \dots$

25. $a_n = \dfrac{(-1)^n}{n}; \quad n = 1, 2, 3, \dots$

26. $a_n = 1 - 10^{-n}; \quad n = 1, 2, 3, \dots$

27. $a_{n+1} = \dfrac{a_n^2}{10}; \quad a_0 = 1$

28. $a_{n+1} = 0.5a_n(1 - a_n); \quad a_0 = 0.8$

29. $a_{n+1} = 0.5a_n + 50; \quad a_0 = 100$

30. $a_{n+1} = 0.9a_n + 100; \quad a_0 = 50$

31–36. Explicit formulas for sequences *Consider the explicit formulas for the following sequences.*

    a. *Find the first four terms of the sequence.*

    b. *Using a calculator, make a table with at least 10 terms and determine a plausible value for the limit of the sequence or state that it does not exist.*

31. $a_n = n + 1; \quad n = 0, 1, 2, \dots$

32. $a_n = 2\tan^{-1}(1000n); \quad n = 1, 2, 3, \dots$

33. $a_n = n^2 - n; \quad n = 1, 2, 3, \dots$

34. $a_n = \dfrac{2n - 3}{n}; \quad n = 1, 2, 3, \dots$

35. $a_n = \dfrac{(n - 1)^2}{(n^2 - 1)}; \quad n = 2, 3, 4, \dots$

36. $a_n = \sin(n\pi/2); \quad n = 0, 1, 2, \dots$

37–38. Limits from graphs *Consider the following sequences.*

    a. *Find the first four terms of the sequence.*

    b. *Based on part (a) and the figure, determine a plausible limit of the sequence.*

37. $a_n = 2 + 2^{-n}; \quad n = 1, 2, 3, \dots$

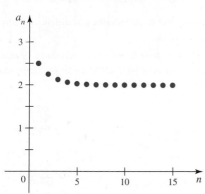

38. $a_n = \dfrac{n^2}{n^2 - 1}; \quad n = 2, 3, 4, \dots$

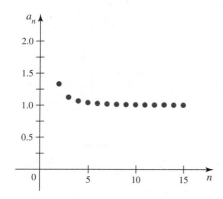

39–44. Recurrence relations to formulas *Consider the following recurrence relations.*

    a. *Find the terms* a_0, a_1, a_2, a_3 *of the sequence.*

    b. *If possible, find an explicit formula for the nth term of the sequence.*

    c. *Using a calculator, make a table with at least 10 terms and determine a plausible value for the limit of the sequence or state that it does not exist.*

39. $a_{n+1} = a_n + 2; \quad a_0 = 3$ **40.** $a_{n+1} = a_n - 4; \quad a_0 = 36$

41. $a_{n+1} = 2a_n + 1; \quad a_0 = 0$ **42.** $a_{n+1} = \dfrac{a_n}{2}; \quad a_0 = 32$

43. $a_{n+1} = \frac{1}{2}a_n + 1; \quad a_0 = 1$ **44.** $a_{n+1} = \sqrt{1 + a_n}; \quad a_0 = 1$

45–48. Heights of bouncing balls *Suppose a ball is thrown upward to a height of* h_0 *meters. Each time the ball bounces, it rebounds to a fraction* r *of its previous height. Let* h_n *be the height after the nth bounce. Consider the following values of* h_0 *and* r.

    a. *Find the first four terms of the sequence of heights* $\{h_n\}$.

    b. *Find a general expression for the nth term of the sequence* $\{h_n\}$.

45. $h_0 = 20, \ r = 0.5$ **46.** $h_0 = 10, \ r = 0.9$

47. $h_0 = 30, \ r = 0.25$ **48.** $h_0 = 20, \ r = 0.75$

49–52. Sequences of partial sums *For the following infinite series, find the first four terms of the sequence of partial sums. Then make a conjecture about the value of the infinite series.*

49. $0.3 + 0.03 + 0.003 + \cdots$ **50.** $0.6 + 0.06 + 0.006 + \cdots$

51. $4 + 0.9 + 0.09 + 0.009 + \cdots$ **52.** $1 + \frac{1}{2} + \frac{1}{4} + \frac{1}{8} + \cdots$

53–56. Formulas for sequences of partial sums *Consider the following infinite series.*

   a. *Find the first four terms of the sequence of partial sums.*
   b. *Use the results of part (a) to propose a formula for S_n.*
   c. *Propose a value of the series.*

53. $\displaystyle\sum_{k=1}^{\infty} \frac{2}{(2k-1)(2k+1)}$ **54.** $\displaystyle\sum_{k=1}^{\infty} \frac{1}{2^k}$

55. $\displaystyle\sum_{k=1}^{\infty} \frac{1}{4k^2 - 1}$ **56.** $\displaystyle\sum_{k=1}^{\infty} \frac{2}{3^k}$

Further Explorations

57. Explain why or why not Determine whether the following statements are true and give an explanation or counterexample.

   a. The sequence of partial sums for the series $1 + 2 + 3 + \cdots$ is $\{1, 3, 6, 10, \dots\}$.
   b. If a sequence of positive numbers converges, then the terms of the sequence must decrease in size.
   c. If the positive terms of the sequence $\{a_n\}$ increase in size, then the sequence of partial sums for the series $\displaystyle\sum_{k=1}^{\infty} a_k$ diverges.

58–59. Distance traveled by bouncing balls *Suppose a ball is thrown upward to a height of h_0 meters. Each time the ball bounces, it rebounds to a fraction r of its previous height. Let h_n be the height after the nth bounce and let S_n be the total distance the ball has traveled at the moment of the nth bounce.*

   a. *Find the first four terms of the sequence $\{S_n\}$.*
   b. *Make a table of 20 terms of the sequence $\{S_n\}$ and determine a plausible value for the limit of $\{S_n\}$.*

58. $h_0 = 20,\ r = 0.5$ **59.** $h_0 = 20,\ r = 0.75$

60–67. Sequences of partial sums *Consider the following infinite series.*

   a. *Write out the first four terms of the sequence of partial sums.*

   b. *Estimate the limit of $\{S_n\}$ or state that it does not exist.*

60. $\displaystyle\sum_{k=1}^{\infty} \cos\left(\frac{\pi k}{2}\right)$ **61.** $\displaystyle\sum_{k=1}^{\infty} 0.5^k$ **62.** $\displaystyle\sum_{k=1}^{\infty} 1.5^k$ **63.** $\displaystyle\sum_{k=1}^{\infty} 3^{-k}$

64. $\displaystyle\sum_{k=1}^{\infty} k$ **65.** $\displaystyle\sum_{k=1}^{\infty} (-1)^k$ **66.** $\displaystyle\sum_{k=1}^{\infty} (-1)^k k$ **67.** $\displaystyle\sum_{k=1}^{\infty} \frac{3}{10^k}$

Applications

68–71. Practical sequences *Consider the following situations that generate a sequence.*

   a. *Write out the first five terms of the sequence.*
   b. *Find an explicit formula for the terms of the sequence.*
   c. *Find a recurrence relation that generates the sequence.*
   d. *Using a calculator or a graphing utility, estimate the limit of the sequence or state that it does not exist.*

68. Population growth When a biologist begins a study, a colony of prairie dogs has a population of 250. Regular measurements reveal that each month the prairie dog population increases by 3%. Let p_n be the population (rounded to whole numbers) at the end of the nth month, where the initial population is $p_0 = 250$.

69. Radioactive decay A material transmutes 50% of its mass to another element every 10 years due to radioactive decay. Let M_n be the mass of the radioactive material at the end of the nth decade, where the initial mass of the material is $M_0 = 20$ g.

70. Consumer Price Index The Consumer Price Index (the CPI is a measure of the U.S. cost of living) is given a base value of 100 in the year 1984. Assume the CPI has increased by an average of 3% per year since 1984. Let c_n be the CPI n years after 1984, where $c_0 = 100$.

71. Drug elimination Jack took a 200-mg dose of a strong pain killer at midnight. Every hour, 5% of the drug is washed out of his bloodstream. Let d_n be the amount of drug in Jack's blood n hours after the drug was taken, where $d_0 = 200$ mg.

72. A square root finder A well-known method for approximating $\sqrt{c}$ for positive real numbers c consists of the following recurrence relation (based on Newton's method; see Guided Projects). Let $a_0 = c$ and

$$a_{n+1} = \frac{1}{2}\left(a_n + \frac{c}{a_n}\right), \qquad \text{for } n = 0, 1, 2, 3, \dots.$$

   a. Use this recurrence relation to approximate $\sqrt{10}$. How many terms of the sequence are needed to approximate $\sqrt{10}$ with an error less than 0.01? How many terms of the sequence are needed to approximate $\sqrt{10}$ with an error less than 0.0001? (To compute the error, assume a calculator gives the exact value.)
   b. Use this recurrence relation to approximate $\sqrt{c}$ for $c = 2$, $3, \dots, 10$. Make a table showing how many terms of the sequence are needed to approximate $\sqrt{c}$ with an error less than 0.01.

Additional Exercises

73–80. Repeating decimals

   a. *Write the following repeating decimals as an infinite series.*

   For example, $0.9999\ldots = \displaystyle\sum_{k=1}^{\infty} 9(0.1^k)$.

   b. *Find the limit of the sequence of partial sums for the infinite series and express it as a fraction.*

73. $0.\overline{3} = 0.333\ldots$ **74.** $0.\overline{6} = 0.666\ldots$

75. $0.\overline{1} = 0.111\ldots$

76. $0.\overline{5} = 0.555\ldots$

77. $0.\overline{09} = 0.090909\ldots$

78. $0.\overline{27} = 0.272727\ldots$

79. $0.\overline{037} = 0.037037\ldots$

80. $0.\overline{027} = 0.027027\ldots$

8.2 Sequences

The overview of the previous section sets the stage for an in-depth investigation of sequences and infinite series. This section is devoted to sequences, and the remainder of the chapter deals with series.

Limit of a Sequence

A fundamental question about sequences concerns the behavior of the terms as we go out farther and farther in the sequence. For example, in the sequence

$$\{a_n\}_{n=0}^{\infty} = \left\{\frac{1}{n^2 + 1}\right\}_{n=0}^{\infty} = \left\{1, \frac{1}{2}, \frac{1}{5}, \frac{1}{10}, \ldots\right\},$$

the terms remain positive and decrease to 0. We say that this sequence **converges** and its **limit** is 0, written $\lim\limits_{n \to \infty} a_n = 0$. Similarly, the terms of the sequence

$$\{b_n\}_{n=1}^{\infty} = \left\{(-1)^n \frac{n(n+1)}{2}\right\}_{n=1}^{\infty} = \{-1, 3, -6, 10, \ldots\}$$

increase in magnitude and do not approach a unique value as n increases. In this case, we say that the sequence **diverges**.

Limits of sequences are really no different from limits at infinity of functions except that the variable n assumes only integer values as $n \to \infty$. This idea works as follows.

Given a sequence $\{a_n\}$, we define a function f such that $f(n) = a_n$ for all indices n. For example, if $\{a_n\} = \{n/(n+1)\}$, then we let $f(x) = x/(x+1)$. By the methods of Section 2.5, we know that $\lim\limits_{x \to \infty} f(x) = 1$; because the terms of the sequence lie on the graph of f, it follows that $\lim\limits_{n \to \infty} a_n = 1$ (Figure 8.11). This reasoning is the basis of the following theorem.

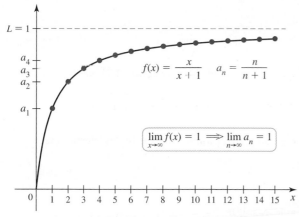

$$f(x) = \frac{x}{x+1} \qquad a_n = \frac{n}{n+1}$$

$$\lim_{x \to \infty} f(x) = 1 \implies \lim_{n \to \infty} a_n = 1$$

FIGURE 8.11

▷ The converse of Theorem 8.1 is not true. For example, if $a_n = \cos 2\pi n$, then $\lim\limits_{n \to \infty} a_n = 1$, but $\lim\limits_{x \to \infty} \cos 2\pi x$ does not exist.

THEOREM 8.1   Limits of Sequences from Limits of Functions
Suppose f is a function such that $f(n) = a_n$ for all positive integers n. If $\lim\limits_{x \to \infty} f(x) = L$, then the limit of the sequence $\{a_n\}$ is also L.

Because of the correspondence between limits of sequences and limits at infinity of functions, we have the following properties that are analogous to those for functions given in Theorem 2.3.

> The limit of a sequence $\{a_n\}$ is determined by the terms in the *tail* of the sequence—the terms with large values of n. If the sequences $\{a_n\}$ and $\{b_n\}$ differ in their first 100 terms but have identical terms for $n > 100$, then they have the same limit. For this reason, the initial index of a sequence (for example, $n = 0$ or $n = 1$) is often not specified.

THEOREM 8.2   Properties of Limits of Sequences

Assume that the sequences $\{a_n\}$ and $\{b_n\}$ have limits A and B, respectively. Then,

1. $\lim\limits_{n\to\infty} (a_n \pm b_n) = A \pm B$

2. $\lim\limits_{n\to\infty} ca_n = cA$,   where c is a real number

3. $\lim\limits_{n\to\infty} a_n b_n = AB$

4. $\lim\limits_{n\to\infty} \dfrac{a_n}{b_n} = \dfrac{A}{B}$,   provided $B \neq 0$.

EXAMPLE 1   Limits of sequences Determine the limits of the following sequences.

a. $\{a_n\}_{n=0}^{\infty} = \left\{ \dfrac{3n^3}{n^3 + 1} \right\}_{n=0}^{\infty}$ **b.** $\{b_n\}_{n=1}^{\infty} = \left\{ \left(\dfrac{5 + n}{n} \right)^n \right\}_{n=1}^{\infty}$

c. $\{c_n\}_{n=1}^{\infty} = \{ e^{-n} n^{10} \}_{n=1}^{\infty}$

SOLUTION

a. A function with the property that $f(n) = a_n$ is $f(x) = \dfrac{3x^3}{x^3 + 1}$. Dividing numerator and denominator by x^3 (Section 2.5), we find that $\lim\limits_{x\to\infty} f(x) = 3$. (Alternatively, we can apply l'Hôpital's Rule and obtain the same result.) Either way, we conclude that $\lim\limits_{n\to\infty} a_n = 3$.

b. The limit

$$\lim_{n\to\infty} b_n = \lim_{n\to\infty} \left(\frac{5 + n}{n} \right)^n = \lim_{n\to\infty} \left(1 + \frac{5}{n} \right)^n$$

> For a review of l'Hôpital's Rule, see Section 4.7, where we showed that
> $$\lim_{x\to\infty} \left(1 + \frac{a}{x} \right)^x = e^a.$$

has the indeterminate form 1^{∞}. Recall that for this limit (Section 4.7), we first evaluate

$$L = \lim_{n\to\infty} \ln \left(1 + \frac{5}{n} \right)^n = \lim_{n\to\infty} n \ln \left(1 + \frac{5}{n} \right)$$

and then, if L exists, $\lim\limits_{n\to\infty} b_n = e^L$. Using l'Hôpital's Rule for the indeterminate form $0/0$, we have

> It is not necessary to convert the terms of a sequence to a function of x, as we did in Example 1a. You can take the limit as $n \to \infty$ of the terms of the sequence directly.

$$L = \lim_{n\to\infty} n \ln \left(1 + \frac{5}{n} \right) = \lim_{n\to\infty} \frac{\ln \left(1 + (5/n) \right)}{1/n} \qquad \text{Indeterminate form } 0/0$$

$$= \lim_{n\to\infty} \frac{\dfrac{1}{1 + (5/n)} \left(-\dfrac{5}{n^2} \right)}{-1/n^2} \qquad \text{L'Hôpital's Rule}$$

$$= \lim_{n\to\infty} \frac{5}{1 + (5/n)} = 5 \qquad \text{Simplify; } 5/n \to 0 \text{ as } n \to \infty.$$

Because $\lim\limits_{n\to\infty} b_n = e^L = e^5$, we have $\lim\limits_{n\to\infty} \left(\dfrac{5 + n}{n} \right)^n = e^5$.

c. Computing the limit $\lim\limits_{n\to\infty} \dfrac{n^{10}}{e^n}$ requires ten applications of l'Hôpital's Rule. Instead, we appeal to the relative growth rates of functions (Section 4.7) and recall that an exponential function grows faster than any power of n as $n \to \infty$. Therefore,

$$\lim_{n\to\infty} \frac{n^{10}}{e^n} = 0.$$

Related Exercises 9–26 ◄

Terminology for Sequences

We now introduce some terminology similar to that used for functions. A sequence $\{a_n\}$ in which each term is greater than or equal to its predecessor $(a_{n+1} \geq a_n)$ is said to be **nondecreasing**. For example, the sequence

$$\left\{1 - \frac{1}{n}\right\}_{n=1}^{\infty} = \left\{0, \frac{1}{2}, \frac{2}{3}, \frac{3}{4}, \ldots\right\}$$

> Nondecreasing sequences include increasing sequences, which satisfy $a_{n+1} > a_n$ (strict inequality). Similarly, nonincreasing sequences include decreasing sequences, which satisfy $a_{n+1} < a_n$. For example, the sequence $\{1, 1, 2, 2, 3, 3, 4, 4, \ldots\}$ is nondecreasing but not increasing.

is nondecreasing (Figure 8.12). A sequence $\{a_n\}$ is **nonincreasing** if each term is less than or equal to its predecessor $(a_{n+1} \leq a_n)$. For example, the sequence

$$\left\{1 + \frac{1}{n}\right\}_{n=1}^{\infty} = \left\{2, \frac{3}{2}, \frac{4}{3}, \frac{5}{4}, \ldots\right\}$$

is nonincreasing (Figure 8.12). A sequence that is either nonincreasing or nondecreasing is said to be **monotonic**; it progresses in only one direction. Finally, a sequence whose terms are all less than or equal to some finite number in magnitude $(|a_n| \leq M$, for some real number $M)$ is said to be **bounded**. For example, the terms of $\left\{1 - \frac{1}{n}\right\}_{n=1}^{\infty}$ satisfy $|a_n| < 1$, and the terms of $\left\{1 + \frac{1}{n}\right\}_{n=1}^{\infty}$ satisfy $|a_n| \leq 2$ (Figure 8.12); so these sequences are bounded.

QUICK CHECK 1 Classify the following sequences as bounded, monotonic, or neither.

a. $\left\{\frac{1}{2}, \frac{3}{4}, \frac{7}{8}, \frac{15}{16}, \ldots\right\}$

b. $\left\{1, -\frac{1}{2}, \frac{1}{4}, -\frac{1}{8}, \frac{1}{16}, \ldots\right\}$

c. $\{1, -2, 3, -4, 5, \ldots\}$

d. $\{1, 1, 1, 1, \ldots\}$ ◄

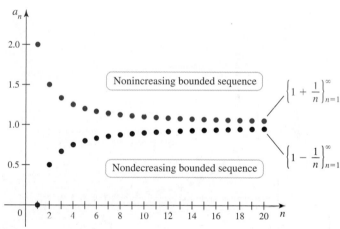

FIGURE 8.12

EXAMPLE 2 **Limits of sequences and graphing** Compare and contrast the behavior of $\{a_n\}$ and $\{b_n\}$ as $n \to \infty$.

a. $a_n = \dfrac{n^{3/2}}{n^{3/2} + 1}$ **b.** $b_n = \dfrac{(-1)^n n^{3/2}}{n^{3/2} + 1}$

SOLUTION

a. The sequence $\{a_n\}$ consists of positive terms. Dividing the numerator and denominator of a_n by $n^{3/2}$, we see that

$$\lim_{n \to \infty} a_n = \lim_{n \to \infty} \frac{n^{3/2}}{n^{3/2} + 1} = \lim_{n \to \infty} \frac{1}{1 + \underbrace{\frac{1}{n^{3/2}}}_{\text{approaches } 0 \text{ as } n \to \infty}} = 1.$$

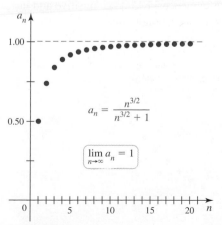

FIGURE 8.13

The terms of this sequence are nondecreasing and bounded (Figure 8.13).

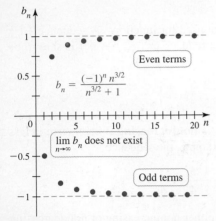

FIGURE 8.14

b. The terms of the bounded sequence $\{b_n\}$ alternate in sign. Using the result of part (a), it follows that the even terms of the sequence approach $+1$ and the odd terms approach -1 (Figure 8.14). Therefore, the sequence diverges, illustrating the fact that the presence of $(-1)^n$ may significantly alter the behavior of the sequence.   *Related Exercises 27–34* ◄

Geometric Sequences

Geometric sequences have the property that each term is obtained by multiplying the previous term by a fixed constant, called the **ratio**. They have the form $\{r^n\}$, where the ratio r is a real number.

EXAMPLE 3   Geometric sequences Graph the following sequences and discuss their behavior.

a. $\{0.75^n\}$ **b.** $\{(-0.75)^n\}$ **c.** $\{1.15^n\}$ **d.** $\{(-1.15)^n\}$

SOLUTION

a. When a number less than 1 in magnitude is raised to increasing powers, the resulting numbers decrease to zero. The sequence $\{0.75^n\}$ converges monotonically to zero (Figure 8.15).

b. Note that $\{(-0.75)^n\} = \{(-1)^n\, 0.75^n\}$. Observe also that $(-1)^n$ oscillates between $+1$ and -1, while 0.75^n decreases to zero as n increases. Therefore, the sequence oscillates and converges to zero (Figure 8.16).

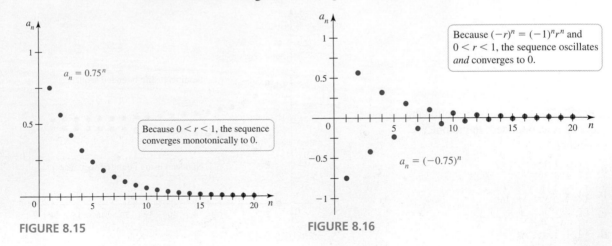

FIGURE 8.15

FIGURE 8.16

c. When a number greater than 1 in magnitude is raised to increasing powers, the resulting numbers increase in magnitude. The terms of the sequence $\{1.15^n\}$ are positive and increase without bound. In this case, the sequence diverges monotonically (Figure 8.17).

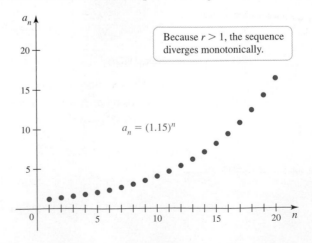

FIGURE 8.17

d. We write $\{(-1.15)^n\} = \{(-1)^n\, 1.15^n\}$ and observe that $(-1)^n$ oscillates between $+1$ and -1, while 1.15^n increases without bound as n increases. The terms of the sequence increase in magnitude without bound and alternate in sign. In this case, the sequence oscillates and diverges (Figure 8.18).

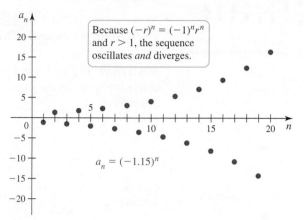

Because $(-r)^n = (-1)^n r^n$ and $r > 1$, the sequence oscillates *and* diverges.

$a_n = (-1.15)^n$

FIGURE 8.18

Related Exercises 35–42 ◄

QUICK CHECK 2 Describe the behavior of $\{r^n\}$ in the cases $r = -1$ and $r = 1$. ◄

The results of Example 3 and Quick Check 2 are summarized in the following theorem.

THEOREM 8.3   Geometric Sequences

Let r be a real number. Then,

$$\lim_{n\to\infty} r^n = \begin{cases} 0 & \text{if } |r| < 1 \\ 1 & \text{if } r = 1 \\ \text{does not exist} & \text{if } r \le -1 \text{ or } r > 1 \end{cases}$$

If $r > 0$, then $\{r^n\}$ converges or diverges monotonically. If $r < 0$, then $\{r^n\}$ converges or diverges by oscillation.

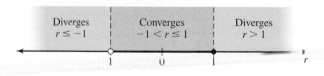

| Diverges $r \le -1$ | Converges $-1 < r \le 1$ | Diverges $r > 1$ |

The previous examples show that a sequence may display any of the following behaviors:

- It may converge to a single value, which is the limit of the sequence.
- Its terms may increase in magnitude without bound (either with one sign or with mixed signs), in which case the sequence diverges.
- Its terms may remain bounded but settle into an oscillating pattern in which the terms approach two or more values; in this case, the sequence diverges.
- Not illustrated in the preceding examples is one other type of behavior: The terms of a sequence may remain bounded, but wander chaotically forever without a pattern. In this case, the sequence also diverges.

The Squeeze Theorem

We cite two theorems that are often useful in either establishing that a sequence has a limit or in finding limits. The first is a direct analog of Theorem 2.5 (the Squeeze Theorem).

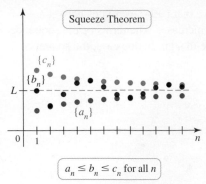

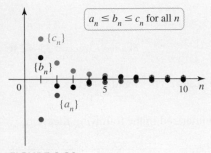

FIGURE 8.19

FIGURE 8.20

THEOREM 8.4 Squeeze Theorem for Sequences
Let $\{a_n\}$, $\{b_n\}$, and $\{c_n\}$ be sequences with $a_n \le b_n \le c_n$ for all integers n greater than some index N. If $\lim_{n \to \infty} a_n = \lim_{n \to \infty} c_n = L$, then $\lim_{n \to \infty} b_n = L$ (Figure 8.19).

EXAMPLE 4 Squeeze Theorem Find the limit of the sequence $b_n = \dfrac{\cos n}{n^2 + 1}$.

SOLUTION The goal is to find two sequences $\{a_n\}$ and $\{c_n\}$ whose terms lie below and above the terms of the given sequence $\{b_n\}$. Note that $-1 \le \cos \pi n \le 1$ for all n. Therefore,

$$\underbrace{-\frac{1}{n^2 + 1}}_{a_n} \le \underbrace{\frac{\cos n}{n^2 + 1}}_{b_n} \le \underbrace{\frac{1}{n^2 + 1}}_{c_n}.$$

Letting $a_n = -\dfrac{1}{n^2 + 1}$ and $c_n = \dfrac{1}{n^2 + 1}$, we have $a_n \le b_n \le c_n$ for $n \ge 1$. Furthermore, $\lim_{n \to \infty} a_n = \lim_{n \to \infty} c_n = 0$. By the Squeeze Theorem, $\lim_{n \to \infty} b_n = 0$ (Figure 8.20).

Related Exercises 43–46 ◄

Bounded Monotonic Sequence Theorem

Suppose a basketball player improves her shooting percentage in every game she plays. Her shooting percentages for each game form an increasing and bounded sequence (the shooting percentage must always be less than or equal to 100%). Therefore, if she plays a *very* large number of games, her sequence of shooting percentages approaches a limit that is less than or equal to 100%. This example illustrates another important theorem that characterizes convergent series in terms of boundedness and monotonicity. This result is easy to believe, but its proof goes beyond the scope of this text and is omitted.

THEOREM 8.5 Bounded Monotonic Sequences
A bounded monotonic sequence converges.

➤ M is called an *upper bound* of the sequence, and N is a *lower bound* of the sequence. If M^* is the smallest of all the upper bounds of an increasing sequence, then a result from advanced calculus tells us that the sequence converges to M^*. Similarly, if N^* is the greatest of all the lower bounds of a decreasing sequence, then the sequence converges to N^*.

Figure 8.21 shows the two cases of this theorem. In the first case, we see a nondecreasing sequence, all of whose terms are less than M. It must converge to a limit less than or equal to M. Similarly, a nonincreasing sequence, all of whose terms are greater than N, must converge to a limit greater than or equal to N.

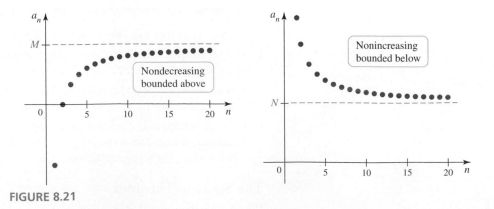

FIGURE 8.21

An Application: Recurrence Relations

> Most drugs decay exponentially in the bloodstream and have a characteristic half-life assuming that the drug is absorbed quickly into the blood.

EXAMPLE 5 **Sequences for drug doses** Suppose your doctor prescribes a 100-mg dose of an antibiotic every 12 hours. Furthermore, the drug is known to have a half-life of 12 hours; that is, every 12 hours half of the drug in your blood is eliminated.

a. Find the sequence that gives the amount of drug in your blood immediately after each dose.

b. Use a graph to propose the limit of this sequence; that is, in the long run, how much drug do you have in your blood?

c. Find the limit of the sequence directly.

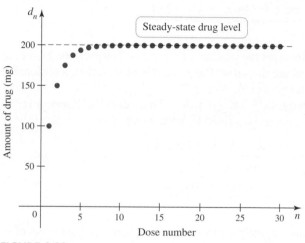

FIGURE 8.22

SOLUTION

a. Let d_n be the amount of drug in the blood immediately following the nth dose, where $n = 1, 2, 3, \ldots$ and $d_1 = 100$ mg. We want to write a recurrence relation that gives the amount of drug in the blood after the $(n + 1)$st dose (d_{n+1}) in terms of the amount of drug after the nth dose (d_n). In the 12 hr between the nth dose and the $(n + 1)$st dose, half of the drug in the blood is eliminated, *and* another 100 mg of drug is added. So, we have

$$d_{n+1} = 0.5\, d_n + 100, \qquad \text{for } n = 1, 2, 3, \ldots \text{ with } d_1 = 100,$$

which is the recurrence relation for the sequence $\{d_n\}$.

b. We see from Figure 8.22 that after about 10 doses (5 days) the amount of antibiotic in the blood is close to 200 mg, and—importantly for your body—it never exceeds 200 mg.

c. The graph of part (b) gives evidence that the terms of the sequence are increasing and bounded (Exercise 80). By the Bounded Monotonic Sequence Theorem, the sequence has a limit; therefore, $\lim\limits_{n \to \infty} d_n = L$, and $\lim\limits_{n \to \infty} d_{n+1} = L$. We now take the limit of both sides of the recurrence relation:

$$d_{n+1} = 0.5\, d_n + 100$$

$$\underbrace{\lim_{n \to \infty} d_{n+1}}_{L} = 0.5 \underbrace{\lim_{n \to \infty} d_n}_{L} + \lim_{n \to \infty} 100$$

$$L = 0.5L + 100$$

Solving for L, the steady-state drug level is $L = 200$.   *Related Exercises 47–50* ◄

QUICK CHECK 3 If a drug had the same half-life as in Example 5, (i) how would the steady-state level of drug in the blood change if the regular dose were 150 mg instead of 100 mg? (ii) How would the steady-state level change if the dosing interval were 6 hr instead of 12 hr? ◄

Growth Rates of Sequences

All the hard work we did in Section 4.7 to establish the relative growth rates of functions is now applied to sequences. Here is the question: Given two nondecreasing sequences of positive terms $\{a_n\}$ and $\{b_n\}$, which sequence grows faster as $n \to \infty$? As with functions, to compare growth rates, we evaluate $\lim\limits_{n \to \infty} a_n/b_n$. If $\lim\limits_{n \to \infty} a_n/b_n = 0$, then $\{b_n\}$ grows faster than $\{a_n\}$. If $\lim\limits_{n \to \infty} a_n/b_n = \infty$, then $\{a_n\}$ grows faster than $\{b_n\}$.

Using the results of Section 4.7, we immediately arrive at the following ranking of growth rates of sequences as $n \to \infty$, with positive real numbers p, q, r, and s and $b > 1$:

$$\{\ln^q n\} \ll \{n^p\} \ll \{n^p \ln^r n\} \ll \{n^{p+s}\} \ll \{b^n\} \ll \{n^n\}.$$

As before, the notation $\{a_n\} \ll \{b_n\}$ means $\{b_n\}$ *grows faster than* $\{a_n\}$ as $n \to \infty$. Another important sequence that should be added to the list is the **factorial sequence** $\{n!\}$, where $n! = n(n-1)(n-2)\cdots 2 \cdot 1$.

Where does the factorial sequence $\{n!\}$ appear in the list? The following argument provides some intuition. Notice that

$$n^n = \underbrace{n \cdot n \cdot n \cdots n}_{n \text{ factors}} \qquad \text{whereas}$$

$$n! = \underbrace{n \cdot (n-1) \cdot (n-2) \cdots 2 \cdot 1}_{n \text{ factors}}.$$

The nth term of both sequences involves the product of n factors; however, the factors of $n!$ decrease, while the factors of n^n are the same. Based on this observation, a reasonable conjecture is that $\{n^n\}$ grows faster than $\{n!\}$.

Therefore, we have the ordering $\{n!\} \ll \{n^n\}$. But where does $\{n!\}$ appear in the list relative to $\{b^n\}$? Again some intuition is gained by noting that

$$b^n = \underbrace{b \cdot b \cdot b \cdots b}_{n \text{ factors}}, \qquad \text{whereas}$$

$$n! = \underbrace{n \cdot (n-1) \cdot (n-2) \cdots 2 \cdot 1}_{n \text{ factors}}.$$

The nth term of both sequences involves the product of n factors; however, the factors of b^n remain constant as n increases, while the factors of $n!$ increase with n. So we claim that $\{n!\}$ grows faster than $\{b^n\}$. This conjecture is supported by computation, although the outcome of the race may not be immediately evident if b is large (Exercise 75).

> 0! = 1 (by definition)
> 1! = 1
> 2! = 2 · 1 = 2
> 3! = 3 · 2! = 6
> 4! = 4 · 3! = 24
> 5! = 5 · 4! = 120
> 6! = 6 · 5! = 720

> **THEOREM 8.6   Growth Rates of Sequences**
> The following sequences are ordered according to increasing growth rates as $n \to \infty$; that is, if $\{a_n\}$ appears before $\{b_n\}$ in the list, then $\lim\limits_{n\to\infty} a_n/b_n = 0$:
>
> $$\{\ln^q n\} \ll \{n^p\} \ll \{n^p \ln^r n\} \ll \{n^{p+s}\} \ll \{b^n\} \ll \{n!\} \ll \{n^n\}$$
>
> The ordering applies for positive real numbers p, q, r, and s and $b > 1$.

QUICK CHECK 4 Which sequence grows faster: $\{\ln n\}$ or $\{n^{1.1}\}$? What is $\lim\limits_{n\to\infty} \dfrac{n^{1,000,000}}{e^n}$? ◄

It is worth noting that the rankings in Theorem 8.6 do not change if a sequence is multiplied by a positive constant (Exercise 88).

EXAMPLE 6   Competing sequences Compare the growth rates of the following pairs of sequences as $n \to \infty$.

a. $\{\ln n^{10}\}$ and $\{0.00001n\}$ **b.** $\{n^8 \ln n\}$ and $\{n^{8.001}\}$ **c.** $\{n!\}$ and $\{10^n\}$

SOLUTION

a. Because $\ln n^{10} = 10 \ln n$, the first sequence is a constant multiple of the sequence $\{\ln n\}$ that appears in Theorem 8.6. Similarly, the second sequence is a constant multiple of the sequence $\{n\}$ that also appears in Theorem 8.6. By Theorem 8.6, $\{n\}$ grows faster than $\{\ln n\}$ as $n \to \infty$; therefore, $\{0.00001n\}$ grows faster than $\{\ln n^{10}\}$ as $n \to \infty$.

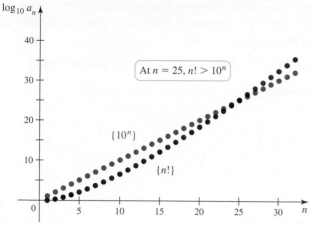

At $n = 25$, $n! > 10^n$

$\{10^n\}$

$\{n!\}$

FIGURE 8.23

b. The sequence $\{n^8 \ln n\}$ is the sequence $\{n^p \ln^r n\}$ of Theorem 8.6 with $p = 8$ and $r = 1$. The sequence $\{n^{8.001}\}$ is the sequence $\{n^{p+s}\}$ of Theorem 8.6 with $p = 8$ and $s = 0.001$. Because $\{n^{p+s}\}$ grows faster than $\{n^p \ln^r n\}$, we conclude that $\{n^{8.001}\}$ grows faster than $\{n^8 \ln n\}$ as $n \to \infty$.

c. Using Theorem 8.6, we see that $n!$ grows faster than any exponential function; therefore, the sequence $\{n!\}$ grows faster than the sequence $\{10^n\}$ (Figure 8.23). Because these sequences grow so quickly, we plot the logarithm of the terms. The exponential sequence $\{10^n\}$ dominates the factorial sequence $\{n!\}$ until $n = 25$ terms. At that point, the factorial sequence overtakes the exponential sequence. *Related Exercises 51–56* ◄

Formal Definition of a Limit of a Sequence

As with limits of functions, there is a formal definition of the limit of a sequence.

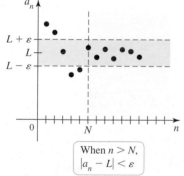

When $n > N$,
$|a_n - L| < \varepsilon$

FIGURE 8.24

> **DEFINITION   Limit of a Sequence**
>
> The sequence $\{a_n\}$ converges to L provided the terms of a_n can be made arbitrarily close to L by taking n sufficiently large. More precisely, $\{a_n\}$ has the unique limit L if given any tolerance $\varepsilon > 0$, it is possible to find a positive integer N (depending only on ε) such that
>
> $$|a_n - L| < \varepsilon \qquad \text{whenever } n > N.$$
>
> If the **limit of a sequence** is L, we say the sequence **converges** to L, written
>
> $$\lim_{n \to \infty} a_n = L.$$
>
> A sequence that does not converge is said to **diverge**.

The formal definition of the limit of a sequence is interpreted in much the same way as the limit at infinity of a function. Given a small tolerance $\varepsilon > 0$, how far out in the sequence must you go so that all succeeding terms are within ε of the limit L (Figure 8.24)? If you are given *any* value of $\varepsilon > 0$ (no matter how small), then you must be able to find a value of N such that all terms beyond a_N are within ε of L.

EXAMPLE 7   Limits using the formal definition Consider the claim that

$$\lim_{n \to \infty} a_n = \lim_{n \to \infty} \frac{n}{n - 1} = 1.$$

a. Given $\varepsilon = 0.01$, find a value of N that satisfies the conditions of the limit definition.

b. Prove that $\lim_{n \to \infty} a_n = 1$.

SOLUTION

a. We must find an integer N such that $|a_n - 1| < \varepsilon = 0.01$, whenever $n > N$. This condition can be written

$$|a_n - 1| = \left| \frac{n}{n - 1} - 1 \right| = \left| \frac{1}{n - 1} \right| < 0.01.$$

Noting that $n > 1$, the absolute value can be removed. The condition on n becomes $n - 1 > 1/0.01 = 100$, or $n > 101$. Thus, we take $N = 101$ or any larger number. This means that $|a_n - 1| < 0.01$ whenever $n > 101$.

b. Given *any* $\varepsilon > 0$, we must find a value of N (depending on ε) that guarantees that
$$|a_n - 1| = \left| \frac{n}{n-1} - 1 \right| < \varepsilon \text{ whenever } n > N. \text{ For } n > 1 \text{ the inequality}$$
$$\left| \frac{n}{n-1} - 1 \right| < \varepsilon \text{ implies that}$$
$$\left| \frac{n}{n-1} - 1 \right| = \frac{1}{n-1} < \varepsilon.$$

▶ In general, $1/\varepsilon + 1$ is not an integer, so N should be the least integer greater than $1/\varepsilon + 1$ or any larger integer.

Solving for n, we find that $\frac{1}{n-1} < \varepsilon$ or $n - 1 > \frac{1}{\varepsilon}$ or $n > \frac{1}{\varepsilon} + 1$. Therefore, given a tolerance $\varepsilon > 0$, we must look beyond a_N in the sequence, where $N \geq \frac{1}{\varepsilon} + 1$, to be sure that the terms of the sequence are within ε of the limit 1. Because we can provide a value of N for *any* $\varepsilon > 0$, the limit exists and equals 1.   *Related Exercises 57–62* ◄

SECTION 8.2 EXERCISES

Review Questions

1. Give an example of a nonincreasing sequence with a limit.

2. Give an example of a nondecreasing sequence without a limit.

3. Give an example of a bounded sequence that has a limit.

4. Give an example of a bounded sequence without a limit.

5. For what values of r does the sequence $\{r^n\}$ converge? Diverge?

6. Explain how the methods used to find the limit of a function as $x \to \infty$ are used to find the limit of a sequence.

7. Explain with a picture the formal definition of the limit of a sequence.

8. Explain how two sequences that differ only in their first ten terms can have the same limit.

Basic Skills

9–26. Limits of sequences *Find the limit of the following sequences or determine that the limit does not exist.*

9. $\left\{ \dfrac{n^3}{n^4 + 1} \right\}$

10. $\left\{ \dfrac{n^{12}}{3n^{12} + 4} \right\}$

11. $\left\{ \dfrac{3n^3 - 1}{2n^3 + 1} \right\}$

12. $\left\{ \dfrac{2e^{n+1}}{e^n} \right\}$

13. $\left\{ \dfrac{\tan^{-1} n}{n} \right\}$

14. $\{ n^{1/n} \}$

15. $\left\{ \left(1 + \dfrac{2}{n} \right)^n \right\}$

16. $\left\{ \left(\dfrac{n}{n + 5} \right)^n \right\}$

17. $\left\{ \sqrt{\left(1 + \dfrac{1}{2n} \right)^n} \right\}$

18. $\left\{ \dfrac{\ln (1/n)}{n} \right\}$

19. $\left\{ \left(\dfrac{1}{n} \right)^{1/n} \right\}$

20. $\left\{ \left(1 - \dfrac{4}{n} \right)^n \right\}$

21. $\{ b_n \}$ if $b_n = \begin{cases} n/(n + 1) & \text{if } n \leq 5000 \\ ne^{-n} & \text{if } n > 5000 \end{cases}$

22. $\{ \ln (n^3 + 1) - \ln (3n^3 + 10n) \}$

23. $\{ \ln \sin (1/n) + \ln n \}$

24. $\left\{ \dfrac{\sin 6n}{5n} \right\}$

25. $\{ n \sin (6/n) \}$

26. $\left\{ \dfrac{n!}{n^n} \right\}$

27–34. Limits of sequences and graphing *Find the limit of the following sequences or determine that the limit does not exist. Verify your result with a graphing utility.*

27. $a_n = \sin \left(\dfrac{n\pi}{2} \right)$

28. $a_n = \dfrac{(-1)^n n}{n + 1}$

29. $a_n = \dfrac{\sin (n\pi/3)}{\sqrt{n}}$

30. $a_n = \dfrac{3^n}{3^n + 4^n}$

31. $a_n = e^{-n} \cos n$

32. $a_n = \dfrac{\ln n}{n^{1.1}}$

33. $a_n = (-1)^n \sqrt[n]{n}$

34. $a_n = \cot \left(\dfrac{n\pi}{2n + 2} \right)$

35–42. Geometric sequences *Determine whether the following sequences converge or diverge and describe whether they do so monotonically or by oscillation. Give the limit when the sequence converges.*

35. $\{ 0.2^n \}$

36. $\{ 1.2^n \}$

37. $\{ (-0.7)^n \}$

38. $\{ (-1.01)^n \}$

39. $\{ 1.00001^n \}$

40. $\{ 2^n 3^{-n} \}$

41. $\{ (-2.5)^n \}$

42. $\{ (-0.003)^n \}$

43–46. Squeeze Theorem *Find the limit of the following sequences or state that they diverge.*

43. $\left\{ \dfrac{\sin n}{2^n} \right\}$

44. $\left\{ \dfrac{\cos (n\pi/2)}{\sqrt{n}} \right\}$

45. $\left\{ \dfrac{2 \tan^{-1} n}{n^3 + 4} \right\}$

46. $\left\{ \dfrac{n \sin^3 n}{n + 1} \right\}$

47. Periodic dosing Many people take aspirin on a regular basis as a preventive measure for heart disease. Suppose a person takes

80 mg of aspirin every 24 hr. Assume also that aspirin has a half-life of 24 hr; that is, every 24 hr half of the drug in the blood is eliminated.

a. Find a recurrence relation for the sequence $\{d_n\}$ that gives the amount of drug in the blood after the nth dose, where $d_1 = 80$.

b. Using a calculator, determine the limit of the sequence. In the long run, how much drug is in the person's blood?

c. Confirm the result of part (b) by finding the limit of $\{d_n\}$ directly.

48. A car loan Marie takes out a $20,000 loan for a new car. The loan has an annual interest rate of 6% or, equivalently, a monthly interest rate of 0.5%. Each month, the bank adds interest to the loan balance (the interest is always 0.5% of the current balance), and then Marie makes a $200 payment to reduce the loan balance. Let B_n be the loan balance immediately after the nth payment, where $B_0 = \$20,000$.

a. Write the first five terms of the sequence $\{B_n\}$.

b. Find a recurrence relation that generates the sequence $\{B_n\}$.

c. Determine how many months are needed to reduce the loan balance to zero.

49. A savings plan James begins a savings plan in which he deposits $100 at the beginning of each month into an account that earns 9% interest annually or, equivalently, 0.75% per month. To be clear, on the first day of each month, the bank adds 0.75% of the current balance as interest, and then James deposits $100. Let B_n be the balance in the account after the nth payment, where $B_0 = \$0$.

a. Write the first five terms of the sequence $\{B_n\}$.

b. Find a recurrence relation that generates the sequence $\{B_n\}$.

c. Determine how many months are needed to reach a balance of $5000.

50. Diluting a solution Suppose a tank is filled with 100 L of a 40% alcohol solution (by volume). You repeatedly perform the following operation: Remove 2 L of the solution from the tank and replace them with 2 L of 10% alcohol solution.

a. Let C_n be the concentration of the solution in the tank after the nth replacement, where $C_0 = 40\%$. Write the first five terms of the sequence $\{C_n\}$.

b. After how many replacements does the alcohol concentration reach 15%?

c. Determine the limiting (steady-state) concentration of the solution that is approached after many replacements.

51–56. Comparing growth rates of sequences *Determine which sequence has the greater growth rate as* $n \to \infty$. *Be sure to justify and explain your work.*

51. $a_n = n^2$; $b_n = n^2 \ln n$

52. $a_n = 3^n$; $b_n = n!$

53. $a_n = 3n^n$; $b_n = 100n!$

54. $a_n = \ln(n^{12})$; $b_n = n^{1/2}$

55. $a_n = n^{1/10}$; $b_n = n^{1/2}$

56. $a_n = e^{n/10}$; $b_n = 2^n$

57–62. Formal proofs of limits *Use the formal definition of the limit of a sequence to prove the following limits.*

57. $\lim_{n\to\infty} \dfrac{1}{n} = 0$

58. $\lim_{n\to\infty} \dfrac{1}{n^2} = 0$

59. $\lim_{n\to\infty} \dfrac{3n^2}{4n^2 + 1} = \dfrac{3}{4}$

60. $\lim_{n\to\infty} b^{-n} = 0$, for $b > 1$

61. $\lim_{n\to\infty} \dfrac{cn}{bn + 1} = \dfrac{c}{b}$, for real numbers $c > 0$ and $b > 0$

62. $\lim_{n\to\infty} \dfrac{n}{n^2 + 1} = 0$

Further Explorations

63. Explain why or why not Determine whether the following statements are true and give an explanation or counterexample.

a. If $\lim_{n\to\infty} a_n = 1$ and $\lim_{n\to\infty} b_n = 3$; then $\lim_{n\to\infty} \dfrac{b_n}{a_n} = 3$.

b. If $\lim_{n\to\infty} a_n = 0$ and $\lim_{n\to\infty} b_n = \infty$; then $\lim_{n\to\infty} a_n b_n = 0$.

c. The convergent sequences $\{a_n\}$ and $\{b_n\}$ differ in their first 100 terms, but $a_n = b_n$ for $n > 100$. It follows that $\lim_{n\to\infty} a_n = \lim_{n\to\infty} b_n$.

d. If $\{a_n\} = \left\{1, \frac{1}{2}, \frac{1}{3}, \frac{1}{4}, \frac{1}{5}, \dots\right\}$ and

$\{b_n\} = \left\{1, 0, \frac{1}{2}, 0, \frac{1}{3}, 0, \frac{1}{4}, 0, \dots\right\}$, then $\lim_{n\to\infty} a_n = \lim_{n\to\infty} b_n$.

e. If the sequence $\{a_n\}$ converges, then the sequence $\{(-1)^n a_n\}$ converges.

f. If the sequence $\{a_n\}$ diverges, then the sequence $\{0.000001 a_n\}$ diverges.

64–65. Reindexing *Express each sequence* $\{a_n\}_{n=1}^{\infty}$ *as an equivalent sequence of the form* $\{b_n\}_{n=3}^{\infty}$.

64. $\{2n + 1\}_{n=1}^{\infty}$

65. $\{n^2 + 6n - 9\}_{n=1}^{\infty}$

66–69. More sequences *Evaluate the limit of the following sequences.*

66. $a_n = \int_1^n x^{-2}\, dx$

67. $a_n = \dfrac{75^{n-1}}{99^n} + \dfrac{5^n \sin n}{8^n}$

68. $a_n = \tan^{-1}\left(\dfrac{10n}{10n + 4}\right)$

69. $a_n = \cos(0.99^n) + \dfrac{7^n + 9^n}{63^n}$

70–74. Sequences by recurrence relations *Consider the following sequences defined by a recurrence relation. Use a calculator, analytical methods, and/or graphing to make a conjecture about the value of the limit or determine that the limit does not exist.*

70. $a_{n+1} = \frac{1}{2} a_n + 2$; $a_0 = 5$, $n = 0, 1, 2, \dots$.

71. $a_{n+1} = 2a_n(1 - a_n)$; $a_0 = 0.3$, $n = 0, 1, 2, \dots$.

72. $a_{n+1} = \frac{1}{2}(a_n + 2/a_n)$; $a_0 = 2$, $n = 0, 1, 2, \dots$.

73. $a_{n+1} = 4a_n(1 - a_n)$; $a_0 = 0.5$, $n = 0, 1, 2, \dots$.

74. $a_{n+1} = \sqrt{2 + a_n}$; $a_0 = 1$, $n = 0, 1, 2, \dots$.

75. Crossover point The sequence $\{n!\}$ ultimately grows faster than the sequence $\{b^n\}$ for any $b > 1$ as $n \to \infty$. However, b^n is generally greater than $n!$ for small values of n. Use a calculator to determine the smallest value of n such that $n! > b^n$ for each of the cases $b = 2$, $b = e$, and $b = 10$.

Applications

76. Fish harvesting A fishery manager knows that her fish population naturally increases at a rate of 1.5% per month while 80 fish are harvested each month. Let F_n be the fish population after the nth month, where $F_0 = 4000$ fish.

a. Write out the first five terms of the sequence $\{F_n\}$.

b. Find a recurrence relation that generates the sequence $\{F_n\}$.

c. Does the fish population decrease or increase in the long run?

d. Determine whether the fish population decreases or increases in the long run if the initial population is 5500 fish.

e. Determine the initial fish population F_0 below which the population decreases.

77. The hungry hippo problem A pet hippopotamus weighing 200 lb today gains 5 lb per day with a food cost of 45¢/day. The price for hippos is 65¢/lb today but is falling 1¢/day.

a. Let h_n be the profit in selling the hippo on the nth day, where $h_0 = (200 \text{ lb}) \times (\$0.65) = \$130$. Write out the first ten terms of the sequence $\{h_n\}$.

b. How many days after today should the hippo be sold to maximize the profit?

78. Sleep model After many nights of observation, you notice that if you oversleep one night you tend to undersleep the following night and vice versa. This pattern of compensation is described by the relationship

$$x_{n+1} = \frac{1}{2}(x_n + x_{n-1}) \quad \text{for } n = 1, 2, 3, \ldots,$$

where x_n is the number of hours of sleep you get on the nth night and $x_0 = 7$ and $x_1 = 6$ are the number of hours of sleep on the first two nights, respectively.

a. Write out the first six terms of the sequence $\{x_n\}$ and confirm that the terms alternately increase and decrease.

b. Find an explicit expression for the nth term of the sequence.

c. What is the limit of the sequence?

79. Calculator algorithm The CORDIC (COordinate Rotation DIgital Calculation) algorithm is used by most calculators to evaluate trigonometric and logarithmic functions. An important number in the CORDIC algorithm, called the *aggregate constant*, is

$$\prod_{n=0}^{\infty} \frac{2^n}{\sqrt{1 + 2^{2n}}}, \text{ where } \prod_{n=0}^{k} a_n \text{ represents the product } a_0 \cdot a_1 \cdots a_k.$$

This infinite product is the limit of the sequence

$$\left\{ \prod_{n=0}^{0} \frac{2^n}{\sqrt{1 + 2^{2n}}}, \prod_{n=0}^{1} \frac{2^n}{\sqrt{1 + 2^{2n}}}, \prod_{n=0}^{2} \frac{2^n}{\sqrt{1 + 2^{2n}}}, \ldots \right\}.$$

(See the Guided Projects.) Estimate the value of the aggregate constant.

Additional Exercises

80. Bounded monotonic proof Prove that the drug dose sequence in Example 5,

$$d_{n+1} = 0.5d_n + 100, \quad \text{for } n = 1, 2, 3, \ldots, d_1 = 100,$$

is bounded and monotonic.

81. Repeated square roots Consider the expression

$$\sqrt{1 + \sqrt{1 + \sqrt{1 + \sqrt{1 + \cdots}}}}, \text{ where the process continues indefinitely.}$$

a. Show that this expression can be built in steps using the recurrence relation $a_0 = 1, a_{n+1} = \sqrt{1 + a_n}$ for $n = 0, 1, 2, 3, \ldots$. Explain why the value of the expression can be interpreted as $\lim_{n \to \infty} a_n$.

b. Evaluate the first five terms of the sequence $\{a_n\}$.

c. Estimate the limit of the sequence. Compare your estimate with $(1 + \sqrt{5})/2$, a number known as the *golden mean*.

d. Assuming the limit exists, use the method of Example 5 to determine the limit exactly.

e. Repeat the above analysis for the expression

$$\sqrt{p + \sqrt{p + \sqrt{p + \sqrt{p + \cdots}}}}, \text{ where } p > 0. \text{ Make}$$

a table showing the approximate value of this expression for various values of p. Does the expression seem to have a limit for all positive values of p?

82. A sequence of products Find the limit of the sequence

$$\{a_n\}_{n=2}^{\infty} = \left\{ \left(1 - \frac{1}{2}\right)\left(1 - \frac{1}{3}\right) \cdots \left(1 - \frac{1}{n}\right) \right\}.$$

83. Continued fractions The expression

$$1 + \cfrac{1}{1 + \cfrac{1}{1 + \cfrac{1}{1 + \cfrac{1}{1 + \cdots}}}}$$

where the process continues indefinitely, is called a *continued fraction*.

a. Show that this expression can be built in steps using the recurrence relation $a_0 = 1, a_{n+1} = 1 + 1/a_n$ for $n = 0, 1, 2, 3, \ldots$. Explain why the value of the expression can be interpreted as $\lim_{n \to \infty} a_n$.

b. Evaluate the first five terms of the sequence $\{a_n\}$.

c. Using computation and/or graphing, estimate the limit of the sequence.

d. Assuming the limit exists, use the method of Example 5 to determine the limit exactly. Compare your estimate with $(1 + \sqrt{5})/2$, a number known as the *golden mean*.

e. Assuming the limit exists, use the same ideas to determine the value of

$$a + \cfrac{b}{a + \cfrac{b}{a + \cfrac{b}{a + \cfrac{b}{a + \cdots}}}}$$

where a and b are positive real numbers.

84. Towers of powers For a positive real number p, how do you interpret $p^{p^{p^{\cdots}}}$, where the tower of exponents continues indefinitely? As it stands, the expression is ambiguous. The tower could be built from the top or from the bottom; that is, it could be evaluated by the recurrence relations

(1) $a_{n+1} = p^{a_n}$ (building from the bottom) or

(2) $a_{n+1} = a_n^p$ (building from the top),

where $a_0 = p$ in either case. The two recurrence relations have very different behaviors that depend on the value of p.

a. Use computations with various values of $p > 0$ to find the values of p such that the recurrence relation (2) has a limit. Find the maximum value of p for which the recurrence relation has a limit.

b. Show that recurrence relation (1) has a limit for certain values of p. Make a table showing the approximate value of the tower for various values of p. Estimate the maximum value of p for which the recurrence relation has a value.

⊤ 85. Fibonacci sequence The famous Fibonacci sequence was proposed by Leonardo Pisano, also known as Fibonacci, in about A.D. 1200 as a model for the growth of rabbit populations. It is given by the recurrence relation $f_{n+1} = f_n + f_{n-1}$, for $n = 1, 2, 3, \ldots$, where $f_0 = f_1 = 1$. Each term of the sequence is the sum of its two predecessors.

a. Write out the first ten terms of the sequence.

b. Is the sequence bounded?

c. Estimate or determine $\varphi = \lim\limits_{n \to \infty} \dfrac{f_{n+1}}{f_n}$, the ratio of successive terms of the sequence. Provide evidence that $\varphi = (1 + \sqrt{5})/2$, a number known as the *golden mean*.

d. Verify the remarkable result that

$$f_n = \frac{1}{\sqrt{5}} \left(\varphi^n - (-1)^n \varphi^{-n} \right)$$

86. Arithmetic-geometric mean Pick two positive numbers a_0 and b_0 with $a_0 > b_0$ and write out the first few terms of the two sequences $\{a_n\}$ and $\{b_n\}$:

$$a_{n+1} = \frac{a_n + b_n}{2}, \qquad b_{n+1} = \sqrt{a_n b_n}, \qquad \text{for } n = 0, 1, 2 \ldots.$$

(Recall that the arithmetic mean $A = (p + q)/2$ and the geometric mean $G = \sqrt{pq}$ of two positive numbers p and q satisfy $A \geq G$.)

a. Show that $a_n > b_n$ for all n.

b. Show that $\{a_n\}$ is a decreasing sequence and $\{b_n\}$ is an increasing sequence.

c. Conclude that $\{a_n\}$ and $\{b_n\}$ converge.

d. Show that $a_{n+1} - b_{n+1} < (a_n - b_n)/2$ and conclude that $\lim\limits_{n \to \infty} a_n = \lim\limits_{n \to \infty} b_n$. The common value of these limits is

called the arithmetic-geometric mean of a_0 and b_0, denoted $\text{AGM}(a_0, b_0)$.

e. Estimate $\text{AGM}(12, 20)$. Estimate Gauss' constant $1/\text{AGM}(1, \sqrt{2})$.

87. The hailstone sequence Here is a fascinating (unsolved) problem known as the hailstone problem (or the Ulam Conjecture or the Collatz Conjecture). It involves sequences in two different ways. First, choose a positive integer N and call it a_0. This is the *seed* of a sequence. The rest of the sequence is generated as follows: For $n = 0, 1, 2, \ldots$

$$a_{n+1} = \begin{cases} a_n/2 & \text{if } a_n \text{ is even} \\ 3a_n + 1 & \text{if } a_n \text{ is odd} \end{cases}$$

However, if $a_n = 1$ for any n, then the sequence terminates.

a. Compute the sequence that results from the seeds $N = 2, 3, 4, \ldots, 10$. You should verify that in all these cases, the sequence eventually terminates. The hailstone conjecture (still unproved) states that for all positive integers N, the sequence terminates after a finite number of terms.

b. Now define the hailstone sequence $\{H_k\}$, which is the number of terms needed for the sequence $\{a_n\}$ to terminate starting with a seed of k. Verify that $H_2 = 1$, $H_3 = 7$, and $H_4 = 2$.

c. Plot as many terms of the hailstone sequence as is feasible. How did the sequence get its name? Does the conjecture appear to be true?

88. Prove that if $\{a_n\} \ll \{b_n\}$ (as used in Theorem 8.6), then $\{ca_n\} \ll \{db_n\}$, where c and d are positive real numbers.

QUICK CHECK ANSWERS

1. (a) bounded, monotonic; (b) bounded, not monotonic; (c) not bounded, not monotonic; (d) bounded, monotonic (both nonincreasing and nondecreasing). **2.** If $r = -1$, the sequence is $\{-1, 1, -1, 1, \ldots\}$, the terms alternate in sign, and the sequence diverges. If $r = 1$, the sequence is $\{1, 1, 1, 1, \ldots\}$, the terms are constant, and the sequence converges. **3.** Both changes would increase the steady-state level of drug. **4.** $\{n^{1.1}\}$ grows faster; the limit is 0. ◄

8.3 Infinite Series

We begin our discussion of infinite series with *geometric series*. These series arise more frequently than any other infinite series, they are used in many practical problems, and they illustrate all the essential features of infinite series in general. First let's summarize some important ideas from Section 8.1.

> The sequence of partial sums may be visualized nicely as follows:
>

Recall that every infinite series $\sum\limits_{k=1}^{\infty} a_k$ has a sequence of partial sums

$$S_1 = a_1, \qquad S_2 = a_1 + a_2, \qquad S_3 = a_1 + a_2 + a_3,$$

where in general $S_n = \sum\limits_{k=1}^{n} a_k$, for $n = 1, 2, 3, \ldots$.

If the sequence of partial sums $\{S_n\}$ converges—that is, if $\lim_{n \to \infty} S_n = L$—then the value of the infinite series is also L. If the sequence of partial sums diverges, then the infinite series also diverges.

In summary, to evaluate an infinite series, it is necessary to determine a formula for the sequence of partial sums $\{S_n\}$ and then find its limit. This procedure can be carried out with the series that we discuss in this section: geometric series and telescoping series.

Geometric Series

> Geometric *sequences* have the form $\{r^k\}$. Geometric *sums* and *series* have the form $\sum_k r^k$ or $\sum_k ar^k$.

As a preliminary step to geometric series, we study **geometric sums**, which are *finite sums* in which each term is a constant multiple of the previous term. A geometric sum with n terms has the form

$$S_n = a + ar + ar^2 + \cdots + ar^{n-1} = \sum_{k=0}^{n-1} ar^k,$$

where a and r are real numbers; r is called the **ratio** of the sum and a is the first term of the series. For example, the geometric sum with $r = 0.1$, $a = 0.9$, and $n = 4$ is

$$0.9 + 0.09 + 0.009 + 0.0009 = 0.9(1 + 0.1 + 0.01 + 0.001)$$

$$= \sum_{k=0}^{3} 0.9(0.1^k).$$

QUICK CHECK 1 Which of the following sums are not geometric sums?

a. $\sum_{k=0}^{10} \left(\frac{1}{2}\right)^k$ **b.** $\sum_{k=0}^{20} \frac{1}{k}$

c. $\sum_{k=0}^{30} (2k + 1)$ ◄

Our goal is to find a formula for the value of the geometric sum

$$S_n = a + ar + ar^2 + \cdots + ar^{n-1} \tag{1}$$

for any values of a, r, and the positive integer n. Doing so requires a clever maneuver: We multiply both sides of equation (1) by the ratio r:

$$rS_n = r(a + ar + ar^2 + ar^3 + \cdots + ar^{n-1})$$
$$= ar + ar^2 + ar^3 + \cdots + ar^{n-1} + ar^n \tag{2}$$

We now subtract equation (2) from equation (1). Notice how most of the terms on the right sides of these equations cancel, leaving

$$S_n - rS_n = a - ar^n.$$

QUICK CHECK 2 Verify that the geometric sum formula gives the correct result for the sums $1 + \frac{1}{2}$ and $\frac{1}{2} + \frac{1}{4} + \frac{1}{8}$. ◄

Solving for S_n gives the formula

$$S_n = a \cdot \frac{1 - r^n}{1 - r}. \tag{3}$$

Having dealt with geometric *sums*, it is a short step to geometric *series*. We simply let the number of terms in the geometric sum $S_n = \sum_{k=0}^{n-1} ar^k$ increase without bound, which results in the geometric series $\sum_{k=0}^{\infty} ar^k$. The value of a geometric series is the limit of its sequence of partial sums (provided it exists). Using equation (3), we have

$$\underbrace{\sum_{k=0}^{\infty} ar^k}_{\text{geometric series}} = \lim_{n \to \infty} \underbrace{\sum_{k=0}^{n-1} ar^k}_{\text{geometric sum } S_n} = \lim_{n \to \infty} a \frac{1 - r^n}{1 - r}.$$

To compute this limit we must examine the behavior of r^n as $n \to \infty$. Recall from our work with geometric sequences (Section 8.2) that

$$\lim_{n \to \infty} r^n = \begin{cases} 0 & \text{if } |r| < 1 \\ 1 & \text{if } r = 1 \\ \text{does not exist} & \text{if } r \leq -1 \text{ or } r > 1 \end{cases}$$

Case 1: $|r| < 1$ Because $\lim\limits_{n \to \infty} r^n = 0$, we have

$$\lim_{n \to \infty} S_n = \lim_{n \to \infty} a \frac{1 - r^n}{1 - r} = a \frac{1 - \overbrace{\lim_{n \to \infty} r^n}^{0}}{1 - r} = \frac{a}{1 - r}.$$

In the case that $|r| < 1$, the geometric series *converges* to $\dfrac{a}{1 - r}$.

Case 2: $|r| > 1$ In this case, $\lim\limits_{n \to \infty} r^n$ does not exist, so $\lim\limits_{n \to \infty} S_n$ does not exist and the series *diverges*.

Case 3: $|r| = 1$ If $r = 1$, then the geometric series is $\sum\limits_{k=0}^{\infty} 1 = 1 + 1 + 1 + \cdots$, which diverges. If $r = -1$, the geometric series is $\sum\limits_{k=0}^{\infty} (-1)^k = 1 - 1 + 1 - \cdots$, which also diverges (because the sequence of partial sums oscillates between 0 and 1). So if $r = \pm 1$, then the geometric series *diverges*.

QUICK CHECK 3 Evaluate $1/2 + 1/4 + 1/8 + 1/16 + \cdots$. ◄

THEOREM 8.7   Geometric Series

Let r and a be real numbers. If $|r| < 1$, then $\sum\limits_{k=0}^{\infty} ar^k = \dfrac{a}{1 - r}$. If $|r| \geq 1$, then the series diverges.

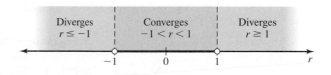

| Diverges $r \leq -1$ | Converges $-1 < r < 1$ | Diverges $r \geq 1$ |

QUICK CHECK 4 Explain why $\sum\limits_{k=0}^{\infty} 0.2^k$ converges and why $\sum\limits_{k=0}^{\infty} 2^k$ diverges. ◄

EXAMPLE 1   Geometric series Evaluate the following geometric series or state that the series diverges.

a. $\sum\limits_{k=0}^{\infty} 1.1^k$ **b.** $\sum\limits_{k=0}^{\infty} e^{-k}$ **c.** $\sum\limits_{k=2}^{\infty} 3(-0.75)^k$

SOLUTION

a. The ratio of this geometric series is $r = 1.1$. Because $|r| \geq 1$, the series diverges.

b. Note that $e^{-k} = \dfrac{1}{e^k} = \left(\dfrac{1}{e}\right)^k$. Therefore, the ratio of the series is $r = \dfrac{1}{e}$, and its first term is $a = 1$. Because $|r| < 1$, the series converges and its value is

$$\sum_{k=0}^{\infty} e^{-k} = \sum_{k=0}^{\infty} \left(\frac{1}{e}\right)^k = \frac{1}{1 - (1/e)} = \frac{e}{e - 1} \approx 1.582.$$

> The series in Example 1c is called an *alternating series* because the terms alternate in sign. Such series are discussed in detail in Section 8.6.

c. Writing out the first few terms of the series is helpful:

$$\sum_{k=2}^{\infty} 3(-0.75)^k = \underbrace{3(-0.75)^2}_{a} + \underbrace{3(-0.75)^3}_{ar} + \underbrace{3(-0.75)^4}_{ar^2} + \cdots.$$

We see that the first term of the series is $a = 3(-0.75)^2$, and the ratio of the series is $r = -0.75$. Because $|r| < 1$, the series converges, and its value is

$$\sum_{k=2}^{\infty} 3(-0.75)^k = \frac{3(-0.75)^2}{1 - (-0.75)} = \frac{27}{28}.$$

Related Exercises 7–40 ◄

EXAMPLE 2   Decimal expansions Write $1.0\overline{35} = 1.0353535\ldots$ as a geometric series and express its value as a fraction.

SOLUTION Notice that the decimal part of this number is a convergent geometric series with $a = 0.035$ and $r = 0.01$:

$$1.0353535\ldots = 1 + \underbrace{0.035 + 0.00035 + 0.0000035 + \cdots}_{\text{geometric series with } a = 0.035 \text{ and } r = 0.01}$$

Evaluating the series, we have

$$1.0353535\ldots = 1 + \frac{a}{1-r} = 1 + \frac{0.035}{1 - 0.01} = 1 + \frac{35}{990} = \frac{205}{198}.$$

Related Exercises 41–46 ◄

Telescoping Series

With geometric series, we carried out the entire evaluation process by finding a formula for the sequence of partial sums and evaluating the limit of the sequence. Not many infinite series can be subjected to this sort of analysis. With another class of series, called **telescoping series**, it can be done. Here is an example.

EXAMPLE 3   Telescoping series Evaluate the following series.

a. $\displaystyle\sum_{k=1}^{\infty} \left(\frac{1}{3^k} - \frac{1}{3^{k+1}} \right)$ **b.** $\displaystyle\sum_{k=1}^{\infty} \frac{1}{k(k+1)}$

SOLUTION

a. The nth term of the sequence of partial sums is

$$S_n = \sum_{k=1}^{n} \left(\frac{1}{3^k} - \frac{1}{3^{k+1}} \right) = \left(\frac{1}{3} - \frac{1}{3^2} \right) + \left(\frac{1}{3^2} - \frac{1}{3^3} \right) + \cdots + \left(\frac{1}{3^n} - \frac{1}{3^{n+1}} \right)$$

$$= \frac{1}{3} + \underbrace{\left(-\frac{1}{3^2} + \frac{1}{3^2} \right)}_{0} + \cdots + \underbrace{\left(-\frac{1}{3^n} + \frac{1}{3^n} \right)}_{0} - \frac{1}{3^{n+1}} \qquad \text{Regroup terms.}$$

> The series in Example 3a is also a geometric series and its value can be found using Theorem 8.7.

$$= \frac{1}{3} - \frac{1}{3^{n+1}} \qquad\qquad\qquad\qquad \text{Simplify.}$$

Observe that the interior terms of the sum cancel (or telescope) leaving a simple expression for S_n. Taking the limit, we find that

$$\sum_{k=1}^{\infty} \left(\frac{1}{3^k} - \frac{1}{3^{k+1}} \right) = \lim_{n\to\infty} S_n = \lim_{n\to\infty} \left(\frac{1}{3} - \underbrace{\frac{1}{3^{n+1}}}_{\to 0} \right) = \frac{1}{3}.$$

▷ See Section 7.4 for a review of partial fractions.

b. Using the method of partial fractions, the sequence of partial sums is

$$S_n = \sum_{k=1}^{n} \frac{1}{k(k+1)} = \sum_{k=1}^{n} \left(\frac{1}{k} - \frac{1}{k+1} \right).$$

Writing out this sum, we see that

$$S_n = \left(1 - \frac{1}{2} \right) + \left(\frac{1}{2} - \frac{1}{3} \right) + \left(\frac{1}{3} - \frac{1}{4} \right) + \cdots + \left(\frac{1}{n} - \frac{1}{n+1} \right)$$

$$= 1 + \underbrace{\left(-\frac{1}{2} + \frac{1}{2} \right)}_{0} + \underbrace{\left(-\frac{1}{3} + \frac{1}{3} \right)}_{0} + \cdots + \underbrace{\left(-\frac{1}{n} + \frac{1}{n} \right)}_{0} - \frac{1}{n+1}$$

$$= 1 - \frac{1}{n+1}.$$

Again, the sum telescopes and all the interior terms cancel. The result is a simple formula for the nth term of the sequence of partial sums. The value of the series is

$$\sum_{k=1}^{\infty} \frac{1}{k(k+1)} = \lim_{n \to \infty} S_n = \lim_{n \to \infty} \left(1 - \frac{1}{n+1} \right) = 1.$$

Related Exercises 47–58 ◀

SECTION 8.3 EXERCISES

Review Questions

1. What is the defining characteristic of a geometric series? Give an example.

2. What is the difference between a geometric sum and a geometric series?

3. What is meant by the *ratio* of a geometric series?

4. Does a geometric sum always have a finite value?

5. Does a geometric series always have a finite value?

6. What is the condition for convergence of the geometric

   series $\sum_{k=0}^{\infty} ar^k$?

Basic Skills

7–18. Geometric sums *Evaluate the following geometric sums.*

7. $\displaystyle\sum_{k=0}^{8} 3^k$

8. $\displaystyle\sum_{k=0}^{10} \left(\frac{1}{4} \right)^k$

9. $\displaystyle\sum_{k=0}^{20} \left(\frac{2}{5} \right)^{2k}$

10. $\displaystyle\sum_{k=4}^{12} 2^k$

11. $\displaystyle\sum_{k=0}^{9} \left(-\frac{3}{4} \right)^k$

12. $\displaystyle\sum_{k=1}^{5} (-2.5)^k$

13. $\displaystyle\sum_{k=0}^{6} \pi^k$

14. $\displaystyle\sum_{k=1}^{10} \left(\frac{4}{7} \right)^k$

15. $\displaystyle\sum_{k=0}^{20} (-1)^k$

16. $1 + \dfrac{2}{3} + \dfrac{4}{9} + \dfrac{8}{27}$

17. $\dfrac{1}{4} + \dfrac{1}{12} + \dfrac{1}{36} + \dfrac{1}{108} + \cdots + \dfrac{1}{2916}$

18. $\dfrac{1}{3} + \dfrac{1}{5} + \dfrac{3}{25} + \dfrac{9}{125} + \cdots + \dfrac{243}{15,625}$

19–34. Geometric series *Evaluate the geometric series or state that it diverges.*

19. $\displaystyle\sum_{k=0}^{\infty} \left(\frac{1}{4} \right)^k$

20. $\displaystyle\sum_{k=0}^{\infty} \left(\frac{3}{5} \right)^k$

21. $\displaystyle\sum_{k=0}^{\infty} 0.9^k$

22. $\displaystyle\sum_{k=0}^{\infty} \frac{2^k}{7^k}$

23. $\displaystyle\sum_{k=0}^{\infty} 1.01^k$

24. $\displaystyle\sum_{j=0}^{\infty} \left(\frac{1}{\pi} \right)^j$

25. $\displaystyle\sum_{k=1}^{\infty} e^{-2k}$

26. $\displaystyle\sum_{m=2}^{\infty} \frac{5}{2^m}$

27. $\displaystyle\sum_{k=1}^{\infty} 2^{-3k}$

28. $\displaystyle\sum_{k=3}^{\infty} \frac{3 \cdot 4^k}{7^k}$

29. $\displaystyle\sum_{k=4}^{\infty} \frac{1}{5^k}$

30. $\displaystyle\sum_{k=0}^{\infty} \left(\frac{4}{3} \right)^{-k}$

31. $\displaystyle\sum_{k=0}^{\infty} \left(\frac{e}{\pi} \right)^k$

32. $\displaystyle\sum_{k=1}^{\infty} \frac{3^{k-1}}{4^{k+1}}$

33. $\displaystyle\sum_{k=0}^{\infty} \left(\frac{1}{4} \right)^k 5^{6-k}$

34. $\displaystyle\sum_{k=2}^{\infty} \left(\frac{3}{8} \right)^{3k}$

35–40. Geometric series with alternating signs *Evaluate the geometric series or state that it diverges.*

35. $\displaystyle\sum_{k=0}^{\infty} \left(-\frac{9}{10} \right)^k$

36. $\displaystyle\sum_{k=1}^{\infty} \left(-\frac{2}{3} \right)^k$

37. $\displaystyle 3\sum_{k=0}^{\infty} \frac{(-1)^k}{\pi^k}$

38. $\displaystyle\sum_{k=1}^{\infty} (-e)^{-k}$

39. $\displaystyle\sum_{k=2}^{\infty} (-0.15)^k$

40. $\displaystyle\sum_{k=1}^{\infty} 3 \left(-\frac{1}{8} \right)^{3k}$

41–46. Decimal expansions *Write each repeating decimal first as a geometric series, then as a fraction (a ratio of two integers).*

41. $0.121212\ldots$ **42.** $1.252525\ldots$ **43.** $0.456456\ldots$

44. $1.00393939\ldots$ **45.** $0.00952952\ldots$ **46.** $5.12838383\ldots$

47–58. Telescoping series *For the following telescoping series, find a formula for the nth term of the sequence of partial sums $\{S_n\}$. Then evaluate $\lim\limits_{n\to\infty} S_n$ to obtain the value of the series or state that the series diverges.*

47. $\displaystyle\sum_{k=1}^{\infty}\left(\frac{1}{k+1}-\frac{1}{k+2}\right)$ **48.** $\displaystyle\sum_{k=1}^{\infty}\left(\frac{1}{k+2}-\frac{1}{k+3}\right)$

49. $\displaystyle\sum_{k=1}^{\infty}\frac{1}{(k+1)(k+2)}$ **50.** $\displaystyle\sum_{k=0}^{\infty}\frac{1}{(3k+1)(3k+4)}$

51. $\displaystyle\sum_{k=1}^{\infty}\ln\left(\frac{k+1}{k}\right)$ **52.** $\displaystyle\sum_{k=1}^{\infty}\left(\sqrt{k+1}-\sqrt{k}\right)$

53. $\displaystyle\sum_{k=1}^{\infty}\frac{1}{(k+p)(k+p+1)}$, where p is a positive integer

54. $\displaystyle\sum_{k=1}^{\infty}\frac{1}{(ak+1)(ak+a+1)}$, where a is a positive integer

55. $\displaystyle\sum_{k=1}^{\infty}\left(\frac{1}{\sqrt{k+1}}-\frac{1}{\sqrt{k+3}}\right)$

56. $\displaystyle\sum_{k=0}^{\infty}\left[\sin\left(\frac{(k+1)\pi}{2k+1}\right)-\sin\left(\frac{k\pi}{2k-1}\right)\right]$

57. $\displaystyle\sum_{k=0}^{\infty}\frac{1}{16k^2+8k-3}$ **58.** $\displaystyle\sum_{k=1}^{\infty}\left[\tan^{-1}(k+1)-\tan^{-1}k\right]$

Further Explorations

59. Explain why or why not Determine whether the following statements are true and give an explanation or counterexample.

a. $\displaystyle\sum_{k=1}^{\infty}\left(\frac{\pi}{e}\right)^{-k}$ is a convergent geometric series.

b. If a is a real number and $\displaystyle\sum_{k=12}^{\infty}a^k$ converges, then $\displaystyle\sum_{k=1}^{\infty}a^k$ converges.

c. If the series $\displaystyle\sum_{k=1}^{\infty}a^k$ converges and $|a|<|b|$, then the series $\displaystyle\sum_{k=1}^{\infty}b^k$ converges.

60. Zeno's paradox The Greek philosopher Zeno of Elea (who lived about 450 B.C.) invented many paradoxes, the most famous of which tells of a race between the swift warrior Achilles and a tortoise. Zeno argued

The slower when running will never be overtaken by the quicker; for that which is pursuing must first reach the point from which that which is fleeing started, so that the slower must necessarily always be some distance ahead.

In other words, giving the tortoise a head start, Achilles will never overtake the tortoise because every time Achilles reaches the point where the tortoise was, the tortoise has moved ahead. Resolve this paradox by assuming that Achilles gives the tortoise a 1-mi head start and runs 5 mi/hr to the tortoise's 1 mi/hr. How far does Achilles run before he overtakes the tortoise, and how long does it take?

61. Archimedes' quadrature of the parabola The Greeks solved several calculus problems almost 2000 years before the discovery of calculus. One example is Archimedes' calculation of the area of the region R bounded by a segment of a parabola, which he did using the "method of exhaustion." As shown in the figure, the idea was to fill R with an infinite sequence of triangles. Archimedes began with one triangle inscribed in the parabola, with area A_1, and proceeded in stages, with the number of new triangles doubling at each stage. He was able to show (the key to the solution) that at each stage, the area of a new triangle is $\frac{1}{8}$ of the area of a triangle at the previous stage; for example, $A_2 = \frac{1}{8}A_1$, and so forth. Show, as Archimedes did, that the area of R is $\frac{4}{3}$ times the area of A_1.

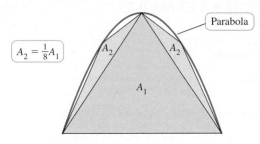

62. Value of a series

a. Find the value of the series

$$\sum_{k=1}^{\infty}\frac{3^k}{(3^{k+1}-1)(3^k-1)}.$$

b. For what value of a does the series

$$\sum_{k=1}^{\infty}\frac{a^k}{(a^{k+1}-1)(a^k-1)}$$

converge, and in those cases, what is its value?

Applications

63. House loan Suppose you take out a home mortgage for $180,000 at a monthly interest rate of 0.5%. If you make payments of $1000 per month, after how many months will the loan balance be zero? Estimate the answer by graphing the sequence of loan balances and then obtain an exact answer.

64. Car loan Suppose you borrow $20,000 for a new car at a monthly interest rate of 0.75%. If you make payments of $600 per month, after how many months will the loan balance be zero? Estimate the answer by graphing the sequence of loan balances and then obtain an exact answer.

65. Fish harvesting A fishery manager knows that her fish population naturally increases at a rate of 1.5% per month. At the end of each month, 120 fish are harvested. Let F_n be the fish population after the nth month, where $F_0 = 4000$ fish. Assuming that this process continues indefinitely, what is the long-term (steady-state) population of the fish?

66. Periodic doses Suppose that you take 200 mg of an antibiotic every 6 hr. The half-life of the drug is 6 hr (the time it takes for half of the drug to be eliminated from your blood). If you continue this regimen indefinitely, what is the long-term (steady-state) amount of antibiotic in your blood?

67. China's one-son policy In 1978, in an effort to reduce population growth, China instituted a policy that allows only one child per family. One unintended consequence has been that, because of a cultural bias toward sons, China now has many more young boys than girls. To solve this problem, some people have suggested replacing the one-child policy with a one-son policy: A family may have children until a boy is born. Suppose that the one-son policy were implemented and that natural birth rates remained the same (half boys and half girls). Using geometric series, compare the total number of children under the two policies.

68. Double glass An insulated window consists of two parallel panes of glass with a small spacing between them. Suppose that each pane reflects a fraction p of the incoming light and transmits the remaining light. Considering all reflections of light between the panes, what fraction of the incoming light is ultimately transmitted by the window? Assume the amount of incoming light is 1.

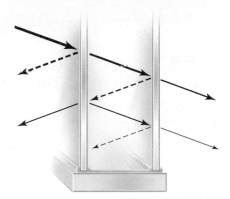

69. Bouncing ball for time Suppose a rubber ball, when dropped from a given height, returns to a fraction p of that height. How long does it take for a ball dropped from 10 m to come to rest? In the absence of air resistance, a ball dropped from a height h requires $\sqrt{2h/g}$ seconds to fall to the ground, where $g \approx 9.8$ m/s^2 is the acceleration due to gravity. The time taken to bounce *up* to a given height equals the time to fall from that height to the ground.

70. Multiplier effect Imagine that the government of a small community decides to give a total of $\$W$, distributed equally, to all of its citizens. Suppose that each month each citizen saves a fraction p of his or her new wealth and spends the remaining $1 - p$ in the community. Assume no money leaves or enters the community, and all of the spent money is redistributed throughout the community.

   a. If this cycle of saving and spending continues for many months, how much money is ultimately spent? Specifically,

by what factor is the initial investment of $\$W$ increased? (Economists refer to this increase in the investment as the multiplier effect.)

   b. Evaluate the limits $p \to 0$ and $p \to 1$ and interpret their meanings.

(See Guided Projects for more on economic stimulus packages.)

71. Snowflake island fractal The fractal called the *snowflake island* (or *Koch island*) is constructed as follows: Let I_0 be an equilateral triangle with sides of length 1. The figure I_1 is obtained by replacing the middle third of each side of I_0 by a new outward equilateral triangle with sides of length $1/3$ (see figure). The process is repeated where I_{n+1} is obtained by replacing the middle third of each side of I_n by a new outward equilateral triangle with sides of length $1/3^{n+1}$. The limiting figure as $n \to \infty$ is called the snowflake island.

   a. Let L_n be the perimeter of I_n. Show that $\lim\limits_{n\to\infty} L_n = \infty$.

   b. Let A_n be the area of I_n. Find $\lim\limits_{n\to\infty} A_n$. It exists!

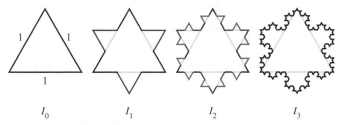

I_0 I_1 I_2 I_3

Additional Exercises

72. Decimal expansions

   a. Consider the number $0.555555\ldots$, which can be viewed as the series $5 \sum\limits_{k=1}^{\infty} 10^{-k}$. Evaluate the geometric series to obtain a rational value of $0.555555\ldots$

   b. Consider the number $0.54545454\ldots$, which can be represented by the series $54 \sum\limits_{k=1}^{\infty} 10^{-2k}$. Evaluate the geometric series to obtain a rational value of the number.

   c. Now generalize parts (a) and (b). Suppose you are given a number with a decimal expansion that repeats in cycles of length p, say, $n_1, n_2 \ldots n_p$, where $n_1, \ldots, n_p$ are integers between 0 and 9. Explain how to use geometric series to obtain a rational form of the number.

   d. Try the method of part (c) on the number $0.123456789123456789\ldots$

   e. Prove that $0.\overline{9} = 1$.

73. Remainder term Consider the geometric series $S = \sum\limits_{k=0}^{\infty} r^k$, which has the value $1/(1 - r)$ provided $|r| < 1$, and let $S_n = \sum\limits_{k=0}^{n-1} r^k = \dfrac{1 - r^n}{1 - r}$ be the sum of the first n terms. The remainder R_n is the error in approximating S by S_n. Show that

$$R_n = |S - S_n| = \left| \frac{r^n}{1 - r} \right|.$$

74–77. Comparing remainder terms *Use Exercise 73 to determine how many terms of each series are needed so that the partial sum is within 10^{-6} of the value of the series (that is, to ensure $R_n < 10^{-6}$).*

74. a. $\displaystyle\sum_{k=0}^{\infty} 0.6^k$ **b.** $\displaystyle\sum_{k=0}^{\infty} 0.15^k$ **75. a.** $\displaystyle\sum_{k=0}^{\infty} (-0.8)^k$ **b.** $\displaystyle\sum_{k=0}^{\infty} 0.2^k$

76. a. $\displaystyle\sum_{k=0}^{\infty} 0.72^k$ **b.** $\displaystyle\sum_{k=0}^{\infty} (-0.25)^k$ **77. a.** $\displaystyle\sum_{k=0}^{\infty} \left(\frac{1}{\pi}\right)^k$ **b.** $\displaystyle\sum_{k=0}^{\infty} \left(\frac{1}{e}\right)^k$

78. Functions defined as series Suppose a function f is defined by the geometric series $f(x) = \displaystyle\sum_{k=0}^{\infty} x^k$.

a. Evaluate $f(0), f(0.2), f(0.5), f(1)$, and $f(1.5)$, if possible.
b. What is the domain of f?

79. Functions defined as series Suppose a function f is defined by the geometric series $f(x) = \displaystyle\sum_{k=0}^{\infty} (-1)^k x^k$.

a. Evaluate $f(0), f(0.2), f(0.5), f(1)$, and $f(1.5)$.
b. What is the domain of f?

80. Functions defined as series Suppose a function f is defined by the geometric series $f(x) = \displaystyle\sum_{k=0}^{\infty} x^{2k}$.

a. Evaluate $f(0), f(0.2), f(0.5), f(1)$, and $f(1.5)$.
b. What is the domain of f?

81. Series in an equation For what values of x does the geometric series

$$f(x) = \sum_{k=0}^{\infty} \left(\frac{1}{1+x}\right)^k$$

converge? Solve $f(x) = 3$.

82. Bubbles Imagine a stack of hemispherical soap bubbles with decreasing radii $r_1 = 1, r_2, r_3, \ldots$ (see figure). Let h_n be the distance between the diameters of bubble n and bubble $n + 1$, and let H_n be the total height of the stack with n bubbles.

a. Use the Pythagorean theorem to show that in a stack with n bubbles, $h_1^2 = r_1^2 - r_2^2, h_2^2 = r_2^2 - r_3^2$, and so forth. Note that $h_n = r_n$.
b. Use part (a) to show that the height of a stack with n bubbles is

$$H_n = \sqrt{r_1^2 - r_2^2} + \sqrt{r_2^2 - r_3^2} + \cdots$$
$$+ \sqrt{r_{n-1}^2 - r_n^2} + r_n.$$

c. The height of a stack of bubbles depends on how the radii decrease. Suppose that $r_1 = 1, r_2 = a, r_3 = a^2, \ldots, r_n = a^n$, where $0 < a < 1$ is a fixed real number. In terms of a, find the height H_n of a stack with n bubbles.
d. Suppose the stack in part (c) is extended indefinitely ($n \to \infty$). In terms of a, how high would the stack be?
e. Challenge problem: Fix n and determine the sequence of radii $r_1, r_2, r_3, \ldots, r_n$ that maximizes H_n, the height of the stack with n bubbles.

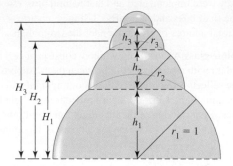

QUICK CHECK ANSWERS

1. b and c   **2.** Using the formula, the values are $\frac{3}{2}$ and $\frac{7}{8}$.

3. 1   **4.** The first converges because $|r| = 0.2 < 1$; the second diverges because $|r| = 2 > 1$. ◄

8.4 The Divergence and Integral Tests

With geometric series and telescoping series, the sequence of partial sums can be found and its limit can be evaluated (when it exists). Unfortunately, it is difficult or impossible to find an explicit formula for the sequence of partial sums for most infinite series. Therefore, it is difficult to obtain the exact value of most convergent series.

In this section, we explore methods to determine whether or not a given infinite series converges, which is simply a *yes/no* question. If the answer is *no*, the series diverges, and there are no more questions to ask. If the answer is *yes*, the series converges and it may be possible to estimate its value.

The Harmonic Series

We begin with an example that has a surprising result. Consider the infinite series

$$\sum_{k=1}^{\infty} \frac{1}{k} = 1 + \frac{1}{2} + \frac{1}{3} + \frac{1}{4} + \frac{1}{5} + \cdots,$$

a famous series known as the **harmonic series**. Does it converge? Suppose you try to answer this question by writing out the terms of the sequence of partial sums:

$$S_1 = 1 \qquad\qquad S_2 = 1 + \frac{1}{2} = \frac{3}{2}$$

$$S_3 = 1 + \frac{1}{2} + \frac{1}{3} = \frac{11}{6} \qquad S_4 = 1 + \frac{1}{2} + \frac{1}{3} + \frac{1}{4} = \frac{25}{12}$$

$$\vdots\ \ \vdots\ \ \vdots$$

$$S_n = \sum_{k=1}^{n} \frac{1}{k} = 1 + \frac{1}{2} + \frac{1}{3} + \frac{1}{4} + \cdots + \frac{1}{n}$$

$$\vdots\ \ \vdots\ \ \vdots$$

> We analyze S_n numerically because an explicit formula for S_n does not exist.

Have a look at the first 200 terms of the sequence of partial sums shown in Figure 8.25. What do you think—does the series converge? The terms of the sequence of partial sums increase, but at a decreasing rate. They could approach a limit or they could increase without bound.

Computing additional terms of the sequence of partial sums does not provide conclusive evidence. Table 8.3 shows that the sum of the first million terms is less than 15; the sum of the first 10^{40} terms—an unimaginably large number of terms—is less than 100. This is a case in which computation alone is not sufficient to determine whether a series converges. We return to this example later with more refined methods.

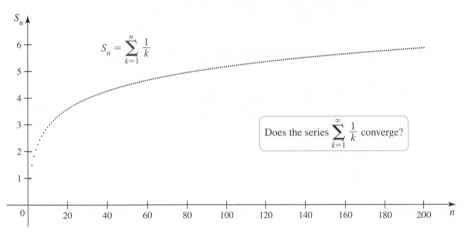

Does the series $\displaystyle\sum_{k=1}^{\infty} \frac{1}{k}$ converge?

FIGURE 8.25

Table 8.3

n	S_n	n	S_n
10^3	≈ 7.49	10^{10}	≈ 23.60
10^4	≈ 9.79	10^{20}	≈ 46.63
10^5	≈ 12.09	10^{30}	≈ 69.65
10^6	≈ 14.39	10^{40}	≈ 92.68

Properties of Convergent Series

For now, we restrict our attention to series with positive terms; that is, series of the form $\sum a_k$, where $a_k > 0$. The notation $\sum a_k$, without initial and final values of k, is used to refer to a general infinite series.

THEOREM 8.8   Properties of Convergent Series

1. Suppose $\sum a_k$ converges to A and let c be a real number. The series $\sum c a_k$ converges and $\sum c a_k = c \sum a_k = cA$.

2. Suppose $\sum a_k$ converges to A and $\sum b_k$ converges to B. The series $\sum (a_k \pm b_k)$ converges and $\sum (a_k \pm b_k) = \sum a_k \pm \sum b_k = A \pm B$.

3. *Whether* a series converges does not depend on a finite number of terms added to or removed from the series. Specifically, if M is a positive integer, then $\displaystyle\sum_{k=1}^{\infty} a_k$

    and $\displaystyle\sum_{k=M}^{\infty} a_k$ both converge or both diverge. However, the *value* of a convergent series does change if nonzero terms are added or deleted.

> The **leading terms** of an infinite series are those at the beginning with a small index. The **tail** of an infinite series consists of the terms at the "end" of the series with a large and increasing index. The convergence or divergence of an infinite series depends on the tail of the series, while the value of a convergent series is determined primarily by the leading terms.

Proof These properties are proved using properties of finite sums and limits of sequences. To prove Property 1, assume that $\sum\limits_{k=1}^{\infty} a_k$ converges and note that

$$\sum_{k=1}^{\infty} ca_k = \lim_{n \to \infty} \sum_{k=1}^{n} ca_k \qquad \text{Definition of infinite series}$$

$$= \lim_{n \to \infty} c \sum_{k=1}^{n} a_k \qquad \text{Property of finite sums}$$

$$= c \lim_{n \to \infty} \sum_{k=1}^{n} a_k \qquad \text{Property of limits}$$

$$= c \sum_{k=1}^{\infty} a_k \qquad \text{Definition of infinite series}$$

$$= cA \qquad \text{Value of the series}$$

Property 2 is proved in a similar way (Exercise 54).

Property 3 follows by noting that for finite sums with $1 < M < n$,

$$\sum_{k=M}^{n} a_k = \sum_{k=1}^{n} a_k - \sum_{k=1}^{M-1} a_k.$$

Letting $n \to \infty$ in this equation and assuming that $\sum\limits_{k=0}^{\infty} a_k = A$, it follows that

$$\sum_{k=M}^{\infty} a_k = \underbrace{\sum_{k=1}^{\infty} a_k}_{A} - \underbrace{\sum_{k=1}^{M-1} a_k}_{\text{finite number}}.$$

QUICK CHECK 1 Explain why if $\sum\limits_{k=1}^{\infty} a_k$ converges, then the series $\sum\limits_{k=5}^{\infty} a_k$ (with a different starting index) also converges. Do the two series have the same value? ◄

Because the right side has a finite value, $\sum\limits_{k=M}^{\infty} a_k$ converges. Similarly, if $\sum\limits_{k=M}^{\infty} a_k$ converges, then $\sum\limits_{k=1}^{\infty} a_k$ converges. By an analogous argument, if one of these series diverges, then the other series diverges. ◄

EXAMPLE 1   Using properties of series Evaluate the infinite series

$$S = \sum_{k=1}^{\infty} \left[5 \left(\frac{2}{3} \right)^k - \frac{2^{k-1}}{7^k} \right].$$

SOLUTION We examine the two series $\sum\limits_{k=1}^{\infty} 5 \left(\frac{2}{3} \right)^k$ and $\sum\limits_{k=1}^{\infty} \frac{2^{k-1}}{7^k}$ individually. The first series is a geometric series and is evaluated using the methods of Section 8.3. Its first few terms are

$$\sum_{k=1}^{\infty} 5 \left(\frac{2}{3} \right)^k = 5 \left(\frac{2}{3} \right) + 5 \left(\frac{2}{3} \right)^2 + 5 \left(\frac{2}{3} \right)^3 + \cdots.$$

The first term of the series is $a = 5 \left(\frac{2}{3} \right)$ and the ratio is $r = \frac{2}{3} < 1$; therefore,

$$\sum_{k=1}^{\infty} 5 \left(\frac{2}{3} \right)^k = \frac{a}{1 - r} = \left[\frac{5 \left(\frac{2}{3} \right)}{1 - \frac{2}{3}} \right] = 10.$$

Writing out the first few terms of the second series, we see that it, too, is geometric:

$$\sum_{k=1}^{\infty} \frac{2^{k-1}}{7^k} = \frac{1}{7} + \frac{2}{7^2} + \frac{2^2}{7^3} + \cdots$$

The first term is $a = \frac{1}{7}$ and the ratio is $r = \frac{2}{7} < 1$; therefore,

$$\sum_{k=1}^{\infty} \frac{2^{k-1}}{7^k} = \frac{a}{1-r} = \frac{\frac{1}{7}}{1 - \frac{2}{7}} = \frac{1}{5}.$$

Both series converge. By Property 2 of Theorem 8.8, we combine them and have
$S = 10 - \frac{1}{5} = \frac{49}{5}.$

Related Exercises 9–14 ◄

QUICK CHECK 2 For a series with positive terms, explain why the sequence of partial sums $\{S_n\}$ is an increasing sequence. ◄

The Divergence Test

The goal of this section is to develop tests to determine whether or not an infinite series converges. One of the simplest and most useful tests determines whether an infinite series *diverges*.

THEOREM 8.9   Divergence Test

If $\sum a_k$ converges, then $\lim_{k \to \infty} a_k = 0$. Equivalently, if $\lim_{k \to \infty} a_k \neq 0$, then the series diverges.

Important note Theorem 8.9 cannot be used to determine convergence.

Proof Let $\{S_k\}$ be the sequence of partial sums for the series $\sum a_k$. Assuming the series converges, it has a finite value, call it S, where

$$S = \lim_{k \to \infty} S_k = \lim_{k \to \infty} S_{k-1}.$$

Note that $S_k - S_{k-1} = a_k$. Therefore,

$$\lim_{k \to \infty} a_k = \lim_{k \to \infty} (S_k - S_{k-1}) = S - S = 0;$$

> If the statement *if p, then q* is true, then its contrapositive, *if (not q), then (not p)*, is also true. However its converse, *if q, then p*, is not necessarily true. Try it out on the true statement, *if I live in Paris, then I live in France.*

that is, $\lim_{k \to \infty} a_k = 0$ (Figure 8.26). The second part of the test follows immediately because it is the *contrapositive* of the first part (see margin note). ◄

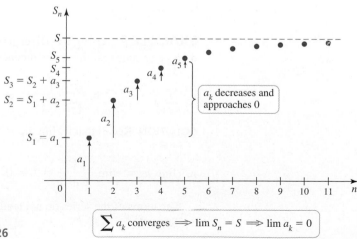

$$\sum a_k \text{ converges} \implies \lim S_n = S \implies \lim a_k = 0$$

FIGURE 8.26

To summarize: If the terms a_k of a given series do *not* tend to zero as $k \to \infty$, then the series diverges. Unfortunately, the test is easy to misuse. It's tempting to conclude that if the terms of the series tend to zero, then the series converges. This is often not true, as illustrated by the harmonic series—to which we now return.

Recall that the harmonic series is

$$\sum_{k=1}^{\infty} \frac{1}{k} = 1 + \frac{1}{2} + \frac{1}{3} + \frac{1}{4} + \frac{1}{5} + \cdots.$$

With $a_k = 1/k$, we have $\lim_{k \to \infty} a_k = 0$, which does *not* imply convergence. We now determine whether the series converges.

The nth term of the sequence of partial sums,

$$S_n = \sum_{k=1}^{n} \frac{1}{k} = 1 + \frac{1}{2} + \frac{1}{3} + \frac{1}{4} + \cdots + \frac{1}{n},$$

is represented geometrically by a left Riemann sum of the function $y = \dfrac{1}{x}$ on the interval $[1, n + 1]$ (Figure 8.27). This fact follows by observing that the areas of the rectangles, from left to right, are $1, \dfrac{1}{2}, \ldots, \dfrac{1}{n}$. By comparing the sum of the areas of these n rectangles with the area under the curve,

which is $\displaystyle\int_{1}^{n+1} \frac{dx}{x}$, we see that $S_n > \displaystyle\int_{1}^{n+1} \frac{dx}{x}$.

We know that $\displaystyle\int_{1}^{n+1} \frac{dx}{x} = \ln(n + 1)$ increases without bound as n increases. Because S_n exceeds $\displaystyle\int_{1}^{n+1} \frac{dx}{x}$, S_n also increases without bound; therefore, $\lim_{n \to \infty} S_n = \infty$ and the harmonic series $\displaystyle\sum_{k=1}^{\infty} \frac{1}{k}$ diverges. This argument justifies the following theorem.

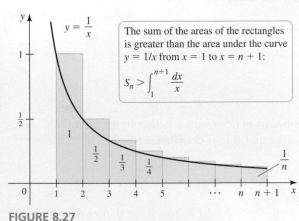

FIGURE 8.27

The sum of the areas of the rectangles is greater than the area under the curve $y = 1/x$ from $x = 1$ to $x = n + 1$:

$$S_n > \int_{1}^{n+1} \frac{dx}{x}$$

$y = \dfrac{1}{x}$

▶ Recall that $\displaystyle\int \frac{dx}{x} = \ln|x| + C$.

In Section 7.7, we showed that $\displaystyle\int_{1}^{\infty} \frac{dx}{x^p}$ diverges for $p \leq 1$. Therefore, $\displaystyle\int_{1}^{\infty} \frac{dx}{x}$ diverges.

THEOREM 8.10   Harmonic Series

The harmonic series $\displaystyle\sum_{k=1}^{\infty} \frac{1}{k} = 1 + \frac{1}{2} + \frac{1}{3} + \frac{1}{4} + \frac{1}{5} + \cdots$ diverges—even though the terms of the series tend to zero.

EXAMPLE 2   Using the Divergence Test Determine whether the following series diverge or state that the test is inconclusive.

a. $\displaystyle\sum_{k=0}^{\infty} \frac{k}{k + 1}$ **b.** $\displaystyle\sum_{k=1}^{\infty} \frac{1}{\sqrt{k}}$ **c.** $\displaystyle\sum_{k=1}^{\infty} \frac{1 + 3^k}{2^k}$

SOLUTION Recall that if $\lim_{k \to \infty} a_k \neq 0$, then the series diverges.

a. $\lim_{k \to \infty} a_k = \lim_{k \to \infty} \dfrac{k}{k + 1} = 1 \neq 0.$

The terms of the series do not tend to zero, so the series diverges by the Divergence Test.

b. $\lim\limits_{k\to\infty} a_k = \lim\limits_{k\to\infty} \dfrac{1}{\sqrt{k}} = 0.$

The terms of the series approach zero, so the Divergence Test is inconclusive. (Remember, the Divergence Test cannot be used to prove that a series converges.)

c. $\lim\limits_{k\to\infty} a_k = \lim\limits_{k\to\infty} \dfrac{1 + 3^k}{2^k}$

$$= \lim_{k\to\infty} \left[\underbrace{2^{-k}}_{\to 0} + \underbrace{\left(\dfrac{3}{2}\right)^k}_{\to\infty} \right] \quad \text{Simplify.}$$

$$= \infty$$

QUICK CHECK 3 Apply the Divergence Test to the geometric series $\sum r^k$. For what values of r does the series diverge? ◄

In this case, $\lim\limits_{k\to\infty} a_k$ does not equal 0, so the corresponding series $\displaystyle\sum_{k=1}^{\infty} \dfrac{1 + 3^k}{2^k}$ diverges by the Divergence Test.

Related Exercises 15–22 ◄

The Integral Test

The fact that infinite series are sums and that integrals are limits of sums suggests a connection between series and integrals. The Integral Test exploits this connection.

> **THEOREM 8.11   Integral Test**
> Suppose f is a continuous, positive, decreasing function for $x \geq 1$ and let $a_k = f(k)$ for $k = 1, 2, 3, \ldots$. Then
>
> $$\sum_{k=1}^{\infty} a_k \quad \text{and} \quad \int_1^{\infty} f(x)\, dx$$
>
> either both converge or both diverge. In the case of convergence, the value of the integral is *not*, in general, equal to the value of the series.

➤ The Integral Test also applies if the terms of the series a_k are decreasing for $k > N$ for some finite $N > 1$. The proof can be modified to account for this situation.

Proof By comparing the shaded regions in Figure 8.28, it follows that

$$\sum_{k=2}^{n} a_k \leq \int_1^n f(x)\, dx \leq \sum_{k=1}^{n-1} a_k. \tag{1}$$

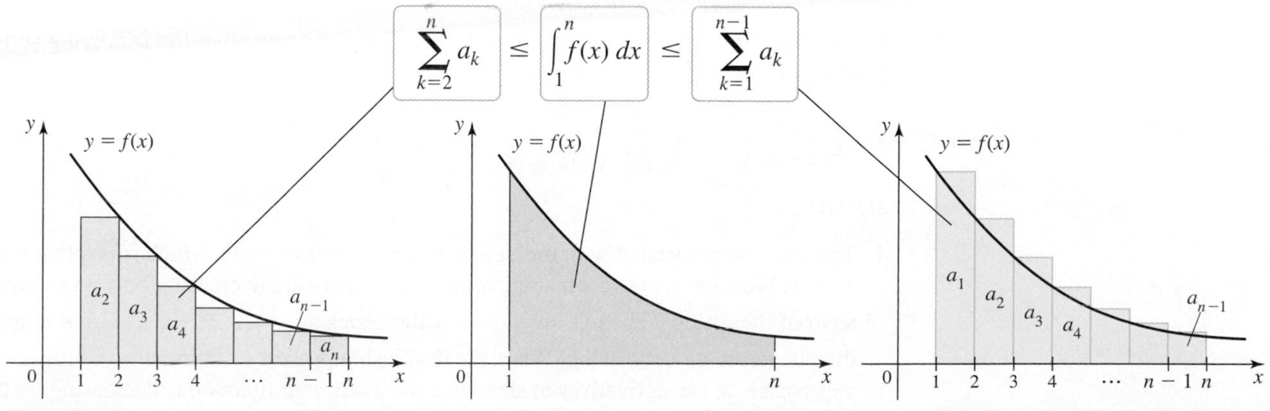

$$\boxed{\sum_{k=2}^{n} a_k} \leq \boxed{\int_1^n f(x)\, dx} \leq \boxed{\sum_{k=1}^{n-1} a_k}$$

FIGURE 8.28

The proof must demonstrate two results: If the improper integral $\int_1^{\infty} f(x)\, dx$ has a finite value, then the infinite series converges, *and* if the infinite series converges, then the improper integral has a finite value. First suppose that the improper integral $\int_1^\infty f(x)\,dx$ has a finite value, say, I. We have

$$\sum_{k=1}^n a_k = a_1 + \sum_{k=2}^n a_k \qquad \text{Separate the first term of the series.}$$

$$\le a_1 + \int_1^n f(x)\,dx \qquad \text{Left inequality in expression (1)}$$

$$< a_1 + \int_1^\infty f(x)\,dx \qquad f \text{ is positive, so } \int_1^n f(x)\,dx < \int_1^\infty f(x)\,dx.$$

$$= a_1 + I.$$

▷ In this proof, we rely twice on the Bounded Monotonic Sequence Theorem of Section 8.2: A bounded monotonic sequence converges.

This argument implies that the terms of the sequence of partial sums $S_n = \sum_{k=1}^n a_k$ are bounded above by $a_1 + I$. Because $\{S_n\}$ is also increasing (the series consists of positive terms), the sequence of partial sums converges, which means the series $\sum_{k=1}^\infty a_k$ converges (to a value less than or equal to $a_1 + I$).

Now suppose the infinite series $\sum_{k=1}^\infty a_k$ converges and has a value S. We have

$$\int_1^n f(x)\,dx \le \sum_{k=1}^{n-1} a_k \qquad \text{Right inequality in expression (1)}$$

$$< \sum_{k=1}^\infty a_k \qquad \text{Terms } a_k \text{ are positive.}$$

$$= S. \qquad \text{Value of infinite series}$$

We see that the sequence $\left\{ \int_1^n f(x)\,dx \right\}$ is increasing (because $f(x) > 0$) and bounded above by a fixed number S. Thus, the improper integral $\int_1^\infty f(x)\,dx = \lim_{n\to\infty} \int_1^n f(x)\,dx$ has a finite value (less than or equal to S).

We have shown that if $\int_1^\infty f(x)\,dx$ is finite, then $\sum a_k$ converges and vice versa. The same inequalities imply that $\int_1^\infty f(x)\,dx$ and $\sum a_k$ also diverge together.   ◄

The Integral Test is used to determine *whether* a series converges or diverges. For this reason, adding or subtracting a few terms in the series *or* changing the lower limit of integration to another finite point does not change the outcome of the test. Therefore, the test does not depend on the lower index in the series or the lower limit of the integral.

EXAMPLE 3 Applying the Integral Test Determine whether the following series converge.

a. $\displaystyle\sum_{k=1}^\infty \frac{k}{k^2 + 1}$ **b.** $\displaystyle\sum_{k=3}^\infty \frac{1}{\sqrt{2k - 5}}$

SOLUTION

a. The function associated with this series is $f(x) = x/(x^2 + 1)$, which is positive for $x \ge 1$. We must also show that the terms of the series are decreasing beyond some fixed term of the series. The first few terms of the series are $\left\{\frac{1}{2}, \frac{2}{5}, \frac{3}{10}, \frac{4}{17}, \dots\right\}$, and it appears that the terms are decreasing. When the decreasing property is difficult to confirm, one approach is to use derivatives to show that the associated function is decreasing. In this case, we have

$$f'(x) = \frac{d}{dx}\left(\frac{x}{x^2 + 1}\right) = \underbrace{\frac{x^2 + 1 - 2x^2}{(x^2 + 1)^2}}_{\text{Quotient Rule}} = \frac{1 - x^2}{(x^2 + 1)^2}.$$

For $x > 1$, $f'(x) < 0$, which implies that the function and the terms of the series are decreasing. The integral that determines convergence is

$$\int_1^\infty \frac{x}{x^2 + 1}\, dx = \lim_{b \to \infty} \int_1^b \frac{x}{x^2 + 1}\, dx \qquad \text{Definition of improper integral}$$

$$= \lim_{b \to \infty} \frac{1}{2} \ln\left(x^2 + 1\right)\Big|_1^b \qquad \text{Evaluate integral.}$$

$$= \frac{1}{2} \lim_{b \to \infty} \left(\ln\left(b^2 + 1\right) - \ln 2\right) \qquad \text{Simplify.}$$

$$= \infty. \qquad \lim_{b \to \infty} \ln\left(b^2 + 1\right) = \infty$$

Because the integral diverges, the series diverges.

b. The Integral Test may be modified to accommodate initial indices other than $k = 1$. The terms of this series decrease for $k \geq 3$. In this case, the relevant integral is

$$\int_3^\infty \frac{dx}{\sqrt{2x - 5}} = \lim_{b \to \infty} \int_3^b \frac{dx}{\sqrt{2x - 5}} \qquad \text{Definition of improper integral}$$

$$= \lim_{b \to \infty} \sqrt{2x - 5}\,\Big|_3^b \qquad \text{Evaluate integral.}$$

$$= \infty \qquad \lim_{b \to \infty} \sqrt{2b - 5} = \infty$$

Because the integral diverges, the series also diverges.   *Related Exercises 23–30* ◄

The *p*-Series

The Integral Test is used to analyze the convergence of an entire family of infinite series $\sum_{k=1}^\infty \frac{1}{k^p}$ known as the *p-series*. For what values of the real number p does the *p*-series converge?

EXAMPLE 4   **The *p*-series** For what values of p does the *p*-series $\sum_{k=1}^\infty \frac{1}{k^p}$ converge?

SOLUTION Notice that $p = 1$ corresponds to the harmonic series, which diverges. To apply the Integral Test, observe that the terms of the given series are positive and decreasing for $p > 0$. The function associated with the series is $f(x) = \frac{1}{x^p}$. The relevant integral is $\int_1^\infty x^{-p}\, dx = \int_1^\infty \frac{dx}{x^p}$. Appealing again to Section 7.7, recall that this integral converges when $p > 1$ and diverges when $p \leq 1$. Therefore, by the Integral Test, the *p*-series $\sum_{k=1}^\infty \frac{1}{k^p}$ converges when $p > 1$ and diverges when $0 < p \leq 1$. For example, the series

$$\sum_{k=1}^\infty \frac{1}{k^3} \quad \text{and} \quad \sum_{k=1}^\infty \frac{1}{\sqrt{k}}$$

converge and diverge, respectively. For $p < 0$, the series diverges by the Divergence Test. This argument justifies the following theorem.   *Related Exercises 31–34* ◄

QUICK CHECK 4 Which of the following series are p-series, and which series converge?

a. $\displaystyle\sum_{k=1}^{\infty} k^{-0.8}$ **b.** $\displaystyle\sum_{k=1}^{\infty} 2^{-k}$ **c.** $\displaystyle\sum_{k=10}^{\infty} k^{-4}$ ◄

> **THEOREM 8.12   Convergence of the p-Series**
>
> The p-series $\displaystyle\sum_{k=1}^{\infty} \frac{1}{k^p}$ converges when $p > 1$ and diverges when $p \leq 1$.

EXAMPLE 5   Using the p-series test Determine whether the following series converge or diverge.

a. $\displaystyle\sum_{k=1}^{\infty} \frac{1}{\sqrt[4]{k^3}}$ **b.** $\displaystyle\sum_{k=4}^{\infty} \frac{1}{(k-1)^2}$

SOLUTION

a. This series is a p-series with $p = \frac{3}{4}$. By Theorem 8.12, it diverges.

b. The series

$$\sum_{k=4}^{\infty} \frac{1}{(k-1)^2} = \sum_{k=3}^{\infty} \frac{1}{k^2} = \frac{1}{3^2} + \frac{1}{4^2} + \frac{1}{5^2} + \cdots$$

is a convergent p-series ($p = 2$) without the first two terms. By property 3 of Theorem 8.8, adding or removing a finite number of terms does not affect the convergence of a series. Therefore, the given series converges.

Related Exercises 31–34 ◄

Estimating the Value of Infinite Series

The Integral Test is powerful in its own right, but it comes with an added bonus. In some cases, it is used to estimate the value of a series. We define the **remainder** to be the error in approximating a convergent infinite series by the sum of its first n terms; that is,

$$R_n = \underbrace{\sum_{k=1}^{\infty} a_k}_{\substack{\text{value of} \\ \text{series}}} - \underbrace{\sum_{k=1}^{n} a_k}_{\substack{\text{approximation based} \\ \text{on first } n \text{ terms}}} = a_{n+1} + a_{n+2} + a_{n+3} + \cdots.$$

QUICK CHECK 5 If $\sum a_k$ is a convergent series of positive terms, why is $R_n \geq 0$? ◄

The remainder consists of the *tail* of the series—those terms beyond a_n.

We now argue much as we did in the proof of the Integral Test. Let f be a continuous, positive, decreasing function such that $f(k) = a_k$ for all relevant k. From Figure 8.29, we see that $\int_{n+1}^{\infty} f(x)\,dx \leq R_n$.

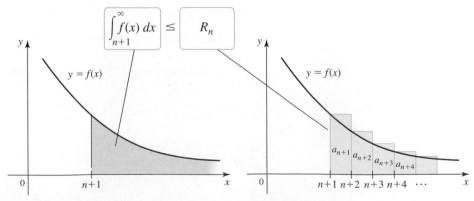

FIGURE 8.29

Similarly, Figure 8.30 shows that $R_n \leq \int_n^\infty f(x)\, dx$.

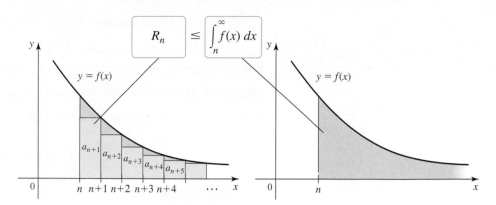

FIGURE 8.30

Combining these two inequalities, the remainder is squeezed between two integrals:

$$\int_{n+1}^\infty f(x)\, dx \leq R_n \leq \int_n^\infty f(x)\, dx \qquad (2)$$

If the integrals can be evaluated, this result provides an estimate of the remainder.

There is, however, another equally useful way to express this result. Notice that the value of the series is

$$S = \sum_{k=1}^\infty a_k = \underbrace{\sum_{k=1}^n a_k}_{S_n} + R_n,$$

which is the sum of the first n terms S_n and the remainder R_n. Adding S_n to each term of (2), we have

$$\underbrace{S_n + \int_{n+1}^\infty f(x)\, dx}_{L_n} \leq \underbrace{\sum_{k=1}^\infty a_k}_{S_n + R_n = S} \leq \underbrace{S_n + \int_n^\infty f(x)\, dx}_{U_n}.$$

These inequalities can be abbreviated as $L_n \leq S \leq U_n$, where S is the exact value of the series, and L_n and U_n are lower and upper bounds for S, respectively. If the integrals can be evaluated, it is straightforward to compute S_n (by summing the first n terms of the series) and to compute both L_n and U_n.

THEOREM 8.13   Estimating Series with Positive Terms
Let f be a continuous, positive, decreasing function for $x \geq 1$ and let $a_k = f(k)$ for $k = 1, 2, 3, \ldots$. Let $S = \sum_{k=1}^\infty a_k$ be a convergent series and let $S_n = \sum_{k=1}^n a_k$ be the sum of the first n terms of the series. The remainder $R_n = S - S_n$ satisfies

$$R_n \leq \int_n^\infty f(x)\, dx.$$

Furthermore, the exact value of the series is bounded as follows:

$$S_n + \int_{n+1}^\infty f(x)\, dx \leq \sum_{k=1}^\infty a_k \leq S_n + \int_n^\infty f(x)\, dx$$

EXAMPLE 6   Approximating a _p_-series

a. How many terms of the series $\displaystyle\sum_{k=1}^{\infty} \frac{1}{k^2}$ must be summed to obtain an approximation that is within 10^{-3} of the exact value of the series?

b. Find an approximation to the series using 50 terms of the series.

SOLUTION The function associated with this series is $f(x) = 1/x^2$.

a. Using the bound on the remainder, we have

$$R_n \le \int_n^\infty f(x)\,dx = \int_n^\infty \frac{dx}{x^2} = \frac{1}{n}.$$

To ensure that $R_n \le 10^{-3}$, we must choose n so that $1/n \le 10^{-3}$, which implies that $n \ge 1000$. In other words, we must sum at least 1000 terms of the series to be sure that the remainder is less than 10^{-3}.

b. Using the bounds on the series itself, we have $L_n \le S \le U_n$, where S is the exact value of the series, and

$$L_n = S_n + \int_{n+1}^\infty \frac{dx}{x^2} = S_n + \frac{1}{n+1} \quad \text{and} \quad U_n = S_n + \int_n^\infty \frac{dx}{x^2} = S_n + \frac{1}{n}.$$

> The values of _p_-series with even values of _p_ are generally known. For example, with $p = 2$ the series converges to $\pi^2/6$ (a proof is outlined in Exercise 58); with $p = 4$, the series converges to $\pi^4/90$. The values of _p_-series with odd values of _p_ are not known.

Therefore, the series is bounded as follows:

$$S_n + \frac{1}{n+1} \le S \le S_n + \frac{1}{n},$$

where S_n is the sum of the first n terms. Using a calculator to sum the first 50 terms of the series, we find that $S_{50} \approx 1.625133$. The exact value of the series is in the interval

$$S_{50} + \frac{1}{50+1} \le S \le S_{50} + \frac{1}{50},$$

or $1.644741 < S < 1.645133$. Taking the average of these two bounds as our approximation of S, we find that $S \approx 1.644937$. This estimate is better than simply using S_{50}. Figure 8.31a shows the lower and upper bounds, L_n and U_n, respectively, for $n = 1, 2, \ldots, 50$. Figure 8.31b shows these bounds on an enlarged scale for $n = 50, 51, \ldots, 100$. These figures illustrate how the exact value of the series is squeezed into a narrowing interval as n increases.

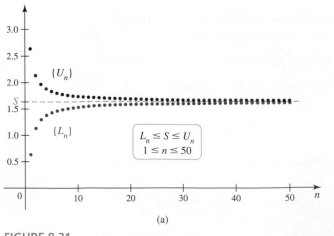

(a)

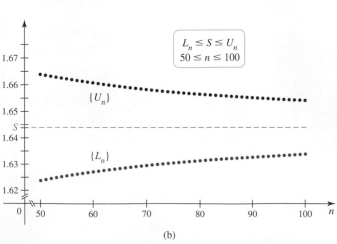

(b)

FIGURE 8.31

SECTION 8.4 EXERCISES

Review Questions

1. Explain why computation alone may not determine whether a series converges.

2. Is it true that if the terms of a series of positive terms decrease to zero, then the series converges? Explain using an example.

3. Can the Integral Test be used to determine whether a series diverges?

4. For what values of p does the series $\sum_{k=1}^{\infty} \frac{1}{k^p}$ converge? For what values of p does it diverge?

5. For what values of p does the series $\sum_{k=10}^{\infty} \frac{1}{k^p}$ converge (initial index is 10)? For what values of p does it diverge?

6. Explain why the sequence of partial sums for a series with positive terms is an increasing sequence.

7. Define the remainder of an infinite series.

8. If a series of positive terms converges, does it follow that the remainder R_n must decrease to zero as $n \to \infty$? Explain.

Basic Skills

9–14. Properties of series *Use the properties of infinite series to evaluate the following series.*

9. $\sum_{k=0}^{\infty} \left[3\left(\frac{2}{5}\right)^k - 2\left(\frac{5}{7}\right)^k \right]$

10. $\sum_{k=1}^{\infty} \left[2\left(\frac{3}{5}\right)^k + 3\left(\frac{4}{9}\right)^k \right]$

11. $\sum_{k=1}^{\infty} \left[\frac{1}{3}\left(\frac{5}{6}\right)^k + \frac{3}{5}\left(\frac{7}{9}\right)^k \right]$

12. $\sum_{k=0}^{\infty} \left[\frac{1}{2}(0.2)^k + \frac{3}{2}(0.8)^k \right]$

13. $\sum_{k=1}^{\infty} \left[\left(\frac{1}{6}\right)^k + \left(\frac{1}{3}\right)^{k-1} \right]$

14. $\sum_{k=0}^{\infty} \frac{2 - 3^k}{6^k}$

15–22. Divergence Test *Use the Divergence Test to determine whether the following series diverge or state that the test is inconclusive.*

15. $\sum_{k=0}^{\infty} \frac{k}{2k + 1}$

16. $\sum_{k=1}^{\infty} \frac{k}{k^2 + 1}$

17. $\sum_{k=2}^{\infty} \frac{k}{\ln k}$

18. $\sum_{k=1}^{\infty} \frac{k^2}{2^k}$

19. $\sum_{k=0}^{\infty} \frac{1}{1000 + k}$

20. $\sum_{k=1}^{\infty} \frac{k^3}{k^3 + 1}$

21. $\sum_{k=2}^{\infty} \frac{\sqrt{k}}{\ln^{10} k}$

22. $\sum_{k=1}^{\infty} \frac{\sqrt{k^2 + 1}}{k}$

23–30. Integral Test *Use the Integral Test to determine the convergence or divergence of the following series. Check that the conditions of the test are satisfied.*

23. $\sum_{k=2}^{\infty} \frac{1}{k \ln k}$

24. $\sum_{k=1}^{\infty} \frac{k}{\sqrt{k^2 + 4}}$

25. $\sum_{k=1}^{\infty} k e^{-2k^2}$

26. $\sum_{k=1}^{\infty} \frac{1}{\sqrt[3]{k + 10}}$

27. $\sum_{k=0}^{\infty} \frac{1}{\sqrt{k + 8}}$

28. $\sum_{k=2}^{\infty} \frac{1}{k(\ln k)^2}$

29. $\sum_{k=1}^{\infty} \frac{k}{e^k}$

30. $\sum_{k=2}^{\infty} \frac{1}{k \ln k \ln (\ln k)}$

31–34. p-series *Determine the convergence or divergence of the following series.*

31. $\sum_{k=1}^{\infty} \frac{1}{k^{10}}$

32. $\sum_{k=2}^{\infty} \frac{k^e}{k^\pi}$

33. $\sum_{k=3}^{\infty} \frac{1}{(k - 2)^4}$

34. $\sum_{k=1}^{\infty} 2k^{-3/2}$

35–42. Remainders and estimates *Consider the following convergent series.*

a. *Find an upper bound for the remainder in terms of n.*
b. *Find how many terms are needed to ensure that the remainder is less than 10^{-3}.*
c. *Find lower and upper bounds (L_n and U_n, respectively) on the exact value of the series.*
d. *Find an interval in which the value of the series must lie if you approximate it using ten terms of the series.*

35. $\sum_{k=1}^{\infty} \frac{1}{k^6}$

36. $\sum_{k=1}^{\infty} \frac{1}{k^8}$

37. $\sum_{k=1}^{\infty} \frac{1}{3^k}$

38. $\sum_{k=2}^{\infty} \frac{1}{k(\ln k)^2}$

39. $\sum_{k=1}^{\infty} \frac{1}{k^{3/2}}$

40. $\sum_{k=1}^{\infty} e^{-k}$

41. $\sum_{k=1}^{\infty} \frac{1}{k^3}$

42. $\sum_{k=1}^{\infty} k e^{-k^2}$

Further Explorations

43. **Explain why or why not** Determine whether the following statements are true and give an explanation or counterexample.

a. If $\sum_{k=1}^{\infty} a_k$ converges, then $\sum_{k=10}^{\infty} a_k$ converges.

b. If $\sum_{k=1}^{\infty} a_k$ diverges, then $\sum_{k=10}^{\infty} a_k$ diverges.

c. If $\sum a_k$ converges, then $\sum (a_k + 0.0001)$ also converges.

d. If $\sum p^k$ diverges, then $\sum (p + 0.001)^k$ diverges, for a fixed real number p.

e. If $\sum k^{-p}$ converges, then $\sum k^{-p+0.001}$ converges.

f. If $\lim_{k \to \infty} a_k = 0$, then $\sum a_k$ converges.

44–49. Choose your test *Determine whether the following series converge or diverge.*

44. $\sum_{k=1}^{\infty} \sqrt{\frac{k + 1}{k}}$

45. $\sum_{k=1}^{\infty} \frac{1}{(3k + 1)(3k + 4)}$

46. $\sum_{k=0}^{\infty} \frac{10}{k^2 + 9}$

47. $\sum_{k=0}^{\infty} \frac{k}{\sqrt{k^2 + 4}}$

48. $\displaystyle\sum_{k=1}^{\infty} \frac{2^k + 3^k}{4^k}$

49. $\displaystyle\sum_{k=2}^{\infty} \frac{4}{k \ln^2 k}$

50. Log p-series Consider the series $\displaystyle\sum_{k=2}^{\infty} \frac{1}{k(\ln k)^p}$, where p is a real number.

    a. Use the Integral Test to determine the values of p for which this series converges.

    b. Does this series converge faster for $p = 2$ or $p = 3$? Explain.

51. Loglog p-series Consider the series $\displaystyle\sum_{k=2}^{\infty} \frac{1}{k \ln k (\ln \ln k)^p}$, where p is a real number.

    a. For what values of p does this series converge?

    b. Which of the following series converges faster? Explain.

$$\sum_{k=2}^{\infty} \frac{1}{k(\ln k)^2} \quad \text{or} \quad \sum_{k=2}^{\infty} \frac{1}{k \ln k (\ln \ln k)^2}?$$

52. Find a series Find a series that ...

    a. converges faster than $\displaystyle\sum \frac{1}{k^2}$ but slower than $\displaystyle\sum \frac{1}{k^3}$.

    b. diverges faster than $\displaystyle\sum \frac{1}{k}$ but slower than $\displaystyle\sum \frac{1}{\sqrt{k}}$.

    c. converges faster than $\displaystyle\sum \frac{1}{k \ln^2 k}$ but slower than $\displaystyle\sum \frac{1}{k^2}$.

Additional Exercises

53. A divergence proof Give an argument, similar to that given in the text for the harmonic series, to show that $\displaystyle\sum_{k=1}^{\infty} \frac{1}{\sqrt{k}}$ diverges.

54. Properties proof Use the ideas in the proof of Property 1 of Theorem 8.8 to prove Property 2 of Theorem 8.8.

55. Property of divergent series Prove that if $\displaystyle\sum a_k$ diverges, then $\displaystyle\sum c a_k$ also diverges, where $c \neq 0$ is a constant.

56. Prime numbers The prime numbers are those positive integers that are divisible by only 1 and themselves (for example, 2, 3, 5, 7, 11, 13, ...). A celebrated theorem states that the sequence of prime numbers $\{p_k\}$ satisfies $\displaystyle\lim_{k \to \infty} p_k/(k \ln k) = 1$. Show that $\displaystyle\sum_{k=2}^{\infty} \frac{1}{k \ln k}$ diverges, which implies that the series $\displaystyle\sum_{k=1}^{\infty} 1/p_k$ diverges.

57. The zeta function The Riemann zeta function is the subject of extensive research and is associated with several renowned unsolved problems. It is defined by $\displaystyle\zeta(x) = \sum_{k=1}^{\infty} \frac{1}{k^x}$. When x is a real number, the zeta function becomes a p-series. For even positive integers p, the value of $\zeta(p)$ is known exactly. For example,

$$\sum_{k=1}^{\infty} \frac{1}{k^2} = \frac{\pi^2}{6}, \quad \sum_{k=1}^{\infty} \frac{1}{k^4} = \frac{\pi^4}{90}, \quad \sum_{k=1}^{\infty} \frac{1}{k^6} = \frac{\pi^6}{945}, \dots$$

Use the estimation techniques described in the text to approximate $\zeta(3)$ and $\zeta(5)$ (whose values are not known exactly) with a remainder less than 10^{-3}.

58. Showing that $\displaystyle\sum_{k=1}^{\infty} \frac{1}{k^2} = \frac{\pi^2}{6}$ In 1734, Leonhard Euler informally proved that $\displaystyle\sum_{k=1}^{\infty} \frac{1}{k^2} = \frac{\pi^2}{6}$. An elegant proof is outlined here that uses the inequality

$$\cot^2 x < \frac{1}{x^2} < 1 + \cot^2 x \quad \text{(provided that } 0 < x < \pi/2\text{)}$$

and the identity

$$\sum_{k=1}^{n} \cot^2 (k\theta) = \frac{n(2n-1)}{3}, \text{ for } n = 1, 2, 3 \dots, \text{ where } \theta = \frac{\pi}{2n+1}.$$

    a. Show that $\displaystyle\sum_{k=1}^{n} \cot^2 (k\theta) < \frac{1}{\theta^2} \sum_{k=1}^{n} \frac{1}{k^2} < n + \sum_{k=1}^{n} \cot^2 (k\theta)$.

    b. Use the inequality in part (a) to show that

$$\frac{n(2n-1)\pi^2}{3(2n+1)^2} < \sum_{k=1}^{n} \frac{1}{k^2} < \frac{n(2n+2)\pi^2}{3(2n+1)^2}.$$

    c. Use the Squeeze Theorem to conclude that $\displaystyle\sum_{k=1}^{\infty} \frac{1}{k^2} = \frac{\pi^2}{6}$.

[*Source: The College Mathematics Journal,* **24**, No. 5 (November, 1993).]

59. Reciprocals of odd squares Given that $\displaystyle\sum_{k=1}^{\infty} \frac{1}{k^2} = \frac{\pi^2}{6}$ (Exercises 57 and 58) and that the terms of this series may be rearranged without changing the value of the series. Determine the sum of the reciprocals of the squares of the odd positive integers.

60. Shifted p-series Consider the sequence $\{F_n\}$ defined by

$$F_n = \sum_{k=1}^{\infty} \frac{1}{k(k+n)},$$

for $n = 0, 1, 2, \dots$. When $n = 0$, the series is a p-series, and we have $F_0 = \pi^2/6$ (Exercises 57 and 58).

    a. Explain why $\{F_n\}$ is a decreasing sequence.

    b. Plot approximations to $\{F_n\}$ for $n = 1, 2, \dots, 20$.

    c. Based on your experiments, make a conjecture about $\displaystyle\lim_{n \to \infty} F_n$.

61. A sequence of sums Consider the sequence $\{x_n\}$ defined for $n = 1, 2, 3, \dots$ by

$$x_n = \sum_{k=n+1}^{2n} \frac{1}{k} = \frac{1}{n+1} + \frac{1}{n+2} + \cdots + \frac{1}{2n}.$$

    a. Write out the terms x_1, x_2, x_3.

    b. Show that $\frac{1}{2} \leq x_n < 1$ for $n = 1, 2, 3, \dots$.

    c. Show that x_n is the right Riemann sum for $\displaystyle\int_1^2 \frac{dx}{x}$ using n subintervals.

    d. Conclude that $\displaystyle\lim_{n \to \infty} x_n = \ln 2$.

62. The harmonic series and Euler's constant

    a. Sketch the function $f(x) = 1/x$ on the interval $[1, n+1]$, where n is a positive integer. Use this graph to verify that

$$\ln(n+1) < 1 + \frac{1}{2} + \frac{1}{3} + \cdots + \frac{1}{n} < 1 + \ln n.$$

b. Let S_n be the sum of the first n terms of the harmonic series, so part (a) says $\ln(n+1) < S_n < 1 + \ln n$. Define the new sequence $\{E_n\}$, where

$$E_n = S_n - \ln(n+1), \quad \text{for } n = 1, 2, 3, \ldots.$$

Show that $E_n > 0$ for $n = 1, 2, 3, \ldots$.

c. Using a figure similar to that used in part (a), show that

$$\frac{1}{n+1} > \ln(n+2) - \ln(n+1).$$

d. Use parts (a) and (c) to show that $\{E_n\}$ is an increasing sequence $(E_{n+1} > E_n)$.

e. Use part (a) to show that $\{E_n\}$ is bounded above by 1.

f. Conclude from parts (d) and (e) that $\{E_n\}$ has a limit less than or equal to 1. This limit is known as **Euler's constant** and is denoted γ (the Greek lowercase letter gamma).

g. By computing terms of $\{E_n\}$, estimate the value of γ and compare it to the value $\gamma \approx 0.5772$. (It has been conjectured, but not proved, that γ is irrational.)

h. The preceding arguments show that the sum of the first n terms of the harmonic series satisfy $S_n \approx 0.5772 + \ln(n+1)$. How many terms must be summed for the sum to exceed 10?

63. Stacking dominoes Consider a set of identical dominoes that are 2 in long. The dominoes are stacked on top of each other with their long edges aligned so that each domino overhangs the one beneath it *as far as possible* (see figure).

a. If there are n dominoes in the stack, what is the *greatest* distance that the top domino can be made to overhang the bottom domino? (*Hint:* Put the nth domino beneath the previous $n-1$ dominoes.)

b. If we allow for infinitely many dominoes in the stack, what is the greatest distance that the top domino can be made to overhang the bottom domino?

QUICK CHECK ANSWERS

1. Adding a finite number of nonzero terms does not change whether the series converges. It does, however, change the value of the series. **2.** Given the nth term of the sequence of partial sums S_n, the next term is obtained by adding a positive number. So $S_{n+1} > S_n$, which means the sequence is increasing. **3.** The series diverges for $|r| \geq 1$. **4. a.** Divergent p-series **b.** Convergent geometric series **c.** Convergent p-series **5.** The remainder is $R_n = a_{n+1} + a_{n+2} + \cdots$, which consists of positive numbers. ◄

8.5 The Ratio, Root, and Comparison Tests

We now consider several more convergence tests: the Ratio Test, the Root Test, and two comparison tests. The Ratio Test will be used frequently throughout the next chapter, and comparison tests are valuable when no other test works. Again, these tests determine *whether* an infinite series converges, but they do not establish the value of the series.

The Ratio Test

The Integral Test is powerful, but limited, because it requires evaluating integrals. For example, the series $\Sigma 1/k!$, with a factorial term, cannot be handled by the Integral Test. The next test significantly enlarges the set of infinite series that we can analyze.

> In words, the Ratio Test says the limit of the ratio of successive terms of the series must be less than 1 for convergence of the series.

THEOREM 8.14   The Ratio Test

Let Σa_k be an infinite series with positive terms and let $r = \lim\limits_{k \to \infty} \dfrac{a_{k+1}}{a_k}$.

1. If $0 \leq r < 1$, the series converges.

2. If $r > 1$ (including $r = \infty$), the series diverges.

3. If $r = 1$, the test is inconclusive.

Proof (outline) We omit the details of the proof, but the idea behind the proof provides insight. Let's assume that the limit r exists. Then, as k gets large and the ratio a_{k+1}/a_k approaches r, we have $a_{k+1} \approx ra_k$. Therefore, as one goes farther and farther out in the series, it behaves like

$$a_k + a_{k+1} + a_{k+2} + \cdots \approx a_k + ra_k + r^2a_k + r^3a_k + \cdots$$
$$= a_k(1 + r + r^2 + r^3 + \cdots)$$

The tail of the series, which determines whether the series converges, behaves like a geometric series with ratio r. We know that if $0 \le r < 1$, the geometric series converges, and if $r > 1$, the series diverges, which is the conclusion of the Ratio Test.   ◄

EXAMPLE 1   Using the Ratio Test Use the Ratio Test to determine whether the following series converge.

a. $\displaystyle\sum_{k=1}^{\infty} \frac{10^k}{k!}$ **b.** $\displaystyle\sum_{k=1}^{\infty} \frac{k^k}{k!}$

SOLUTION In each case, the limit of the ratio of successive terms is determined.

> Recall that
> $$k! = k \cdot (k-1) \cdots 2 \cdot 1.$$
> Therefore,
> $$(k+1)! = (k+1)k!$$

a. $r = \displaystyle\lim_{k \to \infty} \frac{a_{k+1}}{a_k} = \lim_{k \to \infty} \frac{10^{k+1}/(k+1)!}{10^k/k!}$ Substitute a_{k+1} and a_k.

$$= \lim_{k \to \infty} \frac{10^{k+1}}{10^k} \cdot \frac{k!}{(k+1)k!}$$ Invert and multiply.

$$= \lim_{k \to \infty} \frac{10}{k+1} = 0$$ Simplify and evaluate the limit.

Because $r = 0$, the series converges by the Ratio Test.

b. $r = \displaystyle\lim_{k \to \infty} \frac{a_{k+1}}{a_k} = \lim_{k \to \infty} \frac{(k+1)^{k+1}/(k+1)!}{k^k/k!}$ Substitute a_{k+1} and a_k.

$$= \lim_{k \to \infty} \left(\frac{k+1}{k}\right)^k$$ Simplify.

> Recall from Section 4.7 that
> $$\lim_{k \to \infty} \left(1 + \frac{1}{k}\right)^k = e \approx 2.718.$$

$$= \lim_{k \to \infty} \left(1 + \frac{1}{k}\right)^k = e$$ Simplify and evaluate the limit.

Because $r = e > 1$, the series diverges by the Ratio Test. Alternatively, we could have noted that $\displaystyle\lim_{k \to \infty} k^k/k! = \infty$ (Section 8.2) and used the Divergence Test to reach the same conclusion.   *Related Exercises 9–18* ◄

QUICK CHECK 1 Evaluate $10!/9!$, $(k+2)!/k!$, and $k!/(k+1)!$ ◄

The Root Test

Occasionally a series arises for which none of the preceding tests gives a conclusive result. In these situations, the Root Test may be the tool that is needed.

THEOREM 8.15   The Root Test

Let Σa_k be an infinite series with nonnegative terms and let $\rho = \displaystyle\lim_{k \to \infty} \sqrt[k]{a_k}$.

1. If $0 \le \rho < 1$, the series converges.

2. If $\rho > 1$ (including $\rho = \infty$), the series diverges.

3. If $\rho = 1$, the test is inconclusive.

Proof (outline) Assume that the limit ρ exists. If k is large, we have $\rho \approx \sqrt[k]{a_k}$ or $a_k \approx \rho^k$. For large values of k, the tail of the series, which determines whether a series converges, behaves as

$$a_k + a_{k+1} + a_{k+2} + \cdots \approx \rho^k + \rho^{k+1} + \rho^{k+2} + \cdots$$

Therefore, the tail of the series is approximately a geometric series with ratio ρ. If $0 \le \rho < 1$, the geometric series converges, and if $\rho > 1$, the series diverges, which is the conclusion of the Root Test. ◄

EXAMPLE 2   Using the Root Test Use the Root Test to determine whether the following series converge.

a. $\displaystyle\sum_{k=1}^{\infty} \left(\frac{4k^2 - 3}{7k^2 + 6} \right)^k$ **b.** $\displaystyle\sum_{k=1}^{\infty} \frac{2^k}{k^{10}}$

SOLUTION

a. The required limit is

$$\rho = \lim_{k \to \infty} \sqrt[k]{\left(\frac{4k^2 - 3}{7k^2 + 6} \right)^k} = \lim_{k \to \infty} \frac{4k^2 - 3}{7k^2 + 6} = \frac{4}{7}$$

Because $0 \le \rho < 1$, the series converges by the Root Test.

b. In this case,

$$\rho = \lim_{k \to \infty} \sqrt[k]{\frac{2^k}{k^{10}}} = \lim_{k \to \infty} \frac{2}{k^{10/k}} = \lim_{k \to \infty} \frac{2}{(k^{1/k})^{10}} = 2 \qquad \lim_{k \to \infty} k^{1/k} = 1$$

Because $\rho > 1$, the series diverges by the Root Test.

   We could have used the Ratio Test for both series in this example, but the Root Test is easier to apply in each case. In part (b), the Divergence Test leads to the same conclusion.   *Related Exercises 19–26* ◄

The Comparison Test

Tests that use known series to test unknown series are called *comparison tests*. The first test is the Basic Comparison Test or simply the Comparison Test.

> Whether a series converges depends on the behavior of terms in the tail (large values of the index). So the inequalities $0 < a_k \le b_k$ and $0 < b_k \le a_k$ need not hold for all terms of the series. They must hold for all $k > N$ for some positive integer N.

THEOREM 8.16   Comparison Test
Let $\sum a_k$ and $\sum b_k$ be series with positive terms.

1. If $0 < a_k \le b_k$ and $\sum b_k$ converges, then $\sum a_k$ converges.
2. If $0 < b_k \le a_k$ and $\sum b_k$ diverges, then $\sum a_k$ diverges.

Proof Assume that $\sum b_k$ converges, which means that $\sum b_k$ has a finite value B. The sequence of partial sums for $\sum a_k$ satisfies

$$S_n = \sum_{k=1}^{n} a_k \le \sum_{k=1}^{n} b_k \qquad a_k \le b_k$$

$$< \sum_{k=1}^{\infty} b_k \qquad \text{Positive terms are added to a finite sum.}$$

$$= B \qquad \text{Value of series}$$

Therefore, the sequence of partial sums for $\sum a_k$ is increasing and bounded above by B. By the Bounded Monotonic Sequence Theorem (Theorem 8.5), the sequence of partial sums of Σa_k has a limit, which implies that Σa_k converges. The second case of the theorem is proved in a similar way. ◄

The Comparison Test is illustrated with graphs of sequences of partial sums. Consider the series

$$\sum_{k=1}^{\infty} a_k = \sum_{k=1}^{\infty} \frac{1}{k^2 + 10} \quad \text{and} \quad \sum_{k=1}^{\infty} b_k = \sum_{k=1}^{\infty} \frac{1}{k^2}.$$

Because $\dfrac{1}{k^2 + 10} < \dfrac{1}{k^2}$, it follows that $a_k < b_k$ for $k \geq 1$. Furthermore, Σb_k is a convergent p-series. By the Comparison Test, we conclude that Σa_k also converges (Figure 8.32). The second case of the Comparison Test is illustrated with the series

$$\sum_{k=4}^{\infty} a_k = \sum_{k=4}^{\infty} \frac{1}{\sqrt{k} - 3} \quad \text{and} \quad \sum_{k=4}^{\infty} b_k = \sum_{k=4}^{\infty} \frac{1}{\sqrt{k}}.$$

Now $\dfrac{1}{\sqrt{k}} < \dfrac{1}{\sqrt{k} - 3}$ for $k \geq 4$. Therefore, $b_k < a_k$ for $k \geq 4$. Because Σb_k is a divergent p-series, by the Comparison Test, Σa_k also diverges. Figure 8.33 shows that the sequence of partial sums for Σa_k lies above the sequence of partial sums for Σb_k. Because the sequence of partial sums for Σb_k diverges, the sequence of partial sums for Σa_k also diverges.

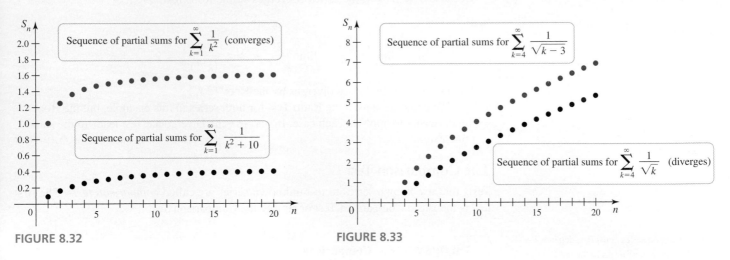

FIGURE 8.32

FIGURE 8.33

The key in using the Comparison Test is finding an appropriate comparison series. Plenty of practice will enable you to spot patterns and choose good comparison series.

EXAMPLE 3 **Using the Comparison Test** Determine whether the following series converge.

a. $\displaystyle\sum_{k=1}^{\infty} \frac{k^3}{2k^4 - 1}$ **b.** $\displaystyle\sum_{k=2}^{\infty} \frac{\ln k}{k^3}$

SOLUTION In using comparison tests, it's helpful to get a feel for how the terms of the given series are decreasing (if they are not decreasing, the series diverges).

a. As we go farther and farther out in this series ($k \to \infty$), the terms behave like

$$\frac{k^3}{2k^4 - 1} \approx \frac{k^3}{2k^4} = \frac{1}{2k}.$$

So a reasonable choice for a comparison series is the divergent series $\displaystyle\sum \frac{1}{2k}$. We must now show that the terms of the given series are *greater* than the terms of the comparison

> If Σa_k diverges, then $\Sigma c a_k$ also diverges for any constant $c \neq 0$ (Exercise 55 of Section 8.4). In Example 3a, one could use either $\displaystyle\sum \frac{1}{2k}$ or $\displaystyle\sum \frac{1}{k}$, both of which diverge, for the comparison series. The first choice makes the required inequality easier to prove.

series. It is done by noting that $2k^4 - 1 < 2k^4$. Inverting both sides, we have

$$\frac{1}{2k^4 - 1} > \frac{1}{2k^4}, \quad \text{which implies that} \quad \frac{k^3}{2k^4 - 1} > \frac{k^3}{2k^4} = \frac{1}{2k}.$$

Because $\sum \dfrac{1}{2k}$ diverges, case (2) of the Comparison Test implies that the given series also diverges.

b. We note that $\ln k < k$ for $k \geq 1$ and then divide by k^3:

$$\frac{\ln k}{k^3} < \frac{k}{k^3} = \frac{1}{k^2}.$$

Therefore, the appropriate comparison series is the convergent p-series $\sum \dfrac{1}{k^2}$. Because $\sum \dfrac{1}{k^2}$ converges, the given series converges.

Related Exercises 27–38 ◀

QUICK CHECK 2 Explain why it is difficult to use the divergent series $\sum 1/k$ as a comparison series to test $\sum 1/(k + 1)$. ◀

The Limit Comparison Test

The Comparison Test should be tried if there is an obvious comparison series and the necessary inequality is easily established. Notice, however, that if the series in Example 3a were $\displaystyle\sum_{k=1}^{\infty} \frac{k^3}{2k^4 + 10}$ instead of $\displaystyle\sum_{k=1}^{\infty} \frac{k^3}{2k^4 - 1}$, then the comparison to the harmonic series would not work. Rather than fiddling with inequalities, it is often easier to use a more refined test called the *Limit Comparison Test*.

THEOREM 8.17 The Limit Comparison Test

Suppose that $\sum a_k$ and $\sum b_k$ are series with positive terms and

$$\lim_{k \to \infty} \frac{a_k}{b_k} = L.$$

1. If $0 < L < \infty$ (that is, L is a finite positive number), then $\sum a_k$ and $\sum b_k$ either both converge or both diverge.
2. If $L = 0$ and $\sum b_k$ converges, then $\sum a_k$ converges.
3. If $L = \infty$ and $\sum b_k$ diverges, then $\sum a_k$ diverges.

> Recall that $|x| < a$ is equivalent to $-a < x < a$.

Proof Recall the definition of $\displaystyle\lim_{k \to \infty} \frac{a_k}{b_k} = L$: Given any $\varepsilon > 0$, $\left| \dfrac{a_k}{b_k} - L \right| < \varepsilon$ provided k is sufficiently large. In this case, let's take $\varepsilon = L/2$. It then follows that for sufficiently large k, $\left| \dfrac{a_k}{b_k} - L \right| < \dfrac{L}{2}$, or (removing the absolute value) $-\dfrac{L}{2} < \dfrac{a_k}{b_k} - L < \dfrac{L}{2}$. Adding L to all terms in these inequalities, we have

$$\frac{L}{2} < \frac{a_k}{b_k} < \frac{3L}{2}.$$

These inequalities imply that, for sufficiently large k,

$$\frac{Lb_k}{2} < a_k < \frac{3Lb_k}{2}.$$

QUICK CHECK 3 For case (1) of the Limit Comparison Test, we must have $0 < L < \infty$. Why can either a_k or b_k be chosen as the known comparison series? That is, why can L be the limit of a_k/b_k or b_k/a_k? ◀

We see that the terms of $\sum a_k$ are sandwiched between multiples of the terms of $\sum b_k$. By the Comparison Test, it follows that the two series converge or diverge together. Cases (2) and (3) ($L = 0$ or $L = \infty$, respectively) are treated in Exercise 69. ◀

EXAMPLE 4   Using the Limit Comparison Test   Determine whether the following series converge.

a. $\displaystyle\sum_{k=1}^{\infty} \frac{k^4 - 2k^2 + 3}{2k^6 - k + 5}$ **b.** $\displaystyle\sum_{k=1}^{\infty} \frac{\ln k}{k^2}$

SOLUTION   In both cases, we must find a comparison series whose terms behave like the terms of the given series as $k \to \infty$.

a. As $k \to \infty$, a rational function behaves like the ratio of the leading (highest-power) terms. In this case, as $k \to \infty$,

$$\frac{k^4 - 2k^2 + 3}{2k^6 - k + 5} \approx \frac{k^4}{2k^6} = \frac{1}{2k^2}.$$

Therefore, a reasonable comparison series is the convergent p-series $\displaystyle\sum_{k=1}^{\infty} \frac{1}{k^2}$ (the factor of 2 does not affect whether the given series converges). Having chosen a comparison series, we compute the limit L:

$$\begin{aligned} L &= \lim_{k\to\infty} \frac{(k^4 - 2k^2 + 3)/(2k^6 - k + 5)}{1/k^2} && \text{Ratio of terms of series} \\[2mm] &= \lim_{k\to\infty} \frac{k^2(k^4 - 2k^2 + 3)}{2k^6 - k + 5} && \text{Simplify.} \\[2mm] &= \lim_{k\to\infty} \frac{k^6 - 2k^4 + 3k^2}{2k^6 - k + 5} = \frac{1}{2} && \text{Simplify and evaluate the limit.} \end{aligned}$$

We see that $0 < L < \infty$; therefore, the given series converges.

b. Why is this series interesting? We know that $\displaystyle\sum_{k=1}^{\infty} \frac{1}{k^2}$ converges and that $\displaystyle\sum_{k=1}^{\infty} \frac{1}{k}$ diverges.

The given series $\displaystyle\sum_{k=1}^{\infty} \frac{\ln k}{k^2}$ is "between" these two series. This observation suggests that we

use either $\displaystyle\sum_{k=1}^{\infty} \frac{1}{k^2}$ or $\displaystyle\sum_{k=1}^{\infty} \frac{1}{k}$ as a comparison series. In the first case, letting $a_k = \ln k/k^2$ and $b_k = 1/k^2$, we find that

$$L = \lim_{k\to\infty} \frac{a_k}{b_k} = \lim_{k\to\infty} \frac{\ln k/k^2}{1/k^2} = \lim_{k\to\infty} \ln k = \infty.$$

Case (3) of the Limit Comparison Test does not apply here because the comparison

series $\displaystyle\sum_{k=1}^{\infty} \frac{1}{k^2}$ converges. So the test is inconclusive.

If, instead, we use the comparison series $\displaystyle\sum b_k = \sum \frac{1}{k}$, then

$$L = \lim_{k\to\infty} \frac{a_k}{b_k} = \lim_{k\to\infty} \frac{\ln k/k^2}{1/k} = \lim_{k\to\infty} \frac{\ln k}{k} = 0.$$

Case (2) of the Limit Comparison Test does not apply here because the comparison

series $\displaystyle\sum_{k=1}^{\infty} \frac{1}{k}$ diverges. Again, the test is inconclusive.

With a bit more cunning, the Limit Comparison Test becomes conclusive. A series that lies "between" $\sum_{k=1}^{\infty} \dfrac{1}{k^2}$ and $\sum_{k=1}^{\infty} \dfrac{1}{k}$ is the convergent p-series $\sum_{k=1}^{\infty} \dfrac{1}{k^{3/2}}$; we try it as a comparison series. Letting $a_k = \ln k / k^2$ and $b_k = 1/k^{3/2}$, we find that

$$L = \lim_{k \to \infty} \frac{a_k}{b_k} = \lim_{k \to \infty} \frac{\ln k / k^2}{1/k^{3/2}} = \lim_{k \to \infty} \frac{\ln k}{\sqrt{k}} = 0.$$

(This limit is evaluated using l'Hôpital's Rule or by recalling that $\ln k$ grows more slowly than any positive power of k.) Now case (2) of the limit comparison test applies; the comparison series $\sum \dfrac{1}{k^{3/2}}$ converges, so the given series converges.

Related Exercises 27–38 ◄

Guidelines

We close by outlining a procedure that puts the various convergence tests in perspective. Here is a reasonable course of action when testing a series of positive terms $\sum a_k$ for convergence.

1. Begin with the Divergence Test. If you show that $\lim\limits_{k \to \infty} a_k \neq 0$, then the series diverges and your work is finished. The order of growth rates of sequences given in Section 8.2 is useful for evaluating $\lim\limits_{k \to \infty} a_k$.

2. Is the series a special series? Recall the convergence properties for the following series.
   - Geometric series: $\sum ar^k$ converges if $|r| < 1$ and diverges for $|r| \geq 1$.
   - p-series: $\sum \dfrac{1}{k^p}$ converges for $p > 1$ and diverges for $p \leq 1$.
   - Check also for a telescoping series.

3. If the general kth term of the series looks like a function you can integrate, then try the Integral Test.

4. If the general kth term of the series involves $k!$, k^k, or a^k, where a is a constant, the Ratio Test is advisable. Series with k in an exponent may yield to the Root Test.

5. If the general kth term of the series is a rational function of k (or a root of a rational function), use the Comparison or the Limit Comparison Test. Use the families of series given in Step 2 as comparison series.

These guidelines will help, but in the end, convergence tests are mastered through practice. It's your turn.

SECTION 8.5 EXERCISES

Review Questions

1. Explain how the Ratio Test works.

2. Explain how the Root Test works.

3. Explain how the Limit Comparison Test works.

4. What is the first test you should use in analyzing the convergence of a series?

5. What tests are advisable if the series involves a factorial term?

6. What tests are best for the series $\sum a_k$ when a_k is a rational function of k?

7. Explain why, with a series of positive terms, the sequence of partial sums is an increasing sequence.

8. Do the tests discussed in this section tell you the value of the series? Explain.

Basic Skills

9–18. The Ratio Test *Use the Ratio Test to determine whether the following series converge.*

9. $\displaystyle\sum_{k=1}^{\infty} \frac{1}{k!}$ **10.** $\displaystyle\sum_{k=1}^{\infty} \frac{2^k}{k!}$ **11.** $\displaystyle\sum_{k=1}^{\infty} \frac{k^2}{4^k}$ **12.** $\displaystyle\sum_{k=1}^{\infty} \frac{2^k}{k^k}$

13. $\displaystyle\sum_{k=1}^{\infty} k e^{-k}$ **14.** $\displaystyle\sum_{k=1}^{\infty} \frac{k!}{k^k}$ **15.** $\displaystyle\sum_{k=1}^{\infty} \frac{2^k}{k^{99}}$ **16.** $\displaystyle\sum_{k=1}^{\infty} \frac{k^6}{k!}$

17. $\displaystyle\sum_{k=1}^{\infty} \frac{(k!)^2}{(2k)!}$ **18.** $\displaystyle\sum_{k=1}^{\infty} k^4 2^{-k}$

19–26. The Root Test *Use the Root Test to determine whether the following series converge.*

19. $\displaystyle\sum_{k=1}^{\infty} \left(\frac{4k^3 + k}{9k^3 + k + 1}\right)^k$ **20.** $\displaystyle\sum_{k=1}^{\infty} \left(\frac{k+1}{2k}\right)^k$

21. $\displaystyle\sum_{k=1}^{\infty} \frac{k^2}{2^k}$ **22.** $\displaystyle\sum_{k=1}^{\infty} \left(1 + \frac{3}{k}\right)^{k^2}$

23. $\displaystyle\sum_{k=1}^{\infty} \left(\frac{k}{k+1}\right)^{2k^2}$ **24.** $\displaystyle\sum_{k=1}^{\infty} \left(\frac{1}{\ln(k+1)}\right)^k$

25. $1 + \left(\dfrac{1}{2}\right)^2 + \left(\dfrac{1}{3}\right)^3 + \left(\dfrac{1}{4}\right)^4 + \cdots$

26. $\left(\dfrac{1}{2}\right)^2 + \left(\dfrac{2}{3}\right)^3 + \left(\dfrac{3}{4}\right)^4 + \cdots$

27–38. Comparison tests *Use the Comparison Test or Limit Comparison Test to determine whether the following series converge.*

27. $\displaystyle\sum_{k=1}^{\infty} \frac{1}{k^2 + 4}$ **28.** $\displaystyle\sum_{k=1}^{\infty} \frac{k^2 + k - 1}{k^4 + 4k^2 - 3}$ **29.** $\displaystyle\sum_{k=1}^{\infty} \frac{k^2 - 1}{k^3 + 4}$

30. $\displaystyle\sum_{k=1}^{\infty} \frac{0.0001}{k + 4}$ **31.** $\displaystyle\sum_{k=1}^{\infty} \frac{1}{k^{3/2} + 1}$ **32.** $\displaystyle\sum_{k=1}^{\infty} \sqrt{\frac{k}{k^3 + 1}}$

33. $\displaystyle\sum_{k=1}^{\infty} \frac{\sin(1/k)}{k^2}$ **34.** $\displaystyle\sum_{k=1}^{\infty} \frac{1}{3^k - 2^k}$ **35.** $\displaystyle\sum_{k=1}^{\infty} \frac{1}{2k - \sqrt{k}}$

36. $\displaystyle\sum_{k=1}^{\infty} \frac{1}{k\sqrt{k+2}}$ **37.** $\displaystyle\sum_{k=1}^{\infty} \frac{\sqrt[3]{k^2 + 1}}{\sqrt{k^3 + 2}}$ **38.** $\displaystyle\sum_{k=2}^{\infty} \frac{1}{(k \ln k)^2}$

Further Explorations

39. Explain why or why not Determine whether the following statements are true and give an explanation or counterexample.

  a. Suppose that $0 < a_k < b_k$. If $\sum a_k$ converges, then $\sum b_k$ converges.

  b. Suppose that $0 < a_k < b_k$. If $\sum a_k$ diverges, then $\sum b_k$ diverges.

  c. Suppose $0 < b_k < c_k < a_k$. If $\sum a_k$ converges, then $\sum b_k$ and $\sum c_k$ converge.

40–57. Choose your test *Use the test of your choice to determine whether the following series converge.*

40. $\displaystyle\sum_{k=1}^{\infty} \frac{(k!)^3}{(3k)!}$ **41.** $\displaystyle\sum_{k=1}^{\infty} \left(\frac{1}{k} + 2^{-k}\right)$ **42.** $\displaystyle\sum_{k=2}^{\infty} \frac{5 \ln k}{k}$

43. $\displaystyle\sum_{k=1}^{\infty} \frac{2^k k!}{k^k}$ **44.** $\displaystyle\sum_{k=1}^{\infty} \left(1 - \frac{1}{k}\right)^{k^2}$ **45.** $\displaystyle\sum_{k=1}^{\infty} \frac{k^8}{k^{11} + 3}$

46. $\displaystyle\sum_{k=1}^{\infty} \frac{1}{(1 + p)^k}, \ p > 0$ **47.** $\displaystyle\sum_{k=1}^{\infty} \frac{1}{k^{1+p}}, \ p > 0$

48. $\displaystyle\sum_{k=2}^{\infty} \frac{1}{k^2 \ln k}$ **49.** $\displaystyle\sum_{k=1}^{\infty} \ln\left(\frac{k+2}{k+1}\right)$ **50.** $\displaystyle\sum_{k=1}^{\infty} k^{-1/k}$

51. $\displaystyle\sum_{k=2}^{\infty} \frac{1}{k^{\ln k}}$ **52.** $\displaystyle\sum_{k=1}^{\infty} \sin^2\left(\frac{1}{k}\right)$ **53.** $\displaystyle\sum_{k=1}^{\infty} \tan\left(\frac{1}{k}\right)$

54. $\displaystyle\sum_{k=2}^{\infty} 100 k^{-k}$ **55.** $\dfrac{1}{1 \cdot 3} + \dfrac{1}{3 \cdot 5} + \dfrac{1}{5 \cdot 7} + \cdots$

56. $\dfrac{1}{2^2} + \dfrac{2}{3^2} + \dfrac{3}{4^2} + \cdots$ **57.** $\dfrac{1}{1!} + \dfrac{4}{2!} + \dfrac{9}{3!} + \dfrac{16}{4!} + \cdots$

58–65. Convergence parameter *Find the values of the parameter p for which the following series converge.*

58. $\displaystyle\sum_{k=2}^{\infty} \frac{1}{(\ln k)^p}$ **59.** $\displaystyle\sum_{k=2}^{\infty} \frac{\ln k}{k^p}$

60. $\displaystyle\sum_{k=2}^{\infty} \frac{1}{k \ln k (\ln \ln k)^p}$ **61.** $\displaystyle\sum_{k=2}^{\infty} \left(\frac{\ln k}{k}\right)^p$

62. $\displaystyle\sum_{k=0}^{\infty} \frac{k! p^k}{(k+1)^k}$ **63.** $\displaystyle\sum_{k=1}^{\infty} \frac{1 \cdot 3 \cdot 5 \cdots (2k - 1)}{k p^{k+1} k!}$

64. $\displaystyle\sum_{k=1}^{\infty} \ln\left(\frac{k}{k+1}\right)^p$ **65.** $\displaystyle\sum_{k=1}^{\infty} \left(1 - \frac{p}{k}\right)^k$

66. Series of squares Prove that if $\sum a_k$ is a convergent series of positive terms, then the series $\sum a_k^2$ also converges.

67. Geometric series revisited We know from Section 8.3 that the geometric series $\sum r^k$ converges if $|r| < 1$ and diverges if $|r| > 1$. Prove these facts using the Integral Test, the Ratio Test, and the Root Test. What can be determined about the geometric series using the Divergence Test?

68. Two sine series Determine whether the following series converge.

  a. $\displaystyle\sum_{k=1}^{\infty} \sin\left(\frac{1}{k}\right)$ **b.** $\displaystyle\sum_{k=1}^{\infty} \frac{1}{k} \sin\left(\frac{1}{k}\right)$

Additional Exercises

69. Limit Comparison Test proof Use the proof of case (1) of the Limit Comparison Test to prove cases (2) and (3).

70–75. A glimpse ahead to power series *Use the Ratio Test to determine the values of $x \geq 0$ for which each series converges.*

70. $\displaystyle\sum_{k=1}^{\infty} \frac{x^k}{k!}$

71. $\displaystyle\sum_{k=0}^{\infty} x^k$

72. $\displaystyle\sum_{k=1}^{\infty} \frac{x^k}{k}$

73. $\displaystyle\sum_{k=1}^{\infty} \frac{x^k}{k^2}$

74. $\displaystyle\sum_{k=1}^{\infty} \frac{x^{2k}}{k^2}$

75. $\displaystyle\sum_{k=1}^{\infty} \frac{x^k}{2^k}$

76. Infinite products An infinite product $P = a_1 a_2 a_3 \ldots$, which is denoted $\displaystyle\prod_{k=1}^{\infty} a_k$, is the limit of the *sequence of partial products* $\{a_1, a_1 a_2, a_1 a_2 a_3, \ldots\}$.

a. Show that the infinite product converges (which means its sequence of partial products converges) provided the series $\displaystyle\sum_{k=1}^{\infty} \ln a_k$ converges.

b. Consider the infinite product

$$P = \prod_{k=2}^{\infty} \left(1 - \frac{1}{k^2} \right) = \frac{3}{4} \cdot \frac{8}{9} \cdot \frac{15}{16} \cdot \frac{24}{25} \cdots.$$

Write out the first few terms of the sequence of partial products,

$$P_n = \prod_{k=2}^{n} \left(1 - \frac{1}{k^2} \right)$$

$\left(\text{for example, } P_2 = \frac{3}{4}, P_3 = \frac{2}{3}\right)$. Write out enough terms to determine the value of the product, which is $\displaystyle\lim_{n \to \infty} P_n$.

c. Use the results of parts (a) and (b) to evaluate the series

$$\sum_{k=2}^{\infty} \ln\left(1 - \frac{1}{k^2}\right).$$

77. Infinite products Use the ideas of Exercise 76 to evaluate the following infinite products.

a. $\displaystyle\prod_{k=0}^{\infty} e^{1/2^k} = 1 \cdot e^{1/2} \cdot e^{1/4} \cdot e^{1/8} \cdots$

b. $\displaystyle\prod_{k=2}^{\infty} \left(1 - \frac{1}{k}\right) = \frac{1}{2} \cdot \frac{2}{3} \cdot \frac{3}{4} \cdot \frac{4}{5} \cdots$

78. An early limit Working in the early 1600s, the mathematicians Wallis, Pascal, and Fermat were attempting to determine the area of the region under the curve $y = x^p$ between $x = 0$ and $x = 1$, where p is a positive integer. Using arguments that predated the Fundamental Theorem of Calculus, they were able to prove that

$$\lim_{n \to \infty} \frac{1}{n} \sum_{k=0}^{n-1} \left(\frac{k}{n}\right)^p = \frac{1}{p+1}.$$

Use what you know about Riemann sums and integrals to verify this limit.

QUICK CHECK ANSWERS

1. $10; (k+2)(k+1); 1/(k+1)$ **2.** To use the Comparison Test, we would need to show that $1/(k+1) > 1/k$, which is not true. **3.** If $\displaystyle\lim_{k \to \infty} \frac{a_k}{b_k} = L$ for $0 < L < \infty$, then $\displaystyle\lim_{k \to \infty} \frac{b_k}{a_k} = \frac{1}{L}$ where $0 < 1/L < \infty$. ◄

8.6 Alternating Series

Our previous discussion focused on infinite series with positive terms, which is certainly an important part of the entire subject. But there are many interesting series with terms of mixed sign. For example, the series

$$1 + \frac{1}{2} - \frac{1}{3} - \frac{1}{4} + \frac{1}{5} + \frac{1}{6} - \frac{1}{7} - \frac{1}{8} + \cdots$$

has the pattern that two positive terms are followed by two negative terms and vice versa. Clearly, infinite series could have a variety of sign patterns, so we need to restrict our attention.

Fortunately, the simplest sign pattern is also the most important. We consider **alternating series** in which the signs strictly alternate, as in the series

$$\sum_{k=1}^{\infty} \frac{(-1)^{k+1}}{k} = 1 - \frac{1}{2} + \frac{1}{3} - \frac{1}{4} + \frac{1}{5} - \frac{1}{6} + \frac{1}{7} - \frac{1}{8} + \cdots.$$

The factor $(-1)^{k+1}$ (or $(-1)^k$) has the pattern $\{\ldots, 1, -1, 1, -1, \ldots\}$ and provides the alternating signs.

Alternating Harmonic Series

Let's see what is different about alternating series by working with the series $\sum_{k=1}^{\infty} \frac{(-1)^{k+1}}{k}$, which is called the **alternating harmonic series**. Recall that this series *without* the alternating signs, $\sum_{k=1}^{\infty} \frac{1}{k}$, is the *divergent* harmonic series. So an immediate question is whether alternating signs change the convergence or divergence of a series.

We investigate this question by looking at the sequence of partial sums for the series. In this case, the first four terms of the sequence of partial sums are

$$S_1 = 1$$

$$S_2 = 1 - \frac{1}{2} = \frac{1}{2}$$

$$S_3 = 1 - \frac{1}{2} + \frac{1}{3} = \frac{5}{6}$$

$$S_4 = 1 - \frac{1}{2} + \frac{1}{3} - \frac{1}{4} = \frac{7}{12}$$

$$\vdots$$

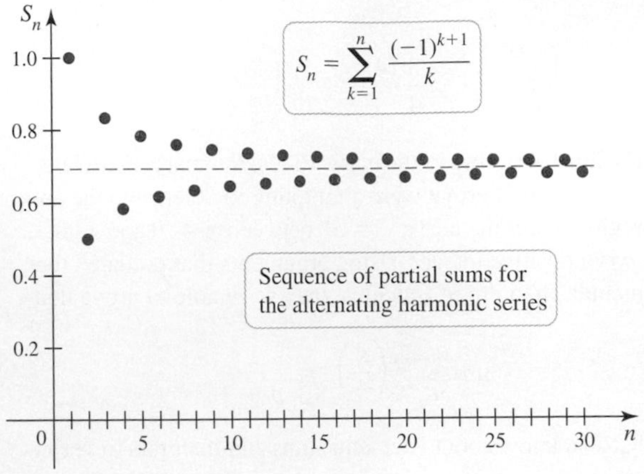

$$S_n = \sum_{k=1}^{n} \frac{(-1)^{k+1}}{k}$$

Sequence of partial sums for the alternating harmonic series

FIGURE 8.34

Plotting the first 30 terms of the sequence of partial sums results in Figure 8.34, which has several noteworthy features.

- The terms of the sequence of partial sums appear to converge to a limit; if they do, it means that, while the harmonic series diverges, the *alternating* harmonic series converges. We will soon learn that taking a divergent series with positive terms and making it an alternating series *may* turn it into a convergent series.

- For series with *positive* terms, the sequence of partial sums is necessarily an increasing sequence. Because the terms of an alternating series alternate in sign, the sequence of partial sums is not increasing.

- Because the sequence of partial sums oscillates, its limit (when it exists) lies between any two consecutive terms.

QUICK CHECK 1 Write out the first few terms of the sequence of partial sums for the alternating series $1 - 2 + 3 - 4 + 5 - 6 + \cdots$. Does this series appear to converge or diverge? ◄

Alternating Series Test

The alternating harmonic series displays many of the properties of all alternating series. We now consider alternating series in general, which are written $\sum (-1)^{k+1} a_k$, where $a_k > 0$. The alternating signs are provided by $(-1)^{k+1}$.

With the exception of the Divergence Test, none of the convergence tests for series with positive terms applies to alternating series. The fortunate news is that only one test needs to be used for alternating series—and it is easy to use.

> Depending on the sign of the first term of the series, an alternating series may be written with $(-1)^k$ or $(-1)^{k+1}$.

> Recall that the Divergence Test of Section 8.4 applies to all series: If the terms of *any* series (including an alternating series) do not tend to zero, then the series diverges.

THEOREM 8.18   The Alternating Series Test

The alternating series $\sum (-1)^{k+1} a_k$ converges provided

1. the terms of the series are nonincreasing in magnitude ($0 < a_{k+1} \le a_k$ for k greater than some index N) and

2. $\lim_{k \to \infty} a_k = 0$.

The first condition is met by most series of interest, so the main job is to show that the terms approach zero. *There is potential for confusion here. For series of positive terms,*

$\lim\limits_{k\to\infty} a_k = 0$ **does** **not** *imply convergence. For alternating series with nonincreasing terms,* $\lim\limits_{k\to\infty} a_k = 0$ **does** *imply convergence.*

Proof The proof is short and instructive; it relies on Figure 8.35. We consider the series in the form

$$\sum_{k=1}^{\infty}(-1)^{k+1}a_k = a_1 - a_2 + a_3 - a_4 + \cdots.$$

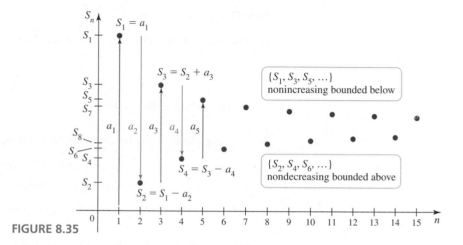

FIGURE 8.35

Because the terms of the series are nonincreasing in magnitude, the even terms of the sequence of partial sums $\{S_{2k}\} = \{S_2, S_4, \dots\}$ form a nondecreasing sequence that is bounded above by S_1. By the Bounded Monotonic Sequence Theorem (Section 8.2), this sequence must have a limit; call it L. Similarly, the odd terms of the sequence of partial sums $\{S_{2k-1}\} = \{S_1, S_3, \dots\}$ form a nonincreasing sequence that is bounded below by S_2. By the Bounded Monotonic Sequence Theorem, this sequence has a limit; call it L'. At the moment, we cannot conclude that $L = L'$. However, notice that $S_{2k} = S_{2k-1} + a_{2k}$. By the condition that $\lim\limits_{k\to\infty} a_k = 0$, it follows that

$$\underbrace{\lim_{k\to\infty} S_{2k}}_{L} = \underbrace{\lim_{k\to\infty} S_{2k-1}}_{L'} + \underbrace{\lim_{k\to\infty} a_{2k}}_{0},$$

or $L = L'$. Thus, the sequence of partial sums converges to a (unique) limit and the corresponding alternating series converges to that limit. ◄

Now we can confirm that the alternating harmonic series $\sum\limits_{k=1}^{\infty}\dfrac{(-1)^{k+1}}{k}$ converges. This fact follows immediately from the Alternating Series Test because the terms $a_k = \dfrac{1}{k}$ decrease and $\lim\limits_{k\to\infty} a_k = 0$.

$\sum\limits_{k=1}^{\infty}\dfrac{1}{k}$

• Diverges
• Partial sums increase

$\sum\limits_{k=1}^{\infty}\dfrac{(-1)^{k+1}}{k}$

• Converges
• Partial sums oscillate

THEOREM 8.19 Alternating Harmonic Series

The alternating harmonic series $\sum\limits_{k=1}^{\infty}\dfrac{(-1)^{k+1}}{k} = 1 - \dfrac{1}{2} + \dfrac{1}{3} - \dfrac{1}{4} + \dfrac{1}{5} - \cdots$

converges (even though the harmonic series $\sum\limits_{k=1}^{\infty}\dfrac{1}{k} = 1 + \dfrac{1}{2} + \dfrac{1}{3} + \dfrac{1}{4} + \dfrac{1}{5} + \cdots$

diverges).

QUICK CHECK 2 Explain why the value of a convergent alternating series is trapped between successive terms of the sequence of partial sums. ◄

EXAMPLE 1   Alternating Series Test Determine whether the following series converge or diverge.

a. $\displaystyle\sum_{k=1}^{\infty} \frac{(-1)^{k+1}}{k^2}$ **b.** $2 - \dfrac{3}{2} + \dfrac{4}{3} - \dfrac{5}{4} + \cdots$ **c.** $\displaystyle\sum_{k=2}^{\infty} \frac{(-1)^k \ln k}{k}$

SOLUTION

a. The terms of this series decrease in magnitude for $k \geq 1$. Furthermore,

$$\lim_{k \to \infty} a_k = \lim_{k \to \infty} \frac{1}{k^2} = 0.$$

Therefore, the series converges.

b. The magnitudes of the terms of this series are $a_k = \dfrac{k+1}{k} = 1 + \dfrac{1}{k}$. While these terms decrease, they approach 1, not 0, as $k \to \infty$. By the Divergence Test, the series diverges.

c. The first step is to show that the terms decrease in magnitude after some fixed term of the series. One way to proceed is to look at the function $f(x) = \dfrac{\ln x}{x}$, which generates the terms of the series. By the Quotient Rule, $f'(x) = \dfrac{1 - \ln x}{x^2}$. The fact that $f'(x) < 0$ for $x > e$ implies that the terms $\dfrac{\ln k}{k}$ decrease for $k \geq 3$. As long as the terms of the series decrease for all k greater than some fixed integer, the first condition of the test is met. Furthermore, using l'Hôpital's Rule or the fact that $\{\ln k\}$ increases more slowly than $\{k\}$ (Section 8.2), we see that

$$\lim_{k \to \infty} a_k = \lim_{k \to \infty} \frac{\ln k}{k} = 0.$$

The conditions of the Alternating Series Test are met and the series converges.

Related Exercises 11–24 ◄

Remainders in Alternating Series

> The absolute value is included in the remainder because with alternating series we have $S > S_n$ for some values of n and $S < S_n$ for other values of n (unlike series with positive terms, in which $S > S_n$ for all n).

Recall that if a series converges to a value S, then the remainder is $R_n = |S - S_n|$, where S_n is the sum of the first n terms of the series. The remainder is the *absolute error* in approximating S by S_n.

An upper bound on the remainder in an alternating series is found by observing that the value of the series is always trapped between successive terms of the sequence of partial sums. Therefore, as shown in Figure 8.36,

$$R_n = |S - S_n| \leq |S_{n+1} - S_n| = a_{n+1}.$$

This argument is a proof of the following theorem.

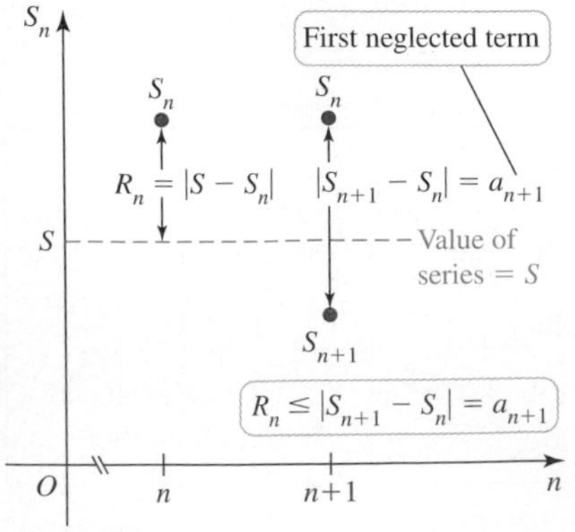

FIGURE 8.36

THEOREM 8.20   Remainder in Alternating Series

Let $R_n = |S - S_n|$ be the remainder in approximating the value of a convergent alternating series $\displaystyle\sum_{k=1}^{\infty} (-1)^{k+1} a_k$ by the sum of its first n terms. Then $R_n \leq a_{n+1}$.

In other words, the remainder is less than or equal to the magnitude of the first neglected term.

EXAMPLE 2 **Remainder in an alternating series** How many terms of the following series are required to approximate the value of the series with a remainder less than 10^{-6}? The exact values of the series are given but are not needed to answer the question (these values are confirmed in Chapter 9).

a. $\ln 2 = 1 - \dfrac{1}{2} + \dfrac{1}{3} - \dfrac{1}{4} + \cdots = \displaystyle\sum_{k=1}^{\infty} \dfrac{(-1)^{k+1}}{k}$

b. $e^{-1} - 1 = -1 + \dfrac{1}{2!} - \dfrac{1}{3!} + \dfrac{1}{4!} - \cdots = \displaystyle\sum_{k=1}^{\infty} \dfrac{(-1)^{k}}{k!}$

SOLUTION

a. The series is expressed as the sum of the first n terms plus the remainder:

$$\sum_{k=1}^{\infty} \frac{(-1)^{k+1}}{k} = \underbrace{1 - \frac{1}{2} + \frac{1}{3} - \frac{1}{4} + \cdots + \frac{(-1)^{n+1}}{n}}_{S_n = \text{ the sum of the first } n \text{ terms}} + \underbrace{\frac{(-1)^{n+2}}{n+1} + \cdots}_{\substack{R_n = |S - S_n| \text{ is less} \\ \text{than the magnitude} \\ \text{of this term}}}.$$

The remainder is less than or equal to the magnitude of the $(n + 1)$st term:

$$R_n = |S - S_n| \le a_{n+1} = \frac{1}{n+1}.$$

To ensure that the remainder is less than 10^{-6}, we require that

$$a_{n+1} = \frac{1}{n+1} < 10^{-6}, \quad \text{or} \quad n + 1 > 10^{6}.$$

Therefore, it takes 1 million terms of the series to approximate ln 2 with a remainder less than 10^{-6}.

b. The series is expressed as the sum of the first n terms plus the remainder:

$$\sum_{k=1}^{\infty} \frac{(-1)^{k}}{k!} = \underbrace{-1 + \frac{1}{2!} - \frac{1}{3!} + \frac{1}{4!} - \cdots + \frac{(-1)^{n}}{n!}}_{S_n = \text{ the sum of the first } n \text{ terms}} + \underbrace{\frac{(-1)^{n+1}}{(n+1)!} + \cdots}_{\substack{R_n = |S - S_n| \text{ is less} \\ \text{than the magnitude} \\ \text{of this term}}}.$$

> The sum of the first nine terms of $\displaystyle\sum_{k=1}^{\infty} \dfrac{(-1)^{k}}{k!}$ is $S_9 = \displaystyle\sum_{k=1}^{9} \dfrac{(-1)^{k}}{k!} \approx$ -0.632120811. A calculator gives $S = e^{-1} - 1 \approx -0.632120559$. Note that the remainder satisfies $R_n = |S - S_n| = 0.000000252$, which is less than 10^{-6}, as claimed in Example 2b.

The remainder satisfies

$$R_n = |S - S_n| \le a_{n+1} = \frac{1}{(n+1)!}.$$

To ensure that the remainder is less than 10^{-6}, we require that

$$a_{n+1} = \frac{1}{(n+1)!} < 10^{-6}, \quad \text{or} \quad (n+1)! > 10^{6}.$$

A bit of experimentation (or a table of factorials) reveals that $9! = 362,880 < 10^{6}$ and $10! = 3,628,800 > 10^{6}$. Therefore, nine terms of the series are needed to approximate $e^{-1} - 1$ with a remainder less than 10^{-6}. *Related Exercises 25–38* ◄

QUICK CHECK 3 Compare and comment on the speed of convergence of the two series in the previous example. Why does one series converge so much more quickly than the other? ◄

Absolute and Conditional Convergence

In this final segment, some terminology is introduced that is needed in Chapter 9. We now let the notation $\sum a_k$ denote any series—a series of positive terms, an alternating series, or even a more general infinite series.

Look again at the alternating harmonic series $\sum(-1)^{k+1}/k$, which converges. The corresponding series of positive terms, $\sum 1/k$, is the harmonic series, which diverges. We also saw in Example 1a that the alternating series $\sum(-1)^{k+1}/k^2$ converges, and the corresponding p-series of positive terms $\sum 1/k^2$ also converges. These examples illustrate that removing the alternating signs in a convergent series may or may not result in a convergent series. The terminology that we now introduce distinguishes these cases.

> **DEFINITION   Absolute and Conditional Convergence**
>
> Assume the infinite series $\sum a_k$ converges. The series $\sum a_k$ **converges absolutely** if the series $\sum |a_k|$ converges. Otherwise, the series $\sum a_k$ **converges conditionally**.

The series $\sum(-1)^{k+1}/k^2$ is an example of an absolutely convergent series because the series of absolute values,

$$\sum_{k=1}^{\infty}\left|\frac{(-1)^{k+1}}{k^2}\right| = \sum_{k=1}^{\infty}\frac{1}{k^2},$$

is a convergent p-series. In this case, removing the alternating signs in the series does *not* affect its convergence.

On the other hand, the convergent alternating harmonic series $\sum(-1)^{k+1}/k$ has the property that the corresponding series of absolute values,

$$\sum_{k=1}^{\infty}\left|\frac{(-1)^{k+1}}{k}\right| = \sum_{k=1}^{\infty}\frac{1}{k},$$

does *not* converge. In this case, removing the alternating signs in the series *does* affect convergence, so this series does not converge absolutely. Instead, we say it converges conditionally. A convergent series (such as $\sum(-1)^{k+1}/k$) may not converge absolutely. It is, however, true that if a series converges absolutely, then it converges.

> **THEOREM 8.21   Absolute Convergence Implies Convergence**
>
> If $\sum |a_k|$ converges, then $\sum a_k$ converges (absolute convergence implies convergence). If $\sum a_k$ diverges, then $\sum |a_k|$ diverges.

Proof Because $|a_k| = a_k$ or $|a_k| = -a_k$, it follows that $0 \leq |a_k| + a_k \leq 2|a_k|$. By assumption $\sum |a_k|$ converges, which, in turn, implies that $2\sum |a_k|$ converges. Using the Comparison Test and the inequality $0 \leq |a_k| + a_k \leq 2|a_k|$, it follows that $\sum(a_k + |a_k|)$ converges. Now note that

$$\sum a_k = \sum(a_k + |a_k| - |a_k|) = \underbrace{\sum(a_k + |a_k|)}_{\text{converges}} - \underbrace{\sum|a_k|}_{\text{converges}}.$$

We see that $\sum a_k$ is the sum of two convergent series, so it also converges. The second statement of the theorem is logically equivalent to the first statement.   ◄

Figure 8.37 gives an overview of absolute and conditional convergence. It shows the universe of all infinite series, split first according to whether they converge or diverge.

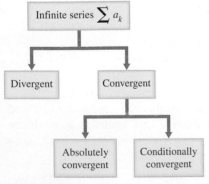

FIGURE 8.37

Convergent series are further divided between absolutely and conditionally convergent series.

Here are a few more consequences of these definitions.

QUICK CHECK 4 Explain why a convergent series of positive terms converges absolutely. ◄

- The distinction between absolute and conditional convergence is relevant only for series of mixed sign, which includes alternating series. If a series of positive terms converges, it converges absolutely; conditional convergence does not apply.

- To test for absolute convergence, we test the series $\sum |a_k|$, which is a series of positive terms. Therefore, the convergence tests of Sections 8.4 and 8.5 (for positive-term series) are used to determine absolute convergence.

EXAMPLE 3   Absolute and conditional convergence Determine whether the following series diverge, converge absolutely, or converge conditionally.

a. $\displaystyle\sum_{k=1}^{\infty} \frac{(-1)^{k+1}}{\sqrt{k}}$ **b.** $\displaystyle\sum_{k=1}^{\infty} \frac{(-1)^{k+1}}{\sqrt{k^3}}$ **c.** $\displaystyle\sum_{k=1}^{\infty} \frac{\sin k}{k^2}$ **d.** $\displaystyle\sum_{k=1}^{\infty} \frac{(-1)^k k}{k+1}$

SOLUTION

a. We examine the series of absolute values,

$$\sum_{k=1}^{\infty} \left| \frac{(-1)^{k+1}}{\sqrt{k}} \right| = \sum_{k=1}^{\infty} \frac{1}{\sqrt{k}},$$

which is a divergent p-series (with $p = \frac{1}{2} < 1$). Therefore, the given alternating series does not converge absolutely. To determine whether the series converges conditionally we look at the original series—with alternating signs. The magnitude of the terms of this series decrease with $\displaystyle\lim_{k\to\infty} 1/\sqrt{k} = 0$, so by the Alternating Series Test, the series converges. Because this series converges, but not absolutely, it converges conditionally.

b. To assess absolute convergence, we look at the series of absolute values,

$$\sum_{k=1}^{\infty} \left| \frac{(-1)^{k+1}}{\sqrt{k^3}} \right| = \sum_{k=1}^{\infty} \frac{1}{k^{3/2}},$$

which is a convergent p-series (with $p = \frac{3}{2} > 1$). Therefore, the original alternating series converges absolutely (and by Theorem 8.21 it converges).

c. The terms of this series do not strictly alternate sign (the first few signs are $+++---$), so the Alternating Series Test does not apply. Because $|\sin k| \le 1$, the terms of the series of absolute values satisfy

$$\left| \frac{\sin k}{k^2} \right| = \frac{|\sin k|}{k^2} \le \frac{1}{k^2}.$$

The series $\sum \dfrac{1}{k^2}$ is a convergent p-series. Therefore, by the Comparison Test, the series $\sum \left| \dfrac{\sin k}{k^2} \right|$ converges, which implies that the series $\sum \dfrac{\sin k}{k^2}$ converges absolutely.

d. Notice that $\displaystyle\lim_{k\to\infty} k/(k+1) = 1$. The terms of the series do not tend to zero and, by the Divergence Test, the series diverges.   *Related Exercises 39–46* ◄

We close the chapter with the summary of tests and series shown in Table 8.4.

Table 8.4 Special Series and Convergence Tests

Series or test	Form of series	Condition for convergence	Condition for divergence	Comments
Geometric series	$\sum_{k=0}^{\infty} ar^k$	$\|r\| < 1$	$\|r\| \geq 1$	If $\|r\| < 1$, then $\sum_{k=0}^{\infty} ar^k = \dfrac{a}{1-r}$.
Divergence Test	$\sum_{k=1}^{\infty} a_k$	Does not apply	$\lim_{k\to\infty} a_k \neq 0$	Cannot be used to prove convergence
Integral Test	$\sum_{k=1}^{\infty} a_k$, where $a_k = f(k)$ and f is continuous, positive, and decreasing	$\displaystyle\int_1^{\infty} f(x)\, dx < \infty$	$\displaystyle\int_1^{\infty} f(x)\, dx$ does not exist.	The value of the integral is not the value of the series.
p-series	$\sum_{k=1}^{\infty} \dfrac{1}{k^p}$	$p > 1$	$p \leq 1$	Useful for comparison tests
Ratio Test	$\sum_{k=1}^{\infty} a_k$, where $a_k > 0$	$\lim_{k\to\infty} \dfrac{a_{k+1}}{a_k} < 1$	$\lim_{k\to\infty} \dfrac{a_{k+1}}{a_k} > 1$	Inconclusive if $\lim_{k\to\infty} \dfrac{a_{k+1}}{a_k} = 1$
Root Test	$\sum_{k=1}^{\infty} a_k$, where $a_k \geq 0$	$\lim_{k\to\infty} \sqrt[k]{a_k} < 1$	$\lim_{k\to\infty} \sqrt[k]{a_k} > 1$	Inconclusive if $\lim_{k\to\infty} \sqrt[k]{a_k} = 1$
Comparison Test	$\sum_{k=1}^{\infty} a_k$, where $a_k > 0$	$0 < a_k \leq b_k$ and $\sum_{k=1}^{\infty} b_k$ converges	$0 < b_k \leq a_k$ and $\sum_{k=1}^{\infty} b_k$ diverges	$\sum_{k=1}^{\infty} a_k$ is given; you supply $\sum_{k=1}^{\infty} b_k$.
Limit Comparison Test	$\sum_{k=1}^{\infty} a_k$, where $a_k > 0, b_k > 0$	$0 \leq \lim_{k\to\infty} \dfrac{a_k}{b_k} < \infty$ and $\sum_{k=1}^{\infty} b_k$ converges.	$\lim_{k\to\infty} \dfrac{a_k}{b_k} > 0$ and $\sum_{k=1}^{\infty} b_k$ diverges.	$\sum_{k=1}^{\infty} a_k$ is given; you supply $\sum_{k=1}^{\infty} b_k$.
Alternating Series Test	$\sum_{k=1}^{\infty} (-1)^k a_k$, where $a_k > 0, 0 < a_{k+1} \leq a_k$	$\lim_{k\to\infty} a_k = 0$	$\lim_{k\to\infty} a_k \neq 0$	Remainder R_n satisfies $R_n < a_{n+1}$

SECTION 8.6 EXERCISES

Review Questions

1. Explain why the sequence of partial sums for an alternating series is not an increasing sequence.

2. Describe how to apply the Alternating Series Test.

3. Why does the value of a converging alternating series lie between any two consecutive terms of its sequence of partial sums?

4. Suppose an alternating series converges to a value L. Explain how to estimate the remainder that occurs when the series is terminated after n terms.

5. Explain why the remainder in terminating an alternating series is less than the first neglected term.

6. Give an example of a convergent alternating series that fails to converge absolutely.

7. Is it possible for a series of positive terms to converge conditionally? Explain.

8. Why does absolute convergence imply convergence?

9. Is it possible for an alternating series to converge absolutely, but not conditionally?

10. Give an example of a series that converges conditionally but not absolutely.

Basic Skills

11–24. Alternating Series Test *Determine whether the following series converge.*

11. $\displaystyle\sum_{k=1}^{\infty} \frac{(-1)^{k+1}}{k^3}$

12. $\displaystyle\sum_{k=0}^{\infty} \frac{(-1)^k}{k^2 + 10}$

13. $\displaystyle\sum_{k=1}^{\infty} (-1)^{k+1} \frac{k^2}{k^3 + 1}$

14. $\displaystyle\sum_{k=2}^{\infty} (-1)^k \frac{\ln k}{k^2}$

15. $\displaystyle\sum_{k=2}^{\infty} (-1)^k \frac{k^2 - 1}{k^2 + 3}$

16. $\displaystyle\sum_{k=0}^{\infty} \left(-\frac{1}{5}\right)^k$

17. $\displaystyle\sum_{k=2}^{\infty} (-1)^k \left(1 + \frac{1}{k}\right)$

18. $\displaystyle\sum_{k=1}^{\infty} \frac{\cos \pi k}{k^2}$

19. $\displaystyle\sum_{k=1}^{\infty} (-1)^{k+1} \frac{k^{10} + 2k^5 + 1}{k(k^{10} + 1)}$

20. $\displaystyle\sum_{k=2}^{\infty} \frac{(-1)^k}{k \ln^2 k}$

21. $\displaystyle\sum_{k=1}^{\infty} (-1)^{k+1} k^{1/k}$

22. $\displaystyle\sum_{k=1}^{\infty} (-1)^{k+1} \frac{k!}{k^k}$

23. $\displaystyle\sum_{k=0}^{\infty} \frac{(-1)^k}{\sqrt{k^2 + 4}}$

24. $\displaystyle\sum_{k=1}^{\infty} (-1)^k k \sin\left(\frac{1}{k}\right)$

⊤ 25–34. Remainders in alternating series *Determine how many terms of the following convergent series must be summed to be sure that the remainder is less than 10^{-4}. Although you do not need it, the exact value of the series is given in each case.*

25. $\displaystyle\ln 2 = \sum_{k=1}^{\infty} \frac{(-1)^{k+1}}{k}$

26. $\displaystyle\frac{1}{e} = \sum_{k=0}^{\infty} \frac{(-1)^k}{k!}$

27. $\displaystyle\frac{\pi}{4} = \sum_{k=0}^{\infty} \frac{(-1)^k}{2k + 1}$

28. $\displaystyle\frac{\pi^2}{12} = \sum_{k=1}^{\infty} \frac{(-1)^{k+1}}{k^2}$

29. $\displaystyle\frac{7\pi^4}{720} = \sum_{k=1}^{\infty} \frac{(-1)^{k+1}}{k^4}$

30. $\displaystyle\frac{\pi^3}{32} = \sum_{k=0}^{\infty} \frac{(-1)^k}{(2k + 1)^3}$

31. $\displaystyle\frac{\pi\sqrt{3}}{9} + \frac{\ln 2}{3} = \sum_{k=0}^{\infty} \frac{(-1)^k}{3k + 1}$

32. $\displaystyle\frac{31\pi^6}{30{,}240} = \sum_{k=1}^{\infty} \frac{(-1)^{k+1}}{k^6}$

33. $\displaystyle\pi = \sum_{k=0}^{\infty} \frac{(-1)^k}{4^k} \left(\frac{2}{4k + 1} + \frac{2}{4k + 2} + \frac{1}{4k + 3}\right)$

34. $\displaystyle\frac{\pi\sqrt{3}}{9} - \frac{\ln 2}{3} = \sum_{k=0}^{\infty} \frac{(-1)^k}{3k + 2}$

⊤ 35–38. Estimating infinite sums *Estimate the value of the following convergent series with an absolute error less than 10^{-3}.*

35. $\displaystyle\sum_{k=1}^{\infty} \frac{(-1)^k}{k^5}$

36. $\displaystyle\sum_{k=1}^{\infty} \frac{(-1)^k}{(2k + 1)^3}$

37. $\displaystyle\sum_{k=1}^{\infty} \frac{(-1)^k}{k^k}$

38. $\displaystyle\sum_{k=1}^{\infty} \frac{(-1)^{k+1}}{(2k + 1)!}$

39–46. Absolute and conditional convergence *Determine whether the following series converge absolutely or conditionally.*

39. $\displaystyle\sum_{k=1}^{\infty} \frac{(-1)^{k+1}}{k^{3/2}}$

40. $\displaystyle\sum_{k=1}^{\infty} \left(-\frac{1}{3}\right)^k$

41. $\displaystyle\sum_{k=1}^{\infty} \frac{\cos k}{k^3}$

42. $\displaystyle\sum_{k=1}^{\infty} \frac{(-1)^k k^2}{\sqrt{k^6 + 1}}$

43. $\displaystyle\sum_{k=1}^{\infty} \frac{(-1)^k k}{2k + 1}$

44. $\displaystyle\sum_{k=2}^{\infty} \frac{(-1)^k}{\ln k}$

45. $\displaystyle\sum_{k=1}^{\infty} \frac{(-1)^k \tan^{-1} k}{k^3}$

46. $\displaystyle\sum_{k=1}^{\infty} \frac{(-1)^{k+1} e^k}{(k + 1)!}$

Further Explorations

47. **Explain why or why not** Determine whether the following statements are true and give an explanation or counterexample.

   a. A series that converges must converge absolutely.
   b. A series that converges absolutely must converge.
   c. A series that converges conditionally must converge.
   d. If $\sum a_k$ diverges, then $\sum |a_k|$ diverges.
   e. If $\sum a_k^2$ converges, then $\sum a_k$ converges.
   f. If $a_k > 0$ and $\sum a_k$ converges, then $\sum a_k^2$ converges.
   g. If $\sum a_k$ converges conditionally, then $\sum |a_k|$ diverges.

48. **Alternating Series Test** Show that the series

$$\frac{1}{3} - \frac{2}{5} + \frac{3}{7} - \frac{4}{9} + \cdots = \sum_{k=1}^{\infty} (-1)^{k+1} \frac{k}{2k + 1}$$

   diverges. Which condition of the Alternating Series Test is not satisfied?

49. **Alternating p-series** Given that $\displaystyle\sum_{k=1}^{\infty} \frac{1}{k^2} = \frac{\pi^2}{6}$, show that

$$\sum_{k=1}^{\infty} \frac{(-1)^{k+1}}{k^2} = \frac{\pi^2}{12}.$$ (Assume the result of Exercise 53.)

50. **Alternating p-series** Given that $\displaystyle\sum_{k=1}^{\infty} \frac{1}{k^4} = \frac{\pi^4}{90}$, show that

$$\sum_{k=1}^{\infty} \frac{(-1)^{k+1}}{k^4} = \frac{7\pi^4}{720}.$$ (Assume the result of Exercise 53.)

51. **Geometric series** In Section 8.3, we established that the geometric series $\sum r^k$ converges provided $|r| < 1$. Notice that if $-1 < r < 0$, the geometric series is also an alternating series. Use the Alternating Series Test to show that for $-1 < r < 0$, the series $\sum r^k$ converges.

⊤ 52. Remainders in alternating series Given any infinite series $\sum a_k$, let $N(r)$ be the number of terms of the series that must be summed to guarantee that the remainder is less than 10^{-r}, where r is a positive integer.

a. Graph the function $N(r)$ for the three alternating p-series

$$\sum_{k=1}^{\infty} \frac{(-1)^{k+1}}{k^p}, \text{ for } p = 1, 2, \text{ and } 3.$$ Compare the three graphs

and discuss what they mean about the rates of convergence of the three series.

b. Carry out the procedure of part (a) for the series $\displaystyle\sum_{k=1}^{\infty} \frac{(-1)^{k+1}}{k!}$

and compare the rates of convergence of all four series.

Additional Exercises

53. Rearranging series It can be proved that if a series converges absolutely, then its terms may be summed in any order without changing the value of the series. However, if a series converges conditionally, then the value of the series depends on the order of summation. For example, the (conditionally convergent) alternating harmonic series has the value

$$1 - \frac{1}{2} + \frac{1}{3} - \frac{1}{4} + \cdots = \ln 2.$$

Show that by rearranging the terms (so the sign pattern is $++-$),

$$1 + \frac{1}{3} - \frac{1}{2} + \frac{1}{5} + \frac{1}{7} - \frac{1}{4} + \cdots = \frac{3}{2} \ln 2.$$

54. A better remainder Suppose an alternating series $\displaystyle\sum_{k=1}^{\infty} (-1)^k a_k$

converges to S and the sum of the first n terms of the series is S_n. Suppose also that the difference between the magnitudes of consecutive terms decreases with k. Then it can be shown that for $n \geq 1$

$$\left| S - \left(S_n + \frac{(-1)^{n+1} a_{n+1}}{2} \right) \right| \leq \frac{1}{2} |a_{n+1} - a_{n+2}|.$$

a. Interpret this inequality and explain why it gives a better approximation to S than simply using S_n to approximate S.

b. For the following series, determine how many terms of the series are needed to approximate its exact value with an error less than 10^{-6} using both S_n and the method explained in part (a).

(i) $\displaystyle\sum_{k=1}^{\infty} \frac{(-1)^k}{k}$ **(ii)** $\displaystyle\sum_{k=2}^{\infty} \frac{(-1)^k}{k \ln k}$ **(iii)** $\displaystyle\sum_{k=2}^{\infty} \frac{(-1)^k}{\sqrt{k}}$

55. A fallacy Explain the fallacy in the following argument. Let

$$x = \frac{1}{1} + \frac{1}{3} + \frac{1}{5} + \frac{1}{7} + \cdots \quad \text{and} \quad y = \frac{1}{2} + \frac{1}{4} + \frac{1}{6} + \frac{1}{8} + \cdots.$$

It follows that $2y = x + y$, which implies that $x = y$. On the other hand,

$$x - y = \underbrace{\left(1 - \frac{1}{2}\right)}_{>0} + \underbrace{\left(\frac{1}{3} - \frac{1}{4}\right)}_{>0} + \underbrace{\left(\frac{1}{5} - \frac{1}{6}\right)}_{>0} + \cdots > 0$$

is a sum of positive terms, so $x > y$. Thus, we have shown that $x = y$ and $x > y$.

QUICK CHECK ANSWERS

1. $1, -1, 2, -2, 3, -3, \ldots$; series diverges.   **2.** The even terms of the sequence of partial sums approach the value of the series from one side; the odd terms of the sequence of partial sums approach the value of the series from the other side.   **3.** The second series with $k!$ in the denominators converges much more quickly than the first series because $k!$ increases much faster than k as $k \to \infty$.   **4.** If a series has positive terms, the series of absolute values is the same as the series itself. ◄

CHAPTER 8   REVIEW EXERCISES

1. **Explain why or why not** Determine whether the following statements are true and give an explanation or counterexample.

   a. The terms of the sequence $\{a_n\}$ increase in magnitude, so the limit of the sequence does not exist.

   b. The terms of the series $\sum 1/\sqrt{k}$ approach zero, so the series converges.

   c. The terms of the sequence of partial sums of the series $\sum a_k$ approach $5/2$, so the infinite series converges to $5/2$.

   d. An alternating series that converges absolutely must converge conditionally.

2–10. Limits of sequences *Evaluate the limit of the sequence or state that it does not exist.*

2. $a_n = \dfrac{n^2 + 4}{\sqrt{4n^4 + 1}}$

3. $a_n = \dfrac{8^n}{n!}$

4. $a_n = \left(1 + \dfrac{3}{n}\right)^{2n}$

5. $a_n = \sqrt[n]{n}$

6. $a_n = n - \sqrt{n^2 - 1}$

7. $a_n = \left(\dfrac{1}{n}\right)^{1/\ln n}$

8. $a_n = \sin\left(\dfrac{\pi n}{6}\right)$

9. $a_n = \dfrac{(-1)^n}{0.9^n}$

10. $a_n = \tan^{-1} n$

11. Sequence of partial sums Consider the series

$$\sum_{k=1}^{\infty} \frac{1}{k(k + 2)} = \frac{1}{2} \sum_{k=1}^{\infty} \left(\frac{1}{k} - \frac{1}{k + 2}\right).$$

   a. Write the first four terms of the sequence of partial sums $S_1, \ldots, S_4$.

   b. Write the nth term of the sequence of partial sums S_n.

   c. Find $\displaystyle\lim_{n \to \infty} S_n$ and evaluate the series.

12–20. Evaluating series *Evaluate the following infinite series or state that the series diverges.*

12. $\displaystyle\sum_{k=1}^{\infty} \left(\frac{9}{10}\right)^k$

13. $\displaystyle\sum_{k=1}^{\infty} 3(1.001)^k$

14. $\displaystyle\sum_{k=0}^{\infty}\left(-\frac{1}{5}\right)^{k}$

15. $\displaystyle\sum_{k=1}^{\infty}\frac{1}{k(k+1)}$

16. $\displaystyle\sum_{k=1}^{\infty}\left(\frac{1}{\sqrt{k}}-\frac{1}{\sqrt{k-1}}\right)$

17. $\displaystyle\sum_{k=1}^{\infty}\left(\frac{3}{3k-2}-\frac{3}{3k+1}\right)$

18. $\displaystyle\sum_{k=1}^{\infty}4^{-3k}$

19. $\displaystyle\sum_{k=1}^{\infty}\frac{2^{k}}{3^{k+2}}$

20. $\displaystyle\sum_{k=0}^{\infty}\left[\left(\frac{1}{3}\right)^{k}-\left(\frac{2}{3}\right)^{k+1}\right]$

21. Sequences of partial sums The sequences of partial sums for three series are shown in the figures below. Assume that the pattern in the sequences continues as $n\to\infty$.

    a. Does it appear that series A converges? If so, what is its (approximate) value?

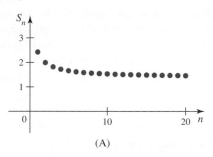

(A)

    b. What can you conclude about the convergence or divergence of series B?

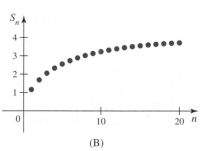

(B)

    c. Does it appear that series C converges? If so, what is its (approximate) value?

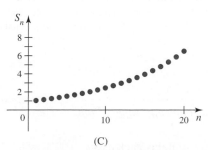

(C)

22–36. Convergence or divergence *Use a convergence test of your choice to determine whether the following series converge or diverge.*

22. $\displaystyle\sum_{k=1}^{\infty}\frac{2}{k^{3/2}}$

23. $\displaystyle\sum_{k=1}^{\infty}k^{-2/3}$

24. $\displaystyle\sum_{k=1}^{\infty}\frac{2k^{2}+1}{\sqrt{k^{3}+2}}$

25. $\displaystyle\sum_{k=1}^{\infty}\frac{2^{k}}{e^{k}}$

26. $\displaystyle\sum_{k=1}^{\infty}\left(\frac{k}{k+3}\right)^{2k}$

27. $\displaystyle\sum_{k=1}^{\infty}\frac{2^{k}k!}{k^{k}}$

28. $\displaystyle\sum_{k=1}^{\infty}\frac{1}{\sqrt{k}\sqrt{k+1}}$

29. $\displaystyle\sum_{k=1}^{\infty}\frac{3}{2+e^{k}}$

30. $\displaystyle\sum_{k=1}^{\infty}k\sin\left(\frac{1}{k}\right)$

31. $\displaystyle\sum_{k=1}^{\infty}\frac{\sqrt[4]{k}}{k^{3}}$

32. $\displaystyle\sum_{k=1}^{\infty}\frac{1}{1+\ln k}$

33. $\displaystyle\sum_{k=1}^{\infty}k^{5}e^{-k}$

34. $\displaystyle\sum_{k=4}^{\infty}\frac{2}{k^{2}-10}$

35. $\displaystyle\sum_{k=1}^{\infty}\frac{\ln k^{2}}{k^{2}}$

36. $\displaystyle\sum_{k=1}^{\infty}ke^{-k}$

37–42. Alternating series *Determine whether the following series converge or diverge. In the case of convergence, state whether the convergence is conditional or absolute.*

37. $\displaystyle\sum_{k=2}^{\infty}\frac{(-1)^{k}}{k^{2}-1}$

38. $\displaystyle\sum_{k=1}^{\infty}\frac{(-1)^{k+1}(k^{2}+4)}{2k^{2}+1}$

39. $\displaystyle\sum_{k=1}^{\infty}(-1)^{k}ke^{-k}$

40. $\displaystyle\sum_{k=1}^{\infty}\frac{(-1)^{k}}{\sqrt{k^{2}+1}}$

41. $\displaystyle\sum_{k=1}^{\infty}\frac{(-1)^{k+1}10^{k}}{k!}$

42. $\displaystyle\sum_{k=2}^{\infty}\frac{(-1)^{k}}{k\ln k}$

43. Sequences vs. series

    a. Find the limit of $\left\{\left(-\frac{4}{5}\right)^{k}\right\}$.

    b. Evaluate $\displaystyle\sum_{k=0}^{\infty}\left(-\frac{4}{5}\right)^{k}$.

44. Sequences vs. series

    a. Find the limit of $\left\{\dfrac{1}{k}-\dfrac{1}{k+1}\right\}$.

    b. Evaluate $\displaystyle\sum_{k=1}^{\infty}\left(\frac{1}{k}-\frac{1}{k+1}\right)$.

45. Partial sums Let S_{n} be the nth partial sum of $\displaystyle\sum_{k=1}^{\infty}a_{k}=8$. Find $\displaystyle\lim_{k\to\infty}a_{k}$ and $\displaystyle\lim_{n\to\infty}S_{n}$.

T 46. Remainder term Let R_{n} be the remainder associated with $\displaystyle\sum_{k=1}^{\infty}\frac{1}{k^{5}}$. Find an upper bound for R_{n} (in terms of n). How many terms of the series must be summed to approximate the series with an error less than 10^{-4}?

47. Conditional p-series Find the values of p for which $\displaystyle\sum_{k=1}^{\infty}\frac{(-1)^{k}}{k^{p}}$ converges conditionally.

48. Logarithmic p-series Show that the series $\displaystyle\sum_{k=2}^{\infty}\frac{1}{k(\ln k)^{p}}$ converges provided $p>1$.

⊤ 49. Error in a finite sum Approximate the series $\sum_{k=1}^{\infty} \frac{1}{5^k}$ by evaluating the first 20 terms. Compute the maximum error in the approximation.

⊤ 50. Error in a finite sum Approximate the series $\sum_{k=1}^{\infty} \frac{1}{k^5}$ by evaluating the first 20 terms. Compute the maximum error in the approximation.

⊤ 51. Error in a finite alternating sum How many terms of the series

$$\sum_{k=1}^{\infty} \frac{(-1)^{k+1}}{k^4}$$

must be summed to ensure that the remainder is less than 10^{-8}?

52. Equations involving series Solve the following equations for x.

a. $\sum_{k=0}^{\infty} e^{kx} = 2$

b. $\sum_{k=0}^{\infty} (3x)^k = 4$

c. $\sum_{k=1}^{\infty} \left(\frac{x}{kx - \frac{x}{2}} - \frac{x}{kx + \frac{x}{2}} \right) = 6$

53. Building a tunnel—first scenario A crew of workers is constructing a tunnel through a mountain. Understandably, the rate of construction decreases because rocks and earth must be removed a greater distance as the tunnel gets longer. Suppose that each week the crew digs 0.95 of the distance it dug the previous week. In the first week, the crew constructed 100 meters of tunnel.

a. How far does the crew dig in 10 weeks? 20 weeks? N weeks?

b. What is the longest tunnel the crew can build at this rate?

54. Building a tunnel—second scenario As in Exercise 53, a crew of workers is constructing a tunnel. The time required to dig 100 m increases by 10% each week, starting with 1 week to dig the first 100 m. Can the crew complete a 1.5-km (1500-m) tunnel in 30 weeks? Explain.

55. Pages of circles On page 1 of a book, there is one circle of radius 1. On page 2, there are two circles of radius $\frac{1}{2}$. On page n there are 2^{n-1} circles of radius 2^{-n+1}.

a. What is the sum of the areas of the circles on page n of the book?

b. Assuming the book continues indefinitely $(n \to \infty)$, what is the sum of the areas of all the circles in the book?

⊤ 56. Sequence on a calculator Let $\{x_n\}$ be generated by the recurrence relation $x_0 = 1$ and $x_{n+1} = x_n + \cos x_n$, for $n = 0, 1, 2, \ldots$. Use a calculator (in radian mode) to generate as many terms of the sequence $\{x_n\}$ needed to find the integer p such that $\lim_{n \to \infty} x_n = \pi/p$.

57. A savings plan Suppose that you open a savings account by depositing \$100. The account earns interest at an annual rate of 3% per year (0.25% per month). At the end of each month, you earn interest on the current balance, and then you deposit \$100. Let B_n be the balance at the beginning of the nth month, where $B_0 = \$100$.

a. Find a recurrence relation for the sequence $\{B_n\}$.

b. Find an explicit formula that gives B_n for $n = 0, 1, 2, 3, \ldots$.

58. Sequences of integrals Find the limits of the sequences $\{a_n\}$ and $\{b_n\}$.

a. $a_n = \int_0^1 x^n \, dx, \ n \geq 1$ **b.** $b_n = \int_1^n \frac{dx}{x^p}, \ p > 1, n \geq 1$

59. Sierpinski triangle The fractal called the *Sierpinski triangle* is the limit of a sequence of figures. Starting with the equilateral triangle with sides of length 1, an inverted equilateral triangle with sides of length $\frac{1}{2}$ is removed. Then, three inverted equilateral triangles with sides of length $\frac{1}{4}$ are removed from this figure (see figure). The process continues in this way. Let T_n be the total area of the removed triangles after stage n of the process. The area of an equilateral triangle with side length L is $A = \sqrt{3}L^2/4$.

a. Find T_1 and T_2, the total area of the removed triangles after stages 1 and 2, respectively.

b. Find T_n for $n = 1, 2, 3, \ldots$.

c. Find $\lim_{n \to \infty} T_n$.

d. What is the area of the original triangle that remains as $n \to \infty$?

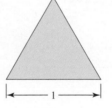
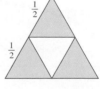
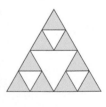

Initial stage   First stage   Second stage

60. Max sine sequence Let $a_n = \max \{\sin 1, \sin 2, \ldots, \sin n\}$, for $n = 1, 2, 3, \ldots$, where $\max \{ \ldots \}$ denotes the maximum element of the set. Does $\{a_n\}$ converge? If so, make a conjecture about the limit.

Chapter 8 Guided Projects

Applications of the material in this chapter and related topics can be found in the following Guided Projects. For additional information, see the Preface.

• Chaos!

• Periodic drug dosing

• The mathematics of loans

• Financial matters

• Economic stimulus packages

• Archimedes' approximation to π

• Exact values of infinite series

• Conditional convergence in a crystal lattice

9

Power Series

Chapter Preview   Until now we have worked with infinite series consisting of real numbers. In this chapter a seemingly small, but significant change is made by considering infinite series whose terms include a variable. With this change, an infinite series becomes a *power series*. Surely one of the most significant ideas in all of calculus is that functions can be represented by power series. As a first step toward this result, we look at approximating functions using polynomials. The transition from polynomials to power series is then straightforward. With these tools, it is possible to represent the familiar functions of mathematics in terms of power series called *Taylor series*. The remainder of the chapter is devoted to the properties and many uses of these series.

9.1 Approximating Functions with Polynomials

Power series—like sets and functions—are among the most fundamental entities of mathematics because they provide a way to represent familiar functions and to define new functions.

What Is a Power Series?

A *power series* is an infinite series of the form

$$\sum_{k=0}^{\infty} c_k x^k = \underbrace{c_0 + c_1 x + c_2 x^2 + \cdots + c_n x^n}_{n\text{th degree polynomial}} + \underbrace{c_{n+1} x^{n+1} + \cdots}_{\text{terms continue}},$$

or, more generally,

$$\sum_{k=0}^{\infty} c_k (x - a)^k = \underbrace{c_0 + c_1 (x - a) + \cdots + c_n (x - a)^n}_{n\text{th degree polynomial}} + \underbrace{c_{n+1}(x - a)^{n+1} + \cdots}_{\text{terms continue}},$$

where the coefficients c_k and the **center** of the series a are constants. This type of series is called a power series because it consists of powers of x or $(x - a)$.

Viewed in another way, a power series is built up from polynomials of increasing degree, as shown in the following progression:

$$
\left.\begin{array}{l}
\text{Degree 0: } c_0 \\
\text{Degree 1: } c_0 + c_1 x \\
\text{Degree 2: } c_0 + c_1 x + c_2 x^2 \\
\quad\vdots \qquad \vdots \qquad \vdots \\
\text{Degree } n: c_0 + c_1 x + c_2 x^2 + \cdots + c_n x^n = \sum_{k=0}^{n} c_k x^k \\
\quad\vdots \qquad \vdots \qquad \vdots
\end{array}\right\} \text{Polynomials}
$$

$$
c_0 + c_1 x + c_2 x^2 + \cdots + c_n x^n + \cdots = \sum_{k=0}^{\infty} c_k x^k \left.\right\} \text{Power series}
$$

We begin our exploration of power series by using polynomials to approximate functions.

Polynomial Approximation

An important observation motivates our work. To evaluate a polynomial $\left(\text{say, } f(x) = x^8 - 4x^5 + \frac{1}{2}\right)$, all we need is arithmetic—addition, subtraction, multiplication, and division. However, algebraic functions $\left(\text{say, } f(x) = \sqrt[3]{x^4 - 1}\right)$ or trigonometric, logarithmic, or exponential functions usually cannot be evaluated exactly using arithmetic. Therefore, it makes practical sense to use the simplest of functions, polynomials, to approximate more complicated functions.

Linear and Quadratic Approximation

Recall that if a function f is differentiable at a point a, it can be approximated near a by its tangent line (Section 4.5); the tangent line provides the linear approximation to f at the point a. The equation of the tangent line at the point $(a, f(a))$ is

$$
y - f(a) = f'(a)(x - a) \quad \text{or} \quad y = f(a) + f'(a)(x - a).
$$

Because the linear approximation function is a first-degree polynomial, we name it p_1:

$$
p_1(x) = f(a) + f'(a)(x - a).
$$

This polynomial has some important properties: it matches f in *value* and in *slope* at a. In other words (Figure 9.1),

$$
p_1(a) = f(a) \quad \text{and} \quad p_1{}'(a) = f'(a).
$$

Linear approximation works well if f has a fairly constant slope near the point a. However if f has a lot of curvature near a, then the tangent line may not provide a good approximation. To remedy this situation, we create a quadratic approximating polynomial by adding a single term to the linear polynomial. Denoting this new polynomial p_2, we have

$$
p_2(x) = \underbrace{f(a) + f'(a)(x - a)}_{p_1(x)} + \underbrace{c_2(x - a)^2}_{\text{quadratic term}}.
$$

The new term consists of a coefficient c_2 that must be determined and a quadratic factor $(x - a)^2$.

To determine c_2 and to ensure that p_2 is a good approximation to f near the point a, we require that p_2 agree with f in value, slope, and concavity at a; that is, p_2 must satisfy the matching conditions

$$
p_2(a) = f(a) \qquad p_2{}'(a) = f'(a) \qquad p_2{}''(a) = f''(a),
$$

where we assume that f and its first and second derivatives exist at a (Figure 9.2).

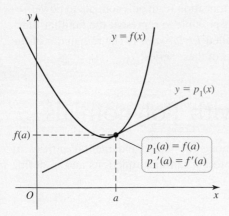

FIGURE 9.1

$p_1(a) = f(a)$
$p_1{}'(a) = f'(a)$

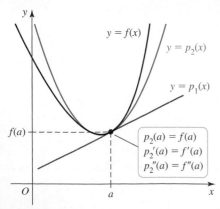

FIGURE 9.2

$p_2(a) = f(a)$
$p_2{}'(a) = f'(a)$
$p_2{}''(a) = f''(a)$

▶ Matching concavity (second derivatives) ensures that the graph of p_2 bends in the same direction as the graph of f at a.

Substituting $x = a$ into p_2, we see immediately that $p_2(a) = f(a)$, so the first matching condition is met. Differentiating p_2 once, we have

$$p_2'(x) = f'(a) + 2c_2(x - a).$$

So, $p_2'(a) = f'(a)$, and the second matching condition is also met. Because $p_2''(a) = 2c_2$, the third matching condition is

$$p_2''(a) = 2c_2 = f''(a).$$

It follows that $c_2 = \frac{1}{2} f''(a)$; therefore, the quadratic approximating polynomial is

$$p_2(x) = \underbrace{f(a) + f'(a)(x - a)}_{p_1(x)} + \frac{1}{2} f''(a)(x - a)^2.$$

EXAMPLE 1 Approximations for $\ln x$

a. Find the linear approximation for $f(x) = \ln x$ at $x = 1$.

b. Find the quadratic approximation for $f(x) = \ln x$ at $x = 1$.

c. Use these approximations to estimate the value of $\ln 1.05$.

SOLUTION

a. Note that $f(1) = 0$, $f'(x) = 1/x$, and $f'(1) = 1$. Therefore, the linear approximation to $f(x) = \ln x$ at $x = 1$ is

$$p_1(x) = f(1) + f'(1)(x - 1) = 0 + 1(x - 1) = x - 1.$$

As shown in Figure 9.3, p_1 matches f in value ($p_1(1) = f(1)$) and in slope ($p_1'(1) = f'(1)$) at $x = 1$.

b. We first compute $f''(x) = -1/x^2$ and $f''(1) = -1$. Building on the linear approximation found in part (a), the quadratic approximation is

$$p_2(x) = \underbrace{x - 1}_{p_1(x)} + \underbrace{\frac{1}{2} f''(1)(x - 1)^2}_{c_2}$$

$$= (x - 1) + \frac{1}{2}(-1)(x - 1)^2$$

$$= (x - 1) - \frac{1}{2}(x - 1)^2.$$

Because p_2 matches f in value, slope, and concavity at $x = 1$, it provides a better approximation to f near $x = 1$ (Figure 9.3).

c. To approximate $\ln 1.05$, we substitute $x = 1.05$ into each polynomial approximation:

$$p_1(1.05) = 1.05 - 1 = 0.05$$

$$p_2(1.05) = (1.05 - 1) - \frac{1}{2}(1.05 - 1)^2 = 0.04875$$

The value of $\ln 1.05$ given by a calculator, rounded to five decimal places, is 0.04879, showing the improvement in quadratic approximation over linear approximation.

Related Exercises 7–12 ◄

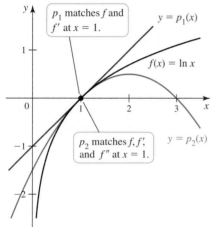

p_1 matches f and f' at $x = 1$.

$y = p_1(x)$

$f(x) = \ln x$

p_2 matches f, f', and f'' at $x = 1$.

$y = p_2(x)$

FIGURE 9.3

▶ Building on ideas that were already circulating in the early 18th century, Brooke Taylor (1685–1731) published Taylor's Theorem in 1715. He is also credited with discovering integration by parts.

▶ Recall that $2! = 2 \cdot 1, 3! = 3 \cdot 2 \cdot 1$, $k! = k \cdot (k - 1)!$, and by definition $0! = 1$.

Taylor Polynomials

The process used to find the approximating polynomial p_2 can be extended to obtain approximating polynomials of higher degree. Assuming that f and its first n derivatives exist at a, we can use p_2 to obtain a cubic polynomial p_3 of the form

$$p_3(x) = p_2(x) + c_3(x - a)^3$$

that satisfies the four matching conditions

$$p_3(a) = f(a), \quad p_3'(a) = f'(a), \quad p_3''(a) = f''(a), \text{ and } p_3'''(a) = f'''(a).$$

Because p_3 is built "on top of" p_2, the first three matching conditions are met. The last condition, $p_3'''(a) = f'''(a)$, is used to determine c_3. A short calculation shows that $p_3'''(x) = 3 \cdot 2c_3 = 3!c_3$, and so the last matching condition becomes $p_3'''(a) = 3!c_3 = f'''(a)$. Solving for c_3, we have $c_3 = \dfrac{f'''(a)}{3!}$. Therefore, the cubic approximating polynomial is

$$p_3(x) = \underbrace{f(a) + f'(a)(x - a) + \frac{f''(a)}{2!}(x - a)^2}_{p_2(x)} + \frac{f'''(a)}{3!}(x - a)^3.$$

QUICK CHECK 1 Verify that p_3 satisfies $p_3^{(k)}(a) = f^{(k)}(a)$, for $k = 0, 1, 2, 3$. ◀

Continuing in this fashion (Exercise 66), building each new polynomial on the previous polynomial, the nth approximating polynomial for f at a is

$$p_n(x) = f(a) + f'(a)(x - a) + \frac{f''(a)}{2!}(x - a)^2 + \cdots + \frac{f^{(n)}(a)}{n!}(x - a)^n.$$

It satisfies the $n + 1$ matching conditions

$$p_n(a) = f(a), \quad p_n'(a) = f'(a), \quad p_n''(a) = f''(a), \ldots, p_n^{(n)}(a) = f^{(n)}(a).$$

These conditions ensure that the graph of p_n conforms as closely as possible to the graph of f near a (Figure 9.4).

▶ Recall that $f^{(n)}$ denotes the nth derivative of f. By convention the zeroth derivative $f^{(0)}$ is f itself.

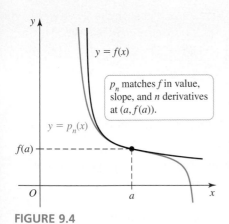

p_n matches f in value, slope, and n derivatives at $(a, f(a))$.

$y = f(x)$

$y = p_n(x)$

FIGURE 9.4

DEFINITION   Taylor Polynomials

Let f be a function with $f', f'', \ldots, f^{(n)}$ defined at a. The **nth-order Taylor polynomial** for f with its **center** at a, denoted p_n, has the property that it matches f in value, slope, and all derivatives up to the nth derivative at a; that is,

$$p_n(a) = f(a), \quad p_n'(a) = f'(a), \ldots, p_n^{(n)}(a) = f^{(n)}(a).$$

The nth-order Taylor polynomial centered at a is

$$p_n(x) = f(a) + f'(a)(x - a) + \frac{f''(a)}{2!}(x - a)^2 + \cdots + \frac{f^{(n)}(a)}{n!}(x - a)^n.$$

More compactly, $p_n(x) = \displaystyle\sum_{k=0}^{n} c_k(x - a)^k$, where the **coefficients** are

$$c_k = \frac{f^{(k)}(a)}{k!}, \quad \text{for } k = 0, 1, 2, \ldots, n.$$

EXAMPLE 2   **Taylor polynomials for sin x**   Find the Taylor polynomials $p_1, \ldots, p_7$ centered at $x = 0$ for $f(x) = \sin x$.

SOLUTION   Differentiating f repeatedly and evaluating the derivatives at 0, a pattern emerges:

$$
\begin{aligned}
f(x) &= \sin x \Rightarrow f(0) = 0 \\
f'(x) &= \cos x \Rightarrow f'(0) = 1 \\
f''(x) &= -\sin x \Rightarrow f''(0) = 0 \\
f'''(x) &= -\cos x \Rightarrow f'''(0) = -1 \\
f^{(4)}(x) &= \sin x \Rightarrow f^{(4)}(0) = 0
\end{aligned}
$$

The derivatives of $\sin x$ at 0 cycle through the values $\{0, 1, 0, -1\}$. Therefore, $f^{(5)}(0) = 1$, $f^{(6)}(0) = 0$, and $f^{(7)}(0) = -1$.

We now construct the polynomials that approximate $f(x) = \sin x$ near 0, beginning with the linear polynomial. The polynomial of order 1 ($n = 1$) is

$$p_1(x) = f(0) + f'(0)x = x,$$

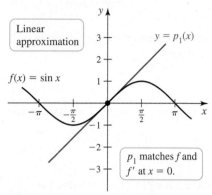

Linear approximation

$y = p_1(x)$

$f(x) = \sin x$

p_1 matches f and f' at $x = 0$.

FIGURE 9.5

whose graph is the line through the origin with slope 1 (Figure 9.5). Notice that f and p_1 agree in value ($f(0) = p_1(0) = 0$) and in slope ($f'(0) = p_1'(0) = 1$) at 0. We see that p_1 provides a good fit to f near 0, but the graphs diverge visibly for $|x| > 0.5$.

The polynomial of order 2 ($n = 2$) is

$$p_2(x) = \underbrace{f(0)}_{0} + \underbrace{f'(0)}_{1}x + \underbrace{\frac{f''(0)}{2!}}_{0}x^2 = x,$$

so p_2 is the same as p_1.

The polynomial of order 3 that approximates f near 0 is

$$p_3(x) = \underbrace{f(0) + f'(0)x + \frac{f''(0)}{2!}x^2}_{p_2(x) = x} + \underbrace{\frac{f'''(0)}{3!}}_{-1/3!}x^3 = x - \frac{x^3}{6}.$$

> It is worth repeating that the next polynomial in the sequence is obtained by adding one new term to the previous polynomial. For example,
>
> $$p_3(x) = p_2(x) + \frac{f'''(a)}{3!}(x - a)^3.$$

QUICK CHECK 2   Verify that $f(0) = p_3(0)$, $f'(0) = p_3'(0)$, $f''(0) = p_3''(0)$, and $f'''(0) = p_3'''(0)$ for $f(x) = \sin x$ and $p_3(x) = x - x^3/6.$ ◄

We have designed p_3 to agree with f in value, slope, concavity, and third derivative at 0 (Figure 9.6). The result is that p_3 provides a better approximation to f over a larger interval than p_1.

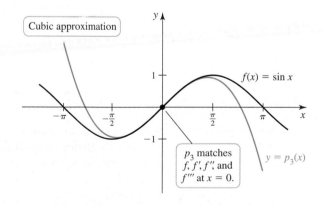

Cubic approximation

$f(x) = \sin x$

p_3 matches f, f', f'', and f''' at $x = 0$.

$y = p_3(x)$

FIGURE 9.6

The procedure for finding Taylor polynomials may be extended to polynomials of any order. Because the even derivatives of $f(x) = \sin x$ are zero, $p_4(x) = p_3(x)$. For the same reason, $p_6(x) = p_5(x)$:

$$p_6(x) = p_5(x) = x - \frac{x^3}{3!} + \frac{x^5}{5!}.$$

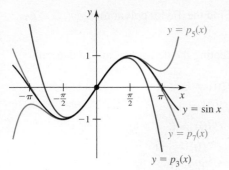

FIGURE 9.7

Finally, it can be shown that the Taylor polynomial of order 7 is

$$p_7(x) = x - \frac{x^3}{3!} + \frac{x^5}{5!} - \frac{x^7}{7!}.$$

From Figure 9.7 we see that as the order of the Taylor polynomials increases, better and better approximations to $f(x) = \sin x$ are obtained over larger and larger intervals centered at 0. For example, p_7 is a good fit to $f(x) = \sin x$ over the interval $[-\pi, \pi]$.

Related Exercises 13–20 ◄

QUICK CHECK 3 Given that $f(x) = \sin x$ is an odd function, why do the Taylor polynomials centered at 0 for f consist only of odd powers of x? ◄

Approximations with Taylor Polynomials

Taylor polynomials find widespread use in approximating functions, as illustrated in the following examples.

EXAMPLE 3 Taylor polynomials for e^x

> Recall that if c is an approximation to x, the absolute error in c is $|c - x|$ and the relative error in c is $|c - x|/|x|$. We use *error* to refer to *absolute error*.

a. Find the Taylor polynomials of order $n = 0, 1, 2,$ and 3 for $f(x) = e^x$ centered at $x = 0$. Graph f and the polynomials.

b. Use the polynomials in part (a) to approximate $e^{0.1}$ and $e^{-0.25}$. Find the absolute errors, $|f(x) - p_n(x)|$, in the approximations. Use calculator values for the exact values of f.

SOLUTION

a. We use the formula for the coefficients in the Taylor polynomials:

$$c_k = \frac{f^{(k)}(0)}{k!}, \qquad \text{for } k = 0, 1, 2, \ldots, n.$$

With $f(x) = e^x$, we have $f^{(k)}(x) = e^x$. Therefore, $f^{(k)}(0) = 1$ and $c_k = 1/k!$, for $k = 0, 1, 2, 3 \ldots$. The first four polynomials are

Taylor polynomials for $f(x) = e^x$ centered at $x = 0$. Approximations improve as n increases.

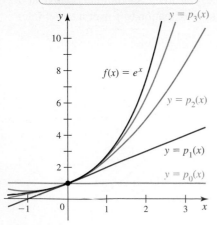

FIGURE 9.8

$$p_0(x) = f(0) = 1$$

$$p_1(x) = \underbrace{f(0)}_{p_0(x) = 1} + \underbrace{f'(0)}_{1}x = 1 + x$$

$$p_2(x) = \underbrace{f(0) + f'(0)x}_{p_1(x) = 1 + x} + \underbrace{\frac{f''(0)}{2!}}_{1/2}x^2 = 1 + x + \frac{x^2}{2}$$

$$p_3(x) = \underbrace{f(0) + f'(0)x + \frac{f''(0)}{2!}x^2}_{p_2(x) = 1 + x + x^2/2} + \underbrace{\frac{f^{(3)}(0)}{3!}}_{1/6}x^3 = 1 + x + \frac{x^2}{2} + \frac{x^3}{6}.$$

Notice that each successive polynomial provides a better fit to $f(x) = e^x$ near 0 (Figure 9.8). Better approximations are obtained with higher-order polynomials. If the pattern in these polynomials is continued, the nth-order Taylor polynomial for e^x centered at 0 is

$$p_n(x) = 1 + x + \frac{x^2}{2!} + \frac{x^3}{3!} + \cdots + \frac{x^n}{n!} = \sum_{k=0}^{n} \frac{x^k}{k!}.$$

b. We evaluate $p_n(0.1)$ and $p_n(-0.25)$ for $n = 0, 1, 2, 3$ and compare these values to the calculator values of $e^{0.1} \approx 1.1051709$ and $e^{-0.25} \approx 0.77880078$. The results are shown in Table 9.1. Observe that the errors in the approximations decrease as n increases. In addition, the errors in approximating $e^{0.1}$ are smaller in magnitude than the errors in approximating $e^{-0.25}$ because $x = 0.1$ is closer to the center of the polynomials than $x = -0.25$. Reasonable approximations based on these calculations are $e^{0.1} \approx 1.105$ and $e^{-0.25} \approx 0.78$.

> A rule of thumb in finding estimates based on several approximations: Keep all of the digits that are common to the last two approximations after rounding.

QUICK CHECK 4 Write out the next two polynomials p_4 and p_5 for $f(x) = e^x$ in Example 3. ◄

Table 9.1

n	Approximations to $e^{0.1}$	Absolute error $\lvert e^{0.1} - p_n(0.1) \rvert$	Approximations to $e^{-0.25}$	Absolute error $\lvert e^{-0.25} - p_n(-0.25) \rvert$
0	1	1.05×10^{-1}	1	2.21×10^{-1}
1	1.1	5.17×10^{-3}	0.75	2.89×10^{-2}
2	1.105	1.71×10^{-4}	0.78125	2.45×10^{-3}
3	1.105167	4.25×10^{-6}	0.778646	1.55×10^{-4}

Related Exercises 21–26 ◄

EXAMPLE 4   Approximating a real number using Taylor polynomials Use polynomials of order $n = 0, 1, 2,$ and 3 to approximate $\sqrt{18}$.

SOLUTION Letting $f(x) = \sqrt{x}$, we choose the center $a = 16$ because it is near 18, and f and its derivatives are easy to evaluate at 16. The Taylor polynomials have the form

$$p_n(x) = f(16) + f'(16)(x - 16) + \frac{f''(16)}{2!}(x - 16)^2 + \cdots + \frac{f^{(n)}(16)}{n!}(x - 16)^n.$$

We now evaluate the required derivatives:

$$f(x) = \sqrt{x} \Rightarrow f(16) = 4$$

$$f'(x) = \frac{1}{2}x^{-1/2} \Rightarrow f'(16) = \frac{1}{8}$$

$$f''(x) = -\frac{1}{4}x^{-3/2} \Rightarrow f''(16) = -\frac{1}{256}$$

$$f'''(x) = \frac{3}{8}x^{-5/2} \Rightarrow f'''(16) = \frac{3}{8192}$$

Therefore, the polynomial p_3 (which includes p_0, p_1, and p_2) is

$$p_3(x) = \underbrace{\underbrace{\underbrace{\underset{p_0}{4} + \frac{1}{8}(x - 16)}_{p_1} - \frac{1}{512}(x - 16)^2}_{p_2} + \frac{1}{16,384}(x - 16)^3}.$$

The graphs of the Taylor polynomials (Figure 9.9) show better approximations to f as the order of the approximation increases.

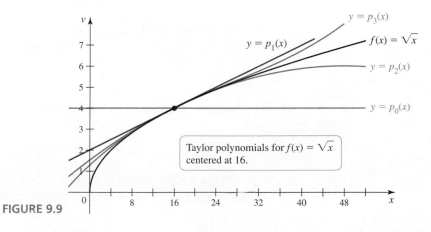

Taylor polynomials for $f(x) = \sqrt{x}$ centered at 16.

FIGURE 9.9

Letting $x = 18$, we obtain the approximations to $\sqrt{18}$ and the associated absolute errors shown in Table 9.2. (A calculator is used for the value of $\sqrt{18}$.) As expected, the errors decrease as n increases. Based on these calculations, a reasonable approximation is $\sqrt{18} \approx 4.24$.

Table 9.2

n	Approximations $p_n(18)$	Absolute Error $|\sqrt{18} - p_n(18)|$
0	4	2.43×10^{-1}
1	4.25	7.36×10^{-3}
2	4.242188	4.53×10^{-4}
3	4.242676	3.51×10^{-5}

Related Exercises 27–40 ◄

QUICK CHECK 5 At what point would you center the Taylor polynomials for $\sqrt{x}$ and $\sqrt[4]{x}$ to approximate $\sqrt{51}$ and $\sqrt[4]{15}$, respectively? ◄

Remainder in a Taylor Polynomial

Taylor polynomials provide good approximations to functions near a specific point. But *how* good are the approximations? To answer this question we define the **remainder** in a Taylor polynomial. If p_n is the Taylor polynomial of order n for f, then the remainder at the point x is

$$R_n(x) = f(x) - p_n(x).$$

The absolute value of the remainder is the error made in approximating $f(x)$ by $p_n(x)$. Equivalently, we have $f(x) = p_n(x) + R_n(x)$, which says that f consists of two components: the polynomial approximation and the associated remainder.

DEFINITION   **Remainder in a Taylor Polynomial**

Let p_n be the Taylor polynomial of order n for f. The **remainder** in using p_n to approximate f at the point x is

$$R_n(x) = f(x) - p_n(x).$$

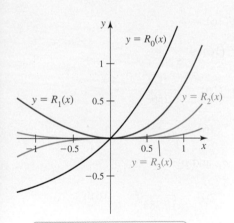

Remainders increase in size as $|x|$ increases. Remainders decrease in size to zero as n increases.

FIGURE 9.10

The idea of a remainder is illustrated in Figure 9.10, where we see the remainder terms associated with the Taylor polynomials for $f(x) = e^x$ centered at 0. For fixed order n, the remainders tend to increase in magnitude as x moves farther from the center of the polynomials (in this case 0). And for fixed x, remainders decrease in magnitude to zero with increasing n.

The remainder term for Taylor polynomials may be written quite concisely, which enables us to estimate remainders. The following result is known as **Taylor's Theorem** (or the **Remainder Theorem**).

➤ The remainder term for a Taylor polynomial can be expressed in several different forms. The form in Theorem 9.1 is called the *Lagrange form* of the remainder.

THEOREM 9.1   Taylor's Theorem
Let f have continuous derivatives up to $f^{(n+1)}$ on an open interval I containing a. For all x in I,

$$f(x) = p_n(x) + R_n(x),$$

where p_n is the nth-order Taylor polynomial for f centered at a, and the remainder is

$$R_n(x) = \frac{f^{(n+1)}(c)}{(n+1)!}(x-a)^{n+1},$$

for some point c between x and a.

Discussion We make two observations here and outline a proof in Exercise 84. First, the case $n = 0$ is the Mean Value Theorem (Section 4.6). This theorem states that the average slope of the curve $y = f(x)$ over an interval equals the slope of the line tangent to the curve at some point on the interval:

$$\frac{f(x) - f(a)}{x - a} = f'(c),$$

where c is between x and a. Rearranging this expression we have

$$f(x) = \underbrace{f(a)}_{p_0(x)} + \underbrace{f'(c)(x - a)}_{R_0(x)}$$

$$= p_0(x) + R_0(x),$$

which is Taylor's Theorem with $n = 0$. Not surprisingly, the term $f^{(n+1)}(c)$ in Taylor's Theorem comes from a Mean Value Theorem argument.

The second observation makes the remainder term easier to remember. If you write the $(n + 1)$st Taylor polynomial p_{n+1}, the highest-degree term is $\dfrac{f^{(n+1)}(a)}{(n + 1)!}(x - a)^{n+1}$. Replacing $f^{(n+1)}(a)$ by $f^{(n+1)}(c)$ results in the remainder term for p_n.

Estimating the Remainder

The remainder has both practical and theoretical importance. We deal with practical matters now and theoretical matters in Section 9.3. The remainder term is used to estimate errors in approximations and to determine the number of terms of a Taylor polynomial needed to achieve a prescribed accuracy.

Because c is generally unknown, the difficulty in estimating of the remainder is finding a bound for $|f^{(n+1)}(c)|$. Assuming this can be done, the following theorem gives a standard estimate for the remainder term.

THEOREM 9.2   Estimate of the Remainder

Let n be a fixed positive integer. Suppose there exists a number M such that $|f^{(n+1)}(c)| \le M$ for all c between a and x inclusive. The remainder in the nth-order Taylor polynomial for f centered at a satisfies

$$|R_n(x)| = |f(x) - p_n(x)| \le M \frac{|x - a|^{n+1}}{(n + 1)!}.$$

Proof The proof requires taking the absolute value of the remainder term in Theorem 9.1, replacing $|f^{(n+1)}(c)|$ by a larger quantity M, and forming an inequality. ◄

EXAMPLE 5   Estimating the remainder for $\cos x$ Find a bound for the magnitude of the remainder term for the Taylor polynomials of $f(x) = \cos x$ centered at 0.

SOLUTION According to Theorem 9.1 with $a = 0$, we have

$$R_n(x) = \frac{f^{(n+1)}(c)}{(n + 1)!} x^{n+1},$$

where c is between 0 and x. Notice that $f^{(n+1)}(c) = \pm \sin c$ or $f^{(n+1)}(c) = \pm \cos c$. In all cases, $|f^{(n+1)}(c)| \le 1$. Therefore, we take $M = 1$ in Theorem 9.2, and the absolute value of the remainder term can be bounded as

$$|R_n(x)| = \left| \frac{f^{(n+1)}(c)}{(n + 1)!} x^{n+1} \right| \le \frac{|x|^{n+1}}{(n + 1)!}.$$

For example, if we approximate $\cos(0.1)$ using the Taylor polynomial p_{10}, the maximum error satisfies

$$|R_{10}(0.1)| \leq \frac{0.1^{11}}{11!} \approx 2.5 \times 10^{-19}.$$

Related Exercises 41–46 ◄

EXAMPLE 6   Estimating the remainder for e^x   Estimate the error in approximating $e^{0.45}$ using the Taylor polynomial of order $n = 6$ for $f(x) = e^x$ centered at 0.

SOLUTION   By Taylor's Theorem with $a = 0$, we have

$$R_n(x) = \frac{f^{(n+1)}(c)}{(n+1)!} x^{n+1},$$

where c is between 0 and x. Because $f^{(k)}(x) = e^x$, for $k = 0, 1, 2, \ldots,$ $f^{(n+1)}(c) = e^c$ and the remainder term is

$$R_n(x) = \frac{e^c}{(n+1)!} x^{n+1}.$$

If we wish to approximate e^x for $x = 0.45$, then $0 < c < x = 0.45$. Because e^c is an increasing function, $e^c < e^{0.45}$. Assuming that $e^{0.45}$ cannot be evaluated exactly (it is the number we are approximating), it must be bounded above by a number M. A conservative bound is obtained by noting that $e^{0.45} < e^{1/2} < 4^{1/2} = 2$. So, if we take $M = 2$, the maximum error satisfies

$$|R_6(0.45)| < 2\frac{0.45^7}{7!} \approx 1.5 \times 10^{-6}.$$

> Recall that if $f(x) = e^x$, then
> $$p_n(x) = \sum_{k=0}^{n} \frac{x^k}{k!}.$$

Using the Taylor polynomial derived in Example 3 with $n = 6$, the resulting approximation to $e^{0.45}$ is

$$p_6(0.45) = \sum_{k=0}^{6} \frac{0.45^k}{k!} \approx 1.5683114;$$

QUICK CHECK 6 In Example 6, give an approximate upper bound for $R_7(0.45)$. ◄

it has an error that does not exceed 1.5×10^{-6}.   *Related Exercises 47–52* ◄

EXAMPLE 7   Maximum error   The nth-order Taylor polynomial for $f(x) = \ln(1 - x)$ centered at 0 is

$$p_n(x) = -\sum_{k=1}^{n} \frac{x^k}{k} = -x - \frac{x^2}{2} - \frac{x^3}{3} - \cdots - \frac{x^n}{n}.$$

a. What is the maximum error in approximating $\ln(1 - x)$ by $p_3(x)$ for values of x in the interval $\left[-\frac{1}{2}, \frac{1}{2}\right]$?

b. How many terms of the Taylor polynomial are needed to approximate values of $f(x) = \ln(1 - x)$ with an error less than 10^{-3} on the interval $\left[-\frac{1}{2}, \frac{1}{2}\right]$?

SOLUTION

a. The remainder for the Taylor polynomial p_3 is $R_3(x) = \dfrac{f^{(4)}(c)}{4!} x^4$, where c is between 0 and x. Computing four derivatives of f, we find that $f^{(4)}(x) = -\dfrac{6}{(1-x)^4}$. On the interval $\left[-\frac{1}{2}, \frac{1}{2}\right]$, the maximum magnitude of this derivative occurs at $x = \frac{1}{2}$ (because the denominator is smallest at $x = \frac{1}{2}$) and is $6/\left(\frac{1}{2}\right)^4 = 96$. Similarly, the factor x^4 has its maximum magnitude at $x = \pm\frac{1}{2}$ and it is $\left(\frac{1}{2}\right)^4 = \frac{1}{16}$. Therefore, $|R_3(x)| \leq \dfrac{96}{4!} \cdot \left(\dfrac{1}{16}\right) = 0.25$ on the interval $\left[-\frac{1}{2}, \frac{1}{2}\right]$. The error in approximating $f(x)$ by $p_3(x)$ for $-\frac{1}{2} \leq x \leq \frac{1}{2}$ does not exceed 0.25.

b. For any positive integer n, the remainder is $R_n(x) = \dfrac{f^{(n+1)}(c)}{(n+1)!}x^{n+1}$. Differentiating f several times reveals that:

$$f^{(n+1)}(x) = -\frac{n!}{(1-x)^{n+1}}.$$

On the interval $\left[-\frac{1}{2}, \frac{1}{2}\right]$, the maximum magnitude of this derivative occurs at $x = \frac{1}{2}$ and is $n!/\left(\frac{1}{2}\right)^{n+1}$. Similarly, x^{n+1} has its maximum magnitude at $x = \pm\frac{1}{2}$, and it is $\left(\frac{1}{2}\right)^{n+1}$. Therefore, a bound on the remainder is

$$|R_n(x)| \le \frac{n!\,2^{n+1}}{(n+1)!}\frac{1}{2^{n+1}} = \frac{1}{n+1}.$$

To ensure that the error is less than 10^{-3} on the entire interval $\left[-\frac{1}{2}, \frac{1}{2}\right]$, n must satisfy $|R_n| \le \dfrac{1}{n+1} < 10^{-3}$ or $n > 999$. This error is likely to be significantly less than 10^{-3} if x is near 0.

Related Exercises 53–64 ◄

SECTION 9.1 EXERCISES

Review Questions

1. Suppose you use a Taylor polynomial with $n = 2$ centered at 0 to approximate a function f. What matching conditions are satisfied by the polynomial?

2. Does the accuracy of a Taylor polynomial generally increase or decrease with the order of the polynomial? Explain.

3. The first three Taylor polynomials for $f(x) = \sqrt{1+x}$ centered at 0 are $p_0 = 1$, $p_1 = 1 + \dfrac{x}{2}$, and $p_2 = 1 + \dfrac{x}{2} - \dfrac{x^2}{8}$. Find three approximations to $\sqrt{1.1}$.

4. In general, how many terms do the Taylor polynomials p_2 and p_3 have in common?

5. How is the remainder in a Taylor polynomial defined?

6. Explain how to estimate the remainder in an approximation given by a Taylor polynomial.

Basic Skills

7–12. Linear and quadratic approximation

a. Find the linear approximating polynomial for the following functions centered at the given point a.

b. Find the quadratic approximating polynomial for the following functions centered at the given point a.

c. Use the polynomials obtained in parts (a) and (b) to approximate the given quantity.

7. $f(x) = e^{-x}, a = 0$; approximate $e^{-0.2}$.

8. $f(x) = \sqrt{x}, a = 4$; approximate $\sqrt{3.9}$.

9. $f(x) = (1 + x)^{-1}, a = 0$; approximate $1/1.05$.

10. $f(x) = \cos x, a = \pi/4$; approximate $\cos(0.24\pi)$.

11. $f(x) = x^{1/3}, a = 8$; approximate $7.5^{1/3}$.

12. $f(x) = \tan^{-1} x, a = 0$; approximate $\tan^{-1} 0.1$.

13–20. Taylor polynomials

a. Find the nth-order Taylor polynomials of the given function centered at 0 for $n = 0, 1$, and 2.

b. Graph the Taylor polynomials and the function.

13. $f(x) = \cos x$

14. $f(x) = e^{-x}$

15. $f(x) = \ln(1 - x)$

16. $f(x) = (1 + x)^{-1/2}$

17. $f(x) = \tan x$

18. $f(x) = (1 + x)^{-2}$

19. $f(x) = (1 + x)^{-3}$

20. $f(x) = \sin^{-1} x$

21–26. Approximations with Taylor polynomials

a. Use the given Taylor polynomial p_2 to approximate the given quantity.

b. Compute the absolute error in the approximation assuming the exact value is given by a calculator.

21. Approximate $\sqrt{1.05}$ using $f(x) = \sqrt{1+x}$ and $p_2(x) = 1 + x/2 - x^2/8$.

22. Approximate $\sqrt[3]{1.1}$ using $f(x) = \sqrt[3]{1+x}$ and $p_2(x) = 1 + x/3 - x^2/9$.

23. Approximate $\dfrac{1}{\sqrt{1.08}}$ using $f(x) = \dfrac{1}{\sqrt{1+x}}$ and $p_2(x) = 1 - x/2 + 3x^2/8$.

24. Approximate $\ln 1.06$ using $f(x) = \ln(1 + x)$ and $p_2(x) = x - x^2/2$.

25. Approximate $e^{-0.15}$ using $f(x) = e^{-x}$ and $p_2(x) = 1 - x + x^2/2$.

26. Approximate $\dfrac{1}{1.12^3}$ using $f(x) = \dfrac{1}{(1+x)^3}$ and $p_2(x) = 1 - 3x + 6x^2$.

27–32. Taylor polynomials centered at $a \neq 0$

    a. *Find the nth-order Taylor polynomials for the given function centered at the given point a for n = 0, 1, and 2.*

    b. *Graph the Taylor polynomials and the function.*

27. $f(x) = \sin x, a = \pi/4$ **28.** $f(x) = \cos x, a = \pi/6$

29. $f(x) = \sqrt{x}, a = 9$ **30.** $f(x) = \sqrt[3]{x}, a = 8$

31. $f(x) = \ln x, a = e$ **32.** $f(x) = \sqrt[4]{x}, a = 16$

33–40. Approximations with Taylor polynomials

    a. *Approximate the given quantities using Taylor polynomials with n = 3.*

    b. *Compute the absolute error in the approximation assuming the exact value is given by a calculator.*

33. $e^{0.12}$ **34.** $\cos(-0.2)$ **35.** $\tan(-0.1)$

36. $\ln(1.05)$ **37.** $\sqrt{1.06}$ **38.** $\sqrt[4]{79}$

39. $\sqrt{101}$ **40.** $\sqrt[3]{126}$

41–46. Remainder terms *Find the remainder term R_n in the nth order Taylor polynomial centered at a for the given functions. Express the result for a general value of n.*

41. $f(x) = \sin x; \ a = 0$ **42.** $f(x) = \cos 2x; \ a = 0$

43. $f(x) = e^{-x}; \ a = 0$ **44.** $f(x) = \cos x; \ a = \pi/2$

45. $f(x) = \sin x; \ a = \pi/2$ **46.** $f(x) = 1/(1-x); \ a = 0$

47–52. Estimating errors *Use the remainder term to estimate the absolute error in approximating the following quantities with the nth-order Taylor polynomial centered at 0. Estimates are not unique.*

47. $\sin 0.3; \ n = 4$ **48.** $\cos 0.45; \ n = 3$

49. $e^{0.25}; \ n = 4$ **50.** $\tan 0.3; \ n = 2$

51. $e^{-0.5}; \ n = 4$ **52.** $\ln 1.04; \ n = 3$

53–58. Maximum error *Use the remainder term to estimate the maximum error in the following approximations on the given interval. Error bounds are not unique.*

53. $\sin x \approx x - x^3/6; \ [-\pi/4, \pi/4]$

54. $\cos x \approx 1 - x^2/2; \ [-\pi/4, \pi/4]$

55. $e^x \approx 1 + x + x^2/2; \ \left[-\frac{1}{2}, \frac{1}{2}\right]$

56. $\tan x \approx x; \ [-\pi/6, \pi/6]$

57. $\ln(1 + x) \approx x - x^2/2; \ [-0.2, 0.2]$

58. $\sqrt{1 + x} \approx 1 + x/2; \ [-0.1, 0.1]$

59–64. Number of terms *What is the minimum order of the Taylor polynomial required to approximate the following quantities with an absolute error no greater than 10^{-3}? (The answer depends on your choice of a center.)*

59. $e^{-0.5}$ **60.** $\sin 0.2$ **61.** $\cos(-0.25)$

62. $\ln 0.85$ **63.** $\sqrt{1.06}$ **64.** $1/\sqrt{0.85}$

Further Explorations

65. Explain why or why not Determine whether the following statements are true and give an explanation or counterexample.

    a. The Taylor polynomials for $f(x) = e^{-2x}$ consist of even powers only.

    b. For $f(x) = x^5 - 1$, the Taylor polynomial of order 10 centered at $x = 0$ is f itself.

    c. The nth-order Taylor polynomial for $f(x) = \sqrt{1 + x^2}$ centered at 0 consists of even powers of x only.

66. Taylor coefficients for $x = a$ Follow the procedure in the text to show that the nth-order Taylor polynomial that matches f and its derivatives up to order n at a has coefficients

$$c_k = \frac{f^{(k)}(a)}{k!} \text{ for } k = 0, 1, 2, \ldots, n.$$

67. Matching functions with polynomials Match the following six functions with the given six Taylor polynomials of order 2. Give reasons for your choices.

    a. $\sqrt{1 + 2x}$ **A.** $p_2(x) = 1 + 2x + 2x^2$

    b. $\dfrac{1}{\sqrt{1 + 2x}}$ **B.** $p_2(x) = 1 - 6x + 24x^2$

    c. e^{2x} **C.** $p_2(x) = 1 + x - \dfrac{x^2}{2}$

    d. $\dfrac{1}{1 + 2x}$ **D.** $p_2(x) = 1 - 2x + 4x^2$

    e. $\dfrac{1}{(1 + 2x)^3}$ **E.** $p_2(x) = 1 - x + \dfrac{3}{2}x^2$

    f. e^{-2x} **F.** $p_2(x) = 1 - 2x + 2x^2$

68. Dependence of errors on x Consider $f(x) = \ln(1 - x)$ and its Taylor polynomials given in Example 7.

    a. Graph $y = |f(x) - p_2(x)|$ and $y = |f(x) - p_3(x)|$ on the interval $\left[-\frac{1}{2}, \frac{1}{2}\right]$ (two curves).

    b. At what points of $\left[-\frac{1}{2}, \frac{1}{2}\right]$ is the error largest? Smallest?

    c. Are these results consistent with the theoretical error bounds obtained in Example 7?

Applications

69–76. Small argument approximations *Consider the following common approximations when x is near zero.*

    a. *Estimate f(0.1) and give the maximum error in the approximation.*

    b. *Estimate f(0.2) and give the maximum error in the approximation.*

69. $f(x) = \sin x \approx x$ **70.** $f(x) = \tan x \approx x$

71. $f(x) = \cos x \approx 1 - x^2/2$

72. $f(x) = \tan^{-1} x \approx x$

73. $f(x) = \sqrt{1 + x} \approx 1 + x/2$

74. $f(x) = \ln(1 + x) \approx x - x^2/2$

75. $f(x) = e^x \approx 1 + x$

76. $f(x) = \sin^{-1} x \approx x$

77. Errors in approximations Suppose you approximate $\sin x$ at the points $x = -0.2, -0.1, 0.0, 0.1, 0.2$ using the Taylor polynomials $p_3 = x - x^3/6$ and $p_5 = x - x^3/6 + x^5/120$. Assume that the exact value of $\sin x$ is given by a calculator.

a. Complete the table showing the absolute errors in the approximations at each point. Show two significant digits.

x	Error $= |\sin x - p_3(x)|$	Error $= |\sin x - p_5(x)|$
-0.2		
-0.1		
0.0		
0.1		
0.2		

b. In each error column, how do the errors vary with x? For what values of x are the errors the largest and smallest in magnitude?

78–81. Errors in approximations *Carry out the procedure described in Exercise 77 with the following functions and approximations.*

78. $f(x) = \cos x$, $p_2(x) = 1 - \dfrac{x^2}{2}$, $p_4(x) = 1 - \dfrac{x^2}{2} + \dfrac{x^4}{24}$

79. $f(x) = e^{-x}$, $p_1(x) = 1 - x$, $p_2(x) = 1 - x + \dfrac{x^2}{2}$

80. $f(x) = \ln(1 + x)$, $p_1(x) = x$, $p_2(x) = x - \dfrac{x^2}{2}$

81. $f(x) = \tan x$, $p_1(x) = x$, $p_3(x) = x + \dfrac{x^3}{3}$

82. Best expansion point Suppose you wish to approximate $\cos(\pi/12)$ using Taylor polynomials. Is the approximation more accurate if you use Taylor polynomials centered at 0 or $\pi/6$? Use a calculator for numerical experiments and check for consistency with Theorem 9.2. Does the answer depend on the order of the polynomial?

83. Best expansion point Suppose you wish to approximate $e^{0.35}$ using Taylor polynomials. Is the approximation more accurate if you use Taylor polynomials centered at 0 or $\ln 2$? Use a calculator for numerical experiments and check for consistency with Theorem 9.2. Does the answer depend on the order of the polynomial?

Additional Exercises

84. Proof of Taylor's Theorem There are several proofs of Taylor's Theorem, which lead to various forms of the remainder. The following proof is instructive because it leads to two different forms of the remainder and it relies on the Fundamental Theorem of Calculus, integration by parts, and the Integral Mean Value Theorem. Assume that f has at least $n + 1$ continuous derivatives on an interval containing a.

a. Show that the Fundamental Theorem of Calculus can be written in the form

$$f(x) = f(a) + \int_a^x f'(t)\, dt.$$

b. Use integration by parts ($u = f'(t)$, $dv = dt$) to show that

$$f(x) = f(a) + (x - a)f'(a) + \int_a^x (x - t)f''(t)\, dt.$$

c. Show that n integrations by parts gives

$$f(x) = f(a) + \frac{f'(a)}{1!}(x - a) + \frac{f''(a)}{2!}(x - a)^2 + \cdots$$
$$+ \frac{f^{(n)}(a)}{n!}(x - a)^n + \underbrace{\int_a^x \frac{f^{(n+1)}(t)}{n!}(x - t)^n\, dt.}_{R_n(x)}$$

d. The result in part (c) looks like $f(x) = p_n(x) + R_n(x)$, where p_n is the nth-order Taylor polynomial and R_n is a new form of the remainder term, known as the integral form of the remainder term. Use the Integral Mean Value Theorem to show that R_n can be expressed in the form

$$R_n(x) = \frac{f^{(n+1)}(c)}{(n + 1)!}(x - a)^{n+1},$$

where c is between a and x.

85. Tangent line is p_1 Let f be differentiable at $x = a$.

a. Find the equation of the line tangent to the curve $y = f(x)$ at $(a, f(a))$.
b. Find the Taylor polynomial p_1 centered at a and confirm that it describes the tangent line found in part (a).

86. Local extreme points and inflection points Suppose that f has two continuous derivatives at a.

a. Show that if f has a local maximum at a, then the Taylor polynomial p_2 centered at a also has a local maximum at a.
b. Show that if f has a local minimum at a, then the Taylor polynomial p_2 centered at a also has a local minimum at a.
c. Is it true that if f has an inflection point at a, then the Taylor polynomial p_2 centered at a also has an inflection point at a?
d. Are the converses to parts (a) and (b) true? If p_2 has a local extreme point at a, does f have the same type of point at a?

QUICK CHECK ANSWERS

3. $f(x) = \sin x$ is an odd function, and its even-ordered derivatives are zero at 0, so its Taylor polynomials are also odd functions. **4.** $p_4(x) = p_3(x) + \dfrac{x^4}{4!}$; $p_5(x) = p_4(x) + \dfrac{x^5}{5!}$

5. $x = 49$ and $x = 16$ are good choices. **6.** Because $e^{0.45} < 2$, $|R_7(0.45)| < 2\dfrac{0.45^8}{8!} \approx 8.3 \times 10^{-8}$. ◄

9.2 Properties of Power Series

The preceding section demonstrated that Taylor polynomials provide accurate approximations to many functions and that, in general, the approximations improve as we let the degree of the polynomials increase. In this section, we take the next step and let the degree of the Taylor polynomials increase without bound to produce a *power series.*

Geometric Series as Power Series

A good way to become familiar with power series is to return to *geometric series*, first encountered in Section 8.3. Recall that for a fixed number r,

$$\sum_{k=0}^{\infty} r^k = 1 + r + r^2 + \cdots = \frac{1}{1-r}, \qquad \text{provided } |r| < 1.$$

It's a small change to replace the real number r by the variable x. In doing so, the geometric series becomes a new representation of a familiar function:

$$\sum_{k=0}^{\infty} x^k = 1 + x + x^2 + \cdots = \frac{1}{1-x}, \qquad \text{provided } |x| < 1.$$

This infinite series is a *power series.* Notice that while $1/(1-x)$ is defined for $\{x: x \neq 1\}$, its power series converges only for $|x| < 1$. The set of values for which a power series converges is called its *interval of convergence.*

Power series are used to represent familiar functions such as trigonometric, exponential, and logarithmic functions. They are also used to define new functions. For example, consider the function defined by

$$g(x) = \sum_{k=1}^{\infty} \frac{(-1)^k k}{4^k} x^{2k}.$$

The term *function* is used advisedly because it's not yet clear whether g really is a function. If so, is it a continuous function? Does it have a derivative? Judging by its graph (Figure 9.11), g appears to be a rather ordinary continuous function (which is identified at the end of the chapter).

In fact, power series satisfy the defining property of all functions: For each value of x, a power series has at most one value. For this reason we refer to a power series as a function, although the domain, properties, and identity of the function may need to be discovered.

▶ Figure 9.11 shows an approximation to the graph of g made by summing the first 500 terms of the power series at selected values of x on the interval $(-2, 2)$.

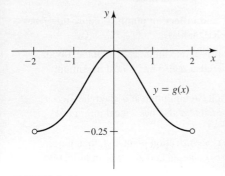

FIGURE 9.11

Convergence of Power Series

We begin by establishing the terminology of power series.

> **QUICK CHECK 1** By substituting $x = 0$ in the power series for g, evaluate $g(0)$ for the function in Figure 9.11. ◀

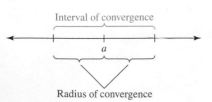

FIGURE 9.12

DEFINITION   Power Series

A **power series** has the general form

$$\sum_{k=0}^{\infty} c_k (x-a)^k,$$

where a and c_k are real numbers, and x is a variable. The c_k's are the **coefficients** of the power series and a is the **center** of the power series. The set of values of x for which the series converges is the **interval of convergence**. The **radius of convergence** of the power series, denoted R, is the distance from the center of the series to the boundary of the interval of convergence (Figure 9.12).

> By the Ratio Test, $\sum |a_k|$ converges if
>
> $$r = \lim_{k \to \infty} \left| \frac{a_{k+1}}{a_k} \right| < 1,$$
>
> it diverges if $r > 1$, and the test is inconclusive if $r = 1$.

> By the Root Test, $\sum |a_k|$ converges if
>
> $\rho = \lim_{k \to \infty} \sqrt[k]{|a_k|} < 1$, it diverges
>
> if $\rho > 1$, and the test is inconclusive if $\rho = 1$.

How do we determine the interval of convergence? The presence of the terms x^k or $(x - a)^k$ in a power series suggests using the Ratio Test or the Root Test. Furthermore, because these terms could be positive or negative, we test a power series for absolute convergence. By Theorem 8.21, if we determine the values of x for which the series converges absolutely, we have a set of values for which the series converges.

Recall that a series $\sum a_k$ converges absolutely if the series $\sum |a_k|$ converges. The following examples illustrate how the Ratio and Root Tests are used to determine the interval and radius of convergence.

EXAMPLE 1   Interval and radius of convergence Find the interval and radius of convergence for each power series.

a. $\displaystyle\sum_{k=0}^{\infty} \frac{x^k}{k!}$ **b.** $\displaystyle\sum_{k=0}^{\infty} \frac{(-1)^k (x - 2)^k}{4^k}$ **c.** $\displaystyle\sum_{k=1}^{\infty} k! \, x^k$

SOLUTION

a. The center of the power series is 0 and the terms of the series are $x^k / k!$. We test the series for absolute convergence using the Ratio Test:

$$\begin{aligned}
r &= \lim_{k \to \infty} \frac{|x^{k+1}/(k+1)!|}{|x^k/k!|} && \text{Ratio Test} \\
&= \lim_{k \to \infty} \frac{|x|^{k+1}}{|x|^k} \cdot \frac{k!}{(k+1)!} && \text{Invert and multiply.} \\
&= |x| \lim_{k \to \infty} \frac{1}{k+1} = 0 && \text{Simplify and take the limit with } x \text{ fixed.}
\end{aligned}$$

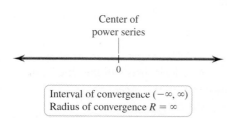

Center of power series

0

Interval of convergence $(-\infty, \infty)$
Radius of convergence $R = \infty$

FIGURE 9.13

Notice that in taking the limit as $k \to \infty$, x is held fixed. Therefore, $r = 0$ for all values of x, which implies that the interval of convergence of the power series is $-\infty < x < \infty$ (Figure 9.13) and the radius of convergence is $R = \infty$.

b. We test for absolute convergence using the Root Test:

$$\rho = \lim_{k \to \infty} \sqrt[k]{\left| \frac{(-1)^k (x - 2)^k}{4^k} \right|} = \frac{|x - 2|}{4}$$

In this case, ρ depends on the value of x. For absolute convergence, x must satisfy

$$\rho = \frac{|x - 2|}{4} < 1,$$

which implies that $|x - 2| < 4$. Using standard techniques for solving inequalities, the solution set is $-4 < x - 2 < 4$, or $-2 < x < 6$. Thus, the interval of convergence includes $(-2, 6)$.

The Root Test does not give information about convergence at the endpoints, $x = -2$ and $x = 6$, because at these points, the Root Test results in $\rho = 1$. To test for convergence at the endpoints, we must substitute each endpoint into the series and carry out separate tests. At $x = -2$, the power series becomes

$$\sum_{k=0}^{\infty} \frac{(-1)^k (x - 2)^k}{4^k} = \sum_{k=0}^{\infty} \frac{4^k}{4^k} \qquad \text{Substitute } x = -2 \text{ and simplify.}$$

$$= \sum_{k=0}^{\infty} 1 \qquad \text{Diverges by Divergence Test.}$$

▶ The Ratio and Root Tests determine the radius of convergence conclusively. However, the interval of convergence is not determined until the endpoints are tested.

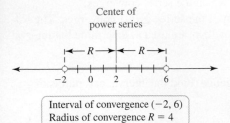

Interval of convergence $(-2, 6)$
Radius of convergence $R = 4$

FIGURE 9.14

The series clearly diverges at the left endpoint. At $x = 6$, the power series is

$$\sum_{k=0}^{\infty} \frac{(-1)^k(x-2)^k}{4^k} = \sum_{k=0}^{\infty}(-1)^k \frac{4^k}{4^k} \quad \text{Substitute } x = 6 \text{ and simplify.}$$

$$= \sum_{k=0}^{\infty}(-1)^k \quad \text{Diverges by Divergence Test.}$$

This series also diverges at the right endpoint. Therefore, the interval of convergence is $(-2, 6)$, excluding the endpoints (Figure 9.14) and the radius of convergence is $R = 4$.

QUICK CHECK 2 Explain why the power series in Example 1b diverges if $x > 6$ or $x < -2$. ◄

c. To test for absolute convergence we use the Ratio Test:

$$r = \lim_{k\to\infty} \frac{|(k+1)! \, x^{k+1}|}{|k! \, x^k|} \quad \text{Ratio Test}$$

$$= |x| \lim_{k\to\infty} \frac{(k+1)!}{k!} \quad \text{Simplify.}$$

$$= |x| \lim_{k\to\infty} (k+1) \quad \text{Simplify.}$$

$$= \infty \quad \text{If } x \neq 0$$

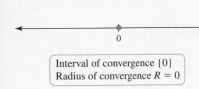

Interval of convergence $\{0\}$
Radius of convergence $R = 0$

FIGURE 9.15

The only way to satisfy $r < 1$ is to take $x = 0$, in which case the power series has a value of 0. The interval of convergence of the power series consists of the single point $x = 0$ (Figure 9.15) and the radius of convergence is $R = 0$.   *Related Exercises 9–20* ◄

Example 1 illustrates the three common types of intervals of convergence, which are summarized in the following theorem (see Appendix B for a proof).

THEOREM 9.3   Convergence of Power Series

A power series $\sum_{k=0}^{\infty} c_k(x-a)^k$ centered at a converges in one of three ways:

1. The series coverges absolutely for all x, in which case the interval of convergence is $(-\infty, \infty)$ and the radius of convergence is $R = \infty$.

2. There is a real number $R > 0$ such that the series converges absolutely for $|x - a| < R$ and diverges for $|x - a| > R$, in which case the radius of convergence is R.

3. The series converges only at a, in which case the radius of convergence is $R = 0$.

▶ Theorem 9.3 says nothing about convergence at the endpoints. For example, the radius of convergence is 2 for the intervals of convergence $(2, 6)$, $(2, 6]$, $[2, 6)$, or $[2, 6]$.

QUICK CHECK 3 What are the interval and radius of convergence of the geometric series $\sum x^k$? ◄

EXAMPLE 2   Interval and radius of convergence Use the Ratio Test to find the radius and interval of convergence of $\sum_{k=1}^{\infty} \frac{(x-2)^k}{\sqrt{k}}$.

SOLUTION
$$r = \lim_{k\to\infty} \frac{|(x-2)^{k+1}/\sqrt{k+1}|}{|(x-2)^k/\sqrt{k}|} \quad \text{Ratio Test}$$

$$= |x-2| \lim_{k\to\infty} \sqrt{\frac{k}{k+1}} \quad \text{Simplify.}$$

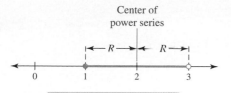

Center of power series

Interval of convergence [1, 3)
Radius of convergence $R = 1$

FIGURE 9.16

$$= |x - 2| \sqrt{\underbrace{\lim_{k \to \infty} \frac{k}{k + 1}}_{1}} \qquad \text{Limit Law}$$

$$= |x - 2| \qquad \text{Limit equals 1.}$$

The series converges absolutely for all x such that $r < 1$, which implies $|x - 2| < 1$, or $1 < x < 3$. Therefore, the radius of convergence is 1 (Figure 9.16).

We now test the endpoints. Substituting $x = 1$ into the power series, we have

$$\sum_{k=1}^{\infty} \frac{(x - 2)^k}{\sqrt{k}} = \sum_{k=1}^{\infty} \frac{(-1)^k}{\sqrt{k}}.$$

This series converges by the Alternating Series Test (the terms of the series decrease in magnitude and approach 0 as $k \to \infty$). Substituting $x = 3$ into the power series, we have

$$\sum_{k=1}^{\infty} \frac{(x - 2)^k}{\sqrt{k}} = \sum_{k=1}^{\infty} \frac{1}{\sqrt{k}},$$

which is a divergent p-series. We conclude that the interval of convergence is $1 \le x < 3$ (Figure 9.16).

Related Exercises 9–20 ◄

Combining Power Series

A power series defines a function on its interval of convergence. When power series are combined algebraically, new functions are defined. The following theorem, stated without proof, gives three common ways to combine power series.

> ➤ New power series can also be defined as the product and quotient of power series. The calculation of the coefficients of such series is more challenging (Exercise 65).

> ➤ Theorem 9.4 also applies to power series centered at points other than $x = 0$. Property 1 applies directly; Properties 2 and 3 apply with slight modifications.

THEOREM 9.4   Combining Power Series

Suppose the power series $\sum c_k x^k$ and $\sum d_k x^k$ converge absolutely to $f(x)$ and $g(x)$, respectively, on an interval I.

1. **Sum and difference:** The power series $\sum (c_k \pm d_k) x^k$ converges absolutely to $f(x) \pm g(x)$ on I.

2. **Multiplication by a power:** The power series $x^m \sum c_k x^k = \sum c_k x^{k+m}$ converges absolutely to $x^m f(x)$ on I, provided m is an integer such that $k + m \ge 0$ for all terms of the series.

3. **Composition:** If $h(x) = bx^m$, where m is a positive integer and b is a real number, the power series $\sum c_k (h(x))^k$ converges absolutely to the composite function $f(h(x))$ for all x such that $h(x)$ is in I.

EXAMPLE 3   Combining power series Given the geometric series

$$\frac{1}{1 - x} = \sum_{k=0}^{\infty} x^k = 1 + x + x^2 + x^3 + \cdots \qquad \text{for } |x| < 1,$$

find the power series and interval of convergence for the following functions.

a. $\dfrac{x^5}{1 - x}$ **b.** $\dfrac{1}{1 - 2x}$ **c.** $\dfrac{1}{1 + x^2}$

SOLUTION

a.
$$\frac{x^5}{1-x} = x^5(1 + x + x^2 + \cdots) \quad \text{Theorem 9.4, Property 2}$$

$$= x^5 + x^6 + x^7 + \cdots$$

$$= \sum_{k=0}^{\infty} x^{k+5}$$

This geometric series has a ratio $r = x$ and converges when $|r| = |x| < 1$. The interval of convergence is $|x| < 1$.

b. We substitute $2x$ for x in the power series for $\dfrac{1}{1-x}$:

$$\frac{1}{1-2x} = 1 + (2x) + (2x)^2 + \cdots \quad \text{Theorem 9.4, Property 3}$$

$$= 1 + 2x + 4x^2 + \cdots$$

$$= \sum_{k=0}^{\infty} (2x)^k$$

This geometric series has a ratio $r = 2x$ and converges provided $|r| = |2x| < 1$ or $|x| < \frac{1}{2}$. The interval of convergence is $|x| < \frac{1}{2}$.

c. We substitute $-x^2$ for x in the power series for $\dfrac{1}{1-x}$:

$$\frac{1}{1+x^2} = 1 + (-x^2) + (-x^2)^2 + \cdots \quad \text{Theorem 9.4, Property 3}$$

$$= 1 - x^2 + x^4 - \cdots$$

$$= \sum_{k=0}^{\infty} (-1)^k x^{2k}$$

This geometric series has a ratio of $r = -x^2$ and converges provided $|r| = |-x^2| = |x^2| < 1$ or $|x| < 1$.   *Related Exercises 21–32* ◄

Differentiating and Integrating Power Series

Some properties of polynomials carry over to power series, but others do not. For example, a polynomial is defined for all values of x, whereas a power series is defined only on its interval of convergence. In general, the properties of polynomials carry over to power series when the power series is restricted to its interval of convergence. The following result illustrates this principle.

> ▶ Theorem 9.5 makes no claim about the convergence of the differentiated or integrated series at the endpoints of the interval of convergence.

THEOREM 9.5 Differentiating and Integrating Power Series

Let the function f be defined by the power series $\sum c_k (x - a)^k$ on its interval of convergence I.

1. f is a continuous function on I.

2. The power series may be differentiated or integrated term by term, and the resulting power series converges to $f'(x)$ or $\int f(x)\, dx + C$, respectively, at all points in the interior of I, where C is an arbitrary constant.

These results are powerful and also deep mathematically. Their proofs require advanced ideas and are omitted. However, some discussion is in order before turning to examples.

The statements in Theorem 9.5 about term-by-term differentiation and integration say two things: The differentiated and integrated power series converge, provided x belongs to the interior of the interval of convergence. But the theorem claims more than convergence. According to the theorem, the differentiated and integrated power series converge to the derivative and indefinite integral of f, respectively, on the interior of the interval of convergence.

EXAMPLE 4 Differentiating and integrating power series Consider the geometric series

$$f(x) = \frac{1}{1-x} = \sum_{k=0}^{\infty} x^k = 1 + x + x^2 + x^3 + \cdots \quad \text{for } |x| < 1.$$

a. Differentiate this series term by term to find the power series for f', and identify the function it represents.

b. Integrate this series term by term and identify the function it represents.

SOLUTION

a. We know that $f'(x) = (1-x)^{-2}$. Differentiating the series, we find that

$$f'(x) = \frac{d}{dx}(1 + x + x^2 + x^3 + \cdots) \qquad \text{Differentiate the power series for } f.$$

$$= 1 + 2x + 3x^2 + \cdots \qquad \text{Differentiate term by term.}$$

$$= \sum_{k=0}^{\infty} (k+1)\,x^k. \qquad \text{Summation notation}$$

Therefore, on the interval $|x| < 1$,

$$f'(x) = (1-x)^{-2} = \sum_{k=0}^{\infty} (k+1)\,x^k.$$

Substituting $x = \pm 1$ into the power series for f' reveals that the series diverges at both endpoints.

b. Integrating f and integrating the power series term by term, we have

$$\int \frac{dx}{1-x} = \int (1 + x + x^2 + x^3 + \cdots)\,dx,$$

which implies that

$$-\ln|1-x| = x + \frac{x^2}{2} + \frac{x^3}{3} + \frac{x^4}{4} + \cdots + C,$$

where C is an arbitrary constant. Notice that the left side is 0 when $x = 0$. The right side is 0 when $x = 0$ provided we choose $C = 0$. Because $|x| < 1$, the absolute value sign on the left side may be removed. Multiplying both sides by -1, we have a representation for $\ln(1-x)$:

$$\ln(1-x) = -x - \frac{x^2}{2} - \frac{x^3}{3} - \frac{x^4}{4} - \cdots = -\sum_{k=1}^{\infty} \frac{x^k}{k}.$$

It is interesting to test the endpoints of the interval $|x| < 1$. When $x = 1$, the series is (a multiple of) the divergent harmonic series, and when $x = -1$, the series is the convergent alternating harmonic series (Section 8.6). So the interval of convergence is $-1 \le x < 1$. Here is a subtle point: Although we know the series converges at $x = -1$, Theorem 9.5 guarantees convergence to $\ln(1 - x)$ only at the interior points. So we cannot use Theorem 9.5 to claim that the series converges to $\ln 2$ at $x = -1$. In fact, it does, as shown in Section 9.3.   *Related Exercises 33–38* ◄

QUICK CHECK 4 Use the result of Example 4 to write a power series representation for $\ln \frac{1}{2} = -\ln 2$. ◄

EXAMPLE 5 **Functions to power series** Find power series representations centered at 0 for the following functions and give their intervals of convergence.

a. $\tan^{-1} x$ **b.** $\ln\left(\dfrac{1 + x}{1 - x}\right)$

SOLUTION In both cases, we work with known power series and use differentiation, integration, and other combinations.

a. The key is to recall that

$$\int \frac{dx}{1 + x^2} = \tan^{-1} x + C$$

and that, by Example 3c,

$$\frac{1}{1 + x^2} = 1 - x^2 + x^4 - \cdots, \qquad \text{provided } |x| < 1.$$

We now integrate both sides of this last expression:

$$\int \frac{dx}{1 + x^2} = \int \left(1 - x^2 + x^4 - \cdots\right) dx,$$

which implies that

$$\tan^{-1} x = x - \frac{x^3}{3} + \frac{x^5}{5} - \cdots + C.$$

Substituting $x = 0$ and noting that $\tan^{-1} 0 = 0$, the two sides of this equation agree provided we choose $C = 0$. Therefore,

$$\tan^{-1} x = x - \frac{x^3}{3} + \frac{x^5}{5} - \cdots = \sum_{k=0}^{\infty} \frac{(-1)^k x^{2k+1}}{2k + 1}.$$

By Theorem 9.5, this power series converges for $|x| < 1$. Testing the endpoints separately, we find that it also converges at $x = \pm 1$. Therefore, the interval of convergence is $[-1, 1]$.

> ➤ Again, Theorem 9.5 does not guarantee that the power series in part (a) converges to $\tan^{-1} x$ at $x = \pm 1$. In fact, it does.

b. We have already seen (Example 4) that

$$\ln(1 - x) = -x - \frac{x^2}{2} - \frac{x^3}{3} - \cdots.$$

> ➤ Nicolaus Mercator (1620–1687) and Sir Isaac Newton (1642–1727) independently derived the power series for $\ln(1 + x)$, which is called the *Mercator series*.

Replacing x by $-x$, we have

$$\ln(1 - (-x)) = \ln(1 + x) = x - \frac{x^2}{2} + \frac{x^3}{3} - \cdots.$$

Subtracting these two power series gives

$$\ln\left(\frac{1+x}{1-x}\right) = \ln(1+x) - \ln(1-x) \quad \text{Properties of logarithms}$$

$$= \underbrace{\left(x - \frac{x^2}{2} + \frac{x^3}{3} - \cdots\right)}_{\ln(1+x)} - \underbrace{\left(-x - \frac{x^2}{2} - \frac{x^3}{3} - \cdots\right)}_{\ln(1-x)}, \quad \text{for } |x| < 1$$

$$= 2\left(x + \frac{x^3}{3} + \frac{x^5}{5} + \cdots\right) \quad \text{Combine, Theorem 9.4.}$$

$$= 2\sum_{k=0}^{\infty} \frac{x^{2k+1}}{2k+1} \quad \text{Summation notation}$$

QUICK CHECK 5 Verify that the power series in Example 5b does not converge at the endpoints $x = \pm 1$. ◄

This power series is the difference of two power series, both of which converge on the interval $|x| < 1$. Therefore, by Theorem 9.4, the new series also converges on $|x| < 1$.

Related Exercises 39–44 ◄

If you look carefully, every example in this section is ultimately based on the geometric series. Using this single series, we were able to develop power series for many other functions. Imagine what we could do with a few more basic power series. The following section accomplishes precisely that end. There, we discover basic power series for all the standard functions of calculus.

SECTION 9.2 EXERCISES

Review Questions

1. Write the first four terms of a power series with coefficients c_0, c_1, c_2, c_3 centered at 0.

2. Write the first four terms of a power series with coefficients c_0, c_1, c_2, c_3 centered at 3.

3. What tests are used to determine the radius of convergence of a power series?

4. Explain why a power series is tested for *absolute* convergence.

5. Do the interval and radius of convergence of a power series change when the series is differentiated or integrated? Explain.

6. What is the radius of convergence of the power series $\sum c_k (x/2)^k$ if the radius of convergence of $\sum c_k x^k$ is R?

7. What is the interval of convergence of the power series $\sum (4x)^k$?

8. How are the radii of convergence of the power series $\sum c_k x^k$ and $\sum (-1)^k c_k x^k$ related?

Basic Skills

9–20. Interval and radius of convergence *Determine the radius of convergence of the following power series. Then test the endpoints to determine the interval of convergence.*

9. $\sum \left(\frac{x}{3}\right)^k$

10. $\sum (-1)^k \frac{x^k}{5^k}$

11. $\sum \frac{x^k}{k^k}$

12. $\sum (-1)^k \frac{k(x-4)^k}{2^k}$

13. $\sum \frac{k^2 x^{2k}}{k!}$

14. $\sum \frac{k^k x^k}{(k+1)!}$

15. $\sum \frac{x^{2k+1}}{3^{k-1}}$

16. $\sum \left(-\frac{x}{10}\right)^{2k}$

17. $\sum \frac{(x-1)^k k^k}{(k+1)^k}$

18. $\sum \frac{(-2)^k (x+3)^k}{3^{k+1}}$

19. $\sum \frac{k^{20} x^k}{(2k+1)!}$

20. $\sum (-1)^k \frac{x^{3k}}{27^k}$

21–26. Combining power series *Use the geometric series*

$$f(x) = \frac{1}{1-x} = \sum_{k=0}^{\infty} x^k, \quad \text{for } |x| < 1,$$

to find the power series representation for the following functions (centered at 0). Give the interval of convergence of the new series.

21. $f(3x) = \dfrac{1}{1-3x}$

22. $g(x) = \dfrac{x^3}{1-x}$

23. $h(x) = \dfrac{2x^3}{1-x}$

24. $f(x^3) = \dfrac{1}{1-x^3}$

25. $p(x) = \dfrac{4x^{12}}{1-x}$

26. $f(-4x) = \dfrac{1}{1+4x}$

27–32. Combining power series *Use the power series representation*

$$f(x) = \ln(1-x) = -\sum_{k=1}^{\infty} \frac{x^k}{k}, \quad \text{for } -1 \le x < 1,$$

to find the power series for the following functions (centered at 0). Give the interval of convergence of the new series.

27. $f(3x) = \ln(1-3x)$

28. $g(x) = x^3 \ln(1-x)$

29. $h(x) = x \ln (1 - x)$

30. $f(x^3) = \ln (1 - x^3)$

31. $p(x) = 2x^6 \ln (1 - x)$

32. $f(-4x) = \ln (1 + 4x)$

33–38. Differentiating and integrating power series *Find the power series representation for g centered at 0 by differentiating or integrating the power series for f (perhaps more than once). Give the interval of convergence for the resulting series.*

33. $g(x) = \dfrac{1}{(1 - x)^2}$ using $f(x) = \dfrac{1}{1 - x}$

34. $g(x) = \dfrac{1}{(1 - x)^3}$ using $f(x) = \dfrac{1}{1 - x}$

35. $g(x) = \dfrac{1}{(1 - x)^4}$ using $f(x) = \dfrac{1}{1 - x}$

36. $g(x) = \dfrac{x}{(1 + x^2)^2}$ using $f(x) = \dfrac{1}{1 + x^2}$

37. $g(x) = \ln (1 - 3x)$ using $f(x) = \dfrac{1}{1 - 3x}$

38. $g(x) = \ln (1 + x^2)$ using $f(x) = \dfrac{x}{1 + x^2}$

39–44. Functions to power series *Find power series representations centered at 0 for the following functions using known power series. Give the interval of convergence for the resulting series.*

39. $f(x) = \dfrac{1}{1 + x^2}$

40. $f(x) = \dfrac{1}{1 - x^4}$

41. $f(x) = \dfrac{3}{3 + x}$

42. $f(x) = \ln \sqrt{1 - x^2}$

43. $f(x) = \ln \sqrt{4 - x^2}$

44. $f(x) = \tan^{-1} (4x^2)$

Further Explorations

45. Explain why or why not Determine whether the following statements are true and give an explanation or counterexample.

  a. The interval of convergence of the power series $\sum c_k (x - 3)^k$ could be $(-2, 8)$.

  b. $\sum (-2x)^k$ converges for $-1/2 < x < 1/2$.

  c. If $f(x) = \sum c_k x^k$ on the interval $|x| < 1$, then $f(x^2) = \sum c_k x^{2k}$ on the interval $|x| < 1$.

  d. If $f(x) = \sum c_k x^k = 0$ for all x on an interval $(-a, a)$, then $c_k = 0$ for all k.

46–49. Summation notation *Write the following power series in summation (sigma) notation.*

46. $1 + \dfrac{x}{2} + \dfrac{x^2}{4} + \dfrac{x^3}{6} + \cdots$

47. $1 - \dfrac{x}{2} + \dfrac{x^2}{3} - \dfrac{x^3}{4} + \cdots$

48. $x - \dfrac{x^3}{4} + \dfrac{x^5}{9} - \dfrac{x^7}{16} + \cdots$

49. $-\dfrac{x^2}{1!} + \dfrac{x^4}{2!} - \dfrac{x^6}{3!} + \dfrac{x^8}{4!} - \cdots$

50. Scaling power series If the power series $f(x) = \sum c_k x^k$ has an interval of convergence of $|x| < R$, what is the interval of convergence of the power series for $f(ax)$, where $a \neq 0$ is a real number?

51. Shifting power series If the power series $f(x) = \sum c_k x^k$ has an interval of convergence of $|x| < R$, what is the interval of convergence of the power series for $f(x - a)$, where $a \neq 0$ is a real number?

52–57. Series to functions *Find the function represented by the following series and find the interval of convergence of the series.*

52. $\displaystyle\sum_{k=0}^{\infty} (x^2 + 1)^{2k}$

53. $\displaystyle\sum_{k=0}^{\infty} (\sqrt{x} - 2)^k$

54. $\displaystyle\sum_{k=1}^{\infty} \dfrac{x^{2k}}{4k}$

55. $\displaystyle\sum_{k=0}^{\infty} e^{-kx}$

56. $\displaystyle\sum_{k=1}^{\infty} \dfrac{(x - 2)^k}{3^{2k}}$

57. $\displaystyle\sum_{k=0}^{\infty} \left(\dfrac{x^2 - 1}{3}\right)^k$

58. A useful substitution Replace x by $x - 1$ in the series

$$\ln (1 + x) = \sum_{k=1}^{\infty} \dfrac{(-1)^{k+1} x^k}{k}$$ to obtain a power series for $\ln x$

centered at $x = 1$. What is the interval of convergence for the new power series?

59–62. Exponential function *In Section 9.3, we show that the power series for the exponential function centered at 0 is*

$$e^x = \sum_{k=0}^{\infty} \dfrac{x^k}{k!}, \qquad for -\infty < x < \infty.$$

Use the methods of this section to find the power series for the following functions. Give the interval of convergence for the resulting series.

59. $f(x) = e^{-x}$

60. $f(x) = e^{2x}$

61. $f(x) = e^{-3x}$

62. $f(x) = x^2 e^x$

Additional Exercises

63. Powers of x multiplied by a power series Prove that if

$$f(x) = \sum_{k=0}^{\infty} c_k x^k$$ converges on the interval I, then the power series

for $x^m f(x)$ also converges on I for positive integers m.

64. Remainders Let

$$f(x) = \sum_{k=0}^{\infty} x^k = \dfrac{1}{1 - x} \quad and \quad S_n(x) = \sum_{k=0}^{n-1} x^k.$$

Then, the remainder in truncating the power series after n terms is $R_n = f(x) - S_n(x)$, which now depends on x.

  a. Show that $R_n(x) = x^n / (1 - x)$.

  b. Graph the remainder function on the interval $|x| < 1$ for $n = 1, 2, 3$. Discuss and interpret the graph. Where on the interval is $|R_n(x)|$ largest? Smallest?

  c. For fixed n, minimize $|R_n(x)|$ with respect to x. Does the result agree with the observations in part (b)?

  d. Let $N(x)$ be the number of terms required to reduce $|R_n(x)|$ to less than 10^{-6}. Graph the function $N(x)$ on the interval $|x| < 1$. Discuss and interpret the graph.

65. Product of power series Let

$$f(x) = \sum_{k=0}^{\infty} c_k x^k \quad \text{and} \quad g(x) = \sum_{k=0}^{\infty} d_k x^k.$$

a. Multiply the power series together as if they were polynomials, collecting all terms that are multiples of 1, x, and x^2. Write the first three terms of the product $f(x)g(x)$.

b. Find a general expression for the coefficient of x^n in the product series, for $n = 0, 1, 2, \ldots$.

66. Inverse sine Given the power series

$$\frac{1}{\sqrt{1 - x^2}} = 1 + \frac{1}{2}x^2 + \frac{1 \cdot 3}{2 \cdot 4}x^4 + \frac{1 \cdot 3 \cdot 5}{2 \cdot 4 \cdot 6}x^6 + \cdots$$

for $-1 < x < 1$, find the power series for $f(x) = \sin^{-1} x$ centered at 0.

⊞ 67. Computing with power series Consider the following function and its power series:

$$f(x) = \frac{1}{(1 - x)^2} = \sum_{k=1}^{\infty} kx^{k-1}, \qquad \text{for } -1 < x < 1.$$

a. Let $S_n(x)$ be the first n terms of the series. With $n = 5$ and $n = 10$, graph $f(x)$ and $S_n(x)$ at the sample points $x = -0.9, -0.8, \ldots, -0.1, 0, 0.1, \ldots, 0.8, 0.9$ (two graphs). Where is the difference in the graphs the greatest?

b. What value of n is needed to guarantee that $|f(x) - S_n(x)| < 0.01$ at all of the sample points?

QUICK CHECK ANSWERS

1. $g(0) = 0$ **2.** For any value of x with $x > 6$ or $x < -2$, the series diverges by the Divergence Test. The Root or Ratio Test gives the same result.   **3.** $|x| < 1, R = 1$ **4.** Substituting $x = 1/2, \ln(1/2) = -\ln 2 = -\sum_{k=1}^{\infty} \frac{1}{2^k k}.$ ◄

9.3 Taylor Series

In the preceding section we saw that a power series represents a function on its interval of convergence. This section explores the opposite question: Given a function, what is its power series representation? We have already made significant progress in answering this question because we know how Taylor polynomials are used to approximate functions. We now extend Taylor polynomials to produce power series—called *Taylor series*—that provide series representations of functions.

Taylor Series for a Function

Suppose a function f has derivatives $f^{(k)}(a)$ of *all* orders at the point a. If we write the Taylor polynomial of degree n for f centered at a and allow n to increase indefinitely, a power series is obtained. The power series consists of a Taylor polynomial of order n plus terms of higher degree called the *remainder*:

$$\underbrace{c_0 + c_1(x - a) + c_2(x - a)^2 + \cdots + c_n(x - a)^n}_{\text{Taylor polynomial of order } n} + \underbrace{c_{n+1}(x - a)^{n+1} + \cdots}_{\text{remainder}}$$

$$= \sum_{k=0}^{\infty} c_k(x - a)^k$$

The coefficients of the Taylor polynomial are given by

$$c_k = \frac{f^{(k)}(a)}{k!}, \qquad \text{for } k = 0, 1, 2, \ldots.$$

> Maclaurin series are named after the Scottish mathematician Colin Maclaurin (1698–1746), who described them (with credit to Taylor) in a textbook in 1742.

These coefficients are also the coefficients of the power series. Furthermore, this power series has the same matching properties as the Taylor polynomials; that is, the function f and the power series agree in *all* of their derivatives at a. This power series is called the **Taylor series for f centered at a**. It is the natural extension of the set of Taylor polynomials for f at a. The special case of a Taylor series centered at 0 is called a **Maclaurin series**.

> **DEFINITION   Taylor/Maclaurin Series for a Function**
>
> Suppose the function f has derivatives of all orders on an interval containing the point a. The **Taylor series for f centered at a** is
>
> $$f(a) + f'(a)(x - a) + \frac{f''(a)}{2!}(x - a)^2 + \frac{f^{(3)}(a)}{3!}(x - a)^3 + \cdots$$
>
> $$= \sum_{k=0}^{\infty} \frac{f^{(k)}(a)}{k!}(x - a)^k.$$
>
> A Taylor series centered at 0 is called a **Maclaurin series**.

For the Taylor series to be useful, we need to know two things:

▸ There are unusual cases in which the Taylor series for a function converges to a different function (Exercise 80).

- The values of x for which the power series converges, which comprise the interval of convergence.
- The values of x for which the power series for f *equals* f. This question is more subtle and is postponed for a few pages. For now, we find the Taylor series for f at a point, but we refrain from saying $f(x)$ equals the power series.

QUICK CHECK 1 Verify that if the Taylor series for f centered at a is evaluated at $x = a$, then the Taylor series equals $f(a)$. ◂

EXAMPLE 1   Maclaurin series and convergence Find the Maclaurin series (which is the Taylor series centered at 0) for the following functions. Give the interval of convergence.

a. $f(x) = \cos x$ **b.** $f(x) = \dfrac{1}{1 - x}$

SOLUTION The procedure for finding the coefficients of a Taylor series is the same as for Taylor polynomials; most of the work is computing the derivatives of f.

a. The Maclaurin series (centered at 0) has the form

$$\sum_{k=0}^{\infty} c_k x^k, \quad \text{where } c_k = \frac{f^{(k)}(0)}{k!}, \quad \text{for } k = 0, 1, 2, \ldots.$$

We evaluate derivatives of $f(x) = \cos x$ at $x = 0$:

$$f(x) = \cos x \ \Rightarrow\ f(0) = 1$$
$$f'(x) = -\sin x \ \Rightarrow\ f'(0) = 0$$
$$f''(x) = -\cos x \ \Rightarrow\ f''(0) = -1$$
$$f'''(x) = \sin x \ \Rightarrow\ f'''(0) = 0$$
$$f^{(4)}(x) = \cos x \ \Rightarrow\ f^{(4)}(0) = 1$$
$$\vdots \qquad\qquad \vdots$$

Because the odd-order derivatives are zero, $c_k = \dfrac{f^{(k)}(0)}{k!} = 0$ when k is odd. Using the even-order derivatives, we have

$$c_0 = f(0) = 1 \qquad\qquad c_2 = \frac{f^{(2)}(0)}{2!} = -\frac{1}{2!}$$

$$c_4 = \frac{f^{(4)}(0)}{4!} = \frac{1}{4!} \qquad\qquad c_6 = \frac{f^{(6)}(0)}{6!} = -\frac{1}{6!}$$

and, in general, $c_{2k} = \dfrac{(-1)^k}{(2k)!}$. Therefore, the Maclaurin series for f is

$$1 - \frac{x^2}{2!} + \frac{x^4}{4!} - \frac{x^6}{6!} + \cdots = \sum_{k=0}^{\infty} \frac{(-1)^k}{(2k)!} x^{2k}.$$

Notice that this series contains all the Taylor polynomials. In this case, it consists only of even powers of x, reflecting the fact that $\cos x$ is an even function.

For what values of x does the series converge? As discussed in Section 9.2, we apply the Ratio Test to $\displaystyle\sum_{k=0}^{\infty} \left| \frac{(-1)^k}{(2k)!} x^{2k} \right|$ to test for absolute convergence:

> **Recall that**
>
> $(2k + 2)! = (2k + 2)(2k + 1)(2k)!$
>
> Therefore, $\dfrac{(2k)!}{(2k + 2)!} = \dfrac{1}{(2k + 2)(2k + 1)}$

$$r = \lim_{k \to \infty} \left| \frac{(-1)^{k+1} x^{2(k+1)}/(2(k + 1))!}{(-1)^k x^{2k}/(2k)!} \right| \qquad \lim_{k \to \infty} \left| \frac{a_{k+1}}{a_k} \right|$$

$$= \lim_{k \to \infty} \left| \frac{x^2}{(2k + 2)(2k + 1)} \right| = 0 \qquad \text{Simplify and take the limit with } x \text{ fixed.}$$

In this case, $r < 1$ for all x, so the Maclaurin series converges absolutely for all x and the interval of convergence is $-\infty < x < \infty$.

b. We proceed in a similar way with $f(x) = 1/(1 - x)$ by evaluating the derivatives of f at 0:

$$f(x) = \frac{1}{1 - x} \quad\Rightarrow\quad f(0) = 1$$

$$f'(x) = \frac{1}{(1 - x)^2} \quad\Rightarrow\quad f'(0) = 1$$

$$f''(x) = \frac{2}{(1 - x)^3} \quad\Rightarrow\quad f''(0) = 2!$$

$$f'''(x) = \frac{3 \cdot 2}{(1 - x)^4} \quad\Rightarrow\quad f'''(0) = 3!$$

$$f^{(4)}(x) = \frac{4 \cdot 3 \cdot 2}{(1 - x)^5} \quad\Rightarrow\quad f^{(4)}(0) = 4!$$

and, in general, $f^{(k)}(0) = k!$. Therefore, the Maclaurin series coefficients are $c_k = \dfrac{f^{(k)}(0)}{k!} = \dfrac{k!}{k!} = 1$, for $k = 0, 1, 2, \ldots$. The series for f centered at 0 is

$$1 + x + x^2 + x^3 + \cdots = \sum_{k=0}^{\infty} x^k.$$

This power series is familiar! The Maclaurin series for $f(x) = 1/(1 - x)$ is a geometric series. We could apply the Ratio Test, but we have already demonstrated that this series converges for $|x| < 1$.

Related Exercises 9–22 ◀

QUICK CHECK 2 Based on Example 1b, what is the Taylor series for $f(x) = (1 + x)^{-1}$? ◀

The preceding example has an important lesson. *There is only one power series representation for a given function about a given point; however, there may be several ways to find it.*

EXAMPLE 2   Manipulating Maclaurin series Let $f(x) = e^x$.

a. Find the Maclaurin series for f (by definition centered at 0).

b. Find its interval of convergence.

c. Use the Maclaurin series for e^x to find the Maclaurin series for the functions $x^4 e^x$, e^{-2x}, and e^{-x^2}.

SOLUTION

a. The coefficients of the Taylor polynomials for $f(x) = e^x$ centered at 0 are $c_k = 1/k!$ (Example 3, Section 9.1). They are also the coefficients of the Maclaurin series. Therefore, the Maclaurin series for f is

$$1 + \frac{x}{1!} + \frac{x^2}{2!} + \cdots + \frac{x^n}{n!} + \cdots = \sum_{k=0}^{\infty} \frac{x^k}{k!}.$$

b. By the Ratio Test,

$$r = \lim_{k \to \infty} \left| \frac{x^{k+1}/(k+1)!}{x^k/k!} \right| \qquad \text{Substitute } (k+1)\text{st and } k\text{th terms.}$$

$$= \lim_{k \to \infty} \left| \frac{x}{k+1} \right| = 0 \qquad \text{Simplify; take the limit with } x \text{ fixed.}$$

Because $r < 1$ for all x, the interval of convergence is $-\infty < x < \infty$.

c. As stated in Theorem 9.4, power series may be added, multiplied by powers of x, or composed with functions on their intervals of convergence. Therefore, the Maclaurin series for $x^4 e^x$ is

$$x^4 \sum_{k=0}^{\infty} \frac{x^k}{k!} = \sum_{k=0}^{\infty} \frac{x^{k+4}}{k!} = x^4 + \frac{x^5}{1!} + \frac{x^6}{2!} + \cdots + \frac{x^{k+4}}{k!} + \cdots.$$

Similarly, e^{-2x} is the composition $f(-2x)$. Replacing x by $-2x$ in the Maclaurin series for f, the series representation for e^{-2x} is

$$\sum_{k=0}^{\infty} \frac{(-2x)^k}{k!} = \sum_{k=0}^{\infty} \frac{(-1)^k (2x)^k}{k!} = 1 - 2x + 2x^2 - \frac{4}{3}x^3 + \cdots.$$

The Maclaurin series for e^{-x^2} is obtained by replacing x by $-x^2$ in the power series for f. The resulting series is

$$\sum_{k=0}^{\infty} \frac{(-x^2)^k}{k!} = \sum_{k=0}^{\infty} \frac{(-1)^k x^{2k}}{k!} = 1 - x^2 + \frac{x^4}{2!} - \frac{x^6}{3!} + \cdots.$$

Because the interval of convergence of $f(x) = e^x$ is $-\infty < x < \infty$, the manipulations used to obtain the series for $x^4 e^x$, e^{-2x}, or e^{-x^2} do not change the interval of convergence. If in doubt about the interval of convergence of a new series, apply the Ratio Test.

Related Exercises 23–28 ◀

QUICK CHECK 3 Find the first three terms of the Taylor series for $2xe^x$ and e^{-x}. ◀

The Binomial Series

We know from algebra that if p is a positive integer then $(1 + x)^p$ is a polynomial of degree p. In fact,

$$(1 + x)^p = \binom{p}{0} + \binom{p}{1}x + \binom{p}{2}x^2 + \cdots + \binom{p}{p}x^p,$$

where the binomial coefficients $\binom{p}{k}$ are defined as follows.

For nonnegative integers p, the binomial coefficients may also be defined as
$$\binom{p}{k} = \frac{p!}{k!(p-k)!}, \text{ where } 0! = 1.$$
The coefficients form the rows of Pascal's triangle. The coefficients of $(1+x)^5$ form the sixth row of the triangle.

$$
\begin{array}{ccccccccccc}
 & & & & & 1 & & & & & \\
 & & & & 1 & & 1 & & & & \\
 & & & 1 & & 2 & & 1 & & & \\
 & & 1 & & 3 & & 3 & & 1 & & \\
 & 1 & & 4 & & 6 & & 4 & & 1 & \\
1 & & 5 & & 10 & & 10 & & 5 & & 1
\end{array}
$$

DEFINITION Binomial Coefficients

For real numbers p and integers $k \geq 1$,
$$\binom{p}{k} = \frac{p(p-1)(p-2)\cdots(p-k+1)}{k!}, \qquad \binom{p}{0} = 1.$$

For example,

$$(1+x)^5 = \underbrace{\binom{5}{0}}_{1} + \underbrace{\binom{5}{1}}_{5}x + \underbrace{\binom{5}{2}}_{10}x^2 + \underbrace{\binom{5}{3}}_{10}x^3 + \underbrace{\binom{5}{4}}_{5}x^4 + \underbrace{\binom{5}{5}}_{1}x^5$$

$$= 1 + 5x + 10x^2 + 10x^3 + 5x^4 + x^5$$

QUICK CHECK 4 Evaluate the binomial coefficients $\binom{-3}{2}$ and $\binom{\frac{1}{2}}{3}$. ◄

Our goal is to extend this idea to the functions $f(x) = (1+x)^p$, where p is a real number other than a nonnegative integer. The result is a Taylor series called the **binomial series**.

THEOREM 9.6 Binomial Series

For real numbers p, the Taylor series for $f(x) = (1+x)^p$ centered at 0 is the **binomial series**

$$\sum_{k=0}^{\infty} \binom{p}{k} x^k = \sum_{k=0}^{\infty} \frac{p(p-1)(p-2)\cdots(p-k+1)}{k!} x^k$$

$$= 1 + px + \frac{p(p-1)}{2!}x^2 + \frac{p(p-1)(p-2)}{3!}x^3 + \cdots.$$

The series converges to f for $|x| < 1$ (and possibly at the endpoints depending on p). If p is a nonnegative integer, the series terminates and results in a polynomial of degree p.

Proof We seek a power series centered at 0 of the form

$$\sum_{k=0}^{\infty} c_k x^k, \qquad \text{where } c_k = \frac{f^{(k)}(0)}{k!}, \qquad \text{for } k = 0, 1, 2, \ldots.$$

To evaluate $\binom{p}{k}$, start with p, successively subtract 1 until k factors are obtained; then take the product of these k factors and divide by $k!$. Recall that $\binom{p}{0} = 1$.

The job is to evaluate the derivatives of f at 0:

$$f(x) = (1+x)^p \Rightarrow f(0) = 1$$
$$f'(x) = p(1+x)^{p-1} \Rightarrow f'(0) = p$$
$$f''(x) = p(p-1)(1+x)^{p-2} \Rightarrow f''(0) = p(p-1)$$
$$f'''(x) = p(p-1)(p-2)(1+x)^{p-3} \Rightarrow f'''(0) = p(p-1)(p-2).$$

A pattern emerges: The kth derivative $f^{(k)}(0)$ involves the k factors $p(p-1)(p-2)\cdots(p-k+1)$. In general, we have

$$f^{(k)}(0) = p(p-1)(p-2)\cdots(p-k+1).$$

Therefore,

$$c_k = \frac{f^{(k)}(0)}{k!} = \frac{p(p-1)(p-2)\cdots(p-k+1)}{k!} = \binom{p}{k} \qquad \text{for } k = 0, 1, 2, \ldots.$$

The Taylor series for $f(x) = (1 + x)^p$ centered at 0 is

$$\binom{p}{0} + \binom{p}{1}x + \binom{p}{2}x^2 + \binom{p}{3}x^3 + \cdots = \sum_{k=0}^{\infty}\binom{p}{k}x^k.$$

This series has the same general form for all values of p. When p is a nonnegative integer, the series terminates and it is a polynomial of degree p.

The interval of convergence for the binomial series is determined by the Ratio Test. Holding p and x fixed, the relevant limit is

$$r = \lim_{k\to\infty}\left|\frac{x^{k+1}p(p-1)\cdots(p-k+1)(p-k)/(k+1)!}{x^k\,p(p-1)\cdots(p-k+1)/k!}\right| \qquad \text{Ratio of } (k+1)\text{st to } k\text{th term}$$

$$= |x| \lim_{k\to\infty}\underbrace{\left|\frac{p-k}{k+1}\right|}_{\text{approaches 1}} \qquad \text{Cancel terms and simplify.}$$

$$= |x| \qquad \qquad \begin{array}{l} \text{With } p \text{ fixed,} \\ \lim_{k\to\infty}\left|\frac{(p-k)}{k+1}\right| = 1. \end{array}$$

Absolute convergence requires that $r = |x| < 1$. Therefore, the series converges absolutely for $|x| < 1$. (Depending on the value of p, the interval of convergence may include the endpoints; they should be tested on a case-by-case basis.) ◄

EXAMPLE 3   Binomial series Consider the function $f(x) = \sqrt{1 + x}$.

> A binomial series is a Taylor series. Because the series in Example 3 is centered at 0, it is also a Maclaurin series.

a. Find the binomial series for f centered at 0.

b. Approximate $\sqrt{1.15}$ to three decimal places.

SOLUTION

a. We use the formula for the binomial coefficients with $p = \frac{1}{2}$ to compute the first four coefficients:

Table 9.3

n	Approximations: $p_n(0.15)$
0	1.0
1	1.075
2	1.0721875
3	1.072398438

$$c_0 = 1 \qquad\qquad c_1 = \binom{\frac{1}{2}}{1} = \frac{\left(\frac{1}{2}\right)}{1!} = \frac{1}{2}$$

$$c_2 = \binom{\frac{1}{2}}{2} = \frac{\frac{1}{2}\left(-\frac{1}{2}\right)}{2!} = -\frac{1}{8} \qquad c_3 = \binom{\frac{1}{2}}{3} = \frac{\frac{1}{2}\left(-\frac{1}{2}\right)\left(-\frac{3}{2}\right)}{3!} = \frac{1}{16}$$

The leading terms of the binomial series are

$$1 + \frac{1}{2}x - \frac{1}{8}x^2 + \frac{1}{16}x^3 - \cdots.$$

> The remainder theorem for alternating series (Section 8.6) could be used to estimate the number of terms of the Taylor series needed to achieve a desired accuracy.

b. Truncating the binomial series in part (a) produces Taylor polynomials that may be used to approximate $f(0.15) = \sqrt{1.15}$. With $x = 0.15$, we find the polynomial approximations shown in Table 9.3. Four terms of the power series ($n = 3$) give $\sqrt{1.15} \approx 1.072$, which is accurate to three decimal places.

Related Exercises 29–34 ◄

QUICK CHECK 5 Use two and three terms of the binomial series in Example 3 to approximate $\sqrt{1.1}$. ◄

EXAMPLE 4   **Working with binomial series**   Consider the functions

$$f(x) = \sqrt[3]{1 + x} \quad \text{and} \quad g(x) = \sqrt[3]{c + x}, \qquad \text{where } c > 0 \text{ is a constant.}$$

a. Find the first four terms of the binomial series for f centered at 0.
b. Use part (a) to find the first four terms of the binomial series for g centered at 0.
c. Use part (b) to approximate $\sqrt[3]{23}, \sqrt[3]{24}, \dots, \sqrt[3]{31}$.

SOLUTION

a. Because $f(x) = (1 + x)^{1/3}$, we find the binomial coefficients with $p = \frac{1}{3}$:

$$c_0 = \binom{\frac{1}{3}}{0} = 1 \qquad\qquad\qquad c_1 = \binom{\frac{1}{3}}{1} = \frac{\left(\frac{1}{3}\right)}{1!} = \frac{1}{3}$$

$$c_2 = \binom{\frac{1}{3}}{2} = \frac{\left(\frac{1}{3}\right)\left(\frac{1}{3} - 1\right)}{2!} = -\frac{1}{9} \qquad c_3 = \binom{\frac{1}{3}}{3} = \frac{\left(\frac{1}{3}\right)\left(\frac{1}{3} - 1\right)\left(\frac{1}{3} - 2\right)}{3!} = \frac{5}{81} \cdots$$

The first four terms of the binomial series are

$$1 + \frac{1}{3}x - \frac{1}{9}x^2 + \frac{5}{81}x^3 - \cdots.$$

b. To avoid deriving a new series for $g(x) = \sqrt[3]{c + x}$, a few steps of algebra allow us to use part (a). Note that

$$g(x) = \sqrt[3]{c + x} = \sqrt[3]{c\left(1 + \frac{x}{c}\right)} = \sqrt[3]{c} \cdot \sqrt[3]{1 + \frac{x}{c}} = \sqrt[3]{c} \cdot f\left(\frac{x}{c}\right).$$

In other words, g can be expressed in terms of f, for which we already have a binomial series. The binomial series for g is obtained by substituting x/c into the binomial series for f and multiplying by $\sqrt[3]{c}$:

$$g(x) = \sqrt[3]{c} \underbrace{\left[1 + \frac{1}{3}\left(\frac{x}{c}\right) - \frac{1}{9}\left(\frac{x}{c}\right)^2 + \frac{5}{81}\left(\frac{x}{c}\right)^3 - \cdots\right]}_{f(x/c)}$$

The series for $f(x/c)$ converges provided $|x/c| < 1$, or, equivalently, for $|x| < c$.

c. The series of part (b) may be truncated (forming a polynomial p_3) to approximate cube roots. For example, note that $\sqrt[3]{29} = \sqrt[3]{\underset{c}{\underbrace{27}} + \underset{x}{\underbrace{2}}}$, so we take $c = 27$ and $x = 2$. The choice $c = 27$ is made because 29 is near 27 and $\sqrt[3]{c} = \sqrt[3]{27} = 3$ is easy to evaluate. Substituting $c = 27$ and $x = 2$, we find that

$$\sqrt[3]{29} \approx \sqrt[3]{27}\left[1 + \frac{1}{3}\left(\frac{2}{27}\right) - \frac{1}{9}\left(\frac{2}{27}\right)^2 + \frac{5}{81}\left(\frac{2}{27}\right)^3\right] \approx 3.0723.$$

The same method is used to approximate the cube roots of $23, 24, \dots, 30, 31$ (Table 9.4). The absolute error is the difference between p_3 and the value given by a calculator. Notice that the errors increase as we move away from 27.

Table 9.4

	Approximation p_3	Absolute error
$\sqrt[3]{23}$	2.8439	6.7×10^{-5}
$\sqrt[3]{24}$	2.8845	2.0×10^{-5}
$\sqrt[3]{25}$	2.9240	3.9×10^{-6}
$\sqrt[3]{26}$	2.9625	2.4×10^{-7}
$\sqrt[3]{27}$	3	0
$\sqrt[3]{28}$	3.0366	2.3×10^{-7}
$\sqrt[3]{29}$	3.0723	3.5×10^{-6}
$\sqrt[3]{30}$	3.1072	1.7×10^{-5}
$\sqrt[3]{31}$	3.1414	5.4×10^{-5}

Related Exercises 35–46 ◄

Convergence of Taylor Series

It may seem that the story of Taylor series is over. But there is a technical point that is easily overlooked. Given a function f, we know how to write its Taylor series centered at a point a. And we know how to find its interval of convergence. We still do not know that the power series actually converges to f. The remaining task is to determine when the Taylor series for f actually converges to f on its interval of convergence. Fortunately, the necessary tools have already been presented in Taylor's Theorem (Theorem 9.1), which gives the remainder for Taylor polynomials.

Assume f has derivatives $f^{(n)}$ of *all* orders on an open interval containing the point a. Taylor's Theorem tells us that

$$f(x) = p_n(x) + R_n(x),$$

where p_n is the nth-order Taylor polynomial for f centered at a,

$$R_n(x) = \frac{f^{(n+1)}(c)}{(n+1)!}(x - a)^{n+1},$$

and c is a point between x and a. We see that the remainder, $R_n(x) = f(x) - p_n(x)$, measures the difference between f and the approximating polynomial p_n. For the Taylor series to converge to f on an interval, the remainder must approach zero at each point of the interval as the order of the Taylor polynomials increases. The following theorem makes these ideas precise.

THEOREM 9.7   Convergence of Taylor Series

Let f have derivatives of all orders on an open interval I containing a. The Taylor series for f centered at a converges to f for all x in I if and only if $\lim_{n\to\infty} R_n(x) = 0$ for all x in I, where

$$R_n(x) = \frac{f^{(n+1)}(c)}{(n+1)!}(x - a)^{n+1}$$

is the remainder at x (with c between x and a).

Proof The theorem requires derivatives of *all* orders. Therefore, by Taylor's Theorem (Theorem 9.1), the remainder term exists in the given form for all n. Let p_n denote the nth-order Taylor polynomial and note that $\lim_{n\to\infty} p_n(x)$ is the Taylor series for f centered at a, evaluated at a point x in I.

First assume that $\lim\limits_{n \to \infty} R_n(x) = 0$ on the interval I and recall that $p_n(x) = f(x) - R_n(x)$. Taking limits of both sides, we have

$$\underbrace{\lim_{n \to \infty} p_n(x)}_{\text{Taylor series}} = \lim_{n \to \infty} (f(x) - R_n(x)) = \underbrace{\lim_{n \to \infty} f(x)}_{f(x)} - \underbrace{\lim_{n \to \infty} R_n(x)}_{0} = f(x).$$

We conclude that the Taylor series $\lim\limits_{n \to \infty} p_n(x)$ equals $f(x)$, for all x in I.

Conversely, if the Taylor series converges to f, then $f(x) = \lim\limits_{n \to \infty} p_n(x)$ and

$$0 = f(x) - \lim_{n \to \infty} p_n(x) = \lim_{n \to \infty} \underbrace{(f(x) - p_n(x))}_{R_n(x)} = \lim_{n \to \infty} R_n(x).$$

It follows that $\lim\limits_{n \to \infty} R_n(x) = 0$ for all x in I. ◀

Even with an expression for the remainder, it may be difficult to show that $\lim\limits_{n \to \infty} R_n(x) = 0$. The following examples illustrate cases in which it is possible.

EXAMPLE 5 Remainder term in the Maclaurin series for e^x Show that the Maclaurin series for $f(x) = e^x$ converges to f for $-\infty < x < \infty$.

SOLUTION As shown in Example 2, the Maclaurin series for $f(x) = e^x$ is

$$\sum_{k=0}^{\infty} \frac{x^k}{k!} = 1 + x + \frac{x^2}{2!} + \cdots + \frac{x^n}{n!} + \cdots,$$

which converges for $-\infty < x < \infty$. In Example 6 of Section 9.1 it was shown that the remainder term is

$$R_n(x) = \frac{e^c}{(n + 1)!} x^{n+1},$$

where c is between 0 and x. Notice that the intermediate point c varies with n, but it is always between 0 and x. Therefore, e^c is between $e^0 = 1$ and e^x; in fact, $e^c \leq e^{|x|}$ for all n. It follows that

$$|R_n(x)| \leq \frac{e^{|x|}}{(n + 1)!} |x|^{n+1}.$$

Holding x fixed, we have

$$\lim_{n \to \infty} |R_n(x)| = \lim_{n \to \infty} \frac{e^{|x|}}{(n + 1)!} |x|^{n+1} = e^{|x|} \lim_{n \to \infty} \frac{|x|^{n+1}}{(n + 1)!} = 0,$$

where we used the fact that $\lim\limits_{n \to \infty} x^n / n! = 0$ for $-\infty < x < \infty$ (Section 8.2). Because $\lim\limits_{n \to \infty} |R_n(x)| = 0$, it follows that for all real numbers x the Taylor series converges to e^x, or

$$e^x = \sum_{k=0}^{\infty} \frac{x^k}{k!} = 1 + x + \frac{x^2}{2!} + \cdots + \frac{x^n}{n!} + \cdots.$$

The convergence of the Taylor series to e^x is illustrated in Figure 9.17, where Taylor polynomials of increasing degree are graphed together with e^x. *Related Exercises 47–50* ◀

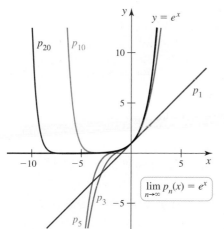

FIGURE 9.17

EXAMPLE 6 Maclaurin series convergence for $\cos x$ Show that the Maclaurin series for $\cos x$,

$$1 - \frac{x^2}{2!} + \frac{x^4}{4!} - \frac{x^6}{6!} + \cdots = \sum_{k=0}^{\infty} (-1)^k \frac{x^{2k}}{(2k)!},$$

converges to $f(x) = \cos x$ for $-\infty < x < \infty$.

SOLUTION To show that the power series converges to f, we must show that $\lim_{n \to \infty} |R_n(x)| = 0$ for $-\infty < x < \infty$. According to Taylor's Theorem with $a = 0$,

$$R_n(x) = \frac{f^{(n+1)}(c)}{(n+1)!} x^{n+1},$$

where c is between 0 and x. Notice that $f^{(n+1)}(c) = \pm\sin c$ or $f^{(n+1)}(c) = \pm\cos c$. In all cases, $|f^{(n+1)}(c)| \le 1$. Therefore, the absolute value of the remainder term is bounded as

$$|R_n(x)| = \left| \frac{f^{(n+1)}(c)}{(n+1)!} x^{n+1} \right| \le \frac{|x|^{n+1}}{(n+1)!}.$$

Holding x fixed and using $\lim_{n \to \infty} x^n/n! = 0$, we see that $\lim_{n \to \infty} R_n(x) = 0$ for all x. Therefore, the given power series converges to $f(x) = \cos x$ for all x; that is, $\cos x = \sum_{k=0}^{\infty} \frac{(-1)^k x^{2k}}{(2k)!}$. The convergence of the Taylor series to $\cos x$ is illustrated in Figure 9.18.   *Related Exercises 47–50* ◀

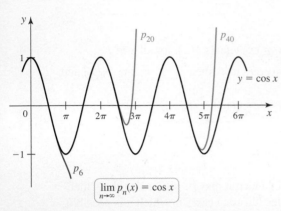

$$\lim_{n \to \infty} p_n(x) = \cos x$$

FIGURE 9.18

The procedure used in Examples 5 and 6 can be carried out for all of the Taylor series we have worked with so far (with varying degrees of difficulty). In each case, the Taylor series converges to the function it represents on the interval of convergence. Table 9.5 summarizes commonly used Taylor series centered at 0 and the functions to which they converge.

▶ Table 9.5 asserts, without proof, that in several cases the Taylor series for f converges to f at the endpoints of the interval of convergence. Proving convergence at the endpoints generally requires advanced techniques. It may also be done using the following theorem:

Suppose the Taylor series for f centered at 0 converges to f on the interval $(-R, R)$. If the series converges at $x = R$, then it converges to $\lim_{x \to R^-} f(x)$. If the series converges at $x = -R$, then it converges to $\lim_{x \to -R^+} f(x)$.

For example, this theorem would allow us to conclude that the series for $\ln(1+x)$ converges to $\ln 2$ at $x = 1$.

Table 9.5

$$\frac{1}{1-x} = 1 + x + x^2 + \cdots + x^k + \cdots = \sum_{k=0}^{\infty} x^k, \quad \text{for } |x| < 1$$

$$\frac{1}{1+x} = 1 - x + x^2 - \cdots + (-1)^k x^k + \cdots = \sum_{k=0}^{\infty} (-1)^k x^k, \quad \text{for } |x| < 1$$

$$e^x = 1 + x + \frac{x^2}{2!} + \cdots + \frac{x^k}{k!} + \cdots = \sum_{k=0}^{\infty} \frac{x^k}{k!}, \quad \text{for } |x| < \infty$$

$$\sin x = x - \frac{x^3}{3!} + \frac{x^5}{5!} - \cdots + \frac{(-1)^k x^{2k+1}}{(2k+1)!} + \cdots = \sum_{k=0}^{\infty} \frac{(-1)^k x^{2k+1}}{(2k+1)!}, \quad \text{for } |x| < \infty$$

$$\cos x = 1 - \frac{x^2}{2!} + \frac{x^4}{4!} - \cdots + \frac{(-1)^k x^{2k}}{(2k)!} + \cdots = \sum_{k=0}^{\infty} \frac{(-1)^k x^{2k}}{(2k)!}, \quad \text{for } |x| < \infty$$

$$\ln(1+x) = x - \frac{x^2}{2} + \frac{x^3}{3} - \cdots + \frac{(-1)^{k+1} x^k}{k} + \cdots = \sum_{k=1}^{\infty} \frac{(-1)^{k+1} x^k}{k}, \quad \text{for } -1 < x \le 1$$

$$-\ln(1-x) = x + \frac{x^2}{2} + \frac{x^3}{3} + \cdots + \frac{x^k}{k} + \cdots = \sum_{k=1}^{\infty} \frac{x^k}{k}, \quad \text{for } -1 \le x < 1$$

$$\tan^{-1} x = x - \frac{x^3}{3} + \frac{x^5}{5} - \cdots + \frac{(-1)^k x^{2k+1}}{2k+1} + \cdots = \sum_{k=0}^{\infty} \frac{(-1)^k x^{2k+1}}{2k+1}, \quad \text{for } |x| \le 1$$

$$(1+x)^p = \sum_{k=0}^{\infty} \binom{p}{k} x^k, \quad \text{for } |x| < 1 \quad \text{and} \quad \binom{p}{k} = \frac{p(p-1)(p-2)\cdots(p-k+1)}{k!}, \binom{p}{0} = 1$$

SECTION 9.3 EXERCISES

Review Questions

1. How are the Taylor polynomials for a function f centered at a related to the Taylor series of the function f centered at a?

2. What conditions must be satisfied by a function f to have a Taylor series centered at a?

3. How do you find the coefficients of the Taylor series for f centered at a?

4. How do you find the interval of convergence of a Taylor series?

5. Suppose you know the Maclaurin series for f and it converges for $|x| < 1$. How do you find the Maclaurin series for $f(x^2)$ and where does it converge?

6. For what values of p does the Taylor series for $f(x) = (1 + x)^p$ centered at 0 terminate?

7. In terms of the remainder, what does it mean for a Taylor series for a function f to converge to f?

8. Write the Maclaurin series for e^{2x}.

Basic Skills

9–16. Maclaurin series

    a. Find the first four nonzero terms of the Maclaurin series for the given function.
    b. Write the power series using summation notation.
    c. Determine the interval of convergence of the series.

9. $f(x) = e^{-x}$ 10. $f(x) = \cos 2x$

11. $f(x) = (1 + x^2)^{-1}$ 12. $f(x) = \ln(1 + x)$

13. $f(x) = e^{2x}$ 14. $f(x) = (1 + 2x)^{-1}$

15. $f(x) = \tan^{-1} x$ 16. $f(x) = \sin 3x$

17–22. Taylor series centered at $a \neq 0$

    a. Find the first four nonzero terms of the Taylor series for the given function centered at a.
    b. Write the power series using summation notation.

17. $f(x) = \sin x, \ a = \pi/2$ 18. $f(x) = \cos x, \ a = \pi$

19. $f(x) = 1/x, \ a = 1$ 20. $f(x) = 1/x, \ a = 2$

21. $f(x) = \ln x, \ a = 3$ 22. $f(x) = e^x, \ a = \ln 2$

23–28. Manipulating Taylor series *Use the Taylor series in Table 9.5 to find the first four nonzero terms of the Taylor series for the following functions centered at 0.*

23. $\ln(1 + x^2)$ 24. $\sin x^2$

25. $\dfrac{e^x - 1}{x}$ 26. $\cos \sqrt{x}$

27. $(1 + x^4)^{-1}$ 28. $x \tan^{-1} x^2$

▇ 29–34. Binomial series

    a. Find the first four nonzero terms of the Taylor series centered at 0 for the given function.
    b. Use the first four terms of the series to approximate the given quantity.

29. $f(x) = (1 + x)^{-2}$; approximate $1/1.21 = 1/1.1^2$.

30. $f(x) = \sqrt{1 + x}$; approximate $\sqrt{1.06}$.

31. $f(x) = \sqrt[4]{1 + x}$; approximate $\sqrt[4]{1.12}$.

32. $f(x) = (1 + x)^{-3}$; approximate $1/1.331 = 1/1.1^3$.

33. $f(x) = (1 + x)^{-2/3}$; approximate $1.18^{-2/3}$.

34. $f(x) = (1 + x)^{2/3}$; approximate $1.02^{2/3}$.

35–40. Working with binomial series *Use properties of power series, substitution, and factoring to find the first four nonzero terms of the Taylor series centered at 0 for the following functions. Give the interval of convergence for the new series. Use the Taylor series*

$$\sqrt{1 + x} = 1 + \frac{x}{2} - \frac{x^2}{8} + \frac{x^3}{16} - \cdots, \qquad \text{for } -1 < x \leq 1.$$

35. $\sqrt{1 + x^2}$ 36. $\sqrt{4 + x}$

37. $\sqrt{9 - 9x}$ 38. $\sqrt{1 - 4x}$

39. $\sqrt{a^2 + x^2}, \ a > 0$ 40. $\sqrt{4 - 16x^2}$

41–46. Working with binomial series *Use properties of power series, substitution, and factoring of constants to find the first four nonzero terms of the Taylor series centered at 0 for the following functions. Use the Taylor series*

$$(1 + x)^{-2} = 1 - 2x + 3x^2 - 4x^3 + \cdots, \qquad \text{for } -1 < x < 1.$$

41. $(1 + 4x)^{-2}$ 42. $\dfrac{1}{(1 - 4x)^2}$

43. $\dfrac{1}{(4 + x^2)^2}$ 44. $(x^2 - 4x + 5)^{-2}$

45. $\dfrac{1}{(3 + 4x)^2}$ 46. $\dfrac{1}{(1 + 4x^2)^2}$

47–50. Remainder terms *Find the remainder in the Taylor series centered at the point a for the following functions. Then show that $\lim\limits_{n \to \infty} R_n(x) = 0$ for all x in the interval of convergence.*

47. $f(x) = \sin x, \ a = 0$ 48. $f(x) = \cos 2x, \ a = 0$

49. $f(x) = e^{-x}, \ a = 0$ 50. $f(x) = \cos x, \ a = \pi/2$

Further Explorations

51. **Explain why or why not** Determine whether the following statements are true and give an explanation or counterexample.

    a. The function $f(x) = \sqrt{x}$ has a Taylor series centered at 0.
    b. The function $f(x) = \csc x$ has a Taylor series centered at $\pi/2$.
    c. If f has a Taylor series that converges only on $(-2, 2)$, then $f(x^2)$ has a Taylor series that also converges only on $(-2, 2)$.
    d. If p is the Taylor series for f centered at 0, then $p(x - 1)$ is the Taylor series for f centered at 1.
    e. The Taylor series for an even function about 0 has only even powers of x.

52–59. Any method

   a. *Use any analytical method to find the first four nonzero terms of the Taylor series centered at 0 for the following functions. In most cases you do not need to use the definition of the Taylor series coefficients.*
   b. *If possible, determine the radius of convergence of the series.*

52. $f(x) = \cos(2x) + 2\sin x$ **53.** $f(x) = \dfrac{e^x + e^{-x}}{2}$

54. $f(x) = \sec x$ **55.** $f(x) = (1 + x^2)^{-2/3}$

56. $f(x) = \tan x$ **57.** $f(x) = \sqrt{1 - x^2}$

58. $f(x) = b^x$ for $b > 0$ **59.** $f(x) = \dfrac{1}{x^4 + 2x^2 + 1}$

60–63. Alternative approach *Compute the coefficients for the Taylor series for the following functions about the given point a and then use the first four terms of the series to approximate the given number.*

60. $f(x) = \sqrt{x}$ with $a = 36$; approximate $\sqrt{39}$.

61. $f(x) = \sqrt[3]{x}$ with $a = 64$; approximate $\sqrt[3]{60}$.

62. $f(x) = 1/\sqrt{x}$ with $a = 4$; approximate $1/\sqrt{3}$.

63. $f(x) = \sqrt[4]{x}$ with $a = 16$; approximate $\sqrt[4]{13}$.

64. Geometric/binomial series Recall that the Taylor series for

$$f(x) = 1/(1 - x) \text{ about 0 is the geometric series } \sum_{k=0}^{\infty} x^k.$$

Show that this series can also be found as a case of the binomial series.

65. Integer coefficients Show that the coefficients in the Taylor series (binomial series) for $f(x) = \sqrt{1 + 4x}$ about 0 are integers.

66. Choosing a good center Suppose you want to approximate $\sqrt{72}$ using four terms of a Taylor series. Compare the accuracy of the approximations obtained using the Taylor series for $\sqrt{x}$ centered at 64 and 81.

67. Alternative means By comparing the first four terms, show that the Maclaurin series for $\sin^2 x$ can be found (a) by squaring the Maclaurin series for $\sin x$ or (b) by using the identity $\sin^2 x = (1 - \cos 2x)/2$.

68. Alternative means By comparing the first four terms, show that the Maclaurin series for $\cos^2 x$ can be found (a) by squaring the Maclaurin series for $\cos x$ or (b) by using the identity $\cos^2 x = (1 + \cos 2x)/2$.

69. Designer series Find a power series that has $(2, 6)$ as an interval of convergence.

70–71. Patterns in coefficients *Find the next two terms of the following Taylor series.*

70. $\sqrt{1 + x}$: $1 + \dfrac{1}{2}x - \dfrac{1}{2 \cdot 4}x^2 + \dfrac{1 \cdot 3}{2 \cdot 4 \cdot 6}x^3 - \cdots$.

71. $\dfrac{1}{\sqrt{1 + x}}$: $1 - \dfrac{1}{2}x + \dfrac{1 \cdot 3}{2 \cdot 4}x^2 - \dfrac{1 \cdot 3 \cdot 5}{2 \cdot 4 \cdot 6}x^3 + \cdots$.

72. Composition of series Use composition of Taylor series to find the first three terms of the Maclaurin series for the following functions.

   a. $e^{\sin x}$ **b.** $e^{\tan x}$ **c.** $\sqrt{1 + \sin^2 x}$

Applications

73–76. Approximations *Choose a Taylor series and a center point a to approximate the following quantities with an accuracy of at least 10^{-4}.*

73. $\cos 40°$ **74.** $\sin(0.98\pi)$

75. $\sqrt[3]{83}$ **76.** $1/\sqrt[4]{17}$

77. Different approximation strategies Suppose you want to approximate $\sqrt[3]{128}$ to within 10^{-4} of the exact value.

   a. Use a Taylor polynomial centered at 0.
   b. Use a Taylor polynomial centered at 125.
   c. Compare the two approaches. Are they equivalent?

Additional Exercises

78. Mean Value Theorem Explain why the Mean Value Theorem (Theorem 4.9 of Section 4.6) is a special case of Taylor's Theorem.

79. Version of the Second Derivative Test Assume that f has at least two continuous derivatives on an interval containing a with $f'(a) = 0$. Use Taylor's Theorem to prove the following version of the Second Derivative Test:

   a. If $f''(x) > 0$ on some interval containing a, then f has a local minimum at a.
   b. If $f''(x) < 0$ on some interval containing a, then f has a local maximum at a.

80. Nonconvergence to f Consider the function

$$f(x) = \begin{cases} e^{-1/x^2} & \text{if } x \neq 0 \\ 0 & \text{if } x = 0 \end{cases}$$

   a. Use the definition of the derivative to show that $f'(0) = 0$.
   b. Assume the fact that $f^{(k)}(0) = 0$, for $k = 1, 2, 3, \ldots$. (You can write a proof using the definition of the derivative.) Write the Taylor series for f centered at 0.
   c. Explain why the Taylor series for f does not converge to f for $x \neq 0$.

QUICK CHECK ANSWERS

1. When evaluated at $x = a$, all terms of the series are zero except for the first term, which is $f(a)$. Therefore the series equals $f(a)$ at this point.
2. $1 - x + x^2 - x^3 + x^4 - \cdots$ **3.** $2x + 2x^2 + x^3$; $1 - x + x^2/2$ **4.** $6, 1/16$ **5.** $1.05, 1.04875$ ◄

9.4 Working with Taylor Series

We now know the Taylor series for many familiar functions and we have tools for working with power series. The goal of this final section is to illustrate additional techniques associated with power series. As you will see, power series cover the entire landscape of calculus from limits and derivatives to integrals and approximation.

Limits by Taylor Series

An important use of Taylor series is evaluating limits. A couple of examples illustrate the essential ideas.

> L'Hôpital's Rule may be impractical when it must be used more than once on the same limit or when derivatives are difficult to compute.

EXAMPLE 1 **A limit by Taylor series** Evaluate $\lim\limits_{x \to 0} \dfrac{x^2 + 2\cos x - 2}{3x^4}$.

SOLUTION Because the limit has the indeterminate form $0/0$, l'Hôpital's Rule can be used, which requires four applications of the rule. Alternatively, because the limit involves values of x near 0, we substitute the Maclaurin series for $\cos x$. Recalling that

$$\cos x = 1 - \frac{x^2}{2} + \frac{x^4}{24} - \frac{x^6}{720} + \cdots, \quad \text{Table 9.5, page 620}$$

we have

> In using a series approach to evaluating limits, it is often not obvious how many terms of the Taylor series to use. When in doubt, include extra (higher-power) terms. The dots in the calculation stand for powers of x greater than the last power that appears.

$$\lim_{x \to 0} \frac{x^2 + 2\cos x - 2}{3x^4} = \lim_{x \to 0} \frac{x^2 + 2\left(1 - \dfrac{x^2}{2} + \dfrac{x^4}{24} - \dfrac{x^6}{720} + \cdots\right) - 2}{3x^4} \quad \begin{array}{l}\text{Substitute}\\ \text{for } \cos x.\end{array}$$

$$= \lim_{x \to 0} \frac{x^2 + \left(2 - x^2 + \dfrac{x^4}{12} - \dfrac{x^6}{360} + \cdots\right) - 2}{3x^4} \quad \text{Simplify.}$$

$$= \lim_{x \to 0} \frac{\dfrac{x^4}{12} - \dfrac{x^6}{360} + \cdots}{3x^4} \quad \text{Simplify.}$$

$$= \lim_{x \to 0} \left(\frac{1}{36} - \frac{x^2}{1080} + \cdots\right) = \frac{1}{36}. \quad \begin{array}{l}\text{Simplify,}\\ \text{evaluate}\\ \text{limit.}\end{array}$$

Related Exercises 7–20 ◄

QUICK CHECK 1 Use the Taylor series $\sin x = x - x^3/6 + \cdots$ to verify that $\lim\limits_{x \to 0} (\sin x)/x = 1.$ ◄

EXAMPLE 2 **A limit by Taylor series** Evaluate

$$\lim_{x \to \infty} \left[6x^5 \sin\left(\frac{1}{x}\right) - 6x^4 + x^2\right].$$

SOLUTION A Taylor series may be centered at any finite point in the domain of the function, but we don't have the tools needed to expand a function about $x = \infty$. Using a technique introduced earlier, we replace x by $1/t$ and note that as $x \to \infty$, $t \to 0^+$. The new limit becomes

$$\lim_{x \to \infty} \left[6x^5 \sin\left(\frac{1}{x}\right) - 6x^4 + x^2\right] = \lim_{t \to 0^+} \left(\frac{6\sin t}{t^5} - \frac{6}{t^4} + \frac{1}{t^2}\right) \quad \text{Replace } x \text{ by } 1/t.$$

$$= \lim_{t \to 0^+} \left(\frac{6\sin t - 6t + t^3}{t^5}\right). \quad \text{Common denominator}$$

This limit has the indeterminate form $0/0$. We now expand $\sin t$ in a Taylor series centered at $t = 0$. Because

$$\sin t = t - \frac{t^3}{6} + \frac{t^5}{120} - \frac{t^7}{5040} + \cdots, \quad \text{Table 9.5, page 620}$$

the value of the original limit is

$$\lim_{t \to 0^+} \left(\frac{6 \sin t - 6t + t^3}{t^5} \right)$$

$$= \lim_{t \to 0^+} \left(\frac{6\left(t - \dfrac{t^3}{6} + \dfrac{t^5}{120} - \dfrac{t^7}{5040} + \cdots \right) - 6t + t^3}{t^5} \right) \qquad \text{Substitute for } \sin t.$$

$$= \lim_{t \to 0^+} \left(\frac{\dfrac{t^5}{20} - \dfrac{t^7}{840} + \cdots}{t^5} \right) \qquad \text{Simplify.}$$

$$= \lim_{t \to 0^+} \left(\frac{1}{20} - \frac{t^2}{840} + \cdots \right) = \frac{1}{20} \qquad \text{Simplify; evaluate limit.}$$

Related Exercises 7–20 ◄

Differentiating Power Series

The following examples illustrate the ways in which term-by-term differention (Theorem 9.5) may be used.

EXAMPLE 3 **Power series for derivatives** Differentiate the Maclaurin series for $f(x) = \sin x$ to verify that $\dfrac{d}{dx}(\sin x) = \cos x$.

SOLUTION The Maclaurin series for $f(x) = \sin x$ is

$$\sin x = x - \frac{x^3}{3!} + \frac{x^5}{5!} - \frac{x^7}{7!} + \cdots,$$

and it converges for $-\infty < x < \infty$. By Theorem 9.5, the differentiated series also converges for $-\infty < x < \infty$ and it converges to $f'(x)$. On differentiating, we have

$$\frac{d}{dx}\left(x - \frac{x^3}{3!} + \frac{x^5}{5!} - \frac{x^7}{7!} + \cdots \right) = 1 - \frac{x^2}{2!} + \frac{x^4}{4!} - \frac{x^6}{6!} + \cdots = \cos x.$$

QUICK CHECK 2 Differentiate the power series for $\cos x$ (given in Example 3) and identify the result. ◄

The differentiated series is the Maclaurin series for $\cos x$, confirming that $f'(x) = \cos x$.

Related Exercises 21–26 ◄

EXAMPLE 4 **A differential equation** Find a power series solution of the differential equation $y'(t) = y(t) + 2$, subject to the initial condition $y(0) = 6$. Identify the function represented by the power series.

SOLUTION Because the initial condition is given at $t = 0$, we expand the solution in a Taylor series about 0 of the form $y(t) = \displaystyle\sum_{k=0}^{\infty} c_k t^k$, where the coefficients c_k must be determined. Recall that the coefficients of the Taylor series are given by

$$c_k = \frac{y^{(k)}(0)}{k!}, \qquad \text{for } k = 0, 1, 2, \ldots.$$

If we can determine $y^{(k)}(0)$, for $k = 0, 1, 2, \ldots$, the coefficients of the series are also determined. The assumption that y has a Taylor series means that y has derivatives of all orders at 0.

Substituting the initial condition $t = 0$ and $y = 6$ into the power series

$$y(t) = c_0 + c_1 t + c_2 t^2 + \cdots,$$

we find that

$$6 = c_0 + c_1(0) + c_2(0)^2 + \cdots.$$

It follows that $c_0 = 6$. To determine $y'(0)$, we substitute $t = 0$ into the differential equation; the result is $y'(0) = y(0) + 2 = 6 + 2 = 8$. Therefore, $c_1 = y'(0)/1! = 8$.

The remaining derivatives are obtained by successively differentiating the differential equation and substituting $t = 0$. We find that $y''(0) = y'(0) = 8$, $y'''(0) = y''(0) = 8$, and, in general, $y^{(k)}(0) = 8$ for $k = 2, 3, 4, \ldots$. Therefore, $c_k = \dfrac{y^{(k)}(0)}{k!} = \dfrac{8}{k!}$, for $k = 1, 2, 3, \ldots$ and the Taylor series for the solution is

$$y(t) = c_0 + c_1 t + c_2 t^2 + \cdots$$

$$= 6 + \frac{8}{1!}t + \frac{8}{2!}t^2 + \frac{8}{3!}t^3 + \cdots.$$

To identify the function represented by this series we write

$$y(t) = \underbrace{-2 + 8}_{6} + \frac{8}{1!}t + \frac{8}{2!}t^2 + \frac{8}{3!}t^3 + \cdots$$

$$= -2 + 8\underbrace{\left(1 + t + \frac{t^2}{2!} + \frac{t^3}{3!} + \cdots\right)}_{e^t}.$$

> You should check that $y(t) = -2 + 8e^t$ satisfies $y'(t) = y(t) + 2$ and $y(0) = 6$.

The power series that appears is the Taylor series for e^t. Therefore, the solution is $y(t) = -2 + 8e^t$.   *Related Exercises 27–30* ◄

Integrating Power Series

The following example illustrates the use of power series in approximating integrals that cannot be evaluated by analytical methods.

EXAMPLE 5   Approximating a definite integral Approximate the value of the integral $\int_0^1 e^{-x^2}\, dx$ with an error no greater than 5×10^{-4}.

SOLUTION The antiderivative of e^{-x^2} cannot be expressed in terms of familiar functions. The strategy is to write the Maclaurin series for e^{-x^2} and integrate it term by term. Recall that integration of a power series is valid within its interval of convergence (Theorem 9.5). Beginning with the Maclaurin series

$$e^x = 1 + x + \frac{x^2}{2!} + \frac{x^3}{3!} + \cdots + \frac{x^n}{n!} + \cdots,$$

which converges for $-\infty < x < \infty$, we replace x by $-x^2$ to obtain

$$e^{-x^2} = 1 - x^2 + \frac{x^4}{2!} - \frac{x^6}{3!} + \cdots + \frac{(-1)^n x^{2n}}{n!} + \cdots,$$

which also converges for $-\infty < x < \infty$. By the Fundamental Theorem of Calculus,

$$\int_0^1 e^{-x^2}\, dx = \left(x - \frac{x^3}{3} + \frac{x^5}{5 \cdot 2!} - \frac{x^7}{7 \cdot 3!} + \cdots + \frac{(-1)^n x^{2n+1}}{(2n+1)n!} + \cdots\right)\Bigg|_0^1$$

$$= 1 - \frac{1}{3} + \frac{1}{5 \cdot 2!} - \frac{1}{7 \cdot 3!} + \cdots + \frac{(-1)^n}{(2n+1)n!} + \cdots.$$

Because the definite integral is expressed as an alternating series, the remainder in truncating the series is less than the first neglected term, which is $\dfrac{(-1)^{n+1}}{(2n+3)(n+1)!}$. By trial and error, we find that the magnitude of this term is less than 5×10^{-4} if $n \geq 5$ (with $n = 5$, we have $\dfrac{1}{13 \cdot 6!} \approx 1.07 \times 10^{-4}$). The sum of the terms of the series up to $n = 5$ gives the approximation

> The integral in Example 5 is important in statistics and probability theory because of its relationship to the *normal distribution*.

$$\int_0^1 e^{-x^2}\, dx \approx 1 - \frac{1}{3} + \frac{1}{5 \cdot 2!} - \frac{1}{7 \cdot 3!} + \frac{1}{9 \cdot 4!} - \frac{1}{11 \cdot 5!} \approx 0.747.$$

Related Exercises 31–38 ◄

Representing Real Numbers

When values of x are substituted into a convergent power series, the result may be a series representation of a familiar real number. The following example illustrates some techniques.

EXAMPLE 6   Evaluating infinite series

a. Use the Maclaurin series for $f(x) = \tan^{-1} x$ to evaluate

$$1 - \frac{1}{3} + \frac{1}{5} - \cdots = \sum_{k=0}^{\infty} \frac{(-1)^k}{2k+1}.$$

b. Let $f(x) = (e^x - 1)/x$ for $x \neq 0$ and $f(0) = 1$. Use the Maclaurin series for f to evaluate $f'(1)$ and $\displaystyle\sum_{k=1}^{\infty} \frac{k}{(k+1)!}$.

SOLUTION

a. From Table 9.5 (page 620), we see that for $|x| \leq 1$,

$$\tan^{-1} x = x - \frac{x^3}{3} + \frac{x^5}{5} - \cdots + \frac{(-1)^k x^{2k+1}}{2k+1} + \cdots = \sum_{k=0}^{\infty} \frac{(-1)^k x^{2k+1}}{2k+1}.$$

Substituting $x = 1$, we have

> This series (known as the *Gregory series*) is one of a multitude of series representations of π. Because this series converges slowly, it does not provide an efficient way to approximate π.

$$\tan^{-1} 1 = 1 - \frac{1^3}{3} + \frac{1^5}{5} - \cdots = \sum_{k=0}^{\infty} \frac{(-1)^k}{2k+1}.$$

Because $\tan^{-1} 1 = \pi/4$, the value of the series is $\pi/4$.

b. Using the Maclaurin series for e^x, the series for $f(x) = (e^x - 1)/x$ is

$$f(x) = \frac{e^x - 1}{x} = \frac{1}{x}\left[\left(1 + x + \frac{x^2}{2!} + \frac{x^3}{3!} + \cdots\right) - 1\right] \quad \text{Substitute series for } e^x.$$

$$= 1 + \frac{x}{2!} + \frac{x^2}{3!} + \frac{x^3}{4!} + \cdots = \sum_{k=1}^{\infty} \frac{x^{k-1}}{k!}, \quad \text{Simplify.}$$

which converges for $-\infty < x < \infty$. By the Quotient Rule,

$$f'(x) = \frac{xe^x - (e^x - 1)}{x^2}.$$

Differentiating the series for f term by term (Theorem 9.5), we find that

$$f'(x) = \frac{d}{dx}\left(1 + \frac{x}{2!} + \frac{x^2}{3!} + \frac{x^3}{4!} + \cdots\right)$$

$$= \frac{1}{2!} + \frac{2x}{3!} + \frac{3x^2}{4!} + \cdots = \sum_{k=1}^{\infty} \frac{kx^{k-1}}{(k+1)!}.$$

We now have two expressions for f'; they are evaluated at $x = 1$ to show that

$$f'(1) = 1 = \sum_{k=1}^{\infty} \frac{k}{(k+1)!}.$$

Related Exercises 39–48 ◄

QUICK CHECK 3 What value of x would you substitute into the Maclaurin series for $\tan^{-1} x$ to obtain a series representation for $\pi/6$? ◄

Representing Functions as Power Series

Power series have a fundamental role in mathematics in defining functions and providing alternative representations of familiar functions. As an overall review, we close this chapter with two illustrations of the many techniques for working with power series.

EXAMPLE 7   Identify the series Identify the function represented by the power series $\sum_{k=0}^{\infty} \frac{(1-2x)^k}{k!}$ and give its interval of convergence.

SOLUTION The Taylor series for the exponential function,

$$e^x = \sum_{k=0}^{\infty} \frac{x^k}{k!},$$

converges for $-\infty < x < \infty$. Replacing x by $1 - 2x$ produces the given series:

$$\sum_{k=0}^{\infty} \frac{(1-2x)^k}{k!} = e^{1-2x}.$$

This replacement is allowed because $1 - 2x$ is within the interval of convergence of the series for e^x; that is, $-\infty < 2x - 1 < \infty$ for all x. Therefore, the given series represents e^{1-2x} for $-\infty < x < \infty$.

Related Exercises 49–58 ◄

EXAMPLE 8   Mystery series The power series $\sum_{k=1}^{\infty} \frac{(-1)^k k}{4^k} x^{2k}$ appeared in the opening of Section 9.2. Determine the interval of convergence of the power series and find the function it represents on this interval.

SOLUTION Applying the Ratio Test to the series, we determine that it converges when $|x^2/4| < 1$, which implies that $|x| < 2$. A quick check of the endpoints of the original series confirms that it diverges at $x = \pm 2$. Therefore, the interval of convergence is $|x| < 2$.

To find the function represented by the series, we apply several maneuvers until we obtain a geometric series. First note that

$$\sum_{k=1}^{\infty} \frac{(-1)^k k}{4^k} x^{2k} = \sum_{k=1}^{\infty} k\left(-\frac{1}{4}\right)^k x^{2k}.$$

The series on the right is not a geometric series because of the presence of the factor k. The key is to realize that k could appear in this way through differentiation; specifically, something like $\dfrac{d}{dx}(x^{2k}) = 2kx^{2k-1}$. To achieve terms of this form, we write

$$\underbrace{\sum_{k=1}^{\infty} \frac{(-1)^k k}{4^k} x^{2k}}_{\text{original series}} = \sum_{k=1}^{\infty} k\left(-\frac{1}{4}\right)^k x^{2k}$$

$$= \frac{1}{2} \sum_{k=1}^{\infty} 2k\left(-\frac{1}{4}\right)^k x^{2k} \qquad \text{Multiply and divide by 2.}$$

$$= \frac{x}{2} \sum_{k=1}^{\infty} 2k\left(-\frac{1}{4}\right)^k x^{2k-1}. \qquad \text{Remove } x \text{ from the series.}$$

Now we identify the last series as the derivative of another series:

$$\underbrace{\sum_{k=1}^{\infty} \frac{(-1)^k k}{4^k} x^{2k}}_{\text{original series}} = \frac{x}{2} \sum_{k=1}^{\infty} \left(-\frac{1}{4}\right)^k 2kx^{2k-1}$$

$$= \frac{x}{2} \sum_{k=1}^{\infty} \left(-\frac{1}{4}\right)^k \frac{d}{dx}(x^{2k}) \qquad \text{Identify a derivative.}$$

$$= \frac{x}{2} \frac{d}{dx} \sum_{k=1}^{\infty} \left(-\frac{x^2}{4}\right)^k \qquad \text{Combine factors; term by term differentiation.}$$

This last series is a geometric series with a ratio $r = -x^2/4$ and first term $-x^2/4$; therefore, its value is $\dfrac{-x^2/4}{1 + (x^2/4)}$, provided $\left|\dfrac{x^2}{4}\right| < 1$. We now have

$$\underbrace{\sum_{k=1}^{\infty} \frac{(-1)^k k}{4^k} x^{2k}}_{\text{original series}} = \frac{x}{2} \frac{d}{dx} \sum_{k=1}^{\infty} \left(-\frac{x^2}{4}\right)^k$$

$$= \frac{x}{2} \frac{d}{dx} \left(\frac{-x^2/4}{1 + (x^2/4)}\right) \qquad \text{Sum of geometric series}$$

$$= \frac{x}{2} \frac{d}{dx} \left(\frac{-x^2}{4 + x^2}\right) \qquad \text{Simplify.}$$

$$= -\frac{4x^2}{(4 + x^2)^2} \qquad \text{Differentiate and simplify.}$$

Therefore, the function represented by the power series on $(-2, 2)$ has been uncovered; it is

$$f(x) = -\frac{4x^2}{(4 + x^2)^2}.$$

Notice that f is defined for $-\infty < x < \infty$ (Figure 9.19), but its power series centered at 0 converges to f only on $(-2, 2)$.   *Related Exercises 49–58* ◄

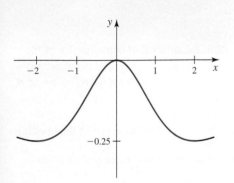

$$\sum_{k=1}^{\infty} \frac{(-1)^k k}{4^k} x^{2k} = -\frac{4x^2}{(4 + x^2)^2} \text{ on } (-2, 2)$$

FIGURE 9.19

SECTION 9.4 EXERCISES

Review Questions

1. Explain the strategy presented in this section for evaluating a limit of the form $\lim_{x \to a} f(x)/g(x)$, where f and g have Taylor series centered at a.

2. Explain the method presented in this section for evaluating $\int_a^b f(x) \, dx$, where f has a Taylor series with an interval of convergence centered at a that includes b.

3. How would you approximate $e^{-0.6}$ using the Taylor series for e^x?

4. Suggest a Taylor series and a method for approximating π.

5. If $f(x) = \sum_{k=0}^{\infty} c_k x^k$ and the series converges for $|x| < b$, what is the power series for $f'(x)$?

6. What condition must be met by a function f for it to have a Taylor series centered at a?

Basic Skills

7–20. Limits *Evaluate the following limits using Taylor series.*

7. $\lim_{x \to 0} \dfrac{e^x - e^{-x}}{x}$

8. $\lim_{x \to 0} \dfrac{1 + x - e^{-x}}{4x^2}$

9. $\lim_{x \to 0} \dfrac{2 \cos 2x - 2 + 4x^2}{2x^4}$

10. $\lim_{x \to \infty} x \sin \left(\dfrac{1}{x} \right)$

11. $\lim_{x \to 0} \dfrac{3 \tan x - 3x - x^3}{x^5}$

12. $\lim_{x \to 4} \dfrac{x^2 - 16}{\ln (x - 3)}$

13. $\lim_{x \to 0} \dfrac{3 \tan^{-1} x - 3x + x^3}{x^5}$

14. $\lim_{x \to 0} \dfrac{\sqrt{1 + x} - 1 - (x/2)}{4x^2}$

15. $\lim_{x \to 0} \dfrac{\sin x - \tan x}{3x^3 \cos x}$

16. $\lim_{x \to 1} \dfrac{x - 1}{\ln x}$

17. $\lim_{x \to 2} \dfrac{x - 2}{\ln (x - 1)}$

18. $\lim_{x \to \infty} \left(x^4 \left(e^{1/x} - 1 \right) - x^3 \right)$

19. $\lim_{x \to 0^+} \dfrac{(1 + x)^{-2} - 4 \cos \sqrt{x} + 3}{2x^2}$

20. $\lim_{x \to 0} \dfrac{(1 - 2x)^{-1/2} - e^x}{8x^2}$

21–26. Power series for derivatives

a. *Differentiate the Taylor series about 0 for the following functions.*

b. *Identify the function represented by the differentiated series.*

c. *Give the interval of convergence of the power series for the derivative.*

21. $f(x) = e^x$

22. $f(x) = \cos x$

23. $f(x) = \ln (1 + x)$

24. $f(x) = \sin (x^2)$

25. $f(x) = e^{-2x}$

26. $f(x) = \sqrt{1 + x}$

27–30. Differential equations

a. *Find a power series for the solution of the following differential equations.*

b. *Identify the function represented by the power series.*

27. $y'(t) - y(t) = 0$, $y(0) = 2$

28. $y'(t) + 4y(t) = 8$, $y(0) = 0$

29. $y'(t) - 3y(t) = 10$, $y(0) = 2$

30. $y'(t) = 6y(t) + 9$, $y(0) = 2$

31–38. Approximating definite integrals *Use a Taylor series to approximate the following definite integrals. Retain as many terms as needed to ensure the error is less than 10^{-4}.*

31. $\int_0^{0.25} e^{-x^2} \, dx$

32. $\int_0^{0.2} \sin x^2 \, dx$

33. $\int_{-0.35}^{0.35} \cos 2x^2 \, dx$

34. $\int_0^{0.2} \sqrt{1 + x^4} \, dx$

35. $\int_0^{0.15} \dfrac{\sin x}{x} \, dx$

36. $\int_0^{0.1} \cos \sqrt{x} \, dx$

37. $\int_0^{0.5} \dfrac{dx}{\sqrt{1 + x^6}}$

38. $\int_0^{0.2} \dfrac{\ln (1 + t)}{t} \, dt$

39–44. Approximating real numbers *Use an appropriate Taylor series to find the first four nonzero terms of an infinite series that is equal to the following numbers.*

39. e^2

40. $\sqrt{e}$

41. $\cos 2$

42. $\sin 1$

43. $\ln \left(\frac{3}{2} \right)$

44. $\tan^{-1} \left(\frac{1}{2} \right)$

45. **Evaluating an infinite series** Let $f(x) = (e^x - 1)/x$ for $x \neq 0$ and $f(0) = 1$. Use the Taylor series for f about 0 and evaluate $f(1)$ to find the value of $\sum_{k=0}^{\infty} \dfrac{1}{(k + 1)!}$.

46. **Evaluating an infinite series** Let $f(x) = (e^x - 1)/x$ for $x \neq 0$ and $f(0) = 1$. Use the Taylor series for f and f' about 0 to evaluate $f'(2)$ and to find the value of $\sum_{k=1}^{\infty} \dfrac{k2^{k-1}}{(k + 1)!}$.

47. **Evaluating an infinite series** Write the Taylor series for $f(x) = \ln (1 + x)$ about 0 and find the interval of convergence. Assume the Taylor series converges to f on the interval of convergence. Evaluate $f(1)$ to find the value of $\sum_{k=1}^{\infty} \dfrac{(-1)^{k+1}}{k}$ (the alternating harmonic series).

48. **Evaluating an infinite series** Write the Taylor series for $f(x) = \ln (1 + x)$ about 0 and find the interval of convergence. Evaluate $f\left(-\frac{1}{2} \right)$ to find the value of $\sum_{k=1}^{\infty} \dfrac{1}{k \cdot 2^k}$.

49–58. Representing functions by power series *Identify the functions represented by the following power series.*

49. $\displaystyle\sum_{k=0}^{\infty} \frac{x^k}{2^k}$

50. $\displaystyle\sum_{k=0}^{\infty} (-1)^k \frac{x^k}{3^k}$

51. $\displaystyle\sum_{k=0}^{\infty} (-1)^k \frac{x^{2k}}{4^k}$

52. $\displaystyle\sum_{k=0}^{\infty} 2^k x^{2k+1}$

53. $\displaystyle\sum_{k=1}^{\infty} \frac{x^k}{k}$

54. $\displaystyle\sum_{k=0}^{\infty} \frac{(-1)^k x^{k+1}}{4^k}$

55. $\displaystyle\sum_{k=1}^{\infty} (-1)^k \frac{kx^{k+1}}{3^k}$

56. $\displaystyle\sum_{k=1}^{\infty} \frac{x^{2k}}{k}$

57. $\displaystyle\sum_{k=2}^{\infty} \frac{k(k-1)x^k}{3^k}$

58. $\displaystyle\sum_{k=2}^{\infty} \frac{x^k}{k(k-1)}$

Further Explorations

59. Explain why or why not Determine whether the following statements are true and give an explanation or counterexample.

a. To evaluate $\displaystyle\int_0^2 \frac{dx}{1-x}$, one could expand the integrand in a Taylor series and integrate term by term.

b. To approximate $\pi/3$, one could substitute $x = \sqrt{3}$ into the Taylor series for $\tan^{-1} x$.

c. $\displaystyle\sum_{k=0}^{\infty} \frac{(\ln 2)^k}{k!} = 2.$

60–62. Limits with a parameter *Use Taylor series to evaluate the following limits. Express the result in terms of the parameter(s).*

60. $\displaystyle\lim_{x \to 0} \frac{e^{ax} - 1}{x}$

61. $\displaystyle\lim_{x \to 0} \frac{\sin ax}{\sin bx}$

62. $\displaystyle\lim_{x \to 0} \frac{\sin ax - \tan ax}{bx^3}$

63. A limit by Taylor series Use Taylor series to evaluate
$$\lim_{x \to 0} \left(\frac{\sin x}{x} \right)^{1/x^2}.$$

64. Inverse hyperbolic sine A function known as the *inverse of the hyperbolic sine* is defined in several ways; among them are
$$\sinh^{-1} x = \ln \left(x + \sqrt{x^2 + 1} \right) = \int_0^x \frac{dt}{\sqrt{1 + t^2}}.$$

Find the first four terms of the Taylor series for $\sinh^{-1} x$ using these two definitions (and be sure they agree).

65–68. Derivative trick *Here is an alternative way to evaluate higher derivatives of a function f that may save time. Suppose you can find the Taylor series for f centered at the point a without evaluating derivatives (for example, from a known series). Explain why $f^{(k)}(a) = k!$ multiplied by the coefficient of $(x - a)^k$.*

Use this idea to evaluate $f^{(3)}(0)$ and $f^{(4)}(0)$ for the following functions. Use known series and do not evaluate derivatives.

65. $f(x) = e^{\cos x}$

66. $f(x) = \dfrac{x^2 + 1}{\sqrt[3]{1 + x}}$

67. $f(x) = \displaystyle\int_0^x \sin(t^2)\, dt$

68. $f(x) = \displaystyle\int_0^x \frac{1}{1 + t^4}\, dt$

Applications

69. Probability: tossing for a head The expected (average) number of tosses of a fair coin required to obtain the first head is

$$\sum_{k=1}^{\infty} k\left(\tfrac{1}{2}\right)^k.$$ Evaluate this series and determine the expected number of tosses. (*Hint:* Differentiate a geometric series.)

70. Probability: sudden death playoff Teams A and B go into sudden death overtime after playing to a tie. The teams alternate possession of the ball and the first team to score wins. Each team has a $\frac{1}{6}$ chance of scoring when it has the ball, with Team A having the ball first.

a. The probability that Team A ultimately wins is $\displaystyle\sum_{k=0}^{\infty} \tfrac{1}{6}\left(\tfrac{5}{6}\right)^{2k}$. Evaluate this series.

b. The expected number of rounds (possessions by either team) required for the overtime to end is $\displaystyle\frac{1}{6}\sum_{k=1}^{\infty} k\left(\tfrac{5}{6}\right)^{k-1}$. Evaluate this series.

71. Elliptic integrals The period of a pendulum is given by
$$T = 4\sqrt{\frac{\ell}{g}} \int_0^{\pi/2} \frac{d\theta}{\sqrt{1 - k^2 \sin^2 \theta}} \equiv 4\sqrt{\frac{\ell}{g}} F(k),$$

where ℓ is the length of the pendulum, $g \approx 9.8$ m/s^2 is the acceleration due to gravity, $k = \sin(\theta_0/2)$, and θ_0 is the initial angular displacement of the pendulum (in radians). The integral in this formula $F(k)$ is called an **elliptic integral** and it cannot be evaluated analytically.

a. Approximate $F(0.1)$ by expanding the integrand in a Taylor (binomial) series and integrating term by term.

b. How many terms of the Taylor series do you suggest using to obtain an approximation to $F(0.1)$ with an error less than 10^{-3}?

c. Would you expect to use fewer or more terms (than in part (b)) to approximate $F(0.2)$ to the same accuracy? Explain.

72. Sine integral function The function $\text{Si}(x) = \displaystyle\int_0^x \frac{\sin t}{t}\, dt$ is called the **sine integral function**.

a. Expand the integrand in a Taylor series about 0.

b. Integrate the series to find a Taylor series for Si.

c. Approximate Si (0.5) and Si (1). Use enough terms of the series so the error in the approximation does not exceed 10^{-3}.

73. Fresnel integrals The theory of optics gives rise to the two **Fresnel integrals**

$$S(x) = \int_0^x \sin(t^2)\, dt \quad \text{and} \quad C(x) = \int_0^x \cos(t^2)\, dt.$$

a. Compute $S'(x)$ and $C'(x)$.
b. Expand $\sin(t^2)$ and $\cos(t^2)$ in a Maclaurin series and then integrate to find the first four nonzero terms of the Maclaurin series for S and C.
c. Use the polynomials in part (b) to approximate $S(0.05)$ and $C(-0.25)$.
d. How many terms of the Maclaurin series are required to approximate $S(0.05)$ with an error no greater than 10^{-4}?
e. How many terms of the Maclaurin series are required to approximate $C(-0.25)$ with an error no greater than 10^{-6}?

74. Error function An essential function in statistics and the study of the normal distribution is the **error function**

$$\text{erf}(x) = \frac{2}{\sqrt{\pi}} \int_0^x e^{-t^2}\, dt.$$

a. Compute the derivative of $\text{erf}(x)$.
b. Expand e^{-t^2} in a Maclaurin series, then integrate to find the first four nonzero terms of the Maclaurin series for erf.
c. Use the polynomial in part (b) to approximate $\text{erf}(0.15)$ and $\text{erf}(-0.09)$.
d. Estimate the error in the approximations of part (c).

75. Bessel functions Bessel functions arise in the study of wave propagation in circular geometries (for example, waves on a circular drum head). They are conveniently defined as power series. One of an infinite family of Bessel functions is

$$J_0(x) = \sum_{k=0}^{\infty} \frac{(-1)^k}{2^{2k}(k!)^2} x^{2k}.$$

a. Write out the first four terms of J_0.
b. Find the radius and interval of convergence of the power series for J_0.
c. Differentiate J_0 twice and show (by keeping terms through x^6) that J_0 satisfies the equation $x^2 y''(x) + x y'(x) + x^2 y(x) = 0$.

Additional Exercises

76. Power series for $\sec x$ Use the identity $\sec x = \dfrac{1}{\cos x}$ and long division to find the first three terms of the Maclaurin series for $\sec x$.

77. Symmetry

a. Use infinite series to show that $\cos x$ is an even function. That is, show $\cos x = \cos(-x)$.
b. Use infinite series to show that $\sin x$ is an odd function. That is, show $\sin x = -\sin(-x)$.

78. Behavior of $\csc x$ We know that $\lim_{x \to 0^+} \csc x = \infty$. Use long division to determine exactly how $\csc x$ grows as $x \to 0^+$. Specifically, find a, b, and c (all positive) in the following sentence:

As $x \to 0^+$, $\csc x \approx \dfrac{a}{x^b} + cx.$

79. L'Hôpital's Rule by Taylor series Suppose f and g have Taylor series about the point a.

a. If $f(a) = g(a) = 0$ and $g'(a) \neq 0$, evaluate $\lim_{x \to a} f(x)/g(x)$ by expanding f and g in their Taylor series. Show that the result is consistent with l'Hôpital's Rule.
b. If $f(a) = g(a) = f'(a) = g'(a) = 0$ and $g''(a) \neq 0$, evaluate $\lim_{x \to a} \dfrac{f(x)}{g(x)}$ by expanding f and g in their Taylor series. Show that the result is consistent with two applications of l'Hôpital's Rule.

80. Newton's derivation of the sine and arcsine series Newton discovered the binomial series and then used it ingeniously to obtain many more results. Here is a case in point.

a. Referring to the figure, show that $x = \sin y$ or $y = \sin^{-1} x$.
b. The area of a circular sector of radius r subtended by an angle θ is $\frac{1}{2} r^2 \theta$. Show that the area of the circular sector APE is $y/2$, which implies that

$$y = 2 \int_0^x \sqrt{1 - t^2}\, dt - x\sqrt{1 - x^2}.$$

c. Use the binomial series for $f(x) = \sqrt{1 - x^2}$ to obtain the first few terms of the Taylor series for $y = \sin^{-1} x$.
d. Newton next inverted the series in part (c) to obtain the Taylor series for $x = \sin y$. He did this by assuming that $\sin y = \sum a_k y^k$ and solving $x = \sin(\sin^{-1} x)$ for the coefficients a_k. Find the first few terms of the Taylor series for $\sin y$ using this idea (a computer algebra system might be helpful as well).

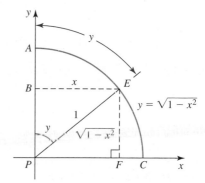

1. $\dfrac{\sin x}{x} = \dfrac{x - x^3/3! + \cdots}{x} = 1 - \dfrac{x^2}{3!} + \cdots \to 1$ as $x \to 0$.

2. The result is the power series for $-\sin x$. **3.** $x = 1/\sqrt{3}$ (which lies in the interval of convergence) ◄

CHAPTER 9   REVIEW EXERCISES

1. Explain why or why not Determine whether the following statements are true and give an explanation or counterexample.

  a. Let p_n be the nth-order Taylor polynomial for f centered at 2. The approximation $p_3(2.1) \approx f(2.1)$ is likely to be more accurate than the approximation $p_2(2.2) \approx f(2.2)$.

  b. If the Taylor series for f centered at 3 has a radius of convergence of 6, then the interval of convergence is $[-3, 9]$.

  c. The interval of convergence of the power series $\sum c_k x^k$ could be $\left(-\frac{7}{3}, \frac{7}{3}\right)$.

  d. The Taylor series for $f(x) = (1 + x)^{12}$ centered at 0 has a finite number of terms.

2–7. Taylor polynomials *Find the nth-order Taylor polynomial for the following functions with the given center point a.*

2. $f(x) = \sin 2x, \ n = 3, \ a = 0$

3. $f(x) = \cos x^2, \ n = 2, \ a = 0$

4. $f(x) = e^{-x}, \ n = 2, \ a = 0$

5. $f(x) = \ln(1 + x), \ n = 3, \ a = 0$

6. $f(x) = \cos x, \ n = 2, \ a = \pi/4$

7. $f(x) = \ln x, \ n = 2, \ a = 1$

8–11. Approximations

  a. *Find the Taylor polynomials of order $n = 0, 1,$ and 2 for the given functions centered at the given point a.*

  b. *Make a table showing the approximations and the absolute error in these approximations using a calculator for the exact function value.*

8. $f(x) = \cos x, \ a = 0$; approximate $\cos(-0.08)$.

9. $f(x) = e^x, \ a = 0$; approximate $e^{-0.08}$.

10. $f(x) = \sqrt{1 + x}, \ a = 0$; approximate $\sqrt{1.08}$.

11. $f(x) = \sin x, \ a = \pi/4$; approximate $\sin(\pi/5)$.

12–14. Estimating remainders *Find the remainder term $R_n(x)$ for the Taylor series centered at 0 for the following functions. Find an upper bound for the magnitude of the remainder on the given interval for the given value of n. (The bound is not unique.)*

12. $f(x) = e^x$; bound $R_3(x)$ for $|x| < 1$.

13. $f(x) = \sin x$; bound $R_3(x)$ for $|x| < \pi$.

14. $f(x) = \ln(1 - x)$; bound $R_3(x)$ for $|x| < 1/2$.

15–20. Radius and interval of convergence *Use the Ratio or Root Test to determine the radius of convergence of the following power series. Test the endpoints to determine the interval of convergence, when appropriate.*

15. $\displaystyle\sum \frac{k^2 x^k}{k!}$

16. $\displaystyle\sum \frac{x^{4k}}{k^2}$

17. $\displaystyle\sum (-1)^k \frac{(x + 1)^{2k}}{k!}$

18. $\displaystyle\sum \frac{(x - 1)^k}{k \cdot 5^k}$

19. $\displaystyle\sum \left(\frac{x}{9}\right)^{3k}$

20. $\displaystyle\sum \frac{(x + 2)^k}{\sqrt{k}}$

21–26. Power series from the geometric series *Use the geometric series $\displaystyle\sum_{k=0}^{\infty} x^k = \frac{1}{1 - x}$, for $|x| < 1$ to determine the Maclaurin series and the interval of convergence for the following functions.*

21. $f(x) = \dfrac{1}{1 - x^2}$

22. $f(x) = \dfrac{1}{1 + x^3}$

23. $f(x) = \dfrac{1}{1 - 3x}$

24. $f(x) = \dfrac{10x}{1 + x}$

25. $f(x) = \dfrac{1}{(1 - x)^2}$

26. $f(x) = \ln(1 + x^2)$

27–32. Taylor series *Write out the first three terms of the Taylor series for the following functions centered at the given point a. Then write the series using summation notation.*

27. $f(x) = e^{3x}, \ a = 0$

28. $f(x) = \dfrac{1}{x}, \ a = 1$

29. $f(x) = \cos x, \ a = \pi/2$

30. $f(x) = -\ln(1 - x), \ a = 0$

31. $f(x) = \tan^{-1} x, \ a = 0$

32. $f(x) = \sin 2x, \ a = -\pi/2$

33–36. Binomial series *Write out the first three terms of the Maclaurin series for the following functions.*

33. $f(x) = (1 + x)^{1/3}$

34. $f(x) = (1 + x)^{-1/2}$

35. $f(x) = (1 + x/2)^{-3}$

36. $f(x) = (1 + 2x)^{-5}$

37–40. Convergence *Write the remainder term $R_n(x)$ for the Taylor series for the following functions centered at the given point a. Then show that $\displaystyle\lim_{n \to \infty} R_n(x) = 0$ for all x in the given interval.*

37. $f(x) = e^{-x}, \ a = 0, \ -\infty < x < \infty$

38. $f(x) = \sin x, \ a = 0, \ -\infty < x < \infty$

39. $f(x) = \ln(1 + x), \ a = 0, \ -\frac{1}{2} \le x \le \frac{1}{2}$

40. $f(x) = \sqrt{1 + x}, \ a = 0, \ -\frac{1}{2} \le x \le \frac{1}{2}$

41–46. Limits by power series *Use Taylor series to evaluate the following limits.*

41. $\displaystyle\lim_{x \to 0} \frac{x^2/2 - 1 + \cos x}{x^4}$

42. $\displaystyle\lim_{x \to 0} \frac{2 \sin x - \tan^{-1} x - x}{2x^5}$

43. $\displaystyle\lim_{x \to 4} \frac{\ln(x - 3)}{x^2 - 16}$

44. $\displaystyle\lim_{x \to 0} \frac{\sqrt{1 + 2x} - 1 - x}{x^2}$

45. $\displaystyle\lim_{x \to 0} \frac{\sec x - \cos x - x^2}{x^4}$

46. $\displaystyle\lim_{x \to 0} \frac{(1 + x)^{-2} - \sqrt[3]{1 - 6x}}{2x^2}$

47. A differential equation Find a power series solution of the differential equation $y'(x) - 4y(x) + 12 = 0$, subject to the condition $y(0) = 4$. Identify the solution in terms of known functions.

T 48. Rejected quarters The probability that a random quarter is *not* rejected by a vending machine is given by the integral $11.4 \int_0^{0.14} e^{-102x^2} \, dx$ (assuming that the weights of quarters are normally distributed with a mean of 5.670 g and a standard deviation of 0.07 g). Expand the integrand in $n = 2$ and $n = 3$ terms of a Taylor series and integrate to find two estimates of the probability. Check for agreement between the two estimates.

T 49. Approximating ln 2 Consider the following three ways to approximate ln 2.

   a. Use the Taylor series for $\ln(1 + x)$ centered at 0 and evaluate it at $x = 1$ (convergence was asserted in Table 9.5). Write the resulting infinite series.

   b. Use the Taylor series for $\ln(1 - x)$ centered at 0 and the identity $\ln 2 = -\ln\left(\frac{1}{2}\right)$. Write the resulting infinite series.

   c. Use the property $\ln(a/b) = \ln a - \ln b$ and the series of parts (a) and (b) to find the Taylor series for $f(x) = \ln\left(\frac{1 + x}{1 - x}\right)$ centered at 0.

   d. At what value of x should the series in part (c) be evaluated to approximate ln 2? Write the resulting infinite series for ln 2.

   e. Using four terms of the series, which of the three series derived in parts (a)–(d) gives the best approximation to ln 2? Which series gives the worst approximation? Can you explain why?

T 50. Graphing Taylor polynomials Consider the function $f(x) = (1 + x)^{-4}$.

   a. Find the Taylor polynomials p_0, p_1, p_2, and p_3 centered at 0.

   b. Use a graphing utility to plot the Taylor polynomials and f for $-1 < x < 1$.

   c. For each Taylor polynomial, give the interval on which its graph appears indistinguishable from the graph of f.

Chapter 9 Guided Projects

Applications of the material in this chapter and related topics can be found in the following Guided Projects. For additional information, see the Preface.

• Euler's formula (Taylor series with complex numbers)

• Fourier Series

• Series approximations to π

• Stirling's formula and $n!$

• Three-sigma quality control

Parametric and Polar Curves

Chapter Preview Until now, all our work has been done in the Cartesian coordinate system with functions of the form $y = f(x)$. There are, however, alternative ways to generate curves and represent functions. We begin by introducing parametric equations, which are featured prominently in Chapter 11 to represent curves and trajectories in three-dimensional space. When working with objects that have circular, cylindrical, or spherical shapes, other coordinate systems are often advantageous. In this chapter, we introduce the polar coordinate system for circular geometries. Cylindrical and spherical coordinate systems appear in Chapter 13. After working with parametric equations and polar coordinates, the next step is to investigate calculus in these settings. How do we find slopes of tangent lines and rates of changes? How are areas of regions bounded by curves in polar coordinates computed? The chapter ends with the related topic of *conic sections*. Ellipses, parabolas, and hyperbolas can be represented in both Cartesian and polar coordinates. These important families of curves have many fascinating properties and appear throughout the remainder of the book.

10.1 Parametric Equations

So far, we have used functions of the form $y = f(x)$ to describe curves in the xy-plane. In this section we look at another way to define curves, known as *parametric equations*. As you will see, parametric curves enable us to describe both common and exotic curves; they are also indispensable for modeling the trajectories of moving objects.

Basic Ideas

A motor boat speeds counterclockwise around a circular course with a radius of 4 mi, completing one lap every hour at a constant speed. Suppose we wish to describe the points on the path of the boat $(x(t), y(t))$ at any time $t \geq 0$, where t is measured in hours. We assume that the boat starts on the positive x-axis at the point $(4, 0)$ (Figure 10.1). Note that the angle θ corresponding to the position of the boat increases by 2π radians every hour beginning with $\theta = 0$ when $t = 0$; therefore, $\theta = 2\pi t$ for $t \geq 0$. As we show in Example 2, the x- and y-coordinates of the boat are

$$x = 4 \cos \theta = 4 \cos 2\pi t \quad \text{and} \quad y = 4 \sin \theta = 4 \sin 2\pi t,$$

where $t \geq 0$. You can confirm that when $t = 0$, the boat is at the starting point $(4, 0)$; when $t = 1$, it returns to the starting point.

The equations $x = 4 \cos 2\pi t$ and $y = 4 \sin 2\pi t$ are examples of **parametric equations**. They specify x and y in terms of a third variable t called a **parameter**, which often represents time (Figure 10.2).

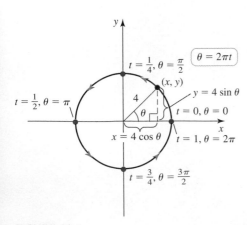

FIGURE 10.1

▶ You can think of the parameter t as the independent variable. There are two dependent variables, x and y.

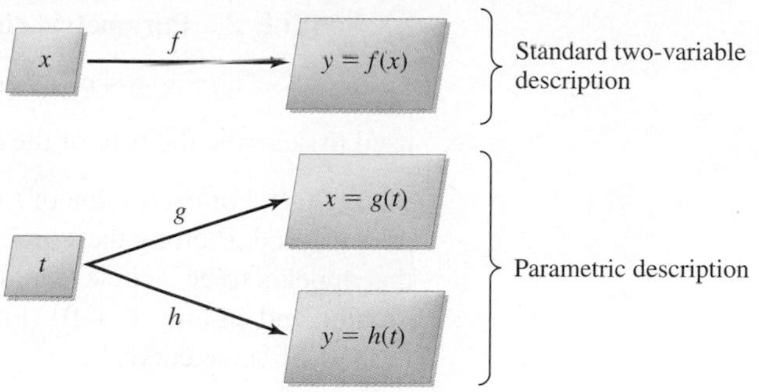

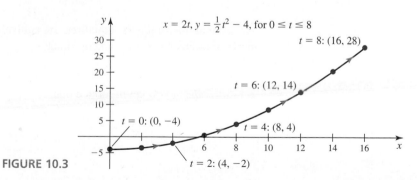

FIGURE 10.2

In general, parametric equations have the form

$$x = g(t), \qquad y = h(t),$$

where g and h are given functions and the parameter t typically varies over a specified interval, such as $a \le t \le b$. The **parametric curve** described by these equations consists of the points in the plane that satisfy

$$(x, y) = (g(t), h(t)), \qquad \text{for } a \le t \le b.$$

EXAMPLE 1 **Parametric parabola** Graph and analyze the parametric equations

$$x = g(t) = 2t, \qquad y = h(t) = \frac{1}{2}t^2 - 4, \qquad \text{for } 0 \le t \le 8.$$

SOLUTION Plotting individual points often helps visualize a parametric curve. Table 10.1 shows the values of x and y corresponding to several values of t on the interval $[0, 8]$. By plotting the (x, y) pairs in Table 10.1 and connecting them with a smooth curve, we obtain the graph shown in Figure 10.3. We see that as t increases from its initial value of $t = 0$ to its final value of $t = 8$, the curve is generated from the initial point $(0, -4)$ to the final point $(16, 28)$. Notice that the values of the parameter do not appear in the graph. The only signature of the parameter is the direction in which the curve is generated: in this case, it unfolds upward and to the right.

Table 10.1

t	x	y	(x, y)
0	0	-4	$(0, -4)$
1	2	$-\frac{7}{2}$	$\left(2, -\frac{7}{2}\right)$
2	4	-2	$(4, -2)$
3	6	$\frac{1}{2}$	$\left(6, \frac{1}{2}\right)$
4	8	4	$(8, 4)$
5	10	$\frac{17}{2}$	$\left(10, \frac{17}{2}\right)$
6	12	14	$(12, 14)$
7	14	$\frac{41}{2}$	$\left(14, \frac{41}{2}\right)$
8	16	28	$(16, 28)$

FIGURE 10.3

Occasionally, it is possible to eliminate the parameter from a set of parametric equations and obtain a description of the curve in terms of x and y. In this case, from the x-equation, we have $t = x/2$, which may be substituted into the y-equation to give

$$y = \frac{1}{2}t^2 - 4 = \frac{1}{2}\left(\frac{x}{2}\right)^2 - 4 = \frac{x^2}{8} - 4.$$

Expressed in this form, we identify the graph as part of a parabola.

Related Exercises 7–14 ◀

QUICK CHECK 1 Identify the graph that is generated by the parametric equations $x = t^2$, $y = t$, for $-10 \le t \le 10$. ◀

EXAMPLE 2 **Parametric circle** Graph and analyze the parametric equations

$$x = 4\cos 2\pi t, \qquad y = 4\sin 2\pi t, \qquad \text{for } 0 \le t \le 1$$

used to describe the path of the motor boat in the opening paragraphs.

SOLUTION For each value of t in Table 10.2, the corresponding ordered pairs (x, y) are recorded. Plotting these points as t increases from $t = 0$ to $t = 1$ results in a graph that appears to be a circle of radius 4; it is generated in a counterclockwise direction, beginning and ending at $(4, 0)$ (Figure 10.4). Letting t increase beyond $t = 1$ would simply retrace the same curve.

Table 10.2

t	(x, y)
0	$(4, 0)$
$\frac{1}{8}$	$(2\sqrt{2}, 2\sqrt{2})$
$\frac{1}{4}$	$(0, 4)$
$\frac{3}{8}$	$(-2\sqrt{2}, 2\sqrt{2})$
$\frac{1}{2}$	$(-4, 0)$
$\frac{3}{4}$	$(0, -4)$
1	$(4, 0)$

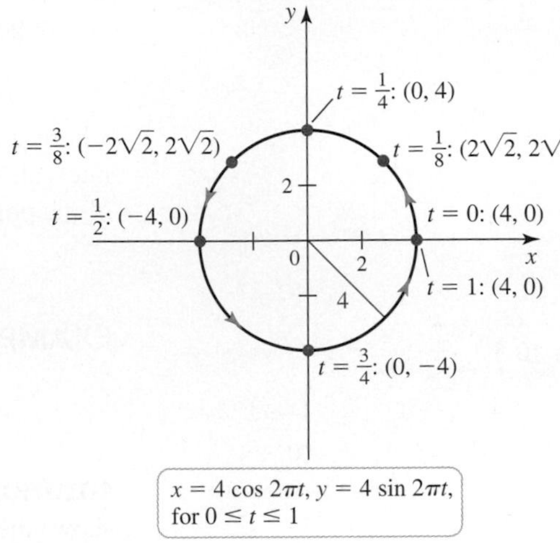

FIGURE 10.4

To identify the curve conclusively, the parameter t is eliminated by writing

$$x^2 + y^2 = (4\cos 2\pi t)^2 + (4\sin 2\pi t)^2$$
$$= 16\underbrace{(\cos^2 2\pi t + \sin^2 2\pi t)}_{1} = 16.$$

We see that the parametric equations are equivalent to $x^2 + y^2 = 16$, whose graph is a circle of radius 4 in Cartesian coordinates.   *Related Exercises 15–22* ◀

Generalizing Example 2 for nonzero real numbers a and b in the parametric equations $x = a\cos bt$, $y = a\sin bt$, notice that

$$x^2 + y^2 = (a\cos bt)^2 + (a\sin bt)^2$$
$$= a^2\underbrace{(\cos^2 bt + \sin^2 bt)}_{1} = a^2.$$

> Recall that the functions $\sin bt$ and $\cos bt$ have period $2\pi/|b|$. The equations $x = a\cos bt$, $y = -a\sin bt$ also describe a circle of radius $|a|$, as do the equations $x = a\sin bt$, $y = \pm a\cos bt$.

Therefore, the parametric equations $x = a\cos bt$, $y = a\sin bt$ describe the circle $x^2 + y^2 = a^2$, centered at the origin with radius $|a|$, for *any* nonzero value of b. The circle is traversed once as t varies over any interval of length $2\pi/|b|$. If t represents time, the circle is traversed in $2\pi/|b|$ time units, which means we can vary the speed at which the curve unfolds by varying b. If $b > 0$, the curve is generated in the counterclockwise direction. If $b < 0$, the curve has a clockwise direction.

More generally, the parametric equations

$$x = x_0 + a\cos bt, \qquad y = y_0 + a\sin bt$$

describe the circle $(x - x_0)^2 + (y - y_0)^2 = a^2$, centered at (x_0, y_0) with radius $|a|$. If $b > 0$, then the circle is generated in the counterclockwise direction.

> Example 3 shows that a single curve—for example, a circle of radius 4—may be parameterized in many different ways.

EXAMPLE 3   Circular path A turtle walks with constant speed in the counterclockwise direction on a circular track of radius 4 ft centered at the origin. Starting from the point $(4, 0)$, the turtle completes one lap in 30 minutes. Find a parametric description of the path of the turtle at any time $t \geq 0$.

SOLUTION Example 2 showed that a circle of radius of 4 may be described by the parametric equations

$$x = 4 \cos bt, \qquad y = 4 \sin bt.$$

> The constant $|b|$ is called the *angular frequency* because it is the number of radians the object moves per unit time. The turtle travels 2π rad every 30 min, so the angular frequency is $2\pi/30 = \pi/15$ rad/min. Because radians have no units, the angular frequency in this case has units *per minute*, sometimes written as min^{-1}.

The *angular frequency* b must be chosen so that, as t varies from 0 to 30, the product bt varies from 0 to 2π. Specifically, when $t = 30$, we must have $30b = 2\pi$, or $b = \pi/15$ rad/min. Therefore, the parametric equations for the turtle's motion are

$$x = 4 \cos\left(\frac{\pi t}{15}\right), \qquad y = 4 \sin\left(\frac{\pi t}{15}\right), \qquad \text{for } 0 \leq t \leq 30.$$

You should check that as t varies from 0 to 30, the points (x, y) make one complete circuit of a circle of radius 4 (Figure 10.5). *Related Exercises 23–26* ◀

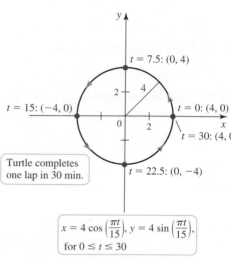

$$x = 4 \cos\left(\frac{\pi t}{15}\right), y = 4 \sin\left(\frac{\pi t}{15}\right),$$ for $0 \leq t \leq 30$

FIGURE 10.5

QUICK CHECK 2 Give the center and radius of the circle generated by the equations $x = 3 \sin t, y = -3 \cos t$, for $0 \leq t \leq 2\pi$. Specify the direction in which the curve is generated. ◀

EXAMPLE 4   Parametric lines Express the curve described by the equations $x = x_0 + at, y = y_0 + bt$ in the form $y = f(x)$. Assume that $x_0, y_0, a \neq 0$, and b are constants and $-\infty < t < \infty$.

SOLUTION The parameter t may be eliminated by solving the x-equation for t, resulting in $t = (x - x_0)/a$. Substituting t into the y-equation, we have

$$y = y_0 + bt = y_0 + b\left(\frac{x - x_0}{a}\right) \quad \text{or} \quad y - y_0 = \frac{b}{a}(x - x_0).$$

This equation describes the line with slope b/a passing through the point (x_0, y_0). Figure 10.6 illustrates the line $x = 2 + 3t, y = 1 + t$, which passes through the point $(2, 1)$ at $t = 0$ with slope $\frac{1}{3}$.

> We can also vary the point on the line that corresponds to $t = 0$. For example, the equations
>
> $$x = -1 + 6t, \qquad y = 2t$$
>
> produce the same line shown in Figure 10.6. However, the point corresponding to $t = 0$ is $(-1, 0)$.

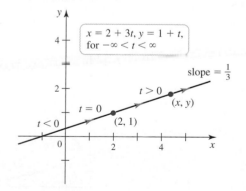

$x = 2 + 3t, y = 1 + t,$ for $-\infty < t < \infty$

slope $= \frac{1}{3}$

FIGURE 10.6

Notice that the parametric description of a given line is not unique: If k is any nonzero constant, the numbers a and b may be replaced by ka and kb, respectively, and the resulting equations describe the same line (although it is traversed at a different speed). If $b = 0$ and $a \neq 0$, the line has zero slope and is horizontal. If $a = 0$ and $b \neq 0$, the line is vertical. *Related Exercises 27–34* ◀

QUICK CHECK 3 Describe the curve generated by $x = 3 + 2t, y = -12 - 6t$, for $-\infty < t < \infty$. ◀

EXAMPLE 5   Parametric equations of curves   A common task (particularly in upcoming chapters) is to parameterize curves given either by Cartesian equations or by graphs. Find a parametric representation of the following curves.

a. The segment of the parabola $y = 9 - x^2$, for $-1 \le x \le 3$

b. The complete curve $x = (y - 5)^2 + \sqrt{y}$

c. The piecewise linear path that connects $P(-2, 0)$ to $Q(0, 3)$ to $R(4, 0)$ (in that order), where the parameter varies over the interval $0 \le t \le 2$.

SOLUTION

a. The simplest way to represent the curve $y = f(x)$ parametrically is to let $x = t$ and $y = f(t)$, where t is the parameter. We must then find the appropriate interval for the parameter. Using this approach, the curve $y = 9 - x^2$ has the parametric representation

$$x = t, \qquad y = 9 - t^2, \qquad \text{for } -1 \le t \le 3.$$

This representation is not unique. You should check that the parametric equations

$$x = 1 - t, \qquad y = 9 - (1 - t)^2, \qquad \text{for } -2 \le t \le 2$$

also do the job, although these equations trace the parabola from right to left, while the original equations trace the curve from left to right (Figure 10.7).

b. In this case, it is easier to let $y = t$. Then, a parametric description of the curve is

$$x = (t - 5)^2 + \sqrt{t}, \qquad y = t.$$

Notice that t can take values only in the interval $[0, \infty)$. As $t \to \infty$, we see that $x \to \infty$ and $y \to \infty$ (Figure 10.8).

c. The path consists of two line segments (Figure 10.9) that can be parameterized separately in the form $x = x_0 + at$ and $y = y_0 + bt$. The line segment PQ originates at $(-2, 0)$ and unfolds in the positive x-direction with slope $\frac{3}{2}$. It can be represented as

$$x = -2 + 2t, \qquad y = 3t, \qquad \text{for } 0 \le t \le 1.$$

The line segment QR originates at $(0, 3)$ and unfolds in the positive x-direction with slope $-\frac{3}{4}$. On the interval $1 \le t \le 2$, the point $(0, 3)$ corresponds to $t = 1$. Therefore, the line segment has the representation

$$x = -4 + 4t, \qquad y = 6 - 3t, \qquad \text{for } 1 \le t \le 2.$$

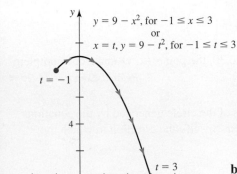

$y = 9 - x^2$, for $-1 \le x \le 3$
or
$x = t, y = 9 - t^2$, for $-1 \le t \le 3$

FIGURE 10.7

➤ In moving from P to Q, y increases as x increases. In moving from Q to R, y decreases as x increases. The parametric equations must reflect these changes. Recall that the line $x = x_0 + at$, $y = y_0 + bt$ has slope b/a.

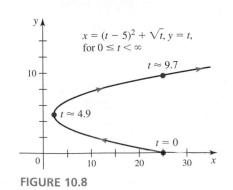

$x = (t - 5)^2 + \sqrt{t}$, $y = t$,
for $0 \le t < \infty$

$t \approx 9.7$

$t \approx 4.9$

$t = 0$

FIGURE 10.8

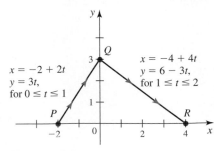

$x = -2 + 2t$
$y = 3t$,
for $0 \le t \le 1$

$x = -4 + 4t$
$y = 6 - 3t$,
for $1 \le t \le 2$

FIGURE 10.9

It is always wise to check the endpoints of the line segments for consistency. As before, this representation is not unique.

Related Exercises 35–38 ◄

QUICK CHECK 4 Find parametric equations for the line segment that goes from $Q(0, 3)$ to $P(-2, 0)$. ◄

EXAMPLE 6 Rolling wheels Many fascinating curves are generated by points on rolling wheels. The path of a light on the rim of a rolling wheel (Figure 10.10) is a **cycloid**, which has the parametric equations

$$x = a(t - \sin t), \qquad y = a(1 - \cos t), \qquad \text{for } t \geq 0,$$

where $a > 0$. Use a graphing utility to graph the cycloid with $a = 1$. On what interval does the parameter generate one arch of the cycloid?

SOLUTION The graph of the cycloid for $0 \leq t \leq 3\pi$ is shown in Figure 10.11. The wheel completes one full revolution on the interval $0 \leq t \leq 2\pi$, which gives one arch of the cycloid.

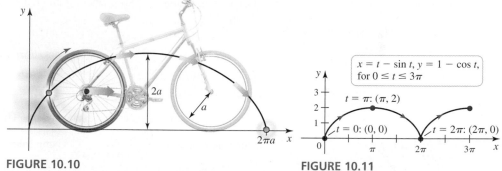

FIGURE 10.10

FIGURE 10.11

Related Exercises 39–44 ◄

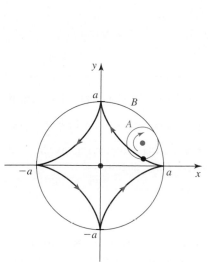

FIGURE 10.12

EXAMPLE 7 More rolling wheels The path of a point on circle A with radius $a/4$ that rolls on the inside of circle B with radius a (Figure 10.12) is an **astroid** or **hypocycloid**. Its parametric equations are

$$x = a \cos^3 t, \qquad y = a \sin^3 t, \qquad \text{for } 0 \leq t \leq 2\pi.$$

Graph the astroid with $a = 1$ and find its equation in terms of x and y.

SOLUTION Because both $\cos^3 t$ and $\sin^3 t$ have a period of 2π, the complete curve is generated on the interval $0 \leq t \leq 2\pi$ (Figure 10.13). To eliminate t from the parametric equations, note that $x^{2/3} = \cos^2 t$ and $y^{2/3} = \sin^2 t$. Therefore,

$$x^{2/3} + y^{2/3} = \cos^2 t + \sin^2 t = 1,$$

where the Pythagorean identity has been used. We see that an alternative description of the astroid is $x^{2/3} + y^{2/3} = 1$.

Related Exercises 39–44 ◄

Given a set of parametric equations, the preceding examples show that as the parameter increases, the corresponding curve unfolds in a particular direction. The following definition captures this fact and is important in upcoming work.

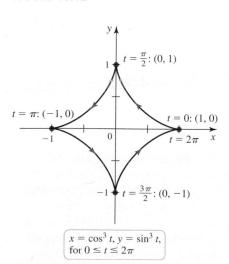

FIGURE 10.13

DEFINITION Forward or Positive Orientation

The direction in which a parametric curve is generated as the parameter increases is called the **forward** or **positive orientation** of the curve.

Derivatives and Parametric Equations

Parametric equations express a relationship between the variables x and y. Therefore, it makes sense to ask about dy/dx, the rate of change of y with respect to x at a point on a parametric curve. Once we know how to compute dy/dx, it can be used to determine slopes of lines tangent to parametric curves.

Consider the parametric equations $x = g(t), y = h(t)$ on an interval on which both g and h are differentiable. The Chain Rule relates the derivatives dy/dt, dx/dt, and dy/dx:

$$\frac{dy}{dt} = \frac{dy}{dx} \frac{dx}{dt}$$

Provided that $dx/dt \neq 0$, we divide both sides of this equation by dx/dt and solve for dy/dx to obtain the following result.

> We will soon interpret $x'(t)$ and $y'(t)$ as the horizontal and vertical velocities, respectively, of an object moving along a curve. The slope of the curve at a point is the ratio of the velocity components at that point.

THEOREM 10.1   Derivative for Parametric Curves

Let $x = g(t)$ and $y = h(t)$, where g and h are differentiable on an interval $[a, b]$. Then

$$\frac{dy}{dx} = \frac{dy/dt}{dx/dt} = \frac{h(t)}{g(t)},$$

provided $dx/dt \neq 0$.

Figure 10.14 gives a geometric explanation of Theorem 10.1. The slope of the line tangent to a curve at a point is $\dfrac{dy}{dx} = \lim\limits_{\Delta x \to 0} \dfrac{\Delta y}{\Delta x}$. Using linear approximation (Section 4.5), we have $\Delta x \approx x'(t)\Delta t$ and $\Delta y \approx y'(t)\Delta t$, with these approximations improving as $\Delta t \to 0$. Notice also that $\Delta t \to 0$ as $\Delta x \to 0$. Therefore, the slope of the tangent line is

$$\frac{dy}{dx} = \lim_{\Delta x \to 0} \frac{\Delta y}{\Delta x} = \lim_{\Delta t \to 0} \frac{y'(t)\Delta t}{x'(t)\Delta t} = \frac{y'(t)}{x'(t)}.$$

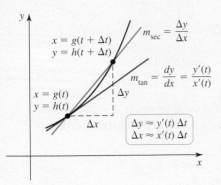

FIGURE 10.14

EXAMPLE 8   Slopes of tangent lines Find dy/dx for the following curves. Interpret the result and determine the points (if any) at which the curve has a horizontal or a vertical tangent line.

a. $x = t, y = 2\sqrt{t}$, for $t \geq 0$

b. $x = 4 \cos t, y = 16 \sin t$, for $0 \leq t \leq 2\pi$

SOLUTION

a. We find that $x'(t) = 1$ and $y'(t) = 1/\sqrt{t}$. Therefore,

$$\frac{dy}{dx} = \frac{y'(t)}{x'(t)} = \frac{1/\sqrt{t}}{1} = \frac{1}{\sqrt{t}},$$

provided $t \neq 0$. Notice that $dy/dx \neq 0$ for $t > 0$, so the curve has no horizontal tangent lines. On the other hand, as $t \to 0^+$, we see that $dy/dx \to \infty$. Therefore, the curve has a vertical tangent line at the point $(0, 0)$. To eliminate t from the parametric equations, we substitute $t = x$ into the y-equation to find that $y = 2\sqrt{x}$, or $x = y^2/4$. Because $y \geq 0$, the curve is the upper half of a parabola (Figure 10.15). Slopes of tangent lines at other points on the curve are found by substituting the corresponding values of t. For example, the point $(4, 4)$ corresponds to $t = 4$ and the slope of the tangent line at that point is $1/\sqrt{4} = \frac{1}{2}$.

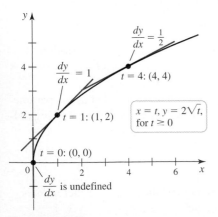

FIGURE 10.15

b. These parametric equations describe an **ellipse** with a long axis of length 32 on the y-axis and a short axis of length 8 on the x-axis (Figure 10.16). In this case, $x'(t) = -4 \sin t$ and $y'(t) = 16 \cos t$. Therefore,

$$\frac{dy}{dx} = \frac{y'(t)}{x'(t)} = \frac{16 \cos t}{-4 \sin t} = -4 \cot t.$$

At $t = 0$ and $t = \pi$, $\cot t$ is undefined, and vertical tangent lines occur at the corresponding points $(\pm 4, 0)$. At $t = \pi/2$ and $t = 3\pi/2$, $\cot t = 0$ and the curve has horizontal tangent lines at the corresponding points $(0, \pm 16)$. Slopes of tangent lines at other points on the curve may be found. For example, the point $(2\sqrt{2}, 8\sqrt{2})$ corresponds to $t = \pi/4$; the slope of the tangent line at that point is $-4 \cot \pi/4 = -4$.

> In general, the equations $x = a \cos t$, $y = b \sin t$, for $0 \le t \le 2\pi$, describe an ellipse. The constants a and b can be seen as horizontal and vertical scalings of the unit circle $x = \cos t$, $y = \sin t$. Ellipses are explored in Exercises 57–62 and in Section 10.4.

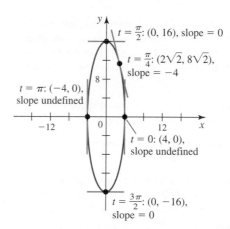

FIGURE 10.16

Related Exercises 45–50 ◄

SECTION 10.1 EXERCISES

Review Questions

1. Explain how a set of parametric equations generates a curve in the xy-plane.

2. Give two sets of parametric equations that generate a circle centered at the origin with radius 6.

3. Give a set of parametric equations that describes a full circle of radius R, where the parameter varies over the interval $[0, 10]$.

4. Give a set of parametric equations that generates the line with slope -2 passing through $(1, 3)$.

5. Find a set of parametric equations for the parabola $y = x^2$.

6. Describe the similarities and differences between the parametric equations $x = t$, $y = t^2$ and $x = -t$, $y = t^2$, where $t \ge 0$ in each case.

Basic Skills

7–10. Working with parametric equations *Consider the following parametric equations.*

   a. *Make a brief table of values of t, x, and y.*
   b. *Plot the points in the table and the full parametric curve, indicating the positive orientation (the direction of increasing t).*
   c. *Eliminate the parameter to obtain an equation in x and y.*
   d. *Describe the curve.*

7. $x = 2t$, $y = 3t - 4$; $-10 \le t \le 10$

8. $x = t^2 + 2$, $y = 4t$; $-4 \le t \le 4$

9. $x = -t + 6$, $y = 3t - 3$; $-5 \le t \le 5$

10. $x = \ln 5t$, $y = \ln t^2$; $1 \le t \le e$

11–14. Working with parametric equations *Consider the following parametric equations.*

   a. *Eliminate the parameter to obtain an equation in x and y.*
   b. *Describe the curve and indicate the positive orientation.*

11. $x = \sqrt{t} + 4$, $y = 3\sqrt{t}$; $0 \le t \le 16$

12. $x = (t + 1)^2$, $y = t + 2$; $-10 \le t \le 10$

13. $x = t - 1$, $y = t^3$; $-4 \le t \le 4$

14. $x = e^{2t}$, $y = e^t + 1$; $0 \le t \le 25$

15–18. Circles and arcs *Eliminate the parameter to find a description of the following circles or circular arcs in terms of x and y. Give the center and radius, and indicate the positive orientation.*

15. $x = 3 \cos t$, $y = 3 \sin t$; $\pi \le t \le 2\pi$

16. $x = 3 \cos t$, $y = 3 \sin t$; $0 \le t \le \pi/2$

17. $x = -7 \cos 2t$, $y = -7 \sin 2t$; $0 \le t \le \pi$

18. $x = 1 - 3 \sin 4\pi t$, $y = 2 + 3 \cos 4\pi t$; $0 \le t \le \frac{1}{2}$

19–22. Parametric equations of circles *Find parametric equations for the following circles (give an interval for the parameter values). Graph the circle and find a description in terms of x and y. There is more than one way to describe any circle.*

19. A circle centered at the origin with radius 4, generated counterclockwise

20. A circle centered at the origin with radius 12, generated clockwise with initial point $(0, 12)$

21. A circle centered at $(-2, -3)$ with radius 8, generated clockwise

22. A circle centered at $(2, -4)$ with radius 3/2, generated counterclockwise with initial point $\left(\frac{7}{2}, -4\right)$

23–26. Circular motion *Find parametric equations that describe the circular path of the following objects. Assume (x, y) denotes the position of the object relative to the origin at the center of the circle. Use the units of time specified in the problem. There is more than one way to describe any circle.*

23. A go-cart moves counterclockwise with constant speed around a circular track of radius 400 m, completing a lap in 1.5 min.

24. The tip of the 15-in second hand of a clock completes one revolution in 60 s.

25. A bicyclist rides counterclockwise with constant speed around a circular velodrome track with a radius of 50 m, completing one lap in 24 s.

26. A Ferris wheel has a radius of 20 m and completes a revolution in the clockwise direction at constant speed in 3 min. Assume that x and y measure the horizontal and vertical positions of a seat on the Ferris wheel relative to a coordinate system whose origin is at the low point of the wheel. Assume the seat begins moving at the origin.

27–30. Parametric lines *Find the slope of each line and a point on the line. Then graph the line.*

27. $x = 3 + t, y = 1 - t$ **28.** $x = 4 - 3t, y = -2 + 6t$

29. $x = 8 + 2t, y = 1$ **30.** $x = 1 + 2t/3, y = -4 - 5t/2$

31–34. Line segments *Find a parametric description of the line segment from the point P to the point Q. The solution is not unique.*

31. $P(0, 0), Q(2, 8)$ **32.** $P(1, 3), Q(-2, 6)$

33. $P(-1, -3), Q(6, -16)$ **34.** $P(-8, 2), Q(1, 2)$

35–38. Curves to parametric equations *Give a set of parametric equations that describes the following curves. Graph the curve and indicate the positive orientation. Be sure to specify the interval over which the parameter varies.*

35. The segment of the parabola $y = 2x^2 - 4$, where $-1 \le x \le 5$

36. The complete curve $x = y^3 - 3y$

37. The piecewise linear path from $P(-2, 3)$ to $Q(2, -3)$ to $R(3, 5)$.

38. The path consisting of the line segment from $(-4, 4)$ to $(0, 8)$, followed by the segment of the parabola $y = 8 - 2x^2$ from $(0, 8)$ to $(2, 0)$

39–44. More parametric curves *Use a graphing utility to graph the following curves. Be sure to choose an interval for the parameter that generates all features of interest.*

39. Spiral $x = t \cos t, y = t \sin t; \ t \ge 0$

40. Witch of Agnesi $x = 2 \cot t, y = 1 - \cos 2t$

41. Folium of Descartes $x = \dfrac{3t}{1 + t^3}, y = \dfrac{3t^2}{1 + t^3}$ (graph the curve in two pieces)

42. Involute of a circle $x = \cos t + t \sin t, y = \sin t - t \cos t$

43. Evolute of an ellipse $x = (a^2 - b^2) \cos^3 t, y = (a^2 - b^2) \sin^3 t$; $a = 4$ and $b = 3$.

44. Cissoid of Diocles $x = 2 \sin 2t, y = \dfrac{2 \sin^3 t}{\cos t}$

45–50. Derivatives *Consider the following parametric curves.*

   a. Determine dy/dx in terms of t and evaluate it at the given value of t.
   b. Make a sketch of the curve showing the tangent line at the point corresponding to the given value of t.

45. $x = 2 + 4t, y = 4 - 8t; \ t = 2$

46. $x = 3 \sin t, y = 3 \cos t; \ t = \pi/2$

47. $x = \cos t, y = 8 \sin t; \ t = \pi/2$

48. $x = 2t, y = t^3; \ t = -1$

49. $x = t + 1/t, y = t - 1/t; \ t = 1$

50. $x = \sqrt{t}, y = 2t; \ t = 4$

Further Explorations

51. **Explain why or why not** Determine whether the following statements are true and give an explanation or counterexample.

   a. The equations $x = -\cos t, y = -\sin t$, for $0 \le t \le 2\pi$ generate a circle in the clockwise direction.
   b. An object following the parametric curve $x = 2 \cos 2\pi t$, $y = 2 \sin 2\pi t$ circles the origin once every 1 time unit.
   c. The parametric equations $x = t, y = t^2$ for $t \ge 0$ describe the complete parabola $y = x^2$.
   d. The parametric equations $x = \cos t, y = \sin t$ for $-\pi/2 \le t \le \pi/2$ describe a semicircle.

52–55. Words to curves *Find parametric equations for the following curves. Include an interval for the parameter values.*

52. The left half of the parabola $y = x^2 + 1$, originating at $(0, 1)$

53. The line that passes through the points $(1, 1)$ and $(3, 5)$, oriented in the direction of increasing x

54. The lower half of the circle centered at $(-2, 2)$ with radius 6, oriented in the counterclockwise direction

55. The upper half of the parabola $x = y^2$, originating at $(0, 0)$

56. **Matching curves and equations** Match the following four equations with the four graphs in the accompanying figure. Explain your reasoning.

   a. $x = t^2 - 2, y = t^3 - t$
   b. $x = \cos (t + \sin 50t), y = \sin (t + \cos 50t)$

c. $x = t + \cos 2t,\ y = t - \sin 4t$

d. $x = 2\cos t + \cos 20t,\ y = 2\sin t + \sin 20t$

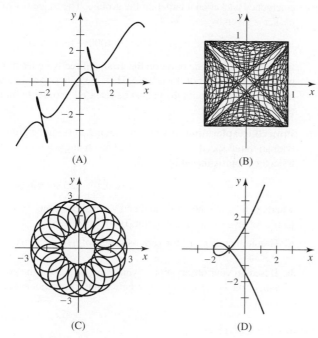

(A)   (B)

(C)   (D)

57–58. Ellipses *An* ***ellipse*** *(discussed in detail in Section 10.4) is generated by the parametric equations* $x = a\cos t,\ y = b\sin t$. *If* $0 < a < b$, *then the long axis (or* ***major axis***) *lies on the y-axis and the short axis (or* ***minor axis***) *lies on the x-axis. If* $0 < b < a$, *the axes are reversed. The lengths of the axes in the x- and y-directions are 2a and 2b, respectively. Sketch the graph of the following ellipses. Specify an interval in t over which the entire curve is generated.*

57. $x = 4\cos t,\ y = 9\sin t.$ **58.** $x = 12\sin 2t,\ y = 3\cos 2t$

59–62. Parametric equations of ellipses *Find parametric equations of the following ellipses (see Exercises 57–58). Graph the ellipse and find a description in terms of x and y. Solutions are not unique.*

59. An ellipse centered at the origin with major axis of length 6 on the *x*-axis and minor axis of length 3 on the *y*-axis, generated counterclockwise

60. An ellipse centered at the origin with major and minor axes of length 12 and 2, on the *x*- and *y*-axes, respectively, generated clockwise

61. An ellipse centered at $(-2, -3)$ with major and minor axes of length 30 and 20, on the *x*- and *y*-axes, respectively, generated counterclockwise (Shift the parametric equations.)

62. An ellipse centered at $(0, -4)$ with major and minor axes of length 10 and 3, on the *x*- and *y*-axes, respectively, generated clockwise (Shift the parametric equations.)

63. Multiple descriptions Which of the following parametric equations describe the same line?

a. $x = 3 + t,\ y = 4 - 2t;\ -\infty < t < \infty$
b. $x = 3 + 4t,\ y = 4 - 8t;\ -\infty < t < \infty$
c. $x = 3 + t^3,\ y = 4 - t^3;\ -\infty < t < \infty$

64. Multiple descriptions Which of the following parametric equations describe the same curve?

a. $x = 2t^2,\ y = 4 + t;\ -4 < t < 4$
b. $x = 2t^4,\ y = 4 + t^2;\ -2 < t < 2$
c. $x = 2t^{2/3},\ y = 4 + t^{1/3};\ -64 < t < 64$

65–70. Eliminating the parameter *Eliminate the parameter to express the following parametric equations as a single equation in x and y.*

65. $x = 2\sin 8t,\ y = 2\cos 8t$ **66.** $x = 3 - t,\ y = 3 + t$

67. $x = t,\ y = \sqrt{4 - t^2}$ **68.** $x = \sqrt{t + 1},\ y = \dfrac{1}{t + 1}$

69. $x = \tan t,\ y = \sec^2 t - 1$

70. $x = a\sin^n t,\ y = b\cos^n t$, where *a* and *b* are real numbers and *n* is a positive integer

71–74. Slopes of tangent lines *Find all the points on the following curves that have the given slope.*

71. $x = 4\cos t,\ y = 4\sin t;\ \text{slope} = \frac{1}{2}$

72. $x = 2\cos t,\ y = 8\sin t;\ \text{slope} = -1$

73. $x = t + 1/t,\ y = t - 1/t;\ \text{slope} = 1$

74. $x = 2 + \sqrt{t},\ y = 2 - 4t;\ \text{slope} = 0$

75–76. Equivalent descriptions *Find real numbers a and b such that equations A and B describe the same curve.*

75. *A*: $x = 10\sin t,\ y = 10\cos t;\ 0 \le t \le 2\pi$
        B: $x = 10\sin 3t,\ y = 10\cos 3t;\ a \le t \le b$

76. *A*: $x = t + t^3,\ y = 3 + t^2;\ -2 \le t \le 2$
        B: $x = t^{1/3} + t,\ y = 3 + t^{2/3};\ a \le t \le b$

77–78. Lissajous curves *Consider the following Lissajous curves. Find all points on the curve at which there is (a) a horizontal tangent line and (b) a vertical tangent line. (See the Guided Projects for more on Lissajous curves.)*

77. $x = \sin 2t,\ y = 2\sin t;$
        $0 \le t \le 2\pi$

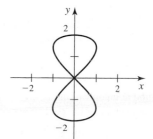

78. $x = \sin 4t,\ y = \sin 3t;$
        $0 \le t \le 2\pi$

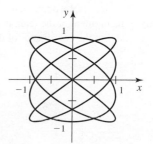

79. Lamé curves The *Lamé curve* described by $\left|\dfrac{x}{a}\right|^n + \left|\dfrac{y}{b}\right|^n = 1$, where *a*, *b*, and *n* are positive real numbers, is a generalization of an ellipse.

a. Express this equation in parametric form (four sets of equations are needed).

b. Graph the curve for $a = 4$ and $b = 2$, for various values of n.

c. Describe how the curves change as n increases.

80. Hyperbolas A family of curves called *hyperbolas* (discussed in Section 10.4) has the parametric equations $x = a \tan t$, $y = b \sec t$, for $-\pi < t < \pi$, where a and b are nonzero real numbers. Graph the hyperbola with $a = b = 1$. Indicate clearly the direction in which the curve is generated as t increases from $t = -\pi$ to $t = \pi$.

81. Trochoid explorations A *trochoid* is the path followed by a point b units from the center of a wheel of radius a as the wheel rolls along the x-axis. Its parametric description is $x = at - b \sin t$, $y = a - b \cos t$. Choose specific values of a and b, and use a graphing utility to plot different trochoids. In particular, explore the difference between the cases $a > b$ and $a < b$.

82. Epitrochoid An *epitrochoid* is the path of a point on a circle of radius b as it rolls on the outside of a circle of radius a. It is described by the equations

$$x = (a + b) \cos t - c \cos \left[\frac{(a + b)t}{b} \right]$$

$$y = (a + b) \sin t - c \sin \left[\frac{(a + b)t}{b} \right]$$

Use a graphing utility to explore the dependence of the curve on the parameters a, b, and c.

83. Hypocycloid A general *hypocycloid* is described by the equations

$$x = (a - b) \cos t + b \cos \left[\frac{(a - b)t}{b} \right]$$

$$y = (a - b) \sin t - b \sin \left[\frac{(a - b)t}{b} \right]$$

Use a graphing utility to explore the dependence of the curve on the parameters a, b, and c.

Applications

84. Air drop A plane traveling horizontally at 80 m/s over flat ground at an elevation of 3000 m releases an emergency packet. The trajectory of the packet is given by

$$x = 80t, y = -4.9t^2 + 3000, \quad \text{for } t \geq 0,$$

where the origin is the point on the ground directly beneath the plane at the moment of the release. Graph the trajectory of the packet and find the coordinates of the point where the packet lands.

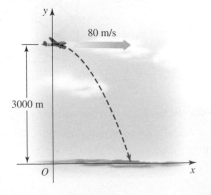

85. Air drop—inverse problem A plane traveling horizontally at 100 m/s over flat ground at an elevation of 4000 m must drop an emergency packet on a target on the ground. The trajectory of the packet is given by

$$x = 100t, \quad y = -4.9t^2 + 4000, \quad \text{for } t \geq 0,$$

where the origin is the point on the ground directly beneath the plane at the moment of the release. How many horizontal meters before the target should the packet be released in order to hit the target?

86. Projectile explorations A projectile launched from the ground with an initial speed of 20 m/s and a launch angle θ follows a trajectory approximated by

$$x = (20 \cos \theta)t, \quad y = -4.9t^2 + (20 \sin \theta)t,$$

where x and y are the horizontal and vertical positions of the projectile relative to the launch point $(0, 0)$.

a. Graph the trajectory for various values of θ in the range $0 < \theta < \pi/2$.

b. Based on your observations, what value of θ gives the greatest range (the horizontal distance between the launch and landing points)?

Additional Exercises

87. Implicit function graph Explain and carry out a method for graphing the curve $x = 1 + \cos^2 y - \sin^2 y$ using parametric equations and a graphing utility.

88. Second derivative Assume a curve is given by the parametric equations $x = g(t)$ and $y = h(t)$, where g and h are twice differentiable. Use the Chain Rule to show that

$$y''(x) = \frac{x'(t)y''(t) - y'(t)x''(t)}{[x'(t)]^3}.$$

89. General equations for a circle Prove that the equations

$$x = a \cos t + b \sin t, \quad y = c \cos t + d \sin t$$

where a, b, c, and d are real numbers, describe a circle of radius R provided $a^2 + c^2 = b^2 + d^2 = R^2$ and $ab + cd = 0$.

90. x^y versus y^x Consider positive real numbers x and y. Notice that $4^3 < 3^4$, while $3^2 > 2^3$, and $4^2 = 2^4$. Describe the regions in the first quadrant of the xy-plane in which $x^y > y^x$ and $x^y < y^x$. (*Hint:* Find a parametric description of the curve that separates the two regions.)

QUICK CHECK ANSWERS

1. A segment of the parabola $x = y^2$ opening to the right with vertex at the origin   **2.** The circle has center $(0, 0)$ and radius 3; it is generated in the counterclockwise direction starting at $(0, -3)$.   **3.** The line $y = -3x - 3$ with slope -3 passing through $(3, -12)$ (when $t = 0$)   **4.** One possibility is $x = -2t, y = 3 - 3t$, for $0 \leq t \leq 1$. ◄

10.2 Polar Coordinates

> Recall that the terms *Cartesian* coordinate system and *rectangular* coordinate system both describe the usual *xy*-coordinate system.

Suppose you work for a company that designs heat shields for space shuttles. The shields are thin plates that are either rectangular or circular in shape. To solve the heat transfer equations for these two shields, you must choose a coordinate system that best fits the geometry of the problem. A Cartesian (rectangular) coordinate system is a natural choice for the rectangular shields (Figure 10.17a). However, it does not provide a good fit for the circular shields (Figure 10.17b). On the other hand, a **polar coordinate** system, in which the coordinates are constant on circles and rays, is much better suited for the circular shields (Figure 10.17c).

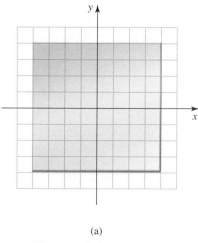

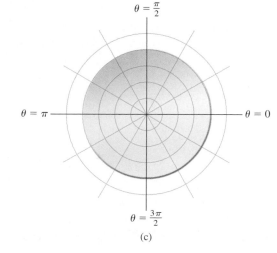

(a)   (b)   (c)

FIGURE 10.17

> Polar points and curves are usually sketched on a rectangular coordinate system, with standard "*x*" and "*y*" labels attached to the axes. However, plotting polar points and curves is often easier using polar graph paper, which has concentric circles centered at the origin and rays emanating from the origin (Figure 10.19).

Defining Polar Coordinates

Like Cartesian coordinates, polar coordinates are used to locate points in the plane. When working in polar coordinates, the origin of the coordinate system is also called the **pole**, and the *x*-axis is called the **polar axis**. The polar coordinates for a point P have the form (r, θ). **The radial coordinate** r describes the *signed*, or *directed*, distance from the origin to P. The **angular coordinate** θ describes an angle whose initial side is the positive *x*-axis and whose terminal side lies on the ray passing through the origin and P (Figure 10.18a). Positive angles are measured counterclockwise from the positive *x*-axis.

With polar coordinates, points have more than one representation for two reasons. First, angles are determined up to multiples of 2π radians, so the coordinates (r, θ) and $(r, \theta \pm 2\pi)$ refer to the same point (Figure 10.18b). Second, the radial coordinate may be negative, which is interpreted as follows: The points (r, θ) and $(-r, \theta)$ are reflections of each other through the origin (Figure 10.18c). This means that (r, θ), $(-r, \theta + \pi)$, and $(-r, \theta - \pi)$ all refer to the same point. The origin is specified as $(0, \theta)$ in polar coordinates, where θ is any angle.

QUICK CHECK 1 Which of the following coordinates represent the same point: $(3, \pi/2)$, $(3, 3\pi/2)$, $(3, 5\pi/2)$, $(-3, -\pi/2)$, and $(-3, 3\pi/2)$? ◄

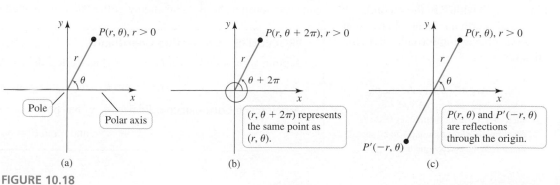

(a)   (b)   (c)

FIGURE 10.18

EXAMPLE 1   Points in polar coordinates Graph the following points in polar coordinates: $Q\left(1, \frac{5\pi}{4}\right)$, $R\left(-1, \frac{7\pi}{4}\right)$, and $S\left(2, -\frac{3\pi}{2}\right)$. Give two alternative representations for each point.

SOLUTION The point $Q\left(1, \frac{5\pi}{4}\right)$ is one unit from the origin on a line OQ that makes an angle of $\frac{5\pi}{4}$ with the positive x-axis (Figure 10.19a). Subtracting 2π from the angle, the point Q can be represented as $\left(1, -\frac{3\pi}{4}\right)$. Subtracting π from the angle and negating the radial coordinate means Q also has the coordinates $\left(-1, \frac{\pi}{4}\right)$.

To locate the point $R\left(-1, \frac{7\pi}{4}\right)$, it is easiest first to find the point $R'\left(1, \frac{7\pi}{4}\right)$ in the fourth quadrant. Then, $R\left(-1, \frac{7\pi}{4}\right)$ is the reflection of R' through the origin (Figure 10.19b). Other representations of R include $\left(-1, -\frac{\pi}{4}\right)$ and $\left(1, \frac{3\pi}{4}\right)$.

The point $S\left(2, -\frac{3\pi}{2}\right)$ is two units from the origin, found by rotating *clockwise* through an angle of $\frac{3\pi}{2}$ (Figure 10.19c). The point S can also be represented as $\left(2, \frac{\pi}{2}\right)$ or $\left(-2, -\frac{\pi}{2}\right)$.

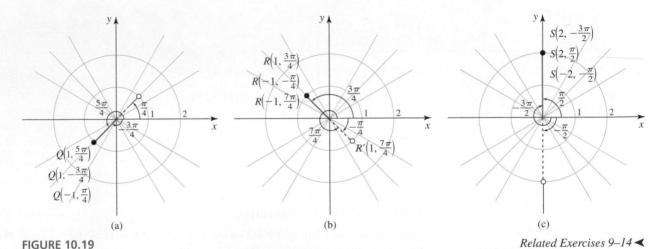

(a)   (b)   (c)

FIGURE 10.19   *Related Exercises 9–14* ◀

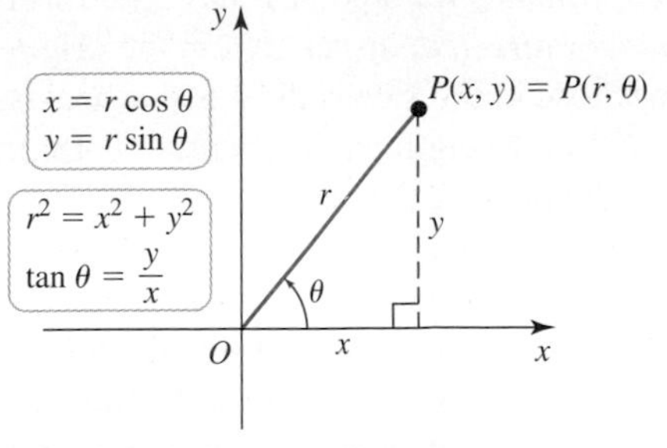

Converting Between Cartesian and Polar Coordinates

$$\boxed{\begin{aligned} x &= r\cos\theta \\ y &= r\sin\theta \end{aligned}}$$

$$\boxed{\begin{aligned} r^2 &= x^2 + y^2 \\ \tan\theta &= \frac{y}{x} \end{aligned}}$$

$P(x, y) = P(r, \theta)$

FIGURE 10.20

We often need to convert between Cartesian and polar coordinates. The conversion equations emerge when we look at a right triangle (Figure 10.20) in which

$$\cos\theta = \frac{x}{r} \quad \text{and} \quad \sin\theta = \frac{y}{r}.$$

Given a point with polar coordinates (r, θ), we see that its Cartesian coordinates are $x = r\cos\theta$ and $y = r\sin\theta$. Conversely, given a point with Cartesian coordinates (x, y), its radial polar coordinate satisfies $r^2 = x^2 + y^2$. The coordinate θ is determined using the relation $\tan\theta = y/x$, where the quadrant in which θ lies is determined by the signs of x and y. Figure 10.20 illustrates the conversion formulas for a point P in the first quadrant. The same relationships hold if P is in any of the other three quadrants.

QUICK CHECK 2 Draw versions of Figure 10.20 with P in the second, third, and fourth quadrants. Verify that the same conversion formulas hold in all cases. ◀

▶ To determine θ, you may also use the relationships $\cos\theta = x/r$ and $\sin\theta = y/r$. Either method requires checking the signs of x and y to be sure that θ is in the correct quadrant.

PROCEDURE   Converting Coordinates

A point with polar coordinates (r, θ) has Cartesian coordinates (x, y), where

$$x = r\cos\theta \quad \text{and} \quad y = r\sin\theta.$$

A point with Cartesian coordinates (x, y) has polar coordinates (r, θ), where

$$r^2 = x^2 + y^2 \quad \text{and} \quad \tan\theta = y/x.$$

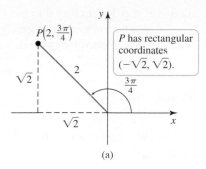

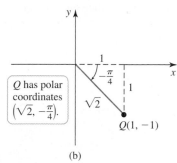

FIGURE 10.21

EXAMPLE 2 Converting coordinates

a. Express the point with polar coordinates $P\left(2, \frac{3\pi}{4}\right)$ in Cartesian coordinates.

b. Express the point with Cartesian coordinates $Q(1, -1)$ in polar coordinates.

SOLUTION

a. The point P has Cartesian coordinates

$$x = r \cos \theta = 2 \cos \left(\frac{3\pi}{4}\right) = -\sqrt{2}$$
$$y = r \sin \theta = 2 \sin \left(\frac{3\pi}{4}\right) = \sqrt{2}$$

As shown in Figure 10.21a, P is in the second quadrant.

b. It's best to locate this point first to be sure that the angle θ is chosen correctly. As shown in Figure 10.21b, the point $Q(1, -1)$ is in the fourth quadrant at a distance $r = \sqrt{1^2 + (-1)^2} = \sqrt{2}$ from the origin. The coordinate θ satisfies

$$\tan \theta = \frac{y}{x} = \frac{-1}{1} = -1.$$

The angle in the fourth quadrant with $\tan \theta = -1$ is $\theta = -\frac{\pi}{4}$ or $\frac{7\pi}{4}$. Therefore, two (of infinitely many) polar representations of Q are $\left(\sqrt{2}, -\frac{\pi}{4}\right)$ and $\left(\sqrt{2}, \frac{7\pi}{4}\right)$.

Related Exercises 15–26 ◄

QUICK CHECK 3 Give two polar coordinate descriptions of the point with Cartesian coordinates $(1, 0)$. What are the Cartesian coordinates of the point with polar coordinates $\left(2, \frac{\pi}{2}\right)$? ◄

Basic Curves in Polar Coordinates

A curve in polar coordinates is the set of points that satisfy an equation in r and θ. Some sets of points are easier to describe in polar coordinates than in Cartesian coordinates. Let's begin with two simple curves.

The polar equation $r = 3$ is satisfied by the set of points whose distance from the origin is 3. The angle θ is arbitrary because it is not specified by the equation, so the graph of $r = 3$ is the circle of radius 3 centered at the origin. In general, the equation $r = a$ describes a circle of radius $|a|$ centered at the origin (Figure 10.22a).

The equation $\theta = \pi/3$ is satisfied by the points whose angle with respect to the positive x-axis is $\pi/3$. Because r is unspecified, it is arbitrary (and can be positive or negative). Therefore, $\theta = \pi/3$ describes the line through the origin making an angle of $\pi/3$ with the positive x-axis. More generally, $\theta = \theta_0$ describes the line through the origin making an angle of θ_0 with the positive x-axis (Figure 10.22b).

> If the equation $\theta = \theta_0$ is accompanied by the condition $r \geq 0$, the resulting set of points is a *ray* emanating from the origin.

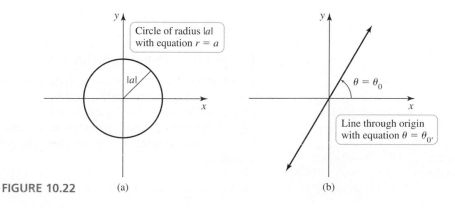

FIGURE 10.22   (a)   (b)

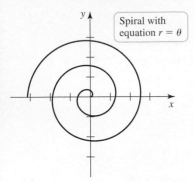

FIGURE 10.23

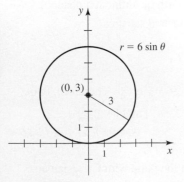

FIGURE 10.24

The simplest polar equation that involves both r and θ is $r = \theta$. Restricting θ to the interval $\theta \geq 0$, we see that as θ increases, r increases. Therefore, as θ increases, the points on the curve move away from the origin as they circle the origin in a counterclockwise direction, generating a spiral (Figure 10.23).

QUICK CHECK 4 Describe the polar curves $r = 12$, $r = 6\theta$, and $r \sin \theta = 10$. ◄

EXAMPLE 3 Polar to Cartesian coordinates Convert the polar equation $r = 6 \sin \theta$ to Cartesian coordinates and describe the corresponding graph.

SOLUTION We first assume that $r \neq 0$ and multiply both sides of the equation by r, which produces the equation $r^2 = 6r \sin \theta$. Using the conversion relations $r^2 = x^2 + y^2$ and $y = r \sin \theta$, the equation

$$\underbrace{r^2}_{x^2 + y^2} = \underbrace{6r \sin \theta}_{6y}$$

becomes $x^2 + y^2 - 6y = 0$. Completing the square gives the equation

$$x^2 + \underbrace{y^2 - 6y + 9}_{(y - 3)^2} - 9 = x^2 + (y - 3)^2 - 9 = 0.$$

We recognize $x^2 + (y - 3)^2 = 9$ as the equation of a circle of radius 3 centered at $(0, 3)$ (Figure 10.24).   *Related Exercises 27–36* ◄

Calculations similar to those in Example 3 lead to the following equations of circles in polar coordinates.

SUMMARY Circles in Polar Coordinates

The equation $r = a$ describes a circle of radius $|a|$ centered at $(0, 0)$.

The equation $r = 2a \sin \theta$ describes a circle of radius $|a|$ centered at $(0, a)$.

The equation $r = 2a \cos \theta$ describes a circle of radius $|a|$ centered at $(a, 0)$.

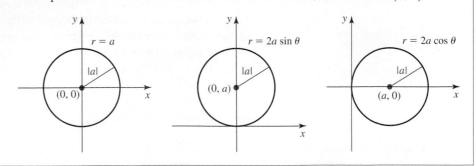

Graphing in Polar Coordinates

Equations in polar coordinates often describe curves that are difficult to represent in Cartesian coordinates. Partly for this reason, curve-sketching methods for polar coordinates differ from those used for curves in Cartesian coordinates. Conceptually, the easiest graphing method is to choose several values of θ, calculate the corresponding r-values, and tabulate the coordinates. The points are then plotted and connected with a smooth curve.

> When a curve is described as $r = f(\theta)$, it is natural to tabulate points in θ-r format, just as we list points in x-y format for $y = f(x)$. Despite this fact, the standard form for an ordered pair in polar coordinates is (r, θ).

EXAMPLE 4 **Plotting a polar curve** Graph the polar equation $r = f(\theta) = 1 + \sin \theta$.

SOLUTION The domain of f consists of all real values of θ; however, the complete curve is generated by letting θ vary over any interval of length 2π. Table 10.3 shows several θ-r pairs, which are plotted in Figure 10.25. The resulting curve, called a **cardioid**, is symmetric about the y-axis.

Table 10.3

θ	$r = 1 + \sin \theta$
0	1
$\pi/6$	$3/2$
$\pi/2$	2
$5\pi/6$	$3/2$
π	1
$7\pi/6$	$1/2$
$3\pi/2$	0
$11\pi/6$	$1/2$
2π	1

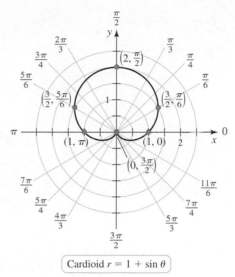

FIGURE 10.25

Cardioid $r = 1 + \sin \theta$

Related Exercises 27–36 ◄

Cartesian-to-Polar Method Plotting polar curves point by point is time consuming, and important details may not be revealed. Here is an alternative procedure for graphing polar curves that is usually quicker and more reliable.

> For some (but not all) curves, it suffices to graph $r = f(\theta)$ over any interval in θ whose length is the period of f. See Examples 6 and 9 for exceptions.

PROCEDURE **Cartesian-to-Polar Method for Graphing $r = f(\theta)$**

1. Graph $r = f(\theta)$ *as if r and θ were Cartesian coordinates* with θ on the horizontal axis and r on the vertical axis. Be sure to choose an interval in θ on which the entire polar curve is produced.

2. Use the Cartesian graph in Step 1 as a guide to sketch the points (r, θ) on the final *polar* curve.

EXAMPLE 5 **Plotting polar graphs** Use the Cartesian-to-polar method to graph the polar equation $r = 1 + \sin \theta$ (Example 4).

SOLUTION Viewing r and θ as Cartesian coordinates, the graph of $r = 1 + \sin \theta$ on the interval $[0, 2\pi]$ is a standard sine curve with amplitude 1 shifted up 1 unit (Figure 10.26). Notice that the graph begins with $r = 1$ at $\theta = 0$, increases to $r = 2$ at $\theta = \pi/2$, decreases to $r = 0$ at $\theta = 3\pi/2$ (which indicates an intersection with the origin on the polar graph), and increases to $r = 1$ at $\theta = 2\pi$. The second row of Figure 10.26 shows the final polar curve (a cardioid) as it is transferred from the Cartesian curve.

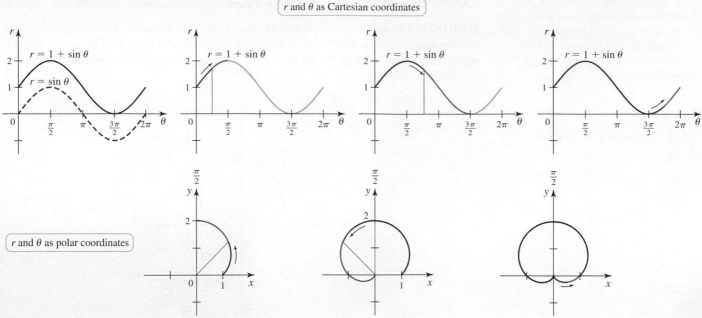

FIGURE 10.26

Related Exercises 37–44 ◄

Symmetry   Given a polar equation in r and θ, three types of symmetry are easy to spot (Figure 10.27).

> **SUMMARY   Symmetry in Polar Equations**
>
> **Symmetry about the x-axis** occurs if the point (r, θ) is on the graph whenever $(r, -\theta)$ is on the graph.
>
> **Symmetry about the y-axis** occurs if the point (r, θ) is on the graph whenever $(r, \pi - \theta) = (-r, -\theta)$ is on the graph.
>
> **Symmetry about the origin** occurs if the point (r, θ) is on the graph whenever $(-r, \theta) = (r, \theta + \pi)$ is on the graph.

➤ Any two of these three symmetries implies the third. For example, if a graph is symmetric about both the x- and y-axes, then it must be symmetric about the origin.

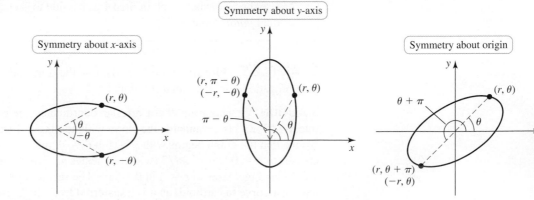

FIGURE 10.27

QUICK CHECK 5 Identify the symmetry in the graph of (a) $r = 4 + 4\cos\theta$ and (b) $r = 4\sin\theta$. ◄

For instance, consider the polar equation $r = 1 + \sin\theta$ in Example 5. If (r, θ) satisfies the equation, then $(r, \pi - \theta)$ also satisfies the equation because $\sin\theta = \sin(\pi - \theta)$. Therefore, the graph is symmetric about the y-axis, as shown in Figure 10.26. Testing for symmetry produces a more accurate graph and often simplifies the task of graphing polar equations.

EXAMPLE 6 **Plotting polar graphs** Graph the polar equation $r = 3 \sin 2\theta$.

SOLUTION The Cartesian graph of $r = 3 \sin 2\theta$ on the interval $[0, 2\pi]$ has amplitude 3 and period π (Figure 10.28). The θ-intercepts occur at $\theta = 0, \pi/2, \pi, 3\pi/2$, and 2π, which correspond to the intersections with the origin on the polar graph. Furthermore, the arches of the Cartesian curve between θ-intercepts correspond to loops in the polar curve. The resulting polar curve is a **four-leaf rose** (Figure 10.28).

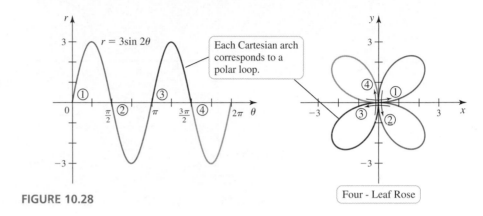

FIGURE 10.28

Four - Leaf Rose

The graph is symmetric about the x-axis, the y-axis, and the origin. It is instructive to see how these symmetries are justified. To prove symmetry about the y-axis, notice that

$$(r, \theta) \text{ on the graph} \Rightarrow r = 3 \sin 2\theta$$
$$\Rightarrow r = -3 \sin 2(-\theta) \qquad \sin(-\theta) = -\sin\theta$$
$$\Rightarrow -r = 3 \sin 2(-\theta) \qquad \text{Simplify.}$$
$$\Rightarrow (-r, -\theta) \text{ on the graph.}$$

We see that if (r, θ) is on the graph, then $(-r, -\theta)$ is also on the graph, which implies symmetry about the y-axis. Similarly, to prove symmetry about the origin, notice that

$$(r, \theta) \text{ on the graph} \Rightarrow r = 3 \sin 2\theta$$
$$\Rightarrow r = 3 \sin(2\theta + 2\pi) \qquad \sin(\theta + 2\pi) = \sin\theta$$
$$\Rightarrow r = 3 \sin[2(\theta + \pi)] \qquad \text{Simplify.}$$
$$\Rightarrow (r, \theta + \pi) \text{ on the graph.}$$

We have shown that if (r, θ) is on the graph, then $(r, \theta + \pi)$ is also on the graph, which implies symmetry about the origin. Symmetry about the y-axis and the origin imply symmetry about the x-axis. Had we proved these symmetries in advance, we could have graphed the curve only in the first quadrant—reflections about the x- and y-axes would produce the full curve.   *Related Exercises 37–44* ◀

EXAMPLE 7 **Plotting polar graphs** Graph the polar equation $r^2 = 9 \cos \theta$. Use a graphing utility to check your work.

SOLUTION The graph of this equation has symmetry about the origin (because of the r^2) and about the x-axis (because of $\cos \theta$). These two symmetries imply symmetry about the y-axis.

A preliminary step is required before using the Cartesian-to-polar method for graphing the curve. Solving the given equation for r, we find that $r = \pm 3\sqrt{\cos \theta}$. Notice that $\cos \theta < 0$ for $\pi/2 < \theta < 3\pi/2$, so the curve does not exist on that interval. Therefore, we plot the curve on the intervals $0 \le \theta \le \pi/2$ and $3\pi/2 \le \theta \le 2\pi$ (the interval $[-\pi/2, \pi/2]$ would also work). Both the positive and negative values of r are included in the Cartesian graph (Figure 10.29a).

Now we are ready to transfer points from the Cartesian graph to the final polar graph (Figure 10.29b). The resulting curve is called a **lemniscate**.

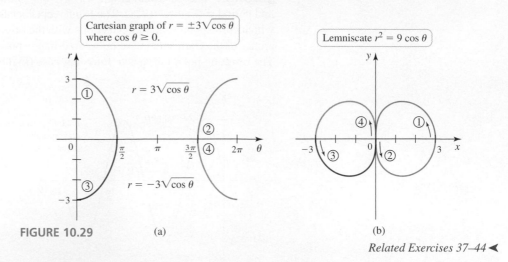

FIGURE 10.29   (a)   (b)

Related Exercises 37–44 ◄

EXAMPLE 8   Matching polar and Cartesian graphs   The butterfly curve

$$r = e^{\sin\theta} - 2\cos 4\theta, \qquad \text{for } 0 \le \theta \le 2\pi$$

is plotted in polar coordinates in Figure 10.30b. The same function, $r = e^{\sin\theta} - 2\cos 4\theta$, is plotted in a Cartesian coordinate system with θ on the horizontal axis and r on the vertical axis (Figure 10.30a). Follow the Cartesian graph through the points $A, B, C, \ldots, N, O$ and mark the corresponding points on the polar curve.

> See Exercise 99 for a spectacular enhancement of the butterfly curve.

SOLUTION   Point A in Figure 10.30a has the Cartesian coordinates $(\theta = 0, r = -1)$. The corresponding point in the polar plot (Figure 10.30b) with polar coordinates $(-1, 0)$ is marked A. Point B in the Cartesian plot is on the θ-axis; therefore, $r = 0$. The corresponding point in the polar plot is the origin. The same argument used to locate B applies to $F, H, J, L,$ and N, all of which appear at the origin in the polar plot. In general, the local and endpoint maxima and minima in the Cartesian graph $(A, C, D, E, G, I, K, M,$ and $O)$ correspond to the extreme points of the loops of the polar plot and are marked accordingly in Figure 10.30b.

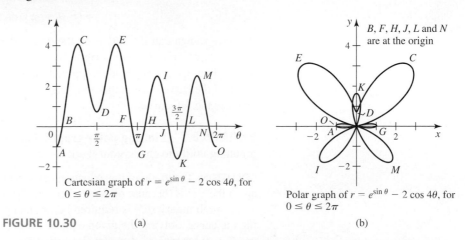

FIGURE 10.30   (a)   (b)

Source: The butterfly curve is due to T. H. Fay *Amer. Math. Monthly* **96** (1989), revived in Wagon and Packel, *Animating Calculus*, Freeman, 1994.

Related Exercises 45–48 ◄

Using Graphing Utilities

With many graphing utilities, it is necessary to specify an interval in θ that generates the entire curve. In some cases, this problem is a challenge in itself.

> **Using a parametric equation plotter to graph polar curves**
>
> To graph $r = f(\theta)$, treat θ as a parameter and define the parametric equations
>
> $$x = r \cos \theta = \underbrace{f(\theta)}_{r} \cos \theta$$
>
> $$y = r \sin \theta = \underbrace{f(\theta)}_{r} \sin \theta$$
>
> Then graph $(x(\theta), y(\theta))$ as a parametric curve with θ as the parameter.

> Once P is found, the complete curve is generated as θ varies over any interval of length P. This choice of P described here ensures that the complete curve is generated. Smaller values of P work in some cases.

EXAMPLE 9 **Plotting complete curves** Consider the curve described by $r = \cos(2\theta/5)$. Give an interval in θ that generates the entire curve and then graph the curve.

SOLUTION Recall that $\cos \theta$ has a period of 2π. Therefore, $\cos(2\theta/5)$ completes one cycle when $2\theta/5$ varies from 0 to 2π, or when θ varies from 0 to 5π. Therefore, it is tempting to conclude that the complete curve $r = \cos(2\theta/5)$ is generated as θ varies from 0 to 5π. But you can check that the point corresponding to $\theta = 0$ is *not* the point corresponding to $\theta = 5\pi$, which means the curve does not close on itself over the interval $[0, 5\pi]$ (Figure 10.31a).

In general, an interval $[0, P]$ over which the complete curve $r = f(\theta)$ is guaranteed to be generated must satisfy two conditions: P is the smallest positive number such that

- P is a multiple of the period of f (so that $f(0) = f(P)$), and
- P is a multiple of 2π (so that the points $(0, f(0))$ and $(P, f(P))$ are the same).

To graph the *complete* curve $r = \cos(2\theta/5)$, we must find an interval $[0, P]$, where P is a multiple of 5π and a multiple of 2π. The smallest number satisfying these conditions is 10π. Graphing $r = \cos(2\theta/5)$ over the interval $[0, 10\pi]$ produces the complete curve (Figure 10.31b).

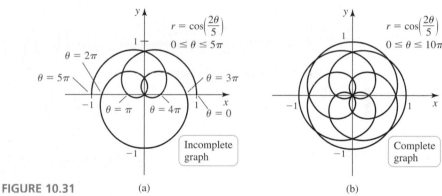

FIGURE 10.31   (a)   (b)

Related Exercises 49–56 ◄

SECTION 10.2 EXERCISES

Review Questions

1. Plot the points with polar coordinates $\left(2, \frac{\pi}{6}\right)$ and $\left(-3, -\frac{\pi}{2}\right)$. Give two alternative sets of coordinate pairs for both points.

2. Write the equations that are used to express a point with polar coordinates (r, θ) in Cartesian coordinates.

3. Write the equations that are used to express a point with Cartesian coordinates (x, y) in polar coordinates.

4. What is the polar equation of a circle of radius $|a|$ centered at the origin?

5. What is the polar equation of the vertical line $x = 5$?

6. What is the polar equation of the horizontal line $y = 5$?

7. Explain three symmetries in polar graphs and how they are detected in equations.

8. Explain the Cartesian-to-polar method for graphing polar curves.

Basic Skills

9–13. *Graph the points with the following polar coordinates. Give two alternative representations of the points in polar coordinates.*

9. $\left(2, \frac{\pi}{4}\right)$
10. $\left(3, \frac{2\pi}{3}\right)$
11. $\left(-1, -\frac{\pi}{3}\right)$
12. $\left(2, \frac{7\pi}{4}\right)$
13. $\left(-4, \frac{3\pi}{2}\right)$

14. Points in polar coordinates Give two sets of polar coordinates for each of the points A–F in the figure.

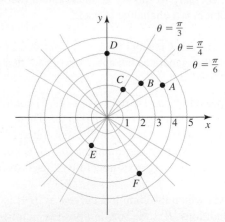

15–20. Converting coordinates *Express the following polar coordinates in Cartesian coordinates.*

15. $\left(3, \frac{\pi}{4}\right)$ **16.** $\left(1, \frac{2\pi}{3}\right)$ **17.** $\left(1, -\frac{\pi}{3}\right)$

18. $\left(2, \frac{7\pi}{4}\right)$ **19.** $\left(-4, \frac{3\pi}{4}\right)$ **20.** $(4, 5\pi)$

⊤ 21–26. Converting coordinates *Express the following Cartesian coordinates in polar coordinates in at least two different ways. Make approximations when necessary.*

21. $(2, 2)$ **22.** $(-1, 0)$ **23.** $(1, \sqrt{3})$

24. $(-9, 0)$ **25.** $(-4, 4\sqrt{3})$ **26.** $(4, 4\sqrt{2})$

27–30. Simple curves *Tabulate and plot enough points to sketch a rough graph of the following equations.*

27. $r = 8 \cos \theta$

28. $r = 4 + 4 \cos \theta$

29. $r(\sin \theta - 2 \cos \theta) = 0$

30. $r = 1 - \cos \theta$

31–36. Polar to Cartesian coordinates *Convert the following equations to Cartesian coordinates. Describe the resulting curve.*

31. $r \cos \theta = -4$ **32.** $r = \cot \theta \csc \theta$

33. $r \cos \theta = \sin 2\theta$ **34.** $r = \sin \theta \sec^2 \theta$

35. $r = 8 \sin \theta$ **36.** $r = \dfrac{1}{2 \cos \theta + 3 \sin \theta}$

⊤ 37–44. Graphing polar curves *Graph the following equations. Use a graphing utility to check your work and produce a final graph.*

37. $r = 1 + \sin \theta$ **38.** $r = 2 - 2 \sin \theta$ **39.** $r = \sin^2 (\theta/2)$

40. $r^2 = 4 \sin \theta$ **41.** $r^2 = 16 \cos \theta$ **42.** $r^2 = 16 \sin 2\theta$

43. $r = \sin 3\theta$ **44.** $r = 2 \sin 5\theta$

45–48. Matching polar and Cartesian curves *A Cartesian and a polar graph of $r = f(\theta)$ are given in the figures. Mark the points on the polar graph that correspond to the points shown on the Cartesian graph.*

45. $r = 1 - 2 \sin 3\theta$

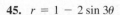
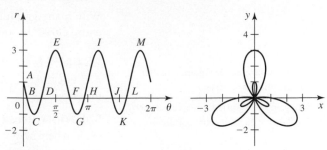

46. $r = \sin (1 + 3 \cos \theta)$

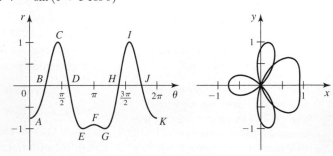

47. $r = \frac{1}{4} - \cos 4\theta$

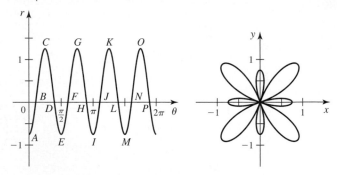

48. $r = \cos \theta + \sin 2\theta$

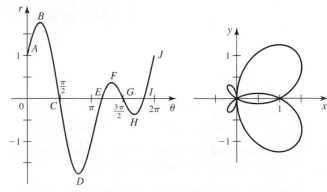

⊤ 49–56. Using a graphing utility *Use a graphing utility to graph the following equations. In each case, give the smallest interval $[0, P]$ that generates the entire curve (if possible).*

49. $r = \theta \sin \theta$ **50.** $r = 2 - 4 \cos 5\theta$

51. $r = \cos 3\theta + \cos^2 2\theta$ **52.** $r = \sin^2 2\theta + 2 \sin 2\theta$

53. $r = \cos (3\theta/5)$ **54.** $r = \sin (3\theta/7)$

55. $r = 1 - 3 \cos 2\theta$ **56.** $r = 1 - 2 \sin 5\theta$

Further Explorations

57. Explain why or why not Determine whether the following statements are true and give an explanation or counterexample.

   a. The point with Cartesian coordinates $(-2, 2)$ has polar coordinates $(2\sqrt{2}, 3\pi/4)$, $(2\sqrt{2}, 11\pi/4)$, $(2\sqrt{2}, -5\pi/4)$, and $(-2\sqrt{2}, -\pi/4)$.

   b. The graphs of $r \cos \theta = 4$ and $r \sin \theta = -2$ intersect exactly once.

   c. The graphs of $r = 2$ and $\theta = \pi/4$ intersect exactly once.

   d. The point $(3, \pi/2)$ lies on the graph of $r = 3 \cos 2\theta$.

58–65. Sets in polar coordinates *Sketch the following sets of points.*

58. $\{(r, \theta): r = 3\}$　　　　　**59.** $\{(r, \theta): \theta = 2\pi/3\}$

60. $\{(r, \theta): 2 \leq r \leq 8\}$　　　**61.** $\{(r, \theta): \pi/2 \leq \theta \leq 3\pi/4\}$

62. $\{(r, \theta): 1 < r < 2 \text{ and } \pi/6 \leq \theta \leq \pi/3\}$

63. $\{(r, \theta): |\theta| \leq \pi/3\}$

64. $\{(r, \theta): |r| < 3 \text{ and } 0 \leq \theta \leq \pi\}$

65. $\{(r, \theta): r \geq 2\}$

66. Circles in general Show that the polar equation

$$r^2 - 2r(a \cos \theta + b \sin \theta) = R^2 - a^2 - b^2$$

describes a circle of radius R centered at (a, b).

67. Circles in general Show that the polar equation

$$r^2 - 2rr_0 \cos(\theta - \theta_0) = R^2 - r_0^2$$

describes a circle of radius R whose center has polar coordinates (r_0, θ_0).

68–73. Equations of circles *Use the results of Exercises 66–67 to describe and graph the following circles.*

68. $r^2 - 6r \cos \theta = 16$

69. $r^2 - 4r \cos(\theta - \pi/3) = 12$

70. $r^2 - 8r \cos(\theta - \pi/2) = 9$

71. $r^2 - 2r(2 \cos \theta + 3 \sin \theta) = 3$

72. $r^2 + 2r(\cos \theta - 3 \sin \theta) = 4$

73. $r^2 - 2r(-\cos \theta + 2 \sin \theta) = 4$

74. Equations of circles Find equations of the circles in the figure. Determine whether the combined area of the circles is greater than or less than the area of the region inside the square but outside the circles.

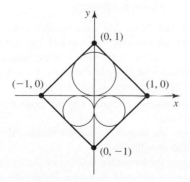

75. Vertical lines Consider the polar curve $r = 2 \sec \theta$.

   a. Graph the curve on the intervals $(\pi/2, 3\pi/2)$, $(3\pi/2, 5\pi/2)$, and $(5\pi/2, 7\pi/2)$. In each case, state the direction in which the curve is generated as θ increases.

   b. Show that on any interval $(n\pi/2, (n + 2)\pi/2)$, where n is an odd integer, the graph is the vertical line $x = 2$.

76. Lines in polar coordinates

   a. Show that an equation of the line $y = mx + b$ in polar coordinate is $r = \dfrac{b}{\sin \theta - m \cos \theta}$.

   b. Use the figure to find an alternate polar equation of a line, $r \cos(\theta_0 - \theta) = r_0$. Note that $Q(r_0, \theta_0)$ is a fixed point on the line such that OQ is perpendicular to the line and $r_0 \geq 0$; $P(r, \theta)$ is an arbitrary point on the line.

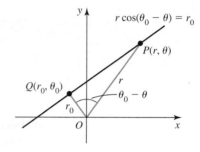

77–80. Equations of lines *Use the result of Exercise 76 to describe and graph the following lines.*

77. $r \cos(\theta - \pi/3) = 3$　　　**78.** $r \cos(\theta + \pi/6) = 4$

79. $r(\sin \theta - 4 \cos \theta) - 3 = 0$　**80.** $r(4 \sin \theta - 3 \cos \theta) = 6$

81. The limaçon family The equations $r = a + b \cos \theta$ and $r = a + b \sin \theta$ describe curves known as *limaçons* (from Latin for *snail*). We have already encountered cardioids, which occur when $|a| = |b|$. The limaçon has an inner loop if $|a| < |b|$. The limaçon has a dent or dimple if $|b| < |a| < 2|b|$. And, the limaçon is oval-shaped if $|a| > 2|b|$. Match the limaçons in the figures A–F with the following equations.

   a. $r = -1 + \sin \theta$　　　　**b.** $r = -1 + 2 \cos \theta$

   c. $r = 2 + \sin \theta$　　　　　**d.** $r = 1 - 2 \cos \theta$

   e. $r = 1 + 2 \sin \theta$　　　　**f.** $r = 1 + (2/3) \sin \theta$

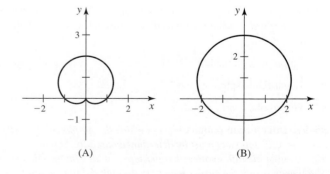

(A)　　　　　　　　　(B)

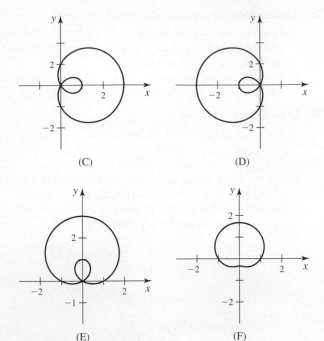

(C)   (D)

(E)   (F)

82. Limiting limaçon Consider the family of limaçons
$r = 1 + b \cos \theta$. Describe the limiting curve as $b \to \infty$.

83–86. The lemniscate family *Equations of the form $r^2 = a \sin 2\theta$ and $r^2 = a \cos 2\theta$ describe lemniscates (see Example 7). Graph the following lemniscates.*

83. $r^2 = \cos 2\theta$ **84.** $r^2 = 4 \sin 2\theta$

85. $r^2 = -2 \sin 2\theta$ **86.** $r^2 = -8 \cos 2\theta$

87–90. The rose family *Equations of the form $r = a \sin m\theta$ or $r = a \cos m\theta$, where a and b are real numbers and m is a positive integer, have graphs known as roses (see Example 6). Graph the following roses.*

87. $r = \sin 2\theta$ **88.** $r = 4 \cos 3\theta$

89. $r = 2 \sin 4\theta$ **90.** $r = 6 \sin 5\theta$

91. Number of rose petals Show that the graph of $r = a \sin m\theta$ or $r = a \cos m\theta$ is a rose with m leaves if m is an odd integer and a rose with $2m$ leaves if m is an even integer.

92–94. Spirals *Graph the following spirals. Indicate the direction in which the spiral winds outward as θ increases, where $\theta > 0$. Let $a = 1$ and $a = -1$.*

92. Spiral of Archimedes: $r = a\theta$

93. Logarithmic spiral: $r = e^{a\theta}$

94. Hyperbolic spiral: $r = a/\theta$

95–98. Intersection points *Points at which the graphs of $r = f(\theta)$ and $r = g(\theta)$ intersect must be determined carefully. Solving $f(\theta) = g(\theta)$ identifies some—but perhaps not all—intersection points. The reason is that the curves may pass through the same point for different values of θ. Use analytical methods and a graphing utility to find all the intersection points between the following curves.*

95. $r = 2 \cos \theta$ and $r = 1 + \cos \theta$

96. $r^2 = 4 \cos \theta$ and $r = 1 + \cos \theta$

97. $r = 1 - \sin \theta$ and $r = 1 + \cos \theta$

98. $r^2 = \cos 2\theta$ and $r^2 = \sin 2\theta$

99. Enhanced butterfly curve The butterfly curve of Example 8 may be enhanced by adding a term:
$$r = e^{\sin \theta} - 2 \cos 4\theta + \sin^5 (\theta/12), \quad \text{for } 0 \le \theta \le 24\pi.$$
   a. Graph the curve.
   b. Explain why the new term produces the observed effect.
   (*Source:* S. Wagon and E. Packel, *Animating Calculus,* Freeman, New York, 1994)

100. Finger curves Consider the curve $r = f(\theta) = \cos (a^\theta) - 1.5$, where $a = (1 + 12\pi)^{1/2\pi} \approx 1.78933$ (see figure).
   a. Show that $f(0) = f(2\pi)$ and find the points on the curve that correspond to $\theta = 0$ and $\theta = 2\pi$.
   b. Is the same curve produced over the intervals $[-\pi, \pi]$ and $[0, 2\pi]$?
   c. Let $f(\theta) = \cos (a^\theta) - b$, where $a = (1 + 2k\pi)^{1/2\pi}$, k is an integer, and b is a real number. Show that $f(0) = f(2\pi)$ and the curve closes on itself.
   d. Plot the curve with various values of k. How many fingers can you produce?

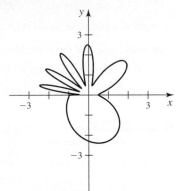

Applications

101. Earth–Mars system A simplified model assumes that the orbits of Earth and Mars are circular with radii of 2 and 3, respectively, and that Earth completes a complete orbit in one year while Mars takes two years. The position of Mars as seen from Earth is given by the parametric equations
$$x = (3 - 4 \cos \pi t) \cos \pi t + 2, \quad y = (3 - 4 \cos \pi t) \sin \pi t.$$
   a. Graph the parametric equations for $0 \le t \le 2$.
   b. Letting $r = (3 - 4 \cos \pi t)$, explain why the path of Mars as seen from Earth is a limaçon.

102. Channel flow Water flows in a shallow semicircular channel with inner and outer radii of 1 m and 2 m (see figure). At a point $P(r, \theta)$ in the channel, the flow is in the tangential direction (counterclockwise along circles), and it depends only on r, the distance from the center of the semicircles.
   a. Express the region formed by the channel as a set in polar coordinates.
   b. Express the inflow and outflow regions of the channel as sets in polar coordinates.
   c. Suppose the tangential velocity of the water in m/s is given by $v(r) = 10r$ for $1 \le r \le 2$. Is the velocity greater at $\left(1.5, \frac{\pi}{4}\right)$ or $\left(1.2, \frac{3\pi}{4}\right)$? Explain.

d. Suppose the tangential velocity of the water is given by
$$v(r) = \frac{20}{r} \text{ for } 1 \le r \le 2.$$ Is the velocity greater at
$\left(1.8, \frac{\pi}{6}\right)$ or $\left(1.3, \frac{2\pi}{3}\right)$? Explain.

e. The total amount of water that flows through the channel (across a cross section of the channel $\theta = \theta_0$) is proportional to $\int_1^2 v(r)\, dr$. Is the total flow through the channel greater for the flow in part (c) or (d)?

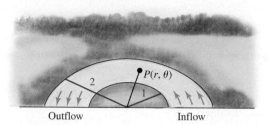

Outflow   Inflow

Additional Exercises

103. **Special circles** Show that the equation $r = a\cos\theta + b\sin\theta$, where a and b are real numbers, describes a circle. Find the center and radius of the circle.

104. **Cartesian lemniscate** Find the equation in Cartesian coordinates of the lemniscate $r^2 = a^2 \cos 2\theta$, where a is a real number.

105. **Subtle symmetry** Without using a graphing utility, determine the symmetries (if any) of the curve $r = 4 - \sin(\theta/2)$.

106. **Complete curves** Consider the polar curve $r = \cos(n\theta/m)$, where n and m are integers.

    a. Graph the complete curve when $n = 2$ and $m = 3$.
    b. Graph the complete curve when $n = 3$ and $m = 7$.
    c. Find a general rule in terms of m and n for determining the least positive number P such that the complete curve is generated over the interval $[0, P]$.

QUICK CHECK ANSWERS

1. All the points are the same except $(3, 3\pi/2)$.   **3.** Polar coordinates: $(1, 0)$, $(1, 2\pi)$; Cartesian coordinates: $(0, 2)$ **4.** A circle centered at the origin with radius 12; a double spiral; the horizontal line $y = 10$ **5.** (a) Symmetric about the x-axis; (b) symmetric about the y-axis ◄

10.3 Calculus in Polar Coordinates

Having learned about the *geometry* of polar coordinates, we now have the groundwork needed to explore *calculus* in polar coordinates. Familiar topics, such as slopes of tangent lines and areas bounded by curves, are now revisited in a different setting.

Slopes of Tangent Lines

Given a function $y = f(x)$, the slope of the line tangent to the graph at a given point is dy/dx or $f'(x)$. So, it may be tempting to conclude that the slope of a curve described by the polar equation $r = f(\theta)$ is $dr/d\theta = f'(\theta)$. Unfortunately, it's not that simple.

The key observation is that the slope of a tangent line—in any coordinate system—is the rate of change of the vertical coordinate y with respect to the horizontal coordinate x, which is dy/dx. We begin by writing the polar equation $r = f(\theta)$ in parametric form with θ as a parameter:

> The slope is the change in the vertical coordinate divided by the change in the horizontal coordinate, independent of the coordinate system. In polar coordinates, neither r nor θ corresponds to a vertical or horizontal coordinate.

$$x = r\cos\theta = f(\theta)\cos\theta \quad \text{and} \quad y = r\sin\theta = f(\theta)\sin\theta \tag{1}$$

From Section 10.1, when x and y are defined parametrically as differentiable functions of θ, the derivative is $\dfrac{dy}{dx} = \dfrac{y'(\theta)}{x'(\theta)}$. Using the Product Rule to compute $y'(\theta)$ and $x'(\theta)$ in equation (1), we have

$$\frac{dy}{dx} = \frac{\overbrace{f'(\theta)\sin\theta + f(\theta)\cos\theta}^{y'(\theta)}}{\underbrace{f'(\theta)\cos\theta - f(\theta)\sin\theta}_{x'(\theta)}}. \tag{2}$$

If the graph passes through the origin for some angle θ_0, then $f(\theta_0) = 0$, and equation (2) simplifies to

$$\frac{dy}{dx} = \frac{\sin\theta_0}{\cos\theta_0} = \tan\theta_0,$$

provided $f'(\theta_0) \neq 0$. However, $\tan\theta_0$ is the slope of the line $\theta = \theta_0$, which also passes through the origin. We conclude that if $f(\theta_0) = 0$, then the tangent line at $(0, \theta_0)$ is simply $\theta = \theta_0$ (Figure 10.32).

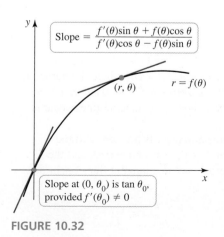

$$\text{Slope} = \frac{f'(\theta)\sin\theta + f(\theta)\cos\theta}{f'(\theta)\cos\theta - f(\theta)\sin\theta}$$

(r, θ) $r = f(\theta)$

Slope at $(0, \theta_0)$ is $\tan\theta_0$, provided $f'(\theta_0) \neq 0$

FIGURE 10.32

QUICK CHECK 1 Verify that if $y = f(\theta) \sin \theta$, then $y'(\theta) = f'(\theta) \sin \theta + f(\theta) \cos \theta$ (which was used earlier to find dy/dx). ◄

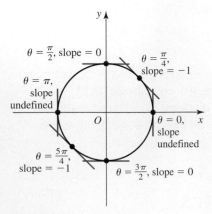

FIGURE 10.33

> **THEOREM 10.2   Slope of a Tangent Line**
> Let f be a differentiable function at θ_0. The slope of the line tangent to the curve $r = f(\theta)$ at the point $(f(\theta_0), \theta_0)$ is
> $$\frac{dy}{dx} = \frac{f'(\theta_0) \sin \theta_0 + f(\theta_0) \cos \theta_0}{f'(\theta_0) \cos \theta_0 - f(\theta_0) \sin \theta_0},$$
> provided the denominator is nonzero at the point. At angles θ_0 for which $f(\theta_0) = 0$ and $f'(\theta_0) \neq 0$, the tangent line is $\theta = \theta_0$ with slope $\tan \theta_0$.

EXAMPLE 1   Slopes on a circle Find the slopes of the lines tangent to the circle $r = f(\theta) = 10$.

SOLUTION In this case, $f(\theta)$ is constant (independent of θ). Therefore, $f'(\theta) = 0$, $f(\theta) \neq 0$, and the slope formula becomes

$$\frac{dy}{dx} = \frac{f'(\theta) \sin \theta + f(\theta) \cos \theta}{f'(\theta) \cos \theta - f(\theta) \sin \theta} = -\frac{\cos \theta}{\sin \theta} = -\cot \theta.$$

We can check a few points to see that this result makes sense. With $\theta = 0$ and $\theta = \pi$, the slope $dy/dx = -\cot \theta$ is undefined, which is correct (Figure 10.33). With $\theta = \pi/2$ and $\theta = 3\pi/2$, the slope is zero; with $\theta = 3\pi/4$ and $\theta = 7\pi/4$, the slope is 1; and with $\theta = \pi/4$ and $\theta = 5\pi/4$, the slope is -1. At all points $P(r, \theta)$ on the circle, the slope of the line OP from the origin to P is $\tan \theta$, which is the negative reciprocal of $-\cot \theta$. Therefore, OP is perpendicular to the tangent line at all points P on the circle.

Related Exercises 5–20 ◄

EXAMPLE 2   Vertical and horizontal tangent lines Find the points on the interval $-\pi \leq \theta \leq \pi$ at which the cardioid $r = f(\theta) = 1 - \cos \theta$ has a vertical or horizontal tangent line.

SOLUTION Applying Theorem 10.2, we find that

$$\frac{dy}{dx} = \frac{f'(\theta) \sin \theta + f(\theta) \cos \theta}{f'(\theta) \cos \theta - f(\theta) \sin \theta}$$

$$= \frac{\overbrace{\sin \theta \sin \theta}^{\sin^2 \theta = 1 - \cos^2 \theta} + (1 - \cos \theta) \cos \theta}{\underbrace{\sin \theta \cos \theta - (1 - \cos \theta) \sin \theta}_{\sin \theta(2 \cos \theta - 1)}} \quad \text{Substitute for } f(\theta) \text{ and } f'(\theta).$$

$$= -\frac{(2 \cos^2 \theta - \cos \theta - 1)}{\sin \theta(2 \cos \theta - 1)} \quad \text{Simplify.}$$

$$= -\frac{(2 \cos \theta + 1)(\cos \theta - 1)}{\sin \theta(2 \cos \theta - 1)} \quad \text{Factor the numerator.}$$

The points with a horizontal tangent line satisfy $dy/dx = 0$ and occur where the numerator is zero and the denominator is nonzero. The numerator is zero when $\theta = 0$ and $\pm 2\pi/3$. Because the denominator is *not* zero when $\theta = \pm 2\pi/3$, horizontal tangent lines occur at $\theta = \pm 2\pi/3$ (Figure 10.34).

Vertical tangent lines occur where the numerator of dy/dx is nonzero and the denominator is zero. The denominator is zero when $\theta = 0$, $\pm \pi$, and $\pm \pi/3$, and the numerator is not zero at $\theta = \pm \pi$ and $\pm \pi/3$. Therefore, vertical tangent lines occur at $\theta = \pm \pi$ and $\pm \pi/3$.

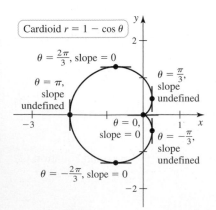

FIGURE 10.34

The point $(0, 0)$ on the curve must be handled carefully because both the numerator and denominator of dy/dx equal 0 at $\theta = 0$. Notice that with $f(\theta) = 1 - \cos\theta$, we have $f(0) = f'(0) = 0$. Therefore, dy/dx may be computed as a limit using l'Hôpital's Rule. As $\theta \to 0^+$, we find that

$$\frac{dy}{dx} = \lim_{\theta \to 0^+}\left[-\frac{(2\cos\theta + 1)(\cos\theta - 1)}{\sin\theta\,(2\cos\theta - 1)}\right]$$

$$= \lim_{\theta \to 0^+}\frac{4\cos\theta\sin\theta - \sin\theta}{-2\sin^2\theta + 2\cos^2\theta - \cos\theta} \qquad \text{L'Hôpital's Rule}$$

$$= \frac{0}{1} = 0. \qquad \text{Evaluate the limit.}$$

QUICK CHECK 2 What is the slope of the line tangent to the cardioid in Example 2 at the point corresponding to $\theta = \pi/4$? ◄

A similar calculation using l'Hôpital's Rule shows that as $\theta \to 0^-$, $dy/dx \to 0$. Therefore, the curve has a slope of 0 at $(0, 0)$.

Related Exercises 5–20 ◄

Area of Regions Bounded by Polar Curves

The problem of finding the area of a region bounded by polar curves brings us back to the slice-and-sum strategy used extensively in Chapters 5 and 6. The objective is to find the area of the region R bounded by the graph of $r = f(\theta)$ between the two rays $\theta = \alpha$ and $\theta = \beta$ (Figure 10.35a). We assume that f is continuous and nonnegative on $[\alpha, \beta]$.

The area of R is found by slicing the region in the radial direction creating wedge-shaped slices. The interval $[\alpha, \beta]$ is partitioned into n subintervals by choosing the grid points

$$\alpha = \theta_0 < \theta_1 < \theta_2 < \cdots < \theta_k < \cdots < \theta_n = \beta.$$

We let $\Delta\theta_k = \theta_k - \theta_{k-1}$, for $k = 1, 2, \ldots, n$, and we let $\bar\theta_k$ be any point of the interval $[\theta_{k-1}, \theta_k]$. The kth slice is approximated by the sector of a circle swept out by an angle $\Delta\theta_k$ with radius $f(\bar\theta_k)$ (Figure 10.35b). Therefore, the area of the kth slice is approximately $\frac{1}{2}f(\bar\theta_k)^2\Delta\theta_k$, for $k = 1, 2, \ldots, n$ (Figure 10.35c). To find the approximate area of R, we sum the areas of these slices:

$$\text{area} \approx \sum_{k=1}^{n}\frac{1}{2}f(\bar\theta_k)^2\,\Delta\theta_k$$

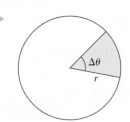

Area of circle $= \pi r^2$

Area of $\Delta\theta/(2\pi)$ of a circle

$= \left(\dfrac{\Delta\theta}{2\pi}\right)\pi r^2 = \dfrac{1}{2}r^2\Delta\theta$

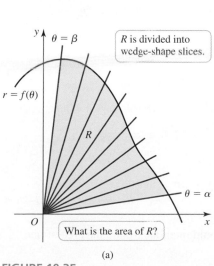

(a)

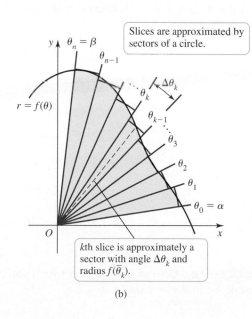

Slices are approximated by sectors of a circle.

kth slice is approximately a sector with angle $\Delta\theta_k$ and radius $f(\bar\theta_k)$.

(b)

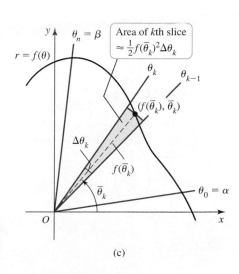

Area of kth slice $\approx \frac{1}{2}f(\bar\theta_k)^2\Delta\theta_k$

(c)

FIGURE 10.35

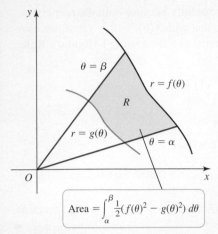

$$\text{Area} = \int_{\alpha}^{\beta} \frac{1}{2}(f(\theta)^2 - g(\theta)^2)\, d\theta$$

FIGURE 10.36

▶ If R is bounded by the graph of $r = f(\theta)$ between $\theta = \alpha$ and $\theta = \beta$, then $g(\theta) = 0$ and the area of R is
$$\int_{\alpha}^{\beta} \frac{1}{2} f(\theta)^2\, d\theta.$$

▶ The equation $r = 2\cos 2\theta$ is unchanged when θ is replaced by $-\theta$ (symmetry about the x-axis) and when θ is replaced by $\pi - \theta$ (symmetry about the y-axis).

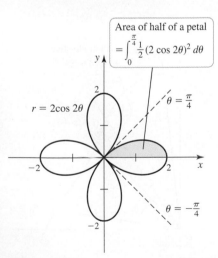

FIGURE 10.37

This approximation is a Riemann sum, and the approximation improves as we take more sectors ($n \to \infty$) and let $\Delta\theta_k \to 0$ for all k. The exact area is given by $\displaystyle\lim_{n \to \infty} \sum_{k=1}^{n} \frac{1}{2} f(\overline{\theta}_k)^2 \Delta\theta_k$, which we identify as a definite integral. Therefore, the area of R is $\displaystyle\int_{\alpha}^{\beta} \frac{1}{2} f(\theta)^2\, d\theta$.

With a slight modification, a more general result is obtained for the area of a region R bounded by two curves, $r = f(\theta)$ and $r = g(\theta)$, between the rays $\theta = \alpha$ and $\theta = \beta$ (Figure 10.36). We assume that f and g are continuous and $f(\theta) \geq g(\theta) \geq 0$ on $[\alpha, \beta]$. To find the area of R, we subtract the area of the region bounded by $r = g(\theta)$ from the area of the entire region bounded by $r = f(\theta)$ (all between $\theta = \alpha$ and $\theta = \beta$); that is,

$$\text{area} = \int_{\alpha}^{\beta} \frac{1}{2} f(\theta)^2\, d\theta - \int_{\alpha}^{\beta} \frac{1}{2} g(\theta)^2\, d\theta = \int_{\alpha}^{\beta} \frac{1}{2}(f(\theta)^2 - g(\theta)^2)\, d\theta.$$

DEFINITION Area of Regions in Polar Coordinates

Let R be the region bounded by the graphs of $r = f(\theta)$ and $r = g(\theta)$, between $\theta = \alpha$ and $\theta = \beta$, where f and g are continuous and $f(\theta) \geq g(\theta) \geq 0$ on $[\alpha, \beta]$. The area of R is

$$\int_{\alpha}^{\beta} \frac{1}{2}(f(\theta)^2 - g(\theta)^2)\, d\theta.$$

QUICK CHECK 3 Find the area of the circle $r = f(\theta) = 8$ (for $0 \leq \theta \leq 2\pi$). ◄

EXAMPLE 3 Area of a polar region Find the area of the four-leaf rose $r = f(\theta) = 2\cos 2\theta$.

SOLUTION The graph of the rose (Figure 10.37) *appears* to be symmetric about the x- and y-axes; in fact, these symmetries can be proved. Appealing to this symmetry, we find the area of one-half of a leaf and then multiply the result by 8 to obtain the area of the full rose. The upper half of the rightmost leaf is generated as θ increases from $\theta = 0$ (when $r = 2$) to $\theta = \pi/4$ (when $r = 0$). Therefore, the area of the entire rose is

$$8\int_{0}^{\pi/4} \frac{1}{2} f(\theta)^2\, d\theta = 4\int_{0}^{\pi/4} (2\cos 2\theta)^2\, d\theta \qquad f(\theta) = 2\cos 2\theta$$

$$= 16\int_{0}^{\pi/4} \cos^2 2\theta\, d\theta \qquad \text{Simplify.}$$

$$= 16\int_{0}^{\pi/4} \frac{1 + \cos 4\theta}{2}\, d\theta \qquad \text{Double-angle formula}$$

$$= (8\theta + 2\sin 4\theta)\Big|_{0}^{\pi/4} \qquad \text{Fundamental Theorem}$$

$$= (2\pi - 0) - (0 - 0) = 2\pi. \qquad \text{Simplify.}$$

Related Exercises 21–28 ◄

QUICK CHECK 4 Give an interval over which you could integrate to find the area of one leaf of the rose $r = 2\sin 3\theta$. ◄

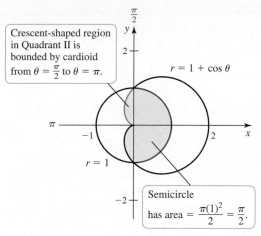

Crescent-shaped region in Quadrant II is bounded by cardioid from $\theta = \frac{\pi}{2}$ to $\theta = \pi$.

$r = 1 + \cos\theta$

$r = 1$

Semicircle has area $= \dfrac{\pi(1)^2}{2} = \dfrac{\pi}{2}$.

FIGURE 10.38

EXAMPLE 4   Areas of polar regions Consider the circle $r = 1$ and the cardioid $r = 1 + \cos\theta$ (Figure 10.38).

a. Find the area of the region inside the circle and inside the cardioid.

b. Find the area of the region inside the circle and outside the cardioid.

SOLUTION

a. The points of intersection of the two curves can be found by solving $1 + \cos\theta = 1$, or $\cos\theta = 0$. The solutions are $\theta = \pm\pi/2$. The region inside the circle and inside the cardioid consists of two subregions:

- A semicircle with radius 1 in the first and fourth quadrants bounded by the circle $r = 1$

- Two crescent-shaped regions in the second and third quadrants bounded by the cardioid $r = 1 + \cos\theta$ and the y-axis

The area of the semicircle is $\pi/2$. To find the area of the upper crescent-shaped region in the second quadrant, notice that it is bounded by $r = 1 + \cos\theta$, as θ varies from $\pi/2$ to π. Therefore, its area is

$$\int_{\pi/2}^{\pi} \frac{1}{2}(1 + \cos\theta)^2\, d\theta = \int_{\pi/2}^{\pi} \frac{1}{2}(1 + 2\cos\theta + \cos^2\theta)\, d\theta \qquad \text{Expand.}$$

$$= \frac{1}{2}\int_{\pi/2}^{\pi}\left(1 + 2\cos\theta + \frac{1 + \cos 2\theta}{2}\right) d\theta \qquad \begin{array}{l}\text{Double-angle}\\ \text{formula}\end{array}$$

$$= \frac{1}{2}\left(\theta + 2\sin\theta + \frac{\theta}{2} + \frac{\sin 2\theta}{4}\right)\Bigg|_{\pi/2}^{\pi} \qquad \begin{array}{l}\text{Fundamental}\\ \text{Theorem}\end{array}$$

$$= \frac{3\pi}{8} - 1. \qquad \text{Simplify.}$$

The area of the entire region (two crescents and a semicircle) is

$$2\left(\frac{3\pi}{8} - 1\right) + \frac{\pi}{2} = \frac{5\pi}{4} - 2.$$

b. The region inside the circle and outside the cardioid is bounded by the outer curve $r = 1$ and the inner curve $r = 1 + \cos\theta$ on the interval $[\pi/2, 3\pi/2]$ (Figure 10.38). Using the symmetry about the x-axis, the area of the region is

$$2\int_{\pi/2}^{\pi} \frac{1}{2}(1^2 - (1 + \cos\theta)^2)\, d\theta = \int_{\pi/2}^{\pi} (-2\cos\theta - \cos^2\theta)\, d\theta \quad \text{Simplify the integrand.}$$

$$= 2 - \frac{\pi}{4}. \qquad \text{Evaluate the integral.}$$

Note that the regions in parts (a) and (b) comprise a circle of radius 1; indeed, their areas have a sum of π.

Related Exercises 21–28 ◄

EXAMPLE 5   Final note of caution Find the points of intersection of the circle $r = 3\cos\theta$ and the cardioid $r = 1 + \cos\theta$ (Figure 10.39).

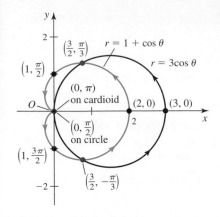

FIGURE 10.39

SOLUTION The fact that a point has multiple representations in polar coordinates may lead to subtle difficulties in finding intersection points. We first proceed algebraically. Equating the two expressions for r and solving for θ, we have

$$3\cos\theta = 1 + \cos\theta \quad \text{or} \quad \cos\theta = \frac{1}{2},$$

which has roots $\theta = \pm\pi/3$. Therefore, two intersection points are $(3/2, \pi/3)$ and $(3/2, -\pi/3)$ (Figure 10.39). Without graphs of the curves, we might be tempted to stop here. Yet, the figure shows another intersection point O that has not been detected. To find the third intersection point, we must investigate the way in which the two curves are generated. As θ increases from 0 to 2π, the cardioid is generated counterclockwise, beginning at $(2, 0)$. The cardioid passes through O when $\theta = \pi$. As θ increases from 0 to π, the circle is generated counterclockwise, beginning at $(3, 0)$. The circle passes through O when $\theta = \pi/2$. Therefore, the intersection point O is $(0, \pi)$ on the cardioid (and these coordinates do not satisfy the equation of the circle), while O is $(0, \pi/2)$ on the circle (and these coordinates do not satisfy the equation of the cardioid). There is no foolproof rule for detecting such "hidden" intersection points. Care must be used.   *Related Exercises 29–32* ◄

SECTION 10.3 EXERCISES

Review Questions

1. Express the polar equation $r = f(\theta)$ in parametric form in Cartesian coordinates, where θ is the parameter.

2. How do you find the slope of the line tangent to the polar graph of $r = f(\theta)$ at a point?

3. Explain why the slope of the line tangent to the polar graph of $r = f(\theta)$ is not $dr/d\theta$.

4. What integral must be evaluated to find the area of the region bounded by the polar graphs of $r = f(\theta)$ and $r = g(\theta)$ on the interval $\alpha \le \theta \le \beta$, where $f(\theta) \ge g(\theta) \ge 0$?

Basic Skills

5–14. Slopes of tangent lines *Find the slope of the line tangent to the following polar curves at the given points. At the points where the curve intersects the origin (when this occurs), find the equation of the tangent line in polar coordinates.*

5. $r = 1 - \sin\theta$; $\left(\frac{1}{2}, \frac{\pi}{6}\right)$

6. $r = 4\cos\theta$; $\left(2, \frac{\pi}{3}\right)$

7. $r = 8\sin\theta$; $\left(4, \frac{5\pi}{6}\right)$

8. $r = 4 + \sin\theta$; $(4, 0)$ and $\left(3, \frac{3\pi}{2}\right)$

9. $r = 6 + 3\cos\theta$; $(3, \pi)$ and $(9, 0)$

10. $r = 2\sin 3\theta$; tips of the leaves

11. $r = 4\cos 2\theta$; tips of the leaves

12. $r^2 = 4\sin 2\theta$; tips of the lobes

13. $r^2 = 4\cos 2\theta$; $\left(0, \pm\frac{\pi}{4}\right)$

14. $r = 2\theta$; $\left(\frac{\pi}{2}, \frac{\pi}{4}\right)$

15–20. Horizontal and vertical tangents *Find the points at which the following polar curves have a horizontal or a vertical tangent line.*

15. $r = 4\cos\theta$

16. $r = 2 + 2\sin\theta$

17. $r = \sin 2\theta$

18. $r = 3 + 6\sin\theta$

19. $r^2 = 4\cos 2\theta$

20. $r = 2\sin 2\theta$

21–28. Areas of regions *Make a sketch of the region and its bounding curves. Find the area of the region.*

21. The region inside the circle $r = 8\sin\theta$

22. The region inside the cardioid $r = 4 + 4\sin\theta$

23. The region inside the limaçon $r = 2 + \cos\theta$

24. The region inside all the leaves of the rose $r = 3\sin 2\theta$

25. The region inside one leaf of the rose $r = \cos 5\theta$

26. The region inside the rose $r = 4\cos 2\theta$ and outside the circle $r = 2$

27. The region inside the rose $r = 4\sin 2\theta$ and inside the circle $r = 2$

28. The region inside the lemniscate $r^2 = 2\sin 2\theta$ and outside the circle $r = 1$

29–32. Intersection points *Use algebraic methods to find as many intersection points of the following curves as possible. Use graphical methods to identify the remaining intersection points.*

29. $r = 3\sin\theta$ and $r = 3\cos\theta$

30. $r = 2 + 2\sin\theta$ and $r = 2 - 2\sin\theta$

31. $r^2 = 4\cos\theta$ and $r = 1 + \cos\theta$

32. $r = 1$ and $r = \sqrt{2}\cos 3\theta$

Further Explorations

33. Explain why or why not Determine whether the following statements are true and give an explanation or counterexample.

    a. The area of the region bounded by the polar graph of $r = f(\theta)$ on the interval $[\alpha, \beta]$ is $\int_\alpha^\beta f(\theta)\,d\theta$.

    b. The slope of the line tangent to the polar curve $r = f(\theta)$ at a point (r, θ) is $f'(\theta)$.

34. Multiple identities Explain why the point $(-1, 3\pi/2)$ is on the polar graph of $r = 1 + \cos\theta$ even though it does not satisfy the equation $r = 1 + \cos\theta$.

35–38. Area of plane regions *Find the area of the following regions.*

35. The region common to the circles $r = 2\sin\theta$ and $r = 1$.

36. The region inside the inner loop of the limaçon $r = 2 + 4\cos\theta$.

37. The region inside the outer loop but outside the inner loop of the limaçon $r = 3 - 6\sin\theta$.

38. The region common to the circle $r = 3\cos\theta$ and the cardioid $r = 1 + \cos\theta$.

39. Spiral tangent lines Use a graphing utility to determine the first three points with $\theta \geq 0$ at which the spiral $r = 2\theta$ has a horizontal tangent line. Find the first three points with $\theta \geq 0$ at which the spiral $r = 2\theta$ has a vertical tangent line.

40. Area of roses

    a. *Even number of leaves*: What is the relationship between the total area enclosed by the $4m$-leaf rose $r = \cos(2m\theta)$ and m?

    b. *Odd number of leaves*: What is the relationship between the total area enclosed by the $(2m + 1)$-leaf rose $r = \cos(2m + 1)\theta$ and m?

41. Regions bounded by a spiral Let R_n be the region bounded by the nth turn and the $(n + 1)$st turn of the spiral $r = e^{-\theta}$ in the first and second quadrants for $\theta \geq 0$ (see figure).

    a. Find the area A_n of R_n.

    b. Evaluate $\lim\limits_{n \to \infty} A_n$.

    c. Evaluate $\lim\limits_{n \to \infty} A_{n+1}/A_n$.

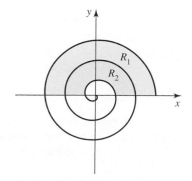

42–45. Area of polar regions *Find the area of the regions bounded by the following curves.*

42. The complete three-leaf rose $r = 2\cos 3\theta$

43. The lemniscate $r^2 = 6\sin 2\theta$

44. The limaçon $r = 2 - 4\sin\theta$

45. The limaçon $r = 4 - 2\cos\theta$

Applications

46. Blood vessel flow A blood vessel with a circular cross section of constant radius R carries blood that flows parallel to the axis of the vessel with a velocity of $v(r) = V(1 - r^2/R^2)$, where V is a constant and r is the distance from the axis of the vessel.

    a. Where is the velocity a maximum? A minimum?

    b. Find the average velocity of the blood over a cross section of the vessel.

    c. Suppose the velocity in the vessel is given by $v(r) = V(1 - r^2/R^2)^{1/p}$, where $p \geq 1$. Graph the velocity profiles for $p = 1, 2$, and 6 on the interval $0 \leq r \leq R$. Find the average velocity in the vessel as a function of p. How does the average velocity behave as $p \to \infty$?

47–49. Grazing goat problems. *Consider the following sequence of problems related to grazing goats tied to a rope. (See The Guided Projects for more grazing goat problems.)*

47. A circular corral of unit radius is enclosed by a fence. A goat inside the corral is tied to the fence with a rope of length $0 \leq a \leq 2$ (see figure). What is the area of the region (inside the corral) that the goat can graze? Check your answer with the special cases $a = 0$ and $a = 2$.

48. A circular concrete slab of unit radius is surrounded by grass. A goat is tied to the edge of the slab with a rope of length $0 \leq a \leq 2$ (see figure). What is the area of the grassy region that the goat can graze? Note that the rope can extend over the concrete slab. Check your answer with the special cases $a = 0$ and $a = 2$.

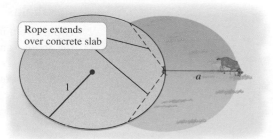

Rope extends over concrete slab

49. A circular corral of unit radius is enclosed by a fence. A goat is outside the corral and tied to the fence with a rope of length $a \geq 0$ (see figure). What is the area of the region (outside the corral) that the goat can reach?

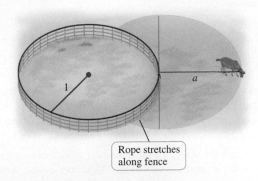

Rope stretches along fence

Additional Exercises

50. Tangents and normals Let a polar curve be described by $r = f(\theta)$ and let ℓ be the line tangent to the curve at the point $P(x, y) = P(r, \theta)$ (see figure).

a. Explain why $\tan \alpha = dy/dx$.

b. Explain why $\tan \theta = y/x$.

c. Let φ be the angle between ℓ and OP. Prove that $\tan \varphi = f(\theta)/f'(\theta)$.

d. Prove that the values of θ for which ℓ is parallel to the x-axis satisfy $\tan \theta = -f(\theta)/f'(\theta)$.

e. Prove that the values of θ for which ℓ is parallel to the y-axis satisfy $\tan \theta = f'(\theta)/f(\theta)$.

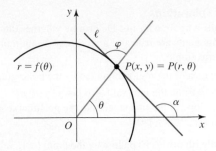

51. Isogonal curves Let a curve be described by $r = f(\theta)$, where $f(\theta) > 0$ on its domain. Referring to the figure of Exercise 50, a curve is **isogonal** provided the angle φ is constant for all θ.

a. Prove that φ is constant for all θ provided $\cot \varphi = f'(\theta)/f(\theta)$ is constant, which implies that $\dfrac{d}{d\theta}[\ln f(\theta)] = k,$ where k is a constant.

b. Use part (a) to prove that the family of logarithmic spirals $r = Ce^{k\theta}$ consists of isogonal curves, where C and k are constants.

c. Graph the curve $r = 2e^{2\theta}$ and confirm the result of part (b).

QUICK CHECK ANSWERS

1. Apply the Product Rule.   **2.** $\sqrt{2} + 1$

3. Area $= \displaystyle\int_0^{2\pi} \tfrac{1}{2}(8)^2 \, d\theta = 64\pi$

4. $\left[0, \dfrac{\pi}{3}\right]$ or $\left[\dfrac{\pi}{3}, \dfrac{2\pi}{3}\right]$ (among others) ◄

10.4 Conic Sections

Conic sections are best visualized as the Greeks did over 2000 years ago by slicing a double cone with a plane (Figure 10.40). Three of the seven different sets of points that arise in this way are *ellipses*, *parabolas*, and *hyperbolas*. These curves have practical applications and broad theoretical importance. For example, celestial bodies travel in orbits that are modeled by ellipses and hyperbolas. Mirrors for telescopes are designed using the properties of conic sections. And architectural structures, such as domes and arches, are sometimes based on these curves.

Parabolas

A **parabola** is the set of points in a plane that are equidistant from a fixed point F (called the **focus**) and a fixed line (called the **directrix**). In the four standard orientations, a parabola may open upward, downward, to the right, or to the left. We derive the equation of the parabola that opens upward.

Suppose the focus F is on the y-axis at $(0, p)$ and the directrix is the horizontal line $y = -p$, where $p > 0$. The parabola is the set of points P that satisfy the defining property

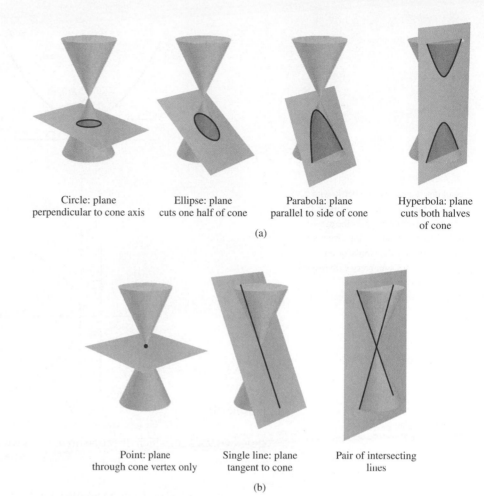

FIGURE 10.40. The standard conic sections (a) are the intersection sets of a double cone and a plane that does not pass through the vertex of the cones. Degenerate conic sections (lines and points) are produced when a plane passes through the vertex of the cone (b).

Circle: plane perpendicular to cone axis   Ellipse: plane cuts one half of cone   Parabola: plane parallel to side of cone   Hyperbola: plane cuts both halves of cone

(a)

Point: plane through cone vertex only   Single line: plane tangent to cone   Pair of intersecting lines

(b)

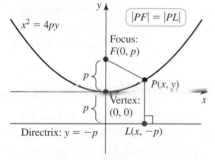

FIGURE 10.41

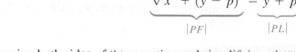

QUICK CHECK 1 Verify that $\sqrt{x^2 + (y - p)^2} = y + p$ is equivalent to $x^2 = 4py$. ◄

$|PF| = |PL|$, where $L(x, -p)$ is the point on the directrix closest to P (Figure 10.41). Consider an arbitrary point $P(x, y)$ that satisfies this condition. Applying the distance formula, we have

$$\underbrace{\sqrt{x^2 + (y - p)^2}}_{|PF|} = \underbrace{y + p}_{|PL|}.$$

Squaring both sides of this equation and simplifying gives the equation $x^2 = 4py$. This is the equation of a parabola that is symmetric about the y-axis and opens upward. The **vertex** of the parabola is the point closest to the directrix; in this case it is $(0, 0)$ (which satisfies $|PF| = |PL| = p$).

The equations of the other three standard parabolas are derived in a similar way.

Equations of Four Standard Parabolas

Let p be a real number. The parabola with focus at $(0, p)$ and directrix $y = -p$ is symmetric about the y-axis and has the equation $x^2 = 4py$. If $p > 0$, then the parabola opens *upward*; if $p < 0$, then the parabola opens *downward*.

The parabola with focus at $(p, 0)$ and directrix $x = -p$ is symmetric about the x-axis and has the equation $y^2 = 4px$. If $p > 0$, then the parabola opens *to the right*; if $p < 0$, then the parabola opens *to the left*.

Each of these parabolas has its vertex at the origin (Figure 10.42).

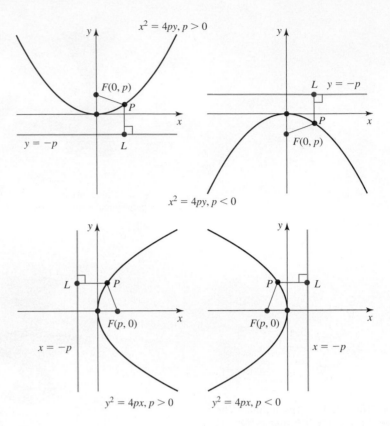

FIGURE 10.42

> Recall that a curve is symmetric with respect to the x-axis if $(x, -y)$ is on the curve whenever (x, y) is on the curve. So, a y^2-term indicates symmetry with respect to the x-axis. Similarly, an x^2-term indicates symmetry with respect to the y-axis.

QUICK CHECK 2 In which direction do the following parabolas open?
a. $y^2 = -4x$ **b.** $x^2 = 4y$ ◄

EXAMPLE 1 Graphing parabolas Find the focus and directrix of the parabola $y^2 = -12x$. Sketch its graph.

SOLUTION The y^2-term indicates that the parabola is symmetric with respect to the x-axis. Rewriting the equation as $x = -y^2/12$, we see that $x \leq 0$ for all y, implying that the parabola opens to the left. Comparing $y^2 = -12x$ to the standard form $y^2 = 4px$, we see that $p = -3$; therefore, the focus is $(-3, 0)$, and the directrix is $x = 3$ (Figure 10.43).
Related Exercises 13–18 ◄

EXAMPLE 2 Equations of parabolas Find the equation of the parabola with vertex $(0, 0)$ that opens downward and passes through the point $(2, -3)$.

SOLUTION The standard parabola that opens downward has the equation $x^2 = 4py$. The point $(2, -3)$ must satisfy this equation. Substituting $x = 2$ and $y = -3$ into $x^2 = 4py$, we find that $p = -\frac{1}{3}$. Therefore, the focus is at $\left(0, -\frac{1}{3}\right)$, the directrix is $y = \frac{1}{3}$, and the equation of the parabola is $x^2 = -4y/3$, or $y = -3x^2/4$ (Figure 10.44).
Related Exercises 19–26 ◄

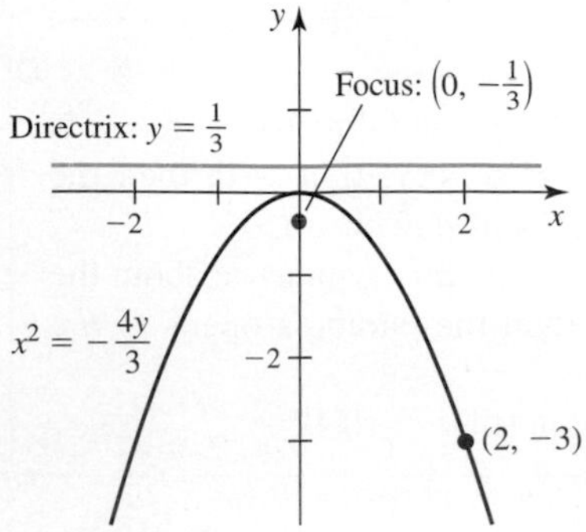

Reflection Property

Parabolas have a property that makes them useful in the design of reflectors and transmitters. A particle approaching a parabola on any line parallel to the axis of the parabola is reflected on a line that passes through the focus (Figure 10.45); this property is used to focus incoming light by a parabolic mirror on a telescope. Alternatively, signals emanating from the focus are reflected on lines parallel to the axis, a property used to design radio transmitters and headlights (Exercise 83).

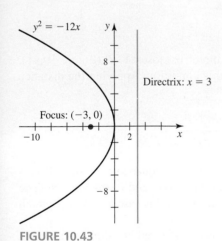

FIGURE 10.43

FIGURE 10.44

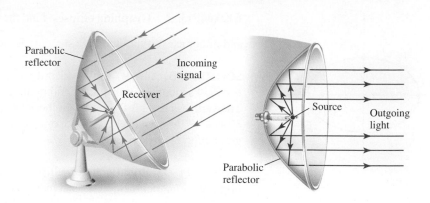

FIGURE 10.45

Ellipses

An **ellipse** is the set of points in a plane whose distances from two fixed points have a constant sum that we denote $2a$ (Figure 10.46). Each of the two fixed points is a **focus** (plural **foci**). The equation of an ellipse is simplest if the foci are on the x-axis at $(\pm c, 0)$ or on the y-axis at $(0, \pm c)$. In either case, the **center** of the ellipse is $(0, 0)$. If the foci are on the x-axis, the points $(\pm a, 0)$ lie on the ellipse and are called **vertices**. If the foci are on the y-axis, the vertices are $(0, \pm a)$ (Figure 10.47). A short calculation (Exercise 85) using the definition of the ellipse results in the following equations for an ellipse.

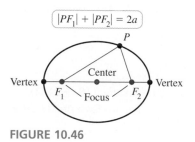

$|PF_1| + |PF_2| = 2a$

FIGURE 10.46

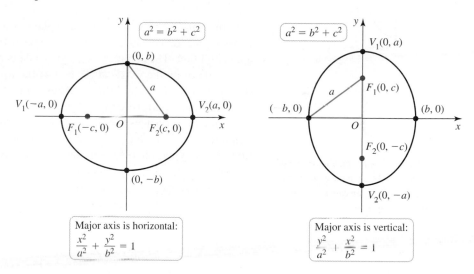

FIGURE 10.47

> When necessary, we may distinguish between the *major-axis vertices* $(\pm a, 0)$ or $(0, \pm a)$, and the *minor-axis vertices* $(\pm b, 0)$ or $(0, \pm b)$. The word *vertices* (without further description) is understood to mean *major-axis vertices*.

QUICK CHECK 3 In the case that the vertices and foci are on the x-axis, show that the length of the minor axis of an ellipse is $2b$. ◄

Equations of Standard Ellipses

An ellipse centered at the origin with foci at $(\pm c, 0)$ and vertices at $(\pm a, 0)$ has the equation

$$\frac{x^2}{a^2} + \frac{y^2}{b^2} = 1, \qquad \text{where } a^2 = b^2 + c^2.$$

An ellipse centered at the origin with foci at $(0, \pm c)$ and vertices at $(0, \pm a)$ has the equation

$$\frac{y^2}{a^2} + \frac{x^2}{b^2} = 1, \qquad \text{where } a^2 = b^2 + c^2.$$

In both cases, $a > b > 0$ and $a > c > 0$, the length of the long axis (called the **major axis**) is $2a$, and the length of the short axis (called the **minor axis**) is $2b$.

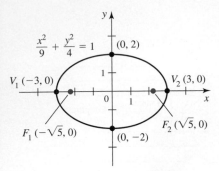

$$\frac{x^2}{9} + \frac{y^2}{4} = 1$$

$V_1(-3, 0)$ $V_2(3, 0)$ $(0, 2)$ $F_1(-\sqrt{5}, 0)$ $F_2(\sqrt{5}, 0)$ $(0, -2)$

FIGURE 10.48

EXAMPLE 3   Graphing ellipses Find the vertices, foci, and the length of the major and minor axes of the ellipse $\dfrac{x^2}{9} + \dfrac{y^2}{4} = 1$. Graph the ellipse.

SOLUTION Because $9 > 4$, we identify $a^2 = 9$ and $b^2 = 4$. Therefore, $a = 3$ and $b = 2$. The lengths of the major and minor axes are $2a = 6$ and $2b = 4$, respectively. The vertices are $(\pm 3, 0)$ and lie on the x-axis, as do the foci. The relationship $c^2 = a^2 - b^2$ implies that $c^2 = 5$, or $c = \sqrt{5}$. Therefore, the foci are $(\pm\sqrt{5}, 0)$. The graph of the ellipse is shown in Figure 10.48.   *Related Exercises 27–32* ◄

EXAMPLE 4   Equation of an ellipse Find the equation of the ellipse centered at the origin with its foci on the y-axis, a major axis of length 8, and a minor axis of length 4. Graph the ellipse.

SOLUTION Because the length of the major axis is 8, the vertices are located at $(0, \pm 4)$, and $a = 4$. Because the length of the minor axis is 4, we have $b = 2$. Therefore, the equation of the ellipse is

$$\frac{y^2}{16} + \frac{x^2}{4} = 1.$$

Using the relation $c^2 = a^2 - b^2$, we find that $c = 2\sqrt{3}$ and the foci are at $(0, \pm 2\sqrt{3})$. The ellipse is shown in Figure 10.49.   *Related Exercises 33–38* ◄

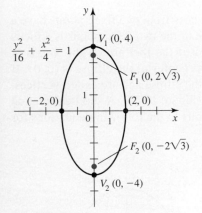

$$\frac{y^2}{16} + \frac{x^2}{4} = 1$$

$V_1(0, 4)$ $F_1(0, 2\sqrt{3})$ $(-2, 0)$ $(2, 0)$ $F_2(0, -2\sqrt{3})$ $V_2(0, -4)$

FIGURE 10.49

Hyperbolas

A **hyperbola** is the set of points in a plane whose distances from two fixed points have a constant difference, either $2a$ or $-2a$ (Figure 10.50). As with ellipses, the two fixed points are called **foci**. The equation of a hyperbola is simplest if the foci are on either the x-axis at $(\pm c, 0)$ or on the y-axis at $(0, \pm c)$. If the foci are on the x-axis, the points $(\pm a, 0)$ on the hyperbola are called the **vertices**. In this case, the hyperbola has no y-intercepts, but it has the **asymptotes** $y = \pm bx/a$, where $b^2 = c^2 - a^2$. Similarly, if the foci are on the y-axis, the vertices are $(0, \pm a)$, the hyperbola has no x-intercepts, and it has the asymptotes $y = \pm ax/b$ (Figure 10.51). A short calculation (Exercise 86) using the definition of the hyperbola results in the following equations for standard hyperbolas.

▶ Asymptotes that are not parallel to one of the coordinate axes, as in the case of the standard hyperbolas, are called **oblique**, or **slant, asymptotes**.

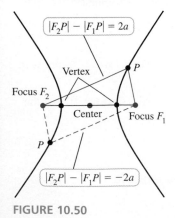

$|F_2P| - |F_1P| = 2a$

Vertex P Focus F_2 Center   Focus F_1 P

$|F_2P| - |F_1P| = -2a$

FIGURE 10.50

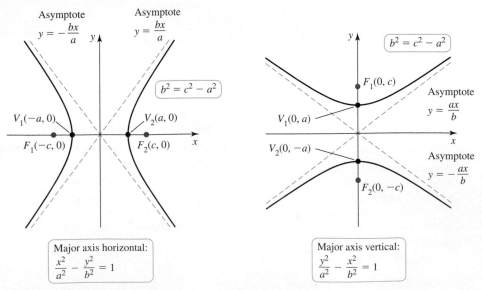

Asymptote $y = -\dfrac{bx}{a}$ Asymptote $y = \dfrac{bx}{a}$

$b^2 = c^2 - a^2$

$V_1(-a, 0)$ $V_2(a, 0)$ $F_1(-c, 0)$ $F_2(c, 0)$

$b^2 = c^2 - a^2$

$F_1(0, c)$ Asymptote $y = \dfrac{ax}{b}$ $V_1(0, a)$ $V_2(0, -a)$ Asymptote $y = -\dfrac{ax}{b}$ $F_2(0, -c)$

Major axis horizontal:
$$\frac{x^2}{a^2} - \frac{y^2}{b^2} = 1$$

Major axis vertical:
$$\frac{y^2}{a^2} - \frac{x^2}{b^2} = 1$$

FIGURE 10.51

> ### Equations of Standard Hyperbolas
>
> A hyperbola centered at the origin with foci at $(\pm c, 0)$ and vertices at $(\pm a, 0)$ has the equation
>
> $$\frac{x^2}{a^2} - \frac{y^2}{b^2} = 1, \quad \text{where} \quad b^2 = c^2 - a^2.$$
>
> The hyperbola has **asymptotes** $y = \pm bx/a$.
>
> A hyperbola centered at the origin with foci at $(0, \pm c)$ and vertices at $(0, \pm a)$ has the equation
>
> $$\frac{y^2}{a^2} - \frac{x^2}{b^2} = 1, \quad \text{where} \quad b^2 = c^2 - a^2.$$
>
> The hyperbola has **asymptotes** $y = \pm ax/b$.
> In both cases, $c > a > 0$ and $c > b > 0$.

▶ Notice that the asymptotes for hyperbolas are $y = \pm bx/a$ when the vertices are on the x-axis and $y = \pm ax/b$ when the vertices are on the y-axis (the roles of a and b are reversed).

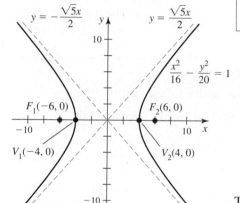

FIGURE 10.52

EXAMPLE 5   Graphing hyperbolas Find the equation of the hyperbola centered at the origin with vertices at $(\pm 4, 0)$ and foci at $(\pm 6, 0)$. Graph the hyperbola.

SOLUTION Because the foci are on the x-axis, the vertices are also on the x-axis, and there are no y-intercepts. With $a = 4$ and $c = 6$, we have $b^2 = c^2 - a^2 = 20$, or $b = 2\sqrt{5}$. Therefore, the equation of the hyperbola is

$$\frac{x^2}{16} - \frac{y^2}{20} = 1.$$

The asymptotes are $y = \pm bx/a = \pm \sqrt{5}x/2$ (Figure 10.52).   *Related Exercises 39–50* ◀

QUICK CHECK 4 Identify the vertices and foci of the hyperbola $y^2 - x^2/4 = 1$. ◀

Eccentricity and Directrix

Parabolas, ellipses, and hyperbolas may also be developed in a single unified way called the *eccentricity-directrix* approach. We let ℓ be a line called the **directrix** and F be a point not on ℓ called a **focus**. The **eccentricity** is a real number $e > 0$. Consider the set C of points P in a plane with the property that the distance $|PF|$ equals e multiplied by the perpendicular distance $|PL|$ from P to ℓ (Figure 10.53); that is,

$$|PF| = e|PL| \quad \text{or} \quad \frac{|PF|}{|PL|} = e = \text{constant}.$$

▶ The conic section lies in the plane formed by the directrix and the focus.

Depending on the value of e, the set C is one of the three standard conic sections, as described in the following theorem.

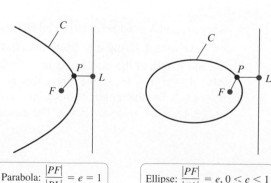

FIGURE 10.53

Parabola: $\dfrac{|PF|}{|PL|} = e = 1$ Ellipse: $\dfrac{|PF|}{|PL|} = e, 0 < e < 1$ Hyperbola: $\dfrac{|PF|}{|PL|} = e > 1$

> Theorem 10.3 for ellipses and hyperbolas describes how the entire curve is generated using just one focus and one directrix. Nevertheless, every ellipse or hyperbola has two foci and two directrices.

THEOREM 10.3   Eccentricity-Directrix Theorem

Let ℓ be a line, F be a point not on ℓ, and $e > 0$ be a real number. Let C be the set of points P in a plane with the property that $\dfrac{|PF|}{|PL|} = e$, where $|PL|$ is the perpendicular distance from P to ℓ.

1. If $e = 1$, C is a **parabola**.

2. If $0 < e < 1$, C is an **ellipse**.

3. If $e > 1$, C is a **hyperbola**.

The proof of the theorem is straightforward; it requires an algebraic calculation that can be found in Appendix B. The proof establishes relationships between five parameters a, b, c, d, and e that are characteristic of any ellipse or hyperbola. The relationships are given in the following summary.

SUMMARY   Properties of Ellipses and Hyperbolas

An ellipse or hyperbola centered at the origin has the following properties.

	Foci on x-axis	Foci on y-axis
Major-axis vertices:	$(\pm a, 0)$	$(0, \pm a)$
Minor-axis vertices (for ellipses):	$(0, \pm b)$	$(\pm b, 0)$
Foci:	$(\pm c, 0)$	$(0, \pm c)$
Directrices:	$x = \pm d$	$y = \pm d$
Eccentricity: $0 < e < 1$ for ellipses, $e > 1$ for hyperbolas.		

Given any two of the five parameters a, b, c, d, and e, the other three are found using the relations

$$c = ae \qquad d = \frac{a}{e}$$

$$b^2 = a^2 - c^2 \quad \text{(for ellipses)} \qquad b^2 = c^2 - a^2 \quad \text{(for hyperbolas)}$$

QUICK CHECK 5 Given an ellipse with $a = 3$ and $e = \frac{1}{2}$, what are the values of b, c, and d? ◄

EXAMPLE 6   Equations of ellipses Find the equation of the ellipse centered at the origin with foci at $(0, \pm 4)$ and eccentricity $e = \frac{1}{2}$. Give the length of the major and minor axes, the location of the vertices, and the directrices. Graph the ellipse.

SOLUTION An ellipse with its major axis along the y-axis has the equation

$$\frac{y^2}{a^2} + \frac{x^2}{b^2} = 1,$$

where a and b must be determined (with $a > b$). Because the foci are at $(0, \pm 4)$, we have $c = 4$. Using $e = \frac{1}{2}$ and the relation $c = ae$, it follows that $a = c/e = 8$. So, the length of the major axis is $2a = 16$, and the major-axis vertices are $(0, \pm 8)$. Also $d = a/e = 16$, so the directrices are $y = \pm 16$. Finally, $b^2 = a^2 - c^2 = 48$, or $b = 4\sqrt{3}$. So, the length of the minor axis is $2b = 8\sqrt{3}$, and the minor-axis vertices are $(\pm 4\sqrt{3}, 0)$ (Figure 10.54). The equation of the ellipse is

$$\frac{y^2}{64} + \frac{x^2}{48} = 1.$$

Related Exercises 51–54 ◄

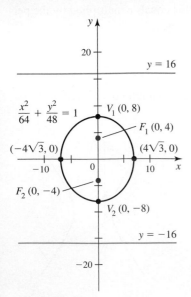

FIGURE 10.54

Polar Equations of Conic Sections

It turns out that conic sections have a natural representation in polar coordinates, provided we use the eccentricity-directrix approach given in Theorem 10.3. Furthermore, a single polar equation covers parabolas, ellipses, and hyperbolas.

When working in polar equations, the key is to place one focus of the conic section at the origin of the coordinate system. We begin by placing one focus F at the origin and taking a directrix perpendicular to the x-axis through $(d, 0)$, where $d > 0$ (Figure 10.55). We now use the definition $\dfrac{|PF|}{|PL|} = e$, where $P(r, \theta)$ is an arbitrary point on the conic. As shown in Figure 10.55, $|PF| = r$ and $|PL| = d - r \cos \theta$. The condition $\dfrac{|PF|}{|PL|} = e$ implies that $r = e(d - r \cos \theta)$. Solving for r, we have

$$r = \frac{ed}{1 + e \cos \theta}.$$

A similar derivation (Exercise 74) with the directrix at $x = -d$, where $d > 0$, results in the equation

$$r = \frac{ed}{1 - e \cos \theta}.$$

For horizontal directrices at $y = \pm d$ (Figure 10.56), a similar argument (Exercise 74) leads to the equations

$$r = \frac{ed}{1 \pm e \sin \theta}.$$

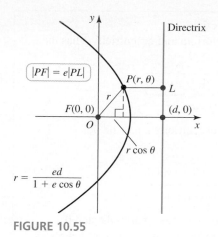

$$r = \frac{ed}{1 + e \cos \theta}$$

FIGURE 10.55

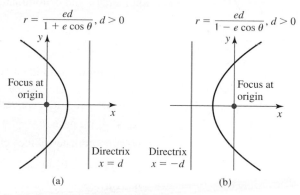

(a)   (b)

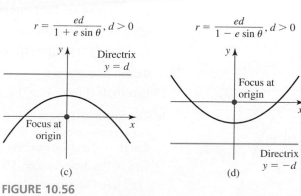

(c)   (d)

FIGURE 10.56

THEOREM 10.4   Polar Equations of Conic Sections

Let $d > 0$. The conic section with a focus at the origin and eccentricity e has the polar equation

$$r = \underbrace{\frac{ed}{1 + e\cos\theta}}_{\text{if one directrix is } x = d} \quad \text{or} \quad r = \underbrace{\frac{ed}{1 - e\cos\theta}}_{\text{if one directrix is } x = -d}.$$

The conic section with a focus at the origin and eccentricity e has the polar equation

$$r = \underbrace{\frac{ed}{1 + e\sin\theta}}_{\text{if one directrix is } y = d} \quad \text{or} \quad r = \underbrace{\frac{ed}{1 - e\sin\theta}}_{\text{if one directrix is } y = -d}.$$

If $0 < e < 1$, the conic section is an ellipse; if $e = 1$, it is a parabola; and if $e > 1$, it is a hyperbola. The curves are defined over any interval in θ of length 2π.

> **QUICK CHECK 6** On which axis do the vertices and foci of the conic section $r = 2/(1 - 2\sin\theta)$ lie? ◄

EXAMPLE 7   Conic sections in polar coordinates Find the vertices, foci, and directrices of the following conic sections. Graph each curve and then check your work with a graphing utility.

a. $r = \dfrac{8}{2 + 3\cos\theta}$ **b.** $r = \dfrac{2}{1 + \sin\theta}$

SOLUTION

a. The equation must be put in the standard polar form for a conic section. Dividing numerator and denominator by 2, we have

$$r = \frac{4}{1 + \frac{3}{2}\cos\theta},$$

which allows us to identify $e = \frac{3}{2}$. Therefore, the equation describes a hyperbola (because $e > 1$) with one focus at the origin.

The directrices are vertical (because $\cos\theta$ appears in the equation). Knowing that $ed = 4$, we have $d = \frac{4}{e} = \frac{8}{3}$, and one directrix is $x = \frac{8}{3}$. Letting $\theta = 0$ and $\theta = \pi$, the polar coordinates of the vertices are $\left(\frac{8}{5}, 0\right)$ and $(-8, \pi)$; equivalently, the vertices are $\left(\frac{8}{5}, 0\right)$ and $(8, 0)$ in Cartesian coordinates (Figure 10.57). The center of the hyperbola is halfway between the vertices; therefore, its Cartesian coordinates are $\left(\frac{24}{5}, 0\right)$. The distance between the focus at $(0, 0)$ and the nearest vertex $\left(\frac{8}{5}, 0\right)$ is $\frac{8}{5}$. Therefore, the other focus is $\frac{8}{5}$ units to the right of the vertex $(8, 0)$. So, the Cartesian coordinates of the foci are $\left(\frac{48}{5}, 0\right)$ and $(0, 0)$. Because the directrices are symmetric about the center and the left directrix is $x = \frac{8}{3}$, the right directrix is $x = \frac{104}{15} \approx 6.9$. The graph of the hyperbola (Figure 10.57) is generated with $0 \le \theta \le 2\pi$ $\left(\text{with } \theta \ne \pm\cos^{-1}\left(-\frac{2}{3}\right)\right)$.

b. The equation is in standard form, and it describes a parabola because $e = 1$. The sole focus is at the origin. The directrix is horizontal (because of the $\sin\theta$ term); $ed = 1$ implies that $d = 2$ and the directrix is $y = 2$. The parabola opens downward because of the plus sign in the denominator. The vertex corresponds to $\theta = \frac{\pi}{2}$ and has polar coordinates $\left(1, \frac{\pi}{2}\right)$, or Cartesian coordinates $(0, 1)$. Setting $\theta = 0$ and $\theta = \pi$, the parabola crosses the x-axis at $(2, 0)$ and $(2, \pi)$ in polar coordinates, or $(\pm 2, 0)$ in Cartesian coordinates. As θ increases from $-\frac{\pi}{2}$ to $\frac{\pi}{2}$, the right branch of the parabola is generated and as θ increases from $\frac{\pi}{2}$ to $\frac{3\pi}{2}$, the left branch of the parabola is generated (Figure 10.58).

Related Exercises 55–64 ◄

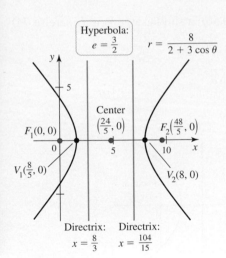

Hyperbola: $e = \frac{3}{2}$ $r = \dfrac{8}{2 + 3\cos\theta}$

Center $\left(\frac{24}{5}, 0\right)$

$F_1(0, 0)$ $F_2\left(\frac{48}{5}, 0\right)$

$V_1\left(\frac{8}{5}, 0\right)$ $V_2(8, 0)$

Directrix: $x = \frac{8}{3}$ Directrix: $x = \frac{104}{15}$

FIGURE 10.57

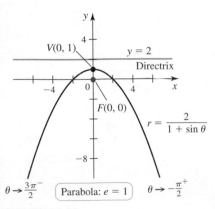

$V(0, 1)$ $y = 2$ Directrix

$F(0, 0)$ $r = \dfrac{2}{1 + \sin\theta}$

$\theta \to \frac{3\pi}{2}^-$ Parabola: $e = 1$ $\theta \to -\frac{\pi}{2}^+$

FIGURE 10.58

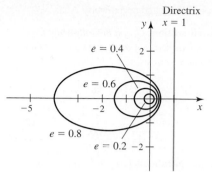

EXAMPLE 8 **Conics in polar coordinates** Use a graphing utility to plot the curves

$$r = \frac{e}{1 + e \cos \theta} \text{ with } e = 0.2, 0.4, 0.6, \text{ and } 0.8. \text{ Comment on the effect of varying the eccentricity, } e.$$

SOLUTION Because $0 < e < 1$, all the curves are ellipses. Notice that the equation is in standard form with $d = 1$; therefore, the curves have the same directrix, $x = d = 1$. As the eccentricity increases, the ellipses becomes more elongated. Small values of e correspond to more circular ellipses (Figure 10.59).   *Related Exercises 65–66* ◄

FIGURE 10.59

SECTION 10.4 EXERCISES

Review Questions

1. Give the property that defines all parabolas.

2. Give the property that defines all ellipses.

3. Give the property that defines all hyperbolas.

4. Sketch the three basic conic sections in standard position with vertices and foci on the x-axis.

5. Sketch the three basic conic sections in standard position with vertices and foci on the y-axis.

6. What is the equation of the standard parabola with its vertex at the origin that opens downward?

7. What is the equation of the standard ellipse with vertices at $(\pm a, 0)$ and foci at $(\pm c, 0)$?

8. What is the equation of the standard hyperbola with vertices at $(0, \pm a)$ and foci at $(0, \pm c)$?

9. Given the vertices $(\pm a, 0)$ and the eccentricity e, what are the coordinates of the foci of an ellipse and a hyperbola?

10. Give the equation in polar coordinates of a conic section with a focus at the origin, eccentricity e, and a directrix $x = d$, where $d > 0$.

11. What are the equations of the asymptotes of a standard hyperbola with vertices on the x-axis?

12. How does the eccentricity determine the type of conic section?

Basic Skills

13–18. Graphing parabolas *Sketch the graph of the following parabolas. Specify the location of the focus and the equation of the directrix. Use a graphing utility to check your work.*

13. $x^2 = 12y$

14. $y^2 = 20x$

15. $x = -y^2/16$

16. $4x = -y^2$

17. $8y = -3x^2$

18. $12x = 5y^2$

19–24. Equations of parabolas *Find an equation of the following parabolas, assuming the vertex is at the origin. Use a graphing utility to check your work.*

19. A parabola that opens to the right with directrix $x = -4$

20. A parabola that opens downward with directrix $y = 6$

21. A parabola with focus at $(3, 0)$

22. A parabola with focus at $(-4, 0)$

23. A parabola symmetric about the y-axis that passes through the point $(2, -6)$

24. A parabola symmetric about the x-axis that passes through the point $(1, -4)$

25–26. From graphs to equations *Write an equation of the following parabolas.*

25.
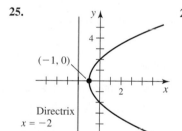

26.
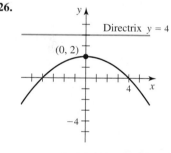

27–32. Graphing ellipses *Sketch the graph of the following ellipses. Plot and label the coordinates of the vertices and foci, and find the lengths of the major and minor axes. Use a graphing utility to check your work.*

27. $\dfrac{x^2}{4} + y^2 = 1$

28. $\dfrac{x^2}{9} + \dfrac{y^2}{4} = 1$

29. $\dfrac{x^2}{4} + \dfrac{y^2}{16} = 1$

30. $x^2 + \dfrac{y^2}{9} = 1$

31. $\dfrac{x^2}{5} + \dfrac{y^2}{7} = 1$

32. $12x^2 + 5y^2 = 60$

33–36. Equations of ellipses *Find an equation of the following ellipses, assuming the center is at the origin. Sketch a graph labeling the vertices and foci. Use a graphing utility to check your work.*

33. An ellipse whose major axis is on the x-axis with length 8 and whose minor axis has length 6

34. An ellipse with vertices $(\pm 6, 0)$ and foci $(\pm 4, 0)$

35. An ellipse with vertices $(\pm 5, 0)$, passing through the point $\left(4, \frac{3}{5}\right)$

36. An ellipse with vertices $(0, \pm 10)$, passing through the point $(\sqrt{3}/2, 5)$

37–38. From graphs to equations *Write an equation of the following ellipses.*

37.

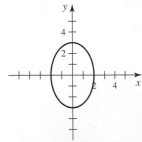

38.

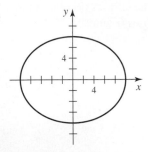

39–44. Graphing hyperbolas *Sketch the graph of the following hyperbolas. Specify the coordinates of the vertices and foci and find the equations of the asymptotes. Use a graphing utility to check your work.*

39. $\dfrac{x^2}{4} - y^2 = 1$

40. $\dfrac{y^2}{16} - \dfrac{x^2}{9} = 1$

41. $4x^2 - y^2 = 16$

42. $25y^2 - 4x^2 = 100$

43. $\dfrac{x^2}{3} - \dfrac{y^2}{5} = 1$

44. $10x^2 - 7y^2 = 140$

45–48. Equations of hyperbolas *Find an equation of the following hyperbolas, assuming the center is at the origin. Sketch a graph labeling the vertices, foci, and asymptotes. Use a graphing utility to check your work.*

45. A hyperbola with vertices $(\pm 4, 0)$ and foci $(\pm 6, 0)$

46. A hyperbola with vertices $(\pm 1, 0)$ that passes through $\left(\frac{5}{3}, 8\right)$

47. A hyperbola with vertices $(\pm 2, 0)$ and asymptotes $y = \pm 3x/2$

48. A hyperbola with vertices $(0, \pm 4)$ and asymptotes $y = \pm 2x$

49–50. From graphs to equations *Write an equation of the following hyperbolas.*

49.

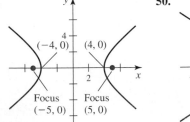

50.

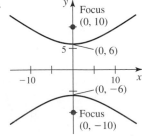

51–54. Eccentricity-directrix approach *Find an equation of the following curves, assuming the center is at the origin. Sketch a graph labeling the vertices, foci, asymptotes, and directrices. Use a graphing utility to check your work.*

51. An ellipse with vertices $(\pm 9, 0)$ and eccentricity $\frac{1}{3}$

52. An ellipse with vertices $(0, \pm 9)$ and eccentricity $\frac{1}{4}$

53. A hyperbola with vertices $(\pm 1, 0)$ and eccentricity 3

54. A hyperbola with vertices $(0, \pm 4)$ and eccentricity 2

55–60. Polar equations for conic sections *Graph the following conic sections, labeling the vertices, foci, directrices, and asymptotes (if they exist). Use a graphing utility to check your work.*

55. $r = \dfrac{4}{1 + \cos \theta}$

56. $r = \dfrac{4}{2 + \cos \theta}$

57. $r = \dfrac{1}{2 - \cos \theta}$

58. $r = \dfrac{6}{3 + 2 \sin \theta}$

59. $r = \dfrac{1}{2 - 2 \sin \theta}$

60. $r = \dfrac{12}{3 - \cos \theta}$

61–64. Tracing hyperbolas and parabolas *Graph the following equations. Then use arrows and labeled points to indicate how the curve is generated as θ increases from 0 to 2π.*

61. $r = \dfrac{1}{1 + \sin \theta}$

62. $r = \dfrac{1}{1 + 2 \cos \theta}$

63. $r = \dfrac{3}{1 - \cos \theta}$

64. $r = \dfrac{1}{1 - 2 \cos \theta}$

65. Parabolas with a graphing utility Use a graphing utility to graph the parabolas $y^2 = 4px$ for $p = -5, -2, -1, 1, 2,$ and 5 on the same set of axes. Explain how the shapes of the curves vary as p changes.

66. Hyperbolas with a graphing utility Use a graphing utility to graph the hyperbolas $r = \dfrac{e}{1 + e \cos \theta}$ for $e = 1.1, 1.3, 1.5, 1.7,$ and 2 on the same set of axes. Explain how the shapes of the curves vary as e changes.

Further Explorations

67. Explain why or why not Determine whether the following statements are true and give an explanation or counterexample.

a. The hyperbola $x^2/4 - y^2/9 = 1$ has no y-intercepts.

b. On every ellipse, there are exactly two points at which the curve has slope s, where s is any real number.

c. Given the directrices and foci of a standard hyperbola, it is possible to find its vertices, eccentricity, and asymptotes.

d. The point on a parabola closest to the focus is the vertex.

68–71. Tangent lines *Find an equation of the line tangent to the following curves at the given point.*

68. $y^2 = 8x;\ (8, -8)$

69. $x^2 = -6y;\ (-6, -6)$

70. $r = \dfrac{1}{1 + \sin \theta};\ \left(\dfrac{2}{3}, \dfrac{\pi}{6}\right)$

71. $y^2 - \dfrac{x^2}{64} = 1;\ \left(6, -\dfrac{5}{4}\right)$

72–73. Graphs to polar equations *Find a polar equation for each conic section.*

72.

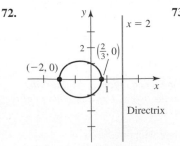

73.

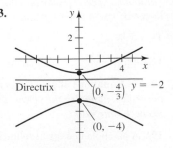

74. Deriving polar equations for conics Modify Figure 10.55 to derive the polar equation of a conic section with a focus at the origin in the following three cases.

    a. Vertical directrix at $x = -d$, where $d > 0$
    b. Horizontal directrix at $y = d$, where $d > 0$
    c. Horizontal directrix at $y = -d$, where $d > 0$

75. Another construction for a hyperbola Suppose two circles are centered at F_1 and F_2, respectively, whose centers are at least $2a$ units apart (see figure). The radius of one circle is $2a + r$ and the radius of the other circle is r, where $r \geq 0$. Show that as r increases, the intersection point P of the two circles describes one branch of a hyperbola with foci at F_1 and F_2.

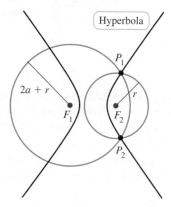

76. The ellipse and the parabola Let R be the region bounded by the upper half of the ellipse $x^2/2 + y^2 = 1$ and the parabola $y = x^2/\sqrt{2}$.

    a. Find the area of R.
    b. Which is greater, the volume of the solid generated when R is revolved about the x-axis or the volume of the solid generated when R is revolved about the y-axis?

77. Tangent lines for an ellipse Show that an equation of the line tangent to the ellipse $x^2/a^2 + y^2/b^2 = 1$ at the point (x_0, y_0) is

$$\frac{xx_0}{a^2} + \frac{yy_0}{b^2} = 1.$$

78. Tangent lines for a hyperbola Find an equation of the line tangent to the hyperbola $x^2/a^2 - y^2/b^2 = 1$ at the point (x_0, y_0).

79. Volume of an ellipsoid Suppose that the ellipse $x^2/a^2 + y^2/b^2 = 1$ is revolved about the x-axis. What is the volume of the *ellipsoid* that is generated? Is the volume different if the same ellipse is revolved about the y-axis?

80. Area of a sector of a hyperbola Consider the region R bounded by the right branch of the hyperbola $x^2/a^2 - y^2/b^2 = 1$ and the vertical line through the right focus.

    a. What is the area of R?
    b. Sketch a graph that shows how the area of R varies with the eccentricity e, for $e > 1$.

81. Volume of a hyperbolic cap Consider the region R bounded by the right branch of the hyperbola $x^2/a^2 - y^2/b^2 = 1$ and the vertical line through the right focus.

    a. What is the volume of the solid that is generated when R is revolved about the x-axis?
    b. What is the volume of the solid that is generated when R is revolved about the y-axis?

82. Volume of a paraboloid (Archimedes) The region bounded by the parabola $y = ax^2$ and the horizontal line $y = h$ is revolved about the y-axis to generate a solid bounded by a surface called a **paraboloid** (where $a > 0$ and $h > 0$). Show that the volume of the solid is $\frac{3}{2}$ the volume of the cone with the same base and vertex.

Applications
(See the Guided Projects for additional applications of conic sections.)

83. Reflection property of parabolas Consider the parabola $y = x^2/(4p)$ with its focus at $F(0, p)$ (see figure). We know that light is reflected from a surface in such a way that the angle of incidence equals the angle of reflection. The goal is to show that the angle between the ray ℓ and the tangent line L (α in the figure) equals the angle between the line PF and L (β in the figure). If these two angles are equal, then the reflection property is proved because ℓ is reflected through F.

    a. Let $P(x_0, y_0)$ be a point on the parabola. Show that the slope of the line tangent to the curve at P is $\tan \theta = x_0/(2p)$.
    b. Show that $\tan \varphi = (p - y_0)/x_0$.
    c. Show that $\alpha = \pi/2 - \theta$; therefore, $\tan \alpha = \cot \theta$.
    d. Note that $\beta = \theta + \varphi$. Use the tangent addition formula

$$\tan(\theta + \varphi) = \frac{\tan \theta + \tan \varphi}{1 - \tan \theta \tan \varphi} \text{ to show that}$$

    $\tan \alpha = \tan \beta = 2p/x_0$.

    e. Conclude that because α and β are acute angles, $\alpha = \beta$.

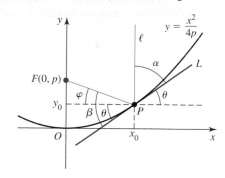

84. Golden Gate Bridge Completed in 1937, San Francisco's Golden Gate Bridge is 2.7 km long and weighs about 890,000 tons. The length of the span between the two central towers is 1280 m; the towers themselves extend 152 m above the roadway. The cables that support the deck of the bridge between the two towers hang in a parabola (see figure). Assuming the origin is midway between the towers on the deck of the bridge, find an equation that describes the cables. How long is a guy wire that hangs vertically from the cables to the roadway 500 m from the center of the bridge?

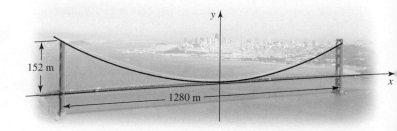

Additional Exercises

85. Equation of an ellipse Consider an ellipse to be the set of points in a plane whose distances from two fixed points have a constant sum $2a$. Derive the equation of an ellipse. Assume the two fixed points are on the x-axis equidistant from the origin.

86. Equation of a hyperbola Consider a hyperbola to be the set of points in a plane whose distances from two fixed points have a constant difference of $2a$ or $-2a$. Derive the equation of a hyperbola. Assume the two fixed points are on the x-axis equidistant from the origin.

87. Equidistant set Show that the set of points equidistant from a circle and a line not passing through the circle is a parabola.

88. Polar equation of a conic Show that the polar equation of an ellipse or hyperbola with one focus at the origin, major axis of length $2a$ on the x-axis, and eccentricity e is

$$r = \frac{a(1 - e^2)}{1 + e \cos \theta}.$$

89. Shared asymptotes Suppose that two hyperbolas with eccentricities e and E have perpendicular major axes and share a set of asymptotes. Show that $e^{-2} + E^{-2} = 1$.

90–94. Focal chords *A **focal chord** of a conic section is a line through a focus joining two points of the curve. The **latus rectum** is the focal chord perpendicular to the major axis of the conic. Prove the following properties.*

90. The lines tangent to the endpoints of any focal chord of a parabola $y^2 = 4px$ intersect on the directrix and are perpendicular.

91. Let L be the latus rectum of the parabola $y^2 = 4px$ for $p > 0$. Let F be the focus of the parabola, P be any point on the parabola to the left of L, and D be the (shortest) distance between P and L. Show that for all P, $D + |FP|$ is a constant. Find the constant.

92. The length of the latus rectum of the parabola $y^2 = 4px$ or $x^2 = 4py$ is $4|p|$.

93. The length of the latus rectum of an ellipse centered at the origin is $2b^2/a = 2b\sqrt{1 - e^2}$.

94. The length of the latus rectum of a hyperbola centered at the origin is $2b^2/a = 2b\sqrt{e^2 - 1}$.

95. Confocal ellipse and hyperbola Show that an ellipse and a hyperbola that have the same two foci intersect at right angles.

96. Approach to asymptotes Show that the vertical distance between a hyperbola $x^2/a^2 - y^2/b^2 = 1$ and its asymptote $y = bx/a$ approaches zero as $x \to \infty$, where $0 < b < a$.

97. Sector of a hyperbola Let H be the right branch of the hyperbola $x^2 - y^2 = 1$ and let ℓ be the line $y = m(x - 2)$ that passes through the point $(2, 0)$ with slope m, where $-\infty < m < \infty$. Let R be the region in the first quadrant bounded by H and ℓ (see figure). Let $A(m)$ be the area of R. Note that for some values of m, $A(m)$ is not defined.

    a. Find the x-coordinates of the intersection points between H and ℓ as functions of m; call them $u(m)$ and $v(m)$ where $v(m) > u(m) > 1$. For what values of m are there two intersection points?

    b. Evaluate $\lim_{m \to 1^+} u(m)$ and $\lim_{m \to 1^+} v(m)$.

    c. Evaluate $\lim_{m \to \infty} u(m)$ and $\lim_{m \to \infty} v(m)$.

    d. Evaluate and interpret $\lim_{m \to \infty} A(m)$.

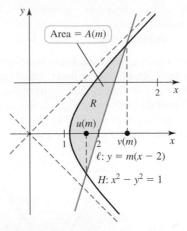

98. The anvil of a hyperbola Let H be the hyperbola $x^2 - y^2 = 1$ and let S be the 2-by-2 square bisected by the asymptotes of H. Let R be the anvil-shaped region bounded by the hyperbola and the horizontal lines $y = \pm p$ (see figure).

    a. For what value of p is the area of R equal to the area of S?

    b. For what value of p is the area of R twice the area of S?

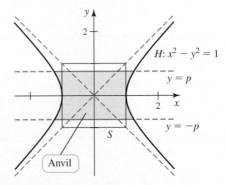

99. Parametric equations for an ellipse Consider the parametric equations

$$x = a \cos t + b \sin t, \qquad y = c \cos t + d \sin t,$$

where a, b, c, and d are real numbers

    a. Show that (apart from a set of special cases) the equations describe an ellipse of the form $Ax^2 + Bxy + Cy^2 = K$, where A, B, C, and K are constants

    b. Show that (apart from a set of special cases), the equations describe an ellipse with its axes aligned with the x- and y-axes provided $ab + cd = 0$.

    c. Show that the equations describe a circle provided $ab + cd = 0$ and $c^2 + d^2 = a^2 + b^2 \neq 0$.

QUICK CHECK ANSWERS

2. a. Left **b.** Up **3.** The minor-axis vertices are $(0, \pm b)$. The distance between them is $2b$, which is the length of the minor axis. **4.** Vertices: $(0, \pm 1)$; foci: $(0, \pm \sqrt{5})$ **5.** $b = 3\sqrt{3}/2, c = 3/2, d = 6$ **6.** y-axis ◄

CHAPTER 10 REVIEW EXERCISES

1. **Explain why or why not** Determine whether the following statements are true and give an explanation or counterexample.

   a. A set of parametric equations for a given curve is always unique.
   b. The equations $x = e^t$, $y = 2e^t$ for $-\infty < t < \infty$ describe a line passing through the origin with slope 2.
   c. The polar coordinates $(3, -3\pi/4)$ and $(-3, \pi/4)$ describe the same point in the plane.
   d. The limaçon $r = f(\theta) = 1 - 4\cos\theta$ has an outer and inner loop. The area of the region between the two loops is
   $$\frac{1}{2}\int_0^{2\pi} (f(\theta))^2\, d\theta.$$
   e. The hyperbola $y^2/2 - x^2/4 = 1$ has no x-intercepts.
   f. The equation $x^2 + 4y^2 - 2x = 3$ describes an ellipse.

2–5. Parametric curves

   a. *Plot the following curves, indicating the positive orientation.*
   b. *Eliminate the parameter to obtain an equation in x and y.*
   c. *Identify or briefly describe the curve.*
   d. *Evaluate dy/dx at the specified point.*

2. $x = t^2 + 4$, $y = 6 - t$, for $-\infty < t < \infty$; evaluate dy/dx at $(5, 5)$.

3. $x = e^t$, $y = 3e^{-2t}$, for $-\infty < t < \infty$; evaluate dy/dx at $(1, 3)$.

4. $x = 10\sin 2t$, $y = 16\cos 2t$, for $0 \le t \le \pi$; evaluate dy/dx at $(5\sqrt{3}, 8)$.

5. $x = \ln t$, $y = 8\ln t^2$, for $1 \le t \le e^2$; evaluate dy/dx at $(1, 16)$.

6. **Circles** For what values of a, b, c, and d do the equations $x = a\cos t + b\sin t$, $y = c\cos t + d\sin t$ describe a circle? What is the radius of the circle?

7. **Tangent lines** Find an equation of the line tangent to the cycloid $x = t - \sin t$, $y = 1 - \cos t$ at the points corresponding to $t = \pi/6$ and $t = 2\pi/3$.

8–9. Sets in polar coordinates *Sketch the following sets of points.*

8. $\{(r, \theta): 4 \le r^2 \le 9\}$

9. $\{(r, \theta): 0 \le r \le 4, -\pi/2 \le \theta \le -\pi/3\}$

10. **Matching polar curves** Match equations a–f with graphs A–F.

   a. $r = 3\sin 4\theta$ b. $r^2 = 4\cos\theta$
   c. $r = 2 - 3\sin\theta$ d. $r = 1 + 2\cos\theta$
   e. $r = 3\cos 3\theta$ f. $r = e^{-\theta/6}$

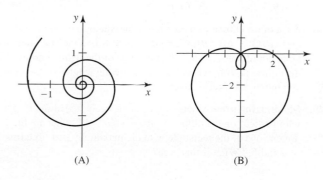

(A)   (B)

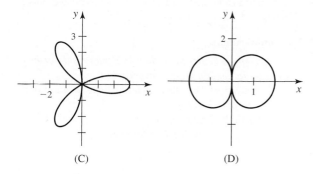

(C)   (D)

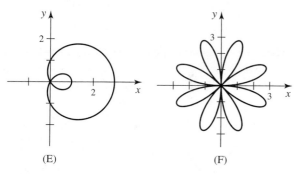

(E)   (F)

11. **Polar conversion** Write the equation
$r^2 + r(2\sin\theta - 6\cos\theta) = 0$ in Cartesian coordinates and identify the corresponding curve.

12. **Polar conversion** Consider the equation $r = 4/(\sin\theta - 6\cos\theta)$.

   a. Convert the equation to Cartesian coordinates and identify the curve it describes.
   b. Graph the curve and indicate the points that correspond to $\theta = 0, \pi/2$, and 2π.
   c. Give an interval in θ on which the entire curve is generated.

13. **Intersection points** Consider the equations $r = 1$ and $r = 2 - 4\cos\theta$.

   a. Graph the curves. How many intersection points do you observe?
   b. Give the approximate polar coordinates of the intersection points.

14–17. Slopes of tangent lines

   a. *Find all points where the following curves have vertical and horizontal tangent lines.*
   b. *Find the slope of the lines tangent to the curve at the origin (when relevant).*
   c. *Sketch the curve and all the tangent lines identified in parts (a) and (b).*

14. $r = 2\cos 2\theta$ 15. $r = 4 + 2\sin\theta$

16. $r = 3 - 6\cos\theta$ 17. $r^2 = 2\cos 2\theta$

18–21. Areas of regions *Find the area of the following regions. In each case, graph the curve(s) and shade the region in question.*

18. The region enclosed by all the leaves of the rose $r = 3\sin 4\theta$

19. The region enclosed by the limaçon $r = 3 - \cos\theta$

20. The region inside the limaçon $r = 2 + \cos \theta$ and outside the circle $r = 2$

21. The region inside the lemniscate $r^2 = 4 \cos 2\theta$ and outside the circle $r = \frac{1}{2}$

22–27. Conic sections

    a. *Determine whether the following equations describe a parabola, an ellipse, or a hyperbola.*

    b. *Use analytical methods to determine the location of the foci, vertices, and directrices.*

    c. *Find the eccentricity of the curve.*

    d. *Make an accurate graph of the curve*

22. $x = 16y^2$ **23.** $x^2 - y^2/2 = 1$

24. $x^2/4 + y^2/25 = 1$ **25.** $y^2 - 4x^2 = 16$

26. $y = 8x^2 + 16x + 8$ **27.** $4x^2 + 8y^2 = 16$

28. Matching equations and curves Match equations a–f with graphs A–F.

    a. $x^2 - y^2 = 4$ **b.** $x^2 + 4y^2 = 4$

    c. $y^2 - 3x = 0$ **d.** $x^2 + 3y = 1$

    e. $x^2/4 + y^2/8 = 1$ **f.** $y^2/8 - x^2/2 = 1$

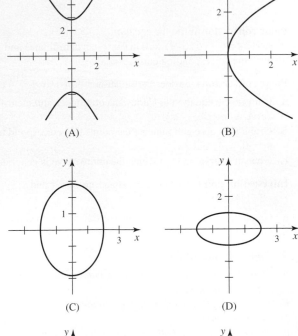

(A)   (B)

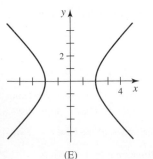 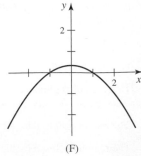

(C)   (D)

(E)   (F)

29–32. Tangent lines *Find an equation of the line tangent to the following curves at the given point. Check your work with a graphing utility.*

29. $y^2 = -12x;\ \left(-\frac{4}{3}, -4\right)$ **30.** $x^2 = 5y;\ \left(-2, \frac{4}{5}\right)$

31. $\dfrac{x^2}{100} + \dfrac{y^2}{64} = 1;\ \left(-6, -\frac{32}{5}\right)$ **32.** $\dfrac{x^2}{16} - \dfrac{y^2}{9} = 1;\ \left(\frac{20}{3}, -4\right)$

33–36. Polar equations for conic sections *Graph the following conic sections, labeling vertices, foci, directrices, and asymptotes (if they exist). Give the eccentricity of the curve. Use a graphing utility to check your work.*

33. $r = \dfrac{2}{1 + \sin \theta}$ **34.** $r = \dfrac{3}{1 - 2 \cos \theta}$

35. $r = \dfrac{4}{2 + \cos \theta}$ **36.** $r = \dfrac{10}{5 + 2 \cos \theta}$

37. A polar conic section Consider the equation $r^2 = \sec 2\theta$.

    a. Convert the equation to Cartesian coordinates and identify the curve.

    b. Find the vertices, foci, directrices, and eccentricity of the curve.

    c. Graph the curve. Explain why the polar equation does not have the form given in the text for conic sections in polar coordinates.

38–41. Eccentricity-directrix approach *Find an equation of the following curves, assuming the center is at the origin. Graph the curve, labeling vertices, foci, asymptotes (if they exist), and directrices.*

38. An ellipse with foci $(\pm 4, 0)$ and directrices $x = \pm 8$

39. An ellipse with vertices $(0, \pm 4)$ and directrices $y = \pm 10$

40. A hyperbola with vertices $(\pm 4, 0)$ and directrices $x = \pm 2$

41. A hyperbola with vertices $(0, \pm 2)$ and directrices $y = \pm 1$

42. Conic parameters A hyperbola has eccentricity $e = 2$ and foci $(0, \pm 2)$. Find the location of the vertices and directrices.

43. Conic parameters An ellipse has vertices $(0, \pm 6)$ and foci $(0, \pm 4)$. Find the eccentricity, the directrices and the minor axis vertices.

44–47. Intersection points *Use analytical methods to find as many intersection points of the following curves as possible. Use methods of your choice to find the remaining intersection points.*

44. $r = 1 - \cos \theta$ and $r = \theta$ **45.** $r^2 = \sin 2\theta$ and $r = \theta$

46. $r^2 = \sin 2\theta$ and $r = 1 - 2 \sin \theta$

47. $r = \theta/2$ and $r = -\theta$, for $\theta \geq 0$

48. Area of an ellipse Consider the polar equation of an ellipse $r = ed/(1 \pm e \cos \theta)$, where $0 < e < 1$. Evaluate an integral in polar coordinates to show that the area of the region enclosed by the ellipse is πab, where $2a$ and $2b$ are the lengths of the major and minor axes, respectively.

49. Maximizing area Among all rectangles centered at the origin with vertices on the ellipse $x^2/a^2 + y^2/b^2 = 1$, what are the dimensions of the rectangle with the maximum area (in terms of a and b)? What is that area?

50. Equidistant set Let S be the square centered at the origin with vertices $(\pm a, \pm a)$ and $(\pm a, \mp a)$. Describe and sketch the set of points that are equidistant from the square and the origin.

51. Bisecting an ellipse Let R be the region in the first quadrant bounded by the ellipse $x^2/a^2 + y^2/b^2 = 1$. Find the value of m (in terms of a and b) such that the line $y = mx$ divides R into two subregions of equal area.

52. Parabola-hyperbola tangency Let P be the parabola $y = px^2$ and H be the right half of the hyperbola $x^2 - y^2 = 1$.

  a. For what value of p is P tangent to H?
  b. At what point does the tangency occur?
  c. Generalize your results for the hyperbola $x^2/a^2 - y^2/b^2 = 1$.

53. Another ellipse construction Start with two circles centered at the origin with radii $0 < a < b$ (see figure). Assume the line ℓ though the origin intersects the smaller circle at Q and the larger circle at R. Let $P(x, y)$ have the y-coordinate of Q and the x-coordinate of R. Show that the set of points $P(x, y)$ generated in this way for all lines ℓ through the origin is an ellipse.

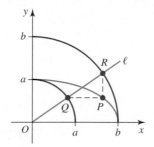

54–55. Graphs to polar equations *Find a polar equation for the coinc sections in the figures.*

54.

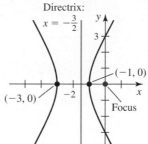

55.

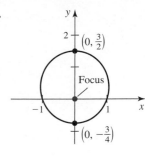

Chapter 10 Guided Projects

Applications of the material in this chapter and related topics can be found in the following Guided Projects. For additional information, see the Preface.

• The amazing cycloid

• Polar art

• Translations and rotations of axes

• Properties of conic sections

• Parametric art

• Grazing goat problems

• Celestial orbits

A

Appendix

The goal of this appendix is to establish the essential notation, terminology, and algebraic skills that are used throughout the book.

Algebra

EXAMPLE 1 **Algebra review**

a. Evaluate $(-32)^{2/5}$.

b. Simplify $\dfrac{1}{x-2} - \dfrac{1}{x+2}$.

c. Solve the equation $\dfrac{x^4 - 5x^2 + 4}{x - 1} = 0$.

SOLUTION

a. Recall that $(-32)^{2/5} = [(-32)^{1/5}]^2$. Because $(-32)^{1/5} = \sqrt[5]{-32} = -2$, we have $(-32)^{2/5} = (-2)^2 = 4$.

Another option is to write $(-32)^{2/5} = [(-32)^2]^{1/5} = 1024^{1/5} = 4$.

b. Finding a common denominator and simplifying leads to

$$\frac{1}{x-2} - \frac{1}{x+2} = \frac{(x+2) - (x-2)}{(x-2)(x+2)} = \frac{4}{x^2 - 4}.$$

c. Notice that $x = 1$ cannot be a solution of the equation because the left side of the equation is undefined at $x = 1$. Because $x \neq 1$, both sides of the equation can be multiplied by $x - 1$ to produce $x^4 - 5x^2 + 4 = 0$. After factoring, this equation becomes $(x^2 - 4)(x^2 - 1) = 0$, which implies $x^2 - 4 = (x - 2)(x + 2) = 0$ or $x^2 - 1 = (x - 1)(x + 1) = 0$. The roots of $x^2 - 4 = 0$ are $x = \pm 2$ and the roots of $x^2 - 1 = 0$ are $x = \pm 1$. Excluding $x = 1$, the roots of the original equation are $x = -1$ and $x = \pm 2$.   *Related Exercises 15–26* ◄

Sets of Real Numbers

Figure A.1 shows the notation for **open intervals**, **closed intervals**, and various **bounded** and **unbounded** intervals. Notice that either interval notation or set notation may be used.

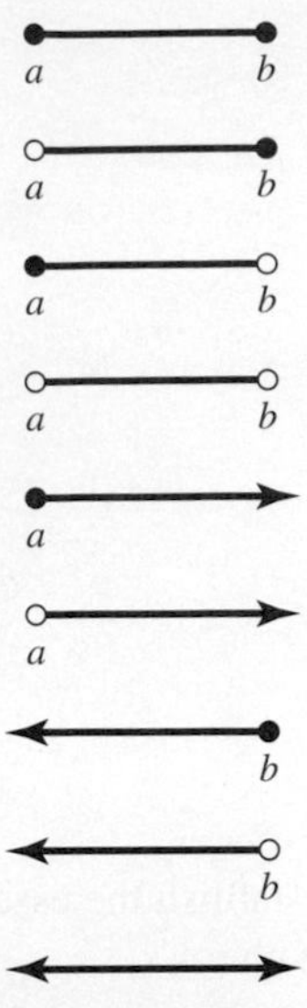

$$[a, b] = \{x : a \le x \le b\} \qquad \text{Closed, bounded interval}$$

$$(a, b] = \{x : a < x \le b\} \qquad \text{Bounded interval}$$

$$[a, b) = \{x : a \le x < b\} \qquad \text{Bounded interval}$$

$$(a, b) = \{x : a < x < b\} \qquad \text{Open, bounded interval}$$

$$[a, \infty) = \{x : x \ge a\} \qquad \text{Unbounded interval}$$

$$(a, \infty) = \{x : x > a\} \qquad \text{Unbounded interval}$$

$$(-\infty, b] = \{x : x \le b\} \qquad \text{Unbounded interval}$$

$$(-\infty, b) = \{x : x < b\} \qquad \text{Unbounded interval}$$

$$(-\infty, \infty) \qquad \text{Unbounded interval}$$

FIGURE A.1

EXAMPLE 2   Solving inequalities   Solve the following inequalities.

a. $-x^2 + 5x - 6 < 0$ **b.** $\dfrac{x^2 - x - 2}{x - 3} \le 0$

SOLUTION

a. We multiply by -1, reverse the inequality, and then factor:

$$x^2 - 5x + 6 > 0 \quad \text{Multiply by } -1.$$
$$(x - 2)(x - 3) > 0 \quad \text{Factor.}$$

The roots of the corresponding equation $(x - 2)(x - 3) = 0$ are $x = 2$ and $x = 3$. These roots partition the number line (Figure A.2) into three intervals: $(-\infty, 2)$, $(2, 3)$, and $(3, \infty)$. On each interval, the product $(x - 2)(x - 3)$ does not change sign. To determine the sign of the product on a given interval, a **test value** x is selected and the sign of $(x - 2)(x - 3)$ is determined at x.

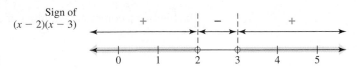

Sign of
$(x - 2)(x - 3)$

FIGURE A.2

A convenient choice for x in $(-\infty, 2)$ is $x = 0$. At this test value,

$$(x - 2)(x - 3) = (-2)(-3) > 0.$$

Using a test value of $x = 2.5$ in the interval $(2, 3)$, we have

$$(x - 2)(x - 3) = (0.5)(-0.5) < 0.$$

A test value of $x = 4$ in $(3, \infty)$ gives

$$(x - 2)(x - 3) = (2)(1) > 0.$$

Therefore, $(x - 2)(x - 3) > 0$ on $(-\infty, 2)$ and $(3, \infty)$. We conclude that the inequality $-x^2 + 5x - 6 < 0$ is satisfied for all x in either $(-\infty, 2)$ or $(3, \infty)$ (Figure A.2).

> The set of numbers
> $\{x : x \text{ is in } (-\infty, 2) \text{ or } (3, \infty)\}$ may also
> be expressed using the union symbol:
>
> $(-\infty, 2) \cup (3, \infty)$

b. The expression $\dfrac{x^2 - x - 2}{x - 3}$ changes sign only at points where the numerator or

denominator of $\dfrac{x^2 - x - 2}{x - 3}$ equals 0. Because

$$\frac{x^2 - x - 2}{x - 3} = \frac{(x + 1)(x - 2)}{x - 3},$$

the numerator is 0 when $x = -1$ and $x = 2$, and the denominator is 0 at $x = 3$.

Therefore, we examine the sign of $\dfrac{(x + 1)(x - 2)}{x - 3}$ on the intervals $(-\infty, -1)$,

$(-1, 2), (2, 3),$ and $(3, \infty)$.

Using test values on these intervals, we see that $\dfrac{(x + 1)(x - 2)}{x - 3} < 0$ on

$(-\infty, -1)$ and $(2, 3)$. Furthermore, the expression is 0 when $x = -1$ and $x = 2$.

Therefore, $\dfrac{x^2 - x - 2}{x - 3} \leq 0$ for all values of x in either $(-\infty, -1]$ or $[2, 3)$

(Figure A.3).

Test Value	$x + 1$	$x - 2$	$x - 3$	Result
-2	$-$	$-$	$-$	$-$
0	$+$	$-$	$-$	$+$
2.5	$+$	$+$	$-$	$-$
4	$+$	$+$	$+$	$+$

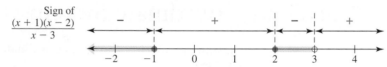

FIGURE A.3

Related Exercises 27–30 ◄

Absolute Value

The **absolute value** of a real number x, denoted $|x|$, is the distance between x and the origin on the number line. (Figure A.4) More generally, $|x - y|$ is the distance between the points x and y on the number line. The absolute value has the following definition and properties:

> The absolute value is useful in simplifying square roots. Because $\sqrt{a}$ is nonnegative, we have $\sqrt{a^2} = |a|$. For example, $\sqrt{3^2} = 3$ and $\sqrt{(-3)^2} = \sqrt{9} = 3$. Note that the solutions of $x^2 = 9$ are $|x| = 3$ or $x = \pm 3$.

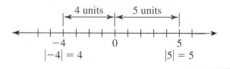

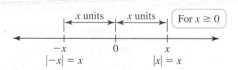

FIGURE A.4

Definition and Properties of the Absolute Value

The absolute value of a real number x is defined as

$$|x| = \begin{cases} x & \text{if } x \geq 0 \\ -x & \text{if } x < 0 \end{cases}$$

Let a be a positive real number.

1. $|x| = a \Leftrightarrow x = \pm a$

2. $|x| < a \Leftrightarrow -a < x < a$

3. $|x| > a \Leftrightarrow x > a$ or $x < -a$

4. $|x| \leq a \Leftrightarrow -a \leq x \leq a$

> Property 6 is called the **triangle inequality**.

5. $|x| \geq a \Leftrightarrow x \geq a$ or $x \leq -a$

6. $|x + y| \leq |x| + |y|$

EXAMPLE 3   Inequalities with absolute values Solve the following inequalities. Then sketch the solution on the number line and express it in interval notation.

a. $|x - 2| < 3$ **b.** $|2x - 6| \geq 10$

SOLUTION

a. Using Property 2 of the absolute value, $|x - 2| < 3$ is written as

$$-3 < x - 2 < 3.$$

Adding 2 to each term of these inequalities results in $-1 < x < 5$ (Figure A.5). This set of numbers is written as $(-1, 5)$ in interval notation.

b. Using Property 5, the inequality $|2x - 6| \geq 10$ implies that

$$2x - 6 \geq 10 \quad \text{or} \quad 2x - 6 \leq -10.$$

We add 6 to both sides of the first inequality to obtain $2x \geq 16$, which implies $x \geq 8$. Similarly, the second inequality yields $x \leq -2$ (Figure A.6). In interval notation, the solution is $(-\infty, -2]$ or $[8, \infty)$.   *Related Exercises 31–34* ◄

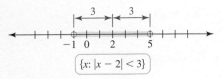

$\{x: |x - 2| < 3\}$

FIGURE A.5

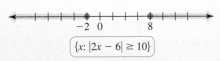

$\{x: |2x - 6| \geq 10\}$

FIGURE A.6

Cartesian Coordinate System

The conventions of the **Cartesian coordinate system** or **xy-coordinate system** are illustrated in Figure A.7.

▷ The familiar (x, y) coordinate system is named after René Descartes (1596–1650). However, it was introduced independently and simultaneously by Pierre de Fermat (1601–1665).

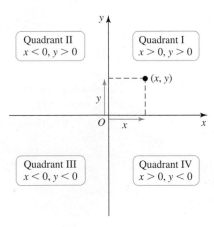

FIGURE A.7

Distance Formula and Circles

By the Pythagorean theorem (Figure A.8), we have the following formula for the distance between two points $P_1(x_1, y_1)$ and $P_2(x_2, y_2)$.

> **Distance Formula**
>
> The distance between the points $P_1(x_1, y_1)$ and $P_2(x_2, y_2)$ is
>
> $$|P_1P_2| = \sqrt{(x_2 - x_1)^2 + (y_2 - y_1)^2}.$$

For any right triangle, $a^2 + b^2 = c^2$.

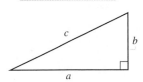

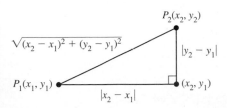

FIGURE A.8

A **circle** is the set of points in the plane whose distance from a fixed point (the **center**) is a constant (the **radius**). This definition leads to the following equations that describe a circle.

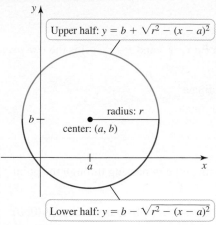

Upper half: $y = b + \sqrt{r^2 - (x - a)^2}$

radius: r

center: (a, b)

Lower half: $y = b - \sqrt{r^2 - (x - a)^2}$

FIGURE A.9

Equation of a Circle

The equation of a circle centered at (a, b) with radius r is

$$(x - a)^2 + (y - b)^2 = r^2.$$

Solving for y, the equations of the upper and lower halves of the circle (Figure A.9) are

$$y = b + \sqrt{r^2 - (x - a)^2} \quad \text{upper half of the circle}$$
$$y = b - \sqrt{r^2 - (x - a)^2} \quad \text{lower half of the circle.}$$

EXAMPLE 4 Sets involving circles

a. Find the equation of the circle with center $(2, 4)$ passing through $(-2, 1)$.

b. Describe the set of points satisfying $x^2 + y^2 - 4x - 6y < 12$.

SOLUTION

a. The radius of the circle is the length of the line segment between the center $(2, 4)$ and the point on the circle $(-2, 1)$, which is

$$\sqrt{(2 - (-2))^2 + (4 - 1)^2} = 5.$$

Therefore, the equation of the circle is

$$(x - 2)^2 + (y - 4)^2 = 25.$$

b. To put this inequality in a recognizable form, we complete the square on the left side of the inequality:

> Recall that the procedure shown here for completing the square works when the coefficient on the quadratic term is 1. When the coefficient is not 1, it must be factored out before completing the square.

$$x^2 + y^2 - 4x - 6y = x^2 - 4x \underbrace{+ 4 - 4}_{\substack{\text{Add and subtract the square} \\ \text{of half the coefficient of } x.}} + y^2 - 6y \underbrace{+ 9 - 9}_{\substack{\text{Add and subtract the square} \\ \text{of half the coefficient of } y.}}$$

$$= \underbrace{x^2 - 4x + 4}_{(x - 2)^2} + \underbrace{y^2 - 6y + 9}_{(y - 3)^2} - 4 - 9$$

$$= (x - 2)^2 + (y - 3)^2 - 13$$

Therefore, the original inequality becomes

$$(x - 2)^2 + (y - 3)^2 - 13 < 12, \quad \text{or} \quad (x - 2)^2 + (y - 3)^2 < 25.$$

> A **circle** is the set of all points whose distance from a fixed point is a constant. A **disk** is the set of all points within and possibly on a circle.

This inequality describes those points that lie within the circle centered at $(2, 3)$ with radius 5 (Figure A.10). Note that a dashed curve is used to indicate that the circle itself is not part of the solution.

The solution to
$(x - 2)^2 + (y - 3)^2 < 25$
is the interior of a circle.

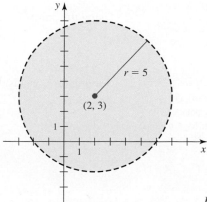

$r = 5$

$(2, 3)$

FIGURE A.10

Related Exercises 35–36 ◄

Equations of Lines

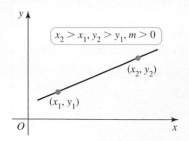

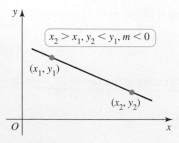

FIGURE A.11

The **slope** m of the line passing through the points $P_1(x_1, y_1)$ and $P_2(x_2, y_2)$ is the *rise over run* (Figure A.11), computed as

$$m = \frac{\text{change in vertical coordinate}}{\text{change in horizontal coordinate}} = \frac{y_2 - y_1}{x_2 - x_1}.$$

Equations of a Line

Point-slope form The equation of the line with slope m passing through the point (x_1, y_1) is $y - y_1 = m(x - x_1)$.

Slope-intercept form The equation of the line with slope m and y-intercept $(0, b)$ is $y = mx + b$ (Figure A.12a).

General linear equation The equation $Ax + By + C = 0$ describes a line in the plane, provided A and B are not both zero.

Vertical and horizontal lines The vertical line that passes through $(a, 0)$ has an equation $x = a$; its slope is undefined. The horizontal line through $(0, b)$ has an equation $y = b$, with slope equal to 0 (Figure A.12b).

> Given a particular line, we often talk about *the* equation of a line. But the equation of a specific line is not unique. Having found one equation, we can multiply it by any nonzero constant to produce another equation of the same line.

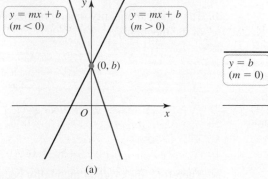

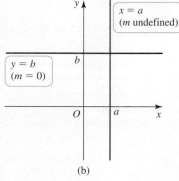

FIGURE A.12

EXAMPLE 5 Working with linear equations Find an equation of the line passing through the points $(1, -2)$ and $(-4, 5)$.

SOLUTION The slope of the line through the points $(1, -2)$ and $(-4, 5)$ is

$$m = \frac{5 - (-2)}{-4 - 1} = \frac{7}{-5} = -\frac{7}{5}.$$

Using the point $(1, -2)$, the point-slope form of the equation is

$$y - (-2) = -\frac{7}{5}(x - 1).$$

> Because both points $(1, -2)$ and $(-4, 5)$ lie on the line and must satisfy the equation of the line, either point can be used to determine an equation of the line.

Solving for y yields the slope-intercept form of the equation:

$$y = -\frac{7}{5}x - \frac{3}{5}$$

Related Exercises 37–40 ◄

Parallel and Perpendicular Lines

Two lines in the plane may have either of two special relationships to each other: they may be parallel or perpendicular.

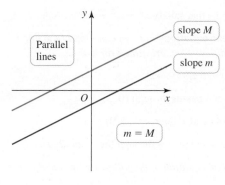

> **Parallel Lines**
>
> Two distinct nonvertical lines are **parallel** if they have the same slope; that is, the lines with equations $y = mx + b$ and $y = Mx + B$ are parallel if and only if $m = M$. Two distinct vertical lines are parallel.

EXAMPLE 6   Parallel lines Find an equation of the line parallel to $3x - 6y + 12 = 0$ that intersects the x-axis at $(4, 0)$.

SOLUTION Solving the equation $3x - 6y + 12 = 0$ for y, we have

$$y = \frac{1}{2}x + 2.$$

This line has a slope of $\frac{1}{2}$ and any line parallel to it has a slope of $\frac{1}{2}$. Therefore, the line that passes through $(4, 0)$ with slope $\frac{1}{2}$ has the point-slope equation $y - 0 = \frac{1}{2}(x - 4)$. After simplifying, an equation of the line is

$$y = \frac{1}{2}x - 2.$$

Notice that the slopes of the two lines are the same; only the y-intercepts differ.

Related Exercises 41–42 ◄

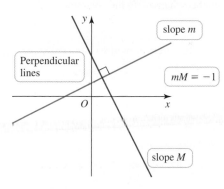

> The slopes of perpendicular lines are *negative reciprocals* of each other.

> **Perpendicular Lines**
>
> Two lines with slopes $m \neq 0$ and $M \neq 0$ are **perpendicular** if and only if $mM = -1$, or equivalently, $m = -1/M$.

EXAMPLE 7   Perpendicular lines Find an equation of the line passing through the point $(-2, 5)$ perpendicular to the line $\ell: 4x - 2y + 7 = 0$.

SOLUTION The equation of ℓ can be written $y = 2x + \frac{7}{2}$, which reveals that its slope is 2. Therefore, the slope of any line perpendicular to ℓ is $-\frac{1}{2}$. The line with slope $-\frac{1}{2}$ passing through the point $(-2, 5)$ is

$$y - 5 = -\frac{1}{2}(x + 2), \quad \text{or} \quad y = -\frac{x}{2} + 4.$$

Related Exercises 43–44 ◄

APPENDIX A EXERCISES

Review Questions

1. State the meaning of $\{x: -4 < x \leq 10\}$ in words. Express the set $\{x: -4 < x \leq 10\}$ using interval notation and draw it on a number line.

2. Write the interval $(-\infty, 2)$ in set notation and draw it on a number line.

3. Give the definition of $|x|$.

4. Write the inequality $|x - 2| \leq 3$ without absolute value symbols.

5. Write the inequality $|2x - 4| \geq 3$ without absolute value symbols.

6. Write an equation of the set of all points that are a distance 5 units from the point $(2, 3)$.

7. Explain how to find the distance between two points whose coordinates are known.

8. Sketch the set of points $\{(x, y): x^2 + (y - 2)^2 > 16\}$.

9. What is the equation of the upper half of the circle centered at the origin with radius 6?

10. What are the possible solution sets of the equation $x^2 + y^2 + Cx + Dy + E = 0$?

11. Give an equation of the line with slope m that passes through the point $(4, -2)$.

12. Give an equation of the line with slope m and y-intercept $(0, 6)$.

13. What is the relationship between the slopes of two parallel lines?

14. What is the relationship between the slopes of two perpendicular lines?

Basic Skills

15–20. Algebra review *Simplify or evaluate the following expressions without a calculator.*

15. $(1/8)^{-2/3}$

16. $\sqrt[3]{-125} + \sqrt{1/25}$

17. $(u + v)^2 - (u - v)^2$

18. $\dfrac{(a + h)^2 - a^2}{h}$

19. $\dfrac{1}{x + h} - \dfrac{1}{x}$

20. $\dfrac{2}{x + 3} - \dfrac{2}{x - 3}$

21–26. Algebra review

21. Factor $y^2 - y^{-2}$.

22. Solve $x^3 - 9x = 0$.

23. Solve $u^4 - 11u^2 + 18 = 0$.

24. Solve $4^x - 6(2^x) = -8$.

25. Simplify $\dfrac{(x + h)^3 - x^3}{h}$ for $h \neq 0$.

26. Rewrite $\dfrac{\sqrt{x + h} - \sqrt{x}}{h}$, where $h \neq 0$, without square roots in the numerator.

27–30. Solving inequalities *Solve the following inequalities and draw the solution on a number line.*

27. $x^2 - 6x + 5 < 0$

28. $\dfrac{x + 1}{x + 2} < 6$

29. $\dfrac{x^2 - 9x + 20}{x - 6} \leq 0$

30. $x\sqrt{x - 1} > 0$

31–34. Inequalities with absolute values *Solve the following inequalities. Then draw the solution on a number line and express it using interval notation.*

31. $|3x - 4| > 8$

32. $1 \leq |x| \leq 10$

33. $3 < |2x - 1| < 5$

34. $2 < \left|\frac{x}{2} - 5\right| < 6$

35–36. Circle calculations *Solve the following problems using the distance formula.*

35. Find the equation of the lower half of the circle with center $(-1, 2)$ and radius 3.

36. Describe the set of points that satisfy $x^2 + y^2 + 6x + 8y \geq 25$.

37–40. Working with linear equations *Find an equation of the line ℓ that satisfies the given condition. Then draw the graph of ℓ.*

37. ℓ has slope $5/3$ and y-intercept $(0, 4)$.

38. ℓ has undefined slope and passes through $(0, 5)$.

39. ℓ has y-intercept $(0, -4)$ and x-intercept $(5, 0)$.

40. ℓ is parallel to the x-axis and passes through the point $(2, 3)$.

41–42. Parallel lines *Find an equation of the following lines and draw their graphs.*

41. the line with y-intercept $(0, 12)$ parallel to the line $x + 2y = 8$

42. the line with x-intercept $(-6, 0)$ parallel to the line $2x - 5 = 0$.

43–44. Perpendicular lines *Find an equation of the following lines.*

43. the line passing through $(3, -6)$ perpendicular to the line $y = -3x + 2$

44. the perpendicular bisector of the line joining the points $(-9, 2)$ and $(3, -5)$

Further Explorations

45. **Explain why or why not** State whether the following statements are true and give an explanation or counterexample.

  a. $\sqrt{16} = \pm 4$

  b. $\sqrt{4^2} = \sqrt{(-4)^2}$

  c. There are two real numbers that satisfy the condition $|x| = -2$.

  d. $|\pi^2 - 9| < 0$.

  e. The point $(1, 1)$ is inside the circle of radius 1 centered at the origin.

  f. $\sqrt{x^4} = x^2$ for all real numbers x.

  g. $\sqrt{a^2} < \sqrt{b^2}$ implies $a < b$ for all real numbers a and b.

46–48. Intervals to sets *Express the following intervals in set notation. Use absolute value notation when possible.*

46. $(-\infty, 12)$

47. $(-\infty, -2]$ or $[4, \infty)$

48. $(2, 3]$ or $[4, 5)$

49–50. Sets in the plane *Graph each set in the xy-plane.*

49. $\{(x, y): |x - y| = 0\}$

50. $\{(x, y): |x| = |y|\}$

B

Appendix

Proofs of Selected Theorems

THEOREM 2.3 Limit Laws

Assume $\lim_{x \to a} f(x)$ and $\lim_{x \to a} g(x)$ exist. The following properties hold, where c is a real number and $m > 0$ and $n > 0$ are integers.

1. **Sum** $\lim_{x \to a} [f(x) + g(x)] = \lim_{x \to a} f(x) + \lim_{x \to a} g(x)$

2. **Difference** $\lim_{x \to a} [f(x) - g(x)] = \lim_{x \to a} f(x) - \lim_{x \to a} g(x)$

3. **Constant multiple** $\lim_{x \to a} [cf(x)] = c \lim_{x \to a} f(x)$

4. **Product** $\lim_{x \to a} [f(x)g(x)] = [\lim_{x \to a} f(x)][\lim_{x \to a} g(x)]$

5. **Quotient** $\lim_{x \to a} \left[\dfrac{f(x)}{g(x)} \right] = \dfrac{\lim\limits_{x \to a} f(x)}{\lim\limits_{x \to a} g(x)}$ provided $\lim_{x \to a} g(x) \neq 0$

6. **Power** $\lim_{x \to a} [f(x)]^n = [\lim_{x \to a} f(x)]^n$

7. **Fractional power** $\lim_{x \to a} [f(x)]^{n/m} = [\lim_{x \to a} f(x)]^{n/m}$ provided $f(x) \geq 0$ for x near a if m is even and n/m is reduced to lowest terms

---

Proof The proof of Law 1 is given in Example 5 of Section 2.7. The proof of Law 2 is analogous to that of Law 1; the triangle inequality in the form $|x - y| \leq |x| + |y|$ is used. The proof of Law 3 is outlined in Exercise 26 of Section 2.7. The proofs of Laws 4 and 5 are given below. The proof of Law 6 involves the repeated use of Law 4. The proof of Law 7 is given in advanced texts. ◄

Proof of Product Law Let $L = \lim_{x \to a} f(x)$ and $M = \lim_{x \to a} g(x)$. Using the definition of a limit, the goal is to show that given any $\varepsilon > 0$, it is possible to specify a $\delta > 0$ such that $|f(x)g(x) - LM| < \varepsilon$ whenever $0 < |x - a| < \delta$. Notice that

$$|f(x)g(x) - LM| = |f(x)g(x) - Lg(x) + Lg(x) - LM| \quad \text{Add and subtract } Lg(x).$$
$$= |(f(x) - L)g(x) + (g(x) - M)L| \quad \text{Group terms.}$$
$$\leq |(f(x) - L)g(x)| + |(g(x) - M)L| \quad \text{Triangle inequality}$$
$$= |f(x) - L||g(x)| + |g(x) - M||L|. \quad |xy| = |x||y|$$

> Real numbers x and y obey the triangle inequality $|x + y| \leq |x| + |y|$.

We now use the definition of the limits of f and g and note that L and M are fixed real numbers. Given $\varepsilon > 0$, there exist $\delta_1 > 0$ and $\delta_2 > 0$ such that

$$|f(x) - L| < \frac{\varepsilon}{2(|M| + 1)} \quad \text{and} \quad |g(x) - M| < \frac{\varepsilon}{2(|L| + 1)}$$

> $|g(x) - M| < 1$ implies that $g(x)$ is less than 1 unit from M. Therefore, whether $g(x)$ and M are positive or negative, $|g(x)| < |M| + 1$.

whenever $0 < |x - a| < \delta_1$ and $0 < |x - a| < \delta_2$, respectively. Furthermore, by the definition of the limit of g, there exits a $\delta_3 > 0$ such that $|g(x) - M| < 1$ whenever $0 < |x - a| < \delta_3$. It follows that $|g(x)| < |M| + 1$ whenever $0 < |x - a| < \delta_3$. Now take δ to be the minimum of $\delta_1, \delta_2,$ and δ_3. Then for $0 < |x - a| < \delta$, we have

$$|f(x)g(x) - LM| \le \underbrace{|f(x) - L|}_{< \frac{\varepsilon}{2(|M| + 1)}} \underbrace{|g(x)|}_{<(|M| + 1)} + \underbrace{|g(x) - M|}_{< \frac{\varepsilon}{2(|L| + 1)}} |L|$$

$$< \frac{\varepsilon}{2} + \frac{\varepsilon}{2} \underbrace{\frac{|L|}{|L| + 1}}_{<1} < \frac{\varepsilon}{2} + \frac{\varepsilon}{2} = \varepsilon.$$

It follows that $\displaystyle\lim_{x \to a} [f(x)g(x)] = LM$.   ◄

Proof of Quotient Law We first prove that if $\displaystyle\lim_{x \to a} g(x) = M$ exists, where $M \ne 0$, then $\displaystyle\lim_{x \to a} \frac{1}{g(x)} = \frac{1}{M}$. The Quotient Law then follows by replacing g by $1/g$ in the Product Law. Therefore, the goal is to show that given any $\varepsilon > 0$, it is possible to specify a $\delta > 0$ such that $\left|\dfrac{1}{g(x)} - \dfrac{1}{M}\right| < \varepsilon$ whenever $0 < |x - a| < \delta$. First note that $M \ne 0$ and $g(x)$ can be made arbitrarily close to M. For this reason, there exists a $\delta_1 > 0$ such that $|g(x)| > |M|/2$ whenever $0 < |x - a| < \delta_1$. Furthermore, using the definition of the limit of g, given any $\varepsilon > 0$, there exists a $\delta_2 > 0$ such that $|g(x) - M| < \dfrac{\varepsilon|M|^2}{2}$ whenever $0 < |x - a| < \delta_2$. Now take δ to be the minimum of δ_1 and δ_2. Then for $0 < |x - a| < \delta$, we have

$$\left|\frac{1}{g(x)} - \frac{1}{M}\right| = \left|\frac{M - g(x)}{Mg(x)}\right| \qquad \text{Common denominator}$$

> Note that if $|g(x)| > |M|/2$, then $1/|g(x)| < 2/|M|$.

$$= \frac{1}{|M|} \underbrace{\frac{1}{|g(x)|}}_{<\frac{2}{|M|}} \underbrace{|g(x) - M|}_{<\frac{\varepsilon|M|^2}{2}} \qquad \text{Rewrite.}$$

$$< \frac{1}{|M|} \frac{2}{|M|} \cdot \frac{\varepsilon|M|^2}{2} = \varepsilon. \qquad \text{Simplify.}$$

By the definition of a limit, we have $\displaystyle\lim_{x \to a} \frac{1}{g(x)} = \frac{1}{M}$. The proof can be completed by applying the Product Rule with g replaced by $1/g$.   ◄

THEOREM 9.3   Convergence of Power Series

A power series $\sum_{k=0}^{\infty} c_k(x-a)^k$ centered at a converges in one of three ways:

1. The series converges absolutely for all x, in which case the interval of convergence is $(-\infty, \infty)$ and the radius of convergence is $R = \infty$.

2. There is a real number $R > 0$ such that the series converges absolutely for $|x - a| < R$ and diverges for $|x - a| > R$, in which case the radius of convergence is R.

3. The series converges only at a, in which case the radius of convergence is $R = 0$.

Proof Without loss of generality, we take $a = 0$. (If $a \neq 0$, the following argument may be shifted so it is centered at $x = a$.) The proof hinges on a preliminary result:

If $\sum_{k=0}^{\infty} c_k x^k$ converges for $x = b \neq 0$, then it converges absolutely for

$|x| < |b|$. If $\sum_{k=0}^{\infty} c_k x^k$ diverges for $x = d$, then it diverges for $|x| > |d|$.

To prove this fact, assume that $\sum_{k=0}^{\infty} c_k b^k$ converges, which implies that $\lim_{k \to \infty} c_k b^k = 0$. Then there exists a real number $M > 0$ such that $|c_k b^k| < M$ for $k = 0, 1, 2, 3, \ldots$. It follows that

$$\sum_{k=0}^{\infty} |c_k x^k| = \sum_{k=0}^{\infty} \underbrace{|c_k b^k|}_{M} \left| \frac{x}{b} \right|^k < M \sum_{k=0}^{\infty} \left| \frac{x}{b} \right|^k.$$

If $|x| < |b|$, then $|x/b| < 1$ and $\sum_{k=0}^{\infty} \left| \frac{x}{b} \right|^k$ is a convergent geometric series. Therefore,

$\sum_{k=0}^{\infty} |c_k x^k|$ converges by the comparison test, which implies that $\sum_{k=0}^{\infty} c_k x^k$ converges absolutely for $|x| < |b|$. The second half of the preliminary result is proved by supposing the series diverges at $x = d$. The series cannot converge at a point x_0 with $|x_0| > |d|$ because by the preceding argument, it would converge for all $|x| < |x_0|$, which includes $x = d$. Therefore, the series diverges for all $|x| > |d|$.

Now we may deal with the three cases in the theorem. Let S be the set of real numbers for which the series converges, which always includes 0. If $S = \{0\}$, then we have Case 3. If S consists of all real numbers, then we have Case 1. For Case 2, assume that $d \neq 0$ is a point at which the series diverges. By the preliminary result, the series diverges for $|x| > |d|$. Therefore, if x is in S, then $|x| < |d|$, which implies that S is bounded. By the Least Upper Bound Property for real numbers, S has a least upper bound R, such that $x \leq R$ for all x in S. If $|x| > R$, then x is not in S and the series diverges. If $|x| < R$, then x is not the least upper bound of S and there exists a number b in S with

$|x| < b \leq R$. Because the series converges at $x = b$, by the preliminary result, $\sum_{k=0}^{\infty} |c_k x^k|$

converges for $|x| < |b|$. Therefore, the series $\sum_{k=0}^{\infty} c_k x^k$ converges absolutely for $|x| < R$

and diverges for $|x| > R$.   ◄

> The Least Upper Bound Property for real numbers states that if a nonempty set S is bounded (that is, there is exists a number M, called an *upper bound*, such that $x \leq M$ for all x in S), then S has a *least upper bound* L, which is the smallest of the upper bounds.

THEOREM 10.3   **Eccentricity-Directrix Theorem**

Let ℓ be a line, F be a point not on ℓ, and $e > 0$ be a real number. Let C be the set of points P in a plane with the property that $\dfrac{|PF|}{|PL|} = e$, where $|PL|$ is the perpendicular distance from P to ℓ.

1. If $e = 1$, C is a **parabola**.
2. If $0 < e < 1$, C is an **ellipse**.
3. If $e > 1$, C is a **hyperbola**.

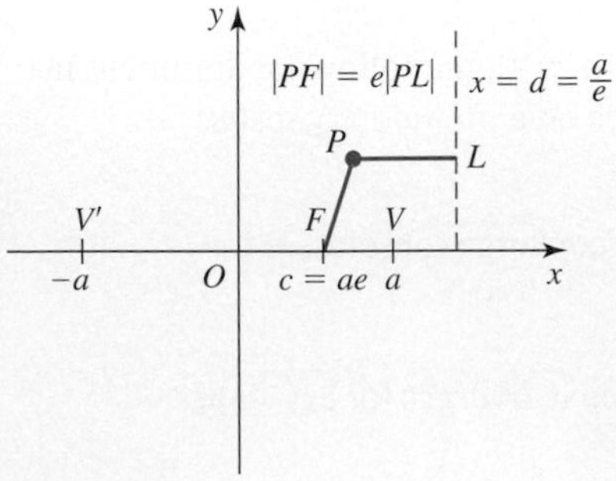

FIGURE B.1

Proof If $e = 1$, then the defining property becomes $|PF| = |PL|$, which is the standard definition of a parabola (Section 10.4). We prove the result for ellipses ($0 < e < 1$) and a small modification handles the case of hyperbolas ($e > 1$).

Let E be the curve whose points satisfy $|PF| = e\,|PL|$; the goal is to show that E is an ellipse. We locate the point F (a *focus*) at $(c, 0)$ and label the line ℓ (a *directrix*) $x = d$, where $c > 0$ and $d > 0$. It can be shown that E intersects the x-axis at the symmetric points (the *vertices*) $V(a, 0)$ and $V'(-a, 0)$ (Figure B.1). These choices place the center of E at the origin. Notice that we have four parameters (a, c, d, and e) that must be related.

Because the vertex $V(a, 0)$ is on E, it satisfies the defining property $|PF| = e\,|PL|$, with $P = V$. This condition implies that $a - c = e(d - a)$. Because the vertex $V'(-a, 0)$ is on the ellipse, it also satisfies the defining property $|PF| = e\,|PL|$, with $P = V'$. This condition implies that $a + c = e(d + a)$. Solving these two equations for c and d, we find that $c = ae$ and $d = a/e$. To summarize, the parameters a, c, d, and e are related by the equations

$$c = ae \quad \text{and} \quad a = de.$$

Because $e < 1$, it follows that $c < a < d$.

We now use the property $|PF| = e\,|PL|$ with an arbitrary point on the ellipse $P(x, y)$. Figure B.1 shows the geometry with the focus $(c, 0) = (ae, 0)$ and the directrix $x = d = a/e$. The condition $|PF| = e\,|PL|$ becomes

$$\sqrt{(x - ae)^2 + y^2} = e\left(\frac{a}{e} - x\right).$$

The goal is to find the simplest possible relationship between x and y. Squaring both sides and collecting terms, we have

$$(1 - e^2)x^2 + y^2 = a^2(1 - e^2).$$

Dividing through by $a^2(1 - e^2)$ gives the equation of the standard ellipse:

$$\frac{x^2}{a^2} + \frac{y^2}{a^2(1 - e^2)} = \frac{x^2}{a^2} + \frac{y^2}{b^2} = 1, \quad \text{where} \quad b^2 = a^2(1 - e^2).$$

This is the equation of an ellipse centered at the origin with vertices and foci on the x-axis.

The preceding proof is now applied with $e > 1$. The argument for ellipses with $0 < e < 1$ led to the equation

$$\frac{x^2}{a^2} + \frac{y^2}{a^2(1 - e^2)} = 1.$$

With $e > 1$, we have $1 - e^2 < 0$, so we write $(1 - e^2) = -(e^2 - 1)$. The resulting equation describes a hyperbola centered at the origin with the foci on the x-axis:

$$\frac{x^2}{a^2} - \frac{y^2}{b^2} = 1, \quad \text{where} \quad b^2 = a^2(e^2 - 1). \qquad \blacktriangleleft$$

THEOREM 12.3   Continuity of Composite Functions

If $u = g(x, y)$ is continuous at (a, b) and $z = f(u)$ is continuous at $g(a, b)$, then the composite function $z = f(g(x, y))$ is continuous at (a, b).

Proof Let P and P_0 represent the points (x, y) and (a, b), respectively. Let $u = g(P)$ and $u_0 = g(P_0)$. The continuity of f at u_0 means that $\lim_{u \to u_0} f(u) = f(u_0)$. This limit implies that given any $\varepsilon > 0$, there exists a $\delta^* > 0$ such that

$$|f(u) - f(u_0)| < \varepsilon \quad \text{whenever} \quad 0 < |u - u_0| < \delta^*.$$

The continuity of g at P_0 means that $\lim_{P \to P_0} g(P) = g(P_0)$. Letting $|P - P_0|$ denote the distance between P and P_0, this limit implies that given any $\delta^* > 0$, there exists a $\delta > 0$ such that

$$|g(P) - g(P_0)| = |u - u_0| < \delta^* \quad \text{whenever} \quad 0 < |P - P_0| < \delta.$$

We now combine these two statements. Given any $\varepsilon > 0$, there exists a $\delta > 0$ such that

$$|f(g(P)) - f(g(P_0))| = |f(u) - f(u_0)| < \varepsilon \quad \text{whenever} \quad 0 < |P - P_0| < \delta.$$

Therefore, $\lim_{(x,y) \to (a,b)} f(g(x, y)) = f(g(a, b))$ and $z = f(g(x, y))$ is continuous at (a, b).

◄

THEOREM 12.5   Conditions for Differentiability

Suppose the function f has partial derivatives f_x and f_y defined in a region containing (a, b) with f_x and f_y continuous at (a, b). Then f is differentiable at (a, b).

Proof Figure B.2 shows a region on which the conditions of the theorem are satisfied containing the points $P_0(a, b), Q(a + \Delta x, b)$, and $P(a + \Delta x, b + \Delta y)$. By the definition of differentiability of f at P_0, we must show that

$$\Delta z = f(P) - f(P_0) = f_x(a, b)\Delta x + f_y(a, b)\Delta y + \varepsilon_1 \Delta x + \varepsilon_2 \Delta y,$$

where ε_1 and ε_2 depend only on $a, b, \Delta x$, and Δy, with $(\varepsilon_1, \varepsilon_2) \to (0, 0)$ as $(\Delta x, \Delta y) \to (0, 0)$. We can view the change Δz taking place in two stages:

- $\Delta z_1 = f(a + \Delta x, b) - f(a, b)$ is the change in z as (x, y) moves from P_0 to Q.
- $\Delta z_2 = f(a + \Delta x, b + \Delta y) - f(a + \Delta x, b)$ is the change in z as (x, y) moves from Q to P.

Applying the Mean Value Theorem to the first variable and noting that f is differentiable with respect to x, we have

$$\Delta z_1 = f(a + \Delta x, b) - f(a, b) = f_x(c, b)\,\Delta x,$$

where c lies in the interval $(a, a + \Delta x)$. Similarly, applying the Mean Value Theorem to the second variable and noting that f is differentiable with respect to y, we have

$$\Delta z_2 = f(a + \Delta x, b + \Delta y) - f(a + \Delta x, b) = f_y(a + \Delta x, d)\,\Delta y,$$

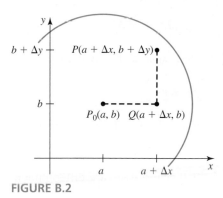

FIGURE B.2

where d lies in the interval $(b, b + \Delta y)$. We now express Δz as the sum of Δz_1 and Δz_2:

$$
\begin{aligned}
\Delta z &= \Delta z_1 + \Delta z_2 \\
&= f_x(c, b)\Delta x + f_y(a + \Delta x, d)\Delta y \\
&= \underbrace{(f_x(c, b) - f_x(a, b)}_{\varepsilon_1} + f_x(a, b))\Delta x \qquad \text{Add and subtract } f_x(a, b). \\
&\quad + \underbrace{(f_y(a + \Delta x, d) - f_y(a, b)}_{\varepsilon_2} + f_y(a, b))\Delta y \quad \text{Add and subtract } f_y(a, b). \\
&= (f_x(a, b) + \varepsilon_1)\Delta x + (f_y(a, b) + \varepsilon_2)\Delta y
\end{aligned}
$$

Note that as $\Delta x \to 0$ and $\Delta y \to 0$, we have $c \to a$ and $d \to b$. Because f_x and f_y are continuous at (a, b) it follows that

$$\varepsilon_1 = f_x(c, b) - f_x(a, b) \to 0 \quad \text{and} \quad \varepsilon_2 = f_y(a + \Delta x, d) - f_y(a, b) \to 0.$$

Therefore, the condition for differentiability of f at (a, b) has been proved. ◄

THEOREM 12.7   Chain Rule (One Independent Variable)

Let $z = f(x, y)$ be a differentiable function of x and y on its domain, where x and y are differentiable functions of t on an interval I. Then

$$\frac{dz}{dt} = \frac{\partial f}{\partial x}\frac{dx}{dt} + \frac{\partial f}{\partial y}\frac{dy}{dt}.$$

Proof Assume $(a, b) = (x(t_0), y(t_0))$ is in the domain of f, where t is in I. Let $\Delta x = x(t + \Delta t) - x(t)$ and $\Delta y = y(t + \Delta t) - y(t)$. Because f is differentiable at (a, b), we know (Section 12.4) that

$$\Delta z = \frac{\partial f}{\partial x}(a, b)\,\Delta x + \frac{\partial f}{\partial y}(a, b)\,\Delta y + \varepsilon_1 \Delta x + \varepsilon_2 \Delta y,$$

where $(\varepsilon_1, \varepsilon_2) \to (0, 0)$ as $(\Delta x, \Delta y) \to (0, 0)$. Dividing this equation by Δt gives

$$\frac{\Delta z}{\Delta t} = \frac{\partial f}{\partial x}\frac{\Delta x}{\Delta t} + \frac{\partial f}{\partial y}\frac{\Delta y}{\Delta t} + \varepsilon_1 \frac{\Delta x}{\Delta t} + \varepsilon_2 \frac{\Delta y}{\Delta t}.$$

As $\Delta t \to 0$, several things occur. First, because $x = g(t)$ and $y = h(t)$ are differentiable on I, $\dfrac{\Delta x}{\Delta t}$ and $\dfrac{\Delta y}{\Delta t}$ approach $\dfrac{dx}{dt}$ and $\dfrac{dy}{dt}$, respectively. Similarly, $\dfrac{\Delta z}{\Delta t}$ approaches $\dfrac{dz}{dt}$ as $\Delta t \to 0$. The fact that x and y are continuous on I (because they are differentiable there), means that $\Delta x \to 0$ and $\Delta y \to 0$ as $\Delta t \to 0$. Therefore, because $(\varepsilon_1, \varepsilon_2) \to (0, 0)$ as $(\Delta x, \Delta y) \to (0, 0)$, it follows that $(\varepsilon_1, \varepsilon_2) \to (0, 0)$ as $\Delta t \to 0$. Letting $\Delta t \to 0$, we have

$$\underbrace{\lim_{\Delta t \to 0} \frac{\Delta z}{\Delta t}}_{\frac{dz}{dt}} = \frac{\partial f}{\partial x} \underbrace{\lim_{\Delta t \to 0} \frac{\Delta x}{\Delta t}}_{\frac{dx}{dt}} + \frac{\partial f}{\partial y} \underbrace{\lim_{\Delta t \to 0} \frac{\Delta y}{\Delta t}}_{\frac{dy}{dt}} + \underbrace{\lim_{\Delta t \to 0} \varepsilon_1}_{\to 0} \underbrace{\frac{\Delta x}{\Delta t}}_{\to \frac{dx}{dt}} + \underbrace{\lim_{\Delta t \to 0} \varepsilon_2}_{\to 0} \underbrace{\frac{\Delta y}{\Delta t}}_{\to \frac{dy}{dt}}$$

or

$$\frac{dz}{dt} = \frac{\partial f}{\partial x}\frac{dx}{dt} + \frac{\partial f}{\partial y}\frac{dy}{dt}.$$

◄

> **THEOREM 12.14   Second Derivative Test**
> Suppose that the second partial derivatives of f are continuous throughout an open disk centered at the point (a, b) where $f_x(a, b) = f_y(a, b) = 0$. Let
> $$D(x, y) = f_{xx}f_{yy} - f_{xy}^2.$$
>
> 1. If $D(a, b) > 0$ and $f_{xx}(a, b) < 0$, then f has a local maximum value at (a, b).
> 2. If $D(a, b) > 0$ and $f_{xx}(a, b) > 0$, then f has a local minimum value at (a, b).
> 3. If $D(a, b) < 0$, then f has a saddle point at (a, b).
> 4. If $D(a, b) = 0$, then the test is inconclusive.

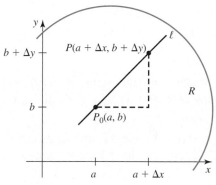

FIGURE B.3

Proof The proof relies on a two-variable version of Taylor's Theorem, which we prove first. Figure B.3 shows the open disk R on which the conditions of the theorem are satisfied; it contains the points $P_0(a, b)$ and $P(a + \Delta x, b + \Delta y)$. The line ℓ through $P_0 P$ has a parametric description

$$\langle x(t), y(t) \rangle = \langle a + t\Delta x, b + t\Delta y \rangle,$$

where $t = 0$ corresponds to P_0 and $t = 1$ corresponds to P.

We now let $F(t) = f(a + t\Delta x, b + t\Delta y)$ be the value of f along that part of ℓ that lies in R. By the Chain Rule we have

$$F'(t) = f_x x'(t) + f_y y'(t) = f_x \underbrace{\Delta x}_{\Delta x} + f_y \underbrace{\Delta y}_{\Delta y}.$$

Differentiating again with respect to t (f_x and f_y are differentiable), we use $f_{xy} = f_{yx}$ to obtain

$$F''(t) = \frac{\partial F'}{\partial x}\underbrace{x'(t)}_{\Delta x} + \frac{\partial F'}{\partial y}\underbrace{y'(t)}_{\Delta y}$$

$$= \frac{\partial}{\partial x}(f_x \Delta x + f_y \Delta y)\Delta x + \frac{\partial}{\partial y}(f_x \Delta x + f_y \Delta y)\Delta y$$

$$= f_{xx}\Delta x^2 + 2f_{xy}\Delta x\Delta y + f_{yy}\Delta y^2.$$

Noting that F meets the conditions of Taylor's Theorem for one variable with $n = 1$, we write

$$F(t) = F(0) + F'(0)(t - 0) + \frac{1}{2}F''(c)(t - 0)^2,$$

where c is between 0 and t. Setting $t = 1$, it follows that

$$F(1) = F(0) + F'(0) + \frac{1}{2}F''(c), \qquad (1)$$

where $0 < c < 1$. Recalling that $F(t) = f(a + t\Delta x, b + t\Delta y)$ and invoking the condition $f_x(a, b) = f_y(a, b) = 0$,

$$F(1) = f(a + \Delta x, b + \Delta y)$$

$$= f(a, b) + \underbrace{f_x(a, b)\Delta x + f_y(a, b)\Delta y}_{F'(0) = 0}$$

$$+ \frac{1}{2}(f_{xx}\Delta x^2 + 2f_{xy}\Delta x\Delta y + f_{yy}\Delta y^2)\Big|_{(a+c\Delta x, b+c\Delta y)}$$

$$= f(a, b) + \frac{1}{2}\underbrace{(f_{xx}\Delta x^2 + 2f_{xy}\Delta x\Delta y + f_{yy}\Delta y^2)\Big|_{(a+c\Delta x, b+c\Delta y)}}_{H(c)}$$

$$= f(a, b) + \frac{1}{2}H(c)$$

The existence and type of extreme point at (a, b) is determined by the sign of $f(a + \Delta x, b + \Delta y) - f(a, b)$ (for example, if $f(a + \Delta x, b + \Delta y) - f(a, b) \geq 0$ for all Δx and Δy near 0, then f has a local minimum at (a, b)). Note that $f(a + \Delta x, b + \Delta y) - f(a, b)$ has the same sign as the quantity we have denoted $H(c)$. Assuming $H(0) \neq 0$, for Δx and Δy sufficiently small and nonzero, the sign of $H(c)$ is the same as the sign of

$$H(0) = \Delta x^2 f_{xx}(a, b) + 2\Delta x \Delta y\, f_{xy}(a, b) + \Delta y^2 f_{yy}(a, b)$$

(because the second partial derivatives are continuous at (a, b) and $(a + c\Delta x, b + c\Delta y)$ can be made arbitrarily close to (a, b)). Multiplying both sides of the previous expression by f_{xx} and rearranging terms leads to

$$\begin{aligned}
f_{xx} H(0) &= f_{xx}{}^2 \Delta x^2 + 2 f_{xy} f_{xx} \Delta x \Delta y + f_{yy} f_{xx} \Delta y^2 \\
&= \underbrace{(\Delta x\, f_{xx} + \Delta y\, f_{xy})^2}_{\geq 0} + (f_{xx} f_{yy} - f_{xy}{}^2)\Delta y^2,
\end{aligned}$$

where all derivatives are evaluated at (a, b). Recall that the signs of $H(0)$ and $f(a + \Delta x, b + \Delta y) - f(a, b)$ are the same. Letting $D(a, b) = (f_{xx} f_{yy} - f_{xy}{}^2)|_{(a,b)}$, we reach the following conclusions:

- If $D(a, b) > 0$ and $f_{xx}(a, b) < 0$, then $H(0) < 0$ (for Δx and Δy sufficiently close to 0) and $f(a + \Delta x, b + \Delta y) - f(a, b) < 0$. Therefore, f has a local maximum value at (a, b).

- If $D(a, b) > 0$ and $f_{xx}(a, b) > 0$, then $H(0) > 0$ (for Δx and Δy sufficiently close to 0) and $f(a + \Delta x, b + \Delta y) - f(a, b) > 0$. Therefore, f has a local minimum value at (a, b).

- If $D(a, b) < 0$, then $H(0) > 0$ for some small nonzero values of Δx and Δy (implying $f(a + \Delta x, b + \Delta y) > f(a, b)$), *and* $H(0) < 0$ for other small nonzero values of Δx and Δy (implying $f(a + \Delta x, b + \Delta y) < f(a, b)$). (The relative sizes of $(f_{xx} \Delta x + f_{xy} \Delta y)^2$ and $(f_{xx} f_{yy} - f_{xy}{}^2)\Delta y^2$ can be adjusted by varying Δx and Δy.) Therefore, f has a saddle point at (a, b).

- If $D(a, b) = 0$, then $H(0)$ may be zero, in which case the sign of $H(c)$ cannot be determined. Therefore, the test is inconclusive.　　◄

C

Appendix

Hyperbolic Functions

When we construct trigonometric-like functions with respect to a hyperbola rather than a circle, the result is a new family of six functions that involve e^x and e^{-x}. They are called the hyperbolic trigonometric functions. We begin with the *hyperbolic sine* and *hyperbolic cosine* functions.

> sinh is pronounced *sinch* and cosh is pronounced *cosh*.

DEFINITION **Hyperbolic sine and cosine**

Hyperbolic sine $\quad \sinh x = \dfrac{e^x - e^{-x}}{2}$

Hyperbolic cosine $\quad \cosh x = \dfrac{e^x + e^{-x}}{2}$

The properties of the exponential function lead immediately to the following properties of the hyperbolic sine and cosine.

1. The domain of $y = \sinh x$ is $\{x: -\infty < x < \infty\}$ and the range is $\{y: -\infty < y < \infty\}$.

2. The domain of $y = \cosh x$ is $\{x: -\infty < x < \infty\}$ and the range is $\{y: y \geq 1\}$.

3. $\sinh 0 = 0$, $\cosh 0 = 1$.

4. $\lim\limits_{x \to \infty} \sinh x = \lim\limits_{x \to \infty} \cosh x = \infty$, $\lim\limits_{x \to -\infty} \sinh x = -\infty$, $\lim\limits_{x \to -\infty} \cosh x = \infty$.

5. $\sinh x$ is an odd function ($\sinh(-x) = -\sinh x$) and $\cosh x$ is an even function ($\cosh(-x) = \cosh x$).

6. $\cosh^2 x - \sinh^2 x = 1$.

7. The graphs of $\sinh x$ and $\cosh x$ are shown in Figure C.1.

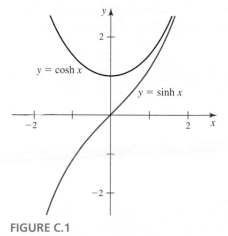

FIGURE C.1

Two quick calculations lead to the derivatives of $\sinh x$ and $\cosh x$:

$$\frac{d}{dx}(\sinh x) = \frac{d}{dx}\left(\frac{e^x - e^{-x}}{2}\right) = \frac{1}{2}(e^x + e^{-x}) = \cosh x$$

$$\frac{d}{dx}(\cosh x) = \frac{d}{dx}\left(\frac{e^x + e^{-x}}{2}\right) = \frac{1}{2}(e^x - e^{-x}) = \sinh x$$

As shown in Figure C.1, $\sinh x$ is increasing for all x (because its derivative $\cosh x$ is positive for all x). Similarly, $\cosh x$ is decreasing for $x < 0$ (because its derivative $\sinh x$ is negative for $x < 0$) and $\cosh x$ is increasing for $x > 0$ (because $\sinh x > 0$ for $x > 0$).

The corresponding indefinite integrals are just as easy to remember:

$$\int \sinh x \, dx = \cosh x + C \quad \text{and} \quad \int \cosh x \, dx = \sinh x + C$$

EXAMPLE 1   Derivatives and integrals of hyperbolic functions

a. Evaluate $\dfrac{d}{dx}(x \sinh 3x)$.   **b.** Evaluate $\int_1^3 x \cosh(x^2 - 1) \, dx$.

SOLUTION

a.

$$\frac{d}{dx}(x \sinh 3x) = \frac{d}{dx}(x) \cdot \sinh 3x + x \frac{d}{dx}(\sinh 3x) \quad \text{Product Rule}$$
$$= \sinh 3x + 3x \cosh 3x \quad \text{Chain Rule}$$

b. Using the substitution $u = x^2 - 1$ and noting that $du = 2x \, dx$, we find that

$$\int_1^3 x \cosh(x^2 - 1) \, dx = \frac{1}{2} \int_0^8 \cosh u \, du \quad u = x^2 - 1, u(3) = 8, u(1) = 0$$

$$= \frac{1}{2} \sinh u \Big|_0^8$$

$$= \frac{1}{2} \sinh 8 \approx 745.24.$$

Related Exercises 3–8 ◄

> To see all the connections between the hyperbolic trigonometric functions and the regular (or circular) trigonometric functions it is necessary to write all these functions in terms of a complex variable. Then all the properties of one family (for example, identities, derivatives, integrals) have mirror properties in the other family.

The derivative results for $\sinh x$ and $\cosh x$ show the parallels between the hyperbolic trigonometric functions and the ordinary trigonometric functions. In fact, the connections are much deeper. So perhaps it is not surprising that we can define four more hyperbolic trigonometric functions.

DEFINITION   Other hyperbolic functions

Hyperbolic tangent $\tanh x = \dfrac{\sinh x}{\cosh x} = \dfrac{e^x - e^{-x}}{e^x + e^{-x}}$

Hyperbolic cotangent $\coth x = \dfrac{\cosh x}{\sinh x} = \dfrac{e^x + e^{-x}}{e^x - e^{-x}}$

Hyperbolic secant $\operatorname{sech} x = \dfrac{1}{\cosh x} = \dfrac{2}{e^x + e^{-x}}$

Hyperbolic cosecant $\operatorname{csch} x = \dfrac{1}{\sinh x} = \dfrac{2}{e^x - e^{-x}}$

The graphs of the other four hyperbolic trigonometric functions are shown in Figure C.2. The derivatives of these functions are given in Example 2 and Exercise 9.

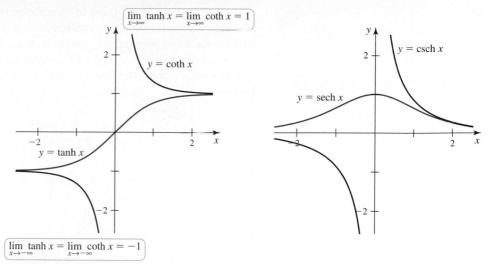

$$\lim_{x\to\infty} \tanh x = \lim_{x\to\infty} \coth x = 1$$

$$\lim_{x\to-\infty} \tanh x = \lim_{x\to-\infty} \coth x = -1$$

FIGURE C.2

EXAMPLE 2   Derivatives and integrals of hyperbolic functions

a. Evaluate $\dfrac{d}{dx}(\tanh x)$.

b. Evaluate $\int \tanh ax\, dx$ for real numbers a.

SOLUTION

a.

$$\frac{d}{dx}(\tanh x) = \frac{d}{dx}\left(\frac{\sinh x}{\cosh x}\right)$$

$$= \frac{\cosh x \dfrac{d}{dx}(\sinh x) - \sinh x \dfrac{d}{dx}(\cosh x)}{\cosh^2 x} \qquad \text{Quotient Rule}$$

$$= \frac{\cosh^2 x - \sinh^2 x}{\cosh^2 x} \qquad \text{Simplify.}$$

$$= \frac{1}{\cosh^2 x} \qquad \cosh^2 x - \sinh^2 x = 1$$

$$= \text{sech}^2 x$$

It follows from this result that $\int \text{sech}^2 x\, dx = \tanh x + C$.

b.

$$\int \tanh ax\, dx = \int \frac{\sinh ax}{\cosh ax}\, dx$$

$$= \frac{1}{a}\int \frac{du}{u} \qquad u = \cosh ax,\ du = a\sinh ax$$

$$= \frac{1}{a}\ln|u| + C \qquad \text{Evaluate integral.}$$

$$= \frac{1}{a}\ln \cosh x + C \qquad u = \cosh x,\ \cosh x > 0$$

Related Exercises 10–18 ◄

A well-known application of hyperbolic trigonometric functions arises in the field of statics. A flexible chain or cable, hanging at its ends and acted on only by the gravitational force (its own weight) takes the shape of a **catenary**, described by the function $y = a \cosh(x/a)$ (Figure C.3). (The term comes from the Latin for *chain*; however, Thomas Jefferson is often credited with introducing the English word.) See Exercise 25 for an example.

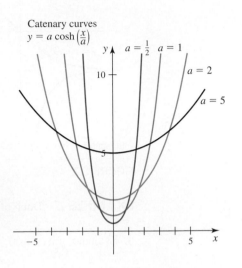

FIGURE C.3

Finally, the hyperbolic trigonometric functions are either one-to-one on $(-\infty, \infty)$ ($\sinh x$ and $\tanh x$), or they can be restricted to be one-to-one on either $(-\infty, 0)$, or $(0, \infty)$ ($\cosh x$, $\coth x$, $\operatorname{sech} x$, and $\operatorname{csch} x$). It follows that inverse hyperbolic trigonometric functions can be defined on the appropriate intervals (Exercises 23–24).

APPENDIX C EXERCISES

1. **Domains** Find the domain of $\tanh x$, $\coth x$, $\operatorname{sech} x$, and $\operatorname{csch} x$.

2. **Identities** Prove each identity. When possible, give the analogous identity for ordinary trigonometric functions.

   a. $\cosh^2 x - \sinh^2 x = 1$
   b. $\operatorname{sech}^2 x + \tanh^2 x = 1$
   c. $\coth^2 x - \operatorname{csch}^2 x = 1$
   d. $\sinh 2x = 2 \sinh x \cosh x$
   e. $\sinh(\ln x) = (x - x^{-1})/2$ for $x > 0$
   f. $\cosh(\ln x) = (x + x^{-1})/2$ for $x > 0$

3–8. Derivatives and integrals with $\sinh x$ and $\cosh x$ *Evaluate the following derivatives and integrals.*

3. $\dfrac{d}{dx}\left(\dfrac{\cosh 3x}{x^2}\right)$

4. $\dfrac{d}{dx}\left(\sinh \sqrt{x^2 + 1}\right)$

5. Find $y'(x)$ when $\sinh(xy) = x + 8$

6. $\displaystyle\int x \sinh(x^2 + 4)\, dx$

7. $\displaystyle\int_0^{\ln 3} \sinh^2 x \cosh x \, dx$

8. $\displaystyle\int x \cosh x \, dx$ (Integrate by parts.)

9. **Derivatives** Prove the following derivative results.

$$\dfrac{d}{dx}(\coth x) = -\operatorname{csch}^2 x \qquad \dfrac{d}{dx}(\operatorname{sech} x) = -\operatorname{sech} x \tanh x$$

$$\dfrac{d}{dx}(\operatorname{csch} x) = -\operatorname{csch} x \coth x$$

10–12. More derivatives *Use Example 2 and Exercise 9 to evaluate the following derivatives.*

10. $\dfrac{d}{dx}(x \tanh x)$

11. $\dfrac{d}{dx}(\sqrt{\tanh 3x})$

12. $\dfrac{d}{dx}(\ln \coth 3x)$

13–18. More integrals *Use Example 2 and Exercise 9 to evaluate the following integrals.*

13. $\displaystyle\int \operatorname{csch}^2 x \, dx$

14. $\displaystyle\int \operatorname{sech} 2x \tanh 2x \, dx$

15. $\displaystyle\int \coth 5x \, dx$

16. $\displaystyle\int \operatorname{sech}^5 x \operatorname{csch} x \, dx$

17. $\displaystyle\int_1^4 \dfrac{\tanh \sqrt{x}}{\sqrt{x}} \, dx$

18. $\displaystyle\int_{\ln 2}^{\ln 3} \dfrac{dx}{\coth x}$

19. Area under sech x Is the area of the region bounded by the graph of $y = \operatorname{sech} x$ and the x-axis on $(-\infty, \infty)$ finite? If so, what is its value? Use the fact that $\int \operatorname{sech} x \, dx = 2 \tan^{-1}(e^x) + C$.

20. Area under cosh x Let R be the region bounded by the graph of $y = \cosh x$ and the line $y = 2$.

  a. Find the area of R.

  b. Find the volume of the solid generated when R is revolved about the x-axis. (Use $\cosh^2 x = (1 + \cosh 2x)/2$.)

21. Arc length Show that the arc length of $y = \cosh x$ on the interval $[0, a]$ is $\sinh a$.

22. Differential equations Hyperbolic trigonometric functions are useful in solving differential equations. Show that the functions $y = A \sinh kx$ and $y = B \cosh kx$, where A, B, and k are constants, satisfy the equation $y''(x) - k^2 y(x) = 0$.

23. Inverse hyperbolic sine Because the hyperbolic sine is one-to-one on $(-\infty, \infty)$, it has an inverse on that interval.

  a. Solve the equation $y = \sinh x = \dfrac{e^x - e^{-x}}{2}$ for x in terms of y to show that the inverse hyperbolic sine function is $\sinh^{-1} x = \ln(x + \sqrt{x^2 + 1})$.

  b. Give the domain and range of $\sinh^{-1} x$.

  c. Graph $\sinh^{-1} x$ (using either a graphing utility or the reflection property of the graphs of a function and its inverse).

  d. Show that $\dfrac{d}{dx}(\sinh^{-1} x) = \dfrac{1}{\sqrt{x^2 + 1}}$.

24. Inverse hyperbolic cosine Because the hyperbolic cosine is one-to-one on $[0, \infty)$, it has an inverse on that interval.

  a. Solve the equation $y = \cosh x = \dfrac{e^x + e^{-x}}{2}$ for x in terms of y to show that the inverse hyperbolic cosine function is $\cosh^{-1} x = \ln(x + \sqrt{x^2 - 1})$.

  b. Give the domain and range of $\cosh^{-1} x$ (as it is defined here).

  c. Graph $\cosh^{-1} x$ (using either a graphing utility or the reflection property of the graphs of a function and its inverse).

  d. Show that $\dfrac{d}{dx}(\cosh^{-1} x) = \dfrac{1}{\sqrt{x^2 - 1}}$, for $x > 1$.

25–26. "Trigonometric" substitutions *Use the change of variables* $x = a \sinh u$ *to evaluate the following integrals.*

25. $\displaystyle \int \frac{dx}{\sqrt{x^2 + 9}}$

26. $\displaystyle \int_0^4 \sqrt{x^2 + 9} \, dx$

27. Catenary A flexible chain hangs from the tops of two poles of equal length whose bases are at $x = \pm L$. The height of the chain above the ground ($y = 0$) is given by $y = a \cosh\left(\dfrac{x}{a}\right)$.

  a. The lowest point of the chain is at $x = 0$, where the chain is 2 m above the ground. Find a.

  b. If the poles are 4 m high, find L.

  c. Graph the resulting catenary.

Answers

CHAPTER 1

Section 1.1 Exercises, pp. 7–9

1. A function is a rule that assigns to each value of the independent variable in the domain a unique value of the dependent variable in the range. **3.** A graph is that of a function provided no vertical line intersects the graph at more than one point. **5.** The first statement is true of a function, by definition. **7.** $f(g(2)) = 2; g(f(-2)) = -2$
9.

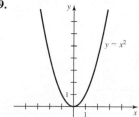
$$f(-x) = f(x)$$

11. B **13.** $D = \mathbf{R}, R = [-10, \infty)$ **15.** $D = [-2, 2], R = [0, 2]$
17. $D = \mathbf{R}, R = \mathbf{R}$ **19.** The independent variable is t; the dependent variable is d. $D = [0, 8]$ **21.** 96 **23.** $1/z^3$

25. $1/(y^3 - 3)$ **27.** $(u^2 - 4)^3$ **29.** $\dfrac{x - 3}{10 - 3x}$

31. $g(x) = x^3 - 5; f(x) = x^{10}; D = \mathbf{R}$
33. $g(x) = x^4 + 2, f(x) = \sqrt{x}; D = \mathbf{R}$
35. $(f \circ g)(x) = |x^2 - 4|; D = \mathbf{R}$

37. $(f \circ G)(x) = \dfrac{1}{|x - 2|}; \quad D = \{x: x \neq 2\}$

39. $(G \circ g \circ f)(x) = \dfrac{1}{x^2 - 6}; \quad D = \{x: x \neq \sqrt{6}, -\sqrt{6}\}$

41. $f(x) = x^2$ **43.** $f(x) = x^2$ **45. a.** 4 **b.** 1 **c.** 3 **d.** 3
e. 7 **f.** 8 **47.** y-axis **49.** no symmetry **51.** x-axis, y-axis, origin **53.** A is even, B is odd, C is even **55. a.** True **b.** False
c. True **d.** False **e.** False **f.** True **g.** True **h.** False **i.** True.
57.

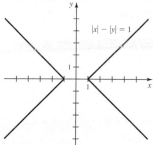

59. $f(x) = 3x - 2$ **61.** $f(x) = x^2 - 6$ **63. a.** $[0, 3 + \sqrt{14}]$

b.

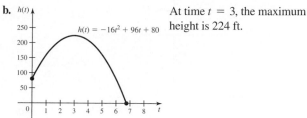

At time $t = 3$, the maximum height is 224 ft.

65. None **67.** Symmetry about the origin **69.** y-axis **71.** y-axis
73. $4, 4$ **75.** $-1/(2ax); -1/(2x(x + h))$

Section 1.2 Exercises, pp. 19–22

1. A formula, a graph, a table, words **3.** Set of all real numbers except points at which the denominator is zero
5.

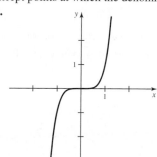

7. Shift the graph to the left 2 units **9.** Compress the graph horizontally by a factor of 3 **11.** $y = -\frac{2}{3}x - 1$
13. $d = -3p/50 + 27; D = [0, 450]$

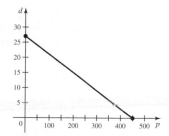

15. $y = \begin{cases} x + 3 & \text{if } x < 0 \\ -\frac{1}{2}x + 3 & \text{if } x \geq 0 \end{cases}$

17.

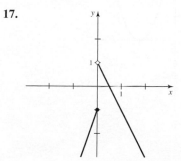

19.

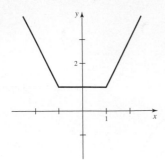

21. a.

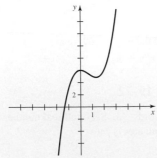

b. Polynomial function; D $=$ **R**   **c.** one peak near $x = 0$; one valley near $x = 4/3$; x-intercept near $x = -1.34$, y-intercept at $(0, 6)$

23. a.

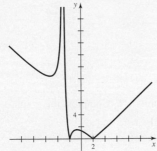

b. Absolute value of a rational function; D $= \{x: x \neq -3\}$
c. Undefined at $x = -3$; a valley near $x = -5.2$; x-intercepts (and valleys) at $x = -2$ and $x = 2$; a peak near $x = -0.8$

25. $g(x) = \begin{cases} 1 & \text{if } x < 0 \\ -\dfrac{1}{2} & \text{if } x > 0 \end{cases}$ **27. a.** 12   **b.** 36   **c.** $A(x) = 6x$

29. $f(x) = |x - 2| + 3; g(x) = -|x + 2| - 1$
31. a. Shift 3 units to the right

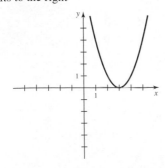

b. Compress horizontally by a factor of 2, then shift 2 units to the right.

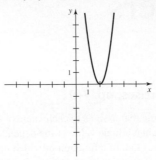

c. Shift to the right 2 units, vertical scaling and flip by a factor of 3, shift up 4 units

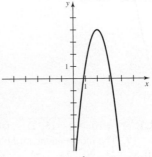

d. Horizontal scaling by a factor of $\frac{1}{3}$, horizontal shift right 2 units, vertical scaling by a factor of 6, vertical shift up 1 unit

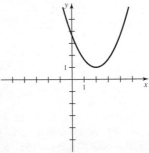

33. Stretch the graph of $y = x^2$ vertically by a factor of 3; then reflect across the x-axis.   **35.** Shift the graph of $y = x^2$ left 3 units and stretch vertically by a factor of 2.   **37.** Shift the graph of $y = x^2$ to the left $\frac{1}{2}$ unit, stretch vertically by a factor of 4, reflect through the x-axis, and then shift up 13 units to obtain the graph of h.   **39. a.** True
b. False   **c.** True   **d.** False   **41.** $(0, 0)$ and $(4, 16)$
43. $y = \sqrt{x} - 1$
45. $y = 5x; D = [0, 4]$ hr

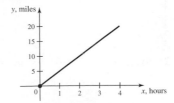

47. $y = 3200/x$; D $= [0, 5)$

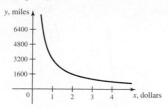

49. $y = \lceil x \rceil$

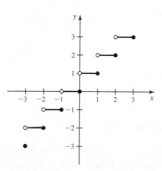

51.

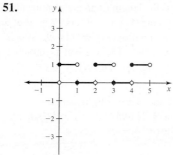

53.

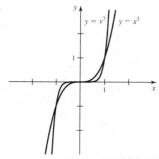

55. **a.** $p(t) = 328.3t + 1875$ **b.** 4830
57. **a.** $f(m) = 350m + 1200$ **b.** Buy
59. $0 \le h \le 2$

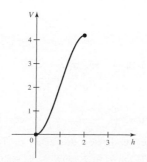

61. **a.** $S(x) = x^2 + \dfrac{500}{x}$ **b.** ≈ 6.30 ft

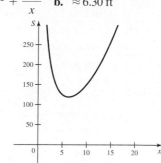

65. **a.**

n	1	2	3	4	5
$f(n)$	1	2	6	24	120

b.   **c.** 10

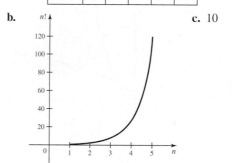

67. **a.**

n	1	2	3	4	5	6	7	8	9	10
$T(n)$	1	5	14	30	55	91	140	204	285	385

b. D $= \{n : n$ is a positive integer$\}$ **c.** 14

Section 1.3 Exercises, pp. 31–34

1. D $= \mathbf{R}$; R $= \{y : y > 0\}$ **3.** If a function f is not one-to-one, there are domain values, x_1 and x_2, such that $x_1 \ne x_2$ but $f(x_1) = f(x_2)$. If f^{-1} were to exist, by definition $f^{-1}(f(x_1)) = x_1$ and $f^{-1}(f(x_2)) = x_2$ so that f^{-1} assigns two different range values to the single domain value of $f(x_1)$.

5.

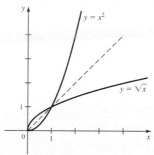

7. The expression $\log_b x$ represents the power to which b must be raised to obtain x.   **9.** D $= (0, \infty)$; R $= \mathbf{R}$
11. $(-\infty, -1], [-1, 1], [1, \infty)$ **13.** $(-\infty, \infty)$
15. $(-\infty, 5), (5, \infty)$ **17.** **a.** $f^{-1}(x) = (6 - x)/4$

19. a. $f^{-1}(x) = (x - 5)/3$ **21. a.** $f^{-1}(x) = x^2 - 2$ for $x \geq 0$
23. a. $f_1(x) = \sqrt{1 - x^2};\ 0 \leq x \leq 1$

$f_2(x) = \sqrt{1 - x^2};\ -1 \leq x \leq 0$

$f_3(x) = -\sqrt{1 - x^2};\ -1 \leq x \leq 0$

$f_4(x) = -\sqrt{1 - x^2};\ 0 \leq x \leq 1$

  b. $f_1^{-1}(x) = \sqrt{1 - x^2};\ 0 \leq x \leq 1$

$f_2^{-1}(x) = -\sqrt{1 - x^2};\ 0 \leq x \leq 1$

$f_3^{-1}(x) = -\sqrt{1 - x^2};\ -1 \leq x \leq 0$

$f_4^{-1}(x) = \sqrt{1 - x^2};\ -1 \leq x \leq 0$

25. $f^{-1}(x) = \dfrac{8 - x}{4}$ **27.** $f^{-1}(x) = x^2$ for $x \geq 0$

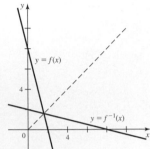   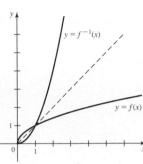

29. $f^{-1}(x) = \sqrt[4]{x - 4}$ **31.**

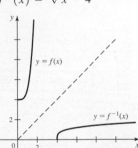   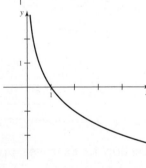

33. 1000 **35.** 2 **37.** 1.19 **39.** $-0.09\overline{6}$ **41.** ± 2 **43.** ± 4
45. $1 + \ln 3/\ln 7$ **47.** $\ln 5/(3 \ln 3) + 5/3$ **49.** $\ln 15/\ln 2 \approx 3.9069$
51. $\ln 40/\ln 4 \approx 2.6610$ **53.** $e^{x \ln 2}$ **55.** $\log_5 |x|/\log_5 e$
57. a. False **b.** False **c.** False **d.** True **e.** False **f.** False
g. True **59.** A is $y = \log_2 x$. B is $y = \log_4 x$. C is $y = \log_{10} x$.
61.

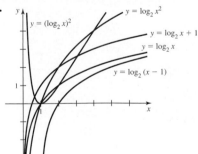

63. $f^{-1}(x) = \sqrt[3]{x} - 1, \mathbf{D} = \mathbf{R}$
65. $f_1^{-1}(x) = \sqrt{2/x - 2}, \mathbf{D}_1 = (0, 1]; f_2^{-1}(x) = -\sqrt{2/x - 2},$

$\mathbf{D}_2 = (0, 1]$ **67. b.** $\dfrac{p(t + 12)}{p(t)} = 2$ **c.** 38,400

d. 19.0 hr **e.** 72.7 hr **69. a.** No   **b.** $f^{-1}(h) = 2 - \sqrt{\dfrac{64 - h}{16}}$

c. $f^{-1}(h) = 2 + \sqrt{\dfrac{64 - h}{16}}$ **d.** 0.5423 s **e.** 3.8371 s

71. Let $y = \log_b x$. Then $b^y = x$ and $(1/b)^y = 1/x$. Hence,
$y = -\log_{1/b} x$. Thus, $\log_{1/b} x = -y = -\log_b x$.
75. a.

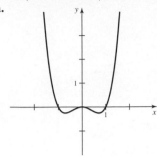

f is one-to-one on the intervals $(-\infty, -1/\sqrt{2}], [-1/\sqrt{2}, 0],$ $[0, 1/\sqrt{2}], [1/\sqrt{2}, \infty]$

b. $x = \pm\sqrt{\dfrac{1 \pm \sqrt{4y + 1}}{2}}$

Section 1.4 Exercises, pp. 43–46

1. $\sin\theta = $ opp/hyp; $\cos\theta = $ adj/hyp; $\tan\theta = $ opp/adj
$\cot\theta = $ adj/opp; $\sec\theta = $ hyp/adj; $\csc\theta = $ hyp/opp
3. The radian measure of an angle θ is the length of an arc s on the unit
circle associated with θ.   **5.** $\sin^2\theta + \cos^2\theta = 1, 1 + \cot^2\theta = \csc^2\theta,$
$\tan^2\theta + 1 = \sec^2\theta$ **7.** $\{x: x$ is an odd multiple of $\pi/2\}$ **9.** Sine is
not one-to-one on its domain.   **11.** Yes; no **13.** Vertical asymptotes
at $x = \pi/2$ and $x = -\pi/2$ **15.** $-\frac{1}{2}$ **17.** 1 **19.** $-1/\sqrt{3}$
21. $1/\sqrt{3}$ **23.** Dividing both sides of $\cos^2\theta + \sin^2\theta = 1$ by $\cos^2\theta$
gives $1 + \tan^2\theta = \sec^2\theta$. **25.** If α and β are complementary
angles, we have seen that $\cos\alpha = \sin\beta$. Thus $1/(\cos\alpha) = 1/(\sin\beta)$.
Letting $\alpha = \pi/2 - \theta$ and $\beta = \theta$, $\sec(\pi/2 - \theta) = \csc\theta$.

27. $\dfrac{\sqrt{2 + \sqrt{3}}}{2}$ or $\dfrac{\sqrt{6} + \sqrt{2}}{4}$ **29.** $\pi/4 + n\pi, n = 0, \pm 1, \pm 2, \ldots$

31. $\pi/4 + 2n\pi, 3\pi/4 + 2n\pi, n = 0, \pm 1, \pm 2, \ldots$
33. $\{\pi/12, 5\pi/12, 3\pi/4, 13\pi/12, 17\pi/12, 7\pi/4\}$ **35.** $\pi/3$

37. $2\pi/3$ **39.** -1 **41.** $\sqrt{1 - x^2}$ **43.** $\dfrac{\sqrt{4 - x^2}}{2}$

45. $2x\sqrt{1 - x^2}$ **47.** $\cos^{-1} x + \cos^{-1}(-x) = \theta + (\pi - \theta) = \pi$

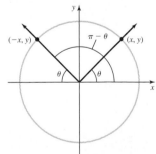

49. $\pi/3$ **51.** $\pi/3$ **53.** $\pi/4$ **55.** $\pi/2 - 2$

57. $\dfrac{1}{\sqrt{x^2 + 1}}$ **59.** $1/x$ **61.** $x/\sqrt{x^2 + 16}$

63. $\theta = \sin^{-1}\left(\dfrac{x}{6}\right) = \tan^{-1}\left(\dfrac{x}{\sqrt{36 - x^2}}\right) = \sec^{-1}\left(\dfrac{6}{\sqrt{36 - x^2}}\right)$
65. a. False **b.** False **c.** False **d.** False **e.** True **f.** False
g. True **h.** False **67.** $\sin\theta = \frac{12}{13}; \tan\theta = \frac{12}{5}; \sec\theta = \frac{13}{5};$
$\csc\theta = \frac{13}{12}; \cot\theta = \frac{5}{12}$ **69.** $\sin\theta = \frac{12}{13}; \cos\theta = \frac{5}{13}; \tan\theta = \frac{12}{5};$
$\sec\theta = \frac{13}{5}; \cot\theta = \frac{5}{12}$ **71.** amp $= 3$; period $= 6\pi$

73. amp = 3.6; period = 48   **75.** Stretch the graph of $y = \cos x$ horizontally by a factor of 3; stretch vertically by a factor of 2; and reflect through the x-axis

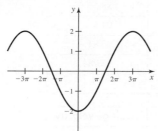

77. Stretch the graph of $y = \cos x$ horizontally by a factor of $24/\pi$; then stretch it vertically by a factor of 3.6 and shift it up 2 units

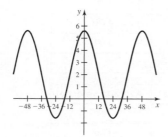

79. $y = 3 \sin(\pi x/12 - 3\pi/4) + 13$ **81.** $d(t) = 10 \cos(4\pi t/3)$
83. $\sqrt{a^2 - h^2} + k$

85. $s(t) = 117.5 - 87.5 \sin\left(\dfrac{\pi}{182.5}(t - 95)\right)$

$S(t) = 843.5 - 87.5 \sin\left(\dfrac{\pi}{182.5}(t - 67)\right)$

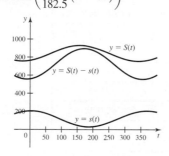

87. Area of circle is πr^2; $\theta/(2\pi)$ represents the proportion of area swept out by a central angle θ. Thus, the area of such a sector is $(\theta/2\pi)\pi r^2 = r^2\theta/2$.

Chapter 1 Review Exercises, pp. 46–49

1. a. True   **b.** False   **c.** False   **d.** True
e. False   **f.** False   **g.** True

3. a.

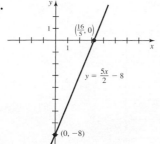

b.

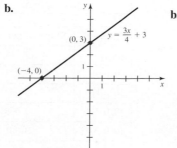

c.

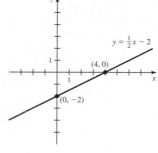

5. $f(x) = \begin{cases} 0 & \text{if } x \geq 0 \\ 4x & \text{if } x < 0 \end{cases}$

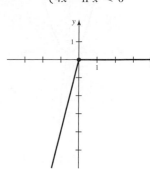

7. a.

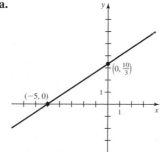

b.

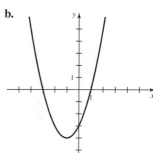

c.

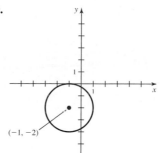

d.

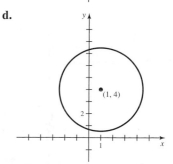

9. $D_f = \mathbf{R}, R_f = \mathbf{R}; D_g = [0, \infty), R_g = [0, \infty)$

11. $B = -\dfrac{2}{375}a + 212$

13. a.

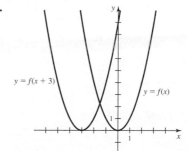

b.

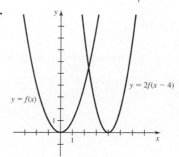

c.

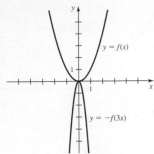

d.

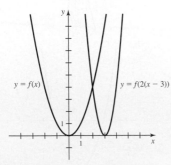

15. a. 1 **b.** $\sqrt{x^3}$ **c.** $\sin^3\sqrt{x}$ **d.** $(-\infty, \infty)$ **e.** $[-1, 1]$
17. a. y-axis **b.** y-axis **c.** x-axis, y-axis, origin **19.** $x = 2$; base does not matter **21.** $(-\infty, 0], [0, 2],$ and $[2, \infty)$
23. $f^{-1}(x) = 2 + \sqrt{x - 1}$

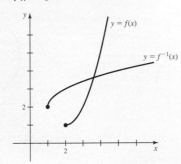

25. a. $3\pi/4$ **b.** $144°$ **c.** $40\pi/3$ **27. a.** $f(t) = -2\cos\left(\dfrac{\pi t}{3}\right)$

b. $f(t) = 5\sin\left(\dfrac{\pi t}{12}\right) + 15$ **29. a.** F **b.** E **c.** D **d.** B

e. C **f.** A **31.** $\pi/6$ **33.** $-\pi/2$ **35.** x **37.** $\cos\theta = \frac{5}{13}$;

$\tan\theta = \frac{12}{5}$; $\cot\theta = \frac{5}{12}$; $\sec\theta = \frac{13}{5}$; $\csc\theta = \frac{13}{12}$

39. $\dfrac{\sqrt{4 - x^2}}{2}$ **41.** $\pi/2 - \theta$ **43.** 0 **45.** $1 - 2x^2$

CHAPTER 2

Section 2.1 Exercises, pp. 55–56

1. $\dfrac{s(b) - s(a)}{b - a}$ **3.** $\dfrac{f(b) - f(a)}{b - a}$ **5.** The instantaneous velocity at
$t = a$ is the slope of the tangent line to the position curve at $t = a$.
7. a. 48 **b.** 64 **c.** 80 **d.** $16(7 - h)$

9.

Time interval	Average velocity
[1, 2]	80
[1, 1.5]	88
[1, 1.1]	94.4
[1, 1.01]	95.84
[1, 1.001]	95.984
$v_{\text{inst}} = 96$	

11.

Time interval	Average velocity
[2, 3]	20
[2.9, 3]	5.60
[2.99, 3]	4.16
[2.999, 3]	4.016
[2.9999, 3]	4.002
$v_{\text{inst}} = 4$	

13.

Time interval	Average velocity
[3, 3.5]	−24
[3, 3.1]	−17.6
[3, 3.01]	−16.16
[3, 3.001]	−16.016
[3, 3.0001]	−16.002
$v_{\text{inst}} = -16$	

15.

Time interval	Average velocity
[0, 1]	36.372
[0, 0.5]	67.318
[0, 0.1]	79.468
[0, 0.01]	79.995
[0, 0.001]	80.000
$v_{\text{inst}} = 80$	

17.

Interval	Slope of secant line
[1, 2]	6
[1.5, 2]	7
[1.9, 2]	7.8
[1.99, 2]	7.98
[1.999, 2]	7.998
$m_{\text{tan}} = 8$	

19.

Interval	Slope of secant line
[0, 1]	1.718
[0, 0.5]	1.297
[0, 0.1]	1.052
[0, 0.01]	1.005
[0, 0.001]	1.001
$m_{\text{tan}} = 1$	

21. a. 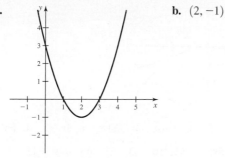 **b.** $(2, -1)$

c.

Interval	Slope of secant line
$[2, 2.5]$	0.5
$[2, 2.1]$	0.1
$[2, 2.01]$	0.01
$[2, 2.001]$	0.001
$[2, 2.0001]$	0.0001
$m_{\tan} = 0$	

23. a. 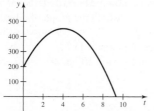 **b.** $t = 4$

c.

Interval	Slope of secant line
$[4, 4.5]$	-8
$[4, 4.1]$	-1.6
$[4, 4.01]$	-0.16
$[4, 4.001]$	-0.016
$[4, 4.0001]$	-0.0016
$v_{\text{inst}} = 0$	

d. $0 < t < 4$ **e.** $4 < t \le 9$ **25.** $0.6366, 0.9589, 0.9996, 1$

Section 2.2 Exercises, pp. 61–64

1. As x approaches a from either side, the values of $f(x)$ approach L.
3. As x approaches a from the right, the values of $f(x)$ approach L.
5. L must equal M. **7. a.** 5 **b.** 3 **c.** Does not exist **d.** 1
e. 2 **9. a.** -1 **b.** 1 **c.** 2 **d.** 2
11. a.

x	$f(x)$	x	$f(x)$
1.9	3.9	2.1	4.1
1.99	3.99	2.01	4.01
1.999	3.999	2.001	4.001
1.9999	3.9999	2.0001	4.0001

b. 4

13. a.

t	$g(t)$	t	$g(t)$
8.9	5.983287	9.1	6.016621
8.99	5.998333	9.01	6.001666
8.999	5.999833	9.001	6.000167

b. 6

15. $\lim\limits_{x \to 5^+} f(x) = 10$; $\lim\limits_{x \to 5^-} f(x) = 10$; $\lim\limits_{x \to 5} f(x) = 10$
17. a. 0 **b.** 1 **c.** 0 **d.** Does not exist; $\lim\limits_{x \to 1^-} f(x) \neq \lim\limits_{x \to 1^+} f(x)$
19. a. 3 **b.** 2 **c.** 2 **d.** 2 **e.** 2 **f.** 4 **g.** 1 **h.** Does not exist;
$\lim\limits_{x \to 3^-} f(x) \neq \lim\limits_{x \to 3^+} f(x)$ **i.** 3 **j.** 3 **k.** 3 **l.** 3
21. a.

x	$\sin\left(\dfrac{1}{x}\right)$
$2/\pi$	1
$2/(3\pi)$	-1
$2/(5\pi)$	1
$2/(7\pi)$	-1
$2/(9\pi)$	1
$2/(11\pi)$	-1

The value alternates between 1 and -1.

b. The alternation between 1 and -1 happens infinitely many times on the interval $(0, h)$ no matter how small $h > 0$ becomes. **c.** $\lim\limits_{x \to 0} \sin(1/x)$

does not exist. **23. a.** False **b.** False **c.** False
25.

27. Approximately 403.43 **29.** 1 **31. a.** $-2, -1, 1, 2$ **b.** 2, 2, 2
c. $\lim\limits_{x \to a^-} \lfloor x \rfloor = a - 1$ and $\lim\limits_{x \to a^+} \lfloor x \rfloor = a$, if a is an integer

d. $\lim\limits_{x \to a^-} \lfloor x \rfloor = \lfloor a \rfloor$ and $\lim\limits_{x \to a^+} \lfloor x \rfloor = \lfloor a \rfloor$, if a is not an integer

e. $\lim\limits_{x \to a} \lfloor x \rfloor$ exists only if a is not an integer. **33.** $\dfrac{16}{9} \approx 1.78$

35. a. **b.** $\$0.93$

c. $\lim\limits_{w \to 1^+} f(w)$ is the cost of a letter that weighs just over 1 oz.

$\lim\limits_{w \to 1^-} f(w)$ is the cost of a letter that weighs just under 1 oz.
d. No; $\lim\limits_{w \to 4^+} f(w) \neq \lim\limits_{w \to 4^-} f(w)$ **39. a.** 8 **b.** 5

41. a. $2; 3; 4$ **b.** n **43.** $\dfrac{n}{m}$

Section 2.3 Exercises, pp. 73–75

1. $\lim\limits_{x \to a} f(x) = f(a)$ **3.** Those values of a for which the denominator

is not zero   **5.** $\dfrac{x^2 - 7x + 12}{x - 3} = x - 4$ for $x \neq 3$.   **7.** 20

9. 4   **11.** 5   **13.** −45   **15.** 4   **17.** 32; Constant Multiple Law

19. 12; Quotient and Product Laws   **21.** 32; Power Law   **23.** 8

25. 3   **27.** 3   **29.** −5   **31. a.** 2   **b.** 0   **c.** Does not exist

33. a. 0   **b.** $\sqrt{x - 2}$ is not defined for $x < 2$.

35. $\lim\limits_{x \to 0^-} |x| = \lim\limits_{x \to 0^-} (-x) = 0$ and $\lim\limits_{x \to 0^+} |x| = \lim\limits_{x \to 0^+} x = 0$

37. 2   **39.** −8   **41.** −1   **43.** −12   **45.** $\frac{1}{6}$ **47.** $\frac{1}{8}$

49. a.

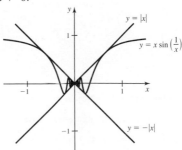

b. $\dfrac{2^x - 1}{x}$

c.

x	$\dfrac{2^x - 1}{x}$
−1	0.5
−0.1	0.6697
−0.01	0.6908
−0.001	0.6929
−0.0001	0.6931
−0.00001	0.6931
Limit ≈ 0.693	

51. a. Because $\left| \sin\left(\dfrac{1}{x}\right) \right| \leq 1$ for all $x \neq 0$, we have that

$$\left| x \right| \left| \sin\left(\dfrac{1}{x}\right) \right| \leq |x|.$$

That is, $\left| x \sin\left(\dfrac{1}{x}\right) \right| \leq |x|$, so that $-|x| \leq x \sin\left(\dfrac{1}{x}\right) \leq |x|$

for all $x \neq 0$.

b.

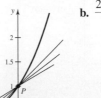

c. $\lim\limits_{x \to 0} -|x| = 0$ and $\lim\limits_{x \to 0} |x| = 0$; by part (a) and the Squeeze

Theorem, $\lim\limits_{x \to 0} x \sin\left(\dfrac{1}{x}\right) = 0$

53. a.

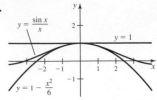

b. $\lim\limits_{x \to 0} \dfrac{\sin x}{x} = 1$ **55. a.** False   **b.** False   **c.** False   **d.** False

e. False   **57.** 8   **59.** 5   **61.** 10   **63.** −3   **65.** $a = -13$;

$\lim\limits_{x \to -1} g(x) = 6$.   **67.** 6   **69.** $5a^4$ **71.** $\frac{1}{3}$ **73.** 2   **75.** −54

77. $f(x) = x - 1$, $g(x) = \dfrac{5}{x - 1}$ **79.** $b = 2$ and $c = -8$; yes

81. $\lim\limits_{S \to 0^+} r(S) = 0$; the radius of the cylinder approaches 0 as the surface

area of the cylinder approaches 0.   **83.** 0.0435 N/C   **85.** 6; 4

Section 2.4 Exercises, pp. 81–84

1. $\lim\limits_{x \to a^+} f(x) = -\infty$ means that as x approaches a from the right,

the values of $f(x)$ are negative and become arbitrarily large in

magnitude.   **3.** A vertical line $x = a$, which the graph of a function

approaches as x approaches a **5.** $-\infty$ **7.** ∞ **9. a.** ∞ **b.** ∞

c. ∞ **d.** ∞ **e.** $-\infty$ **f.** Does not exist   **11. a.** $-\infty$ **b.** $-\infty$

c. $-\infty$ **d.** ∞ **e.** $-\infty$ **f.** Does not exist   **13. a.** ∞ **b.** $-\infty$

c. $-\infty$ **d.** ∞

15.

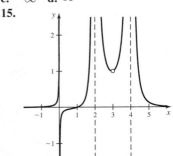

17. a. ∞ **b.** $-\infty$ **c.** Does not exist   **19.** −5   **21.** ∞

23. $x = 3$; $\lim\limits_{x \to 3^+} f(x) = -\infty$; $\lim\limits_{x \to 3^-} f(x) = \infty$; $\lim\limits_{x \to 3} f(x)$ does not

exist.   **25.** $x = 0$ and $x = 2$; $\lim\limits_{x \to 0^+} f(x) = \infty$; $\lim\limits_{x \to 0^-} f(x) = -\infty$;

$\lim\limits_{x \to 0} f(x)$ does not exist; $\lim\limits_{x \to 2^+} f(x) = \infty$; $\lim\limits_{x \to 2^-} f(x) = \infty$;

$\lim\limits_{x \to 2} f(x) = \infty$ **27.** ∞ **29.** $-\infty$ **31. a.** $-\infty$ **b.** ∞

c. $-\infty$ **d.** ∞ **33. a.** False   **b.** True   **c.** False

35. $f(x) = \dfrac{1}{x - 6}$ **37.** $x = 0$ **39.** $x = -1$ **41.** $\theta = (2k + 1)5$

for any integer k **43.** $x = 0$ or any positive odd multiple of $\pi/2$

45. a. $a = 4$ or $a = 3$ **b.** Either $a > 4$ or $a < 3$ **c.** $3 < a < 4$

47. a. $\dfrac{1}{\sqrt[3]{h}}$, regardless of the sign of h **b.** $\lim\limits_{h \to 0^+} \dfrac{1}{\sqrt[3]{h}} = \infty$;

$\lim\limits_{h \to 0^-} \dfrac{1}{\sqrt[3]{h}} = -\infty$; the tangent line at $(0, 0)$ is vertical.

Section 2.5 Exercises, pp. 92–94

1. The values of $f(x)$ approach 10 as x increases without bound negatively. **3.** 0 **5.** $\lim_{x \to \infty} f(x) = -\infty$; $\lim_{x \to -\infty} f(x) = \infty$

7. $\infty, 0, 0$ **9.** 3 **11.** 0 **13.** 0 **15.** ∞ **17.** $-\infty$ **19.** 0

21. $\lim_{x \to \infty} f(x) = 2$; $\lim_{x \to -\infty} f(x) = 2$; $y = 2$

23. $\lim_{x \to \infty} f(x) = \lim_{x \to -\infty} f(x) = 0$; $y = 0$

25. $\lim_{x \to \infty} f(x) = \infty$; $\lim_{x \to -\infty} f(x) = -\infty$; none

27. $\lim_{x \to \infty} f(x) = \frac{2}{3}$; $\lim_{x \to -\infty} f(x) = -2$; $y = \frac{2}{3}$; $y = -2$

29. $\lim_{x \to \infty} f(x) = \lim_{x \to -\infty} f(x) = \dfrac{1}{4 + \sqrt{3}}$; $y = \dfrac{1}{4 + \sqrt{3}}$

31. $\lim_{x \to \infty} (-3e^{-x}) = 0$; $\lim_{x \to -\infty} (-3e^{-x}) = -\infty$

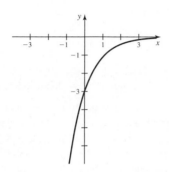

33. $\lim_{x \to \infty} (1 - \ln x) = -\infty$; $\lim_{x \to 0^+} (1 - \ln x) = \infty$

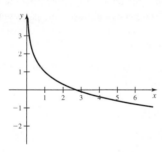

35. $\lim_{x \to \infty} \sin x$ does not exist; $\lim_{x \to -\infty} \sin x$ does not exist

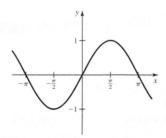

37. a. False **b.** False **c.** True

39. a. $\lim_{x \to \infty} f(x) = 2$; $\lim_{x \to -\infty} f(x) = 2$; $y = 2$

b. $x = 0$; $\lim_{x \to 0^+} f(x) = \infty$; $\lim_{x \to 0^-} f(x) = -\infty$

41. a. $\lim_{x \to \infty} f(x) = 3$; $\lim_{x \to -\infty} f(x) = 3$; $y = 3$

b. $x = -3$ and $x = 4$; $\lim_{x \to -3^-} f(x) = \infty$; $\lim_{x \to -3^+} f(x) = -\infty$; $\lim_{x \to 4^-} f(x) = -\infty$; $\lim_{x \to 4^+} f(x) = \infty$

43. a. $\lim_{x \to \infty} f(x) = 1$; $\lim_{x \to -\infty} f(x) = 1$; $y = 1$

b. $x = 0$; $\lim_{x \to 0^+} f(x) = \infty$; $\lim_{x \to 0^-} f(x) = -\infty$

45. a. $\lim_{x \to \infty} f(x) = 1$; $\lim_{x \to -\infty} f(x) = -1$; $y = 1$ and $y = -1$

b. No vertical asymptotes **47. a.** $\lim_{x \to \infty} f(x) = 0$; $\lim_{x \to -\infty} f(x) = 0$; $y = 0$ **b.** No vertical asymptotes **49. a.** $\dfrac{\pi}{2}$ **b.** $\dfrac{\pi}{2}$

51. a. $\lim_{x \to \infty} \sinh x = \infty$; $\lim_{x \to -\infty} \sinh x = -\infty$

b. $\sinh 0 = 0$

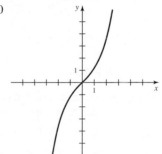
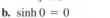

53.

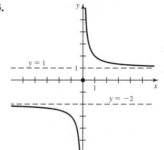
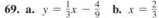

55. $x = 0$ is a vertical asymptote; $y = 2$ is a horizontal asymptote **57.** 3500 **59.** No steady state

61. 2 **63.** 1 **65.** 0

67. a. $y = x - 6$ **b.** $x = -6$

c.

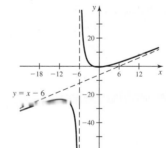

69. a. $y = \frac{1}{3}x - \frac{4}{9}$ **b.** $x = \frac{2}{3}$

c.

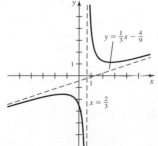

71. a. $y = 4x + 4$ **b.** No vertical asymptotes

c.

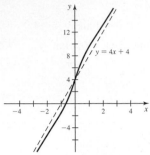

Section 2.6 Exercises, pp. 103–106

1. a, c   **3.** A function is continuous on an interval if it is continuous at each point of the interval.   **5. a.** $\lim\limits_{x \to a^-} f(x) = f(a)$

b. $\lim\limits_{x \to a^+} f(x) = f(a)$ **7.** $\{x : x \neq 0\}, \{x : x \neq 0\}$ **9.** $a = 2$,

item 3; $a = 3$, items 2 and 3; $a = 1$, item 1   **11.** $a = 1$, item 1; $a = 2$, items 2 and 3; $a = 3$, item 1   **13.** No; $f(1)$ is undefined.   **15.** No; $\lim\limits_{x \to 1} f(x) = 2$ but $f(1) = 3$ **17.** No; $f(4)$ is undefined.

19. $(-\infty, \infty)$ **21.** $\{x : x \neq 3 \text{ and } x \neq -3\}$
23. $\{x : x \neq 2 \text{ and } x \neq -2\}$ **25.** 1   **27.** 16
29. $[0, 1), (1, 2), (2, 3], (3, 4]$
31. $[0, 1), (1, 2), [2, 3), (3, 5]$ **33. a.** $\lim\limits_{x \to 1} f(x)$ does not exist.

b. From the right   **c.** $(-\infty, 1), [1, \infty)$
35. $(-\infty, -2\sqrt{2}]; [2\sqrt{2}, \infty)$ **37.** $(-\infty, \infty)$
39. $(-\infty, \infty)$ **41.** 3   **43.** 4
45. $\{x : x \neq n\pi, \text{ where } n \text{ is an integer}\}; \sqrt{2}, -\infty$
47. $\{x : x \neq \dfrac{k\pi}{2}, \text{ where } k \text{ is an odd integer}\}; \infty, \sqrt{3} - 2$
49. a. A is continuous on $[0, 0.08]$ and 7000 is between $A(0) = 5000$ and $A(0.08) = 11{,}098.20$. So, by the Intermediate Value Theorem, there is at least one c in $(0, 0.08)$ such that $A(c) = 7000$.

b.

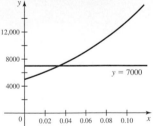

$c \approx 0.034$ or 3.4%

51. a. $2x^3 + x - 2$ is continuous on $[-1, 1]$ and takes on the values -5 and 1 at the endpoints of this interval. Since $-5 < 0 < 1$, there exists c in $(-1, 1)$ such that $2c^3 + c - 2 = 0$.   **b.** $x \approx 0.835$
53. a. $x^3 - 5x^2 + 2x$ is continuous on $[-1, 5]$ and takes on the values -8 and 10 at the endpoints. Since $-8 < -1 < 10$, there exists c in $(-1, 5)$ such that $c^3 - 5c^2 + 2c = -1$.   **b.** $x \approx -0.285$; $x \approx 0.778; x \approx 4.507$ **55. a.** True   **b.** True   **c.** False
d. False   **57.** $(-\infty, \infty)$ **59.** $[0, 16), (16, \infty)$ **61.** 1   **63.** 2

65. $-\dfrac{1}{2}$ **67.** 0   **69.** $-\infty$ **71.** The vertical line segments should not appear.

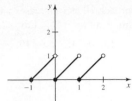

73. a, b.

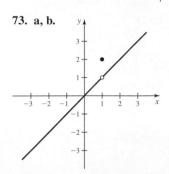

75. a. 2   **b.** 8   **c.** No; $\lim\limits_{x \to 1^-} g(x) = 2$ and $\lim\limits_{x \to 1^+} g(x) = 8$
77. $c_1 = \dfrac{1}{7}; c_2 = \dfrac{1}{2}; c_3 = \dfrac{3}{5}$ **79. a.** $A(r)$ is continuous on $[0.01, 0.10]$ and $A(0.01) = 2615.55$, while $A(0.10) = 3984.36$. Thus, $A(0.01) < 3500 < A(0.10)$. So, by the Intermediate Value Theorem, there exists c in $(0.01, 0.10)$ such that $A(c) = 3500$. This is the desired interest rate needed to achieve the goal of \$3500 in 10 years.   **b.** $r \approx 7.28\%$ **81.** Yes. Imagine there is a clone of the monk who walks down the path at the same time the monk walks up the path. The monk and his clone must cross paths with his clone at some time between dawn and dusk.   **83.** No; f cannot be made continuous at $x = a$ by redefining $f(a)$.   **85.** $\lim\limits_{x \to 2} f(x) = -3$; define $f(2)$ to be -3.   **87. a.** Yes   **b.** No   **89.** $a = 0$ removable discontinuity; $a = 1$ infinite discontinuity.

Section 2.7 Exercises, pp. 115–118

1. 1   **3.** c   **5.** Given any $\varepsilon > 0$, there exists a $\delta > 0$ such that $|f(x) - L| < \varepsilon$ whenever $0 < |x - a| < \delta$.   **7.** $0 < \delta < 2$
9. a. $\delta = 1$ **b.** $\delta = \dfrac{1}{2}$ **11. a.** $\delta = 2$ **b.** $\delta = \dfrac{1}{2}$ **13. a.** $\delta = 1$
b. $\delta = 0.79$ **15. a.** $\delta = 1$ **b.** $\delta = \dfrac{1}{2}$ **c.** $\delta = \varepsilon$
17. a. $\delta = 0.23$ **b.** $\delta = 0.12$ **c.** $\delta = \varepsilon/5$ **19.** $\delta = \varepsilon/8$
21. $\delta = \varepsilon$ **23.** $\delta = \sqrt{\varepsilon}$ **27. a.** Use any $\delta > 0$ **b.** $\delta = \varepsilon$
29. $\delta = 1/\sqrt{N}$ **31.** $\delta = 1/\sqrt{N-1}$ **33. a.** False   **b.** False
c. True   **d.** True   **35.** $\delta = \min\{1, 6\varepsilon\}$
37. $\delta = \min\{1/20, \varepsilon/200\}$ **39.** For $x > a, |x - a| = x - a$.
41. a. $\delta = \varepsilon/2$ **b.** $\delta = \varepsilon/3$
c. Since $\lim\limits_{x \to 0^+} f(x) = \lim\limits_{x \to 0^-} f(x) = -4, \lim\limits_{x \to 0} f(x) = -4$.
43. $\delta = \varepsilon^2$ **45. a.** For each $N > 0$, there is a corresponding $\delta > 0$ such that $f(x) > N$ whenever $a < x < a + \delta$.   **b.** For each $N < 0$, there is a corresponding $\delta > 0$ such that $f(x) < N$ whenever $a - \delta < x < a$.   **c.** For each $N > 0$, there is a corresponding $\delta > 0$ such that $f(x) > N$ whenever $a - \delta < x < a$.
47. $\delta = 1/N$ **49.** $\delta = (-1/M)^{1/4}$ **51.** $N = 1/\varepsilon$
53. $N = M - 1$

Chapter 2 Review Exercises, pp. 118–120

1. a. False **b.** False **c.** False **d.** True **e.** False **f.** False
g. True **3.** $x = -1$; $\lim_{x \to -1} f(x)$ does not exist; $x = 1$; $\lim_{x \to 1} f(x) \ne$
$f(1)$; $x = 3$; $f(3)$ is undefined. **5. a.** 1.4142 **b.** $\sqrt{2}$
7.

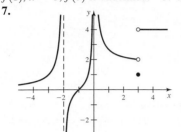

9. $\sqrt{11}$ **11.** 2 **13.** $\frac{1}{3}$ **15.** $-\frac{1}{16}$ **17.** 108 **19.** $\frac{1}{108}$ **21.** 0
23. a.

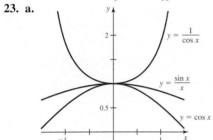

b. $\lim_{x \to 0} \cos x \le \lim_{x \to 0} \frac{\sin x}{x} \le \lim_{x \to 0} \frac{1}{\cos x}$;

$1 \le \lim_{x \to 0} \frac{\sin x}{x} \le 1$;

$\lim_{x \to 0} \frac{\sin x}{x} = 1$

25. $-\infty$ **27.** ∞ **29.** $-\infty$ **31.** $\frac{1}{2}$ **33.** ∞
35. $3\pi/2 + 2$ **37.** $\lim_{x \to \infty} f(x) = -4$; $\lim_{x \to -\infty} f(x) = -4$
39. $\lim_{x \to \infty} f(x) = 1$; $\lim_{x \to -\infty} f(x) = -\infty$ **41.** Horizontal
asymptotes at $y = 2/\pi$ and $y = -2/\pi$; vertical asymptote at $x = 0$
43. No; $f(5)$ does not exist. **45.** Yes; $\lim_{x \to 3^+} h(x) = h(3) = 0$
47. $(-\infty, -\sqrt{5}]$ and $[\sqrt{5}, \infty)$; left-continuous at $-\sqrt{5}$ and right-
continuous at $\sqrt{5}$ **49.** $(-\infty, -5), (-5, 0), (0, 5),$ and $(5, \infty)$
51. $a = 3, b = 0$
53.

55. a. $m(0) < 30 < m(5)$ and $m(5) > 30 > m(15)$.
b. $m = 30$ when $t \approx 2.4$ and $t \approx 10.8$. **c.** No; the maximum
amount is approximately $m(5.5) \approx 38.5$. **57.** $\delta = \varepsilon$
59. $\delta = 1/\sqrt[4]{N}$.

CHAPTER 3

Section 3.1 Exercises, pp. 131–135

1. Given the point $(a, f(a))$ and any point $(x, f(x))$ near $(a, f(a))$,
the slope of the secant line joining these points is $\frac{f(x) - f(a)}{x - a}$. The
limit of this quotient as x approaches a is the slope of the tangent line
at the point. **3.** The average rate of change over the interval $[a, x]$ is
$\frac{f(x) - f(a)}{x - a}$. The limit $\lim_{x \to a} \frac{f(x) - f(a)}{x - a}$ is the slope of the tangent
line; it is also the limit of average rates of change, which is the instan-
taneous rate of change at $x = a$. **5.** $f'(a)$ is the slope of the tangent
line at $(a, f(a))$ or the instantaneous rate of change of f at a. **7.** $\frac{dy}{dx}$ is
the limit of $\frac{\Delta y}{\Delta x}$ and is the rate of change of y with respect to x. **9.** No.
11. a. $m_{\tan} = 6$ **b.** $y = 6x - 14$ **c.**

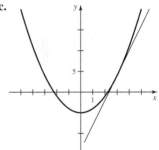

13. a. $m_{\tan} = -5$ **b.** $y = -5x + 1$
c.

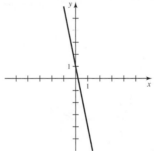

15. a. $m_{\tan} = -1$ **b.** $y = -x - 2$
c.

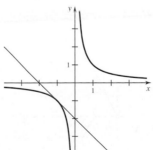

17. a. $m_{\tan} = 2$ **b.** $y = 2x + 1$ **19. a.** $m_{\tan} = -4$
b. $y = -4x - 3$ **21. a.** $m_{\tan} = \frac{2}{25}$ **b.** $y = \frac{2}{25}x + \frac{7}{25}$
23. a. $f'(-3) = 8$ **b.** $y = 8x$ **25. a.** $f'(-2) = -14$
b. $y = -14x - 16$ **27. a.** $f'\left(\frac{1}{4}\right) = -4$ **b.** $y = -4x + 3$
29. a. $f'(x) = 6x + 2$ **b.** $y = 8x - 13$

c.

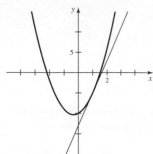

31. a. $f'(x) = 10x - 6$ **b.** $y = 14x - 19$

c.

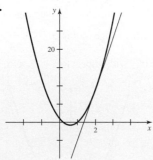

33. a. $2ax + b$ **b.** $8x - 3$ **35.** $-\frac{1}{4}$ **37.** $\frac{1}{5}$

39.

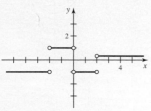

41. a–D; b–C; c–B; d–A   **43.**

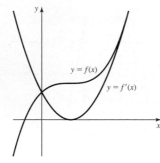

45. a. $x = 1$ **b.** $x = 1, x = 2$ **c.**

47. a. True   **b.** False   **c.** True   **d.** True

49. a. $f'(x) = \dfrac{3}{2\sqrt{3x + 1}}$ **b.** $y = 3x/10 + 13/5$

51. a. $f'(x) = \dfrac{-6}{(3x + 1)^2}$ **b.** $y = -3x/2 - 5/2$

53. a. C, D   **b.** A, B, E   **c.** A, B, E, D, C
55. Yes.

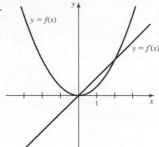

57. a. Approximately 10 kW; approximately -5 kW   **b.** $t = 6$ and $t = 18$ **c.** $t = 12$ **59. b.** $f'_+(2) = 1, f'_-(2) = -1$ **c.** f is continuous but not differentiable.

61. a.

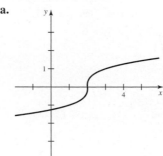

Vertical tangent line $x = 2$

b.

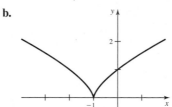

Vertical tangent line $x = -1$.

c.

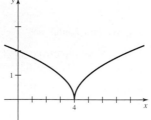

Vertical tangent line $x = 4$.

d.

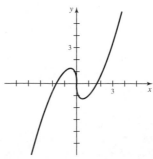

Vertical tangent line $x = 0$

63. $f'(x) = \dfrac{1}{3}x^{-2/3}$ and $\lim\limits_{x \to 0^-} |f'(0)| = \lim\limits_{x \to 0^+} |f'(0)| = \infty$

65. $f(x) = \dfrac{1}{x + 1}$; $a = 2; -\dfrac{1}{9}$ **67.** $f(x) = x^4$; $a = 2; 32$

69. No; f is not continuous at $x = 2$.   **71.** $a = 4$.

Section 3.2 Exercises, pp. 142–145

1. Using the definition can be tedious.   **3.** $f(x) = e^x$ **5.** Take the product of the constant and the derivative of the function.
7. $5x^4$ **9.** 0   **11.** 1   **13.** $15x^2$ **15.** 8   **17.** $200t$ **19.** $12x^3 + 7$
21. $40x^3 - 32$ **23.** $6w^2 + 3 + e^w$ **25.** $18x^2 + 6x + 4$
27. $4x^3 + 4x$ **29.** $2w$ for $w \neq 0$ **31.** 1   for $x \neq 1$

33. $\dfrac{1}{2\sqrt{x}}$ for $x \neq a$ **35. a.** $y = -6x + 5$

b.

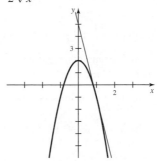

37. a. $y = 3x + (3 - 3\ln 3)$ **b.**

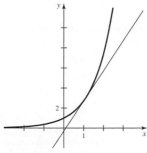

39. a. $x = 3$ **b.** $x = 4$ **41. a.** $(-1, 11), (2, -16)$
b. $(-3, -41), (4, 36)$ **43.** $f'(x) - 20x^3 + 30x^2 + 3$;
$f''(x) = 60x^2 + 60x; f^{(3)}(x) - 120x + 60$
45. $f'(x) = 1; f''(x) = f^{(3)}(x) = 0$ for $x \neq -1$
47. a. False   **b.** True   **c.** False   **d.** False   **e.** False
49. a. $y = 7x - 1$ **b.** $y = -2x + 5$ **c.** $y = 16x + 4$

51. -10 **53.** 4   **55.** 7.5   **57. a.** $f(x) = \sqrt{x}; a = 9$ **b.** $f'(9) = \dfrac{1}{6}$

59. a. $f(x) = x^{100}; a = 1$ **b.** $f'(1) = 100$ **61.** 3   **63.** 1
65. $f(x) = e^x;\ a = 0;\ f'(0) = 1$ **67. a.** $d'(t) = 32\,t$; ft/s;
the velocity of the stone.   **b.** 576 ft; ≈ 131 mi/hr

69. a. $\dfrac{dD}{dg} - 0.10g + 35$, mi/gal; the rate of change of mi driven per gal of gas consumed   **b.** 35 mi/gal, 35.5 mi/gal, 36 mi/gal; the gas mileage improves when driving longer distances.   **c.** ≈ 427 mi

Section 3.3 Exercises, pp. 152–154

1. $\dfrac{d}{dx}[f(x) \cdot g(x)] = f'(x)\,g(x) + f(x)\,g'(x)$ **3.** $\dfrac{d}{dx}(x^n) = nx^{n-1}$
for any integer n **5.** $y' = ke^{kx}$ for any real number k
7. $36x^5 - 12x^3$ **9.** $300x^9 + 135x^8 + 105x^6 + 120x^3 + 45x^2 + 15$
11. $e^w(w^3 + 3w^2 - 1)$ **13. a.** $6x + 1$ **15. a.** $18y^5 - 52y^3 + 8y$

17. $\dfrac{1}{(x+1)^2}$ **19.** $\dfrac{-1}{(t-1)^2}$ **21.** $\dfrac{e^x(x^2 - 2x - 1)}{(x^2 - 1)^2}$

23. a. $2w$ for $w \neq 0$ **25. a.** $\dfrac{1}{2\sqrt{x}}$

27. a. $y = -3x/2 + 17/2$ **b.**

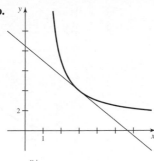

29. a. $y = -1/e$ **b.**

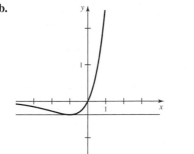

31. $-27x^{-10}$ **33.** $6t - 42/t^8$ **35.** $-3/t^2 - 2/t^3$
37. $45e^{3x}$ **39.** $(1 - x)/e^x$ **41.** $\frac{2}{3}e^x - e^{-x}$

43. a. $p'(t) = \left(\dfrac{20}{t+2}\right)^2$ **b.** $p'(5) \approx 8.16$ **c.** $t = 0$

d. $\lim\limits_{t \to \infty} p'(t) = 0$; the population reaches a steady state.

e.

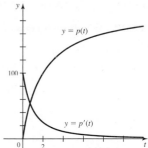

45. a. $Q'(t) = -1.386e^{-0.0693t}$ **b.** -1.386 mg/hr; -1.207 mg/hr
c. $\lim\limits_{t \to \infty} Q(t) = 0$—eventually none of the drug remains in the
bloodstream; $\lim\limits_{t \to \infty} Q'(t) = 0$—the rate at which the body excretes
the drug goes to zero over time   **47. a.** $x = -\frac{1}{2}$ **b.** The line tangent
to the graph of $f(x)$ at $x = -\frac{1}{2}$ is horizontal.   **49.** $g'(x) = -3/(2x^2)$

51. $g'(x) = \dfrac{-e^{x/4}(x^3 - 12x^2 - x + 4)}{(x - x^3)^2}$ **53. a.** False   **b.** False
c. False   **d.** True
55. $f'(x) = xe^{3x}(3x + 2)$
  $f''(x) = e^{3x}(9x^2 + 12x + 2)$
  $f^{(3)}(x) = 9e^{3x}(3x^2 + 6x + 2)$
57. $f'(x) = \dfrac{x^2 + 2x - 7}{(x + 1)^2}$

  $f''(x) = \dfrac{16}{(x + 1)^3}$

  $f^{(3)}(x) = \dfrac{-48}{(x + 1)^4}$

59. $8x - \dfrac{2}{(5x + 1)^2}$ **61.** $\dfrac{r - 6\sqrt{r} - 1}{2\sqrt{r}(r + 1)^2}$

63. a. $y = -\dfrac{108}{169}x + \dfrac{567}{169}$ **b.**

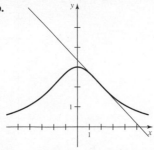

65. $-\dfrac{3}{2}$ **67.** $\dfrac{1}{9}$ **69.** $\dfrac{7}{8}$ **71. a.** $F'(x) = -\dfrac{1.8 \times 10^{10}\,Qq}{x^3}$ N/m

b. 1.8×10^{19} N/m **c.** $|F'(x)|$ decreases and $F'(x)$ increases as x increases. **73.** One possible pair: $f(x) = e^{ax}$ and $g(x) = e^{bx}$,

where $b = \dfrac{a}{a-1}$, $a \neq 1$. **77.** $f''g + 2f'g' + fg''$

81. a. $f'gh + fg'h + fgh'$ **b.** $2e^{2x}(x^2 + 3x - 2)$

Section 3.4 Exercises, pp. 161–163

1. $\dfrac{\sin x}{x}$ is undefined at $x = 0$. **3.** The tangent and cotangent

functions are defined as ratios of the sine and cosine functions. **5.** -1
7. 3 **9.** 5 **11.** 7 **13.** $\dfrac{1}{4}$ **15.** $\cos x - \sin x$
17. $e^{-x}(\cos x - \sin x)$ **19.** $\cos^2 x - \sin^2 x = \cos 2x$
21. $-2\sin x \cos x = -\sin 2x$ **27.** $\sec x \tan x - \csc x \cot x$
29. $\dfrac{-\csc x}{1 + \csc x}$ **31.** $\cos^2 z - \sin^2 z = \cos 2z$ **33.** $2\csc^2 x \cot x$
35. $2(\sec^2 x \tan x + \csc^2 x \cot x)$ **37. a.** False **b.** False
c. True **d.** True **39.** a/b **41.** $\dfrac{3}{4}$ **43.** 0 **45.** $x\cos 2x + \dfrac{1}{2}\sin 2x$
47. $\dfrac{-2}{1 + \sin x}$ **49.** $\dfrac{2\sin x}{(1 + \cos x)^2}$ **51. a.** $y = \sqrt{3}x + 2 - \pi\sqrt{3}/6$
b.

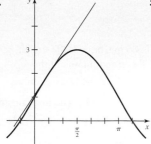

53. a. $y = -2\sqrt{3}x + 2\sqrt{3}\pi/3 + 1$

b.

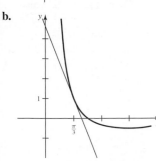

55. $x = 7\pi/6 + 2k\pi$ and $x = 11\pi/6 + 2k\pi$, where k is any integer

57. a.

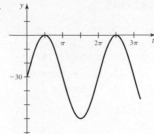

b. $v(t) = 30\cos t$

c.

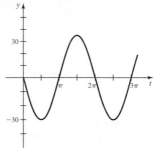

d. $v(t) = 0$ for $t = (2k+1)\dfrac{\pi}{2}$,
where k is any nonnegative
integer and the position is
$\left((2k+1)\dfrac{\pi}{2}, 0\right)$ if k is even or
$\left((2k+1)\dfrac{\pi}{2}, -60\right)$ if k is odd.

e. $v(t)$ is at a maximum at $t = 2k\pi$, where k is a nonnegative
integer; the position is $(2k\pi, -30)$.
f. $a(t) = -30\sin t$

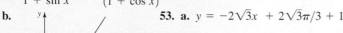

67. a. $2\sin x \cos x$ **b.** $3\sin^2 x \cos x$ **c.** $4\sin^3 x \cos x$

d. $n\sin^{n-1} x \cos x$ The conjecture is true for $n = 1$. If it holds for
$n = k$, then when $n = k + 1$, we have $\dfrac{d}{dx}(\sin^{k+1} x) =$

$\dfrac{d}{dx}(\sin^k x \cdot \sin x) = \sin^k x \cos x + \sin x \cdot k \sin^{k-1} x \cos x =$
$(k + 1)\sin^k x \cos x$. **69. a.** $f(x) = \sin x; a = \pi/6$ **b.** $\sqrt{3}/2$
71. a. $f(x) = \cot x; a = \pi/4$ **b.** -2

Section 3.5 Exercises, pp. 171–175

1. The average rate of change is $\dfrac{f(x + \Delta x) - f(x)}{\Delta x}$; whereas the
instantaneous rate of change is the limit as Δx goes to zero in this
quotient. **3.** Small **5.** If the position of the object at time t is $s(t)$,
then the acceleration at time t is $a(t) = d^2s/dt^2$. **7.** Each of the first
200 stoves costs, on average, \$70 to produce. When 200 stoves have
already been produced, the 201st stove costs \$65 to produce.
9. a. 40 mi/hr **b.** 40 mi/hr; yes **c.** -60 mi/hr; -60 mi/hr;
south **d.** The police car drives away from the police station going
north until about 10:08, when it turns around and heads south, toward
the police station. It continues south until it passes the police station
at about 11:02 and keeps going south until about 11:40, when it turns
around and heads north.

11. a.

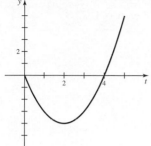

b. $v(t) = 2t - 4$; stationary at $t = 2$, to the right on $(2, 5]$, to the left on $[0, 2)$

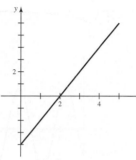

c. $v(1) = -2$ ft/s; $a(1) = 2$ ft/s^2 **d.** $a(2) = 2$ ft/s^2

13. a.

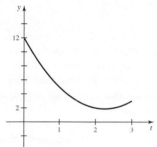

b. $v(t) = 4t - 9$; stationary at $t = \frac{9}{4}$ to the right on $\left(\frac{9}{4}, 3\right]$, to the left on $\left[0, \frac{9}{4}\right)$

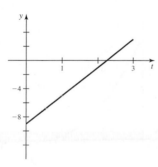

c. $v(1) = -5$ ft/s; $a(1) = 4$ ft/s^2 **d.** $a\left(\frac{9}{4}\right) = 4$ ft/s^2

15. a.

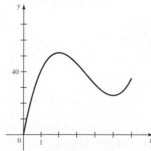

b. $v(t) = 6t^2 - 42t + 60$; stationary at $t = 2$ and $t = 5$, to the right on $[0, 2)$ and $(5, 6]$, to the left on $(2, 5)$

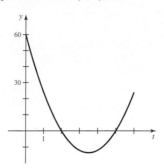

c. $v(1) = 24$ ft/s; $a(1) = -30$ ft/s^2 **d.** $a(2) = -18$ ft/s; $a(5) = 18$ ft/s^2 **17. a.** $v(t) = -32t + 64$ ft/s   **b.** At $t = 2$ s
c. 96 ft   **d.** At $2 + \sqrt{6}$ s   **e.** $-32\sqrt{6}$ ft/s
19. a. 98,300 people/year   **b.** 99,920 people/year in 1997; 95,600 people/year in 2005   **c.** $p'(t) = -0.54t + 101$; population increased, growth rate is positive but decreasing.
21. a. $\overline{C}(x) = \dfrac{1000}{x} + 0.1$; $C'(x) = 0.1$

b. $\overline{C}(2000) = \$0.60$/item; $C'(2000) = \$0.10$/item
c. The average cost per item when 2000 items are produced is \$0.60/item. The cost of producing the 2001st item is \$0.10.
23. a. $\overline{C}(x) = -0.01x + 40 + 100/x$; $C'(x) = -0.02x + 40$
b. $\overline{C}(1000) = \$30.10$/item; $C'(1000) = \$20$/item   **c.** The average cost per item is about \$30.10 when 1000 items are produced. The cost of producing the 1001st item is \$20.
25. a. False   **b.** True   **c.** False   **d.** True   **27.** 37,500 ft
29. a. $t = 1, 2, 3$ **b.** It is moving in the positive direction for t in $(0, 1)$ and $(2, 3)$; it is moving in the negative direction for t in $(1, 2)$ and $t > 3$.   **c.**

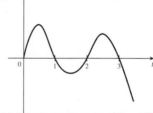

31. a. $P(x) = 0.02x^2 + 50x - 100$

b. $\dfrac{P(x)}{x} = 0.02x + 50 - \dfrac{100}{x}$; $\dfrac{dP}{dx} = 0.04x + 50$

c. $\dfrac{P(500)}{500} = 59.8$; $\dfrac{dp}{dx}(500) = 70$ **d.** The profit, on average, for each of the first 500 items produced is 59.8; the profit for the 501st item produced is 70.   **33. a.** $P(x) = 0.04x^2 + 100x - 800$

b. $\dfrac{P(x)}{x} = 0.04x + 100 - \dfrac{800}{x}$; $\dfrac{dp}{dx} = 0.08x + 100$

c. $\dfrac{P(1000)}{1000} = 139.2$; $\dfrac{dp}{dx}(1000) = 180$ **d.** The average profit per item for each of the first 1000 items produced is \$139.20. The profit for the 1001st item produced is \$180.   **35. a.** 1930, 1.1 million people/yr
b. 1960, 2.9 million people/yr   **c.** The population did not decrease.
d. $[1905, 1915], [1930, 1960], [1980, 1990]$

37. a.

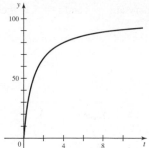

b. $v = \dfrac{100}{(t + 1)^2}$

c.

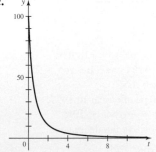

The marble moves fastest at the beginning and slows considerably over the first 5 s. It continues to slow but never actually stops.
d. $t = 4$ s   **e.** $t = -1 + \sqrt{2} \approx 0.414$ s

39. a. $C'(x) = \dfrac{-125,000,000}{x^2} + 1.5$;

$\overline{C}(x) = \dfrac{C(x)}{25,000} = 50 + \dfrac{5000}{x} + 0.00006x$

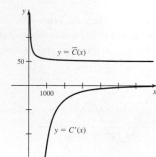

b. $C'(5000) = -3.5$; $\overline{C}(5000) = 51.3$ **c.** Marginal cost: If the batch size is increased from 5000 to 5001, then the cost of producing 25,000 gadgets would *decrease* by about $3.50.   Average cost: When batch size is 5000, it costs $51.30 *per item* to produce all 25,000 gadgets.

41. a.

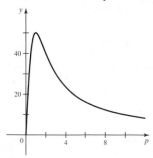

b. $v = \dfrac{100}{(t + 1)^2}$ **b.** $R'(p) = \dfrac{100(1 - p^2)}{(p^2 + 1)^2}$ **c.** $p = 1$

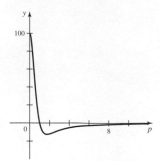

43. a.

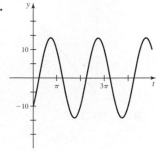

b. $dx/dt = 10 \cos t + 10 \sin t$ **c.** $t = 3\pi/4 + k\pi$, where k is any positive integer.   **d.** The graph implies that the spring never stops oscillating. In reality, the weight would eventually come to rest.
45. a. Juan starts faster than Jean and opens up a big lead. Then, Juan slows down while Jean speeds up. Jean catches up, and the race finishes in a tie.   **b.** Same average velocity   **c.** Tie   **d.** At $t = 2$, $\theta'(2) = \pi/2$ rad/min; $\theta'(4) = \pi =$ Jean's greatest velocity   **e.** At $t = 2$, $\varphi'(2) = \pi/2$ rad/min; $\varphi'(0) = \pi =$ Juan's greatest velocity
47. a. $v(0) = 40,000$ m³

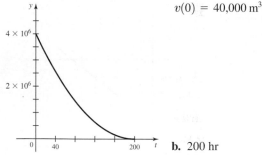

b. 200 hr
c.

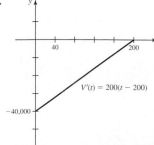

d. The magnitude of the flow rate is greatest (most negative) at $t = 0$ and least (zero) at $t = 200$.

49. a. $v(t) = -15e^{-t}(\sin t + \cos t)$; $v(1) \approx -7.6$ m/s, $v(3) \approx 0.63$ m/s   **b.** Down $(0, 2.4)$ and $(5.5, 8.6)$; up $(2.4, 5.5)$ and $(8.6, 10)$ **c.** ≈ 0.65 m/s   **51. a.** $-T'(1) = -80, -T'(3) = 80$
b. $-T'(x) < 0$ for $0 \le x < 2$; $-T'(x) > 0$ for $2 < x \le 4$
c. $-T'(0) = -160 < 0$ and $-T'(4) = 160 > 0$

Section 3.6 Exercises, pp. 180–183

1. $\dfrac{dy}{dx} = \dfrac{dy}{du} \cdot \dfrac{du}{dx}$; $\dfrac{d}{dx}(f(g(x)) = f'(g(x)) \cdot g'(x)$ **3.** $g(x), x$

5. Outer: $f(x) = x^{-5}$; inner: $u = x^2 + 10$ **7.** $30(3x + 7)^9$

9. $\dfrac{x}{\sqrt{x^2 + 1}}$ **11.** $10x \sec^2 5x^2$ **13.** $e^x \sec e^x \tan e^x$ **15.** $-2xe^{-x^2}$

17. $10(6x + 7)(3x^2 + 7x)^9$ **19.** $\dfrac{-315x^2}{(7x^3 + 1)^4}$ **21.** $e^x \sec^2 e^x$

23. $(12x^2 + 3)\cos(4x^3 + 3x + 1)$

25. $50^2 \sec 5\theta \tan 5\theta + 2\theta \sec 5\theta$ **27.** $5 \sec x (\sec x + \tan x)^5$

29. a. $u = \cos x, y = u^3; \dfrac{dy}{dx} = -3\cos^2 x \sin x$ **b.** $u = x^3$,

$y = \cos u; \dfrac{dy}{dx} = -3x^2 \sin x^3$ **31. a.** 100   **b.** -100 **c.** -16

d. 40   **e.** 40   **33.** $y' = 25(12x^5 - 9x^2)(2x^6 - 3x^3 + 3)^{24}$

35. $y' = 30(1 + 2\tan x)^{14}\sec^2 x$ **37.** $y' = \dfrac{-\cot x \csc^2 x}{\sqrt{1 + \cot^2 x}}$

39. $y' = -15 \sin^4(\cos 3x)(\sin 3x)[\cos(\cos 3x)]$

41. $y' = \dfrac{3e^{\sqrt{3x}}}{2\sqrt{3x}} \sec^2(e^{\sqrt{3x}})$ **43.** $y' = \dfrac{1}{2\sqrt{x + \sqrt{x}}}\left(1 + \dfrac{1}{2\sqrt{x}}\right)$

45. $y' = f'(g(x^2))g'(x^2)2x$ **47. a.** True   **b.** True   **c.** True

d. False   **49.** $\dfrac{d^2y}{dx^2} = 2\cos x^2 - 4x^2 \sin x^2$

51. $\dfrac{d^2y}{dx^2} = 4e^{-2x^2}(4x^2 - 1)$ **53.** $y' = \dfrac{f'(x)}{2\sqrt{f(x)}}$

55. $y = -9x + 35$

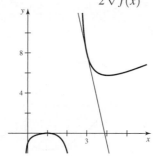

57. a. $h(4) = 9, h'(4) = -6$ **b.** $y = -6x + 33$
59. $y = 6x + 3 - 3\ln 3$

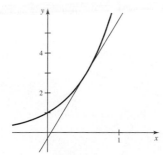

61. a. -3π **b.** -5π **63. a.** $\dfrac{d^2y}{dt^2} = \dfrac{-y_0 k}{m}\cos\left(t\sqrt{\dfrac{k}{m}}\right)$

65. a.

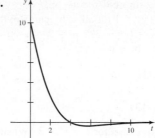

b. $v(t) = -5e^{-t/2}\left[\dfrac{\pi}{4}\sin\left(\dfrac{\pi t}{8}\right) + \cos\left(\dfrac{\pi t}{8}\right)\right]$

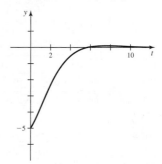

67. a. 10.88 hr   **b.** $D'(t) = \dfrac{6\pi}{365}\sin\left(\dfrac{2\pi(t + 10)}{365}\right)$

c. 2.87 min/day; on March 1, the length of day is increasing at a rate of about 2.87 min/day

d.

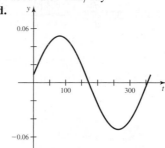

e. Most rapidly: Approximately March 22 and September 22; least rapidly: approximately December 21 and June 21

69. a. $E'(t) = 400 + 200\cos\left(\dfrac{\pi t}{12}\right)$ MW   **b.** At noon;

$E'(0) = 600$ MW   **c.** At midnight; $E'(12) = 200$ MW
d.

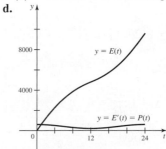

71. a. $f'(x) = -2\cos x \sin x + 2\sin x \cos x = 0$
b. $f(0) = \cos^2 0 + \sin^2 0 = 1$; $f(x) = 1$ for all x, by part (b); that is,
$\cos^2 x + \sin^2 x = 1$ **75. a.** $h(x) = (x^2 - 3)^5$; $a = 2$ **b.** 20
77. a. $h(x) = \sin(x^2)$; $a = \pi/2$ **b.** $\pi \cos(\pi^2/4)$

79. $\lim\limits_{x \to 5} \dfrac{f(x)^2 - f(25)}{x - 5} = 10 f'(25)$

Section 3.7 Exercises, pp. 188–191

1. There may be more than one expression for y or y'.   **3.** When derived implicitly, dy/dx is usually given in terms of both x and y.

5. a. $\dfrac{dy}{dx} = \dfrac{2}{y}$ **b.** 1   **7. a.** $\dfrac{dy}{dx} = \dfrac{20x^3}{\cos y}$ **b.** -20

9. a. $\dfrac{dy}{dx} = -\dfrac{1}{\sin y}$ **b.** -1 **11.** $\dfrac{dy}{dx} = \dfrac{1 - y \cos(xy)}{x \cos(xy) - 1}$

13. $\dfrac{dy}{dx} = \dfrac{1}{2y \sin(y^2) + e^y}$ **15.** $\dfrac{dy}{dx} = \dfrac{3x^2(x - y)^2 + 2y}{2x}$

17. $\dfrac{dy}{dx} = \dfrac{13y - 18x^2}{21y^2 - 13x}$ **19.** $\dfrac{dy}{dx} = \dfrac{5\sqrt{x^4 + y^2} - 2x^3}{y - 6y^2\sqrt{x^4 + y^2}}$

21. a. $2^2 + 2 \cdot 1 + 1^2 = 7$ **b.** $y = -5x/4 + 7/2$

23. a. $\sin \pi + 5\left(\dfrac{\pi^2}{5}\right) = \pi^2$ **b.** $y = \dfrac{\pi(1 + \pi)}{1 + 2\pi} + \dfrac{5}{1 + 2\pi}x$

25. a. $\cos\left(\dfrac{\pi}{2} - \dfrac{\pi}{4}\right) + \sin\dfrac{\pi}{4} = \sqrt{2}$ **b.** $y = \dfrac{x}{2}$

27. $\dfrac{d^2 y}{dx^2} = \dfrac{-1}{4y^3}$ **29.** $\dfrac{d^2 y}{dx^2} = \dfrac{2y^2(5 + 8x\sqrt{y})}{(1 + 2x\sqrt{y})^3}$

31. $\dfrac{d^2 y}{dx^2} = \dfrac{4e^{2y}}{(1 - 2e^{2y})^3}$ **33.** $\dfrac{dy}{dx} = \dfrac{5}{4}x^{1/4}$ **35.** $\dfrac{dy}{dx} = \dfrac{10}{3(5x + 1)^{1/3}}$

37. $\dfrac{dy}{dx} = \dfrac{-3}{2^{7/4} x^{3/4}(4x - 3)^{5/4}}$ **39.** $\dfrac{dy}{dx} = \dfrac{5x^2 + 20x + 3}{3(x^2 + 5x + 1)^{2/3}}$

41. $\dfrac{-1}{4}$ **43.** $\dfrac{-24}{13}$ **45.** -5 **47. a.** False   **b.** True   **c.** False
d. False   **49. a.** $y = x - 1$ and $y = -x + 2$
b.

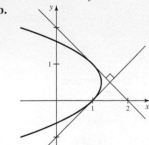

51. a. $y' = \dfrac{-2xy}{x^2 + 4}$ **b.** $y = \frac{1}{2}x + 2, y = -\frac{1}{2}x + 2$

c. $-\dfrac{16x}{(x^2 + 4)^2}$ **53. a.** $\left(\frac{5}{4}, \frac{1}{2}\right)$ **b.** No

55. a. $\dfrac{dy}{dx} = \dfrac{y - 1}{3y^2 - x} = \dfrac{1}{2y + 1}$ if $y \neq 1$; $\dfrac{dy}{dx} = 0$ if $y = 1$

b. $f_1(x) = 1, f_2(x) = \dfrac{-1 + \sqrt{4x - 3}}{2}, f_3(x) = \dfrac{-1 - \sqrt{4x - 3}}{2}$

c.

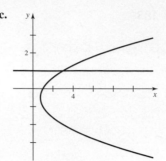

d. $\dfrac{d}{dx}[f_1(x)] = 0, \dfrac{d}{dx}[f_2(x)] = \dfrac{1}{\sqrt{4x - 3}}, \dfrac{d}{dx}[f_3(x)] = \dfrac{-1}{\sqrt{4x - 3}}$

57. a. $\dfrac{dy}{dx} = \dfrac{x - x^3}{y}$ **b.** $f_1(x) = \sqrt{x^2 - \dfrac{x^4}{2}}; f_2(x) = -\sqrt{x^2 - \dfrac{x^4}{2}}$

c.

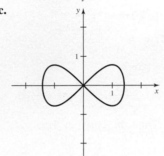

d. $\dfrac{d}{dx}[f_1(x)] = \dfrac{x - x^3}{\sqrt{x^2 - \dfrac{x^4}{2}}}; \dfrac{d}{dx}[f_2(x)] = \dfrac{x^3 - x}{\sqrt{x^2 - \dfrac{x^4}{2}}}$

59. $y = \dfrac{4x}{5} - \dfrac{3}{5}$

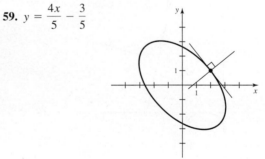

61. $y = -\dfrac{1 + 2\pi}{5}x + \pi\left(\dfrac{25 + \pi + 2\pi^2}{25}\right)$

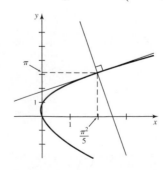

63. $y = -2x + \dfrac{5\pi}{4}$

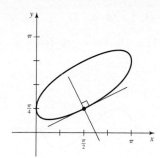

65. a. $y = -\dfrac{9x}{11} + \dfrac{20}{11}$ and $y = \dfrac{11x}{9} - \dfrac{2}{9}$

b.

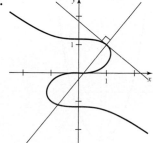

67. a. $y = -\dfrac{x}{3} + \dfrac{8}{3}$ and $y = 3x - 4$

b.

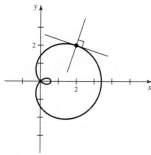

69. a. $\dfrac{dK}{dL} = \dfrac{-K}{2L}$ **b.** -4 **71.** $\dfrac{dr}{dh} = \dfrac{h - 2r}{h}; -3$ **73.** Note that for $y = mx, \, dy/dx = m$; for $x^2 + y^2 = a^2, \, dy/dx = -x/y$.
75. For $xy = a, \, dy/dx = -y/x$. For $x^2 - y^2 = b, \, dy/dx = x/y$. Since $(-y/x) \cdot (x/y) = -1$, the families of curves are orthogonal trajectories.

Section 3.8 Exercises, pp. 199–202

1. $x = e^y \Rightarrow 1 = e^y y'(x) \Rightarrow y'(x) = 1/e^y = 1/x$.
3. $\dfrac{d}{dx}(\ln kx) = \dfrac{d}{dx}(\ln k + \ln x) = \dfrac{d}{dx}(\ln x)$
5. $f'(x) = \dfrac{1}{x \ln b}$. If $b = e$, then $f'(x) = \dfrac{1}{x}$. **7.** $f(x) = e^{h(x) \ln g(x)}$
9. $2/x, \, x \neq 0$ **11.** $-2/(x^2 - 1)$ for $|x| > 1$
13. $(x^2 + 1)/x + 2x \ln x, \, x > 0$ **15.** $1/(x \ln x), \, x > 1$
17. $y' = 5 \cdot 4^x \ln 4$ **19.** $y' = 3^x \cdot x^2 (x \ln 3 + 3)$
21. $A' = 1000(1.045)^{4t} \ln (1.045)$ **23. a.** About 28.7 s
b. -46.512 s/1000 ft **c.** $dT/da = -2.74 \cdot 2^{-0.274a} \ln 2$
At $t = 8, \dfrac{dT}{da} = -0.4150$ min/1000 ft
$= -24.938$ s/1000 ft.

If a plane is traveling at 30,000 feet and it increases its altitude by 1,000 feet, the time of useful consciousness in the event of a sudden loss of pressure would decrease by about 25 seconds. **25. a.** About 67.19 hr
b. $Q'(12) = -9.815 \, \mu\text{Ci/hr}$
$Q'(24) = -5.201 \, \mu\text{Ci/hr}$
$Q'(48) = -1.461 \, \mu\text{Ci/hr}$
The rate at which iodine-123 leaves the body decreases with time.
27. $g'(y) = e^y y^{e-1}(y + e)$ **29.** $r' = 2e^{2\theta}$
31. $f'(x) = \dfrac{\sqrt{x}}{2}(10x - 9)$ **33.** $y = x \sin 1 + 1 - \sin 1$
35. $y = e^{2/e}$ and $y = e^{-2/e}$ **37.** $y' = \dfrac{8x}{(x^2 - 1) \ln 3}$
39. $y' = -\sin x (\ln (\cos^2 x) + 2)$ **41.** $y' = -\dfrac{\ln 4}{x \ln^2 x}$
43. $f'(x) = \dfrac{(x + 1)^{10}}{(2x - 4)^8}\left[\dfrac{10}{x + 1} - \dfrac{8}{x - 2}\right]$
45. $f'(x) = 2x^{(\ln x)-1} \ln x$
47. $f'(x) = \dfrac{(x + 1)^{3/2}(x - 4)^{5/2}}{(5x + 3)^{2/3}} \cdot$
$\left[\dfrac{3}{2(x + 1)} + \dfrac{5}{2(x - 4)} - \dfrac{10}{3(5x + 3)}\right]$
49. $f'(x) = (\sin x)^{\tan x}[1 + \sec^2 x \ln (\sin x)]; \, 0 < x < \pi, x \neq \pi/2$
51. a. False **b.** True **c.** False **d.** False **e.** True
53. $\dfrac{d^2}{dx^2}[\log x] = \dfrac{-1}{x^2 \ln 10}$ **55.** $d^3/dx^3 = 2/x$ **57.** $y' = 3^x \ln 3$
59. $f'(x) = 12/(3x + 1)$ **61.** $f'(x) = 1/(2x \ln 10)$
63. $f'(x) = \dfrac{2}{2x - 1} + \dfrac{3}{x + 2} + \dfrac{8}{1 - 4x}$
65. $y = 2$

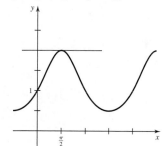

67. $\dfrac{d}{dx}(x^{10x}) = 10x^{10x}(1 + \ln x)$
69. $\dfrac{d}{dx}(x^{\cos x}) = x^{\cos x}\left(\dfrac{\cos x}{x} - \ln x \sin x\right)$
71. $\dfrac{d}{dx}\left(1 + \dfrac{1}{x}\right)^x = \left(1 + \dfrac{1}{x}\right)^x\left[\ln\left(1 + \dfrac{1}{x}\right) - \dfrac{1}{x + 1}\right]$
73. $\dfrac{d}{dx}\left(x^{(x^{10})}\right) = x^{9+x^{10}}(1 + 10 \ln x)$
75. a.

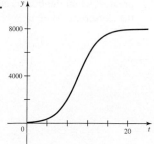

b. $t = 2 \ln{(265)} \approx 11.2$ years; about 14.5 years
c. $P'(0) \approx 25$ fish/year; $P'(5) \approx 264$ fish/year
d.

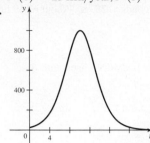

The population is growing fastest after about 10 years.

77. b. $r(11) \approx 0.0133$; $r(21) \approx 0.0118$; the relative growth rate is decreasing. **c.** $\lim\limits_{t\to\infty} r(t) = 0$; as the population gets close to carrying capacity, the rate of population growth vanishes.
79. a. $A(5) = \$17{,}443$
$\quad A(15) = \$72{,}705$
$\quad A(25) = \$173{,}248$
$\quad A(35) = \$356{,}178$
$\$5526.20$/year, $\$10{,}054.30$/year, $\$18{,}293$/year
b. $A(40) = \$497{,}873$

c. $\dfrac{dA}{dt} = 600{,}000 \ln{(1.005)}[(1.005)^{12t}]$

$\quad \approx (2992.5)(1.005)^{12t}$

A increases at an increasing rate.   **81.** $p = e^{1/e}; (e, e)$
83. $1/e$ **85.** $27(1 + \ln 3)$

Section 3.9 Exercises, pp. 208–211

1. $\dfrac{d}{dx}(\sin^{-1} x) = \dfrac{1}{\sqrt{1 - x^2}}$; $\dfrac{d}{dx}(\tan^{-1} x) = \dfrac{1}{1 + x^2}$;

$\dfrac{d}{dx}(\sec^{-1} x) = \dfrac{1}{|x|\sqrt{x^2 - 1}}$ **3.** $\frac{1}{5}$ **5.** $\frac{1}{4}$ **7.** $\dfrac{2}{\sqrt{1 - 4x^2}}$

9. $\dfrac{-4w}{\sqrt{1 - 4w^2}}$ **11.** $\dfrac{-2e^{-2x}}{\sqrt{1 - e^{-4x}}}$ **13.** $\dfrac{4y}{1 + (2y^2 - 4)^2}$

15. $\dfrac{-1}{2\sqrt{z}(1 + z)}$ **17.** $\dfrac{1}{|x|\sqrt{x^2 - 1}}$ **19.** $\dfrac{-1}{|2u + 1|\sqrt{u^2 + u}}$

21. $\dfrac{2y}{(y^2 + 1)^2 + 1}$ **23.** $\dfrac{1}{x|\ln x|\sqrt{(\ln x)^2 - 1}}$

25. $\dfrac{-e^x\sec^2{(e^x)}}{|\tan e^x|\sqrt{\tan^2 e^x - 1}}$ **27.** $\dfrac{-e^s}{1 + e^{2s}}$

29. a. ≈ -0.00055 rad/m
b.

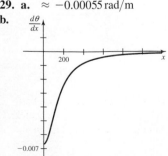

The magnitude of the change in angular size, $|d\theta/dx|$, is greatest when the boat is at the skyscraper (i.e., at $x = 0$).

31. $\frac{1}{3}$ **33.** $\frac{1}{2}$ **35.** 4   **37.** $\frac{1}{12}$ **39.** $\frac{1}{4}$ **41.** $\frac{5}{4}$

43. a. True   **b.** False   **c.** True   **d.** True   **e.** True
45. a.

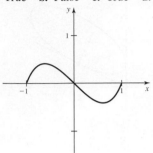

b. $f'(x) = 2x\sin^{-1}(x) + \dfrac{x^2 - 1}{\sqrt{1 - x^2}}$

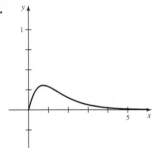

47. a.

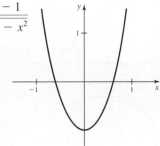

b. $f'(x) = \dfrac{e^{-x}}{1 + x^2} - e^{-x}\tan^{-1}(x)$

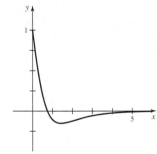

49. $(f^{-1})'(x) = \frac{1}{3}$ **51.** $(f^{-1})'(x) = 1/(2\sqrt{x + 4})$
53. $(f^{-1})'(x) = 2x, \ x \geq 0$ **55.** $(f^{-1})'(x) = -2/x^3, \ x > 0$
57. a. $\sin\theta = \dfrac{10}{\ell}$ implies $\theta = \sin^{-1}\left(\dfrac{10}{\ell}\right)$.

Thus, $\dfrac{d\theta}{d\ell} = \dfrac{1}{\sqrt{1 - \left(\dfrac{10}{\ell}\right)^2}} \cdot (-10\ell^{-2}) = \dfrac{-10}{\ell\sqrt{\ell^2 - 100}}$.

b. $d\theta/d\ell = -0.0041, -0.0289,$ and -0.1984 **c.** $\lim\limits_{\ell \to 10^+} d\theta/d\ell = -\infty$

d. The length ℓ is decreasing.   **59. a.** $d\theta/dc = 1/\sqrt{R^2 - c^2}$
b. $1/R$ **63.** Use the identity $\cot^{-1}(x) + \tan^{-1}(x) = \pi/2$.

Section 3.10 Exercises, pp. 214–218

1. As the side length s of a cube changes, the surface area $6s^2$ changes as well.   **3.** The other two opposite sides decrease in length.
5. a. $40 \text{ m}^2/\text{s}$ **b.** $80 \text{ m}^2/\text{s}$
c.

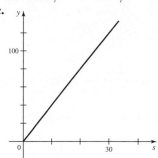

7. $-40\pi \text{ ft}^2/\text{min}$ **9.** $\dfrac{3}{80\pi} \text{ in/min}$ **13.** $\dfrac{1}{500} \text{ m/min}$; 2000 min

15. $10 \tan 20° \text{ km/hr} \approx 3.6 \text{ km/hr}$ **17.** $\dfrac{5}{24} \text{ ft/s}$ **19.** $-\dfrac{8}{3} \text{ ft/s}$,

$-\dfrac{32}{3} \text{ ft/s}$ **21.** $2592\pi \text{ cm}^3/\text{s}$ **23.** $\dfrac{-8}{9\pi} \text{ ft/s}$ **25. a.** $\dfrac{-\sqrt{3}}{10} \text{ m/hr}$

b. $-1 \text{ m}^2/\text{hr}$ **27.** 57.89 ft/s **29.** 4.66 in/s **31.** $\dfrac{3\sqrt{5}}{2} \text{ ft/s}$

33. $\approx 720.3 \text{ mi/hr}$ **35.** 11.06 m/hr **37. a.** 187.5 ft/s **b.** 0.938 rad/s
39. $\dfrac{d\theta}{dt} = 0.543 \text{ rad/hr}$ **41.** $\dfrac{d\theta}{dt} = \dfrac{1}{5} \text{ rad/s}, \dfrac{d\theta}{dt} = \dfrac{1}{8} \text{ rad/s}$

43. $\dfrac{d\theta}{dt} = 0 \text{ rad/s}$ for all $t \geq 0$ **45.** -0.0201 rad/s

Review Exercises, pp. 219–222

1. a. False   **b.** False   **c.** False   **d.** False   **e.** True
3. a. 16   **b.** $y = 16x - 10$

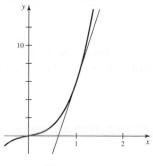

5. a. $\dfrac{-3}{4}$ **b.** $y = -\dfrac{3x}{4} + \dfrac{1}{2}$

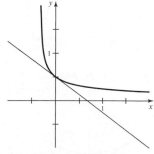

7. a. 2.70 million people/year   **b.** The slope of the secant line through the two points is approximately equal to the slope of

that tangent line at $t = 55$.   **c.** 2.217 million people/year
9. a. $\approx 40 \text{ m/s}$ **b.** $\approx 7 \text{ m/s}$ **c.** $\approx 18 \text{ m/s}$
d.

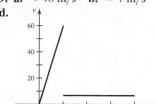

e. The skydiver deployed the parachute.

13.

15. $2x^2 + 2\pi x + 7$ **17.** $5t^2 \cos t + 10t \sin t$
19. $(8\theta + 12) \sec^2(\theta^2 + 3\theta + 2)$ **21.** $\dfrac{32u^2 + 8u + 1}{(8u + 1)^2}$

23. $\sec^2(\sin \theta) \cdot \cos \theta$ **25.** $\dfrac{9x \sin x - 2 \sin x + 6x^2 \cos x - 2x \cos x}{\sqrt{3x - 1}}$

27. $(2 + \ln x) \ln x$ **29.** $(2x - 1) 2^{x^2 - x} \ln 2$ **31.** $\dfrac{-1}{|x| \sqrt{x^2 - 1}}$

33. 1   **35.** $\sqrt{3} + \pi/6$ **37.** $\dfrac{dy}{dx} = \dfrac{y \cos x}{e^y - 1 - \sin x}$

39. $\dfrac{dy}{dx} = -\dfrac{xy}{x^2 + 2y^2}$ **41.** $y = x$ **43.** $y = -\dfrac{4x}{5} + \dfrac{24}{5}$

45. $x = 4$ **47.** $y' = \dfrac{\cos \sqrt{x}}{2\sqrt{x}}, y'' = \dfrac{-(\sqrt{x} \sin \sqrt{x} + \cos \sqrt{x})}{4x^{3/2}}$,

$y''' = \dfrac{3\sqrt{x} \sin \sqrt{x} + (3 - x) \cos \sqrt{x}}{8x^{5/2}}$ **49.** $x^2 f'(x) + 2x f(x)$

51. $\dfrac{g(x)[xf'(x) + f(x)] - xf(x)g'(x)}{g^2(x)}$ **53. a.** 27   **b.** $\frac{25}{27}$ **c.** 294
55. $f(x) = \tan(\pi\sqrt{3x - 11})$, $a = 5$; $f'(5) = 3\pi/4$ **57.** -1
59. $(f^{-1})'(x) = -3/x^4$ on $(-\infty, 0)$ and $(0, \infty)$
61. a. $(f^{-1})'(1/\sqrt{2}) = \sqrt{2}$ **63. a.** $\overline{C}(3000) = \$341.67$;
$C'(3000) = \$280$ **b.** The average cost of producing the first 3000 lawnmowers is $341.67 per mower. The cost of producing the 3001st lawnmower is $280.   **65. a.** 6550 people/year
b. $p'(40) = 4800$ people/year   **67.** 50 mi/hr
69. $-5 \sin(65°)$ ft/s or ≈ -4.5 ft/s   **71.** 0.166 rad/s

CHAPTER 4

Section 4.1 Exercises, pp. 229–232

1. f has an absolute maximum at c in $[a, b]$ if $f(x) \leq f(c)$ for all x in $[a, b]$. f has an absolute minimum at c in $[a, b]$ if $f(x) \geq f(c)$ for all x in $[a, b]$.   **3.** The function must be continuous on a closed interval.

5.

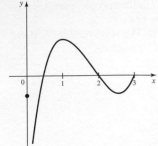

7.

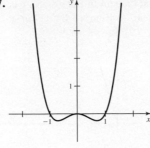

9. Evaluate the function at the critical points and at the endpoints of the interval.   **11.** Abs. min at $x = c_2$; abs. max at $x = b$
13. Abs. min at $x = a$; no abs. max   **15.** Local min at $x = q, s$; local max at $x = p, r$; abs. min at $x = a$; abs. max at $x = b$
17. Local max at $x = p$ and $x = r$; local min at $x = q$; abs. max at $x = p$; abs. min at $x = b$

19.

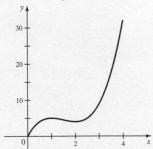

21.

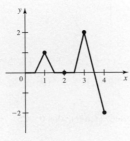

23. a. $x = \frac{2}{3}$ **b.** Local min   **25. a.** $x = \pm 1$ **b.** $x = -1$
local min; $x = 1$ local max   **27. a.** $x = 0$ **b.** Local min
29. a. No critical points   **31. a.** $x = 0$ **b.** Abs. max: -1 at $x = 3$; abs. min: -10 at $x = 0$ **c.**

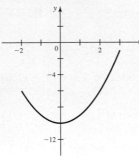

33. a. $x = \pi/2$ **b.** Abs. max: 1 at $x = 0, \pi$; abs. min: 0 at $x = \pi/2$ **c.**

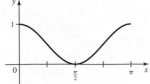

35. a. $x = \pm\pi/6$ **b.** Abs. max: 1 at $x = \pi/6$; abs. min: -1 at $x = -\pi/6$ **c.**

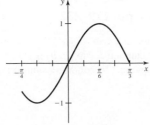

37. a. $x = 1/(2e)$ **b.** Abs. min: $(\sqrt{1/e})^{1/e}$ at $x = 1/(2e)$; abs. max: 2 at $x = 1$ **c.**

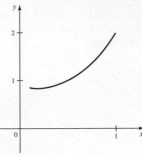

39. a. $x = 1/\sqrt{2}$ **b.** Abs. max: $1 + \pi$ at $x = -1$; abs. min: 1 at $x = 1$ **c.**

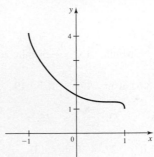

41. a. No critical points   **b.** No abs. max or min
c.

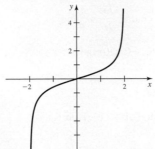

43. $t = 2$ s   **45. a.** 50   **b.** 45   **47. a.** False   **b.** False   **c.** False
d. True   **e.** False   **49. a.** $x = -0.96, 2.18, 5.32$ **b.** Abs. max: 3.72 at $x = 2.18$; abs. min: -32.80 at $x = 5.32$
c.

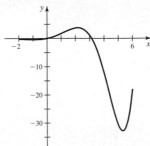

51. a. $x = 0$ **b.** Abs. max: $\sqrt{2}$ at $x = \pm\pi/4$; abs. min: 1 at $x = 0$
c.

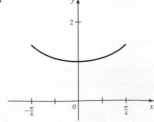

53. a. $x = 0$ and $x = 3$ **b.** Abs. max: $27/e^3$ at $x = 3$; abs. min: $-e$ at $x = -1$ **c.**

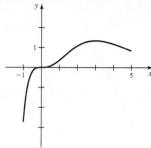

55. a. $x = 8$ **b.** Abs. max: $3\sqrt{2}$ at $x = 6$ and $x = 12$; abs. min: 4 at $x = 8$ **c.**

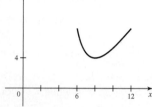

57. If $a \geq 0$, there is no critical point. If $a < 0$, $x = 2a/3$ is the only critical point. **59.** $x = \pm a$ **61. a.** $x = \tan^{-1}2 + k\pi$ for $k = -2, -1, 0, 1$ **b.** $x = \tan^{-1}2 + k\pi$ for $k = -2, 0$ correspond to local max; $x = \tan^{-1}2 + k\pi$ for $k = -1, 1$ correspond to local min. **c.** Abs. max: 2.24; abs. min: -2.24 **63. a.** $x = -\frac{1}{8}$ and $x = 3$ **b.** $x = -\frac{1}{8}$ corresponds to a local min; $x = 3$ is neither **c.** Abs. max: 51.23; abs. min: -12.52 **65. a.** $x = 5 - 4\sqrt{2}$ **b.** $x = 5 - 4\sqrt{2}$ corresponds to a local max. **c.** No abs. max or min **67.** Abs. max: 4 at $x = -1$; abs. min: -8 at $x = 3$

69. a. $T(x) = \dfrac{\sqrt{2500 + x^2}}{2} + \dfrac{50 - x}{4}$ **b.** $x = 50/\sqrt{3}$

c. $T(50/\sqrt{3}) = 34.15$, $T(0) = 37.50$, $T(50) = 35.36$

d.

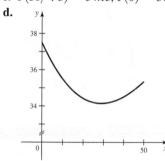

71. a. 1, 3, 0, 1 **b.** Since $g'(2) = 0$, g could have a local extreme value at $x = 2$. Since $h'(2) \neq 0$, h does not have a local extreme value at $x = 2$. **73. a.** A local min at $x = -c$ **b.** A local max at $x = -c$ **75. a.** $f(x) - f(c) \leq 0$ for all x near c

b. $\lim\limits_{x \to c^+} \dfrac{f(x) - f(c)}{x - c} \leq 0$ **c.** $\lim\limits_{x \to c^-} \dfrac{f(x) - f(c)}{x - c} \geq 0$

d. Since $f'(c)$ exists, $\lim\limits_{x \to c^+} \dfrac{f(x) - f(c)}{x - c} = \lim\limits_{x \to c^-} \dfrac{f(x) - f(c)}{x - c}$.

By parts (b) and (c), we must have that $f'(c) = 0$.

Section 4.2 Exercises pp. 243–247

1. f is increasing on I if $f'(x) > 0$ for all x in I; f decreasing on I if $f'(x) < 0$ for all x in I. **3.**

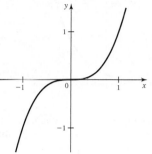

5. Because f has a local maximum at c, $f'(x) > 0$ for $x < c$ and $f'(x) < 0$ for $x > c$. Therefore, f' is decreasing near c and $f''(c) < 0$.
7. A point in the domain at which f changes concavity. **9.** Yes; consider $f(x) = x^2$ on the interval $[1, 2]$.
11.

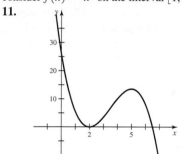

13. **15.**

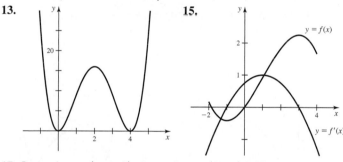

17. Increasing on $(-\infty, 0)$; decreasing on $(0, \infty)$ **19.** Decreasing on $(-\infty, 1)$; increasing on $(1, \infty)$ **21.** Increasing on $(-\infty, 1/2)$; decreasing on $(1/2, \infty)$ **23.** Increasing on the intervals $(-\pi, -2\pi/3)$, $(-\pi/3, 0)$, $(\pi/3, 2\pi/3)$; decreasing on the intervals $(-2\pi/3, -\pi/3)$, $(0, \pi/3)$, $(2\pi/3, \pi)$ **25.** Increasing on $(0, \infty)$; decreasing on $(-\infty, 0)$ **27.** Increasing on $(-\infty, \infty)$ **29.** Decreasing on $(-\infty, 1)$, $(4, \infty)$; increasing on $(1, 4)$ **31. a.** $x = 0$ **b.** Local min at $x = 0$ **c.** Abs. min: 3 at $x = 0$; abs. max: 12 at $x = -3$ **33. a.** $x = \pm 3/\sqrt{2}$ **b.** Local min at $x = -3/\sqrt{2}$; local max at $x = 3/\sqrt{2}$ **c.** Abs. max: $9/2$ at $x = 3/\sqrt{2}$; abs. min: $-9/2$ at $x = -3/\sqrt{2}$ **35. a.** $x = 8/5$ and $x = 0$ **b.** Local max at $x = 0$; local min at $x = 8/5$ **c.** Abs. min: -26.32 at $x = -5$; abs. max: 2.92 at $x = 5$ **37. a.** $x = e^{-2}$ **b.** Local min at $x = e^{-2}$ **c.** Abs. min: $-2/e$ at $x = e^{-2}$; no abs. max **39.** Abs. max: $1/e$ at $x = 1$ **41.** Abs. min: $36\sqrt[3]{\pi/6}$ at $x = \sqrt[3]{6/\pi}$.

43.
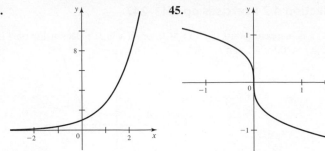

45.
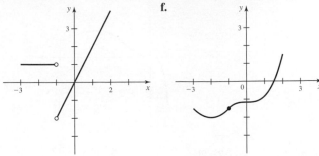

47. Concave up on $(-\infty, 0)$ and $(2, \infty)$; concave down on $(0, 2)$; inflection points at $x = 0$ and $x = 2$ **49.** Concave up on $(-1/\sqrt{3}, 1/\sqrt{3})$; concave down on $(-\infty, -1/\sqrt{3})$ and $(1/\sqrt{3}, \infty)$; inflection points at $x = \pm 1/\sqrt{3}$ **51.** Concave up on $(-\infty, -1)$ and $(1, \infty)$; concave down on $(-1, 1)$; inflection points at $x = \pm 1$
53. Concave up on $(0, 1)$; concave down on $(1, \infty)$; inflection point at $x = 1$ **55.** Concave up on $(0, 2)$ and $(4, \infty)$; concave down on $(-\infty, 0)$ and $(2, 4)$; inflection points at $x = 0, 2, 4$ **57.** Critical pt. at $x = 0$; local max at $x = 0$ **59.** Critical pt. at $x = 0$ and $x = 1$; local max at $x = 0$; local min at $x = 1$ **61.** Critical pt. at $x = 0$ and $x = 2$; local min at $x = 0$; local max at $x = 2$
63. a. True   **b.** False   **c.** True   **d.** False   **e.** False
65.

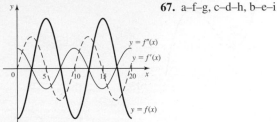

67. a–f–g, c–d–h, b–e–i

69.
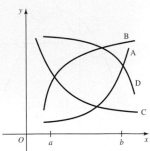

71.
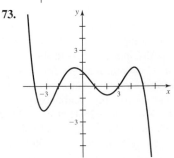

73.
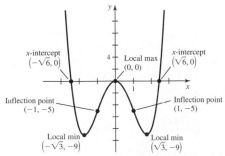

75. a. Increasing on $(-2, 2)$; decreasing on $(-3, -2)$
b. Critical pt. at $x = -2$ and $x = 0$; local min at $x = -2$; neither a

local min or max at $x = 0$ **c.** Inflection pts. at $x = -1$ and $x = 0$
d. Concave up on $(-3, -1)$ and $(0, 2)$; concave down on $(-1, 0)$
e.   **f.**

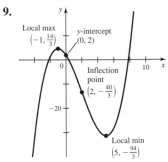

77. Critical pt. at $x = -3$ and $x = 4$; local min at $x = -3$; inconclusive at $x = 4$ **79.** No critical pts.   **81. a.** $E = \dfrac{10p}{10p - 500}$ **b.** -1.4%
c. $E'(p) = -\dfrac{ab}{(a - bp)^2} < 0$ for $p \geq 0, p \neq a/b$ **d.** $E(p) = -b$ for $p \geq 0$ **83. a.** 300   **b.** $t = \sqrt{10}$ **c.** $t = \sqrt{b/3}$
85. a. $f''(x) = 6x + 2a = 0$ when $x = -a/3$
b. $f(-a/3) - f(-a/3 + x) = (a^2/3)x - bx - x^3$; also, $f(-a/3 - x) - f(-a/3) = (a^2/3)x - bx - x^3$

Section 4.3 Exercises, pp. 254–257

1. We need to know over which interval(s) to graph f.   **3.** No; the domain of any polynomial is $(-\infty, \infty)$; there are no vertical asymptotes. Also, $\lim\limits_{x \to \pm\infty} p(x) = \pm\infty$ where p is any polynomial; there are no horizontal asymptotes.   **5.** Evaluate the function at the critical points and at the endpoints. Then find the largest and smallest values among those candidates.
7.   **9.**

11.

13.

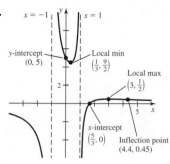

x-intercept
$(-2.77, 0)$

Local max
$(0, 0)$

x intercept
$(1.44, 0)$

Local min
$(1, -5)$

Inflection point
$(0.55, -2.68)$

Inflection point
$(-1.21, -18.36)$

Local min
$(-2, -32)$

15.

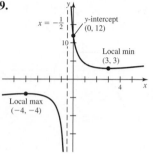

Local min
$(4, 8)$

Local max
$(0, 0)$

$x = 2$

27.

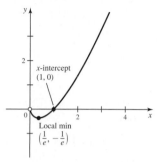

Inflection point
$(0, 0)$

Inflection point
(π, π)

Inflection point
$(-\pi, -\pi)$

17.

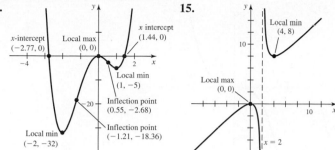

$x = -1$ $x = 1$

y-intercept
$(0, 5)$

Local min
$\left(\frac{1}{3}, \frac{9}{2}\right)$

Local max
$\left(3, \frac{1}{2}\right)$

x-intercept
$\left(\frac{5}{3}, 0\right)$

Inflection point
$(4.4, 0.45)$

19.

$x = -\frac{1}{2}$

y-intercept
$(0, 12)$

Local min
$(3, 3)$

Local max
$(-4, -4)$

29.

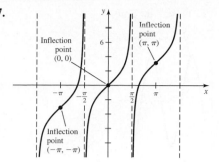

x-intercept
$(1, 0)$

Local min
$\left(\frac{1}{e}, -\frac{1}{e}\right)$

21.

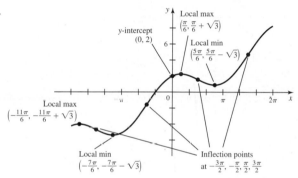

Local max
$\left(\frac{\pi}{6}, \frac{\pi}{6} + \sqrt{3}\right)$

y-intercept
$(0, 2)$

Local min
$\left(\frac{5\pi}{6}, \frac{5\pi}{6} - \sqrt{3}\right)$

Local max
$\left(-\frac{11\pi}{6}, -\frac{11\pi}{6} + \sqrt{3}\right)$

Local min
$\left(-\frac{7\pi}{6}, -\frac{7\pi}{6} - \sqrt{3}\right)$

Inflection points
at $-\frac{3\pi}{2}, -\frac{\pi}{2}, \frac{\pi}{2}, \frac{3\pi}{2}$

31.

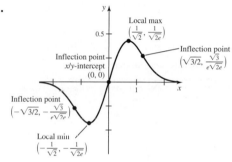

Local max
$\left(\frac{1}{\sqrt{2}}, \frac{1}{\sqrt{2e}}\right)$

Inflection point
x/y-intercept
$(0, 0)$

Inflection point
$\left(\sqrt{3/2}, \frac{\sqrt{3}}{e\sqrt{2e}}\right)$

Inflection point
$\left(-\sqrt{3/2}, -\frac{\sqrt{3}}{e\sqrt{2e}}\right)$

Local min
$\left(-\frac{1}{\sqrt{2}}, -\frac{1}{\sqrt{2e}}\right)$

33.

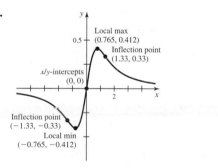

Local max
$(0.765, 0.412)$

Inflection point
$(1.33, 0.33)$

x/y-intercepts
$(0, 0)$

Inflection point
$(-1.33, -0.33)$

Local min
$(-0.765, -0.412)$

23.

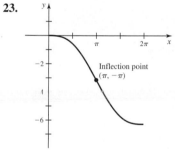

Inflection point
$(\pi, -\pi)$

25.

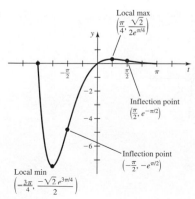

Local max
$\left(\frac{\pi}{4}, \frac{\sqrt{2}}{2e^{\pi/4}}\right)$

Inflection point
$\left(\frac{\pi}{2}, e^{-\pi/2}\right)$

Inflection point
$\left(-\frac{\pi}{2}, -e^{\pi/2}\right)$

Local min
$\left(-\frac{3\pi}{4}, \frac{-\sqrt{2}}{2}e^{3\pi/4}\right)$

35.

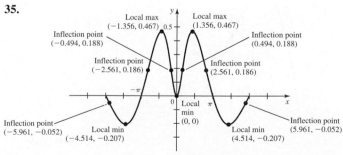

Local max
$(-1.356, 0.467)$

Local max
$(1.356, 0.467)$

Inflection point
$(-0.494, 0.188)$

Inflection point
$(0.494, 0.188)$

Inflection point
$(-2.561, 0.186)$

Inflection point
$(2.561, 0.186)$

Inflection point
$(-5.961, -0.052)$

Inflection point
$(5.961, -0.052)$

Local min
$(-4.514, -0.207)$

Local min
$(4.514, -0.207)$

Local min
$(0, 0)$

37. a. False   **b.** False   **c.** False   **d.** True

39.

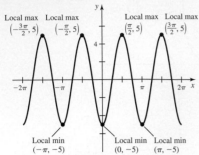

41.

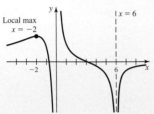

43. Critical pt. at $x = 1, 3$; local max at $x = 1$; local min at $x = 3$; inflection pt. at $x = 2$; increasing on $(0, 1), (3, 4)$; decreasing on $(1, 3)$; concave up on $(2, 4)$; concave down on $(0, 2)$

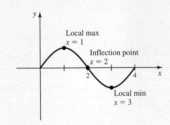

45.

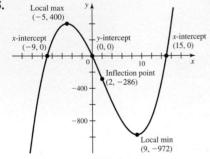

47.

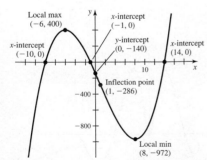

49. Local max at $(e, e^{1/e})$

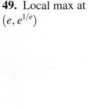

51.

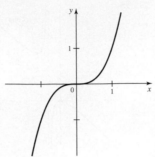

53.

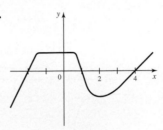

55.

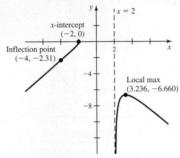

57. a.

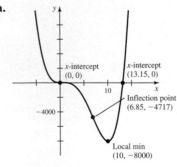

b.

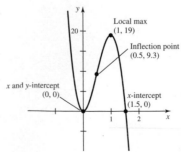

59.

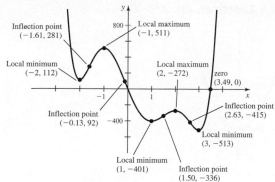

61.

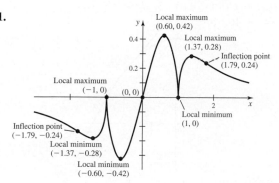

63. a.

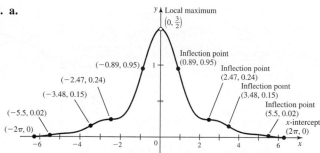

b.

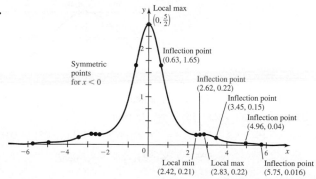

65. (a) a.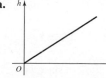

b. Water is being added at all times
c. No concavity   **d.** h' has an abs. max at all points of $[0, 10]$.

(b) a. 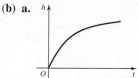   **c.** Concave down
d. h' has abs. max at $t = 0$.

(c) a. 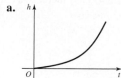   **c.** Concave up
d. h' has abs. max at $t = 10$.

(d) a. 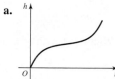   **c.** Concave up on $(0, 5)$, then concave down on $(5, 10)$; inflection pt. at $t = 5$
d. h' has abs. max at $t = 0$ and $t = 10$.

(e) a. 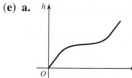   **c.** First, no concavity; then, concave down, no concavity, concave up, and, finally, no concavity   **d.** h' has abs. max at all points of an interval $[0, a]$ and $[b, 10]$.

(f) a. 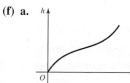   **c.** Concave down on $(0, 5)$; concave up on $(5, 10)$; inflection pt. at $t = 5$ **d.** h' has abs. max at $t = 0$ and $t = 10$.

67. $f'(0)$ does not exist.

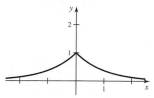

69. 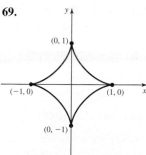   **71.**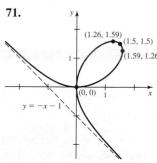

73. 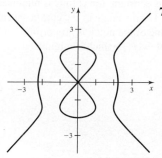   **75.**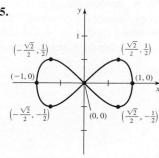

77.

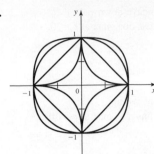

79. a. $\{x: x < a\}$
b. $f(a) = 0$, $\lim_{x \to -\infty} f(x) = 0$
c. $f'(x) = (a - x)^{x-1}$
     $[(a - x) \ln(a - x) - x]$
d. See part (c).
e. z and $f(z)$ increase as a increases.

Section 4.4 Exercises, pp. 261–267

1. Objective function, constraints
3. $Q = x^2(10 - x); Q = (10 - y)^2 y$ **5.** Width = length = $\frac{5}{2}$ m
7. $\frac{23}{2}$ and $\frac{23}{2}$ **9.** $5\sqrt{2}$ and $5\sqrt{2}$
11. Length = width = height = $\sqrt[3]{100}$
13. $\dfrac{4}{\sqrt[3]{5}}$ ft by $\dfrac{4}{\sqrt[3]{5}}$ ft by $5^{2/3}$ ft   **15. a.** A point $8/\sqrt{5}$ mi from
the point on the shore nearest the woman in the direction of the
restaurant   **b.** $9/\sqrt{13}$ mi/hr   **17.** 18.2 ft
19. $\dfrac{10}{\sqrt{2}}$ cm by $\dfrac{5}{\sqrt{2}}$ cm   **21.** $\theta = 2\pi\left(1 - \dfrac{\sqrt{6}}{3}\right)$
23. $\sqrt{15}$ m by $2\sqrt{15}$ m   **25.** $r/h = \sqrt{2}$ **27.** $r = h = \sqrt[3]{450/\pi}$ m
29. The point $12/\left(\sqrt[3]{2} + 1\right) \approx 5.3$ m from the weaker source
31. A point $7\sqrt{3}/6$ mi from the point on shore nearest the island,
in the direction of the power station   **33. a.** $P = 2/\sqrt{3}$ units
from the midpoint of the base   **35.** $r = \sqrt{6}, h = \sqrt{3}$
37. For $L \leq 4r$ max at $\theta = 0$ and $\theta = 2\pi$; min at
$\theta = \cos^{-1}(-L/(4r))$ and $\theta = 2\pi - \cos^{-1}(-L/(4r))$
39. a. $r = \sqrt[3]{177/\pi} \approx 3.83$ cm; $h = 2\sqrt[3]{177/\pi} \approx 7.67$ cm
b. $r = \sqrt[3]{177/2\pi} \approx 3.04$ cm; $h = 2\sqrt[3]{708/\pi} \approx 12.17$ cm.
Part (b) is closer to the real can.   **41.** $\sqrt{30} \approx 5.5$ ft   **43.** When the
seat is at its lowest point   **45.** $r = \sqrt{2}\,R/\sqrt{3}; h = 2R/\sqrt{3}$
47. a. $r = 2R/3; h = \frac{1}{3}H$ **b.** $r = R/2; h = H/2$ **49.** 3:1
51. $(1 + \sqrt{3})$ mi ≈ 2.732 mi   **53.** You can run 12 mi/hr if you run
toward the point $3/16$ mi ahead of the locomotive (when it passes the
point nearest you).   **55. a.** $(-6/5, 2/5)$ **b.** Approx $(0.59, 0.65)$
c. (i) $\left(p - \frac{1}{2}, \sqrt{p - \frac{1}{2}}\right)$ **(ii)** $(0, 0)$ **57. a.** 0, 30, 25
b. 42.5 mi/hr   **c.** The units of $p/g(v)$ are \$/mi and so are the units
of W/V. Thus, $L\left(\dfrac{p}{g(v)} + \dfrac{w}{v}\right)$ gives the total cost of a trip of L miles.
d. ≈ 62.9 mi/hr   **e.** Neither; the zeros of $C'(v)$ are independent of L.
f. Decreased slightly, to 62.5 mi/hr   **g.** Decreased to 60.8 mi/hr
59. b. Because the speed of light is constant, travel time is minimized
when distance is minimized.   **61.** Let the angle of the cuts be ϕ_1 and
ϕ_2, where $\phi_1 + \phi_2 = \theta$. The volume of the notch is proportional to
$\tan \phi_1 + \tan \phi_2 = \tan \phi_1 + \tan(\theta - \phi_1)$, which is minimized
when $\phi_1 = \phi_2 = \dfrac{\theta}{2}$.

Section 4.5 Exercises, pp. 273–274

1.

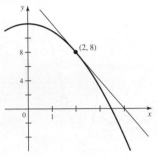

3. $f(x) \approx f(a) + f'(a)(x - a)$ **5.** $dy = f'(x)\,dx$
7. a. $y = -4x + 16$ **b.**

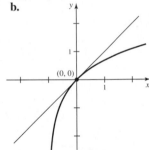

c. 7.6   **d.** 0.13% error   **9. a.** $y = x$ **b.**

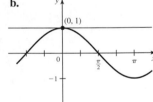

c. 0.9   **d.** 40% error   **11. a.** $y = 1$ **b.**

c. 1   **d.** 0.005% error   **13.** $y = 1/x$ near
$a = 200; \frac{1}{203} \approx 0.004925$ **15.** $y = \sqrt{x}$ near
$a = 144; \sqrt{146} \approx 12\frac{1}{12}$ **17.** $y = \ln x$ near $a = 1$; $\ln(0.05) \approx 0.05$
19. $y = e^x$ near $a = 0$; $e^{0.06} \approx 1.06$ **21.** $y = \dfrac{1}{\sqrt[3]{x}}$ near
$a = 512; \dfrac{1}{\sqrt[3]{510}} \approx \dfrac{769}{6144} \approx 0.125$ $f(x) = f(a) + f'(a)(x - a)$
23. $\Delta V \approx 10\pi$ ft^3 **25.** $\Delta V \approx -40\pi$ cm^3 **27.** $\Delta S \approx \dfrac{-59\pi}{5\sqrt{34}}$ m^3
29. $dy = 2\,dx$ **31.** $dy = \dfrac{-3}{x^4}\,dx$ **33.** $dy = a \sin x\,dx$
35. $dy = (9x^2 - 4)\,dx$ **37.** $dy = \sec^2 x\,dx$ **39. a.** True
b. False   **c.** True

41. $y = 1 - x$;   **a.**

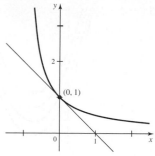

b. $1/1.1 \approx 0.9$ **c.** 1% error   **43.** $y = 1 - x$

a.

b. $e^{-0.03} \approx 0.97$ **c.** 0.05% error

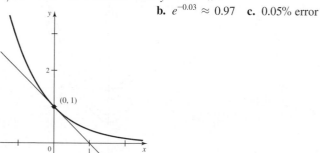

45. $L(x) = 2 + (x - 8)/12$

x	Linear Approximation	Exact Value	Percent Error
8.1	$2.008\overline{3}$	2.00829885	1.717×10^{-3}
8.01	$2.0008\overline{3}$	2.000832986	1.734×10^{-5}
8.001	$2.00008\overline{3}$	2.00008333	1.736×10^{-7}
8.0001	$2.0000083\overline{3}$	2.000008333	1.735×10^{-9}
7.9	$1.991\overline{6}$	1.991631701	1.736×10^{-3}
7.99	$1.99916\overline{6}$	1.999166319	1.736×10^{-5}
7.999	$1.999916\overline{6}$	1.999916663	1.738×10^{-7}
7.9999	$1.9999916\overline{6}$	1.999991667	1.738×10^{-9}

47. a. f; the rate at which f' is changing at 1 is smaller than the rate at which g' is changing at 1. The graph of f bends away from the linear function more slowly than the graph of g.   **b.** The larger the value of $|f''(a)|$, the greater the deviation of the curve $y = f(x)$ from the tangent line at points near $x = a$.

Section 4.6 Exercises, pp. 279–280

1. If f is a continuous function on the closed interval $[a, b]$ and is differentiable on (a, b) and the slope of the secant line that joins $(a, f(a))$ to $(b, f(b))$ is zero, then there is at least one value c in (a, b) at which the slope of the line tangent to f at $(c, f(c))$ is also zero.

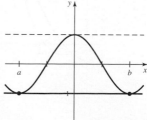

3. $f(x) = |x|$ is not differentiable at 0.

5.

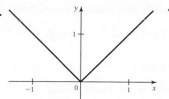

7. $x = \frac{1}{3}$ **9.** $x = \pi/4$

11. Does not apply   **13.** Average lapse rate $= -6.3°/\text{km}$. You cannot conclude that the lapse rate at a point exceeds the critical value.

15. a. Yes   **b.** $c = \frac{1}{2}$ **c.**

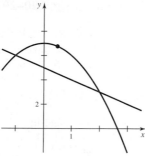

17. a. Yes   **b.** $c = \ln\left(\dfrac{3}{\ln 4}\right)$ **c.**

19. a. Yes   **b.** $c = \sqrt{1 - 9/\pi^2}$ **c.**

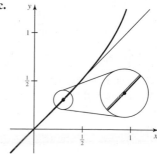

21. a. Does not apply   **23. a.** False   **b.** True   **c.** False   **25.** h and p

27.

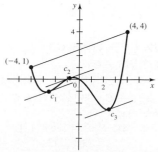

29. The car's average velocity is $(30 - 0)/(28/60) = 64.3$ mi/hr. By the MVT, the car's instantaneous velocity was 64.3 mi/hr at some time.   **31.** Average speed $= 11.6$ mi/hr. By MVT, the speed was exactly 11.6 mi/hr at least once. By the Intermediate Value Theorem, all speeds between 0 and 11.6 mi/hr were reached. Because the initial and final speed was 0 mi/hr, the speed of 11 mi/hr was reached at least twice.

33. $\dfrac{f(b) - f(a)}{b - a} = A(a + b) + B$ and $f'(x) = 2Ax + B$;

$2Ax + B = A(a + b) + B$ implies that $x = \dfrac{a + b}{2}$, the midpoint

of $[a, b]$.   **35.** $\tan^2 x$ and $\sec^2 x$ differ by a constant; in fact,
$\tan^2 x - \sec^2 x = -1$.   **37.** Bolt's average speed was 37.58 km/hr,
so he exceeded 37 km/hr during the race.   **39. b.** $c = \frac{1}{2}$

Section 4.7 Exercises, pp. 290–292

1. If $\lim\limits_{x\to a} f(x) = 0$ and $\lim\limits_{x\to a} g(x) = 0$, then we say $\lim\limits_{x\to a} f(x)/g(x)$ is
of indeterminate form $0/0$.   **3.** Take the limit of the quotient of the
derivatives of the functions.   **5.** If $\lim\limits_{x\to a} f(x)g(x)$ has the indeterminate

form $0 \cdot \infty$, then $\lim\limits_{x\to a}\left(\dfrac{f(x)}{1/g(x)}\right)$ has the indeterminate form $0/0$ or

∞/∞.   **7.** If $\lim\limits_{x\to a} f(x) = 1$ and $\lim\limits_{x\to a} g(x) = \infty$, then

$f(x)^{g(x)} \to 1^\infty$ as $x \to a$, which is meaningless; so direct

substitution does not work.   **9.** $\lim\limits_{x\to\infty} \dfrac{g(x)}{f(x)} = 0$ **11.** $\ln x, x^3, 2^x, x^x$

13. -1 **15.** $\frac{12}{5}$ **17.** 4   **19.** $\frac{9}{16}$ **21.** 4   **23.** $-\frac{1}{2}$ **25.** $\cos x$
27. $\frac{1}{2}$ **29.** 0   **31.** 1   **33.** 1   **35.** 0   **37.** 0   **39.** 1   **41.** 1   **43.** e
45. 1   **47.** e **49.** $e^{0.01x}$ **51.** Comparable growth rates   **53.** x^x
55. 1.00001^x **57.** x^x **59.** e^{x^2} **61. a.** False   **b.** False   **c.** False
d. False   **e.** True   **f.** True   **63.** $\frac{2}{5}$ **65.** $-\frac{9}{4}$ **67.** 0   **69.** $\frac{1}{6}$
71. ∞ **73.** $(\ln 3)/(\ln 2)$ **75.** $\frac{1}{2}$ **77. a.** Approx. 3.44×10^{15}
b. Approx. 3536   **c.** e^{100} **d.** Approx. 163   **79.** 1   **81.** $\ln a - \ln b$

83. b. $\lim\limits_{m\to\infty}(1 + r/m)^m = \lim\limits_{m\to\infty}\left(1 + \dfrac{1}{(m/r)}\right)^{(m/r)r} = e^r$

85. $\sqrt{a/c}$ **87.** $\lim\limits_{x\to\infty}\dfrac{x^p}{b^x} = \lim\limits_{t\to\infty}\dfrac{\ln^p t}{t \ln^p b} = 0$ (let $t = b^x$, see

Example 8).   **89.** Show $\lim\limits_{x\to\infty}\dfrac{\log_a x}{\log_b x} = \dfrac{\ln b}{\ln a} \neq 0$.   **91.** $1/3$

95. a. $b > e$ **b.** e^{ax} grows faster than e^x as $x \to \infty$ for $a > 1$; e^{ax}
grows slower than e^x as $x \to \infty$ for $0 < a < 1$.

Section 4.8 Exercises, pp. 301–303

1. Derivative, antiderivative   **3.** $x + C$, where C is any real number

5. $\dfrac{x^{p+1}}{p + 1} + C$, where C is any real number and $p \neq -1$
7. $\ln x + C$ **9.** 0   **11.** $x^5 + C$
13. $-\frac{1}{2}\cos 2x + C$
15. $3\tan x + C$ **17.** $y^{-2} + C$ **19.** $\frac{1}{2}x^6 - \frac{1}{2}x^{10} + C$
21. $\frac{8}{3}x^{3/2} - 8x^{1/2} + C$ **23.** $(5s + 3)^3/15 + C$
25. $\frac{9}{4}x^{4/3} + 6x^{2/3} + 6x + C$ **27.** $-\frac{1}{2}\cos 2y + \frac{1}{3}\sin 3y + C$
29. $\tan x - x + C$ **31.** $\tan\theta + \sec\theta + C$ **33.** $\frac{1}{2}\ln|y| + C$
35. $6\sin^{-1}(x/5) + C$ **37.** $1/10 \sec^{-1}|x/10| + C$
39. $F(x) = x^6/6 + 2/x + x$ **19/6**   **41.** $F(v) = \sec v + 1$
43. $f(x) = x^2 - 3x + 4$ **45.** $g(x) = \dfrac{7}{8}x^8 - \dfrac{x^2}{2} + \dfrac{13}{8}$
47. $f(u) = 4\sin u + 2\cos 2u - 3$

49. $f(x) = x^2 - 5x + 4$

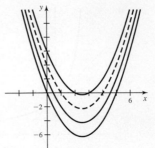

51. $f(x) = \dfrac{3x^2}{2} - \dfrac{\cos(\pi x)}{\pi} + \dfrac{1 - 3\pi}{\pi}$

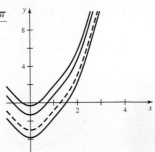

53. $f(t) = \ln t + 4$

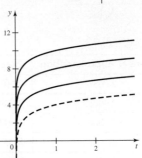

55. $s(t) = t^2 + 4t$

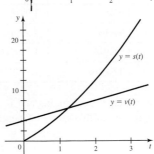

57. $s(t) = \frac{4}{3}t^{3/2} + 1$

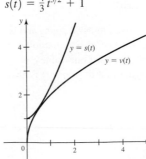

59. $s(t) = 2t^3 + 2t^2 - 10t$

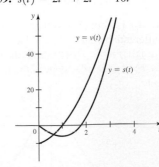

61. Runner A overtakes runner B at $t = \pi/2$ s.

63. a. $v(t) = -9.8t + 30$ **b.** $s(t) = -4.9t^2 + 30t$ **c.** 45.92 m
at time $t = 3.06$ **d.** $t = 6.12$ s   **65. a.** $v(t) = -9.8t + 10$

b. $s(t) = -4.9t^2 + 10t + 400$ **c.** 405.10 m at time $t = 1.02$

d. $t = 10.11$ s   **67. a.** True   **b.** False   **c.** True   **d.** False   **e.** False

69. $(e^{2x} + e^{-2x})/4 + C$ **71.** $-\cot\theta + 2\theta^3/3 - 3\theta^2/2 + C$

73. $\ln|x| + 2\sqrt{x} + C$ **75.** $\frac{4}{15}x^{15/2} - \frac{24}{11}x^{11/6} + C$

77. $F(x) = -\cos x + 3x + 3 - 3\pi$

79. $F(x) = 2x^8 + x^4 + 2x + 1$ **81. a.** $Q(t) = 10t - t^3/30$ gal

b.

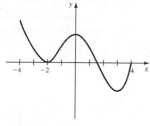

c. $\dfrac{200}{3}$ gal

83. $\displaystyle\int \sin^2 x\, dx = x/2 - (\sin 2x)/4 + C;$

$\displaystyle\int \cos^2 x\, dx = x/2 + (\sin 2x)/4 + C$

Review Exercises, pp. 303–305

1. a. False   **b.** False   **c.** True   **d.** True

3.

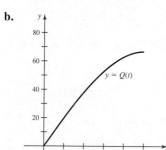

5.

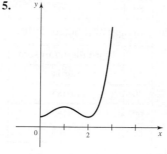

7. $x = 3$ and $x = -2$; no abs. max or min.

9. $x = 1/e$; abs. min at $(1/e, 10 - 2/e)$

11.

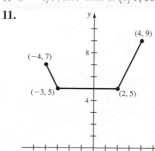

Critical pts.: x in the interval $[-3, 2]$; abs. max: $(4, 9)$; abs. and local min at $(x, 5)$ for all x in $[-3, 2]$; local max at $(x, 5)$ for all x in $(-3, 2)$.

13.

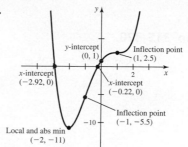

15.

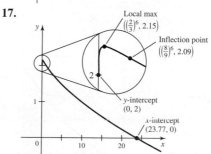

17.

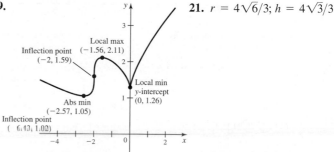

19.

21. $r = 4\sqrt{6}/3;\ h = 4\sqrt{3}/3$

23. a. $a = b = \frac{23}{2}$ **b.** $a = 0, b = 23$ or $a = 23, b = 0$

25. a. $\frac{100}{9}$ cells/week   **b.** $t = 2$ weeks   **27.** 0   **29.** $2\sqrt{3} - \frac{4}{3}$

31. 2/3   **33.** ∞ **35.** 0   **37.** 1   **39.** 0   **41.** $1/e^3$ **43.** $x^{1/2}$ **45.** $\sqrt{x}$

47. 3^x **49.** Comparable growth rate   **51.** $\dfrac{-1}{x} + \dfrac{4}{3}x^{-3/2} + C$

53. $\theta + \frac{1}{3}\sin 3\theta + C$ **55.** $\frac{1}{2}\sec 2x + C$ **57.** $12\ln|x| + C$

59. $\tan^{-1} x + C$ **61.** $\frac{4}{7}x^{7/4} + \frac{2}{7}x^{7/2} + C$

63. $f(t) = -\cos t + t^2 + 6$

65. $h(x) = \dfrac{x}{2} - \dfrac{1}{4}\sin 2x + \left(\dfrac{1}{2} + \dfrac{\sin 2}{4}\right)$

67. $v(t) = -9.8t + 120;\ s(t) = -4.9t^2 + 120t + 125$
The rocket reaches a height of 859.69 m at time $t = 12.24$ s and then falls to the ground, hitting at time $t = 25.49$ s.   **69.** 1; 1   **71.** 0

73. $\displaystyle\lim_{x\to 0^+} f(x) = 1;\ \lim_{x\to 0^+} g(x) = 0$

CHAPTER 5

Section 5.1 Exercises, pp. 315–320

1.
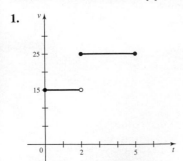
Displacement $= 105$ m

3. We can subdivide the interval $[0, \pi/2]$ into several segments, which will be the bases of rectangles that fit under the curve. The heights of the rectangles can be computed by taking the value of $\cos x$ at the right-hand value of each base. We can calculate the area of each rectangle and add them to get a lower bound on the area.

5. $\frac{1}{2}$; 1, 1.5, 2, 2.5, 3; 1, 1.5, 2, 2.5; 1.5, 2, 2.5, 3; 1.25, 1.75, 2.25, 2.75

7. Underestimate; the rectangles all fit under the curve.

9. a. 67 ft **b.** 67.75 ft **11.** Displacement ≈ 2.78 m

13. Displacement ≈ 148.96 mi **15.** 20; 25

17. a. c.
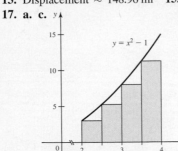

Left Riemann sum underestimates.   Right Riemann sum overestimates.

b. $\Delta x = \frac{1}{2}$; 2, 2.5, 3, 3.5, 4 **d.** 13.75; 19.75

19. a. c.
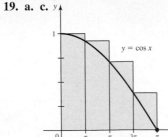

Left Riemann sum overestimates.   Right Riemann sum underestimates.

b. $\Delta x = \pi/8$; 0, $\pi/8$, $\pi/4$, $3\pi/8$, $\pi/2$ **d.** 1.18; 0.79 **21.** 670

23. a. c.
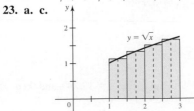
b. $\Delta x = \frac{1}{2}$; 1, $\frac{3}{2}$, 2, $\frac{5}{2}$, 3
d. 2.80

25. a. c.
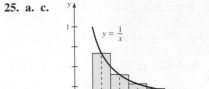
b. $\Delta x = 1$; 1, 2, 3, 4, 5, 6
d. 1.76

27. 5.5, 3.5 **29. b.** 110, 117.5 **31. a.** $\sum_{k=1}^{5} k$ **b.** $\sum_{k=1}^{6} (k + 3)$

c. $\sum_{k=1}^{4} k^2$ **d.** $\sum_{k=1}^{4} \frac{1}{k}$ **33. a.** 55 **b.** 48 **c.** 30 **d.** 60 **e.** 6 **f.** 6

g. 85 **h.** 0 **35. a.** $\frac{1}{10} \sum_{k=1}^{40} \sqrt{\frac{k-1}{10}} \approx 5.227$; $\frac{1}{10} \sum_{k=1}^{40} \sqrt{\frac{k}{10}} \approx 5.427$;

$\frac{1}{10} \sum_{k=1}^{40} \sqrt{\frac{2k-1}{20}} \approx 5.3$ **b.** $\frac{16}{3}$ **37. a.** $\frac{1}{15} \sum_{k=1}^{75} \left[\left(\frac{k+29}{15} \right)^2 - 1 \right] =$

$\frac{14{,}198}{135} \approx 105.17$; $\frac{1}{15} \sum_{k=1}^{75} \left[\left(\frac{k+30}{15} \right)^2 - 1 \right] = \frac{14{,}603}{135} \approx 108.17$;

$\frac{1}{15} \sum_{k=1}^{75} \left[\left(\frac{2k+59}{30} \right)^2 - 1 \right] = \frac{57{,}599}{540} \approx 106.66$ **b.** 106.7

39.

n	Right Riemann sum
10	10.56
30	10.65
60	10.664
80	10.665

The sums appear to approach $10\frac{2}{3}$.

41.

n	Right Riemann sum
10	5.655
30	6.074
60	6.178
80	6.205

The sums appear to approach 2π.

43.

n	Right Riemann sum
10	1.0844
30	1.0285
60	1.0143
80	1.0108

The sums appear to approach 1.

45. a. True **b.** False
c. True

47.
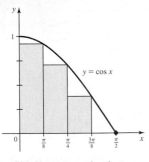
sum ≈ 1.1375

49. $\sum_{k=1}^{50} \left(\frac{4k}{50} + 1 \right) \cdot \frac{4}{50} = \frac{304}{25} = 12.16$

51. $\sum_{k=1}^{32} \left(3 + \frac{2k-1}{8} \right)^3 \cdot \frac{1}{4} \approx 3639.1$

53. Left; $[2, 6]$; 4 or Right; $[1, 5]$; 4 **55.** Midpoint; $[2, 6]$; 4

57. a.

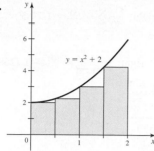

Left Riemann sum is
$$\frac{23}{4} = 5.75.$$

b.

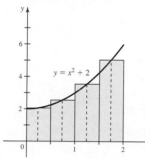

Midpoint Riemann sum is
$$\frac{53}{8} = 6.625.$$

c.

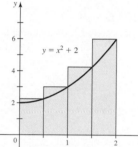

Right Riemann sum is
$$\frac{31}{4} = 7.75.$$

59. Left sum: 34; right sum: 24   **61. a.** The object is speeding up on the interval $[0, 1]$, moving at a constant rate on $[1, 3]$, slowing down on $[3, 5]$, and maintaining a constant velocity on $[5, 6]$.
b. 30 m   **c.** 50 m   **d.** $s(t) = 30 + 10t$ **63. a.** 14.5 g   **b.** 29.5 g
c. 44 g   **d.** $x = 6\frac{1}{3}$ cm   **65.** 107 mi
67.

n	Midpoint Riemann sum
16	0.503906
32	0.500977
64	0.500244

The sums appear to approach 0.5.

69.

n	Midpoint Riemann sum
16	4.7257
32	4.7437
64	4.7485

The sums appear to approach 4.75.

Section 5.2 Exercises, pp. 331–334

1. The area of the regions above the x-axis minus the area of the regions below the x-axis.   **3.** When the function is nonnegative on the entire interval; when the function has negative values on the interval
5. Both integrals = 0.   **7.** The length of the interval $[a, a]$ is
$a - a = 0$, so the net area is 0.   **9.** $\dfrac{a^2}{2}$

11. a.

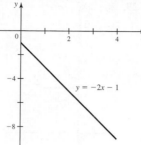

b. $-16, -24, -20$

13. a.

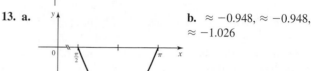

b. $\approx -0.948, \approx -0.948,$
≈ -1.026

15. a.

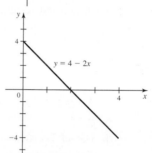

b. $4, -4, 0$ **c.** Positive contributions on $[0, 2]$; negative contributions on $[2, 4]$.

17. a.

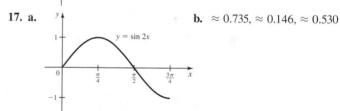

b. $\approx 0.735, \approx 0.146, \approx 0.530$

c. Positive contribution on $[0, \pi/2]$, negative contribution on $[\pi/2, 3\pi/4]$.   **19.** $\displaystyle\int_0^2 (x^2 + 1)\, dx$ **21.** $\displaystyle\int_1^2 x \ln x\, dx$

23. 16

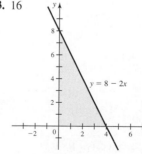

25. $-\dfrac{5}{2}$

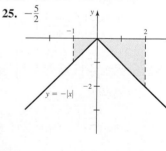

27. 4π

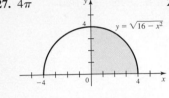

29. 26

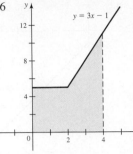

Left: $\dfrac{1}{50}\displaystyle\sum_{k=1}^{50}e^{(k-1)/50} = 1.70116$; **right:** $\dfrac{1}{50}\displaystyle\sum_{k=1}^{50}e^{k/50} = 1.73552$

$n = 100$

Left: $\dfrac{1}{100}\displaystyle\sum_{k=1}^{100}e^{(k-1)/100} = 1.70970$; **right:** $\dfrac{1}{100}\displaystyle\sum_{k=1}^{100}e^{k/100} = 1.72689$

b. 1.71   **61. a.** $\displaystyle\sum_{k=1}^{n}\dfrac{6}{n}\sqrt{\dfrac{2n+6k-3}{2n}}$

b.

Estimate: 9.33

n	Midpoint Riemann sum
20	9.33380
50	9.33341
100	9.33335

31. 16   **33.** 6   **35.** π **37.** -2π **39. a.** -32 **b.** $\frac{32}{3}$ **c.** -64
d. Not possible   **41. a.** 10   **b.** -3 **c.** -16 **d.** 3   **43. a.** $\frac{3}{2}$
b. $-\frac{3}{4}$ **45.** 6   **47.** 104   **49.** 18   **51. a.** True   **b.** True
c. True   **d.** False   **e.** False

53. a.

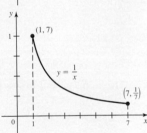

b. $\Delta x = \frac{1}{2}$; 3, 3.5, 4, 4.5, 5, 5.5, 6   **c.** -22.5; -25.5
d. The left Riemann sum overestimates; the right Riemann sum underestimates.

63. a. $\displaystyle\sum_{k=1}^{n}(2k-1)(2n+1-2k)\cdot\dfrac{16}{n^3}$

b.

Estimate: 10.67

n	Midpoint Riemann sum
20	10.6800
50	10.6688
100	10.6672

65. a. 15   **b.** 5   **c.** 3   **d.** -2 **e.** 24   **f.** -10

67.

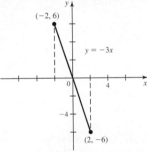

The area is 12; the net area is 0.

55. a.

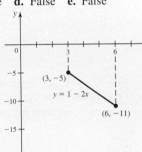

b. $\Delta x = 1$; 1, 2, 3, 4, 5, 6, 7
c. $\frac{49}{20}$, $\frac{223}{140}$
d. The left Riemann sum overestimates; The right Riemann sum underestimates.

69.

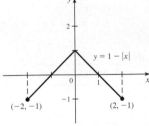

The area is 2; the net area is 0.

57. a. $n = 20$

Left: $\displaystyle\sum_{k=1}^{20}\left[\left(\dfrac{k-1}{20}\right)^2+1\right]\cdot\dfrac{1}{20} = 1.30875$;

right: $\displaystyle\sum_{k=1}^{20}\left[\left(\dfrac{k}{20}\right)^2+1\right]\cdot\dfrac{1}{20} = 1.35875$

$n = 50$

Left: $\displaystyle\sum_{k=1}^{50}\left[\left(\dfrac{k-1}{50}\right)^2+1\right]\cdot\dfrac{1}{50} = 1.3234$;

right: $\displaystyle\sum_{k=1}^{50}\left[\left(\dfrac{k}{50}\right)^2+1\right]\cdot\dfrac{1}{50} = 1.3434$

$n = 100$

Left: $\displaystyle\sum_{k=1}^{100}\left[\left(\dfrac{k-1}{100}\right)^2+1\right]\cdot\dfrac{1}{100} = 1.32835$;

right: $\displaystyle\sum_{k=1}^{100}\left[\left(\dfrac{k}{100}\right)^2+1\right]\cdot\dfrac{1}{100} = 1.33835$ **b.** 1.33

59. a. $n = 20$

Left: $\dfrac{1}{20}\displaystyle\sum_{k=1}^{20}e^{(k-1)/20} = 1.67568$; **right:** $\dfrac{1}{20}\displaystyle\sum_{k=1}^{20}e^{k/20} = 1.76160$

$n = 50$

71. 17   **73.** $25\pi/2$ **75.** 25   **79.** For any such partition on the interval $[0,1]$, the grid points are $x_k = k/n$ for $k = 0, 1, \ldots, n$. That is, x_k is rational for each k so that $f(x_k) = 1$ for $k = 0, 1, \ldots, n$. Thus, the left, right, and midpoint Riemann sums are $\displaystyle\sum_{k=1}^{n}1\cdot(1/n) = 1$.

Section 5.3 Exercises, pp. 345–349

1. A is an antiderivative of f; $A'(x) = f(x)$ **3.** Let f be continuous on $[a,b]$. Then $\displaystyle\int_{a}^{b}f(x)\,dx = F(b) - F(a)$, where F is any antiderivative of f.   **5.** Increasing   **7.** The derivative of the integral of f is f, or $\dfrac{d}{dx}\left(\displaystyle\int_{a}^{x}f(t)\,dt\right) = f(x)$.   **9.** $f(x)$, 0   **11. a.** 0
b. -9 **c.** 25   **d.** 0   **e.** 16

13. a. 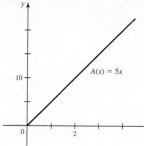   **b.** $A'(x) = 5$

$A(x) = 5x$

15. a. 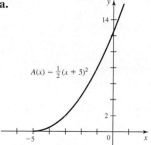   **b.** $A'(x) = 5$

$A(x) = 5x + 25$

17. a. $A(2) = 2$, $A(4) = 8$; $A(x) = \frac{1}{2}x^2$ **b.** $F(4) = 6$, $F(6) = 16$;

$F(x) = \frac{1}{2}x^2 - 2$ **c.** $A(x) - F(x) = \frac{1}{2}x^2 - \left(\frac{1}{2}x^2 - 2\right) = 2$

19. a.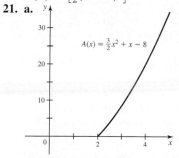

$A(x) = \frac{1}{2}(x + 5)^2$

b. $A'(x) = \left[\frac{1}{2}(x + 5)^2\right]' = x + 5 = f(x)$

21. a.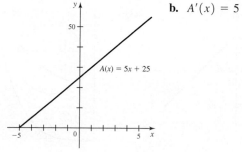

$A(x) = \frac{3}{2}x^2 + x - 8$

b. $A'(x) = \left(\frac{3}{2}x^2 + x - 8\right)' = 3x + 1 = f(x)$

23. $\frac{7}{3}$ **25.** $\frac{9}{2}$

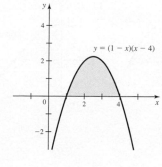

$y = (1 - x)(x - 4)$

27. $-\frac{125}{6}$ **29.** $-\frac{10}{3}$

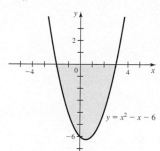
$y = x^2 - x - 6$

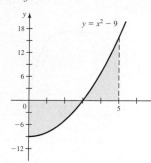
$y = x^2 - 9$

31. $-\frac{32}{3}$ **33.** $-\frac{5}{2}$ **35.** 1 **37.** $-\frac{3}{8}$ **39.** 3 ln 2

41. (i) $\frac{14}{3}$ **(ii)** $\frac{14}{3}$ **43. (i)** -51.2 **(ii)** 51.2

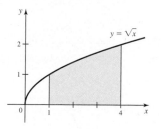
$y = \sqrt{x}$

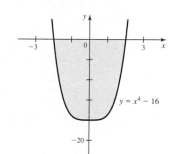
$y = x^4 - 16$

45. 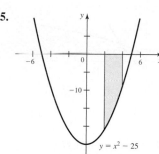   **47.**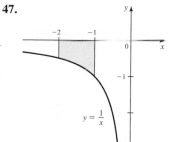

$y = x^2 - 25$

$y = \frac{1}{x}$

Area $= \frac{94}{3}$ Area $=$ ln 2

49. 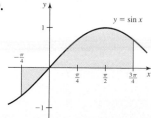   Area $= 2$

$y = \sin x$

51. $x^2 + x + 1$ **53.** $3/x^4$ **55.** $-\sqrt{x^4 + 1}$ **57.** a–C, b–B, c–D,
d–A   **59. a.** $x \approx 4.5$ **b.** Local min at $x \approx 2$; local max at $x \approx 8$
c.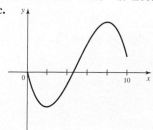

61. a. $x = 10$ **b.** Local max at $x = 5$
c.

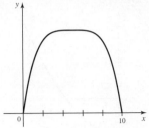

63. $-\pi, -\pi + \frac{9}{2}, -\pi + 9, 5 - \pi$ **65. a.** $A(x) = e^x - 1$
b.   **c.** $A(b) = 1; A(c) = 3$

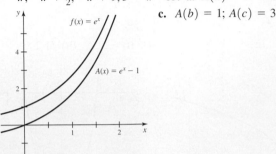

$f(x) = e^x$
$A(x) = e^x - 1$

67. a. $A(x) = \frac{2}{3}(x^{3/2} - 1)$ **b.**

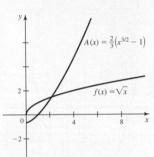

$A(x) = \frac{2}{3}(x^{3/2} - 1)$
$f(x) = \sqrt{x}$

c. $A(b) = \frac{14}{3}; A(c) = \frac{52}{3}$
69. a.   **b.** $g'(x) = \sin^2(x)$

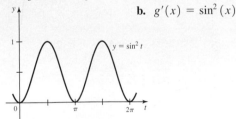

$y = \sin^2 t$

c.

71. a.   **b.** $g'(x) = \sin(\pi x^2)$

$y = \sin(\pi t^2)$

c.

73. a. True **b.** True **c.** False **d.** True **75.** $\frac{2}{3}$ **77.** 1 **79.** $\frac{45}{4}$
81. $\frac{3}{2} + 4\ln 2$
83.   **85.**

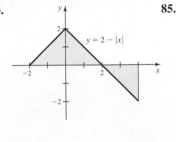

$y = 2 - |x|$

$y = x^4 - 4$

Area = 6   Area ≈ 194.05
87. $f(8) - f(3)$ **89.** $-(\cos^4 x + 6)\sin x$
91. a.   **b.** $b = 6$
c. $b = \dfrac{3a}{2}$

$y = x^2 - 4x$

93. 3 **95.** $f(x) = -2\sin x + 3$ **97.** $\pi/2 \approx 1.57$
99. $[s'(x)]^2 + \left[\dfrac{s''(x)}{2x}\right]^2 = [\sin(x^2)]^2 + \left[\dfrac{2x\cos(x^2)}{2x}\right]^2$
$\qquad = \sin^2(x^2) + \cos^2(x^2) = 1$

Section 5.4 Exercises, pp. 354–357

1. If f is odd, the region between f and the positive x-axis and between f and the negative x-axis are reflections of each other through the origin. Thus, on $[-a, a]$, the areas cancel each other out. **3.** Even; even
5. If f is continuous on $[a, b]$, then there is a c in (a, b) such that
$$f(c) = \frac{1}{b - a}\int_a^b f(x)\, dx. \quad \textbf{7. } \frac{1000}{3} \quad \textbf{9. } -\frac{88}{3} \quad \textbf{11. } 0 \quad \textbf{13. } 0$$
15. 0   **17.** 0

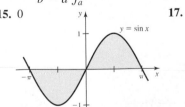

$y = \sin x$

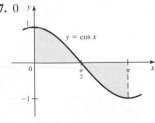

$y = \cos x$

19. $1/(e - 1)$

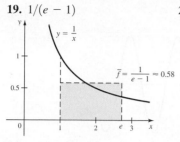

21. $2/\pi$

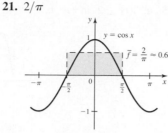

23. $1/(n + 1)$ **25.** $2000/3$ **27.** $20/\pi$ **29.** $\frac{4}{\pi}$ km **31.** $c = 2$
33. $c = a/\sqrt{3}$ **35.** $c = \pm\frac{1}{2}$ **37. a.** True **b.** True **c.** True
d. False **39.** 2 **41.** 0 **43.** 420 ft **47.** $f(g(-x)) = f(g(x)) \Rightarrow$

the integrand is even; $\displaystyle\int_{-a}^{a} f(g(x))\,dx = 2\int_{0}^{a} f(g(x))\,dx$

49. $p(g(-x)) = p(g(x)) \Rightarrow$ the integrand is even;

$\displaystyle\int_{-a}^{a} p(g(x))\,dx = 2\int_{0}^{2} p(g(x))\,dx$ **51. a.** $a/6$

b. $(3 \pm \sqrt{3})/6$, independent of a

55. $c = \sqrt[4]{12}$ **59.**

Even	Even
Even	Odd

Section 5.5 Exercises, pp. 363–366

1. The Chain Rule **3.** $u = g(x)$ **5.** We let a become $g(a)$ and b
become $g(b)$. **7.** $\dfrac{x}{2} + \dfrac{\sin 2x}{4} + C$ **9.** $\dfrac{(x + 1)^{13}}{13} + C$

11. $\dfrac{(2x + 1)^{3/2}}{3} + C$ **13.** $\dfrac{(x^2 + 1)^5}{5} + C$ **15.** $\frac{1}{4}\sin^4 x + C$

17. $\dfrac{(x^2 - 1)^{100}}{100} + C$ **19.** $\dfrac{-(1 - 4x^3)^{1/2}}{3} + C$ **21.** $\dfrac{(x^2 + x)^{11}}{11} + C$

23. $\dfrac{(x^4 + 16)^7}{28} + C$ **25.** $\dfrac{\sin^{-1}(3x)}{3} + C$ **27.** $\dfrac{(x^6 - 3x^2)^5}{30} + C$

29. $\frac{2}{3}(x - 4)^{1/2}(x + 8) + C$ **31.** $\frac{3}{5}(x + 4)^{2/3}(x - 6) + C$
33. $\frac{3}{112}(2x + 1)^{4/3}(8x - 3) + C$ **35.** $\frac{7}{2}$ **37.** $\frac{1}{3}$ **39.** $(e^9 - 1)/3$

41. $\sqrt{2} - 1$ **43.** $\pi/6$ **45.** π **47.** $\dfrac{\theta}{2} - \dfrac{1}{4}\sin\left(\dfrac{6\theta + \pi}{3}\right) + C$

49. $\dfrac{\pi}{4}$ **51. a.** True **b.** True **c.** False **d.** False **e.** False

53. $\frac{1}{10}\tan(10x) + C$ **55.** $\frac{1}{2}\tan^2 x + C$ **57.** $\frac{1}{7}\sec^7 x + C$ **59.** $\frac{1}{3}$

61. $\frac{3}{4}(4 - 3^{2/3})$ **63.** $\frac{32}{3}$ **65.** 1 **67.** $\dfrac{64}{5}$ **69.** $\frac{2}{3}$; constant

71. a. 160 **b.** $\dfrac{4800}{49} \approx 98$ **c.** $\Delta p = \displaystyle\int_{0}^{T} \dfrac{200}{(t + 1)^r}\,dt$; decreases as

r increases **d.** $r \approx 1.28$ **e.** As $t \to \infty$, the population approaches

100. **73.** $2/\pi$ **77.** One area is $\displaystyle\int_{4}^{9} \dfrac{(\sqrt{x} - 1)^2}{2\sqrt{x}}\,dx$. Changing

variables by letting $u = (\sqrt{x} - 1)$ yields $\int_{1}^{2} u^2\,du$, which is the

other area. **79.** 7297/12 **81.** $\dfrac{[f^{(p)}(x)]^{n+1}}{n + 1} + C$

83. $\frac{2}{15}(3 - 2a)(1 + a)^{3/2} + \frac{4}{15}a^{5/2}$ **85.** $\frac{1}{3}\sec^3\theta + C$
87. a. $I = \int\left(\frac{1}{2}\sin 2x\right)^2 dx = \frac{1}{8}x - \frac{1}{32}\sin 4x + C$
b. $I = \int(\sin^2 x - \sin^4 x)\,dx = \frac{1}{8}x - \frac{1}{32}\sin 4x + C$
91. $\frac{4}{3}(-2 + \sqrt{1 + x})\sqrt{1 + \sqrt{1 + x}}$ **93.** $-4 + \sqrt{17}$

Chapter 5 Review Exercises, pp. 366–369

1. a. True **b.** False **c.** True **d.** True **e.** False **f.** True **g.** True
3. $\frac{23}{2}$ **5.** 4π **7. a.** $1[(3\cdot 2 - 2) + (3\cdot 3 - 2) + (3\cdot 4 - 2)] = 21$

b. $\displaystyle\sum_{k=1}^{n} \frac{3}{n}\left[3\left(1 + \frac{3k}{n}\right) - 2\right]$ **c.** $\frac{33}{2}$ **9.** $\displaystyle\int_{0}^{4}(1 + x^8)\,dx = \frac{36 + 4^9}{9}$

11. $\frac{212}{5}$ **13.** 20 **15.** $x^9 - x^7 + C$ **17.** $\frac{7}{6}$ **19.** $\dfrac{\pi}{6}$ **21.** 1

23. $\frac{1}{2}\theta - \frac{1}{20}\sin 10\theta + C$ **25.** $\frac{1}{3}\ln|x^3 + 3x^2 - 6x| + C$
27. a. 20 **b.** 0 **c.** 80 **29.** 18 **31.** 10 **33.** Not enough
information **35.** Displacement = 0; distance = $20/\pi$ **37. a.** $\frac{5}{2}$
b. 3 **39.** 24 **41.** $f(1) = 0; f'(x) > 0$ on $[1, \infty); f''(x) < 0$
on $[1, \infty)$

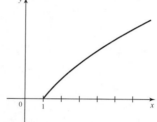

47. $\cos\left(\dfrac{1}{x}\right) + C$ **49.** $\ln|\tan^{-1} x| + C$ **51.** $\ln(e^x + e^{-x}) + C$
53. Differentiating the first equation gives the second equation.
55. a. Increasing on $(-\infty, 1)$ and $(2, \infty)$; decreasing on $(1, 2)$
b. Concave up on $\left(\frac{13}{8}, \infty\right)$; concave down on $\left(-\infty, \frac{13}{8}\right)$
c. Local max at $x = 1$; local min at $x = 2$ **d.** Inflection point
at $x = \frac{13}{8}$

CHAPTER 6

Section 6.1 Exercises, pp. 378–383

1. The position, $s(t)$, is the location of the object relative to the origin.
The displacement between time $t = a$ and $t = b$ is $s(b) - s(a)$.

The distance traveled between $t = a$ and $t = b$ is $\displaystyle\int_{a}^{b}|v(t)|\,dt$,

where $v(t)$ is the velocity at time t. **3.** The displacement between

$t = a$ and $t = b$ is $\displaystyle\int_{a}^{b} v(t)\,dt$. **5.** $Q(t) = Q(0) + \displaystyle\int_{0}^{t} Q'(x)\,dx$

7. a.

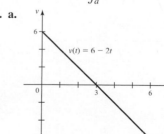

Positive direction for
$0 \le t < 3$; negative direction
for $3 < t \le 6$
b. 0 **c.** 18 m

9. a. 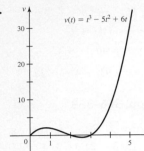 Positive direction for $0 < t < 2, 3 < t < 5$; negative direction for $2 < t < 3$

b. $\dfrac{275}{12}$ m   **c.** 23.75 m

11. a. 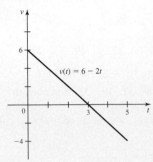 Positive direction for $0 \le t < 3$; negative direction for $3 < t \le 5$

b. $s(t) = 6t - t^2$ **c.**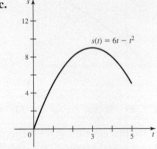

13. a. 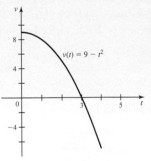 Positive direction for $0 < t < 3$; negative direction for $3 < t < 5$

b. $s(t) = 9t - \dfrac{t^3}{3} - 2$ **c.**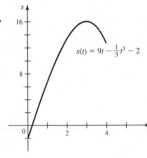

15. a. $s(t) = 2 \sin \pi t$ **b.** 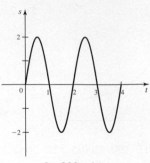   **c.** $\dfrac{3}{2}, \dfrac{7}{2}, \dfrac{11}{2}$ **d.** $\dfrac{1}{2}, \dfrac{5}{2}, \dfrac{9}{2}$

17. a. $s(t) = 10t(48 - t^2)$ **b.** 880 mi   **c.** $\dfrac{2720\sqrt{6}}{9} \approx 740.29$ mi

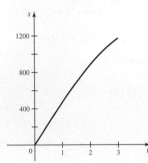

19. a. 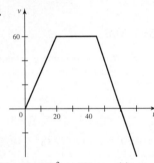 Velocity is a maximum for $20 \le t \le 45$; $v = 0$ at $t = 0$ and $t = 60$ **b.** 1200 m   **c.** 2550 m   **d.** 2100 m in the positive direction from $s(0)$

21. $s(t) = -4.9t^2 + 20t$ m; $v(t) = -9.8t + 20$ m/s

23. $s(t) = \dfrac{-0.005}{3}t^3 + 10t$ m; $v(t) = -0.005t^2 + 10$ m/s

25. a. $s(t) = 44t^2$ ft   **b.** 704 ft   **c.** $\sqrt{30} \approx 5.477$ s   **d.** $\dfrac{5\sqrt{33}}{11} \approx 2.611$ s   **e.** $\dfrac{89^2}{44} \approx 180.023$ ft

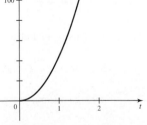

27. 6.154 mi; 1.465 mi   **29. a.** 27,250 barrels   **b.** 31,000 barrels **c.** 4000 barrels   **31. a.** ≈ 2639 people   **b.** $P(t) = 250 + 20t^{3/2} + 30t$ people   **33. a.** 1897 cells; 1900 cells   **b.** $N(t) = -400e^{-0.25t} + 1900$ cells   **35. a.** \$96,875   **b.** \$86,875   **37. a.** \$69,583.33 **b.** \$139,583.33   **39. a.** False   **b.** True   **c.** True   **d.** True

41. a. 3   **b.** $\dfrac{13}{3}$ **c.** 3   **d.** $s(t) = \begin{cases} \dfrac{-t^2}{2} + 2t, & 0 \le t \le 3 \\[2mm] \dfrac{3t^2}{2} - 10t + 18, & 3 < t \le 4 \\[2mm] -t^2 + 10t - 22, & 4 < t \le 5 \end{cases}$

43. $\frac{2}{3}$ **45.** $\frac{25}{3}$ **47. a.**

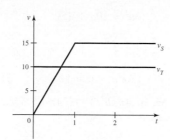

b. Theo   **c.** Sasha   **d.** Theo hits the 10-mi mark before Sasha; Sasha and Theo hit the 15-mi mark at the same time; Sasha hits the 20-mi mark before Theo.   **e.** Sasha   **f.** Theo
49. a. Abe initially runs into a headwind; Bess initially runs with a tailwind.

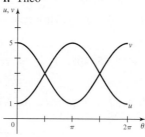

b. Both runners have an average speed of 3 mi/hr.   **c.** $\pi\sqrt{5}/25$ s.
51. a. $\frac{1}{200}$ year^{-1} **b.** $200 \ln 2 \approx 138.63$ years
53. a. $\frac{120}{\pi} + 40 \approx 78.20$ m^3 **b.** $Q(t) = 20\left(t + \frac{12}{\pi}\sin\left(\frac{\pi}{12}t\right)\right)$

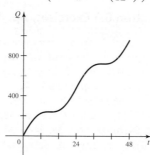

c. After ≈ 122.6 hr   **55. a.**

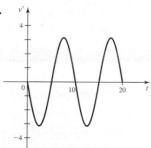

b. $V(t) = 5\cos\left(\frac{\pi t}{5}\right) + 5$

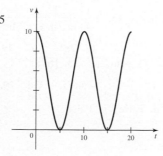

c. 6 breaths/min   **57. a.** 7200 MWh or 2.592×10^{13} J   **b.** 16,000 kg; 5,840,000 kg   **c.** 450 g; 164,250 g   **d.** About 1500 turbines

Section 6.2 Exercises, pp. 388–392

1.

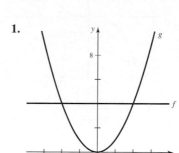

$\int_{-2}^{2}[f(x) - g(x)]\,dx$ represents the area between these curves.

3. See solution to Exercise 1.   **5.** $\frac{9}{2}$ **7.** $\frac{5}{2} - \frac{1}{\ln 2}$ **9.** $\frac{25}{2}$ **11.** $\frac{81}{32}$
13. $8\pi/3 - 2\sqrt{3}$ **15.** $2 - \sqrt{2}$ **17.** $\frac{1}{2} + \ln 2$ **19.** 1   **21.** 3
23. 48   **25.** $\frac{9}{2}$ **27. a.** $\int_{-\sqrt{2}}^{-1}(2 - x^2)\,dx + \int_{-1}^{0}(-x)\,dx$

b. $\int_{-1}^{0}(y + \sqrt{y + 2})\,dy$

29. a. $2\int_{-3}^{-2}\sqrt{x + 3}\,dx + \int_{-2}^{6}\left(\sqrt{x + 3} - \frac{x}{2}\right)dx$

b. $\int_{-1}^{3}[2y - (y^2 - 3)]\,dy$ **31. a.** $\frac{63}{4}$

b. $\frac{63}{4}$ **33.** $\frac{64}{5}$ **35.** $\ln 2$ **37.** $\frac{5}{24}$ **39. a.** False   **b.** False
c. True   **41.** $\frac{1}{6}$ **43.** $\frac{9}{2}$ **45.** $\frac{32}{3}$ **47.** $\frac{63}{4}$ **49.** $\frac{15}{8} - 2\ln 2$
51. a. Area $(R_1) = \dfrac{p - 1}{2(p + 1)}$ for all positive integers p;

area $(R_2) = \dfrac{q - 1}{2(q + 1)}$ for all positive integers q; they are the same.
b. R_1 has greater area.   **c.** R_2 has greater area.
53. $\dfrac{135 + 17\sqrt{17} - 128\sqrt{2}}{96}$ **55.** $\dfrac{81}{2}$ **57.** $\dfrac{n - 1}{2(n + 1)}$
59. $A_n = \dfrac{n - 1}{n + 1}$; $\lim\limits_{n\to\infty} A_n = 1$; the region approximates a square with side length of 1.   **61. a.** The lowest $p\%$ of households owns exactly $p\%$ of the wealth for $0 \le p \le 100$.   **b.** The function must be increasing and concave up because the poorest $p\%$ cannot own more than $p\%$ of the wealth.   **c.** $p = 1.1$ is most equitable; $p = 4$ is least equitable.
e. $G(p) = 1 - \dfrac{2}{p + 1}$ **f.** $0 \le G < 1$ for $p \ge 1$.   **g.** $\dfrac{5}{18}$ **63.** -1
65. $\frac{4}{9}$ **67. a.** $F(a) = a(b^3/6) - b^4/12$; $F(a) = 0$ if $a = b/2$
b. Since $A'(b/2) = 0$ and $A''(b/2) > 0$, A has a minimum at $a = b/2$. The maximum value of $b^4/12$ occurs if $a = 0$ or $a = b$.

69. a.

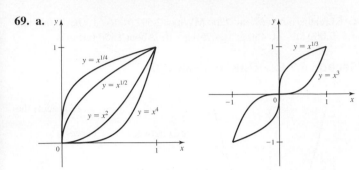

b. $A_n(x)$ is the net area of the region between the graphs of f and g from 0 to x.   **c.** $x = n^{n/(n^2-1)}$; the roots decrease with n.

Section 6.3 Exercises, pp. 399–403

1. $A(x)$ is the area of the cross section through the solid at the point x.

3. $V = \int_0^2 \pi(4x^2 - x^4)\,dx$ **5.** The cross sections are disks and $A(x)$ is the area of a disk.   **7.** 30   **9.** $\frac{1000}{3}$ **11.** $\frac{\pi}{3}$ **13.** $\frac{16\sqrt{2}}{3}$

15. 36π **17.** $15\pi/32$ **19.** $\pi^2/2$ **21.** $\pi^2/6$ **23.** $32\pi/3$

25. $\frac{753\pi}{128}$ **27.** $\frac{256\pi}{35}$ **29.** $(4\pi - \pi^2)/4$ **31.** 54π **33.** $64\pi/5$

35. $\frac{256\pi}{3}$ **37.** Volumes are equal.   **39.** x-axis   **41. a.** False   **b.** True

c. True   **43.** $\pi \ln 3$ **45.** $\frac{\pi}{2}(e^4 - 1)$ **47.** $49\pi/2$

49. Volume $S = 8\pi a^{5/2}/15$; volume $T = \pi a^{5/2}/3$

51. a.   **b.**

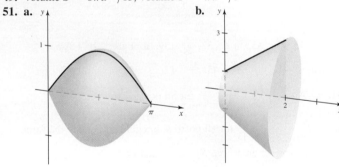

53. a. $\frac{1}{3}V_C$ **b.** $\frac{2}{3}V_C$ **55.** $24\pi^2$ **57.** $\frac{704\pi}{15}$ **59.** $\frac{192\pi}{5}$

63. b. $2/\sqrt{\pi}$ m

Section 6.4 Exercises, pp. 410–413

1. $\int_a^b 2\pi x(f(x) - g(x))\,dx$ **3.** x; y **5.** $\pi \ln 5$ **7.** π **9.** π

11. 8π **13.** $\frac{32\pi}{3}$ **15.** 90π **17.** $11\pi/6$ **19.** $23\pi/15$

21. 24π **23.** 54π **25.** $16\sqrt{2}\,\pi/3$ **27.** $4\pi/15$; shell method
29. $8\pi/27$; shell method   **31.** $\pi(\sqrt{e} - 1)^2$; shell method
33. a. True   **b.** False   **c.** True   **35.** $4\pi \ln 2$ **37.** $2\pi e(e - 1)$
39. $16\pi/3$ **41.** $608\pi/3$ **43.** $\pi/4$ **45.** $\pi/3$

47. a. $V_1 = \dfrac{\pi}{15}(3a^2 + 10a + 15)$

$V_2 = \dfrac{\pi}{2}(a + 2)$

b. $V(S_1) = V(S_2)$ for $a = 0$ and $a = -\frac{5}{6}$ **51.** $\dfrac{\pi h^2}{3}(24 - h)$

53. $24\pi^2$ **55. a.** $4\pi ab^2/3$ **b.** $4\pi a^2 b/3$ **c.** No; they will agree only in the case $a = b$.

Section 6.5 Exercises, pp. 418–420

1. Determine if f has a continuous derivative on $[a, b]$. If so, calculate $f'(x)$ and $[f'(x)]^2$. Then evaluate the integral $\int_a^b \sqrt{1 + [f'(x)]^2}\,dx$.

3. $4\sqrt{5}$ **5.** 168   **7.** $\frac{4}{3}$ **9.** $\frac{123}{32}$ **11. a.** $\int_{-1}^1 \sqrt{1 + 4x^2}\,dx$ **b.** 2.96

13. a. $\int_1^4 \sqrt{1 + \dfrac{1}{x^2}}\,dx$ **b.** 3.34   **15. a.** $\int_3^4 \sqrt{\dfrac{4x - 7}{4x - 8}}\,dx$ **b.** 1.08

17. a. $\int_0^\pi \sqrt{1 + 4\sin^2(2x)}\,dx$ **b.** 5.27   **19. a.** $\int_1^{10} \sqrt{1 + 1/x^4}\,dx$

b. 9.15   **21.** $7\sqrt{5}$ **23.** $\frac{123}{32}$ **25. a.** False   **b.** True   **c.** False
27. a. $f(x) = \pm 4x^3/3 + C$ **b.** $f(x) = \pm 3\sin(2x) + C$
29. $y = 1 - x^2$ **31.** Approximately 1326 m   **33. a.** $L/2$ **b.** L/c

Section 6.6 Exercises, pp. 428–431

1. 150 g   **3.** Work is the force times the distance moved.
5. Different volumes of water are moved different distances.
7. $39,200\ \text{N/m}^2$ **9.** $\pi + 2$ **11.** 3   **13.** $(2\sqrt{2} - 1)/3$ **15.** 10
17. 25 J   **19. a.** 112.5 J   **b.** 12.5 J   **21.** 525 J   **23.** 11,484,375 J
25. a. $66,150\pi$ J   **b.** No   **27. a.** $200,704,000\ \pi/3$ J
b. $120,422,400\ \pi$ J   **29. a.** 32,667 J   **b.** Yes   **31.** 14,700,000 N
33. 29,400,000 N   **35.** 800,000 N   **37.** 6737.5 N   **39. a.** True
b. True   **c.** True   **d.** False   **41. a.** Compared to a linear spring
$F(x) = 16x$, the restoring force is less for large displacements.
b. 17.87 J   **c.** 31.6 J   **43.** 0.28 J   **45. a.** 8.87×10^9 J
b. $500\,GMx/(R(x + R)) = (2 \times 10^{17})x/(R(x + R))$ J   **c.** GMm/R
d. $v = \sqrt{2GM/R}$ **47. a.** 2250g J   **b.** 3750g J
51. The left-hand plate   **53. a.** Yes   **b.** 4.296 m

Section 6.7 Exercises, pp. 439–440

1. $D = (0, \infty)$, $R = (-\infty, \infty)$ **3.** $\dfrac{4^x}{\ln 4} + C$

5. $e^{x \ln 3}, e^{\pi \ln x}, e^{(\sin x)(\ln x)}$ **7.** 3   **9.** $\cos(\ln x)/x$, $x \in (0, \infty)$

11. $6(1 - \ln 2)$ **13.** $-\dfrac{1}{4}\left[\dfrac{1}{\ln^2(\ln 4)} - \dfrac{1}{\ln^2(\ln 3)}\right]$

15. $\ln|e^x - e^{-x}| + C$ **17.** $2e^{\sqrt{x}} + C$ **19.** $99/(10 \ln 10)$ **21.** 3
23. $4^{?x+1} x^{4x}(1 + \ln 2x)$ **25.** $(\ln 2)\,2^{x^2+1}\,x$

27. $2(x + 1)^{2x}\left[\dfrac{x}{x + 1} + \ln(x + 1)\right]$ **29. a.** True   **b.** False
c. False   **d.** False   **e.** False

31.

h	$(1 + 2h)^{1/h}$	h	$(1 + 2h)^{1/h}$
10^{-1}	6.1917	-10^{-1}	9.3132
10^{-2}	7.2446	-10^{-2}	7.5404
10^{-3}	7.3743	-10^{-3}	7.4039
10^{-4}	7.3876	-10^{-4}	7.3905
10^{-5}	7.3889	-10^{-5}	7.3892
10^{-6}	7.3890	-10^{-6}	7.3891

$\lim_{h \to 0} (1 + 2h)^{1/h} = e^2$

33.

x	$\dfrac{2^x - 1}{x}$	x	$\dfrac{2^x - 1}{x}$
10^{-1}	0.71773	-10^{-1}	0.66967
10^{-2}	0.69556	-10^{-2}	0.69075
10^{-3}	0.69339	-10^{-3}	0.69291
10^{-4}	0.69317	-10^{-4}	0.69312
10^{-5}	0.69315	-10^{-5}	0.69314
10^{-6}	0.69315	-10^{-6}	0.69315

$\lim_{x \to 0} \dfrac{2^x - 1}{x} = \ln 2$

35. a. No **b.** No **37.** $\dfrac{\ln p}{p - 1}, 0$ **39.** $-20xe^{-10x^2}$

41. $-(1/x)^x(1 + \ln x)$ **43.** $\left[-\dfrac{4}{x + 4} + \ln\left(\dfrac{x + 4}{x}\right) \right]\left(1 + \dfrac{4}{x} \right)^x$

45. $-\sin(x^{2 \sin x}) x^{2 \sin x}\left(\dfrac{2 \sin x}{x} + 2 \cos x \ln x \right)$ **47.** $-\dfrac{1}{9^x \ln 9} + C$

49. $\dfrac{10^{x^3}}{3 \ln 10} + C$ **51.** $\dfrac{3 \cdot 3^{\ln 2} - 1}{\ln 3}$ **53.** $\dfrac{32}{3}$ **55.** $\dfrac{1}{2}(\ln 2 + 1) \approx 0.85$

59. $\ln 2 = \displaystyle\int_1^2 \dfrac{dt}{t} < L_2 = \dfrac{5}{6} < 1$

$\ln 3 = \displaystyle\int_1^3 \dfrac{dt}{t} > R_7$

$= 2\left(\dfrac{1}{9} + \dfrac{1}{11} + \dfrac{1}{13} + \dfrac{1}{15} + \dfrac{1}{17} + \dfrac{1}{19} + \dfrac{1}{21} \right) > 1$

Section 6.8 Exercises, pp. 447–449

1. The relative growth is constant. **3.** The time it takes for a function to double in value **5.** $T_2 = \ln 2/k$ **7.** Compound interest, world population **9.** $\dfrac{df}{dt} = 10.5; \dfrac{dg}{dt} \cdot \dfrac{1}{g} = \dfrac{10e^{t/10}}{100e^{t/10}} = \dfrac{1}{10}$

11. $P(t) = 90{,}000e^{0.024t}$ people with $t = 0$ in 2010; in 2039

13. \$134.39 **15. a.** 99.367 yr; $P(100) \approx 564$ million **b.** 139 yr and 77 yr; $P(100) \approx 463$ million and $P(100) \approx 688$ million

c. The projections are highly sensitive to the growth rate.

17. $H(t) = 800e^{-0.030t}$ homicides/yr with $t = 0$ in 2010; in 2019

19. 18,928 ft; 125,754 ft **21. a.** 15.87 mg **b.** after 119.59 hr $\approx$ 5 days **23.** ≈ 1.055 billion yr **25. a.** False

b. False **c.** True **d.** True **e.** True **27.** If $A(t) = A_0e^{kt}$ and $A(T) = 2A_0$, then $e^{kt} = 2$ and $T = (\ln 2)/k$. Thus the doubling time is a constant **29. a.** Bob; Abe

b. $y = 4 \ln(t + 1)$ and $y = 8 - 8e^{-t/2}$; Bob

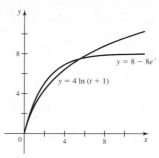

31. $\approx 10.034\%$; no **33.** ≈ 1.2643 s **35.** ≈ 1044 days **37.** \$50

39. $k = \ln(1 + r); r = 2^{(1/T_2)} - 1; T_2 = (\ln 2)/k$

Chapter 6 Review Exercises, pp. 450–452

1. a. True **b.** True **c.** True **d.** False **e.** False **f.** False

3. $s(t) = 20t - 5t^2$; displacement $(t) = 20t - 5t^2$;

$D(t) = \begin{cases} 20t - 5t^2 & 0 \le t < 2 \\ 5t^2 - 20t + 40 & 2 \le t \le 4 \end{cases}$

5. a. $v(t) = -\dfrac{8}{\pi} \cos \dfrac{\pi t}{4}$

$s(t) = -\dfrac{32}{\pi^2} \sin \dfrac{\pi t}{4}$

b. min value $= -\dfrac{32}{\pi^2}$; max value $= \dfrac{32}{\pi^2}$ **c.** 0; 0 **7. a.** $R(t) = 3t^{4/3}$

b. $R(t) = \begin{cases} 3t^{4/3} & \text{if } 0 \le t \le 8 \\ 2t + 32 & \text{if } t > 8 \end{cases}$ **c.** $t = 59$ min

9. a.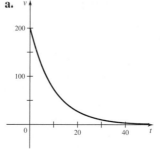

b. $10 \ln 4 \approx 13.86$ s

c. $s(t) = 2000[1 - e^{-t/10}]$ **d.** No

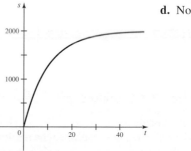

11. a. $s_{\text{Tom}}(t) = -10e^{-2t} + 10$
$s_{\text{Sue}}(t) = -15e^{-t} + 15$

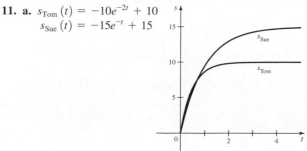

b. $t = 0$ and $t = \ln 2$ **c.** Sue   **13.** $\dfrac{21\pi}{4}$ **15.** 8   **17.** 1   **19.** 1/3

21. 16   **23.** $\dfrac{8\pi}{5}$ **25.** $\dfrac{\pi r^2 h}{3}$ **27. a.** V_y **b.** V_y

c. $V_x = \begin{cases} \pi\left(\dfrac{a^{1-2p} - 1}{1 - 2p}\right) & \text{if } p \neq 1/2 \\ \pi \ln a & \text{if } p = 1/2 \end{cases}$

d. $V_y = \begin{cases} 2\pi\left(\dfrac{a^{2-p} - 1}{2 - p}\right) & \text{if } p \neq 2 \\ 2\pi \ln a & \text{if } p = 2 \end{cases}$

29. $2\sqrt{3} - \dfrac{4}{3}$ **31.** $\sqrt{b^2 + 1} - \sqrt{2} + \ln\left(\dfrac{(\sqrt{b^2 + 1} - 1)(1 + \sqrt{2})}{b}\right)$;

$b \approx 2.715$ **33.** $450 - \dfrac{450}{e}\,g$ **35.** 56.25   **37.** $5.2 \times 10^7\,N$

39. $\ln 4$ **41.** $\dfrac{1}{2}\ln(x^2 + 8x + 25) + C$ **43.** 48.37 yr

45. Local max at $x = -\dfrac{1}{2}(\sqrt{5} + 1)$;
local min at $x = \dfrac{1}{2}(\sqrt{5} - 1)$;
inflection points at $x = -3$ and $x = 0$;
$\displaystyle \lim_{x \to -\infty} f(x) = 0$; $\displaystyle \lim_{x \to \infty} f(x) = \infty$.

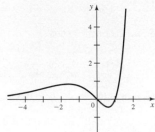

47. a.

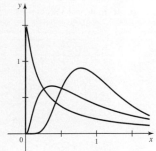

b. $\displaystyle\lim_{x \to 0} f(x) = 0$ **c.** $f'(x^*) = 0$

d. $f(x^*) = \dfrac{1}{\sqrt{2\pi}}\dfrac{e^{\sigma^2/2}}{\sigma}$ **e.** $\sigma = 1$

CHAPTER 7

Section 7.1 Exercises, pp. 458–460

1. The Product Rule   **3.** $u = x^n$ **5.** Those for which the choice for dv is easily integrated and when the resulting new integral is no more difficult than the original   **7.** $x \sin x + \cos x + C$ **9.** $te^t - e^t + C$

11. $\left(\dfrac{2x^2 - 1}{4}\right)\cos 2x + \dfrac{x}{2}\sin 2x + C$

13. $\dfrac{e^{4x}}{32}(8x^2 - 4x + 1) + C$ **15.** $\dfrac{x^3}{9}(3\ln x - 1) + C$

17. $-\dfrac{1}{9x^9}\left(\ln x + \dfrac{1}{9}\right) + C$ **19.** $x\tan^{-1} x - \dfrac{1}{2}\ln(x^2 + 1) + C$

21. $\dfrac{1}{8}\sin 2x - \dfrac{x}{4}\cos 2x + C$ **23.** $\dfrac{e^x}{2}(\sin x + \cos x) + C$

25. $-\dfrac{e^{-x}}{17}(\sin 4x + 4\cos 4x) + C$

27. $-e^{-t}(t^3 + 3t^2 + 6t + 6) + C$ **29.** π **31.** $-\dfrac{1}{2}$

33. $\dfrac{1}{9}(5e^6 + 1)$ **35.** $\left(\dfrac{2\sqrt{3} - 1}{12}\right)\pi + \dfrac{1 - \sqrt{3}}{2}$ **37.** $\pi(1 - \ln 2)$

39. $\dfrac{2\pi}{27}(13e^6 - 1)$ **41. a.** False   **b.** True   **43.** Let $u = x^n$
and $dv = \cos(ax)\,dx$.   **45.** Let $u = \ln^n x$ and $dv = dx$.

47. $\dfrac{x^2 \sin(5x)}{5} + \dfrac{2x\cos(5x)}{25} - \dfrac{2\sin(5x)}{125} + C$

49. $x\ln^4 x - 4x\ln^3 x + 12x\ln^2 x - 24x\ln x + 24x + C$

51. $(\tan x + 2)\ln(\tan x + 2) - \tan x + C$

53. $\displaystyle\int \log_b x\,dx = \int \dfrac{\ln x}{\ln b}\,dx = \dfrac{1}{\ln b}(x\ln x - x) + C$

55. $2\sqrt{x}\sin\sqrt{x} + 2\cos\sqrt{x} + C$ **57.** $2e^3$ **59.** $\pi(\pi - 2)$

61. x-axis $\dfrac{\pi^2}{2}$; y-axis $2\pi^2$ **63. a.** Let $u = x$ and $v = f(x)$.

b. $\dfrac{e^{3x}}{9}(3x - 1) + C$ **65.** Use $u = \sec x$ and $dv = \sec^2 x\,dx$.

67. a.

$t = k\pi$ for $k = 0, 1, 2, \ldots$

b. $\dfrac{e^{-\pi} + 1}{2\pi}$

c. $(-1)^n\left(\dfrac{e^{\pi} + 1}{2\pi e^{(n+1)\pi}}\right)$

d. $a_n = a_{n-1}\cdot\dfrac{1}{e^\pi}$

69. $\displaystyle\int_a^b u\,dv + \int_a^b v\,du = A + B = f(b)\,g(b) - f(a)\,g(a) = uv\Big]_a^b$

71. a. $I_1 = -\dfrac{1}{2}e^{-x^2} + C$ **b.** $I_3 = -\dfrac{1}{2}e^{-x^2}(x^2 + 1) + C$
c. $I_5 = -\dfrac{1}{2}e^{-x^2}(x^4 + 2x^2 + 2) + C$
d. $I_{2n+1} = -\dfrac{1}{2}e^{-x^2}x^{2n} + n\,I_{2n-1}$

Section 7.2 Exercises, pp. 466–467

1. $\sin^2 x = \dfrac{1}{2}(1 - \cos 2x); \cos^2 x = \dfrac{1}{2}(1 + \cos 2x)$ **3.** Rewrite $\sin^3 x$ as $(1 - \cos^2 x)\sin x$.   **5.** A reduction formula expresses an integral with a power in the integrand in terms of another integral with a smaller power in the integrand.   **7.** Let $u = \tan x$.

9. $\dfrac{x}{2} - \dfrac{1}{4}\sin 2x + C$ **11.** $-\cos x + \dfrac{2}{3}\cos^3 x - \dfrac{\cos^5 x}{5} + C$

13. $\dfrac{1}{8}x - \dfrac{1}{32}\sin 4x + C$ **15.** $\sec x + 2\cos x - \dfrac{\cos^3 x}{3} + C$

17. $\dfrac{\sin^3 x \cos^3 x}{6} + \dfrac{1}{16}x - \dfrac{1}{64}\sin 4x + C$ **19.** $\tan x - x + C$

21. $\dfrac{1}{8}\tan^2 4x + \dfrac{1}{4}\ln|\cos 4x| + C$

23. $4\tan^5 x - \dfrac{20}{3}\tan^3 x + 20\tan x - 20x + C$

25. $\dfrac{2}{3}\tan^{3/2} x + C$ **27.** $\tan x - \cot x + C$ **29.** $\dfrac{4}{3}$ **31.** $\dfrac{4}{3} - \ln\sqrt{3}$

33. a. True   **b.** False   **37.** $\dfrac{1}{a}\ln|\sec ax| + C,$

$\dfrac{1}{a}\ln|\sec ax + \tan ax| + C$ **39.** $\dfrac{1}{2}\ln\left(\sqrt{2} + \dfrac{3}{2}\right)$

41. $\frac{1}{3}\tan(\ln\theta)\sec^2(\ln\theta) + \frac{2}{3}\tan(\ln\theta) + C$ **43.** $\ln 4$ **45.** $8\sqrt{2}/3$

47. $\sqrt{2}$ **49.** $2\sqrt{2}/3$ **51.** $\ln(\sqrt{2}+1)$ **53.** $\frac{1}{2} - \ln\sqrt{2}$

55. $\dfrac{\cos 4x}{8} - \dfrac{\cos 10x}{20} + C$ **57.** $\dfrac{\sin x}{2} - \dfrac{\sin 5x}{10} + C$

61. $\displaystyle\int_0^\pi \sin^2 nx\,dx = \int_0^\pi \cos^2 nx\,dx = \pi/2 \quad n = 1, 2, 3, \ldots$

$\displaystyle\int_0^\pi \sin^4 nx\,dx = \frac{3\pi}{8}, \quad n = 1, 2, 3, \ldots$

Section 7.3 Exercises, pp. 473–476

1. $x = 3\sec\theta$ **3.** $x = 10\sin\theta$ **5.** $\sqrt{4-x^2}/x$ **7.** $\pi/6$

9. $25(2\pi/3 - \sqrt{3}/2)$ **11.** $\sin^{-1}(x/4) + C$

13. $3\ln\left(\dfrac{\sqrt{9-x^2}-3}{x}\right) + \sqrt{9-x^2} + C$

15. $\dfrac{x}{2}\sqrt{64-x^2} + 32\sin^{-1}\left(\dfrac{x}{8}\right) + C$ **17.** $\sin^{-1}(x/6) + C$

19. $\ln\left(\sqrt{x^2-81}+x\right) + C$ **21.** $x/\sqrt{1+4x^2} + C$

23. $8\sin^{-1}(x/4) - x\sqrt{16-x^2}/2 + C$

25. $\sqrt{x^2-9} - 3\sec^{-1}(x/3) + C$

27. $\dfrac{x}{2}\sqrt{4+x^2} - 2\ln\left(x + \sqrt{4+x^2}\right) + C$

29. $\sin^{-1}\left(\dfrac{x+1}{2}\right) + C$ **31.** $\dfrac{9}{10}\cos^{-1}\left(\dfrac{5}{3x}\right) - \dfrac{45\sqrt{9x^2-25}}{90x^2} + C$

33. $\dfrac{1}{10}\left[\tan^{-1}\left(\dfrac{x}{5}\right) - \dfrac{5x}{25+x^2}\right] + C$

35. $x/\sqrt{100-x^2} - \sin^{-1}(x/10) + C$

37. $81/(2(81-x^2)) + \ln\left(\sqrt{81-x^2}\right) + C$

39. $-1/\sqrt{x^2-1} - \sec^{-1}(x) + C$ **41.** $\ln\left(\dfrac{1+\sqrt{17}}{4}\right)$

43. $\sqrt{2}/6$ **45.** $\frac{1}{16}[1 - \sqrt{3} - \ln(21 - 12\sqrt{3})]$ **47. a.** False

b. True **c.** False **d.** False **49.** $\dfrac{1}{3}\tan^{-1}\left(\dfrac{x+3}{3}\right) + C$

51. $\left(\dfrac{x-1}{2}\right)\sqrt{x^2-2x+10}$

$\qquad - \dfrac{9}{2}\ln\left(x - 1 + \sqrt{x^2-2x+10}\right) + C$

53. $\dfrac{x-4}{\sqrt{9+8x-x^2}} - \sin^{-1}\left(\dfrac{x-4}{5}\right) + C$ **55.** $\dfrac{\pi\sqrt{2}}{48}$

57. a. $A_{\text{seg}} = A_{\text{sector}} - A_{\text{triangle}} = \dfrac{\theta r^2}{2} - \dfrac{r^2\sin\theta}{2} = \dfrac{r^2}{2}(\theta - \sin\theta)$

59. a. $\ln 3$ **b.** $\dfrac{\pi}{3}\tan^{-1}\left(\dfrac{4}{3}\right)$ **c.** 4π

61. $\dfrac{1}{4a}\left[20a\sqrt{1+400a^2} + \ln\left(20a + \sqrt{1+400a^2}\right)\right]$

63.

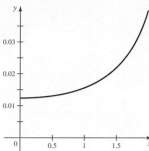

$\dfrac{1}{81} + \dfrac{\ln 3}{108}$

65.

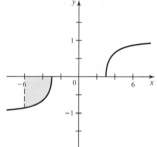

$25\left(\sqrt{3} - \ln\sqrt{2+\sqrt{3}}\right)$

67. $\ln[(2+\sqrt{3})(\sqrt{2}-1)]$ **69.** $192\pi^2$

71. b. $\displaystyle\lim_{L\to\infty}\dfrac{kQ}{a\sqrt{a^2+L^2}} = \lim_{L\to\infty} 2\rho k\dfrac{1}{a\sqrt{\left(\dfrac{a}{L}\right)^2+1}} = \dfrac{2\rho k}{a}$

73. a. $\dfrac{1}{\sqrt{g}}\left[\dfrac{\pi}{2} - \sin^{-1}\left(\dfrac{2\cos b - \cos a + 1}{\cos a + 1}\right)\right]$ **b.** For $b = \pi$,

the descent time is $\dfrac{\pi}{\sqrt{g}}$, a constant.

77.

$\pi - 3\sqrt{3}$

Section 7.4 Exercises, pp. 483–485

1. Rational functions **3. a.** $\dfrac{A}{x-3}$ **b.** $\dfrac{A_1}{(x-4)}, \dfrac{A_2}{(x-4)^2}, \dfrac{A_3}{(x-4)^3}$

c. $\dfrac{Ax+B}{x^2+2x+6}$ **5.** $\dfrac{\frac{1}{3}}{x-4} + \dfrac{-\frac{1}{3}}{x+2}$ **7.** $\dfrac{\frac{1}{2}}{x-4} + \dfrac{\frac{1}{2}}{x+4}$

9. $\ln\left|\dfrac{x-1}{x+2}\right|^{1/3} + C$ **11.** $\ln\left|\dfrac{x-1}{x+1}\right|^{3/2} + C$ **13.** $\ln\left|\dfrac{x-3}{x+2}\right|^{2/5} + C$

15. $\ln\left|\dfrac{x-6}{x+4}\right|^{1/10} + C$ **17.** $\ln\left|\dfrac{(x-3)^{1/3}(x+1)}{(x+3)^{1/3}(x-1)}\right|^{1/16} + C$

19. $\ln\left|\dfrac{x-9}{x}\right|^{1/27} + \dfrac{1}{3x} + C$ **21.** $\ln|x+3| + \dfrac{3}{x+3} + C$

23. $-\dfrac{2}{x} + \ln\left|\dfrac{x+1}{x}\right|^2 + C$ **25.** $\dfrac{5}{x} + \ln\left|\dfrac{x}{x+1}\right|^6 + C$

27. $\dfrac{A}{x-1} + \dfrac{B}{(x-1)^2} + \dfrac{Cx+D}{x^2+1}$

29. $\dfrac{A}{x-4} + \dfrac{B}{(x-4)^2} + \dfrac{Cx+D}{x^2+3x+4}$

31. $\ln\left(\dfrac{|x-4|}{\sqrt{x^2+2x+6}}\right)^{1/15} - \dfrac{\sqrt{5}}{15}\tan^{-1}\left(\dfrac{x+1}{\sqrt{5}}\right) + C$

33. $\ln\left|(x-1)(x^2+4x+5)^{9/2}\right|^{1/10} - \dfrac{13}{10}\tan^{-1}(x+2) + C$

35. $\ln\left|(x-1)^{1/5}(x^2+4)^{2/5}\right| + \dfrac{2}{5}\tan^{-1}\left(\dfrac{x}{2}\right) + C$

37. a. False **b.** False **c.** False **d.** True **39.** $\ln 6$

41. $4\sqrt{2} + \ln(17 - 12\sqrt{2})^{1/3}$ **43.** $\left(\dfrac{24}{5} - 2\ln 5\right)\pi$

45. $\dfrac{2}{3}\pi \ln 2$ **47.** $2\pi\left(3 + \ln\frac{2}{5}\right)$ **49.** $x - \ln(1 + e^x) + C$

51. $3x + \ln\dfrac{(x-2)^{14}}{|x-1|} + C$ **53.** $\ln\sqrt{2e^x + 1} + C$

55. $\dfrac{1}{2}\left(\sec x \tan x - \sec^2 x + \ln|\sec x + \tan x|\right) + C$

57. $\ln\left|\dfrac{e^x-1}{e^x+2}\right|^{1/3} + C$ **59.** $-\dfrac{1}{2(e^{2x}+1)} + C$

61. $\dfrac{4}{3}(x+2)^{3/4} - 2(x+2)^{1/2} + 4(x+2)^{1/4} - \ln\left[(x+2)^{1/4} + 1\right]^4 + C$

63. $2\sqrt{x} - 3\sqrt[3]{x} + 6\sqrt[6]{x} - \ln(\sqrt[6]{x}+1)^6 + C$

65. $\dfrac{4}{3}\sqrt{1 + \sqrt{x}}\left(\sqrt{x} - 2\right) + C$ **67.** $\ln\left(\dfrac{x^2}{x^2+1}\right) + \dfrac{1}{x^2+1} + C$

69. $\dfrac{1}{50}\left[\dfrac{5(3x+4)}{x^2+2x+2} + 11\tan^{-1}(1+x) + \ln\left|\dfrac{(x-1)^2}{x^2+2x+2}\right|\right] + C$

71. $\ln\sqrt{\left|\dfrac{x-1}{x+1}\right|}$ **73.** $\tan x - \sec x + C$ **75.** $-\cot x - \csc x + C$

77. $\dfrac{\sqrt{2}}{2}\ln\left(\dfrac{\sqrt{2}+1+\tan(\theta/2)}{\sqrt{2}-1-\tan(\theta/2)}\right) + C$ **79. a.** Car A **b.** Car C

c. $S_A(t) = 88t - 88\ln|t+1|;$

$S_B(t) = 88\left[t - \ln(t+1)^2 - \dfrac{1}{t+1} + 1\right];$

$S_C(t) = 88(t - \tan^{-1}t)$

d. Car C **81.** Since $\dfrac{x^4(1-x)^4}{1+x^2} > 0$ on $(0,1);$

$\displaystyle\int_0^1 \dfrac{x^4(1-x^4)}{1+x^2}\,dx > 0;$ thus, $\dfrac{22}{7} > \pi$.

Section 7.5 Exercises, pp. 489–491

1. Substitutions, integration by parts, partial fractions **3.** The CAS may not include the constant of integration and it may use a trigonometric identity or other algebraic simplification.

5. $\ln(x + \sqrt{16 + x^2}) + C$ **7.** $\dfrac{3}{4}(7 + 2u - 7\ln|7 + 2u|) + C$

9. $-\dfrac{1}{4}\cot 2x + C$ **11.** $\dfrac{\sqrt{4x+1}}{2} + C$

13. $\dfrac{1}{3}\ln\left|x + \sqrt{x^2 - \left(\frac{10}{3}\right)^2}\right| + C$ **15.** $\dfrac{x}{16\sqrt{16+9x^2}} + C$

17. $-\dfrac{1}{12}\ln\left|\dfrac{12 + \sqrt{144 - x^2}}{x}\right| + C$

19. $\ln x - \dfrac{1}{10}\ln(x^{10} + 1) + C$

21. $\dfrac{1}{3}\tan^{-1}\left(\dfrac{x+1}{3}\right) + C$ **23.** $2\ln\left(\sqrt{x-6} + \sqrt{x}\right) + C$

25. $\ln(e^x + \sqrt{4 + e^{2x}}) + C$ **27.** $-\dfrac{1}{2}\ln\left|\dfrac{2 + \sin x}{\sin x}\right| + C$

29. $-\dfrac{\tan^{-1}(x^3)}{3x^3} + \ln\left|\dfrac{x}{(x^6+1)^{1/6}}\right| + C$

31. $\dfrac{2(\ln x)^2 - 1}{4}\sin^{-1}(\ln x) + \dfrac{\ln x\sqrt{1 - (\ln x)^2}}{4} + C$

33. $4\sqrt{17} + \ln(4 + \sqrt{17})$ **35.** $\sqrt{5} - \sqrt{2} + \ln\left(\dfrac{2 + 2\sqrt{2}}{1 + \sqrt{5}}\right)$

37. $\dfrac{128\pi}{3}$ **39.** $\dfrac{\pi^2}{4}$ **41.** $\dfrac{(x-3)\sqrt{3+2x}}{3} + C$

43. $\dfrac{1}{3}\tan(3x) - x + C$

45. $\dfrac{(x^2-a^2)^{3/2}}{3} - a^2\sqrt{x^2 - a^2} + a^3\cos^{-1}\left(\dfrac{a}{x}\right) + C$

47. $-\dfrac{x}{8}(2x^2 - 5a^2)\sqrt{a^2 - x^2} + \dfrac{3a^4}{8}\sin^{-1}\left(\dfrac{x}{a}\right) + C$

49. $\dfrac{\left(\frac{4}{5}\right)^9 - \left(\frac{2}{3}\right)^9}{9}$ **51.** $\dfrac{1540 + 243\ln 3}{8}$ **53.** $\dfrac{\pi}{4}$ **55.** $2 - \dfrac{\pi^2}{12} - \ln 4$

57. a. True **b.** True **61.** $\dfrac{1}{8}e^{2x}(4x^3 - 6x^2 + 6x - 3) + C$

63. $\dfrac{\tan^3(3y)}{9} - \dfrac{\tan(3y)}{3} + y + C$

65. $\dfrac{1}{16}\left[(8x^2 - 1)\sin^{-1}(2x) + 2x\sqrt{1 - 4x^2}\right] + C$

67. $-\dfrac{\tan^{-1}x}{x} + \ln\left(\dfrac{|x|}{\sqrt{x^2+1}}\right) + C$ **69.** $\sin^{-1}\left(\dfrac{x-a}{a}\right) + C$

71. a.

θ_0	T
0.10	6.27927
0.20	6.26762
0.30	6.24854
0.40	6.22253
0.50	6.19021
0.60	6.15236
0.70	6.10979
0.80	6.06338
0.90	6.01399
1.00	5.96247

b. All are within 10%.

73. $\dfrac{1}{a^2}\left[ax - b\ln|b + ax|\right] + C$

75. $\dfrac{1}{a^2}\left[\dfrac{(ax+b)^{n+2}}{n+2} - \dfrac{b(ax+b)^{n+1}}{n+1}\right] + C$ **77. b.** $\dfrac{63\pi}{512}$

c. Decrease

Section 7.6 Exercises, pp. 499–501

1. $\dfrac{1}{2}$ **3.** The Trapezoid Rule approximates areas under curves using trapezoids. **5.** $-1, 1, 3, 5, 7, 9$ **7.** $1.59 \times 10^{-3}; 5.04 \times 10^{-4}$
9. $1.72 \times 10^{-3}; 6.32 \times 10^{-4}$ **11.** 576; 640; 656 **13.** 0.643950551

15. 704; 672; 664   **17.** 0.622   **19.** $M(25) = 0.63703884$, $T(25) = 0.63578179$; 6.58×10^{-4}, 1.32×10^{-3}

21.

n	$M(n)$	$T(n)$	Abs. Error $M(n)$	Abs. Error $T(n)$
4	99	102	1.00	2.00
8	99.75	100.5	0.250	0.500
16	99.9375	100.125	0.0625	0.125
32	99.984375	100.03125	0.0156	0.0313

23.

n	$M(n)$	$T(n)$	Abs. Error $M(n)$	Abs. Error $T(n)$
4	1.50968181	1.48067370	9.68×10^{-3}	1.93×10^{-2}
8	1.50241228	1.49517776	2.41×10^{-3}	4.82×10^{-3}
16	1.50060256	1.49879502	6.03×10^{-4}	1.20×10^{-3}
32	1.50015061	1.49969879	1.51×10^{-4}	3.01×10^{-4}

25.

n	$M(n)$	$T(n)$	Abs. Error $M(n)$	Abs. Error $T(n)$
4	-1.96×10^{-16}	0	1.96×10^{-16}	0
8	7.63×10^{-17}	-1.41×10^{-16}	7.63×10^{-17}	1.42×10^{-16}
16	1.61×10^{-16}	1.09×10^{-17}	1.61×10^{-16}	1.09×10^{-17}
32	6.27×10^{-17}	-4.77×10^{-17}	6.27×10^{-17}	4.77×10^{-17}

27. $\dfrac{164}{3} \approx 54.7$ **29.** $\dfrac{421}{12} \approx 35.1$

31. a. $T(25) = 3.19623162$
$T(50) = 3.19495398$
b. $S(50) = 3.19452809$
c. $e_T(50) = 4.26 \times 10^{-4}$
$e_S(50) = 4.05 \times 10^{-8}$

33. a. $T(50) = 1.00008509$
$T(100) = 1.00002127$
b. $S(100) = 1.00000000$
c. $e_T(100) = 2.13 \times 10^{-5}$
$e_S(100) = 4.57 \times 10^{-9}$

35.

n	$T(n)$	$S(n)$	Error $T(n)$	Error $S(n)$
4	1820.0000	—	284	—
8	1607.7500	1537.0000	71.8	1
16	1553.9844	1536.0625	18.0	6.25×10^{-2}
32	1540.4990	1536.0039	4.50	3.90×10^{-3}

37.

n	$T(n)$	$S(n)$	Error $T(n)$	Error $S(n)$
4	0.46911538	—	5.25×10^{-2}	—
8	0.50826998	0.52132152	1.33×10^{-2}	2.85×10^{-4}
16	0.51825968	0.52158957	3.35×10^{-3}	1.74×10^{-5}
32	0.52076933	0.52160588	8.38×10^{-4}	1.08×10^{-6}

39. a. True   **b.** False   **c.** True

41.

n	$M(n)$	$T(n)$	Abs. Error $M(n)$	Abs. Error $T(n)$
4	0.40635058	0.40634782	1.38×10^{-6}	1.38×10^{-6}
8	0.40634920	0.40634920	7.6×10^{-10}	7.62×10^{-10}
16	0.40634920	0.40634920	6.55×10^{-13}	6.56×10^{-13}
32	0.40634920	0.40634920	8.88×10^{-16}	7.77×10^{-16}

43.

n	$M(n)$	$T(n)$	Abs. Error $M(n)$	Abs. Error $T(n)$
4	4.72531819	4.72507878	0.00012	0.00012
8	4.72519850	4.72519849	9.12×10^{-9}	9.12×10^{-9}
16	4.72519850	4.72519850	0.	8.88×10^{-16}
32	4.72519850	4.72519850	0.	8.88×10^{-16}

49. Approximations will vary; exact value is 38.753792 . . . .
51. Approximations will vary; exact value is 68.26894921 . . . .
53. a. Approximately 1.6×10^{11} barrels   **b.** Approximately 6.8×10^{10} barrels   **55. a.** $T(40) = 0.874799972 \ldots$
b. $f''(x) = e^x \cos(e^x) - e^{2x} \sin(e^x)$ **d.** $E_T \leq \dfrac{1}{3200}$
59. Overestimate

Section 7.7 Exercises, pp. 510–513

1. The interval of integration is infinite or the integrand is unbounded on the interval of integration.   **3.** $\displaystyle\int_0^1 \frac{1}{\sqrt{x}}\,dx = \lim_{b \to 0^+} \int_b^1 \frac{1}{\sqrt{x}}\,dx$

5. 1   **7.** Diverges   **9.** $\frac{1}{2}$ **11.** $\dfrac{1}{(p-1)\,2^{p-1}}$ **13.** $\frac{1}{2}$ **15.** $1/\pi$

17. Diverges   **19.** Diverges   **21.** $\dfrac{\pi}{3}$ **23.** $3\pi/2$ **25.** $\pi/(\ln 2)$

27. 6   **29.** Diverges   **31.** $4 \cdot 10^{3/4}/3$ **33.** -2 **35.** π **37.** 2π
39. $\dfrac{72 \cdot 2^{1/3}\,\pi}{5}$ **41.** 48   **43.** 0.76   **45.** 10 mi   **47. a.** True
b. False   **c.** False   **d.** True   **e.** True   **49. a.** 2;   **b.** 0
51. $\displaystyle\int_0^\infty e^{-x^2}\,dx \approx 0.886227$ **53.** $-\frac{1}{4}$
55. $\displaystyle\int_0^\infty xe^{-x^2}\,dx = \frac{1}{2}$; $\displaystyle\int_0^\infty x^2 e^{-x^2}\,dx = \sqrt{\pi}/4 \approx 0.443$
57. $1/b - 1/a$ **59. a.** $A(a, b) = \dfrac{e^{-ab}}{a}$ for $a > 0$
b. $b = g(a) = \dfrac{-1}{a}\ln(2a)$ **c.** $b^* = -2/e$ **61. a.** $p < \frac{1}{2}$ **b.** $p < 2$
67. $41,666.67   **71.** 20,000 hr   **73. a.** $6.28 \times 10^7 m$ J
b. 11.2 km/s   **c.** ≤ 9 mm   **75. a.**

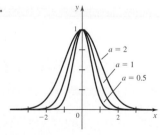

b. $\sqrt{2\pi}$, $\sqrt{\pi}$, $\sqrt{\pi/2}$ **c.** $e^{(b^2-4ac)/(4a)}\sqrt{\pi/a}$ **81. a.** π **b.** $\pi/(4e^2)$
83. $p > 1$

Section 7.8 Exercises, pp. 520–523

1. Second order   **3.** Two constants   **5.** A separable equation can be written in the form $g(y)\,y'(t) = h(t)$.   **7.** Integrate both sides with respect to t and convert the integral on the left side to an integral with respect to y.   **9.** $y = t^3 - 2t^2 + 10t + 20$ **11.** $y = t^2 + 4\ln t + 1$

13. $y = Ce^{3t} + \frac{4}{3}$ **15.** $y = Ce^{-2x} - 2$ **17.** $y = 7e^{3t} + 2$
19. $y = 2e^{-2t} - 2$ **21. a.** $y = 150(1 - e^{-0.02t})$ **b.** 150 mg
c. $t = \dfrac{\ln 10}{0.02}$ hr ≈ 115 hr

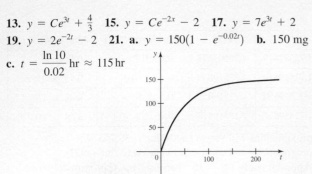

23. $y = \pm\sqrt{2t^3 + C}$ **25.** $y = -2\ln\left(\frac{1}{2}\cos t + C\right)$
27. Not separable   **29.** $y = \sqrt{e^t - 1}$ **31.** $y = \ln(e^x + 2)$
33. a. $P = \dfrac{200}{3e^{-0.08t} + 1}$ **b.** 200

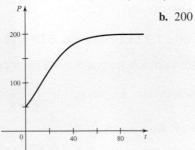

35.

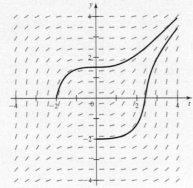

37. A–c, B–b, C–d, D–a   **39.**

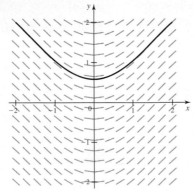

41. a. False   **b.** False   **c.** False   **d.** True
43. a. $y = 0$

b.

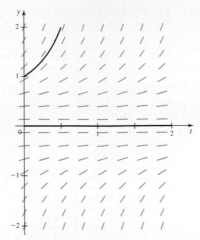

45. a. Equilibrium solutions $y = 0$ and $y = 3$
b.

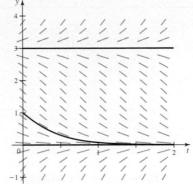

47. a. Equilibrium points $y = 0$, $y = 3$, and $y = -2$
b.

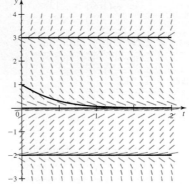

49. $p = 4e^{1 - 1/t} - 1$ **51.** $w = \tan^{-1}(t^2 + 1)$
53. a. $y = \dfrac{y_0}{(1 - y_0)e^{-kt} + y_0}$ **b.**

c. For any $0 < y_0 < 1$, $\lim\limits_{t \to \infty} y(t) = 1$. Eventually everyone knows the rumor.   **55. b.** $v = \dfrac{mg}{R}$ **c.** $v = \dfrac{g}{b}(1 - e^{-bt})$

d.

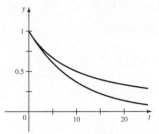

57. a. General solution $y = Ce^{-kt}$ **b.** $y = \dfrac{1}{kt + 1/y_0}$

c.

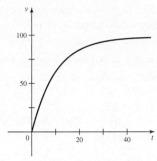

59. a. $B = 20,000 - 5000e^{0.05t}$; balance decreases
b. $m = \$2500$; constant balance $= \$50,000$

Chapter 7 Review Exercises, pp. 523–525

1. a. True   **b.** False   **c.** False   **3.** $\dfrac{e^x}{2}(\sin x - \cos x) + C$

5. $\theta/2 + (1/16)\sin(8\theta) + C$ **7.** $(\sec^5 z)/5 + C$

9. $(256 - 147\sqrt{3})/480$ **11.** $\sin^{-1}(x/2) + C$

13. $-\dfrac{1}{9y}\sqrt{9 - y^2} + C$ **15.** $\pi/9$ **17.** $\dfrac{1}{8}\ln\left|\dfrac{x - 5}{x + 3}\right| + C$

19. $(1/4)\ln 2 + \pi/8$ **21.** $\dfrac{\sqrt{6}}{3}\tan^{-1}\left(\sqrt{\dfrac{2x - 3}{3}}\right) + C$

23. 1.196288 **25. a.** $T(6) = 9.125$, $M(6) = 8.9375$
b. $T(12) = 9.03125$, $M(12) = 8.984375$ **27.** 0.4054651 **29.** 1

31. $\pi/2$ **33.** $\dfrac{1}{3}\ln\left|\dfrac{x - 2}{x + 1}\right| + C$ **35.** $2(x - 2\ln|x + 2|) + C$

37. $e^{2t}/2\sqrt{1 + e^{4t}} + C$ **39.** $\pi(e - 2)$ **41.** $\dfrac{\pi}{2}(e^2 - 3)$

43. y-axis   **45. a.** 1.603 **b.** 1.870 **c.** $b \ln b - b = a \ln a - a$
d. Decreasing   **47.** $20/(3\pi)$ **49.** 1901 cars

51. a. $I(p) = \dfrac{1}{(p - 1)^2}(1 - pe^{1-p})$ if $p \neq 1$, $I(1) = \dfrac{1}{2}$ **b.** $0, \infty$

c. $I(0) = 1$ **53.** $n = 2$ **55.** $y = 10e^{2t} - 2$
57. $y = \sqrt{t + \ln t} + 15$ **59.** $\pi/2$

61.

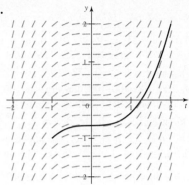

63. $t = \dfrac{-s - 5\ln s + 50 + 5\ln 50}{10}$

$\lim\limits_{t \to \infty} s(t) = 0$

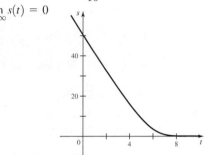

65. a. $V_1(a) = \pi[a \ln^2 a - 2a \ln a + 2(a - 1)]$

b. $V_2(a) = \dfrac{\pi}{2}(2a^2 \ln a - a^2 + 1)$

c. $V_2(a) > V_1(a)$ for all $a > 1$

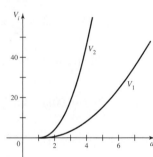

67. $a = \ln 2/(2b)$

CHAPTER 8

Section 8.1 Exercises, pp. 534–537

1. A sequence is an ordered list of numbers. Example: $1, \dfrac{1}{3}, \dfrac{1}{9}, \dfrac{1}{27}, \ldots$

3. $1, 1, 2, 6, 24$ **5.** Given a sequence $\{a_1, a_2, \ldots\}$, an infinite series is

the sum $a_1 + a_2 + a_3 + \ldots$. Example: $\displaystyle\sum_{k=1}^{\infty} \dfrac{1}{k^2}$ **7.** $1, 5, 14, 30$ **9.** $\dfrac{1}{10}, \dfrac{1}{100}$,

$\dfrac{1}{1000}, \dfrac{1}{10,000}$ **11.** $2, 1, 0, 1$ **13.** $10, 18, 42, 114$ **15.** $0, 2, 15, 679$

17. a. $1/32, 1/64$ **b.** $a_1 = 1$, $a_{n+1} = \dfrac{1}{2}a_n$ for $n \geq 1$

c. $a_n = \dfrac{1}{2^{n-1}}$ for $n \geq 1$ **19. a.** $32, 64$ **b.** $a_1 = 1$, $a_{n+1} = 2a_n$ for $n \geq 1$ **c.** $a_n = 2^{n-1}$ for $n \geq 1$ **21. a.** $243, 729$ **b.** $a_1 = 1$, $a_{n+1} = 3a_n$ for $n \geq 1$ **c.** $a_n = 3^{n-1}$ for $n \geq 1$ **23.** $9, 99, 999$, 9999; diverges   **25.** $-1, \frac{1}{2}, -\frac{1}{3}, \frac{1}{4}$; converges to 0   **27.** $10^{-1}, 10^{-3}$, $10^{-7}, 10^{-15}$; converges to 0   **29.** $100, 100, 100, 100$; converges to 100   **31. a.** $1, 2, 3, 4$ **b.** The limit does not exist.   **33. a.** $0, 2, 6, 12$ **b.** The limit does not exist.   **35. a.** $\frac{1}{1}, \frac{1}{2}, \frac{3}{5}, \frac{2}{3}$ **b.** 1   **37. a.** $\frac{5}{2}, \frac{9}{4}, \frac{17}{8}, \frac{33}{16}$ **b.** 2   **39. a.** $3, 5, 7, 9$ **b.** $a_n = 2n + 3$, $n \geq 0$ **c.** The limit does not exist.   **41. a.** $0, 1, 3, 7$ **b.** $a_n = 2^n - 1$, $n \geq 0$ **c.** The limit does not exist.   **43. a.** $1, \frac{3}{2}, \frac{7}{4}, \frac{15}{8}$ **b.** $a_n = 2 - \dfrac{1}{2^n}$ **c.** Limit is 2.

45. a. $20, 10, 5, \frac{5}{2}$ **b.** $h_n = 20\left(\frac{1}{2}\right)^n$, $n \geq 0$ **47. a.** $30, \frac{15}{2}, \frac{15}{8}, \frac{15}{32}$ **b.** $h_n = 30\left(\frac{1}{4}\right)^n$, $n \geq 0$ **49.** $S_1 = 0.3, S_2 = 0.33, S_3 = 0.333$, $S_4 = 0.3333$; $\frac{1}{3}$ **51.** $S_1 = 4, S_2 = 4.9, S_3 = 4.99, S_4 = 4.999$;   5   **53. a.** $\frac{2}{3}, \frac{4}{5}, \frac{6}{7}, \frac{8}{9}$ **b.** $S_n = \dfrac{2n}{2n + 1}$ **c.** $\lim\limits_{n \to \infty} S_n = 1$ **55. a.** $\frac{1}{3}, \frac{2}{5}, \frac{3}{7}, \frac{4}{9}$ **b.** $S_n = \dfrac{n}{2n + 1}$ **c.** $\lim\limits_{n \to \infty} S_n = \frac{1}{2}$ **57. a.** True   **b.** False   **c.** True   **59. a.** $40, 70, \frac{185}{2}, \frac{875}{8}$ **b.** 160   **61. a.** $\frac{1}{2}, \frac{3}{4}, \frac{7}{8}, \frac{15}{16}$ **b.** 1   **63. a.** $\frac{1}{3}, \frac{4}{9}, \frac{13}{27}, \frac{40}{81}$ **b.** $\frac{1}{2}$ **65. a.** $-1, 0, -1, 0$ **b.** Diverges   **67. a.** $0.3, 0.33, 0.333, 0.3333$ **b.** $\frac{1}{3}$ **69. a.** $20, 10, 5, \frac{5}{2}, \frac{5}{4}$ **b.** $M_n = 20\left(\frac{1}{2}\right)^n$, $n \geq 0$ **c.** $M_0 = 20, M_{n+1} = \frac{1}{2}M_n$ for $n \geq 0$ **d.** $\lim\limits_{n \to \infty} a_n = 0$ **71. a.** $200, 190, 180.5, 171.475, 162.90125$ **b.** $d_n = 200(0.95)^n$, $n \geq 0$ **c.** $d_0 = 200, d_{n+1} = (0.95)d_n$ for $n \geq 0$ **d.** $\lim\limits_{n \to \infty} d_n = 0$.   **73. a.** $0.\overline{3} = \sum\limits_{k=1}^{\infty} 3(0.1)^k$ **b.** $\frac{1}{3}$

75. a. $0.\overline{1} = \sum\limits_{k=1}^{\infty} (0.1)^k$ **b.** $\frac{1}{9}$ **77. a.** $0.\overline{09} = \sum\limits_{k=1}^{\infty} 9(0.01)^k$ **b.** $\frac{1}{11}$

79. a. $0.\overline{037} = \sum\limits_{k=1}^{\infty} 37(0.001)^k$ **b.** $\frac{1}{27}$

Section 8.2 Exercises, pp. 546–549

1. $a_n = \dfrac{1}{n}$, $n \geq 1$ **3.** $a_n = \dfrac{n}{n+1}$, $n \geq 1$ **5.** Converges for $-1 < r \leq 1$, diverges otherwise   **7.** A sequence $\{a_n\}_{n=1}^{\infty}$ converges to L if, given any $\varepsilon > 0$ there exists a positive integer N such that whenever $n > N$, $|a_n - L| < \varepsilon$

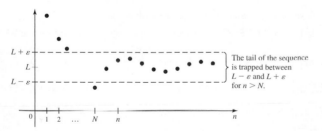

The tail of the sequence is trapped between $L - \varepsilon$ and $L + \varepsilon$ for $n > N$.

9. 0   **11.** 3/2   **13.** 0   **15.** e^2 **17.** $e^{1/4}$ **19.** 1   **21.** 0   **23.** 0   **25.** 6   **27.** Limit doesn't exist.   **29.** 0   **31.** 0   **33.** The limit doesn't exist.   **35.** Converges monotonically; 0   **37.** Converges by oscillation; 0   **39.** Diverges monotonically   **41.** Diverges by oscillation   **43.** 0   **45.** 0

47. a. $d_{n+1} = \dfrac{1}{2}d_n + 80$, $n \geq 1$ **b.** 160 mg   **49. a.** $\$0, \$100$, $\$200.75, \$302.26, \$404.53$ **b.** $B_{n+1} = 1.0075B_n + 100$, $n \geq 0$ **c.** During the 43rd month   **51.** $\{b_n\}$ grows faster than $\{a_n\}$.   **53.** $\{a_n\}$ grows faster than $\{b_n\}$.   **55.** $\{b_n\}$ grows faster than $\{a_n\}$.   **57.** Given a tolerance $\varepsilon > 0$, look beyond a_N where $N > 1/\varepsilon$.   **59.** Given a tolerance $\varepsilon > 0$, look beyond a_N where $N > \frac{1}{4}\sqrt{3/\varepsilon}$, provided $\varepsilon < \frac{3}{4}$ **61.** Given a tolerance $\varepsilon > 0$, look beyond a_N where $N > c/(\varepsilon b^2)$.   **63. a.** True   **b.** False   **c.** True   **d.** True   **e.** False   **f.** True   **65.** $\{n^2 + 2n - 17\}$ **67.** 0   **69.** 1   **71.** 1/2   **73.** 0   **75.** $n = 4$, $n = 6, n = 25$ **77. a.** $\{h_n\} = \{(200 + 5n)(0.65 - 0.01n) - 0.45n\}$ **b.** The profit is maximized after 8 days.   **79.** 0.607   **81. b.** $1, \sqrt{2} \approx 1.4142$, $1.5538, 1.5981, 1.6119$ **c.** Limit ≈ 1.618 **e.** $\dfrac{1 + \sqrt{1 + 4p}}{2}$ **83. b.** $1, 2, 1.5, 1.6667, 1.6$ **c.** Limit ≈ 1.618 **e.** $\dfrac{a + \sqrt{a^2 + 4b}}{2}$ **85. a.** $1, 1, 2, 3, 5, 8, 13, 21, 34, 55$ **b.** No

Section 8.3 Exercises, pp. 553–556

1. Consecutive terms differ by a constant ratio. Example: $2 + 1 + \frac{1}{2} + \frac{1}{4} + \cdots$ **3.** The constant r in the series $\sum\limits_{k=0}^{\infty} r^k$.

5. No   **7.** 9841   **9.** ≈ 1.1905 **11.** ≈ 0.5392 **13.** $\dfrac{1 - \pi^7}{1 - \pi}$ **15.** 1

17. $\frac{1093}{2916}$ **19.** $\frac{4}{3}$ **21.** 10   **23.** Diverges   **25.** $\dfrac{1}{e^2 - 1}$ **27.** $\frac{1}{7}$

29. $\dfrac{1}{500}$ **31.** $\dfrac{\pi}{\pi - e}$ **33.** $\frac{312,500}{19}$ **35.** $\frac{10}{19}$ **37.** $\dfrac{3\pi}{\pi + 1}$ **39.** $\dfrac{9}{460}$

41. $0.12 + 0.0012 + \cdots = \dfrac{4}{33}$ **43.** $0.456 + 0.000456 + \cdots = \dfrac{152}{333}$

45. $0.00952 + 0.00000952 + \cdots = \dfrac{952}{99,900}$ **47.** $S_n = \dfrac{1}{2} - \dfrac{1}{n + 2}$;

$\lim\limits_{n \to \infty} S_n = \dfrac{1}{2}$ **49.** $S_n = \dfrac{1}{2} - \dfrac{1}{n + 2}$; $\lim\limits_{n \to \infty} S_n = \dfrac{1}{2}$

51. $S_n = \ln(n + 1)$; $\lim\limits_{n \to \infty} S_n$ diverges

53. $S_n = \dfrac{1}{p + 1} - \dfrac{1}{n + p + 1}$; $\lim\limits_{n \to \infty} S_n = \dfrac{1}{p + 1}$

55. $S_n = \left(\dfrac{1}{\sqrt{2}} + \dfrac{1}{\sqrt{3}}\right) - \left(\dfrac{1}{\sqrt{n + 2}} + \dfrac{1}{\sqrt{n + 3}}\right)$;

$\lim\limits_{n \to \infty} S_n = \dfrac{1}{\sqrt{2}} + \dfrac{1}{\sqrt{3}}$ **57.** $S_n = -\dfrac{n + 1}{4n + 3}$; $\lim\limits_{n \to \infty} S_n = -\dfrac{1}{4}$

59. a. True   **b.** True   **c.** False   **61.** $\sum\limits_{k=0}^{\infty}\left(\dfrac{1}{4}\right)^k A_1 = \dfrac{A_1}{1 - 1/4} = \dfrac{4}{3}A_1$

63. 462 months   **65.** 0   **67.** There will be twice as many children.

69. $\sqrt{\dfrac{20}{g}\dfrac{1 + \sqrt{p}}{1 - \sqrt{p}}}$ s   **71. a.** $L_n = 3 \cdot \left(\dfrac{4}{3}\right)^n$, so $\lim\limits_{n \to \infty} L_n = \infty$

b. $\lim\limits_{n \to \infty} A_n = \dfrac{2\sqrt{3}}{5}$ **73.** $R_n = |S - S_n| = \left|\dfrac{1}{1 - r} - \left(\dfrac{1 - r^n}{1 - r}\right)\right| = \left|\dfrac{r^n}{1 - r}\right|$ **75. a.** 60   **b.** 9   **77. a.** 13   **b.** 15   **79. a.** $1, \frac{5}{6}, \frac{2}{3}$, undefined, undefined   **b.** $(-1, 1)$ **81.** Converges for x in $(-\infty, -2)$ or $(0, \infty)$; $f(x) = 3$ for $x = \frac{1}{2}$

Section 8.4 Exercises, pp. 567–569

1. Computation may not show whether the sequence of partial sums converges.   **3.** Yes, if the terms are positive and decreasing.

5. Converges for $p > 1$ and diverges for $p \leq 1$.   **9.** -2 **11.** $\frac{113}{30}$
13. $\frac{17}{10}$ **15.** Diverges   **17.** Diverges   **19.** Inconclusive   **21.** Diverges
23. Diverges   **25.** Converges   **27.** Diverges   **29.** Converges

31. Converges   **33.** Converges   **35. a.** $\dfrac{1}{5n^5}$ **b.** 3

c. $L_n = \sum_{k=1}^{n} \dfrac{1}{k^6} + \dfrac{1}{5(n+1)^5}$ $U_n = \sum_{k=1}^{n} \dfrac{1}{k^6} + \dfrac{1}{5n^2}$

d. (1.017342754, 1.017343512)   **37. a.** $\dfrac{3^{-n}}{\ln 3}$ **b.** 7

c. $L_n = \sum_{k=1}^{n} 3^{-k} + \dfrac{3^{-n-1}}{\ln 3}$ $U_n = \sum_{k=1}^{n} 3^{-k} + \dfrac{3^{-n}}{\ln 3}$

d. (0.499996671, 0.500006947)   **39. a.** $\dfrac{2}{\sqrt{n}}$ **b.** $4 \cdot 10^6$

c. $L_n = \sum_{k=1}^{n} \dfrac{1}{k^{3/2}} + \dfrac{2}{\sqrt{n+1}}$ $U_n = \sum_{k=1}^{n} \dfrac{1}{k^{3/2}} + \dfrac{2}{\sqrt{n}}$

d. (2.598359183, 2.627792025)   **41. a.** $\dfrac{1}{2n^2}$ **b.** 23

c. $L_n = \sum_{k=1}^{n} \dfrac{1}{k^3} + \dfrac{1}{2(n+1)^2}$ $U_n = \sum_{k=1}^{n} \dfrac{1}{k^3} + \dfrac{1}{2n^2}$

d. (1.201664217, 1.202531986)   **43. a.** True   **b.** True   **c.** False
d. False   **e.** False   **f.** False   **45.** Converges   **47.** Diverges

49. Converges   **51. a.** $p > 1$ **b.** $\sum_{k=2}^{\infty} \dfrac{1}{k(\ln k)^2}$ converges more quickly.

57. $\zeta(3) \approx 1.202, \zeta(5) \approx 1.037$ **59.** $\frac{\pi^2}{8}$ **61. a.** $\frac{1}{2}, \frac{7}{12}, \frac{37}{60}$

63. a. $\sum_{k=2}^{n} \dfrac{1}{k}$ **b.** Infinitely many

Section 8.5 Exercises, pp. 575–577

5. Ratio Test   **7.** $S_{n+1} - S_n = a_{n+1} > 0$ thus $S_{n+1} > S_n$
9. Converges   **11.** Converges   **13.** Converges   **15.** Diverges
17. Converges   **19.** Converges   **21.** Converges   **23.** Converges
25. Converges   **27.** Converges   **29.** Diverges   **31.** Converges
33. Converges   **35.** Diverges   **37.** Diverges   **39. a.** False
b. True   **c.** True   **41.** Diverges   **43.** Converges
45. Converges   **47.** Converges   **49.** Diverges   **51.** Converges
53. Diverges   **55.** Converges   **57.** Converges   **59.** $p > 1$
61. $p > 1$ **63.** $p > 2$ **65.** Diverges for all p **67.** Diverges if
$|r| \geq 1$ **71.** $x < 1$ **73.** $x \leq 1$ **75.** $x < 2$ **77. a.** e^2 **b.** 0

Section 8.6 Exercises, pp. 584–586

1. Because $S_{n+1} - S_n = a_{n+1}$ alternates sign.   **3.** Because
$\lim_{k \to \infty} a_k = 0$ and the terms $\{a_k\}$ alternate in sign.
5. $R_n = |S - S_n| \leq |S_{n+1} - S_n| = a_{n+1}$ **7.** No; if a series of
positive terms converges, if does so absolutely and not conditionally.

9. Yes, $\sum_{k=1}^{\infty} \dfrac{(-1)^k}{k^2}$ has this property.   **11.** Converges   **13.** Converges

15. Diverges   **17.** Diverges   **19.** Converges   **21.** Diverges
23. Converges   **25.** 10,000   **27.** 5000   **29.** 10   **31.** 3334   **33.** 6
35. -0.973 **37.** -0.783 **39.** Converges absolutely   **41.** Converges
absolutely   **43.** Diverges   **45.** Converges absolutely   **47. a.** False
b. True   **c.** True   **d.** True   **e.** False   **f.** True   **g.** True   **51.** The

conditions of the Alternating Series Test are met; thus $\sum_{k=1}^{\infty} r^k$ converges

for $-1 < r < 0$.   **55.** x and y are divergent series.

Chapter 8 Review Exercises, pp. 586–588

1. a. False   **b.** False   **c.** True   **d.** False   **3.** 0
5. 1   **7.** $1/e$ **9.** Diverges   **11. a.** $\frac{1}{3}, \frac{11}{24}, \frac{21}{40}, \frac{17}{30}$

b. $S_1 = \dfrac{1}{3}, S_n = \dfrac{1}{2}\left(\dfrac{3}{2} - \dfrac{1}{n+1} - \dfrac{1}{n+2}\right), n \geq 2$ **c.** $3/4$

13. Diverges   **15.** 1   **17.** 3   **19.** 2/9   **21. a.** Yes; 1.5
b. Convergence uncertain.   **c.** Appears to diverge   **23.** Diverges
25. Converges   **27.** Converges   **29.** Converges   **31.** Converges
33. Converges   **35.** Converges   **37.** Converges absolutely
39. Converges absolutely   **41.** Converges absolutely   **43. a.** 0

b. $\dfrac{5}{9}$ **45.** $\lim_{k \to \infty} a_k = 0, \lim_{n \to \infty} S_n = 8$ **47.** $0 < p \leq 1$
49. 0.25 (to 14 digits); 6.5×10^{-15} **51.** 100   **53. a.** 803 m, 1283 m,
$2000(1 - 0.95^N)$ m   **b.** 2000 m   **55. a.** $\dfrac{\pi}{2^{n-1}}$ **b.** 2π
57. a. $B_{n+1} = 1.0025 B_n + 100, B_0 = 100$
b. $B_n = 40,000(1.0025^{n+1} - 1)$
59. a. $T_1 = \dfrac{\sqrt{3}}{16}, T_2 = \dfrac{7\sqrt{3}}{64}$ **b.** $T_n = \dfrac{\sqrt{3}}{4}\left(1 - \left(\dfrac{3}{4}\right)^n\right)$
c. $\lim_{n \to \infty} T_n = \dfrac{\sqrt{3}}{4}$ **d.** 0

CHAPTER 9

Section 9.1 Exercises, pp. 599–601

1. $f(0) = p(0), f'(0) = p'(0),$ and $f''(0) = p''(0)$
3. 1, 1.05, 1.04875   **5.** $R_n(x) = f(x) - p_n(x)$
7. a. $p_1(x) = 1 - x$ **b.** $p_2(x) = 1 - x + \dfrac{x^2}{2}$ **c.** 0.8, 0.82
9. a. $p_1(x) = 1 - x$ **b.** $p_2(x) = 1 - x + x^2$ **c.** 0.95, 0.9525
11. a. $p_1(x) = 2 + \dfrac{1}{12}(x - 8)$
b. $p_2(x) = 2 + \dfrac{1}{12}(x - 8) - \dfrac{1}{288}(x - 8)^2$ **c.** $1.958\overline{3}, 1.95747$
13. a. $p_0(x) = 1, p_1(x) = 1, p_2(x) = 1 - \dfrac{x^2}{2}$
b.

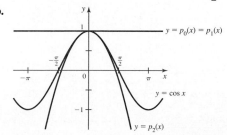

15. a. $p_0(x) = 0$, $p_1(x) = -x$, $p_2(x) = -x - \dfrac{x^2}{2}$

b.

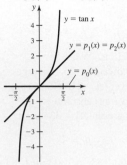

17. a. $p_0(x) = 0$, $p_1(x) = x$, $p_2(x) = x$

b.

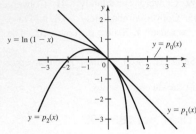

19. a. $p_0(x) = 1$, $p_1(x) = 1 - 3x$, $p_2(x) = 1 - 3x + 6x^2$

b.

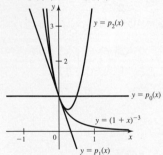

21. a. 1.0247   **b.** 7.58×10^{-6} **23. a.** 0.9624   **b.** 1.50×10^{-4}
25. a. 0.8613   **b.** 5.42×10^{-4}

27. a. $p_0(x) = \dfrac{\sqrt{2}}{2}$, $p_1(x) = p_0(x) + \dfrac{\sqrt{2}}{2}\left(x - \dfrac{\pi}{4}\right)$,

$p_2(x) = p_1(x) - \dfrac{\sqrt{2}}{4}\left(x - \dfrac{\pi}{4}\right)^2$

b.

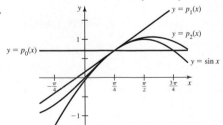

29. a. $p_0(x) = 3$, $p_1(x) = p_0(x) + \dfrac{(x-9)}{6}$,

$p_2(x) = p_1(x) - \dfrac{(x-9)^2}{216}$

b.

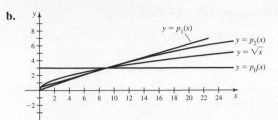

31. a. $p_0(x) = 1$, $p_1(x) = p_0(x) + \dfrac{x - e}{e}$,

$p_2(x) = p_1(x) - \dfrac{(x - e)^2}{2e^2}$

b.

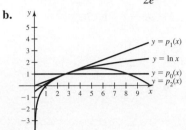

33. a. 1.12749   **b.** 8.85×10^{-6} **35. a.** -0.100333
b. 1.34×10^{-6} **37. a.** 1.029564   **b.** 4.86×10^{-7}

39. a. 10.04987563   **b.** 3.88×10^{-9} **41.** $R_n(x) = \dfrac{\sin^{(n+1)}(c)}{(n+1)!}x^{n+1}$

for some c between x and 0.   **43.** $R_n(x) = \dfrac{(-1)^{n+1}e^{-c}}{(n+1)!}x^{n+1}$ for some

c between x and 0.   **45.** $R_n(x) = \dfrac{\sin^{(n+1)}(c)}{(n+1)!}\left(x - \dfrac{\pi}{2}\right)^{n+1}$ for some

c between x and $\dfrac{\pi}{2}$.   **47.** 2.03×10^{-5} **49.** $1.63 \times 10^{-5}\,(e^{0.25} < 2)$

51. 2.60×10^{-4} **53.** With $n = 4$, max error $= 2.49 \times 10^{-3}$
55. With $n = 2$, max error $= 4.17 \times 10^{-2}\,(e^{0.5} < 2)$ **57.** With
$n = 2$, max error $= 2.67 \times 10^{-3}$ **59.** 4   **61.** 3   **63.** 1
65. a. False   **b.** True   **c.** True   **67. a.** C   **b.** E   **c.** A   **d.** D
e. B   **f.** F   **69. a.** 0.1; with $n = 2$, 1.67×10^{-4} **b.** 0.2; 1.33×10^{-3}
71. a. 0.995; with $n = 3$, 4.17×10^{-6} **b.** 0.98; 6.67×10^{-5}
73. a. 1.05; $\dfrac{1}{800}$ **b.** 1.1; $\dfrac{1}{200}$ **75.**   **a.** 1.1; $\dfrac{1}{100}$ **b.** 1.2; $\dfrac{1}{25}$
77. a.

x	$\lvert \sin x - p_3(x) \rvert$	$\lvert \sin x - p_5(x) \rvert$
-0.2	2.7×10^{-6}	2.5×10^{-9}
-0.1	8.3×10^{-8}	2.0×10^{-11}
0.0	0	0
0.1	8.3×10^{-8}	2.0×10^{-11}
0.2	2.7×10^{-6}	2.5×10^{-9}

b. The error increases as $\lvert x \rvert$ increases.
79. a.

x	$\lvert e^{-x} - p_1(x) \rvert$	$\lvert e^{-x} - p_2(x) \rvert$
-0.2	2.1×10^{-2}	1.4×10^{-3}
-0.1	5.2×10^{-3}	1.7×10^{-4}
0.0	0	0
0.1	4.8×10^{-3}	1.6×10^{-4}
0.2	1.9×10^{-2}	1.3×10^{-3}

b. The error increases as $|x|$ increases.

81. a.

x	$|\tan x - p_1(x)|$	$|\tan x - p_3(x)|$
-0.2	2.7×10^{-3}	4.3×10^{-5}
-0.1	3.3×10^{-4}	1.3×10^{-6}
0.0	0	0
0.1	3.3×10^{-4}	1.3×10^{-6}
0.2	2.7×10^{-3}	4.3×10^{-5}

b. The error increases as $|x|$ increases.   **83.** Centered at $x = 0$ for all n **85. a.** $y = f(a) + f'(a)(x - a)$

Section 9.2 Exercises, pp. 609–611

1. $c_0 + c_1 x + c_2 x^2 + c_3 x^3$ **3.** Ratio and Root Test   **5.** The radius of convergence does not change. The interval of convergence may change.
7. $|x| < \frac{1}{4}$ **9.** $r = 3; (-3, 3)$ **11.** $R = \infty; (-\infty, \infty)$
13. $R = \infty; (-\infty, \infty)$ **15.** $R = \sqrt{3}; (-\sqrt{3}, \sqrt{3})$

17. $R = 1; (0, 2)$ **19.** $R = \infty; (-\infty, \infty)$ **21.** $\sum_{k=0}^{\infty} (3x)^k; \left(-\frac{1}{3}, \frac{1}{3}\right)$

23. $2\sum_{k=0}^{\infty} x^{k+3}; (-1,1)$ **25.** $4\sum_{k=0}^{\infty} x^{k+12}; (-1,1)$ **27.** $-\sum_{k=1}^{\infty} \frac{(3x)^k}{k};$

$\left[-\frac{1}{3}, \frac{1}{3}\right)$ **29.** $-\sum_{k=1}^{\infty} \frac{x^{k+1}}{k}; [-1,1)$ **31.** $-2\sum_{k=1}^{\infty} \frac{x^{k+6}}{k}; [-1,1)$

33. $g(x) = \sum_{k=1}^{\infty} kx^{k-1}; (-1,1)$ **35.** $g(x) = \sum_{k=3}^{\infty} \frac{k(k-1)(k-2)}{6} x^k$ ³;

$(-1,1)$ **37.** $g(x) = -\sum_{k=1}^{\infty} \frac{3^k x^k}{k}; \left[-\frac{1}{3}, \frac{1}{3}\right)$ **39.** $\sum_{k=0}^{\infty} (-x^2)^k; (-1,1)$

41. $\sum_{k=0}^{\infty} \left(-\frac{x}{3}\right)^k; (-3, 3)$ **43.** $\ln 2 - \frac{1}{2}\sum_{k=1}^{\infty} \frac{x^{2k}}{k4^k}; (-2, 2)$ **45. a.** True

b. True   **c.** True   **d.** True   **47.** $\sum_{k=0}^{\infty} \frac{(-1)^k x^k}{k+1}$ **49.** $\sum_{k=1}^{\infty} \frac{(-x^2)^k}{k!}$

51. $|x - a| < R$ **53.** $f(x) = \frac{1}{3 - \sqrt{x}}; 1 < x < 9$

55. $f(x) = \frac{e^x}{e^x - 1}; 0 < x < \infty$ **57.** $f(x) = \frac{3}{4 - x^2}; -2 < x < 2$

59. $\sum_{k=0}^{\infty} \frac{(-x)^k}{k!}; -\infty < x < \infty$ **61.** $\sum_{k=0}^{\infty} \frac{(-3x)^k}{k!}; -\infty < x < \infty$

63. $\lim_{k \to \infty} \left| \frac{c_{k+1} x^{k+1}}{c_k x^k} \right| = \lim_{k \to \infty} \left| \frac{c_{k+1} x^{k+m+1}}{c_k x^{k+m}} \right|$, so by the Ratio Test the two series converge on the same interval.
65. a. $f(x) \cdot g(x) = c_0 d_0 + (c_0 d_1 + c_1 d_0)x$
$\qquad + (c_0 d_2 + c_1 d_1 + c_2 d_0)x^2 + \cdots.$

b. $\sum_{k=0}^{n} c_k d_{n-k}$ **67. b.** $n = 112$

Section 9.3 Exercises, pp. 621–622

1. The nth Taylor polynomial is the nth partial sum of the corresponding Taylor series.   **3.** Calculate $c_k = \frac{f^{(k)}(a)}{k!}$ for $k = 0, 1, 2, \ldots$.

5. Replace x by x^2 in the Taylor series for $f(x); |x| < 1$.   **7.** The Taylor series for a function f converges to f on an interval if, for all x in the interval, $\lim_{n \to \infty} R_n(x) = 0$, where $R_n(x)$ is the remainder at x.

9. a. $1 - x + \frac{x^2}{2!} - \frac{x^3}{3!}$ **b.** $\sum_{k=0}^{\infty} \frac{(-1)^k x^k}{k!}$ **c.** $(-\infty, \infty)$

11. a. $1 - x^2 + x^4 - x^6$ **b.** $\sum_{k=0}^{n} (-1)^k x^{2k}$ **c.** $(-1, 1)$

13. a. $1 + 2x + \frac{(2x)^2}{2!} + \frac{(2x)^3}{3!}$ **b.** $\sum_{k=0}^{\infty} \frac{(2x)^k}{k!}$ **c.** $(-\infty, \infty)$

15. a. $x - \frac{x^3}{3} + \frac{x^5}{5} - \frac{x^7}{7}$ **b.** $\sum_{k=0}^{\infty} \frac{(-1)^k x^{2k+1}}{2k+1}$ **c.** $[-1, 1]$

17. a. $1 - \frac{(x - \pi/2)^2}{2!} + \frac{(x - \pi/2)^4}{4!} - \frac{(x - \pi/2)^6}{6!}$

b. $\sum_{k=0}^{\infty} \frac{(-1)^k}{(2k)!} (x - \pi/2)^{2k}$

19. a. $1 - (x - 1) + (x - 1)^2 - (x - 1)^3$ **b.** $\sum_{k=0}^{\infty} (-1)^k (x - 1)^k$

21. a. $\ln 3 + \frac{(x - 3)}{3} - \frac{(x - 3)^2}{3^2 \cdot 2} + \frac{(x - 3)^3}{3^3 \cdot 3}$

b. $\ln 3 + \sum_{k=1}^{\infty} \frac{(-1)^{k+1}(x - 3)^k}{k3^k}$ **23.** $x^2 - \frac{x^4}{2} + \frac{x^6}{3} - \frac{x^8}{4} + \cdots$

25. $1 + \frac{x}{2} + \frac{x^2}{6} + \frac{x^3}{24} + \cdots$ **27.** $1 - x^4 + x^8 - x^{12} + \cdots$

29. a. $1 - 2x + 3x^2 - 4x^3$ **b.** 0.826
31. a. $1 + \frac{1}{4}x - \frac{3}{32}x^2 + \frac{7}{128}x^3$ **b.** 1.029
33. a. $1 - \frac{2}{3}x + \frac{5}{9}x^2 - \frac{40}{81}x^3$ **b.** 0.895

35. $1 + \frac{x^2}{2} - \frac{x^4}{8} + \frac{x^6}{16} - \cdots; [-1, 1]$

37. $3 - \frac{3x}{2} - \frac{3x^2}{8} - \frac{3x^3}{16} - \cdots; [-1, 1)$

39. $a + \frac{x^2}{2a} - \frac{x^4}{8a^3} + \frac{x^6}{16a^5} - \cdots; |x| \le a$

41. $1 - 8x + 48x^2 - 256x^3 + \cdots$

43. $\frac{1}{16} - \frac{x^2}{32} + \frac{3x^4}{256} - \frac{x^6}{256} + \cdots$

45. $\frac{1}{9} - \frac{2}{9}\left(\frac{4x}{3}\right) + \frac{3}{9}\left(\frac{4x}{3}\right)^2 - \frac{4}{9}\left(\frac{4x}{3}\right)^3 + \cdots$

47. $R_n(x) = \frac{f^{(n+1)}(c)}{(n+1)!} x^{n+1}$, where c is between 0 and x and $f^{(n+1)}(c) = \pm \sin c$ or $\pm \cos c$. Thus, $|R_n(x)| \le \frac{|x|^{n+1}}{(n+1)!} \to 0$ as $n \to \infty$, for $-\infty < x < \infty$.

49. $R_n(x) = \frac{f^{(n+1)}(c)}{(n+1)!} x^{n+1}$, where c is between 0 and x and $f^{(n+1)}(c) = \begin{cases} e^{-c} & \text{if } n \text{ even} \\ -e^{-c} & \text{if } n \text{ odd} \end{cases}$

Thus, $\lim_{n \to \infty} |R_n(x)| = \lim_{n \to \infty} \left| \frac{x^{n+1}}{e^c(n+1)!} \right| = 0$ and so $\lim_{n \to \infty} R_n(x) = 0$, for $-\infty < x < \infty$.   **51. a.** False   **b.** True   **c.** False   **d.** False
e. True   **53. a.** $1 + \frac{x^2}{2!} + \frac{x^4}{4!} + \frac{x^6}{6!} + \cdots$ **b.** $R = \infty$

55. a. $1 - \frac{2}{3}x^2 + \frac{5}{9}x^4 - \frac{40}{81}x^6 + \cdots$ **b.** $R = 1$

57. a. $1 - \frac{1}{2}x^2 - \frac{1}{8}x^4 - \frac{1}{16}x^6 - \cdots$ **b.** $R = 1$

59. a. $1 - 2x^2 + 3x^4 - 4x^6 + \cdots$ **b.** $R = 1$ **61.** $\sqrt[3]{60} \approx 3.9149$
using the first four terms **63.** $\sqrt[4]{13} \approx 1.8989$ using the first four terms

69. $\displaystyle\sum_{k=0}^{\infty}\left(\frac{x-4}{2}\right)^k$ **71.** $\dfrac{1 \cdot 3 \cdot 5 \cdot 7}{2 \cdot 4 \cdot 6 \cdot 8}x^4, \dfrac{-1 \cdot 3 \cdot 5 \cdot 7 \cdot 9}{2 \cdot 4 \cdot 6 \cdot 8 \cdot 10}x^5$

73. Use three terms of the Taylor series for $\cos x$ centered at $a = \pi/4$;
$\cos 40° = \cos(40\pi/180) \approx 0.766$ **75.** Use six terms of the Taylor
series for $\sqrt[3]{x}$ centered at $a = 64$; $\sqrt[3]{83} \approx 4.362$ **77. a.** Use three
terms of the Taylor series for $\sqrt[3]{125 + x}$ centered at $a = 0$;
$\sqrt[3]{128} \approx 5.03968$ **b.** Use three terms of the Taylor series for $\sqrt[3]{x}$
centered at $a = 0$; $\sqrt[3]{128} \approx 5.03968$ **c.** Yes.

Section 9.4 Exercises, pp. 629–631

1. Replace f and g by their Taylor series centered at a and evaluate the
limit. **3.** Substitute $x = -0.6$ into the Taylor series for e^x centered at
0. Because the resulting series is an alternating series, the error can
be estimated. **5.** $f'(x) = \displaystyle\sum_{k=1}^{\infty} kc_k x^{k-1}$ **7.** 2 **9.** $\frac{2}{3}$ **11.** $\frac{2}{5}$ **13.** $\frac{3}{5}$

15. $-\frac{1}{6}$ **17.** 1 **19.** $\frac{17}{12}$ **21. a.** $1 + x + \dfrac{x^2}{2!} + \cdots + \dfrac{x^n}{n!} + \cdots$

b. e^x **c.** $-\infty < x < \infty$

23. a. $1 - x + x^2 - \cdots (-1)^{n-1}x^{n-1} + \cdots$ **b.** $\dfrac{1}{1+x}$ **c.** $|x| < 1$

25. a. $-2 + 4x - 8 \cdot \dfrac{x^2}{2!} + \cdots + (-2)^n \dfrac{x^{n-1}}{(n-1)!} + \cdots$ **b.** $-2e^{-2x}$

c. $-\infty < x < \infty$ **27. a.** $2 + 2t + \dfrac{2t^2}{2!} + \cdots + \dfrac{2t^n}{n!} + \cdots$

b. $y(t) = 2e^t$

29. a. $2 + 16t + 24t^2 + 24t^3 + \cdots + \dfrac{3^{n-1} \cdot 16}{n!}t^n + \cdots$

b. $y(t) = \frac{16}{3}e^{3t} - \frac{10}{3}$ **31.** 0.2448 **33.** 0.6958 **35.** 0.1498

37. 0.4994 **39.** $e^2 = \displaystyle\sum_{k=0}^{\infty} \dfrac{2^k}{k!} = 1 + 2 + \dfrac{2^2}{2!} + \dfrac{2^3}{3!} + \cdots$

41. $\cos 2 = \displaystyle\sum_{k=0}^{\infty} \dfrac{(-1)^k 2^{2k}}{(2k)!} = 1 - 2 + \dfrac{2}{3} - \dfrac{4}{45} + \cdots$

43. $\ln(3/2) = \displaystyle\sum_{k=1}^{\infty} \dfrac{(-1)^{k+1}}{k2^k}$

$= \frac{1}{2} - \frac{1}{8} + \frac{1}{24} - \frac{1}{64} + \cdots$

45. $\dfrac{e^x - 1}{x} = \displaystyle\sum_{k=0}^{\infty} \dfrac{x^k}{(k+1)!}$. Therefore, $\displaystyle\sum_{k=0}^{\infty} \dfrac{1}{(k+1)!} = e - 1$.

47. $\displaystyle\sum_{k=1}^{\infty} \dfrac{(-1)^{k+1}x^k}{k}$ for $-1 < x \leq 1$. At $x = 1$, $\displaystyle\sum_{k=1}^{\infty} \dfrac{(-1)^{k+1}}{k} = \ln 2$.

49. $f(x) = \dfrac{2}{2-x}$ **51.** $f(x) = \dfrac{4}{4+x^2}$ **53.** $f(x) = -\ln(1-x)$

55. $f(x) = \dfrac{-3x^2}{(3+x)^2}$ **57.** $f(x) = \dfrac{6x^2}{(3-x)^3}$ **59. a.** False

b. False **c.** True **61.** $\frac{a}{b}$ **63.** $e^{-1/6}$ **65.** $f^{(3)}(0) = 0$;
$f^{(4)}(0) = 4e$ **67.** $f^{(3)}(0) = 2; f^{(4)}(0) = 0$ **69.** 2 **71. a.** 1.5741
using four terms **b.** At least three **c.** More terms would be

needed. **73. a.** $S'(x) = \sin(x^2); C'(x) = \cos(x^2)$

b. $\dfrac{x^3}{3} - \dfrac{x^7}{7 \cdot 3!} + \dfrac{x^{11}}{11 \cdot 5!} - \dfrac{x^{15}}{15 \cdot 7!}; x - \dfrac{x^5}{5 \cdot 2!} + \dfrac{x^9}{9 \cdot 4!} - \dfrac{x^{13}}{13 \cdot 6!}$

c. $S(0.05) \approx 0.00004166664807$ $C(-0.25) \approx -0.2499023614$ **d.** 1

e. 2 **75. a.** $1 - \dfrac{x^2}{4} + \dfrac{x^4}{64} - \dfrac{x^6}{2304}$ **b.** $-\infty < x < \infty, R = \infty$

c. $\left(-\dfrac{x^2}{2} + \dfrac{3x^4}{16} - \dfrac{5x^6}{384}\right) + \left(-\dfrac{x^2}{2} + \dfrac{x^4}{16} - \dfrac{x^6}{384}\right) +$

$\left(x^2 - \dfrac{x^4}{4} + \dfrac{x^6}{64}\right) = 0$. **77. a.** The Maclaurin series for $\cos x$
consists of even powers of x, which are even functions. **b.** The
Maclaurin series for $\sin x$ consists of odd powers of x, which are odd
functions.

Chapter 9 Review Exercises, pp. 632–633

1. a. True **b.** False **c.** True **d.** True **3.** $p_2(x) = 1$

5. $p_3(x) = x - \dfrac{x^2}{2} + \dfrac{x^3}{3}$ **7.** $p_2(x) = (x-1) - \dfrac{(x-1)^2}{2}$

9. a. $p_2(x) = 1 + x + \dfrac{x^2}{2}$ **b.**

n	$p_n(x)$	error
0	1	7.7×10^{-2}
1	0.92	3.1×10^{-3}
2	0.9232	8.4×10^{-5}

11. a. $p_2(x) = \dfrac{\sqrt{2}}{2} + \dfrac{\sqrt{2}}{2}\left(x - \dfrac{\pi}{4}\right) - \dfrac{\sqrt{2}}{4}\left(x - \dfrac{\pi}{4}\right)^2$

b.

n	$p_n(x)$	error
0	0.7071	1.2×10^{-1}
1	0.5960	8.2×10^{-3}
2	0.5873	4.7×10^{-4}

13. $R_3(x) = \dfrac{\sin c}{4!}x^4, |c| < \pi; |R_3| < \dfrac{\pi^4}{4!}$ **15.** $(-\infty, \infty), R = \infty$

17. $(-\infty, \infty), R = \infty$ **19.** $(-9, 9), R = 9$ **21.** $\displaystyle\sum_{k=0}^{\infty} x^{2k}; (-1, 1)$

23. $\displaystyle\sum_{k=0}^{\infty} 3^k x^k; \left(-\dfrac{1}{3}, \dfrac{1}{3}\right)$ **25.** $\displaystyle\sum_{k=1}^{\infty} kx^{k-1}; (-1, 1)$

27. $1 + 3x + \dfrac{9x^2}{2!}; \displaystyle\sum_{k=0}^{\infty} \dfrac{(3x)^k}{k!}$

29. $-(x - \pi/2) + \dfrac{(x - \pi/2)^3}{3!} - \dfrac{(x - \pi/2)^5}{5!};$

$\displaystyle\sum_{k=0}^{\infty} (-1)^{k+1} \dfrac{(x - \pi/2)^{2k+1}}{(2k+1)!}$

31. $x - \dfrac{x^3}{3} + \dfrac{x^5}{5}; \displaystyle\sum_{k=0}^{\infty} \dfrac{(-1)^k x^{2k+1}}{2k+1}$

33. $1 + \dfrac{x}{3} - \dfrac{x^2}{9} + \cdots$ **35.** $1 - \dfrac{3}{2}x + \dfrac{3}{2}x^2 - \cdots$

37. $R_n(x) = \dfrac{(-1)^{n+1}e^{-c}}{(n+1)!}x^{n+1}$, where c is between 0 and x.

$\displaystyle\lim_{n \to \infty} |R_n(x)| = \lim_{n \to \infty} \dfrac{|x^{n+1}|}{e^{|x|}} \cdot \dfrac{1}{(n+1)!} = 0$ for $-\infty < x < \infty$.

39. $R_n(x) = \dfrac{(-1)^n(1 + c)^{-(n+1)}}{n + 1} x^{n+1}$ where c is between 0 and x.

$\lim_{n\to\infty} |R_n(x)| = \lim_{n\to\infty}\left(\dfrac{|x|}{1 + c}\right)^{n+1} \cdot \dfrac{1}{n + 1} < \lim_{n\to\infty} 1^{n+1} \cdot \dfrac{1}{n + 1} = 0$

for $|x| \le \frac{1}{2}$.   **41.** $\frac{1}{24}$ **43.** $\frac{1}{8}$ **45.** $\frac{1}{6}$

47. $y(x) = 4 + 4x + \dfrac{4^2}{2!}x^2 + \dfrac{4^3}{3!}x^3 + \cdots + \dfrac{4^n}{n!}x^n + \cdots$

$= 3 + e^{4x}$.

49. a. $\displaystyle\sum_{k=1}^{\infty} \dfrac{(-1)^{k+1}}{k}$ **b.** $\displaystyle\sum_{k=1}^{\infty} \dfrac{1}{k2^k}$ **c.** $2\displaystyle\sum_{k=0}^{\infty} \dfrac{x^{2k+1}}{2k + 1}$

d. $x = \dfrac{1}{3}; 2\displaystyle\sum_{k=0}^{\infty} \dfrac{1}{3^{2k+1}(2k + 1)}$ **e.** Series in part (d)

CHAPTER 10

Section 10.1 Exercises, pp. 641–644

1. If $x = g(t)$ and $y = h(t)$ for $a \le t \le b$, then plotting the set $\{(g(t), (h(t)): a \le t \le b\}$ results in a graph in the xy-plane.
3. $x = R\cos(\pi t/5), y = R\sin(\pi t/5)$
5. $x = t; y = t^2, -\infty < t < \infty$
7. a.

t	-10	-8	-6	-4	-2	0	2	4	6	8	10
x	-20	-16	-12	-8	-4	0	4	8	12	16	20
y	-34	-28	-22	-16	-10	-4	2	8	14	20	26

b. 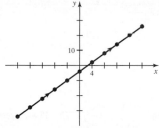   **c.** $y = \frac{3}{2}x - 4$

d. A line rising up and to the right as t increases
9. a.

t	-5	-4	-3	-2	-1	0	1	2	3	4	5
x	11	10	9	8	7	6	5	4	3	2	1
y	-18	-15	-12	-9	-6	-3	0	3	6	9	12

b. 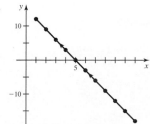   **c.** $y = -3x + 15$

d. A line rising up and to the left as t increases
11. a. $y = 3x - 12$ **b.** A line rising up and to the right as t increases   **13. a.** $y = (x + 1)^3$ **b.** A cubic function rising up and to the right as t increases   **15.** Center $(0, 0)$; radius 3; lower half of circle generated counterclockwise   **17.** Center $(0, 0)$; radius 7; full circle generated counterclockwise

19. $x = 4\cos t, y = 4\sin t, 0 \le t \le 2\pi$: The circle has equation $x^2 + y^2 = 16$.

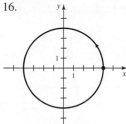

21. $x = 8\sin t - 2, y = 8\cos t - 3, 0 \le t \le 2\pi$: The circle has equation $(x + 2)^2 + (y + 3)^2 = 64$.

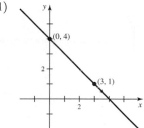

23. $x = 400\cos\left(\dfrac{4\pi t}{3}\right), y = 400\sin\left(\dfrac{4\pi t}{3}\right),$

$0 \le t \le 1.5$ (min)   **25.** $x = 50\cos\left(\dfrac{\pi t}{12}\right), y = 50\sin\left(\dfrac{\pi t}{12}\right),$

$0 \le t \le 24$ (s)
27. Slope: -1; point: $(3, 1)$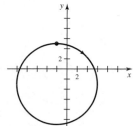

29. Slope: 0; point: $(8, 1)$

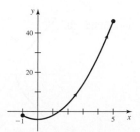

31. $x = 2t, y = 8t, 0 \le t \le 1$
33. $x = -1 + 7t, y = -3 - 13t, 0 \le t \le 1$
35. $x = t, y = 2t^2 - 4, -1 \le t \le 5$ (not unique)

37. $x = 4t - 2, y = -6t + 3, 0 \le t \le 1$;
$x = t + 1, y = 8t - 11, 1 \le t \le 2$ (not unique)

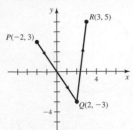

39. 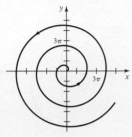   **41.**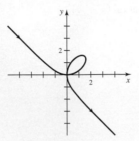

43.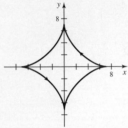

45. a. $\dfrac{dy}{dx} = -2; -2$ **b.**

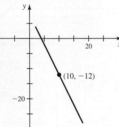

47. a. $\dfrac{dy}{dx} = -8 \cot t; 0$ **b.**

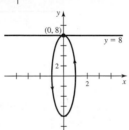

49. a. $\dfrac{dy}{dx} = \dfrac{t^2 + 1}{t^2 - 1}, t \ne 0$; undefined   **b.**

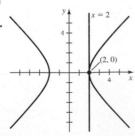

51. a. False   **b.** True   **c.** False   **d.** True
53. $x = 1 + 2t, y = 1 + 4t, -\infty < t < \infty$
55. $x = t^2, y = t, t \ge 0$

57. 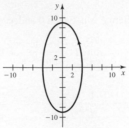 $0 \le t \le 2\pi$

59. $x = 3 \cos t, y = \frac{3}{2} \sin t, 0 \le t \le 2\pi; \left(\dfrac{x}{3}\right)^2 + \left(\dfrac{2y}{3}\right)^2 = 1$;

in the counterclockwise direction

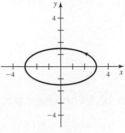

61. $x = 15 \cos t - 2, y = 10 \sin t - 3, 0 \le t \le 2\pi$;
$\left(\dfrac{x + 2}{15}\right)^2 + \left(\dfrac{y + 3}{10}\right)^2 = 1$; in the counterclockwise direction

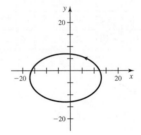

63. a and b   **65.** $x^2 + y^2 = 4$ **67.** $y = \sqrt{4 - x^2}$
69. $y = x^2$ **71.** $\left(\dfrac{-4}{\sqrt{5}}, \dfrac{8}{\sqrt{5}}\right)$ and $\left(\dfrac{4}{\sqrt{5}}, \dfrac{-8}{\sqrt{5}}\right)$ **73.** There are no

such points.   **75.** $a = p, b = p + \dfrac{2\pi}{3}$, for all real p **77. a.** $(0, 2)$
and $(0, -2)$ **b.** $(1, \sqrt{2}), (1, -\sqrt{2}), (-1, \sqrt{2}), (-1, -\sqrt{2})$
79. a. $x = \pm a \cos^{2/n}(t), y = \pm b \sin^{2/n}(t)$ **c.** The curves become
more square as n increases.   **85.** ≈ 2857 m

Section 10.2 Exercises, pp. 653–657

1. 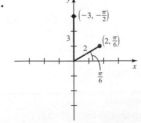 $(-2, -5\pi/6), (2, 13\pi/6)$;
$(3, \pi/2), (3, 5\pi/2)$

3. $r^2 = x^2 + y^2$, $\tan \theta = \dfrac{y}{x}$ **5.** $r \cos \theta = 5$
7. x-axis symmetry occurs if (r, θ) on the graph implies $(r, -\theta)$ is on
the graph. y-axis symmetry occurs if (r, θ) on the graph implies
$(r, \pi - \theta) = (-r, -\theta)$ is on the graph. Symmetry about the origin occurs if (r, θ) on the graph implies $(-r, \theta) = (r, \theta + \pi)$ is on the graph.

9.

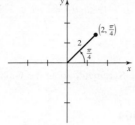

$(-2, -3\pi/4), (2, 9\pi/4)$

11.

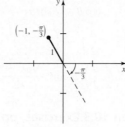

$(1, 2\pi/3), (1, 8\pi/3)$

$(4, \pi/2), (4, 5\pi/2)$

49.

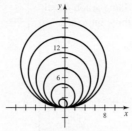

No interval $[0, P]$ generates the entire curve; $-\infty < \theta < \infty$

13.

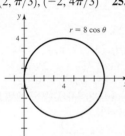

15. $(3\sqrt{2}/2, 3\sqrt{2}/2)$ **17.** $(1/2, -\sqrt{3}/2)$
19. $(2\sqrt{2}, -2\sqrt{2})$ **21.** $(2\sqrt{2}, \pi/4), (-2\sqrt{2}, 5\pi/4)$
23. $(2, \pi/3), (-2, 4\pi/3)$ **25.** $(8, 2\pi/3), (-8, -\pi/3)$
27.

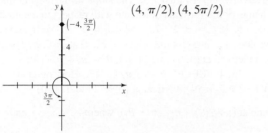

29.

31. $x = -4$; vertical line passing through $(-4, 0)$
33. $x^2 + (y - 1)^2 = 1$; circle of radius 1 centered at $(0, 1)$ and $x = 0$;
y-axis   **35.** $x^2 + (y - 4)^2 = 16$; circle of radius 4 centered at $(0, 4)$
37.

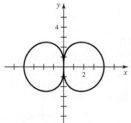

39.

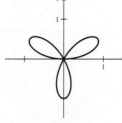

41.

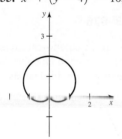

43.

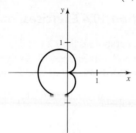

45.

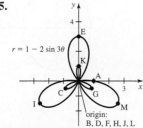

47.

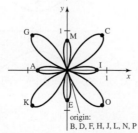

51.

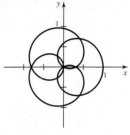

$[0, 2\pi]$

53.

$[0, 5\pi]$
$[0, 2\pi]$

55.

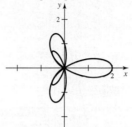

57. a. True   **b.** True   **c.** False   **d.** True

59.

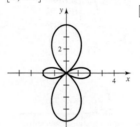

61.

63.

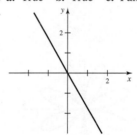

65.

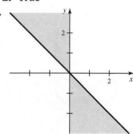

69. A circle of radius 4 and
center $(2, \pi/3)$ (polar
coordinates)

71. A circle of radius 4
centered at $(2, 3)$
(Cartesian coordinates)

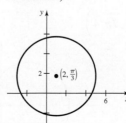

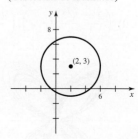

73. A circle of radius 3 centered at $(-1, 2)$ (Cartesian coordinates)

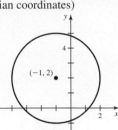

75. Same graph on all three intervals.

77.

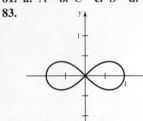

$y = -\dfrac{x}{\sqrt{3}} + 2\sqrt{3}$

79.

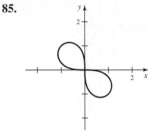

$y = 4x + 3$

81. a. A **b.** C **c.** B **d.** D **e.** E **f.** F

83.

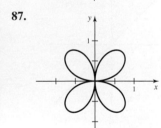

85.

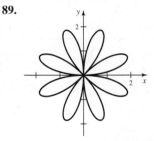

87.

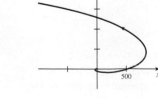

89.

93.

For $a = -1$, the spiral winds inward toward the origin.

95. $(2, 0)$ and $(0, 0)$

97. $(0, 0), \left(\dfrac{2 - \sqrt{2}}{2}, 3\pi/4\right), \left(\dfrac{2 + \sqrt{2}}{2}, 7\pi/4\right)$

99. a.

101. a.

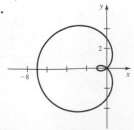

103. $r = a\cos\theta + b\sin\theta = \dfrac{a}{r}(r\cos\theta) + \dfrac{b}{r}(r\sin\theta) = \dfrac{a}{r}x + \dfrac{b}{r}y$

Thus, $\left(x - \dfrac{a}{2}\right)^2 + \left(y - \dfrac{b}{2}\right)^2 = \dfrac{a^2 + b^2}{4}$.

Center: $\left(\dfrac{a}{2}, \dfrac{b}{2}\right)$; radius: $\dfrac{\sqrt{a^2 + b^2}}{2}$ **105.** Symmetry about the x-axis

Section 10.3 Exercises, pp. 662–664

1. $x = f(\theta)\cos\theta, y = f(\theta)\sin\theta$ **3.** The slope of the tangent line is the rate of change of the vertical coordinate with respect to the horizontal coordinate.   **5.** $0; \theta = \pi/2$ **7.** $-\sqrt{3}; \theta = 0$ **9.** Vertical, vertical; the curve does not intersect the origin.   **11.** 0 at $(-4, \pi/2)$ and $(-4, 3\pi/2)$, undefined at $(4, 0)$ and $(4, \pi)$; $\theta = \pi/4, \theta = 3\pi/4, \theta = 5\pi/4, \theta = 7\pi/4$ **13.** $\pm 1; \theta = \pm\pi/4$ **15.** Horizontal at $(2\sqrt{2}, \pi/4), (-2\sqrt{2}, 3\pi/4)$; vertical at $(0, \pi/2)$ and $(4, 0)$

17. Horizontal at $(\pm r_1, \theta_1), (\pm r_1, \pi - \theta_1)$, where $r_1 = \dfrac{2\sqrt{2}}{3}$ and $\tan\theta_1 = \sqrt{2}$; vertical at $(\pm r_1, \theta_2), (\pm r_1, \pi - \theta_2)$, where $r_1 = \dfrac{2\sqrt{2}}{2}$ and $\tan\theta_2 = \dfrac{1}{\sqrt{2}}$ **19.** Horizontal at $\left(\pm\sqrt{2}, \frac{\pi}{6}\right), \left(\pm\sqrt{2}, -\frac{\pi}{6}\right)$; vertical at $(2, 0)$ and $(2, \pi)$ **21.** 16π **23.** $9\pi/2$

25. $\pi/20$ **27.** $2\left(\frac{8\pi}{3} - 2\sqrt{3}\right)$ **29.** $(0, 0), (3/\sqrt{2}, \pi/4)$

31. $(2, 0), (0, 0), (r, \theta) \approx (-2 + 2\sqrt{2}, \pm\cos^{-1}(-3 + 2\sqrt{2})) \approx (0.828, \pm 1.74)$ **33. a.** False   **b.** False   **35.** $2\pi/3 - \sqrt{3}/2$

37. $9\pi + 27\sqrt{3}$ **39.** Horizontal: $(0, 0), (4.05, 2.03), (9.83, 4.91)$; vertical: $(1.72, 0.86), (6.85, 3.43), (12.87, 6.44)$

41. a. $A_n = \dfrac{1}{4e^{(4n+2)\pi}} - \dfrac{1}{4e^{4n\pi}} - \dfrac{1}{4e^{(4n-2)\pi}} + \dfrac{1}{4e^{(4n-4)\pi}}$ **b.** 0

c. $e^{-4\pi}$ **43.** 6 **45.** 18π **47.** $(a^2 - 2)\theta^* + \pi - \sin 2\theta^*$, where $\theta^* = \cos^{-1}(a/2)$.   **49.** $a^2(\pi/2 + a/3)$

Section 10.4 Exercises, pp. 673–676

1. A parabola is the set of all points in a plane equidistant from a fixed point and a fixed line.   **3.** A hyperbola is the set of all points in a plane, the difference of whose distances from two fixed points is constant.

5. Parabola:

Hyperbola: 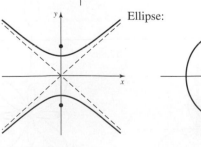   Ellipse: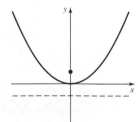

7. $\left(\dfrac{x}{a}\right)^2 + \dfrac{y^2}{a^2 - c^2} = 1$ **9.** $(\pm ae, 0)$ **11.** $y = \pm\dfrac{b}{a}x$

13.

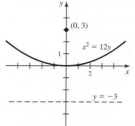

15.

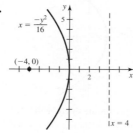

17.

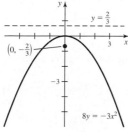

19. $y^2 = 16x$ **21.** $y^2 = 12x$

23. $x^2 = -\dfrac{2}{3}y$ **25.** $y^2 = 4(x + 1)$

27.

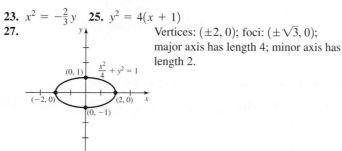

Vertices: $(\pm 2, 0)$; foci: $(\pm\sqrt{3}, 0)$; major axis has length 4; minor axis has length 2.

29.

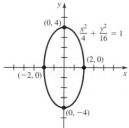

Vertices: $(0, \pm 4)$; foci: $(0, \pm 2\sqrt{3})$; major axis has length 8; minor axis has length 4.

31.

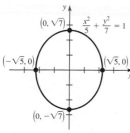

Vertices: $(0, \pm\sqrt{7})$; foci: $(0, \pm\sqrt{2})$; major axis has length $2\sqrt{7}$; minor axis has length $2\sqrt{5}$.

33.

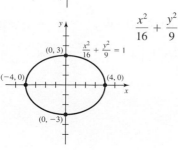

$\dfrac{x^2}{16} + \dfrac{y^2}{9} = 1$

35.

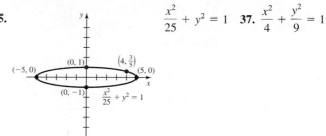

$\dfrac{x^2}{25} + y^2 = 1$ **37.** $\dfrac{x^2}{4} + \dfrac{y^2}{9} = 1$

39.

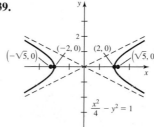

Vertices: $(\pm 2, 0)$; foci: $(\pm\sqrt{5}, 0)$; asymptotes: $y = \pm\dfrac{1}{2}x$

41.

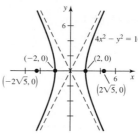

Vertices: $(\pm 2, 0)$; foci: $(\pm 2\sqrt{5}, 0)$; asymptotes: $y = \pm 2x$

43.

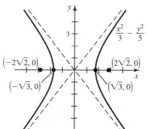

Vertices: $(\pm\sqrt{3}, 0)$; foci: $(\pm 2\sqrt{2}, 0)$; asymptotes: $y = \pm\sqrt{\dfrac{5}{3}}x$

45.

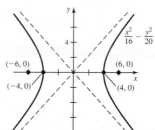

Vertices: $(\pm 4, 0)$; foci: $(\pm 6, 0)$; asymptotes: $y = \pm\dfrac{\sqrt{5}}{2}x$

47.

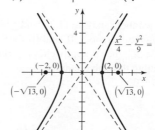

Vertices: $(\pm 2, 0)$; foci: $(\pm\sqrt{13}, 0)$; asymptotes: $y = \pm\dfrac{3}{2}x$

49. $\dfrac{x^2}{16} - \dfrac{y^2}{9} = 1$ **51.** $\dfrac{x^2}{81} + \dfrac{y^2}{72} = 1$
Directrices:
$x = \pm 27$

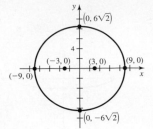

53. $x^2 - \dfrac{y^2}{8} = 1$

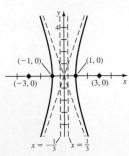

55.

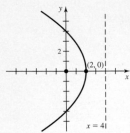

Vertex: $(2, 0)$; focus: $(0, 0)$; directrix: $x = 4$

57.

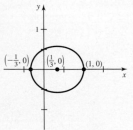

Vertices: $(1, 0)$, $\left(-\frac{1}{3}, 0\right)$; center: $\left(\frac{1}{3}, 0\right)$; foci: $(0, 0)$, $\left(\frac{2}{3}, 0\right)$; directrices: $x = -1$, $x = \frac{5}{3}$

59.

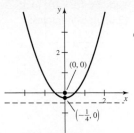

Vertex: $\left(0, -\frac{1}{4}\right)$; focus: $(0, 0)$; directrix: $y = -\frac{1}{2}$

61.

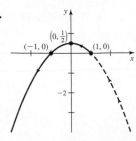

The parabola starts at $(1, 0)$ and goes through quadrants I, II, and III for θ in $[0, 3\pi/2]$; then it approaches $(1, 0)$ by traveling through quadrant IV on $(3\pi/2, 2\pi)$.

63.

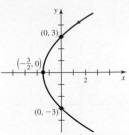

The parabola begins in the first quadrant and passes through the points $(0, 3)$ and then $\left(-\frac{3}{2}, 0\right)$ and $(0, -3)$ as θ ranges from 0 to 2π.

65. The parabolas open to the right if $p > 0$, open to the left if $p < 0$, and are more vertically compressed as $|p|$ decreases.   **67. a.** True
b. True   **c.** True   **d.** True   **69.** $y = 2x + 6$ **71.** $y = \frac{-3}{40}x - \frac{4}{5}$

73. $r = \dfrac{-4}{1 + 2\sin\theta}$ **77.** $\dfrac{dy}{dx} = \left(\dfrac{-b^2}{a^2}\right)\left(\dfrac{x}{y}\right)$, so
$\dfrac{y - y_0}{x - x_0} = \left(\dfrac{-b^2}{a^2}\right)\left(\dfrac{x_0}{y_0}\right)$, which is equivalent to the given equation.

79. $\dfrac{4\pi b^2 a}{3}$; $\dfrac{4\pi a^2 b}{3}$; yes, if $a \neq b$ **81. a.** $\dfrac{\pi b^2}{3a^2}\cdot(a - c)^2(2a + c)$
b. $\dfrac{4\pi b^4}{3a}$ **91.** $2p$ **97. a.** $u(m) = \dfrac{2m^2 - \sqrt{3m^2 + 1}}{m^2 - 1}$;
$v(m) = \dfrac{2m^2 + \sqrt{3m^2 + 1}}{m^2 - 1}$; 2 intersection points for $|m| > 1$
b. $\frac{5}{4}, \infty$ **c.** $2, 2$ **d.** $2\sqrt{3} - \ln(\sqrt{3} + 2)$

Chapter 10 Review Exercises, pp. 677–679

1. a. False   **b.** False   **c.** True   **d.** False   **e.** True   **f.** True
3. a.

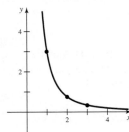

b. $y = 3/x^2$

c. The right branch of the function $y = 3/x^2$.   **d.** $\dfrac{dy}{dx} = -6$

5. a.

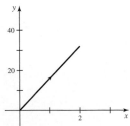

b. $y = 16x$

c. A line segment from $(0, 0)$ to $(2, 32)$ **d.** $\dfrac{dy}{dx} = 16$

7. At $t = \pi/6$: $y = (2 + \sqrt{3})x + \left(2 - \dfrac{\pi}{3} - \dfrac{\pi\sqrt{3}}{6}\right)$; at
$t = \dfrac{2\pi}{3}$: $y = \dfrac{x}{\sqrt{3}} + 2 - \dfrac{2\pi}{3\sqrt{3}}$

9.

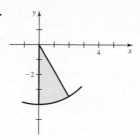

11. $(x - 3)^2 + (y + 1)^2 = 10$; a circle of radius $\sqrt{10}$ centered at $(3, -1)$ **13. a.**

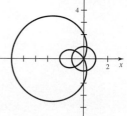

4 intersection points

b. $(1, 1.32), (1, 4.97), (-1, 0.7), (-1, 5.56)$

15. a. $(4.73, 2.77), (4.73, 0.38); (6, \pi/2), (2, 3\pi/2)$ **b.** There is no point at the origin.   **c.**

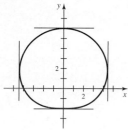

17. a. Horizontal tangent lines at $(1, \pi/6)$ $(1, 5\pi/6)$ $(1, 7\pi/6)$ and $(1, 11\pi/6)$; vertical tangent lines at $(\sqrt{2}, 0)$ and $(\sqrt{2}, \pi)$
b. Tangent lines at the origin have slopes ± 1.
c.

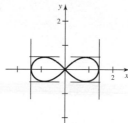

19. $\dfrac{19\pi}{2}$

21. $\frac{1}{4}(\sqrt{255} - \cos^{-1}(1/16))$

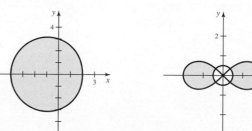

23. a. Hyperbola   **b.** Foci $(\pm\sqrt{3}, 0)$, vertices $(\pm 1, 0)$, directrices $x = \pm\dfrac{1}{\sqrt{3}}$ **c.** $e = \sqrt{3}$ **d.**

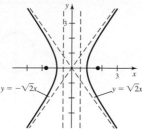

25. a. Hyperbola   **b.** Foci $(0, \pm 2\sqrt{5})$, vertices $(0, \pm 4)$, directrices $y = \pm = \dfrac{8}{\sqrt{5}}$ **c.** $e = \dfrac{\sqrt{5}}{2}$ **d.**

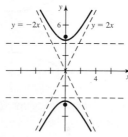

27. a. Ellipse   **b.** Foci $(\pm\sqrt{2}, 0)$, vertices $(\pm 2, 0)$, directrices $x = \pm 2\sqrt{2}$ **c.** $e = \dfrac{\sqrt{2}}{2}$ **d.**

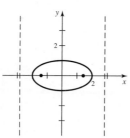

29. $y = \frac{3}{2}x - 2$ **31.** $y = -\frac{3}{5}x - 10$ **33.**

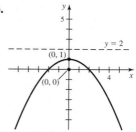

35.

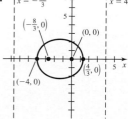

37. a. $x^2 - y^2 = 1$; hyperbola

b. $(\pm 1, 0), (\pm\sqrt{2}, 0); x = \pm\dfrac{1}{\sqrt{2}}; e = \sqrt{2}$

c.

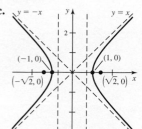

39. $\dfrac{y^2}{16} + \dfrac{25x^2}{336} = 1$

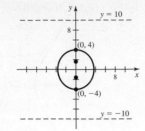

41. $\dfrac{y^2}{4} - \dfrac{x^2}{12} = 1$;

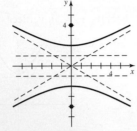

43. $e = 2/3, y = \pm 9, (\pm 2\sqrt{5}, 0)$ **45.** $(0, 0), (0.97, 0.97)$

47. $(0, 0)$ and $(r, \theta) = ((2n - 1)\pi, 0), n = 1, 2, 3, \ldots$

49. $\dfrac{2a}{\sqrt{2}} \cdot \dfrac{2b}{\sqrt{2}}; 2ab$ **51.** $m = \dfrac{b}{a}$ **55.** $r = \dfrac{3}{3 - \sin\theta}$

Index

TABLE OF INTEGRALS

Substitution Rule	Integration by Parts
$\displaystyle\int f(g(x))g'(x)\,dx = \int f(u)\,du \quad (u = g(x))$	$\displaystyle\int u\,dv = uv - \int v\,du$
$\displaystyle\int_a^b f(g(x))g'(x)\,dx = \int_{g(a)}^{g(b)} f(u)\,du$	$\displaystyle\int_a^b uv'\,dx = uv\Big|_a^b - \int_a^b vu'\,dx$

Basic Integrals

1. $\displaystyle\int x^n\,dx = \frac{1}{n+1}x^{n+1} + C;\ n \neq -1$

2. $\displaystyle\int \frac{dx}{x} = \ln|x| + C$

3. $\displaystyle\int \cos ax\,dx = \frac{1}{a}\sin ax + C$

4. $\displaystyle\int \sin ax\,dx = -\frac{1}{a}\cos ax + C$

5. $\displaystyle\int \tan x\,dx = \ln|\sec x| + C$

6. $\displaystyle\int \cot x\,dx = \ln|\sin x| + C$

7. $\displaystyle\int \sec x\,dx = \ln|\sec x + \tan x| + C$

8. $\displaystyle\int \csc x\,dx = -\ln|\csc x + \cot x| + C$

9. $\displaystyle\int e^{ax}\,dx = \frac{1}{a}e^{ax} + C$

10. $\displaystyle\int b^{ax}\,dx = \frac{1}{a\ln b}b^{ax} + C;\ b > 0, b \neq 1$

11. $\displaystyle\int \ln x\,dx = x\ln x - x + C$

12. $\displaystyle\int \log_b x\,dx = \frac{1}{\ln b}(x\ln x - x) + C$

13. $\displaystyle\int \frac{dx}{\sqrt{a^2 - x^2}} = \sin^{-1}\frac{x}{a} + C$

14. $\displaystyle\int \frac{dx}{x^2 + a^2} = \frac{1}{a}\tan^{-1}\frac{x}{a} + C$

15. $\displaystyle\int \frac{dx}{x\sqrt{x^2 - a^2}} = \frac{1}{a}\sec^{-1}\left|\frac{x}{a}\right| + C$

16. $\displaystyle\int \sin^{-1} x\,dx = x\sin^{-1}x + \sqrt{1 - x^2} + C$

17. $\displaystyle\int \cos^{-1} x\,dx = x\cos^{-1}x - \sqrt{1 - x^2} + C$

18. $\displaystyle\int \tan^{-1} x\,dx = x\tan^{-1}x - \frac{1}{2}\ln(1 + x^2) + C$

19. $\displaystyle\int \sec^{-1} x\,dx = x\sec^{-1}x - \ln\left(x + \sqrt{x^2 - 1}\right) + C;\ x \geq 1$

Trigonometric Integrals

20. $\displaystyle\int \cos^2 x\,dx = \frac{x}{2} + \frac{\sin 2x}{4} + C$

21. $\displaystyle\int \sin^2 x\,dx = \frac{x}{2} - \frac{\sin 2x}{4} + C$

22. $\displaystyle\int \sec^2 ax\,dx = \frac{1}{a}\tan ax + C$

23. $\displaystyle\int \csc^2 ax\,dx = -\frac{1}{a}\cot ax + C$

24. $\displaystyle\int \tan^2 x\,dx = \tan x - x + C$

25. $\displaystyle\int \cot^2 x\,dx = -\cot x - x + C$

26. $\displaystyle\int \cos^3 x = -\frac{1}{3}\sin^3 x + \sin x + C$

27. $\displaystyle\int \sin^3 x = \frac{1}{3}\cos^3 x - \cos x + C$

28. $\displaystyle\int \sec^3 x = \frac{1}{2}\sec x\tan x + \frac{1}{2}\ln|\sec x + \tan x| + C$

29. $\displaystyle\int \csc^3 x = -\frac{1}{2}\csc x\cot x - \frac{1}{2}\ln|\csc x + \cot x| + C$

30. $\displaystyle\int \tan^3 x\,dx = \frac{1}{2}\tan^2 x - \ln|\sec x| + C$

31. $\displaystyle\int \cot^3 x\,dx = -\frac{1}{2}\cot^2 x - \ln|\sin x| + C$

32. $\displaystyle\int \sec^n ax\tan ax\,dx = \frac{1}{na}\sec^n ax + C;\ n \neq 0$

33. $\displaystyle\int \csc^n ax\cot ax\,dx = -\frac{1}{na}\csc^n ax + C;\ n \neq 0$

34. $\displaystyle\int \frac{dx}{1 + \sin ax} = -\frac{1}{a}\tan\left(\frac{\pi}{4} - \frac{ax}{2}\right) + C$

35. $\displaystyle\int \frac{dx}{1 - \sin ax} = \frac{1}{a}\tan\left(\frac{\pi}{4} + \frac{ax}{2}\right) + C$

36. $\displaystyle\int \frac{dx}{1 + \cos ax} = \frac{1}{a}\tan\frac{ax}{2} + C$

37. $\displaystyle\int \frac{dx}{1 - \cos ax} = -\frac{1}{a}\cot\frac{ax}{2} + C$

38. $\int \sin mx \cos nx \, dx = -\dfrac{\cos (m+n)x}{2(m+n)} - \dfrac{\cos (m-n)x}{2(m-n)} + C; \ m^2 \neq n^2$

39. $\int \sin mx \sin nx \, dx = \dfrac{\sin (m-n)x}{2(m-n)} - \dfrac{\sin (m+n)x}{2(m+n)} + C; \ m^2 \neq n^2$

40. $\int \cos mx \cos nx \, dx = \dfrac{\sin (m-n)x}{2(m-n)} + \dfrac{\sin (m+n)x}{2(m+n)} + C; \ m^2 \neq n^2$

Reduction Formulas for Trigonometric Functions

41. $\int \cos^n x \, dx = \dfrac{1}{n} \cos^{n-1} x \sin x + \dfrac{n-1}{n} \int \cos^{n-2} x \, dx$

42. $\int \sin^n x \, dx = -\dfrac{1}{n} \sin^{n-1} x \cos x + \dfrac{n-1}{n} \int \sin^{n-2} x \, dx$

43. $\int \tan^n x \, dx = \dfrac{\tan^{n-1} x}{n-1} - \int \tan^{n-2} x \, dx; \ n \neq 1$

44. $\int \cot^n x \, dx = -\dfrac{\cot^{n-1} x}{n-1} - \int \cot^{n-2} x \, dx; \ n \neq 1$

45. $\int \sec^n x \, dx = \dfrac{\sec^{n-2} x \tan x}{n-1} + \dfrac{n-2}{n-1} \int \sec^{n-2} x \, dx; \ n \neq 1$

46. $\int \csc^n x \, dx = -\dfrac{\csc^{n-2} x \cot x}{n-1} + \dfrac{n-2}{n-1} \int \csc^{n-2} x \, dx; \ n \neq 1$

47. $\int \sin^m x \cos^n x \, dx = -\dfrac{\sin^{m-1} x \cos^{n+1} x}{m+n} + \dfrac{m-1}{m+n} \int \sin^{m-2} x \cos^n x \, dx; \ m \neq -n$

48. $\int \sin^m x \cos^n x \, dx = \dfrac{\sin^{m+1} x \cos^{n-1} x}{m+n} + \dfrac{n-1}{m+n} \int \sin^m x \cos^{n-2} x \, dx; \ m \neq -n$

49. $\int x^n \sin ax \, dx = -\dfrac{x^n \cos ax}{a} + \dfrac{n}{a} \int x^{n-1} \cos ax \, dx; \ a \neq 0$

50. $\int x^n \cos ax \, dx = \dfrac{x^n \sin ax}{a} - \dfrac{n}{a} \int x^{n-1} \sin ax \, dx; \ a \neq 0$

Integrals Involving $a^2 - x^2$; $a > 0$

51. $\int \sqrt{a^2 - x^2} \, dx = \dfrac{x}{2} \sqrt{a^2 - x^2} + \dfrac{a^2}{2} \sin^{-1} \dfrac{x}{a} + C$

52. $\int \dfrac{dx}{x \sqrt{a^2 - x^2}} = -\dfrac{1}{a} \ln \left| \dfrac{a + \sqrt{a^2 - x^2}}{x} \right| + C$

53. $\int \dfrac{dx}{x^2 \sqrt{a^2 - x^2}} = -\dfrac{\sqrt{a^2 - x^2}}{a^2 x} + C$

54. $\int x^2 \sqrt{a^2 - x^2} \, dx = \dfrac{x}{8} (2x^2 - a^2) \sqrt{a^2 - x^2} + \dfrac{a^4}{8} \sin^{-1} \dfrac{x}{a} + C$

55. $\int \dfrac{\sqrt{a^2 - x^2}}{x^2} \, dx = -\dfrac{1}{x} \sqrt{a^2 - x^2} - \sin^{-1} \dfrac{x}{a} + C$

56. $\int \dfrac{x^2}{\sqrt{a^2 - x^2}} \, dx = -\dfrac{x}{2} \sqrt{a^2 - x^2} + \dfrac{a^2}{2} \sin^{-1} \dfrac{x}{a} + C$

57. $\int \dfrac{dx}{a^2 - x^2} = \dfrac{1}{2a} \ln \left| \dfrac{x+a}{x-a} \right| + C$

Integrals Involving $x^2 - a^2$; $a > 0$

58. $\int \sqrt{x^2 - a^2} \, dx = \dfrac{x}{2} \sqrt{x^2 - a^2} - \dfrac{a^2}{2} \ln \left| x + \sqrt{x^2 - a^2} \right| + C$

59. $\int \dfrac{dx}{\sqrt{x^2 - a^2}} = \ln \left| x + \sqrt{x^2 - a^2} \right| + C$

60. $\int \dfrac{dx}{x^2 \sqrt{x^2 - a^2}} = \dfrac{\sqrt{x^2 - a^2}}{a^2 x} + C$

61. $\int x^2 \sqrt{x^2 - a^2} \, dx = \dfrac{x}{8} (2x^2 - a^2) \sqrt{x^2 - a^2} - \dfrac{a^4}{8} \ln \left| x + \sqrt{x^2 - a^2} \right| + C$

62. $\int \dfrac{\sqrt{x^2 - a^2}}{x^2} \, dx = \ln \left| x + \sqrt{x^2 - a^2} \right| - \dfrac{\sqrt{x^2 - a^2}}{x} + C$

63. $\int \dfrac{x^2}{\sqrt{x^2 - a^2}} \, dx = \dfrac{a^2}{2} \ln \left| x + \sqrt{x^2 - a^2} \right| + \dfrac{x}{2} \sqrt{x^2 - a^2} + C$

64. $\int \dfrac{dx}{x^2 - a^2} = \dfrac{1}{2a} \ln \left| \dfrac{x-a}{x+a} \right| + C$

65. $\int \dfrac{dx}{x(x^2 - a^2)} = \dfrac{1}{2a^2} \ln \left| \dfrac{x^2 - a^2}{x^2} \right| + C$

Integrals Involving $a^2 + x^2$; $a > 0$

66. $\int \sqrt{a^2 + x^2} \, dx = \dfrac{x}{2} \sqrt{a^2 + x^2} + \dfrac{a^2}{2} \ln \left(x + \sqrt{a^2 + x^2} \right) + C$

67. $\int \dfrac{dx}{\sqrt{a^2 + x^2}} = \ln \left(x + \sqrt{a^2 + x^2} \right) + C$

68. $\int \dfrac{dx}{x^2 \sqrt{a^2 + x^2}} = -\dfrac{\sqrt{a^2 + x^2}}{a^2 x} + C$

69. $\int x^2 \sqrt{a^2 + x^2} \, dx = \dfrac{x}{8} (a^2 + 2x^2) \sqrt{a^2 + x^2} - \dfrac{a^4}{8} \ln \left(x + \sqrt{a^2 + x^2} \right) + C$

70. $\displaystyle \int \frac{\sqrt{a^2 + x^2}}{x^2}\,dx = \ln\left|x + \sqrt{a^2 + x^2}\right| - \frac{\sqrt{a^2 + x^2}}{x} + C$

71. $\displaystyle \int \frac{x^2}{\sqrt{a^2 + x^2}}\,dx = -\frac{a^2}{2}\ln\left(x + \sqrt{a^2 + x^2}\right) + \frac{x\sqrt{a^2 + x^2}}{2} + C$

72. $\displaystyle \int \frac{\sqrt{a^2 + x^2}}{x}\,dx = \sqrt{a^2 + x^2} - a\ln\left|\frac{a + \sqrt{a^2 + x^2}}{x}\right| + C$

73. $\displaystyle \int \frac{dx}{(a^2 + x^2)^{3/2}} = \frac{x}{a^2\sqrt{a^2 + x^2}} + C$

74. $\displaystyle \int \frac{dx}{x(a^2 + x^2)} = \frac{1}{2a^2}\ln\left(\frac{x^2}{a^2 + x^2}\right) + C$

Integrals Involving $ax \pm b$; $a \neq 0, b > 0$

75. $\displaystyle \int (ax + b)^n\,dx = \frac{(ax + b)^{n+1}}{a(n + 1)} + C; \; n \neq -1$

76. $\displaystyle \int \left(\sqrt{ax + b}\right)^n\,dx = \frac{2}{a}\frac{\left(\sqrt{ax + b}\right)^{n+2}}{n + 2} + C; \; n \neq -2$

77. $\displaystyle \int \frac{dx}{x\sqrt{ax - b}} = \frac{2}{\sqrt{b}}\tan^{-1}\sqrt{\frac{ax - b}{b}} + C; \; b > 0$

78. $\displaystyle \int \frac{dx}{x\sqrt{ax + b}} = \frac{1}{\sqrt{b}}\ln\left|\frac{\sqrt{ax + b} - \sqrt{b}}{\sqrt{ax + b} + \sqrt{b}}\right| + C; \; b > 0$

79. $\displaystyle \int \frac{x}{ax + b}\,dx = \frac{x}{a} - \frac{b}{a^2}\ln|ax + b| + C$

80. $\displaystyle \int \frac{x^2}{ax + b}\,dx = \frac{1}{2a^3}\left((ax + b)^2 - 4b(ax + b) + 2b^2\ln|ax + b|\right) + C$

81. $\displaystyle \int \frac{dx}{x^2(ax + b)} = -\frac{1}{bx} + \frac{a}{b^2}\ln\left|\frac{ax + b}{x}\right| + C$

82. $\displaystyle \int x\sqrt{ax + b}\,dx = \frac{2}{15a^2}(3ax - 2b)(ax + b)^{3/2} + C$

83. $\displaystyle \int \frac{x}{\sqrt{ax + b}}\,dx = \frac{2}{3a^2}(ax - 2b)\sqrt{ax + b} + C$

84. $\displaystyle \int x(ax + b)^n\,dx = \frac{(ax + b)^{n+1}}{a^2}\left(\frac{ax + b}{n + 2} - \frac{b}{n + 1}\right) + C; \; n \neq -1, -2$

85. $\displaystyle \int \frac{dx}{x(ax + b)} = \frac{1}{b}\ln\left|\frac{x}{ax + b}\right| + C$

Integrals with Exponential and Trigonometric Functions

86. $\displaystyle \int e^{ax}\sin bx\,dx = \frac{e^{ax}(a\sin bx - b\cos bx)}{a^2 + b^2} + C$

87. $\displaystyle \int e^{ax}\cos bx\,dx = \frac{e^{ax}(a\cos bx + b\sin bx)}{a^2 + b^2} + C$

Integrals with Exponential and Logarithmic Functions

88. $\displaystyle \int \frac{dx}{x\ln x} = \ln|\ln x| + C$

89. $\displaystyle \int x^n\ln x\,dx = \frac{x^{n+1}}{n + 1}\left(\ln x - \frac{1}{n + 1}\right) + C; \; n \neq -1$

90. $\displaystyle \int xe^x\,dx = xe^x - e^x + C$

91. $\displaystyle \int x^n e^{ax}\,dx = \frac{1}{a}x^n e^{ax} - \frac{n}{a}\int x^{n-1}e^{ax}\,dx; \; a \neq 0$

92. $\displaystyle \int \ln^n x\,dx = x\ln^n x - n\int \ln^{n-1} x\,dx$

Miscellaneous Formulas

93. $\displaystyle \int x^n\cos^{-1}x\,dx = \frac{1}{n + 1}\left(x^{n+1}\cos^{-1}x + \int \frac{x^{n+1}dx}{\sqrt{1 - x^2}}\right); \; n \neq -1$

94. $\displaystyle \int x^n\sin^{-1}x\,dx = \frac{1}{n + 1}\left(x^{n+1}\sin^{-1}x - \int \frac{x^{n+1}dx}{\sqrt{1 - x^2}}\right); \; n \neq -1$

95. $\displaystyle \int x^n\tan^{-1}x\,dx = \frac{1}{n + 1}\left(x^{n+1}\tan^{-1}x - \int \frac{x^{n+1}dx}{x^2 + 1}\right); \; n \neq -1$

96. $\displaystyle \int \sqrt{2ax - x^2}\,dx = \frac{x - a}{2}\sqrt{2ax - x^2} + \frac{a^2}{2}\sin^{-1}\left(\frac{x - a}{a}\right) + C; \; a > 0$

97. $\displaystyle \int \frac{dx}{\sqrt{2ax - x^2}} = \sin^{-1}\left(\frac{x - a}{a}\right) + C; \; a > 0$

GRAPHS OF ELEMENTARY FUNCTIONS

Lincar functions

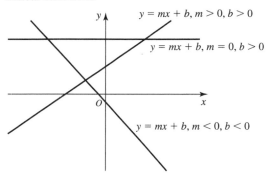

$y = mx + b, m > 0, b > 0$

$y = mx + b, m = 0, b > 0$

$y = mx + b, m < 0, b < 0$

Quadratic functions

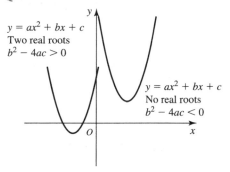

$y = ax^2 + bx + c$
Two real roots
$b^2 - 4ac > 0$

$y = ax^2 + bx + c$
No real roots
$b^2 - 4ac < 0$

Positive even powers

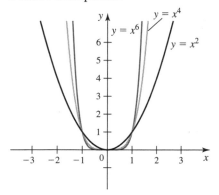

$y = x^6$ $y = x^4$ $y = x^2$

Positive odd powers

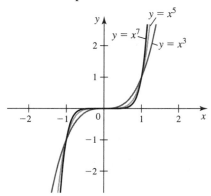

$y = x^5$ $y = x^7$ $y = x^3$

Negative even powers

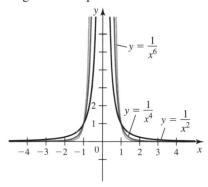

$y = \frac{1}{x^6}$ $y = \frac{1}{x^4}$ $y = \frac{1}{x^2}$

Negative odd powers

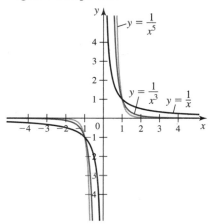

$y = \frac{1}{x^5}$ $y = \frac{1}{x^3}$ $y = \frac{1}{x}$

Exponential functions

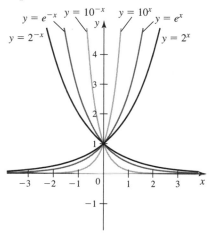

$y = e^{-x}$ $y = 10^{-x}$ $y = 10^x$ $y = e^x$
$y = 2^{-x}$ $y = 2^x$

Natural logarithmic and exponential functions

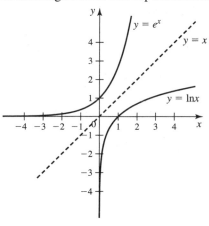

$y = e^x$ $y = x$ $y = \ln x$

DERIVATIVES

General Formulas

$$\frac{d}{dx}(c) = 0$$

$$\frac{d}{dx}(cf(x)) = cf'(x)$$

$$\frac{d}{dx}(f(x) + g(x)) = f'(x) + g'(x)$$

$$\frac{d}{dx}(f(x) - g(x)) = f'(x) - g'(x)$$

$$\frac{d}{dx}(f(x)g(x)) = f'(x)g(x) + f(x)g'(x)$$

$$\frac{d}{dx}\left(\frac{f(x)}{g(x)}\right) = \frac{g(x)f'(x) - f(x)g'(x)}{(g(x))^2}$$

$$\frac{d}{dx}(x^n) = nx^{n-1}, \text{ for real numbers } n$$

$$\frac{d}{dx}[f(g(x))] = f'(g(x)) \cdot g'(x) \text{ (Chain Rule)}$$

Trigonometric Functions

$$\frac{d}{dx}(\sin x) = \cos x$$

$$\frac{d}{dx}(\cos x) = -\sin x$$

$$\frac{d}{dx}(\tan x) = \sec^2 x$$

$$\frac{d}{dx}(\cot x) = -\csc^2 x$$

$$\frac{d}{dx}(\sec x) = \sec x \tan x$$

$$\frac{d}{dx}(\csc x) = -\csc x \cot x$$

Inverse Trigonometric Functions

$$\frac{d}{dx}(\sin^{-1} x) = \frac{1}{\sqrt{1 - x^2}}$$

$$\frac{d}{dx}(\cos^{-1} x) = -\frac{1}{\sqrt{1 - x^2}}$$

$$\frac{d}{dx}(\tan^{-1} x) = \frac{1}{1 + x^2}$$

$$\frac{d}{dx}(\cot^{-1} x) = -\frac{1}{1 + x^2}$$

$$\frac{d}{dx}(\sec^{-1} x) = \frac{1}{|x|\sqrt{x^2 - 1}}$$

$$\frac{d}{dx}(\csc^{-1} x) = -\frac{1}{|x|\sqrt{x^2 - 1}}$$

Exponential and Logarithmic Functions

$$\frac{d}{dx}(e^x) = e^x$$

$$\frac{d}{dx}(b^x) = b^x \ln b$$

$$\frac{d}{dx}(\ln |x|) = \frac{1}{x}$$

$$\frac{d}{dx}(\log_b x) = \frac{1}{x \ln b}$$

Hyperbolic Functions

$$\frac{d}{dx}(\sinh x) = \cosh x$$

$$\frac{d}{dx}(\cosh x) = \sinh x$$

$$\frac{d}{dx}(\tanh x) = \text{sech}^2 x$$

$$\frac{d}{dx}(\coth x) = -\text{csch}^2 x$$

$$\frac{d}{dx}(\text{sech } x) = -\text{sech } x \tanh x$$

$$\frac{d}{dx}(\text{csch } x) = -\text{csch } x \coth x$$

FORMS OF THE FUNDAMENTAL THEOREM OF CALCULUS

Fundamental Theorem of Calculus	$$\int_a^b f'(x)\,dx = f(b) - f(a)$$
Fundamental Theorem of Line Integrals	$$\int_C \nabla f \cdot d\mathbf{r} = f(B) - f(A)$$ (A and B are the initial and final points of C.)
Green's Theorem	$$\iint_R (g_x - f_y)\,dA = \oint_C f\,dx + g\,dy$$ $$\iint_R (f_x + g_y)\,dA = \oint_C f\,dy - g\,dx$$
Stokes' Theorem	$$\iint_S (\nabla \times \mathbf{F}) \cdot \mathbf{n}\,dS = \oint_C \mathbf{F} \cdot d\mathbf{r}$$
Divergence Theorem	$$\iint_S \mathbf{F} \cdot \mathbf{n}\,dS = \iiint_D \nabla \cdot \mathbf{F}\,dV$$

FORMULAS FROM VECTOR CALCULUS

Assume $\mathbf{F}(x, y, z) = f(x, y, z)\mathbf{i} + g(x, y, z)\mathbf{j} + h(x, y, z)\mathbf{k}$, where f, g, and h are differentiable on a region D of $\mathbf{R}^3$.

Gradient: $\nabla f(x, y, z) = \dfrac{\partial f}{\partial x}\mathbf{i} + \dfrac{\partial f}{\partial y}\mathbf{j} + \dfrac{\partial f}{\partial z}\mathbf{k}$

Divergence: $\nabla \cdot \mathbf{F}(x, y, z) = \dfrac{\partial f}{\partial x} + \dfrac{\partial g}{\partial y} + \dfrac{\partial h}{\partial z}$

Curl: $\nabla \times \mathbf{F}(x, y, z) = \begin{vmatrix} \mathbf{i} & \mathbf{j} & \mathbf{k} \\ \dfrac{\partial}{\partial x} & \dfrac{\partial}{\partial y} & \dfrac{\partial}{\partial z} \\ f & g & h \end{vmatrix}$

$\nabla \times (\nabla f) = \mathbf{0}$ $\qquad \nabla \cdot (\nabla \times \mathbf{F}) = 0$

$\mathbf{F}$ conservative on $D \Leftrightarrow \mathbf{F} = \nabla\varphi$ for some potential function φ

$\qquad\qquad \Leftrightarrow \oint_C \mathbf{F} \cdot d\mathbf{r} = 0$ over closed paths C in D

$\qquad\qquad \Leftrightarrow \int_C \mathbf{F} \cdot d\mathbf{r}$ is independent of path for C in D.

$\qquad\qquad \Leftrightarrow \nabla \times \mathbf{F} = \mathbf{0}$ on D